STATISTICS

STATISTICS
William L. Hays

University of Texas at Austin

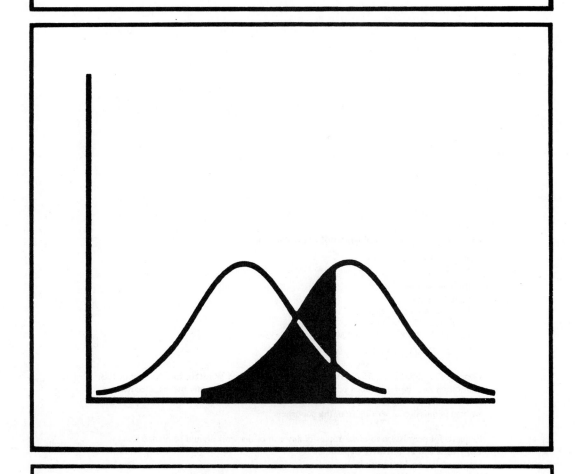

Fourth Edition

Harcourt Brace Jovanovich College Publishers

Fort Worth   Philadelphia   San Diego   New York   Orlando   Austin   San Antonio

Toronto   Montreal   London   Sydney   Tokyo

Publisher: Susan Meyers
Acquisitions Editor: Susan Arellano
Senior Project Manager: Sondra Greenfield
Production Manager: Annette Mayeski
Design Supervisor: Robert Kopelman
Cover Design: Caliber Design Planning, Inc.

Library of Congress Cataloging-in-Publication Data

Hays, William Lee, 1926–
  Statistics.

  Bibliography: p.
  Includes index.
  1. Social sciences—Statistical methods.
  2. Statistics.   3. Psychometrics.   I. Title.
  HA29.H352   1988   519.5′0243   87-32115

ISBN 0-03-002464-1

Requests for permission to make copies of any part of the work should be mailed to:
Copyrights and Permissions Department
Harcourt Brace Jovanovich, Inc.
Orlando, Florida 32887
PRINTED IN THE UNITED STATES OF AMERICA

    2 3   039   9 8 7 6 5

Harcourt Brace Jovanovich, Inc.
The Dryden Press
Saunders College Publishing

PREFACE

This book represents an attempt to give the elements of modern statistics in a relatively nonmathematical form, but in somewhat more detail than is customary in texts designed for psychologists, and with considerably more emphasis on the theoretical rather than the applied aspects of the subject. It is designed as a text for at least an intermediate level of difficulty. I have felt for some time that the serious student in experimental psychology both needs and wishes to know something more of the language and concepts of theoretical statistics than is provided in the usual "cookbook" of statistical methods. Granted that a real understanding of statistical methods requires a considerable degree of mathematical training and sophistication, a great deal of statistical theory can be got across to the serious student familiar only with elementary algebra, provided that a relatively detailed exposition of the concepts accompanies the mathematical treatment, and provided that the student is genuinely interested in acquiring a grasp of this subject.

A quarter century has now passed since these words appeared in the preface to the first edition of this text. Obviously, a great deal has happened in the meantime, affecting both how one uses and how one teaches statistical methods. Nevertheless, I tend to believe that the aims set forth in the first edition remain relevant even in our modern circumstances.

The biggest change, of course, is the wide accessibility and the ease of use of the computer in the solution of statistical problems. When this book was first written the computer age had but recently dawned, and working with the computer was the privilege of only a few experts with some very special skills. Now everyone in a research setting has access to some sort of computation facility, and generally to a computer with speed, power, and sophistication undreamed of only a few years ago. Furthermore, the availability of the statistical packages has made it possible to use the computer for data analyses of almost any size and level of complexity, even when one has little or no knowledge of the computer per se.

In addition, almost every graduate student in the social and behavioral sciences either owns or has easy access to a microcomputer, and these too are often equipped with one or more packages of statistical programs. Even hand-held calculators, still at a primitive stage of development when this book was first written, now often possess a considerable amount of computational power of their own, especially through their programming capabilities.

In short, "doing statistics" is now easier than ever before. On the other hand, "using statistics" remains as big a problem as ever for the researcher. Indeed, given the wealth of methods available for little or no effort on the user's part, the problem of choice and correct application of a statistical method is perhaps even more difficult now than in the old days. Hence I, at least, believe that the approach represented by this text is still timely.

Since computer packages are now such a ubiquitous part of applying statistics, I have incorporated sections giving limited information about the popular packages and how they relate to the methods this book covers. However, this information is confined to very brief descriptions of the capabilities of the most popular packages. This reflects my very strong belief that a student using this text should at the same time be learning to use at least one of the standard packages, and actually studying the relevant manuals which contain far more information about each method than it would be appropriate to include in any text. I hope that teachers of courses for which this text is appropriate will use this strategy, supplementing this text with the manual for the package in use. I have not, however, ventured to suggest which of the packages should be employed, as this must depend on the instructor's own preferences and background, as well as on the hardware and software available. Given these assumptions, I have also resisted the current trend for showing extensive computer input and output in the text itself. Such items are now far too familiar and ephemeral to occupy expensive text space, and I think it far more important to suggest to the student where he or she should look in the program manual, where all the relevant detail is already provided.

I also believe that it is extremely educational for the student to create statistical programs for a microcomputer, using BASIC or a similar simple language, while studying the methods. This is not all that difficult to do, and can be a good deal of fun. It is my experience that nothing teaches one more about the innards of a statistical method than trying to write and debug a program for carrying out the method. I would hope that some enterprising teachers will give students this experience while leading them through a text such as this.

The universality of computers has also affected the coverage of this book in other ways. For example, multiple regression and its variants, at one time a considerable computational challenge for most researchers, has now become such a staple of research reports that a greatly expanded coverage seemed required. The same is true of the analysis of covariance, which, since it seems to come as a sort of bonus in a number of the programs for analysis of variance, appears as a feature of an ever-increasing number of studies. Such a popular and important technique now deserves a far fuller treatment than it has received heretofore. The same can be said for a number of other methods and concerns that are increasingly prominent in the research literature, largely because of their prominence in the statistical packages. This comment could also be made, perhaps with more force, about multivariate methods such as MANOVA, discriminant function, and canonical correlation. The researcher is beginning to need a much earlier and more thorough introduction to these methods since they have become far more commonplace than before. Although it was tempting to expand the coverage of univariate statistical methods a bit in the direction of MANOVA, and the first draft of this edition actually contained such a section, prudence (and problems of length) finally prevailed, and so I will leave these matters to other, more capable, hands. Much the same is true of other important developments such as structural equations analysis and metanalysis, to name but two prominent examples. Students need to be aware of these methods early on in their research careers, but space priorities prevented even their mention here.

Given the increased coverage of this edition, and especially if, as I have urged above, the teacher incorporates at least one of the statistical packages as a parallel feature of the course, the material included should now be adequate for a two-quarter or even a two-semester course. On the other hand, the chapter sections are relatively self-contained, especially in the later portions of the text, and so the teacher should be able to cut and tailor the content fairly freely to fit his or her own needs. In addition, the teacher who wants to introduce matrix algebra as a way of approaching some of the more complex topics may be aided by the material in the new Appendix D, devoted to matrix terminology and some matrix versions of important topics covered in the text. However, the vector and matrix terminology is certainly not a required feature of the text itself.

A good deal of effort has been put forth in the editorial preparation of this new edition. Furthermore, many of the exercises have been reworked for accuracy and/or clarity, and the later problems have been rearranged so that the odd-numbered exercises will not tend always to be alike. A new solutions manual is available for the even-numbered exercises.

As in the prefaces to previous editions, I wish to thank all the many people, colleagues and students alike, who shared their experiences of this text with me, and have made valuable suggestions for its improvement. I am especially grateful to my colleagues at the University of Texas at Austin, including (but not limited to) Earl Jennings, William Koch, Ed Emmer, and H. Paul Kelley, for the many things they have taught me over the years about statistics and about good teaching. I am grateful to the reviewers whose comments were of critical importance: David Schlundt, Vanderbilt University; James Stevens, University of Cincinnati; Robert Seibel, Penn State University; James Whipple, Washington State University; J. E. Keith Smith, University of Michigan; Greg C. Oden, University of Wisconsin (Madison); Nancy Anderson, University of Maryland; Charles Reichardt, University of Denver; James Terwilliger, University of Minnesota. My special thanks go to Mrs. Joy Bohmfalk for her splendid assistance, especially in the preparation of the solutions manual.

Renewed thanks are due to the Biometrika Trustees for their permission to reproduce portions of tables from the *Biometrika Tables for Statisticians* (3rd ed.) by E. S. Pearson and H. O. Hartley, to the McGraw-Hill Company for permission to use the binomial table from *Handbook of Probability and Statistics with Tables* (2nd ed.) by R. S. Burington and D. C. May, and to the American Statistical Association for permission to reproduce the table of percentage points for the Kolmogorov-Smirnov test.

I also wish to express my gratitude to Susan Arellano, Acquisitions Editor, and to Sondra Greenfield, Senior Project Manager at Holt, Rinehart and Winston. Both have been highly professional, helpful, and sympathetic. Linda Davoli has contributed greatly as copy editor, proofreader, and indexer. Finally, my thanks must go to my dear family, Palma, Leeann, and Scott. Each has taken several more steps toward sainthood during this latest long ordeal with Daddy's book!

W.L.H.
Austin, Texas
January 1988

CONTENTS

Chapter *8*   Inferences About Population Means   *287*

Chapter *9*   The Chi-Square and the *F* Distributions   *322*

Chapter 14   Problems in Regression and Correlation   544

STATISTICS

INTRODUCTION

0.1 ON THE NATURE AND THE ROLE OF INFERENTIAL STATISTICS

The word *statistics* came into English from Latin and German, and ultimately derives from the same Indo-European root which gave us *standing, status, state,* and even *understand*. In the minds of most people, *statistics* has a lot in common with these related words, meaning roughly, a description of "how things are." It is, of course, true that a part of the theory of statistics concerns effective ways of summarizing and communicating masses of information which describe some situation. This part of the overall theory and set of methods is usually known as *descriptive statistics*.

Although descriptive statistics form an important basis for dealing with data, a major part of the theory of statistics is concerned with another question: How does one go beyond a given set of data, and make general statements about the large body of potential observations, of which the data collected represent but a sample? This is the theory of *inferential statistics,* with which this book is mainly concerned.

Applications of inferential statistics occur in virtually all fields of research—the physical sciences, the biological sciences, the social sciences, engineering, market and consumer research, quality control in industry, and so on, almost without end. Although the actual methods differ somewhat in the different fields, the applications all rest on the same general theory of statistics. By examining what the fields have in common in their applications of statistics we can gain a picture of the basic problem studied in mathematical statistics. The major applications of statistics in any field all rest on the possibility of *repeated observations* or experiments made under essentially the *same conditions*. That is, either the researcher actually can observe the same process repeated many times, as in industrial quality control, or there is the *conceptual* possibility of repeated observation, as in a scientific experiment that might, in principle, be repeated under identical conditions. However, in any circumstance where repeated observations are made, even though every precaution is taken to make conditions exactly the same, the results of observations will vary, or tend to be different, from trial to trial. The researcher has control over some, but not all, of the factors that make the outcomes of observations tend to differ from one another.

When observations are made under the same conditions in one or more respects, but they give outcomes differing in other ways, then there is some *uncertainty* connected with observation of any given object or phenomenon. Even though some things are known to be true about that object in advance of the observation, the experimenter cannot predict with complete certainty what its other characteristics will be. Given enough repeated observations of the same object or kind of object a good bet may be formulated about what the other characteristics are likely to be, but one cannot be completely sure of the status of any given object.

This fact leads us to the central problem of inferential statistics: in one sense, inferential statistics is a *theory about uncertainty,* the tendency of outcomes to vary when repeated observations are made under identical conditions. Granted that certain conditions are fulfilled, theoretical statistics permits deductions about the *likelihood* of the various possible outcomes of observation. The essential concepts in statistics derive from the theory of probability, and the deductions made within the theory of statistics are, by and large, statements about the probability of particular kinds of outcomes, given that initial, mathematical, conditions are met.

Mathematical statistics is a formal mathematical system. Any mathematical system consists of these basic parts:

1. A collection of undefined **things** or **elements,** considered only as abstract entities;
2. A set of undefined **operations,** or possible relations among the abstract elements;
3. A set of **postulates** and **definitions,** each asserting that some specific relation holds among the various elements, the various operations, or both.

In any mathematical system the application of logic to combinations of the postulates and definitions leads to new statements, or *theorems,* about the undefined elements of the system. *Given* that the original postulates and definitions are true, then the new statements *must* be true. Mathematical systems are purely abstract, and essentially undefined, **deductive** structures. In the first chapter we will see that the abstract system known as the theory of probability has this character.

Mathematical systems are not really "about" anything in particular. They are systems of statements about "things" having the formal properties given by the postulates. No one may know what the original mathematician really had in mind to call these abstract elements. Indeed, they may represent absolutely nothing that exists in the real world of experience, and the sole concern may be in what one can derive about the other necessary relations among abstract elements given particular sets of postulates. It is perfectly true, of course, that many mathematical systems originated from attempts to describe real objects or phenomena and their interrelationships: historically, the abstract systems of geometry, school algebra, and the calculus grew out of problems where something very practical and concrete was in the back of the mathematician's mind. However, as *mathematics* these systems deal with completely abstract entities.

When a mathematical system is interpreted in terms of real objects or events, then the system is said to be a **mathematical model** for those objects or events. Somewhat more precisely, the undefined terms in the mathematical system are identified with particular, relevant **properties** of objects or events; thus, in applications of arithmetic, the number symbols are identified with magnitudes or amounts of some particular property that objects possess, such as weight, or extent, or numerosity. The system of arithmetic need not apply to other characteristics of the same objects, as, for example, their colors. Once this identification can be made between the mathematical system and the relevant properties of objects, then anything that is a logical consequence in the system is a true statement about objects in the model, *provided,* of course, that the formal characteristics of the system actually *parallel* the real characteristics of objects in terms of the particular properties considered. In short, in order to be useful as a mathematical model, a mathematical system must have a formal structure that ''fits'' at least one aspect of a real situation.

Probability theory and statistics are each both mathematical systems *and* mathematical models. Probability theory deals with elements called *events,* which are completely abstract. Furthermore, these abstract things are paired with numbers called *probabilities*. The theory itself is the system of logical relations among these essentially undefined things. The experimenter uses this abstract system as a mathematical model: the experiment produces a real outcome, which is called an event, and the model of probability theory provides a value which is interpreted as the relative frequency of occurrence for the outcome. If the requirements of the model are met, this is a true, and perhaps useful, result. If the experiment really does not fit the requirements of probability theory as a system, then the statement made about the actual result need not be true. (This point must not be overstressed, however. We will find that often a statistical method can yield practically useful results even when its requirements are not fully satisfied. Much of the art in applying statistical methods lies in understanding when and how this is true.)

Mathematical systems such as probability theory and the theory of statistics are, by their very nature, deductive. That is, formal assertions are postulated as true, and then by logical argument, true conclusions are reached. All well-developed theories have this formal, logico-deductive character.

On the other hand, the problem of the empirical scientist is essentially different from that of the logician or mathematician. Scientists search for general relations among events; these general relations are those which can be expected to hold whenever the appropriate set of circumstances exists. The very name *empirical science* asserts that these laws shall be discovered and verified by the actual observation of what happens in the real world of experience. However, no mortal scientist ever observes all the phenomena about which a generalization must be made. Scientific conclusions about what would happen for *all* of a certain class of phenomena always come from observations of only a very few particular cases of that phenomenon.

The student acquainted with logic will recognize that this is a problem of **induction.** The rules of logical *deduction* are rules for arriving at true consequences from true premises. Scientific theories are, for the most part, systems of deductions from basic principles held to be true. If the basic principles are true, then the deductions must be true. However, how does one go about arriving at and checking the truth of the initial propositions? The answer is, for an empirical science, observation and inductive generalization—going from what is true of some observations to a statement that this is true for *all possible* observations made under the same conditions. Any empirical science begins with observation and generalization.

Furthermore, even after deductive theories exist in a science, experimentation is used to check on the truth of these theories. Observations that contradict deductions made within the theory are prima facie evidence against the truth of the theory itself. Yet, how does the scientist know that the results are not an accident, the product of some chance variation in procedure or conditions over which there is no control? Would the result be the same in the long run if the experiment could be repeated many times?

It takes only a little imagination to see that this process of going from the specific to the general is a very risky one. Each observation the scientist makes is different in some way from the next. Innumerable influences are at work altering—sometimes minutely, sometimes radically—the similarities and differences the scientist observes among events. Controlled experimentation in any science is an attempt to minimize at least part of the accidental variation or *error* in observation. Precise techniques of measurement are aids to scientists in sharpening their own rather dull powers of observation and comparison among events. So-called exact sciences, such as physics and chemistry, have thus been able to remove a substantial amount of the unwanted variation among observations from time to time, place to place, observer to observer, and hence are often able to make general statements about physical phenomena with great assurance from the observation of quite limited numbers of events. Observations in these sciences can often be made in such a way that the generality of conclusions is not a major point at issue. Here, there is relatively little reliance on probability and statistics. (However, as even these scientists delve into the molecular, atomic, and subatomic domain, negligible differences turn into enormous unpredictabilities and statistical theories become an important adjunct to their work.)

In the biological, behavioral, and social sciences, however, the situation is radically different. In these sciences the variations between observations are not subject to the precise experimental controls that are possible in the physical sciences. Refined measurement techniques have not reached the stage of development that they have attained in physics and chemistry. Consequently, the drawing of general conclusions is a much more dangerous business in these fields, where the sources of variability among living things are extremely difficult to identify, measure, and control. And yet the aim of the social or biological scientist is precisely the same as that of the physical scientist— arriving at general statements about the phenomena under study.

Faced with only a limited number of observations or with an experiment that can be conducted only once, the scientist can reach general conclusions only in the form of a "bet" about what the true, long-run situation actually is like. Given only sample evidence, the scientist is always unsure of the "goodness" of any assertion made about the true state of affairs. The theory of statistics provides ways to assess this uncertainty and to calculate the probability of being wrong in deciding in a particular way. Provided that the experimenter can make some assumptions about what is true, then the deductive theory of statistics tells us *how likely* particular results should be. Armed with this information, the experimenter is in a better position to decide what to say about the true situation. Regardless of what one decides from evidence, it *could* be wrong; but deductive statistical theory can at least determine the probabilities of error in a particular decision.

In the last half century, a branch of mathematics has been developed around this problem of decision making under uncertain conditions. This is sometimes called "statistical decision theory." One of the main problems treated in decision theory is the choice of a decision rule, or "deciding how to decide" from evidence. Decision theory evaluates rules for deciding from evidence in the light of what the decision maker wants to accomplish. As we shall see in later chapters, mathematics can tell us wise ways to decide how to decide under some circumstances, but it can never tell the experimenter how a decision must be reached in any particular situation. The theory of statistics supplies one very important piece of information to the experimenter: the probability of sample results *given* certain conditions. Decision theory can supply another ingredient: optimal ways of using this and other information to accomplish certain ends. Nevertheless, neither theory tells the experimenter *exactly* how to decide—how to make the inductive leap from observation to what is true in general. This is the experimenter's problem, and the answer must be sought outside of deductive mathematics, and in the light of what the experimenter is trying to do.

0.2 ABOUT THIS BOOK

This book is addressed to upper-division or graduate students in the social and behavioral sciences. Such students are the people who will produce the significant social and behavioral science research in years to come and those who will make up the audience for much of this research. As a part of their professional equipment, these students need to know statistics, at a level beyond an undergraduate course, and just short of the specialized research design and methodology courses needed to round out their graduate programs. Such students are the "you" in this book.

You will soon discover that the main concern in this book is with the theory underlying inferential methods, rather than with a detailed exposition of all the different methods social scientists and others find useful. The author had no intention of writing a "cookbook" that would equip students to meet every possible situation they might encounter. Many methods will be introduced, it is true, and we will, in fact, discuss most of the elementary techniques for statistical inference currently in use. However, in the past few years the concerns of the social scientist have begun to grow increasingly complicated. Theory is growing, and social scientists are turning their attention to new problems and techniques for data analysis that are becoming much more sophisticated than in the past. The statistical analyses required in many such studies are simply not in the cookbooks. From all indications, this trend will continue, and by the time that you, the student, are in the midst of your professional career, it may well be the case that entirely new statistical methods will be required, replacing many of the methods currently found useful.

Furthermore, a true revolution has occurred in the past two decades, deeply affecting the application and teaching of statistical methodology. This has been brought about by the new generations of computers, which are faster, more flexible, and cheaper to use than anyone would have dreamed only a few years back. Most statistical analysis is now done by computer in almost all research settings. Even the beginner in research turns to the computer for all but the simplest computational routines. Techniques that were once viewed as impracticable because of the amount of computation involved are now quite routine parts of the social scientist's methodology. Prepackaged programs for all of the common statistical techniques are now available in virtually all computing centers. Some of the most popular of these statistical packages will be mentioned in Section 0.4.

Prior to this development, when most computations were done by hand with the assistance of a desk calculator, it was necessary to devote a large part of statistical instruction to computational methods, with emphasis on shortcuts to lessen the computational burden. Now all has changed, and it is no longer necessary to dwell so long on the how-to-do-it aspects. Even inexpensive pocket calculators are influencing how statistics can be taught and learned. These pocket, or hand-held, calculators provide an amazing range of functions in addition to basic arithmetic. Pocket calculators have taken us light-years away from even the most sophisticated desk calculators, and cost a tiny fraction of what one paid for the old machines. One of the best investments a student of statistics can make is in a really good hand-held calculator of the scientific type, which includes such functions as square roots, reciprocals, logarithms, powers, etc. Many calculators have common statistical analyses built in, which can be a big labor-saving device, whereas others can be programmed for unusual operations required for repeated, short-term use.

For all of these reasons, the process of learning (and of teaching) statistics is changing rapidly. With many new techniques at the disposal of the social scientist, and with the computer ready to give almost any analysis we request, the process is shifting from "how-to-do-it" to "what can be done."

As social and behavioral research becomes more sophisticated and more and more methods are made available for use in particular situations, a point is rapidly being reached where the research worker simply cannot be familiar with all the statistical methods that might pertain to a given problem. It seems unfair to demand that each competent researcher must also be a competent statistician as well, although a few gifted individuals (not including the author) have somehow found time and brainpower to be both. In short, the days when each researcher was his own "do it yourself" statistician, relying on a handy cookbook of set methods, are about over.

Instead, it is becoming increasingly important for the research worker to understand some of the principles that unite all methods for statistical inference and some of the general models that are capable of almost limitless adaptation and modification to fit particular research requirements. Secondly, it is important for the serious worker in research to form consultative relationships with statisticians, who can give guidance when problems begin to require solutions that extend beyond familiar methods. A large part of the work of most applied statisticians consists of consultation on problems of design and analysis of experiments, and many are available for such consultation on a professional basis. The statistician can usually provide answers to the research worker's questions, provided that the statistician is asked about the problem *before* the data are collected, and can participate in the efficient and logical *planning* of the experiment. It is most unreasonable to expect the statistician to reach in a hat and pull out a method that will extract meaning from a poorly designed or executed study.

The statistician does not expect the scientist to know all about theoretical statistics, nor does the scientist expect the statistician to know all about a particular problem. But to work together effectively, each must have some idea of the basic concepts the other uses. This is one reason for the theoretical emphasis in this book. At the very outset, the student needs to know something about the points of view and concepts of theoretical statistics in order to appreciate its resources and not become lost in the complexities of using the statistical language.

This book is not, nor does it pretend to be, a first course in mathematical statistics. Ideally, the serious student in the social or behavioral sciences should take at least one such course, although there are two practical difficulties: The content and the organization of courses in mathematical statistics are framed for the training of statisticians, not behavioral scientists, and the peculiar problems of these research areas are not emphasized in such courses; this is as it should be. In the second place, to become a really good researcher is a full-time job, and the student may not have the time to devote to the mathematical statistics courses and their prerequisites in order to gain the essential background needed.

Thus, this book contains some of the concepts, results, and theoretical arguments that come from mathematical statistics, but these results and arguments are given at a far more intuitive and informal level than would be the case for a student in mathematical statistics. Only very seldom will the level of mathematics used rise above the high-school level, although the mathematical concepts used will occasionally be unfamiliar to some students. Occasionally we will use some results coming from the application of the calculus, especially results having to do with the idea of a ''limit''; these ideas really cannot be treated adequately at an elementary level. Furthermore, use of the mathematical vocabulary of vectors and matrices, and of the algebra for manipulating such arrays of numbers, can often shed light on the more advanced topics that we will be treating. Although a knowledge of vector and matrix theory is not a prerequisite for following the mathematical arguments that this text contains, the material on vector manipulations in Appendix C and the elementary discussions of matrix theory in Appendix D may provide useful supplements to certain of the discussions in the text. This should be true especially for the student who has a better than usual background in mathematics.

From a mathematician's technical point of view, many of our statements are incomplete, poorly framed, or imprecise. On the other hand, many of these ideas can be grasped intuitively by the serious student, and the author feels that this intuitive understanding is better than no understanding at all, provided that the student understands the *limitations* of a presentation such as this.

0.3 THE ORGANIZATION OF THE TEXT

A glance at the table of contents reveals the topics covered, and there is little point in a detailed listing here. However, it should be pointed out that the chapters in this book fall roughly into two sections: Chapter 1 through 6 deal largely with the essential ideas of probability and of distributions, the two central notions of theoretical statistics. The first chapter lays a foundation for these topics by introducing very fundamental concepts of probability. A clear idea of these concepts can do a great deal to clarify the remainder of the book. Chapters 1 through 6 are very closely related in the topics covered, and each succeeding chapter builds on the concepts introduced in the preceding ones.

Chapter 7 develops some of the issues connected with the actual use of results from statistics, particularly the problem of making up one's mind from data. Chapters 10 through 17 discuss methods for various kinds of inferences to be made in different research situations. The methods are closely linked, as each is a special instance of a general linear model relating an observed value to factors that may influence it. Chapter 18 deals with qualitative data, and Chapter 19 gives some of the basic ideas of order statistics, an alternative approach to many problems.

A theme that runs throughout this book is the search for relationships. A statistical relation will be said to exist when knowledge of one property of an object or event *reduces our uncertainty* about another property that object or event will show. A statistical relation occurs when things tend to "go together" in a systematic way. This theme will recur repeatedly in the chapters to follow, but it is an important one.

Finally, Appendixes A and B, rules for the manipulation of summations and of expectations, are very important, since we will use these rules to considerable extent in our simple derivations of results. Appendix C concerns the principles underlying bivariate distributions, linear combinations of variables, and ways of handling arrays, or vectors, of numbers. These are more advanced matters that will be useful to know about in the later chapters of this book. Appendix D is devoted to matrix algebra, and Appendix E contains useful tables.

Many mathematical expressions occur throughout this text. These are of three kinds: one, algebraic equivalences serving as steps in some derivation; two, actual definitions or principles stated mathematically; and three, computational formulas useful in some method. Some of the mathematical expressions are numbered; ordinarily this occurs when some reference will be made to that expression at a later point. If the number for any expression is followed by an asterisk (*), then this is an important definition or relationship that is worthy of your special attention.

Every chapter is followed by a set of problems which cover most, though not necessarily all, of the concepts and the methods introduced in that chapter. An effort has been made to make these problems relatively interesting in content and to keep them at a level which will not be beyond the ability of a student who has read the text carefully. Solutions to odd-numbered problems may be found in Appendix F.

0.4 STATISTICAL PACKAGES FOR THE COMPUTER

For all but the very smallest sets of data, statistical analysis is now carried out by computer, and for individuals working in or near universities or other large research settings, this has come to mean the use of one or another of the comprehensive statistical packages for the computer. These packages are sets of computer programs, developed and marketed by companies on a proprietary basis. Such a package permits the user to create files of data, to manage and manipulate those files, and to analyze the file contents (or any set of data) through any of a wide variety of statistical techniques. These packages also have graphics capabilities, and some even offer report-writing among their options.

These statistical packages provide the advantage of high-speed analysis of large amounts of data, using very sophisticated modern methods, at very modest cost. Best of all, these packages require very little skill at using the computer. When you stop to consider that only a short time ago one had either to know a lot about computers or be prepared to spend a lot of money to gain these advantages of computing power, the true importance of the revolution brought about by the computer packages begins to emerge.

If you are using this text as part of a statistics course, particularly at the graduate level, you will almost certainly be expected to learn to use one or more of these packages as preparation for doing research on your own. On the other hand, this is a text about statistics, not about computer usage, and so our references to the packages can only be superficial and suggestive at best. Fortunately, most of the computer packages are fairly easy to use, and most have excellent user's manuals, which greatly clarify and simplify their use. In addition, university computation centers generally have consultants who are familiar with the packages and who are prepared to help the novice user with any problems.

There are a number of such packages designed for the large, or mainframe, computers generally found on university campuses and research centers, and these packages are widely available for use. However, just as there is, to date, no uniformity across campuses in terms of the kinds and capabilities of the computers that are in use, so too the statistical packages differ somewhat in their availability and in the particulars of their capacity and operation in different locations. You should familiarize yourself with such limitations, if any, as you start to learn to use a particular package.

Among the packages currently available, special mention must be made of MINITAB (Ryan, Joiner, and Ryan, 1976) since it is probably the simplest to learn to use. For this reason MINITAB is now very popular as an adjunct to statistics courses at both the graduate and undergraduate levels. Indeed, a number of textbooks exist that are organized around the MINITAB programs. However, MINITAB does not seem to offer quite the breadth of coverage nor reach the level of sophistication that is desirable for someone training to do doctoral level research. Thus, it seems more appropriate here to emphasize those packages that appear likely to be most servicable to the future researcher over a long span of his or her professional career and that will still prove useful when harder problems in data analysis present themselves.

There are three statistical packages that will be referred to in the remainder of this text: SPSS (and SPSS^x), BMDP, and SAS. Table 0.4.1 may be useful in giving you some idea about what these packages contain, and how the programs making up these packages coordinate with the material in this text.

Although each of these packages was first developed at a time when most users employed punched cards to give instructions and enter data to the computer, it is now more common to communicate with the computer through a CRT terminal and keyboard. The packages work in the same way, however, and one just substitutes a direct entry through the terminal for the information on a punched card. Furthermore, in most computation centers it is more economical to work in the so-called batch mode. In this mode the computer enters the request for an analysis into its own internal queue of work to be done; the output is then produced at some time (usually very quickly) after data entry. On the other hand, some of the statistical packages also lend themselves to an interactive mode, where the results of part of an analysis can be inspected before the next part is requested. This is generally a more expensive approach, however.

A very popular and widely available statistical package goes by the acronym SPSS,

Table 0.4.1

Computer procedures arranged by chapter topics

Book chapter	Topic	SPSS	BMDP	SAS
2	distributions graphs	FREQUENCIES GRAPHICS	2D 5D	FREQ CHART
4	central tendency variability	FREQUENCIES CONDESCRIPTIVE	2D	TTEST
8	inferences about means	T-TEST	3D	TTEST
10	analysis of variance	ONEWAY, ANOVA	1V	ANOVA
11	comparisons of means	ONEWAY	1V	ANOVA
12	multiway ANOVA	ANOVA	2V	ANOVA
13	random and mixed model ANOVA	MANOVA	2V,3V	ANOVA
14	regression and corre- lation	PEARSON CORR REGRESSION	8D,1R	REG
15	partial and multiple regression	PARTIAL CORR REGRESSION	1R,2R 6R	REG
16	unbalanced designs polynomial regression	ANOVA MANOVA	2V 5R	GLM
17	ANCOVA	ANOVA	2V	GLM
18	chi-square	CROSSTABS LOGLINEAR (SPSS[x])	4F	FREQ CATMOD
19	order methods	NONPAR TESTS NONPAR CORR	3S	NPAR1WAY

standing for *Statistical package for the social sciences* (Nie, Hull, Jenkins, Steinbrenner, and Bent 1975; Hull and Nie, 1981). A newer version, known as SPSS[x] (SPSS, Inc., 1986), contains a number of improvements on the original and is compatible with some computer configurations not suitable for the earlier SPSS package. Table 0.4.1 lists the major programs (or, more correctly, subprograms) in SPSS anad SPSS[x], and indicates how they coordinate with the coverage in this text. Each of the subprograms carries a name that is descriptive of the general type of analysis that it provides. Thus, for example, the SPSS subprogram *Frequencies* arranges your raw data into frequency distributions, and so on for other subprograms as outlined in Table 0.4.1. (The subprograms listed for SPSS and for the other packages are only a small selection of those actually available. These are included here simply because they happen to correspond fairly closely to the coverage in this text.)

Another very popular and useful statistical package for the computer is known as BMDP (Dixon, 1985). The acronym is derived from *Biomedical data programs,* and the package was formerly referred to simply as BMD. At present, BMDP consists of 40 programs that cover most of the well-known methods of statistical analysis, as well as many other techniques including scaling, factor analysis, and time-series analysis. The BMDP programs that have some relation to the contents of this text are listed in Table 0.4.1. Although the BMDP programs are distinguished by codes consisting of a number followed by a letter, it is fairly easy to tell the type of analysis that a given program yields. For example, the letter D symbolizes a program for relatively simple data management and data description; the letter R is used for methods that establish relationships among variables through correlation and regression, and so on.

The third of the trio of popular and useful statistical packages discussed here is SAS (1982). SAS is the most flexible and probably the most sophisticated of the widely used packages. It is especially strong in methods based on the general linear model, a concept which will be of major importance in our discussions, starting in Chapter 10. Table 0.4.1 shows a few of the general procedures that are available in SAS and how these could coordinate with the chapters to follow in this text.

In addition to programs for data analysis, such as those mentioned in Table 0.4.1, each of these packages has elaborate subprograms or procedures for file creation, data sorting, and data management. We will not be able to give any of the details of these uses of the computer programs here, but it should be realized that these are very important aspects of the statistics packages for the computer, which the researcher will want to learn to utilize to the fullest.

The statistical packages also differ widely from each other in how the data are to be entered, or files of data requested, how the variables are to be listed and formatted for input, and in many other operating details. Other operating details depend on the type of computer system you are using to run the package. It is not appropriate or even possible to go into these aspects of the packages and computer usage here; the manuals for use of the packages are helpful in some of these matters, but you will almost certainly require some instruction in the use of a given package in the available computer configuration before trying to strike out on your own.

Although the exercises that this book contains do not *demand* the use of a computer for their solution, and you can, with a certain amount of effort, solve all of these problems by hand, you are well advised to learn to use one or another of the computer packages while studying the text, and actually to solve some of the numerical exercises by computer. Computers are usually at their most efficient when handling large sets of data, but there is nothing to prevent their use for small sets such as those contained in the exercises in this book. The methods of input and the form of output will be just the same and just as instructive as if you were dealing with a great many more cases measured in a great many more ways.

0.5 STATISTICAL PACKAGES FOR MICROCOMPUTERS

A recent development is the widespread availability of personal computers or microcomputers that are powerful enough to handle fairly large problems in statistical analysis. There are now small statistical packages available for all personal computers that have at least 64K of memory. If you own such a computer, it is worthwhile checking software catalogs and with vendors to see what is available that may already be tailored to your special microcomputer set-up. Unless your computer has been upgraded to 256K or 512K, your packaged program may not handle large sets of data, and the speed of execution of the programs may be a little slow compared with the mainframes. Nevertheless, the real convenience of being able to use your own microcomputer can offset these disadvantages.

Because of the recent history of competition in the microcomputer industry, some of the best and most sophisticated of these smaller statistical packages have been developed for the IBM line of microcomputers, such as the IBM-PC and AT, and for other computers that are compatible with these. Indeed, as of this writing, the most recent development has been the release of personal computer versions of the three major packages discussed above: SPSS, BMDP, SAS, suitable for IBM/AT models with 512K memory. This means that in future the researcher will not necessarily be dependent on a mainframe installation to make use of the tremendous statistical computing power and data management capabilities that these packages give, provided that he or she has a relatively powerful microcomputer available. Since the future seems to be pointing toward the dominance of the microcomputer for handling data sets other than the largest, this seems another reason for emphasizing those packages which have already realized this capability.

0.6 WHY LEARN STATISTICS IN THE COMPUTER AGE?

Given that computers have now taken over much of the drudgery of computation from statistics and can provide answers to our questions far faster and with more accuracy than any mere human being, why then should we study statistics? In particular, why is it necessary to learn to do statistical analyses by hand, when in all likelihood you will have these done routinely on the computer?

The answer to the first part of this question is that the statistical packages are designed for use in a wide variety of situations and apply to all sorts of problems. This implies that a great many options and choices are open to the user of any of these packaged programs. There is no way that a person can use these packaged programs correctly and effectively without a pretty thorough understanding of the theoretical basis for the statistics being requested. In the absence of such knowledge, the choices presented and the options available may simply make no sense to the user, and can consequently be invitations to disaster.

The second part of this question can be answered by noting that statistics is a branch of *applied* mathematics. For a very large portion of the statistical methods used in research, the theory is important mostly as a justification for *doing* something in a certain way. If you really want to understand how the numbers fed into the computer as data are changed into the results listed on the computer printout, then you must have some idea about the mathematical and logical machinery that produces these conclusions. If you have no idea about the mechanics of the statistical method, then you are totally at the mercy of some anonymous computer programmer or of some statistical consultant when it comes to reaching conclusions from your data, just as you are at the mercy of the auto repairman if you know nothing about how an internal-combustion engine works. In a real sense, the meaning of a statistical technique actually *is* how it works, and thus we will continually emphasize this aspect of statistics, at least when it is practically feasible to carry out the steps that the computer will, in effect, also follow. With few exceptions, we will be able to do this for simple sets of data, and these data sets will usually be sufficient to give a feel for what the method does.

Finally, the computer sometimes gives a false sense of security to the user, who now finds it very easy to carry out quite complex analyses that would have been prohibitively laborious to do only a few years ago. However, just because the computer appears to do this simply and without visible effort does *not* mean that the analysis is conceptually simple, nor that the data necessarily fit the analysis chosen for use. You must be in a position to be thoughtful and *critical* in the wide choice of methods that will be offered. For any given set of data, the methods available cannot all be appropriate. You must acquire the background and experience to be able to pick and choose.

0.7 A WORD ON ROUNDING

In the social and behavioral sciences it is usual to round statistical results to a maximum of two decimal places beyond the number to be found in the original measurements. Thus, for measurements in whole units rounding is done to the nearest hundredth, for measurements in tenths rounding occurs to the nearest thousandth, and so on. Examples in this text will generally follow this practice. Typically, we too will round *final* results to two decimal places, unless the raw data definitely reflect a greater degree of accuracy, or there are specific reasons for doing otherwise.

On the other hand, statistical calculations frequently require several steps in which rounding could occur. Unfortunately, the practice of rounding during such intermediate steps in computation can make a decided difference in the final answer that one gets, depending on when and how the rounding takes place. For this reason it is usually good policy to carry a good number of decimal places (say, four or more) during the intermediate steps of statistical calculations, and then to round to one or two decimals at the end. Modern calculators, and certainly computers, make it easy to carry a fairly large number of decimals, without the need for rounding until the final answer is reached. This produces results that tend to be more accurate numerically than those found back in the old days of pencil and paper calculations.

Nevertheless, computers (and even calculators) are also having the bad effect of breaking down some of the standard conventions about accuracy in reporting statistical results. The fact that a computer output presents a result to an impressive six or eight decimal places makes us forget that such apparent accuracy is usually meaningless. As statistical analysis on the computer becomes almost universal, these long strings of decimals are finding their way into research reports, implying a degree of precision in measurement that is simply not present. As users of statistics, we should take care to round these results to reflect the accuracy present in the original numbers and not fill the research literature with these impressive, but really specious, strings of decimals.

To repeat: When you are calculating, avoid rounding during intermediate steps whenever possible, as you easily can on a good modern calculator, and as a computer does routinely. However, when you reach the final answer, try to round the numbers to about two decimal places, as appropriate.

Chapter *1*

Elementary Probability Theory

Statistical inference involves statements about probability. Everyone's vocabulary includes the words *probable* and *likely,* and most of us have some notion of the meaning of statements such as "The probability is one-half that the coin will come up heads." However, in order to understand the methods of statistical inference and use them correctly, one must have some grasp of probability theory. This is a mathematical system, closely related to the algebra of sets now taught in most elementary schools. Like many mathematical systems, probability theory becomes a useful model when its elements are identified with particular things, in this case the outcomes of real or conceptual experiments. Then the theory lets us deduce propositions about the likelihood of various outcomes, if certain conditions are true.

Originally, of course, the theory of probability was developed to serve as a model of games of chance—in this case the "experiment" in question was rolling a die or spinning a roulette wheel or dealing a hand from a deck of playing cards. As this theory developed, it became apparent that it could also serve as a model for many other kinds of things having little obvious connection with games, such as results in the sciences. One feature is common to most applications of this theory, however: the observer is uncertain about what the outcome of some observation will be, and must eventually infer or guess what will happen. In particular, the observer needs to know what will happen *in the long run* if observations could be made indefinitely.

In this respect the scientist is like the gambler keeping track of the numbers coming up on a roulette wheel. At any given opportunity only the tiniest part of the total set of things the scientist would like to know about can be observed. Given our human frailty as observers we must fall back on logic; *given* that certain things are true, we can make deductions about what *should* be true in the long run. The logical machinery of "in the long run" is formalized in the theory of probability. The scientist's statements about *all* observations of such and such phenomena are on a par with the gambler's; both are deductions about what should be true, if the initial conditions are met. Furthermore, the gambler, the businessman, the engineer, the scientist, and, indeed, every person must make decisions based on incomplete evidence. Each does so in the face of risk about how good those decisions will turn out to be. Probability theory alone does not tell any of these people how they should decide, but it does give ways of evaluating the degree of risk one takes for some decisions.

The theory of probability is erected on a few very simple concepts. Indeed, the main terms in the theory, the "events," are simply sets of possibilities. However, before we develop this idea further, we need to talk about the ways that events are made to happen: simple experiments.

1.1 SIMPLE EXPERIMENTS

Modern probability theory starts with the notion of a *simple experiment.* We shall mean nothing very fancy by the term *simple experiment,* and there is no implication that a simple experiment need be anything even remotely resembling a laboratory experiment. **A simple experiment is some well-defined act of process that leads to a single well-defined outcome.** Some simple experiments are tossing a coin to see whether heads or tails comes up, cutting a deck of cards and noting the particular card that appears on the bottom of the cut, opening a book at random and noting the first word on the right-hand page, running a rat through a **T** maze and noting whether it turns to the right or the left, lifting a telephone receiver and recording the time until the dial tone is heard, asking an individual's political preference, giving a person an intelligence test and computing the score, counting the number of colonies of bacteria seen through a microscope, and so on, literally without end. The simple experiment may be real (actually carried out) or conceptual (completely idealized), but it must be capable of being described and be repeatable (at least in principle). We also require that each performance of the simple experiment have one and only one outcome, that we can know when it occurs, and that the set of all possible outcomes can be specified. Any single performance, or trial, of the experiment must eventuate in exactly one of these possibilities.

Obviously, this concept of a simple experiment is a very broad one, and almost any describable act or process can be called a simple experiment. There is no implication that the act or process even be initiated by the experimenter, who needs function only in the role of an observer. On the other hand, it is essential that the outcome, whatever it is, be unambiguous and capable of being categorized among all possibilities.

1.2 EVENTS

The basic elements of probability theory are the possible distinct outcomes of some idealized simple experiment. The set of all possible distinct outcomes for a simple experiment will be called the **sample space** for that experiment. Any member of the sample space is called a **sample point,** or an **elementary event.** Every separate and "thinkable" result of an experiment is represented by *one and only one* sample point or elementary event. Any elementary event is *one possible result* of a single trial of the experiment.

For example, we have a standard deck of playing cards. The simple experiment is drawing out one card, haphazardly, from the shuffled deck. The sample space consists of the 52 separate and distinct cards that we might draw. If the experiment is stopping a person on the street, then the sample space consists of all the different individuals we might possibly stop. If the experiment is reading a thermometer under particular conditions, then the sample space is all the different numerical readings that the thermometer might show.

Seldom, however, is the *particular* elementary event that occurs on a trial of any special interest. Rather the actual outcome takes on importance only in relation to all possible outcomes. Ordinarily we are interested in the *kind* or *class* of outcome an observation represents. The outcome of an experiment is measured at least by allotting it to some qualitative class. For this reason, the main concern of probability theory is with *sets* of elementary events. **Any set of elementary events is known simply as an event, or an event class.**

Imagine, once again, that the experiment is carried out by drawing playing cards from a standard pack. The 52 sample points (the distinct cards) can be grouped into sets in a variety of ways. The suits of the cards make up four sets of 13 cards each. The event "spades" is the set of all card possibilities showing this suit, the event "hearts" is another set, and so on. The event "ace" consists of four different elementary events, as does the event "king," and so on.

If the experiment involves noting the eye color of some person stopped on the street, the experimenter may designate seven or eight different eye-color classes. Each such set is an event. The event "blue eyes" is said to "occur" when we encounter a person who is a member of the class "blue eyes." If, on the other hand, we find the weight of each person we stop, then there may be a vast number of "weight events," different weight numbers standing for classes into which people may fall. If a person is stopped who weighs exactly 168 pounds, then the event "168 pounds" is said to occur.

In short, **events are sets, or classes, having the elementary events as members.** The elementary events are the raw materials that make up event classes. The occurrence of any member of event class *A* makes us say that event *A* has occurred. Since *any* subset of the sample space is an event, then some event *must* occur on each and every trial of the experiment.

1.3 EVENTS AS SETS OF POSSIBILITIES

The elementary events making up any sample space are, as we have seen, those separate, distinct, and indentifiable outcomes to a given simple experiment. However, as we have also seen, interest is usually focused on groupings or sets of elementary events, according to one or more characteristics that the elementary events show. Thus, it is convenient to use some of the language of sets in discussing events.

In the following discussion the symbol S will be used to stand for the sample space, which is the set of all possible elementary events for the simple experiment. Then capital letters such as *A, B,* and so on will represent events, each of which is a subset, or subgrouping, of elementary events in S. For a given simple experiment, every event discussed must be a subset of S.

The set S may contain either a finite or an infinite number of elementary events. In this initial discussion the number of elements in S may be assumed to be finite. Nevertheless, in many of the most important parts of the theory of probability the sample space is assumed to be infinitely large.

What are the possible events making up a sample space S? First of all, the set S is an event: S is the "sure" event, since it is bound to occur (that is, some elementary event must occur on every trial of the simple experiment).

The event $\varnothing$ is called "the impossible event," since it cannot occur. However, the "null set" $\varnothing$ is defined to be a subset of every other set. Hence, it is quite possible to think even of $\varnothing$ as an event, or subset of S.

Now suppose that there are two events, A and B, each composed of elementary events in S, and thus each subsets of S. Then we can imagine an outcome to the simple experiment which is a member *both* of A and of B. In this instance we say that the event "A and B" has occurred. The outcome to the experiment qualifies both as an event A and as an event B.

Similarly, if A is an event and B is an event, we can imagine an outcome which is *either* an event A *or* an event B. Such an outcome represents an occurrence of the event "A or B." This event occurs when the elementary event qualifies as an A, or it qualifies as a B, or when the elementary event qualifies *both* as an A and a B. Notice that the occurrence of "A and B" is thus automatically an occurrence of the event "A or B." However, the reverse is not true. In discussions of such *joint events*, a little of the notation of set theory becomes useful. We shall use the symbol ∩, read as *intersection*, to stand for the word *and* in discussing an event such as "A and B." Thus,

(A and B) = (A ∩ B).

The symbol ∪, read *union*, will stand for the word *or* when we discuss events such as "A or B." Therefore,

(A or B) = (A ∪ B).

Suppose that there is some simple experiment with sample space S; also suppose that A is some event in that sample space. Now an elementary event occurs which does *not* qualify for the event A. Then we say that the event "not A" has occurred. The event "not A" is often symbolized by $\overline{A}$, the letter standing for the event, but with a horizontal bar above it. Given any event such as A, then the event $\overline{A}$ is called the *complement* of the event A. Every event in any sample space S will have a complement

A very useful way to discuss a sample space S and the events in that space is by use of a group of figures know as Venn diagrams, after the logician J. Venn (1834–1923). Figure 1.3.1 pictures the sample space S as a rectangle. Let us imagine that each and every point within the rectangle is a possible elementary event in S. Now within the rectangle S there is a circle marked A. This circle symbolizes the event A, and every point within A represents an elementary event which is an occurrence of the event A as well. Now look at Figure 1.3.2. Here there is once again the sample space S, shown as a rectangle, and also two events, A and B. Every point in circle A represents an elementary event qualifying as an A event, and every point within B is a possible ele-

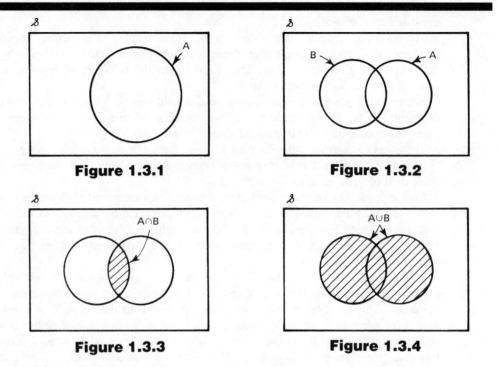

Figure 1.3.1   **Figure 1.3.2**

Figure 1.3.3   **Figure 1.3.4**

mentary event qualifying as a *B*. Notice that some events qualify *both* for event *A* and for event *B*. This is the set making up the event *A* ∩ *B*. This event *A* ∩ *B* is shown as the shaded area in Figure 1.3.3.

Recall that the event *A* ∪ *B* includes all elementary events that qualify for *A*, or for *B*, or for both. The Venn diagram of Figure 1.3.4 shows the event *A* ∪ *B* as a shaded area. Notice that when an elementary event qualifies for *A* ∩ *B* it automatically qualifies for *A* ∪ *B* but that the reverse is not true, since it is entirely possible for an elementary event to qualify for *A* ∪ *B* even though it does not qualify for *A* ∩ *B*.

The complementary event $\overline{A}$ for event *A* is shown as the shaded area in Figure 1.3.5. An elementary event qualifying as a member of $\overline{A}$ is a member of $\mathcal{S}$, of course, but not a member of *A*. Figure 1.3.6 shows the event $A \cap \overline{B}$, or "*A* and not *B*." This event includes all elementary events that qualify for *A* but do not qualify for *B*.

An important situation exists when it is impossible for an elementary event to qualify both for event *A* and for event *B*. In this case, *A* ∩ *B* = ∅, as shown in Figure 1.3.7. Two sets such as *A* and *B* in this figure, for which *A* ∩ *B* = ∅, are said to be *mutually exclusive:* if one of these events occurs, the other event cannot occur.

It is possible to have three or more different events in the same sample space $\mathcal{S}$. For example, the three circles in Figure 1.3.8 stand for the three events *A*, *B*, and *C* in the same sample space $\mathcal{S}$. The shaded portion of this figure represents the event *A* ∪ *B* ∪ *C*, which means "*A* or *B* or *C*." Notice that in this example an elementary event might qualify for events such as *A* ∩ *B*, *A* ∩ *C*, *B* ∩ *C*, *A* ∩ *B* ∩ *C*, and so on for all of the events *A*, *B*, *C*, as well as their complementary events.

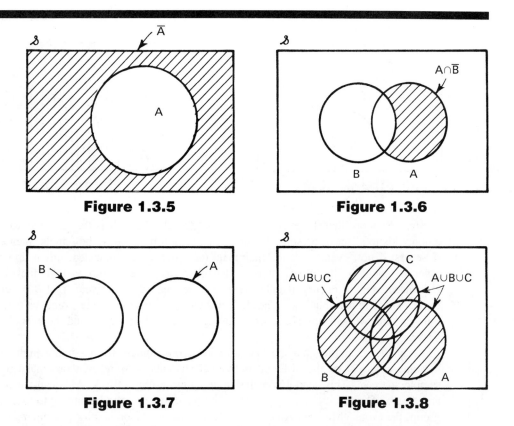

Figure 1.3.5

Figure 1.3.6

Figure 1.3.7

Figure 1.3.8

The use of Venn diagrams such as those shown above can sometimes be helpful in thinking about probabilities and even in simple probability calculations. It is good practice to learn to visualize events within a sample space in this way, especially when one is dealing with a number of different possible events.

A Venn diagram can also help to explain another pair of concepts that occur frequently in probability theory. We have just seen that two events are said to be mutually exclusive if no elementary event can qualify for their intersection, so that $A \cap B = \varnothing$. More than two events such as A, B, and C may also be mutually exclusive. This means simply that any *pair* of these events, such as A and B, or B and C, are themselves mutually exclusive. This situation is illustrated by the shaded circles in Figure 1.3.9. In addition, two or more events may be described as *exhaustive*, which means that each elementary event in S must qualify for one (or more) of these events. Three *mutually exclusive and exhaustive* events $\{A, B, C\}$ are shown in Figure 1.3.10. Such a set of mutually exclusive and exhaustive events is called a *partition* of the sample space S. A glance at Figure 1.3.10 should show you why this name is appropriate.

Let us take a concrete example of some events. Suppose that we had a list containing the name of every living person in the United States. We close our eyes, point a finger at some spot on the list, and choose one person to observe. The elementary event is the actual person we see as the result of that choice, and the set S is the total set of possible

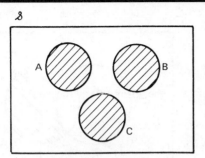

Figure 1.3.9

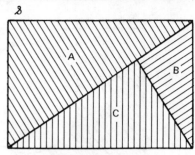

Figure 1.3.10

persons. Suppose that the event A is the set "female," and B is the set "redheaded," among this total set of persons. If our chosen person turns out to be female, then event A occurs; if not, event $\overline{A}$. If the individual turns out to be redheaded, this is an occurrence of event B; if not, event $\overline{B}$. If the observation shows up as both female and redheaded, $A \cap B$ occurs; if either female or redheaded, then event $A \cup B$ occurs. If the person is female but not redheaded, this is an occurrence of the event $A \cap \overline{B}$. One thing is sure: we will observe a person living in the United States, and the event S must occur.

As another example, imagine that we choose one student from a junior high school in some city. All of the different individual students we might choose make up the sample space S. This school contains three ethnic groups: Black, Mexican-American, and Anglo. The event "Black" occurs when we draw a person belonging to that ethnic group, the event "Mexican-American" occurs when a person is drawn belonging to that classification, and so on. Let us suppose that the three ethnic groups are mutually exclusive and exhaustive, so that for instance, a person cannot be both Black and Mexican-American, and each student must belong to one of the three groups. Then the three events form a partition of the sample space S. (Try to visualize both of these examples in terms of a Venn diagram such as those in the preceding figures.)

In short, each and every subset that can be formed from the elementary events in a sample space S is an event. How many events then are there for a given sample space? It can be shown that if a sample space contains exactly N elementary events (meaning that there are exactly N discernibly different outcomes to the simple experiment), then there are exactly 2^N different events, or subsets, in S, including both $\emptyset$ and S itself. This group of all possible subsets of the sample space S is known as the *family of events* in S. This concept of the family of events in some sample space S is useful when we wish to talk about some event that is possible in this S but without having to specify exactly which one. We will use this idea in the next section.

1.4 PROBABILITIES

Formally, probabilities are just numbers assigned to each and every event in some sample space. That is, each possible outcome or set of outcomes for a simple experiment is assigned a number, the probability of that particular outcome or event. As we have seen, sets of outcomes are called events, and so each possible event, such as A, is given a number $p(A)$, the probability of event A. Thus,

$p(A) = $ probability of event A.

Recall that the entire sample space, the set of all possible distinct outcomes of the simple experiment, is denoted by the symbol S. Even the set S itself can be thought of as an event ("some outcome occurred"), and the event S is associated with probability $p(S)$.

On the other hand, some outcomes are simply impossible in any given simple experiment. These impossible outcomes are symbolized by the event $\varnothing$, with probability $p(\varnothing)$.

In exactly this same way any other event which is a subset of S is associated with a probability. Thus, if there are two events A and B, both subsets of S, then we can speak of probabilities such as $p(A)$, $p(B)$, $p(A \cap B)$, $p(A \cup B)$, and so on. Exactly this same idea applies to any event which is a part of the family of events that can be constructed from the elementary events of S. Each and every such event has a number, its probability.

We are now ready for a somewhat more formal definition of these numbers known as probabilities. Modern probability theory is constructed from a small group of basic statements, or axioms, giving the properties that these probability numbers must exhibit. Although a truly sophisticated treatment of this topic is beyond our scope, a somewhat simplified version of these axioms can be stated as follows:

DEFINITION: Given the sample space S, and the family of events in S, a probability function associates with each event A a real number $p(A)$, the probability of event A, such that the following axioms are true:

1. $p(A) \geq 0$ for every event A.
2. $p(S) = 1.00$
3. If there exists some countable set of events, $\{A_1, A_2, \cdots, A_N\}$, and if these events are all mutually exclusive, then $p(A_1 \cup A_2 \cup \cdots \cup A_N) = p(A_1) + p(A_2) + \cdots + p(A_N)$. The probability of the union of mutually exclusive events is the sum of their separate probabilities.

(In this connection, the word *countable* means simply that it is possible to associate each and every distinct event in the set with one and only one of the "natural," or counting, numbers. Later, when we deal with uncountably infinite sets of events, the definition will be extended accordingly.)

In essence, this definition states that paired with each event A is a nonnegative number, probability $p(A)$, and that the probability of the sure event S, or $p(S)$, is always 1.00. Furthermore, if A and B are any two *mutually exclusive* events in the sample space, the probability of their union $p(A \cup B)$ is simply the sum of their two probabilities $p(A) + p(B)$. In the same way, if A, B, and C are all mutually exclusive, then $p(A \cup B \cup C)$ is equal to $p(A) + p(B) + p(C)$, and so on for any number of mutually exclusive events.

It is important to remember at this stage that this is a purely formal definition of probability, in terms of numbers associated with sets of possibilities making up a sample space $\mathcal{S}$. *Events* have probabilities, and in order to discuss probabilities we must always specify the events to which these probabilities belong. When we speak of the probability that a citizen of the United States has red hair, we are referring to the probability number associated with the event "red hair" in the sample space, "all citizens of the United States at a given moment in time." Similarly when we say that the probability that a coin will come up heads is .50, we are stating that the number .50 is paired with the event "heads" in the sample space "all possible results of tossing a particular coin."

At first this may seem an extremely unmotivated and arbitrary way to discuss probability. Everyone knows that the word *probability* means more than a mere number assigned to a set, and we shall certainly give these probability numbers additional meaning in sections to come. For the present, however, let us accept this formal definition at face value.

1.5 SOME SIMPLE RULES OF PROBABILITY

The concepts of simple experiment, sample space, and events, taken together with the three axioms given above, can be used to deduce any number of new consequences. These consequences, supplemented by new definitions and further consequences or theorems, are what is meant by the theory of probability. Each of these mathematical results gives some relationship that probability numbers must exhibit. Along with the three basic axioms, these derived relationships can be thought of as the rules of probability. A few of these rules, which follow directly from the axioms, will now be discussed.

Recall that for any event A there will also be a complementary event "not A," or $\overline{A}$. Then for any event A in $\mathcal{S}$,

---

Probability rule 1: $p(\overline{A}) = 1 - p(A)$

(This is sometimes called *the rule of complementary probability*.) This simply says that the probability of the event $\overline{A}$ (not A) is one minus the probability of the event A. Why should this be true? First notice that the two events A and $\overline{A}$ are exhaustive, since logically any elementary event must be a member of one of these classifications. Consequently, $A \cup \overline{A} = \mathcal{S}$. Note further that A and $\overline{A}$ are mutually exclusive, since an elementary event cannot be a member of both A and $\overline{A}$. On putting these ideas together with Axioms 3 and 2 we have

$$p(A \cup \overline{A}) = p(A) + p(\overline{A}),$$

and
$$p(A \cup \overline{A}) = p(\mathcal{S}),$$

so that $p(A) + p(\overline{A}) = 1.00,$

and thus $p(\overline{A}) = 1 - p(A).$

Within what range of values must any probability lie? The answer is given by

---

Probability rule 2: $0 \leq p(A) \leq 1.00$ for any event A.

(this may be thought of as *the probability range*.) In other words, only positive numbers lying between 0 and 1, inclusive, may be used to signify probabilities.

This too is easy to deduce by use of the axioms and Rule 1. We have just seen that the probability of an event A plus the probability of its complement $\overline{A}$ must sum to 1.00. Now, just suppose that Rule 2 were not true, and that some event A *could* have a probability greater than 1.00. However, then event $\overline{A}$ would have to have a probability less than 0 in order for Rule 1 to be true. This would contradict Axiom 1, which says that no event may have a probability less than 0. Since axioms may not be contradicted, then Rule 2 must be true.

We already know from Axiom 2 that $p(\mathcal{S})$ must have the value 1.00. However, what event has probability of zero? This is given by

---

Probability rule 3: $p(\varnothing) = 0$, for any $\mathcal{S}$.

(You may wish to think of this as the *rule of the impossible event*.) That is, the impossible event $\varnothing$ always receives probability 0.

This is also easy to demonstrate. For any simple experiment, the sample space $\mathcal{S}$ contains all of the possible outcomes or elementary events. For an outcome to occur which is not in $\mathcal{S}$ is impossible. Therefore, the event "not $\mathcal{S}$ ", or $\overline{\mathcal{S}}$, is the same as $\varnothing$. From Rule 1 it is known that

$$p(\mathcal{S}) + p(\overline{\mathcal{S}}) = 1.00.$$

It is also known from Axiom 2 that $p(\mathcal{S}) = 1.00$. Hence $p(\varnothing) = 0$.

A very important rule applies for any two events A and B, whether or not they are mutually exclusive. In other words, this rule applies even if $p(A \cap B)$ is not equal to zero. Thus,

---

Probability rule 4: For any two events A and B in $\mathcal{S}$,
   $p(A \cup B) = p(A) + p(B) - p(A \cap B)$.

(This is often called the *"or" rule of probability*.) That is, the probability of the event "A or B" is always equal to the probability of event A plus the probability of event B, minus the probability of the event "A and B." Look at Figure 1.3.4 and see if you can tell why this must be true.

Notice that Axiom 3 provides a special case of the "or" rule: When A and B are mutually exclusive, then $p(A \cap B) = 0$, and $p(A \cup B) = p(A) + p(B)$.

(It is also possible to generalize Rule 4 to three or more events and we shall do so in Section 1.11.)

Finally, a useful rule deals with a *partition* of the sample space $\mathcal{S}$. Recall that a partition of the sample space $\mathcal{S}$ consists of two or more events, which are all mutually exclusive, and which exhaust the possibilities in $\mathcal{S}$. Suppose there were a set of two or more events in $\mathcal{S}$, represented by A for the first event, and by L for the last event. Then

---

Probability rule 5: If the set of events $A, \cdots, L$ constitute a partition of $\mathcal{S}$, then
$$p(A \cup \cdots \cup L) = p(A) + \cdots + p(L) = 1.00.$$

(If you like, this may be described as the *partition rule*.) In other words, when any set of events are mutually exclusive and exhaustive, and thus form a partition of $\mathcal{S}$, the sum of their probabilities must be equal to 1.00.

It is a simple matter to illustrate these elementary rules of probability. Take once again the example of drawing playing cards one at a time from a standard deck which is well shuffled before each draw. The sample space $\mathcal{S}$ here consists of all of the 52 different cards that we might draw. Some of the possible events are "the card has the number 10," "the card is a diamond," "the card is a queen of spades," "the suit of the card is red," "the card is an odd number in a black suit, but not a picture card," and so on. Some outcomes cannot occur: the event "card is a gorilla" is (or surely ought to be) impossible, and thus is the same as $\varnothing$.

Suppose that the event A in this example is "card is a club," and that we happen to know in this example that $p(A) = 1/4$. What is the event $\overline{A}$? This would be the event "card is not a club," and we could find its probability by use of Rule 1 by taking $p(\overline{A}) = 1 - 1/4 = 3/4$.

If the event A is taken to be "card is a king," and the event B is "card is a jack," then we could also consider the event $A \cup B$, "the card is a king or a jack." For this playing-card experiment it happens that $p(A) = 1/13$ and that $p(B) = 1/13$. However, a card cannot be both a king and a jack, so that $p(A \cap B) = 0$. Then by Axiom 3,

$$p(\text{king or jack}) = \frac{1}{13} + \frac{1}{13} = \frac{2}{13}.$$

On the other hand, suppose that event A is "card is a king," and that event C is "card is a heart." Here it happens that $p(A) = 1/13$, and $p(C) = 1/4$. Furthermore, $p(A \cap C)$ is not zero here, but rather is 1/52. Then, in order to find $p(A \cup C)$, we use Rule 4 to obtain

$$p(\text{king or heart}) = \frac{1}{13} + \frac{1}{4} - \frac{1}{52} = \frac{16}{52}.$$

What is the probability of observing a "seven," an "eight," or a "nine" for the card drawn? Among playing cards each of these events has probability of 1/13, and, since these are mutually exclusive events, we know from Axiom 3 that

$$p(\text{seven or eight or nine}) = \frac{1}{13} + \frac{1}{13} + \frac{1}{13} = \frac{3}{13}.$$

For the simple experiment of drawing playing cards, the four suits, (spades, hearts, diamonds, and clubs) form a partition of the sample space. That is, these suit events are mutually exclusive, since if a card belongs to one suit it cannot belong to any other. Furthermore, the suits are exhaustive: each card drawn from a standard deck must belong to one and only one of these four suits. Then, by Rule 5,

$$p(\text{spades}) + p(\text{hearts}) + p(\text{clubs}) + p(\text{diamonds}) = 1.00.$$

Many more examples of these simple manipulations are given in the exercises at the end of this chapter.

1.6 EQUALLY PROBABLE ELEMENTARY EVENTS

So far probabilities of events have been discussed only in a rather formal sense, and we have not really gone into how these probabilities are to be calculated nor what these probability numbers mean. Now we turn to the problem of calculating probabilities for a given simple experiment.

Bear in mind that each distinct outcome making up the sample space S can itself be thought of as an event and that each of these elementary events is associated with a probability. Any pair of elementary events must be mutually exclusive since each is a distinct outcome. Now suppose that for some simple experiment we were actually given the probability for each elementary event. Then we would be able to calculate the probability for any event A: The probability of any event A in the sample space S is simply the *sum* of the probabilities of all of the elementary events in S qualifying for the event A. In other words, the probability of event A is nothing more than the sum of the probabilities of the elementary events that A contains. This follows directly from Axiom 3.

This principle for finding the probability of an event is certainly simple enough, but there is one obvious hitch: How does one know the probabilities of the elementary events in the first place? Fortunately, a great many of the simple experiments for which we need to calculate probabilities, and particularly games of chance, have a feature that does away with this problem. A great many simple experiments are conducted in such a way that it is reasonable to assume that each and every distinct elementary event has the *same probability*. When tickets are drawn in a lottery, for example, great pains are taken to have the tickets well shaken up in a tumbler before each one is drawn; this mixing operation makes it reasonable to believe that any particular ticket has the same chance of being drawn as any other. Cards are thoroughly shuffled and cut, perfectly balanced dice are thrown in a dice-cage, and, in fact, almost all gambling situations have some feature which makes this equal-chances assumption reasonable. As we shall see presently, even experiments that are not games of chance also are carried out in such a way that each and every elementary event should have the same probability.

When there is some finite number of elementary events in a sample space, where each elementary event has exactly the same probability, then the probability of any event is particularly easy to compute. Imagine a sample space containing N distinct elementary events. For any event A, let the number of elementary events that are members of A be denoted by $n(A)$. The following principle applies:

If all elementary events in the sample space $\mathcal{S}$ have exactly the same probability, then the probability of any event A is given by

$$p(A) = \frac{\text{number of elementary events in } A}{\text{total number of elementary events in } \mathcal{S}}$$

$$= \frac{n(A)}{N}.$$

For equally likely elementary events, the probability of an event A is its relative frequency in the sample space.

As an illustration, suppose that a box contains 10 marbles. Five of these marbles are white, three are red, and two are black. We perform the simple experiment of drawing a marble out of the box (without looking) in such a way that the marbles are well mixed up, and that there is no reason for any given marble to be favored in our drawing. Now we can identify our experiment with the model of probability we have been discussing. An outcome is the result of our drawing a marble and looking at it. An elementary event is a particular marble, in this instance, and there are exactly 10 distinct marbles; hence there are 10 elementary events making up the sample space $\mathcal{S}$. The marble events are mutually exclusive, and each has probability 1/10.

We are concerned with the three events: "white," "red," and "black." Notice that these three events are subsets of $\mathcal{S}$, containing five, three, and two elementary events (marbles), respectively. What is the probability of our drawing a red marble? The answer is given by

$$p(\text{red}) = \frac{\text{number of red marbles}}{\text{total number of marbles}}$$

$$= \frac{3}{10}$$

so that we may say correctly that the probability of the event "red" in this experiment is .30. In the same way, one can find

$p(\text{white}) = .50$, and $p(\text{black}) = .20$.

Furthermore, it is easy to see, by applying the rules of Section 1.5 that

$p(\text{no color}) = .00$

$p(\text{red or white}) = .30 + .50 = .80$

$p(\text{red and white}) = .00$

$p(\text{red or black}) = .30 + .20 = .50$

$p(\text{red or white or black}) = .30 + .20 + .50 = 1.00$

and so on, for any other event.

Consider another example. A teacher has 30 children in a classroom. In some completely "random" and unsystematic way, the teacher chooses a child and notes the father's occupation. The children are each equally likely to be chosen by the teacher. In this example the elementary events are the different children who might be chosen: there are 30 such elementary events possible, making up the sample space $\mathcal{S}$. Now suppose that there are only four classes of occupation represented in the room: professional, white-collar, skilled labor, and unskilled labor. These four classes will be labeled events A, B, C, and D. If 3 children represent group A, 15 group B, 10 group C, and 2 group D, what is the probability that a child chosen will fall into a given group? In other words, what are the probabilities of the four different events? Once again, these probabilities are given by the relative frequencies:

$$p(A) = \frac{\text{number of elementary events in } A}{\text{total number of elementary events}}$$

$$= \frac{3}{30} = .10.$$

In the same way we find:

$$p(B) = \frac{15}{30} = .50$$

$$p(C) = \frac{10}{30} = .33$$

$$p(D) = \frac{2}{30} = .07$$

Finally, suppose that the simple experiment consisted of giving some poor unprepared graduate student a choice of six essay topics on an examination. These topics are numbered in order from 1 through 6. Suppose further that the student knows nothing about any of these topics, so that they are all equally likely to be picked. Then the probability that item 1 is picked should be 1/6, item 2 has probability of 1/6, and so on. What then is the probability that the student picks an odd-numbered topic? This event occurs when the choice falls on topic 1, topic 3, or topic 5. Consequently, the desired probability is

$$p(1 \text{ or } 3 \text{ or } 5) = p(\text{odd number}).$$

Since these events are mutually exclusive, the probability is found from Axiom 3 to be

$$p(1 \text{ or } 3 \text{ or } 5) = \frac{1}{6} + \frac{1}{6} + \frac{1}{6} = .50.$$

What is the probability that the student chooses the first or the last topic? This event involves the choice of topic 1 with probability of 1/6, or topic 6, also with a probability of 1/6. Thus

$$p(\text{first or last topic}) = \frac{1}{3}.$$

In summary, the simplest approach to the calculation of probabilities rests on the assumption that all of the elementary events in a given sample space S have the *same* probability. However, there is nothing in the formal theory of probability that makes this assumption necessary, and one may talk perfectly well about probabilities even in situations where this assumption may not be true. Even so, the calculation of probabilities of events is made very much simpler when this assumption is made, and in virtually all that follows we will confine our discussion to this case.

1.7 "IN THE LONG RUN"

So far we have seen how the outcomes of an experiment are identified with the elementary events in a sample space, and how probability values may be assigned to each event. In the situation where the elementary events are themselves equally probable, then the probability of any event is just its relative frequency in the space. However, we still have not reached an answer to the question of what these probabilities mean. How are probabilities related to the things one observes?

In the first example of the preceding section, if one were to keep drawing marbles out of the box and after each draw were to replace the marble in the box before drawing again, then in the *long run,* after very many observations, what proportion of the marbles drawn should be red? It should take only a very little thought to arrive at the conclusion that one should *expect* about 3 in 10 such observations to be red. In the same way, in the second example, suppose that the teacher keeps choosing children in this unsystematic way, and any given child can be chosen over and over again. If each child is equally likely to be chosen at any time, then it seems reasonable that in the long run, over a large number of such observations, about .10 or 1 in 10 of the children chosen should have fathers in occupation group A.

In the third example, suppose that many students were asked to choose one essay from the six topics, and that the topics were equally likely to be chosen by any student. Then the first topic should be chosen by about 1/6 of the students, the second by 1/6, and so on. Similarly, we should expect about half of the students to choose an odd-numbered topic, and about 1/3 to choose either the first or last topic.

Despite the different contexts of these three examples, they all illustrate the same principle:

The probability of an event denotes the relative frequency of occurrence of that event to be expected in the long run.

Simple examples of this sort all involve equally likely elementary events, and in many applications of probability theory, especially to games of chance, this assumption is made. On the other hand, the same idea applies even when elementary events are *not* equally likely: *in the long run, the relative frequency of occurrence approaches the probability for any event.*

Thus, the idea of relative frequency is connected with probability in two ways:

1. If elementary events are equally probable, the probability of an event is its relative frequency in the sample space.
2. The *long-run* relative frequency of occurrences of event A over trials of the experiment *should approach* p(A). This is true regardless of whether elementary events are equally probable or not, provided that observations are made independently and at random.

(Shortly we will give more formal meanings to the terms *independently* and *at random*. However, for the moment think of *independent* as meaning "each has nothing to do with any other," and *at random* as meaning, "by chance alone.")

The connection between probability and long-run relative frequency is both a simple and an appealing one, and it does form a tie between the purely formal notion, "the probability of an event," and something we can actually observe, a relative frequency of occurrence. A statement of probability tells us what to *expect* about the relative frequency of occurrence of an event given that enough sample observations are made at random: In the long run, the relative frequency of occurrence of event X should approach the probability of this event, if independent trials are made at random over an indefinitely long sequence. As it stands, this idea is a familiar and reasonable one, but it will be useful to have a more exact statement of this principle for use in future work. The principle was first formulated and proved by James Bernoulli in the early 18th century, and it often goes by the name "Bernoulli's theorem." A more or less precise statement of Bernoulli's theorem goes as follows:

If the probability of occurrence of the event X is p(X), and if N trials are made, independently and under exactly the same conditions, then the probability that the relative frequency of occurrence of X differs from p(X) by any amount, however small, approaches zero as the number of trials grows indefinitely large.

In effect, the theorem says this: Imagine some N trials made at random and in such a way that the outcome on any one trial cannot possibly influence the probability of outcomes on any other trial; that is, N independent trials are made. This act of taking N independent trials of the same simple experiment may itself be regarded as another experiment in which the possible outcomes are the numbers of times event X occurs out of N trials. Each possible such outcome has a probability. Then as N becomes very large, the probability that the relative frequency of X differs from p(X) in any way becomes very small. This does not mean, however, that the proportion of X occurences among any N trials *must* be p(X); the proportion actually observed might be any number between 0 and 1. Nevertheless, given more and more trials, the relative frequency of X occurrences may be expected to come closer and closer to p(X).

In practical terms, Bernoulli's theorem says that even if we only have a limited number of trials, we should expect the probability of any event to be reflected in the relative frequency we actually observe for that event. In the long run, such an observed relative frequency should approach the true probability. Although Bernoulli's theorem does work for us in this way, it does so not because of any necessary compensation for early "misses" by "hits" later on. Rather, the theorem holds because departures from what one expects to occur are simply "swamped out" as the total number of trials or observations becomes very large.

Please do not fall into the error of thinking that an event is ever due to occur *on any given trial*. If you toss a coin 100 times, you expect about 50% of the tosses to show the event "heads," since the theoretical probability of that event for a fair coin is .50. This does not mean, however, that for any given 100 trials (or one thousand, or million, or billion trials) the coin must show 50% heads. This need not be true at all. Every one of your 100 tosses *could* result in the event heads, or none of them might result in this event—the coin does not ever have to come up heads in any finite number of tosses. Only in an infinite number of tosses must the relative frequency equal .50. The only thing which we can say with assurance is that .50 is the relative frequency of heads we should *expect* to observe in any given number N of tosses, and that it is increasingly probable that we observe close to 50% heads as the N grows larger. But on any finite number of tosses of the coin, the relative frequency of heads observed can be anything. This same is true for the occurrence of any event in any simple experiment in which observations are made at random and in which the probability $p(X)$ of the event X is other than 1 or 0.

Although it is true that the relative frequency of occurrence of any event must exactly equal the probability only for an infinite number of independent trials, this point must not be overstressed. Even with relatively small numbers of trials we have very good reason to expect the observed relative frequency to be quite close to the probability. The rate of convergence of the relative frequency to the probability is very rapid, even at the lower levels of sample size, although the probability of a small discrepancy between relative frequency and probability is much smaller for extremely large than for extremely small samples. A probability is not a curiosity requiring unattainable conditions but rather a value that can be estimated with considerable accuracy from a sample. Our best bet about the probability of an event is the actual relative frequency we have observed from some N trials, and the larger N is, the better the bet.

1.8 AN EXAMPLE OF SIMPLE STATISTICAL INFERENCE

Imagine that you have a friend who claims to possess the power of extrasensory perception (ESP). You have some doubts about this, but wish to check out this claim. As it happens, there is a standard deck of cards often used in such experiments. This deck is composed of cards printed with circles, squares, and triangles. There are, lets say, 60 cards in all, with 20 of each kind.

The experiment goes as follows: You shuffle the cards thoroughly, and draw one card unsystematically from the deck, completely out of sight of the subject. You then concentrate intently on the figure for the card drawn, and your friend uses the alleged ESP to guess the figure represented. Any card drawn is then replaced, and the experiment repeated for as many trials as desired.

The sample space of elementary events can be represented by all combinations of card drawn and figure guessed, as shown below:

		Figure guessed		
		Circle	Square	Triangle
	circle	1/9	1/9	1/9
card	square	1/9	1/9	1/9
	triangle	1/9	1/9	1/9

If nothing but chance is going on in this experiment, it is reasonable to suppose that each of these card-guess events has the same probability, or 1/9. Furthermore, the event "correct" since it involves three of these possibilities, has probability p(correct) = 1/3.

On the other hand, suppose that ESP actually is operating here. This should lead to more correct guesses than otherwise, or more than 1/3 correct. In other words, there are really two theories being compared: "no ESP," represented by the situation with p(correct) = 1/3, and "some ESP," representeed by p(correct) > 1/3.

Now let us say that you make a series of trials of the experiment, where a card is drawn and a guess made on each trial. Suppose that your friend is correct on the first trial. Does this say anything about ESP? Not very much, since we expect one out of three trials to be correct anyway, even without ESP. However, suppose that you keep making trials, so that after 12 cards have been drawn, there are eight correct guesses. Here, one expects 1/3 of 12, of only four correct guesses, so that this result of eight correct is somewhat above the expected number. Such a result is not very likely to have occurred when p(correct) = 1/3, but is more likely when p(correct) > 1/3. The data so far accord better with the theory "some ESP" than with "no ESP." Still, you keep going until, perhaps, 900 trials have been completed. Here imagine that 600 correct guesses out of 900 trials were recorded. This is *very* far from the (900)1/3 or 300 correct expected by chance alone, and is very unlikely to have occurred if the true probability of a correct guess is 1/3. Your friend's claim begins to look pretty good at this point. Your best guess about p(correct) is 2/3, not 1/3.

Nevertheless, 600 correct out of 900 trials is not impossible even when the probability of a correct guess is only 1/3. Perhaps your friend simply got lucky. Regardless of how many trials are made, short of an infinite number, the number of correct guesses could be anything, from none correct to all correct. Even so, large departures from expectation are themselves unlikely, and make us suspect that we were not expecting the right thing in the first place.

By the same token, even though your friend actually has ESP which is slightly less than perfect, most or even all of the guesses *could* have been wrong, in any limited series of trials. Nevertheless, this is not what one expects to happen if the ESP theory is true. The mere fact of many failures would make you very skeptical of the ESP claim, and the more these failures piled up over many trials, the more skeptical you would tend to become.

Please understand that this little experiment was not described in order to show how to do experiments on ESP. Rather, it merely illustrates, in fairly simple form, the process that is often applied in statistical inference. In such situations, empirical evidence is to be used to check the validity of some theoretical state of affairs. The theory can itself be used to determine the probability with which a certain kind of event should be observed. Thereupon an experiment is conducted in which the event or events in question are possible outcomes, and some N observations are made at random and independently. The theoretical probabilities tell what to expect with regard to the events of interest, and the relative frequencies are compared with those that are expected. To the extent that the obtained relative frequencies *depart widely* from the theoretical probabilities, then there is evidence that the theory is *not true*. Moreover, the larger the number of observations N the more weight is given to any extent of departure from expectation as evidence against the theory. On the other hand, one can never be completely sure that the theory is false unless an infinite number of observations have been made.

By the same token, even though the evidence appears to agree well with a theory, this in no way implies that the theory must be true. Only after an infinite number of trials, when the relative frequency of occurrence of events matches the theoretical probability exactly, can one assert with complete confidence that the theory is true. Increased numbers of trials lend confidence to our judgments about the true state of affairs, but we can always be wrong, short of an infinite number of observations.

The principle embodied in Bernoulli's theorem can be used not only to compare empirical results with theoretical probabilities, but also to *estimate* the true probabilities of events by observing their relative frequencies over some limited numbers of trials. For example, if 67% of your friend's guesses are correct, the best estimate of the actual probability is 2/3, not 1/3.

As another example, visualize a box containing marbles of different solid colors. The number of marbles in the box, and how many different colors they have are both unknown, but the task is to find out the probability of each color when the marbles are drawn individually from the box. The simple experiment consists of mixing the marbles well, drawing one out, observing its color, and then putting it back in the box (random sampling with replacement). Now suppose that the first marble is white. Then the observed relative frequencies of colors are white 1.00, other .00. We wish to decide about the probabilities in such a way that our estimate would, if correct, have made the occurrence of this particular sample result have the greatest prior probability. What probability of the event "white" would make drawing a white marble most likely? The answer is $p(\text{white}) = 1.00$, and $p(\text{other}) = .00$, so that this is our first estimate of the probabilities of the colors of the marbles.

However, we now draw a second marble at random with replacement. This marble turns out to be red, and our best guess about the probabilities is now white .50, red .50, other .00, as these probabilities would make the occurrence of the obtained sample of two marbles most likely. If we kept on drawing, observing, and replacing marbles for 100 trials, shaking the box so that the marbles (elementary events) are equally likely to be chosen on a trial, we might begin to get a reasonably clear picture of the probabilities of the colors:

Color	Relative frequency
white	.24
red	.50
blue	.26

By the time we had made 10,000 observations, we could be even more confident that the probabilities are close to:

Color	Relative frequency
white	.24
red	.50
blue	.24
green	.02

and so on. Given an infinite number of trials, we could specify the relative frequencies (that is, the probabilities) of the marbles in the box precisely. The larger the number of observations, the less do we expect to "miss" in our estimates of what the box contains. However, for fewer than an infinite number of observations, the observed relative frequencies *need not* reflect the probabilities of the colors exactly, though they may well do so.

It is important to note that since we are drawing samples *with replacement* in this situation, the fact that there may be only some finite numbers of distinct elementary events (marbles in this case) in no way prevents us from taking a very large or even an infinite number of sample observations with replacement after each trial. Thus Bernoulli's theorem applies perfectly well to situations in which there may be a very small number of distinct elementary events possible, so long as one can draw an unlimited number of samples or make an unlimited number of trials of the experiment (as in coin tossing or card drawing).

The scientist making observations is doing something quite similar to drawing marbles from a box. Because of limited powers of observation, the scientist cannot "see into the box" and generalize from observing all such phenomena. Nevertheless, sample observations can be made and generalizations can be drawn from what is actually observed. Such a generalization is much like a bet, and how good the bet is largely depends on how many observations the scientist is able to make. The scientist can never be sure that the generalizations drawn are the correct ones. Yet, the risk of being wrong can be made quite small if a sufficiently large number of observations are made.

1.9 PROBABILITIES AND BETTING ODDS

One common way of expressing the probabilities of two mutually exclusive events is in terms of betting odds. The formal connection between betting odds and probabilities is as follows:

If the probability of an event is p, then the odds in favor of the event are p to $(1 - p)$.

Thus, if some event A has a probability $p(A) = 3/4$, then the odds in favor of A are 3/4 to 1/4, or 3 to 1. If $p(A) = 1/8$, then the odds in favor of A are only 1/8 to 7/8, or

1 to 7. When two fair dice are tossed, the probability of the event "7" is 1/6, so that the odds in favor of a "7" are 1 to 5. The odds against the occurrence of a "7" are 5 to 1. As another example, suppose that a class contains 15 girls and 11 boys. A student is selected at random from the class. What are the odds in favor of the selection of a girl? Since the probability of the event "girl" is 15/26, the odds in favor of the selection of a girl are 15 to 11.

It is equally simple to convert statements of odds into statements of probability:

If the odds in favor of some event A are x to y, then the probability of that event is given by $p(A) = x/(x + y)$.

If the odds for some event are 9 to 2, then the probability of that event is $9/(9 + 2)$ or 9/11. If the odds for some event are 1 to 1, or "even," then the probability of that event must be 1/2.

In betting situations individuals often give or accept odds. This is true not only in games of chance but also in situations where the "objective" probability of the event in question would be difficult to determine. In such situations, a statement of acceptable odds can sometimes be taken as a reflection of an individual's judgment of probability. When someone says that he believes the odds are 2 to 1 that the Pittsburgh Steelers will beat the Los Angeles Rams in their impending game, he is saying that the probability is, for him, $p(\text{Steelers beat Rams}) = 2/3$. When the weather forecaster says that the odds are 5 to 2 against rain tomorrow, she is saying that her judged probability $p(\text{no rain tomorrow}) = 5/7$, or $p(\text{rain tomorrow}) = 2/7$.

Odds are often stated in monetary terms as well. Suppose that Bettor I and Bettor II agree that the odds are 2 to 1 in favor of the Pittsburgh Steelers against the Los Angeles Rams (there will be a sudden-death playoff, so that no tie is possible in this game). Bettor I picks the Steelers and II picks the Rams. How much should each put up in order to make this a fair bet? The first bettor has chosen the event he believes to be the more likely, and thus he should stand to gain less if he wins than Bettor II, who has the less likely event. The bet becomes fair when Bettor I agrees to put up $2 for every $1 from Bettor II. In this way Bettor I gets only $1 if he wins, whereas Bettor II gets $2. In general, a bet is fair when the ratio of the moneys put up is the same as the odds. In this case, the bettor on the more likely event puts up $2 for each $1 on the less likely event, since the odds are 2 to 1. Looked at in another way, a bet is fair if the following relationship holds:

(amount won if A occurs) $p(A) = $ (amount lost if $\overline{A}$ occurs) $[1 - p(A)]$.

When a bet is fair, it is also true that

$$\frac{\text{amount bet on event } A}{\text{amount bet against event } A} = \text{odds in favor of } A.$$

The amounts bet in a fair bet thus give the odds and, consequently, the probabilities involved. When someone says "Five dollars will get you ten dollars that such and such will occur," he is actually saying that he wants to bet on an event with odds of 2 to 1, so that he is willing to put up $2 for every $1 you put up. This is another way of saying that the probability is, in his judgment, 2/3.

Strictly speaking, in a bet such as we have been describing, the two bettors agree only that true odds are *at least* as favorable as those accepted by the person who bets on the event, and, *at most* as favorable as those accepted by the person who bets against it. If Bettor I accepts odds of 2 to 1 in the example given above, but really believes that the odds are, say, 4 to 1, he should also believe that the bet is biased in his favor. On the other hand, if Bettor II believes that the odds are less in favor of the event, say 1 to 1 rather than 2 to 1, and he nevertheless accepts the 2-to-1 bet, he should also believe the bet to be biased in his favor. In short, when two people agree on a bet with odds of *x* to *y,* we can say only that the judged probability of the person who bets on event *A* is $p(A) \geq x/(x + y)$, and that the judged probability of the person who bets against event *A* is $p(A) \leq x/(x + y)$. Even so, the fact that two individuals can agree on a bet gives information on the degree of belief each holds about the occurrence of an event.

The expression of probabilities in terms of odds is a quite common feature of daily life. The fact that people can make these judgments about events, and act on the basis of such judgments, is one of the bases for an alternative interpretation, which we will now explore.

1.10 OTHER INTERPRETATIONS OF PROBABILITY

Long-run relative frequency is but one interpretation that can be given to the formal notion of probability. It is important to remember that this is an *interpretation* of the abstract model. The model per se is a system of relations among and rules for calculating with numbers that happen to be called probabilities. There is no "true meaning" of probability, any more than there is a true meaning of the symbol *x* used as a variable in school algebra. Probability is an abstract mathematical concept that takes on meaning in the ordinary sense only when it is identified with something in our experience, such as relative frequency of real events.

The probability concept acquired its interpretation as relative frequency because it was originally developed to describe certain games of chance where plays (such as spinning a roulette wheel or tossing dice or dealing cards) are indeed repeated for very many trials. Similarly, there are situations in which the scientist makes many observations under the same conditions, and so the mathematical theory of probability is given a relative frequency interpretation here as well.

On the other hand, there are some students of the matter who object to this as the exclusive interpretation of probability, and who have shown that quite different interpretations can be advanced that do not identify formal probability exclusively with relative frequency. Indeed, an everyday use of the probability concept does not have this relative frequency connotation at all. We say "It will probably rain tomorrow," or "The Yankees will probably win the pennant," or "I am unlikely to pass this test"; our hearers have no difficulty in understanding what we mean in each instance, but it is very difficult to see how these statements describe long-run relative frequencies of outcomes of simple experiments repeated over and over again. Each such statement describes the speaker's *certainty* or *degree of belief* about an event that will occur once and once only. Our inclination to use naive notions of probability in this way is one of the reasons theorists have sought other interpretations.

For many years some theorists of probability and statistics have studied the implications of a subjective, or personalistic, approach to the interpretation of probability. Such an interpretation is not so much an alternative to, as an extension of, the usual relative frequency interpretation of probability. Under this subjective approach, a probability is seen as a measure of individual degree of belief, the quantified judgment of a particular individual with respect to the occurrence of a particular event. An individual is assumed to hold, with some measurable degree of confidence, the belief that a particular event will or will not occur. Then the measure of that degree of belief about the event is represented as a probability. Under this conception, it is perfectly reasonable to assign a probability value to an event that can occur only once. The event "rain tomorrow" for a particular day is such a nonrepetitive event. That particular "tomorrow" will occur once and only once, and it will or will not rain at the location in question. When an individual says that "the probability is 1/5 that it will rain at this place tomorrow," this is a value reflecting the strength of this person's belief that the event in question will occur. It is very difficult to interpret such a statement in terms of relative frequency, but it does make intuitive sense when viewed as an individual's subjective assessment of a situation. We make similar statements all the time without a moment's second thought.

Degree of belief, and thus subjective probability, can be inferred from the choice behavior of an individual. It is not from what one says, but rather from the choices one makes, that a measure of subjective probability can be gained. It is true that in order to find these measures, we must make certain assumptions about behavior. Certain reasonable "axioms of choice" or "axioms of coherence" must be satisfied if the values inferred from individual behavior are to be treated as probabilities satisfying the mathematical properties given above. In this rapid overview we cannot take the time to go into these axioms. Suffice it to say that if an individual's behavior is consistent in the ways specified by these axioms, then probabilities inferred from individual behavior *can* be treated by the full machinery of probability theory.

It is also quite possible to apply the subjective interpretation of probability to events that are repetitive, and which thus lend themselves to a relative frequency interpretation as well. Again consider a game of dice. A player might very well have a definite degree of belief that the dice will turn up "7" on the next play. Given certain assumptions, such that each die is "fair," that the method of tossing does not influence the outcome, and so on, an intelligent player might well behave as though the chances were 1/6 that the dice would total 7 on the next throw. This is the same value that simple probability calculations dictate under these assumptions. After all, remember that the probability computations are absolutely neutral with respect to the interpretation to be placed on the probabilities. The difference between the subjective and the relative-frequency approaches lies only in the way the probability values are interpreted, not in the way they are treated mathematically. In one view, probability calculations can be thought of as ways of settling on an appropriate degree of belief, given certain assumptions and information about the circumstances of the event in question.

Critics of this approach are quick to point out that because they stand for degrees of belief, subjective probabilities may vary from individual to individual for the same event. Since individuals vary in their backgrounds, knowledge, and available information, it is quite reasonable to suppose that they will vary in the degree of belief they hold for the occurrence of a given event. If this were not the case, professional gamblers, as well as thousands of perfectly respectable businesspeople, would soon be out of work. However, such individual differences in degrees of belief do not unduly trouble advocates of the subjective approach, provided that differing probability values accurately reflect differences in judgment among individuals. Students of the subjective approach to probability are especially interested in the choice, or preference behavior, of an individual among bets or lotteries. Furthermore, they are concerned with changes in degree of belief by individuals, as new information is gained and as information is shared. Subjective probabilities can be assessed from such individual choice behavior, and in this sense they are just as "objective" as probabilities interpreted in other ways.

In an elementary discussion such as this, probability can usually be interpreted as relative frequency. Most applications of statistics deal with situations were sampling can be repeated many times, at least in theory, and in these situations the relative-frequency interpretation does make sense. However, when we deal with decision making based on statistical information, we will have occasion to discuss subjective probabilities once again.

Generally, in the material to follow, we will work with probabilities without specifying whether these are to be interpreted in one way or the other. We can do this because, essentially, probability is an abstract mathematical concept, for which certain consequences must follow from certain premises, regardless of what the "real" interpretation of the concept may be.

1.11 MORE ABOUT JOINT EVENTS

As we have seen, the elementary events making up a sample space may be grouped into events of various kinds. Given some event A, then the occurrence of some elementary event either does or does not qualify as an occurrence of A. Furthermore, given that an elementary event is an A, it may also qualify for some event B as well. In that instance, when the elementary event occurs we say that the *joint event $A \cap B$* occurs.

Provided that the elementary events in S are equally likely to occur, then the probability of $A \cap B$ is found just as for any other event:

$$p(A \cap B) = \frac{\text{number of elementary events in } A \cap B}{\text{total number of elementary events}}. \qquad [1.11.1^*]$$

Furthermore, Bernoulli's theorem holds for such joint events just as it does for any events: the long-run relative frequency of any joint event should approach the probability of that event. The best estimate of the probability of a joint event is the relative frequency we actually observe in some N number of random and independent trials of the simple experiment.

Given the joint event "A and B" for some sample space S, it is also possible to define three other joint events in terms of "not A" and "not B." The relations among these four events and their probabilities are given by the "or" rule for probabilities

(Rule 4) of Section 1.5. Recall that for any event A there is also an event $\overline{A}$, consisting of all of the elementary events not in the set A, so that

$$p(A) + p(\overline{A}) = 1.$$

Then along with $A \cap B$ we can discuss the other joint events:

$A \cap \overline{B}$. consisting of all elementary events in A and not in B;

$\overline{A} \cap B$. consisting of all elementary events not in A but in B;

$\overline{A} \cap \overline{B}$. consisting of all elementary events neither in A nor in B.

These four events are all mutually exclusive, so that

$$p(A \cap B) + p(A \cap \overline{B}) = p(A)$$

$$p(A \cap B) + p(\overline{A} \cap B) = p(B)$$

$$p(\overline{A} \cap B) + p(\overline{A} \cap \overline{B}) = p(\overline{A})$$

$$p(A \cap \overline{B}) + p(\overline{A} \cap \overline{B}) = p(\overline{B}).$$

[1.11.2]

Rule 4 of Section 1.5 also covers the situation where two events A and B are not mutually exclusive:

$$p(A \cup B) = p(A) + p(B) - p(A \cap B).$$

By use of the complementary events $\overline{A}$ and $\overline{B}$ and the relationships given above, we can also write the "or" rule in still another way. Since, from Eq. 1.11.2,

$$p(A) + p(B) = p(A \cap \overline{B}) + p(\overline{A} \cap B) + 2p(A \cap B),$$

then

$$p(A \cup B) = p(A \cap \overline{B}) + p(\overline{A} \cap B) + p(A \cap B).$$

[1.11.3]

This says that the probability of A or B or both is the sum of the probabilities of the events "A and not B," "not A and B," and "A and B."

As an illustration of the use of the "or" rule for combining probabilities, think of a city school system, where we are going to sample one pupil at random. In this school system, we know that 35% of the pupils are left-handed, so that we know also that the probability is .35 for the event "left-handed." In the same way we know that the probability is .51 of our observing a girl, and that the probability is .10 of observing a girl who is left-handed. What is the probability of observing *either* a left-hand student *or* a girl (or both)?

Let the event A be "girl," and let the event B be "left-handed." Then we know that

$$p(A) = .51, \quad p(B) = .35, \quad p(A \cap B) = .10.$$

Thus.

$$p(A \cup B) = p(A) + p(B) - p(A \cap B) = .51 + .35 - .10 = .76.$$

What is the probability of a girl who is right-handed? We know from Eq. 1.11.2 that

$$p(A \cap \overline{B}) + p(A \cap B) = p(A),$$

so that

$$p(A \cap \overline{B}) = p(A) - p(A \cap B) = .51 - .10 = .41.$$

In a similar way we can find the probability of a boy who is left-handed:

$$p(\overline{A} \cap B) + p(A \cap B) = p(B),$$

so that

$$p(\overline{A} \cap B) = p(B) - p(A \cap B) = .35 - .10 = .25.$$

What is the probability of a boy who is right-handed? Since

$$p(\overline{A} \cap \overline{B}) + p(\overline{A} \cap B) = p(\overline{A}) = 1 - p(A),$$

then

$$p(\overline{A} \cap \overline{B}) = 1 - p(A) - p(\overline{A} \cap B) = 1 - .51 - .25 = .24.$$

Finally, what is the probability of a boy *or* right-handed?

$$p(\overline{A} \cup \overline{B}) = p(\overline{A}) + p(\overline{B}) - p(\overline{A} \cap \overline{B})$$
$$= 1 - p(A) + 1 - p(B) - p(\overline{A} \cap \overline{B})$$
$$= .49 + .65 - .24 = .90.$$

This example illustrates that given the probabilities of some of the joint events, it may be possible to find the probabilities for other joint events, because of the relationships shown in Eq. 1.11.2, and because of the "or" rule of probability. This can be very important in problems in which the probabilities of only some of the joint events can be assessed, and one wishes to deduce the probabilities that other joint events must show. When one is faced with this sort of problem, drawing a Venn diagram is often extremely helpful.

These notions about joint events extend quite readily to more than two events. Perhaps there are three events A, B, and C. Then we might be interested in the joint event "A and B and C," symbolized by $A \cap B \cap C$. Thus, suppose that students on a large university campus were sampled at random. For each person observed, a record is made of age, sex, and year in college. Furthermore, suppose that the event A is "21 years or older," the event B is "female," and the event C is "senior." Then the event "21 years or older and female and senior" is the joint event $A \cap B \cap C$. The probability that such an event will occur when we sample a student at random is

$$p(A \cap B \cap C) = \frac{\text{number of students in } A \cap B \cap C}{\text{total number of students}}.$$

There are, of course, only four possibilities for the joint occurrence of two sets and their complements: $A \cap B$, $A \cap \overline{B}$, $\overline{A} \cap B$, and $\overline{A} \cap \overline{B}$. However, when one is dealing with three events, A, B, and C, there are eight possible joint events involving all three. These are diagrammed in Figure 1.11.1.

Notice the relationships among these triple events and the pairwise joint events. Since the eight joint events shown in Figure 1.11.1 are all mutually exclusive, it must be true,

Figure 1.11.1

Eight joint events in a sample space.

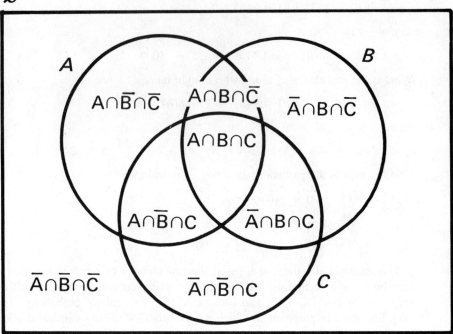

for example, that

$$p(A) = p(A \cap B \cap C) + p(A \cap B \cap \overline{C}) + p(A \cap \overline{B} \cap C) + p(A \cap \overline{B} \cap \overline{C})$$

and that

$$p(A \cap B) = p(A \cap B \cap C) + p(A \cap B \cap \overline{C}),$$

and so on for events B and C.

Such relationships may also be used to deduce the probabilities of joint events other than the ones we already know.

Again using the college example begun above, suppose that we know that

$$p(A) = .60, p(B) = .51, p(C) = .44.$$

In addition we know that

$$p(A \cap B) = .30, p(A \cap B \cap C) = .05, p(\overline{A} \cap \overline{B} \cap C) = .09,$$

$$p(A \cap \overline{B} \cap \overline{C}) = .10.$$

What then is the probability of a female nonsenior who is 21 or over? Since

$$p(A \cap B \cap C) + p(A \cap B \cap \overline{C}) = p(A \cap B)$$

then

$$p(A \cap B \cap \overline{C}) = p(A \cap B) - p(A \cap B \cap C) = .30 - .05 = .25.$$

What is the probability of a male senior who is over 21? We note from the figure that it must be true that

$$p(A \cap \overline{B} \cap C) = p(A) - p(A \cap B \cap C) - p(A \cap B \cap \overline{C}) - p(A \cap \overline{B} \cap \overline{C})$$

$$= .60 - .05 - .25 - .10 = .20.$$

We can, of course, proceed in this same way for any other combination for which the relevant information is given.

In the Figure 1.11.1, an elementary event qualifies for the event $A \cup B \cup C$ if it is represented by a point within any of the three circles. The events A, B, and C thus include among them seven possible joint events involving all three. Hence

$$p(A \cup B \cup C) = p(A \cap B \cap C) + p(A \cap B \cap \overline{C}) + p(A \cap \overline{B} \cap C)$$

$$+ p(\overline{A} \cap B \cap C) + (A \cap \overline{B} \cap \overline{C}) + p(\overline{A} \cap B \cap \overline{C}) \qquad [1.11.4]$$

$$+ p(\overline{A} \cap \overline{B} \cap C).$$

This is equivalent to

$$p(A \cup B \cup C) = p(A) + p(B) + p(C) - p(A \cap B) - p(A \cap C)$$
$$\qquad [1.11.5]$$
$$- p(B \cap C) + p(A \cap B \cap C).$$

Equation 1.11.5 is the generalization of the ''or'' rule of Section 1.5 for three events.

By exactly this same method we could show the probability of $A \cup B \cup C \cup D$ to be

$$p(A \cup B \cup C \cup D) = p(A) + p(B) + p(C) + p(D) - p(A \cap B)$$

$$- p(A \cap C) - p(A \cap D) - p(B \cap C)$$

$$- p(B \cap D) - p(C \cap D) + p(A \cap B \cap C)$$

$$+ p(A \cap B \cap D) + p(A \cap C \cap D) + p(B \cap C \cap D)$$

$$- p(A \cap B \cap C \cap D), \qquad [1.11.6]$$

and so on for any number of events.

Once again, given part of the information about the probabilities of joint events, it is possible to deduce other probabilities by the relationships given above.

So far, we have spoken of joint events where the simple experiment produces elementary events that qualify simultaneously for two or more events in $\mathcal{S}$. However, there is another situation that also produces joint events. This is when two (or more) simple experiments are done at the same time, with each one producing an outcome. Then we may be interested in a joint outcome, ''what happens on Experiment I *and* what happens on Experiment II.''

For example a simple experiment (Experiment I) here consists of rolling a die of six sides. At the same time, a simple experiment (Experiment II) consists of tossing a penny. Now we are interested in the joint outcome of rolling the die *and* tossing the penny. For this joint experiment each elementary event consists of the number that comes up on the die and the side that comes up on the penny. The sample space is shown below:

(II) Side of penny

$$\begin{array}{c} H \\ T \end{array}$$
$$\begin{array}{cccccc} \bullet & \bullet & \bullet & \bullet & \bullet & \bullet \\ \bullet & \bullet & \bullet & \bullet & \bullet & \bullet \\ 1 & 2 & 3 & 4 & 5 & 6 \end{array}$$

(I) Number on die

Each point in the diagram is one elementary event, a possible outcome of this experiment, and there are exactly (2)(6) = 12 such outcomes.

Consider the joint event "coin comes up heads, and die comes up with an odd number." There are exactly three elementary events that qualify: (heads, 1) (heads, 3), and (heads, 5). Since there are 12 elementary events in all,

$$p(\text{H and odd}) = \frac{3}{12} = \frac{1}{4}.$$

This simple example illustrates that sometimes an elementary event can also be a joint event: the event (head, 1) is the intersection of the event "heads" and the event "1," and so on.

1.12 CONDITIONAL PROBABILITY

Suppose that in sampling children in a school system, we restrict our observations only to girls. Here, there is a new sample space including only part of the elementary events in the original set. In this new sample space, consisting only of girls, what is the probability of observing a left-handed person? If all girls in the school system are equally likely to be observed,

$$p(\text{left-handed given girl}) = \frac{\text{number left-handed girls}}{\text{total number of girls}}.$$

This probability can be found from the *original* probabilities for the total sample space: first of all we know from the example in Section 1.11 that

$$p(\text{left-handed and girl}) = \frac{\text{number left-handed girls}}{\text{total number of pupils}} = .10$$

and that

$$p(\text{girl}) = \frac{\text{total number of girls}}{\text{total number of pupils}} = .51.$$

It follows that

$$p(\text{left-handed given girl}) = \frac{p(\text{left-handed and girl})}{p(\text{girl})} = \frac{.10}{.51} = .196$$

since the two probabilities put into ratio each have the same denominator.

The probability just found, based on only a part of the total sample space, is called *a conditional probability*. A more formal definition follows:

Let *A* and *B* be events in a sample space made up of a finite number of elementary events. Then the conditional probability of *B* given *A*, denoted by $p(B|A)$, is

$$p(B|A) = \frac{p(A \cap B)}{p(A)}, \qquad [1.12.1^*]$$

provided that $p(A)$ is not zero.

Notice that the conditional probability symbol $p(B|A)$ is read as "the probability of *B* given *A*." In the example it was given that the observation would be a girl; the value desired was the probability of "left-handed," given that a girl were observed.

For any two events *A* and *B*, there are two conditional probabilities that may be calculated:

$$p(B|A) = \frac{p(A \cap B)}{p(A)} \qquad [1.12.2]$$

and

$$p(A|B) = \frac{p(A \cap B)}{p(B)}. \qquad [1.12.3]$$

We might find the probability of left-handed, given girl ($.10/.51 = .20$) or the conditional probability of girl, given left-handed ($.10/.35 = .29$). In general, these two conditional probabilities will not be equal, since they represent quite different sets of elementary events.

As another example, take the probabilities found in the example for a coin and a die in Section 1.11. Here, the probability of a tail coming up on the coin is .50, and the probability of an odd number on the die *and* a tail on the coin is .25. The probability that the die comes up with an odd number, *given* that the coin comes up tails is

$$p(\text{odd}|\text{tails}) = \frac{.25}{.50} = .50.$$

The probability that the coin comes up tails and the die comes up with a 1 or a 2 can be found from the example to be 1/6 or .167. In order to know the probability that the die comes up with a 1 or a 2 given that the coin comes up tails, we take

$$p(1 \text{ or } 2|\text{tails}) = \frac{.167}{.50} = .33.$$

What are some of the properties that conditional probabilities must exhibit? In order to answer this question, think once again of two events *A* and *B* for which none of the probabilities $p(A)$, $p(B)$, and $p(A \cap B)$ is equal to zero. Then the conditional probability $p(B|A)$ can either be larger or smaller than $p(B)$; there is no necessary relation between the size of $p(B)$ and the size of $p(B|A)$, other than that they both must lie between zero and one.

Now consider the complementary event to B, or $\overline{B}$. What can we say about its conditional probability, given A. For any event B and its complement $\overline{B}$,

$$p(B|A) + p(\overline{B}|A) = 1. \qquad\qquad [1.12.4]$$

That is, the sum of the probability of B given A and the probability of not-B given A must be equal to 1.00. This is true because

$$p(A \cap B) + p(A \cap \overline{B}) = p(A)$$

so that

$$\frac{p(A \cap B)}{p(A)} + \frac{p(A \cap \overline{B})}{p(A)} = 1. \qquad\qquad [1.12.5]$$

If we know the conditional probability $p(B|A)$ and also $p(A)$, then we can also find the joint probability $p(A \cap B)$:

$$p(A)p(B|A) = p(A \cap B) \qquad\qquad [1.12.6]$$

and

$$p(B)p(A|B) = p(A \cap B),$$

since, by definiton.

$$p(B|A) = \frac{p(A \cap B)}{p(A)}$$

and

$$p(A|B) = \frac{p(A \cap B)}{p(B)}.$$

It follows that

$$p(A)p(B|A) = p(B)p(A|B)$$

and that

$$\frac{p(B|A)}{p(A|B)} = \frac{p(B)}{p(A)}.$$

Why should we be concerned with conditional probabilities? In the first place, in a sense all probabilities are conditional probabilities. Whenever we specify the circumstances surrounding, and the assumptions underlying, a given simple experiment, we are laying down certain conditions. These conditions could, if we wished, be represented through the use of conditional probabilites. For example, many times in preceding sections the experiment of drawing one card from a well-shuffled standard deck of playing cards has been used. In view of the stated condition "well-shuffled deck of standard playing cards," an event such as "king of spades" actually has a conditional probability given by

p(king of spades|well-shuffled standard deck of playing cards.)

The sample space is limited by the condition "well-shuffled deck." Usually, however, we simply assume that a well-shuffled standard deck has been chosen, and then proceed to ignore this condition in the remainder of the discussion.

Secondly, most of the information we deal with in everyday life, as the basis for the choices we must make, has a conditional character. We must constantly make choices and decisions, from the most trivial all the way to some that are of tremendous importance. These decisions are all predicated upon things that may or may not be true, on things that may or may not happen. The very best information we have to go on is usually no more than a probability. These probabilities are conditional, since virtually all of our information is of an "if-then" character. "If so and so is true, then the probability of this event must be such and such." It is not surprising that a concern with conditional probability, and with decisions based upon such probabilities, has been an important part of probability theory since its very beginning.

In fact, the theorem about conditional probabilities to be discussed in the next section has had great influence on the way people think about conditional probabilities, and especially how probabilities can change, given additional information.

1.13 BAYES'S THEOREM

The relations among various conditional probabilities are embodied in the theorem named for Thomas Bayes, an English clergyman who did early work in probability and decision theory. The simplest version of this theorem is: For two events A and B, where none of the probabilities $p(A)$, $p(B)$, and $p(A \cap B)$ is either 1.00 or 0, then the relation must hold:

$$p(A|B) = \frac{p(B|A)p(A)}{p(B|A)p(A) + p(B|\overline{A})p(\overline{A})}. \qquad [1.13.1]$$

Bayes's theorem gives a way to find the conditional probability of event A given event B, provided that you know the probability of A, the conditional probability of B given A, and the conditional probability of B given $\overline{A}$ [once you know the probability of A you already know the probability of $\overline{A}$, or $1 - p(A)$]. This theorem, which must be true for any pair of events with probabilities as specified above, summarizes the connection between the conditional and unconditional probabilities. In addition, Bayes's theorem has a number of important practical uses.

Consider this example: At a university it is decided to try out a new placement test for admitting students to a special mathematics class. Experience has shown that in general only 60% of students applying for admission actually can pass this course. Heretofore, each student applying for the course was admitted. Of the students who passed the course, some 80% passed the placement test beforehand, whereas only 40% of those who failed the course could pass the placement test initially. If this test is to be used for placement, and only students who pass the test are to be admitted to the course, what is the probability that such a student will pass the course?

First of all, event A is defined to be "passes the course," and it is assumed that $p(A) = .60$. The event B is "passes the test." The conditional probabilities are taken to be

$$p(B|A) = .80, \; p(\overline{B}|A) = .20,$$

$$p(B|\overline{A}) = .40, \; p(\overline{B}|\overline{A}) = .60.$$

However, we want to know the conditional probability, $p(A|B)$, that a student will pass the course, *given* that he has passed the test. Bayes's theorem shows

$$p(A|B) = \frac{(.80)\,(.60)}{(.80)\,(.60) + (.40)\,(.40)}$$

$$= \frac{.48}{.48 + .16} = .75.$$

Thus, the probability is .75 for a student's passing the course *given* that the test was passed.

If we wanted to find other conditional probabilities, we could also apply the theorem. For example,

$$p(\overline{A}|\overline{B}) = \frac{(.60)\,(.40)}{(.60)\,(.40) + (.20)\,(.60)}$$

$$= \frac{.24}{.24 + .12} = .67.$$

Notice that the probability of a student's passing the course given a passing score on the test is greater than the probability in general of passing the course. Any student is a better bet to pass the course given that the placement test was passed. What is the probability of being *right* if a student is admitted or refused the course strictly on the basis of this test? The probability of a correct decision can be found by

$$p(A \cap B) + p(\overline{A} \cap \overline{B}) = p(\text{correct}).$$

By Eq. 1.12.6, this is

$$p(\text{correct}) = p(B|A)p(A) + p(\overline{B}|\overline{A})p(\overline{A})$$

$$= (.80)\,(.60) + (.60)\,(.40) = .72.$$

In other words, the administrator will have a probability of .72 of being right about a student's proper placement by use of the test. If the administrator does not use the test and simply admits all students to the course, the probability of being right (of the student passing the course) is only .60. Thus, the test does something for the administrator; its use allows an increase of the probability of being right about a given student selected at random.

Bayes's theorem and other calculations with conditional probability are often used in this way, especially in questions of selection or diagnosis of subjects where initial probabilities are known. Good selection or diagnostic procedures are those permitting an increase in the probability of being right about an individual given some prior information, and such *conditional* probabilities can often be calculated by using Bayes's theorem.

As a mathematical result, Bayes's theorem is necessarily true for conditional probabilities satisfying the basic axioms of probability theory. In and of itself Bayes's theorem is in no sense controversial. However, the question of its appropriate use has, in years past, been a focal point in the controversy between those who favor a strict relative-frequency interpretation of probability and those who would admit a subjective interpretation as well. The issue emerges quite clearly when some of the probabilities figuring in Bayes's theorem are associated with ''states of nature'' or with nonrepetitive events. As we have seen, it is usually quite difficult to give meaningful relative-frequency interpretations to probabilities for such states or events. The difficulty was compounded by the fact that in some past applications, Bayes's theorem yielded rather ridiculous results. This had the effect of casting many applications of the theorem into dispute. The history of the development of statistics in the late 19th and early 20th centuries demonstrates some rather elaborate attempts to ignore or to ''finesse'' the problems of prior information and subjective probabilities.

In recent years, some theorists have shown that subjective probabilities, including those for nonrepetitive events, can be given an axiomatic status on a par with probabilities subject only to relative-frequency interpretations. Furthermore, a basis exists for the experimental determination of subjective probabilities through preferences among betting odds. Consequently, there has been a renewed interest in the application of Bayes's theorem. In particular, Bayes's theorem occupies a central place in the theory of how an individual's subjective probabilities change in the face of accumulating information.

Not all statisticians agree that some of these applications are proper, just as not all agree that a subjective interpretation of probability is meaningful. Nevertheless, the use of Bayes's theorem is becoming a feature of a considerable portion of the modern theory of statistics and of decision making, as these theories begin to allow once again for the subjective interpretation of probabilities. We will examine this approach more closely in Chapter 7.

1.14 INDEPENDENCE OF EVENTS

Now we are ready to take up the topic of independence of events and give it a more precise formulation. The general idea used heretofore is that independent events are those having nothing to do with each other; the occurrence of one event in no wise affects the probability of the other event. But how does one know if events A and B are independent? If the occurrence of event A has nothing whatever to do with the occurrence of event B, then we should expect the conditional probability of B given A to be exactly the same as the probability of B, $p(B|A) = p(B)$. Likewise, the conditional probability of A given B should be equal to the probability of A, or $p(A|B) = p(A)$. The information that one event has occurred does not affect the probability of the other event, when the events are independent.

The condition of independence of two events may also be stated in another form: if $p(B|A) = p(B)$, then

$$\frac{p(A \cap B)}{p(A)} = p(B)$$

so that

$$p(A \cap B) = p(A)p(B).$$

In the same way,

$$p(A|B) = p(A)$$

leads to the statement that

$$p(A \cap B) = p(A)p(B).$$

This fact leads to the usual definition of independence:

If events A and B are independent, then the joint probability $p(A \cap B)$ is equal to the probability of A times the probability of B.

$$p(A \cap B) = p(A)p(B). \hspace{3cm} [1.14.1^*]$$

For example, suppose that you go into a library and at random select one book (each book having an equal likelihood of being drawn). The sample space consists of all distinct books that a person might select. Suppose that the proportion of books then on the shelves and classified as fiction is exactly .15, so that the probability of selecting such a book is also .15. Furthermore, suppose that the proportion of books having red covers is exactly .30. If the event "fiction" is independent of the event "red cover," the probability of the joint event is found very easily;

$$p(\text{fiction and red cover}) = p(\text{fiction})p(\text{red cover}) = .15 \times .30 = .045.$$

You should come up with a red-covered piece of fiction in about 45 out of 1,000 random selections.

It is certainly not true that all events must be independent, and it is very easy to give examples where the joint probability of two events is not equal to the product of their separate probabilities. For example, suppose that children were selected at random. A child is observed and hair color noted; also noted is whether or not this child is freckled. For our purposes, event A consists of the set of all children having red hair, and event B is the set of all children who are freckled. Is it reasonable that event A will be independent of event B? The answer is no; everyone knows that among red-heads freckles are much more common than among children in general. One would expect in this case $p(B|A)$ to be *greater* than $p(B)$, so that it should *not* be true that $p(A \cap B) = p(A)p(B)$. The events "red hair" and "freckled" *do* tend to occur together and are not ordinarily regarded as independent.

This example suggests one of the uses of the concept of independent events. The definition of independence permits us to decide whether or not events *are associated* or *dependent* in some way:

If, for two events A and B, $p(A \cap B)$ is not equal to $p(A)p(B)$, then A and B are said to be associated or dependent.

For example, consider a sample space with elementary events consisting of all the employed adult persons in the United States. We wish to answer the question, "Is making over \$30,000 a year independent of having a college education?" Let us call event A "makes over \$30,000 a year," and event B "has a college education." Now suppose that the proportion of employed adults in the United States who make more than \$30,000 a year is .36, so that if each such person is equally likely to be selected, the probability of event A is .36. Suppose also that the proportion of employed adults with a college education is .41, so that $p(B) = .41$. Finally, the probability of the event $A \cap B$, our observing an employed adult making over \$30,000 a year who has a college education is .22.

Now if the events A and B were independent, this probability $p(A \cap B)$ should be .36 × .41 = .15. The actual probability is, say, .22. This leads immediately to the conclusion that the two events are *not* independent, or *are* associated. We can also see that the probability of occurrence of event A given event B is much larger than it should be if A and B were independent: That is,

$$p(A|B) = \frac{p(A \cap B)}{p(B)} = \frac{.22}{.41} = .54.$$

If A and B were independent, it should have been true that

$$p(A|B) = p(A) = .36.$$

Incidentally, some students tend to confuse independent with mutually exclusive events. These are by no means the same! In fact, **two events A and B that are mutually exclusive cannot be independent unless one or both events have zero probability.** This is easy to show: Consider two mutually exclusive events, A and B. Since they are mutually exclusive,

$$A \cap B = \varnothing,$$

so that $p(A \cap B) = 0$. Now if A and B were independent, it would be true that $p(A \cap B) = p(A)p(B)$, but since $p(A \cap B) = 0$, this cannot be unless $p(A)$ or $p(B)$ or both are zero. Even at a commonsense level the difference between mutually exclusive and idependent events is easy to show. Suppose that all men are either members of the class "balding" or the class "full head of hair." These two classes are mutually exclusive. Then if the events "balding" and "full head of hair" were independent, among those who are balding there would be the same proportion of men with a full head of hair, as the proportion of men with full heads of hair generally [i.e., $p(B|A) = p(B)$]. You must admit that this is a little hard to visualize.

The idea of independence is very important to many of the statistical techniques to be discussed in later chapters. The question of association or dependence of events is central to the scientific question of relationships among the phenomena we observe (literally, the question of "what goes with what"). Many techniques for studying the presence and degree of relatedness among observations depend directly upon this idea of comparing probabilities for joint events with the probabilities when events are independent.

Even at this early stage of the game, a little warning is in order: The concept of association of events must not be confused with the idea of causation. Association, as used in statistics, means simply that the events are not independent, and that a certain correspondence between their joint and separate probabilities does not hold. When events A and B are associated, it need not mean that A causes B or that B causes A, but only that the events occur together with probability different from the product of their separate probabilities. This warning carries force whenever we talk of the association of events; association may be a consequence of causation, but this need not be true.

1.15 REPRESENTING JOINT EVENTS IN TABLES

Sometimes it is convenient to list joint events by means of a table, with each cell representing one joint possibility. For example, consider a sample space consisting of all individual students at a particular college campus. Each student is either male or female, of course, and each responds either yes or no to an opinion question. The possible joint events are as follows:

Male	(male and yes)	(male and no)
Female	(female and yes)	(female and no)
	Yes	No

It might be that on this particular campus, the probability of our observing a male student is .55 and of observing a female .45. Futhermore, suppose that the probability of obtaining a yes answer is .40, and .60 for obtaining a no. Then the table of probabilities follows this general pattern:

p(Male) = .55	p(male and yes)	p(male and no)
p(Female) = .45	p(female and yes)	p(female and no)
	p(Yes) = .40	p(No) = .60

The values of p(male), p(female), p(yes,) and p(no) are called the **marginal probabilities** (they get the name from the obvious circumstance of appearing in the *margin* of the table). Each marginal probability is a sum of all the joint probabilities in some particular row or column of the table:

$$p(\text{male}) = p(\text{male and yes}) + p(\text{male and no});$$

$$p(\text{female}) = p(\text{female and yes}) + p(\text{female and no});$$

$$p(\text{yes}) = p(\text{male and yes}) + p(\text{female and yes});$$

$$p(\text{no}) = p(\text{male and no}) + p(\text{female and no}).$$

Ordinarily the event classes that appear together along any margin of the table are *mutually exclusive and exhaustive*. That is, the set of event classes forms a partition of the total sample space. Thus, the events "yes" and "no" are mutually exclusive and exhaustive and so are the events "male" and "female." Any set of mutually exclusive and exhaustive event classes that make up one margin of the table can be called an **attribute** or a **dimension.** This table embodies two attributes, the sex of the student and the response of the student, respectively.

Suppose that the two attributes of the table were independent: this means that any event along one margin must be independent of every event along the other margin. Here this is tantamount to saying that the sex of the student has absolutely nothing to do with how the question was answered. **If independence exists, the probability of each joint event** (as given in a cell of the table) **must be equal to the product of the probabilities of the corresponding marginal events.**

For example, if the two attributes are independent, then the probability $p(\text{male and yes})$ must be equal to the product of the two marginal probabilities,

$$p(\text{male and yes}) = p(\text{male})p(\text{yes}) = .55 \times .40 = .22.$$

In the same way the joint probabilities for each of the other cells may be found from products of marginal probabilities, and the following table should be correct:

p(Male) = .55 p(Female) = .45	$p(\text{male})p(\text{yes})$ = .55 × .40 = .22 $p(\text{female})p(\text{yes})$ = .45 × .40 = .18	$p(\text{male})p(\text{no})$ = .55 × .60 = .33 $p(\text{female})p(\text{no})$ = .45 × .60 = .27
	p(Yes) = .40	p(No) = .60

It is important to remember, however, that these will be the correct joint probabilities only if the attributes are independent. It might be that the following table is the true one:

p(Male) = .55 p(Female) = .45	$p(\text{male and yes})$ = .10 $p(\text{female and yes})$ = .30	$p(\text{male and no})$ = .45 $p(\text{female and no})$ = .15
	p(Yes) = .40	p(No) = .60

When this is true it is safe to say that for this sample space the sex of the student *is* associated with the answer to the question. Look at the conditional probabilities: here, the probability of a male students' answering yes is

$$p(\text{yes}|\text{male}) = \frac{.10}{.55} = .18$$

and the probability of a yes answer if the student is a female is

$$p(\text{yes}|\text{female}) = \frac{.30}{.45} = .67.$$

Thus, a female is much more likely to answer yes to this question than is a male. On the other hand, had the two attributes been independent, these two conditional probabilities *should* have been the same:

$$p(\text{yes}|\text{male}) = \frac{.22}{.55} = .40$$

$$p(\text{yes}|\text{female}) = \frac{.18}{.45} = .40,$$

indicating that sex gives no information about how a person tends to respond to the question.

Given such a table of joint events and their relative frequencies, we can always calculate the relative frequencies (or probabilities) that *should* have appeared in the various cells *if* the two attributes actually had been independent. We can also examine the relative frequencies or probabilities that actually did appear in each cell. Then, if the relative frequencies, *given* independence, are different from the relative frequencies obtained for each cell, we have a basis for saying that the two attributes are associated, or are dependent, to some extent. Many methods for assessing the relationship between attributes, and particularly those to be discussed in Chapter 18, are based directly on this idea.

1.16 RANDOM SAMPLES AND RANDOM SAMPLING

Given some simple experiment and the sample space of elementary events, the set of outcomes of N separate trials is a **sample.** When the sample is drawn *with replacement,* the same elementary event can occur more than once. When the sample is drawn *without replacement,* the same elementary event can occur no more than once in a given sample. Ordinarily, probability and statistics deal with **random samples.** The word *random* has been used rather loosely in the foregoing discussion, and now the time has come to give it a more restricted meaning in connection with random samples.

It has already been suggested that probability calculations can be made quite simple when the elementary events in a sample space have equal probabilities. The theory of statistics deals with samples of size N from a specified sample space, and here, too, great simplification is introduced if each distinct sample of a particular size can be assumed to have equal probability of selection. For this reason, the elementary theory of statistics is based on the idea of simple random sampling:

A method of drawing samples such that each and every distinct sample of the same size N has exactly the same probability of being selected is called simple random sampling.

In most social science applications of sampling, the sample space consists simply of some large group of individuals that we might select and observe. Random sampling of one individual at a time means that every possible individual in the large group has an equal probability of being drawn. If we take N such individuals, this is simple random sampling only if each possible set of N has the same probability of being selected on such an occasion. Most of what follows will deal with simple random samples. In other words, our discussion will be confined to the situation where all possible samples of the same size have exactly the same probability. For us, sampling "at random" will always mean simple random sampling, as defined above. This does not mean, however, that the theory of statistics does not apply to situations where samples have unequal probabilities of occurrence; in more advanced work the theory and methods can be extended to any sampling scheme where the probabilities of the various samples are *known,* even though they are unequal.

An alternative definition of simple random sampling in terms of elementary events can also be given: "Simple random sampling is a process of selecting elementary events for observation in such a way that each and every elementary event has precisely the same probability of being included in any sample of N observations." In *random sampling with replacement,* each elementary event has exactly the same probability of occurring on each trial. In random sampling *without replacement,* the composition of sample space $\mathcal{S}$ changes with each trial since an elementary event can occur only once in N trials. However, we shall assume that among the elementary events available for selection on a given trial, the probabilities are equal.

We shall also have many occasions to require **independent random sampling.** You may recall that independence was specified in Bernoulli's theorem. Elementary events are sampled independently when the occurrence of one elementary event has *absolutely no connection* with the probability of occurrence of the same or another elementary event on a subsequent trial. A series of tosses of a coin can be thought of as a random and independent sampling of events: what happens on one toss has no conceivable connection with what happens on any other. Each trial of a simple experiment is a sampling of elementary events, and the trials are independent when no connection exists between the particular outcomes of different trials. In independent random sampling of individuals for observation from some large "population" of such individuals, the inclusion of one individual in the sample has absolutely nothing to do with the possible inclusion or noninclusion of anyone else.

Take care to notice that our assuming samples of size N thus to be equally probable does not imply that *events* must be equally probable. Thus, in the marble example above, samples of N marbles were assumed equally probable, but the probabilities of the colors of the marbles were not equal. The assumption of equal probabilities is simply a way of saying that any elementary event has just as good an opportunity as any other to serve as a sample on any given trial.

1.17 RANDOM NUMBERS

In practical situations there are a number of schemes for making sure that each unit drawn for observation, or each elementary event in the sample space, has equal probability. Until recently, the most common was use of a table of random numbers. Random number tables consist of many pages filled with digits, from 0 through 9. These tables have been composed in such a way that each digit is approximately equally likely to occur in any spot in the table, and there is no systematic connection between the occurrence of any digit in the sequence and any other. Many books on the design of experiments contain pages of random numbers, and very extensive tables of random numbers may be found in a book prepared by the RAND Corporation (1955).

At present, computers are most commonly used to generate random numbers. Most computing centers have programs available for the generation of as many random numbers as may be needed on a given occasion; computer packages also provide this feature. Quite often a file of the potential units for observation is placed in the computer memory storage, and the computer itself selects the sample in a random way. These methods have pretty well replaced the use of random number tables in other than the smallest research settings.

These methods share a common drawback: A listing of the members of the sample space (all potential units for observation) must be possible. Except in some very restricted situations this is very difficult, or even impossible, to do. What usually happens in the social and behavioral sciences is that the investigator utilizes a sample space much more restricted than the one actually of interest. For example, an experimenter might like to make statements about behaviors of all people of a certain age, but the only group in reach for sampling consists of college sophomores in a particular locale. Then one of two things must occur, either it is assumed that the sample space employed is itself a random sample from some larger space, or inferences are confined to the group that can be sampled at random.

Random numbers are used in much the same way to achieve randomization in experiments. For example, an experimenter is going to administer three different treatments to a sample of rats, with one-third of the rats getting each treatment. The members of the total sample are listed and then assigned to the different groups by random numbers. This randomization is an extremely important part of experimental procedure, as we shall see, even though the sample itself is a random selection from some sample space.

Quite apart from their utility in drawing random samples and in randomization, random numbers are interesting as another instance of the notion of *randomness* that is idealized in the theory of probability. A random process is one in which only chance factors determine the exact outcome of any particular trial of an experiment or the result of an observation. Although the possible outcomes may be known in advance, the particular outcome of a given trial is not. Nevertheless, built into the process is some regularity, so that each class of outcome (that is, each event) can reasonably be assigned a probability, representing its long-run relative frequency. Perhaps the simplest example of a random phenomenon is the result of tossing a coin. Only chance factors determine whether heads or tails will come up on a particular toss, but a fair coin is so constructed that there is just as much physical reason for heads to appear as for tails, and so we have the justification we need to say that, in the long run, heads will appear just as often as tails. This property of the coin is idealized when we say that the probability of heads is .50. In practice, of course, random numbers are generated electronically, but the general idea is the same: the physical process is such that no single number is favored for any particular outcome, and chance alone actually dictates the number that does occur in any place in a sequence. When we use random numbers to select samples, we are using this property of randomness, which was inherent in the process by which the numbers themselves were generated. The point is that the ideas of randomness and of probability are not just misty abstractions; we know how to create processes and devise operations that are approximately random and we *do* know how to manufacture events with particular probabilities, at least with probabilities that are approximately known.

EXERCISES

1. Describe an elementary event for each of the following simple experiments:

(a) Choosing a telephone number from the Moline, Illinois directory for the current year.
(b) Selecting a student currently enrolled in a college or university in the United States.
(c) Opening a box of cookies taken down from a shelf of a certain grocery store and noting the number of cookies it contains.
(d) Selecting a current U.S. Senator to call for an interview.
(e) Counting the number of fast-food restaurants in a given 50-mile stretch of a certain highway.
(f) Picking a particular day of a given year, and noting by how many points the New York stock market changed on that day, according to the Dow-Jones index.
(g) Choosing a particular area of India and noting the percentage of persons infected with leprosy.

2. For which of the simple experiments listed in Exercise 1, might the following events reasonably occur?

(a) The individual selected comes from the state of Wisconsin.
(b) The number is 28.5.
(c) The number is -10.2
(d) The person chosen is 20 years old.
(e) The number represents a hospital.
(f) The number is 0.
(g) The number begins with the digit 3.

3. Write out in symbols and words the events represented by the following Venn diagrams:

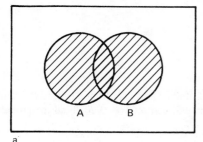

a

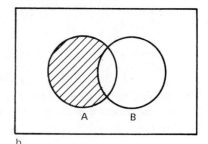

b

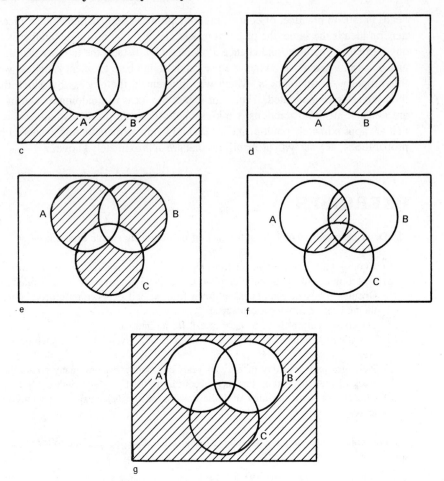

4. A bag contains pennies, nickels, and dimes. You draw one coin from the bag in such a way that any coin is equally likely to be selected. Let A be the event "penny," let B be the event "nickel," and let C be the event "dime." These events are mutually exclusive, of course. Give a verbal description of the following events:

(a) S
(b) $A \cup B$
(c) $A \cap B$
(d) $A \cup C$
(e) not A
(f) $B \cup$ not C
(g) not $B \cap$ not C
(h) $A \cup B \cup C$

5. Suppose that for the simple experiment of drawing a coin from a bag, as in Exercise 4 above, the probabilities were $p(A) = 1/4$, $p(B) = 3/5$, and $p(C) = 3/20$. Find the probabilities of the events.

(a) S

(b) $\varnothing$

(c) $A \cup B$

(d) $B \cup C$

(e) $A \cap B$

(f) not B

(g) $A \cap$ not B (**Hint:** Are A and not B mutually exclusive?)

(h) $C \cap$ not B

(i) $C \cup$ not B

6. A simple experiment consists of drawing exactly one playing card from a well-shuffled standard deck. The suit and value of the card are then noted. How many elements qualify for the following events?

(a) Ace

(b) Red suit

(c) A face card (i.e., king, queen, jack)

(d) An even-valued card, not a face card

(e) A spade or a diamond

(f) A 10 in a red suit

7. If the cards in the deck in Exercise 6 are each equally likely to be drawn, so that each has probability of 1/52, find the probabilities of the events listed in Exercise 6.

8. A person is selected at random from the list of registered voters in some community. Let the event A be "votes Democratic," let the event B be "is 60 years of age or older," and let the event C be "female." Give a verbal description of the following events:

(a) $A \cup B$

(b) $A \cap C$

(c) $A \cap$ not B

(d) $A \cup$ not $B \cap$ not C

(e) $A \cap B \cap C$

9. A part-time secretarial service has nine persons on its roster, as follows:

Name	Sex	Hair color	Height in inches
Mary	F	blonde	63
Susan	F	brunette	66
Jane	F	blonde	65
Alice	F	red	68
Eleanor	F	brunette	67
Bill	M	brunette	71
Lorna	F	blonde	63
John	M	red	70
Henry	M	blonde	73

Each individual is equally likely to be sent out on assignment. What is the probability that an individual is sent who is

(a) Female
(b) Brunette
(c) Red-haired
(d) A red-haired male
(e) 68 or more inches tall
(f) A blonde female less than 65 inches tall

10. Suppose that over a period of time the secretarial service in Exercise 9 filled 150 different assignments, and that each person was equally likely to be sent on any assignment. How often should we expect each of these events to have occurred?

(a) A male was sent
(b) A person over 65 inches tall was sent
(c) A brunette female was sent
(d) A brunette female or a blonde male was sent
(e) Either Mary or Alice was sent, but not both
(f) A redhead over 70 inches tall was sent

11. Two dice are rolled simultaneously. One die is white with black spots, and the other die is black with white spots. Let an elementary event be the occurrence of a pair of numbers of (x,y), where x is the number of spots coming up on the white die, and y is the number of spots coming up on the black die. List all of the possible elementary events in the sample space   . How many elementary events are there? Find the number of elements or (x,y) pairs that make up the following events:

(a) $x + y = 7$
(b) $x + y > 7$
(c) $x = y$
(d) $x \neq y$
(e) x is odd and y is even
(f) x is even *or* y is even
(g) $x \cdot y \geq 8$

12. Given that each of the elementary events found in Exercise 11 is equally likely, find the probabilities of the events listed there.

13. Suppose that one letter is chosen from among the 26 letters making up the English alphabet. This is done is such a way that each letter is equally likely to be chosen. Find the probabilities for the following events:

(a) The letter is a vowel (include y as a vowel).
(b) The letter is a consonant other than s or t.
(c) The letter is a consonant from the first half of the alphabet.
(d) The letter chosen appears in the word *exodus*.
(e) The letter chosen appears either in the word *born* or in the word *film*.
(f) The letter appears either in the word *lead* or the word *load*.

14. Suppose that a letter is drawn from the word *sensation* in such a way that any of the letters are equally probable. Then that letter is replaced and a second letter is drawn. Find the probabilities of the events listed below. (**Hints:** Treat two occurences of the letter s in *sensation* as though they were two separate elementary events, and the same for *n*.) A simple graphic plot of all of the possible joint events (first letter, second letter) will be helpful.

    (a) The two letters drawn are the same.
    (b) The two letters drawn are both consonants.
    (c) The letters drawn are both vowels.
    (d) The letters drawn are two different consonants.
    (e) The letters drawn in order form any of the words *so, as,* or *is.*
    (f) The letters drawn in order form any of the words, *in, on, at,* or *an.*
    (g) If this simple experiment were repeated 162 times, how many times would you expect to get a consonant followed by a vowel?

15. Thirty-six poker chips are placed in a box. One-third of the chips are red, one-third blue, and one-third white. The chips of each color are numbered from 1 through 12. If one chip is drawn from the box, and if any chip is equally likely to be drawn, find the probabilities of the following events:

    (a) The number drawn is a 3.
    (b) The chip is red with an even number.
    (c) The chip is blue with a number of 5 or more.
    (d) The chip is red or the number is 8.
    (e) The chip is white and the number is odd.
    (f) The chip is white or the number is odd.

16. Suppose that the simple experiment described in Exercise 15 is carried out 360 times, with the poker chip being replaced in the box after each drawing. How many times should you expect the following events to occur?

    (a) The chip is not blue.
    (b) The number is 3 or 5.
    (c) The chip is red and the number is 12.
    (d) The number drawn is less than 5.
    (e) The number drawn is less than 2 or greater than 11.
    (f) The chip is white with the number 5, 6, or 7.
    (g) The chip is red with a number which is a perfect square.

17. The graduating class of a certain high school contains 52% boys and 48% girls. Among the boys, 37% are 19 years old or older, whereas among the girls only 12% are 19 years old or older. Suppose that a student is drawn at random from this class.

    (a) What is the probability that he or she is less than 19 years of age?
    (b) Are sex and age of student independent in this class? How do you know?
    (c) Given that a student is 19 years old or older, what is the conditional probability that this is a boy?
    (d) Given that a student has an age under 19 years, what is the conditional probability that this is a girl?

18. From the information provided in Exercise 9, find the following conditional probabilities:

    (a) p(male|blonde).
    (b) p(red-headed|female).

(c) p(female|height under 70 inches).

(d) p(height under 70 inches|female).

(e) p(Alice|red hair).

19. Each subject in a behavioral experiment was given one of three experimental tasks. Let these three tasks by symbolized by A_1, A_2, and A_3. The number of errors each subject made in performing the task was recorded by use of the categories "0 errors" "1 error," and "2 or more errors." Thus, any subject was classified in two ways: by task, and number of errors. The following table shows the probabilities for these joint events, for any subject drawn at random from the experimental group. Use these probabilities to find

(a) the marginal probabilities of the task events A_1, A_2, and A_3.

(b) The marginal probabilities for numbers of errors.

(c) The conditional probability for 0 errors given task A_1.

(d) The conditional probability for 2 or more errors given task A_2.

(e) The probability of task A_3, given no errors.

	Task		
errors	A_1	A_2	A_3
0	.05	.02	.13
1	.08	.17	.10
2 or more	.20	.15	.10

20. A child's game has a metal spinner on a card. This card consists of a circular area divided into six colored sectors of equal area: black, white, red, yellow, green, and blue. The spinner is attached to the center of the card and is free to come to rest at any point along the circle. If the spinner is equally likely to come to rest in any sector, then what is the probability of the following:

(a) Black or white.

(b) Yellow.

(c) Neither white nor yellow.

(d) Either green or blue.

(e) Blue or not black.

21. Suppose that the card described in the exercise above were such that black, white, and yellow sectors were each twice the size of the individual sectors devoted to red, blue, and green. Calculate the probabilities of the events listed above under these circumstances, assuming that the pointer has equal probability of stopping at any point on the circle.

22. An archeologist is interested in three geographical areas containing ancient village sites. The area below shown as the circle came under the influence of culture A, the rectangular area under the influence of culture B, and the triangular area was influenced by culture C. The black dots show the possible sites. Suppose that the archeologist picks one of these sites at random to excavate. What is the probability that the site selected came under the influence of

(a) Culture A?

(b) C?

(c) B?

(d) *A* and *B?*
(e) *A* or *B?*
(f) *B* and *C?*
(g) *B* or *C?*
(h) *A* and *B* and *C?*
(i) *B* and *C* but not *A?*
(j) *A* and not *B* and not *C?*

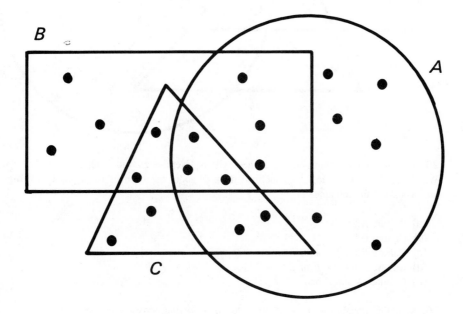

23. In Exercise 22, what is the probability that the site selected came under the influence of exactly one culture? Exactly two cultures? Three cultures? More than one culture? Less than three culutres?

24. The archeologist in Exercise 22 knows that some of the sites that might be explored are productive in terms of artifacts, whereas others that might be explored will be nonproductive. The chart below shows the productive sites with an *X* and the nonproducive with an *O*. If a site is productive, what is the probability that it was influenced by

   (a) Culture *A?*
   (b) *B?*
   (c) *A* and *B?*
   (d) *C* but not *A?*
   (e) Two cultures only?
   (f) *A* or *B?*
   (g) *B* and *C* but not *A?*

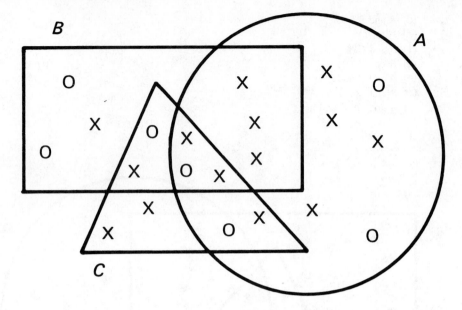

25. A native worker returns to the archeologist in Exercise 24 with a report of finding artifacts. What would the archeologist be inclined to infer about the chances for a productive dig if the artifacts show the influence of

(a) Culture *A?*
(b) Culture *C?*
(c) Culture *B* but not *A* or *C?*
(d) Culture *A* but not *B* or *C?*
(e) Cultures *A* and *B* and *C?*

26. State the probabilities corresponding to the following odds and bets:

(a) 5 to 3 in favor of an event *a.*
(b) 2 to 9 against event *a.*
(c) 6 to 4 in favor of event *a.*
(d) 17 to 15 in favor of *a.*
(e) $5 to $2 bet on *a.*
(f) $9 to $4 bet against *a.*
(g) $1 to $49 bet on *a.*

27. The following events can occur only once. If an individual regards the bets listed as fair bets, estimate the subjective probability held for these events.

(a) A bet of $.50 to $.10 that it will rain tomorrow.
(b) A bet of $3 to $7 that the Big Ten representative will win or tie in the Rose Bowl game next New Year's Day.
(c) A bet of $1 to $1 that this individual will fail a psychology course this term.

28. A sociologist identified 270 distinct neighborhoods in a large American city. In terms of ethnic balance, these neighborhoods were classified as being "mostly minority," "evenly mixed," and "mostly nonminority," Each neighborhood was also classified according to average income level, as "low income," "medium income," and "high income." The numbers of neighborhoods in each combination of classifications is shown in the following table.

		Income		
		low	medium	high
ethnic balance	minority	79	9	3
	mixed	22	20	10
	nonminority	18	63	46

Suppose that a neighborhood is selected at random from this city. What is the probability that it will have these characteristics:

(a) Low income.
(b) Minority.
(c) Mixed.
(d) Low income or minority.
(e) Not minority.
(f) Not minority and high income.
(g) Low income and mixed.
(h) Not low income and not minority.

29. If a neighborhood is selected, as in Exercise 28, what is the conditional probability that it will have the characteristic of

(a) Low income, given that it is minority.
(b) High income, given that it is minority.
(c) Nonminority, given high income.
(d) Medium income, given mixed.
(e) Not middle income, given mixed.

30. Assume for the moment that the ethnic balance of a neighborhood in the city in Exercise 28 is independent of the neighborhood income. What then should the probabilities for the nine possible joint events in the table have been? How many neighborhoods of each ethnic-balance and income combination should there have been? How do these compare with the actual numbers given? What does this suggest about the relation between ethnic balance and income in these neighborhoods?

31. The neighborhoods in Exercise 28 were also classified according to whether they fell in a three-mile radius of the city center, or whether they were beyond three miles. The table showing all three classifications is as follows:

	Within three miles			Beyond three miles		
	low	medium	high	low	medium	high
minority	69	5	0	10	4	3
mixed	14	4	1	8	16	9
no minority	2	1	0	16	62	46

Once again, consider a neighborhood selected at random. What is the probability that it is:

(a) Within three miles of the city center.
(b) Nonminority and beyond three miles.
(c) Minority and within three miles.
(d) High income and beyond three miles.
(e) Low income and minority, given beyond three miles.
(f) Within three miles, given minority.
(g) Low income and within three miles, given minority.
(h) Low income, given minority and within three miles.

32. A certain test of vocational interests provides one of five interest patterns for any individual taking the test. Call these patterns *A, B, C, D,* and *E.* The test was given to 1,000 persons, consisting of 294 engineers, 264 physicians, 187 writers, and 255 accountants. The table below gives the numbers in each occupational group giving each of the possible patterns.

	Pattern				
	A	B	C	D	E
engineer	117	58	70	33	16
physician	52	125	20	38	29
writer	13	24	12	37	101
accountant	79	33	94	35	14

If one individual is selected at random from this group, state the probability of observing

(a) Pattern *A.*
(b) Pattern *A* or *B.*
(c) An engineer.
(d) A writer or an accountant.
(e) A physician with Pattern *D.*
(f) A writer with Pattern *A.*
(g) An accountant with Pattern *C.*
(h) An engineer with Pattern *A* or *C.*
(i) A physician with Pattern *B* or an accountant with Pattern *C.*

33. For the situation in Exercise 32, what is the probability of observing

(a) An engineer, given Pattern *A.*
(b) A physician, given Pattern *B.*
(c) A writer, given Pattern *E.*
(d) Pattern *D,* given an accountant.
(e) Pattern *A* or Pattern *B,* given a physician.
(f) An engineer, given Pattern *A* or Pattern *B.*
(g) A writer, given Pattern *A.*
(h) An accountant, given any Pattern except *E.*

34. Once again, an individual is drawn at random from the group in Exercise 32. Under each of the following conditions, what is your best bet about the pattern that individual will exhibit?

(a) There is no information given about the person's occupation.
(b) The person is an accountant.
(c) The person is an engineer.
(d) The occupation is writer.
(e) The individual is a physician.
(f) Occupation is engineer or accountant.
(g) The person is not a writer.

35. An individual is selected at random from the group in Exercise 32. Find out what the odds are that this person is

(a) A writer (as opposed to ''not a writer'').
(b) A physician.
(c) An engineer.
(d) An accountant.
(e) An engineer or a physician.

36. Answer the questions in Exercise 35 once again, but this time suppose that we know that the individual selected exhibits Pattern B. What happens to the odds when one is provided with some information about the pattern, as opposed to no information? What would the addition of information about pattern do to the odds for occupations if pattern and occupation were actually independent? Is there evidence that pattern and occupation are independent for this group of people?

37. Now suppose that two individuals are drawn at random and independently (with replacement) from the group in Exercise 32. What is the probability that

(a) The first is an engineer and the second a physician?
(b) Both are accountants?
(c) Both are writers?
(d) The first shows Pattern A and the second Pattern B?
(e) Both are physicians showing Pattern B?
(f) Both are writers showing Pattern E or both are accountants showing Pattern C?
(g) Neither is a physician?

Chapter *2*

Frequency and Probability Distributions

Given the definitions of *event* and of *probability* in the last chapter, we are now ready to take up the first major set of concepts in statistics.

The idea of a frequency distribution for sets of observations will be introduced, together with some of the mechanics for constructing distributions of data. Then this idea of a frequency distribution will be paralleled by that of a probability distribution. The important special case of numerical events will be introduced under the heading *random variables*.

Prior to any concerns with summarizing the data or with comparing the obtained results with some theoretical state of affairs, the investigator must first measure the phenomena under study. It therefore seems appropriate at this point to mention the process of measurement and to indicate the role that measurement considerations will play in the remainder of this text. A thorough discussion of this topic will not even be attempted, but we will try to examine its relevance to statistics.

2.1 MEASUREMENT SCALES

Whenever the scientist makes observations of any kind, some classifying and recording scheme must be used. Any phenomenon or "thing" will have many distinguishable characteristics or attributes, but the scientist must first single out those properties *relevant to the question* being studied. For example, a scientist tries out a new serum on laboratory rats infected with a disease. Any single rat differs from any other rat in innumerable ways: coat markings, heart rate, length of tail, exact age, and so on. The scientist is interested, however, in only one thing: did the rat recover or did it die of the disease? Or, perhaps, the intelligence level of a group of boys is being compared with that of a group of girls. Individual boys and girls differ and same-sex groups differ among themselves in ways as diverse as body temperature, size of head, weight, color of hair, father's income, name, and so on, ad infinitum. All these properties are ignored by the experimenter as immaterial to the immediate purpose, which is giving each child a number representing intelligence.

The classifying and recording scheme the scientist uses is based upon differences in some *particular* property or attribute that objects of observation exhibit. The scientist simply cannot pay attention to all the ways in which things differ from one another. Many of these differences obviously are irrelevant to the purpose, and other potential differences are controlled by the scientist in making the observations: only rats (not all kinds of animals) are given the serum, and each child is give the same intelligence test in the same way. Still other differences that are germane to the conclusions, but not specifically controlled, are treated as statistical "error."

Once the scientist has singled out the property or properties to be studied and established controls for the others, the classifying scheme is applied to each observation. Such a scheme is essential in order to record, categorize, and communicate the things observed. At its very simplest, this scheme is a rule for arranging observations into *equivalence classes,* so that observations falling into the same set are thought of as *qualitatively the same* and those in different classes as *qualitatively different* in some respect. In general, each observation is placed *in one and only one* class, making the classes mutually exclusive and exhaustive.

The process of grouping individual observations into qualitative classes is measurement at its most primitive level. Sometimes this is called **categorical or nominal scaling.** The set of equivalence classes itself is called a **nominal scale.** There are many areas in science where the best one can do is to group observations into classes, each given a distinguishing symbol or name. For example, taxonomy in biology consists of grouping living things into phyla, genera, species, and so on, which are simply sets or classes. Psychiatric nosology contains classes such as schizophrenic, manic-depressive, paretic, and so on; individual patients are classified on the basis of symptoms and the history of their disease.

At this level of measurement, what the classes happen to be called is quite irrelevant. Any set of equivalence classes can be transformed into another set of equivalence classes (that is, renamed), provided that common class membership, or equivalence, is strictly preserved. Thus, consider a set of seniors at a given university in a given year. Each might be classified as "liberal arts major," "engineering major," "education major," "business major," and "other major." However, each could just as well be classified as belonging to "group alpha," "group beta," "group gamma," "group delta," and "group omega" so far as the measurement process is concerned, provided that all members of a group in the first instance wind up as members of the same group in the second instance. Only the names of the classifications are changed. Any such equivalence-preserving transformation is permissible in nominal measurement.

The word *measurement* is usually reserved, however, to refer to the situation in which each individual is assigned a *number;* this number reflects a magnitude of some *quantitative* property. A definite rule must exist associating one and only one number with each individual, and the measurement procedure fills the role of this rule.

There are at least three kinds of numerical measurement that can be distinguished:

these are often called *ordinal scaling, interval scaling,* and *ratio scaling.* A really accurate description of each kind of measurement is beyond the scope of this book, but we can gain some idea of these basic distinctions.

Imagine a set O consisting of N distinct objects. These objects may be given the labels o_1, o_2, and so on up to o_N. Now imagine any distinct pair of these objects, which we will call o_i and o_j.

Suppose that there is some property that pertains to each object in the set O, such as temperature, weight, length, age, intelligence, or motivation. Each object has a certain amount or degree of that property. In principle, to any object o_i we could assign a number $t(o_i)$, standing for the amount that o_i actually "has" of that characteristic. In the same way, any other object o_j in the set could also be given its value $t(o_j)$.

Ideally, in measuring an object o_i we should be able to determine its $t(o_i)$ value directly. Unfortunately, this is not always (or even usually) possible. Rather, what we must do is to devise a procedure for pairing each o_i with another number, say $m(o_i)$, that we call its numerical measurement. The actual procedure we use to assign the $m(o_i)$ value constitutes a measurement rule. However, just any old procedure, regardless of how precise and "scientific" it appears, will not do: we want the various values of $m(o_i)$ assigned to the various possible objects o_i at least to reflect the $t(o_i)$ values showing the different degrees of the property. A measurement rule would surely be nonsense if it gave numbers having no connection at all with the true amounts of some property that different objects possess. Even though we may never be able to determine the $t(o_i)$ value for any object o_i exactly, we at least hope to find numbers $m(o_i)$ that will be related to these true vaules in a systematic way. The measurement numbers we obtain must be good reflections of the true quantities, so that information about magnitudes or amounts of the property can be at least inferred from the values observed.

Measurement operations or procedures differ in the information that the numerical measurements themselves provide about the *true magnitudes*. Some ways of measuring permit us to make very strong statements about what the differences or ratios among the true magnitudes must be, and thus about the actual differences in, or proportional amounts of, some property that different objects possess. On the other hand, some measurement operations permit only the roughest inferences to be made about true magnitudes from the measurement numbers themselves.

Now suppose that we have a measurement procedure that gives a number $m(o_i)$ to any object o_i, and also gives a number $m(o_j)$ to a different object o_j. Then we say that this is measurement at the *ordinal level* if the following statements are true:

---

1. $m(o_i) \neq m(o_j)$ implies that $t(o_i) \neq t(o_j)$.
2. $m(o_i) > m(o_j)$ implies that $t(o_i) > t(o_j)$.

In other words, when we measure the ordinal level, we can say at least that if two measurements are unequal, the true magnitudes are unequal, and if one measurement is larger than another, one magnitude exceeds another. However, we really cannot say by *how much* the objects truly differ on the property in question.

For example, suppose that the objects in question were minerals of various kinds. Each mineral has a certain degree of *hardness,* represented by the quantity $t(o)$. We have no way of knowing these quantities directly, and so we devise the following measurement rule: take each pair of minerals and find if one *scratches* the other. Presumably, the harder mineral will scratch the softer in each case. When this has been done for each pair of minerals, give the mineral scratched by everything some number, the mineral scratched by all but the first a higher number, and so on, until the mineral scratching all others but scratched by nothing else gets the highest number of all. In each pair the "scratcher" gets a higher number than the "scratchee." (Here we are assuming tacitly that "*a* scratches *b*" and "*b* scratches *c*" implies that "*a* scratches *c*," and that we will get a *simple ordering* of "what scratches what.")

This measurement procedure gives an example of *ordinal* scaling. The possible numerical measurements themselves might be some set of numbers, such as (1, 2, 3, 4, . . .) or (10, 17, 24, . . .) or even other arbitrary symbols having some conventional order, as (*A, B, C,* . . .). In any case, the assignment of numbers or symbols to objects is only a form of ranking, showing which is "more" something. In the example, if one mineral gets a higher number than another, then we can say that the first mineral is harder than the second. Notice, however, that this is really all we can say about the degree or amount of hardness each possesses. Although the numbers standing for ordinal measurements may be manipulated by arithmetic, the answer cannot necessarily be interpreted as a statement about the true *magnitudes* of objects, nor about the true *amounts* of some property.

Other measurement procedures give functions pairing objects with numbers where much stronger statements can be made about the true magnitudes from the numerical measurements. Suppose that the following statement, in addition to Statements 1 and 2, were true:

---

3. For any object o_i, $t(o_i) = x$ if and only if $m(o_i) = ax + b$, where $a \neq 0$.

That is, the measurement number $m(o)$ is some *linear function* of the true magnitude x (the rule for a linear function is x multiplied by some constant a and added to some constant b). When the statements 1, 2, and 3 are all true, the measurement operation is called **interval scaling,** or measurement at the **interval scale level.**

Much stronger inferences about magnitudes can be made from interval scale measurements than from ordinal measurements. In particular, we can say something precise about *differences* in objects in terms of magnitude. Consider two objects, o_i and o_j, once again. Then if we find

$$m(o_i) - m(o_j) = 4,$$

we can conclude that

$$t(o_i) - t(o_j) = \frac{4}{a},$$

the difference between the magnitudes of the two objects is 4 units, where a is simply some constant changing measurement units into "real" units, whatever they may be.

For example, finding temperature in Fahrenheit units is measurement on an interval scale. If object o_i has a reading of 180° and o_j has 160°, the difference (180 − 160) times a constant actually *is* the difference in temperature magnitude between the objects. It is perfectly meaningful to say that the first object has 20 units more temperature than the second.

Given some measurement operation yielding an interval scale of some property, then any other measurement operation for the same property *also* gives an interval scale, provided that the second way of measuring yields numbers that are a *linear function* of the first.

A familiar example of this principle is temperature measurement in Fahrenheit and in Celsius degrees; each is an interval scale measurement procedure, and the reading on one scale is a linear function of the reading on the other:

$$C° = \frac{5}{9} (F° - 32).$$

When measurement is at the interval scale level, any of the ordinary operations of arithmetic may be applied to the differences between numerical measurements, and the result interpreted as a statement about *magnitudes* of the underlying property. The important part is this interpretation of a numerical result as a *quantitative statement* about the property shown by the objects. This is not generally possible for ordinal scale measurement numbers, but it *can* be done for differences between interval scale numbers. In very simple language: you can do arithmetic to your heart's content on any set of numbers, but your results are not necessarily true statements about amounts of some property that objects possess unless interval scale requirements are met by the procedure for obtaining those numbers.

Interval scaling is about the best one can do in most scientific work, and even this level of measurement is all too rare in social and behavioral sciences. However, especially in physical science, it is sometimes possible to find measurement operations making the following statement true:

---

4. For any object o_i, $t(o_i) = x$ if and only if $m(o_i) = ax$, where $a > 0$.

When the measurement operation defines a function such that Statements 1 through 4 are all true, then measurement is said to be at the **ratio scale level.** For such scales, ratios of numerical measurements can be interpreted directly as ratios of magnitudes of objects:

$$\frac{m(o_i)}{m(o_j)} = \frac{t(o_i)}{t(o_j)}.$$

For example, the usual procedure for finding the length of objects provides a ratio scale. If one object has a measurement value of 10 ft, and another a value of 20 ft, then it is quite legitimate to say that the second object has twice as much length as the first. Notice that this is not a statement one ordinarily makes about the *temperatures* of objects (on an interval scale): if the first object has a temperature reading of 10° and the second 20°, we do not ordinarily say that the second has twice the temperature of the first. Only when scaling is at the ratio level can the full force of ordinary arithmetic be applied directly to the measurements themselves, and the results reinterpreted as statements about magnitudes of objects. An important connection exists between interval scaling and ratio scaling. **When objects are measured on an interval scale, then differences between objects are measured on a ratio scale.** The concept of zero difference, or zero "distance," does have a fixed and nonarbitrary definition, and differences between objects can be treated by any of the methods available for ratio scale values, provided that the original objects were measured on an interval scale. This accounts, in part, for the considerable preoccupation with differences among measurements that one encounters not only in the physical sciences but also in statistics.

Only one sort of transformation is permissible for values measured on a ratio scale. This is multiplication of each value by the same positive constant. In other words, the only permissible change in a ratio scale is in the unit of measurement employed.

As a related point, observe that an interval scale has an *arbitrary* origin or zero point, which is determined by the constant b in Statement 3. However, any ratio scale has a *fixed* zero point, which cannot be modified. One can alter the point on a thermometer that represents zero degrees of temperature (except on the Kelvin scale), but one cannot change the point on a ruler that stands for zero units of length.

There are any number of examples that could be adduced to illustrate the differences among these levels of measurement and other possible intermediate levels as well. The student who is interested in this topic enough to explore it further is urged to look into the initial chapters of the books by Stevens (1951), by Thrall, Coombs, and Davis (1954), and by Torgerson (1958); each gives a slightly different perspective on psychological measurement. Our immediate concern, however, is to see the implications of different levels of measurement for probability and statistics.

The problem of measurement, and especially of attaining interval scales, is an extremely serious one for the social and behavioral sciences. It is unfortunate that in their search for quantitative methods researchers sometimes overlook the question of level of measurement and tend to read quite unjustified meanings into their results. This has brought about a reaction in some quarters, where some people insist that such investigators are not justified in using most of the machinery of mathematics on their generally low-level measurements.

However, the core problem of level of measurement really lies outside the province of mathematics and statistics. Statistics deals with numbers, and statistical methods yield conclusions based on numbers, but there is absolutely nothing in the mathematical machinery that whistles and waves a flag to show that the numbers supplied are not really interval scale measurements of some property of objects or human beings. The machinery works the same way and gives the same result regardless of whether the numbers are made up from the whole cloth or are the product of the most refined measurement procedures imaginable. Only the users of the statistical result, the investigators and their readers, can judge the reinterpretability of the numerical result into a valid statement about the properties of things. Sometimes the investigator regards the numerical conclusion as a statement about *scores* and does not go beyond this statement to a conclusion about the real magnitudes of some property. In other instances the whole investigation makes no sense at all unless the numerical conclusion can refer directly to magnitudes of some property. Surely the individual researcher knows his problem and his measurement procedures better than a stranger dogmatizing about hypothetical situations in a textbook! The scientist must face the problem of the interpretation of statistical results *within* the context of the scientific discipline and on *extramathematical* grounds.

Thus, it seems to this author that statistics as a discipline is quite neutral on this issue. In developing procedures, mathematical statisticians have assumed that techniques involving numerical scores, orderings, or categorization are to be applied where these numbers or classes are appropriate and meaningful within the experimenter's problem. If the statistical method involves the procedures of arithmetic used on numerical scores, then the numerical answer is formally correct. Even if the numbers are the purest non-sense, having no relation to real magnitudes or the properties of real things, the answers are still right *as numbers*. The difficulty comes with the interpretation of these numbers back into statements about the real world. If nonsense is put into the mathematical system, nonsense is sure to come out.

For these reasons, very little more will be said about scales of measurement in this book. Everything depends on what is being measured in whatever way, what is to be found out, and, most of all, what the investigator wishes to say about the "real prop-erties" underlying the numerical measurements. From time to time, statistical tech-niques will be examined which seem especially appropriate to one or another of the scales of measurement, and this will, of course, be pointed out. Thus, in Chapter 18 techniques will be examined which lend themselves quite naturally to nominally scaled data, and in Chapter 19 the emphasis will be on data at the ordinal level of measure-ment. For many other methods, interval scale measurement may be tacitly assumed, even though it is completely obvious that this level of measurement cannot be attained for some things. In these cases the user of the statistical technique should understand the limitations imposed on the conclusions by the measuring device used (*not* by statis-tics).

2.2 FREQUENCY DISTRIBUTIONS

As we have just seen, even in the simplest instances of meaurement the experimenter makes some N observations and classifies them into a set of qualitative measurement classes. These qualitative classes are mutually exclusive and exhaustive, so that each observation falls into one and only one class.

As a summary of these observations, the experimenter often reports the various pos-sible classes, together with the number of observations falling into each. This may be done by a simple listing of classes, each paired with its frequency number, the number of cases observed in that category. The same information may be displayed as a graph, perhaps with the different classes represented by points or segments of a horizontal axis, the frequency shown by a point or a vertical bar above each class. Regardless of how this information is displayed, such a listing of classes and their frequencies is called a **frequency distribution.** Any representation of the relation between a set of mutually exclusive and exhaustive measurement classes and the frequency of each is a frequency distribution.

A frequency distribution is simply a function in which each of a set of classes is paired with a number, its frequency. Thus, in principle, a frequency distribution of real or theoretical data can be shown by any of the three ways one specifies any function: an explicit listing of class and frequency pairs, graphically, or by the statement of a rule for pairing a class of observations with its frequency. In describing actual data, the first two methods are almost always used alone, but there are circumstances where the experimenter wants to describe some theoretical frequency distribution, and in this case the mathematical rule for the function sometimes is stated.

The set of measurement classes may correspond to a nominal, an ordinal, an interval, or a ratio scale. Although the various possible classes will be qualitative in some instances and quantitative in others, a frequency distribution can always be constructed, provided that each and every observation goes into one and only one class. Thus, for example, suppose that some N native U.S. citizens are observed, and the state in which each was born is noted. This is like nominal measurement, where the measurement classes consist of the 50 states (and the District of Columbia) into which subjects could be categorized. On the other hand, the subjects might be students in a course and graded according to A, B, C, and so on. Here the frequency distribution would show how many got A, B, and so on. Notice that in this case the measurement actually is ordinal: A is better than B, B better than C. Nevertheless, the frequency distribution is constructed in the same way, except that the various classes are displayed in their proper order.

Even when measurement is at the interval or ratio scale level, one reports frequency distributions in this general way. Suppose that N college students were each weighed, and the weights noted to the nearest pound. The set of weight classes into which students are placed would perhaps consist of 50 or 60 different numbers; a weight class is the set of students getting the same weight number. There may be only one, or even no students, in a particular weight class. Nevertheless, the frequency distribution would show the pairing of each possible class with some number from zero to N, its frequency.

As we shall see, when the number of possible classes is very large, or even potentially infinite, the task of constructing a frequency distribution is made easier by a process of combining classes. The principle is, however, always the same: a display of the relation between classes and their frequencies.

2.3 FREQUENCY DISTRIBUTIONS WITH A SMALL NUMBER OF MEASUREMENT CLASSES

The reasons for dealing with frequency distributions rather than raw data are not hard to see. Raw data almost always take the form of some sort of listing of pairs; each consisting of an object and its measurement class or number. Thus, if we were noting the hair color of 11 persons, our raw data might be something like this:

John Jones	black
Mary Smith	blonde
Jim Hardy	brown
Horace Goodman	brown
Alice Adams	blonde
Ann Wilk	red
William Thomas	brown
Bert Fox	red
Homer Giddens	black
Shirley Snider	brown
Richard Rowan	blonde

This set of pairs does contain all the relevant information, but if there are a great many objects being measured such a listing is not only laborious but also very confusing to anyone who is trying to get a picture of the set of observations as a whole. If we are interested only in the "pattern" of hair color in the group, the names of the persons are irrelevant, and all that is necessary is the number of individuals having each hair color.

There are only four measurement classes in this example that show one or more cases, and so the frequency distribution can be displayed in this way:

Hair color	f*
black	2
red	2
blonde	3
brown	4
	11 = N

*The letter f symbolizes *frequency* throughout.

It is possible to illustrate this idea of a frequency distribution in any number of ways: consider a study done on a group of 25 males in order to determine their blood types. The subjects were classified variously as having blood types A, B, AB, or O. Table 2.3.1 lists the measurements of these 25 men. These data condense into the frequency distribution that follows the table.

In another study 2,000 families in some city were interviewed about their perferences in art, music, literature, recreation, and so forth. After each interview, the family was characterized as having "highbrow," "middlebrow," or "lowbrow" tastes. This we can consider as a case of ordinal scale measurement since these three "taste" categories do seem to be ordered; "highbrow" and "lowbrow" should be more different from each other than "middlebrow" is from either. A listing of families in terms of these classes would be confusing if not grounds for libel suits. However, just as before, we can construct a frequency distribution summarizing these data:

Class	f
highbrow	50
middlebrow	990
lowbrow	960
	2,000 = N

Notice how much more clearly the characteristics of the group as a whole emerge from a frequency distribution such as this than they possibly could from a listing of 2,000 names each paired with a rating.

It is appropriate to point out that a frequency distribution provides clarity at the expense of some information in the data. It is not possible to know from the frequency distribution alone whether the family of John Jones at 2193 Spring Street is high-, middle-, or lowbrowed. Such information about *particular* objects is sacrificed in a frequency distribution to gain a picture of the group of measurements *as a whole*. This

Man number	Blood type
1	A
2	B
3	A
4	A
5	A
6	AB
7	O
8	A
9	A
10	A
11	O
12	B
13	O
14	B
15	A
16	B
17	O
18	B
19	O
20	A
21	B
22	B
23	A
24	A
25	O

Table 2.3.1

Blood types of 25 men.

Class	f
A	11
B	7
AB	1
O	6
	25 = N

is true of all descriptive statistics; we want clear pictures of large numbers of measurements, and we can do this only by losing detail about particular objects. The process of weeding out particular qualities of the object that happen to be irrelevant to our purpose, begun whenever we measure, is continued when we summarize a set of measurements.

2.4 GROUPED DISTRIBUTIONS

In the distributions shown above, there were a few "natural" categories into which all the data fit. It often happens, however, that data are measured in some way giving a great many categories into which a given observation might fall. Indeed, the number of potential categories of description often vastly exceeds the number of cases observed, so that little or no economy of description would be gained by constructing a distribution showing each *possible* class. The most usual such situation occurs when the data are measured in numerical terms. For instance, in the measurement of height or weight, there are, in principle, an infinite number of numerical classes into which an observation might fall: all the positive real numbers. Even when the measurements result in fewer than an infinite number of possibilities, it is still quite common for the number of available categories to be very large.

For this reason, it is necessary to form **grouped** frequency distributions. Here, frequencies are assigned not to each possible measurement category, but rather to intervals or groupings of categories.

For example, consider a number of people who have been given an intelligence test. The data of interest are the numerical IQ scores. Immediately, we run into a problem of procedure. It could very well turn out that for, say 100 individuals, the IQ score would be different for each case, and thus a frequency distribution using each different IQ number as a measurement class would not condense that data any more than a simple listing.

The solution to this problem lies in grouping the possible measurement classes into new classes, called **class intervals,** each including several score possibilities. Proceeding in this way, if the IQs 105, 104, 100, 101, and 102 should turn up in the data, instead of listing each in a different class, we might put them all into a single class, 100–105, along with any other IQ measurements that fall between these limits. Similarly, we group other sets of numbers into class intervals. On our doing this, one way that the frequency distribution might look for a group of 150 persons is shown in Table 2.4.1.

Forming class intervals has enabled us to condense the data so that a simple statement of the frequency distribution can be made in terms of only a few classes. In this distribution, the various class intervals are shown in order in the first column on the left. The extreme right column lists the frequencies, each paired with the class in the row of the table. The sum of the frequencies must be N, the total number of observations. The middle column contains the midpoint of each class interval; these will be discussed further in Section 2.7

Class interval	Midpoint	f	**Table 2.4.1**
124–129	126.5	8	Intelligence quotients for a group of 150 persons, arranged in class intervals.
118–123	120.5	0	
112–117	114.5	10	
106–111	108.5	20	
100–105	102.5	65	
94– 99	96.5	22	
88– 93	90.5	23	
82– 87	84.5	2	
		150 = N	

2.5 CLASS INTERVAL SIZE AND CLASS LIMITS

The first problem in constructing a grouped frequency distribution is deciding how big the intervals shall be. What are the largest and the smallest numbers that may go into an interval? The difference between the largest and the smallest number that may go into any class interval is called the *interval size*. We will let the symbol i denote this interval size.

In the example above, as in most examples to follow, the size i is the same for each class interval. This is the accepted convention for most work in the social sciences. (Some exceptions will be mentioned in Section 2.9, however.) For the example, $i = 6$, which means that the largest and the smallest score going into a class interval will differ by six units. Take the class interval labeled 100–105. It may seem that the smallest number going into this interval is 100, and the largest 105; this is not, however, true. Actually, the numbers 99.5, 99.6, 99.7 would also go in this interval, should they occur in the data. Also, the numbers 105.1, 105.2 are included. On the other hand, 99.2 or 105.8 would be excluded. The interval actually includes any number *greater than or equal to* 99.5 *and less than* 105.5. These are called the **real limits** of the interval, in contrast to those actually listed, which are the **apparent limits.** Thus, the real limits of the interval 100–105 are

99.5 to 105.5

and

$i = 105.5 - 99.5 = 6.$

In general,

real lower limit = apparent lower limit − .5 (unit difference)

real upper limit = apparent upper limit + .5 (unit difference).

The term *unit difference* demands some explanation. In measuring something we usually find some limitation on the accuracy of the measurement, and seldom can one measure with *any* desired degree of accuracy. For this reason, measurement in numerical terms is always rounded, either during the measurement operation itself, or after the measurement has taken place. In measuring weight, for example, we obtain accuracy only within the nearest pound, or the nearest one tenth of a pound, or one hundredth of a pound, and so on. If one were constructing a frequency distribution where weight had been rounded to the nearest pound, then a unit difference is 1lb, and the real limits of an interval such as 150–190 would be 149.5–190.5. The i here would be 41 lbs.

On the other hand, suppose that weights were accuate to the nearest one tenth of a pound; then the unit difference would be .1 lb, and the real class interval limits would be

149.95–190.05.

The way that rounding has been carried out will have a real bearing on how the successive class intervals will be constructed. Suppose that we wanted to construct the distribution of annual income for a group of American men, where the income figures have been rounded to the nearest thousand dollars. We have decided that the apparent upper limit of the top interval shall be $25,000, and that i should equal $5,000. Then the classes would have the following limits:

Real	Apparent
20,500–25,500	21,000–25,000
15,500–20,500	16,000–20,000
10,500–15,500	11,000–15,000
5,500–10,500	6,000–10,000
500– 5,500	1,000– 5,000

Here one half of a unit difference is $500.

Now suppose that each man's income has been rounded to the nearest hundred dollars. Once again, with $i = $5,000 and with the top apparent limit $25,000, we form class intervals, but this time the class limits are:

Real	Apparent
20,050–25,050	20,100–25,000
15,050–20,050	15,100–20,000
10,050–15,050	10,100–15,000
5,050–10,050	5,100–10,000
50– 5,050	100– 5,000

This difference in unit for the two distributions could make a difference in the picture we get of the data. For example, a man making exactly $20,100 would fall into the top interval in the second distribution, but into the second from top interval in the first. It is important to decide upon the accuracy represented in the data before the real limits for the class intervals are chosen. Is every digit recorded in the raw data regarded as significant, or will the data be rounded to the "nearest" unit?

2.6 INTERVAL SIZE AND THE NUMBER OF CLASS INTERVALS

One can use any number for i in setting up a distribution. However, it should be obvious that there is very little point in choosing i smaller than the unit difference (for example, letting $i = .1$ lb when the data are in nearest whole pounds), or in choosing i so large that all observations fall into the same class interval. Furthermore, i is usually chosen to be a *whole number* of units, whatever the unit may be. Even within these restrictions, there is considerable flexibility of choice. For example, the data in Section 2.4 could have been put into a distribution with $i = 3$, as shown in Table 2.6.1. Notice that this distribution with $i = 3$ is somewhat different from the distribution with $i = 6$, even though they are based on the same set of data. For one thing, here there are 16 class intervals, whereas there were 8 before. This second distribution also gives somewhat more detail then the first about the original set of data. We now know, for example,

Class interval	Midpoint	f
126–128	127	5
123–125	124	3
120–122	121	0
117–119	118	0
114–116	115	7
111–113	112	3
108–110	109	8
105–107	106	13
102–104	103	44
99–101	100	20
96– 98	97	10
93– 95	94	12
90– 92	91	14
87– 89	88	9
84– 86	85	0
81– 83	82	2
		$150 = N$

Table 2.6.1

Intelligence quotient scores of 150 persons, arranged into intervals of three units.

that there was no one in the group of persons who got an IQ score between the real limits of 83.5 and 86.5, whereas we could not have told this from the first distribution. We can also tell that more cases showed IQs between 101.5 and 104.5 than between 98.5 and 101.5; conceivably this could be a fact of some importance. On the other hand, the first distribution gives a simple and, in a sense, a neater picture of the group than does the second.

The decision facing the maker of frequency distributions is, "Shall I use a small class-interval size and thus get more detail, or shall I use a large class-interval size and get more condensation of the data?" There is no fixed answer to this question; it depends on the kind of data and the uses to which they will be put. However, a convention does exist which says that for most purposes *10 to 20 class intervals* give a good balance between condensation and necessary detail, and in practice one usually chooses class-interval size to make about that number of intervals.

A handy rule of thumb for deciding about the size of class intervals is given by

$$i = \frac{\text{highest score in data } - \text{ lowest score}}{\text{number of class intervals}}.$$

After deciding on some convenient number of class intervals, you divide this number into the difference between the highest and lowest scores, or the **range** of scores, and find the size that i will have to be. In practice, this may give an i that is not a whole number, in which case you simply use the nearest whole number for i. In the example just given, the data showed 128 as the highest and 82 as the lowest scores, with a range of 46. To find i giving about 16 class intervals we would divide the range by 16:

$$\frac{128 - 82}{16} = 2.87.$$

Since this is a decimal number, we choose the nearest whole number, or 3.

Note that in the last distribution some intervals show a frequency of 0; it is not absolutely necessary that such intervals be listed. However, intervals with zero frequency are usually listed when they fall between other intervals that do not have a zero frequency. We could have had an interval 129–131 having a zero frequency in the last distribution; however, we did not list this interval because, unlike the interval 120–122, it did not fall between other nonzero intervals. In addition, it is important to take the total N into account when deciding upon the number and size of class intervals. It is really not very interesting to look at a distribution with fewer than five or so observations per class interval on the average, although in some exceptional situations this might be allowable. Usually, then, when fairly small numbers of observations are involved, it will pay to examine the average number of cases per class interval. This can be found from N/C, where C is the number of class intervals, or from iN/range. If this number comes out to be as small as 5 or less, consideration should be given to a larger interval size, or perhaps even to an unequal interval size over some intervals of values.

Two other conventions are useful in making distributions from data. The first is that, in the social sciences, the class intervals are usually listed starting with low numbers at the bottom of the list and with high numbers at the top, as shown in the example. Another convention followed here is to start figuring the class intervals by listing the *highest score* in the data as the *apparent upper limit of the top interval;* then it is a simple matter to find the other apparent upper limits by subtracting the class-interval size successively from this number. The apparent *lower* limit of each interval is then *one unit more* than the *upper* apparent limit of the interval below it. For example, suppose 128 were the top score in the data, and that i were 3. Taking 128 as the highest apparent limit, the upper limit of the next interval would be $128 - 3$, or 125, next would be $125 - 3$ or 122, and so on, until all upper limits were found. Then, given an interval with an apparent upper limit of 122, the lower limit of the next interval up would be 123, and given the interval with upper limit 125, the lower limit of the next interval up would be 126, and so on.

2.7 MIDPOINTS OF CLASS INTERVALS

There is one more feature of frequency distributions that has not yet been discussed. On putting a set of measurements into a frequency distribution, one loses the power to say exactly what the original numbers were. For the example of Section 2.6 a person in the group of 150 cases who has an IQ of 102 is simply counted as one of the 44 individuals who make up the frequency for the interval 102–104. You cannot tell from the distribution exactly what the IQs of those 44 individuals were, but only that they fell within the limits 102–104. What do we call the IQs of these 44 individuals? We call them all 103, the midpoint of the interval. **The midpoint of any class interval is that number which would fall exactly halfway among the possible numbers included in the interval.** A moment's thought will convince you that 103 falls halfway between the real limits of 101.5 and 104.5. In a like fashion, all 5 cases falling into the class interval 126–128 will be called by the midpoint 127; all 3 cases in the interval 123–125 will be called 124, and so on for each of the other class intervals.

The real limits can also be defined in terms of the midpoint:

real limits $=$ midpoint $\pm$ $.5i$.

Thus, given only the midpoints of the distribution, one can find the class limits.

In computations using grouped distributions the midpoint is used to substitute for each raw score in the interval. For this reason it is convenient to choose an odd number for i whenever possible; this makes the midpoint a whole number of units and simplifies computations. (Actually, this is a far less important consideration than in former times. Efficient calculators and computers have almost eliminated the need for calculations from such grouped data, even for a large N value.)

2.8 ANOTHER EXAMPLE OF A GROUPED FREQUENCY DISTRIBUTION

Suppose that 75 students in high school had been used as subjects in an experiment on verbal memory. Each student was given a list containing 48 pairs of words and allowed to study the list as a whole for 10 mins. The first member of each pair was then shown in order to the individual student, and his task was to recall the second word. The raw data are shown in Table 2.8.1. The largest number of words recalled by any student was 35, and the smallest number was 10. The range was thus $35 - 10$ or 25.

Student	Score	Student	Score	Student	Score
1	19	26	18	51	15
2	16	27	14	52	15
3	11	28	26	53	14
4	13	29	16	54	11
5	12	30	19	55	23
6	20	31	21	56	13
7	11	32	17	57	19
8	24	33	20	58	17
9	19	34	12	59	21
10	16	35	22	60	15
11	12	36	11	61	20
12	24	37	13	62	16
13	19	38	16	63	12
14	17	39	10	64	10
15	18	40	17	65	10
16	25	41	17	66	15
17	16	42	19	67	18
18	20	43	11	68	10
19	17	44	10	69	19
20	19	45	15	70	14
21	15	46	10	71	12
22	35	47	13	72	18
23	16	48	13	73	14
24	11	49	14	74	15
25	26	50	10	75	20

Table 2.8.1

Scores of 75 students in one trial of paired-associate learning.

Now we want to make a frequency distribution having about 10 class intervals. According to our rule of thumb, given above:

$$i = \frac{\text{range}}{\text{number of intervals}} = \frac{25}{10} = 2.5.$$

Thus $i = 3$ should provide us with a convenient class-interval size.

We start with the largest number, 35, and make it the upper apparent limit of the highest class interval. That interval must then have real limits of 32.5 and 35.5 with apparent limits of 33 to 35. The real limits of the second from highest interval must be 29.5–32.5, and so on. Each time, the difference between the real limits to an interval must be i, or 3, and the differences between the successive lower apparent limits (and also successive upper limits) must also be 3. Proceeding in this way, we find the class intervals shown in Table 2.8.2.

The table is completed by inspecting the raw data to find the number of cases that fall into each class interval. Only one case falls between the real limits 32.5–35.5, and so the frequency for the highest interval is 1. The frequency for the lowest interval is 13, since exactly 13 individuals showed scores between the real limits 8.5 and 11.5. Just as in the preceding examples, the midpoint for each interval stands exactly midway between the two real limits.

What can the experimenter tell from looking at this distribution that would not have been obvious from the raw data? First of all the "typical" range of performance is really a fairly short interval of scores; the large majority of students scored in the range of 11 points from 9 through 20. The most "popular" score interval is 15–17. Note that this concentration of cases lies at the low end of the *conceivable range* of scores 0 through 48. If the experimenter thinks of a score as reflecting the student's ability to memorize paired associates, then the task is, by and large, a hard one for the students. The single individual scoring in the interval 33–35 is really most atypical. This student's score falls very far from the bulk of the cases in the distribution.

Naturally, there are other, more precise statements that the experimenter can make about what these data show. We will discuss these summary indices in Chapter 4. For the moment, however, it should be clear that a distribution does communicate information about a set of observations *as a whole* even without further analysis of the data.

Class interval	Midpoint	f
33–35	34	1
30–32	31	0
27–29	28	0
24–26	25	5
21–23	22	4
18–20	19	17
15–17	16	20
12–14	13	15
9–11	10	13
		75 = N

Table 2.8.2

Number of paired associates recalled on the first trial by a sample of 75 high-school students.

2.9 FREQUENCY DISTRIBUTIONS WITH OPEN OR UNEQUAL CLASS INTERVALS

Quite often the data are such that it is not possible to make a frequency distribution with intervals of a constant size. This most commonly occurs when exact scores are not known for some of the cases. For example, in a study of the trials that it takes an animal to learn a discrimination problem, the experimenter found that out of 100 animals, 5 could not learn the problem in 60 trials. It was felt that some of the animals would never learn the problem at all, and running these animals ceased at the 60th trial. This means, however, that the five animals could not be given an exact score, and could only be put in the top class of the score distribution, in an interval called "60 or more." This interval is **open,** since there is no way to determine its upper real limit. Furthermore, there is no way to give such an open interval a midpoint.

In other instances, it may be that there are extreme scores in a distribution that are widely separated from the bulk of the cases. When this is true, a class-interval size that is small enough to show detail in the more concentrated part of the distribution will eventuate in very many classes with zero frequency before the extreme scores are included. Enlarging the class-interval size will reduce the number of unnecessary intervals but will also sacrifice detail. When this situation arises, it is often wise to have a varying class-interval size, so that the class intervals are narrow where the detail is desired, but rather broad toward the extreme or extremes of the distribution. In this case, class-interval limits and midpoints can be found in the usual way, provided that one takes care to notice where the interval size changes.

2.10 GRAPHS OF DISTRIBUTIONS: HISTOGRAMS

Now we direct our attention for a time to the problem of putting a frequency distribution into graphic form. Often a graph shows that the old bromide, "a picture is worth a thousand words," is really true. If the purpose is to provide an easily grasped summary of data, nothing is so effective as a graph of a distribution.

There are undoubtedly all sorts of ways any frequency distribution might be graphed. A glance at any national news magazine will show many examples of graphs which have been made striking by some ingenious artist. However, we shall deal with only three "graden varieties": the histogram, the frequency polygon, and the cumulative frequency polygon.

The histogram is really a version of the familiar bar graph. In a histogram, each class or class interval is represented by a portion of the horizontal axis. Then over each class interval is erected a bar or rectangle equal in height to the frequency for that particular class or class interval. The histograms representing the frequency distributions in Sections 2.3 and 2.6 are shown in Figures 2.10.1 to 2.10.3. It is customary to label both the horizontal and the vertical axes as shown in these figures, and to give a label to the graph as a whole. In the case where class intervals are employed to group numerical measurements, a saving of time and space may be accomplished by labeling the class intervals by their midpoints (as in Figure 2.10.3) rather than by their apparent or real limits. With categorical measurements, this problem does not arise, of course, since each class can only be given its name or symbol.

Figures 2.10.1/2.10.2

Blood types of 25 American males *(left)*. Taste ratings for 2,000 American families *(right)*.

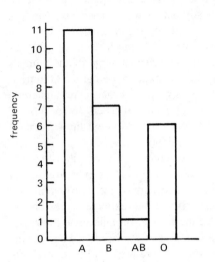 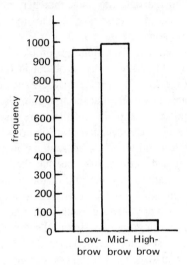

If you will look at the two historgrams shown in Figures 2.10.4 and 2.10.5 you will get very different first impressions, even though they represent the same frequency distribution. These two figures illustrate the effect that proportion can have on the viewer's first impressions of graphed distributions. Various conventions about graphs of distributions exist in different fields, of course. In fact, those who make a business of using statistics for persuasion often choose a proportion designed to give a certain impression. However, it is most usual to find graphs arranged so that the vertical axis is about three-

Figure 2.10.3

Intelligence quotients of a sample of 150 subjects, with $i = 3$.

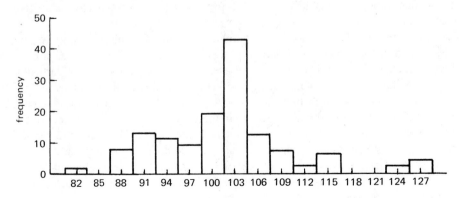

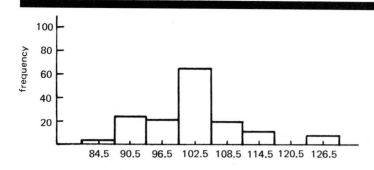

Figure 2.10.4

Intelligence quotients of a sample of 150 subjects, with *i* = 6.

fourths as long as the horizontal. This usually will give a clear and esthetically pleasing picture of the distribution. (An amusing and instructive recital of ways to "adjust" graphs and other statistical presentations so as to get a desired effect is given in the little book by Huff, *How to Lie with Statistics,* 1954)

A word must also be said about pie charts and bar charts, which are alternative ways to display the information contained in one or more frequency distributions. These figures are especially common as computer-generated graphics. A pie chart, like that in Figure 2.10.6 and based on the data of Table 2.3.1, is especially useful when the distribution describes a few qualitative classes, and one wishes to emphasize the relative frequencies or proportions that fell into each class. Each class is shown by one wedge of the total "pie," and distinctive hatchings, shadings, or colors are used to point up the different groupings. Sometimes, when it is desired to emphasize one particular class, and especially a class with small frequency, the pie is "exploded" to bring out and emphasize the wedge in question. Clear and adequate labeling of pie charts can be a problem when there are a number of classes to be shown. In addition, pies are not as

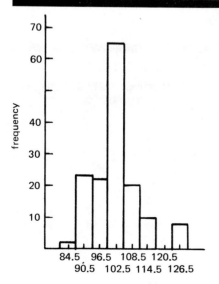

Figure 2.10.5

Another version of Figure 2.10.4.

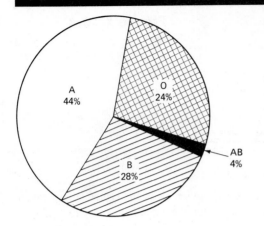

Figure 2.10.6

Blood types of 25 American males.

effective for numerical data as for qualitative classes because the circle is confusing when ordered classes are represented. Pies are also not very effective for displaying two or more sets of data at once.

The problem of showing two or more distributions for comparison or contrast is better handled by so-called bar charts. These are, essentially, histograms in which there is a spatial separation between adjacent classes. For displaying two or more distrubtions at the same time, especially when the basic measurement classes are qualitative, the frequency bars for the different data sets are lined up side by side, as in Figure 2.10.7. However, when the classes are actually grouped into class intervals, the frequency bars for the different sets of data are often superimposed. The bars may be oriented vertically or horizontally. Shadings, hatchings, or different colors are used to distinguish among the data sets being compared.

Microcomputer graphics and sophisticated printers are capable of an incredible variety of graphs, although a good bit of planning and a certain amount of taste on the part of

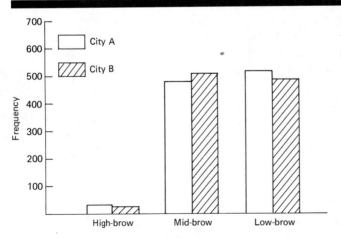

Figure 2.10.7

Taste ratings for 1,000 families in each of two American cities.

the user must be exercised if these graphs are to be effective. One of the dangers of computer-generated graphs such as these, in which many different line patterns or shadings are possible, is that visually the graph will be so "busy" that it fails to communicate the information intended. A good discussion of general principles for creating graphs that avoid such pitfalls is given in Tufte (1983).

2.11 FREQUENCY POLYGONS

The histogram is a useful way to picture any sort of frequency distribution, regardless of the scale of measuement of the data. However, a second kind of graph is often used to show frequency distributions, particularly those that are based upon numerical data: this is the frequency polygon. In order to construct the frequency polygon from a frequency distribution one proceeds exactly as though one were making a histogram; that is, the horizontal axis is marked off into class intervals and the vertical into numbers representing frequencies. However, instead of using a bar to show the frequency for each class, a point on the graph corresponding to the midpoint of the interval and the frequency of the interval is found. These points are then joined by straight lines, each being connected to the point immediately preceding and the point immediately following, as in Figure 2.11.1.

The frequency polygon is especially useful when there are a great many potential class intervals, and it thus finds its chief use with numerical data that could, in principle, be shown as a smooth curve, given enough observations. This would be true of a distribution based on a very large number of numerical measurements on a potentially continuous scale. If we maintained the same proportions in our graph but employed a much greater number of class intervals, the frequency polygon would provide a picture more like a smooth curve, as in Figures 2.11.2 to 2.11.4, based on several thousand cases. Although the function rule describing the frequency polygon may be extremely complicated to state, the function rule for a smooth curve approximately the same as that of the distribution may be relatively easy to find. For this reason, a frequency polygon based upon interval or ratio scale measurement with a relatively large number

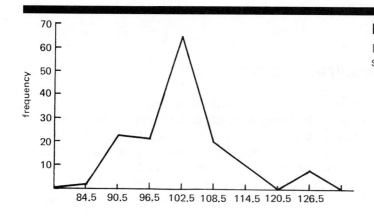

Figure 2.11.1

Intelligence quotients of a sample of 150 subjects.

Figures 2.11.2–2.11.4

Frequency polygons for a large number of cases arranged into successively smaller class intervals.

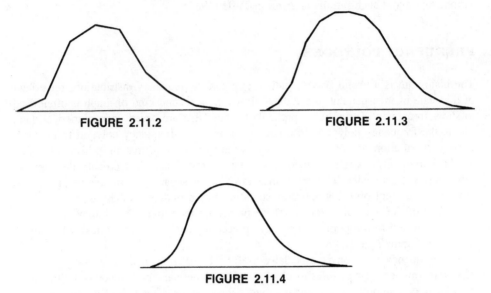

FIGURE 2.11.2   FIGURE 2.11.3

FIGURE 2.11.4

of classes is sometimes "smoothed" by creating a curve that approximates the shape of the frequency polygon. This new curve may have a rule that is simple to state and that may serve as an approximation of the rule describing the frequency distribution itself. Although the procedures for smoothing such a frequency polygon and finding a function rule that comes close to describing the distribution are outside the scope of this book, they may be found in advanced texts on the analysis of experimental data. Suffice it to say, for the moment, that when we come to a discussion of the "shapes" of distributions, we shall often refer to smooth curves as examples of types of distributions, when actually the frequency distributions themselves could only be represented by "jagged" polygons composed of straight lines. Nevertheless we may use the smooth curves as good approximations of the picture that the frequency polygon presents of the distribution.

2.12. CUMULATIVE FREQUENCY DISTRIBUTIONS

For some purposes it is convenient to make a different arrangement of data into a distribution, called a cumulative frequency distribution. For example, when it is desired to find where certain scores rank relative to all of the others ("percentile ranks," as discussed in Section 4.19), a cumulative distribution may be used. Often a learning curve may be shown as a form of cumulative distribution as well.

Instead of showing the relation that exists between a class interval and its frequency, **a cumulative distribution shows the relation between a class interval and the fre-**

Class interval	Cumulative frequency
124–129	150
118–123	142
112–117	142
106–111	132
100–105	112
94– 99	47
88– 93	25
82– 87	2

Table 2.12.1

Cumulative frequency distribution for intelligence score data of Section 2.4.

quency at or below its real upper limit. In other words, the cumulative frequency shows how many cases fall into *this interval or below*. Thus, in the distribution in Section 2.4 the lowest interval shows two cases, and its cumulative frequency is 2. Now the next class interval has a frequency of 23, so that its cumulative frequency is 23 + 2, or 25. The third interval has a frequency of 22, and its cumulative frequency is 22 + 23 + 2, or 47, and so on. The cumulative frequency of the very top interval must always be N since all cases must lie either in the top interval or below. The cumulative frequency distribution for the example in Section 2.4 is given by Table 2.12.1.

Cumulative frequency distributions are often graphed as polygons. The axes of the graph are laid out just as for a frequency polygon, except that the class intervals are labeled by their *real upper limits* rather than by their midpoints. The numbers on the vertical axis are now cumulative frequencies, of course. The graph of the example from Section 2.4 is shown in Figure 2.12.1.

Estimates of percentile ranks may be read from a cumulative frequency polygon such as Figure 2.12.1 by finding the cumulative frequency for any number and then multiplying this cumulative frequency by $100/N$. Thus, for example, the percentile rank corresponding to 99.5 is $47(100)/150$ or 31.33. The cumulative frequency for a score of 92 is about 19, so that its percentile rank is about $19(100)/150 = 12.7$, and so on. Methods for determining exact value for percentile ranks are discussed in Section 4.19.

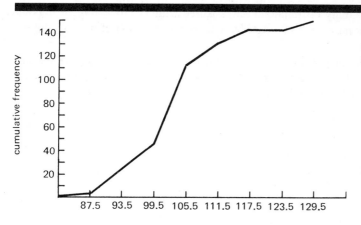

Figure 2.12.1

Cumulative frequency polygon for data of Table 2.12.1.

Cumulative frequency distributions often have more or less this characteristic **S** shape, although the slope and the size of the "tails" on the **S** will vary greatly from distribution to distribution. Like a frequency polygon, a cumulative frequency polygon is sometimes smoothed into a curve as similar as possible to the polygon. You sometimes encounter a reference to such a smoothed cumulative distribution curve under the term **ogive.**

2.13 PROBABILITY DISTRIBUTIONS

Suppose that there were some finite set of objects, and that we sampled these objects at random with replacement. Each object sampled is measured in some way. The measurement may be at the nominal, ordinal, interval, or ratio scale level, so long as each object is measured by the same procedure. Now notice that the occurrence of any object in a sample is an elementary event, and that the *actual measurement class or number* assigned any object is an event. Since many different elementary events can be allotted the same number or classification, each measurement class is an event class. These measurement events are mutually exclusive and exhaustive, each object being given one and only one number or category.

Furthermore, since the different possible measurement classes define events, there is a probability that can be assigned to each; there is, for example, some probabilty that the score of an object will be 10, or that the object will be assigned to the class "high need for achievement." In short, given the sample space S and a set of measurement events, we can think of a probability function pairing some probability with each possible measurement class.

Any statement of a function associating each of a set of mutually exclusive and exhaustive events with its probability is a probability distribution.

Notice how the idea of probability distribution exactly parallels that of frequency distribution: each type is based on a set of mutually exclusive and exhaustive measurement classes. In a frequency distribution, each measurement class is paired with a frequency, and in a probability distribution each is paired with its probability. Just as the sum of the frequencies must be N in a distribution, so must the sum of the probabilities be 1.00.

As a simple example of probability distribution, imagine a sample space of all girls of high-school age. Such a girl is selected at random and classified as "right-handed," "left-handed," or "ambidextrous." The probability distribution might be:

Class	p
right	.60
left	.30
ambidextrous	.10
	1.00

Or, perhaps, the height of a boy drawn at random from all U.S. boys is measured. Some seven class intervals are used to record the height of any boy observed, and the probability distribution might be:

Height in inches	p
78–82	.002
73–77	.021
68–72	.136
63–67	.682
58–62	.136
53–57	.021
48–52	.002
	1.000

Here, each class interval is an event.

Corresponding to any frequency distribution of N cases there is a probability distribution, if single cases are selected from the total group of N cases at random with replacement. Thus, imagine a set of 50 persons, whose scores on some test formed the following distribution:

Scores	f
90–99	3
80–89	8
70–79	15
60–69	14
50–59	10
	50

One person is drawn at random from among this set. What are the probabilities associated with the various intervals he might fall into? If the individuals are all assumed equally likely to be observed, then for any class interval A.

$$pA = \frac{\text{frequency of } A}{\text{total frequency}}.$$

Thus, the class interval 90–99 has a probability of 3/50, or .06, and the other probabilities can be found in the same way, giving

Scores	p
90–99	.06
80–89	.16
70–79	.30
60–69	.28
50–59	.20
	1.00

In short, any group of cases summarized in a frequency distribution can also be regarded as a sample space of *potential* observations. If a single case is sampled at random, then the probabilities of the various classes are the same as the relative frequencies of those classes.

Furthermore, given some theoretical probability distribution, and N observations made independently and at random with replacement, then there is a *theoretical frequency distribution* as well, where for any event class A

theoretical frequency $= Np(A)$.

This is the frequency distribution we *expect* to observe, given the theoretical probability distribution. For example, suppose that the probability distribution of heights of boys given above were true for some specific population of boys. What is the theoretical frequency distribution of heights for 1,000 boys sampled independently and at random? The answer is

Height in inches	$f = Np(A)$
78–82	2
73–77	21
68–72	136
63–67	682
58–62	136
53–57	21
43–52	2
	1,000

Given that the probability distribution is correct, then for a sample of 1,000 boys we should expect our obtained frequency distribution to show 682 boys in the interval 63–67 in. only 2 boys in the interval 78–82, and so on. This distribution might never actually occur in practice, but it is our best guess about what will occur if the probability distribution is a correct statement. Whereas the frequencies in real distributions must always be whole numbers, in theoretical distributions it is possible to have fractional frequencies; if only 100 boys were observed in this example, then the frequencies for the top and bottom intervals would be, in theory, .2 each.

Like frequency distributions, a probability distribution may be specified either by a listing such as we have shown here, or in graphic form. However, unlike actual obtained distributions of data, the most important probability distributions are theoretical ones that can be given a mathematical rule. In order to discuss such distributions we need the concept of a "random variable."

2.14 RANDOM VARIABLES

A special terminology is useful for discussing probability distributions of numerical scores. Imagine that each and every possible elementary event in some S is assigned a number. That is, various elementary events are paired with various values of a variable.

Thus, an elementary event may be a person, with some height in inches; or the elementary event may be the result of tossing a pair of dice, with the assigned number being the total of the spots that come up; or the elementary event may be a rat, with the number standing for the trials taken to learn a maze. Each and every elementary event is thought of as getting one and only one such number.

Let X represent a function that associates a real number with each and every elementary event in some sample space $\mathcal{S}$. Then X is called a random variable on the sample space $\mathcal{S}$.

Bear in mind the distinction between the elementary events themselves, which may or may not be numbers, and the values associated with the various elementary events, which are the values of the random variable. Thus, for example, consider the set of American males 21 years of age or older. An individual is drawn at random from this set. The sample space here consists of the entire set of American males of specified age, and each such male is an elementary event in this sample space. Now we can associate with each elementary event a real value, the income of the man during the current calendar year. The values that the random variable X can thus assume are the various income values associated with the men. The particular value x occurs when a man is chosen who has income x.

On the other hand, suppose that a box were filled with slips of paper, each inscribed with a real number. Here the elementary events are the slips bearing the numbers. The value x is associated with a particular slip, so that x occurs when the particular slip is drawn at random. However, there are many different assignments of numbers to slips that might be chosen for X. Thus, we might define X as the square of the number on a slip, so that $X = 4$ occurs when a slip is drawn bearing the number 2. The point is that a random variable represents values that are associated in some way with elementary events, so that particular values of X occur when the appropriate elementary events occur. The way in which the numerical values get associated with the elementary events is open to the widest latitude of definition, however.

Although the term *random variable* is rather awkward, it will be used here because of its popularity in statistical writing. The terms *chance variable* or *stochastic variable* are sometimes used. These all mean precisely the same thing, however; a symbol for number events each having a probability.

Given the random variable X, other events may be defined. Thus, given two numbers a and b,

$$a \leq X \leq b$$

is the event of some value of X lying between the numbers a and b (inclusive). This event has some probability $p(a \leq X \leq b)$. Furthermore, there is the event

$$a \leq X,$$

a value greater than or equal to a; this has probability $p(a \leq X)$.

As another example of a random variable, suppose that X symbolizes the height of an American man, measured to the nearest inch. Here, there is some probability that X

≤ 60, or $p(X \leq 60)$, since there is presumably some set of American men with heights less than or equal to 60 ins. Furthermore, there is some probability $p(70 \leq X \leq 72)$, since there is a set of American men having heights between 5 ft 10 ins. and 6 ft, inclusive. Here, X symbolizes the numerical value (height in inches) assigned to an American man (an elementary event). This symbol X represents any one of many different such values, and for any arbitrary number a there is some probability that the particular value x is less than or equal to the value of a.

Although the notation employed for random variables varies considerably among various authors, we will find it convenient to use capital letters, such as X, Y, Z, to denote random variables, and lowercase letters, such as x, y, z and $a, b, c,$ to denote *particular values* that the random variable in question can take on. (Occasionally, this convention becomes awkward, and it will be violated from time to time in future sections. However, the context will usually make the random variable notation clear in these instances.) The expression $p(X = x)$ symbolizes the probability that the random variable X takes on the particular value x. Often, for convenience, this will be written simply as $p(x)$. In the same way, $p(a \leq X)$ stands for the probability that X takes on some particular value x greater than or equal to the value symbolized by a, $p(a \leq X \leq b)$ symbolizes the event in which X takes on some value lying between the two values symbolized by a and b, and so on.

2.15 DISCRETE RANDOM VARIABLES

In a great many situations, only a limited set of numbers can occur as values of a random variable. That is, some values have zero probability, even though there may be values to either side that have probability greater than zero. Quite often, the set of numbers that can occur as values of the random variable is relatively small, or at least finite in extent. As an example of a limited set of values assumed by a random variable, suppose that a volume of the *Encyclopedia Britannica* is opened at random, and the page number noted. In this instance, the values of the random variable are all of the different page numbers that might occur. These would be all of the whole numbers between 1 and, say, 1009 (perhaps with a few Roman numbers thrown in). The number of possibilities is thus finite, and, indeed, quite limited with respect to all possible numbers. Real numbers such as -34.6, 1.1103, and the square root of 2 would not occur, of course.

Some random variables can assume what is called a "countably infinite" set of values. One example of a countably infinite set would be the ordinary counting numbers themselves, where the count goes on without end. A simple experiment in which one counts the number of trials until a particular event occurs would give a random variable taking on these counting values.

In either of these situations, the random variable is said to be discrete. **If a random variable X can assume only a particular finite or countably infinite set of values, it is said to be a discrete random variable.** As we shall see, not all random variables are discrete, but a large number of random variables of practical and theoretical interest to us will have this property.

If cases are sampled from any frequency distribution that has been grouped into class intervals, then the random variable X is usually regarded as symbolizing the midpoints of the intervals. For a finite number of class intervals, such an X is a discrete random variable. Also, when measurements have been rounded to the *nearest* unit, as in height to the nearest inch, this is like forming class intervals, and the random variable standing for such a rounded measurement is often thought of as discrete. In fact, almost any practical situation involving numerical data can be thought of in terms of a discrete random variable.

Probability calculations are often very simple when one is dealing with a discrete random variable where only very few values can occur. For example, take the simple experiment of rolling a pair of fair dice. Each die has six sides, and each side contains one to six spots.

Table 2.15.1 shows the sample space S for the simple experiment of rolling two fair dice. Each dot is an elementary event. Thus the dot corresponding to 5 for die 1 and 4 for die 2, is the elementary event "die 1 comes up 5 and die 2 comes up 4." You will notice that there are exactly 36 possiblities, or 36 elementary events, in this sample space.

Now let us define the random variable X to be the sum of the spots on the two dice. That is,

X = (spots on die 1) + (spots on die 2).

Table 2.15.2 then gives the value of X associated with each elementary event in S .

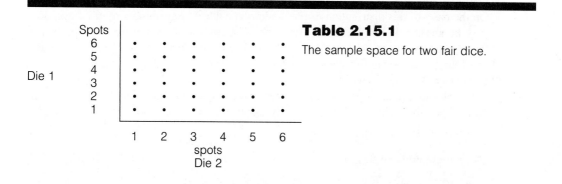

Table 2.15.1

The sample space for two fair dice.

Table 2.15.2

A random variable X defined on the sample space of Table 2.15.1.

x	p
12	1/36
11	2/36
10	3/36
9	4/36
8	5/36
7	6/36
6	5/36
5	4/36
4	3/36
3	2/36
2	1/36
	36/36

Table 2.15.3

Probability distribution for the number of spots appearing on two fair dice.

Here, we are assuming that the dice are both fair, so that the probability of each elementary event is the same, or 1/36. Then what are the probabilites of the various values of the random variable associated with these elementary events? Notice that the value "2" is associated with exactly one elementary event, so that its relative frequency, or probability, is 1/36. The value "3" is associated with two elementary events, so that the probability of a "3" is 2/36. The value of "7" has the highest relative frequency, so that the probability of a "7" is 6/36. Proceeding in this way we can work out the following probability distribution of the particular values x of the random variable X, shown in Table 2.15.3.

On the basis of this distribution of the random variable X, any number of other questions may be answered. For example, often one is interested in finding the probability that the obtained value of some random variable X will fall *between* two particular values, or in some interval. In this instance, what is the probability that X, the number of spots, is between 3 and 5 inclusive?

In order to find such probabilities, we rely once again on Rule 5 of Section 1.5. The various possible values of X are mutually exclusive events, and so

$$p(3 \leq X \leq 5) = p(3 \cup 4 \cup 5) = p(3) + p(4) + p(5).$$

Using these probabilities, we find

$$p(3 \leq X \leq 5) = p(X = 3) + p(X = 4) + p(X = 5)$$

$$= \frac{2 + 3 + 4}{36} = \frac{1}{4}.$$

Similarly,

$$p(9 \leq X \leq 10) = p(X = 9) + p(X = 10) = \frac{7}{36}.$$

In general, the probability that X falls in the interval between any two numbers a and b, inclusive, is found by the sum of probabilities for X over *all possible* values between a and b, inclusive:

$p(a \leq X \leq b) = $ sum of $p(X = c)$ for all c such that $a \leq c \leq b$.

By the same argument,

$p(a \leq X) = $ sum of $p(X = c)$ for all c such that $a \leq c$.

Furthermore,

$p(X < a) = 1 - p(a \leq X)$.

For this example,

$$p(5 \leq X) = p(X = 5) + \cdots + p(X = 12) = \frac{4}{36} + \cdots + \frac{1}{36} = \frac{30}{36} = \frac{5}{6}.$$

On the other hand;

$$p(X < 5) = p(X = 4) + p(X = 3) + p(X = 2) = \frac{3}{36} + \frac{2}{36} + \frac{1}{36} = \frac{1}{6}.$$

This example, simple as it is, illustrates the main concepts underlying any discrete random variable. A simple experiment generates a sample space of possible outcomes or elementary events, such as that illustrated in Table 2.15.1. Then, each elementary event is thought of as associated with a number, such as the sum of the number of spots on the dice as in the example. Such numerical values are themselves events, and have probabilites. The value events, or values of the random variable, can then be displayed along with their probabilities in the form of a probability distribution. The probabilites for other numerical events, such as intervals of values, can be calculated from the distribution.

Notice that this example is purely theoretical. It rests on the theory that the two dice in question are fair, so that each side of each die is equally likely to come up. This need not be true, of course. However, we could use this theoretical distribution of a discrete random variable to check on the fairness of an actual pair of dice. Thus, if we toss a real pair of dice 100 times, we expect that $100 \times (6/36)$ or about 17% of the tosses should show the event "7," and in the same way we could work out the other frequencies we expect, if the dice are fair. A comparison of the frequencies we obtain with the frequencies we expect then gives us evidence about the fairness or unfairness of the dice.

Once again, this is an illustration of very simple statistical inference. We have a question, "Are these particular dice fair, or are they loaded?" That question is to be answered by actually examining the behavior of the pair of dice. The theory of probability tells us the probabilities of the various outcomes for tossing a pair of fair dice, and the idea of a random variable lets us construct the theoretical distribution of values that a pair of fair dice should show, along with their probabilities. These probabilities tell the long-run relative frequencies that we should expect any pair of fair dice to show. For any N actual trials, we also know the expected frequency of any value by taking N times the probability of that value. Thus, we can construct a distribution of expected frequencies from the probability distribution. Now we actually toss the dice N times. For any N we can construct a frequency distribution of the results. A comparison of the frequency distribution we *obtain* with the frequency distribution we *expect* gives evidence in answer to our original question. If the obtained frequencies are very different from those that we expect, then we have evidence that the dice are not fair, and we reject our theory that they are. On the other hand, if the expected and the obtained frequencies are very close, we have no good basis for saying that the dice are anything but fair.

Most of the things to be discussed in succeeding sections of this book are simply ways of formalizing this basic set of procedures, and extending them to more complex situations. Moreover, this general idea will underlie almost all of the methods of statistical inference still to be introduced.

2.16 GRAPHS OF PROBABILITY DISTRIBUTIONS

The same general procedure is used for graphing either a frequency distribution or the probability distribution of a discrete random variable. The two most common forms are histograms and probability mass functions.

First, let us deal with the histogram. When the random variable is discrete and can take on only a few values, it is customary to show each value as an interval on the X axis, with the height of the bar above that interval indicating the probability. For example, the distribution in Table 2.15.3 gives the histogram of Figure 2.16.1. Notice that each of the blocks making up the column above a value of X has exactly the same area, and the number of blocks divided by the total number gives the probability. In other words, if each block had an area of 1/36, the total area would be 1.00, and the area of any column would be the probability of that value of X.

This idea extends to the histogram of any probability distribution: the total area covered by all columns is regarded as 1.00, and the area of any single column representing an interval or class is thought of as a probability for that interval or class. Another example is given by Fig. 2.16.2.

The use of a histogram for the distribution of a discrete random variable emphasizes the analogy between probabilities and relative areas; with total probability corresponding to total area in the graph. The values on the X axis to either side of any midpoint of an interval are treated as though they could occur with the same probability as the midpoint itself. This is, of course, not generally the case with a discrete random variable, al-

Figure 2.16.1

Relative frequencies for the sum of the spots on two fair dice.

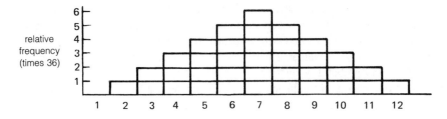

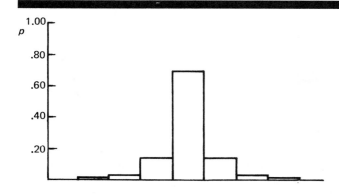

Figure 2.16.2

A hypothetical probability distribution graphed as a histogram.

though this convenient fiction is adopted in order that the connection between areas and probabilities can be seen.

An alternative method of graphing the probability distribution for a discrete random variable emphasizes the discreteness of the random variable at the expense of the analogy between probability and area. This alternative form of graph is called a **probability mass function.** An example of a probability mass function is shown in Figure 2.16.3.

Note that in the probability mass function graphed in Figure 2.16.3 the same basic features exist as in a histogram. The X axis of the graph is marked off with the values that the random variable can actually assume. The Y axis, which may or may not be explicitly shown, is a measure of units of probability. Then above each possible value x of X a vertical line is raised to the height corresponding to $p(X = x)$. The discreteness of the random variable is emphasized by the gaps between the vertical lines standing for the various possible values of X. A value corresponding to a point in one of these gaps has zero probability.

The probabilities, or masses, in any such probability mass function must be nonnegative, of course, and the sum of the probabilities over all possible values of X must be equal to 1.00.

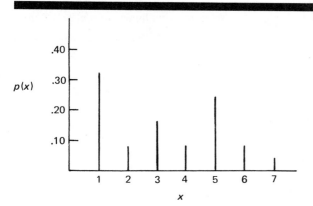

Figure 2.16.3

Probability mass function for a discrete random variable.

2.17 FUNCTION RULES FOR DISCRETE RANDOM VARIABLES

Recall that a probability function is simply a way of pairing each of a set of events with its probability. When the events are themselves numerical values, we call the function a random variable. Random variables can be specified in at least three ways: by listing each possible numerical event along with its probability, by graphing this relationship and by stating a rule that tells how to find the probability for each possible value.

In many instances, particularly in theoretical statistics, it will be much more convenient to specify the distribution of a discrete random variable by its rule, rather than by a simple listing or by a graph. Some very simple examples will show the form that such function rules often take. Later we will deal with much more complicated function rules, but the general ideas presented here will still obtain.

Suppose that we are interested in a discrete random variable X, which can take on one only of six different values on any occurrence, the values 1, 2, . . . , 6. Furthermore, the probability for the occurrence of any particular value from among these six possible values is exactly the same as for any other. How would one symbolize the rule for this random variable? The answer is

$$p(x) = \begin{cases} 1/6 & \text{if } x = 1, 2, 3, \ldots, 6, \\ 0 & \text{otherwise.} \end{cases}$$

You can check for yourself to see that the requirements of a probability function are met for this random variable: the probability of any value is nonegative, and the sum of probabilities over all possible values is 1.00. Figure 2.17.1 shows the probability mass function for this discrete random variable.

As another example, consider the discrete random variable X, which, once again, can assume only the values 1, 2, 3, 4, 5, and 6. This time, however, the probability function for the random variable obeys the following rule:

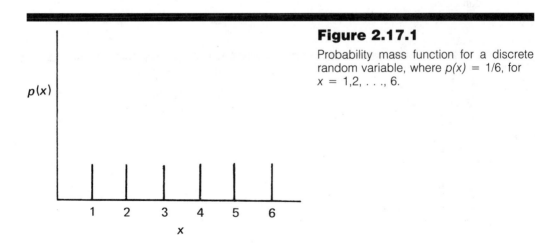

Figure 2.17.1

Probability mass function for a discrete random variable, where $p(x) = 1/6$, for $x = 1, 2, \ldots, 6$.

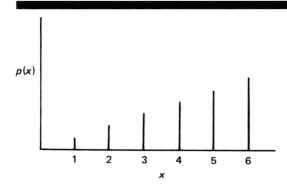

Figure 2.17.2

Probability mass function where $p(x) = x/21$
for $x = 1, 2, 3, 4, 5, 6$.

$$p(x) = \begin{cases} x/21 & \text{if } x = 1, 2, 3, 4, 5, 6, \\ 0 & \text{otherwise.} \end{cases}$$

Thus $p(X = 1) = 1/21$, $p(X = 2) = 2/21$, and so on. The probability mass function is graphed as Figure 2.17.2

Notice that the sum of the probabilities over all of the different values of the random variable is

$$\frac{1 + 2 + 3 + 4 + 5 + 6}{21} = \frac{21}{21} = 1.00$$

just as it should be for any discrete random variable.

Finally, once agan consider a random variable X that can take on the values 1, 2, 3, 4, 5, and 6. This time, however, let the function rule be as follows:

$$p(x) = \begin{cases} x/12 & \text{if } x = 1, 2, 3, \\ (7 - x)/12 & \text{if } x = 4, 5, 6, \\ 0 & \text{otherwise.} \end{cases}$$

This rule gives the following probabilities to values of the random variable:

$p(X = 1) = 1/12$

$p(X = 2) = 2/12$

$p(X = 3) = 3/12$

$p(X = 4) = (7 - 4)/12 = 3/12$

$p(X = 5) = (7 - 5)/12 = 2/12$

$p(X = 6) = (7 - 6)/12 = 1/12$

Once again, as they should, these probabilities sum to 1.00. This probability mass function is shown as Figure 2.17.3. Note how this rule produces a very different picture from those given by the first two rules.

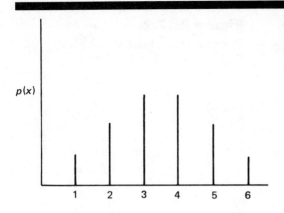

Figure 2.17.3

Probability mass function where $p(x) = x/12$ for $x = 1, 2, 3$ and $p(x) = (7 - x)/12$ for $x = 4, 5, 6$.

These are among the simplest of all function rules for discrete random variables, of course. Those of ultimate intrest for us will tend to be more elaborate. However, even these simple examples show how hypothetical random variables can be described in terms of their rules, even when the rules are really quite arbitrary, and completely hypothetical, as these examples happen to be. Even though the possible function rules for discrete random variables are limitless in their variety, they all show the same basic features. These are rules for pairing with each possible value of the random variable, $X = x$, with a probability $p(X = x)$ that the random variable takes on that value. Each probability provided by the rule is nonnegative, and the sum of the probabilities over all possible values of X must be equal to 1.00.

2.18 CONTINUOUS RANDOM VARIABLES

As we have just seen, the graph for a discrete random variable shows sudden jumps or breaks in the plot of the values and their probabilities. However, the idea of a continuous random variable should suggest a smooth curve. Moreover, the distribution of a continous random variable has a slightly different interpretation from that of a discrete varible, and thus demands a somewhat different terminology.

We can set the stage for the discussion of continous random variables by considering a variable X whose values are grouped into intervals. Consider once again the examples of heights from Section 2.13. In Figure 2.18.1 each interval of values is shown as the corresponding interval on the X axis. If the total area of the histogram is equal to 1.00, as it should be, then the area of any column is the probability of the corresponding interval.

Now suppose that we were able to measure height to any degree of precision, regardless of how many decimal places this might involve. In other words, suppose that our measurements, and the population being measured, were such that we could choose any class interval size i for the random variable X, and still have a nonzero probability that values in the interval would occur, so long as the interval lay between two reasonable extreme values, say u and v. In this instance we could arrange measurements of height

Figure 2.18.1

Probability distribution of height in inches for a specific popuation of boys.

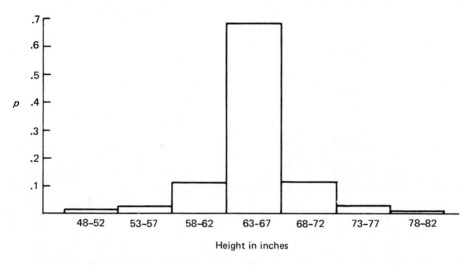

according to a scheme of class intervals of any size, however small, and still have a chance of observing a case in any interval.

For an interval size smaller than the $i = 5$ used above, say $i = 1$, our histogram might look like that shown in Figure 2.18.2. Here the number of intervals is larger and the histogram area for each is smaller. Let us use the symbol ΔX to stand for a very small positive number, representing a very small difference in the X values. Then suppose that the class interval size were made *extremely* small, $i = \Delta X$, still under the assumption that any class interval within the limits u and v will have a probability greater than zero. In this circumstance each class interval will have an extremely small corresponding area, and the area for the histogram as a whole would be very nearly the same as that enclosed in a smooth curve (Fig. 2.18.3). The width of the bar shown on the graph is a representative class interval. Notice that the area cut off by the interval with width ΔX under the curve is almost the same as the area of the histogram bar itself; that is, the area under the curve for that interval is almost the same as the probability.

Finally, imagine that the class-interval size is made *very nearly zero*. As the class-interval size ΔX approaches zero, the probability associated with any class interval and the area cut off by the interval under the curve should agree increasingly well. A random variable that can be represented in terms of arbitrarily small class intervals of size ΔX, with the probability of any interval corresponding *exactly* to the area cut off by the interval under a smooth curve, is called **continuous.** Notice, however, that as the class-interval size ΔX approaches zero, the probability associated with any class interval must also approach zero, since the corresponding area under the curve is steadily reduced. Next we will discuss some of the ramifications of this property of a continuous random variable.

In Section 2.15 it was possible for us to discuss discrete random variables in terms

Figures 2.18.2/2.18.3

The approximation of a histogram by a smooth curve, when the number of cases is very large and the interval size is made very small.

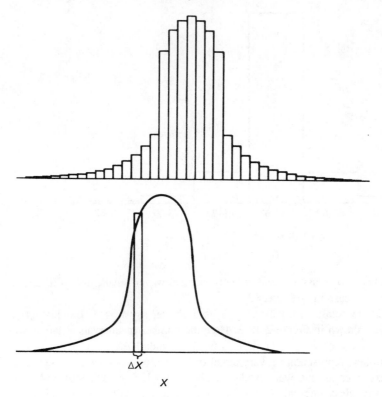

of probabilities $p(x)$, the probability that the random variable takes on exactly some value. However, the fact we have just illustrated, that the area in any interval under a smooth curve approaches zero when interval size goes toward zero, makes it necessary to discuss continuous random variables in a special way.

Consider the example of an infinite set of men weighed with any degree of accuracy. What is the probability that a man will be observed at random weighing *exactly* 160 lbs? This event is an interval of values smaller than 159.5–160.5, or 159.95–160.05, or any other interval $160 \pm .5i$, where i is not exactly zero. Since the smaller the interval, the smaller the probability, the probability of exactly 160 lbs is, in effect, zero. **In general, for continuous random variables, the occurrence of any exact value of X may be regarded as having zero probability.**

For this reason, one does not usually discuss the probability per se for a value of a continuous random variable; probabilities are discussed only for intervals of X values in a continuous distribution. Instead of the probability that X takes on some value a, we deal with so-called **probability density** of X at a, symbolized by

$f(a)$ = probability density of X at a.

(Notice that this use of the symbol $f(a)$ to denote a probability density is different from the use of the letter f alone, which stands for *frequency*.)

Loosely speaking, one can say that the probability density at a is the *rate of change* in the probability of an interval with lower limit a, for minute changes in the size of the interval. This rate of change will depend on two things: the function rule assigning probabilities to intervals such as $(X \leq a)$ and the particular "region" of X values we happen to be talking about.

Rather than talk about probabilities of X values per se, for continuous random variables it is mathematically far more convenient to discuss the probability density and reserve the idea of probability only for the discussion of *intervals* of X values. This need not trouble us especially, as usually we will be interested only in intervals of values in the first place. We will even continue to speak loosely of the probabilities for different X values. However, distribution functions plotted as smooth curves are really plots of probability *densities,* and the only probabilites represented are the areas cut off by nonzero intervals under these curves. Furthermore, when we come to look at a function rule for a continuous random variable, we will actually be looking at the rule for a density function.

If the random variable is discrete, then we can say that

$p(X = a) = f(a)$ = the probability density of X at value a.

The terms *probability at value x* and *density at value x* can be used as though they were synonomous in the case of discrete random variables. On the other hand, these terms are not interchangeable in the case of continuous random variables. For a continuous random variable probabilities are defined only for intervals of values: e.g., $p(X \leq a)$.

Although probability densities such as $f(x)$ are not probabilities, **intervals of values can always be assigned probabilities,** regardless of whether the variable is discrete or continuous.

We have already seen that for discrete variables, the probability of an interval, say $a \leq X \leq b$, is simply the sum of the probabilities for all values of X such that $a \leq X$ and $X \leq b$. For continuous random variables, the probability of any interval depends on the probability density associated with each value in the interval. The probability of any continuous interval is given by

$$p(a \leq X \leq b) = \int_a^b f(x)\, dx.$$

The mathematically unsophisticated reader need not worry over the symbolism used here; it suffices to say that **the probability of an interval is the same as the area cut off by that interval under the curve for the probability densities, when the random variable is continuous and the total area is equal to 1.00.** The expression $f(x)$ may be thought of as the height, and dx can be thought of as the width, of the area created by a minute interval with midpoint $X = x$, somewhere between a and b. When the number of such intervals approaches an infinite number and their size approaches zero, the sum of all these areas is the *entire* area cut off by the limits a and b. Since there is an infinite number of such intervals, this sum is expressed by the definite integral sign $\int_a^b$. This agrees in general form with the definition of the probability of an interval for a discrete variable, which also is a sum, though of probabilites rather than probability densities. Furthermore in a histogram for a discrete variable, the area in an interval is a sum of areas, and this too is like the summing of areas under a smooth curve, yielding the probability of an interval for a continuous variable.

A continuous distribution of a random variable is a theoretical state of affairs. Continuous distributions can never be observed, but are only idealizations. In the first place, no set of potential observations is actually infinite in number; the mortal scientist can deal only with finite numbers of real things, and so obtained data distributions are always discrete. Regardless of the size of the sample, it will never be large enough to permit each of the possible real numbers in any interval to be observed as values of the random variable. In the second place, measurements are imprecise. Not even in the most precise work known can accuracy be obtained to *any* number of decimals. This puts a limit to the actual possibility of encountering a continuous distribution in practice.

Why, then, do statisticians so often deal with these idealizations? The answer is that, mathematically, continuous distributions are far more tractable than discrete distributions. The function rules for continuous distributions are relatively easy to state and to study using the full power of mathematical analysis. This is not usually true for discrete distributions. On the other hand, continuous distributions are very good approximations to many truly discrete distributions. This fact makes it possible to organize statistical theory about a few such idealized distributions and find methods that are good approximations to results for the more complicated discrete situations. Nevertheless, the student should realize that these continuous distributions are mathematical abstractions that happen to be quite useful; they do not necessarily describe "real" situations.

2.19 CUMULATIVE DISTRIBUTION FUNCTIONS

As we have seen, the distribution of a discrete random variable may be represented by a histogram or by a probability mass function, and that of a continuous random variable by the smooth curve of its probability density function. There is also another important way to describe the distribution of a random variable. This is through use of the so-called *cumulative distribution function*. Any frequency distribution can be converted into a cumulative frequency distribution, which shows the relation between a class interval and the frequency of cases falling at or below the interval's upper limit. In much the same way, a probability or density function can be converted into a cumulative probability distribution. A cumulative probability distribution shows the relation between the possible values a of a random variable X, and the probability that the value of X is less than or equal to a. That is, the cumulative probability distribution is a function relating all possible values a to the probabilities $p(X \leq a)$.

The probability that a random variable X takes on a value less than or equal to some particular value a is often written as $F(a)$:

$$F(a) = p(X \leq a).$$

The symbol $F(a)$ denotes the particular probability for the interval $X \leq a$; the general symbol $F(X)$ is sometimes used to represent the existence of the function relating the various values of X to the corresponding *cumulative* probabilities.

The probability that a continuous random variable takes on any value betwen limits a and b can be found from

$$p(a \leq X \leq b) = F(b) - F(a).$$

This is seen easily if it is recalled that $F(b)$ is the probability that X takes on value b or below, $F(a)$ is the probability that X takes on value a or below; their difference must be the probability of a value between a and b.

In the inequalities such as $a \leq X \leq b$, the less-than-or-equal-to sign is used. These could just as well be written with less-than signs alone, however, and the statements would still be true for a *continuous* distribution. The reason is that for a continuous random variable the probability that X equals any exact number is regarded as zero, and thus the probabilities remain the same whether or not a or b or both are considered inside or outside of the interval. However, for discrete variables $<$ and $\leq$ signs may lead to different probabilities.

The symbol $F(a)$ can be used to represent the cumulative probability that X is less than or equal to a either for a continuous or a discrete random variable. All random variables have cumulative distribution functions. Occasionally in mathematical statistics the terms *distribution function* or *probability function* are used to refer only to the cumulative distribution of a random variable, and *density function* is used where we have used *probability distribution*. However, the author believes the terms used here are simpler for the beginning student to learn and use.

Given the probabilities associated with the possible values of a discrete random variable, it is merely a matter of addition to find the $F(a)$ values.

As an example of a cumulative distribution function, consider a discrete random variable with the following rule:

$$p(x) = \begin{cases} 1/4 & \text{for } x = 2 \\ 1/2 & x = 1, \\ 1/4 & x = 0, \\ 0 & \text{otherwise.} \end{cases}$$

Now to determine the cumulative distribution function, we need only the values $F(0)$, $F(1)$, and $F(2)$, since these are the only values of X with nonzero probabilities. For $X < 0$, the probability is zero, of course, and for $X = 0$, the probability is 1/4. Hence

$$F(0) = 0 + 1/4 = 1/4.$$

For $F(1)$ we have

$$F(1) = F(0) + p(1) = 1/4 + 1/2 = 3/4.$$

For $F(2)$, we find

$$F(2) = F(1) + p(2) = 3/4 + 1/4 = 1.00.$$

For any value of X above 2, the value of $F(x) = 1.00$.

The cumulative distribution function for this random variable is now fully determined. We can plot this function as in Figure 2.19.1. Note that in this plot the same $F(x)$ value occurs for all points in the interval $0 \leq X < 1$, the same value holds for all points in the interval $1 \leq X < 2$, and so on. This gives the graph the distinctive appearance of a

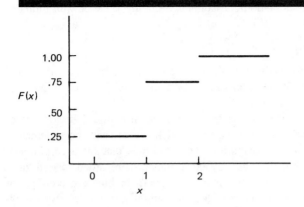

Figure 2.19.1

A step function, showing the cumulative probabilities for values of a discrete random variable.

set of plateaus, or "steps," and for this reason such a graph is called a *step function*. Graphing the cumulative distribution function of any discrete variable will produce such a step function.

2.20 GRAPHIC REPRESENTATIONS OF CONTINUOUS DISTRIBUTIONS

A continuous probability distribution is always represented as a smooth curve erected over the horizontal axis, which itself represents the possible values of the random variable X. A curve for some hypothetical distribution is shown in Figure 2.20.1. The two points a and b marked off in the horizontal axis represent limits to some interval. The shaded portion between a and b shows $p(a \leq X \leq b)$, the probability that X takes on a value between the limits a and b. Remember that this probability corresponds to an *area* under the curve: in any continuous distribution, the probability of an interval can be represented by an area under the distribution curve. The total area under the curve represents 1.00, the probability that X takes on *some* value.

In Figure 2.20.1 the shaded portion of the curve to the right of the point c is the probability $p(c < X)$. This is found by first taking

$$F(c) = p(X \leq c).$$

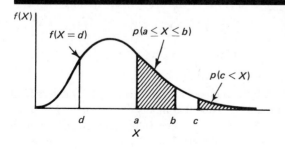

Figure 2.20.1

Probability densities, areas, and probabilities in a continuous distribution.

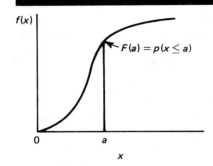

Figure 2.20.2

A continuous cumulative function, with $F(a)$ value.

In figure: $f(x)$, $F(a) = p(x \leq a)$, 0, a, x

Since the total probability is 1.00, and since $(X \leq c)$ and $(c < X)$ are mutually exclusive events, then

$$p(c < X) = 1.00 - F(c).$$

Cumulative distribution curves often have more or less the characteristics **S** shape shown in Figure 2.20.2. Here, the horizontal axis shows possible values of X; for any point on the axis, a, the *height* of the curve, $F(a)$, is the probability that X is less than or equal to a.

Tables of theoretical probability distributions are most often given in terms of the cumulative distribution or $F(X)$, for a random variable. Given these cumulative probabilites for various values of X it is easy to find the probability for any interval by subtraction. When we come to use the table of the so-called normal distribution, we will find that cumulative probabilities are shown.

$F(X)$ - cumulative distribution

2.21 JOINT DISTRIBUTIONS OF RANDOM VARIABLES

Just as the basic idea of an event was extended to the concept of joint events in Chapter 1, it is perfectly possible to extend the idea of a random variable and its distribution to that of *joint random variables* having a joint probability or probability density function. This means that instead of identifying an elementary event with a single numerical value, we identify each elementary event with two or more such values, each representing a random variable. The former situation is said to be *univariate,* and the latter *bivariate* when there are two variables, or *multivariate* when there are more than two variables.

The theory of joint random variables, both discrete and continuous, plays a very large role in the theory of statistics. On a number of occasions in future sections we will have to draw on principles from this theory. A few of the elementary ideas about joint random variables are given in the first section of Appendix C. Since we will be appealing to one or another of these principles from time to time in sections to come, the student may at this point find it worthwhile to start becoming familiar with the materials in the first section of this appendix. Most of it is a direct translation of the ideas of single-variable distributions to distributions involving joint variables.

2.22 FREQUENCY AND PROBABILITY DISTRIBUTIONS IN USE

Both frequency and probability distributions exhibit the same basic features: each is a statement of a relation between the various possible measurement classes into which observations may fall and a number attached to each class. In the case of frequency distributions, this number is the frequency of cases actually observed in a particular class or class interval. For probability distributions, each class or interval is accompanied by its probability. In future chapters we will see that the same language is applied in the further summarization of either kind of distribution.

However, there is an important difference between frequency distributions of data and probability distributions. The latter almost invariably represent *hypothetical* or *theoretical* distributions. A probability distribution is some idealization of the way things might be if we only had all of the information. The frequency distribution represents what we have actually seen to be true from some limited number of observations.

The connection between frequency and probability distributions should be obvious by now: the probability distribution dictates what we should *expect* to observe in a frequency distribution, if the given state of affairs is true. Thus, if theoretically the random variable X has a probability of .30 of taking on a value in the interval between 100 and 102, then given a frequency distribution of 500 sample observations we should expect the class interval 100–102 to contain $500 \times .30$ or 150 cases. This does not mean that this will be true for any given sample; we *might* observe any frequency between 0 and 500. Nevertheless, in the long run, if we sample indefinitely, 30% of all cases sampled should show scores in that interval, provided that our hypothetical probability distribution is right.

This is the reason for the parallel development of frequency and probability distributions carried on here. Each form of distribution plays a role in inference: the probability distribution specifies what to expect when the data are put into distribution form, and the obtained frequency distribution supplies us with our best evidence about the probability distribution (the *relative* frequencies of occurence being estimates of the true probabilities). This parallel discussion will be continued in other chapters. However, the immediate goal is to show how theoretical probability distributions can be constructed. In order to do this we must pick up a few more rudiments of probability theory. These will be presented in the next chapter.

2.23 FINDING FREQUENCY DISTRIBUTIONS WITH THE COMPUTER PACKAGES

Each of the widely available packages of statistical programs for the computer has one or more programs providing frequency distributions for the variables included in a study. These programs also include a wide variety of statistics calculated on the data, but we will defer description of these options until later chapters.

The SPSS and SPSS[x] packages include the subprogram FREQUENCIES, which operates in one of two modes: GENERAL, which permits one to find distributions either of nominal or of numerical data; and INTEGER, which applies only to numerical "scores," and which is faster and, therefore, more economical to use than the GENERAL program. The frequency distributions can be obtained either in a normal or "default'' format, with 15 to 20 class intervals, or a condensed format. This last option is used when the variable in question can take on many values, and a distribution without grouping is desired. When there are a number of variables to be examined in the same set of data, a mixture of these options may be specified. The output routinely includes the frequency for each category or value, the percentage of the total that this frequency represents, and the cumulative percentage. In addition, each frequency distribution may also be requested in the form of a histogram.

The GRAPHICS option ins SPSS may also be used to obtain various other forms of frequency graph, either for displaying a single set of data or for comparing two or more data sets. The available procedures include PIECHART, BARCHART, and LINE-CHART, each of which produces one or more graphs of the type implied by the name. Options include fancy shadings and crosshatch patterns for the pie and bar charts, choice of line patterns of markers for line graphs, and groupings of two or more bar charts or line graphs, along with many other features. (SPSS[x] simply lumps these procedures under FREQUENCIES).

Probably the most useful of the BMDP programs for finding frequency distributions of data is P2D. In addition to ordinary frequency distributions of the variables, including raw frequencies, relative frequencies, and cumulative frequencies, the program also has options such as the ''stem and leaf'' displays proposed by John Tukey (1977). Small histograms are routinely produced for the frequency distributions, but if large histograms are desired, they can be found through program P5D, which also gives other graphics options.

Simple frequency distributions can be achieved with the SAS package through use of the PROC FREQ procedure. However, the user must dictate the format if a variable is to be grouped into class intervals. The simplest situation, where the distribution of a variable named x is desired, requires these statements:

```
PROC FREQ;
TABLES X;
```

These statements are sufficient to produce the frequencies, cumulative frequencies, percentages of the total frequency, and cumulative percentages for each value or level of the variable X. Any number of frequency distributions can be requested in a single TABLES statement by naming different variables. However, if PROC FREQ is listed without a TABLES statement, then frequency distributions are produced for *all* of the variables identified in the data set.

EXERCISES

1. Given data of the following types, state the scale of measurement which each type appears most clearly to represent.

(a) The nationality of an individual's father.
(b) Hand pressure, as applied to a flexible bulb (i.e., on a dynamometer).
(c) Memory ability, as measured by the number of words recalled from an initially memorized list.

(d) The excellence of baseball teams, as determined by their won–lost records at the end of the season.

(e) The time to clotting of samples of human blood.

(f) Air distance between New York and other cities in the United States.

(g) Time, as measured before and after the discovery of the New World by Columbus in 1492.

(h) Reading ability of fifth-grade children, as shown by their test performance relative to a national norm group.

(i) U.S. Department of Agriculture classifications of fresh meats ("choice," etc.).

(j) Social Security numbers.

2. For which sorts of measurement scales are the following manipulations most appropriate:

(a) Taking the arithmetic difference between two values.

(b) Changing the labels assigned to a set of objects to another arbitrarily chosen set of labels.

(c) Stating that one value represents a higher level of some property than does another value.

(d) Taking the ratio of two values.

(e) Taking the ratio of two differences between values.

(f) Multiplying each value by a constant and then adding a constant to each.

3. A well-known magazine provides evaluative reports to consumers on the quality of various products. In a report on television sets of different brands, the ratings were as follows:

Brand	Rating
A	Acceptable
B	Good
C	Good
D	Acceptable
E	Poor
F	Acceptable
G	Acceptable
H	Poor
I	Very Good
J	Good
K	Acceptable
L	Acceptable
M	Poor
N	Very Good
O	Acceptable
P	Poor

Construct a frequency table displaying these data.

4. The mental health clinic of a university uses the following codes for the major types of problems that bring students in for assistance.

(A) General anxiety.

(B) General depression.

(C) Sexually related problems.

(D) Alcohol or drug-related problems.

(E) Problems of social adjustment.

(F) Family problems.

(G) Other problems.

During one day, 54 students visted the clinic, and the classifications used (one per student) were as follows:

A	B	B	E	B	D
B	C	E	B	B	A
C	F	D	G	G	D
G	A	G	A	F	G
B	B	B	G	G	G
G	G	B	D	B	B
F	B	G	C	C	F
E	G	B	G	G	B
B	B	D	G	A	B

Construct a frequency distribution showing the numbers receiving each code.

5. A questionnarire consisted of only three items. On each item the respondent was asked to circle one of the numbers from 1 through 5, where 1 indicated "strongly agree," 2 indicated "agree somewhat," 3 was "neither agree nor disagree," 4 stood for "disagree somewhat," and 5 represented "disagree strongly." The responses for 30 people are listed below for all three items. Form a frequency distribution of responses to Item I, and then do the same thing for Items II and III. On the basis of these distributions, what would one tend to conclude about the responses of these subjects?

Individual	Item I	Item II	Item III
1	1	2	2
2	2	4	4
3	2	1	1
4	1	3	4
5	3	5	3
6	3	3	3
7	4	2	5
8	1	5	4
9	4	3	3
10	2	4	4
11	4	2	2
12	1	3	4
13	2	4	4
14	5	3	3
15	3	1	1
16	2	3	4
17	2	4	5
18	2	2	2
19	1	4	1
20	3	2	2
21	5	3	5
22	2	3	3
23	5	4	4
24	1	2	5
25	4	3	5
26	3	3	3
27	3	3	5
28	2	4	4
29	4	2	2
30	3	3	3

N = 30

6. Put each of the distributions found in Exercise 5 above into the form of a histogram. Are the histograms of similar appearance? Why?

7. In the calendar year 1979, a group of elderly people had the following incomes. Form a frequency distribution of these data.

3881	3350	3509	3427	3169
3813	3563	3587	3275	3308
2808	3474	3710	3352	2955
3650	3456	3457	3137	2978
3547	3568	3913	3133	3093

(Suggestion: Employ a class-interval size of 121.)

8. For the distribution in Exercise 7, draw a frequency polygon.

9. On a test of mechanical aptitude, the score is the number of seconds required to finish a certain task. A group of 52 students in the second grade received the following scores:

76.3	90.1	57.5	57.8
66.0	92.0	55.0	59.0
80.0	87.4	61.9	55.0
82.0	91.0	59.2	61.1
76.2	94.8	60.0	59.9
80.0	95.0	61.3	60.0
84.7	91.9	74.0	61.7
86.0	94.5	90.4	59.0
80.6	93.2	48.0	52.0
86.5	57.1	57.2	53.0
85.0	53.6	53.5	44.0
87.0	55.7	55.6	51.5
74.0	56.0	56.0	45.0

Construct a frequency distribution table for these data.

10. Suppose that a certan state contains 84 counties. For a given year the numbers of residents by county (shown in thousands) was as follows. Make a frequency distribution for these data, using $i = 10,000$. (**Hint:** Allow for one or more open or unequal intervals.)

3.2	415.8	23.5	91.6	19.0	4.1
29.1	22.1	27.2	5.1	40.9	23.8
31.1	57.5	14.1	32.5	24.4	15.6
13.2	42.3	52.2	14.1	32.1	28.0
41.2	104.6	27.7	43.3	1,001.2	13.3
31.7	48.1	52.0	12.5	56.0	44.4
6.7	22.3	18.8	47.9	26.0	16.5
55.5	30.5	31.4	7.6	53.4	35.0
26.7	12.2	84.7	25.7	57.1	2.3
11.9	34.6	25.9	40.1	110.2	66.6
22.6	18.3	70.7	27.9	78.2	42.7
8.3	43.3	30.4	11.7	57.2	34.2
58.8	23.6	88.8	28.3	28.1	11.8
40.8	7.2	14.3	75.1	53.7	51.3

11. Plot the distribution in Exercise 10 as a frequency polygon.

12. Plot the cumulative frequency polygon for the data in Exercise 9. Then see if you can estimate the frequencies corresponding to the intervals in (a)–(c). (**Hint:** Treat the cases in any given class interval as though they were evenly spread across that interval.

(a) $x \le 70$
(b) $83 \le x$
(c) $50 \le x \le 60$
(d) Find the x value that cuts off the bottom 50% of cases in the distribution.
(e) Between what two values do the middle 50% of cases seem to fall?

13. A student of the history of the English language noted the dates of the first recorded appearance in written English of 1,000 words of foreign origin. The dates were as follows:

Dates	Frequency	Dates	Frequency
Before 1050	2	1451–1500	76
1051–1100	2	1501–1550	84
1101–1150	1	1551–1600	91
1151–1200	15	1601–1650	69
1201–1250	64	1651–1700	34
1251–1300	127	1701–1750	24
1301–1350	120	1751–1800	16
1351–1400	180	1801–1850	23
1401–1450	70	1851–1900	2

Plot the cumulative frequency distribution for these data.

14. Suppose that a word were drawn at random from among the 1,000 words given in the distribution in Exercise 13. Find the probability that the following events occur:

(a) The word drawn is one which appeared in English before 1301.
(b) The word drawn appeared in English after 1601.
(c) The word drawn appeared in English between 1501 and 1700 inclusive.
(d) The word drawn appeared in English either before 1201 or after 1801.

15. Suppose that one of the group of students in Exercise 4 was selected at random. Determine the probability that this individual would represent

(a) Code A.
(b) Code B.
(c) Code E or Code F.
(d) Any code except G.

16. Suppose that 10 students had been drawn from the group in Exercise 4, with each student being replaced before the next student is drawn. Give the frequencies that one should expect over the seven categories.

17. The following data represent weights, to the nearest pound, of 78 boys in junior high school. Arrange these data into a frequency polygon with $i = 5$.

122	122	110	118	120	111
117	122	107	127	146	113
116	119	108	118	127	116
114	118	153	125	138	126
110	117	148	119	133	113
109	112	134	125	134	106
107	108	140	119	128	118
105	117	108	124	144	115
103	117	126	120	132	119
102	126	103	121	137	133
134	123	136	128	136	148
126	118	112	116	146	137
123	113	118	127	152	124

18. Imagine that the 78 boys in Exercise 17 represent but a sample from a much larger group of junior high school boys. If this sample had been drawn at random, what is our best guess about the probability of each of the intervals of values as used above?

19. Suppose that a discrete random variable can take on only the values 1, 2, 3, 4, and 5. The rule giving the probability for each of these values is:

$$p(x) = \begin{cases} \dfrac{x^2}{55} & \text{for } x = 1, 2, 3, 4, 5 \\ 0 & \text{otherwise} \end{cases}$$

(a) Plot the probability mass function for this random variable.
(b) Plot the cumulative distribution function (step function) for this variable.
(c) What is the probability that the variable takes on the values 2, 3, or 4?
(d) What is the probability that the random variable takes on *some* value?

20. A theoretical random variable can take on only the whole number values between -3 and 3, inclusive. The probability for each value is given by the rule:

$$p(x) = \begin{cases} \dfrac{4 + x}{16} & \text{for } x = -3, -2, -1, 0 \\ \dfrac{4 - x}{16} & \text{for } x = 1, 2, 3 \\ 0 & \text{otherwise} \end{cases}$$

(a) Find the probability mass function for this variable.
(b) Plot the cumulative distribution function, or step function, for this variable.
(c) If 32 independent values of this random variable should be obtained, what should we expect the frequency distribution to be like?

21. A random variable has a probability function defined as follows:

$$f(x) = \begin{cases} \dfrac{1}{10} & -3 \leq x \leq -1 \\ \dfrac{2}{5} & -1 < x \leq 0 \\ \dfrac{1}{5} & 0 < x \leq 1 \\ \dfrac{3}{10} & 1 < x \leq 3 \\ 0 & \text{elsewhere} \end{cases}$$

(a) Plot this probability function.
(b) Indicate the probability $p(-3 \le x \le 1)$.
(c) Is this a continuous distribution? How can one tell?

22. A carnival game uses a large wheel which is spun and which then stops at a number. Suppose that the wheel is so constructed that it can stop at any of the real numbers between 0 and 6, inclusive. Suppose further that each of these numbers has the same likelihood of occuring (i.e., the wheel is ''fair''). Then the number where the wheel stops on any trial is the value of a continuous random variable, with a rule given by:

$$f(x) = \begin{cases} \dfrac{1}{6} & \text{for } 0 \le x \le 6 \\ 0 & \text{otherwise} \end{cases}$$

Graph the distribution of this random variable. Then, use the appropriate areas under this curve to find the following probabilities:

(a) $p(x \le 1.5)$

(b) $p(x \le 5)$

(c) $p(3 \le x \le 4)$

(d) $p(5 \le x)$

23. In another carnival operation a wheel is used similar to that in Exercise 22. However, this wheel is biased, with numbers close to 3 much more likely to occur than numbers far from 3. This might be represented by a random variable following the rule:

$$f(x) = \begin{cases} \dfrac{x(6 - x)}{36} & 0 \le x \le 6 \\ 0 & \text{otherwise} \end{cases}$$

(a) Draw a graph of the distribution of this random variable.
(b) On this graph mark off areas corresponding to the probabilities:

$p(x \le 1.5)$
$p(x \le 5)$
$p(3 \le x \le 4)$
$p(5 \le x)$.

24. Suppose that you really observed a wheel at a carnival, which might be working in the way described in Exercise 22 or in the manner of Exercise 23. Describe how you might be able to decide which of these two models of the wheel's behavior seems to provide the best fit to what is actually observed.

25. Consider the following cumulative distribution function for the discrete random variable X:

a	$p(X \le a)$
20	1.00
18	.98
16	.83
14	.74
12	.63
10	.52
8	.37
6	.21
4	.10
2	.03
0	0

Suppose that 48 observations were made, independently and at random, from the process generating this random variable. Construct the frequency distribution of observations which one should *expect* to obtain.

Chapter **3**

A Discrete Random Variable: The Binomial

The main burden of this chapter is to show how the theoretical distribution of a discrete random variable can be constructed. We will develop the distribution known as the *binomial,* which will play a very important role in much that follows. However, first of all we will need ways to make elementary probability calculations and to know some "counting rules" that underlie the development to follow.

3.1 CALCULATING PROBABILITIES

As we have seen, almost any simple problem in probability can be reduced to a problem in counting. Especially for a sample space containing equally probable elementary events, the calculation of probability involves two quantities, both of which are *counts of possibilities:* the total number of elementary events, and the number qualifying for a particular event class. The key to solving probability problems is to learn to ask: "How many distinctly different ways can this event happen?" It may be possible simply to list the number of different elementary events that make up the event class in question, but it is often much more convenient to use a rule for finding this number. Once the probabilities of events are found, the rules of Chapter 1 permit deductions of other probabilities.

There is really no way to become expert in probability computations except by practice in the application of these various counting procedures. Naturally, problems differ widely in the particular principles they involve, but most problems should be approached by these steps:

1. Determine exactly the sample space of elementary events with which this problem deals. Draw a graph or picture such as Table 2.15.1 if necessary.

2. Find out how many elementary events make up the sample space. What are all the distinct outcomes that might conceivably occur? If the elementary events are equally probable, then the probability of any single elementary event is one over the total number of elementary events.

3. Decide on the particular events for which probabilities are to be found. How many elementary events qualify for each event class? If no other counting method is available, *list* the elements of these different events and count them. Be sure to remember that if event A can occur in $n(A)$ ways and an independent event B in $n(B)$ ways, then event $(A \cap B)$ might occur in $n(A)n(B)$ ways.

4. For equally probable elementary events the probability of any event A is simply the ratio of the number of members of A to the total number of elementary events. Even though elementary events are not equally probable, we can still use the fact that elementary events are mutually exclusive to find the probability of any event A by taking the *sum* of the probabilities of all members of that set.

A game of chance such as roulette shows how probabilities may be computed simply by listing elementary events. A standard roulette wheel has 37 equally spaced slots into which a ball may come to rest after the wheel is spun. These slots are numbered from 0 through 36. One half of the numbers from 1 through 36 are red, the others are black, and the zero is generally green.

In playing roulette, you may bet on any single number, on certain groups of numbers, on colors, etc. One such bet might be on the odd numbers (excluding 0 of course). The only elementary events in the sample space are the numbers 0 through 36 with their respective colors. Which numbers qualify for the event "odd"? It is easy to count these numbers: they form the set

$$\{1,3,5,7,9,11,13,15,17,19,21,23,25,27,29,31,33,35\}.$$

Since there are exactly 18 elementary events qualifying as odd numbers, if each slot on the roulette wheel is equally likely to receive the ball, the probability of "odd" is 18/37, slightly less than 1/2.

As a more complicated example that can be solved by listing, let us find the probability that the number contains a 3 as one of the digits. The elementary events qualifying are:

$$\{3,13,23,30,31,32,33,34,35,36\}.$$

Here the probability is 10/37, if the 37 different elementary events are equally likely.

As a final example, consider the probability of a number that is "even and red" on the roulette wheel. These are

$$\{12,14,16,18,30,32,34,36\}$$

and the probability is 8/37.

There is no end to the examples that might be brought in at this point to show this counting principle in operation, but there is little point in spending more time with it. Just remember that almost all simple problems in probability can be solved in this same way. The important thing to understand is, "When in doubt, make a list." This will usually work, although, as we shall see, other counting methods are often more efficient.

3.2 SEQUENCES OF EVENTS

In many problems, an elementary event may be the set of outcomes of a *series or sequence* of observations. Such sequences can also be considered joint events, but it is convenient to regard an entire sequence as an elementary event. Suppose that any trial of some simple experiment must result in one of K mutually exclusive and exhaustive events, $\{A_1, \cdots, A_K\}$. Now the experiment is *repeated N times*. This leads to a sequence of events: the outcome of the first trial, the outcome of the second, and so on in order through the outcome of the Nth trial. The outcome of the whole series of trials might be the sequence $(A_3, A_1, A_2, \cdots, A_3)$. This denotes that the event A_3 occurred on the first trial, A_1 on the second trial, A_2 on the third, and so on. The place in order gives the trial on which the event occurred, and the symbol occupying that place shows the event occurring for that trial. For example, it might be that the simple experiment is drawing one marble at a time from a box, with replacement after each drawing. The marbles observed may be red, white, or black. For four drawings a sequence of outcomes might be (R, W, B, W), or perhaps (W, W, B, B).

Remember that each possible sequence can be thought of in two ways; it is a joint event from the standpoint of a series of simple experiments, but it is also an elementary event if one thinks of the *experiment itself* as the series of trials with a sample space consisting of *n*-tuples, or possible sequences of outcomes. Each *whole* or *compound experiment* has as its outcome one and only one sequence. Such experiments producing sequences as outcomes are very important to statistics, and so we will begin our study of counting rules by finding how many different sequences a given series of trials could produce.

3.3 COUNTING RULE 1: NUMBER OF POSSIBLE SEQUENCES FOR *N* TRIALS

Suppose that a series of N trials were carried out, and that on each trial any of K events might occur. Then the following rule holds:

Counting rule 1: If any one of K mutually exclusive and exhaustive events can occur on each of N trials, then there are K^N different sequences that may result from a set of such trials.

As an example of this rule, consider a coin's being tossed. Each toss can result only in a H or a T event ($K = 2$). Now the coin is tossed five times, so that $N = 5$. The total number of possible results of tossing the coin five times is $K^N = 2^5 = 32$ sequences. Exactly the same number is obtained for the possible outcomes of tossing five coins simultaneously, if the coins are thought of as numbered and a sequence describes what happens to Coin 1, to Coin 2, and so on.

As another example, the outcome of tossing two dice is a sequence: what number comes up on the first die, and what number comes up on the second. Here $K = 6$ (six different numbers per die), and $N = 2$. There are exactly $6^2 = 36$ different sequences possible as results of this experiment, as we saw in Section 2.15.

3.4 COUNTING RULE 2: FOR SEQUENCES

Sometimes the number of possible events in the first trial of a series is different from the number possible in the second, the second different from the third, and so on. It is obvious that if there are different numbers $K_1, K_2, \ldots, K_N$ of events possible on the respective trials, then the total number of sequences will not be given by rule 1. Instead, the following rule holds:

Counting rule 2: If $K_1, \cdots, K_N$ are the numbers of distinct events that can occur on trials $1, \cdots, N$ in a series, then the number of different sequences of N events that can occur is $(K_1)(K_2) \cdots (K_N)$.

For example, suppose that for the first trial you toss a coin (two possible outcomes) and for the second you roll a die (six possible outcomes). Then the total number of different sequences would be $(2)(6) = 12$.

Notice that counting rule 1 is actually a special case of rule 2. If the same number K of events can occur on any trial, then the total number of sequences is K multiplied by itself N times, or K^N. We saw this illustrated in Section 2.15.

3.5 COUNTING RULE 3: PERMUTATIONS

A rule of extreme importance in probability computations concerns the number of ways that objects may be arranged in order. The rule will be given for arrangements of objects, but it is equally applicable to sequences of events:

Counting rule 3: The number of different ways that N distinct things may be arranged in order is $N! = (1)(2)(3) \cdots (N - 1)(N)$, (where $0! = 1$). An arrangement in order is called a permutation, so that the total number of permutations of N objects is $N!$ The symbol $N!$ is called N *factorial*.

A useful table of factorials of integers may be found as Table VII, Appendix E. In addition, many calculators provide values of $N!$. As an illustration of this rule, suppose that a classroom contained exactly 10 seats for 10 students. How many ways could the students be assigned to the chairs? Any of the students could be put into the first chair, making 10 possibilities for chair 1. But, given the occupancy of chair 1, there are only 9 students for chair 2; the total number of ways chair 1 and chair 2 could be filled is $(10)(9) = 90$ ways. Now consider chair 3. With chairs 1 and 2 occupied, 8 students remain, so that there are $(10)(9)(8) = 720$ ways to fill chairs 1, 2, and 3. Finally, when 9 chairs have been filled, there remains only 1 student to fill the remaining place, so that there are

$$(10)(9)(8)(7)(6)(5)(4)(3)(2)(1) = (10)! = 3{,}628{,}800$$

ways of arranging the 10 students into the 10 chairs.

Now suppose that there are only N different events that can be observed in N trials. Imagine that the occurrence of any given event "uses up" that event for the sequence, so that each event may occur *once and only once* in the sequence. In this case there must be $N!$ different orders in which these events might occur in sequence. Each sequence is a permutation of the N possible events.

To take a homely example, suppose that a man is observed dressing. At each point in our observation there are three articles of clothing he might put on: his shoes, his pants, or his shirt. However each article can be put on only once. We might observe the following sequences:

(shoes, pants, shirt)
(shoes, shirt, pants)
(shirt, shoes, pants)
(shirt, pants, shoes)
(pants, shoes, shirt)
(pants, shirt, shoes)

The man's putting on any one of these articles of clothing literally uses up that outcome for any sequence of observations, so that there are $N! = (1)(2)(3) = 6$ possible permutation sequences.

The procedure of counting sequences as permutations in order is especially important for sampling without replacement. For example, suppose that a teacher has the names of five students in a hat. She draws the names out at random one at a time, *without replacement*. Then there are exactly 5! or 120 different sequences of names that she might observe. If all sequences of names are equally likely to be drawn, the the probability for any one sequence is 1/120.

What is the probability that the name of any given child will be drawn first? Given that a certain child is drawn first, the order of the remainder of the sequence is still unspecified. There are $(N - 1)! = 4! = 24$ different orders in which the other children can appear. Thus the probability of any given child being first in sequence is $24/120 = .20$ or 1/5.

Suppose that exactly *two* of the names in the hat are girls'. What is the probability that the first two names drawn belong to the girls? Given the first two names are girls', there are still $3! = 6$ ways three boys may be arranged. The girls' names may themselves be ordered in two ways. Thus, the probability of the girls' names being drawn first and second is $(2)(6)/120 = 1/10$.

3.6 COUNTING RULE 4: ORDERED COMBINATIONS

Sometimes it is necessary to count the number of ways that r objects might be selected from among some N objects in all, $(r \leq N)$. Furthermore, each different *arrangement* of the r objects is considered separately. Then the following rule applies:

Counting rule 4: The number of ways of selecting and arranging r objects from among N distinct objects is $\dfrac{N!}{(N - r)!}$.

The reasoning underlying this rule becomes clear if a simple example is taken: Consider a classroom teacher, once again, who has 10 students to be assigned to seats. This time, however, imagine that there are only 5 seats. How many different ways could the teacher select 5 students and arrange them into the available seats? Notice that there are 10 ways that the first seat might be filled, 9 ways for the second, and so on, until seat 5 could be filled in 6 ways. Thus there are

$$(10)(9)(8)(7)(6)$$

ways to select students to fill the 5 seats. This number is equivalent to

$$\frac{10!}{5!} = \frac{N!}{(N-r)!} = 30{,}240,$$

the number of ways that 10 students out of 10 may be selected and arranged, but divided by the number of arrangements of the 5 *un*selected students in the 5 *missing* seats.

As an example of the use of this principle in probability calculations, consider the following: In lotteries it is usual for the first person whose name is drawn to receive a large amount, the second some smaller amount, and so on, until some r prizes are awarded. This means that some r names are drawn in all, and the *order* in which those names are drawn determines the size of the prizes awarded to individuals. Suppose that in a rather small lottery 40 tickets had been sold, each to a different person, and only 3 were to be drawn for first, second, and third prizes. Here $N = 40$ and $r = 3$. How many different assignments of prizes to persons could there be? The answer, by counting principle 4, is

$$\frac{40!}{(40-3)!} = \frac{40!}{37!} = (38)(39)(40) = 59{,}280.$$

On how many of these possible sequences of winners would a given person, John Doe, appear as first, second, *or* third prize winner? If he were drawn first, the number of possible selections for second and third prize would be $39!/37! = 1{,}482$. Similarly, there would be 1,482 sequences in which he could appear second, and a like number where he could appear third. Thus, the probability that he appears in a sequence of three drawings, winning first, second, or third prize is (by rule 4, Section 1.5.)

$$p(\text{first, second, or third}) = \frac{3(1{,}482)}{59{,}280} = \frac{3}{40} = .075.$$

Calculators often give values for ordered combinations of N things taken r at a time, under the symbol $_NP_r$.

3.7 COUNTING RULE 5: COMBINATIONS

In a very large class of probability problems, we are not interested in the *order* of events, but only in the number of ways that r things could be selected from among N things, *irrespective of order*. We have just seen that the total number of ways of selecting r things from N and ordering them is $N!/(N - r)!$, by rule 4. Each set of r objects has $r!$ possible orderings, by rule 3. A combination of these two facts gives us counting rule 5.

Counting rule 5: The total number of ways of selecting r distinct combinations of N objects, irrespective of order, is

$$\frac{N!}{r!(N-r)!} = \binom{N}{r}.$$

The symbol $\binom{N}{r}$ is *not* a symbol for a fraction, but instead denotes the number of combinations of N things, taken r at a time. Sometimes the number of combinations is known as a *binomial coefficient,* and occasionally $\binom{N}{r}$ is replaced by the symbols $^{N}C_r$ or $_{N}C_r$. However, the name and symbol introduced in rule 5 will be used here.

It is helpful to note that

$$\binom{N}{r} = \binom{N}{N-r}.$$

Thus, $\binom{10}{3} = \binom{10}{7}$, $\binom{50}{49} = \binom{50}{1}$, and so on.

As an example of the use of this rule, suppose that a total of 33 persons were candidates for the board of supervisors in some community. Three supervisors are to be elected at large; how many ways could 3 persons be selected from among these 33 candidates? Here, $r = 3$, $N = 33$, so that

$$\binom{N}{r} = \frac{(33)!}{3!\,(30)!} = \frac{(1 \times 2 \cdots 32 \times 33)}{(1 \times 2 \times 3)(1 \times 2 \cdots 29 \times 30)}.$$

Canceling in numerator and denominator, we get

$$\binom{33}{3} = \frac{(31)(32)(33)}{6} = (31)(16)(11) = 5{,}456.$$

If all sets of three are equally likely to be chosen, the probability for any given set is $1/5{,}456$.

Because of their utility in probability calculations, a table of $\binom{N}{r}$ values for various values of N and r is included as Table IX Appendix E. Although this table shows values of N only up to 20, and values of r up to 10, other values can be found by the relation given above, and also by the relation known as Pascal's Rule:

$$\binom{N}{r} = \binom{N-1}{r-1} + \binom{N-1}{r}.$$

Still other values may be worked out from the table of factorials also included in Appendix E, in terms of rule 5.

3.8 SOME EXAMPLES: POKER HANDS

The five counting rules provide the basis for many calculations of probability when the number of elementary events is finite, especially when sampling is without replacement. One of the easiest examples of their application is the calculation of probabilitites for poker hands.

The game of poker as discussed here will be highly simplified and not very exciting to play: the player simply deals five cards to himself from a well-shuffled standard deck of 52 cards. Nevertheless, the probabilities of the various hands are interesting, and can be computed quite easily.

The particular hands we will examine are the following:

  a.   one pair, with three different remaining cards.
  b.   full house (three of a kind, and one pair)
  c.   flush (all cards of the same suit).

First of all we need to know how many different hands may be drawn. A given hand of five cards will be thought of as an event; notice that the order in which the cards appear in a hand is immaterial, so that a number of sequences of cards can correspond to a given hand. By rule 5, since there are 52 different cards in all and only 5 are selected, then there are

$$\binom{52}{5} = \frac{52!}{5!47!}$$

different hands that might be drawn. If all hands are equally likely, then the probability of any given one is $1 \Big/ \binom{52}{5}$, or about 4 in 10 million.

There are 13 numbers that a pair of cards may show (counting the picture cards), and each member of a pair must have a suit. Thus by rule 5 there are $13 \binom{4}{2}$ different pairs that might be observed. The remaining cards must show 3 of the 12 remaining numbers and each of the 3 cards may be of any suit. By rules 1 and 5 there are $\binom{12}{3}$ (4)(4)(4) ways of filling out the hand in this way. Finally, we find that there are

$$13 \binom{4}{2} \binom{12}{3} 4^3$$

different ways for this event to occur. Thus,

$$p(\text{one pair}) = \frac{13\binom{4}{2}\binom{12}{3}4^3}{\binom{52}{5}}.$$

This number can be worked out by writing out the factorials, canceling in numerator and denominator, and dividing. It is approximately equal to .42. The chances are roughly 4 in 10 of drawing a single pair in five cards, if all possible hands are equally likely to be drawn.

This same scheme can be followed to find the probability of a full house. There are 13 numbers that the three-of-a-kind may have, and then 12 numbers possible for the pair. The three-of-a-kind must represent three of four suits, and the pair two of four suits. This gives

$$13\binom{4}{3}\,12\binom{4}{2}$$

different ways to get a full house. The probability is

$$p(\text{full house}) = \frac{13\binom{4}{3}\,12\binom{4}{2}}{\binom{52}{5}}$$

which is about .0014.

Finally, a flush is a hand of five cards all of which are in the same suit. There are exactly four suit possibilities, and a selection of five out of 13 numbers that the cards may show. Hence, the number of different flushes is $4\binom{13}{5}$, and

$$p(\text{flush of five cards}) = \frac{4\binom{13}{5}}{\binom{52}{5}}$$

or about .00198.

The probabilities of the various other hands can be worked out in a similar way. The point of this illustration is that it typifies the use of these counting rules for actually figuring probabilities of complicated events, such as particular poker hands. Naturally, a great deal of practice is usually necessary before one can visualize and carry out probability calculations from "scratch" with any facility. Nevertheless, such probability calculations usually depend on counting how many ways events of a certain kind can occur, if the elementary events are finite in number.

Now we will turn to a special use of these counting rules in finding the distribution of an important discrete random variable.

3.9 BERNOULLI TRIALS

The very simplest probability distribution is one with only two event classes. For example, a coin is tossed and one of two events, heads or tails, must occur, each with some probability. Or a normal human being is selected at random and his or her sex recorded: the outcome can be only male or female. Such an experiment or process that can eventuate in only one of two outcomes is usually called a Bernoulli trial, and we will call the two event classes and their associated probabilities a Bernoulli process.

In general, one of the two events is called a ''success'' and the other a ''failure'' or ''nonsuccess.'' These names serve only to tell the events apart, and are not meant to bear any connotation of ''goodness'' of the event. In the discussion to follow, the symbol p will stand for the probability of a success, and $q = 1 - p$ for the probability of a failure. Thus, in tossing a fair coin, let a head be a success. Then $p = 1/2$, $q = 1 - p = 1/2$. If the coin is biased, so that heads are twice as likely to come up as tails, then $p = 2/3$, $q = 1/3$. In the following, take care to distinguish between p, standing for the probability of a success, and $p(x)$, standing for the probability of some value of a random variable.

3.10 SAMPLING FROM A BERNOULLI PROCESS

Suppose that some sample space exists fitting a Bernoulli process. Furthermore, suppose that either we sample independently with replacement, or that an infinite number of elementary events exist, so that for each sample observation out of N trials, p and q are unchanged. When the outcome is generated by the same process on every trial, so that p and q remain constant over the trials, the process is said to be **stationary.** If the p and q values change from trial to trial, as may well be the case in some practical situations, then the Bernoulli process is said to be nonstationary. For the time being, we will confine our attention to stationary processes and to independent trials made from such a process.

Now we proceed to make N independent observations. Let N be, say, 5. How many different sequences of five outcomes could be observed? The answer, by rule 1, is $2^5 = 32$. However, here it is not necessarily true that all sequences will be *equally probable*. The probability of a given sequence depends upon p and q, the probabilities of the two events. Fortunately, since trials are independent, one can compute the probability of any sequence by the application of Eq. 1.14.1.

We want to find the probability of the particular sequence of events:

(S, S, F, F, S)

where S stands for a success and F for failure. The probability of first observing an S is p. If the second observation is independent of the first, then by Eq. 1.14.1,

probability of (S, S) $= p \times p = p^2$.

The probability of an F on the third trial is q, so that the probability of (S, S) followed by F is p^2q. In the same way the probability of (S, S, F, F) $= p^2q^2$, and that of the entire sequence is $p^2q^2p = p^3q^2$.

The same argument shows that the probability of the sequence (S, F, F, F, F) is pq^4, that of (S, S, S, S, S) is p^5, of (F, S, S, S, F) is p^3q^2, and so on.

Now if we write out all of the possible sequences and their probabilities, an interesting fact emerges: **the probability of any given sequence of N independent Bernoulli trials depends only on the number of successes and p, the probability of a success.** That is, regardless of the *order* in which successes and failures occur in a sequence, the probability is

$$p^r q^{N-r}$$

[3.10.1]

where r is the number of successes, and $N - r$ is the number of failures. Suppose that in a sequence of 10 trials, exactly 4 successes occur. Then the probability of that particular sequence is

$$p^4 q^6.$$

If $p = 2/3$, then the probability can be worked out from

$$\left(\frac{2}{3}\right)^4 \left(\frac{1}{3}\right)^6.$$

The same procedure would be followed for any r successes out of N trials for any p.

For example, if we toss a fair coin ($p = 1/2$) six times, what is the probability of observing three heads followed in order by three tails? The answer is

$$p^3 q^3 = \left(\frac{1}{2}\right)^3 \left(\frac{1}{2}\right)^3 = \frac{1}{64}.$$

This is also the probability of the sequence (H, T, H, T, H, T), of the sequence (H, T, T, T, H, H), and of any other sequence containing exactly three successes or heads.

The probabilities just found are for *particular sequences*, arrangements of r successes and $N - r$ failures in a certain order. What we have found is that if we want to know the probability of a particular sequence of outcomes of independent Bernoulli trials, that sequence will have the same probability as any other sequence with exactly the same number of successes, given N and p.

In most instances, however, one is not especially interested in particular sequences *in order*. We would like to know probabilities of given numbers of successes *regardless* of the order in which they occur. For example, when a coin is tossed five times, there are several sequences of outcomes where exactly two heads occur:

(H, H, T, T, T)
(H, T, H, T, T)
(H, T, T, H, T)
(H, T, T, T, H)
(T, H, T, T, H)
(T, H, H, T, T)
(T, H, T, H, T)
(T, T, H, H, T)
(T, T, H, T, H)
(T, T, T, H, H).

Each and every one of these different sequences must have the same probability, p^2q^3, since each shows exactly two successes and three failures. Notice that there are

$$\binom{N}{r} = \binom{5}{2} = 10 \text{ different such sequences, exactly as counting rule 5 gives for the}$$

number of ways 5 things can be taken 2 at a time.

What we want now is the probability that $r = 2$ successes will occur *regardless* of order. This could be paraphrased as "the probability of the sequence (H, H, T, T, T) *or* the sequence (H, T, H, T, T) *or* any other sequence showing exactly 2 successes in 5 trials." Such "or" statements for mutually exclusive events recall Axiom 3 of Section 1.4: whenever A and B are mutually exclusive events, then $p(A \cup B) = p(A) + p(B)$. Thus, the probability of 2 successes in any sequence of 5 trials is

$$p(2 \text{ successes in 5 trials}) = p^2q^3 + p^2q^3 + \cdots + p^2q^3 = \binom{5}{2} p^2q^3$$

since each of these sequences has the same probability and there are $\binom{5}{2}$ of them.

Generalizing this idea for any r, N, and p, we have the following principle: **in sampling from a stationary Bernoulli process, with the probability of a success equal to p, the probability of observing exactly r successes in N independent trials is**

$$p(r \text{ successes}; N, p) = \binom{N}{r} p^r q^{N-r}. \tag{3.10.2}$$

In understanding the basis for this rule, the thing to keep in mind is that $p^r q^{N-r}$ is the probability of *any* of the events consisting of a specific sequence showing exactly r successes out of N trials. Then $\binom{N}{r}$ is the number of such sequence events that qualify for the event "exactly r successes in N trials." It is important to notice that there is an exact correspondence between the binomial coefficients $\binom{N}{r}$ and the number of sequences possible in which exactly r successes occur out of N trials.

An experiment carried out in such a way that N independent trials are made from a stationary Bernoulli process is known as **binomial sampling.** In binomial sampling the value of N is predetermined, and it is the value of r, the number of successes, that is left to chance.

For example, imagine that in some very large population of animals 80% of the individuals have normal coloration, and only 20% are albino (no skin and hair pigmentation). This may be regarded as a Bernoulli process, with "albino" being a success and "normal" a failure. Suppose that the probabilities are $.20 = p$ and $.80 = q$. A biologist manages to sample this population at random, catching three animals. What is the probability of catching one albino? Here, $N = 3$, $r = 1$, so that

$$p(1 \text{ albino in 3 animals}) = \binom{3}{1} (.20)^1 (.80)^2$$

$$= .384.$$

If the sampling is random, and if the population is so large that sampling without replacement still permits one to regard the results of the successive trials as independent, then the biologist has about 38 chances in 100 of observing exactly 1 albino in the sample of 3 animals.

3.11 NUMBER OF SUCCESSES AS A RANDOM VARIABLE: THE BINOMIAL DISTRIBUTION

When samples of N trials are taken from a Bernoulli process, the number of successes is a discrete random variable. Since the various values are counts of successes out of N observations, the random variable can take on only the whole values from 0 through N. We have just seen how the probability for any given number of successes can be found. Now we will discuss the distribution pairing each possible number of successes with its probability. This distribution of number of successes in N trials is called the **binomial distribution.** This is the first distribution we have studied that can most readily be described by its mathematical rule.

The general definition of the binomial distribution can be stated as follows:

Any random variable X with probability function given by

$$p(X = r; N,p) = \binom{N}{r} p^r q^{N-r}, \quad X = 0, 1, \cdot \cdot \cdot , N,$$ [3.11.1]

is said to have a binomial distribution with parameters N and p.

The unspecified mathematical constants such as N and p that enter into the rules for probability or density functions indicate **parameters.** By changing the particular values of p and N that enter into the rule, one can produce different binomial distributions, each following the same mathematical rule, but differing in the probabilities that particular values of the random variable receive. Thus, what we are calling *the binomial distribution* is actually a whole *family* of binomial distributions. Families of distributions share the same mathematical rule for assigning probabilities or probability densities to values of X; in these rules the parameters are simply symbolized as constants. Actually assigning values to the parameters, such as we did for p and N in the distributions above, gives some particular distribution belonging to the family. Thus, *the* binomial distribution usually refers to the family of distributions having the same rule, and *a* binomial distribution is a particular one of this family found by fixing N and p.

A binomial distribution can be illustrated most simply as follows: Consider a simple experiment repeated independently five times. Each trial must result in only one of two outcomes and the result of five trials is a sequence of outcomes like those in the preceding sections. However, we are not at all interested in the order of the outcomes, but only in the number of successes in the set of trials.

In order to find the probability for each value of the discrete random variable, given N and p, let us begin with the largest value, $X = 5$. By counting rule 5, there must be $\binom{5}{5} = 1$ possible sequence where all of the outcomes are successes. The probability of this sequence is $p^5 q^0 = p^5$, so that we have

$$p(X = 5; N = 5, p) = \binom{5}{5} p^5 = p^5.$$

For four successes, we find that by counting rule 5,

$$\binom{5}{4} = \frac{5!}{4!1!} = \frac{(1)(2)(3)(4)(5)}{(1)(2)(3)(4)(1)} = 5,$$

so that four successes can appear in five different sequences. Each sequence has probability $p^4 q^1$. Thus,

$$p(X = 4; N = 5, p) = \binom{5}{4} p^4 q^1 = 5p^4 q^1.$$

Going on in this way, we find

$$p(X = 3; N = 5, p) = \binom{5}{3} p^3 q^2 = 10p^3 q^2$$

$$p(X = 2; N = 5, p) = \binom{5}{2} p^2 q^3 = 10p^2 q^3$$

$$p(X = 1; N = 5, p) = \binom{5}{1} pq^4 = 5pq^4$$

$$p(X = 0; N = 5, p) = \binom{5}{0} q^5 = q^5.$$

(Note that $\binom{5}{0} = \dfrac{5!}{0!5!} = 1$, since 0! is 1 by definition.)

Now, to take a concrete instance of this binomial distribution, let the experiment be that of tossing a fair coin five times. Then $p = 1/2$, and the binomial distribution for the number of heads is

x	p(x)		
5	$(1/2)^5$	=	1/32
4	$5(1/2)^4(1/2)$	=	5/32
3	$10(1/2)^3(1/2)^2$	=	10/32
2	$10(1/2)^2(1/2)^3$	=	10/32
1	$5(1/2)(1/2)^4$	=	5/32
0	$(1/2)^5$	=	1/32
			32/32

Notice that the probabilities over all values of X sum to 1.00, just as they must for any probability distribution.

Now consider another example that is similar to this last one but provides different probability values. Suppose that among American women college students who are undergraduates, only 1 in 10 is married. A sample of five female students is drawn at random. Let X be the number of married students observed. (We will assume the total set of students to be large enough that sampling can be without replacement without affecting the probabilities and that observations are independent.) Hence, $p = .10$ and the distribution is

x	$p(x)$		
5	$(1/10)^5$	$=$	1/100,000
4	$5(1/10)^4(9/10)$	$=$	45/100,000
3	$10(1/10)^3(9/10)^2$	$=$	810/100,000
2	$10(1/10)^2(9/10)^3$	$=$	7,290/100,000
1	$5(1/10)(9/10)^4$	$=$	32,805/100,000
0	$(9/10)^5$	$=$	59,049/100,000
			100,000/100,000

Contrast this distribution with the preceding one: When p was 1/2, the distribution showed the greatest probability for $X = 2$ and $X = 3$, with the probabilities diminishing gradually both toward $X = 0$ and toward $X = 5$. On the other hand, in the second distribution, the most probable value of X is 0, with a steady decrease in probability for the values 1 through 5. The distribution over such values of X is very different in these two situations, even though the probabilities are found by exactly the same *formal* rule. This illustrates once again that the binomial is actually a family of theoretical distributions, each following the same mathematical rule for associating probabilities with values of the random variable but differing in particular probabilities, depending on the values of p and N.

Almost all theoretical distributions of interest in statistics can be specified by stating the function rule. The way this simplifies the discussion of distributions will be obvious as we go along; indeed, continuous distributions cannot really be discussed at all except in terms of their function rule. Typically, any one of these theoretical distributions will be a member of a family of distributions, each of which follows the same mathematical rule as the others, but which differs according to the values of the constants, or parameters, appearing in the rule.

3.12 THE BINOMIAL DISTRIBUTION AND THE BINOMIAL EXPANSION

In school algebra you were very likely taught how to expand an expression such as $(a + b)^n$ by the following rule:

$$(a + b)^n = a^n + \frac{n!}{(n - 1)!1!} a^{n-1}b + \frac{n!}{(n - 2)!2!} a^{n-2}b^2$$

$$+ \cdots + \frac{n!}{1!(n - 1)}ab^{n-1} + b^n.$$

For example, $(a + b)^3 = a^3 + 3a^2b + 3ab^2 + b^3$ according to this rule. This is the familiar *binomial theorem* for expanding a sum of two terms raised to a power.

Notice that the various probabilities in the binomial distribution are simply terms in such a binomial expansion. Thus, if we take $a = p$, $b = q$, and $n = N$,

$$(p + q)^N = p^N + \binom{N}{N-1} p^{N-1}q + \binom{N}{N-2} p^{N-2}q^2 + \cdots + q^N.$$

Since $p + q$ must equal 1.00, then $(p + q)^N = 1.00$, and the sum of all of the probabilities in a binomial distribution is 1.00.

3.13 PROBABILITIES OF INTERVALS IN THE BINOMIAL DISTRIBUTION

In Chapter 2 we saw how to find a probability that a value of a random variable lies in an interval, such as $p(2 \leq X \leq 8)$, the probability that the random variable X takes on some value between 2 and 8 inclusive. This idea is easy to extend to a binomial variable.

Consider the binomial distribution shown in Table 3.13.1, with $p = .3$ and $N = 10$. First of all we will find the probability $p(1 \leq X \leq 7)$, that X lies between the values 1 and 7 inclusive. This is given by the sum

$p(X = 1)$	.12106
$+\ p(X = 2)$	.23347
$+\ p(X = 3)$	.26683
$+\ p(X = 4)$	.20012
$+\ p(X = 5)$	.10292
$+\ p(X = 6)$	.03676
$+\ p(X = 7)$	.00900

$$.97016 = p(1 \leq X \leq 7).$$

The probability is about .97 that an observed value of X will lie between 1 and 7 inclusive. Thus, if we were drawing random samples of 10 observations from a Ber-

Table 3.13.1
A binomial distribution with $p = .3$ and $N = 10$.

r	$\binom{N}{r} p^r q^{N-r} = p(X = r)$
10	.00001
9	.00014
8	.00145
7	.00900
6	.03676
5	.10292
4	.20012
3	.26683
2	.23347
1	.12106
0	.02824
	1.00000

noulli process where the probability of a success was .3, we should be very likely to observe a number of successes between 1 and 7 inclusive.

By the same token, we can find

$$p(8 \leq X) = p(X = 8) + p(X = 9) + p(X = 10) = .00160.$$

The chances are less than two in a thousand of observing eight or more successes, if $p = .30$.

Notice that

$$p(X = 0, \text{ or } 8 \leq X) = .02824 + .00160 = .02984,$$

which is the same as

$$1 - p(1 \leq X \leq 7) = 1 - .97016 = .02984,$$

so that this is also the probability that X falls *outside* the interval bounded by 1 and 7.

Binomial distributions can also be put into cumulative form. Thus, the probability that X falls at or below a certain value a is the probability of the interval $X \leq a$. For this particular distribution, the corresponding cumulative distribution is given in Table 3.13.2.

In this distribution, we see that about 65% of samples of 10 should show 3 or fewer successes, about 85% should have 4 or fewer, 99.8% should have 7 or fewer. Every sample (100%) must have 10 or fewer successes, of course, since $N = 10$.

It must be reemphasized that the binomial distribution is *theoretical*. It shows the probabilities for various numbers of successes out of N trials *if* independent random samplings are carried out from a stationary Bernoulli process and *if* p is the probability of a success. Given a different value of p or of N (or of both) the probabilities will be different. Nevertheless, regardless of N or p, the probabilities are found by the same binomial rule.

Table II of Appendix E gives binomial probabilities for $0 \leq r \leq N$ and $1 \leq N \leq 20$, for selected values of p. For p values which are greater than .50, one simply deals with $q = 1 - p$ and with the number of failures, or $N - r$. Thus, for $p = .70$ and $N = 10$, the probability of 6 successes is the same as the probability of 4 failures given $q = .30$. In the table under $N = 10$, $r = 4$, and $p = .30$, this is .2001. Cumulative

r	$p(x \leq r)$	**Table 3.13.2**
		The cumulative distribution for Table 3.13.1
10	1.00000	
9	.99999	
8	.99985	
7	.99840	
6	.98940	
5	.95264	
4	.84972	
3	.64960	
2	.38277	
1	.14930	
0	.02824	

probabilities such as $p(X \leq r; N, p)$ are found from the sum of the values of $p(x)$ for $x = 0, 1, \cdots, r$, given N and p. Probabilities for larger values of N can be found directly by use of the factorials provided in Appendix E, Table VIII. Provided that only a few probabilities are needed, these can also be found quickly on a calculator.

3.14 THE BINOMIAL DISTRIBUTION OF PROPORTIONS

Quite often researchers are interested not in the number of successes that occur for some N trial observations, but rather in the *proportion* of successes, r/N. The proportion of successes is also a random variable, taking on fractional (or decimal) values between 0 and 1.00. Such sample proportions will be designated by the capital letter P to distinguish them from p, the probability of a success.

The probability of any given proportion P of successes among N cases sampled from a given Bernoulli process is exactly the same as the probability of the number of successes; that is

$$p\left(P = \frac{r}{N}\right) = p(X = r).$$ [3.14.1]

For instance, if $N = 6$ and $r = 4$

$$p(X = 4) = p\left(P = \frac{4}{6}\right) = \binom{6}{4} p^4 q^2.$$

The distribution of sample proportions is therefore given by the binomial distribution, the only difference being that each possible value of X becomes a value of $P = X/N$, so that $p(P = a) = p(X = Na)$ for any particular value a. In the further discussion of the binomial distribution in Chapter 6, care will be taken to specify whether the random variable is regarded as X or P, since the arithmetic is slightly different in the two situations. Nevertheless, the probability of any given P is the same as for the corresponding X, given the sample size N.

3.15 A PREVIEW OF A USE OF THE BINOMIAL DISTRIBUTION

It is now possible to point ahead to an important use of the binomial distribution. This example will deal with an experiment in psychology where we must use the data to decide whether or not a particular hypothesis actually seems to fit the data we obtain.

The context of the experiment is this: a psychophysical threshold or limen is that value on some physical measurement of a stimulus object at which a human subject is just capable of responding—in somewhat inexact terms, if the stimulus is a point of light in a darkened room, the threshold might be the physical intensity the light would have to have so that the subject would just be able to "see" the stimulus. It has, however, been suspected for some time that subjects may be able to respond in certain ways to stimuli which are actually below their known threshold of awareness. Such stimuli are said to be subliminal; the subject may not really be conscious that he "sees" the light—nevertheless, the subject may be able to respond *as though* capable of seeing the light.

Think of a hypothetical study of this question: "If a human is subjected to a stimulus below the threshold of conscious awareness, can the behavior somehow still be influenced by the presence of the stimulus?" The experimental task is as follows: the subject is seated in a room in front of a square screen divided into four equal parts and instructed that the task is to guess in which part of the screen a small, very faint, spot of light is thrown. He is to do this for many trials, and is told the light will be projected on the screen in a completely "haphazard," "random" manner over the trials. The light projected is made to be so faint that the subject cannot in any conscious sense actually "see" the light. However, unknown to the subject, the spot is always projected into the same one of the four parts of the screen over the various trials. For our computational convenience, suppose that only 10 trials are taken for this one subject.

Our hypothesis goes like this: if the subject really is in no way being influenced by the small "invisible" spot of light, then guesses should be random, haphazard affairs themselves, so that the guess should be right only 1/4 of the time by accident. Thus, under this hypothesis of "only guessing," the sample space of the subject's response to this situation should be distributed in this way:

Class	p
right	1/4
wrong	3/4
	4/4 = 1.00

The explicit assumption made is that the various trials for the subject are independent of each other. Now what sampling distribution holds for 10 trials for this subject if this hypothesis is true? The number of correct guesses *could* range from 10 right to 0 right, and the distribution would be a binomial distribution with $p(\text{right}) = 1/4$, $N = 10$. If we apply the rule for the binomial, we find the distribution that appears in Table 3.15.1. This binomial distribution gives all the possible numbers of correct guesses that this subject *could* make in the 10 trials, and the probability for each, by guessing in a truly haphazard manner.

Given this theoretical distribution of possible outcomes, we turn to the actual result of the experiment. It is found that the subject guessed correctly on 7 out of the 10 trials. What then is the probability that exactly this result *should* have come up by chance? The binomial probability for 7 correct is 3,240 out of 1,048, 576, or about .0031. Thus the probability of getting exactly this number correct by random guessing alone is about 31 chances in 10,000. In other words, if we repeated the experiment, giving the subject 10,000 independent sets of 10 trials, about 31 of these repetitions should give us exactly 7 correct guesses. The exact sample result obtained is clearly not very likely to occur if the hypothesis is correct.

However, we should be interested in the probability not only of getting exactly 7 correct, but also in the probability of getting *this many or more* correct, since we are really interested in this result as evidence of whether or not the subject is guessing, or

x	p(x)
10 right	$\binom{10}{10}(1/4)^{10}(3/4)^0 = .0000(+)$
9	$\binom{10}{9}(1/4)^9(3/4)^1 = .0000(+)$
8	$\binom{10}{8}(1/4)^8(3/4)^2 = .0004$
7	$\binom{10}{7}(1/4)^7(3/4)^3 = .0031$
6	$\binom{10}{6}(1/4)^6(3/4)^4 = .0162$
5	$\binom{10}{5}(1/4)^5(3/4)^5 = .0584$
4	$\binom{10}{4}(1/4)^4(3/4)^6 = .1460$
3	$\binom{10}{3}(1/4)^3(3/4)^7 = .2503$
2	$\binom{10}{2}(1/4)^2(3/4)^8 = .2816$
1	$\binom{10}{1}(1/4)^1(3/4)^9 = .1877$
0	$\binom{10}{0}(1/4)^0(3/4)^{10} = .0563$

Table 3.15.1

A binomial distribution for $p = .25$ and $N = 10$.

doing something else which gives more correct responses than should simple, haphazard guessing. What is the probability of 7 or more correct trials? The answer is readily seen if we remember that this is asking for the probability of an interval:

$$p(7 \le X) = p(7) + p(8) + p(9) + p(10) = .0035(+).$$

Seven or more correct guesses should occur only about 35 times in 10,000 independent replications of this experiment, if guessing alone is responsible for the subject's behavior. Does this unlikely result cast any doubt on the theory that the stimulus has no effect on guessing behavior? The answer is yes. For a theory to be "good," it should forecast results that agree with what we actually obtain. If the subject had come up with 2, 3, or 4 correct responses, then we would have little reason to doubt the "guessing" hypothesis, since these results fall among those that are quite probable according to the sampling distribution. However, the results obtained are quite unlikely to occur if the hypothesis is true, and so the evidence does not seem to favor this hypothesis.

On the other hand, are we completely safe in inferring that the subject was not just guessing? The answer is, of course, no. Even though this many or more correct responses obtained is improbable if the hypothesis is true, it is still not impossible that the subject was only guessing. We need some way to state just how much of a chance we are taking in saying that the hypothesis is not true, and that the subject is in some way being influenced by the stimulus. The best measure of the amount of risk we run by abandoning this hypothesis on the evidence is given by the probability of sample results as *extreme or more extreme* than those actually obtained, if the hypothesis were true. The evidence apparently diverges considerably from what we should expect (about 1/4 correct) if the subject were guessing, and the probability of such a divergent sample is only about 35 in 10,000. The probability of our being *wrong* in rejecting the guessing-hypothesis equals the probability of the divergent result, or .0035. We *could* be wrong in rejecting the hypothesis that the subject is guessing, but the probability of our being wrong is not very great. The more our sample result departs from what we expect given a hypothetical situation, and the more improbable that departure is, the less credence is given the hypothesis. In a later chapter we shall discuss in detail the rules for (and precautions in) evaluating hypotheses in the light of obtained results. However, this illustration exhibits the general logic underlying all "tests" of hypotheses: from the hypothetical population distribution one obtains a theoretical sampling distribution. Then the obtained results are compared with the sampling distribution probabilities. If the probability of samples such as the one obtained is high, the hypothesis is regarded as tenable. On the other hand, if the probability of such a sample (or one in more extreme disagreement with what is expected) is quite small, then doubt is cast on the hypothesis.

Although this little example should not be taken as a model of sophisticated scientific practice, it does suggest the use of a theoretical distribution of sample results as an aid in making inferences about a hypothetical situation. The binomial distribution is only one of a number of families of distributions that are employed in this general way.

Since we have been rather heavily theoretical in our discussion up to this point, the next section will deal with one method based on the binomial distribution that has some immediate practical use in experimental situations. Although not all of the qualifications involved in the use of such methods can be explained at this time, perhaps it will prove interesting and useful nevertheless. Then, some close relatives of the binomial distribution will be introduced, along with some indications of how these distributions are used.

3.16 THE SIGN TEST FOR TWO GROUPS

We have already seen that the binomial distribution can be used to test a hypothesis about a proportion on the basis of a random sample of N independent observations. This is but one of a very large number of uses of the idea of a binomial variable in applied statistics. Now, a method based on the binomial distribution will be introduced, which is designed for the comparison of two groups of observations.

This simple method based on the binomial distribution is the so-called sign test. The sign test is used in situations in which N *pairs of matched* observations are made. The first member of each pair is an observation of some type, A, and the second an observation of another type, B. Thus, husband-wife pairs may be observed, in which each husband (type A) is paired with a wife (type B). Now each A individual has a numerical score X_A, and this is paired with a numerical score for the B individual, X_B. The question to be answered is, "Is the distribution of X_A scores identical to the distribution of X_B scores in the long run, if all possible pairs could be observed?" We approach this question by noting the difference $(x_A - x_B)$ for each of the N pairs. If $(x_A - x_B)$ is positive in value, then a "+" or a "success" is recorded. If $(x_A - x_B)$ is negative, then a "−" or "failure" is recorded. When $(x_A - x_B) = 0$, a fair coin is tossed, and the result is noted as a success or a failure depending on the outcome of the coin.

If the distribution of X_A values and the distribution of the X_B values really are identical, in the long run for all possible pairs, then it should be true that $p(+) = p(-) = 1/2$. Furthermore, the occurrence of successes and failures should correspond to a Bernoulli process, with N equal to the number of pairs observed, and $p = 1/2$. In order to reach a conclusion about the tenability of the hypothesis that $p = 1/2$, one calculates the probability that results as extreme, or more extreme, should occur, given that $p = 1/2$, and given the value of N that is actually used. If this probability is sufficiently small (say .05 or less), then the hypothesis that $p = 1/2$ is rejected, meaning that the hypothesis of equal distributions is rejected as well. If a number of successes or failures equally or more extreme than the number actually obtained does not have a sufficiently small probability, then the hypothesis is not rejected.

For example, suppose that the basic question had been, "Are women actually better drivers than their husbands?" To shed light on this question, we take 20 married couples at random, and separately give each wife and each husband a driving test. If a wife scores higher than a husband, this is a "success" or a "+," and if the husband scores higher, this is a "failure," or "−." We decide that we will reject the hypothesis of "no difference in driving ability" only if the number of successes observed should be equaled or exceeded with a probability no larger than .05. That is, we need to find the probability of the number of successes we actually observed, *or more,* in a binomial distribution with $N = 20$ and $p = 1/2$. If this probability is .05 or less, then we will reject the hypothesis of equal ability.

Suppose that exactly 15 successes were observed. Then the probability of 15 successes or more can be found quite simply in Table II in Appendix E. Here, $N = 20$ and $p = 1/2$, so that we find the column in the table labeled $p = .50$, and the row section for $N = 20$. Adding up the probabilities for 15 or more, we find p (15 or more) $= .0148 + .0046 + .0011 + .0002 = .0207$. Then, since this value is less than the probability of .05 we had decided to employ in making the decision, we reject the hypothesis that husbands and wives are equal in driving ability. The evidence obtained suggests that wives are actually better drivers.

On the other hand, if we were able to observe the entire population of husbands and wives and give the members of each pair the driving test, it might well be true that there is no long-run difference, or even that husbands turn out to be the better drivers. After all, this is only a sample of 20 from among millions of couples, and by chance it could be that we have drawn a most unrepresentative group. However, we have a bit of insurance: since the probability of 15 or more successes is less than .05, the chances of our having got such an extreme sample (when husbands and wives are not truly different) is also less than .05. We are thus running a small risk in saying that wives are indeed better drivers than their husbands, provided this sample was indeed random.

The sign test is useful in a variety of problems where pairs of observations are to be compared on some characteristic. It can also be extended to situations where the sample size is too large to permit use of Table II; this extension is discussed in Chapter 19.

3.17 THE GEOMETRIC AND PASCAL DISTRIBUTIONS

A close relative of the binomial distribution is the Pascal distribution, named for the French mathematician-philosopher Blaise Pascal (1623–1662). Whereas a random variable following the binomial rule corresponds to the number of successes r out of a fixed number of trials N, the random variable following the Pascal rule has a different interpretation. Here one is interested in the number of trials N necessary to achieve a given number of successes r. Thus N is the random variable and r is a constant in a geometric, or Pascal, distribution. Unlike binomial sampling, in geometric and Pascal sampling, r is fixed and N is left to chance.

Consider a stationary and independent Bernoulli process in which the probability of a success on any trial is p. Then for any sequence of trials we might ask, "What is the probability that the first success occurs on the first trial?" or "What is the probability that the first success occurs on the sixth trial?" and so on for a trial of any number. Let us study the first success. Since the first success never *has* to occur at all, short of an indefinitely large number of trials, then the possible trials on which the first success *might* occur are countably infinite in number. The distribution of N, the trial number on which the first success occurs, given trials from a stable Bernoulli process, is known as the **geometric distribution.** In a geometric distribution, the random variable N can take on any value that is one of the counting numbers $1, 2, 3, \cdots$. This means that, unlike a binomial variable, a geometric variable takes on a countably infinite set of values. However, such a geometric variable is still discrete, since it can take on only whole-number values.

Now, the probability that the first success occurs on the first trial is

$$p(N = 1;p) = p.$$

The probability that the first success occurs on the second trial must be

$$p(N = 2;p) = (1 - p)p,$$

since the first trial must be a failure if the first success occurs on the second trial. In the same way, we can show that

$$p(N = 3;p) = (1 - p)^2 p$$

and so on, until for the probability that the first success occurs on trial $N = n$ we have

$$p(N = n;p) = (1 - p)^{n-1} p.$$

This is the rule for the geometric distribution, in which the random variable is the trial number on which the first success occurs, in trials from a stable, independent Bernoulli process with probability p.

The Pascal distribution can be thought of as a generalization of the geometric distribution. That is, the random variable in the Pascal distribution is the trial number on which the rth success occurs, where r can be any whole number $r = 1, 2, \cdots$, and where $r \le N$. The geometric distribution is a Pascal distribution in which $r = 1$. For a Pascal distribution, the probability that N equals a given value n depends on the fixed value r, or number of successes, and p, the probability of a success, as follows:

$$p(N = n; r, p) = \frac{(n - 1)!}{(r - 1)!(n - r)!} p^r (1 - p)^{n-r}, N \ge r. \qquad [3.17.1]$$

(Notice that there is no upper limit to N; in principle the rth success might *never* occur. Thus a Pascal variable is discrete, but *countably infinite* in the number of values.)

For an example of how the idea of a Pascal variable can be used, consider the following: A team of investigators are choosing items for inclusion in a test. They believe that one item is moderately easy, so that two out of three schoolchildren of a given age should pass the item. They interpret this to mean that the probability that any given child will pass is 2/3. However, they want to test this belief against the possibility that the item may be harder. Time and expense are factors, so they do not want to sample more children than absolutely necessary in order to arrive at a judgment. Thus the team decides to administer the item to individual children, one at a time, selected at random. They also decide to test only until four children have successfully passed the item. Then, on the basis of the total number of children tested in order to achieve four successes, they will reach a conclusion about the notion that the probability of a given child's passing is 2/3. The more children it is necessary to test in order to achieve four successes, the more doubt will be cast upon 2/3 as the value of p. For reasons that we will elaborate at length later on, the team decides to reject the idea that $p = 2/3$ only if the number of children required to reach four successes is in some sense "excessive." In particular, $p = 2/3$ will be rejected as the true probability only if N, the number of children actually tested, should be equaled or exceeded with probability of only .05 or less, given that $p = 2/3$ is the true probability of passing the test. Otherwise, $p = 2/3$ will be retained as a tenable hypothesis about the true value.

Now suppose that the investigators begin to test children in the order of their random selection. Sure enough, the fourth child to pass the item is actually the eighth child tested. Thus, $N = 8$, and the fixed number of successes $r = 4$, whereas the value of p is believed to be 2/3. The critical question then is the value of $p(N \ge 8; 4, 2/3)$, which is the same as $1 - F(7; 4, 2/3)$. Is this, or is it not, .05 or less?

The value of this probability is determined by the Pascal rule. Since

$$F(7; 4, 2/3) = p(4) + p(5) + P(6) + p(7),$$

it is necessary to find the probability for each of these values of the random variable. (Remember that one cannot have four successes in fewer than four trials.) Proceeding, we find

$$p(4) = \frac{(4 - 1)!}{(4 - 1)!(0)!} p^4 = (2/3)^4 = 16/81,$$

$$p(5) = \frac{(5 - 1)!}{(4 - 1)!(1)!} p^4 (1 - p) = 4(2/3)^4 (1/3) = 64/243,$$

$$p(6) = \frac{(6-1)!}{(4-1)!(6-4)!} p^4(1-p)^2 = 10(2/3)^4(1/3)^2 = 160/729,$$

$$p(7) = \frac{(7-1)!}{(4-1)!(7-4)!} p^4(1-p)^3 = 20(2/3)^4(1/3)^3 = 320/2,187.$$

Then

$$F(7;4, 2/3) = 16/81 + 64/243 + 160/729 + 320/2,187 = 1,808/2,187,$$

so that $F(7;4, 2/3)$ is about .83. Thus, a number of trials equal to 8 or more can occur with probability of about .17. This is far in excess of the probability of .05 that the investigators had decided upon. On this evidence they decide not to reject the hypothesis that $p = 2/3$.

On the other hand, had it been true that 10 trials were required in order to reach four successes, then $F(9;4, 2/3)$ would have been approximately .96. This in turn implies that the probability of 10 or more trials required to reach four successes would have been .04, which is less than the criterion agreed upon. In this case, the investigators would have decided to reject the hypothesis that $p = 2/3$. Either it is true that $p = 2/3$, and a rare event has occurred, or 2/3 is not the correct value of p. The investigators have already decided to conclude the latter, should such a rare event occur.

Notice what has been assumed here. First of all, we have assumed that $p = 2/3$ for each and every one of the children tested, so that observation of each child represents a trial in a stationary Bernoulli process. Futhermore, we have assumed that the results for each child, or trial, are independent of those for any other child. Finally, we have assumed that the order of selection of the children was completely random. The failure of any of these assumptions to be true could, of course, make a difference in the final conclusions.

Problems sometimes arise in which interest focuses not on the trial number on which the rth success occurs, but rather on the number of failures that occur before the rth success. However, when $y = N - r$ (the number of failures before the rth success) is the random variable, it is customary to refer to the distribution as the **negative binomial distribution,** and this formulation will sometimes be encountered in the statistical literature in place of the Pascal distribution. For all practical purposes these two kinds of distributions are the same, and problems can be solved in terms of either the number of failures before the rth success, $(N - r)$ or the trial number of the rth success, N.

Like the binomial distribution, the Pascal distribution can be used as the basis for a variety of statistical techniques. Ordinarily, the kind of situation producing a random variable following a Pascal rule will be one involving **sequential sampling,** as opposed to simple random sampling. In sequential sampling no predetermined limit is set upon the number of observations. Rather a series of observations are carried out until sufficient data are accumulated to enable a decision to be made, according to a predetermined criterion. Since a Pascal variable is interpreted as the number of trials required in order to reach a given number of successes, this conception lends itself readily to the simplest kinds of sequential experiments.

Other methods of sequential sampling, not depending on the Pascal distribution, are also prominently used in statistics. Unfortunately, space does not permit going further with this topic. However, you should know this approach is a valid alternative to some of the sampling methods and statistical techniques we will be employing. Details may be found in many standard texts, such as Mood, Graybill, and Boes (1974).

3.18 THE POISSON DISTRIBUTION

Still other relatives of the binomial family of distributions play a large role in theoretical and applied statistics. One is the family of Poisson distributions, named after the 19th century French mathematician S. Poisson. A random variable following this rule is referred to as a Poisson variable, and the process generating values of such a random variable is known as a Poisson process. The probability function for a Poisson variable X follows the rule

$$p(x;m) = \begin{cases} \dfrac{e^{-m}m^x}{x!}, & x = 0, 1, 2, 3, \cdots; m > 0; \\ \\ 0, & \text{otherwise,} \end{cases}$$

[3.18.1]

where e is a mathematical constant equal approximately to 2.718, and m is a constant known as the *intensity* of the Poisson process.

Like a Pascal variable, a Poisson variable X can take on only integral or "whole" values—in this case from zero to an indefinitely large value. This random variable thus can assume any of a countably infinite set of values. It is thus a discrete variable.

Although Poisson variables can be given a variety of useful interpretations, perhaps the simplest approach to the study of the Poisson is to regard it as a special case of the binomial, where N is thought of as very large, and p is very small.

Not only does this connection with the binomial distribution permit one interpretation of the Poisson distribution; there are practical consequences as well. In cases in which N is large and p is relatively small, the binomial probabilities may be very laborious to calculate. In this instance, it is a much simpler matter to approximate the exact binomial probabilities through use of Poisson probabilities for the various values of x.

When a Poisson probability for some specific x value is desired, one simply calculates $m = Np$, and then applies the Poisson rule

$$p(x; m) = \frac{e^{-m}m^x}{x!}$$

for the value $X = x$ of interest. Table X in Appendix E gives selected values of e^{-m}, and Table VIII gives various values of factorials. For example, suppose that for $N = 3,000$ and $p = .001$, we wished to find the probability for $X = 5$. Then, $m = (3,000)(.001) = 3$, and we calculate the probability by taking

$$p(5;3) = \frac{e^{-3}3^5}{5!}$$

$$= \frac{(.0498)(243)}{120}$$

$$= .1008.$$

This is an exact Poisson probability, $p(5;3)$. It is also an approximation of the binomial probability $p(5;3,000, .001)$. The Poisson probability will be approximately equal to the actual binomial probability only for very large N and very small p, of course. Nevertheless, the approximation is good enough to be useful even when N is only moderately large (say $N \geq 50$) and p only relatively small ($p \leq .2$).

Poisson probabilities and cumulative probabilities can be determined relatively easily in the way just shown, given Table VIII for the values of factorials and Table X for values of e^{-m}. Most calculators of the "scientific" type also give values for e^{-m} and for $x!$. Many books on advanced statistics also give extensive tables of Poisson probabilities, particularly when they are designed to be used in fields such as the physical sciences or industry. However, since our use of the Poisson will not be extensive, space will not be given to such tables here.

A great many illustrations of Poisson processes occur in the physical and the biological sciences, as well as in everyday life. For example, the degeneration of a radioactive substance can be regarded as a Poisson process. At any given instant the probability is very small that an alpha particle will be emitted, whereas there are vast numbers of opportunities for such an event to occur. The distribution of bacteria on a petri plate can be viewed as a Poisson process. Each tiny area on the plate can be viewed as a trial, and a bacterium may or may not occur on such an area. The probability of such an occurrence on any given area is very small indeed, but there are very many areas on such a plate. The distribution of misprints in a book can be studied as a Poisson process, as can the occurrence of accidents of a certain kind in a manufacturing plant. The Poisson distribution is thus important in its own right, quite apart from its connection with the binomial distribution.

As a simple example of a problem involving the Poisson distribution consider the following. Suppose it is known that the annual rate of suicides on U.S. college campuses is 5 in 50,000 students, or about .0001. Now a campus of 25,000 students is studied, which is typical in all respects of such campuses, so that its students might be considered a random sample of such students. However, on this campus last year there were four suicides. What is the probability of four or more suicides if the true rate is actually .0001?

Here, we take $p = .0001$ and $N = 25,000$, so that $m = (.0001)(25,000) = 2.5$. Then,

$$p(x \geq 4;2.5) = 1 - p(x = 0) - p(x = 1) - p(x = 2) - p(x = 3)$$

$$= 1 - \frac{e^{-2.5}(2.5)^0}{0!} - \frac{e^{-2.5}(2.5)}{1!} - \frac{e^{-2.5}(2.5)^2}{2!} - \frac{e^{-2.5}(2.5)^3}{3!}$$

so that, by Table X (or from any reasonably sophisticated hand calculator) we obtain

$$p(x \geq 4;2.5) = 1 - .7576 = .2424.$$

The probability of four or more suicides on this campus, if it *is* typical, is about .24. However, if we figure the corresponding probability for six or more suicides, the probability falls to only .04. Thus, for six or more such occurrences, we might begin to have real doubts about the typicality of this campus.

3.19 THE MULTINOMIAL DISTRIBUTION

The basic rationale underlying the binomial distribution can also be generalized to situations with more than two event classes. This generalization is known as the multinomial distribution, having the following rule:

Consider K classes, mutually exclusive and exhaustive, and with probabilities $p_1, p_2, \cdots, p_K$, If N observations are made independently and at random, then the probability that exactly n_1 will be of kind 1, n_2 of kind 2, $\cdots$, and n_K of kind K, where $n_1 + n_2 + \cdots + n_K = N$, is given by

$$\frac{N!}{n_1! n_2! \cdots n_K!} (p_1)^{n_1}(p_2)^{n_2} \cdots (p_K)^{n_K}.$$ [3.19.1]

Think once again of colored marbles mixed together in a box, where the following probability distribution holds:

Color	p
Black	.40
Red	.30
White	.20
Blue	.10
	1.00

Now suppose that 10 marbles were drawn at random and with replacement. The sample shows 2 black, 3 red, 5 white, and 0 blue. What is the probability of a sample distribution such as this? On substituting into the multinomial rule, we have

$$\frac{10!}{(2!)(3!)(5!)(0!)}(.4)^2(.3)^3(.2)^5(.10)^0$$

$$= \frac{1 \cdot 2 \cdot 3 \cdot 4 \cdot 5 \cdot 6 \cdot 7 \cdot 8 \cdot 9 \cdot 10}{(1 \cdot 2)(1 \cdot 2 \cdot 3)(1 \cdot 2 \cdot 3 \cdot 4 \cdot 5)(1)}(.4)^2(.3)^3(.2)^5$$

since $0!$ and $(.1)^0$ are both equal to 1. Working out this number, we find that .0035 is the probability of the *sample* distribution

Black	2
Red	3
White	5
Blue	0
	10

if the probability distribution given above is the true one. Using this multinomial rule, one could work out a probability for each *possible sample distribution*.

Although the multinomial rule is relatively easy to state, a tabulation or graph of this distribution is very complicated: here a sample result is not a single number, as for the binomial, but rather an *entire* frequency distribution. In principle, once a given discrete probability distribution is specified, then the probability of every conceivable sample distribution of size N could be worked out by the multinomial rule. Even though this possibility exists, the multinomial distribution plays a relatively modest role in statistics. Most often, the probability of obtaining an entire frequency distribution is not so much of concern as the probability of one or more indices summarizing the sample distribution. This problem will occupy most of our attention in the sections to come. Nevertheless, some of the methods we will consider in Chapter 18 will be founded directly on the multinomial distribution.

3.20 THE HYPERGEOMETRIC DISTRIBUTION

Another theoretical probability distribution deserves a passing mention for the same reason as the multinomial. The multinomial rule (and the binomial, of course) assumes either that the sampling is done with replacement, or that the sample space is infinite, so that the basic probabilities do not change over the trials made. However, suppose that one were sampling from a finite space *without* replacement; then the probabilities would change for each observation made. By a series of arguments very similar to those used for finding probabilities of poker hands in Section 3.8, we could arrive at a new rule for finding the probabilities of sample results.

This rule describes the hypergeometric distribution, and can be stated as follows:

Given a sample space containing a finite number T of elements, suppose that the elements are divided into K mutually exclusive and exhaustive classes, with T_1 in class 1, T_2 in class 2, $\cdots$, T_K in class K. A sample of N observations is drawn at random without replacement, and is found to contain n_1 of class 1, n_2 of class 2, $\cdots$, n_K of class K. Then the probability of occurrence of such a sample is given by

$$\frac{\binom{T_1}{n_1}\binom{T_2}{n_2}\cdots\binom{T_K}{n_K}}{\binom{T}{N}} \qquad\qquad [3.20.1]$$

where $n_1 + n_2 + \cdots + n_K = N$ and $T_1 + T_2 + \cdots + T_K = T$.

For an illustration of the use of the hypergeometric rule, let us return to the problem of drawing marbles at random from a box (Section 3.19), but this time *without* replacement. Suppose there had been 30 marbles in the box originally, with the following distribution of colors:

Color	f
Black	12
Red	9
White	6
Blue	3
	30

Notice that the relative frequencies are the same as the probabilities in the previous example. Now 10 marbles are drawn at random without replacement. We want the probability that the *sample* distribution is

Color	f
Black	2
Red	3
White	5
Blue	0
	10

Using the hypergeometric rule, we get

$$\frac{\binom{12}{2}\binom{9}{3}\binom{6}{5}\binom{3}{0}}{\binom{30}{10}}$$

which works out to be about .0011. This is not, of course, the same probability as we found for this sample result using the multinomial rule, since the entire sampling scheme is assumed to be different in this second example. This illustrates that the *sampling scheme* adopted makes a real difference in the probability of a given result.

Ordinarily, this is a practical consideration only when the basic sample space contains a small number of elementary events. When there is a very large number of elementary events in the sample space, the selection and nonreplacement of a particular unit for observation has negligible effect on the probabilities of events for successive samplings. For this reason, the hypergeometric probabilities are very closely approximated by the binomial or multinomial probabilities when T, the total number of elements in the sample space, is extremely large. The distinction between these different distributions becomes practically important only when samples are taken from relatively small sets of potential units for observation. We will have more to say on sampling without replacement from small sample spaces in Section 5.11. The hypergeometric distribution will also be applied in Section 18.6.

EXERCISES

1. A well-known intelligence test item consists of arranging cartoon pictures in a meaningful order. If there are six pictures in all, into how many orders might a subject arrange these cards?

2. In the test item in Exercise 1, only two sequences or orders are scored as correct. If a subject is operating purely by chance, so that each sequence is equally likely to occur, what is the probability of a correct response?

3. Suppose that a child is seated before a toy piano with 12 keys. The child plays four notes. If any note is equally likely to be struck on any attempt, how many different sequences of notes are possible (including repetitions of the same note)? What is the probability of a sequence consisting of the same note struck four times? What is the probability that the sequence consists of at least two different notes? What is the probability of a four-note, ascending scale?

4. Each day a cafeteria offers four types of salad, three types of meat, eight kinds of vegetables, two types of bread, and six varieties of dessert. If a person's meal consists of one selection from each of these choices, how many different meals are possible? (consider two meals to be different if they differ on any item.) Suppose that two people choose their meals independently on the same day. If each is equally likely to select any possible meal, what is the probability that they will come up with exactly the same choices? (**Hint:** Bear in mind how many possible meals there are on which the people could agree.)

5. In arranging a series of interviews, a social scientist wished to assign seven subjects to seven available time periods. If all subjects were available at all times, in how many ways could the assignments of subjects to times be made?

6. Suppose that the researcher in Exercise 5 actually had 15 time slots available for interviewing the seven subjects. In how many ways could the subjects be assigned to time periods? (Give your answer in symbolic form.)

7. An experimenter has 20 subjects available. It is desirable to form two groups of 10 subjects each. In how many ways can the first group be formed? Given that the first group has been selected, how many ways exist for forming the second group? Given the assumption that all assignments of subjects to groups are equally likely, what is the probability of any given assignment? (Give your answers in symbolic form.)

8. In still another experiment, 50 subjects are available. Six groups of five subjects each are to be formed. In how many different ways can these assignments of subjects to groups be made? If all assignments are equally probable, what is the probability of any given arrangement of subjects into groups? (Give your answers in symbolic form.)

9. In the situation in Exercise 8, exactly 20 of the 50 subjects are black, and the remainder are white. If all assignments of subjects to groups are equally likely, what is the probability of exactly one all-black group, with the remainder of the groups all white? (Remember that there are six possibilities for the all-black group.) What is the probability of exactly two all-black groups? (Give your answers in symbolic form.)

10. In a fraternity house, three boys share a room with a single closet. Each boy can wear items of each of the other boys' clothing, and they share freely. The closet contains three pairs of shoes, seven shirts, five pairs of pants, eight pairs of socks, and four coats. If each boy dresses in shoes, shirt, pants, socks, and coat, in how many combinations of clothing may the boys appear together? (Give your answer in symbolic form.)

11. A college contains three departments, A, B, and C. Department A has 10 faculty members, B has 15 members, and C, 20 members. It is decided to form a college committee consisting of 1/5 of the members of each department. How many possible such committees could be formed? (Give your answer in symbolic form.)

12. Suppose that two people work in the same office. Given that their birthdays were independent of each other, and that births are equally likely to occur in any month, what is the probability that they have a birthday in the same month?

13. Let there be four people working in an office. Under the same assumptions made in Exercise 12, find the probability that no two or more of these people's birthdays fall in the same month. Then find the probability that two or more do have a birthday in the same month. (Give your answers in symbolic form.)

14. The legislature of a certain state wants to make automobile license plates carry three letters followed by three digits. Will there by enough possible license plates under this system to provide different plates for all of the five million automobiles the state registers? What if the plates contain two letters and three digits? Will there still be enough different plates? How about with three letters and two digits? (Count 0 through 9 as the possible digits, and include all 26 letters of the alphabet.)

15. In the circumstances in Exercise 14, where three letters and three digits are used, what is the probability that a car owner will get a plate with three repeated letters and three repeated digits, such as BBB-222, if all combinations are equally probable? What is the probability of three repeated letters and three different digits? Three different letters and three different digits? (Leave answers in symbolic form.)

16. Suppose that in the college mentioned in Exercise 11, it was decided to form a nine-member committee by selecting faculty members completely at random. What is the probability that such a committee would wind up containing:

(a) Exactly 1/5 of Department A?
(b) Exactly 1/5 of Department B?
(c) Exactly 1/5 of Department C?
(d) Fewer than 1/5 of the members of Department A?
(e) *Only* members from Department C?

(Again, give your answers in symbolic form.)

17. A very old machine that manufactures automobile parts is now believed to produce defective parts with a frequency of about 30 in 100. We will assume that this process is stationary and independent. Given that 15 parts produced by this machine are sampled at random, find the probability that one or fewer parts will turn out to be defective. Two or more?

18. In a test of possible side effects of a new medication, a physician matched 20 pairs of persons on their physical charcteristics. One member of each pair was given the medicine, and the pair-mate was given a placebo. A success was recorded when the member receiving the medication showed more of the side effect than the pair-mate, and a failure was recorded otherwise. There were 13 successes and 7 failures. If medication and placebo were equal in their tendency to produce the effect, how likely is a result that is deviant or more deviant from what one should expect?

19. On a multiple-choice examination each item has exactly five options, of which the student must pick one. Only one of the options is correct for each item. If a student is merely guessing at the answer, each option should be equally likely to be chosen. Furthermore, the student's answer on a given item is believed to be independent of the answer on any other. Given that the test has 18 items, and that a student is just guessing on each item, find the probability of the following events:

(a) Three items correct.
(b) Seven items correct.
(c) Fourteen items correct.
(d) More than five items correct.
(e) Fewer than two items correct.
(f) Between two and seven items correct (inclusive).

20. In a certain lottery, 40% of the tickets were purchased by men, and 60% by women. Each person purchased only one ticket. Ten tickets were drawn at random and with replacement. What is the probability that:

(a) Four or more winners were women?
(b) Two or fewer winners were women?
(c) The winners were all of the same sex?
(d) Exactly four men and six women were winners?

21. A public health officer in a certain area suspects that 25% of children in that area are severely undernourished. When a sample of 20 children is taken at random, it is found that 9 show severe malnutrition. What is the probability of nine or more such children in the sample if the true proportion in the population is .25? What would the officer be inclined to conclude?

22. In a test of reaction time, husband and wife pairs were studied. Sixteen pairs chosen at random were given the same reaction-time test, and the individual time noted. In nine of the pairs, the wife showed the faster reaction time, and in seven pairs, the husband showed the shorter time. If husbands and wives tend to be about equal in reaction time, what is the probability of a result this much or more in favor of the wives?

23. In an experiment two matched groups of 10 individuals each were employed. Each individual in each group was selected independently and at random. The score of each individual on a certain perceptual test was determined. The results were as indicated in the table on the next page. Before the experiment, the experimenter expected group II to have the larger scores. Use the sign test to examine the hypothesis that the groups represented by these samples are actually equal in the scores one should expect.

Group I	Group II
5	14
8	21
7	23
6	6
21	11
13	5
20	10
17	18
10	21
17	25

24. In an experiment on extrasensory perception (ESP), three shells are used, one of which covers a pea. On each trial the shells are rearranged into a row and the subject's task is to locate the shell under which the pea lies. Suppose that the first, second, and third shells are equally likely to cover the pea on any trial. Suppose also that without ESP a subject is equally likely to guess any of the three shells on a trial. Under these conditions, assuming 10 independent trials, how probable is it that out of 10 trials the subject guesses the correct shell two or fewer times? Eight or more times?

25. An expert typist tends, on the average, to make just one error for five business letters typed, about .2 of an error per letter. What is the probability that the typist will make no errors at all on a letter? (**Hint:** Use the Poisson distribution, letting $m = .2$ and $x = 0$.) What is the probability that the typist will make at least one error?

26. A suicide-prevention unit in a city receives calls for help at a rate of about 2.3 per day. If this is a stationary and independent Bernoulli process, what is the probability of one day without a call? Only one call? Two or more calls in a day? (**Hint:** Poisson.)

27. On a college football team, suppose that a passer has a probability of .60 of a completion on any attempt. If his passing performance reflects an independent and stable random process, what is the probability that in a game he gets his first completion on his fourth attempt? What is the probability of his first completion on his first attempt?

28. Suppose that a certain door-to-door salesman believes that he has a probability of .30 of making a sale on any given call. If his sales can be regarded as corresponding to events in a stable and independent Bernoulli process, what is the probability that he makes his fifth sale on his 10th call, given that $p = .30$? What is the probability that he makes his first sale on his fifth call? (Give your answers in symbolic form.)

29. Students at a very large university campus show the following distribution of religious preferences: Protestant, 32%; Catholic, 14%; Jewish, 19%; other religion, 4%; no preference stated, 31%. Suppose that a given group of 20 students could be regarded as a random selection of students on this campus. What then is the probability of the following set of religious preferences? (**Hint:** Treat the campus as large enough that sampling can be thought of as with replacement. Give the answer in symbolic form.)

Preference	Frequency
Protestant	5
Catholic	3
Jewish	4
Other	0
None reported	8

30. An item in a questionnaire asked for a response on a five-point scale extending from 1 for "strongly agree," to 5 for "strongly disagree." If all respondents are actually equally likely to use any of the categories, how probable is the following distribution of 10 responses? (Give your answer in symbolic form.)

Response	Frequency
1	0
2	2
3	6
4	2
5	0

31. Suppose that 10 students out of a class of 50 undergraduates are chosen for a behavioral experiment. The total class consists of 15 freshmen, 18 sophomores, 10 juniors, and 7 seniors. If all students are equally likely to be picked, what is the probability of the following? (Give your answers in symbolic form.)

(a) The group chosen consists solely of sophomores.
(b) The group contains exactly seven seniors and three juniors.
(c) The gorup contains three freshmen, four sophomores, two juniors, and one senior.

32. A large metropolitan area reports 2.4 traffic fatalities a day, or .1 per hour, on the average. If traffic fatalities represent a stable, independent Poisson process, what is the probability of no traffic fatalities in a given hour? Of two or more in an hour? (**Hint:** use $m = .1$)

Chapter 4

Central Tendency and Variability

Any frequency distribution is a summary of data, but for many purposes it is necessary to summarize still further. Rather than compare entire distributions of data with each other or with hypothetical distributions, it is generally more efficient to compare only certain characteristics of distributions. Two such general characteristics of any distribution, whether frequency or probability, obtained or theoretical, are its measures of central tendency and variability. Indices of central tendency are ways of describing the "typical" or the "average" value in the distribution. Indices of variability, on the other hand, describe the "spread" or the extent of difference among the observations making up the distribution.

Perhaps the more basic concern is the description of central tendency. This will be treated first for obtained frequency distributions, and then the ideas will be extended to probability distributions as well. Next, the measurement of the dispersion or spread of a frequency distribution will be taken up, and a parallel treatment will once again be given for probability distributions. Finally, an attempt will be made to show how two indices, the mean and standard deviation, form the cornerstone of most statistical inference.

4.1 THE SUMMATION NOTATION

In this chapter it will be necessary to employ the summation symbol Σ (capital Greek sigma). This is read as "the sum of," and tells us to take the sum of the values represented by the expressions following the symbol. Thus, for example, Σx stands for the sum over all the different values that the variable X can assume. Most simple statistical derivations involve various sums, and the use of this symbol introduces considerable economy of statement into these formulations.

(Strictly speaking, when we wish to indicate summation over a set of values which are labeled as x_1, x_2, and so on up to x_N, this should be indicated by writing

$$\sum_{i=1}^{N} x_i.$$

However, often, when the context is clear, this will be abbreviated to

$$\sum_{i} x_i, \text{ or, occasionally, } \sum x_i, \text{ or even } \sum x.$$

Nevertheless, the $i = 1$ at the bottom and the N on top of the summation should still be understood. Printed summation signs also vary in size, from very large to very small, depending on the space requirements of the expression that follows the sigma or the available space for Σ in some formulas. However, the size is irrelevant: it always means to sum the values given by the expression immediately following.)

There are a number of simple rules for the algebraic manipulation of the summation sign. These rules are given and illustrated in Appendix A. The student who has not already encountered summation notation in school algebra or elsewhere is urged to study these rules until thoroughly familiar with them. The exercises following Appendix A should also be done. Actually, the rules themselves are easy, and a little practice at writing out the sums symbolized can familiarize one very quickly with the various ways sums can be manipulated. A little time spent in this way will greatly increase your ability to follow the simple mathematical arguments used in later sections.

4.2 MEASURES OF CENTRAL TENDENCY

Imagine an obtained distribution of numerical scores. If you were asked to state *one value* that would best "capture" and communicate the distribution as a whole, which value should you choose? One answer is to find that score or value which is a good "bet" about any randomly selected case from this distribution. Such a score may not be exactly correct for any given case, but it should be a good guess about the obtained score for that case. However, there are at least three ways to specify what we mean by a good bet about any case: (1) the most frequent (most probable) measurement class, (2) the point exactly midway between the top and bottom halves of the distribution, and (3) the arithmetic average of the distribution. The first of these ways of defining the central tendency leads to the measure known as the **mode**; the second leads to the **median** of the distribution; the third is merely the familiar average, or **mean.**

The mode is the most easily computed and the simplest to interpret of all the measures of central tendency. The mode of a frequency distribution is merely the *midpoint* or class name of the *most frequent* measurement class. If a case were drawn at random from the distribution, then that case is more likely to fall in the *modal class* than any other. So it is that in the graph of any distribution, the modal class shows the highest "peak" or "hump" in the graph or interval. The graph of a continuous distribution reaches its maximum density at the mode. The mode may be used to describe any distribution, regardless of whether the events are categorized or numerical.

On the other hand, there are some disadvantages to the mode. One is that there may be more than one modal class: it is perfectly possible for two or more measurement classes to show frequencies that are equal to each other and higher than the frequency shown by any other class. In this case, there is ambiguity about which class gives *the* mode of the distribution, as two or more values, or intervals are "most popular." Fortunately, this does not happen very often; even though there are two or more "humps" in the graph of a distribution, the midpoint of the interval having the highest frequency or probability is still taken as *the* mode. Nevertheless, in such a distribution, where one measurement class is not clearly most popular, the mode loses its effectiveness as a characterization of the distribution as a whole.

Another disadvantage is that the mode is very sensitive to the size and number of class intervals employed when events are numerical; the value of the mode may be made to "jump around" considerably by changing the class intervals for a distribution.

Finally, the mode of a sample distribution forms a very undependable source of information about the mode of the basic probability distribution. For these reasons, our use of the mode will be restricted to situations where: (1) the data are truly nominal scale in nature; or (2) only the simplest, most easily computed, measure of central tendency is needed.

The median score in any set of observations or in a distribution is also a good bet about any case in the total set represented. Just as its name implies, the median is the score corresponding to the middle case when all individual cases are arranged in order by scores, or the median reflects the score that divides the cases into two intervals having equal frequency. Thus, if you drew a case at random from any set of N observations, and guessed that this case showed the median score, you are just as likely to be guessing too high as too low.

The median for a set of observed data is defined in slightly different ways depending upon whether N is odd or even, and upon whether raw data or a grouped frequency distribution is to be described. For a set of raw scores **when N is odd, the median corresponds to the score of individual number $(N + 1)/2$, when all individuals are arranged in order by scores; when N is even, the median is defined as the score midway between the scores for individual $(N/2)$ and individual $(N/2) + 1$ in order.** Then either for odd or even N it will be true that exactly as many cases fall above as fall below the median score.

On the other hand, this way of computing a median is not usually applicable if the data have been arranged into a grouped frequency distribution. **For any such grouped distribution, the median is defined as the point at or below which exactly 50% of the cases fall.** Consequently the first step in finding the median of a grouped distribution is to construct the *cumulative* frequency distribution. Such a cumulative frequency distribution is illustrated in Table 4.2.1.

The last column in Table 4.2.1 shows the cumulative frequencies for the class intervals. Since, by definition, the median will be that point in the distribution at or below which 50% of the cases fall, the cumulative frequency *at* the median score should be $.50N$. Thus, the cumulative frequency is $.50(200) = 100$ at the median for this example. Where would such a score fall? It can be seen that it could not fall in any interval

Class	f	cf	**Table 4.2.1**
			A cumulative frequency distribution for a sample of 200 cases.
74–78	10	200	
69–73	18	190	
64–68	16	172	
59–63	16	156	
54–58	11	140	
49–53	27	129	
44–48	17	102	
39–43	49	85	
34–38	22	36	
29–33	6	14	
24–28	8	8	
	200		

below the real limit of 43.5, since only 85 cases fall at or below that point. However, it does fall below the real limit 48.5, since 102 cases fall at or below that point. We have thus located the median as being in the class interval with real limits 43.5 and 48.5.

We must still ascertain the *score* that corresponds to the median, and this is where the process of interpolation comes in. The median score is somewhere in the interval 44–48. We assume that the 17 cases in that interval are evenly scattered over the interval width of five units, as in Figure 4.2.1.

The median score, which is greater than or equal to exactly 100 cases, must then exceed not only the 85 cases below 43.5, but also equal or exceed 15 cases above 43.5. In other words, the median is 15/17 of the way *up* the interval from the lower limit. Next, what score is exactly 15/17 of the way between 43.5 and 48.5? Since the difference between these two limits is 5, the class interval size, we take (15/17)5, or 4.4 as the amount that must be added to the lower real limit to find the median score, so that the median must be 43.5 + 4.4 or 47.9.

A little computational formula that summarizes the steps just described is given by:

$$\text{median} = \text{lower real limit} + i\frac{(.50N - cf \text{ below lower limit})}{f \text{ in interval}} \qquad [4.2.1]$$

where the lower real limit used belongs to the interval containing the median, and the *cf* refers to the cumulative frequency *up to* the lower limit of the interval. For the example, we find

$$\text{median} = 43.5 + 5\frac{[.50(200) - 85]}{17} = 47.9.$$

If the median can fall only in some class interval with nonzero frequency, this method of interpolation gives a unique value. However, if the interval frequency happens to be zero, then any point in the interval serves equally well as the distribution median; here one usually takes the midpoint as the median.

Even when the raw data are available, it is often worthwhile to define and compute

Figure 4.2.1

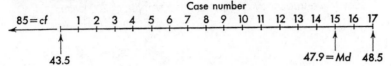

the median as for a grouped frequency distribution. In the first place, for a sizable N, ordering all of the cases by their score magnitudes can be a considerable chore, and it may be simpler to construct a grouped distribution. Secondly, a troublesome problem arises when two or more cases in the raw data have the same order at the median position; here it often makes sense to calculate the median by interpolation as for a grouped distribution.

In principle a median may be found for any distribution in which the variable represents an interval or even an ordinal scale; it is not applied to a distribution in which the measurement classes are purely categorical, since such classes are unordered.

The median is considerably less sensitive to the distribution's grouping into class intervals than is the mode. Furthermore, when one is making inferences about a large "population" of potential observations from a sample, the median is generally more useful and informative than the mode, although the median itself is not ordinarily so useful as the mean, to be discussed next. We will have more to say about characteristics of the median in future sections.

By far the most used and familiar index of central tendency for a set of raw data or a distribution is the *mean*, or *simple arithmetic average*. Surely everyone knows that to take the average of a set of raw scores you simply add them all up and divide by the total number, N:

$$\bar{x} = \frac{\sum\limits_i x_i}{N}.$$

[4.2.2*]

(Here, x_i stands for the score of the observation labeled i, and the sum is taken over all of the N for different observations i.) Thus, Eq. 4.2.2 defines the mean for any set of raw data in the form of numerical scores.

Since expressions representing means will occur so frequently in all of the later sections, it is well to point out that we might also represent the arithmetic mean by

$$\bar{x} = \sum_i \frac{x_i}{N},$$

standing for each value x_i first divided by N and then summed over the individual observations i. The value $\bar{x}$ represented by either of these expressions is precisely the same; this accords with Rule 1 of Appendix A, that the sum of N observations each multiplied by a constant number is the same as the sum itself multiplied by that number. Therefore, in succeeding sections sometimes one and sometimes the other way of expressing the mean will be used, depending on the algebraic and typographical requirements of the particular discussion in which these expressions occur.

Incidentally, it should be mentioned that various texts in statistics use various symbols for the sample mean. Frequently, just as here, the sample mean is shown as $\overline{X}$ or as $\overline{x}$, when X is the variable of interest. On the other hand, there are situations where another symbol for the sample mean, such as M, is more convenient. This is true especially when the sample mean figures as part of a subscript. Thus, from time to time we will employ both $\overline{x}$ and M as ways of symbolizing the sample mean.

The definition and computation of the arithmetic mean for raw data is simple enough, but the situation is slightly more complicated when one wishes to find the mean of a grouped distribution of scores. You will recall (Section 2.7) that when a distribution was grouped in class intervals the midpoint x_j of each class interval j was taken to represent the score of each of the cases in the interval. Thus, in an interval 59–73 with midpoint 66 and frequency 16, the sum of the scores of the 16 cases falling into this interval is taken to be 66 summed 16 times, or $66 \times 16 = x_j f_j$. Similarly, when all of the scores in any interval are assumed to be the same, their sum is the midpoint of that particular interval times the frequency for that interval, or $x_j f_j$. Then the sum of *all* of the scores in the distribution is taken to be the sum of the values of x times f over all of the respective intervals

$$\overline{x} = \frac{1}{N} \sum_{j=1}^{J} x_j f_j$$

[4.2.3*]

(Note that here x_j is the midpoint of any interval, f_j is the frequency corresponding to that interval, and the sum is taken over *all intervals*.)

For example, consider once again the distribution shown in Tables 4.2.1 and 4.2.2. The frequency of each class interval is multipled by its midpoint, and these are then summed and divided by N to give the distribution mean.

The mean calculated from a distribution with grouped class intervals need not agree exactly with that calculated from raw scores. Information is lost and a certain amount

Class	x	f	xf
74–78	76	10	760
69–73	71	18	1,278
64–68	66	16	1,056
59–63	61	16	976
54–58	56	11	616
49–53	51	27	1,377
44–48	46	17	782
39–43	41	49	2,009
34–38	36	22	792
29–33	31	6	186
24–28	26	8	208
		200	10,040 = Σxf

$$\overline{x} = \frac{\sum_{j} x_j f_j}{N} = \frac{10,040}{200} = 50.2$$

Table 4.2.2

Computation of a mean from a grouped distribution

of inaccuracy introduced when scores are grouped and treated as though each corresponded to the midpoint of some interval. The coarser the grouping, in general, the
more likely is the distribution mean to differ from the raw-score mean. For most practical work the rule of 10 to 20 class intervals gives relatively good agreement, however.
Nevertheless, it is useful to think of the mean calculated from any given distribution as
the mean of that *particular* distribution, a particular set of groupings with their associated frequencies.

Actually, in modern statistical work it is somewhat unusual to find sample means
calculated from grouped frequency distributions. Any extensive statistical analysis is
done by computer these days, and the computer is perfectly capable of working directly
with the raw data input, almost regardless of the number of cases involved. Even manual computations of means are much simpler than in former days. Pocket calculators
are easy to use for computing means and other statistical indices, and a great many of
these calculators have built-in programs for computing means and other statistical values. These are nice features to look for if you are shopping for a calculator.

There are also two "minor means" which are encountered, though rarely, in statistics. When the influence of extreme values is to be minimized, sometimes one employs
the *harmonic mean,* defined by

$$\bar{x}_H = \frac{N}{\sum_i \left(\frac{1}{x_i}\right)}$$

The harmonic mean makes a single appearance in this text (Eq. 11.17.1 and the comment following). The other is the *geometric mean,* defined by

$$\bar{x}_G = (x_1 x_2 \cdots x_N)^{\frac{1}{N}},$$

the product of all of the x_i values with the Nth root taken. Then the logarithm of $\bar{x}_G$ is
the average logarithm of the x values. (This mean is often used to find averages of
ratios. The geometric mean will be used in Eq. 14.19.4.)

4.3 THE MEAN AS THE "CENTER OF GRAVITY" OF A DISTRIBUTION

The mean of a distribution parallels the physical idea of a center of gravity, or balance
point, of ideal objects arranged in a straight line. For example, imagine an ideal board
having zero weight. Along this board are arranged stacks of objects at various positions.
The objects have uniform weight and differ from each other only in their positions on
the board. The board is marked off in equal units of some kind, and each object is
assigned a number according to its position. This is shown in Figure 4.3.1. Now given
this idealized situation, at what point would a fulcrum placed under the board create a
state of balance? That is, what is the point at which the "push" of objects on one side
of the board is exactly equal to the push exerted by objects on the other side? This is
found from the mean of the positions of the various objects:

$$\bar{x} = \frac{2 + 2 + 8 + 10 + 10 + 10 + 15 + 15 + 18 + 20}{10} = 11.$$

Figure 4.3.1

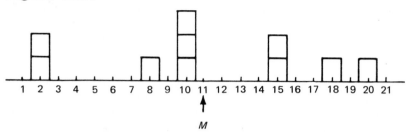

Here, the board would exacly balance if a fulcrum were placed at the position marked 11. Note that since there were piles of uniform objects at various positions on the board, this center of gravity was found in exactly the same way as for the mean of a distribution, since the position (midpoint of an interval) was in effect multiplied by the number of objects at that position (the class frequency), and then these values were summed and divided by the number of objects (the total frequency).

In short, the position of any object on the board is analogous to the score of a case, and each case is treated as having equal "weight" in our computations. The arithmetic mean is then like the center of gravity, or balance point. The mean is that score about which deviations in one direction exactly equal deviations in the other.

This property of the mean is bound up in the statement that **the sum of the signed deviations about the mean is zero in any distribution.** A deviation from the mean is simply the signed difference between the score for any case and the mean score:

$$d_i = (x_i - \bar{x}) \tag{4.3.1}$$

Then it is easy to show that

$$\sum_i d_i = \sum_i x_i - \sum_i \bar{x} = \sum_i x_i - N\bar{x}$$

$$= N\bar{x} - N\bar{x} = 0, \tag{4.3.2*}$$

that is, the sum of the signed deviations about the mean is always zero.

A simple consequence of this fact is that *the mean signed deviation from the mean is zero:*

$$\frac{\sum_i d_i}{N} = \frac{0}{N} = 0. \tag{4.3.3*}$$

4.4 "BEST GUESS" INTERPRETATIONS OF CENTRAL TENDENCY

We have just seen that the tendency for cases in a distribution to differ from the mean in one way is exactly balanced by the tendency to differ in the opposite way.

Suppose once again that you were told to *guess* the score of some case picked at random from a distribution. If you guessed the mean for each case, you might be in error to some extent on each trial since it need not be true that the mean is exactly the same as any obtained score. The *extent* of error for a given case is d, the departure of the true score from the mean. Over all possible cases that might be drawn from the distribution, the average signed error would be $\sum_i d_i/N$, the mean signed deviation. But we have just seen that the mean signed deviation is always zero. Hence, the following statement is true: **if the mean is guessed as the score for any case drawn at random from the distribution, on the average the amount of signed error will be zero.** This is a most important interpretation of the mean: the mean is the best guess about the score of any case in the distribution, if one wishes the *average signed error* to be zero.

Now suppose there were some distribution where you had to guess the score of a case picked at random, and you wished to be *absolutely right* with the highest possible probability. Then you should guess the mode rather than the mean; since it is the most frequent score, guessing the mode guarantees the greatest likelihood of hitting the score "on the nose" for a case drawn at random.

On the other hand, it might be that in guessing the score drawn at random you are not interested in being exactly right most often, nor in making signed error zero on the average, but rather wanted to make the smallest absolute error on the average. Here, the sign of the error is unimportant, but the size of the error is what matters. Then you should guess the median for any score. By doing so you would make the *smallest absolute error* on the average. The median is the typical score in this sense: it is closest on the average to all of the scores in the distribution.

There is really no way to say which is the best measure of central tendency in general terms. This depends very much on what one is trying to do and what one wants to communicate in summary form about the distribution. Each of the measures of central tendency is, in its way, a best guess about any score, but the sense of "best" differs with the way error is regarded. If both the size of the errors and their signs are considered, and we want zero error in the long run, then the mean serves as a best guess. If a miss is as good as a mile, and one wants to be exactly right as often as possible, the mode is indicated. If one wants to come as close as possible on the average, irrespective of sign of error, then the median is a best guess.

From the point of view of purely descriptive statistics, as apart from inferential work, the median is a most servicable measure. Its property of representing the typical (most nearly like) score makes it fit the requirements of simple and effective communication better than the mean in many contexts.

On the other hand, the median is usually inferior to the mean when our purpose is to make inferences beyond the sample. The median has mathematical properties making it difficult to work with, whereas the mean is mathematically tractable. For this and other reasons, mathematical statistics has taken the mean as the focus of most of its inferential methods, and the median is relatively unimportant in inferential statistics. Nevertheless, as a description of a given set of data, the median is extremely useful in communicating the typical score.

4.5 CENTRAL TENDENCY IN DISCRETE PROBABILITY DISTRIBUTIONS

The ideas of mean, median, and mode apply to distributions of discrete random variables just as they do to frequency distributions. However, as we shall see, in probability distributions measures of central tendency such as the mean often play a much more important role. Not only do such measures summarize the distribution, they may also serve as parameters entering into the mathematical rule giving the probability for any value or interval of values of the random variable.

For a discrete random variable, the mode is simply the *most probable* value. For example, in the discrete distribution shown in Table 4.5.1 the mode is 41, since this is the midpoint of the most probable class (recall that each case in any interval of a grouped distribution is ordinarily treated as though it had the value of the midpoint).

A median value for a discrete random variable need not be unique. Any value qualifies if the probability for X less than or equal to that value equals .50, $p(X \le Md) = .5$. The median is any value that evenly divides the distribution. The median of a grouped probability distribution is found in exactly the same way as for a frequency distribution, except that probabilities take the place of frequencies in the computations:

$$Md = \text{lower real limit} + i\left[\frac{.50 - p(X \le \text{lower real limit})}{p(\text{lower limit} \le X \le \text{upper limit})}\right] \qquad [4.5.1]$$

where *lower limit* and *upper limit* refer to the interval containing the median. The example below shows the computation of the median in such a distribution.

Class interval	x	p(X in interval)	xp(X in interval)
74–78	76	.050	3.80
69–73	71	.090	6.39
64–68	66	.080	5.28
59–63	61	.080	4.88
54–58	56	.055	3.08
49–53	51	.135	6.89
44–48	46	.085	3.91
39–43	41	.245	10.05
34–38	36	.110	3.96
29–33	31	.030	.93
24–28	26	.040	1.04
			50.21

Table 4.5.1

Computation of central tendency measures in a discrete probability distribution.

Mode = 41 (midpoint of interval with probability = .245)

Median = $43.5 + \dfrac{5(.50 - .425)}{.085} = 47.91$

(where $p(X \le 43.5) = .425$)

Mean = $\Sigma\, xp(X \text{ in interval}) = 50.21$

Finally, the mean of a discrete distribution is found much as for a grouped frequency distribution: each distinct value of X, or midpoint of an interval, is multiplied by the *probability* that X takes on that value (or that X lies in the interval). Suppose there are J intervals, any one of which might be called j, with midpoint x_j. Then the sum of these products is found:

$$\text{mean} = \sum_{j=1}^{J} x_j p(x_j) \qquad\qquad [4.5.2^*]$$

summed over all possible values of x_j. Alternatively,

$$\text{mean} = \sum_x x p(x), \qquad\qquad [4.5.3^*]$$

where the sum is taken over all possible values of x.

In general, **the mean of a discrete random variable is the sum of the products of the different values of X, each times the probability that X takes on that value.**

All three central tendency indices are found in Table 4.5.1, an example of a probability distribution grouped into class intervals.

4.6 THE MEAN OF A RANDOM VARIABLE AS THE EXPECTATION

A special term is often used to denote the mean of a probability distribution. This is the *expectation* or the *expected value* of a random variable X. The symbol $E(X)$ simply represents the mean of the probability distribution of X, and if X is discrete,

$$E(X) = \sum_x x p(x) = \text{mean of } X, \qquad\qquad [4.6.1^*]$$

the sum being taken over all values that X can assume.

The mean or expectation of a continuous random variable is defined in a way very similar to that for a discrete variable. However, since X may assume any of an infinite set of particular values, and because, for the reasons outlined in Section 2.18, it is necessary to discuss the *probability density* associated with any particular value of X, the actual definition of the mean is somewhat different. If we let $f(x)$ symbolize the probability density associated with any particular value of X, then

$$E(X) = \int_{-\infty}^{\infty} x f(x) \, dx. \qquad\qquad [4.6.2]$$

Here, the integral sign indicates the infinite sum of x times a factor $f(x) \, dx$ for all real number values between the ultimate limits of $-\infty$ and $+\infty$. Notice that, much as in the discrete case, the expectation of the continuous random variable X is actually a sum of products. In the former the products were of values of X, each times the probability of that value, but for the continuous case each value of X is weighted by a factor depending on the probability density at that value.

Since the idea of expectation is so pervasive in theoretical statistics, it is very convenient to have available some list of the formal rules for dealing with expectations mathematically. These rules are summarized in Appendix B and should clarify the ways that expectations will be treated algebraically in other sections. As with Appendix A, exercises are available. The student is advised to become familiar with these rules, as with the rules of summation, which they greatly resemble. A student who does so should have very little trouble in following the simple derivations in mathematical statistics such as those contained in this book.

The rules for expectations and their manipulations given in Appendix B are valid either for continuous or for discrete random variables, with a few minor exceptions that need not bother us in an elementary discussion. This is extremely convenient, since it makes it possible to demonstrate certain general features of statistics without having to qualify the result as pertaining to discrete or continuous variables.

4.7 EXPECTATION AS EXPECTED VALUE

The idea of expectation of a random variable is closely connected with the origin of statistics in games of chance. Gamblers were interested in how much they could expect to win in the long run in a game, and in how much they should wager in certain games if the game was to be "fair." Thus, expected value originally meant the expected long-run winnings (or losings) over repeated play; this term has been retained in mathematical statistics to mean the long-run average for any random variable over an indefinite number of samplings.

The use of expectation in a game of chance is easy to illustrate. For example, suppose that someone were setting up a lottery, selling 1,000 tickets at $1 per ticket. A prize of $750 will go to the winner of the first draw. Suppose now that you buy a ticket. How *good* is this ticket in the sense of *how much you should expect to gain?* Should you have bought it in the first place? You can think of your chances of winning and losing as represented in a probability distribution where the outcome of any drawing falls into one of two event categories:

Class	Prob.
win	1/1,000
don't win	999/1,000

Translated into the amount of money gained (the random variable X), and with a loss regarded as a negative gain, this distribution becomes

x	p(x)
$749	1/1,000
−$ 1	999/1,000

Since this is the distribution of a discrete random variable, the mean of the distribution can be found by the methods in Section 4.5; that is,

$$E(X) = \Sigma\, xp(x) = 749(1/1{,}000) + (-1)(999/1{,}000)$$

$$= .749 - .999 = -.25$$

This amount, a *minus* 25 cents, is the amount which you can expect to gain by buying the ticket, meaning that if the lottery were run over and over indefinitely and you bought a ticket each time, in the long run you should be poorer by a quarter. Should you buy the lottery ticket? Probably not, if you are going to be strictly rational about it; the mean winnings (the expected value) is certainly not in your favor.

On the other hand, suppose that the prize offered were $1,000, so that the gain in winning is $999. Now the expected value is

$$E(X) = 999(1/1,000) + (-1)(999/1,000) = .00.$$

Here the lottery is more worth your while, as there is at least no amount of money to be lost *or* gained in the long run. Such a game is often called "fair." Obviously, truly fair lotteries and other games of chance are hard to find, since their purpose is to make money for the proprietors, not to break even or lose money. If the prize were $2,000, you would likely take the opportunity, as in this case the expected value would be exactly the *gain* of one dollar.

In figuring odds in gambling situations one uses the expected value to find what constitutes a fair bet: that is a bet where the mean of the probability distribution of gains and losses is zero. For instance, it is known that in a particular game the odds are 4 to 1 *against* winning. This means that the probability of winning is 1/5 and that of losing is 4/5. It costs the player exactly $1 to play the game once. How much should the amount he gains by winning be in order to make this a fair game with expectation of zero? The expectation is

$$E(X) = (\text{gain value})p(\text{win}) + (\text{loss value})p(\text{lose})$$

$$= (\text{gain value})(1/5) - \$1(4/5).$$

Setting $E(X)$ equal to zero and solving gives

$$(\text{gain value})(1/5) - \$1(4/5) = 0$$

or

$$\text{gain value} = \$4.$$

In short, one should stand to gain $4 for $1 put up if the game is to be fair. In general, if odds are A to B against winning, the game or bet is fair when B dollars put up gains A dollars.

In betting situations, the random variable is, of course, gains or losses of amounts of money, or of other things having utility value for the person. Nevertheless, the same general idea applies to any random variable; the expectation is the long-run average value that one should observe.

4.8 THEORETICAL EXPECTATIONS: THE MEAN OF THE BINOMIAL DISTRIBUTION

As an example of how the mean of a theoretical probability distribution may be deduced mathematically, we will consider the binomial distribution once again. Here we will see that the distribution rule alone dictates what the expectation of a binomial variable must be. What we will show is that if X is a binomial variable, then

$$E(X) = Np, \qquad\qquad\qquad\qquad\qquad\qquad [4.8.1^*]$$

the expectation is the number of observations times the probability of a success.

We start off with the definition of expectation for any discrete random variable

$$E(X) = \sum_x xp(x).$$

For the binomial distribution, the probability that the number of successes X takes on any value r is $\binom{N}{r} p^r q^{N-r}$ for $0 \le r \le N$. Thus, any value r multiplied by the probability $p(X = r)$ is

$$r\left[\frac{N!}{r!\,(N-r)!} p^r q^{N-r} \right]. \qquad\qquad\qquad\qquad [4.8.2]$$

Now notice that we could factor this expression somewhat, canceling r in the numerator and denominator and bringing an N and a p outside the brackets:

$$Np\left[\frac{(N-1)!}{(r-1)!(N-r)!} p^{r-1} q^{N-r} \right]. \qquad\qquad\qquad\qquad [4.8.3]$$

For $r = 0$, Eq. 4.8.2 is equal to zero, and so there is no equivalent Eq. 4.8.3. On substituting Eq. 4.8.3 into the equation for $E(X)$ we have

$$E(X) = \sum_{r=1}^{N} Np\left[\frac{(N-1)!}{(r-1)!(N-r)!} p^{r-1} q^{N-r} \right] \qquad\qquad\qquad [4.8.4]$$

with the sum going from $r = 1$ to $r = N$, since the term is zero for $r = 0$. By rule 1 in Appendix A, this is the same as

$$E(X) = Np \sum_{r=1}^{N} \frac{(N-1)!}{(r-1)!(N-r)!} p^{r-1} q^{N-r}. \qquad\qquad\qquad [4.8.5]$$

However, if we wrote out the various terms represented in Eq. 4.8.5 beyond the summation sign, we would find that each is a binomial probability for a distribution with parameters $N - 1$ and p, and thus their sum must be 1.00. Thus

$$E(X) = Np.$$

We have just proved that the mean of a binomial distribution depends only on the two parameters, N and p. If $N = 10$ and p is 1/2, then the expectation or mean of the distribution is $(10)(1/2) = 5$. If $N = 25$ and $p = .3$, the mean is $(25)(.3) = 7.5$, and so on. Notice that the mean *can* be some value that X cannot take on, as in this last example. Nevertheless, the mean or expectation is a perfectly good statement about the best guess for any set of N observations, provided that we want our long-run error to be zero in guessing.

In a very similar way, the mean for a Poisson distribution can be found. It turns out that

$$E(X) = m. \qquad\qquad\qquad\qquad\qquad\qquad [4.8.6]$$

The mean for a Poisson distribution is its intensity parameter, m.

Furthermore, in a closely related fashion one can show that the expectation of a Pascal variable N, with probability given by

$$p(N = n; r, p) = \binom{n-1}{r-1} p^r q^{n-r},$$

is

$$E(N) = \frac{r}{p}. \qquad [4.8.7]$$

That is, the expected number of trials required in order to achieve the rth success from a Bernoulli process is r/p. Hence if $p = 1/2$ and $r = 5$, we expect that 10 trials will be required to reach the fifth success, and so on for any other r and p.

4.9 THE MEAN AS A PARAMETER OF A PROBABILITY DISTRIBUTION

In many important instances of probability distributions, the mean or expectation is a parameter that enters into the function rule assigning a probability or probability density to each possible value of X. A simple example is the binomial distribution of sample proportions (or, more properly, the distribution of sample proportions P, which can be found from the binomial distribution). Here, it is easy to show that $E(P) = p$, one of the two parameters figuring in the function rule for the binomial distribution.

In the preceding section we showed that, for a binomial distribution with parameters p and N,

$$E(X) = Np.$$

Now suppose the random variable were

$$P = \frac{X}{N}. \qquad [4.9.1^*]$$

By Rule 1 for expectations (Appendix B),

$$E(P) = E\left(\frac{X}{N}\right) = \frac{E(X)}{N} = \frac{Np}{N} = p. \qquad [4.9.2^*]$$

For the binomial distribution of proportions P, the expectation is the parameter p.

We have also seen that for a Poisson distribution, the mean of the distribution is the parameter m. For many, although not all, distributions, the mean is one of the parameters of the distribution rule. (One exception is the Pascal distribution, where the mean involves both the parameters r and p.)

For this reason, it will sometimes be convenient to use still another symbol for the mean of a random variable, or of a distribution, especially when we are thinking of the mean in its role as a parameter. In these instances, the lowercase Greek letter mu, or μ, will stand for the mean of a random variable or of a probability distribution. That is,

$$E(X) = \mu = \text{mean of the distribution of } X. \qquad [4.9.3^*]$$

In much that follows, small Greek letters will be used to indicate parameters of probability distributions, whereas Roman letters will stand for sample values. The word *parameter* will always indicate a characteristic of a probability distribution, and the word *statistic* will denote a summary value calculated from a sample. Thus, given some sample space and the random variable X, the mean of the distribution of X is a parameter, μ. However, the mean $\bar{x}$ of any given sample of N values of X is a statistic.

4.10 RELATIONS BETWEEN CENTRAL TENDENCY MEASURES AND THE "SHAPES" OF DISTRIBUTIONS

In discussions either of obtained frequency or of theoretical probability distributions, it is often expedient to describe the general "shape" of the distribution curve. Although the terms used here can be applied to any distribution, it will be convenient to illustrate them by referring to graphs of continuous distributions.

First of all, a distribution may be described by the number of relative maximum points it exhibits, its *modality*. This usually refers to the number of "humps" apparent in the graph of the distribution. Strictly speaking, if the density (or probability or frequency) is greatest at one point, then that value is *the* mode, regardless of whether other relative maxima occur in the distribution or not. Nevertheless, it is common to find a distribution described as **bimodal** or **multimodal** whenever there are two or more pronounced humps in the curve, even though there is only one distinct mode. Thus, a distribution may have no modes (Fig. 4.10.1), may be unimodal (Fig. 4.10.2 and 4.10.4), or may be multimodal (Fig. 4.10.3). Once again, notice that the possibility of multimodal distributions lowers the effectiveness of the mode as a description of central tendency.

Another characteristic of a distribution is its symmetry, or conversely, its skewness. A distribution is **symmetric** only if it is possible to divide its graph into two "mirror-image halves," as illustrated in Figures 4.10.3 and 4.10.4. Note that in the graph of Figure 4.10.2 there is no point at which the distribution may be divided into two similar parts, as in the other examples. When a distribution is symmetric, it will be true that the mean and the median are equal in value. It is not necessarily true, however, that the mode(s) will equal either the mean or the median; witness the example of Figure 4.10.5. On the other hand, a nonsymmetric distribution is sometimes described as **skewed,** which means that the length of one of the **tails** of the distribution, relative to the central section, is disproportionate to the other. For example, the distribution in Figure 4.10.6 is skewed to the right, or **skewed positively.** In a positively skewed distribution, the bulk of the cases fall into the lower part of the range of scores, and relatively few show extremely high values. This is reflected by the relation

Mean > Median

usually found in a positively skewed distribution.

On the other hand, it is possible to find distributions skewed to the left, or **negatively skewed.** In such a distribution, the long tail of the distribution occurs among the low values of the variable. That is, the bulk of the distribution shows relatively high scores, although there are a few quite low scores (Fig. 4.10.7). Generally in a negatively

Figures 4.10.1—4.10.4

Idealized distributions of various "shapes."

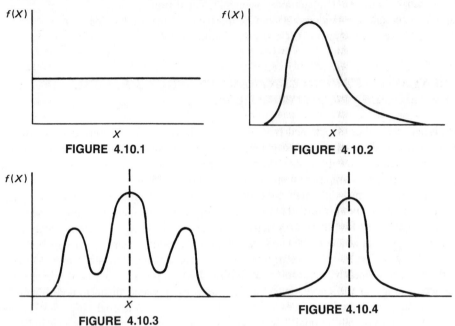

FIGURE 4.10.1

FIGURE 4.10.2

FIGURE 4.10.3

FIGURE 4.10.4

skewed distribution Median > Mean. Thus, a rough and ready way to describe the skewness of a distribution is to find the mean and median; if Mean > Median, then you can conclude that the distribution is skewed to the right (positively). If Median > Mean, then you may conclude that negative skewness exists. If a more accurate determination is needed, other indices reflecting skewness are available, although space will not be devoted to these here. A word of warning: if a distribution is symmetric, then Mean = Median, but the fact that Mean = Median does not necessarily imply that the distribution is symmetric.

Describing the skewness of a distribution in terms of measures of central tendency again points up the contrast between the mean and the median as measures of central tendency. The mean is much more affected by the extreme cases in the distribution than is the median. Any alteration of the scores of cases at the extreme ends of a distribution will have no effect at all on the median so long as the rank order of all of the scores is roughly preserved; only when scores near the center of the distribution are altered is there a good chance of altering the median. This is not true of the mean, which is very sensitive to score changes at the extremes of the distribution. The alteration of the score for a single extreme case in a distribution may have a profound effect on the mean. It is evident that the mean follows the skewed tail in the distribution, but the median does so to a lesser extent. The occurrence of *even a few* very high or very low cases can seriously distort the impression of the distribution given by the mean, provided that one mistakenly interprets the mean as the typical value. If you are dealing with a nonsym-

Figures 4.10.5–4.10.7

Relationships of central tendency measures in symmetric and skewed distributions.

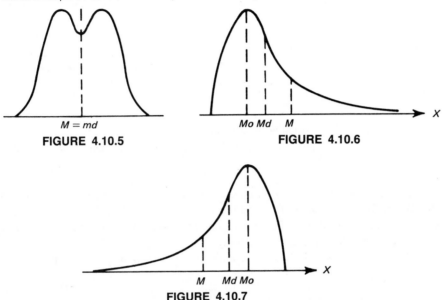

$M = md$

FIGURE 4.10.5

Mo Md M

FIGURE 4.10.6

M   Md Mo

FIGURE 4.10.7

metric distribution, and you want to communicate the typical value, you must report the median. On the other hand, in spite of the distribution's shape, the mean always communicates the same thing: the point about which the sum of deviations is zero. The choice of an index must depend on what the user is trying to *communicate* about the distribution.

Differences in shape are only rough, qualitative ways of distinguishing among distributions. The only adequate description of a theoretical distribution is its function rule. Distributions that look similar in their graphic form may be very different functions. Conversely, distributions that appear quite different when graphed actually may belong to the same family.

A description in terms of modality and skewness is sometimes useful for giving a general impression of what a distribution is like, but it does not communicate its essential character. Similarly, for obtained frequency distributions, an actual statement of the distribution contains far more information than is ever given by any report of central tendency or shape alone.

4.11 MEASURES OF DISPERSION IN FREQUENCY DISTRIBUTIONS

An index of central tendency summarizes only one special aspect of a distribution, be it mode, median, or mean. Any distribution has at least one more feature that must be summarized in some way. Distributions exhibit **spread** or **dispersion,** the tendency for observations to depart from central tendency. If central tendency measures are thought of as good bets about observations in a distribution, then measures of spread represent the other side of the question: dispersion reflects the "poorness" of central tendency as a description of a randomly selected case, the tendency of observations *not* to be like the average.

The mean is a good bet about the score of any observation sampled at random from a distribution, but *no observed* case need be exactly like the mean. A **deviation** from the mean expresses how "off" the mean is as a bet about a particular case, or how much *in error* is the mean as a description of this case:

$$d_i = (x_i - \bar{x})$$

where x_i is the value of a particular observation.

In the same way, we could talk about a deviation from the median, if we wished,

$$d'_i = (x_i - Md)$$

or even perhaps from the mode. It is quite obvious that the larger such deviations are, on the whole, from a measure of central tendency, the more do cases differ from each other and the more spread does the distribution show. What we need is an index (or set of indices) to reflect this spread or variability.

First of all, why not simply take the average of the deviations about the mean as our measure of variability:

$$\frac{\sum_i (x_i - \bar{x})}{N}$$

or

$$\sum_j \frac{(x_j - \bar{x}) \, \text{freq}(x_j)}{N} \, ?$$

This will not work, however, because in Section 4.3 it was shown that in any frequency distribution the average signed deviation from the mean must be zero.

The device used to get around this difficulty is to take the *square* of each deviation from the mean and then to find the average of these squared deviations:

$$S^2 = \frac{\sum_i (x_i - \bar{x})^2}{N} = \frac{\sum_i d_i^2}{N} .$$ [4.11.1*]

For any distribution the index S^2, equal to the average of the squared deviations from the mean, is called the **variance** of the distribution. The variance reflects the *degree of spread*, since S^2 will be zero if and only if each and every case in the distribution shows exactly the same score, the mean. The more that the cases tend to differ from each other and the mean, the larger will the variance be.

The variance is defined in Eq. 4.11.1 *as though* the raw score of each of N cases were known, and can always be computed by this formula when the raw data have not been grouped into a distribution. However, when data are in a grouped distribution, an equivalent definition of the variance is given by

$$S^2 = \frac{\sum_j (x_j - \bar{x})^2 \text{ freq } (x_j)}{N}$$ [4.11.2*]

where x_j is, as usual, the midpoint of an interval j. Here, for each interval, the deviation of the midpoint from the mean is squared and multiplied by the frequency for that interval. When this has been done for each interval, the average of these products is the variance S^2. Just as with the mean, the value of S^2 calculated for a grouped distribution need not agree exactly with that based on the raw scores; nevertheless, if a relatively large number of class intervals is used, these two values should agree very closely.

4.12 THE STANDARD DEVIATION

Although the variance is an adequate way of describing the degree of variability in a distribution, it does have one important drawback. The variance is a quantity in squared units of measurement. For instance, if measurements of height are made in inches, then the mean is some number of inches, and a deviation from the mean is a difference in inches. However, the square of a deviation is in *square-inch units,* and thus the variance, being a mean squared deviation, must also be in square inches. Naturally, this is not an insurmountable problem: taking the positive square root of the variance gives an index of variability in the original units. **The square root of the variance for a distribution is called the standard deviation, and is an index of variability in the original measurement units.**

A letter S will be used to denote the standard deviation for a frequency distribution:

$$S = \sqrt{S^2} = \sqrt{\frac{\sum_i (x_i - \bar{x})^2}{N}}$$ [4.12.1*]

or

$$S = \sqrt{\frac{\sum_j (x_j - \bar{x})^2 \text{ freq}(x_j)}{N}}.$$ [4.12.2*]

4.13 THE COMPUTATION OF THE VARIANCE AND STANDARD DEVIATION

The variance and standard deviation can be computed from a list of raw scores by using Eqs. 4.11.1 and 4.12.1. This deviation method entails finding the mean, subtracting it successively from each score, squaring each result, adding the squared deviations together, and dividing by N to find the variance. Obviously, this can be relatively laborious for a sizable number of cases. However, it is possible to simplify the computations by a few algebraic manipulations of the original formulas. This may be shown as follows: For any single deviation, expanding the square gives

$$(x_i - \bar{x})^2 = x_i^2 - 2 x_i \bar{x} + \bar{x}^2$$

so that, on averaging these squares, we have

$$\sum_i \frac{(x_i - \bar{x})^2}{N} = \sum_i \frac{(x_i^2 - 2 x_i \bar{x} + \bar{x}^2)}{N}.$$

By Rule 4 in Appendix A, the summation may be distributed so that

$$\sum_i \frac{(x_i - \bar{x})^2}{N} = \sum_i \frac{x_i^2}{N} - 2 \sum_i \frac{x_i \bar{x}}{N} + \sum_i \frac{\bar{x}^2}{N}.$$

However, wherever $\bar{x}$ appears it is a constant over the sum, and so, by Rules 1 and 2 in Appendix A,

$$S^2 = \sum_i \frac{x_i^2}{N} - 2 \bar{x} \sum_i \frac{x_i}{N} + N \left(\frac{\bar{x}^2}{N} \right)$$

or

$$S^2 = \sum_i \frac{x_i^2}{N} - \bar{x}^2. \qquad [4.13.1^*]$$

Finally,

$$S = \sqrt{\sum_i \frac{x_i^2}{N} - \bar{x}^2}. \qquad [4.13.2^*]$$

These last formulas, 4.13.1 and 4.13.2, give a way to calculate the indices S^2 and S with some saving in steps. These formulas will be referred to as the "raw-score computing forms" for the variance and standard deviation. For an example of the use of these forms, study the example shown in Table 4.13.1, based on seven cases. These two methods must always agree exactly with the values obtained by the deviation method, as in the example.

It is also possible to find similar computing forms for grouped distributions. Starting with the definition of the variance of a grouped distribution,

$$S^2 = \frac{\sum_j (x_j - \bar{x})^2 \, \text{freq}(x_j)}{N}$$

we could develop the following computing forms:

$$S^2 = \frac{\sum_j x_j^2 \, \text{freq}(x_j)}{N} - \bar{x}^2$$

$$= \frac{\sum_j x_j^2 \, \text{freq}(x_j)}{N} - \left[\frac{\sum_j x_j \, \text{freq}(x_j)}{N} \right]^2. \qquad [4.13.3^*]$$

Then, as usual,

$$S = \sqrt{S^2}.$$

This little derivation is left for you as an exercise.

Scores x_i	Deviation method		Raw-score method x_i^2
	$d_i = (x_i - \bar{x})$	d_i^2	
11	6	36	121
10	5	25	100
9	4	16	81
8	3	9	64
6	1	1	36
-4	-9	81	16
-5	-10	100	25
$\overline{35} = \sum_i x_i$	$\overline{0}$	$\overline{268} = \sum_i d_i^2$	$\overline{443} = \sum_i x_i^2$

Table 4.13.1

Computation of the variance and the standard deviation from raw data.

$$\bar{x} = 5 \qquad S^2 = \frac{\sum_i d_1^2}{N} = \frac{268}{7} = 38.29 \qquad S^2 = \frac{\sum_i x_i^2}{N} - \bar{x}^2$$

$$= \frac{443}{7} - 25 = 38.29$$

$$S = \sqrt{38.29} = 6.19$$

An example of the computation of the mean and standard deviation from a frequency distribution is given in Table 4.13.2.

Since the operations for squaring and summing a set of numbers are very simple on most pocket calculators, the variance and standard deviation are almost as easy to compute as the mean on these little devices, particularly if only a small set of numbers is involved. In addition, many small calculators are preprogrammed to yield the variance and the standard deviation directly (as well as the mean, of course) once the raw data are entered. Naturally, this is extremely convenient in many statistical problems.

In times past, grouping large sets of data into frequency distributions was the main labor-saving device for computing things such as the variance from such data. These days, when the data set is even moderately large, or when many variances or standard deviations are to be calculated, the work is almost always done on a computer.

Class	x	f	xf	x^2	x^2f
46–50	48	6	288	2,304	13,824
41–45	43	8	344	1,849	14,792
36–40	38	10	380	1,444	14,440
31–35	33	5	165	1,089	5,445
26–30	28	3	84	784	2,352
21–25	23	1	23	529	529
		$\overline{33}$	$\overline{1,284} = \Sigma\, xf$		$\overline{51,382} = \Sigma\, x^2f$

Table 4.13.2

Variance and standard deviation from a grouped distribution.

$$\bar{x} = \frac{1,284}{33} = 38.91$$

$$S^2 = \frac{51,382}{33} - 1,513.99 = 43.04 \quad S = \sqrt{43.04} = 6.56$$

4.14 SOME MEANINGS OF THE VARIANCE AND STANDARD DEVIATION

Since the variance and standard deviation will figure very largely in our subsequent work, it is important to gain some intuition about what they represent. One interpretation is provided by the fact that **the variance is directly proportional to the average squared difference between all pairs of observations:**

$$\sum_{(i,j)} \frac{(x_i - x_j)^2}{\binom{N}{2}} = \frac{2N}{N-1} S^2. \tag{4.14.1}$$

(Here (i,j) indicates summation over all possible *pairs* of scores.) The variance summarizes how different the various cases are from *each other*, just as it reflects how different each case is from the mean. Given some N cases, the more that pairs of cases tend to be unlike in their scores, the larger the variance and standard deviation.

Still another way to think of the variance and standard deviation is by a physical analogy. A deviation from the mean can be identified with a certain amount of *force* exerted by a variety of factors making this case different from others in its group. Think of any score as composed of the mean, plus a deviation from the mean,

$$x_i = \bar{x} + d_i,$$

where the deviation is the resultant force of these "influences." Picturing a deviation from the mean as we would show a physical force away from a point, we have:

$$\underbrace{\bar{x} \longrightarrow x_i.}_{d_i}$$

Now suppose that we think of two cases from the larger group, each of which has a deviation from the mean, or d_1 and d_2 repectively. These two cases are independent, so that we can represent their deviation as forces acting at right angles (Fig. 4.14.1). How would you find the net force away from the mean for these two cases? The rule of the parallelogram of forces shows the resultant force to be the diagonal length in Figure 4.14.1. This diagonal has length

$$\sqrt{\sum d^2} = \sqrt{(x_1 - \bar{x})^2 + (x_2 - \bar{x})^2}$$

by the Pythagorean theorem. If we divide by N before taking the square root, this looks much like the standard deviation.

Suppose that there were three independent cases. When their deviations are interpreted as forces away from the mean, the diagram shown in Figure 4.14.2 holds, and the resultant force is $\sqrt{d_1^2 + d_2^2 + d_3^2}$. The resultant force away from the mean per observation would again be calculated much like a standard deviation. In short, a physical analogy to the standard deviation is a resultant force away from the mean per unit observation. A large standard deviation is analogous to a large "push" away from the mean per observation, due to all the factors making observations heterogeneous. In statistics, "error" is often viewed as such a resultant force away from homogeneity,

Figure 4.14.1/4.14.2

Physical analogies to the standard deviation.

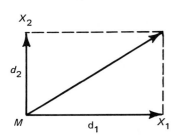
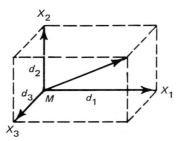

and in terms of this physical analogy, the standard deviation should reflect the net effect of such forces per observation.

The reader acquainted with elementary physics may recognize not only that the mean is the center of gravity of a physical distribution of objects, but also that the variance is the moment of inertia of a distribution of mass. Furthermore, the standard deviation corresponds to the radius of gyration of a mass distribution; this is the real basis for regarding the standard deviation as analogous to a resultant force away from the mean. These physical conceptions and their associated mathematical formulations have influenced the course of theoretical statistics very strongly and have helped to shape the form of statistical inference as we will encounter it.

4.15 THE MEAN AS THE "ORIGIN" FOR THE VARIANCE

This question may already have occurred to the reader: "Why is the variance, and hence the standard deviation, always calculated in terms of deviations from the *mean*? Why couldn't one of the other measures of central tendency be used as well?" The answer lies in the fact that **average squared deviation (i.e., the variance) is smallest when calculated from the mean.** That is, if the disagreement of any score with the mean is indicated by the square of its difference from the mean, then, on the average, the mean agrees better with the scores than any other single value one might choose.

This may be demonstrated as follows: Suppose that we chose some arbitrary real number C, and calculated a "pseudo-variance" S_C^2 by subtracting C from each score, squaring, and averaging:

$$S_C^2 = \sum_i \frac{(x_1 - C)^2}{N}.$$

Adding and subtracting $\bar{x}$ for each score would not change the value of S_C^2 at all. However, if we did so, we could expand each squared deviation as follows:

$$(x_i - C)^2 = (x_i - \bar{x} + \bar{x} - C)^2$$
$$= (x_i - \bar{x})^2 + 2(x_i - \bar{x})(\bar{x} - C) + (\bar{x} - C)^2.$$

Substituting into the expression for S_C^2 and distributing the summation by Rule 4, Appendix A, we have

$$S_C^2 = \frac{\sum_i (x_i - \bar{x})^2}{N} + 2\frac{\sum_i (x_i - \bar{x})(\bar{x} - C)}{N} + \frac{\sum_i (\bar{x} - C)^2}{N}.$$

Notice that $(\bar{x} - C)$ is the same for every score summed, so that by Rules 1 and 2 in Appendix A

$$S_C^2 = \sum_i \frac{(x_i - \bar{x})^2}{N} + 2(\bar{x} - C)\sum_i \frac{(x_i - \bar{x})}{N} + (\bar{x} - C)^2.$$

However, the first term to the right of the equals sign in the equation above is simply S^2, the variance about the mean, and the second term is zero, since the average deviation from $\bar{x}$ is zero. On making these substitutions, we find

$$S_C^2 = S^2 + (\bar{x} - C)^2.$$

Since $(\bar{x} - C)^2$ is a squared real number it can be only positive or zero, and so S_C^2 must be greater than or equal to S^2. The value of S_C^2 can be equal to S^2 only when $\bar{x}$ and C are the same. In short, we have shown that the variance calculated about the mean will always be smaller than about any other point. This method of finding a value that has the property of minimizing the sum of squared deviations is an application of the so-called *principle of least squares*. We will encounter many examples of values determined by this principle in later chapters.

On the other hand, if we appraise error by taking the absolute difference (disregarding sign) between a score and a measure of central tendency, then **the average absolute deviation is smallest when the median is used.** This is one of the reasons that squared deviations rather than absolute deviations figure in the indices of variability when the mean is used to express central tendency. When the median is used to express central tendency it is often accompanied by the average absolute deviation from the median to indicate dispersion, rather than by the standard deviation which is more appropriate to the use of the mean. The average absolute deviation is simply

$$\text{A.D.} = \sum_i \frac{|x_i - Md|}{N} \tag{4.15.1}$$

where the vertical bars indicate a disregarding of sign. Analogically speaking this measure is to the median as the standard deviation is to the mean.

4.16 THE VARIANCE AND OTHER MOMENTS OF A PROBABILITY DISTRIBUTION

The variance and standard deviation of a probability distribution have exactly the same interpretations as do the corresponding indices for a frequency distribution: each is a measure of the variability or spread, the former in squared units and the latter in the original units of the random variable. However, like the mean of a probability distribution, the variance (or standard deviation) often figures as a parameter entering into the function rule for the distribution.

The small Greek letter sigma, σ, is generally used to denote the standard deviation of a random variable, and σ^2 its variance. When the random variable is discrete, the variance is defined by

$$\sigma^2 = \Sigma \, [x - E(X)]^2 p(x) \qquad\qquad [4.16.1^*]$$

which is equivalent to

$$\sigma^2 = E(X^2) - [E(X)]^2, \qquad\qquad [4.16.2^*]$$

the expectation of the square of X minus the square of the expectation of X. Then the standard deviation σ of the random variable X is just the square root of the variance, exactly as in a frequency distribution. (See Appendix B.)

By methods almost identical to those used in Section 4.8 to find the mean of the binomial distribution, it can be shown that the variance of a binomial distribution of number of "successes" is

$$\sigma^2 = Npq. \qquad\qquad [4.16.3^*]$$

Thus, the standard deviation for a binomial distribution is

$$\sigma = \sqrt{Npq}. \qquad\qquad [4.16.4^*]$$

If the random variable is P, the proportion of successes out of N observations, then the variance becomes

$$\sigma_P^2 = \frac{pq}{N}, \qquad\qquad [4.16.5^*]$$

and the standard deviation is

$$\sigma_P = \sqrt{\sigma_P^2} = \sqrt{\frac{pq}{N}}. \qquad\qquad [4.16.6^*]$$

Similarly, the Poisson distribution has a variance

$$\sigma^2 = m. \qquad\qquad [4.16.7]$$

(Note this interesting property of a Poisson distribution: its mean must have the same value as its variance.)

Finally, our other prime example of a discrete random variable, the Pascal, has a variance given by

$$\sigma^2 = \frac{r(1 - p)}{p^2}. \qquad\qquad [4.16.8]$$

The variance for any continuous distribution has the same form as for any other probability distribution:

$$\sigma^2 = E(X^2) - [E(X)]^2, \qquad\qquad [4.16.9^*]$$

the expectation of the square of the variable, minus the square of the expectation. This means that for a random variable X with probability density $f(x)$ at any point $X = x$, the variance is

$$\sigma^2 = \int_{-\infty}^{\infty} x^2 f(x) \, dx - [E(X)]^2. \qquad [4.16.10]$$

A truly mathematical treatment of distributions would introduce not only the mean and the variance, but also a number of other summary characteristics. These are the so-called moments of a distribution, which are simply **the expectations of different powers of the random variable.** Thus, the first moment about the origin of a random variable X is $E(X) = $ the mean. The second moment about the origin is $E(X^2)$, the third is $E(X^3)$, and so on. When the mean is subtracted from X before the power is taken, then the moment is said to be **about the mean;** the variance $E[X - E(X)]^2$ is the second moment about the mean; $E[X - E(X)]^3$ is the third moment about the mean, and so on.

Just as the mean describes the "location" of the distribution on the X axis, and the variance describes its dispersion, so do the higher moments reflect other features of the distribution. For example, the third moment about the mean is used in certain measures of degree of **skewness:** the third moment will be zero for a symmetric distribution, negative for skewness to the left, positive for skewness to the right. The fourth moment indicates the degree of "peakedness" or **kurtosis** of the distribution, and so on. These higher moments have relatively little use in elementary applications of statistics, but they are important for mathematical statisticians in the study of the properties of distributions and in arriving at theoretical distributions fitting observed data. The entire set of moments for a distribution will ordinarily determine the distribution exactly, and distributions are sometimes specified in this way when their general function rules are unknown or difficult to state.

4.17 STANDARDIZED SCORES

A major use of the mean and standard deviation is in transforming a raw score into a **standardized score,** showing the *relative status* of that score in a distribution. If you are given the information, "John Doe has a score of 60," you really know very little about what this score means. Is this score high, low, middling, or what? However, if you know something about the distribution of scores for the group including John Doe, you can judge the location of the score in the distribution. The score of 60 gives quite a different picture when the mean is 30 and the standard deviation 10 than when the mean is 65 and the standard deviation is 20.

Each value in any distribution can be converted into a standardized score, or **z score,** expressing the deviation from the mean in standard deviation units:

$$z = \frac{x - \bar{x}}{S}. \qquad [4.17.1^*]$$

The z score tells how many standard deviations away from the mean is x. The two distributions mentioned above give quite different z scores to the score of 60:

$$z_1 = \frac{60 - 30}{10} = 3$$

$$z_2 = \frac{60 - 65}{20} = -.25.$$

The conversion of raw scores to z scores is handy when one wishes to emphasize the *location or status* of a score in the distribution, and in future sections we will deal with standardized scores when this aspect of any score is to be discussed.

Changing each of the scores in a distribution to a standardized score creates a distribution having a "standard" mean and standard deviation: **the mean of a distribution of standardized scores is always 0, and the standard deviation is always 1.** This is easily shown as follows:

$$\text{mean of } z \text{ scores} = \bar{z} = \sum_i \frac{z_i}{N} = \sum_i \frac{(x_i - \bar{x})}{NS}.$$

Since N and S are constant over the summation, this becomes

$$\bar{z} = \frac{1}{NS} \sum_i (x_i - \bar{x}) = 0 \qquad [4.17.2^*]$$

since the sum of deviations about the mean is always zero.

In a similar fashion,

$$\text{variance of } z \text{ scores} = S_z^2 = \sum_i \frac{z_i^2}{N} = \sum_i \frac{(x_i - \bar{x})^2}{NS^2} \qquad [4.17.3^*]$$

$$= \frac{S^2}{S^2} = 1,$$

so that

$$\text{standard deviation of } z \text{ scores} = S_z = 1. \qquad [4.17.4^*]$$

Although it is true that the mean and standard deviation of a distribution of z scores will always be 0 and 1 respectively, changing the scores in any distribution to z scores *does not alter* the form of the distribution. The frequency of any given z score *is exactly that* of the x score corresponding to it in the original distribution.

Standardized scores have exactly the same format in probability distributions as in frequency distributions. If x is a value of a random variable, then the corresponding standardized score, relative to the probability distribution, is

$$z = \frac{x - E(X)}{\sigma} = \frac{x - \mu}{\sigma}. \qquad [4.17.5^*]$$

Furthermore, the mean of the standardized scores is zero for any probability distribution:

$$E(z) = E\left[\frac{X - E(X)}{\sigma}\right] = \frac{E(X) - E(X)}{\sigma} = 0 \qquad [4.17.6^*]$$

since $E(X)$ and σ are constants over the various possible values of X. The standard deviation of standardized scores is always 1 in a probability distribution:

$$\sigma_z^2 = E(z^2) - [E(z)]^2 = E\left[\frac{X - E(X)}{\sigma}\right]^2 = E\frac{[X - E(X)]^2}{\sigma^2} \qquad [4.17.7^*]$$

$$= \frac{\sigma^2}{\sigma^2} = 1$$

and

$$\sigma_z = 1. \qquad\qquad [4.17.8^*]$$

Again, the form of the probability distribution is not changed at all by the transformation to z scores, in the sense that the probability (or probability density) of any value of z is simply the probability (or probability density) of the corresponding value of X.

4.18 TCHEBYCHEFF'S INEQUALITY

There is a very close connection between the size of deviations from the mean and probability, holding for distributions having finite expectation and variance. The following relation is called Tchebycheff's inequality, after the Russian mathematician who first proved this very general principle:

$$\text{prob}(|X - \mu| \geq b) \leq \frac{\sigma^2}{b^2}, \qquad\qquad [4.18.1]$$

the probability that a random variable X will differ absolutely from expectation by b or more units ($b > 0$) is *always* less than or equal to the ratio of σ^2 to b^2. Any deviation from expectation of b or more units can be *no more probable* than σ^2/b^2.

This relation can be clarified somewhat by dealing with the deviation in σ units, making the random variable a standardized score. If we let $b = k\sigma$, then the following version of the Tchebycheff inequality is true:

$$\text{prob}\left(\frac{|X - \mu|}{\sigma} \geq k\right) \leq \frac{1}{k^2}; \qquad\qquad [4.18.2]$$

the probability that a standardized score drawn at random from the distribution has *absolute* magnitude greater than or equal to some positive number k is *always* less than or equal to $1/k^2$. Thus, given a distribution with some mean and variance, the probability of drawing a case having a standardized score of 2 or more (disregarding sign) must be *at most* 1/4. The probability of a standardized score of 3 or more must be no more than 1/9, the probability of 10 or more can be no more than 1/100, and so on.

Although this principle is very important theoretically, it is not very powerful as a tool in applied problems. The Tchebycheff inequality can be strengthened somewhat if we are willing to make assumptions about the general form of the distribution, however. For example, if we assume that the distribution of the random variable is both *symmetric and unimodal,* then the relation becomes

$$p(|z| \geq k) \leq \frac{4}{9}\left(\frac{1}{k^2}\right). \qquad\qquad [4.18.3]$$

For such distributions, we can make somewhat "tighter" statements about how large the probability of a given amount of deviation may be. For example, how probable is a score value falling three or more standard deviations from the mean in some distribution? By Tchebycheff's inequality, we know that regardless of what the true distribution of scores may be like, a score three or more standard deviations away from the expected value in either direction can be no more probable than 1/9, or about .11. However, if the basic distribution of scores can be assumed to be unimodal–symmetric, then Eq.

4.18.3 above tells that a value three or more standard deviations from the mean should be observed with probability *no greater than* (4/9) (1/9), or about .05. Similarly, we could find that at least 1 − 4/9 or 5/9 of such scores fall within one standard deviation to either side of the mean. Furthermore, 1 − (4/9) (1/4) or 8/9 must fall within two standard deviations, and so on.

By adding assumptions about the form of the distribution to the general principle relating standardized values to probabilites, we are able to make stronger and stronger probability statements about a sample result's departures from the mean. In order to make very precise statements one has to make even stronger assumptions about the distribution of the random variable, unless dealing with very large samples of cases, as we shall see. This chapter has concluded with the introduction of the Tchebycheff inequality to suggest that the mean and the standard deviation play a key role in the theory of statistical inference. The mean and standard deviation are, of course, useful devices for summarizing data if that is our purpose, although there are situations where other measures of central tendency and variability may do equally well or better. The really paramount importance of mean and standard deviation does not emerge until one is interested in making *inferences,* involving the estimation of parameters, or assigning probabilities to sample results. Here, we shall find that most "classical" statistical theory is erected around these two indices, together with their combination in standardized scores.

So far in this chapter we have been concerned with measures of central tendency and of variability. Such indices are summary measures of an entire distribution. However, another type of question concerns the relative location of a particular score value within the distribution. Next we will discuss percentiles and percentile ranks, which provide one way to answer this question of location of an individual or a score within any given distribution.

4.19 PERCENTILES AND PERCENTILE RANKS

In any frequency distribution of numerical scores, the percentile rank of any specific value x is the percent of cases out of the total that fall at or below x in value. In a probability distribution, the percentile rank of a given value x is simply $100F(x)$, or 100 times the cumulative probability associated with $X = x$. Note especially that the percentile rank associated with the median in any distribution must be 50, since 50% of all observed values in a frequency distribution must lie at or below the median, and $F(X = Md) = .50$ in a probability distribution.

One use of the idea of percentile ranks is in finding the relative locations of values within given distributions. The value need not actually have occurred in the frequency distribution in order for a percentile rank to be determined. One asks where that value would have fallen relative to the other cases if it actually had occurred. This is done by linear interpolation:

$$\frac{(x - \text{value below } x)}{(\text{value above } x - \text{value below } x)} (\%\text{ile rank above } x - \%\text{ile rank below } x)$$
$$+ \%\text{ile rank below} = \%\text{ile rank of } x,$$

where "%ile" is simply an abbreviation for "percentile."

Such a use of percentile ranks is probably already familiar to most students, since this typically is the way that the test performance of an individual is compared to the performance of some reference group, or norm group.

The percent at or below a given value of x is the percentile rank of that value. Often, however, we are interested in the reverse problem: What score x cuts off the bottom G percent of the distribution? That score is then referred to the G percentile of the distribution. To find the value corresponding to the Gth percentile, we first look to see if there is any exact value that occurred in the distribution such that the number of cases falling at or below that value is $GN/100$. If so, the value is the Gth percentile. However, it may happen that no value can be located that corresponds exactly to the Gth percentile, among those values of X that actually occurred. In that case, linear interpolation is employed once again:

$$\frac{[((GN/100) - cf \text{ below})(\text{value above} - \text{value below})]}{(cf \text{ above} - cf \text{ below})} + \text{value below}$$

$$= G\text{th percentile,}$$

where "cf below" symbolizes the cumulative frequency nearest below the number $GN/100$ among the values of X that occurred, and "value below" is the actual X value associated with that cumulative frequency. Similarly, "cf above" stands for the cumulative frequency just above $GN/100$ for one of the values of X that occurred, and "value above" is the actual value associated with that cumulative frequency.

A similar set of problems exists when one is trying to find percentile ranks and percentiles in a grouped frequency distribution. Suppose that N is the total number in the distribution, i is the class-interval size, and we want to find the percentile rank for some value of X. First of all we locate the class interval in which $X = x$ must fall. That done, we take

$$\text{percentile rank} = \frac{(x - \text{lower limit})}{i}\left(\frac{100(\text{frequency in interval})}{N}\right)$$

$$+ \frac{100(\text{cumulative frequency below interval})}{N},$$

using the lower limit of the interval and i, the class-interval size, or (upper real limit − lower real limit).

To find the Gth percentile value, we find the interval in which this percentile value must fall by taking $NG/100$, and seeing that the interval chosen has a cumulative frequency greater than or equal to $NG/100$, with the interval just below having a cumulative frequency less than $NG/100$. (Remember that the cumulative frequency of an interval is the number of cases in the total distribution falling at or below the *upper* real limit of the interval.) Then we take

$$\frac{(NG/100 - cf \text{ for interval below})}{\text{frequency in interval}}(i) + \text{lower real limit of interval}$$

$$= G\text{th percentile.}$$

Since we are using linear interpolation to find percentile ranks and percentiles in such a grouped distribution, it follows that we are assuming that within any given class interval, the scores falling into the interval are evenly spaced in value across the extent of the interval.

Occasionally one runs into reports of deciles, quartiles, or other "fractiles" of a distribution. Deciles are simply the values corresponding to the 10th, 20th, 30th, · · · percentiles: those values which divide the distribution into 10ths. Similarly, quartiles divide the distribution into fourths and are the values of the 25th, 50th, 75th, and 100th percentiles. Any other arbitrary division of a distribution might be worked out by percentiles as the occasion demands.

4.20 MEANS AND VARIANCES IN STATISTICS PACKAGES

In Section 2.23 some of the programs that produce frequency distributions were mentioned. Each of those programs also yields the sample mean, the sample variance (or standard deviation), and various other summary values describing the data on one or more variables. Thus, for example, for each variable the FREQUENCIES subprogram in SPSS and SPSS^X will give (on request) the mean, the standard deviation, the variance, the standard error of the mean (to be discussed in Chapter 5) the median, the mode, the range, and measures of skewness and kurtosis. Percentiles may also be requested in SPSS^X. Since, in principle, the FREQUENCIES procedure can be used with hundreds of variables, this can produce a staggering amount of output, and one will generally wish to be very selective in what output is requested. A good rule of thumb both here and elsewhere is to ask, "What do I *really* need to know at the end of this analysis?" and then to request only the information that is high on your priority list.

The SPSS and SPSS^X subprogram CONDESCRIPTIVE handles variables that can be treated *as though* they were measured at the interval scale level. Then a variety of descriptive statistics can be requested for any or all of a set of variables specified in the input data. These include the set of statistics listed above under FREQUENCIES, except for the median and mode. The CONDESCRIPTIVE subprogram also does not provide the frequency distributions. On the other hand, one may request the z scores or standardized values for any variable represented among the N data cases, and these will be produced and read into a file that can be used in a variety of ways.

The BMDP program P2D has already been mentioned in connection with frequency distributions. Summary statistics produced for each variable include the minimum and maximum values, the range, the variance and standard deviation, the interquartile range (which is half the difference between the 75th and 25th percentiles), the mean and the standard error of the mean, the median and its standard error, the mode, measures of skewness and kurtosis, and other specialized measures. Approximately 200 variables may be analyzed at one time in the P2D program.

In the SAS package the means for a set of variables may be examined by use of the PROC MEANS command. In addition means, variances, and other summary statistics are provided by a number of the procedures in this comprehensive package.

EXERCISES

1. Chooose and calculate the appropriate index of central tendency for the data in Exercise 4, Chapter 2. Why must you choose this particular index? If you guess that any individual drawn at random from this group shows this status, how likely are you to be wrong?

2. A social psychologist determined the number of verbal exchanges between pairs of people attempting to solve a puzzle in a given time period. For 45 pairs, the numbers were as follows. Calculate the mean and the median numbers of exchanges observed for this group.

27	31	37	15	31	8	39	33	24
26	23	43	25	47	22	26	42	20
41	7	33	21	33	61	50	27	52
38	32	25	46	54	22	45	63	17
13	21	48	26	40	51	24	53	34

3. A sample was taken of newly divorced women over 40. Each gave the age (in nearest years) at which she had been first married. The results were as follows. Calculate the mean and the median ages.

20	24	20	27	18	22	22	35	26	25	30	18	25	26	30	18
19	25	27	24	26	20	22	28	24	21	24	32	26	34	28	16

4. Calculate the mean and the median from the raw (i.e., ungrouped) data presented in Exercise 17, Chapter 2.

5. From the raw data in Exercise 7, Chapter 2, calculate the mean income of the group.

6. Using the data in Exercise 5, Chapter 2, find the mean, median, and mode for each of the three questionnaire items. What do the relative sizes of mean, median, and mode reflect about the distributions of responses?

7. A teacher was interested in the difference in the time it took students to complete each of two different tests of simple arithmetic. It was believed that practice effects should make the time shorter on the second test. A group of 25 students showed the following differences in time taken on the first and second tests (in minutes). Find the mean difference in time. Does the notion that the second test took less time seem to be supported by the data?

−2.3	− .4	− .6	1.7	1.9
− .3	.7	1.4	.8	−1.3
1.6	1.1	.2	.5	−1.8
− .5	.4	1.5	2.4	.8
.9	−1.5	−1.2	.5	1.2

8. Using the data in Exercise 3 above, demonstrate that the sum of the deviations about the mean is indeed zero.

9. A nutritionist was studying the eating patterns of early adolescent boys and girls. Two randomly selected groups were used, the first consisting of 50 thirteen-year-old boys, and the second of 50 thirteen-year-old girls. The data follow in the form of two frequency distributions, showing the calories (in hundreds) consumed daily by the members of each group. Compute the median and the modal number of calories consumed by the boys and by the girls. (Leave the answer in hundreds of calories.)

Class interval	Boys	Girls
50.00–54.99	1	1
45.00–49.99	2	0
40.00–44.99	4	2
35.00–39.99	16	5
30.00–34.99	12	10
25.00–29.99	7	20
20.00–24.99	5	8
15.00–19.99	2	2
10.00–14.99	1	2
	50	50

10. For the distributions in Exercise 9 calculate the mean number of calories consumed by each group. What do these means reveal about the two groups?

11. Suppose that an individual were drawn at random from 150 cases with a distribution like that in Table 2.6.1. Find the expected value for the score of this individual. What is the most likely value for the individual's score?

12. Compute the mean for the distribution in Table 2.4.1. How does this differ from the mean of the distribution in Table 2.6.1, based on the same cases? To what can you attribute the difference, if any?

13. Find the mean, median, and mode for the distribution shown in Table 2.8.2. What do the relative sizes of these three indices reflect about the distribution's form?

14. Choose and calculate the index of central tendency which seems to be most appropriate for the data in Exercise 10, Chapter 2. Why did you choose this particular index?

15. Find the expected value for each of the three theoretical distributions described in Section 2.17.

16. Find the expected value of the random variable in Exercise 19, Chapter 2.

17. Find the expected value for the random variable in Exercise 20, Chapter 2.

18. Calculate the variance and standard deviation for the data in Exercise 2 above.

19. What is the value of the standard deviation for the data in Exercise 3 above?

20. In the same study referred to in Exercise 3 above, another sample was taken of married women over 40 who had never been divorced. These also reported the age at which they had been married, with the following results. Do the samples of newly divorced and married women differ in mean age of marriage and in the variability of age at marriage? If so, how?

```
18   27   30   26   28   26   34   22   26   24   23   31   26   29   26
22   20   24   28   32   27   32   20   28   20   26   35   24   29   26
```

21. Find the variance and standard deviation for the data in Exercise 7.

22. What is the standard deviation of the raw income data in Exercise 7, Chapter 2?

23. Compare the two groups in Exercise 9 on their means and standard deviations. Which group has the higher caloric consumption, on the average? Which is the more variable in terms of calories consumed?

24. Calculate the variance and standard deviation for the distribution in Table 2.8.2.

25. Two fair dice are tossed independently and at random, yielding the number of "spots" on each trial ranging from 2 to 12. Find the variance and standard deviation of the random variable "number of spots coming up on the dice." (Refer to Section 2.15).

26. Find the mean and standard deviations of the following theoretical distributions:

(a) Binomial, with $N = 17$ and $p = .35$.
(b) Binomial, with $N = 75$ and $p = .80$.
(c) Poisson, with $m = 1.25$.
(d) Pascal, with $r = 3$ and $p = .40$.
(e) Geometric, with $p = .19$.
(f) Binomial, with $N = 2,000$ and $p = .01$.
(g) Poisson, with $m = 20$.

27. Find the variance and standard deviation for each of the three theoretical distributions in Section 2.17.

28. What is the standard deviation of the random variable in Exercise 19, Chapter 2?

29. Find the standard deviation of the random variable in Exercise 20, Chapter 2.

30. Determine the mean and standard deviation of the random variable, "number of items correct," based on the test described in Exercise 19, Chapter 3. Assume that the student in question is only guessing throughout the test.

31. What is the mean and standard deviation of the random variable "number of women winners" as defined in Exercise 20, Chapter 3.

32. Convert the ages in Exercise 20 above to the corresponding standardized scores. Then show that (within a small rounding error) these new scores must have a mean of zero and a standard deviation of 1.00.

33. Describe the relative position of each of the following ages in the group of divorced women in Exercise 3, using standardized scores. Then show where these same ages fall in the married group in Exercise 20.

(a) 25
(b) 30
(c) 17
(d) 37
(e) 20

34. Convert the two distributions in Exercise 9 into distributions of standardized or z scores. Determine the following:

(a) What is the percentile rank corresponding to a z value of -1 among the boys? Among the girls?
(b) An individual consumes 3,300 calories per day. Where does this place him among the boys, in standardized score terms? Among the girls?
(c) Between what two standardized scores do the middle 90% of the girls lie? Between what two z values do the middle 90% of the boys lie?
(d) Is a caloric intake of 4,200 relatively more deviant for a boy or for a girl?

(Hint: Any grouped distribution can be converted into a z-score distribution by simply converting the limits of the intervals into z scores.)

35. A certain probability distribution has a mean μ of 68 and a standard deviation σ of 11. What is the maximum probability that a case drawn at random from this distribution will show a value greater than or equal to 115.5, or less than or equal to 20.5? In this distribution, what is the maximum probability of observing a case which is more than 1.7 standard deviations from the mean? What is the smallest proportion of cases that we should expect to fall in the interval 53.7–82.3 in this distribution?

Chapter 5

Sampling Distributions and Point Estimation

5.1 POPULATIONS, PARAMETERS, AND STATISTICS

So far the entire set of elementary events has been called the *sample space,* since this term is useful and current in probability theory. However, in many fields using statistics it is common to find the word **population** used to mean the totality of potential units for observation. These potential units for observation are very often real or hypothetical sets of people, plants, or animals, and *population* provides a very appropriate alternative to *sample space* in such instances. Nevertheless, whenever the term *population* is used in the following, we shall mean only the sample space of elementary events from which samples are drawn.

Given a population of potential observations, the particular numerical score assigned to any particular unit observation is a value of a random variable; the distribution of this random variable is the **population distribution.** This distribution will have some mathematical form, with a mean μ, a variance σ^2, and all the other characteristic features of any distribution. If you like, you may think of the population distribution as a frequency distribution based on some large but finite number of cases. However, population distributions are almost always discussed as though they were theoretical probability distributions; the process of random sampling of single units with replacement ensures that the long-run relative frequency of any value of the random variable is the same as the probability of that value. Later we shall have occasion to idealize the population distribution and treat it as though the random variable were continuous. This is impossible for real-world observations, but we shall assume that it is "true enough" as an approximation to the population state of affairs.

Population values such as μ and σ^2 will be called **parameters of the population** (or sometimes, **true values**). Strictly speaking, a parameter is a value entering as an arbitrary constant in the particular function rule for a probability distribution, although the term is used more loosely to mean any value summarizing the population distribution. Just as parameters are characteristic of populations, so are **statistics** associated with samples.

Some amplification of this idea of a statistic is called for, however. We have already seen that the same sample of data may be used as the basis for a wide variety of statistics. Thus, even in samples from a Bernoulli process, we have examined both the number of successes out of N trials and the proportion of successes, as well as the trial number on which the rth success occurred (as in Pascal sampling). Each of these ways of regarding the same basic set of data provides a useful statistic. However, these are not the only statistics that might have been formed. We might, for example, have taken the product of the proportion of successes and the proportion of failures, or perhaps the logarithm of the proportion of successes. In some contexts we might have been interested in the reciprocal of the proportion of successes. In still other contexts we might have chosen to form a statistic in any of a variety of other ways.

The point is that there is no limit to the number of ways in which statistics can be constructed and associated with samples, even for samples as simple as binomial sequences. Not all of these statistics would be very useful perhaps, but we are perfectly free to define them. **A statistic is simply a function on samples, such that any sample is paired with a value of that statistic.** For samples of numerical data we ordinarily construct and use familiar statistics such as means, variances, medians, percentile ranks, and the like because they happen to be simple and useful. However, in some situations we would want to use still other statistics, such as the sum of the logarithms of all of the values, or perhaps the sum of the reciprocals of the values observed, or the difference between the highest and lowest values, and so on for any other way of combining or transforming the values in the sample. Each way of relating samples to new "summarizing" values is legitimate.

Moreover, a statistic need not use all of the information in a sample. Certainly the median, like the other percentiles, appears to be based on less of the information in a sample than is the mean or the variance. If we wished, we could use even less of the information in a sample in defining a statistic; thus, we might define our statistic as merely the value of the fifth observation made, for example, and ignore the rest of the sample values. Indeed, we might even let the value of a statistic be a constant, having no relationship to the sample values themselves, so that none of the information in the sample is used in the formation of the statistic.

This point has been stressed to bring out a related point: Since there are no "natural" statistics among the wide variety possible, and if an unlimited variety of statistics might be associated with any given sample, we need criteria for choosing among such statistics. In the last chapter some descriptive statistics were compared on the basis of what they tell and what we wish to communicate about a given sample distribution of values. Now, however, we will examine sample statistics against criteria of what they tell about population distributions. Although it is often true that the best way to gain information about a population parameter is through the use of the analogous sample statistic (that is, population mean and sample mean, population variance and sample variance), this is not always or necessarily true. Thus, in Chapters 8 and 9 we will use some statistics that have no direct parallels in the population. It may be that some other statistic with a different form contains more, or more useful, information about the population from which the sample came than does a familiar statistic. The main business of this chapter is, then, to examine how population distributions determine (or induce) distributions of sample statistics, and then to outline some of the criteria that have been developed for the choice of statistics to be used in inferences about populations.

5.2 SAMPLING DISTRIBUTIONS

In actual practice, random samples seldom consist of single observations. Almost always some N observations are drawn from the same population. Furthermore, the value of some statistic is associated with the sample. Interest then lies in the distribution of values of this statistic *across all possible samples* of N observations from this population. Accordingly, we must distinguish still another kind of theoretical distribution, called a sampling distribution.

A sampling distribution is a theoretical probability distribution that shows the functional relation between the possible values of a given statistic based on a sample of N cases and the probability (density) associated with each value, for all possible samples of size N drawn from a particular population.

In general, the sampling distribution of values for a particular sample statistic will not be the same as the distribution of the random variable for the population. However, the sampling distribution always depends in some specifiable way upon the population distribution, provided the probability structure underlying the occurrence of samples is known.

Notice that this definition is not confined to simple random samples, even though in most applications it will be assumed that samples are drawn at random from the population. Nevertheless, some probability structure linking the occurrence of the possible samples with the population must exist and be known if the population distribution is to be related to the sampling distribution of any statistic. For our elementary purposes this probability structure will be that of simple random sampling, in which each possible sample of size N has exactly the same probability of occurrence as any other. However, in more advanced work, assumptions other than simple random sampling are sometimes made.

Actually, we have already used sampling distributions in the preceding chapters. For example, a binomial distribution is a sampling distribution. Recall that a binomial distribution is based on a two-category population distribution, or Bernoulli process. A sample of N independent cases is drawn at random from such a distribution, and the number (or proportion) of successes is calculated for each sample. Then the binomial distribution is the sampling distribution showing the relation between each possible sample result and the theoretical probability of occurrence. The binomial distribution is *not* the same as the Bernoulli process unless N is 1; however, given the Bernoulli process and the size of the sample N, the binomial distribution may be worked out.

Other examples of sampling distributions will now be given. A most important distribution we shall employ is the sampling distribution of the mean. Here, samples of N cases are drawn independently and at random from some population and each observation is measured numerically. For each sample drawn the sample mean $\bar{x}$ is calculated. **The theoretical distribution that relates the possible values of the sample mean to the probability (density) of each over all possible samples of size N is called the sampling distribution of the mean.**

Furthermore, for each sample of size N drawn, the sample variance S^2 may be found. The theoretical distribution of sample variances in relation to the probability of each is the sampling distribution of the variance. By the same token, the sampling distribution of any *summary characteristic* (mode, median, range, etc.) of samples of N cases may be found, given the population distribution and the sample size N.

5.3 CHARACTERISTICS OF SINGLE-VARIATE SAMPLING DISTRIBUTIONS

A sampling distribution is a theoretical probability distribution, and like any such distribution, is a statement of the functional relation between the values or intervals of values of some random variable and probabilities. Sampling distributions differ from population distributions in that the random variable is always the *value of some statistic* based on a sample of N cases, such as the sample mean, sample variance, or sample median, etc. Thus, a plot of a sampling distribution, such as that in Figure 5.3.1, always has for the abscissa (or horizontal axis) the different sample statistic values that might occur. Figure 5.3.1, for example, shows a theoretical distribution for sample variances for all possible samples of size 7 drawn from a particular population. Any point on the horizontal axis is a possible value of a sample variance, and the height of the curve on the vertical axis gives the probability density $f(S^2)$, for that particular value.

Like population distributions, sampling distributions may be either continuous or discrete. The binomial distribution is discrete, although in applied problems it is sometimes treated as though it were continuous. Most of the commonly encountered sampling distributions based on a continuous population distribution will be continuous.

Since a sample statistic is a random variable, the mean and variance of any sampling distribution are defined in the usual way. That is, let G be any sample statistic; then if the sampling distribution of G is discrete, its expectation or mean over all values g is

$$E(G) = \mu_G = \sum_g gp(g) \qquad [5.3.1]$$

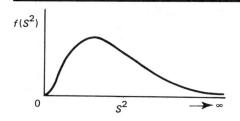

Figure 5.3.1

A theoretical sampling distribution of S^2.

(see Section 4.6). If the variable G is continuous, then

$$E(G) = \mu_G = \int_{-\infty}^{\infty} gf(g)\, dg,$$

just as for any other continuous random variable (Section 4.6).

In the same way, the variance of a sampling distribution for any statistic G can be defined:

$$\sigma_G^2 = E(G - \mu_G)^2 \qquad\qquad [5.3.2]$$

or

$$\sigma_G^2 = E(G^2) - [E(G)]^2. \qquad\qquad [5.3.3^*]$$

The standard deviation σ_G of the sampling distribution reflects the extent to which sample G values tend to be unlike the expectation, or are *in error*. To aid in distinguishing the standard deviation of a sampling distribution from the standard deviation of a population distribution, a standard deviation such as σ_G is usually called the *standard error* of the statistic G:

standard error of $G = \sigma_G$. $\qquad\qquad [5.3.4^*]$

Thus, the standard error of the mean, σ_M is identical to the standard deviation of the distribution of possible sample means, for all possible samples of size N drawn from a specified population. Similar meanings hold for the standard error of the median, the standard error of the standard deviation, the standard error of the range, and so on.

5.4 SAMPLE STATISTICS AS ESTIMATORS

Some population parameters have obvious parallels in sample statistics. The population mean μ has its sample counterpart in $\bar{x}$, the variance σ^2 in the sample variance S^2, the population proportion p in the sample proportion P, and so on. On the other hand, the relationship between population parameters and sample statistics is not necessarily 1 to 1. It is entirely possible for a population distribution to depend upon parameters that are not directly paralleled in one of the common sample statistics, and for sample statistics to be used that are not direct parallels of population parameters.

It is true, however, that a sample of cases drawn from a population contains information about the population distribution and its parameters. Furthermore, a statistic computed from the data in the sample contains some of that information. Some statistics contain more information than others, and some statistics may contain more information about certain parameters than about others.

A central problem of inferential statistics is **point estimation,** the use of the value of some statistic to infer the value of a population parameter. The value of some statistic (or point in the "space" of all possible values) is taken as the "best estimate" of the value of some parameter of the population distribution.

How does one go from a sample statistic to an inference about the population parameter? In particular, which sample statistic does one use, if it is to give an estimate that is in some sense "best"?

The fact that the sample represents only a small subset of observations drawn from a much larger set of potential observations makes it nearly impossible to say that any estimate is exactly like the population value. As a matter of fact they very probably will not be the same, as all sorts of different factors of which we are in ignorance may make the sample a poor representation of the population. Such factors we lump together under the general rubrics *chance* or *random effects*. In the long run such samples should reflect the population characteristics. However, practical action can seldom wait for "in the long run"; things must be decided here and now in the face of limited evidence. We need to know how to use the available evidence in the best possible ways to infer the characteristics of the population.

Various statistics differ in the information they provide about population parameters. They also differ in the extent to which this is "good" information, that can be used to estimate the value of the parameter in question. We are now going to examine some statistics in terms of their properties as estimators.

In the following, it is good to keep in mind the difference between an "estimator" of some population parameter and an "estimate" of the value of that parameter. An estimator is a formula or method for combining the values occurring in the data. Hence an estimator is a random variable, which takes on values dependent upon the sample data. On the other hand, the particular value that results from the application of that formula is an estimate of the population parameter in question. Viewed as a random variable, whose value is arrived at in a certain way from the sample data, the sample mean is an estimator. On the other hand, a particular value of the sample mean, based on a particular sample, is an estimate. Our immediate interest will focus on the properties of statistics as estimators—that is, as random variables with certain characteristics.

5.5 THE PRINCIPLE OF MAXIMUM LIKELIHOOD

Statisticians have established a number of criteria for choosing among the statistics that might be used as estimators. One of these is the very important **principle of maximum likelihood,** which will be discussed next. Even though a formal presentation of the maximum-likelihood principle is far beyond our scope, some intuitive feel for this idea may help you understand the methods that follow.

Basically, the problem of the person using statistics is, "Given several possible population situations, any one of which might be true, how shall I bet on the basis of the evidence so as to be as confident as possible of being right?" The user of statistical inference knows that any number of things might be true of the population. Fortunately, one can "snoop" on the population to a certain extent by taking a sample, but the evidence gained will almost certainly be faulty. Since this evidence is all there is, the person must use it nevertheless to form an opinion or make a decision. How does one decide on the basis of evidence that is probably erroneous? The principle of maximum likelihood gives a general strategy for such decisions, and may be paraphrased as follows.

Suppose that a random variable X has a distribution that depends only upon some population parameter θ (symbolized by lowercase Greek theta). The form of the density function will be assumed known, but not the value of θ. A sample of N independent observations is drawn, producing a set of values $(x_1, x_2 \cdots, x_N)$. Let $L(x_1, \cdots, x_N; \theta)$ represent the likelihood, or probability (density), of this particular sample result *given* θ. For each possible value of θ, the likelihood of the sample result will be different, perhaps. Then, **the principle of maximum likelihood requires us to choose as our estimate that possible value of θ making $L(x_1 \cdots x_N; \theta)$ take on its largest value.**

In effect this principle says that when faced with several parameter values, any of which might be the true one for a population, the best "bet" is that parameter value which *would* have made the sample actually obtained have the highest prior probability. When in doubt, place your bet on that parameter value which would have made the obtained result most likely.

This principle may be illustrated very simply for the binomial distribution. Suppose that a sample is drawn at random from a population of college graduates. Each graduate is classified either as a "liberal arts major" or as a graduate from some other university college. Now three possibilities or hypotheses are entertained about the proportion of college graduates from liberal arts colleges. Hypothesis 1 states that .5 of all graduates are from such colleges, Hypothesis 2 states that .4 of the graduates are from such schools, and Hypothesis 3 states that .6 are liberal arts graduates.

Now suppose that some 15 college graduates are drawn at random and classified according to "liberal arts degree" versus "other degree." The result is that 9 out of the 15 hold a liberal arts college degree (we are supposing that no one in the population holds more than one degree). What decision about the three available hypotheses should we reach? This is a simple binomial sampling problem, so that the probability of each sample result may be calculated for each of the three hypothetical values of the parameter p: $p = .5$, $p = .4$, or $p = .6$.

If $p = .5$ the prior probability of a sample result such as that obtained would be

$$\binom{15}{9} (.5)^9 (.5)^6 = .153.$$

For $p = .4$ the prior probability becomes

$$\binom{15}{9} (.4)^9 (.6)^6 = .061.$$

Finally, if p were .6 the prior probability of 9 successes out of 15 would be

$$\binom{15}{9} (.6)^9 (.4)^6 = .207.$$

The use of the principle of maximum likelihood to decide among these three possibilities leads to the choice of hypothesis 3, that .6 is the population proportion, since this is the parameter value among the possibilities considered that would have made the obtained sample result most likely, a priori.

This principle has a great deal of use in theoretical statistics, since general methods exist for finding the value of θ that maximizes the likelihood of a sample result. Statistics chosen as estimators because their values substituted for the parameter maximize the likelihood of the sample result are called **maximum-likelihood estimators.** For example, if there were no prior information at all about the number of liberal arts graduates, so that any number between 0 and 1.00 might be entertained as a hypothesis about the value of p, the maximum-likelihood estimate of p would be the sample P, or 9/15, since among all possible values of p this value makes the occurrence of the actual result have greatest a priori likelihood (the student may check a few values of p for himself to see that this is true).

The principle of maximum likelihood also provides a fairly routine way of finding estimators having "good" properties. However, the principle of maximum likelihood is introduced here not only because of its importance in estimation, but also because of the general point of view it represents about inference. This point of view is that true population situations should be those making our empirical results *likely;* if a theoretical situation makes our obtained data have *very low prior likelihood* of occurrence, then *doubt* is cast on the truth of the theoretical situation. Theoretical propositions are believable to the extent that they accord with actual observation. If a particular result should be very unlikely given a certain theoretical state of affairs, and we do get this result nevertheless, then we are led back to examine our theory. Good theoretical statements accord with observation by giving predictions having a high probability of being observed.

Naturally, the results of a single experiment, or even of any number of experiments, cannot prove or disprove a theory. Replications, variants, different ways of measuring the phenomena, must all be brought into play. Even then, proof or disproof is never absolute. Nevertheless, the principle of maximum likelihood is in the spirit of empirical science, and it runs throughout the methods of statistical inference.

5.6 OTHER DESIRABLE PROPERTIES OF ESTIMATORS

Since there are many ways for devising a sample statistic for estimating a population parameter's value, several criteria are used for judging how effectively a given statistic serves this purpose. As we have just seen, some statistics have the desirable property of being the maximum-likelihood estimator of a particular parameter. In addition, good estimators should be *unbiased, consistent,* and *relatively efficient,* and a set of estimators used for estimating a set of parameters should be *sufficient.* A brief description of each of these criteria follows.

Imagine, once again, a population distribution depending only on some parameter θ. Some sample statistic G is to be used as an estimator of θ. Then G is said to be an *unbiased* estimator of θ if the expected value of G is *exactly* θ:

$$E(G) = \theta.$$

That is, when the value of G is an *unbiased* estimate of θ, then in the long run the average value of G over all possible random samples is exactly the value of θ.

As we shall see in subsequent sections, the sample mean $\bar{x}$ is an unbiased estimator of the population mean μ. Furthermore, under binomial sampling, the sample proportion P is an unbiased estimator of the population proportion p. On the other hand, the sample variance S^2 is an example of a biased estimator, since $E(S^2)$ is not, in general, equal to the population variance σ^2. (A simple correction exists to solve this problem, as we shall see.)

Another desirable property of an estimator is *consistency*. Roughly speaking, this means that the larger the sample size N, the higher the probability that G comes *close* to the population value θ. Statistics that have this property are called *consistent estimators*. The sample mean, the sample variance, and many other common statistics are consistent estimators, as they tend in likelihood to be closer to the population value as the sample size increases. However, not all possible statistics meet this criterion. For example, suppose that we wished to estimate a population mean, and that we chose to use the score of the second case observed in any sample as our estimate of the population mean. Obviously, by the criterion of consistency this is not a good way to make an estimate. For an N of 2 or more, the probability of coming close to the population value does not increase as we increase the sample size. (Notice that this method does give an unbiased estimate, however.)

A third criterion for choosing an estimator is called *relative efficiency*. Imagine the population depending on the parameter θ, once again. This time, however, suppose that there were two statistics, G and H, that might be computed from any sample of size N, and that both G and H are unbiased estimators of θ. Given a constant N, the sampling distribution of G will have some variance of σ_G^2 and the sampling distribution of H will have some variance σ_H^2. Then

$$\frac{\sigma_H^2}{\sigma_G^2} = \text{efficiency of } G \text{ relative to } H.$$

The *more efficient* estimator has the *smaller sampling variance*. If G is more efficient then, on the average, the sampling error of G is smaller than it is for H, and thus one tends to make a smaller error in using G rather than H in estimating θ. As we shall see, one of the reasons for preferring the mean to the median is that when the population is of a "normal" type, and both are unbiased estimates of the population mean μ, the mean is relatively more efficient than the median, given the same sample size N.

Still another concept of major importance in the modern theory of statistics is that of *sufficiency*. Once again consider a random variable X which depends only on one parameter θ, and let G be a statistic which is computed from N sample values of X. Then we say that G is a *sufficient* estimator (or a sufficient statistic) if G contains *all* of the information in the data about the parameter θ. That is, if G is a sufficient statistic, our estimate of θ cannot be improved by considering any other aspect of the data not already included in G itself.

In some population distributions where there may be more than one parameter required to specify the distribution, then two or more statistics may be required for sufficiency. In these instances one refers to the set of sufficient statistics, rather than to a single sufficient estimator.

Sufficient statistics do not always even exist, and situations can be constructed in which no sufficient set of estimators can be found for a set of parameters. Nevertheless, sets of sufficient estimators, when they do exist, are important, since if one can find a set of sufficient estimators, then it is ordinarily possible to find unbiased and efficient estimators based on that sufficient set. In particular, when a set of suffcent statistics exists, then the maximum-likelihood estimators will be based upon that set.

In most of the work to follow, the estimators will be the sample mean and the (corrected) sample variance, both of which will fulfill these criteria for good estimators in the particular situations where they will be used. This does not imply, however, that other estimators failing to meet one or more of these criteria are useless. In particular, special situations exist where other methods of estimating central tendency may be better than either mean or median, and variance estimates other than S^2 are called for. Other statistics are useful on occasion, but we will focus most attention on the mean and the corrected variance, since they occupy a central place in the classical statistical methods we will be treating.

5.7 THE SAMPLING DISTRIBUTION OF THE MEAN

By the criteria just listed, the sample mean comes off very well as an estimator of its population counterpart μ. In general, as an estimate of μ, the sample mean is unbiased, is consistent, and, in an important set of circumstances, is both efficient relative to other statistics and, taken with S^2, sufficient.

It is simple to show that the sample mean must be an unbiased estimator, or that

$$E(\bar{x}) = \mu.$$

By definition, for any N independent observations,

$$\bar{x} = (x_1 + x_2 + \cdots + x_N)/N,$$

so that

$$E(\bar{x}) = E(x_1 + x_2 \cdots + x_N)/N.$$

By Rule 5 of Appendix B, it must be true that

$$E(\bar{x}) = [E(x_1) + E(x_2) + \cdots + E(x_N)]/N.$$

For any given observation, $E(x) = \mu$, so that

$$E(\bar{x}) = N\mu/N = \mu.$$

The expected value of the sample mean is the population mean μ. This same statement can be interpreted in another way: **the mean of the sampling distribution of means is the same as the population mean.**

However, it is not true that the sampling distribution of the mean will be the same as the population distribution; in fact, these distributions will ordinarily be quite different, depending particularly on sample size. In the first place, the variance of the sampling distribution of means will not be the same as σ^2 but instead will be smaller than σ^2 for samples of size 2 or larger. This will be shown next.

Intuitively it seems quite reasonable that the larger the sample size the more confident we may be that the sample mean is a close estimate of μ. When it was said that the mean is a consistent estimator, this was another way of saying that the mean is a better estimate of μ for large rather than for small samples. Now we can put that intuition on a firm basis by looking into the effect of sample size on the variance and standard deviation of the distribution of sample means.

By definition, the variance of the distribution of means, from samples drawn at random from a population with mean μ and variance σ^2, is

$$\sigma_M^2 = E(\bar{x}^2) - [E(\bar{x})]^2$$

$$= E(\bar{x}^2) - \mu^2.$$

[5.7.1*]

Bear in mind that we are assuming that the sample mean is based on N *independent* observations. Let us call any pair of these observations i and j, with x_i, and x_j. The square of the sample mean is

$$\bar{x}^2 = \frac{(x_1 + x_2 + \cdots + x_N)^2}{N^2}$$

$$= \frac{1}{N^2} (x_1^2 + \cdots + x_N^2 + 2 \sum_{i<j} x_i x_j),$$

the sum of the squared scores, plus twice the sum of the products of all pairs of scores, all divided by N^2. For any single observation i,

$$E(x_i^2) = \sigma^2 + \mu^2$$

[5.7.2]

since, from Section 4.16,

$$\sigma^2 = E(x_i^2) - \mu^2.$$

[5.7.3]

For a pair of *independent* observations, i and j, Rule 9 in Appendix B gives

$$E(x_i x_j) = E(x_i)E(x_j) = \mu^2.$$

[5.7.4]

Thus, putting these two facts (5.7.2 and 5.7.4) together, we have

$$E(\bar{x}^2) = \frac{1}{N^2} [E(x_1^2) + E(x_2^2) + \cdots + 2 \sum_{i<j} E(x_i x_j)]$$

$$= \frac{N\sigma^2 + N\mu^2 + (N)(N - 1)\mu^2}{N^2}$$

$$= \frac{\sigma^2}{N} + \mu^2.$$

Making this substitution in Eq. 5.7.1 we find

$$\sigma_M^2 = E(\bar{x}^2) - \mu^2 = \frac{\sigma^2}{N}.$$

The variance of the sampling distribution of means for independent samples of size N is always the population variance divided by the sample size, σ^2/N.

This is a most important fact, and gives direct support to our feeling that large samples produce better estimates of the population mean than do small. When the sample size is only 1, then the variance of the sampling distribution is exactly the same as the population variance. If however, the sample mean is based on two cases, $N = 2$, then the sampling variance is only 1/2 as large as σ^2. Ten cases give a sampling distribution with variance only 1/10 of σ^2, $N = 500$ gives a sampling distribution with variance 1/500 of σ^2, and so on. If the sample size approaches infinity, then σ_M^2 approaches zero.

If the sample is large enough to embrace the entire population, there is no difference between the sample mean and μ.

In general, the larger the sample size, the more probable it is that the sample mean comes arbitrarily close to the population mean. This fact is often called **the law of large numbers.**

5.8 STANDARDIZED SCORES CORRESPONDING TO SAMPLE MEANS

The standard error of the mean is

$$\sigma_M = \sqrt{\sigma_M^2} = \frac{\sigma}{\sqrt{N}}, \qquad [5.8.1*]$$

so that when a sample mean is put into standard form, we have

$$z_M = \frac{\bar{x} - \mu}{\sigma_M} = \frac{\bar{x} - \mu}{\sigma/\sqrt{N}}. \qquad [5.8.2*]$$

The larger the absolute value of the standard score of a mean, relative to μ and the standard error, the *less likely* is one to observe a mean this much or more deviant from μ. Any given degree of departure of $\bar{x}$ from μ corresponds to a larger absolute standard score value, and hence a less probable class of result, the larger the sample size N.

For example, suppose that in sampling from some population a sample mean was found differing by 10 points from the population mean. Suppose that σ were 5 and the sample size were 2. Then the standard score z_M would be

$$z_M = \frac{10}{5/\sqrt{2}} = 2.83,$$

disregarding sign. The Tchebycheff inequality (Section 4.18) tells us that means that are this deviant or more so from expectation can occur with probability no greater than about $1/(2.83)^2$ or .125.

Suppose, however, that the sample size had been 200. Here

$$z_M = \frac{10}{5/\sqrt{200}} = 28.28$$

so that a sample $\bar{x}$ deviating this much or more from expectation could occur only with probability no greater than about $1/(28.28)^2$ or .0013. Notice that knowing the standard error of the mean is essential if we are going to judge the agreement between a sample and a population mean in terms of the probability of a given extent of deviation. For the moment we are assuming that this information is simply given to us.

5.9 CORRECTING THE BIAS IN THE SAMPLE VARIANCE AS AN ESTIMATOR

It has already been mentioned that the sample variance S^2 is a biased estimate of the population variance σ^2, since

$$E(S^2) \neq \sigma^2. \qquad [5.9.1]$$

This can be demonstrated as follows: the expectation of a sample variance is

$$E(S^2) = E\left(\frac{\sum_i x_i^2}{N} - \bar{x}^2\right) = E\left(\frac{\sum_i x_i^2}{N}\right) - E(\bar{x}^2).$$

Let us consider the two terms on the extreme right separately. By Rules 5 and 2 for expectations,

$$E\left(\frac{\sum_i x_i^2}{N}\right) = \frac{\sum_i E(x_i^2)}{N}.$$

From the definition of the variance of the population,

$$\sigma^2 = E(x^2) - \mu^2$$

so that

$$E(x_i^2) = \sigma^2 + \mu^2 \qquad\qquad [5.9.2]$$

for any observation i. Thus

$$E\left(\frac{\sum_i x_i^2}{N}\right) = \frac{\sum_i (\sigma^2 + \mu^2)}{N} = \sigma^2 + \mu^2. \qquad\qquad [5.9.3]$$

Now the variance of the sampling distribution of means is, from Eq. 5.7.1,

$$\sigma_M^2 = E(\bar{x}^2) - \mu^2$$

so that

$$E(\bar{x}^2) = \sigma_M^2 + \mu^2. \qquad\qquad [5.9.4]$$

Putting these two results (Eqs. 5.9.3 and 5.9.4) together, we have

$$E(S^2) = \sigma^2 - \sigma_M^2, \qquad\qquad [5.9.5^*]$$

the expectation of the sample variance is the *difference* between the population variance σ^2 and the variance of the sampling distribution of means σ_M^2. In general this difference will not be the same as σ^2, since ordinarily the variance σ_M^2 will not be zero, and so the sample variance is biased as an estimator of the population variance. In particular, the sample variance is, on the average, smaller than the population variance σ^2.

We have just shown in Section 5.7 that

$$\sigma_M^2 = \frac{\sigma^2}{N}.$$

On making this substitution into Eq. 5.9.5, we have

$$E(S^2) = \sigma^2 - \frac{\sigma^2}{N} = \left(\frac{N-1}{N}\right)\sigma^2, \qquad\qquad [5.9.6]$$

so that on the average the sample variance is *too small* by a factor of $\dfrac{N-1}{N}$.

Since this is true, a way emerges for correcting the variance of a sample to make it an unbiased estimator. **The unbiased estimate of the variance based on any sample of N independent cases is**

$$s^2 = \frac{N}{N-1} S^2. \tag{5.9.7*}$$

(In all that follows, we will reserve the symbol S^2 to stand for the uncorrected variance of a sample, as originally defined, and use the lowercase letter s^2 to indicate the corrected variance. Similarly, S will stand for the standard deviation based upon S^2, and s for the standard deviation based upon s^2.)

It is simple to show that s^2 is indeed unbiased:

$$E(s^2) = \frac{N}{N-1} E(S^2) = \left(\frac{N}{(N-1)} \right) \left(\frac{(N-1)}{N} \right) \sigma^2 = \sigma^2. \tag{5.9.8*}$$

Quite often it is convenient to calculate the unbiased variance estimate s^2 directly, without the intermediate step of calculating S^2. This is done either by the formula

$$s^2 = \frac{\sum_i x_i^2 - N \bar{x}^2}{N-1} \tag{5.9.9*}$$

or by

$$s^2 = \frac{\sum_i x_i^2}{N-1} - \frac{\left(\sum_i x_i \right)^2}{N(N-1)}. \tag{5.9.10*}$$

Incidentally, some modern statistics texts completely abandon the idea of the sample variance S^2 as used here, and introduce only the unbiased estimate s^2 as *the* variance of a sample. However, this is apt to be confusing in some work, and so we will follow the older practice of distinguishing between the sample variance S^2 as a descriptive statistic, and s^2 as the unbiased estimate of σ^2.

Even though we will be using the square root of s^2, or s, to estimate the population σ, it should be noted that s is not *itself* an unbiased estimate of σ, and that

$$E(s) \neq \sigma$$

in general. The correction factor used to make s an unbiased estimate of σ depends on the form of the population distribution; thus, for the unimodal symmetric, normal, distribution (Chapter 6) an unbiased estimate for large N is provided by

$$\text{unbiased estimate of } \sigma = \left[1 + \frac{1}{4(N-1)} \right] s. \tag{5.9.11}$$

Furthermore, special tables exist for correcting the estimate of σ for relatively small samples from such populations (Dixon & Massey, 1969). However, the problem of estimating σ from s is bypassed, in part, by the methods we will use in Chapter 9 and elsewhere, and if the sample size is reasonably large, the amount of bias in s as an estimator of σ is ordinarily rather small. For these reasons, we will not trouble to correct for the bias in s found from the *unbiased* estimate s^2 of σ^2.

One estimates the standard error of the mean by using the unbiased estimate s^2.

$$\text{estimated } \sigma_M = \frac{\text{estimated } \sigma}{\sqrt{N}} = \sqrt{\frac{s^2}{N}}.$$

[5.9.12*]

On the other hand, if the sample variance (not the unbiased estimate) is used, then the estimate may be arrived at as follows:

$$\text{estimated } \sigma_M = \sqrt{\frac{s^2}{N}} = \sqrt{\left(\frac{N}{(N-1)}\right)\frac{s^2}{N}} = \frac{S}{\sqrt{N-1}}.$$

[5.9.13*]

5.10 PARAMETER ESTIMATES BASED ON POOLED SAMPLES

Sometimes it happens that one has several independent samples where each provides an estimate of the same parameter or set of parameters. The most usual situation occurs when one is estimating μ or σ^2, or both. When this happens there is a real advantage in pooling the sample values to get an unbiased estimate. These pooled estimates are actually weighted averages of the estimates from the different samples. The big advantage lies in the fact that the sampling error will tend to be smaller for the pooled estimate than that for any single sample's value taken alone.

Suppose that there were two independent samples, based on N_1 and N_2 observations respectively. From each sample a mean $\bar{x}$ is calculated, each of which estimates the same value μ. Then the pooled estimate of the mean is

$$\text{pooled } \bar{x} = \text{est. } \mu = \frac{N_1\bar{x}_1 + N_2\bar{x}_2}{N_1 + N_2}.$$

[5.10.1*]

It is easy to see that this must give an unbiased estimate of μ:

$$E\left[\frac{N_1\bar{x}_1 + N_2\bar{x}_2}{N_1 + N_2}\right] = \frac{N_1E(\bar{x}_1) + N_2E(\bar{x}_2)}{N_1 + N_2}$$

$$= \frac{\mu(N_1 + N_2)}{N_1 + N_2} = \mu.$$

If there were three samples of sizes N_1, N_2, and N_3, giving means $\bar{x}_1$, $\bar{x}_2$, and $\bar{x}_3$, then

$$\text{pooled } \bar{x} = \text{est. } \mu = \frac{N_1\bar{x}_1 + N_2\bar{x}_2 + N_3\bar{x}_3}{N_1 + N_2 + N_3}.$$

This same idea applies to any number of samples. The fact that the pooled estimate is likely to be better than the single sample estimates taken alone is shown by the standard error of the pooled mean. For two independent samples, each composed of independent observations drawn from the same population,

$$\sigma_M = \frac{\sigma}{\sqrt{N_1 + N_2}},$$

[5.10.2*]

which *must* be smaller than the standard error either of $\bar{x}_1$ or of $\bar{x}_2$. Similarly, for a pooled value of $\bar{x}$ found from three independent samples,

$$\sigma_M = \frac{\sigma}{\sqrt{N_1 + N_2 + N_3}}, \qquad [5.10.3]$$

and so on for any number of independent samples.

Naturally, these standard errors refer to the situation where either all of the respective samples are drawn from the same population, or where they all come from populations having the same mean and the same population variance σ^2.

In the same general way, it is possible to find pooled estimates of σ^2, for populations with the same variance σ^2, even though the population means may be different. Thus for two independent samples from normal distributions

$$\text{est. } \sigma^2 = \frac{(N_1 - 1)s_1^2 + (N_2 - 1)s_2^2}{(N_1 - 1) + (N_2 - 1)}, \qquad [5.10.4^*]$$

which is an unbiased estimate of the variance σ^2 of each of the populations.

5.11 SAMPLING FROM RELATIVELY SMALL POPULATIONS

It has been mentioned repeatedly that most sampling problems deal with populations so large that the fact that samples are taken without replacement of single cases can safely be ignored. However, it may happen that the population under study is not only finite, but relatively small, so that the process of sampling without replacement has a real effect on the sampling distribution.

Even in this situation the sample mean is still an unbiased estimate regardless of the size of the population sampled. Hence no change in procedure for mean estimation is needed.

However, for finite populations the sample variance is biased as an estimator of σ^2 in a way somewhat different from that for an infinite population. When samples are drawn without replacement of individuals, the unbiased estimate of σ^2 is

$$\text{est. } \sigma^2 = \left(\frac{N(T - 1)}{(N - 1)T}\right)s^2$$

$$= \left(\frac{(T - 1)}{T}\right)s^2, \qquad [5.11.1]$$

where T is the *total number* of elements or individuals in the population.

Another difference from an infinite population comes with the variance of the sampling distribution of means. For a population with T cases in all, from which samples of size N are drawn, the sampling variance of the mean is

$$\sigma_M^2 = \left(\frac{T - N}{T - 1}\right)\frac{\sigma^2}{N}. \qquad [5.11.2]$$

The variance of the mean tends to be *somewhat smaller* for a fixed value of N when sampling is from a finite population than when it is from an infinite population. Note that here the size of σ_M^2 depends both on T, the total number in the population, and on N, the sample size. An unbiased estimate of the variance of the mean is thus given by

$$\text{est. } \sigma_M^2 = \left(\frac{T - N}{T}\right) \frac{s^2}{N}. \tag{5.11.3}$$

The square root of this value gives the estimate of σ_M, for sampling from a finite population.

Once this adjustment to the estimated value of σ_M has been made, any of the inferential methods calling for this value may be employed in the usual way.

5.12 THE IDEA OF INTERVAL ESTIMATION

So far, our discussion of estimation of parameters from statistics has been confined to the subject of "point estimation," making an estimate of a parameter value in terms of the value of a sample statistic. Thus, if a sample of 50 second-grade schoolchildren drawn at random from a large city school system shows a mean arithmetic achievement score of 92.7, we estimate the true value of μ, the mean of all second-grade schoolchildren in that school system, also to be 92.7.

Nevertheless, it is clear that the sample mean ordinarily will not be exactly equal to the true population value because of sampling error. It is necessary to qualify our estimate in some way to indicate the general magnitude of this error. Usually this is done by showing a **confidence interval,** which is **an estimated range of values with a given high probability of covering the true population value.** When there is a large degree of sampling error the confidence interval calculated from any sample will be large; the range of values likely to cover the population mean is wide. On the other hand, if sampling error is small the true value is likely to be covered by a small range of values; in this case one can feel confident that the population value has been "trapped" within a small range of values calculated from the sample.

The general idea is to find two values, say a and b, such that the chances are very good that the interval of values from a to b actually does include the population value. Such an interval of values is called a confidence interval.

One rather primitive way to establish an approximate confidence interval for the mean of a population, given a sample mean, is by use of the Tchebycheff inequality (Section 4.18). We know from Eq. 4.18.2 that for any probability distribution it must be true that for any positive number k

$$\text{prob}\left\{\left|\frac{(X - \mu)}{\sigma}\right| \geq k\right\} \leq \frac{1}{k^2}.$$

Since the sampling distribution of the mean is such a probability distribution of the random variable $\bar{x}$, it must be true that

$$\text{prob}\left\{\left|\frac{(\bar{x} - \mu)}{\sigma_M}\right| \geq k\right\} \leq \frac{1}{k^2}.$$

Then, it must also be true that

$$\text{prob}\left\{-k \le \frac{(\bar{x} - \mu)}{\sigma_M} \le k\right\} \ge 1 - \frac{1}{k^2}$$

which is equivalent to

$$\text{prob}\,(\bar{x} - k\sigma_M \le \mu \le \bar{x} + k\sigma_M) \ge 1 - \frac{1}{k^2}. \qquad [5.12.1]$$

For any sampling distribution of means, the probability is at least $1 - (1/k^2)$ that the interval of values $\bar{x} - k\sigma_M$ to $\bar{x} + k\sigma_M$ covers the true population value μ, where k is any arbitrary positive value.

For example, what is the probability that the true mean μ lies in an interval $\bar{x} - 3\sigma_M$ to $\bar{x} + 3\sigma_M$, as found from any sample mean $\bar{x}$? Here, $k = 3$, so that the probability must be at least $1 - (1/9)$ or .89 that this interval covers the true value of μ.

To go further, suppose that we are sampling from a population where μ is unknown, but where we know that the population standard deviation is 20. There are $N = 50$ cases in each sample. Now we draw a sample, where the mean turns out to be 124. Then, since the standard error of the mean must be

$$\sigma_M = \frac{\sigma}{\sqrt{N}}$$

we find that

$$\sigma_M = \frac{20}{\sqrt{50}} = \frac{20}{7.07} = 2.83.$$

Now let us find the probability that the true mean is covered by the interval of values extending from 3 standard errors below the sample mean to 3 standard errors above. That is, we want the probability that the interval from

$$\bar{x} - 3\sigma_M = 124 - 3(2.83) = 115.51$$

to

$$\bar{x} + 3\sigma_M = 124 + 3(2.83) = 132.49$$

covers the value of μ. From Eq. 5.11.1 we know this to be greater than or equal to .89. We can then make the following statement: The probability is at least .89 that the interval 115.51–132.49 covers the value of the population mean.

Although the best estimate of the population value μ is, of course, the obtained sample mean $\bar{x}$, the confidence interval based on any sample can be thought of as a range of ''good'' estimates to either side of $\bar{x}$. Be sure to notice that the probability statement made in connection with a confidence interval *does not* refer to the probability of μ, but rather to the probability of a *sample*. In drawing samples, one may decide beforehand to compute a confidence interval for each one. The actual range of numbers making up the confidence interval on a given occasion will depend on the sample mean, $\bar{x}$. For some samples, the confidence interval actually will cover the value of μ; for others, it will not. However, the Tchebycheff inequality lets us say what the minimum probability must be that a given confidence interval covers the value of μ.

Now suppose that instead of drawing samples of 50 observations, we take a sample of 600. Once again, imagine that the population has a standard deviation of 20, and that the actual sample obtained gave a mean of 124. In this circumstance, if we calculate the standard error of the mean we get

$$\sigma_M = \frac{20}{\sqrt{600}} = \frac{20}{24.49} = .82.$$

Now, if we take 3 standard errors to either side of the sample mean we find an interval

$$124 - 3(.82) \text{ to } 124 + 3(.82)$$

or

$$121.54 \text{ to } 126.46.$$

We can state that the probability is at least .89 that the interval 121.54–126.46 covers the true value of μ.

Notice in this second example that we have pinned the value of the population mean within a much narrower range of values. This is because the sample size is so much larger in the second example, so that the standard error of the mean is much smaller. The use of the Tchebycheff inequality gives only a very "rough and ready" sort of confidence interval. This method can be improved to a certain extent if we can assume that the sampling distribution of means is unimodal and symmetric. As we shall see in the next chapter, this is an entirely reasonable assumption in the case of means based on samples of at least moderate size ($N \geq 20$), almost regardless of the population distribution. Then the following statement is true:

$$p\,(\bar{x} - 3\sigma_M \leq \mu \leq \bar{x} + 3\sigma_M) \geq .95. \qquad [5.11.2]$$

That is, the probability is at least .95 that the interval $\bar{x} - 3\sigma_M$ to $\bar{x} + 3\sigma_M$ covers the true value of the population mean, for random samples of N independent observations drawn from some population. An approximate "95% confidence interval" for the mean can be constructed for moderately large samples by substracting and adding $3\sigma_M$ to the obtained value of $\bar{x}$. Notice that here we assume nothing about the form of the population distribution, and for the sampling distribution of $\bar{x}$ we assume only that it is unimodal and symmetric. Later we will employ a principle permitting a much narrower confidence interval to be found.

In this situation, where the population σ is known, an analogy to the idea of finding confidence intervals for the mean is tossing rings of a certain size at a post. The size of the confidence interval is like the diameter of a ring, and random sampling is like random tosses at the post. The predetermined size of the ring determines the chances that the ring will cover the post on a given try, just as the size of the confidence interval governs the probability that the true value μ will be covered by the range of values in the estimated interval.

The width of the confidence interval calculated from any sample depends on three things: how many standard errors to either side of the sample mean one chooses to go (i.e., the value of k), the population standard deviation, and the size of the sample.

If, instead of going 3 standard errors to each side of $\bar{x}$ in this last example, we had chosen to take 4 standard errors in either direction, then the following statement would have been true:

$$p\ (\bar{x}\ -\ 4\sigma_M \leq \mu \leq \bar{x}\ +\ 4\sigma_M) \geq .97.$$

That is, instead of being at least .95, the probability would be at least .97 that the interval calculated covers the true population mean. This interval would have been $124\ -\ 4.(.82)$ to $124\ +\ 4(.82)$, or $120.72–127.28$. This interval is a bit bigger than the interval found before. A larger confidence interval permits one to be more certain that the population value is covered; however, a larger interval introduces more uncertainty about what the population value actually is, since there are more possibilities in the interval.

The larger the standard deviation of the population, the larger will the confidence interval be, other things being equal. The spread of possibilities in the population is reflected in the range of possibilities estimated for the population mean.

Sample size works to reduce the size of the confidence interval; the larger the sample, the smaller the confidence interval, other things being equal. In the extreme situation, where the sample is infinite in size, any confidence interval for the mean must be zero in width. Here, we can be sure that the sample mean is exactly right as an estimate of the population. Even in practical situations, the size of the confidence interval can be made very small by taking a large number of cases.

In the next chapter we will begin to use confidence intervals where the probability of covering the population value can be stated exactly. There, because of other things we will know about the sampling distribution of the mean, we will no longer have to rely on the Tchebycheff inequality in order to make a probability statement.

Even so, the main features of confidence intervals are as they have been introduced here. Any confidence interval consists of a set of values constructed from a sample, which has a known, and generally high, probability of covering the population value of interest.

5.13 OTHER KINDS OF SAMPLING

The model of simple random sampling is the basis of almost all the discussion to follow. The classical procedures of statistical inference rest on the sampling scheme in which each and every population element sampled is independent of every other and is equally likely to be included in any sample.

However, statistical inference is not at all limited to such equally probable samples. Samples may be drawn according to any other probability structure, just so long as the probability of occurrence of any particular sample is known or can be calculated. Samples are drawn by some method consistent with these probabilities, and the probabilities are then taken into account in the treatment of any information gained from the sample. The generic term for the process of drawing samples according to some known probability structure is, simply, **probability sampling.** Simple random sampling is then a special case of probability sampling; the probability structure in simple random sampling dictates equal probability for each possible sample. Many methods exist for treating samples drawn according to other probability structures, and often these methods differ from those we will discuss.

One common scheme is called **stratified sampling.** Here, the population is divided into a number of parts, called *strata*. A sample is drawn independently and at random in each part. Given the sizes of the various strata, one can make inferences about the total population represented. Such a scheme is very good for ensuring a representative sample, and it may reduce the error in estimation, although it does require special methods over and above those given here.

Another approach to sampling is called *two-stage sampling*. Here, the population may break naturally into relatively large units, and then within these large units one wishes to examine specific individuals or objects. For example, suppose that social workers as a group were being studied. Most social workers are employed by large governmental units such as states, and then within the governmental unit by some agency. Thus, an effective approach to the sampling of social workers might be first to choose a set of states. Then, within the states selected a random sample of agencies could be chosen, with the social workers making up those agencies used as the ultimate sample.

This idea can be extended to sampling in three or more stages, of course. Thus in the social worker example, one might first sample states, then agencies, then social workers employed by the agencies selected.

Staged sampling of this sort is often the only practical approach when a large and complicated population must be sampled. However, this approach also requires some specialized techniques for statistical inference.

Mention has already been made (Section 3.17) of a sequential approach to sampling. That is, instead of regarding the sample of N independent observations all made at essentially the same time, sampling is done on an individual unit basis. Then the decision to proceed or not to proceed with sampling depends on statistical criteria the investigator has set up ahead of time. Ordinarily these criteria depend on such things as how precise an estimate of the population parameter has been gained, or how much risk remains in making a particular decision in the light of the evidence so far.

The student interested in some of these other sampling strategies might look into Mood, Graybill, Boes (1974), or Snedecor and Cochran (1980). Unfortunately, space does not permit a further discussion of them here.

5.14 TO WHAT POPULATIONS DO OUR INFERENCES REFER?

Most who use inferential statistics in research rely on the model of simple random sampling. Yet how does one go about getting such a *truly* random sample? It is not easy to do, unless, as in all probability sampling, each and every potential member of the population may somehow be listed. Then, by means of a device such as random numbers, individuals may be assigned to the sample with approximately equal probabilities.

However, in social and behavioral sciences, interest often lies in experimental or natural effects that, presumably, should apply to very large populations of people or things. Such a listing procedure is simply not possible. Still other experiments may refer to all possible measurements that *might* be made of some phenomenon under various experimental conditions, where estimated true values may be sought from the experimental observation of a few instances. Here, the population is not only infinite, it is hypothetical, since it includes all *future* or *potential* observations of that phenomenon under the different conditions. In sampling from such experimental populations, where there is no possibility of listing the elements for random assignment to the sample, the only recourse of the experimenter is to draw basic experimental units in some more or less random, or haphazard, way, and then make sure that in the experiment *only random factors* determine *which unit gets which experimental treatment.* In other words, there are two ways in which randomness is important in an experiment: the first is in the selection of the sample as a whole, and the second is in the allotment of individuals to experimental treatments. Each kind of randomness is important for the ''generalizability'' of the experimental results, so that in doing an experiment one takes pains to see that both kinds of randomness are present. However, even given that individual cases are assigned to experimental manipulations at random, the possible inferences are still limited by the fundamental population from which the total sample is drawn.

How does one know the population to which statistical inferences drawn from a sample apply? If random sampling is to be assumed, then the population is defined by the manner in which the sample is drawn. The only population to which the inferences strictly apply is that for which the *units* sampled have equal (and nonzero) likelihood of appearing in any sample. In some contexts a unit may be one thing, and in other contexts quite another: thus, for example, if the units sampled are families inhabiting the same dwelling, it may be that each such family unit has equal probability of appearing in a sample of N such units. However, it would not then necessarily be true that individual persons have equal probability for appearing in any sample. On the other hand, if individual persons are sampled at random, it will not necessarily be true that family units have equal probability of appearing, since families may contain different numbers of people. It is most important to have a clear definition of the unit that is to be sampled, since in random sampling the totality of such possible units with equal probability of appearing in any given sample must make up the population.

It should be obvious that simple random samples from one population may not be random samples of another population. For example, suppose that someone wishes to sample American college students. He or she obtains a directory of college students from a midwestern university and, using a random number table, takes a sample of these students. One is not, however, justified in calling this a random sample of the population of American college students, although it may well be that this is a random sample of students at *that* university. The population is defined not by what is said, but rather by what was *done* to get the sample. For any sample, one should always ask the question, ''What is the set of potential units that could have appeared in any sample with equal probability?'' If there is some well-defined set of units that fits this qualification, then inferences may be made to that population. However, if there is some population whose members could not have been represented in the sample with equal probability, then inferences do not *necessarily* apply to that population when methods based on simple random sampling are used. Any generalization beyond the population actually sampled at random must rest on extrastatistical, scientific considerations.

From a mathematical-statistical point of view the assumption of random sampling makes it relatively simple to determine the sampling distribution of a particular statistic given some particular population distribution. As we noted above, it is possible to use other probability structures for the selection of samples, and in some situations there are advantages in doing so. The point is that *some* probability structure must be known or assumed to underlie the occurrence of samples if statistical inference is to proceed. This point is belabored only because it is so often overlooked, and statistical inferences are so often made with only the most casual attention to the process by which the sample was generated. The assumption of some probability structure underlying the sampling is a little "price tag" attached to a statistical inference. It is a sad fact that if one knows nothing about the probability of occurrence for particular samples of units for observation, very little of the machinery we are describing here applies. This is why our assumption of random sampling is not to be taken lightly. All the techniques and theory that we will discuss apply to random samples, and do not necessarily hold for any old data collected in any old way. In practical situations, the experimenter may be hard put to show that a given sample is truly random, but he must realize that he is acting *as if* this were a random sample from some well-defined population when he applies methods of statistical inference. Unless this assumption is at least reasonable, the probability results of inferential methods mean very little, and these methods might as well be omitted. Although there may be plenty of meaning in the data, and this may be worth studying for its own sake, the probability statements we attach to such data depend on random sampling methods.

In the social sciences, studies made "in the field" are often interesting and valuable, even though random sampling may be out of the question in such studies. In addition, the social scientist is often interested in natural experiments, where a set of highly relevant factors or conditions come together in a society or a culture, providing a unique opportunity to observe the responses of individuals or groups of people. To rule such investigations off limits simply because random sampling models do not happen to fit would be the sheerest folly. Such data should be collected and studied with as much scientific and intellectual rigor as we can muster. However, merely prettifying these studies with the ordinary methods of statistical inference adds little or nothing to their meaning. This is a most serious problem in the development of the social sciences and has not been given the attention that it deserves. One thoughtful treatment of this subject may be found in Cook and Campbell (1979).

Certainly, the application of some statistical method does not somehow magically make a sample random, and the statistical conclusions therefore valid. Inferential methods apply to probability samples, drawn either by simple random sampling or in accordance with some other probability structure. There is no guarantee of their validity in other circumstances.

5.15 LINEAR COMBINATIONS OF RANDOM VARIABLES

So far in our discussion of sampling distributions, we have assumed that there is a single variable of interest, and that a sample quantity such as a mean or a variance is to be studied. Then, the population distribution dictates the sampling distribution of that particular statistic.

However, in other circumstances, the sampling process may produce values of *two or more* random variables. Our interest may not be in the usual statistics such as a sample mean at all, but rather in a particular *weighted sum* of the values of the random variables. Thus, each sample may give us the value of one random variable X and the value of another random variable Y. We might be interested in still another variable W, which is formed by weighting and summing the values of X and Y as they appear in each sample. For example, we might take

$$w = 3x + 4y,$$

or perhaps

$$w = x - y.$$

Over all samples producing values of X and of Y, there will then be a sampling distribution of W as well.

Any weighted sum of two or more variables is known as a *linear combination*. In later chapters we will have occasion to form linear combinations of random variables, and we will need to know something about their distributions over samples. Fortunately, such linear combinations of random variables follow a few simple rules and principles that can be readily summarized. A summary of these is provided in Appendix C. These rules are worthy of study, particularly when references to them are made in later sections. As you will see, the same general principles apply to such linear combinations as to other probability or sampling distributions.

An especially useful notation and concept that is also introduced in Appendix C is that of a *vector*, which is simply an ordered set of values. Thus, as in the example above, when we wish to represent one value $X = x$, and another value $Y = y$, we can actually do so with a pair of values, or a vector, (x,y). This principle extends easily to several variables or values, and gives a simple way to discuss several sets of values at the same time. Some principles involving vectors are given in Appendix C.4.

EXERCISES

1. A psychologist kept a large population of rats. Each rat was to be observed individually running a maze. If a rat had a tendency to turn right in the maze, it was given a score $X = 1$. On the other hand, if the rat had a tendency to turn left it got a score $X = 2$. If any of the population of rats had equal probability of being a "right-turner" or "left-turner," find the population distribution of X. Find the mean and variance of this distribution.

2. Now suppose that the experimenter in Exercise 1 chooses two rats independently and at random to run the maze. On the basis of its performance, each is given a value of X. However, a new variable G is formed by summing the X values for the two rats, or $G = x_1 + x_2$. Given that the population of rats is as described in Exercise 1, find the theoretical sampling distribution of G over all possible pairs of rats. (**Hint:** Simply list all possible pairs of outcomes and the resulting G values, and find the probabilities.)

3. Using the sampling distribution of G found in Exercise 2, find the expected value of G, or $E(G)$. How does this compare with the expected value for the population, or $E(X)$? Show that $E(G)$ in this instance is exactly twice the value of $E(X)$.

4. Again using the sampling distribution of G in Exercise 2, find the variance of this sampling distribution, or σ_G^2. How does this value compare with the variance of the population, or σ_X^2? Show that in this instance the variance of the sampling distribution is precisely twice the variance of the population.

5. Using the results of Exercises 1 through 4 above, make a verbal comparison of the sampling distribution of G to the population distribution on which it is based. How does this sampling distribution differ from the population distribution on which it is based?

6. Suppose that the experimenter in Exercise 1 selected three rats at random and ran each through the maze. Now the score G was formed by taking the X values for all three rats and adding them up and dividing by 3 or $G = (x_1 + x_2 + x_3)/3$. If the population distribution is as found in Exercise 1, what now is the distribution of G?

7. Find the mean and the variance for the sampling distribution of G found in Exercise 6. How do these compare with the mean and variance of the population distribution? Show that $E(G) = E(X)$ and $\sigma_G^2 = \sigma_X^2/3$ in this instance.

8. See if you can generalize from the results of Exercises 1 through 7. That is, if the experimenter selected any number N of rats from the population and formed a statistic G from the sum of their X values divided by N, what would the mean of the sampling distribution of G turn out to be? The variance σ_G^2? What would the sampling distribution be like relative to this population distribution?

9. An educational researcher has devised a measure of schoolchildren's motivation to achieve. Children are assigned a score $x = -1$ if they are classified as "underachievers," $x = 0$ for "parachievers," and $x = 1$ for "overachievers." The researcher theorizes that about 1/4 of children in the school system are overachievers, another 1/4 are underachievers, and the rest achieve about at par for their ability. Find the population distribution for the random variable X. What is the value of $E(X)$? What is the value of σ_X^2?

10. Suppose that the researcher in Exercise 9 selected two children at random from this school system, and formed the average of their two scores, $\bar{x} = (x_1 + x_2)/2$. Given the population situation assumed in Exercise 9, work out the exact sampling distribution of the statistic $\bar{x}$ (**Hint:** Again work out all possible pairs of scores, the corresponding $\bar{x}$ value, and the probability of each value.)

11. Work out the expected value and the variance for the sampling distribution found in Exercise 10. How do these compare with the expected value and variance for the population, as found in Exercise 9? Show that the variance for the sampling distribution is here exactly one-half of the variance for the population.

12. Suppose that the two children selected at random in Exercise 10 showed one of the average scores listed below. How probable is such an average score, given the specified population?

(a) exactly 0
(b) 1.00
(c) −1.00
(d) .5 or less
(e) −.5 or less

13. Imagine now that the researcher in Exercise 9 sampled three children independently and at random from the population of students as given above, and again computed their average score, $\bar{x}$. Find the sampling distribution of $\bar{x}$ under these conditions. How does this distribution compare in its general shape to that of the population distribution? To that of Exercise 10?

14. Work out the expected value and the variance of the sampling distribution of $\bar{x}$ for $N = 3$, as found in Exercise 13. Demonstrate that $E(\bar{x}) = E(X)$, and that $\sigma_M^2 = (\sigma_X^2/3)$.

15. In the sampling distribution found in Exercise 13, determine the following probabilities:

   (a) $p(\bar{x} = 1)$
   (b) $p(\bar{x} = 0)$
   (c) $p(\bar{x} \leq 1/3)$
   (d) $p(\bar{x} \leq 0)$
   (e) $p(\bar{x} \leq 2/3)$
   (f) $p(\bar{x} < -1/3)$

16. Let us suppose that instead of the theory about the population stated in Exercise 9, there is good reason to believe that 60% of the school population consists of underachievers, 30% of parachievers, and only 10% of overachievers. Repeat Exercise 10 under this new assumption. How do the mean and variance of the sampling distribution now relate to the mean and variance of the population? Describe the "form" of this sampling distribution relative to that of the population.

17. A research worker is faced with a large stack of computer cards, each containing data for one individual. Unfortunately, these cards are not labeled, and they could belong either to Study I or to Study II in this particular research project. As it happens, Study I contained 60% female subjects and Study II contained only 40%. The worker draws 15 cards at random and notes the code for sex that each contains. It turns out that there are eight females out of 15 cards. What should the worker conclude? Why? (**Hint:** Regard "female" as a success in a Bernoulli process.)

18. Consider the table given in Exercise 32 of Chapter 1. Suppose that one individual were drawn at random from the 1,000 represented in this table. According to the principle of maximum likelihood, what would you guess the occupation of this individual to be, given each of the different patterns possible?

19. A public health officer believes that something between 5 and 20% of three-year-old children in a certain area have been exposed to tuberculosis. A random sample of 20 such children is taken and a test for tuberculosis given to each. Three children show a positive test. Is 5% or 20% a better bet for the proportion actually exposed in this population? What is the *best* bet about the true proportion of children who would give a positive response to the tuberculosis test? Why?

20. An archeologist has assembled a collection of 18 potshards (pieces of pottery) found at the same level of an ancient site. It is already known that only about 20% of pottery produced in this area before 3000 B.C. was decorated, although about 40% of later pottery had decoration. Four of the potshards in the present group are decorated. What will the archeologist tend to conclude about the age of these potshards? Why? What is being assumed here?

21. There are three typists is a business office. Typist I is not very accurate, and tends to average one error per business letter. Typist II is a little better, and makes about three errors per five letters. Typist III is quite expert, and makes only one error for each 10 letters on the average. The boss reads five letters all typed by the same person, and discovers two errors. Who would you guess typed the letters? (**Hint:** Use the rule for the Poisson distribution.)

22. Four interviewers were sent door to door in a neighborhood. On a given try, the occupant of the dwelling might or might not agree to give the interview. One of the interviewers (A) had a past record of 60% success in getting interviews, whereas interviewer B had a record of only 40% success. Interviewer C had an even worse record of 30%, and D was worst of all with 10%. For one of these interviewers, the first success occurred at the third house. If each interviewer was operating true to form, which interviewer do you think this was? Why? (**Hint:** Look at Section 3.17.)

23. One of the interviewers in Exercise 22 had the third successful interview at the fifth house contacted. Again assuming that the true proportion of successes for each interviewer remains constant and that each interview is independent of every other, which interviewer do you think this was? Why?

24. A candidate for governor of a certain state feels confident in receiving 52% of the votes in the impending election. A political scientist takes a random sample of 250 of this state's registered voters, and asks each person whether or not he or she intends to vote for this candidate. The proportion P who gave an affirmative response was recorded. Find the expected value of P and its sampling variance and standard deviation, *if* the candidate's idea is actually correct. What is the form of this sampling distribution?

25. Suppose that the political scientist in Exercise 24 found that in the sample of 250 voters, only $P = .40$ expressed the intention of voting for this candidate. If the candidate's theory is right, what z score does this result represent in the sampling distribution? What is the probability of getting a sample result as deviant as this (in absolute terms) if the expected value of P really is .52? Use Tchebycheff's inequality in a unimodal symmetric distribution.

26. Assume that the researcher in Exercise 9 sampled a total of 500 children and scored each for achievement level. The mean of the scores was taken. What is the expected value of the mean score for any such sample? What is the standard deviation of the sampling distribution of such means? If a sample mean of .25 is obtained, what z score in the sampling distribution does this represent? Assuming a unimodal symmetric sampling distribution, about how likely is a z value this much or more deviant from 0 if the researcher's theory about the population is true?

27. Let us assume that the psychologist in Exercise 1 took a random sample of 50 rats and ran each one through the maze. The average score for left–right tendency these rats achieved was 1.55. If the population of rats actually has a mean of 1.5 and a variance of .25, where does this result fall (in standardized score terms) in the sampling distribution of means? Again assuming a unimodal symmetric distribution, does this appear to be a relatively likely or a relatively unlikely result, given the population situation specified? Why?

28. Based on the sample result in Exercise 27, find an interval of possible values for the population mean that has probability of at least .95 of covering the true value.

29. A personnel manager of an industry wished to see if the introduction of a new training method reduced the hours lost by injury by a group of employees. In the past such employees lost an average of 17.4 hrs per year through injury, with a standard deviation of 4.9. A group of 150 employees were given the new training method and then placed on the job. After a year's time, this group had lost only 12.3 hrs on the average. If the training method actually made no difference, and these employees truly represent a sample from the larger group with old mean and standard deviation, how likely is a sample deviating this much or more from 17.4? (Assume a unimodal, symmetric distribution). How safe is it to say that a trained population will be different from the old group of employees?

30. Form a confidence interval for the mean of the new population of employees in Exercise 29, such that the probability that this interval covers the true mean is at least .90. Does this interval include the old value of 17.4? What does this say about the new training?

31. Ten observations were drawn at random from some very large population. The values in the sample were as follows. What are the best estimates of the population's mean and variance?

26 25 28 33 42 27 31 37 39 20

32. A sample of 25 employed American men, 40 years of age, was drawn at random. Each was asked how many hours he had watched television during the preceding week. The answers are listed below. On the basis of these data, estimate the mean and standard deviation of the population of such men, with respect to number of television hours watched that week. According to these estimates, how deviant (in z score terms) was a man who watched no television at all? What is the maximum percentage of such cases there could possibly be in this population?

9	8	7	14	15
2	11	9	7	5
3	7	10	11	10
6	8	0	14	20
6	10	7	8	9

33. A medical school enrolls 208 students during a year. At random, a sample of 49 students was taken, and it was determined how much each one had spent for books during that school year. The data are given below, in whole dollars spent. On the basis of this sample, estimate the mean and standard deviation for the entire set of students. Estimate the *total* amount spent for books.

512	650	817	791	526	525	602
715	871	592	725	622	551	617
606	540	720	643	681	708	910
545	819	648	677	715	552	619
811	611	814	519	682	644	560
950	646	772	518	754	624	850
632	575	710	731	816	881	939

34. Consider the medical school described in Exercise 33 once again. Suppose that all possible samples of size 49 were taken from this medical school student body, and that the true mean expenditure for books is $600 a year. Where does the sample mean found in Exercise 34 appear to lie in the sampling distribution of means in terms of its z score value? (Use the estimate found in Exercise 33 in place of the unknown population variance.)

35. Suppose that the same population was sampled three times. On the first occasion 10 observations were made independently and at random, on the second, 20 observations were made independently and at random, and on the third occasion, 15 observations were similarly made. The results were

Sample 1	Sample 2	Sample 3
$\bar{x}_1 = 96$	$\bar{x}_2 = 105$	$\bar{x}_3 = 103$
$S_1^2 = 22$	$S_2^2 = 29$	$S_3^2 = 31$

Estimate the mean and the variance of the population. What is the estimated standard error of this estimate of the population mean?

36. A population is known to have a variance of 18.5. Two samples of 90 observations each are made independently and at random from this population. One sample produces a mean of 65 and a variance of 16 and the second independent sample a mean of 70 and a variance of 20. Estimate the population mean, and find the standard error of this estimate.

37. In a certain sampling distribution of means based on 44 cases each, a sample mean of 438 is known to correspond to a z value of 1.75 in the sampling distribution. If the standard error of the mean is 9, what is the population mean? What is the population variance?

38. In a certain sampling distribution of the mean based on samples of size N_1, a sample mean of 100 corresponds to a z value of 1.5. However, when samples of size N_2 are taken, a sample mean of 100 corresponds to a z value of 2.00. How large is N_2 relative to N_1?

39. Two independent samples were drawn from the same population with a standard deviation of 9. The first sample contained 10 cases and the second sample contained 20 cases. The mean of the first sample corresponded to a z score of 1.8 in its sampling distribution of means, whereas the second sample mean corresponded to a z score of .5. How much larger was the first mean than the second? What z score would correspond to the mean of the pooled samples?

40. A random variable X is generated by a stable and independent Poisson process with intensity 10. Five independent observations of values of X are made and a mean calculated. Find the mean and standard deviation of the sampling distribution of these means.

Chapter *6*

Normal Population and Sampling Distributions

Heretofore, the only theoretical distribution we have considered in any detail has been the discrete binomial distribution. Now we consider a type of distribution having a domain which is *all* of the real numbers. This is the so-called normal or "Gaussian" distribution. The normal distribution is but one of a vast number of mathematical functions one might invent for a distribution; it is purely theoretical. At the very outset let it be clear that, like binomial or other probability functions, the normal distribution is not a fact of nature that one actually observes to be exactly true. Rather, the normal distribution is a theory about what might be true of the relation between intervals of values and probabilities for some variable.

6.1 NORMAL DISTRIBUTIONS

Like most theoretical functions, a normal distribution is completely specified only by its mathematical rule. Quite often, the distribution is symbolized by a graph of the functional relation generated by that rule, and the general picture that a normal distribution presents is the familiar bell-shaped curve of Figure 6.1.1. The horizontal axis represents all the different values x of some random variable X, and the vertical axis $f(x)$ their densities. The normal distribution is continuous for all values of X between $-\infty$ and ∞, so that each conceivable nonzero *interval* of real numbers has a probability other than zero. For this reason, the curve is shown as never quite touching the horizontal axis; the tails of the curve show decreasing probability densities as values grow extreme in either direction from the mode, but any interval representing *any* degree of deviation from central tendency is possible in this theoretical distribution. The distribution is absolutely symmetric and unimodal, and mean, median, and mode all have the same

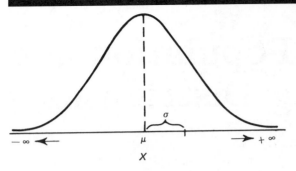

Figure 6.1.1

The graph of a normal distribution.

value. Bear in mind that since the normal distribution is continuous, the height of the curve shows the probability *density* for each value x of X. However, as for any continuous distribution, **the area cut off beneath the curve by any interval is its probability,** and the entire area under the curve is 1.00.

The mathematical rule for a normal density function is as follows:

$$f(x; \mu, \sigma^2) = \frac{1}{\sqrt{2\pi\sigma^2}} e^{-(x-\mu)^2/2\sigma^2}. \qquad [6.1.1^*]$$

This rule pairs a probability density $f(x)$ with each and every possible value x. This rule looks somewhat forbidding to the mathematically uninitiated, but actually it is not difficult to understand. The π, the e, and the 2, of course, are simply positive numbers acting as mathematical constants. The "working" part of the rule is the exponent

$$- \frac{(x - \mu)^2}{2\sigma^2}$$

where the particular value of the variable X appears, along with two parameters μ and σ^2.

It is important to notice that the precise density value assigned to any x by this rule cannot be found unless the two parameters μ and σ are specified. The parameter μ can be any finite number, and σ can be any finite *positive* number. Thus, like the binomial, the normal distribution rule actually specifies a family of distributions. Although each distribution in the family has a density value paired with each specific value x by this same general rule, the *particular* density that is paired with a given x value differs with different assignments of μ and σ. Thus, normal distributions may differ in their means (Fig. 6.1.2); in their standard deviations (Fig. 6.1.3); or in both means and standard deviations (Fig. 6.1.4). Nevertheless, given the mean and standard deviation of the distribution, the rule for finding the probability density of any value of the variable is the same.

Since the normal distribution is really a family of distributions, statisticians constructing tables of probabilities found by the normal rule find it convenient to think of the variable in terms of standardized, or z, scores. That is to say, if the random variable is a standardized score z, so that $\mu = 0$ and $\sigma = 1.00$, then the rule becomes simpler:

$$f(z) = \frac{1}{\sqrt{2\pi}} e^{-z^2/2} \qquad [6.1.2^*]$$

Figures 6.1.2–6.1.4

Normal distributions differing in μ and σ.

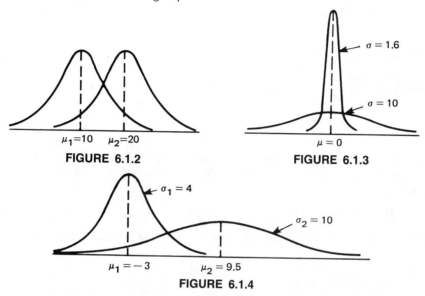

FIGURE 6.1.2

FIGURE 6.1.3

FIGURE 6.1.4

For standardized normal variables, the density depends only on the *absolute* value of z; since both z and $-z$ give the same value z^2, they both have the same density. The higher the z in absolute value, the less the associated density. The standardized-score form of the distribution makes it possible to use one table of densities for any normal distribution, regardless of its particular parameters. Thus, for the density of a score x in a normal distribution, with mean equal to 80 and standard deviation equal to 5, one can look up the density associated with a standardized score $(x - 80)/5$, and this gives the desired number. In the same way, cumulative probabilities are usually found from a table of the standardized normal distribution, such as Table I in Appendix E.

6.2 CUMULATIVE PROBABILITIES AND AREAS FOR THE NORMAL DISTRIBUTION

In our use of the normal distribution, we will be concerned almost exclusively with cumulative probabilities and with the probabilities of intervals of values. The cumulative probability

$$F(a) = p(X \le a) \tag{6.2.1*}$$

can be thought of as the area under the normal curve in the interval bounded by $-\infty$ and the value a (Fig. 6.2.1).

These cumulative probabilities can be used to find the probability of any interval. For example, suppose that in some normal distribution we want to find the probability that X lies between 5 and 10, given some exact values for μ and σ. This is the probability

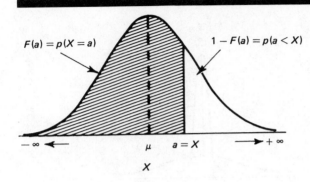

$F(a) = p(X = a)$

$1 - F(a) = p(a < X)$

Figure 6.2.1

A cumulative probability.

$-\infty \longleftarrow$ μ $a = X$ $\longrightarrow +\infty$

X

represented by the area shown in Figure 6.2.2. The cumulative probability *up to and including* 10, or $F(10)$, minus the cumulative probability up to and including 5, or $F(5)$, gives the probability in the interval:

$$p(5 \le X \le 10) = F(10) - F(5).$$

In the same way, the probability of any other interval with limits a and b can be found from cumulative probabilities:

$$P(a \le X \le b) = F(b) - F(a). \qquad [6.2.2^*]$$

As mentioned above, most tables of the normal distribution give the cumulative probabilities for various *standardized* values. That is, for a given z score the table provides the cumulative probability *up to and including that standardized score* in a normal distribution; for example,

$$F(2) = p(z \le 2)$$

$$F(-1) = p(z \le -1)$$

and so on.

If the cumulative probability is to be found for a positive z, then Table I in Appendix E can be read directly. For example, suppose that a normal distribution is known to have a mean of 50 and a standard deviation of 5. What is the cumulative probability of a score of 57.5? The corresponding standardized score is

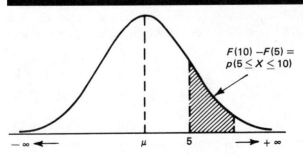

Figure 6.2.2

The probability of an interval in a normal distribution.

$F(10) - F(5) =$
$p(5 \le X \le 10)$

$-\infty \longleftarrow$ μ 5 $\longrightarrow +\infty$

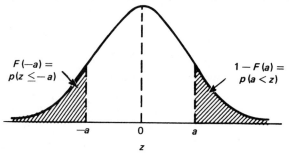

Figure 6.2.3

Probabilities on the extreme tails of a normal distribution.

$$z = \frac{57.5 - 50}{5} = 1.5.$$

A look at Table I shows that for 1.5 in the first column, the corresponding cumulative probability in the column labeled $F(z)$ is .933, approximately. This is the probability of observing a z value *less than or equal to* 1.5 in a standardized normal distribution. It is thus also the probability of observing a score less than or equal to 57.5 in a normal distribution with a mean of 50 and a standard deviation of 5.

If the z score is negative, this procedure is changed slightly. Since the normal distribution is symmetric, the density associated with any z is the same as for the corresponding value with a negative sign, or $-z$. However, the cumulative probability for a negative standardized score is 1 minus the cumulative probability for the z value with a positive sign:

$$F(-z) = 1 - F(z)$$

where z is the positive standardized score of the same magnitude. Figure 6.2.3 will perhaps clarify this point. The shaded area on the left is the cumulative probability for a value $z = -a$. The table gives only the area falling below the point $z = a$, the positive number of the same absolute value as $-a$. However, the shaded area on the *right*, which is $1 - F(a)$, is the same as the shaded area on the *left*, $F(-a)$.

For example, suppose that in a distribution with mean equal to 107 and standard deviation of 70, we want the cumulative probability of the score 100. In standardized form,

$$z = \frac{100 - 107}{70} = -.1.$$

We look in the table for z equal to *positive* .1, which has a cumulative probability of approximately .5398. The cumulative probability of a z equal to $-.1$ must be approximately

$$F(-.1) = 1 - .5398 = .4602.$$

We may also answer questions about the probabilities of various intervals from the table. For example, what proportion of cases in a normal distribution must lie within one standard deviation of the mean? This is the same as the probability of the interval

$(-1 \le z \le 1)$. The table shows that $F(1) = .8413$, and we know that $F(-1)$ must be equal to $1 - .8413$ or $.1587$.

Thus,

$$p(-1 \le z \le 1) = F(1) - F(-1) = .8413 - .1587 = .6826.$$

About 68% of all cases in a normal distribution must lie within one standard deviation of the mean.

To find the proportion lying between one and two standard deviations above the mean, we take

$$p(1 \le z \le 2) = F(2) - F(1).$$

From Table I, these numbers are $.9772$ and $.8413$, so that

$$p(1 \le z \le 2) = .9772 - .8413 = .1359.$$

About 13.6% of cases in a normal distribution lie in the interval between 1σ and 2σ above the mean. By the symmetry of the distribution we know immediately that

$$p(-2 \le z \le -1) = .1359$$

as well.

Beyond what number must only 5% of all standardized scores fall? That is, we want a number b such that the following statement is true for a normal distribution:

$$p(b < z) = .05.$$

This is equivalent to saying that

$$p(z \le b) = .95.$$

A look at the table shows that roughly $.95$ of all observations must have z scores at or below 1.65. Thus,

$$p(1.65 < z) = .05, \text{ approximately.}$$

On the other hand, we can see from Table I that $p(z \ge 1.96) = .025$. In addition, we know that $p(z \le -1.96) = .025$. It must therefore be true that the probability of a z value in a normal distribution exceeding 1.96 in absolute value is $.025 + .025 = .05$. Similarly, two intervals $(z \le -2.58)$ and $(2.58 \le z)$ define

$$p(z \le -2.58 \text{ or } 2.58 \le z) = .005 + .005 = .01.$$

Put the other way,

$$p(2.58 \le |z|) = .01,$$

the probability that z exceeds 2.58 in absolute value is about 1 in 100.

6.3 THE IMPORTANCE OF THE NORMAL DISTRIBUTION

The normal is by far the most used distribution in inferential statistics. There are at least three very good reasons why this is true.

A population may be assumed to follow a normal law because of what is known

or presumed to be true of the measurements themselves. There are two rather broad instances in which the random variable is a measurement of some kind, and the distribution is conceived as normal. The first is when one is considering a hypothetical distribution of "errors," such as errors in reading a dial, in discriminating between stimuli, or in test performance. Any observation can be assumed to represent a "true" component plus an error. Each error has a magnitude, and this number is thought of as a reflection of "pure chance," the result of a vast constellation of circumstances operative at the moment. Any factor influencing performance at the moment contributes a tiny amount to the size of the error and to its direction. Furthermore, such errors are appropriately considered to push the observed measurement up or down with equal likelihood, to be independent of each other over samplings, and to cancel out in the long run. Thus, in theory, it makes sense that errors of measurement or errors of discrimination follow something like the normal rule.

In other instances, the distribution of *true magnitudes* of some trait may be thought of as normal. For example, human heights form approximately a normal distribution, and so, we believe, does human intelligence. Indeed, there are many examples, especially in biology, where the distribution of measurements of a natural trait seems to follow something closely resembling a normal rule. No obtained distribution is ever exactly normal, of course, but some distributions of measurements have relative frequencies very close to normal probabilities. However, the view that the distribution of almost any trait of living things should be normal, although prominent in the 19th century, has been discredited. The normal distribution is not, in any sense, "nature's rule," and all sorts of distributions having little resemblance to the normal distribution may occur in all fields.

It may be convenient, on mathematical grounds alone, to assume a normally distributed population. Mathematical statisticians do not devote so much attention to normal distributions just because they think bell-shaped curves are pretty! The truth of the matter is that the normal function has important mathematical properties shared by no other theoretical distribution. Assuming a normal distribution gives the statistician an extremely rich set of mathematical consequences that are useful in developing methods. Very many problems in mathematical statistics either are solved, or can be solved, *only* in terms of a normal population distribution. We will find this true especially when we come to methods for making inferences about a population variance. The normal distribution is the "parent" of several other important theoretical distributions that figure in statistics. In some practical applications the methods developed using normal theory work quite well even when this assumption is not met, despite the fact that the problem can be given a *formal* solution only when a normal population distribution is assumed. In other instances, there just is not any simple way to solve the problem when the normal rule does not hold for the population, at least approximately.

There is a very intimate connection between the size of sample N and the extent to which a sampling distribution approaches the normal form. Many sampling distributions based on large N can be approximated by the normal distribution even though the population distribution itself is definitely not normal. This is the extremely important principle that we will call the **central limit theorem.** The normal distribution is the *limiting form* for large N for a very large variety of sampling distributions. This is one of the most remarkable and useful principles to come out of theoretical statistics.

Thus, you see, it is no accident that the normal distribution is the workhorse of inferential statistics. The assumption of normal population distributions or the use of the normal distribution as an approximation device is not as arbitrary as it sometimes appears; this distribution is part of the very fabric of inferential statistics. These features of the normal distribution will be illustrated in the succeeding sections.

6.4 THE NORMAL APPROXIMATION TO THE BINOMIAL

One of the interpretations given to the normal distribution is that it is the limiting form of a binomial distribution as $N \to \infty$ for a fixed p.

Imagine samples of N from a Bernoulli distribution with the probabilities of the two categories being p and q, respectively. Then we should expect a binomial distribution of the number of successes. Suppose that we do this for all possible samples of size 10. Next we draw all possible samples with $N = 10{,}000$, then repeat the sampling with an N of 10,000,000, and so on. For each sample size, there will be a different binomial distribution we should observe when *all possible* samples of that size have been drawn.

How does the binomial distribution change with increasing sample size? In the first place, the actual range of the possible number of successes grows larger since the whole numbers 0 through N form a larger set with each increase in N. Second, the expectation Np is larger for each increase in N for a fixed p, and the standard deviation $\sqrt{Npq}$ also increases with N. Finally, the probability associated with any given exact value x of X tends to *decrease* with *increase* in N, for $0 \le x \le N$.

However, suppose that for any value of N, the obtained number of successes x were put into standardized form:

$$z = \frac{x - Np}{\sqrt{Npq}} = \frac{x - E(X)}{\sigma}. \qquad [6.4.1^*]$$

Then regardless of the size of N, the mean of the z scores for a binomial distribution will be 0, and the standard deviation 1. The probability of any z score is the same as for the corresponding x. Any z-score interval can be given a probability in the particular binomial distribution, and the same interval can also be given a probability for a normal distribution. We can compare these two probabilities for any interval; if the two probabilities are quite close over all intervals, the normal distribution gives a good approximation to the binomial probabilities. On making this comparison of binomial and normal interval probabilities, we should find that the larger the sample size N the better is the

x	$p(x)$
5	.0312
4	.1563
3	.3125
2	.3125
1	.1563
0	.0312

Table 6.4.1

A binomial distribution with $p = .5$ and $N = 5$.

fit between the two kinds of probabilities. **As *N* grows infinitely large, the normal and binomial probabilities become identical for any standardized interval.**

For example, imagine that $p = .5$, and $N = 5$. Here the expectation is $(5)(.5)$, or 2.5, and the standard deviation is $\sqrt{5(.5)(.5)} = \sqrt{1.25}$ or about 1.12. The binomial distribution for these values of p and N is given in Table 6.4.1.

For the moment, let us pretend that this is actually a continuous distribution, and that the possible numbers of successes are midpoints of class intervals. Furthermore, let us turn each specific value x into its standardized value z, by taking

$$z = \frac{x - E(X)}{\sigma} = \frac{x - 2.5}{1.12}.$$

Then the distribution is as shown in Table 6.4.2.

Here, the z intervals were found by converting each real limit for the distribution of successes (regarded as continuous) into its equivalent z score: thus, $1.79 = (4.5 - 2.5)/1.12$, and so on. Note that the probabilities are the same for these intervals as for the original x values.

The last column gives the probabilities that these same intervals would have in a normal distribution, found by using Table I in Appendix E. Thus, the interval 0.89–1.79 has probability

$$F(1.79) - F(0.89) = .96327 - .81327 = .1500$$

and the other probabilities are found in the same way. The top and bottom intervals are given the probabilities

$$1 - F(1.79) = .0367$$

and

$$F(-1.79) = .0367$$

in order to make the probabilities over all intervals sum to 1.00.

Observe how closely the normal probabilities approximate their binomial counterparts: each normal probability is correct to two decimal places as an approximation to a binomial probability. The difference in the two probabilities is smallest for the middle intervals, and is larger for the extremes. Thus, even when N is only 5, the normal distribution gives a respectable approximation of binomial probabilities for $p = .5$.

Now suppose that for this same example, N had been 15. Here the expectation is 7.5,

z real limits	*z midpoint*	*p(x)*	*Normal*
(1.79 to 2.68)	2.23	.0312	.0367
(0.89 to 1.79)	1.34	.1563	.1500
(0.00 to 0.89)	0.45	.3125	.3133
(−0.89 to 0.00)	−0.45	.3125	.3133
(−1.79 to −0.89)	−1.34	.1563	.1500
(−2.68 to −1.79)	−2.23	.0312	.0367

Table 6.4.2

The normal approximation to a binomial with $p = .5$ and $N = 5$.

x	z intervals	p(x)	Normal probabilities
15	3.608 to 4.124	.00003	.0002
14	3.092 to 3.608	.0005	.0008
13	2.577 to 3.092	.0032	.0040
12	2.061 to 2.577	.0139	.0146
11	1.546 to 2.061	.0416	.0414
10	1.030 to 1.546	.0916	.0905
9	0.515 to 1.030	.1527	.1518
8	0.000 to 0.515	.1964	.1967
7	−0.515 to 0.000	.1964	.1967
6	−1.030 to −0.515	.1527	.1518
5	−1.546 to −1.030	.0916	.0905
4	−2.061 to −1.546	.0416	.0414
3	−2.577 to −2.061	.0139	.0146
2	−3.092 to −2.577	.0032	.0040
1	−3.608 to −3.092	.0005	.0008
0	−4.124 to −3.608	.00003	.0002

Table 6.4.3

The normal approximation to a binomial with $p = .5$ and $N = 15$.

and the standard deviation is $\sqrt{(15)(.25)} = 1.94$. The distribution, with z-score intervals and both binomial and normal probabilities, is shown in Table 6.4.3.

For this distribution the normal approximation gives an even better fit to the exact binomial probabilities; the average absolute difference in probability over the intervals is about .0006, whereas when N was only 5, the average absolute difference was about .004. In general, as N is made larger, the fit between normal and binomial probabilities grows increasingly good. In the limit, when N approaches infinite size, the binomial probabilities are exactly the same as the normal probabilities for any interval, and thus one can say that the normal distribution is the limit to the binomial. This is true regardless of the value of p.

On the other hand, for any finite N, the more p departs from .5, the less well does the normal distribution approximate the binomial. When p is not exactly equal to .5, the binomial distribution is somewhat skewed for any finite sample size N, and for this reason the normal probabilities will tend to fit less well than for $p = .5$, which always gives a symmetric distribution.

In any practical use of the normal approximation to binomial probabilities, it is important to remember that here we regard the binomial distribution as though it were continuous, and actually find the normal probability associated with an interval with real lower limit of $(x - .5)$ and real upper limit of $(x + .5)$, where x is any given number of successes. Thus, in order to find the normal approximation to the binomial probability of x, we take

$$\text{prob}\left(\frac{x - NP - .5}{\sqrt{Npq}} \leq z \leq \frac{x - Np + .5}{\sqrt{Npq}}\right) \qquad [6.4.2]$$

by use of the standardized normal tables. Similarly, in terms of sample $P = x/N$, we have

$$\text{prob}\left(\frac{P - p - .5/N}{\sqrt{pq/N}} \leq z \leq \frac{P - p + .5/N}{\sqrt{pq/N}}\right). \qquad [6.4.3]$$

This adjustment, by which the probability of an interval of values is taken in place of the exact value of x or P, gives rise to the so-called **correction for continuity.** That is, in later sections when the normal distribution is to be used to approximate a binomial probability, we will deal with the z value

$$z = \frac{x - Np - .5}{\sqrt{Npq}}$$ [6.4.4*]

when x is larger than Np, and with

$$z = \frac{x - Np + .5}{\sqrt{Npq}}$$ [6.4.5*]

when x is less than Np. This will allow for the fact that the continuous normal distribution is being used in order to approximate the probability in a discrete distribution. Only when the sample size N is relatively large does this correction become unimportant enough to ignore.

As noted above, the normal distribution need not be particularly good as an approximation to the binomial if either p or q is quite small; and if either p or q is extremely small, the normal approximation may not be satisfactory even when N is quite large. In these circumstances another theoretical distribution, the Poisson, provides a better approximation to binomial probabilities. This approximation of binomial probabilities by use of Poisson probabilities was outlined in Chapter 3.

6.5 THE THEORY OF THE NORMAL DISTRIBUTION OF ERROR

The fact that the limiting form of the binomial is the normal distribution actually provides a rationale for thinking of random error as distributed in this normal way. Consider an object measured over and over again independently and in exactly the same way. Imagine that the value Y obtained on any occasion is a sum of two independent parts:

$Y = T + e$.

That is, the obtained score Y is a sum of a constant *true* part T plus a random and independent *error* component. However, the error portion can also be thought of as a sum:

$e = g[e_1 + e_2 + e_3 + \cdots + e_N]$

Here, e_1 is a random variable that can take on only two values:

$e_1 = 1$, when factor 1 is operating

$e_1 = -1$, when factor 1 is not operating.

The g is merely a constant, reflecting the "weight" of the error in Y. Similarly, the other random errors are attributable to different factors, and take on only the values 1 and -1. Now imagine a vast number of influences at work at the moment of any measurement. Each of these factors operates independently of each of the others, and whether any factor exerts an influence at any given moment is purely a chance matter.

If you want to be a little anthropomorphic in your thinking about this, visualize old Dame Fortune tossing a vast number of coins on any occasion and, from the result of each, deciding on the pattern of factors that will operate. If the first coin comes up heads, e_1 gets value 1; if it comes up tails, value -1, and the same principle determines the value for every other error portion of the observed score.

Now under this conception, the number of factors operating at the moment one observes Y is a number of successes in N independent trials of a Bernoulli experiment. If this number of successes is X, then

$$e = gX - g[N - X] = g[2X - N].$$

The value of e is exactly determined by X, the number of "influences" in operation at the moment, and the probability associated with any value of e must be the same as for the corresponding value of X. **If N is very large, then the distribution of e must approach a normal distribution.** Furthermore, if any factor is equally likely to operate or not operate at a given moment, so that $p = 1/2$, then

$$E(e) = g[2E(X) - N],$$

so that, since $E(X) = Np = N/2$,

$$E(e) = 0.$$

In the long run, over all possible measurement occasions the errors all cancel out. This makes it true that

$$E(Y) = E(T) + E(e) = T.$$

The long-run expectation of a measurement Y is the true value T, provided that error really behaves in this random way as an additive component of any score.

This is a highly simplified version of the argument for the normal distribution of errors in measurement. Much more sophisticated rationales can be invented, but they all partake of this general idea. Moreover, the same kind of reasoning is sometimes used to explain why distributions of natural traits, such as height, weight, and size of head, follow a more or less normal rule. Here, the mean of some population is thought of as the "true" value, or the "norm." However, associated with each individual is some departure from the norm, or error, representing the culmination of all the billions of chance factors that operate on him, quite independently of other individuals. Then by regarding these factors as generating a binomial distribution, we can deduce that the whole population should take on a form like the hypothetical normal distribution. However, this is only a theory about how errors might operate and there is no reason at all why errors must behave in the sample additive way assumed here. If they do not, then the distribution need not be normal in form at all.

6.6 TWO IMPORTANT PROPERTIES OF NORMAL POPULATION DISTRIBUTIONS

As suggested earlier in this chapter, the normal distribution has mathematical properties that are most important for theoretical statistics. For the moment, we are going to discuss only two of these general properties, both of which will be useful to know in later sections. The first has to do with the sampling distribution of means for samples drawn from a normally distributed population. The second deals with the independence of sample mean and sample variance.

Suppose that samples of N independent observations are being drawn from a population with mean μ and standard deviation σ. Furthermore, suppose that it is known that this population shows a normal distribution of the random variable of interest, X. Can we then say anything special about the sampling distribution of means $\bar{x}$ drawn from this normal population?

As it happens, we can say a great deal:

Given random samples of N independent observations, each drawn from a normal population, the distribution of the sample means is normal, irrespective of the size of N, (N>0).

In other words, when one can assume that the population itself is normally distributed with respect to the random variable X, the problem of the sampling distribution of $\bar{x}$ is solved: this sampling distribution will itself be normal, with expectation equal to μ and standard error equal to $\sigma/\sqrt{N}$. This is true *regardless* of the size of the sample, or N (so long as N is at least 1, of course).

A great deal of use is made of this principle in statistical inference, especially in those situations where one is forced to deal with small samples. As we shall see in subsequent chapters, a normal population distribution is almost always assumed when the sample size is small since this assumption permits one to say exactly what the sampling distribution of $\bar{x}$ must be like.

Any sample consisting of N independent observations of the same random variable provides both a sample mean $\bar{x}$ and a sample variance. The sample mean estimates μ and the unbiased sample variance s^2 estimates σ^2. These two values ($\bar{x}$, s^2) obtained from any sample can be thought of as a joint event. But are these two estimates independent? Or, does the value found for one of these statistics somehow depend on the value found for the other?

The answer to these questions is provided by the following key principle:

Given random and independent observations, the sample mean $\bar{x}$ and the sample variance (either S^2 or s^2) are independent if and only if the population distribution is normal.

The information contained in the sample mean in no way dictates the value of the sample variance, and vice versa, when a normal population is sampled. Furthermore, **unless the population actually is normal, these two sample statistics are not independent across samples.**

This is a most important principle, since a great many problems concerned with a population mean can be solved only if one knows something about the value of the population variance. At least an estimate of the population variance is required before particular sorts of inferences about the value of μ can be made. Unless the estimate of the population variance is statistically independent of the estimate of μ made from the sample, no simple way to make these inferences may exist. This question will be considered in more detail in Chapters 10 through 12. For the moment, suffice it to say that this principle is one of the main reasons for statisticians' assuming normal population distributions.

6.7 THE CENTRAL LIMIT THEOREM

As we have just seen, there are some very good reasons for the user of statistical methods to be interested in populations having a normal distribution of the random variable under study. Sometimes there are compelling reasons to believe that the population really would show such a distribution, could all of the observations actually be made. On other occasions, no very good reasons can be advanced why the population should *not* be normal, and so one assumes this to be the case, even though evidence to support this assumption may be lacking.

Nevertheless, it is quite common for us to be concerned with populations where the distribution should *definitely not* be normal. We may know this either from empirical evidence about the distribution or because some theoretical issue makes it impossible for scores to represent this kind of random variable. Simple illustrations are distributions of intelligence scores among graduate engineers, and of measures of socially nonconforming behavior among normal adults. In principle both these distributions should be extremely skewed, though for quite different reasons: in the first, there is a selection in the education process making it rather unlikely for anyone in the middle or low intelligence ranges to qualify as an engineer; in the second, social nonconformity is something that most people must show in very small degree. The assumption of a normal distribution does not make sense in such instances.

Very often an inference must nonetheless be made about the mean of such a population. To do this effectively the experimenter needs to know the sampling distribution of the mean, and to know this exactly, one has to be able to specify the particular form of the population distribution. However, if we had enough evidence to permit this, we would likely have an extremely good estimate of the population mean in the first place, and would not need any other statistical methods!

The way out of this apparent impasse in provided by the **central limit theorem,** which is stated approximately as follows:

If a population has a finite variance σ^2 and mean μ, then the distribution of sample means from samples of N independent observations approaches a normal distribution with variance σ^2/N and mean μ as the sample size N increases. When N is very large, the sampling distribution of $\bar{x}$ is approximately normal.

The central limit theorem may be the only example in the universe where one almost gets something for nothing! Absolutely nothing is said in this theorem about the *form* of the population distribution. Regardless of the population distribution, if sample size

Figures 6.7.1/6.7.2

A negatively skewed population distribution (*left*) and a sampling distribution of $\bar{x}$ for $N = 2$ (*right*), with comparable normal distributions.

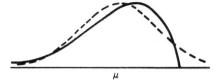
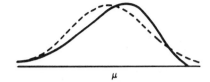

N is large enough, the normal distribution is a good approximation of the sampling distribution of the mean. This is the heart of the theorem since, as we have already seen, the sampling distribution will have a variance σ^2/N and mean μ for any N.

The sense of the central limit theorem is illustrated by Figures 6.7.1 through 6.7.4. The solid curve in Figure 6.7.1 is a very skewed population distribution in z-score form, and the broken curve is a standardized normal distribution. The other figures show the standardized form of the sampling distribution of means for samples of size 2, 4, and 10 respectively from this population together with the corresponding standardized normal distributions. Notice that it is *not* the number of samples drawn, but rather the *size* of each sample, that is the effective part of the central limit theorem. A vast number of samples of any size N may be drawn; the larger N is for each sample, the more nearly is the distribution of means normal.

Even with relatively small sample sizes, it is obvious that each increase in sample size gives a sampling distribution that is more nearly symmetric and more closely resembling the normal distribution, with the same mean and variance. This symmetry increases with increasing sample size until, in the limit, the normal distribution is reached.

It must be emphasized that in most instances the tendency for the sampling distribution of $\bar{x}$ to be like the normal distribution is very strong, even for samples of moderate size. Naturally, the more similar to a normal distribution the original population distribution, the more nearly will the sampling distribution of $\bar{x}$ be like the normal distribution for any given sample size. However, even extremely skewed or other nonnormal distributions may yield sampling distributions of $\bar{x}$ that can be approximated quite well by the normal distribution, for samples of at least moderate size. In the examples shown in Figure 6.7.3 and 6.7.4, the correspondence between the exact probabilities of intervals in the sampling distribution and intervals in the normal distribution is fairly good even for samples of only 10 observations; for a rough approximation, the normal distribution probabilities might be useful even here in some statistical work. In a great many instances in social research, a sample size of 30 or more is considered large enough to permit a satisfactory use of normal probabilities to approximate the unknown exact probabilities associated with the sampling distribution of $\bar{x}$. Thus, even though the central limit theorem is actually a statement about what happens *in the limit* as N approaches an infinite value, the principle at work is so strong that in many instances the theorem is practically useful even for only moderately large samples.

Figures 6.7.3/6.7.4

Sampling distributions of $\bar{x}$ for $N = 4$ (*left*) and $N = 10$ (*right*), with comparable normal distributions.

 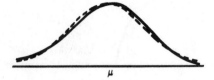

The mathematical proof of this theorem is extremely advanced, and many eminent mathematicians over the centuries contributed to its development before it was finally proved in full generality. However, some intuitive feel for why it should be true can be gained from the following example. Here we will actually work out the sampling distribution of the mean for a special and very simple population distribution.

Imagine a random variable with the following distribution:

x	$p(x)$
5	1/3
4	1/3
3	1/3

The μ of this little distribution is 4. However, instead of dealing directly with this random variable, let us consider the deviation d from μ, or

$$d = (X - \mu).$$

The distribution of d has a form identical to the distribution of X itself, although the mean d is zero. Since X can assume only three values, each equally probable, there are three possible deviation values of d, also equally probable.

Suppose that sample observations were taken two at a time, independently and at random. The deviation value of the first observation made is d_1 and that of the second is d_2. Corresponding to each joint event (d_1, d_2) there is a mean deviation $(d_1 + d_2)/2$, which is equal to $\bar{x} - \mu$. The following table shows the mean deviation value that is produced by each possible joint event:

d_2

1	0	.5	1
0	$-.5$	0	.5
-1	-1	$-.5$	0

$\qquad -1 \qquad 0 \qquad 1 \quad d_1$

Since X_1 and X_2 are independent, then d_1 and d_2 are also independent.

The probability associated with each cell in the table is thus $p(d_1)p(d_2)$ or 1/9.

Now let us find the probabilities of the various values of mean d. The value -1 can occur in only one way, and so its probability is 1/9. On the other hand, $-.5$ can occur in two mutually exclusive ways, giving a probability of 2/9. In this same way we can find the other probabilities and form the distribution of the mean d and the mean X values, as follows:

$\bar{x}$	Mean d	$p(\bar{x}) = p(mean\ d)$
5.0	1.0	1/9
4.5	0.5	2/9
4.0	0	3/9
3.5	−0.5	2/9
3.0	−1.0	1/9

where $\bar{x}$ = mean d + μ. Now notice that whereas in the original distribution there was no distinct mode, in this sampling distribution there *is* a distinct mode at 4, with exact symmetry about this point. What causes the sampling distribution to differ from the population distribution in this way? Look at the table of mean deviations corresponding to joint events once again: there are simply more possible ways for a sample to show a small, rather than an extreme, deviation from μ. There are more ways for a sample joint event to occur where the deviations tend to ''cancel out'' rather than an event where deviations tend to cumulate in the same direction.

Exactly the same idea could be illustrated for any discrete population distribution, whatever its form. The small and middling deviations of $\bar{x}$ from μ always have a numerical advantage over the more extreme deviations. This superiority in number of possibilities for a small mean deviation increases as N is made larger. Regardless of how skewed or otherwise irregular a population distribution is, by increasing sample size one can make the advantage given to small deviations so big that it will overcome any initial advantage given to extreme deviations by the original population distribution. The numerical advantage given to relatively small deviations from μ effectively swamps any initial advantage given to other deviations by the form of the population distribution, as N becomes large.

Perhaps this simple example gives you some feel for why it is that the sampling distribution of the mean approaches a unimodal, symmetric form. Of course, this is a far cry from showing that the sampling distribution must approach a *normal* distribution, which is the heart of the central limit theorem. Nevertheless, the basic operation of chance embodied in this theorem is of this general nature: for large N, it is relatively much easier for the deviation of $\bar{x}$ from μ to be small than to be large in absolute value.

6.8 CONFIDENCE INTERVALS FOR THE MEAN

When the idea of a confidence interval was introduced in the last chapter, it was illustrated in terms of Tchebycheff's inequality. This was done without our having to specify the exact form of the sampling distribution of a statistic such as $\bar{x}$. On the other hand, these confidence intervals were associated with quite general statements such as "the probability is at least .89" or the "probability is at least .95" that the obtained interval actually covers the population mean value.

Now we are ready to construct confidence intervals for the mean, where more precise probability statements may be made. This can be done because two principles can be applied: the sampling distribution of $\bar{x}$ is normal when the population is normal, and, when N is large, the sampling distribution of $\bar{x}$ is normal without respect to the population distribution.

In any normal distribution of sample means, with population mean μ and standard deviation σ_M, the following statement is true: **Over all samples of size N, the probability is .95 for the event**

$$-1.96\sigma_M \leq \bar{x} - \mu \leq +1.96\sigma_M. \tag{6.8.1}$$

That is, very nearly 95% of all possible sample means from the population in question must lie within 1.96 standard errors to either side of the true mean. You can check this for yourself from Table I in Appendix E.

However, this is still the same event even if we alter the inequality by changing all the signs and adding $\bar{x}$ to each term: **Over all sample values of $\bar{x}$ the probability is .95 for the event**

$$\bar{x} - 1.96\sigma_M \leq \mu \leq \bar{x} + 1.96\sigma_M. \tag{6.8.2*}$$

That is, over all possible samples, the probability is about .95 that the range between $\bar{x} - 1.96\sigma_M$ and $\bar{x} + 1.96\sigma_M$ will include the true mean, μ.

This range of values between $\bar{x} \pm 1.96\sigma_M$ is called the **95% confidence interval for μ.** The two boundaries of the interval, or $\bar{x} + 1.96\sigma_M$ and $\bar{x} - 1.96\sigma_M$, are called **the 95% confidence limits.**

Be sure to notice that μ is not a random variable and the probability statement is really not about μ, but about *samples*. Before any sample mean is seen, one may decide to compute the 95% confidence interval for each and every sample. The actual range of numbers obtained for a sample will depend on the value of $\bar{x}$. In short, over all possible samples, there will be many possible 95% confidence intervals. Some of these confidence intervals will represent the event "covers the true mean," and others will not. If one such confidence interval were sampled at random, then the probability is .95 that it covers the true mean.

The 99% confidence interval for the mean is given by

$$\bar{x} - 2.58\sigma_M \leq \mu \leq \bar{x} + 2.58\sigma_M. \tag{6.8.3*}$$

Notice that the 99% confidence interval is larger than the 95% interval. More possible values are included, and thus one can assert with more confidence that the true mean is covered. A 100% confidence interval would approach

$$\bar{x} - \infty\sigma_M < \mu < \bar{x} + \infty\sigma_M.$$

Here, one can be completely confident that the true value is covered, since the range includes all possible values of μ. On the other hand, the 0% interval is simply $\bar{x}$; for a continuous sampling distribution of $\bar{x}$, the probability is, in effect, zero that $\bar{x}$ exactly equals μ. Any other confidence interval may be found by finding the z values in a normal distribution that cut off the desired percent of cases in the middle of the distribution. These new z values then take the place of the values of ± 1.96 and ± 2.58 used above.

It is interesting to note that, in principle, the 95%, or 99%, or any other confidence interval for a mean might be defined in many other ways. For example, for a normal sampling distribution of $\bar{x}$ it is also true that the probability is approximately .95 that

$$-1.75\sigma_M \leq \bar{x} - \mu \leq 2.33\sigma_M.$$

This implies that a 95% confidence interval might possibly be found with limits

$$\bar{x} - 2.33\sigma_M \text{ and } \bar{x} + 1.75\sigma_M.$$

If many different ways of defining the 95% confidence interval are possible, what is the advantage of defining this confidence interval in terms of the *middle* portion of a normal distribution? In the particular case of the mean, this way of defining the confidence interval gives the *shortest* possible range of values such that the probability statement holds. Naturally, there is an advantage in pinning the population parameter within the narrowest possible range with a given probability, and this dictates the form of the confidence interval we use.

This method for finding a confidence interval applies either when the population is normally distributed or when the sample size is sufficiently large to make the central limit theorem apply. For many practical purposes, a sample size of 30 or more cases seems to be large enough to permit this method to be used.

A more serious drawback is the assumption that the population value of σ is known. Strictly speaking, these confidence intervals do not have the correct probabilities unless a known and constant value of σ is used in figuring the value of σ_M for each sample. However, this too is not an insurmountable problem. When σ is unknown, and the sample size N is large, confidence intervals that are approximately correct may still be found. We do this by using estimated σ_M or $s/\sqrt{N}$ in place of the unknown value of σ_M. That is, if the sample size is large (in this instance certainly 30 or more) the 95% confidence interval may be found by taking the approximate limits

$$\bar{x} - (1.96) (s/\sqrt{N}) \text{ and } \bar{x} + (1.96) (s/\sqrt{N}) \qquad [6.8.4^*]$$

and interpreting this interval in the usual way. Although a full rationale for our doing this will not be developed until a later chapter, this is the way that confidence intervals for the mean are usually calculated.

6.9 SAMPLE SIZE AND THE ACCURACY OF ESTIMATION OF THE MEAN

The width of any confidence interval for the mean μ depends upon σ_M, the standard error, and anything that makes σ_M proportionately smaller reduces the width of the interval. Thus, any increase in sample size, operating to reduce σ_M, makes the confidence interval shorter.

A practical result of this relation between the standard error of the mean and sample size is that the population mean may be estimated within *any desired degree of preci-*

sion, given a large enough sample size. Especially if the desired accuracy can be stated in population σ units, the required N is easy to find. For instance, how many cases should one sample to make the probability .99 that the sample mean lies within .1σ of the true mean? That is, the experimenter wants

$$\text{prob}(|\bar{x} - \mu| \leq .1\sigma) = .99.$$

Assuming that the sampling distribution is nearly normal, which it should be if sample size is large enough, this is equivalent to requiring that the 99% confidence interval should have limits

$$\bar{x} - .1\sigma$$

and

$$\bar{x} + .1\sigma.$$

However, this is the same as saying that

$$.1\sigma = 2.58\sigma_M$$

so that

$$.1\sigma = 2.58 \frac{\sigma}{\sqrt{N}}.$$

Solving for N, we find

$$\sqrt{N} = 25.8$$

$$N = 665.64.$$

In short, if the experimenter makes 666 independent observations, he or she can be sure that the probability of the estimate's being wrong by more than .1σ is only 1 in 100. Notice that we do not have to say exactly what σ is in order to specify the desired accuracy in σ units, and to find the required sample size.

Obviously, the question of adequate sample size could be settled this way in many social science studies. Once we have some idea of the desirable accuracy of estimation, stated in population standard deviation units, then we can set the sample size so as to give that degree of accuracy a high probability. Although this is a very sensible and easy way to determine sample size, it is not used as often as it should be in social science research. The main reason is that social scientists usually have not paid much attention to the question of the desirable accuracy of estimation. Although this is a complicated scientific question, other disciplines routinely concern themselves with such issues of accuracy in estimation. Perhaps as the social sciences continue to develop and mature, these questions will become as routine as they are in other sciences.

6.10 USE OF A CONFIDENCE INTERVAL IN A QUESTION OF INFERENCE

To see a simple example of how a confidence interval might be used in a problem concerning a population mean, imagine the following: A gerontologist is engaged in a study of the eating habits of women aged seventy or older. In particular this scientist is concerned that, due to inflation and other factors, the elderly may not be eating as well today as in years past. A study was done 10 years ago, in which it was found that healthy women over age seventy, living independently and not in a nursing or retirement home, had an average daily intake of 2,032 calories. Now the gerontologist wishes to estimate the current daily calorie intake of such a population. In particular, the question to be asked is, "Is the average daily calorie intake of this population the same as it was 10 years ago?"

A sample of 100 women over the age of seventy is drawn, where each woman sampled meets the criteria listed above. The daily caloric intake of each individual sampled is measured. The sample mean turns out to be 1,847, with a standard deviation s equal to 310. On the face of it, this sample suggests that elderly women are consuming fewer daily calories on the average than they did 10 years ago.

This is, after all, just a sample result, and it could be true that the population of elderly women still has the same, or even a higher, average calorie intake than before. Can the investigator say with assurance that the present calorie intake is not the same? This question may be answered by calculating confidence limits for the population mean. Here, since $s = 310$ the estimated standard error of the mean is

$$\text{est } \sigma_M = 310/\sqrt{100} = 31.$$

Then the approximate 95% confidence interval has limits given by

$$1,847 - (1.96)(31) \text{ and } 1,847 + (1.96)(31)$$

or

$$1,786.24 \text{ to } 1,907.76.$$

Now notice that the interval between these two values *does not* contain the former population value of 2,032. Since the probability is .95 that the true population value of the mean is covered in this interval, then the probability is only .05 or less that the sample represents a population with a mean outside this range of values. In other words, the investigator rejects the hypothesis that the new population mean is the same as the old. However, this could be the wrong conclusion. What is the probability of the investigtor's being wrong in this conclusion? It is no more than .05—the probability of the sample confidence interval's not covering the true value of μ.

This exemplifies one use to which a confidence interval for a mean may be applied: checking on a hypothetical population situation in the light of sample evidence. Such hypothesis testing forms an important part of statistical inference. The construction of a confidence interval is one way that a hypothesis about an exact value of some population's mean may be tested in the light of a sample. If the confidence interval does not cover the hypothetical value of the mean, then that hypothesis may be rejected. If the K percent confidence interval is used, one says that the hypothesis is rejected at the $(100 - K)$ percent level. We will have a great deal more to say about procedures and conventions for testing hypotheses in subsequent chapters. For the moment, let it suffice that this is one highly useful application of the idea of a confidence interval.

6.11 CONFIDENCE INTERVALS FOR PROPORTIONS

It was seen in Section 6.4 that as N is made larger, the binomial distribution can be approximated quite well by the normal distribution. In consequence, it is possible to construct confidence intervals for population proportions in much the same way as confidence intervals for the mean, given a large N.

When a sample of N independent observations is taken, the sample proportion P is found by taking $P = x/N$, where x is the number of successes, or events of a particular kind. When N is reasonably large (say $N \geq 20$ when P is about .5, but $N \geq 50$ for $p \leq .4$), the distribution of values of P should be close to a normal distribution, where

$$E(P) = p$$

and

$$\sigma_P = \sqrt{pq/N}.$$

We would like to be able to form a confidence interval that is highly likely to cover the value of p, the population proportion. At first glance it appears that this could be done simply by taking $z = 1.96$ or 2.58 and substituting into

$$P - z\sqrt{\frac{pq}{N}} \quad \text{and} \quad P + z\sqrt{\frac{pq}{N}}.$$

However, there is a hitch in doing this: the value of p that we are trying to estimate is, of course, unknown. Nevertheless, the possible values of p corresponding to these limits must satisfy the relation

$$\frac{N(P - p)^2}{(p - p^2)} = z^2.$$

Simply solving this quadratic equation for p by methods familiar from high-school algebra yields the following expression for the confidence interval:

$$\frac{N}{N + z^2}\left[P + \frac{z^2}{2N} \pm z\sqrt{\frac{PQ}{N} + \frac{z^2}{4N^2}} \right] \qquad [6.11.1]$$

where $Q = 1 - P$ and z is the standard score in a normal distribution cutting off the *upper (K/2)* proportion of cases. These limits define the $100 (1 - K)$ percent confidence interval for p.

If N is very large, (certainly $N \geq 100$), this confidence interval may be replaced by the simpler approximation

$$P - z\sqrt{\frac{PQ}{N - 1}} \leq p \leq P + z\sqrt{\frac{PQ}{N - 1}}. \qquad [6.11.2]$$

For example, suppose that we were interested in the question of how many full professors at state universities prefer teaching activities and how many prefer research and writing. We decide to take a random sample from among all such professors, and ask each his or her preference. Here, quite arbitrarily, let us say that

"success" = prefers teaching
"failure" = prefers research

We will not allow a response of indifference or "no prefernce," so that by eliminating such responses from our sample, we are restricting the population of interest to "professors who have a definite preference." Now suppose that we are successful in drawing a random sample of full professors at state universities, providing $N = 132$ usable responses.

From this sample, the proportion of successes, or P, turns out to be .62. In order to find the 95% confidence interval for p, the population proportion, we use Eq. 6.11.1 and take

$$\frac{132}{132 + (1.96)^2} \left[.62 + \frac{(1.96)^2}{2(132)} \pm 1.96 \sqrt{\frac{.62 \times .38}{132} + \frac{(1.96)^2}{4(132)^2}} \right]$$

which works out to provide the confidence limits of approximately

.535 and .698.

Then the probability is about .95 that in the population of college professors the proportion preferring teaching to research is covered by the range of values .535 to .698. Can one say it is reasonable to suppose that just as many professors prefer teaching as prefer research, on the basis of this evidence? If the preferences for the two activities were exactly equal, then the population proportion would be .50. Notice, however, that this value is not covered by the 95% confidence interval just constructed. Thus, on the basis of this evidence, the hypothesis of equal preference may be rejected at the .05, or 5%, level.

EXERCISES

1. A teacher believes that the class scores on a final examination should be approximately normally distributed, provided the class is large enough. If this belief is correct, what proportion of the class should fall at or below the following z values?

(a) -1.2 (d) -1.78
(b) 0.96 (e) -0.43
(c) 1.88 (f) 2.15

2. Another teacher wishes to curve class grades to make them correspond to values in a normal distribution with a mean of 100 and a standard deviation of 10. To do this, she first gives each score a cumulative relative frequency. These are then equated with corresponding z values in a normal distribution, and thus into the scores desired. Carry out this process with the following set of scores:

18, 21, 23, 25, 30, 31, 36, 38, 40, 42, 45, 46, 48, 50, 52, 57, 60, 75

3. A mean from a sample of 36 cases has a value of 100. How probable is a sample mean of 100 or more when the population being sampled is normal with the following parameters?

(a) Mean 103, standard deviation 10
(b) Mean 99, standard deviation 4
(c) Mean 80, standard deviation 50
(d) Mean 98, standard deviation 24
(e) Mean 110, standard deviation 80

4. A normal distribution has a mean of 500 and a standard deviation of 10. The researcher desires to divide this distribution into five intervals of values such that the probability of each interval will be exactly .20. List the desired intervals, starting with the highest.

5. A study was concerened with the age at which American mothers have their first children. Twenty years ago the age at first delivery was known to be 23.6. However, it is also known that this distribution is not symmetric, since mothers tend to have their first children at early rather that at late ages. A sample of 200 first births was drawn at random from records for the past year, and the age of each mother recorded. This sample yielded a mean of 24.1 years, with a standard deviation S of 5.6. How likely is a sample mean this much or more deviant from expectation (in either direction) if the population mean is truly 23.6? Is it reasonable to conclude that the average age has changed? What does the presumed nonnormal distribution of this population do to our inference in this situation?

6. Under each of the conditions outlined for the population in Exercise 3, how probable is it that a sample mean based upon 36 independent observations will fall in the interval of values 95–105?

7. Using the method illustrated in Section 6.4 in the text, compare the binomial probabilities for $N = 8$, $p = .5$ with the corresponding normal probabilities. Then compare $N = 8$, $p = .4$ with normal probabilities. Is the latter appreciably the poorer?

8. Prior to a presidential primary in a certain state, a candidate claimed approximately 45% of the votes of the party. A newspaper took a random sample of 500 registered voters in that party and found that only 37% indicated they would vote for the candidate. If the true proportion of voters favoring the candidate is .45, how probable is a sample result showing 37% or fewer in favor? Would you say that there is good reason to doubt the candidate's assertion?

9. A study found that a large group of children judging the radius of a circle made errors which were approximately normal in their distribution, with a mean of .1 in., and a standard deviation of .03 in. A new study focussed on the extent to which these judgments might change if the circle were embedded in a larger circle. To examine this, 10 children were drawn at random and asked to judge the radius of such a circle. They showed a mean error of $-.2$ in. If embedding the circle in a larger figure is having no effect, and these children can be thought of simply as a sample from the former population, how likely is a sample mean differing this much from expectation (in either direction)?

10. A known population has an average height of 68.2 in. The population distribution is normal, and it is known that the middle 50% of the population have heights between 66.5 and 69.9 in. If a sample of 25 individuals were drawn independently and at random from this population, how likely is it that their mean height would exceed 72 in.? How likely is it that their mean height would be less than 60 in.? Beyond what mean height should only about 5% of sample means fall?

11. A random sample of 3,000 income tax returns was taken from a given tax year. Each was examined for the number of exemptions claimed for that year. The mean was found to be 3.78, with a standard deviation of .97. Find the 99% confidence interval for the true mean based on these data.

12. A standardized test was designed so that it should produce an approximately normal distribution for normal adults. The testmaker stated that the middle 50% of the scores for all such adults should lie between 294 and 306. What is the mean and standard deviation for this population?

13. Suppose that a sample of 40 adults is taken at random from the population described in Exercise 12. Find the probability that the mean of this sample will fall between the limits 303 and 304, and between the limits 304 and 305. What are the odds that the mean will fall into the first interval as opposed to the second?

14. In a study of reading, subjects were asked to identify 100 words flashed on a screen under varying background conditions. Subjects often confused these words with other words with similar appearance. A sample of 269 subjects showed an average of 7.91 such confusions out of the 100 words presented to each. The standard deviation S for these subjects was .47. Find the 99% confidence limits for the average number of confusions in a population of such subjects.

15. Suppose that a sample of 20 normal adults is taken at random and each person given the test described in Exercise 12. What is the probability that this sample will have a mean of 294 or less? What is the probability of a mean of 306 or more?

16. As a part of an experiment, four distinct random samples are drawn from the same population of subjects. The 99% confidence interval for the mean is found for each sample. What is the probability that every single confidence interval found will cover the population mean? What is the probability that at least one confidence interval will fail to do so? (**Hint:**Regard the event ''covers the population mean'' as a success in a Bernoulli process.)

17. In a very large experiment, 200 random samples were taken independently from the same population. The experimenter believed that the population mean was actually 67.9 units. When the 90% confidence interval was found for each of the 200 samples, some failed to cover the value of 67.9. If the population mean actually was 67.9, how many such ''noncovering'' intervals should have been expected?

18. For the situation described in Exercise 17, show (in symbols) the probability that all 200 intervals cover the population value. Is this probability large or small? Is it therefore likely or unlikely that one or more intervals failed to cover the true value?

19. A dress manufacturer suspected that about 25% of the bolts of cloth that were being purchased from a certain wholesaler contained defects. Accordingly, a random sample of 75 from among a large number of bolts obtained from this source were examined. It was found that 15 had defects. Find the 95% confidence limits for the true proportion of defective bolts.

20. In Exercise 19, how likely is the manufacturer to obtain 15 or fewer defective bolts if the true proportion is actually .25? Is the manufacturer running a large risk in concluding that 25% or more are defective on the basis of this evidence?

21. In the past the city council of a large municipal government has proposed numerous bond issues, only to have them defeated. Consistently, 54% of the voters are against any such issue. A new bond issue is about to come up for vote. A random poll of 1,000 voters shows that 51% favor this issue, and 49% are against. Form the 99% confidence interval for the true proportion of voters favoring this issue. Does this include the value .46? Does the council have any reason to believe that 50% or more of voters might favor this issue?

22. Prior to the further development of a small electric automobile, a company conducted a survey to see how many middle-income families would be interested in such a car if it were offered for sale. The results showed that 37% of 5,000 such families expressed an interest. Form the 99% confidence interval for the true precentage of such families, given that the sample was drawn at random from among all middle-income families in the United States. How likely was the company to have received respones at least this favorable if the true percentage were only 30?

23. In a study of the effects of a medication on the body temperature of normal adults a scientist wishes to be 95% sure that the estimates made from a sample are within .01 °F of the population mean. The population under study is believed to have a standard deviation in body temperature of .07 °F. How many subjects should be used in the sample if these conditions are to be met?

24. Suppose that the experimenter in Exercise 23 did not know the standard deviation of the population. Nevertheless, it was desired to estimate mean body temperature to within 1/4 of a standard deviation, with 99% confidence. How many subjects were needed?

25. The public health officer in Exercise 19, Chapter 5, drew a random sample of 900 children from the area under study. Some 135 children showed a positive test for tuberculosis. What is the probability that no more children than this should show this reaction if the true population proportion is .18? How much risk is the officer running in stating that the true proportion is less than .18?

26. The psychologist in Exercise 27, Chapter 5 took a sample of 127 rats and ran each through the maze. The average score was 1.52. If, as the psychologist suspects, the average score of the population of rats is truly 1.5, how probable is a sample mean this far or farther from expectation in either direction?

27. For Exercise 29 of Chapter 5, find the 90% confidence interval for the mean of the population of employees under the new training (using 4.9 as the standard deviation once again). How do these exact limits differ from those found in Exercise 30 of Chapter 5?

28. Use the data in Exercise 31, Chapter 5, and find the 95% confidence limits for the mean, given that the population is normal with a standard deviation of 7.

29. An experimenter wishes to be 95% certain that the mean of a sample will fall within K standard deviations of the true mean. Make a table showing how the size of K changes with different sample sizes. Use the following values for N: 1, 5, 10, 20, 30, 40, 50, and 100.

30. Four independent random samples were taken from the same population. The means and variances of the samples are given below. Establish the 95% confidence interval based upon the mean and variance from the pooled samples.

Samples			
1	2	3	4
$N_1 = 83$	$N_2 = 46$	$N_3 = 74$	$N_4 = 88$
$\bar{X}_1 = 106.2$	$\bar{X}_2 = 112.9$	$\bar{X}_3 = 107.8$	$\bar{X}_4 = 109.3$
$S_1^2 = 10.11$	$S_2^2 = 15.26$	$S_3^2 = 13.19$	$S_4^2 = 14.19$

Is this a "better" confidence interval than that figured from any single one of the samples? Explain why.

31. For Exercise 33, Chapter 5, find the 99% confidence interval for the mean amount spent for books.

32. For Exercise 32, Chapter 5, what are the odds that a 40-year old American watched between 6 and 10 h of television during the week, as opposed to some other number of hours? (Use the unbiased estimate of the variance figured from the sample here.)

33. For the data in Exercise 35, Chapter 5, find a pooled estimate of the population mean and standard deviation. How probable is a mean value this large or larger if the mean of the population is actually 102?

34. Imagine that the archeologist in Exercise 20, Chapter 5 had assembled 720 potshards from the same level of some dig, where each potshard appeared to represent a different pot. Suppose that 27% of these potshards showed decoration. What is the probability of this few or fewer such potshards if this site is later than 3000 B.C.?

35. In Chapter 2, Section 2.17 discusses three discrete random variables following very simple rules. Suppose that a sample of 50 observations were made in turn from each of these three distributions. A sample mean is computed for each. For which of these theoretical distributions would you anticipate that the sampling distribution of means would be most nearly normal? Why? Which sampling distribution would have the smallest standard error?

Chapter 7

Hypothesis Testing

The title of this chapter might well be "How to decide how to decide," or "How much evidence is enough?" The social scientist or anyone else who samples from a population is trying to decide, or at least form an opinion on, something about the population. Quite often the investigator wants to decide if some hypothetical population situation appears reasonable in the light of the sample evidence. Sometimes the problem is to judge which of several possible population situations is best supported by the evidence at hand. In either instance, one is trying to make up his or her mind from evidence. However, as we shall see, there are many possible ways of making this kind of decision, depending on what information in the sample is actually used, the various possibly true population situations being compared, and especially the risk one is willing to take of being wrong in the decision.

Granting for the moment that we know the information we need from the sample and how to extract it, how do we judge the tenability of a particular hypothesis about the population? It seems reasonable that a tenable or "good" hypothesis about the population should provide us with a good expectation about the sample result in terms of the relevant statistic. That is, the hypothesis gives a good fit to the data if, as a consequence of that hypothesis' being true, we can deduce an expected value of the statistic that agrees quite well with the value actually observed; if the hypothesis is true, there is some value that we should expect the appropriate sample statistic to show. If the hypothesis is a good one, our obtained sample result should fall into a region of values relatively close to the expected value the hypothesis dictates for our statistic.

A substantial deviation between the obtained value of the statistic and its expected value under the hypothesis implies one of two things: either the hypothesis is right, and the difference in value between the statistic and its expectation is the product of chance, or the hypothesis is wrong and we were not led to expect the right thing of our sample statistic. In particular, if the sample value falls so far from expectation that it lies in an interval of values very improbable given the hypothesis, but this sort of sample result would be made probable if some alternative hypothesis were true, then doubt is cast on the original hypothesis and we reject it in favor of the alternative. Obviously, however, these notions of "disagree with expectation," "improbable," and "probable" are relative matters; when do we say that a sample result disagrees *enough* with expectation under some hypothesis to warrant rejection of the hypothesis itself?

What we need is a decision rule, a guide giving the conclusion we will reach depending on how the data turn out, and which can be formulated even before the data themselves are seen. However, there is literally no end to the number of decision rules that one might formulate for a particular problem. Some of these ways of deciding might be very good in a particular circumstance but not so good in others, and so we need to decide how to decide on the basis of what we wish to accomplish in a specific situation. The branch of mathematics known as *decision theory* treats this problem of choosing a decision rule, and we are going to apply some of its elementary principles here. We will find that there is no rigid formula supplying the right way to decide in all situations, and, indeed, we often lack the information necessary for choosing an optimal decision rule. For this reason social scientists and others using statistical inference most often fall back on accepted conventions for evaluating evidence, and these conventions may have very little to do with the principles of decision theory. Nevertheless, the application of principles from decision theory to the problem of statistical inference does shed some light on the ideas underlying tests of statistical hypotheses.

Although we will speak of deciding about some single hypothesis, in practice the experimenter always behaves as if the choice lies between *two* hypotheses. A great many issues connected with the use of statistical inference can be understood only if this decision-making task is presupposed.

The process of comparing two hypotheses in the light of sample evidence is usually called a *statistical test* or a *significance test*. The competing hypotheses are stated in terms of the form, or of one or more parameters, of the distribution for the population. Then the statistic containing the relevant information is chosen, having a sampling distribution dictated by the population distribution specified in the particular hypotheses. The hypothesis that dictates the particular sampling distribution against which the obtained sample value is compared is said to be "tested." The significance test is based on the sampling distribution of the statistic, given that the particular hypothesis is true. If the observed value of the statistic falls into an interval representing a kind and degree of deviation from expectation which is improbable given the hypothesis, but which would be relatively probable given the other, alternative hypothesis, then the first hypothesis is said to be rejected. The actual decision to entertain or to reject any hypothesis is thus based on whether or not the sample statistic falls into a particular region of values in the sampling distribution dictated by that hypothesis.

As we look into this problem of the choice of a decision rule, we will be using the mean $\bar{x}$ and the proportion P as examples of statistics on which decisions are based. However, the issues raised apply to the testing of any hypothesis. Before we can consider these ideas more closely, some standard terminology must be given.

7.1 STATISTICAL HYPOTHESES

A statistical hypothesis is usually a statement about one or more population distributions, and specifically about one or more parameters of such population distributions. It is always a statement about the population, not about the sample. The statement is called a hypothesis because it refers to a situation that *might* be true. Statistical hypotheses are almost never equivalent to the hypotheses of science, which are usually statements about phenomena or their underlying bases. Quite commonly, statistical hypotheses grow out of or are implied by scientific hypotheses, but the two are seldom identical. The statistical hypothesis is usually a concrete discription of one or more summary aspects of one or more populations; there is no implication of *why* populations have these characteristics.

We shall use the following scheme to indicate a statistical hypothesis: a letter H followed by a statement about parameters, the form of the distribution, or both, for one or more specific populations. For example, one hypothesis about a population could be written

H: the population in question is normally distributed with $\mu = 48$ and $\sigma = 13$, and another

H: the population in question has a Bernoulli (two-class) distribution with $p = .5$.

Hypotheses that completely specify a population distribution are known as **simple hypotheses.** In general, the sampling distribution of any statistic also is completely specified given a simple hypothesis and N, the sample size.

We will also encounter hypotheses such as

H: the population is normal with $\mu = 48$.

Here, the exact population distribution is not specified, since no requirement was put on σ, the population standard deviation. When the population distribution is not determined completely, the hypothesis is known as **composite.**

Hypotheses may also be classified by whether they specify *exact* parameter values, or merely *a range* or *interval* of such values. For example, the hypothesis

H: $\mu = 100$

would be an **exact hypothesis,** although

H: $\mu \geq 100$

would be **inexact.**

The hypothesis actually to be tested will be given the symbol H_0. Such a hypothesis is commonly referred to as the *null hypothesis,* meaning that this is the hypothesis that is assumed true in generating the sampling distribution to be used in the test. The other hypothesis, which is assumed to be true when the null hypothesis is false, is referred to as the *alternative hypothesis,* and is often symbolized by H_1. Both the null and the alternative hypothesis should be stated before any statistical test of significance is attempted.

Incidentally, there is an impression in some quarters that the term "null hypothesis" refers to the fact that in experimental work the parameter value specified in H_0 is very often zero. Thus, in many experiments the hypothetical situation "no experimental effect" is represented by a statement that some mean or difference between means is exactly zero. However, as we have seen, the tested hypothesis can specify any of the possible values for one or more parameters, and this use of the word *null* is only incidental. It is far better for the student to think of the null hypothesis H_0 as simply designating that hypothesis actually being tested, the one which, if true, determines the sampling distribution referred to in the test.

In applied situations very seldom will a hypothesis specify the sampling distribution completely. Simple hypotheses are just not of interest or are not available in most practical situations. On the other hand, it *is* necessary that the sampling distribution be completely specified before any precise probability statement can be made about the sample results. For this reason assumptions usually are made, which, taken together with the hypothesis itself, determine the relevant statistic and its sampling distribution and justify a test of the hypothesis. These assumptions differ from hypotheses in that they are rarely or never tested against the sample data. They are assertions that are simply assumed true, or the evidence is collected in such a way that they must be true (e.g., random sampling).

These assumptions are necessary for the formal justification of many of the methods to be discussed. It is not true, however, that these assumptions are always or even ever realized in practice. The results of any data analysis leading to statistical inference can always be prefaced by "If such and such assumptions are true, then. . . ." The effects of the violation of these assumptions on the conclusions reached can be very serious in some circumstances, and only minor in others. In later sections the importance of the most common assumptions will be discussed. Nevertheless, these assumptions should always be kept in mind, along with the conditional character of any result subject to the assumptions being true.

7.2 TESTING A HYPOTHESIS IN THE LIGHT OF SAMPLE EVIDENCE

Every test of a hypothesis involves the following features:

1. The hypothesis to be tested is stated, together with an alternative hypothesis.
2. Additional assumptions are made, permitting one to specify the sample statistic that is most relevant and appropriate to the test, as well as the sampling distribution of that statistic.
3. Given the sampling distribution of the test statistic when the hypothesis to be tested is true, a **region of rejection** is decided upon. This is an interval of possible values deviant from expectation if the hypothesis were true, but which more or less accord with expectation if the alternative were true. This region of rejection contains values relatively improbable of occurrence if the first hypothesis were true, but relatively probable given the alternative. The risk one is willing to take in rejecting the tested hypothesis *falsely* determines the size of the region of rejection, and the alternative hypothesis determines its location.

4. The sample itself is obtained. If the computed value of the test statistic falls into the region of rejection, then doubt is cast on the hypothesis, and it is said to be rejected in favor of the alternative. If the result falls out of the region of rejection, then the hypothesis is not rejected, and the experimenter may choose either to accept the hypothesis or to suspend judgment, depending on the circumstances. A sample result falling into the region of rejection is said to be **statistically signifi-cant, or to depart significantly from expectation** under the hypothesis.

As an example, suppose that an experimenter entertains the hypothesis that the mean of some population is 75:

H_0: $\mu = 75$.

This is actually a composite hypothesis, since nothing whatever is said about the form of the population distribution nor about other parameters such as the standard deviation. The experimenter is, however, prepared to assume that whatever is true about the mean, the population has a normal distribution and the standard deviation for the population is 10; for the moment, let us imagine that both of these assumptions are reasonable in the light of what the experimenter already knows about this problem. In effect, the addition of these two assumptions to the original hypothesis makes this hypothesis simple:

H_0: $\mu = 75$, $\sigma = 10$, normal distribution.

Now, what of the alternative hypothesis? It might be that the alternative hypothesis is

H_1: $\mu \neq 75$.

Note that this alternative happens to be inexact, and really asserts that some (unspeci-fied) value of μ other than 75 is true. However, even if H_1 is true and H_0 false, the experimenter will still assume that the distribution is normal with a standard deviation of 10.

It has already been decided that 25 independent observations will be made, Given this sample size N, the information in the hypothesis H_0, and the assumptions, the sampling distribution of the mean is known completely: this must be a normal distribu-tion with a mean of 75 and a standard error of $10/\sqrt{25}$ or 2, if the hypothesis H_0 is true.

Now in accordance with H_1, the experimenter decides to reject this hypothesis H_0 only if the sample mean falls among the extremely deviant ones in either direction from the value 75; in fact, H_0 will be rejected only if the probability is .05 or less for the occurrence of a sample mean as deviant or more so from 75. Thus, the region of rejec-tion for this hypothesis contains only 5% of all possible sample results when the hy-pothesis H_0 is actually true. Such sample values depart widely from expectation and are relatively improbable given that H_0 is true; on the other hand, samples falling into these regions are relatively more probable if a situation covered by H_1 is true. These regions of rejection in the sampling distribution are shown in Figure 7.2.1.

Tables of the normal distribution show that a z score of 1.96 has a cumulative prob-ability $F(1.96)$ of about .975. Hence only .025 of all sample means should lie *above* 1.96 standard errors from μ. Similarly, it must be true that about .025 of all sample

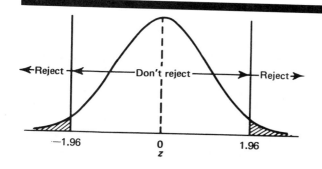

Figure 7.2.1

Possible rejection regions for a normal sampling distribution.

means will lie below -1.96 standard errors from μ. In short, the region of rejection includes all sample means such that, in this sampling distribution,

$$1.96 \leq z_M$$

or

$$z_M \leq -1.96$$

giving a total probability of .025 + .025 or .05 for the combined intervals. A value of the sample mean that lies exactly on the boundary of a region of rejection is called a **critical value** of $\bar{x}$; the critical z_M values here are -1.96 and 1.96, so that the critical values of $\bar{x}$, given $\sigma_M = 2$, are

$$-1.96\sigma_M + \mu = -1.96(2) + 75 = 71.08$$

and

$$1.96\sigma_M + \mu = 1.96(2) + 75 = 78.92.$$

Now the sample is drawn and turns out to have a mean of 79. The corresponding z_M score is

$$z = \frac{79 - \mu}{\sigma_M} = \frac{79 - 75}{2} = 2.$$

Notice that this z_M score *does* fall into the region of rejection since it is greater than a critical value of 1.96. Then the experimenter says that the sample result is **significant beyond the 5% level.** Less than 5% of all samples should show results this deviant (or more so) from expectation under H_0, if H_0 is actually true. Since the experimenter has decided in advance that samples falling into this region show sufficient departure from expectation to be called *improbable* results, then doubt is cast on the truth of the hypothesis, and H_0 is said to be rejected.

In its essentials this simple problem exemplifies all significance tests. Naturally, the details vary with the particular problem. Hypotheses may be about variances or any other characteristic feature of one or more populations. Other statistics may be appropriate to a test of the particular hypothesis. Other assumptions may be made to permit the specification of the sampling distribution of the relevant statistic. Other regions of rejection may be chosen in the sampling distribution. But *once these specifications are made,* the experimenter knows, even before drawing the sample, how the decision will be made in the light of what the sample shows.

Here the region of rejection was simply specified without any particular explanation. In this step, however, lies a key problem of significance tests. In the choice of the region of rejection, the experimenter is deciding how to decide from the data. It is perfectly obvious that there are very many ways that the experimenter could have chosen a rejection region. It could have been decided to reject the hypothesis if the sample result fell 1.65 standard errors below the mean specified by the hypothesis, or if the sample result fell between 3 and 4 standard errors above the hypothetical mean. Each of these intervals has small probability in a normal sampling distribution. There is no law against using any decision rule for rejection, although on intuitive grounds alone, the one actually used here seems somehow more sensible than the others suggested. We now examine this question in more detail.

7.3 A PROBLEM IN DECISION THEORY

Next to breathing, perhaps the most common human activity is decision making. Every moment of existence is filled with choices that the human being must make. From the time that the caveman pondered ''Shall I leave the campfire to go out into the dark for a moment?,'' to the investor asking, ''Shall I buy this common stock while its value is low?,'' life has remained full of choices among actions. Each of these decisions carries with it an element of risk. There might have been a tiger waiting in the dark for the caveman, even though our ancestor hoped that he might catch a small animal for his own family to eat. The investor might buy the stock and see it streak upward in value; or see it hit rock bottom immediately after purchase.

Decision theory, which actually grew out of problems of economic decision making, has established some general principles which extend to decision situations of many kinds. Any number of choices of action are usually open to the person who must make a decision, and once a decision to act has been made, some event or chain of events is going to occur. Sometimes the outcome of a choice of action is almost certain: if I decide to place my hand in a roaring fire, the outcome of a burn is almost sure to occur. However, in most instances the decision maker does not know exactly what the outcome will be be. Such decisions are called *risky* for this reason. In most decision settings there are various ''states of the world'' that might or might not be true (e.g., the tiger in the bushes, the health of the stock market). Then the outcome of any decision depends both on what is decided and what turns out to be true.

In statistical inference these possible states of the world are formulated as hypotheses. We try to form an advance judgment on which state of the world is true by drawing a sample. Then on the basis of that evidence we try to make a decision about the hypothesis under study. Since our knowledge is necessarily imperfect, these decisions too are risky.

To illustrate this point, along with some of the principles of decision theory, we may use a very simple example. Even this example, homely as it is, illustrates the problems of how to decide how to decide in the presence of evidence.

Suppose that you were placed in the following decision situation. You are seated before a table, on which rests a bag full of paper currency. You are told that there are exactly 50 bills in all in the bag. Furthermore, one of two situations is possible: either 80% of the bills in the bag are $20 bills, with the remainder being $10 bills, or 40% of the bills are $20 bills, with the remainder in tens. Your task is to guess which bag is actually on the table. If the guess is right, you will get the bag of money; should your guess be wrong, you will get nothing.

To make your decision a little easier, the experimenter is willing to allow you to extract 10 bills from the bag at random, with replacement after each draw. You are expected to tell the experimenter ahead of time how you will decide after having seen the sample of bills.

The choice open to you is equivalent to a decision between the exact hypotheses

$$H_0: p = .80$$

$$H_1: p = .40.$$

That is, either the true proportion in the bag for $20 bills is .80 or it is .40.

Being a good statistician, you know that drawing a $20 bill from the bag is a success, and the distribution of P, the proportion of successes among 10 draws, is binomial, with $N = 10$. However, the actual sampling distribution depends on p, which in turns depends on which bag is actually on the table. The binomial distribution under each of the two hypotheses is shown in Table 7.3.1.

Now how should you decide on the basis of the sample evidence? You are going to need a decision rule, telling you how to proceed on the basis of the number of $20 bills in the sample. There are any number of decision rules that you might use, but for the sake of simplicity we are going to evaluate the efficacy of only three rules that you might possibly use:

Decision Rule 1: If P is greater than or equal to .8, choose H_0; otherwise H_1.
Decision Rule 2: If P is greater than or equal to .6, choose H_0; otherwise H_1.
Decision Rule 3: If P falls between .2 and .8 inclusive, choose H_0; otherwise H_1.

Table 7.3.1

Binomial distributions of P for $p = .8$ and $p = .4$, with $N = 10$.

P	$p(P)$ if $p = .8$	$p(P)$ if $p = .4$
1.0	.107	.0001(+)
.9	.268	.002
.8	.302	.011
.7	.201	.042
.6	.088	.111
.5	.026	.200
.4	.006	.251
.3	.001	.215
.2	.000(+)	.121
.1	.0000(+)	.040
.0	.0000(+)	.006

Regardless of which decision rule is selected there are two ways that you could be right, and two ways of making an error. This is diagrammed below:

True situation

		H_0	H_1
	H_0	correct	error
Decision			
	H_1	error	correct

If you decide on H_0, and H_1 is actually true, an error is made. Furthermore, you make an error by deciding on H_1 and H_0 is actually true. The other possibilities lead, of course, to correct decisions.

Given the sampling distributions under each of the two hypotheses and any decision rule, we can find the probabilities of these two kinds of error.

Suppose that you have no idea which hypothesis is true, and all you have to go by is the occurrence of a P value leading you to reject one or the other of the hypotheses. Suppose that Decision Rule 1 were adopted. This rule requires that the occurrence of any P of .8 or more leads automatically to a decision that H_0 is true. What is the probability of observing such a sample result *if H_1 is true?* Notice that whenever this happens, there will be a wrong decision and so this probability is that of *one kind* of error. The binomial distribution for $p = .4$ shows a probability of about .013 for P greater than or equal to .8, and so the probability is .013 of *wrongly* deciding that H_0 is true.

In the same way, we can find the probability of the error made in choosing H_1 when H_0 is true. By Decision Rule 1, H_1 is chosen whenever P is less than or equal to .7; in the distribution under H_0, this interval of values has a probability of about .323. Thus, the probabilities of error and correct decisions, under Decision Rule 1, are as shown in Table 7.3.2.

The probability of a correct decision is $1 - p(\text{error})$ for either of the possibly true situations (the two error probabilities appear in italics). Notice that you are far more likely to make an erroneous judgment by using this rule when H_0 is true than for a true H_1.

Exactly the same procedure gives the erroneous and correct decision probabilities under Rule 2, as shown in Table 7.3.3.

By this second rule, the probability of error is relatively smaller when H_0 is true than it is for Rule 1. However, look what happens to the probability of the other error. This illustrates a general principle in the choice of decision rules: Any change in a decision

Table 7.3.2

		True situation	
		H_0	H_1
	H_0	.677	.013
Decision			
	H_1	.323	.987

	H_0	H_1	**Table 7.3.3**
H_0	.966	.166
H_1	.034	.834

	H_0	H_1	**Table 7.3.4**
H_0	.625	.952
H_1	.375	.048

rule that makes the probability of one kind of error *smaller* will ordinarily make the other error probability *larger* (other things, such as sample size, being equal).

Before trying to choose between Rules 1 and 2, let us write down the probabilities for Decision Rule 3 (See Table 7.3.4).

Even on the face of it this decision rule does not look sensible. The probability of error is large when H_0 is true, and the experimenter is almost sure to make an error if H_1 is the true situation! This illustrates that not all decision rules are reasonable if the experimenter has any concern at all with making an error; here regardless of what is true, the experimenter has a larger chance of making an error using this rule than in using either Rules 1 or 2. Rules such as this are called **inadmissible** by decision theorists. We need confine our attention only to the two relatively good rules, 1 and 2.

There is a real problem in deciding between Rules 1 and 2; Rule 1 is good for making error probability small when H_1 is true, but risky when H_0 represents the true state of affairs. On the other hand, Rule 2 makes for small chance of error when H_0 is true, but gives relatively large probability of error when H_1 holds. Obviously, any choice between these two rules must have something to do with the relative importance of the two kinds of errors. It might be that making an error is a minor matter when H_0 is true, but very serious given H_1. In this case Rule 1 is preferable. On the other hand, were error very serious given H_0, Rule 1 could be disastrous. In short, any rational way of choosing among rules must involve some notion of the loss involved in making an error. Thus, before completing this example, we need to examine the role that possible losses to the decision maker can play in the choice of a decision rule.

7.4 EXPECTED LOSS AS A CRITERION FOR CHOOSING A DECISION RULE

In the everyday world, a decision followed by an outcome always involves some payoff to the decision maker. That is, the combination of a decision and the outcome of that decision carrries some positive or negative value for the person affected. The exact nature of these values need not concern us here; they may reflect safety, dollars, time, prestige, self-respect, pleasure, or any of a variety of other things of importance to a human being. We will simply assume that it is possible to assign a value to any combination of what was decided and what turned out to be true. When the choice is between H_0 and H_1, then that value can be represented by

$$u(H_0|H_1),$$

standing for the value of choosing H_0 when H_1 turns out to be true. Similar meaning attaches to $u(H_1|H_0)$ and so on, for the other combinations of decision and outcome.

For our purposes we need consider these values only as *losses*. Each loss value is some positive number or zero given to an action–outcome combination. If the combination represents the best available action under the true circumstances, then the decision is *right* and the loss value is zero. On the other hand, if the action is less desirable than the best available action, then this is an *error* and the loss value is positive. (A more technical term for such a value is *opportunity loss*.)

Now let us return to your choice between the two hypotheses of Section 7.3. In your present decision-making situation, when H_0 is true the money you receive from the bag is \$900. On the other hand, when H_1 is true the amount you receive is only \$700. Then the possible losses associated with the four possible action–outcome possibilities are as shown in Table 7.4.1. Now let us examine the *expected loss,* given that you use a particular decision rule, and given that H_0 is actually true. This value is given by

$$E(u|H_0) = u(H_0|H_0)p(\text{decide } H_0|H_0) + u(H_1|H_0)p(\text{decide } H_1|H_0).$$

Similarly, we can find $E(u|H_1)$ by taking

$$E(u|H_1) = u(H_0|H_1)p(\text{decide } H_0|H_1) + u(H_1|H_1)p(\text{decide } H_1|H_1).$$

These expected losses can be computed for each of the three decision rules we have been considering. For Decision Rule 1 the probabilities are as shown in Table 7.3.2, and the u values as given in Table 7.4.1. Thus,

$$E(u|H_0) = (0 \times .677) + (900 \times .323) = 290.70$$

and

$$E(u|H_1) = (700 \times .013) + (0 \times .987) = 9.10.$$

In other words, when the bag of money on the table actually represents H_0, then the expected loss to you in using Decision Rule 1 is just over \$290. If, on the other hand, the bag represents H_1, then the expected loss is only \$9.10. Clearly, Decision Rule 1 is much better when H_1 is actually true than when H_0 happens to be true.

True situation **Table 7.4.1**

		H_0	H_1
	H_0	0	\$700
Choose			
	H_1	\$900	0

Table 7.4.2

		True	
		H_0	H_1
	1	290.70	9.10
Decision rule	2	30.60	116.20
	3	337.50	666.40

Now in the same way the expected losses under H_0 and H_1 can be calculated for Decision Rule 2, and for Decision Rule 3. These expected losses are summarized in Table 7.4.2.

Decision Rule 1 gives a small loss when H_1 is true, but a high loss under H_0. On the contrary, Decision Rule 2 gives a relatively small loss when H_0 is true, but a larger loss under H_1. The table again shows the absurdity of Rule 3, which gives the highest loss of the three rules regardless of which hypothesis is true.

Can we choose between Rules 1 and 2, however, on the basis of expected loss? A device within decision theory for choosing between rules is known as the *minimax principle*. Applying this principle leads us to choose that rule showing the *minimum maximum-expected-loss* over all possible true situations. For expected-loss tables such as 7.4.2, a minimax decision rule is one having the largest entry in its row that is smaller than the largest entry in any other row of the table. Notice that in this situation, Rule 2 has 116.20 as the largest value in its row; this is smaller than the largest value for Rule 1, or 290.70. Thus, applying the minimax principle, we find that Rule 2 is the one to use. Using Rule 2, the largest that we expect long-run loss to be is smaller than for either Rule 1 or Rule 3.

This minimax criterion for choosing among decision rules is historically important in the theory of games and in decision theory, and it does give a way to compare decision rules on the basis of their expected losses. However, this criterion is no longer taken very seriously by statisticians. Among other things, it is unduly conservative, especially in scientific decision making, since it focuses only on the extreme potential loss in making errors and not on the very large gains one can make through increased knowledge, despite the risk of being wrong on occasion. An alternative to the minimax criterion will be discussed next.

One thing that has been left completely out of this example so far is the fact that you may know something, or at least believe something, about the bag on the table before a sample is even drawn. That is, you might be thought of as having a subjective probability, say $\mathscr{P}(H_0)$, that the bag on the table actually does represent H_0, and another subjective probability, $\mathscr{P}(H_1)$, that the bag represents H_1. The experimenter might even have gone so far as to create these subjective probabilities for you, by saying right at the outset that "the chances are about one in four that H_0 is true, and the chances are about three in four that H_1 is true." Then, another way to choose among the decision rules exists.

We can define *subjective expected loss* for Rule 1 by

$$SE(u|\text{rule } 1) = E(u|H_0)\mathcal{P}(H_0) + E(u|H_1)\mathcal{P}(H_1).$$

In the same way, the subjective expected loss for any other rule can be defined, given the values of $\mathcal{P}(H_0)$ and $\mathcal{P}(H_1)$.

Suppose that you choose the decision rule having the lowest subjective expected loss. It turns out that, since we have established that $\mathcal{P}(H_0) = 1/4$ and $\mathcal{P}(H_1) = 3/4$,

$$SE(u|\text{rule } 1) = 290.70(1/4) + (9.10)(3/4) = 79.50$$

$$SE(u|\text{rule } 2) = 30.60(1/4) + (116.20)(3/4) = 94.80$$

$$SE(u|\text{rule } 3) = (337.50)(1/4) + (666.40)(3/4) = 584.18$$

If you seek to mimimize subjective expected loss in this experiment, then Decision Rule 1 should be chosen. Given that $\mathcal{P}(H_0) = 1/4$ and $\mathcal{P}(H_1) = 3/4$, then on the average Decision Rule 1 is the best way to decide.

This illustrates how decision rules may also be compared by their subjective expected losses; the lower the subjective expected loss, the better the rule. In this case, given the loss values and given the personal probabilities, Rule 1 is clearly superior to Rule 2. Even though error has a fairly high probability by this rule when H_0 is true, your own weighing of what you expect to be true of the situation tends to discount the likelihood of such errors.

Your little problem in decision theory is now solved: provided that you can attach loss values to the action–outcome combinations and provided that personal probabilities are associated with the competing hypotheses, a subjective expected loss can be calculated for each possible decision rule and the best one adopted. This does not guarantee that you will *make* the right decision, but only that you will have chosen a *good way* to make the decision. Essentially, that is the basic concern of decision theory.

7.5 FACTORS IN SCIENTIFIC DECISION MAKING

It requires no great logical leap to see that your problem in trying to decide how much money is in a bag is very like the dilemma of the business executive trying to decide which of the two lots of merchandise to buy, or that of the scientist confronted with two competing hypotheses about human behavior. In all such situations, the decision maker will rely on a sample to help him or her make a judgment between the two possible states of affairs. There will be a variety of decision rules that might be used, and a choice needs to be made among these possible ways of deciding. Just as some of the criteria of decision theory could be applied in the money-bag problem, so these same principles can be extended to scientific or practical decision making.

However, particularly in the case of scientific decision making, the choice of a decision rule is not nearly so simple as it may be in a business situation. In business, it is frequently true that a given action must be followed by one of a fairly small and specifiable set of outcomes. The decision maker does not know what the particular outcome of an action will be, but at least one can state the possible outcomes. The actual outcome may depend on economic conditions in general, the quality of the product, competition, and so on. Furthermore, each of these possible outcomes to any action can be assigned a profit or loss value, again in principle.

On the other hand, the possible outcomes of actions taken by the scientist are far more difficult to specify. It is clear that some definite courses of action are usually open to the experimenter: to decide for or against a hypothesis, to publish or withhold findings, to collect or not to collect more data, to pursue or abandon the line of investigation, to ask for funds for further research, or to forget the whole matter. By stretching a point, we might say that finding out what the true situation is constitutes an outcome following any action. However, this outcome may not apply to the experimenter at all, but rather may serve as an outcome for the science or for humanity.

It is important to recognize that the scientist as a mere human being may be capable of foreseeing only the short-term monetary, prestige, or other outcome possibilities of the actions, which are insignificant compared to the long-range effect of the action on knowledge and human welfare.

For the sake of argument, let us nonetheless suppose that the set of actions and outcomes could be specified for the scientific decision maker. Then, in principle, criteria such as minimax expected loss or minimum subjective expected loss could be applied in the choice of a scientific decision rule, provided that values could be assigned to the action–outcome combinations.

In the sections to follow, various procedures will be given that are in common use in such research. Most such procedures use a rule for deciding when a result is significant that *might* be justified by a decision-making argument very similar to that just given. The experimenter wants to avoid errors in inference, presumably because such errors lead to losses, at least in time and effort. Decision rules differ in the expected losses they give, and it behooves the experimenter to choose a decision rule that minimizes expected loss. The act of deciding on a region of rejection in the test of a hypothesis is the choice of a decision rule, and it should be subject to the same considerations.

Unfortunately, the cost of erroneous decisions is almost never considered in social science research. Indeed, in most instances in such research it seems very unlikely that a numerical value could ever be assigned to the loss incurred in an erroneous decision. Just how bad is it to be wrong in a scientific inference? This problem is not quite so difficult in many other fields using statistical inference, especially in research on applied problems. In business decisions it is often possible to give a value in dollars to the outcomes that might result from various decisions, and statistical decision theory then provides guides for choosing an effective decision rule. Even in some applied social science areas, losses involved in errors might be reckoned in the same general way. Thus, in studying methods for selecting people for various jobs, the cost of an error may be a calculable thing, and the decision rule may be chosen on that basis. Similarly, this possibility may exist for studies of diagnostic methods, of training effectiveness, and so on. By and large, however, most social or behavioral scientists would not know how to assign loss values to decision errors. Even in a relatively clear-cut situation such as diagnosis or selection, the costs to the people involved or even to the community may be extremely important. For example, consider a diagnostic method for potential suicides. Errors of the two different types (false positives and false negatives) have enormously different consequences for the person and for society. Such costs are most difficult to assess, and the effort required is seldom expended. Even so, the decision rules obtained by omitting such incalculable costs could be disastrous. Such an omission is sometimes called the "accountant's error," although that term fails to convey the importance of the problem. In still other circumstances the loss values of errors may be so small that it really does not make much difference how or what the experimenter decides.

In the same way, even if we admit the existence of personal probabilities or $\mathscr{P}$ values for the experimenter, typically one is not in a position to assess objectively what these values should be, even though this is theoretically possible. Perhaps in the future a standard part of statistical analysis will be the statement of prior personal probabilities before the collection of the data. Personal probabilities can be found by the odds the experimenter is willing to accept in betting on the truth of a hypothesis, and it might be possible to make this a standard part of statistical practice. This "Bayesian" approach to inference is already common in some applied fields.

In short, we use much of the terminology of statistical decision theory without its main feature, the choice of a decision rule having optimal properties for a given purpose. Instead, we use conventional decision rules, completely ignoring questions of the loss involved in errors and the degree of prior certainty of the experimenter. These conventional rules can be justified by decision theory in some contexts, but they are surely not appropriate to every situation.

Must the scientist make a decision about what is true from a given set of data? Naturally, choosing to suspend judgment and wait for more evidence is a decision to adopt a course of action. However, why cannot this time-honored strategy of the scientist be used more often than apparently it is? As we have seen, one is usually cut off from the possibility of choosing an appropriate decision rule, using the minimax principle or minimum subjective-expected-loss. There is no recourse except to fall back on some conventional rule. No convention can possibly be ideal or even sensible for all the problems to which it may be applied, as we shall illustrate below.

It is entirely possible that the course of action represented by "no decision, suspend judgment" will be best in a given situation, especially if the penalty involved in waiting to decide is small compared to the loss involved in an incorrect decision.

Why, then, do social scientists bother with significance tests at all? Regardless of what one is going to do with the information—change an opinion, adopt a course of action, or what not—one needs to know *relatively how probable* is a result like that obtained, given a hypothetical true situation. Basically, a significance test gives this information, and *that is all*. The conventions about significance level and regions of rejection can be regarded as ways of defining *improbable*. The occurrence of a significant result in terms of these conventions is really a signal: "Here is a direction and degree of deviation which falls among those relatively unlikely to occur given that the tested hypothesis is true, but which is relatively more likely given the truth of some other hypothesis." On deciding to reject the original hypothesis on the basis of a significant result, then at least one knows the probability of error in doing so. This does not mean that one must decide against the hypothesis simply because some conventional level of significance was met. Other options, such as suspending judgment, may actually be better actions under the circumstances regardless of the result of the conventional signficance test. Even more emphatically, the occurrence of a nonsignificant result does not mean that you must accept the hypothesis as true. As we shall see, one often has not the foggiest idea of the error probability in saying that the tested hypothesis is true; here, making a decision to accept the tested hypothesis is absurd in the light of the unknown, and perhaps very large, probability of such an error. On occasion, however, the probability of such an error can be assessed, and here one can feel safe in asserting that the tested hypothesis is true when this error probability is small.

To sum up: it is perfectly all right to discuss conventional hypothesis testing in the language of decision theory, but the user is speaking as though the conventional rule were a good way to decide, which it may not be. There is *no* God-given rule about when and how to make up your mind in general.

7.6 TYPE I AND TYPE II ERRORS

As we have noted already, in any choice between two hypotheses, there are two ways of being correct in the decision, and two ways of making an error. Ordinarily, these action–outcome possibilities and their associated probabilities are symbolized as follows:

$$
\begin{array}{c}
& \text{True} \\
& \begin{array}{cc} H_0 & H_1 \end{array} \\
\text{Decide} \begin{array}{c} H_0 \\ H_1 \end{array} & \left| \begin{array}{cc} 1 - \alpha & \beta \\ \alpha & 1 - \beta \end{array} \right.
\end{array}
$$

The probability of making an error of the type "decide H_1 and H_0 is true" is symbolized by α, the lowercase Greek letter alpha. Then the probability of being correct, given that H_0 is true, is symbolized by $1 - \alpha$, since given that H_0 is true, we must have decided either on H_0 or H_1.

In a similar way, the probability of making an error of the type "decide H_0 and H_1 is true" is symbolized by the lowercase Greek letter beta, β. Then the probability of making a correct decision, given that H_1 is true must be $1 - \beta$.

The time has also come to dignify with names the two kinds of errors we have been discussing. **Type I error is that made when H_0 (the tested hypothesis) is falsely rejected.** One makes a Type I error whenever the sample result falls into the rejection region even though H_0 is true. Thus the probability α gives the risk run of making a Type I error:

$$\alpha = \text{probability of Type I error.} \tag{7.6.1*}$$

The errors of Type II are those made by not rejecting H_0 when it is false. When a sample result does not fall into the rejection region, even though some H_1 is true, we are led to make a Type II error. Thus for a given true alternative H_1, β is the probability of Type II error

$$\beta = \text{probability of Type II error.} \tag{7.6.2*}$$

7.7 CONVENTIONAL DECISION RULES

In social science research, given some hypothesis H_0 to be tested, the region of rejection is usually found as follows:

CONVENTION: Set α, the probability of falsely rejecting H_0, equal to some small value. Then, in accordance with the alternative H_1, choose a region of rejection such that the probability of observing a sample value in that region is equal to α when H_0 is true. The obtained result is significant beyond the α level if the sample statistic falls within that region.

Ordinarily, the values of α used are .05 to .01, although, on occasion, larger or smaller values are employed. Even though only one hypothesis, H_0, may be exactly specified and this determines the sampling distribution employed in the test, in choosing the region of rejection one acts as though there were two hypotheses, H_0 and H_1. The alternative hypothesis H_1 dictates what portion or **tail** of the sampling distribution contains the rejection region for H_0. In some problems, the region of rejection is contained in only one tail of the distribution, so that only extreme deviations in a given direction from expectation lead to rejection of H_0. In other problems, big deviations of either sign are candidates for the rejection region, so that the region of rejection lies in both tails of the sampling distribution.

In all that follows, these conventional rules will be used for finding a rejection region. Some hypothesis will be designated as H_0, and an arbitrary α value will be chosen, which, taken together with H_1, determines the sample values that lead to a possible rejection of H_0.

Although the conventional rules usually involve α values such as .05 and .01, the choice of α is not always totally arbitrary. The choice of α may also be dictated by a balance that the expermenter strikes between the two kinds of errors possible in drawing conclusions from a test of a hypothesis, as indicated in the section to follow.

The conventions about the size of the α probability of Type I error actually grew out of a particular sort of experimental setting. Here it is known in advance that one kind of error is extremely important and is to be avoided. In this kind of experiment these conventional procedures do make sense when viewed from the decision-making point of view. Furthermore, designation of the hypothesis H_0 as the null hypothesis and the arbitrary setting of the level of α can best be understood within this context.

As an example of an experimental setting where Type I error is clearly to be avoided, imagine that one is testing a new medicine, with the goal of deciding if the medicine is safe for the normal adult population. By *safe* we will mean that the medicine fails to produce a particular set of undesirable reactions on all but a very few normal adults. Now in this instance, deciding that the medicine is safe when actually it tends to produce reactions in a relatively large proportion of adults is certainly an error to be avoided. Such an error might be abhorrent to the experimenter and the interests he or she represents. Therefore, the hypothesis ''medicine unsafe'' or its statistical equivalent is cast in the role of the null hypothesis, H_0, and the value of α chosen to be extremely small, so that the abhorrent Type I error is very unlikely to be committed. A great deal of evidence against the null hypothesis is required before H_0 is to be rejected. The experimenter has complete control over Type I error, and regardless of any other feature this study of the medicine may have, one can be confident of taking very little risk of asserting that H_1, or "medicine safe," is true when actually H_0, or "medicine unsafe," is true.

On the other hand, the experimenter always has some control over the values of β under the various possibly true alternatives. As we shall see, by the choice of an appropriate sample size and by exercise of control over the size of σ^2 in the population considered, the value of β for any possibly true alternative to H_0 may be made as small as desired. The essential point is that the experimenter is absolutely free to set α at any level and that the conventional levels are dictated by the notion that Type I errors are bad and must be avoided. The experimenter does not have this same freedom with respect to Type II errors; low risk of Type II error must be bought in terms of sample size and other features of the test procedure. The value of β can always be made small against any given alternative, but only at some cost to the experimenter. An inappropriately small value of α makes it more difficult than otherwise to avoid Type II error.

Within contexts such as the test of a new medication where Type I error is abhorrent, setting α extremely small is manifestly appropriate. Here, considerations of Type II error are actually secondary. In some instances in a social science as well, Type I error is clearly to be avoided, and from the outset the experimenter wants to be sure that this kind of error is very improbable. The designation of one hypothesis as H_0 should rest on which kind of error is to have the small probability α.

On the other hand, in some social science research it is very hard to see exactly why the particular hypothesis tested, or H_0, should be the one we are loath to abandon, and why Type I errors necessarily have this drastic character. Granting that scientific discretion is commendable, the mistaken conclusion that "something really happened" is not *necessarily* worse than overlooking a real experimental phenomenon. In some situations, perhaps, we should be far more attentive to Type II errors, and less attentive to setting α at one of the conventional levels. Furthermore, if the conventional α levels are to be used, a little more thought might be given to deciding exactly what *is* the null hypothesis we want to be so careful not to reject falsely.

7.8 THE POWER OF A STATISTICAL TEST

Given only the one exact hypothesis H_0, and the required value of α, one may still think of an inference involving two hypotheses:

H_0, the hypothesis actually being tested
H_1, the alternative hypothesis, against which H_0 is to be compared.

Once any true H_1 is specified, then we can determine β, the probability of an error in decision given the true H_1. Similarly, one may compute the probability $1 - \beta$, which is the probability of being right in rejecting H_0 given that H_1 is true. This probability, $1 - \beta$, is often called the **power** of the statistical test. It is literally the probability of finding out that H_0 is wrong, given the decision rule and the true value under H_1.

The idea of power can be discussed most simply in terms of the choice between two exact hypotheses. Admittedly, practical problems very seldom present us with this sort of choice and so any example will appear somewhat contrived. Nevertheless, any power calculation always begins as though this were the choice available to the experimenter.

For example, a psychologist working in industry is assigned to study the possibility of using only blind persons in a certain job requiring unusually fine hearing and touch discrimination but no use of sight at all. Heretofore, only sighted persons have been used, and past experience has shown that for the sighted population the average performance score is 138, with a standard deviation of 20. If the blind perform exactly like the sighted, then their population should have this same mean and standard deviation. On the other hand, if the mean performance of blind persons is at least 142, then a considerable benefit will accrue to the company by employing them. From what is already known about the requirments of the job and blind performance in other settings, there is good reason to believe that one of these two situations will prove true.

Accordingly, the experimenter frames two hypotheses, both of which deal with a population of blind persons who might be placed on this job:

H_0: $\mu = 138$

H_1: $\mu = 142.$

It is assumed that the population distribution in either situation will have a standard deviation of 20.

Now a sample of 100 blind persons is to be drawn at random and put into this job situation on an experimental basis. The psychologist feels that this sample is large enough to permit using the normal approximation to the sampling distribution of the mean. Note that here $\sigma_M = 2$.

The following conventional rule is used: If the sample result falls among the highest 5% of means in a normal distribution given H_0, then reject H_0; otherwise, reject H_1.

This region of rejection for H_0 is shown in Figure 7.8.1.

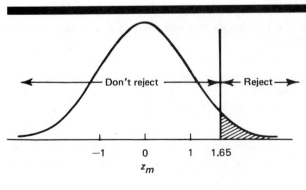

Figure 7.8.1

A region of rejection on the right tail of a normal distribution.

In this instance, there are once again two kinds of errors that can be made. The probabilities for each kind are labeled α and β respectively:

$$\alpha = \text{probability (reject } H_0 | \mu = 138)$$

$$\beta = \text{probability (accept } H_0 | \mu = 142).$$

In this situation with two *exact* hypotheses, it does make sense to talk about accepting or deciding in favor of H_0 since both probabilities can be known. In essence, the conventional rule says to fix α at .05, and the region of rejection chosen does just that. Given that H_0 is true, then the region of rejection must be bounded by a z_M score such that

$$F(z_M) = .95, \text{ or } 1 - F(z_M) = .05.$$

The normal tables show that this z_M score is 1.65. In terms of a sample mean

$$z_M = \frac{\bar{x} - 138}{\sigma_M} = \frac{\bar{x} - 138}{2}.$$

Thus, the *critical* value of $\bar{x}$, forming the boundary of the rejection region, is $\bar{x} = 138 + 1.65\sigma_M = 138 + 3.30 = 141.30$.

However, notice that assigning the value of .05 to the α probability of error automatically fixes β as well, as shown in Figure 7.8.2.

What would the z_M score for this critical value of 141.30 be if H_1 were true? This is found from

$$z_M = \frac{141.3 - 142}{2} = -.35.$$

In a normal distribution, $F(-.35) = .36$, approximately, and so we can see that $\beta = .36$. The two error probabilities are then

$$\alpha = .05$$

$$\beta = .36.$$

Thus by the rule used, H_0 is not rejected unless a mean exceeds 141.3. This gives a probability of error β shown by the shaded region in Figure 7.8.2, if H_1 is true.

We see that here the β probability of making an error is much greater than α. Our experimenter is rather unlikely to decide on H_1 when H_0 is true but has considerable chance of concluding H_0 when H_1 is true.

The unshaded area under the curve to the right is $1 - \beta$, or the power of the test, given H_1. In this instance the power is about .64, so that the chances of correctly rejecting H_0 are about 6 in 10. Notice that the power of a test of H_0 cannot be found until some *true situation* H_1 is specified, as in this example.

What factors determine the power of a statistical test in a situation, such as we have just examined? The power of a test of a mean always depends on four things:

1. The particular alternative hypothesis H_1 that is to be assumed true if H_0 is false.
2. The value of α chosen by the experimenter and the selection of a region of rejection corresponding to α.

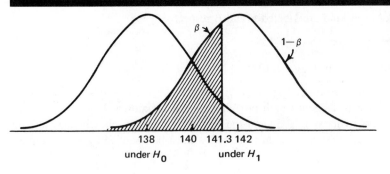

Figure 7.8.2

Type II error (β) and power 1 − β when H_1 is true.

3. The size of the sample.
4. The variability of the population under study.

We shall examine each of these factors in turn.

7.9 POWER OF TESTS AGAINST VARIOUS TRUE ALTERNATIVES

The power of a test of H_0 is not unlike the power of a microscope. It reflects the ability of a decision rule to detect from evidence that the true situation differs from a hypothetical one. Just as a high-powered microscope lets us distinguish gaps in an apparently solid material that we would miss with low power or the naked eye, so does a high-powered test of H_0 almost ensure us of detecting when H_0 is false. Pursuing the analogy further, any microscope will reveal "gaps" with more clarity the larger these gaps are; the larger the departure of H_0 from the true situation H_1, the more powerful is the test of H_0, other things being equal.

For example, suppose that the hypothesis to be tested is

$$H_0: \mu = \mu_0$$

but that the true hypothesis is

$$H_1: \mu = \mu_1, \text{ where } \mu_1 > \mu_0.$$

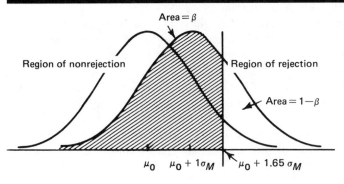

Figure 7.9.1

The region of rejection under H_0, and power under H_1.

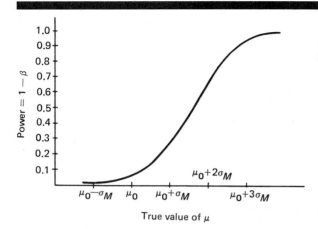

Figure 7.9.2

Power curve for a one-sided test of mean $\alpha = .05$.

Here, μ_0 and μ_1 symbolize two different possible numerical values for μ. It will be assumed that under either hypothesis the sampling distribution of the mean is normal with σ known. Suppose now that the α probability of error is set at .05, and the rejection region is on the *right tail* of the distribution. The decision rule is such that the critical value of $\bar{x}$ (the smallest $\bar{x}$ leading to the rejection of H_0) is $\mu_0 + 1.65\sigma_M$. The power of the test can then be figured for any possible true H_1.

First of all, suppose that μ_1 were actually $\mu_0 + \sigma_M$. Then in the distribution under H_1, the critical value of $\bar{x}$ has a z score given by

$$z_M = \frac{\bar{x} - \mu_1}{\sigma_M}$$

$$= \frac{(\mu_0 + 1.65\sigma_M) - (\mu_0 + 1\sigma_M)}{\sigma_M} = .65.$$

The probability of a sample's falling into the region of rejection for H_0 is found from Table I in Appendix E to be .26. This probability is $1 - \beta$, the power of the test.

This may be clarified by Figure 7.9.1. The region of rejection for H_0 is the segment of the horizontal line to the right of $\bar{x} = \mu_0 + 1.65\sigma_M$. The shaded portion under the curve for $\mu = \mu_0 + 1\sigma_M$ denotes β, the probability of *failing* to reject H_0, and the power is the unshaded part of the righthand curve.

If another alternative hypothesis were true, say,

$$H_1: \mu_1 = \mu_0 + 3\sigma_M,$$

then the power would be much larger. Here, relative to the sampling distribution based on the *true* mean or $\mu_0 + 3\sigma_M$, the critical value of $\bar{x}$ would correspond to a z_M score of

$$z_M = \frac{(\mu_0 + 1.65\sigma_M) - (\mu_0 + 3\sigma_M)}{\sigma_M} = -1.35.$$

Table I shows us that above a z score of -1.35 lie .91 of sample means in a normal sampling distribution, so that the power is .91.

The power of the test for any true value of μ_1 can be found in the same way. Often, to show the relation of power to the true value μ_1, so called **power functions** or **power curves** are plotted. One such curve is given in Figure 7.9.2, where the horizontal axis gives the possible values of true μ_1 in terms of μ_0 and σ_M, and the vertical axis the value of $1 - \beta$, the power for that alternative. Notice that for this particular decision rule, the power curve rises for increasing values of μ_1, and approaches 1.00 for very large values. On the other hand, for this decision rule, when true μ_1 is less than μ_0, the power is very small and approaches 0 for decreasing μ_1 values. In any statistical test where the region of rejection is in the direction of the true value covered by H_1, the greater the discrepancy between the tested hypothesis and the true situation, the greater the power.

7.10 POWER AND THE SIZE OF α

Since β will ordinarily be small for a large α, as we have seen in the discussion above, it follows that setting α larger makes for relatively more powerful tests of H_0. For example, the two power curves given below for the same decision rule show that if α is set at .10 rather than at .05, the test with $\alpha = .10$ will be more powerful than that for $\alpha = .05$ over all possibly true values under alternative H_1 (note Figure 7.10.1). Making the probability of error in rejecting H_0 larger has the effect of making the test more powerful, other things being equal.

In principle, if it is very costly to make the mistake of overlooking a true departure from H_0, but not very costly to reject H_0 falsely, one could (and perhaps should) make the test more powerful by setting the value of α at .10, .20, or more. This is not ordinarily done in social science research, however. There are at least two reasons why α is seldom taken to be greater than .05: In the first place, as observed in Section 7.7 the problem of relative losses incurred by making errors is seldom solved in such research; hence conventions about the size of α are adopted. The other important reason is that given some fixed α the power of the test can be increased either by increasing sample size or by reducing the standard error of the test statistic in some other way.

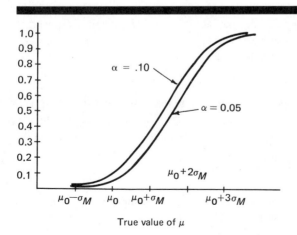

Figure 7.10.1

Power curves for two tests with equal sample sizes but with different values of α.

True value of μ

7.11 THE EFFECT OF SAMPLE SIZE ON POWER

Given a population with true standard deviation σ, the standard error of the mean depends inversely on the square root of sample size N. That is,

$$\sigma_M = \frac{\sigma}{\sqrt{N}}.$$

When N is large, then the standard error is smaller than when N is small. Provided that $1 - \beta > \alpha$ to begin with, *increasing the sample size increases the power* of a given test of H_0 against a true alternative H_1.

For example, suppose that

$$H_0\text{: } \mu = 50$$

$$H_1\text{: } \mu = 60,$$

where H_1 is true. Let us assume that true $\sigma = 20$. If samples of size 25 are taken, then

$$\sigma_M = \frac{20}{5} = 4.$$

Now the α probability for this test is fixed at .01, making the critical value be

$$\bar{x} = \mu_0 + 2.33\sigma_M = 50 + (2.33)(4) = 59.32.$$

In the true sampling distribution (under H_1) this amounts to a z score of

$$z_M = \frac{59.32 - 60}{4} = \frac{-.68}{4} = -.17$$

so that .57 of all sample means should fall into the rejection region for H_0. The power here is thus .57.

Now let the sample size be increased to 100. This changes the standard error of the mean to

$$\sigma_M = \frac{20}{10} = 2$$

and the critical value of $\bar{x}$ to

$$\bar{x} = 50 + (2.33)(2) = 54.66.$$

The corresponding z score when $\mu = 60$ is

$$z_M = \frac{54.66 - 60}{2} = -2.67$$

making the power now in excess of .99. With this sample size we would be almost certain to detect correctly that the H_0 is false when this particular H_1 is true. With only 25 cases, we are quite likely not to do so.

The disadvantages of an arbitrary setting of α can thus be offset, in part, by the choice of a large sample size. Other things being equal and regardless of the size chosen for α, the test may be made powerful against any given alternative H_1 in the direction

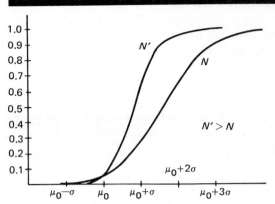

Figure 7.11.1

Power curves for two tests with $\alpha = .05$ but with different sample sizes $N' > N$.

True value of μ in terms of μ_0 and population standard deviation units

of the rejection region, provided that sample N can be made very large. (See Figure 7.11.1.)

Once again, however, it is not always feasible to obtain very large samples. In much social science research, samples of substantial size are costly for the experimenter—if not in money, then in effort. Our ability to attain power through large samples not only partly offsets the failure to choose a decision rule according to cost of error in such research; large samples may not really be necessary in some research using them, especially if the error thereby made improbable is actually not very important. The matter of sample size and power will be considered once again in Chapter 8.

7.12 POWER AND "ERROR VARIANCE"

Even given a fixed sample size, the experimenter has one more device for attaining power in tests of hypotheses. *Anything that makes σ, the population standard deviation, small will increase power,* other things being equal. This is one of the reasons for the careful control of conditions in good experimentation. By making conditions constant, the experimenter rules out many of the factors that contribute to variation in the observations. Statistically this amounts to a relative reduction in the size of σ for some experimental population. By ruling out some of the error variance from the observations one is *decreasing* the standard error of the mean, and thus *increasing* the power of the test against whatever hypothesis H_1 is true. Experiments in which the variability attributable to experimental or sampling error is small are said to be **precise;** the result of such precision is that the experimenter is quite likely to be able to detect when something of interest is happening. The application of experimental controls is like restricting inferences to populations with smaller values of σ^2 than otherwise, and thus control over error variation through careful experimentation implies powerful statistical tests. It follows that controlled experiments in which there is little "natural" variation in the materials observed can attain statistical power with relatively few observations, whereas those involving extremely variable material may require many observations to attain the same degree of power.

7.13 ONE-TAILED REJECTION REGIONS

The primary notions of errors in inference and the power of a statistical test have just been illustrated for an extremely artificial situation, in which a decision must be made between two exact alternatives. Such situations are almost nonexistent in social science research. Instead, the experimenter is far more likely to be called on to evaluate inexact hypotheses, each of which encompasses a whole range of possibly true values. What relevance, then, does the discussion of the exact two-alternative case have to what researchers actually do? The answer is that researchers make inferences as though deciding between two exact alternatives, even though actual interest lies in judging between inexact hypotheses. Thus, the mechanism we have been using for decisions between exact hypotheses is exactly the same as for any other set of alternatives.

In some examples two inexact hypotheses are compared, having the form

H_0: $\mu \leq \mu_0$

H_1: $\mu > \mu_0$.

Here the entire range of possible values for the parameter under study (in this case, the population mean) was divided into two parts, that above and that below (or equal to) an exact value μ_0. The interest of the experimenter is in placing the true value of μ either above or below μ_0.

In this instance the appropriate region of rejection for H_0 consists of values of $\bar{x}$ relatively much larger than μ_0. Such values have a rather small probability of representing true means covered by the hypothesis H_0, but are more likely to represent true means in the range of H_1. Such rejection regions consisting of sample values in a particular direction from the expectation given by the exact value included in H_0 are called **directional** or **one-tailed** rejection regions. For the particular hypotheses compared in Section 7.8, the rejection region was one-tailed, since only the right or "high-value" tail of the sampling distribution under H_0 was considered in deciding between the hypotheses.

For other questions, the two inexact hypotheses are of the form

H_0: $\mu \geq \mu_0$

H_1: $\mu < \mu_0$.

Once again the region of rejection is one-tailed, but this time the lower or left tail of the sampling distribution contains the region of rejection for H_0. The choice of the particular rejection region thus depends both on α and the alternative hypothesis H_1.

Tests of hypotheses using one-tailed rejection regions are also called **directional.** The direction or sign of the value of the statistic (such as z_M) is important in directional tests since the sample result must show not only an extreme departure from expectation under H_0 but also a departure in the right direction to be considered strong evidence against H_0 and for H_1.

Directional hypotheses are implied when the basic question involves terms like *more than, better than, increased, declined*. The essential question to be answered by the data has a clear implication of a difference or change in a specific direction.

An example will probably clarify this point. As a fairly plausible situation, imagine a psychologist who has constructed a new test of verbal fluency. This test has been very carefully standardized on a population of American adults. The standardization has been so carried out that the mean score is 100 and the standard deviation of the test is 15. Furthermore, the distribution of scores in the standardization population is approximately normal.

The psychologist is now interested in the application of the test to residents of England. From knowledge of the skills required by the test, there is some reason to suspect that English adults may, on the average, score somewhat higher than American adults. So, in order to arrive at some idea if this may be the case, and especially to see if the test may require some scoring modification for administration in England, it is decided to try the test on a sample of English adults. Basically, the question to be answered is "Do English adults tend to score higher on this test than Americans?" We will suppose that there is no reason to question the standard deviation of scores among the English nor the general form of the distribution of scores.

The answer to this question is tantamount to a decision between two *inexact* hypotheses:

H_0: $\mu \le 100$

H_1: $\mu > 100$

i.e., the English population has either a mean score less than or equal to the American mean, or a mean greater than the American mean.

In choosing the decision rule to be used, the experimenter decides to set α equal to .01. A result greater than 100 tends to favor H_1, and so the region of rejection includes all z scores greater than or equal to 2.33, since this value cuts off the highest 1% of sample means in a normal sampling distribution.

What, however, is the hypothesis actually being tested? As written, H_0 is inexact, since it states a whole region of possible values for μ. One exact value is specified, however: this is $\mu = 100$. Actually, then, the hypothesis tested is $\mu = 100$ against some unspecified alternative greater than 100. In effect, the decision rule can be put in the following form:

		True situation	
		$\mu = 100$	$\mu > 100$
Decide	$\mu = 100$	.99	(β)?
	$\mu > 100$	.01	$(1 - \beta)$?

The α probability of error can be specified in advance as .01, but β and the power are unknown, depending, as they do, on the true situation. The experimenter has no real interest in the hypothesis that $\mu = 100$ and may even feel extremely confident that

Figure 7.13.1

Representation of α and β for some other hypothesis covered by H_0, in a one-tailed test.

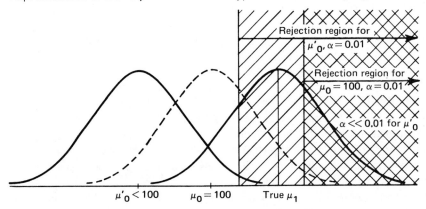

the true mean is not precisely 100. Nevertheless, this is a useful dummy hypothesis, in the sense that if one can reject $\mu = 100$ with $\alpha = .01$, then one can reject *any other hypothesis* that $\mu < 100$ with $\alpha < .01$. In other words, by this decision rule, if the researcher can be confident of not making an error in rejecting the hypothesis actually tested, then there is even more confidence in rejecting any other hypothesis covered by H_0.

But what does this do to β? Given some true mean μ_1 covered by H_1, the power of the test of $\mu = 100$ *is less than* the power for any other hypothesis covered by H_0 with fixed α. If the test is of any other exact hypothesis embodied in H_0, such as $\mu = 90$, then neither α nor β exceed those for $\mu = 100$. Testing the exact hypothesis with given α and β probabilities can be regarded as testing *all* hypotheses covered by H_0, with *at most* α and β probabilities of error (the β depending on the *true* mean, of course). This is illustrated below and in Figure 7.13.1.

		True situation			
		$\mu = 100$	$\mu < 100$	$\mu = \mu_1 > 100$	$\mu > \mu_1$
Decide	H_0	$1 - \alpha$	$> 1 - \alpha$	β	$< \beta$
	H_1	α	$< \alpha$	$1 - \beta$	$> 1 - \beta$

In other words by testing the exact value $\mu = 100$ and using a one-tailed rejection region corresponding to H_1, we can be sure that the probability of wrongly rejecting any of the values covered by H_0 is *no more than* the .05 or .01 we pick for α. Furthermore, we can be sure that the power of the test is *at least* as great for the other hypotheses under H_0 as it is for 100, given some true alternative value greater than 100.

7.14 TWO-TAILED REJECTION REGIONS

Many times an experimenter goes into a problem without a clearly defined notion of the direction of difference to expect if H_0 is false. We ask "Did something happen?" "Is there a difference?" or "Was there a change?" without any specification of expected direction. Next we will examine techniques for nondirectional hypothesis testing.

Imagine a study carried out on the "optical dominance" of human subjects. There is interest in whether or not the dominant eye and the dominant hand of a subject tend to be on the same or different sides. Subjects are to be tested for both kinds of dominance, and then classified as "same side" or "different side" in this respect. We will use the letters S and D to denote these two classes of subjects.

The experimenter knows that in 70% of subjects in this population, the right hand is dominant. It is also known that the right eye is dominant for 70% of subjects. From such past knowledge about the relative frequency of each kind of dominance, the experimenter reasons that if there actually is no tendency for eye dominance to be associated with hand dominance, then in a particular population of subjects one should expect 58% S and 42% D. (Why?) However, little is known about what to expect if there is some connection between the two kinds of dominance. To try to answer this question, our experimenter draws a random sample of 100 subjects, each with a full set of eyes and hands, and classifies them.

The present question may be put into the form of two hypotheses, one exact and one inexact:

$$H_0: p = .58$$

$$H_1: p \neq .58.$$

The first hypothesis represents the possibility of no connection between the two kinds of dominance, and the inexact alternative is simply a statement that H_0 is not true, since H_1 does not specify an exact value of population proportion p as true.

The α chosen is .05. The experimenter then is faced with the choice of a rejection region for the hypothesis H_0. Either a very high percentage of S subjects in the sample or a very low percentage would tend to discount the credibility of the hypothesis that $p = .58$ and would lend support to H_1. Thus the rejection region is chosen so that the H_0 will be rejected when extreme departures from expectation of *either* sign occur. This calls for a rejection region on *both* tails of the sampling distribution of sample P when $p = .58$. Since the sample size is relatively large, the binomial distribution of sample P may be approximated by a normal distribution (Section 6.4), and the two rejection regions may be diagrammed as shown in Figure 7.14.1.

Since the total probability of Type I error has been set at .05, the rejection region on the upper tail of the distribution will contain the highest 2.5% and the lower tail the lowest 2.5% of sample proportions, given that $p = .05$. In short, each region should contain exactly $\alpha/2$ proportion of all samples under H_0. Consequently, either a very large or very small sample P will lead to a rejection of H_0, and the total probability of Type I error is .05. The z score cutting off the upper rejection region is 1.96, and that for the lower is -1.96. Any sample giving a z beyond these two limits will lead to a rejection of H_0.

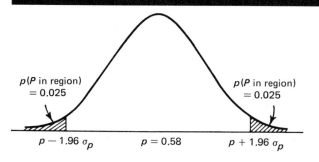

Figure 7.14.1

A two-tailed rejection region, for $\alpha = .05$.

The basic sampling distribution is binomial, and the standard error of P is

$$\sigma_P = \sqrt{\frac{pq}{N}} = \sqrt{\frac{.58 \times .42}{100}}$$

$$= .049$$

when H_0 is true (Eq. 4.16.4). Using the normal approximation to the binomial we find

$$z = \frac{P - p}{\sigma_P}.$$

Now suppose that the proportion of S subjects is .69. Then

$$z = \frac{.69 - .58}{.049} = 2.24.$$

This value exceeds the critical z score of 1.96, and so the result is said to be significant beyond the 5% level. By concluding that eye and hand dominance do tend to be related, the experimenter runs a risk of less than .05 of being wrong.

Suppose, however, that the sample result had come out to be only .53. Then

$$z = \frac{.53 - .58}{.049} = -1.02$$

making the result nonsignificant. Here, the experimenter should most likely suspend judgment pending further evidence and not be willing to assert that H_0 is true: the risk in that assertion is unknown. Indeed, the best guess about the true value of p is .53, not .58, and given enough sample observations, a P value of .53 might be enough to warrant rejection. Thus, for a nonsignificant result such as this, the experimenter cannot wisely accept H_0 as true. The best available choice usually is to suspend judgment and look for more evidence.

One additional refinement of this test must be pointed out here. The actual sampling distribution of concern here is the binomial, which is, of course, discrete. On the other hand, we are approximating this discrete distribution by a continuous normal distribution. In order that the fit between binomial probabilities and the probabilities for intervals under a normal curve be as good as possible, the "correction for continuity" (Eq. 6.4.4) is generally made. Such a correction consists in regarding a given P value as simply the midpoint in an interval of values with limits $P - .5/N$ and $P + .5/N$. Then the z value is computed, not for the difference between P and p, but for the difference between the limiting value of the interval and p. That is, we take

$$z = \frac{P - p - (.5/N)}{\sigma_P}$$

if P happens to be greater in value than p, or we take

$$z = \frac{P - p + (.5/N)}{\sigma_P}$$

if P happens to be smaller in value than p. For the example above, the correction for continuity amounts to .5/100 or .005. This has the result of making the z value in the test 2.14 rather than the 2.24 found previously.

Obviously, the correction for continuity can make a difference in the conclusions reached from a test, although it happened in this example not to do so. In general, the correction really should be used, especially when the sample size N is relatively small.

Although the example just concluded dealt with a hypothesis about a proportion, much the same procedure applies to two-tailed tests of means. An exact and an inexact hypothesis are opposed:

H_0: $\mu = \mu_0$

H_1: $\mu \neq \mu_0$.

The exact hypothesis is tested by forming two regions of rejection, each containing exactly $\alpha/2$ proportion of sample results when H_0 is true, and lying in the higher and lower tails of the distribution. A sample result falling into either of these regions of rejection is said to be significant at the α level.

7.15 RELATIVE MERITS OF ONE- AND TWO-TAILED TESTS

In deciding whether a hypothesis should be tested with a one- or two-tailed rejection region, the primary guide to the experimenter must be the original question. Is one looking for a directional difference between populations, or a difference only in kind or degree? By and large, most significance tests done in social science research are nondirectional, simply because research questions tend to be framed this way. However, there are situations where one-tailed tests are clearly indicated by the question posed.

The powers of one- and two-tailed tests of the same hypothesis will be different, given the same α level and the same true alternative. If a one-tailed test is used, and the true alternative is in the direction of the rejection region, then the one-tailed test is more powerful than the two-tailed over all such possibly true values of α. In a way, we get a little statistical credit in the one-tailed test for asking a more searching question. Conversely, if the true alternative happens to be on the tail opposite the rejection region in a one-tailed test, the power is very low; in fact, the power will be no more than α in the test of a mean. In other words, we are penalized for framing a stupid question. Power curves for one- and two-tailed tests for means are compared in Figure 7.15.1.

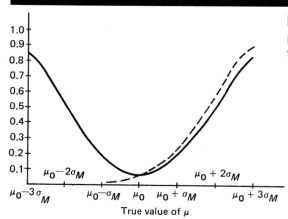

Figure 7.15.1

Power curves for a two-tailed and a one-tailed test of a mean, $\alpha = .05$.

In many circumstances calling for one-tailed tests the form of the question or of the sampling distribution makes it clear that the only alternative of logical or practical consequence must lie in a certain direction. For example, when we turn to the problem of simultaneously testing many means for equality (Chapter 10), it will turn out that the only rejection region making sense lies on one tail of the particular sampling distribution employed. In other circumstances the question involves considerations of "which is better" or "which is more," where the discovery of a difference from H_0 in one direction may have real consequences for practical action, although a mean in another direction from H_0 may indicate nothing. For example, consider a treatment for some disease. The cure rate for the disease is known, and we want to see if the treatment *improves* the cure rate. We have not the slightest practical interest in a possible decrease in cures; this, like no change, leads to nonadoption of the treatment. The thing we want to be sure to detect is whether or not the new treatment is really better than what we have. Thus, we can safely ignore the low power for detecting a poorer treatment, so long as we have high power for detecting a really better treatment. In many such problems where different practical actions depend on the sign of the deviation from expectation, a one-tailed test is clearly called for. Otherwise, the two-tailed test is safest for the experimenter asking only "What happens?"

7.16 INTERVAL ESTIMATION AND HYPOTHESIS TESTING

We have already examined the concept of a confidence interval for a population parameter such as μ and p. These are, of course, constructed by finding sample confidence limits a certain number of standard errors to either side of the sample mean or proportion. Then we are justified in saying that the probability has some value such as .95 or .99 that the true value of μ or p is covered by the interval between those limits. By using confidence intervals, we are able to "hedge our bet" about the value of μ or p estimated from $\bar{x}$ or P.

However, it is also true that confidence limits have a close connection with the procedures we use in two-tailed tests of hypotheses about μ or p. That is, if one forms a 95% confidence interval for the value of μ, based on a sample of N cases drawn independently and at random, then 95 out of 100 such confidence intervals will cover the true value of μ. Any hypothetical value you might have for μ that is covered by the confidence interval *could not* be rejected at the $\alpha = .05$ level in a two-tailed test. That is, if μ_0 is the value you have in a null hypothesis

$$H_0: \mu = \mu_0$$

and it is true that

$$\bar{x} - 1.96\sigma_M \leq \mu_0 \leq \bar{x} + 1.96\,\sigma_M\,,$$

then it must also be true that

$$-1.96 \leq \frac{\bar{x} - \mu_0}{\sigma_M} \leq 1.96\,,$$

meaning that the hypothesis cannot be rejected for $\alpha = .05$ in a two-tailed test.

This fact leads to another interpretation and use for a confidence interval: a $100\,(1 - \alpha)$ percent confidence interval contains all of the values of μ (or of p) that *cannot* be rejected at the $100\,(\alpha)$ percent level of significance. A value not covered by the confidence interval would be significant beyond the $100\,(\alpha)$ level. Therefore, the reporting of a confidence interval along with a point estimate of a population's μ or p value immediately establishes a range of hypotheses that would not be rejected at a particular level. Any hypothesis lying outside that range of values would prove to be significant if tested (two tails).

7.17 EVIDENCE AND CHANGE IN PERSONAL PROBABILITY

In some research settings, a very good case can be made that the purpose of data collection is neither to permit an immediate decision to be made nor to obtain point estimates of parameters. Instead, the purpose is to modify the experimenter's *degrees of belief* in the various situations that may exist. The experimenter starts out with various hypotheses about the true situation. Experimentation is never begun in a state of total ignorance, however. For various theoretical and empirical reasons, there are grounds to believe in some of these hypotheses more strongly than others *before* the data are seen. Then the data alter these beliefs, so that some may be strengthened, some weakened, and still others unchanged. In short, some statisticians and researchers would argue that scientific investigation does not usually deal with decision making, but rather with alteration in personal probabilities. As evidence accumulates, some beliefs become so strong that these propositions are regarded as proved and become part of the body of empirical science. Other beliefs fall away as evidence fails to support them.

The idea of personal probability has already been introduced in the discussion of decision rules. Let us suppose once again that it is possible to assign numerical values to various hypotheses, each standing for the experimenter's personal probability that a given proposition is true.

For example, in the situation described in Section 7.3, the true proportion is either

$$H_0: \ p \ = \ .80$$

or

$$H_1: \ p \ = \ .40,$$

so that we may regard the personal probability for any other H to be zero. Suppose that in the light of what you already know, there is good reason to believe much more strongly in H_0 than in H_1. Imagine the personal probabilities assigned to these hypotheses are

$$\mathcal{P}(H_0) \ = \ .75$$

$$\mathcal{P}(H_1) \ = \ .25.$$

In other words, you are willing to give odds of three to one that H_0 is true. These personal probabilities of the experimenter are *prior* since they reflect the degrees of belief in the two alternatives before the evidence. After the data are observed, something should happen to these personal probabilities; the degree of belief in each of the hypotheses should change somewhat.

The personal probability for H_0, given the evidence y, is designated by $\mathcal{P}(H_0|y)$, and that for H_1 by $\mathcal{P}(H_1|y)$. These are *posterior* values, since they *do* depend upon what is observed.

How should evidence affect the degree of belief in a given hypothesis? The key is given by Bayes's theorem, discussed in Section 1.13. If personal probabilities operate like ordinary, relative-frequency probabilities, then by Bayes's theorem,

$$\mathcal{P}(H_0|y) \ = \ \frac{p(y|H_0)\mathcal{P}(H_0)}{p(y|H_0)\mathcal{P}(H_0) \ + \ p(y|H_1)\mathcal{P}(H_1)} \ .$$

The $p(y|H_0)$ and $p(y|H_1)$ are the usual probabilities dealt with in statistical inference: the conditional probability of evidence given that the particular hypothesis is true. Notice that, other things being equal, the more likely is y given H_0, and the less likely is y given H_1, the larger is $\mathcal{P}(H_0|y)$. Under this conception, evidence made likely by one hypothesis and not by the other always strengthens belief in the first and weakens belief in the second.

However, the effective *strength* of the evidence also depends upon the *prior personal probabilities*. The amount that evidence changes degree of belief in H_0 and H_1 depends on the *prior* $\mathcal{P}(H_0)$ and $\mathcal{P}(H_1)$ values. For example, suppose that the experimenter observes a sample proportion P of .6 among 10 cases. From the sampling distributions given in Section 7.3, we can see that

$$p(.6|H_0) \ = \ .088$$

$$p(.6|H_1) \ = \ .111.$$

Applying Bayes's theorem, we find

$$\mathcal{P}(H_0|.6) \ = \ \frac{p(.6|H_0)\mathcal{P}(H_0)}{p(.6|H_0)\mathcal{P}(H_0) \ + \ p(.6|H_1)\mathcal{P}(H_1)}$$

$$= \frac{.066}{.066 + .028}$$

$$= .70$$

$$\mathcal{P}(H_1|.6) = .30.$$

This evidence lowers the personal probability for H_0 somewhat, since it is less probable under H_0 than H_1. However, the posterior personal probability for H_0 is *still* much greater than for H_1, reflecting the original discrepancy in the degree of belief the experimenter holds for each.

On the other hand, suppose that the sample evidence is a P of .3. Here,

$$p(.3|H_0) = .001; p(.3|H_1) = .215.$$

Notice that this value of P is *very* improbable if H_0 were true. Now,

$$\mathcal{P}(H_0|.3) = \frac{.001 \times .75}{(.001 \times .75) + (.215 \times .25)}$$

$$= \frac{.00075}{.00075 + .05375} = .01$$

$$\mathcal{P}(H_1|.3) = 1 - (H_0|.3) = .99.$$

In spite of the original weighting given the evidence by the prior personal probabilities, this time the P obtained is so unlikely given H_0 that the experimenter is almost certain that H_1 is true in the light of the evidence.

This way of evaluating evidence actually uses the raw materials of ordinary tests of hypotheses; as we have seen, the net result of any significance test is a statement of the likelihood of the sample result given that H_0 is true. However, a new ingredient is added: the prior personal probabilities of the experimenter. Intuitively, this approach has much to recommend it. In many ways it seems a far better description of what the scientist actually does with the statistical result than is the usual discussion of a significance test, burdened as it is with irrelevant decision terminology. In this conception, scientists are viewed as changing their minds in the light of evidence, but they are not thought of as using some phony decision rule.

On the other hand, mathematical statisticians are not always enthusiastic about this approach. The general idea is very old and, at one time, was carried to extreme and unreasonable lengths. Even metaphysical questions were treated by this general method. Much of modern theoretical statistics grew out of attempts to get away from the idea of prior probabilities, and it is not surprising that some statisticians look at such uses of Bayes's theorem with a jaundiced eye.

This is not an ideal place to go further into this approach to inferential statistics. However, the student who would like to become acquainted with this approach is urged to look into the books by Phillips (1974); Winkler (1972); or Novick and Jackson (1974). Each gives a relatively simple introduction to statistics as seen from a Bayesian point of view.

7.18 SIGNIFICANCE TESTS AND COMMON SENSE

Stripped of the language of decision theory and of concern with personal probabilities, all that a significant result implies is that one has observed something relatively unlikely given the hypothetical situation, but relatively more likely given some alternative situation. Everything else is a matter of what one does with this information. Statistical significance is a statement about the likelihood of the observed result, nothing else. It does not guarantee that something important, or even meaningful, has been found.

By shifting our point of view slightly, we can regard the statistical significance or nonsignificance of a result as a measure of the "surprisal value" of that result. That is, the surprisal value of a result can be thought of as the reciprocal of the probability (or likelihood) of the result given H_0. If a result is relatively likely given H_0, the surprisal value of this result is quite small. On the other hand, if the result falls among unlikely events given H_0, then it has great surprisal value. An impossible result should lead to infinite surprise. Then, since we tend to favor those hypotheses that lead to unsurprising results, H_0 is rejected only when the surprisal value of our result, given H_0, is quite high.

The simple matter of parsimony should lead one to be cautious in rejecting H_0. If trivial deviations from expectation are to lead us to rejection of a given hypothesis, our ideas about what is true might flit all over the place. The conservatism of the conventional rules for hypothesis testing act as a kind of brake on our tendency to follow trivial or ephemeral tendencies in the data.

These are very good arguments indeed for the traditional procedures of hypothesis testing. However, these arguments do not apply with equal force to all situations, nor do they engage the full conceptual machinery of decision making.

Any social scientist who is seriously concerned with making an important decision from data should pay attention to the possible losses the various outcomes represent. There is no guarantee that the use of any of the conventional rules for deciding significance is appropriate for this particular purpose. On the other hand, so long as the real costs of decision errors in scientific research remain as obscure as they are at present, then these conventions may not be a bad thing. In a young science, still mapping out its area of study, the error of pursuing a "phantom" result *is* costly, perhaps even more so than failing to recognize a real experimental effect when we see one. If these conventions are actually used to determine which lines of research are pursued and which are abandoned, then at least we tend to follow up the big departures from expectation that our data show.

However, conventions about significant results should not be turned into canons of good scientific practice. Even more emphatically, a convention must not be made a superstition. It is interesting to speculate how many of the early discoveries in physical science would have been statistically significant in the experiments where they were first observed. Even in the crude and poorly controlled experiment, some departures from expectation stand out simply because they are interesting and suggest things to us that we might not be able to explain. These are matters that warrant looking into further regardless of what the conventional rule says to decide. Statistics cannot do the scientist's basic job—looking and wondering and looking again.

It is a grave error to evaluate the "goodness" of an experiment only in terms of the significance level of its results. Regardless of the scientist's convictions about personal probability, even a little bit of evidence coming out of a careful experiment is far more persuasive than a great deal from a sloppy one. To what population of subjects do the conclusions refer? Has a significant result been "bought" by restricting the population so much that the result fails to have any generality at all? In an effort to achieve a large sample N has the experimenter sacrificed all claim that the sample is random? Finally, how potent is the finding in a *predictive* sense: how much do these results permit us to reduce our uncertainty about the status of a given individual in this situation? This is by no means the same as statistical significance! As we shall see in later chapters, it is entirely possible for a highly significant result to contribute little to our ability to predict behavior, and for a nonsignificant result to mask an important gain in predictive ability.

It is very easy for research persons, particularly young persons, to become overly concerned with statistical method. Sometimes the problem itself seems almost secondary to some elegant method of data analysis. Easy access to computers has made it possible even for the novice in research to apply highly sophisticated methods quite routinely. Highly analyzed and significant results are often confused with good results. But over-emphasizing the role of statistical analysis and statistical significance in research is like confusing the paintbrush with the painting. This form of statistical inference is a valuable tool in research, but it is never the arbiter of good research.

The remainder of this book is about statistical inference and largely about significance tests. Since a text in statistics must necessarily deal mostly with statistics, there just is not room to discuss all the *if*s, *and*s, and *but*s that accompany the use of this tool in research. The skills of doing good research are acquired slowly and often painfully, although a good share both of native curiosity and of common sense helps matters along. Statistical texts can display the tools, but they cannot give a guide to their use in every conceivable situation. If the user has some idea of how the tool works, perhaps its uses and limitations in a given situation will become clearer. Expert help is available and should always be sought. But if there is ever a conflict between the use of a statistical technique and common sense, then common sense comes first. Careful observation is the main business of empirical science, and statistical methods are useful only so long as they help, not hinder, the systematic exploration of data and the cumulation and coordination of results.

EXERCISES

1. A study was concerned with the tendency of alcoholic middle-aged men to relapse after having undergone a certain new course of treatment. In the past, under an old treatment, 39% of such men relapsed into alcoholism within six months. It was felt that the six-month relapse rate may be different following the new treatment. A random sample of 150 male alcoholics was thus given the new treatment. After its conclusion, the number that relapsed within six months was noted. Formulate the null and the alternative hypotheses for this example. Sketch the relative locations of the rejection and nonrejection regions in a test of the null hypothesis, given α.

2. In the study in Exercise 1, interest actually focused on the possibility that the new treatment was better, as there is little practical value in a treatment that *increases* relapse rate. Given this interest, state the null and alternative hypotheses, and sketch the rejection and nonrejection regions that appear to be appropriate.

3. There was concern in a public school system that the method of teaching reading then in use might be inferior to other methods. A standardized test was available, giving national norms on reading achievement. For fifth-graders, this test showed a national average of 172 with a standard deviation of 16. A random sample was taken consisting of 250 fifth-graders taught by the method in question, who were each given the test. Frame the null and the alternative hypotheses for this problem, and indicate the regions of rejection and nonrejection that would be appropriate here. Let the probability of incorrectly rejecting the null hypothesis be equal to .01 in this instance.

4. Actually test the hypothesis stated in Exercise 3, given that the random sample of students showed a mean of 170. What conclusions would you draw?

5. Suppose that a child transferring into a school must be placed either in a sixth-grade-level or a seventh-grade-level mathematics class. A standardized test of arithmetic achievement is used. It has been found that people who succeed in the seventh-grade-level class have an average score of 148 on entering, with a standard deviation of 20. On the other hand, those who belong in the lower class have an average entering score of 132, with a standard deviation of 15. Frame this as a decision problem, showing the different correct decisions and errors that can occur.

6. Assume that the two distributions of scores described in Exercise 5 are both normal. Then work out the probabilities of erroneous and correct decisions about this child if the following rules are applied:

(a) Assign to sixth grade if score ≤ 132; seventh grade otherwise.
(b) Assign to seventh grade if score ≥ 148; sixth otherwise.
(c) Assign to sixth if score ≤ 139; otherwise seventh.
(d) Assign to sixth if score ≤ 140; otherwise seventh.
(e) Assign to sixth if score ≤ 141; otherwise seventh.

If the losses associated with each kind of error are equal, which is the best decision rule?

7. Let us assume that the losses associated with errors in Exercises 5 and 6 are known. Let A represent the loss associated with placing a child in a grade which is too high, and let B represent the loss involved in the other kind of error. Decide which of the five rules listed above should be used if the errors have the loss values

(a) $A = 10; B = 10$
(b) $A = 20; B = 10$
(c) $A = 10; B = 20$
(d) $A = 10; B = 12$

If only one of the last three rules may be used, which one is best given each of the loss values indicated above?

8. A social science study is to be based on the use of telephone interviews. A new city is about to be included in the study. If 40% of the persons contacted agree to be interviewed, it will be worthwhile to go on with the study in this city. On the other hand, if 20% (or fewer) will be interviewed, it is not worth proceeding. Based on past experience, it is believed that one of these situations (20% or 40%) is likely to be true. Therefore, a random selection of 10 residential calls are made. Depending on the results, the study will either be completed or abandoned for this city. Find the probabilities of errors and of correct decisions under the following rules:

(a) Continue if 6 or more calls are successful, and not otherwise.
(b) Continue if 2 or more calls are successful; not otherwise.
(c) Continue if 4 or more calls are successful; not otherwise.
(d) Continue if 3 or more calls are successful; not otherwise.

(**Hint:** Use Table II, Appendix E.)

9. Let A symbolize the loss involved in the study above if erroneously continued. Let B stand for the loss if the study is wrongly abandoned. Then find the best decision rule to use under each pair of these values for A and B.

(a) $A = \$1,000; B = \$1,000$
(b) $A = \$1,500; B = \$1,000$
(c) $A = \$1,000; B = \$5,000$
(d) $A = \$10,000; B = \$1,000$

10. In the past, about 60% of cities such as those in Exercises 8 and 9 have given a satisfactory number of interviews, and about 40% have not. The study director thus tends to believe that the odds are about 3 to 2 that this city will prove productive. Under these circumstance, and given equal loss values for the two kinds of errors, which decision rule should the study director prefer?

11. A certain random variable is normally distributed and has a standard deviation of 4.2. Twenty-six observations were made independently and at random, and yielded a sample mean value of 31 for this random variable. Test the hypothesis that the mean of the random variable is 28.6 against the alternative hypothesis that the mean has some other value. (Use the .05 level for significance.)

12. A psychological test was standardized for the population of 10th-grade students in such a way that the mean must be 500 and the standard deviation 100. A sample of 90 twelfth-grade students was selected independently and at random, and each given the test. The sample mean turned out to be 506.7. On this basis, can one say that the population distribution for 12th-grade students would differ from that for 10th-graders?

13. For a particular task given to subjects in an experiment, the researcher theorized that about 25% of this population of subjects should be able to complete the task within the allotted time. To test this hunch, he took 20 subjects chosen at random and gave each the task independently. Of this group, 45% actually did finish within the time allotted. Test the hypothesis that the true proportion is .25 or less against the alternative hypothesis that the proportion is greater than .25.

14. Suppose that in Exercise 13 the experimenter had taken 200 subjects in order to test the hypothesis, and that the sample proportion had come out to be .28. Test the hypothesis that the true proportion is .25 against the alternative that the true proportion is not .25.

15. Are young minority children judged to be more or less cooperative than nonminority peers, when the teacher doing the rating is nonminority? To answer this question, a random sample of 123 nonminority first-grade teachers were asked to rate the cooperation of each of their students. Each had about the same number of minority and nonminority students. The score then assigned to the teacher was the average difference in ratings (i.e., minority average − nonminority) given to the two groups of children in that classroom. Over the entire sample of teachers the mean score was +.23, with a standard deviation of 1.27. Test the hypothesis that there is no difference in the way teachers tend to rate these two groups of children, using a two-tailed test and $\alpha = .05$.

16. A nine-hole golf course was supposed to have a par of 30, but over a long period the population of golfers who played this course had a mean of 38.2, with a standard deviation of

3.3. A designer was called in to extend this course to 18 holes, on the understanding that the last nine would have the same difficulty level as the first nine. After the course was finished a random sample of 121 golfers played the course and produced an average score of 42.6 on the last nine holes. Given that this sample was drawn from the population who played the old nine holes, test the hypothesis that the two sets of nine hole are truly equal in difficulty. Use an α value of .05.

17. Some 80 rats selected at random were taught to run a maze. All of them finally succeeded in learning the maze, and the average number of trials to perfect performance was 15.91. However, long experience with a population of rats trained to run a similar maze shows that the average number of trials to success is 15, with a standard deviation of 2. Would you say that the new maze appears to be harder for rats to learn than the older, more extensively used maze?

18. A teacher wished to study the change in student attitude toward the federal government produced by a political science course. At the beginning of the class, therefore, a specially constructed attitude test was given, and then repeated at the end of the course. The student's score was the difference between scores on the second and the first test. A total of 368 students took the tests on both occasions. The mean change score was 2.3 points, and the sample standard deviation s was 10.5. Assuming that the students in the course constitute a random sample, would you say that there was a significant change in attitude? (**Hint:** since the sample is large, use the unbiased estimate of the population variance as taken from the sample.)

19. Given a normal distribution with a standard deviation of 10, suppose that one of the following two hypotheses must be true:

H_0: $\mu = 100$, H_1: $\mu = 105$

If the probability of Type I error is to be .10, and the probability of Type II is also to be .10, what sample size should be used, and what is the critical value of the sample mean leading one to choose H_0 or H_1 as true?

20. Given the conditions in Exercise 19, suppose that Type I error probability is to be fixed at .01 and Type II error probability at .10. Now what is the sample size that should be used, and what is the critical value of $\bar{x}$?

21. In Exercise 11, what is the power of the test against the alternative hypothesis that the mean is 32? Against the hypothesis that the mean is 25? (Use .05 as the probability of Type I error.)

22. In Exercise 12, what is the power of the test against the hypothesis that the mean is 510? Against the hypothesis that the mean is 480? Against the hypothesis that the mean is 520? (Use .05 as the probability of Type I error.)

23. In Exercise 17, what is the power of the test against the hypotheis that the mean is 16? That the mean is 20?

24. In Exercise 17, if the experimenter had wished to have a power of .95 against the alternative hypothesis that the true mean is 16, what number of subjects would have been sufficient? (Use .05 as the probability of Type I error.)

25. Sketch the power function for a test of hypothesis about a mean which is two-tailed for α = .10 and for α = .05. (Assume a normal population distribution and, for simplicity, let $\sigma/\sqrt{N} = 1.00$. Divide the X axis of the plot into units of .5 or less.) What does this plot show about the relationship of power to the probability of Type I error?

26. For Exercise 29 of Chapter 5, test the hypothesis that the new training method is no better than the old, against the alternative that the new method does cut down on injuries. Use the stated value of 4.9 for the standard deviation of the population, and take α = .01.

27. Imagine that an experimenter draws a random sample and uses it to test a hypothesis. The α value is set at .05. Then another, independent random sample is drawn and the same hypothesis is tested once again, with the same α level. What is the probability that a Type I error is made on both tests? On neither test? On at least one test?

28. Suppose that the experimenter in Exercise 27 takes a whole series of independent random samples, J in all. With each sample the same null hypothesis is tested. What is the probability of no Type I errors in this set of J tests? What is the probability of at least one Type I error? Work out this probability of at least one Type I error for $\alpha = .05$ and for $J = 3, 4, 5,$ and 6. What happens to the probability of at least one Type I error as the total number of tests is increased?

29. Suppose that an experimenter were trying to decide among three hypotheses:

$$H_0: \mu = 200, \quad H_1: \mu = 210, \quad H_2: \mu = 220$$

One and only one of these hypotheses must be true. Drawing a sample of data, it is determined that if H_0 is true, such a result has a likelihood of .15; if H_1 is true, the sample result has a likelihood of .23; and if H_2 is true, the sample result has a likelihood of .20. Prior to the experiment, the experimenter believes that the probability that H_0 is true is 1/2, and that H_1 and H_2 each has probability of only .25. Describe the experimenter's probabilities for these three hypotheses *after* the sample results are in.

30. Suppose that an experimenter is entertaining three hypotheses, one and only one of which must be true:

$$H_0: \mu = 200, \quad H_1: \mu = 210, \quad H_2: \mu = 220$$

It is known that the population is normal and that the population standard deviation is 20. A sample of 25 cases is to be drawn and the sample mean computed. Decision Rule A is formulated: if $\bar{x} \le 205$, accept H_0; if $205 < \bar{x} < 215$, accept H_1; if $215 \le \bar{x}$ accept H_2. Find the probabilities of being correct and of being in error under this rule, and display these probabilities in a table similar to that in Section 7.3.

31. Suppose that the experimenter in Exercise 30 formulates a new rule, Decision Rule B: if $\bar{x} \le 207$ accept H_0; if $207 < \bar{x} < 214$ accept H_1; if $\bar{x} \ge 214$ accept H_2. Calculate the probabilities of correct decisions and of errors under Decision Rule B.

32. The experimenter in Exercises 29, 30, and 31 learns that losses are connected with erroneous decisions in the following way:

		True state		
		H_0	H_1	H_2
	Accept H_0	0	5	20
Decision	Accept H_1	10	0	10
	Accept H_2	20	5	0

Under the criterion of making the expected loss as small as possible, is Decision Rule A or Rule B the better rule? Why?

Chapter *8*

Inferences About Population Means

This chapter deals with ways to make inferences about means. The procedures for point and interval estimation and for hypothesis testing described in the past three chapters will be applied, first to questions about the mean of a single population, and then to questions about the difference in the means of two populations. A new distribution, that of the test statistic t, will be the basis for most of these methods.

8.1 LARGE-SAMPLE PROBLEMS WITH UNKNOWN POPULATION VARIANCE

In some of the examples of hypothesis testing up to this point we have actually "fudged" a bit on the usual situation: we have assumed that σ^2 is somehow known, so that the standard error of the mean is also known exactly. In these examples the author did not explain how σ^2 became known, largely because he could not think up a good reason. Now we must face the cold facts of the matter: for inferences about the population mean, σ^2 is seldom known. Instead, we must use the only substitute available for σ^2, which is our unbiased estimate s^2, calculated from the sample.

Notice that this problem does not exist for hypotheses about a population proportion p, since the existence of an exact hypothesis about p specifies what the value of the standard error of P, the sample proportion, must be. Therefore, the special techniques of this chapter apply only to inferences about means, and not to inferences about proportions.

From what we have already seen of the relation between sample size and accuracy of estimation, it makes sense that for large samples s^2 should be a very good estimate of σ^2. In general, for large samples, there is rather little risk of a sizable error when one uses s in place of σ in estimating the standard error of the mean.

Hence, when the sample size is quite large, tests of hypotheses about a single mean may be carried out in the same way as when σ is known, except that the standard error of the mean is estimated from the sample:

$$\text{est. } \sigma_M = \frac{s}{\sqrt{N}} = \frac{S}{\sqrt{N-1}} \qquad [8.1.1^*]$$

The standardized score corresponding to the sample mean is then referred to the normal distribution. This step is justified by the central limit theorem when N is large, regardless of the population distribution's form.

For example, consider the following problem. A small rodent characteristically shows hoarding behavior for certain kinds of foodstuffs when the environmental temperature drops to a certain point. Numerous previous experiments have shown that in a fixed period of time, and given a fixed food supply, the mean amount of food hoarded by an animal is 9 g. The experimenter is currently interested in possible effects that early food deprivation may have on such hoarding behavior in the animal as an adult. So, the experimenter takes a random sample of 175 infant animals and keeps them on survival rations for a fixed period while they are at a certain age, and on regular rations thereafter. When the animals are adults each one is placed in an experimental situation where the lowered temperature condition is introduced. The amount of food each hoards is recorded, and a score is assigned to each animal. (Here and in the following we are going to adopt the convention that the dependent variable in an experiment is symbolized by Y, with any value shown by y. Thus, the mean of the sample values of the dependent variable will be symbolized by $\bar{y}$.)

What is the null hypothesis implied here? The basic experimental question is "Does the experimental treatment (deprivation) tend to affect Y, the amount of food hoarded?" The experimenter has no special reason to expect either an increase or a decrease in amount, but is interested only in finding out if a difference from normal behavior occurs. This question may be put into the form of a null and an alternative hypothesis:

$$H_0: \mu_0 = 9 \text{ g}$$

$$H_1: \mu \neq 9 \text{ g} .$$

Suppose that the conventional level chosen for α is .01, so that the experimenter will say that the result is significant only if the sample mean falls among either the upper .005 or the lower. 005 of all possible results, given H_0. Reference to Table I in Appendix E shows that .005 is the probability of a z score in a normal distribution falling at or below -2.58, and the probability is likewise .005 for a z equal to or exceeding $+2.58$. Accordingly, the sample result will be significant only if

$$\frac{\bar{y} - E(\bar{y})}{\text{est. } \sigma_M} \qquad [8.1.2]$$

equals or exceeds 2.58 in absolute magnitude (disregarding sign). When the null hypothesis is true, $E(\bar{y}) = 9$, and for a sample this large, the value of the standard error of the mean should be reasonably close to $\frac{s}{\sqrt{N}}$ or $\frac{S}{\sqrt{N-1}}$, the value of the sample estimate.

Everything is now set for a significance test except for the sample results. The sample shows a mean of 8.8 g of food hoarded, with a standard deviation S of 2.30. The estimated standard error of the mean is thus

$$\text{est. } \sigma_M = \frac{2.30}{\sqrt{175 - 1}} = \frac{2.30}{13.19} = .174.$$

The standardized score of the mean is found to be

$$\frac{8.80 - 9.00}{.174} = -1.15.$$

This result does not qualify for the region of rejection for $\alpha = .01$. Since the experimenter feels able to reject H_0 only if the α probability of error is no more than .01, then H_0 will not be rejected on the basis of this sample. On the other hand, the risk run in accepting H_0 is unknown, so that "suspend judgment, pending more evidence" seems the appropriate choice here.

Confidence intervals may also be found by the methods in Chapter 6. However, either when σ^2 is unknown or when the population distribution has unknown form, a normal sampling distribution is assumed only for large samples. Just as in significance tests, the estimated standard error of the mean can be used in place of σ_M in finding confidence limits when the sample is relatively large (say $N \geq 100$).

For example, the experimenter studying hoarding behavior computes the approximate 99% confidence limits in the following way:

$$\bar{y} - 2.58 \text{ (est. } \sigma_M)$$

and

$$\bar{y} + 2.58 \text{ (est. } \sigma_M)$$

so that for this problem, the numerical confidence limits are

$$8.8 - 2.58(.174) \text{ or } 8.35$$

and

$$8.8 + 2.58(.174) \text{ or } 9.25.$$

The experimenter can say that the probability is approximately .99 that the true value of μ is covered by an interval such as that between 8.35 and 9.25.

8.2 THE DISTRIBUTION OF *t*

In inferences about μ, the ratio we would like to evaluate and refer to a normal sampling distribution is the standardized score

$$z_M = \frac{\bar{y} - E(\bar{y})}{\sigma_M}. \qquad [8.2.1]$$

However, when we have only an *estimate* of σ_M, then the ratio we really compute and use is not a normal standardized score at all, although it has much the same form. The ratio actually used is

$$t = \frac{\bar{y} - E(\bar{y})}{\text{est. } \sigma_M} = \frac{\bar{y} - E(\bar{y})}{s/\sqrt{N}}.$$ [8.2.2*]

There is an extremely important difference between the two ratios, z_M and t. For z_M, the numerator $(\bar{y} - E(\bar{y}))$ is a random variable, the value of which depends on the particular sample drawn from a given population situation; on the other hand, the denominator is a constant σ_M, which is the same regardless of the particular sample of size N we observe. Now contrast this ratio with the ratio t: just as before, the numerator of t is a random variable, but the denominator is also a random variable, since the particular value of s—and hence the estimate of σ_M—is a sample quantity. Over several different samples, the same value of $\bar{y}$ must give us precisely the same value of z_M; however, over different samples, the same value of $\bar{y}$ will give us different t values. Similar intervals of t and z_M values should therefore have different probabilities of occurrence. For this reason it is risky to use the ratio t as though it were z_M unless the sample size is very large.

The solution to the problem of the nonequivalence of t and z_M rests on the study of t itself as a random variable. That is, suppose that the t ratio were computed for every conceivable sample of N independent observations drawn from some normal population distribution with true mean μ. Each sample would have some t value,

$$t = \frac{\bar{y} - E(\bar{y})}{\text{est. } \sigma_M} = \frac{\bar{y} - \mu}{s/\sqrt{N}}.$$ [8.2.3*]

Over the different samples the value of t would vary, of course, and the different possible values would each have some probability density. A random variable such as t is an example of a **test statistic,** so called to distinguish it from an ordinary descriptive statistic or estimator, such as $\bar{y}$ or s^2. The t value depends on other sample statistics, but is not itself an estimate of a population value. Nevertheless, such test statistics have sampling distributions just as ordinary sample statistics do, and these sampling distributions have been studied extensively.

To find the exact distribution of t, one must assume that the *basic population distribution is normal*. There are two reasons for this assumption. The first reason is that it permits us to specify the distribution of the numerator of t, or $(\bar{y} - \mu)$, without regard to sample size. The second is that only for a normal distribution will the basic random variables in numerator and denominator, sample $\bar{y}$ and s, be statistically independent; this is a use of the important fact mentioned in Section 6.6. Unless $\bar{y}$ and s are independent, the sampling distribution of t is extremely difficult to specify exactly. On the other hand, for the special case of normal populations, the distribution of the ratio t is quite well known. To learn what this distribution is like, let us take a look at the rule for the density function associated with this random variable.

The density function for t is given by the rule:

$$f(t; \nu) = G(\nu) \left[1 + \frac{t^2}{\nu} \right]^{-(\nu+1)/2} \quad \begin{array}{c} -\infty < t < \infty \\ 0 < \nu \end{array}.$$ [8.2.4]

Here, G(ν) stands for a constant number which depends *only* on the parameter ν (lowercase Greek nu), and how this number is found need not really concern us.

The "working part" of the rule involves only ν and the value of *t*. This looks very different from the normal distribution function rule in Section 6.1. As with the normal function rule, however, a quick look at Figure 8.2.1, the graph corresponding to this mathematical expression, tells us much about the distribution of *t* (the unbroken curve).

First of all, notice that the distribution of sample *t* values must be symmetric, since a positive and a negative value having the same absolute size must be assigned the same probability density by this rule. Second, the largest possible density value is assigned to $t = 0$. Thus $t = 0$ is the distribution mode. Observe also that the distribution is unimodal and bell-shaped. If we inferred from the symmetry and unimodality of this distribution that the mean of *t* is also 0, we should be quite correct. In short, the *t* distribution is a unimodal, symmetric, bell-shaped distribution having a graphic form much like a normal distribution, even though the two function rules are quite dissimilar. Loosely speaking, the curve for a *t* distribution differs from the standardized normal in being "plumper" in extreme regions and "flatter" in the central region, as Figure 8.2.1 shows. (Note that both *t* and the standardized normal distribution have a mean of zero, $\nu > 1$.)

The most important feature of the *t* distribution will appear if we return for a look at the function rule. Notice that the only unspecified constants in the rule are those represented by ν and G(ν), which depends on only ν. This is a one-parameter distribution: the single parameter is ν, called *the degrees of freedom*. Ordinarily, in most applications of the *t* distribution to problems involving a single sample, *ν is equal to N − 1*, one less than the number of independent observations in the sample. For samples of *N* independent observations from any normal population distribution, the exact distribution of sample *t* values depends only on the degrees of freedom, $N - 1$. Remember, however, that the value of $E(\bar{y})$ or μ must be specified when a *t* ratio is computed, although the true value of σ need not be known.

In principle, the value of ν can be any positive number, and it just happens that $\nu = N - 1$ is the value for the degrees of freedom for the particular *t* distributions we will use first. Later we will encounter problems calling for *t* distributions with other numbers of degrees of freedom. Like most theoretical distributions, the *t* distribution is actually a family of distributions, with general form determined by the function rule, but with

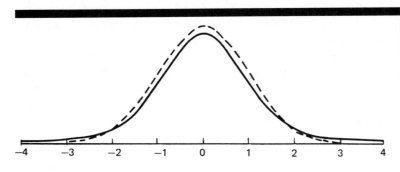

Figure 8.2.1

Distribution of *t* with ν = 4, and standardized normal distribution.

particular probabilities dictated by the parameter. For $v > 2$ the variance of the t distribution is $v/(v - 2)$, so that the smaller the value of v the larger the variance. As v becomes large the variance of the t distribution approaches 1.00, which is the variance of the standardized normal distribution.

Incidentally, the random variable t is often called Student's t, and the distribution of t, Student's distribution. This names comes from the statistician W. S. Gosset, who was the first to use this distribution in an important problem, and who first published his results in 1908 under the pen name Student. Distributions of the general Student form have a number of important applications in statistics.

8.3 THE t AND THE STANDARDIZED NORMAL DISTRIBUTION

As we have seen, the shape of the t distribution is not unlike that of the normal distribution. Just as for the standardized normal, the mean of the distribution of t is 0 for $v > 1$ although the variance of t is greater than 1.00 for finite $v > 2$. Given any extreme interval of fixed size on either tail of the t distribution, the probability associated with this interval in the t distribution is larger than that for the corresponding normal distribution of z_M. The smaller the value of v, the larger is this discrepancy between t and normal probabilities at the extreme ends of each distribution. This reflects and partly explains the danger of using a t ratio as though it were a z ratio except when v is quite large: for small v, extreme values of t are relatively more likely than comparable values of z_M. A small sample size corresponds to a small value of v, or $N - 1$, and thus there is serious danger of underestimating the probability of an extreme deviation from expectation when sample size is small. This is apparent in the illustration (Figure 8.2.1) showing the distribution of t together with the standardized normal function.

On the other hand, what happens to the distribution of t as v becomes large (sample size grows large) is suggested both by Figure 8.2.1 and by the variance of a t distribution. As sample size N grows large, the distribution of t approaches the standardized normal distribution. For *large numbers of degrees of freedom,* the exact probabilities of intervals in the t distribution can be approximated closely by *normal probabilities*.

The practical result of this convergence of the t and the normal probabilities is that the t ratio *can* be treated as a z_M ratio, provided that the sample size is substantial. The normal probabilities are quite close to—though not identical with—the exact t probabilities for large v. On the other hand, when sample size is small the normal probabilities cannot safely be used, and instead one uses a special table based on the t distribution.

How large is "large enough" to permit use of the normal tables? If the population distribution is truly normal, even 40 or so cases permit a fairly accurate use of the normal tables in confidence intervals or tests for a mean. If really good accuracy is desired in determining interval probabilities, the t distribution should be used even when the sample size is around 100 cases. Beyond this number of cases, the normal probabilities are extremely close to the exact t probabilities. For example, in the "hoarding" experiment just discussed, use of t rather than z values would have given confidence limits of about $\bar{y} \pm 2.61(\text{est } \sigma_M)$ instead of $\bar{y} \pm 2.58(\text{est } \sigma_M)$, a very slight difference.

Recall that the stipulation is made that the population distribution be *normal* when the t distribution is used, even when the normal approximations are substituted for the exact *t*-distribution probabilities. Clearly, this requirement limits the usefulness of the *t* distribution, since this is an assumption often hard to justify in practical situations. Fortunately, when sample size is fairly large, and provided that the parent distribution is roughly unimodal and symmetric, the *t* distribution still gives an adequate approximation to the exact (and often unknown) probabilities under these circumstances. However, you should insist on a relatively *larger* sample size the *less* confident you are that the normal rule holds for the population. This is especially true when one-tailed tests of hypotheses are made, since a very skewed population distribution can make the *t* probabilities for one-tailed tests considerably in error. Once again, it is wise to plan on somewhat larger samples when one is considering a one-tailed test using the *t* distribution and when the population is not assumed to be normal.

8.4 TABLES OF THE *t* DISTRIBUTION

Unlike the table of the standardized normal function, which suffices for all possible normal distributions, tables of the *t* distribution must show many different distributions each depending on the value of ν, the degrees of freedom. Consequently, tables of *t* are usually given only in abbreviated form.

Table III in Appendix E shows selected percentage points of the distribution of *t*, in terms of the value of ν. Different ν values appear along the left-hand margin of the table. The top margin gives values of Q, which is $1 - p(t \leq a)$, one minus the cumulative probability that *t* is less than or equal to a specific value a, for a distribution within the given value for ν. A cell of the table then shows the value of *t* cutting off the upper Q proportion of cases in a distribution for ν degrees of freedom.

This sounds rather complicated, but an example will clarify matters considerably: suppose that $N = 10$, and we want to know the value *beyond which* only 10% of all sample *t* values should lie. That is, for the distribution of *t* shown in Figure 8.4.1 we want the *t* value that cuts off the shaded area in the curve, the upper 10%:

First of all, since $N = 10$, $\nu = N - 1 = 9$. So, we enter the table for the row marked 9. Now since we want the upper 10%, we find the column for which $Q = .1$. The corresponding cell in the table is the value of *t* we are looking for, $t = 1.383$. We can say that in a *t* distribution with 9 degrees of freedom, the probability is .10 that a *t* value *equals or exceeds* 1.383. Since the distribution is symmetric, we also know that the probability is .10 that a *t* value *equals or falls below* -1.383. If we wanted to know

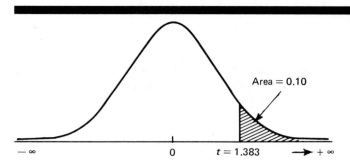

Figure 8.4.1

The upper 10% area in a *t* distribution with 9 degrees of freedom.

Area = 0.10

$-\infty$ 0 $t = 1.383$ $\longrightarrow$ $+\infty$

the probability that t equals or exceeds 1.383 in absolute value, then this must be .10 + .10 = .20, or $2Q$.

Suppose that in a sample of 21 cases, we get a t value of 1.98. We want to see if this value falls into the upper .05 of all values in the distribution. We enter row ν = 21 − 1 = 20, and column Q = .05; the t value in the cell is 1.725. Our obtained value of t is larger than this, and so the obtained value does fall among the top 5% of all such values. On the other hand, suppose that the obtained t had been −3. Does this fall either in the top .001 or the bottom .001 of all such sample values? Again with ν = 20, but this time with Q = .001, we find a t value of 3.552. This means that at or above 3.552 lie .001 of all sample values, and also at or below −3.552 lie .001 of all sample values. Hence our sample value does not fall into either of these intervals; we can say that the sample value does not fall into the rejection region for α = .002.

The very last row in Table III, marked ∞, shows the z scores that cut off various areas in a normal distribution curve. If you trace down any given column, you find that as ν gets larger the t value bounding the area specified by the column comes closer and closer to this normal deviate value, until finally, for an infinite sample size, the required value of t is the same as that for z.

For one-tailed tests of hypotheses, the column Q values are used to find the t value which exactly bounds the rejection region. If the region of rejection is on the upper tail of the distribution, then Q is the probability of a sample value's falling into the region greater than or equal to the tabled value of t. If the region is on the lower tail, the t value in the table is given a negative sign, and Q is the probability of a sample's falling at or below the negative t value. If a two-tailed region is to be used, then the total α probability of error is $2Q$, and the number in the table shows the absolute value of t that bounds the rejection region on *either* tail.

8.5 THE CONCEPT OF DEGREES OF FREEDOM

Before we proceed to the uses of the t distribution, it is well to examine the notion of degrees of freedom. The degrees-of-freedom parameter reflects the fact that a t ratio involves a sample standard deviation as the basis for estimating σ_M. Recall the basic definitions of the sample variance and the sample standard deviation:

$$s^2 = \frac{\Sigma(y - \bar{y})^2}{N}$$

and

$$S = \sqrt{\frac{\Sigma(y - \bar{y})^2}{N}}.$$

The sample variance and standard deviation are both based on a sum of squared deviations from the sample mean $\bar{y}$. However, recall another fact of importance about deviations from a mean: in Section 4.3 it was shown that

$$\sum_i (y_i - \bar{y}) = 0;$$

that is, the sum of all of the deviations about the mean must be zero.

These two facts have an important consequence: Suppose that you are told that $N = 4$ in some sample, and that you are to guess the four deviations from the mean. For the first deviation you can guess any number, and suppose you say

$d_1 = 6.$

Similarly, quite at will, you could assign values to two more deviations, say

$d_2 = -9$

$d_3 = -7.$

However, when you come to the fourth deviation value, you are *no longer free* to guess any number you please. The value of d_4 *must* be

$d_4 = 0 - d_1 - d_2 - d_3$

or

$d_4 = 0 - 6 + 9 + 7 = 10.$

In short, given the values of any $N - 1$ deviations from the mean, which could be any set of numbers, the value of the last deviation is completely determined. Thus we say that there are $N - 1$ degrees of freedom for a sample variance, reflecting the fact that only $N - 1$ deviations are "free" to be any number, but that given these free values, the last deviation is completely determined. It is not the sample size per se that dictates the distribution of t, but rather the number of degrees of freedom in the variance (and standard deviation) estimate.

Perhaps this little argument will give you some intuition about the concept of degrees of freedom and why it is important. However, it must be said that the concept is a good bit more sophisticated than it may appear here, and that in future sections a better explanation will indeed be required. For the moment, however, this interpretation will have to suffice.

8.6 SIGNIFICANCE TESTS AND CONFIDENCE LIMITS FOR MEANS USING THE t DISTRIBUTION

For the moment you can relax; there is really nothing new to learn! When the null hypothesis concerns a single mean, then the test is carried out just as before, except that the table of t (Appendix E, Table III) is used instead of the normal table (Appendix E, Table I). The α level is chosen, and the value (or values) of t corresponding to the region of rejection can be determined from the t table. The number of degrees of freedom used is simply $\nu = N - 1$. The t ratio is

$$t = \frac{\bar{y} - E(\bar{y})}{\text{est. } \sigma_M} \qquad [8.6.1]$$

where

$$\text{est. } \sigma_M = S/\sqrt{N - 1} \qquad [8.6.2]$$

and

$$E(\bar{y}) = \text{expected value of } \bar{y} \text{ if } H_0 \text{ is true.}$$

Then the value of t which is obtained from the sample is compared with values in the rejection region specified by Table III. If the obtained t ratio falls into the rejection region chosen, the sample result is said to be significant beyond the α level.

Naturally, all the considerations hitherto discussed, especially the assumed normal distribution of the population, should be faced before the sample size and rejection region are decided on. If large samples are available, then the assumption of a normal population is relatively unimportant; on the other hand, this matter should be given some serious thought if you are limited to a very small sample size.

The t distribution may also be used to establish confidence limits for the mean. For some fixed percentage representing the confidence level, $100(1 - \alpha)$ percent, the sample confidence limits depend on three things: the sample value of $\bar{y}$; the estimated standard error, or est. σ_M; and the number of degrees of freedom ν. For some specified value of ν, then the $100(1 - \alpha)$ percent confidence limits are found from

$$\bar{y} - t_{(\alpha/2;\nu)} (\text{est. } \sigma_M)$$

[8.6.3*]

$$\bar{y} + t_{(\alpha/2;\nu)} (\text{est. } \sigma_M).$$

Here, $t_{(\alpha/2;\nu)}$ represents the value of t that bounds the upper $\alpha/2$ proportion of cases in a t distribution with ν degrees of freedom. In Table III this is the value listed for $Q = \alpha/2$ and ν. Thus, if one wants the 99% confidence limits, the value of $\alpha = .01$, and one looks in the table for $Q = .005$.

For example, imagine a study using eight independent observations drawn from a normal population. The sample mean is 49 and the estimated standard error of the mean is 3.70. Now we want to find the 95% confidence limits. First of all, $\alpha = .05$, so that $Q = .025$. The value of ν is $N - 1$, or 7. The table shows a t value of 2.365 for $Q = .025$ and $\nu = 7$, so that $t_{(\alpha/2;\nu)} = 2.365$. The confidence limits are

$$49 - (2.365 \times 3.70) = 40.25$$

and

$$49 + (2.365 \times 3.70) = 57.75.$$

Over all random samples, the probability is .95 that the true value of μ is covered by an interval such as that between 40.25 and 57.75, the confidence interval calculated for this sample.

8.7 QUESTIONS ABOUT DIFFERENCES BETWEEN POPULATION MEANS

Examples of hypotheses about single means often sound rather "phony" in their experimental contexts, and the reason is not hard to find. In most experimental work it is not true that the experimenter knows about one particular population in advance and then draws a single sample for the purpose of comparing some experimental population to the known population. Rather, it is far more common to draw two samples, to only one of which the experimental treatment is applied; the other sample is given no treatment and stands as a control group for comparsion with the treated group. In other situations, two different treatments may be compared. The treatments may be something done by the experimenter, or nature may have already made the groups different in some respect. The advantages of this method over the single-sample procedure are obvious; the experimenter can exercise the same experimental controls on both samples, making sure that insofar as possible they are treated in exactly the same way, with the only systematic experimental difference being in the fact that something was done to representatives of one sample which was not done to members of the other. Then, if a very large difference appears between the two samples, this difference should be a product of the experimental treatments and not just a peculiarity introduced by the way in which the data were gathered.

The best available estimate of the difference between the means of two populations is the obtained difference between the means of the two samples. As always, this estimate of the population difference is in error to some unknown extent, and although the obtained difference between the sample means is the best guess the experimenter can make, there is absolutely no guarantee that this estimate is exactly correct. It could well be true that the difference the experimenter observes has no real connection with the treatment administered and is purely a chance result.

What is needed is a way of applying statistical inference to differences between means of samples representing two populations. First, large sample distributions of *differences* between sample means will be studied. Then, the application of the t distribution to small sample differences will be introduced.

Suppose that we wished to test a hypothesis that two populations have means which differ by some specified amount, say 20 points. This is tested against the hypothesis that the population means do not differ by that amount. In our more formal notation:

H_0: $\mu_1 - \mu_2 = 20$

H_1: $\mu_1 - \mu_2 \neq 20$.

We draw a sample size N_1 from Population 1 and an *independent* sample of size N_2 from Population 2, and consider the difference between their means, $\bar{y}_1 - \bar{y}_2$. Now suppose that we kept on drawing pairs of independent samples of these sizes from these populations. For each pair of samples drawn, the difference $\bar{y}_1 - \bar{y}_2$ is recorded. What is the distribution of such sample *differences* that we should expect in the long run? In other words, what is the sampling distribution of the difference between two means?

In the first place,

$$E(\bar{y}_1 - \bar{y}_2) = E(\bar{y}_1) - E(\bar{y}_2) = \mu_1 - \mu_2 , \qquad [8.7.1^*]$$

by Rule 4 of Appendix B. Thus, the *expected difference* between the two sample means is the *difference between the population means*.

Furthermore, since these are two independent samples, of N_1 and N_2 independent cases respectively, then

$$\text{var}(\bar{y}_1 - \bar{y}_2) = \sigma_{M_1}^2 + \sigma_{M_2}^2 . \qquad [8.7.2^*]$$

by Eq. C2.6 of Appendix C.

Hence, the standard error of the difference, $\sigma_{\text{diff.}}$, is

$$\sigma_{\text{diff}} = \sqrt{\sigma_{M_1}^2 + \sigma_{M_2}^2} = \sqrt{\frac{\sigma_1^2}{N_1} + \frac{\sigma_2^2}{N_2}} \qquad [8.7.3^*]$$

provided that Samples 1 and 2 are completely independent. Then, for two *large* samples drawn from populations having different variances σ_1^2 and σ_2^2, the standard error of the difference is estimated by

$$\text{est. } \sigma_{\text{diff.}} = \sqrt{(\text{est } \sigma_{M_1})^2 + (\text{est. } \sigma_{M_2})^2} . \qquad [8.7.4^*]$$

Notice that there is no requirement at all that the samples be of equal size. Regardless of the sample sizes, the expectation of the difference between two means is always the difference between their expectations, and the variance of the difference between two *independent* means is the *sum* of the separate sampling variances.

Furthermore, these statements about the mean and the standard error of a difference between means are true regardless of the form of the parent distributions. However, the form of the sampling distribution can also be specified under either of two conditions:

If the distribution for each of the two populations is normal, then the distribution of differences between sample means is normal.

This is a direct consequence of Principle C2.8, Appendix C. When we can assume both populations normal, the form of the sampling distribution is known to be *exactly* normal.

On the other hand, one or both of the original distributions may not be normal; in this case the central limit theorem comes to our aid:

As both N_1 and N_2 grow infinitely large, the sampling distribution of the difference between means approaches a normal distribution, regardless of the form of the original distributions.

In short, when we are dealing with two very large samples, then the question of the form of the original distributions becomes irrelevant, and we can approximate the sampling distribution of the difference between means by a normal distribution.

Thus, when sample size is large, the test statistic for a difference between means is

$$z_{\text{diff.}} = \frac{(\bar{y}_1 - \bar{y}_2) - E(\bar{y}_1 - \bar{y}_2)}{\text{est. } \sigma_{\text{diff.}}} . \qquad [8.7.5^*]$$

This value may be referred to a normal distribution. The expected difference depends on the hypothesis tested, and the estimated $\sigma_{\text{diff.}}$ is found directly from the estimate of σ_M^2 for each sample by Eq. 8.7.4.

The exact hypothesis actually tested is of the form

$$H_0: \mu_1 - \mu_2 = k,$$

where k is any difference of interest. Quite often, the experimenter is interested only in $k = 0$, but it is entirely possible to test any other meaningful difference value. The alternative hypothesis may be directional,

H_1: $\mu_1 - \mu_2 > k$

or

H_1: $\mu_1 - \mu_2 < k$,

or nondirectional,

H_1: $\mu_1 - \mu_2 \neq k$,

depending on the form of the original question.

When both samples are large, confidence limits are found exactly as for a single mean, except that $(\bar{y}_1 - \bar{y}_2)$ and est. $\sigma_{\text{diff.}}$ are substituted for $\bar{y}$ and est. σ_M respectively. Thus, 95% confidence limits for a difference with large samples are

$\bar{y}_1 - \bar{y}_2 - 1.96$ (est. $\sigma_{\text{diff.}}$) [8.7.6*]

$\bar{y}_1 - \bar{y}_2 + 1.96$ (est. $\sigma_{\text{diff.}}$) .

8.8 AN EXAMPLE OF A LARGE-SAMPLE SIGNIFICANCE TEST FOR A DIFFERENCE BETWEEN MEANS

An experimenter working in the area of motivational factors in perception was interested in the effects of deprivation on the perceived size of objects. Among the studies carried out was one done with orphans, who were compared with nonorphaned children on the basis of the judged size of parental figures viewed at a distance. Each child was seated at a viewing apparatus in which cutout figures appeared. Each figure was actually of the same size and at the same distance from the viewer, although the children were unaware all the figures had the same size. A device was provided on which the child could actually judge the apparent sizes of the different figures in numerical terms. Several of the figures in the set viewed were obviously parents, whereas others were more or less neutral, such as milkmen, postmen, nurses, and so on. Each child was given a score, which was itself a difference in average judged size of parental and nonparental figures.

Now two independent, randomly selected groups were used. Sample 1 was a group of orphaned children without foster parents. Sample 2 was a group of children having a normal family with both parents. Both populations of children sampled showed the same age level, sex distribution, educational level, and so forth.

The question asked by the experimenter was, "Do deprived children tend to judge the parental figures relatively larger than do the nondeprived?" In terms of a null and alternative hypothesis,

H_0: $\mu_1 - \mu_2 \leq 0$

H_1: $\mu_1 - \mu_2 > 0$.

The α level for significance decided on was .05. The actual results were

Sample 1   *Sample 2*

$\bar{y}_1 = 1.82$ $\bar{y}_2 = 1.61$

$s_1 = .7$ $s_2 = .9$

$N_1 = 125$ $N_2 = 150$

These sample sizes are rather large, and the experimenter felt safe in using the normal approximation to the sampling distribution, even though she had no idea about the distribution form for the two populations sampled. The ratio used was

$$\frac{(\bar{y}_1 - \bar{y}_2) - E(\bar{y}_1 - \bar{y}_2)}{\text{est. } \sigma_{\text{diff.}}}$$

In this problem, $E(\bar{y}_1 - \bar{y}_2) = 0$, under the hypothesis tested. It was obviously necessary for the experimenter to estimate the standard error of the difference, since both σ_1 and σ_2 were unknown to her. This estimate was found by first estimating $\sigma_{M_1}^2$ and $\sigma_{M_2}^2$:

$$\text{est. } \sigma_{M_1}^2 = \frac{s_1^2}{N_1} = \frac{.49}{125} = .0039$$

$$\text{est. } \sigma_{M_2}^2 = \frac{s_2^2}{N_2} = \frac{.81}{150} = .0054.$$

Then,

$$\text{est. } \sigma_{\text{diff.}} = \sqrt{\text{est. } \sigma_{M_1}^2 + \text{est. } \sigma_{M_2}^2} = \sqrt{.0039 + .0054} = .0964.$$

On making these substitutions, the experimenter found

$$t = \frac{1.82 - 1.61}{.0964} = 2.18.$$

The rejection region implied by the alternative hypothesis is on the *upper* tail of the sampling distribution. For a normal distribution the upper 5% is bounded by $z = 1.65$. Thus, the result is significant; deviations this far from zero have a probability of less than .05 of occurring by chance alone when the true difference is zero.

The experimenter may conclude that a difference exists between these two populations, *if* an α value less than .05 is a small enough probability of error to warrant this decision. However, the experimenter does *not necessarily* conclude that parental deprivation causes an increase in perceived size. The statistical conclusion suggests that it *might* be safe to assert that a particular direction of numerical difference exists between the mean scores of the two populations of children, but the statistical result is absolutely noncommittal about the reason for this difference, if such exists. The experimenter here takes the step of advancing a reason at her own peril. The statistical test as a mathematical tool is absolutely neutral about what these numbers measure, the level of measurement, what was or was not represented in the experiment, and, most of all the cause of the experimenter's particular finding. As always, the test takes the numerical values as given, and cranks out a conclusion about the conditional probability of such numbers, given certain statistical conditions.

For this example; the 95% limits are

.21 $-$ 1.96(.0964)

.21 $+$ 1.96(.0964)

or .021 and .399. Notice that since the value $\mu_1 - \mu_2 = 0$ does not fall within these values, this value can be rejected as a hypothesis beyond the .05 level (two-tailed).

8.9 USING THE t DISTRIBUTION TO TEST HYPOTHESES ABOUT DIFFERENCES

Given the assumption that both populations sampled have normal distributions, any hypothesis about a difference of means can be tested using the t distribution, regardless of sample size (where $N_1 \geq 2$ and $N_2 \geq 2$ of course). Here, for the null hypothesis

$$H_0: \mu_1 - \mu_2 = k$$

the test is

$$t = \frac{(\bar{y}_1 - \bar{y}_2) - k}{\text{est. } \sigma_{\text{diff}}}. \tag{8.9.1*}$$

However, one additional assumption becomes necessary: to use the t distribution for tests based on *two (or more) samples,* one must assume that the standard deviations of both (or all) populations are *equal.* The basis for this assumption will be discussed in the next chapter.

Given these assumptions, then the distribution of t for a difference has the same form as for a single mean, except that the degrees of freedom are

$$\nu = N_1 - 1 + N_2 - 1 = N_1 + N_2 - 2.$$

When samples are drawn from populations with equal variance, then the estimated standard error of a difference takes a somewhat different form. First of all, when $\sigma_1 = \sigma_2 = \sigma$,

$$\sigma_{\text{diff}} = \sqrt{\frac{\sigma^2}{N_1} + \frac{\sigma^2}{N_2}} = \sqrt{\sigma^2 \left(\frac{1}{N_1} + \frac{1}{N_2} \right)}. \tag{8.9.2*}$$

Now, as we showed in Section 5.10, when one has two or more estimates of the same parameter σ^2 the *pooled* estimate is actually better than either one taken separately. From Eq. 5.10.4 it follows that

$$\text{est.} \sigma^2 = \frac{(N_1 - 1)s_1^2 + (N_2 - 1)s_2^2}{N_1 + N_2 - 2} \tag{8.9.3}$$

is our best estimate of σ^2 based on the two samples. Hence

$$\begin{aligned} \text{est. } \sigma_{\text{diff}} &= \sqrt{\text{est. } \sigma^2 \left(\frac{1}{N_1} + \frac{1}{N_2} \right)} \\ &= \sqrt{\left(\frac{(N_1 - 1)s_1^2 + (N_2 - 1)s_2^2}{N_1 + N_2 - 2} \right) \left(\frac{N_1 + N_2}{N_1 N_2} \right)}. \end{aligned} \tag{8.9.4*}$$

This estimate of the standard error of the difference ordinarily forms the denominator of the t ratio when the t distribution is used for hypotheses about a difference.

Let us imagine that two random samples of subjects are being compared on the basis of their scores on a motor learning task. The subjects are allotted to two experimental groups, with 5 in the first group and 7 in the other. In Group 1 a subject is rewarded for each correct move made, and in the second each incorrect move is punished. The score is the number of errors made in a fixed set of trials. The experimenter wishes to find evidence for the question, "Does the kind of motivation employed, reward or punishment, affect the performance?" This question implies the null and alternative hypotheses:

$$H_0: \mu_1 - \mu_2 = 0$$

$$H_1: \mu_1 - \mu_2 \neq 0.$$

The experimenter is willing to assume that the population distributions of scores are normal, and that the population variances are equal. The probability of Type I error decided upon is .01. Since this is a two-tailed test, a glance at Table III in Appendix E shows that for $N_1 + N_2 - 2$ or $5 + 7 - 2 = 10$ degrees of freedom, and for $2Q = .01$, the required t value is 3.169. Thus an obtained t ratio equaling or exceeding 3.169 in absolute value is grounds for rejecting the hypothesis of no difference between population means.

The sample results are

$$\bar{y}_1 = 18 \qquad \bar{y}_2 = 20$$

$$s_1^2 = 7.00 \qquad s_2^2 = 5.83.$$

The estimated standard error of the difference is found by the pooling procedure given by Eq. 8.9.4

$$\text{est. } \sigma_{\text{diff.}} = \sqrt{\text{est. } \sigma^2 \left(\frac{1}{N_1} + \frac{1}{N_2} \right)}$$

$$= \sqrt{\left(\frac{(4 \times 7) + (6 \times 5.83)}{10} \right) \left(\frac{12}{35} \right)}$$

$$= \sqrt{2.16} = 1.47.$$

Thus, the t ratio is

$$t = \frac{(\bar{y}_1 - \bar{y}_2) - E(\bar{y}_1 - \bar{y}_2)}{\text{est. } \sigma_{\text{diff.}}} = \frac{-2}{1.47} = -1.36.$$

This value comes nowhere close to that required for rejection, and thus if α must be no more than .01, the experimenter does not reject the null hypothesis. The best choice may be to suspend judgment, pending more evidence.

Confidence intervals are found just as for a single small-sample mean: the limits are

$$(\bar{y}_1 - \bar{y}_2) - t_{(\alpha/2;\nu)} \text{ (est. } \sigma_{\text{diff.}})$$

$$(\bar{y}_1 - \bar{y}_2) + t_{(\alpha/2;\nu)} \text{ (est. } \sigma_{\text{diff.}}).$$

[8.9.5*]

For this example, the 99% limits are

$$-2 - (3.169)(1.47)$$

and

$$-2 + (3.169)(1.47)$$

or approximately -6.66 to 2.66. The probability is .99 that true $\mu_1 - \mu_2$ is covered by an interval such as this. Once again, notice that this interval *does* contain the value 0, indicating the hypothesis entertained above is not rejected.

8.10 THE IMPORTANCE OF THE ASSUMPTIONS IN A t TEST OF A DIFFERENCE BETWEEN MEANS

The rationale for a t test of a difference between the means of two groups rests on two assumptions: the populations each have a normal distribution, and each population has the same variance, σ^2. On the other hand, in practical situations these assumptions are sometimes violated with rather small effect on the conclusions.

The first assumption, that of a normal distribution in the populations, is apparently the less important of the two. So long as the sample size is even moderate for each group, quite severe departures from normality seem to make little practical difference in the conclusions reached. Naturally, the results are more accurate the more nearly unimodal and symmetric the population distributions are, and thus if one suspects radical departures from a generally normal form, then he should plan on larger samples. Furthermore, the departure from normality can make more difference in a one-tailed than in a two-tailed result, and once again some special thought should be given to sample size when one-tailed tests are contemplated for such populations. By and large, however, this assumption may be violated almost with impunity provided that sample size is not extremely small.

The assumption of homogeneity of variance is more important. In older work it was often suggested that a separate test for homogeneity of variance be carried out before the t test itself, in order to see if this assumption were at all reasonable. However, the most modern authorities suggest that this is not really worth the trouble involved. In circumstances where they are needed most (small samples), the tests for homogeneity are poorest. Furthermore, for samples of equal size, relatively big differences in the population variances seem to have relatively small consequences for the conclusions derived from a t test. Even so, when the variances are quite unequal using very different sample sizes can have serious effects on the conclusions. The moral should be plain: given the usual freedom about sample size in experimental work, *when in doubt use samples of the same size,* or very nearly so.

However, sometimes it is not possible to obtain an equal number in each group. Then one way out of this problem is by the use of a correction in the value for degrees of freedom. This is useful when one cannot assume equal population variances and samples are of different size. In this situation, the t ratio is calculated as in Section 8.7, where the separate standard errors are computed from each sample and the pooled estimate is not made. Then the corrected number of degrees of freedom is found from

$$\nu = \frac{(\text{est. } \sigma_{M_1}^2 + \text{est. } \sigma_{M_2}^2)^2}{(\text{est. } \sigma_{M_1}^2)^2/(N_1 + 1) + (\text{est. } \sigma_{M_2}^2)^2/(N_2 + 1)} - 2. \qquad [8.10.1]$$

This need not result in a whole value for ν, in which case the use of the nearest whole value for ν is sufficiently accurate for most purposes. When somewhat greater accuracy is desired, the approximate formula for critical values of t given by Eq. 11.14.2 is useful. When both samples are quite large, then both the assumptions of normality and of homogeneous variances become relatively unimportant, and the method of Section 8.7 can be used.

8.11 THE POWER OF t TESTS

The idea of the power of a statistical test was discussed in the preceding chapter only in terms of the normal distribution. Nevertheless, the same general considerations apply to the power of tests based on the t distribution. Thus, the power of a t test increases with sample size, increases with the discrepancy between the null hypothesis value and the true value of a mean or a difference, increases with any reduction in the true value of σ, and increases with any increase in the size of α, given a true value covered by H_1.

Unfortunately, the actual determination of the power for a t test against any given true alternative is more complicated than for the normal distribution. The reason is that when the null hypothesis is false, each t ratio computed involves $E(\bar{y})$ or $E(\bar{y}_1 - \bar{y}_2)$, which is the exact value given by the null (and false) hypothesis. If the true value of the expectation could be calculated into each t ratio, then the distribution would follow the t function tabled in Appendix E. However, when H_0 is false, each t value involves a false expectation; this results in a somewhat different distribution, called the **noncentral t distribution**. The probabilities of the various t's cannot be known unless one more parameter, δ, is specified besides ν. This is the so-called noncentrality parameter, defined by

$$\delta = \left| \frac{\mu_1 - \mu_0}{\sigma_M} \right|. \qquad [8.11.1]$$

The parameter δ expresses the absolute difference between the true expectation μ_1 and that given by the null hypothesis, or μ_0, in terms of σ_M. For a hypothesis about a difference, tested by independent samples,

$$\delta = \left| \frac{(\mu_1 - \mu_2) - (\mu_{0_1} - \mu_{0_2})}{\sigma_{\text{diff.}}} \right|. \qquad [8.11.2]$$

Unlike the regular, central t distribution, the distribution of noncentral t is rather difficult to show in a concise table, depending as it does on the two parameters ν and δ. However, this distribution is the basis for the power graphs for t that appear in some advanced texts in statistics. The power for a t test can also be approximated. One simple way to approximate the power of a t test is as follows: first, some alternative value of the population mean, or μ_1, must be specified. Then if the number of degrees of freedom is ten or more, the power of a t test of $\mu_0 = k$ may be approximated from

$$\text{power} = 1 - \text{prob.} \left\{ z \leq \left(\frac{t' - \delta}{\sqrt{1 + \frac{(t')^2}{2v}}} \right) \right\} \qquad [8.11.3]$$

where δ is estimated by taking

$$\text{est. } \delta = |(\mu_1 - \mu_0)/\text{est. } \sigma_M|$$

and where t' represents the upper critical value of t for v degrees of freedom, in a two-tailed test where α has been specified. After calcualting the ratio to the right of the inequality, this value is referred to a normal distribution, where the probability of a z value less than or equal to this number is found. This probability is the approximate value of β. Then the power is found by subtracting this probability from 1.00.

This same procedure can be applied to estimating the power of t for a hypothesis about a difference as well. Here we take

$$\text{est. } \delta = \left| \frac{\text{difference under } H_0 - \text{difference under } H_1}{\text{est. } \sigma_{\text{diff.}}} \right|. \qquad [8.11.4]$$

The critical value of t, or t', must be found for $N_1 + N_2 - 2$ degrees of freedom, of course.

As an example of such a power calculation involving a difference between means, suppose that there were two independent samples, each containing 35 independent observations drawn at random.

The hypothesis to be tested is

$$H_0: \mu_1 - \mu_2 = 0$$

against the alternative

$$H_1: \mu_1 - \mu_2 \neq 0.$$

The .01 level is to be used for α.

What is the power of this test if the alternative hypothesis, $\mu_1 - \mu_2 = 4$, turns out to be true? Suppose that it is found that $s_1^2 = 19$ and $s_2^2 = 17$. Then

$$\text{est. } \sigma_{\text{diff.}} = \sqrt{\left(\frac{(34 \times 19) + (34 \times 17)}{34 + 34} \right) \left(\frac{2}{35} \right)} = 1.01.$$

Then we estimate δ by taking

$$\delta = \left| \frac{4}{1.01} \right| = 3.96.$$

The critical value of t in this test is read from Table III, Appendix E, to be about

$$t' = 2.66,$$

since 68 degrees of freedom is closest to the tabled value for 60. Then

$$\text{power} = 1 - \text{prob}\left\{ z \leq \left(\frac{2.66 - 3.96}{\sqrt{1 + \frac{(2.66)^2}{2 \times 68}}} \right) \right\}$$

$$= 1 - \text{prob}\{z \leq -1.27\}$$

$$= .898.$$

The power of this test, when the true difference is 4, should be about .90. The power against any other true alternative can be evaluated in this same way, of course.

The big difficulty with this or any similar power calculation is in knowing the value of σ_M or of $\sigma_{\text{diff.}}$. Of course, when sample size is fairly large, one can use estimates of these values, as we have done here. On the other hand, it is common to want a power calculation before a sample is drawn, as an aid in determining sample size itself. Power calculations or the use of power tables become difficult under these circumstances, if one does not already have a value for the standard error. An alternative approach will now be explored.

8.12 STRENGTH OF ASSOCIATION

When an experimenter assigns subjects to two experimental groups and proceeds to give each a different treatment, he or she is generally looking for evidence of a statistical relation. That is, the question asked is basically of the form, "If I do something different to these two groups in one way, is this reflected in a difference between the groups in another way?" The manipulation that is under the control of the experimenter is called the *independent variable*. The performance of the two groups that is not under the experimenter's control is called the *dependent variable*. For notational convenience, let us call the independent variable (which may be either a number, or a qualitative label) the variable X.

We symbolize the dependent variable, which usually is a number, by Y. The question then becomes, "Is Y related to X?"

Evidence for a statistical relation exists when different X treatments lead to different *expectations* about Y. When the expectation of Y is different, depending on which X treatment the group in question receives, then we can say that a *statistical relation*, or *association*, exists between X and Y. On the other hand, if we are led to expect the same Y value, regardless of which X any group receives, then we say that there is *no association*.

Now suppose that you were asked to guess the Y score of an individual drawn at random from one of the two groups given different X treatments. Obviously, you would not have a very good idea about what to guess. However, as we have seen before, when in doubt it is probably wisest to guess the mean of all of the Y values. Then at least your error will be zero on the average.

If you are guessing the mean of all Y scores for any single individual, how "off" should your guess tend to be? Recall that the measure of spread about a mean is the variance. Let us designate this variance about the mean Y score for all of the cases as σ_Y^2. We can say that this value, σ_Y^2, is a measure of our *uncertainty* about the Y scores in general.

Now suppose that you were asked to guess a Y score once again, but this time you are given the X group to which that individual belongs. Then what should you guess? The best guess is again a mean, but this time it is the *mean Y value* of all individuals within that *same X* grouping. Very likely you will still not be exactly right in your guess about the score of an individual in this X group, and the spread about the mean Y value, given X, will be a reflection of how "off" your guess will tend to be. Let us designate this variance about the mean Y score within an X group by $\sigma^2_{Y|X}$, the variance of Y *given* X. Then $\sigma^2_{Y|X}$ can be said to reflect your uncertainty about the Y scores even after you are given the information about X.

In general, we will say that the strength of a statistical relation is reflected in the extent to which *knowing X reduces our uncertainty about Y*. That is, since our uncertainty about Y not knowing X is proportional to σ^2_Y, and our uncertainty about Y *given* X is proportional to $\sigma^2_{Y|X}$, then the *reduction in uncertainty* about Y afforded by X should be based on

$$\sigma^2_Y - \sigma^2_{Y|X}\ .$$

We can say that a statistical relation exists when the *variability of Y given X is smaller than the variability of Y* in general.

It is convenient to turn this reduction in uncertainty into a **relative reduction** by dividing by σ^2_Y, giving the population value

$$\omega^2 = \frac{\sigma^2_Y - \sigma^2_{Y|X}}{\sigma^2_Y}\ . \tag{8.12.1*}$$

The relative reduction in uncertainty about Y given by X is shown by the index ω^2 (Greek omega, squared). Sometimes the value ω^2 is called **the proportion of variance in Y accounted for by X.** Viewed either as a relative reduction in uncertainty or as a proportion of variance accounted for, the index ω^2 represents the strength of association between independent and dependent variables. (The index ω^2 is almost identical to two other indices to be introduced later, the *intraclass correlation* and the *correlation ratio,* usually represented by the symbols ρ_I and η^2 respectively. However, since these indices were developed for and are used in somewhat different contexts, it seems better to use the relatively neutral symbol ω^2 here, to avoid later confusion.)

This index reflects the predictive power afforded by a relationship: when ω^2 is zero, then X does not aid us at all in predicting the value of Y. On the other hand, when ω^2 is 1.00, this tells us that X lets us know Y exactly. All intermediate values of the index represent different degrees of predictive ability. Notice that for any precise functional relation, $\omega^2 = 1.00$ since there can be only one Y for each possible X. A value less than unity tells us that precise prediction is not possible, although X nevertheless gives *some* information about Y unless $\omega^2 = 0$.

Let us return to the problem of guessing scores drawn from two different X groups or populations. Suppose that you are equally likely to have to guess a Y value from Group 1 as from Group 2. Furthermore, suppose that the value of the variance within Group 1, σ^2_1, is exactly the same as the variance within Group 2, or σ^2_2. Let us also designate the mean of Group 1 by μ_1 and the mean of Group 2 by μ_2. Then it will be true that

$$\sigma^2_Y = \sigma^2_{Y|X} + \frac{(\mu_1 - \mu_2)^2}{4} \tag{8.12.2}$$

where μ_1 is the mean of Population 1, μ_2 that of Population 2, and

$$\frac{\mu_1 + \mu_2}{2} = \mu,$$

the mean of the marginal distribution (Appendix C).

On substituting into Eq. 8.12.1 we find

$$\omega^2 = \frac{(\mu_1 - \mu_2)^2}{4\sigma_Y^2}.$$

[8.12.3*]

For two treatment populations with equal variances, the strength of the statistical association between treatment and dependent variable varies directly with the squared difference between the population means, relative to the unconditional, total variance of Y. This connection between the strength-of-association index ω^2 and the difference between population means provides another approach to estimating the power of a t test. This will be examined in the next section.

8.13 STRENGTH OF ASSOCIATION, POWER, AND SAMPLE SIZE

As we have just seen, there is a very close connection between the value of ω^2 in the populations and the difference between the two populations means. Thus, it is possible to rewrite the usual hypothesis about a difference between two means

$$H_0: \mu_1 - \mu_2 = 0$$

in terms of a hypothesis about strength of association:

$$H_0: \omega^2 = 0.$$

Furthermore, the alternative hypothesis

$$H_1: \mu_1 - \mu_2 \neq 0$$

is equivalent to

$$H_1: \omega^2 > 0.$$

In addition, the parameter δ, necessary for us to specify before we can calculate the power of a t test, can be stated in terms of ω^2. That is, for n per group,

$$\delta = \sqrt{\frac{2n\omega^2}{1 - \omega^2}}.$$

[8.13.1]

This can be a distinct advantage, since we seldom know values such as σ_Y^2 and $\sigma_{Y|X}^2$, especially in advance of a sample, when power calculations are often most useful. However, we can preset the power of a t test in terms of that ω^2 value which we want to be sure to detect as a significant result.

If you wish to calculate the power of a t test, given that a certain value of ω^2 is true, the methods of the preceding section give a reasonable approximation. However, the value of δ is found from Eq. 8.13.1, rather than from Eq. 8.11.4.

An even more important situation occurs when we wish to choose a sample size that will afford our test a certain power, given that a particular ω^2 value is true in the population. Actually, a rough and ready way exists to determine sample size, or n, once power against a certain ω^2 value is specified. This method will now be described.

Suppose that two experimental groups of equal size n are to be chosen, in order to test the hypothesis that the means of two populations are equal (or, alternatively, that ω^2 is equal to zero). A two-tailed test is to be carried out, with α equal to one of the conventional values such as .05 or .01.

Suppose further that we want to be very sure that if the true strength of association is at least some value of ω^2, we will get a significant result some sizable proportion (or $1 - \beta$) of the time. In other words, we want to set the power at some given level, given that some specific value of ω^2 is true.

Then a first approximation to the total size of the sample required *in each group* can be obtained from

$$n \geq \frac{[z_{(1-\beta)} - z_{(\alpha/2)}]^2 (1 - \omega^2)}{2\omega^2}, \qquad\qquad [8.13.2]$$

where $z_{(1-\beta)}$ cuts off the lower $(1 - \beta)$ proportion of a normal distribution, and $z_{(\alpha/2)}$ cuts off the lower $\alpha/2$ proportion. Then $2n$ is a rough approximation to N, the total number of cases required. If n is a decimal number, the next largest whole number is taken. Furthermore, since this estimate of n will tend to be slightly small, it is wise to add one more case to each group, or two more cases to the total size of the sample.

The ease of applying this rule can be illustrated by an example. Imagine that we plan to test the difference between the means of two independent groups by use of t, where the α value of .01 will be employed. We would like, say, the probability to be about .90 of rejecting H_0 of no difference if the true value of ω^2 is at least .25. Then, from the normal table we find

$$z_{(1-\beta)} = z_{(.90)} = 1.28$$

$$z_{(\alpha/2)} = z_{(.005)} = -2.58$$

and

$$n \geq \frac{(1.28 + 2.58)^2 (.75)}{2(.25)}$$

or

$$n \geq 22.35.$$

Since the estimate is 22.35, we then know that 23 cases in each group should be in the right ballpark. However, 24 cases in each group will be used. This should make the power be about .90 when the true value of ω^2 is .25 or more.

The required sample size can be estimated in this way for any other combination of α, power, and ω^2. For a one-tailed test, z_α may be substituted for $z_{(\alpha/2)}$, of course. Such a rough and ready procedure should be adequate for most practical purposes. However, if greater precision in finding n is required, one can explore various powers obtained for a given ω^2 by use of Eq. 8.11.3 above, or one can look into power tables such as those given in Winer (1971) and other books on experimental design. One such table is Table XII, which will be discussed in Section 10.21.

8.14 STRENGTH OF ASSOCIATION AND SIGNIFICANCE

When the difference $\mu_1 - \mu_2$ is zero, then ω^2 must be zero. In the usual t test for a difference, the hypothesis of no difference between means is equivalent to the hypothesis that $\omega^2 = 0$. On the other hand, when there is any difference at all between population means, the value of ω^2 must be greater than 0.

A true difference is "big" in the sense of predictive power only if the square of that difference is large relative to σ_Y^2. However, in significance tests such as t, we compare the difference we get with an estimate of $\sigma_{\text{diff.}}$. The standard error of the difference can be made almost as small as we choose if we are given a free choice of sample size. Unless sample size is specified, there is no *necessary* connection between significance and the true strength of association.

Virtually any study can be made to show significant results if one uses enough subjects, *regardless* of how small ω^2 may be. There is surely nothing on earth that is completely independent of anything else. The strength of an association may approach zero, but it should seldom or never be exactly zero. If one applies a large enough sample of the study of any relation, trivial or meaningless as it may be, sooner or later a significant result will almost certainly be achieved. Such a result may be a valid finding, but only in the sense that one can say with assurance that some association is not exactly zero. The degree to which such a finding enhances our knowledge is debatable. If the criterion of strength of association is applied to such a result, it becomes obvious that little or nothing is actually contributed to our ability to predict one thing from another.

This kind of problem occurs when people pay too much attention to the significance test and too little to the degree of statistical association the finding represents. This clutters up the literature with findings that are often not worth pursuing and which serve only to obscure the really important predictive relations that occasionally appear. The serious scientist should ask not only, "Is there any association between X and Y?" but also, "How much does my finding suggest about the power to predict Y from X?" Much too much emphasis is paid to the former, at the expense of the latter, question.

Can sample size be too large? In one sense, even posing this question sounds like heresy! Social scientists are usually trained to think that large samples are good things, and there is much to support this notion. As we have seen, the most elegant features within theoretical statistics are the limit theorems, each implying a connection between sample size and the goodness of inference.

On the other hand, samples are expensive in terms of time, effort, and money. The experimenter purchases information through the sample, and it may be that too high a price is paid for the information received.

It seems reasonable that sample size, and the consequent investment in the experiment or study, can never really be discussed apart from what the experimenter is trying to do and the stakes involved. As long as the experimenter's primary purpose is in precise estimation, then the larger the sample the better. When we want to come as close as possible to the true parameter values, we can always do better by increasing sample size.

This is not, however, the purpose of many studies. These studies are, in the strict sense, exploratory, where the main relationships in some area are to be mapped out. This kind of study serves as a guide for directions that will be pursued in other, perhaps more refined, explorations of the problem. Of main concern are statistical relations that are relatively large and that give considerable promise that a more or less precise relationship is there to be discovered and refined. The experimenter does not want to waste time, effort, and funds by concluding that an association exists when the degree of prediction afforded by that association is negligible.

When this is the situation, it is advisable to look into the effects of sample size on the probability of finding a significant result given a *weak* association. Trivial associations may well show up as significant results when the sample size is very large. If the experimenter wants significance to reflect a sizable association in the data and also wants to be sure not to be led by a significant result into some blind alley, then attention should be paid to both aspects of sample size. Is the sample size *large* enough to give confidence that the big associations will indeed show up, while being *small* enough so that trivial associations will be excluded from significance?

8.15 ESTIMATING THE STRENGTH OF A STATISTICAL ASSOCIATION FROM DATA

It is quite possible to estimate the amount of statistical association implied by any obtained difference between two independent means. The ingredients for this kind of estimation are essentially those used in a t test.

In the first place, a serviceable statistic for *the proportion of variance accounted for in a particular sample* may be had by taking

$$\frac{t^2}{t^2 + N_1 + N_2 - 2} = \text{proportion of variance in } Y \text{ accounted for by } X. \qquad [8.15.1]$$

By taking the square root of this value, and assigning the same sign as for t, we have the measure of association known as the *point biserial correlation,* symbolized by r_{pb}:

$$r_{pb} = \sqrt{t^2/(t^2 + N_1 + N_2 - 2)} \, .$$

The point biserial correlation shows the extent of the relation between the group to which a case belongs, and the dependent variable value y that case exhibits. Use of the point biserial correlation assumes that the two groups represent a genuine, qualitatively defined dichotomy, rather than having been formed by an arbitrary cutoff point on some underlying continuous variable. In the latter case a different index of association, known simply as a *biserial correlation,* is employed. Details may be found in McNemar (1975) and in Guilford and Fruchter, (1978).

Sample values such as those given by Eq. 8.15.1 tend to be somewhat biased as representations of ω^2, the proportion of variance accounted for in the population. In an effort to get around this problem, a number of ways have been proposed for estimating the strength of a statistical association from obtained differences between means. For reasons to be elaborated later, none of these methods is entirely satisfactory. The method to be introduced here is thus only one of the ways that may be encountered in the statistical literature, but it seems to have as much to recommend it as any other.

For samples from two populations, each of which has the same true variance, $\sigma^2_{Y|X}$ a rough estimate of ω^2 is provided by

$$\text{est. } \omega^2 = \frac{t^2 - 1}{t^2 + N_1 + N_2 - 1} \,. \tag{8.15.2}$$

(A more general form for estimating ω^2 will be given in Chapter 10.) Notice that if t^2 is less than 1.00, then this estimate is negative, although ω^2 cannot assume negative values. In this situation the estimate of ω^2 is set equal to zero.

Let us consider an example using this estimate. Imagine a study involving two groups of 30 cases each. Subjects are assigned at random to these two groups, and each set of subjects is given a different treatment. The results are

Group 1	Group 2
$\bar{y}_1 = 65.5$	$\bar{y}_2 = 69$
$s_1^2 = 20.69$	$s_2^2 = 28.96$
$N_1 = 30$	$N_2 = 30$

First of all the t ratio is computed in the usual way (Section 8.9):

$$\text{est. } \sigma^2 = \frac{(29)(20.69 + 28.96)}{58} = 24.83$$

and

$$\text{est. } \sigma_{\text{diff.}} = \sqrt{\frac{24.83(2)}{30}} = 1.29.$$

Thus,

$$t = \frac{65.5 - 69}{1.29} = -2.71 \,.$$

For a two tailed test with 58 degrees of freedom, this value is significant beyond the .01 level. Thus, we are fairly safe in concluding that some association exists.

What do we estimate the true degree of association to be? Substituting into Eq. 8.15.2, we find

$$\text{est. } \omega^2 = \frac{(2.71)^2 - 1}{(2.71)^2 + (60 - 1)} = .096 \,.$$

Our rough estimate is that X (the treatment administered) accounts for about 10% of the variance of Y (the obtained score). By way of contrast, we find that the proportion of variance accounted for *in this sample* is slightly larger, or .11.

Suppose, however, that the groups had contained only 10 cases each, and that the results had been:

Group 1	Group 2
$\bar{y}_1 = 65.5$	$\bar{y}_2 = 69$
$s_1^2 = 5.55$	$s_2^2 = 7.78$
$N_1 = 10$	$N_2 = 10$

Here

$$\text{est. } \sigma^2 = \frac{9(5.55 + 7.78)}{18} = 6.67$$

$$\text{est. } \sigma_{\text{diff.}} = \sqrt{6.67 \left(\frac{2}{10}\right)} = 1.15$$

so that

$$t = \frac{-3.5}{1.15} = -3.04 .$$

For 18 degrees of freedom, this value is also significant beyond the .01 level (two-tailed), and once again we can assert with confidence that some association exists.

Again, we estimate the degree of association represented by this finding:

$$\text{est. } \omega^2 = \frac{(3.04)^2 - 1}{(3.04)^2 + 19} = .29.$$

Here, our rough estimate is that X accounts for about 29% of the variance in Y. Even though the difference between the sample means is the same in these two examples and both results are significant beyond the .01 level, the second experiment gives a much higher estimate of the true association than the first.

8.16 PAIRED OBSERVATIONS

Sometimes it happens that subjects are actually sampled in pairs. Even though each subject is experimentally different in one respect (nominally, the independent variable) from the pair-mate and each has some distinct dependent variable score, the scores of the members of a pair are not necessarily independent. For instance, one may be comparing scores of husbands and wives; a husband is "naturally" matched with his wife, and it makes sense that knowing the husband's score gives us some information about his wife's, and vice versa. Or individuals may be matched on some basis by the experimenter, and within each matched pair the members are assigned at random to experimental treatments. This matching of pairs is one form of experimental control since each member of each experimental group must be identical (or nearly so) to the pair-mate in the other group with respect to the matching factor or factors, and thus the factor or factors used to match pairs is less likely to be responsible for any observed difference in the groups than if two unmatched groups are used.

Given two groups matched in this pairwise way, either by the experimenter or otherwise, it is still true that the difference between the means is an unbiased estimate of the population difference (in two matched populations):

$$E(\bar{y}_1 - \bar{y}_2) = \mu_1 - \mu_2. \tag{8.16.1}$$

However, the matching, and the consequent *dependence* within the pairs, changes the standard error of the difference. Thus, for matched pairs

$$\sigma_{\text{diff.}}^2 = \sigma_{M_1}^2 + \sigma_{M_2}^2 - 2 \text{ cov.}(\bar{y}_1, \bar{y}_2). \tag{8.16.2*}$$

The first of these terms is just $\sigma_{M_1}^2$, and the second is $\sigma_{M_2}^2$. However, what of the third term? Let us denote this last term above as $\text{cov}(\bar{y}_1, \bar{y}_2)$, the **covariance** of the means. (The covariance of two random variables is defined in Appendix B, following Rule 9.)

From Rule 9, Appendix B, we find that the value of the covariance must be zero when the variables are independent. On the other hand, when variables are dependent the expectation is *not* ordinarily zero. In general, for groups matched by pairs, this covariance is a positive number, and thus the variance and standard error of a difference between means will usually be *less* for matched than for unmatched groups. This fact accords with the experimenter's purpose in matching in the first place: to remove one or more sources of variability, and thus to lower the sampling error.

The unknown value of $\text{cov}(\bar{y}_1, \bar{y}_2)$ could be something of a problem, but actually it is quite easy to bypass this difficulty altogether. Instead of regarding this as two samples, we simply think of the data coming from one sample of *pairs*. Associated with each pair i is a difference

$$D_i = (y_{i1} - y_{i2}),$$ [8.16.3]

where y_{i1} is the score of the member of pair i who is in group 1, and y_{i2} is the score of the member of pair i who is in group 2. Then an ordinary t test for a *single* mean is carried out using the scores D_i. That is, for all N pairs,

$$\bar{D} = \frac{\sum_i D_I}{N}$$ [8.16.4]

and

$$s_D^2 = \frac{\sum_i D_i^2}{N-1} - \frac{N(\bar{D})^2}{N-1}.$$ [8.16.5]

Then

$$\text{est. } \sigma_{M_D} = \frac{s_D}{\sqrt{N}}$$ [8.16.6]

and t is found from

$$t = \frac{\bar{D} - E(\bar{D})}{\text{est. } \sigma_{M_D}}$$ [8.16.7*]

with $N - 1$ degrees of freedom. Be sure to notice that here N stands for the number of *differences*, which is the number of *pairs*.

This method turns the test of a hypothesis about two matched populations into the test of an exact hypothesis about a population of pairs. Any of the usual hypotheses about a single mean may be tested. Although the value stated in the hypothesis usually amounts to $E(\bar{D}) = \mu_D = 0$, any other exact value could be used. The test may be either one- or two-tailed.

Matching pairs for an experiment or sampling a population of pairs is very common. Some caution must be exercised in this matching process, however. In the first place, it can happen that the factor on which subject pairs are matched is such that the means are *negatively* related. Thus, for example, suppose that one had an effective measure of the dominance of personality of an individual. It just might be that highly dominant women tend to marry men with low dominance, and vice versa, so that among husband–wife pairs, dominance scores are negatively related. Then, if our interest is basically that of comparing men and women generally on such scores, it would be a mistake to match, since the negative relationship would lead to a larger, rather than a smaller, standard error of the difference than would a comparison of unmatched groups.

Furthermore, such matching may be less efficient than the comparison of unmatched random groups, unless the factor used in matching introduces a relatively strong positive relationship between the means. Although a positive relationship, reflected in a positive covariance term, does reduce the standard error of the difference, this procedure also *halves* the number of degrees of freedom. Dealing with a sample of N pairs gives only half the number of degrees of freedom available when we deal with two independent groups of N cases each. Thus, if the factor entering into the matching is only slightly relevant to the differences between the groups or is even irrelevant to such differences, matching is not a desirable procedure. The experimenter should have quite good reasons for matching before adopting this procedure in preference to the simple comparison of two randomly selected independent groups.

8.17 COMPARING MORE THAN TWO MEANS

In the social sciences it is quite common for an investigator to be interested in an independent variable X that is represented by more than two groupings. Then the dependent variable Y is to be compared among the several X groups. The question is, as always, ''Is a difference in X status reflected in a difference in Y means?''

For example, a study might involve the marital status of women between the ages of 30 and 40 as the independent variable X. Three groupings are used, ''never been married,'' ''married,'' and ''divorced or widowed.'' The dependent variable Y might be a measure such as ''satisfaction with amount of personal independence.'' The idea is to see how these three marital groupings of women vary in their averages on this measure.

Obviously, a simple difference cannot be calculated for all three groups considered simultaneously. However, each pair of groups could be compared, and their difference tested by means of a statistic such as t. There would then be three such differences to be examined.

There is a serious problem in this approach, however. Suppose that the t test between Groups 1 and 2 were carried out, using $\alpha = .05$. Then the probability of a Type I error should be .05, or about 5 in 100 pairs of samples. Similarly the difference between Groups 1 and 3 could be tested at the same α level, and so could the difference between Groups 2 and 3.

In making t tests for all three differences, a Type I error *might* be made for none, for one, for two, or even for all three of the tests. What is the probability that we would be making *at least one* Type I error? If each test can be thought of as an independent trial of a Bernoulli process where $p = \alpha = .05$, then the binomial distribution shows that the probability of making *no* Type I errors in the three tests is

$$\text{prob(no errors)} = \binom{3}{0}(.05)^0(.95)^3 = .86.$$

Hence the probability of making at least one Type I error in these three tests is

$$\text{prob(one or more)} = 1.00 - .86 = .14.$$

In other words, in making three t tests, the probability of making at least one Type I error is almost three times as large as the probability of such an error for a single test, or $\alpha = .05$, even assuming that the tests are independent of each other, as we have done. In general, if we make some J independent t tests, using $\alpha = .05$ for each, then the probability of making at least one Type I error is

$$\text{probability one or more Type I errors} = 1 - (1 - \alpha)^J \qquad [8.17.1]$$

When J is large, this can be a large probability. For example, suppose that we have eight groups, and we examine all 28 differences among them by means of t tests, for $\alpha = .05$. Given that the tests are independent of each other, there will be a probability of .76 for one or more Type I errors. For 20 groups this probability is almost 1.00.

The problem is complicated further by the fact that tests of all differences among a set of means cannot be independent of each other, for reasons we will examine later. Thus, the actual probability of one or more Type I errors might be even worse than the values we calculate.

There is a principle from probability theory that gives us an upper bound for the probability that one or more such tests will lead to a Type I error. This is one of the so-called Bonferroni inequalities (Feller, 1968).

Essentially this says that if there are some J confidence intervals constructed simultaneously, we may think of the event C_j as meaning "the jth confidence interval covers the true mean." Then the event that all confidence intervals simultaneously cover their true means can be symbolized by $(C_1 \cap C_2 \cap \cdots \cap C_J)$ with probability $p(C_1 \cap C_2 \cap \cdots \cap C_J)$. The Bonferroni inequality says that if $(1 - \alpha)$ is the probability for each of the events C_j,

$$p(C_1 \cap C_2 \cap \cdots \cap C_J) \geq 1 - J\alpha.$$

This also implies that the probability that at least one interval *fails* to cover the true mean is

$$1 - p(C_1 \cap C_2 \cap \cdots \cap C_J) \leq J\alpha.$$

(If $J\alpha \geq 1.00$, then the upper bound is set equal to 1.00, of course.) In terms of tests, this is equivalent to saying that the maximum probability for at least one Type I error is $J\alpha$. There is also a practical implication: If we have a series of confidence intervals or corresponding tests, and if we want the probability of making at least one Type I error to be *no larger* than α, then we can set the level of each test at α/J. This will make the maximum probability of a Type I error among all the tests be exactly α. We will have more to say about this problem in Chapter 11.

8.18 t TESTS IN THE COMPUTER PACKAGES

All three of the major computer packages contain easy to use routines for carrying out t tests. In SPSS and SPSS^x the subprogram T-TEST lets one carry out such tests both for the independent-sample and the matched-sample situations. In this procedure applied to two independent groups, as specified in any of several ways by use of the keyword GROUPS, a *t* test is provided for each of the variables listed following the keyword VARIABLES.

For example, think of a situation where there are two groups, with the first designated by a variable name such as "FIRST," and with the other group consisting of the remainder of the cases. Then if *t* tests are desired on all of the variables identified as *Y*1, *Y*2, and *Y*3, the control statements would be:

T-TEST   GROUPS = FIRST/VARIABLE = *Y*1,*Y*2,*Y*3/

On the other hand, when one is dealing with matched pairs, and a *t* test of the difference between the members of the pairs is desired, then values for one group of pair members is identified with a variable name, and values for their pair mates with another variable name. Then a *t* test of the mean difference of pair mates is controlled by the statement

T-TEST   PAIRS = variable name 1 WITH variable name 2/

For example, suppose that father and mature eldest son pairs exist in the data, where the height of a father bears the variable name FATHER and that of his eldest son by SON. The data must of course *list* pairs of FATHER and SON values. The control statement for the *t* test of these matched-pair differences is then

T-TEST   PAIRS = FATHER WITH SON/

In the BMDP series, *t* tests for two groups may be carried out easily through program P3D. Three tests are routinely reported: the ordinary *t* test based on pooled variances, assuming that population variances are equal; the *t* test with adjusted degrees of freedom, assuming that the variances are not equal; and a test of equality of the population variances (Levene's test).

The SAS procedure PROC TTEST is somewhat similar in format to the SPSS procedure. Thus, for example, to compare the daily calories consumed by men and women, coded on the variable SEX, the control statements might look like this:

> PROC TTEST;
> > CLASS SEX;
> > VAR CALORIES;

This gives a variety of statistical results, including *t* with adjusted degrees of freedom under the assumption of unequal variances, and a test of equality of variances.

EXERCISES

1. In a study of leadership, a sample of 186 U.S. military officers was taken, and each was given a test designed to measure "inner direction" of actions. Previously, this test had been given to a large group of business executives, who produced a mean value of 83.4. The random sample of military officers had a mean score of 82.8, with standard deviation $S = 6.5$. Does the population of military officers appear to differ from the business population in this respect? Use $\alpha = .05$.

2. A standardized test for college entry has a mean of 500 and a standard deviation of 10, with a roughly normal distribution of scores when applied to high-school seniors across the United States. A coaching service promises greatly improved scores on this test. Suppose that the scores of 223 seniors who had used this service were examined and found to have a mean of 493 with standard deviation $S = 18.9$. If this group of coached students actually is a random sample of high-school seniors, what can one say about the claim the coaching service makes? Is it reasonable to assume that the students who were coached represent a random sample? Why or why not?

3. Over a long period of time a public library system in a city has charged 5¢ a day for overdue books. At the beginning of this year it was found that the population of library cardholders owed an average of 17.34¢. In an effort to reduce the total debt, the library mailed out notices to everyone with a record of any charges unpaid. Since that time, a random sample of cardholders has been taken. This sample of 392 individuals showed a mean of 17.02¢ due, with standard deviation $S = 2.53$. Do the notices seem to be having the desired effect? Use $\alpha = .01$.

4. A random sample of 175 American women were asked to record their body temperatures twice a day for a full month. From their records an average value was found for each woman. The mean of these values was 98.7 with a standard deviation S of .95. Test the hypothesis that the mean body temperature of such American women is 98.6 against the alternative that the mean is some other value.

5. Find the 99% confidence interval for the mean in Exercise 4.

6. In a study of truth in advertising, a government agency opened 500 boxes selected at random of a well-known brand of raisin bran. For each box the actual number of raisins was counted. The mean number of raisins was 32.4, with a standard deviation $S = 4.1$. Evaluate the company's claim that each box contains 34 raisins on the average, against the alternative of fewer raisins than claimed.

7. Find the 95% confidence interval for the mean in Exercise 6.

8. Suppose that the body weight at birth of normal children (single births) within the United States is approximately normally distributed and has a mean of 115.2 oz. A pediatrician believes that the birth weights of normal children born of mothers who are habitual smokers may be lower on the average than for the population as a whole. To test this hypothesis, records were taken of the birth weights of a random sample of 20 children from mothers who are still smokers. The mean of this sample is 114.0 with $S = 4.3$. Evaluate the pediatrician's theory.

9. Reevaluate the data in Exercise 8 on the assumption that a sample of 80 children had been used.

10. For the results in Exercise 8 find the 99% confidence interval for the mean birth weight of normal children from smoking mothers.

11. Suppose that in a certain large community the number of hours that a TV set is turned on in a given home during a given week is approximately normally distributed. A sample of 26 homes was selected, and careful logs were kept of how many hours per week the TV set was on. The mean number of hours per week in the sample turned out to be 36.1 with a standard deviation S of 3.3 h. Find the 95% confidence interval for the mean number of hours that TV sets are played in the homes of this community.

12. For the data in Exercise 11, test the hypothesis that the true mean number of hours is 35. Test the hypothesis that the mean number of hours is 30.

13. The same government agency referred to in Exercise 6 has decided to compare two well-known brands of raisin bran with respect to the numbers of raisins each contains on the average. Some 100 boxes of Brand A were taken at random, and the same number of boxes of Brand B were randomly selected. On the average the Brand A boxes contained 38.7 raisins, with $S = 3.9$, and Brand B contained an average of 36 raisins with $S = 4$. Test the hypothesis that the two brands are actually identical in the average number of raisins that their boxes contain. Let H_1 be "not H_0."

14. For Exercise 13 find the 99% confidence interval for the difference in average number of raisins for Brands A and B.

15. As editor of a journal in psychology you tend to believe that the contributors to that journal now use shorter sentences on the average than they did a few years ago. To test this hunch, you take a random sample of 150 sentences from journal articles published 10 years ago and take a random sample of 150 sentences from articles published within the last two years. The first sample shows a mean length of 127 type spaces per sentence, whereas the second sample shows a mean length of 113 type spaces. The first standard deviation $S = 41$, and the second standard deviation $S = 45$. Should you conclude that the recent articles do tend to have shorter sentences?

16. Find the 95% confidence interval for difference in sentence length from Exercise 15.

17. In an experiment, subjects were assigned at random between two conditions, five to each. Their scores turned out as follows:

Condition A	Condition B
128	123
115	115
120	130
110	135
103	113

Can one say that there is a significant difference between these two conditions? What must one assume in carrying out this test?

18. Find the 99% confidence interval for the difference between Conditions A and B in Exercise 17. On the evidence of this confidence interval, could one reject the hypothesis that the true mean of Condition B is five points higher than that of Condition A?

19. In an experiment, the null hypothesis is that two means will be equal. The variance of each population is believed to be equal to 16. If $\alpha = .05$, two-tailed, and the test is to have a power of .90 against the alternative that $\bar{y}_1 - \bar{y}_2 = 3$, about how many cases should one take in each experimental group?

20. Suppose that two brands of gasoline were being compared for mileage. Samples of each brand were taken and used in identical cars under identical conditions. Nine tests were made of Brand I and six tests of Brand II. The following miles per gallon were found.

Brand I	Brand II
16	13
18	15
15	11
23	17
17	12
14	13
19	
21	
16	

Are the two brands significantly different? What must be assumed here in order to carry out the test?

21. In a study of the effects of author attribution upon the perceived political tone of otherwise neutral quotations, 50 unfamiliar quotations on a variety of subjects were chosen. A sample of 40 college students was drawn and assigned at random between two groups, 20 to each. Group I received the quotations attributed to authors such as Lenin, Marx, and Mao. Group II received the same quotations but attributed to authors such as Jefferson, Franklin Roosevelt, and Lincoln. Each student rated each statement on a scale running from 1, for "democratic" to 5 for "totalitarian." Total ratings were then calculated for each student, and the mean and standard deviation figured for each group. The results were as follows:

Group I	Group II
$\bar{y}_1 = 155.0$	$\bar{y}_2 = 130.00$
$S_1 = 44.6$	$S_2 = 30.5$

Was there an effect of author attribution on the perception of the quotations? Use $\alpha = .05$.

22. For the data in Exercise 21 above, find the 95% confidence interval for the true difference in the means.

23. Consider the data in Exercise 9, Chapter 4, as two independent random samples of 50 cases each. Test the hypothesis that boys and girls differ on the average by 1,000 calories, against the alternative that the difference is greater than this in the direction of the boys. Use $\alpha = .01$.

24. Calculate the 99% confidence limits for the difference between the means in Exercise 23. Does this interval include the value 1,000? What does this reflect about the significance test carried out in Exercise 23?

25. As experimenter was interested in dieting and weight losses among men and among women. It was believed that in the first two weeks of a standard dieting program, women would tend to lose more weight than men. As a check on this notion, a random sample of 15 husband–wife pairs were put on the same strenuous diet. Their weight losses after two weeks showed the following:

Pair	Husands	Wives
1	5.0 lb	2.7 lb
2	3.3	4.4
3	4.3	3.5
4	6.1	3.7
5	2.5	5.6
6	1.9	5.1
7	3.2	3.8
8	4.1	3.5
9	4.5	5.6
10	2.7	4.2
11	7.0	6.3
12	1.5	4.4
13	3.7	3.9
14	5.2	5.1
15	1.9	3.4

Did wives lose significantly more ($\alpha = .05$) than husbands? What are we assuming here?

26. In the comparison of two brands of gasoline, mentioned in Exercise 20, the investigator felt that the type of automobile in which the gasoline was tried would make a difference in the mileage. Therefore it was decided to take 10 different makes of automobile, and to draw a pair at random from within each make. One member of the pair of automobiles was assigned gasoline Brand I and the other member gasoline Brand II. Then the mileages produced by the pairs of cars was as follows:

Pair	Brand I	Brand II
1	19	16
2	24	22
3	21	20
4	23	15
5	14	13
6	16	16
7	15	14
8	17	18
9	16	15
10	19	20

Do the two brands appear to be significantly different ($\alpha = .05$)?

27. Find the 95% confidence limits for the difference in mileage between the two types of gasoline, given the data in Exercise 26.

28. An experiment was concerned with the possible effect of a small lesion in a particular area of a hamster's brain upon the activity level of the animal. A total of 30 hamsters were selected at random, and randomly divided into two groups of 15 animals each. Group I was used as a control group, and given no lesion, whereas Group II was given a lesion in the designated area. Since 3 animals died in Group II, the actual number turned out to be 12 for that group. After full recovery by the remaining hamsters, each was given access to a running wheel, and the number of revolutions per fixed unit of time recorded. The following data show activity in hundreds of revolutions per unit time. Did the lesion significantly reduce the activity of the hamsters ($\alpha = .05$)?

Group I	Group II
$\bar{y}_1 = 9.26$	$\bar{y}_2 = 5.14$
$S_1 = 1.45$	$S_2 = 2.81$

29. Find the 95% confidence limits for the mean difference in wheel revolutions for the two populations of hamsters in Exercise 28.

30. What is the estimated value of ω^2 from the data in Exercise 28? What should we be inclined to say about the likely strength of association between presence or absence of a lesion and the activity of a hamster?

Chapter 9

The Chi-Square and the *F* Distributions

The essential ideas of inferential statistics are most easily discussed in terms of inferences about means or proportions, and so attention has been focused almost exclusively on these matters in the preceding chapters. However, population distributions can be compared in terms of variability as well as central tendency, and it is important to have inferential methods for the variance at our disposal.

The three basic sampling distributions used so far (the binomial, the normal, and the *t* distribution) no longer apply directly when the variance of a population is under study. Rather, we must turn to two new theoretical distributions. The first of these is called the **chi-square distribution,** or the distribution of the random variable χ^2 (small Greek chi, squared). We will use this distribution first in making inferences about a single population variance, although it has many other applications. The second distribution we will consider is usually called the **F distribution,** or the distribution of the random variable *F* (after Sir Ronald Fisher, who developed the main applications of this distribution). The study of this theoretical distribution grows out of the problem of comparing two population variances. The uses of both of these distributions extend far beyond the problems for which they were originally developed, since, like the normal distribution, they provide good approximations to a large class of other sampling distributions that are not easy to determine exactly. These five theoretical distributions, the three already studied plus the two to be introduced in this chapter, make up the arsenal of theoretical functions from which the statistician draws most heavily; almost all the elementary methods of statistical inference rest on one or more of these theoretical distributions. Furthermore, these five theoretical functions have very close connections with each other, and after we conclude our discussion of the chi-square and *F* distributions, some of these relationships will be pointed out.

Finally, brief mention will be made of three distributions that are close relatives of chi-square or *F* distributions. These are the *exponential,* the *gamma,* and the *beta dis-*

tributions. Although not as important in applied statistics as the five "major" distributions mentioned above, they do have some special interpretations that are worth knowing about.

9.1 THE CHI-SQUARE DISTRIBUTION

Suppose that there exists a population having a normal distribution of scores Y. The mean of this distribution is $E(Y) = \mu$, and the variance is

$$E(Y - \mu)^2 = \sigma^2.$$

Now cases are sampled from this distribution *one* at a time, $N = 1$. For each case sampled the **squared standardized score**

$$z^2 = \frac{(y - \mu)^2}{\sigma^2}$$

is computed. Let us call this squared standardized score $\chi^2_{(1)}$ so that

$$\chi^2_{(1)} = z^2.$$

Now we will look into the sampling distribution of this variable $\chi^2_{(1)}$.

First of all, what is the range of values that $\chi^2_{(1)}$ might take on? The orignial normal variable Y ranges over all real numbers, and this is also the range of the standardized variable z. However, $\chi^2_{(1)}$ is always a squared quantity, and so its range must be all the *nonnegative* real numbers, from zero to a positive infinity. We can also infer something about the form of this distribution of $\chi^2_{(1)}$; the bulk of the cases (about 68%) in a normal distribution of standard scores must lie between -1 and 1. Given a z between -1 and 1, the corresponding $\chi^2_{(1)}$ value lies between 0 and 1, so that the bulk of this sampling distribution will fall in the interval between 0 and 1. This implies that the form of the distribution of $\chi^2_{(1)}$ will be very skewed, with a high probability for a value in the interval from 0 to 1, and relatively low probability in the interval with lower bound 1 and approaching positive ∞ as its upper bound. The graph of the distribution of $\chi^2_{(1)}$ is represented in Figure 9.1.1.

Figure 9.1.1 pictures **the chi-square distribution with one degree of freedom. The distribution of the random variable $\chi^2_{(1)}$ where**

$$\chi^2_{(1)} = \frac{(y - \mu)^2}{\sigma^2} \tag{9.1.1}$$

and Y is normally distributed with mean μ and variance σ^2 is a chi-square distribution with 1 degree of freedom.

Now let us go a little further. Suppose that samples of *two* cases are drawn independently and at random from a normal distribution. We find the squared standardized score corresponding to each observation:

$$z_1^2 = \frac{(y_1 - \mu)^2}{\sigma^2}$$

$$z_2^2 = \frac{(y_2 - \mu)^2}{\sigma^2}.$$

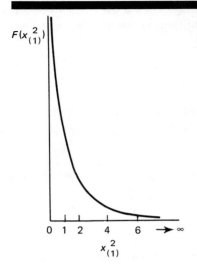

Figure 9.1.1

The distribution of χ^2 for $\nu = 1$.

If the *sum* of these two squared standardized scores is found over repeated independent samplings, the resulting random variable is designated as $\chi^2_{(2)}$:

$$\chi^2_{(2)} = \frac{(y_1 - \mu)^2}{\sigma^2} + \frac{(y_2 - \mu)^2}{\sigma^2} = z_1^2 + z_2^2 \ .$$

[9.1.2]

If we look into the distribution of the random variable $\chi^2_{(2)}$ we find that the range of possible values extends over all nonnegative real numbers. However, since the random variable is based on *two* independent observations, the distribution is somewhat less skewed than for $\chi^2_{(1)}$; the probability is not so high that the sum of two squared standardized scores should fall between 0 and 1. This is illustrated in Figure 9.1.2. This illustrates a **chi-square distribution with two degrees of freedom**.

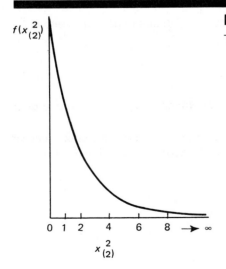

Figure 9.1.2

The distribution of χ^2 for $\nu = 2$.

Finally, suppose that we took N independent observations at random from a normal distribution with mean μ and variance σ^2 and defined the random variable

$$\chi^2_{(N)} = \frac{\sum_{i}^{N} (y_i - \mu)^2}{\sigma^2} = \sum_{i} z_i^2 . \qquad [9.1.3]$$

The distribution of this random variable will have a form that depends on the number of independent observations taken at one time. **In general, for N independent observations from a normal population, the sum of the squared standardized scores for the observations has a chi-square distribution with N degrees of freedom.** Notice that the standardized scores must be relative to the *population mean* and the *population standard deviation*. We say that here $N = v$, the degrees of freedom.

The function rule assigning a probability density to each possible value of χ^2 is given by

$$f(\chi^2;v) = h(v)e^{-\chi^2/2}(\chi^2)^{(v/2)-1}, \quad \text{for } \chi^2 \geq 0 \qquad [9.1.4]$$
$$v > 0.$$

As shown in Figures 9.1.1 through 9.1.3, the plot of this density function always presents the picture of a very positively skewed distribution, at least for relatively small values of v, with a distinct mode at the point $v - 2$ for $v > 2$. However, as v is increased, the form of the distribution appears to be less skewed to the right. In this function rule the value $h(v)$ is a constant depending only on the parameter v. Consequently, there is only one value other than χ^2 that must be specified in order to find the density: this is the parameter v. Like the t distribution, the distribution of χ^2 depends only on the degrees of freedom, the parameter v. The family of chi-square distributions all follow this general rule, but the exact form of the distribution depends on the number of degrees of freedom, v. In principle, the value of v can be any positive number, but in the applications we will make of this distribution in this chapter, v will depend only on the sample size, N.

The mean of a chi-square distribution is simply the value of the parameter v,

$$E(\chi^2_{(v)}) = v$$

and the variance is

$$\text{var}(\chi^2_{(v)}) = 2v.$$

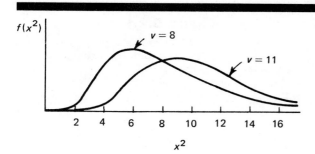

Figure 9.1.3

The general form of the distribution of χ^2 for larger numbers of degrees of freedom.

When several independent random variables each have a chi-square distribution, then the distribution of the sum of the variables is also known. This is a most important and useful property of variables of this kind, which can be stated more precisely as follows:

If a random variable $\chi^2_{(v_1)}$ has a chi-square distribution with v_1 degrees of freedom, and an independent random variable $\chi^2_{(v_2)}$ a chi-square distribution with v_2 degrees of freedom, then the new random variable formed from the sum of these variables

$$\chi^2_{(v_2 + v_1)} = \chi^2_{(v_1)} + \chi^2_{(v_2)} \tag{9.1.5}$$

has a chi-square distribution with $v_1 + v_2$ degrees of freedom.

In short, the new random variable formed by taking the sum of two independent chi-square variables is itself distributed as χ^2, with degrees of freedom equal to the sum of those for the original distributions.

9.2 TABLES OF THE CHI-SQUARE DISTRIBUTION

Like the t distribution, the particular distribution of χ^2 depends on the parameter v, and it is difficult to give tables of the distribution for all values of v that one might need. Thus, Table IV in Appendix E, like Table III, is a condensed table, showing values of χ^2 that correspond to percentage points in various distributions specified by v. The rows of Table IV list various degrees of freedom v and the column headings are probabilities Q, just as in the t table. The numbers in the body of the table give the values of χ^2 such that in a distribution with v degrees of freedom, the probability of a sample chi-square value *this large or larger* is Q (Fig. 9.2.1).

For example, suppose that we are dealing with a chi-square distribution with 5 degrees of freedom. Find the row v equal to 5 and the column headed .05. The value in this row and column is approximately 11.071, showing that for 5 degrees of freedom, random samples showing a chi-square value of 11.071 *or more* should in the long run occur about 5 times in 100. As another example, look at the row labeled $v = 2$ and the

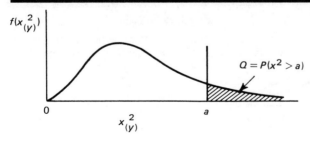

$Q = P(x^2 > a)$

Figure 9.2.1

A right-tail area in a chi-square distribution with v degrees of freedom.

column headed .10. The entry here indicates that in a distribution with 2 degrees of freedom, chi-square values of 4.605 or more should occur with probability of about 1 in 10, under random sampling. Finally, look at the row for $v = 24$ and the column for .990; the entry indicates that chi-square values of 10.856 or more occur with probability of about .990 in a distribution with 24 degrees of freedom.

9.3 THE DISTRIBUTION OF THE SAMPLE VARIANCE FROM A NORMAL POPULATION

In this section, the sampling distribution of the sample variance will be studied. It will be assumed that the population actually sampled is normal, and the results in this section apply, strictly speaking, only when this assumption is true.

Recall that in Section 4.11 the sample variance was defined by

$$S^2 = \frac{\sum\limits_{i=1}^{N} (y_i - \bar{y})^2}{N} .$$

Then, since

$$\frac{NS^2}{\sigma^2} = \sum_{i=1}^{N} \frac{(y_i - \bar{y})^2}{\sigma^2}$$

it can be shown that for N independent observations from a normal population,

$$\frac{NS^2}{\sigma^2} = \chi^2_{(N-1)} , \qquad\qquad [9.3.1^*]$$

the ratio NS^2/σ^2 is a chi-square variable with $N - 1$ degrees of freedom.

If, instead of S^2, we used the unbiased estimate s^2, we find that again for N independent observations from a normal population,

$$\frac{(N - 1)s^2}{\sigma^2} = \chi^2_{(N-1)} \qquad\qquad [9.3.2^*]$$

meaning that the ratio of $(N - 1)s^2$ to σ^2 is a random variable distributed as chi-square with $N - 1$ degrees of freedom. This fact is the basis for inferences about the variance of a single normally distributed population.

This same idea applies to the sampling distribution of the estimate of a variance based on the *pooling* of independent estimates, provided that the basic population sampled is normal. Thus, it is also true that when the unbiased pooled estimate of σ^2 is made from two independent samples of N_1 and N_2 cases, as in Eq. 5.10.4,

$$\frac{(N_1 + N_2 - 2) \text{ est. } \sigma^2}{\sigma^2} = \chi^2_{(N_1 + N_2 - 2)} . \qquad\qquad [9.3.3^*]$$

The pooled unbiased sample estimate of σ^2, multiplied by the degrees of freedom and divided by the true value of σ^2, is a chi-square variable with $v = N_1 + N_2 - 2$.

9.4 TESTING EXACT HYPOTHESES ABOUT A SINGLE VARIANCE

Just as for the mean, when the population is normal, it is possible to test exact hypotheses about a single population variance (and, of course, a standard deviation). The exact hypothesis tested is

H_0: $\sigma^2 = \sigma_0^2$

where σ_0^2 is some specific positive number. The alternative hypothesis may be either directional or nondirectional, depending, as always, on the original question.

As usual, some value of α is decided on and a region of rejection adopted, depending both on α and the alternative H_1. The test statistic itself is

$$\chi_{(N-1)}^2 = \frac{(N-1)s^2}{\sigma_0^2} .$$ [9.4.1*]

The value of this test statistic, computed from the s^2 actually obtained and the σ_0^2 dictated by the null hypothesis, is referred to the distribution of χ^2 for $N - 1$ degrees of freedom.

For example, one might ask this question about height: "It is well known that men and women in the United States differ in terms of their mean height; is it true, however, that women show less variability in height than do men?" Now let us assume that from the past records of the Selective Service System we actually know the mean and standard deviation of height for American men between the ages of 20 and 25. However, such evidence is lacking for women. Assume that the standard deviation of height for the population of men 20 to 25 is known to be 2.5 in. For this same age range, we want to ask if the population of women shows this same standard deviation, or if women are *less* variable, with their distribution having a smaller σ. The null and alternative hypotheses can be framed as

H_0: $\sigma^2 \geq 6.25$ (or $[2.5]^2$)

H_1: $\sigma^2 < 6.25$.

Imagine that we plan to draw a sample of 30 women at random, each between the ages of 20 and 25, and measure the height of each. The test statistic will be

$$\chi_{(29)}^2 = \frac{(29)s^2}{6.25}.$$

What, however, is the region of rejection? Here, *small* values of χ^2 tend to favor H_1, that women actually are less variable than men. Hence, we want to use a region of rejection on the left (or small-value) tail of the chi-square distribution with 29 degrees of freedom. If $\alpha = .01$, then from Table IV in Appendix E the value of χ^2 leading to rejection of H_0 should be less than the value given by the row with $\nu = 29$, and the column for $Q = .99$. This value is 14.257.

Now the actual value of s^2 obtained turns out to be 4.55, so that

$$\chi_{(29)}^2 = \frac{(29 \times 4.55)}{6.25} = 21.11.$$

This value is larger than the critical value decided on, and we cannot reject H_0 if α is to be .01.

This example illustrates that, as with the t and the normal distribution, either or both tails of the chi-square distribution can be used in testing a hypothesis about a variance. Had the alternative hypothesis in this problem been

$$H_1: \sigma^2 \neq 6.25,$$

then the rejection region would lie in both tails of the chi-square distribution. The rejection region on the lower tail of the distribution would be bounded by a chi-square value corresponding to $v = 29$ and $Q = .995$, which is 13.121; the rejection region on the upper tail would be bounded by the value for $v = 29$ and $Q = .005$, which is 52.336. Any obtained χ^2 value falling *below* 13.121 or *above* 52.336 would let one reject H_0 beyond the .01 level.

9.5 CONFIDENCE INTERVALS FOR THE VARIANCE AND STANDARD DEVIATION

Finding confidence intervals for the variance is quite simple, provided that the normal distribution rule holds for the population.

Suppose that we had a sample of some N independent observations, and we wanted the 95% confidence limits for σ^2. For samples from a normal distribution, it must be true that

$$\text{prob}\left[\chi^2_{(N-1;.975)} \leq \frac{(N-1)s^2}{\sigma^2} \leq \chi^2_{(N-1;.025)} \right] = .95 \qquad [9.5.1]$$

where $\chi^2_{(N-1;.975)}$ is the value in a chi-square distribution with $N-1$ degrees of freedom cutting off the upper .975 of samples values, and $\chi^2_{(N-1;.025)}$ the value cutting off the upper .025 of sample values. This inequality can be manipulated to show that it is also true that

$$\text{prob}\left[\frac{(N-1)s^2}{\chi^2_{(N-1;.025)}} \leq \sigma^2 \leq \frac{(N-1)s^2}{\chi^2_{(N-1;.975)}} \right] = .95. \qquad [9.5.2]$$

That is, the probability is .95 that the true value of σ^2 will be covered by an interval with limits found by

$$\frac{(N-1)s^2}{\chi^2_{(N-1;.025)}}$$

and $\qquad\qquad\qquad\qquad\qquad\qquad\qquad\qquad\qquad\qquad\qquad\qquad [9.5.3]$

$$\frac{(N-1)s^2}{\chi^2_{(N-1;.975)}}.$$

Suppose, for example, that a sample of 15 cases is drawn from a normal distribution. We want the 95% confidence limits for σ^2. The value of s^2 is 10; this is our best single estimate of σ^2, of course. For $v = 14$ and for $Q = .025$, $\chi^2_{(14;.025)}$ is found from Table

IV to be 26.12, and the value of $\chi^2_{(14;.975)}$ is 5.63. Thus the confidence limits for σ^2 are

$$\frac{(14 \times 10)}{26.12} \text{ or } 5.36$$

and

$$\frac{(14 \times 10)}{5.63} \text{ or } 24.87.$$

We can say that the probability is .95 that an interval such as 5.36 to 24.87 covers the true value of σ^2.

Unfortunately, because of the skewness of a chi-square distribution, confidence intervals for σ^2 are not necessarily "shortest" in the same sense that holds for confidence intervals for the mean. Thus, these intervals are not as useful as they might be, particularly when our interest is in finding the best possible estimates of σ^2. More advanced methods do exist for finding the shortest possible confidence interval for σ^2, and tables for this purpose are given in Tate and Klett (1959).

9.6 THE NORMAL APPROXIMATION TO THE CHI-SQUARE DISTRIBUTION

As the number of degrees of freedom grows infinitely large, the distribution of χ^2 approaches the normal distribution. You should hardly find this principle surprising by now! Actually the same mechanism is at work here as in the central limit theorem: for any ν greater than 1, the chi-square variable is equivalent to a *sum* of ν independent random variables. Thus, like the mean, given enough summed terms the sampling distribution approaches the normal form in spite of the fact that each component of the sum does not have a normal distribution when sampled alone.

This fact is of more than theoretical interest when very large samples are used. For very large ν, the probability for any interval of values of $\chi^2_{(\nu)}$ can be found from the normal standardized scores

$$z = \frac{\chi^2_{(\nu)} - \nu}{\sqrt{2\nu}} \qquad\qquad [9.6.1]$$

since the mean and variance of a chi-square distribution are ν and 2ν respectively. However, this approximation is not good unless ν is extremely large. A somewhat better approximation procedure is to find the value of $\chi^2_{(\nu)}$ that cuts off the *upper Q* proportion of cases by taking

$$\chi^2_{(\nu;Q)} = \frac{1}{2} (z_Q + \sqrt{2\nu - 1})^2. \qquad\qquad [9.6.2]$$

This can be used to find chi-square values for ν greater than 100, which are not given in Table IV, Appendix E. The necessary z_Q values are listed at the bottom of Table IV and can also be found from Table I.

An even better approximation, giving high accuracy even for small numbers of degrees of freedom, is that devised by Wilson and Hilferty (1931):

$$\chi^2_{(v,Q)} = [z_Q \sqrt{2/(9v)} + 1 - (2/9v)]^3 \times v \tag{9.6.3}$$

9.7 THE IMPORTANCE OF THE NORMALITY ASSUMPTION IN INFERENCES ABOUT σ^2

In the preceding chapter it was pointed out that although the rationale for the *t* test demands the assumption of a normal population distribution of scores, in practice the *t* test may be applied when the parent distribution is not normal, provided that sample *N* is at least moderately large. This is *not* the case for inferences about the variance, however. One runs a considerable risk of error in using the chi-square distribution either to test a hypothesis about a variance or to find confidence limits unless the population distribution is normal, or approximately so. The effect of the violation of the normality assumption is usually minor for large *N*, but can be quite serious when inferences are made about variances, even for moderate *N*. Indeed for some population forms, the effect of using the chi-square statistics may actually grow more serious for larger samples. Thus, the assumption of a normal distribution is important in inferences about the variance.

9.8 THE *F* DISTRIBUTION

It is rather rare to find a problem in research that centers on the value of a single population variance. More common is the situation where the variances of two populations are compared for equality.

Imagine two distinct populations, each showing a normal distribution of the variable *Y*. The means of the two populations may be different, but each population shows the same variance, σ^2. We draw two independent random samples: the first sample, from Population 1, contains N_1 cases, and that from Population 2 consists of N_2 cases.

For each possible pair of samples, one from Population 1, and one from Population 2, we take the *ratio* of s_1^2 to s_2^2 and call this variance ratio the random variable *F*:

$$F = \frac{s_1^2}{s_2^2} = \frac{\text{est. } \sigma_1^2}{\text{est. } \sigma_2^2} . \tag{9.8.1*}$$

The sampling distribution of this ratio of two independent unbiased estimates of the population variance, each based on a sample of independent observations from a normal distribution, is called the *F distribution*.

Notice that when we put the two variance estimates in ratio, this is actually the ratio of two independent chi-square variables, each divided by its degrees of freedom:

$$F = \frac{[\chi^2_{(v_1)}/v_1]}{[\chi^2_{(v_2)}/v^2]} \tag{9.8.2}$$

since

$$\frac{s_1^2}{s_2^2} = \frac{(s_1^2/\sigma_1^2)}{(s_2^2/\sigma_2^2)}$$

provided that $\sigma_1^2 = \sigma_2^2$ (that is to say, the hypothesis of the equality of the population variances is true).

Showing F as a ratio of two independent chi-square variables each divided by its ν is actually a way of defining the F variable:

A random variable formed from the ratio of two independent chi-square variables, each divided by its degrees of freedom, is said to be an F ratio, and to follow the rule for the F distribution.

When this definition is satisfied, meaning that both parent populations are normal, have the same variance, and that the samples drawn are independent, then the theoretical distribution of F values can be found.

Mathematically, the F distribution is rather complicated. However, for our purposes it will suffice to remember that the density function for F depends only on two parameters, ν_1 and ν_2, which can be thought of as the degrees of freedom associated with the *numerator* and the *denominator* of the F ratio. The range for F is *nonnegative real numbers*. The expectation of F is $\nu_2/(\nu_2 - 2)$ for $\nu_2 > 2$. In general form, the distribution for any fixed ν_1 and ν_2 is nonsymmetric, although the particular shape of the function curve varies considerably with changes in ν_1 and ν_2. However, it is worth nothing that for the values of ν_1 and ν_2 that will usually concern us (that is, integral values, with $\nu_1 < \nu_2$) the F distribution is skewed positively and is unimodal for $\nu_1 > 2$. On the other hand, the distribution can take on other forms when other conditions are set for ν_1 and ν_2.

Before we can apply this theoretical distribution to an actual problem of comparing two variances, the use of F tables must be discussed.

9.9 THE USE OF F TABLES

Since the distribution of F depends on two parameters ν_1 and ν_2, it is even more difficult to present tables of F distributions than those of χ^2 or t. Tables of F are usually encountered only in drastically condensed form.

Such tables give only those values of F that cut off the *upper* proportion Q in an F distribution with ν_1 and ν_2 degrees of freedom. The only values of Q given are those that are commonly used as α in a test of significance (that is, the .25, .10, .05, .025, .01, and the .005 values). Table V in Appendix E shows F values required for significance at the $\alpha = Q$ level, given ν_1 and ν_2. The columns of each table give values of ν_1, the degrees of freedom for the numerator, and the rows give values of ν_2, the degrees of freedom for the denominator. Each separate table represents one value of Q.

The entries in the body of the table are the values of F required for significance at this level.

The use of Table V can be illustrated in the following way: Suppose that two independent samples are drawn, containing $N_1 = 10$ and $N_2 = 6$ cases, respectively. The degrees of freedom associated with the two variances are $v_1 = 10 - 1 = 9$ and $v_2 = 6 - 1 = 5$. For the ratio

$$F = \frac{s_1^2}{s_2^2}$$

the degrees of freedom must be 9 for the numerator, and 5 for the denominator. Now suppose that the obtained $F = 7.00$. Does this fall into the upper .05 of values in an F distribution, with 9 and 5 degrees of freedom? We turn to Table V and find the page for .05. Then we look at the column for $v_1 = 9$ and the row for $v_2 = 5$. The tabled value is 4.77. Our obtained F value of 7.00 exceeds 4.77, and thus our sample result falls among the upper 5% in an F distribution.

As another example, suppose that $v_1 = 1$, and $v_2 = 45$. For the problem at hand, the rejection region contains the *upper* .01 proportion in an F distribution. What value must our sample result equal or exceed to be significant? Table V shows that for the .01 level, the required F for this number of degrees of freedom is between 7.31 (the F for $v_2 = 40$) and 7.08 (the F for $v_2 = 60$). It is difficult to find the exact value of F required, but we are reasonably sure that the required F is somewhere below 7.3.

In both of these illustrations we have dealt only with the upper tail of the F distribution. This would be appropriate if we were doing a one-tailed test, where H_1: $\sigma_1^2 > \sigma_2^2$. However, it is possible, although somewhat more complicated, to find F values corresponding to two-tailed rejection regions.

Let $F_{(v_1;v_2)}$ stand for an F ratio with v_1 and v_2 degrees of freedom, and let $F_{(v_2;v_1)}$ be an F ratio with v_2 degrees of freedom in the *numerator*, and v_1 degrees of freedom in the *denominator*. The numbers of degrees of freedom in numerator and denominator are simply reversed for these two F ratios. Then it is true that, for any positive number C,

$$p[C \le F_{(v_1;v_2)}] = p\left[F_{(v_2;v_1)} \le \frac{1}{C}\right] \qquad [9.9.1]$$

the probability that $F_{(v_1;v_2)}$ is greater than or equal to some number C is the same as the probability that the reciprocal of $F_{(v_1;v_2)}$ is *less than or equal to the reciprocal* of C. Practically, this means that the value required for F on the lower tail of some particular distribution can always be found by finding the corresponding value required on the upper tail of a distribution with numerator and denominator degrees of freedom *reversed* and then taking the *reciprocal*.

In practical situations, a simpler way of proceeding is this: For a one-tailed test, take the sample variance value that *should* be the larger if the alternative hypothesis were true. Base the numerator of F on this sample variance and the denominator on the other sample variance. Furthermore, let the numerator degrees of freedom correspond to the cases in the numerator sample, minus one, and the denominator degrees of freedom to the second sample cases, minus one. If the F is less than 1.00 then it is definitely not significant. However, if the *F* is greater than 1.00, use the table to determine the significance level.

For a two-tailed test, simply select the large sample variance and use it in the numerator, with the smaller one in the denominator. Calculate the numerator degrees of freedom to correspond to the sample variance used in the numerator, and do likewise for the denominator. However, to be significant at the α level, the *F* value should reach or exceed that required for the $\alpha/2$ level.

9.10 USING THE *F* DISTRIBUTION TO TEST HYPOTHESES ABOUT TWO VARIANCES

In an investigation of the effect of stress on children's performance on a reasoning test, it is felt that competition with peers represents one form of stressful situation to a child. However, the experimenter suspects that competition has different effects on different children. The conjecture is that bright children might be stimulated to do even better than otherwise by the competitive atmosphere, but that the relatively dull child will appear even more at a disadvantage. One implication of this notion is that if groups of children are sampled from a population having a normal distribution of ability, but the groups are given different amounts of "competitive stress," the group subjected to the greater stress should show a relatively *greater variance* among the scores.

Letting Population 1 stand for the potential group of children tested under stress, and 2 for the control, or nonstressed, population, we frame the following hypotheses:

$$H_0: \sigma_1^2 \le \sigma_2^2$$

$$H_1: \sigma_1^2 > \sigma_2^2 .$$

The rejection region decided on is one-tailed, reflecting the experimenter's prior hunch about the effects of stress. The α level to be used is .05.

Thirty-two children are selected at random, and assigned at random to two experimental groups of 16 cases each. Group 1 is given the stress treatment, and Group 2 is not. The results are

$$s_1^2 = 5.8, s_2^2 = 1.7.$$

If the exact null hypothesis is true, so that both of these values are estimates of the same population value σ^2, the ratio

$$F = \frac{s_1^2}{s_2^2}$$

should be distributed as the *F* distribution with $N_1 - 1 = 15$ and $N_2 - 1 = 15$ degrees of freedom. The value required for significance at the .05 level, one-tailed, is found from Table V to be 2.40. However, the obtained value of *F* is

$$F = \frac{5.8}{1.7} = 3.41,$$

which exceeds the value required. On this evidence we can reject the null hypothesis of equal variances at the .05 level. We are fairly safe in saying that the experimental increase in stress seems to increase the variability of scores.

Had the alternative hypothesis been two-tailed, then we would have had to consider the required value of F on both tails of the distribution, of course. In this instance, the procedure given in Section 9.9 would be used.

Like the chi-square distribution, the use of F depends on the assumption that the population distributions are themselves normal. The failure of this assumption to hold true can make a difference in the conclusions reached from an F test of the equality of variances, *unless* the sample sizes are rather large. (About 30 in each sample should be relatively safe, however.)

9.11 RELATIONSHIPS AMONG THE THEORETICAL DISTRIBUTIONS

Now that the major sampling distributions have been introduced, some of the connections among them can be examined in more detail. Over and over again, the binomial, normal, t, chi-square, and F distributions have proved their utility in the solution of problems in statistical inference. Remember, however, that none of these distributions is empirical in the sense that someone has taken a vast number of samples and found that the sample values actually do occur with exactly the relative frequencies given by the function rule. Rather, it follows mathematically (read *logically*) that if we are drawing random samples from certain kinds of populations, various sample statistics *must* have distributions given by the several function rules. Like any other theory, the theory of sampling distributions deals with "if–then" statements. This is why the assumptions we have introduced are important; if we wish to apply the theory of statistics to making inferences from samples, then we cannot expect the theory necessarily to provide us with correct results unless the conditions specified in the theory hold true. As we have seen, from a practical standpoint these assumptions may be violated to some extent in our use of these theoretical distributions as approximations, especially for large samples. However, these assumptions are quite important for the *mathematical* justification of our methods, in spite of the possible applicability of the methods to situations where the assumptions are not met.

Apart from the general requirement of simple random sampling of independent observations, the most usual assumption made in deriving sampling distributions is that the population distribution is normal. The chi-square, the t, and the F distributions all rest on this assumption. The normal distribution is, in a real sense, the "parent" distribution to these others. This, as mentioned in Chapter 6, is one of the main reasons for the importance of the normal distribution: the normal function rule not only provides probabilities that are often excellent approximations to other probability (density) functions, it also has highly convenient mathematical properties for deriving other distribution functions based on normal populations.

The chi-square distribution rests directly on the assumption that the population is normal. As you will recall from Section 9.1, the chi-square variable is basically a sum of squares of independent *normal* variables, each with mean 0 and with variance 1. At the elementary level, the problem of the distribution of the sample variance can be solved explicitly only for normal populations; this sampling distribution depends on the distribution of χ^2, which in turn rests on the assumption of a normal distribution of single observations. Furthermore, in the limit, the distribution of χ^2 approaches a normal form.

There are close connections in theory between the F distribution and both the normal and chi-square distributions. Basically, the F variable is a ratio of two independent chi-square variables, each divided by its degrees of freedom. Since a chi-square variable is itself defined in terms of the normal distribution, then the F distribution also rests on the assumption of two (or more) normal populations.

In general, a t variable is a standardized normal variable z_M in ratio to the square root of a chi-square variable divided by v. Let us look at t^2 for a single mean in the light of this definition:

$$t^2 = \frac{z_M^2}{\chi^2/(N-1)}.$$ [9.11.1]

The numerator of t^2 is, by definition, a chi-square variable with 1 degree of freedom, and the denominator is a chi-square variable divided by its degrees of freedom, $v = N - 1$. Furthermore, these two chi-square variables are independent, by the principle stating that for a normal population, s^2 is independent of $\bar{y}$ (Section 6.6). Thus, for a single mean, t^2 qualifies as an F ratio, with 1 and $N - 1$ degrees of freedom. In general,

$$t_{(v)}^2 = F_{(1,v_2)}, \quad v = v_2;$$ [9.11.2]

the square of t with v degrees of freedom is an F variable with 1 and v degrees of freedom.

You can check this for yourself by examining the column for 1 and v_2 degrees of freedom in the F table. In the table of F values required for significance at the .05 level, each of the entries in this column is simply the square of the entry in the t table for $v = v_2$ and $t_{.025}$. Similarly, in the table for $\alpha = .01$, each entry in the column for 1 and v_2 degrees of freedom is the square of the corresponding entry for $t_{.005}$ in the distribution of t.

This relation between t^2 and F lets us illustrate the importance of the assumption that the two populations have the same true σ^2 when a difference between means is tested. For n cases in each of two independent groups, the value of z^2 represented by the numerator of t^2 must be

$$z_{\text{diff.}}^2 = \frac{(\bar{y}_1 - \bar{y}_2)^2}{\sigma_{\text{diff.}}^2}.$$ [9.11.3]

When $\mu_1 = \mu_2$, this is a chi-square variable with one degree of freedom. Furthermore, the square of the denominator term in the t ratio must correspond to

$$\frac{\text{est. } \sigma_{\text{diff.}}^2}{\sigma_{\text{diff.}}^2}$$ [9.11.4]

for this to be a chi-square variable divided by its degrees of freedom. When each population has the same true variance, σ^2, the denominator term we compute is equivalent to such a variable, and the value t^2 is then the square of the ratio we actually find:

$$t^2 = \frac{(\bar{y}_1 - \bar{y}_2)^2}{\text{est. } \sigma_{\text{diff.}}^2} = \frac{(\chi_{(1)}^2/1)}{(\chi_{(v_2)}^2/v_2)} = F.$$ [9.11.5]

Table 9.11.1

Connections among the theoretical distributions.

Distribution	Chi-square	F	t
Normal	Parent, and limiting form as $\nu \to \infty$. Defined as sum of normal and independent z^2 values.	Parent, making values in numerator and denominator independent χ^2/ν values.	Parent, and limiting form as $\nu \to \infty$. Numerator is normal z.
Chi-square		Variables in numerator and denominator are independent χ^2/ν.	Denominator is $\sqrt{\chi^2/\nu}$.
F			$t^2_{(\nu)} = F_{(1,\nu)}$

The ratio one calculates, not knowing the true $\sigma^2_{\text{diff.}}$, actually is distributed exactly as t (or its square distributed as F) when the true variances are equal. However, if the values of the two variances are not the same, the ratio of the estimated $\sigma^2_{\text{diff.}}$ to the true $\sigma^2_{\text{diff.}}$ is equivalent to

$$\frac{\text{est. } \sigma^2_{\text{diff.}}}{\text{true } \sigma^2_{\text{diff.}}} = \frac{\chi^2_a \sigma^2_1 + \chi^2_b \sigma^2_2}{(n-1)(\sigma^2_1 + \sigma^2_2)} \qquad [9.11.6]$$

where χ^2_a and χ^2_b symbolize two possibly different values of a chi-square variable with $n - 1$ degrees of freedom. This ratio is not necessarily distributed as a chi-square variable divided by its degrees of freedom. Thus, when variances are unequal for the two populations, the ratio we compute is not really distributed exactly like the random variable t, since the square of the ratio we compute cannot be equivalent to an F ratio. For this reason, a correction procedure, such as that in Section 8.10, is required when the variances are unequal and sample size is small.

The important relationships among these theoretical distributions are summarized in Table 9.11.1, showing how the distribution represented in the column depends for its derivation upon the distribution represented in the row.

All four of these theoretical distributions will figure in the next three chapters as we work though the rationale for the analysis of variance and comparisons among means. Having some idea of the interrelations of these distributions will be of help in understanding how the F distribution can be used to test a hypothesis about *several* means.

9.12 THE EXPONENTIAL, GAMMA, AND BETA DISTRIBUTIONS

Our introduction to the more important continuous distributions will conclude with a brief discription of three families of distributions which are very closely related to the chi-square and the F distributions. In addition, these distributions have special uses in their own right, some of which are of potential value in the social and behavioral sciences.

The exponential distribution has close conceptual ties to the geometric distribution of Section 3.17, and to the Poisson distribution of Section 3.18. Suppose that a random process follows a Poisson law producing successes at the rate of m per unit of time. The time is recorded until the *first* (or the *next*) success occurs. Then the distribution of time until the first success occurs follows the rule of the *exponential distribution:*

$$f(x; m) = \begin{cases} e^{-mx}m & \text{for } x \geq 0, m > 0 \\ 0 & \text{otherwise.} \end{cases}$$

An exponential distribution of the random variable X has a mean $E(X) = 1/m$, and a variance of $1/m^2$.

Thus, for example, a pigeon has been trained to peck at a bar to secure a reward of grain. The trained pigeon has been observed to peck at a stable rate of 5 times a minute. What then is the average time that will elapse before the pigeon will give its first peck after having been placed in front of the bar? The answer is $E(X) = 1/5 = .2$ min, or 12 s. These times to the first peck will have a variance of $1/25 = .04$, or a standard deviation of .2 min. The probability density for a given time such as $X = 6$ s, or .1 min, is $e^{-.5}(.5)$, and so on for any other value of X. (The random variable does not necessarily have to represent only time until the first success; any other continuous dimension such as the distance to the first success may also be used with this model.)

A generalization of the exponential distribution is the *gamma distribution*. One interpretation of a gamma variable is as follows: Again suppose that a Poisson process produces successes at the stable rate of m per unit of time. Then if X is the elapsed time until the rth sucess, the random variable X follows the gamma law

$$(x; r, m) = \begin{cases} \dfrac{e^{-mx}(mx)^{r-1}m}{(r-1)!} & \text{for } x \geq 0, r > 0, m > 0 \\ 0 & \text{otherwise.} \end{cases}$$

Here, when the value of the parameter r is an integer, $(r-1)!$ is an ordinary factorial. However, for noninteger values, $(r-1)!$ stands for a value of the so-called gamma function, studied in the advanced calculus. (Here, as well, X may represent time, distance, or any other continuous dimension relevant to the context, and m is a rate of success events per unit interval on this dimension.) Then the expected value of X for the gamma distribution with parameters r and m is $E(X) = r/m$, and the variance of X is $\text{var}(X) = r/(m^2)$.

For example, the police department in a very large city knows from past records that murders tend to be reported at the rate of 2.6 per week. On a certain January 1, the police department will begin keeping the records for that year. How many weeks should it take them to record exactly 50 murders? Notice that the time it could take to reach this number could be zero weeks (i.e., a very bad early New Year's morning), or it could be some large, or even infinite number of weeks (miraculously, there are no more murders in this city). However, for any starting point, the number of weeks until 50 murders occur should follow a gamma distribution, with $m = 2.6$ and with $r = 50$. The expected number of weeks should be

$$E(X) = 50/2.6 = 19.23 \text{ weeks}$$

and the variance should be

$$E(X) = 50/(2.6)^2 = 7.40.$$

The murder rate in that year could be compared with past years, or with any hypothetical rate, in terms of the time it takes to reach some arbitrary number of cases reported.

A random variable following a *beta distribution* has a probability density for any value $X = x$ given by

$$f(x; r, N) = \begin{cases} \dfrac{(N-1)!}{(r-1)!(N-r-1)!} x^{r-1}(1-x)^{N-r-1} & \text{for } 0 \le x \le 1,\, 0 < r < N \\ 0 & \text{otherwise.} \end{cases}$$

Once again, for integer values of r and N, where $0 < r < N$, the factorial expressions are found in the usual way. However, for noninteger r or N, each factorial is replaced by a value from the gamma function, as mentioned above. The expected value of a beta variable X is given by $E(X) = r/N$, and the variance is $\text{var}(X) = r(N - r)/N^2(N + 1)$.

One of the uses of the beta distribution is when two independent Poisson-like processes are occurring, each producing successes at the same rate m. Then if A stands for the time taken to the rth success in the first process, and B the time taken to the $(N - r)$th success in the second process, the variable $X = A/(A + B)$ has a beta distribution with parameters r and N. Thus, a psychologist might be interested in comparing two groups performing the same task, to see if they come up with errors (successes) at the same rate. Suppose that the time to the *tenth error* is recorded for each group. This time is called A for the first group, and B for the second. Then, $X = A/(A + B)$ and is distributed as a beta variable with $r = 10$, $N - r = 10$, if the occurrence of an error in either group is the outcome of a stable, independent Poission process with the same rate m.

Formally, any chi-square variable may also be considered a gamma variable where $r = v/2$ and $m = 1/2$. Furthermore, for an F variable with v_1 and v_2 degrees of freedom, if one takes

$$X = \frac{F}{F + (v_2/v_1)}$$

then X is a beta variable with $r = v_1/2$ and $(N - r) = v_2/2$. Thus, all of the theoretical continuous variables we have studied in this chapter are related.

Any intermediate text in mathematical statistics will give more details on the gamma and beta distributions, and tables are to be found for each type of distribution in the *Biometrika Tables for Statisticians* (1966).

EXERCISES

1. A normal distribution has a certain mean μ and a certain variance σ^2. What is the probability that a value drawn from this distribution represents a z value of 1.15 or more? What is the probability that $z = -1.15$ or less? What is the probability that *either* $z \ge 1.15$ *or* $z \le -1.15$? What then is the probability that $z^2 \ge (1.15)^2$? Show by use of Table IV in Appendix E that this is also the probability that a value of $\chi^2 \ge 1.3230$, when $v = 1$. Why should this be true?

2. Using the same reasoning as in Exercise 1, find the probability $p(\chi^2 \geq h)$ where h represents in turn each of the values given below, in a chi-square distribution with one degree of freedom.

(a) $h = 1.00$
(b) $h = 2.00$
(c) $h = 3.00$
(d) $h = 0.50$
(e) $h = 0.33$

3. A normal distribution has mean $\mu = 100$ and variance $\sigma^2 = 10$. Five values are taken independently and at random from this distribution. The z value is computed for each value drawn, and then the sum of the squared z values found. What should one expect this sum to be? What is the approximate probability that this sum is greater than 15? Greater than 11? Less than 1.6?

4. A sample of seven observations drawn independently and at random from a normal distribution gave the following results:

22,   2,   0,   30,   28,   26,   32

Test the hypothesis that the population standard deviation is 10 against the alternative that it has some other value. Use $\alpha = .05$. Then test the same hypothesis against the alternative $\sigma < 10$.

5. For the data in Exercise 4 find the 95% confidence limits for the population standard deviation.

6. A study was concerned with the effects of anxiety on test performance. It was theorized that very anxious people might not only show a lower average performance relative to their nonanxious peers, but also a higher degree of variability in performance. A random sample of 22 college students showing normal general ability but high test-anxiety was taken. Each student was given a simple mathematics test. In general, the scores of college students on this test follow an approximately normal distribution with a mean of 55 and a standard deviation of 6. The sample of anxious students had a mean of 51 and a standard deviation $s = 10$. Test both of the hypotheses implied here, using $\alpha = .05$ for each.

7. For Exercise 6 establish both a 95% confidence interval for the mean of the population of anxious students, and a 95% confidence interval for their variance.

8. Suppose that a sample of 20 independent observations is drawn at random from a normal population with standard deviation of 30. What is the probability that the sample value S will turn out to be less than or equal to 25?

9. A certain achievement test is standardized in such a way that a score value of 80 lies one standard deviation below the mean score value of 100 in the normally distributed population. A sample of 30 scores is drawn at random and independently from this population. What is the approximate probability that the sample standard deviation S will be at least 1.3 times larger than the population standard deviation?

10. In Exercise 9, what is the expected value of the ratio of the sample variance S^2 to the population variance σ^2? What is the *most likely* value of that ratio? (**Hint:** Look carefully at Section 9.1, following Eq. 9.1.4.)

11. Six observations are drawn independently and at random from a normal population. The values obtained are

106,   ·98,   97,   103,   101,   105

Find the 95% confidence interval for the population variance.

12. Assume that the sample given in Exercise 32, Chapter 5 represents a normally distributed population. Then test the hypothesis that the variance of this population is 20, against the alternative that the variance is greater than 20, using $\alpha = .05$.

13. From Exercise 14, Chapter 6, test the hypothesis that the population standard deviation is .50, as opposed to the hypothesis that $\sigma \neq .50$. Assume that the population in question is normal. (**Hint:**See Section 9.6.)

14. Find the 95% confidence interval for σ^2 in Exercise 13.

15. In a study of the size of the vocabularies of six-year-old normal children, the experimenter entertained the notion that size of vocabulary should increase from ages five to six, but that the variability for these two age groups should be about the same in this respect. Previous studies had shown that the standard deviation for five-year-olds was about 24 words, and that size of vocabulary is approximately normally distributed. A sample of 25 six-year-old children showed that $S^2 = 945$. Does the standard deviation of the population of six-year-olds seem to be equal to about 24?

16. An experimenter drew random samples from two normal distributions, with different means but with the same variance. The first sample contained 18 independent observations, and the second contained 11. If the two sample variances turned out to be $S_1^2 = 92$ for the first sample, and $S_2^2 = 86$ for the second, what are the 99% confidence limits for the variance of either population?

17. Suppose that the experimenter in Exercise 15 had taken both a sample of 25 six-year-old children, and a similar random sample of the same number of nine-year-olds. The six-year-olds again showed $S^2 = 945$, whereas the nine-year-old children had $S^2 = 919$. Is it reasonable to conclude that six-year-olds are more variable than nine-year-olds? Use $\alpha = .05$.

18. The test carried out in Exercise 17, Chapter 8 involves the assumption that the two populations have exactly the same variance. Test this assumption using $\alpha = .01$.

19. Using the two samples in Exercise 21, Chapter 8, test the hypothesis that the two populations have the same variance, against the alternative that the variance for Population I is greater. Again use $\alpha = .05$.

20. Use the data in Exercise 20, Chapter 8 to test the hypothesis that the two brands of gasoline have the same variance in terms of mileage, against the hypothesis that Brand I is the more variable. The .05 level for α may be used.

21. Two normally distributed populations are being compared to see if they have the same values for their variances. The alternative hypothesis is that Population I has the larger variance. A sample of 13 independent observations are drawn from Population I and a separate random sample of 9 independent observations are made of Population II. Sample I shows that $S_I^2 = 141$, and Sample II shows $S_{II}^2 = 123$. What can one conclude about the variances of the two populations?

22. Suppose that independent random samples each consisting of 16 cases were drawn at random from two normal populations. The first sample produced a sample standard deviation $S = 38.2$ and the second sample a standard deviation $S = 89.2$. The null hypothesis was H_0: $\sigma_I^2 = \sigma_{II}^2$ and the alternative was "not H_0." Test the null hypothesis.

23. Is it possible to test a hypothesis of the form H_0: $\sigma_I^2 = k\sigma_{II}^2$? Explain why it should or should not be possible, and, if possible, how one would go about doing so.

24. An airport has an information counter at which people tend to arrive and ask questions at the rate of four per minute. The process seems to be stable throughout the day, and the arrivals and departures at the counter may be assumed to be independent. A random sample of 100 different times and days is taken. On each occasion the time is recorded between the departure of a person at the counter and the arrival of the next person with a question. The mean of this sample of 100 occasions is taken. What is the expected value and the standard error of such a mean? (**Hint:** Use the exponential distribution.)

25. Suppose that in Exercise 24 the time is recorded until the next 10 people have asked for information. Once again the mean of 100 occasions is taken. What is the expected value and the standard error of such means? (**Hint:** Use the gamma distribution.)

Chapter *10*

The General Linear Model and the Analysis of Variance

In Chapter 8 and elsewhere we discussed the idea of a statistical relation between some independent variable X and a dependent variable Y. The variable X may stand for some set of qualitatively different groupings, or treatments, under the control of the experimenter. The variable Y, however, is not under the direct control of the experimenter, although its value may depend in some way on the X value.

In other contexts, the independent variable X may be quantitative, representing groups or treatments that differ in amount or degree of the thing that the values of X symbolize. Such an independent variable is still thought of as under the direct control of the experimenter, whereas the Y value is thought of as free to vary in some way as the X values are varied.

Furthermore, our discussion so far has been confined to the situation where there are only two independent-variable groups, whereas most studies employ a number of qualitatively different groups, or a number of different values of the variable X. Thus, we need ways to extend the ideas underlying the comparison of two groups to the situation where any number of different groups may be compared on the basis of their Y, or dependent-variable, values.

To do this, we must place the notion of a statistical relation between X and Y on a more concrete basis. This calls for a *mathematical model* of how the variable X influences or is associated with the variable Y. This model should be simple, but it should also be general enough to cover both the situations of qualitative and of quantitative groupings in terms of the independent variable. It should also allow expansion to many independent variables and to variables of mixed types. The model must also distinguish between systematic connections between X and Y, and chance or "error" variations in Y.

10.1 LINEAR MODELS

There are many different kinds of mathematical formulations, or models, that one could invent to represent the relationship between any set of independent variables X and any dependent variable Y. However, the simplest and most flexible of such models is known as a *linear model*. In essence, any linear model of data states that an observed value of the dependent variable is *equal to a weighted sum of values* associated with one or more independent variables, *plus a term standing for error*. Although for the moment we will consider only one of the most elementary versions of this linear model, the concept is broad enough to cover much more complicated situations, as we shall see.

First we will formulate a simple linear model as it applies to three experimental groups. There is nothing special about the number 3 here; we could do the same thing for 2 or for any number of groups. The use of three groups simply lets us show some features most clearly.

Suppose that there are three *qualitatively different* groups or experimental treatments. In all, these groups contain some N subjects or observations. Any subject i will be in exactly one of the groups, and the actual membership of a subject in a group will be shown by the values of one or more variables X. Each subject i will also exhibit some value of the dependent variable, or y_i. Then a linear model showing how y_i depends on the X variable would be.

$$y_i = a_0 x_0 + a_1 x_{i1} + a_2 x_{i2} + a_3 x_{i3} + e_i. \qquad [10.1.1]$$

Here, x_0 is a constant value that enters into the y_i score for each subject. The symbol x_{i1} stands for the value of X as it applies to observation i depending on whether or not that observation is in Group 1. The value of x_{i2} pertains to whether or not observation i is in Group 2, and similarly for x_{i3}. The symbols a_0, a_1, a_2, and a_3 stand for weights that the x values receive in making up the y_i score value. Finally, the term e_i represents the error component of the observed y_i score.

Now you may well ask how we talk about the values of x_{i1}, x_{i2}, and x_{i3} when it was clearly stated that the three treatments or groups are qualitatively different. Here, the values of x are defined in such a way as to indicate the membership of an individual subject or observation in a qualitatively distinct group, as follows:

$x_0 \quad = 1$, for all i

$x_{i1} \begin{cases} = 1, \text{ if } i \text{ is in Group 1} \\ = 0, \text{ otherwise} \end{cases}$

$x_{i2} \begin{cases} = 1, \text{ if } i \text{ is in Group 2} \\ = 0, \text{ otherwise} \end{cases}$

$x_{i3} \begin{cases} = 1, \text{ if } i \text{ is in Group 3} \\ = 0, \text{ otherwise.} \end{cases}$

This device, known as the use of ''indicator'' or ''dummy variables,'' lets us sort out the subjects into the qualitative groupings, according to the x values assigned to particular individuals i.

Then, according to this linear model, the y value of any individual i in Group 1 is found from

$$y_i = a_0 + a_1 + e_i.$$

For a subject in Group 2, the linear model states that

$$y_i = a_0 + a_2 + e_i,$$

and for Group 3,

$$y_i = a_0 + a_3 + e_i.$$

The linear model thus lets us state in detail the makeup of any y_i value, in terms of the overall level of the Y values, the ''effect'' of the treatment group, or a_j, and error, e_i.

In another study, the three groups may be *quantitatively* different with respect to the independent variable. Thus, Group 1 might receive 3 h of practice on some task, Group 2, 6 h, and Group 3, 9 h. Then the linear model for an individual in Group 1 would be

$$y_i = a_0 + a_1(3) + e_i.$$

For Group 2 the model would be

$$y_i = a_0 + a_2(6) + e_i$$

and for Group 3,

$$y_i = a_0 + a_3(9) + e_i.$$

In other words, now the X values are no longer merely indicators, as before, but numerical values standing for amount or degree of treatment applied.

In the next section we will examine a broad version of such a model, known as the general linear model, and point out some of the variations that will prove useful in the remainder of this text.

10.2 A GENERAL LINEAR MODEL

In an experiment or other research study, there are many different things that are going to influence the outcome that we observe for any individual subject. Thus, suppose that we wished to investigate the reading achievement score of fifth-grade boys (variable Y). Then an immediate decision has to be made that some things are going to be held constant. We have already decided to study human children, and not animals; we have decided to study children in the fifth grade, not children of all ages; we have decided to study boys, not boys and girls. Furthermore, when we decide to use the elementary school down the street, we may tacitly have limited the observations to one socio-economic and ethnic mix, and a particular staff of fifth-grade teachers. When we decide to give all the tests next Tuesday, we have also imposed certain constant constraints. In short, when a group of observations are made, many things must be limited or held constant, all of which will go into determining a general level that the obtained values of Y will exhibit. Our model needs to reflect this, and so we will enter a constant value *to baseline*

x_0 to stand for this general level we have established by defining our sample as we have. This value receives some weight a_0 in any observation.

However, a great many things are still going to make fifth-grade boys differ in their reading achievement. Some of these things may be under our control, as when we can ask the different teachers in the school to use different reading texts of our choice. Still other things that may create differences in reading achievement are things we know about and can measure, although they may not be under our direct control. Proper nutrition, or its lack, conceivably affects reading achievement, as does general ability, and so on for a host of other things. All of these known factors or variables in the study, that can operate to make the Y values different, may be thought of as the study's *systematic* features. Each of these features can be symbolized as a variable, such as X_j. Perhaps we can identify a set of some J of the variables, $(X_1, \cdots, X_J)$ which we can either control or observe systematically for any individual.

Finally there are the things that make boys different from each other in reading achievement, but which we can neither control nor measure nor perhaps even adequately explain. Such factors, peculiar to a particular individual in a particular time and place, we identify as "error" in our study. This too must have a place in our underlying model.

Without going further toward specifying the exact nature of the constant, and the systematic and error components that underlie the Y values we observe for particular individuals, we can represent these things in a model:

$$y_i = a_0x_0 + a_1x_{i1} + a_2x_{i2} + \cdots + a_Jx_{iJ} + e_i \qquad [10.2.1^*]$$

Here, x_0 stands for the general level of the group of observations, and a_0 is the weight that the general level carries in determining the score of individual i. Then x_{i1} symbolizes the value of some variable X_1 as it relates to individual i, and a_1 represents the weight or "effect" of that variable in determining y_i. Similarly, for all the other variables that we can identify and observe systematically in the experiment, each variable X_j has some influence x_{ij} on the value y_i. The weight of that influence is given by a_j. Finally, there is the component of the y_i value that entails all of the factors that we either do not or cannot know about. This is error, or e_i. In short, we can envision the actual y_i score for some individual i as being a weighted sum of values of a number of variables X_j, plus a term standing for error.

This is a *general linear model*. It is general in the sense that we really have not specified the precise nature of the X variables, exactly how many there are, nor the weights which they receive. Rather, these specifics of the model may be tailored to fit a given experimental or research situation. Such a conception is enormously flexible in lending itself to all sorts of data-collection situations. By adapting and interpreting the linear model in various ways, we are able to use it as a basis for describing situations which, on their surface, appear quite disparate, but which nevertheless fit under this general conception of how an observed score or value is composed of the weighted contributions of many factors.

Most of the techniques to be discussed in this and subsequent chapters presuppose that such a linear model underlies the data. However, there are many variations that can be applied to this general linear model to fit the particular kind of situation or experiment that is being modeled. As suggested in the previous section, two variations have to do with whether the x values in the model are simply "indicators" or "markers" taking on the values 0 to 1 and showing qualitatively different treatments or groups, or whether any x stands for a numerical value (not necessarily 0 or 1) representing an amount of the experimental treatment in question.

When the x values in a linear model are indicators, taking on only the values of 0 and 1, the model is sometimes said to be an *experimental design model,* or an *analysis of variance model.* We will use the latter term, since the statistical techniques known collectively as *the analysis of variance* will provide our chief use of such models.

More generally, when the x variables in the linear model can take on any real number values, this is called a *multiple regression model.* As we shall see in Chapter 16, analysis of variance models such as we will study in this chapter are actually special cases of the multiple regression model. Various kinds of *problems in correlation* and *problems in regression,* as they will be outlined in Chapters 14 to 17, are also special instances of this overarching model.

In every single instance in which we will be using a linear model, or special case of the general linear model, we will have to identify the values or weights represented by a_0 through a_J. In other words, one of the key problems will be deciding on how much weight each variable actually does have in determining variations among the Y values we observe. To find the values of these weights any of several criteria might, in principle, be employed. However, we will always rely on the *criterion of least squares,* a simple version of which we first encountered in Chapter 4. Although it will not be possible for us to illustrate this principle each time it occurs or to show how it applies in complex situations, you should remember that this is the main avenue for turning the formal statement of the relation between Y and the various X variables into a numerical form that we can actually observe and use. (A detailed and moderately simple treatment of linear models and their variations may be found in Ward and Jennings (1973). This would make good material for study following Chapter 17 of this book for the student seriously interested in getting an overview of methods for the analysis of experiments.)

Incidentally, the topic of the general linear model is the first we have encountered in which it is necessary to think of a number of variables simultaneously. That is, in the general linear model each individual i is conceived as having scores or values on each of a number of X variables. These are combined as a weighted sum to give the value on still another variable Y. Dealing with many variables at once can become quite complicated and awkward when ordinary algebra is used. For this reason, modern statistics often employs the branch of mathematics known as *matrix algebra* in dealing with situations involving many variables. This matrix language permits economical and elegant descriptions and solutions to problems which can be very hard to handle in more conventional mathematical notation.

Obviously, a text such as this must keep its purely mathematical demands at a reasonably low level, and it did not seem appropriate to introduce matrix algebra as an actual part of the text, even though doing so might have helped to clarify the remaining topics we will be discussing, at least for some students. Therefore, we will not be using martix notation and matrix algebra as an explicit part of the discussions to follow. However, from time to time terms and concepts from the language of matrix algebra will be required as we go into more complicated areas. In addition, a few of you may wish to see what some of our results look like in the much "neater" terminology that matrix theory provides. For that reason, a part of Appendix C has been devoted to some basic ideas from vector theory, and Appendix D has been included to give some of the rudiments of matrix notation and matrix algebra. Both Appendix C and Appendix D give some special applications of vector and matrix principles that we will find of use in some later sections, as well as matrix versions of some of the more important principles that will appear in later chapters. Sections C.4 and D.1 can be useful preparation for the material to follow.

10.3 ANALYSIS OF VARIANCE MODELS

In the remainder of this chapter, we will be considering a controlled data-gathering situation, or experiment, of the following type. Some N subjects or cases are available to participate in the experiment, and these will be allocated at random to J experimental groups. Each group will be given a different experimental manipulation or "treatment" (including, of course, the possibility of "no treatment" as applied to a control group.) A useful term for indicating the set of experimental treatments employed is *the experimental factor*. The particular treatments or manipulations are then called the *levels* of that factor. Thus, an experiment on the influence of diet on blood pressure among adults might use four different, randomly formed groups, each receiving a different diet. "Diet" would be the experimental factor, with four levels in this particular experiment. The term *randomized groups* is used to describe such an experimental layout. For the time being, we are going to confine our attention to analysis of variance models, or experimental design models, as special cases of the general linear model. Recall that for an experimental design model, the x_{ij} values entering into the model of Eq. 10.2.1 are indicator variables, taking on only the value 1 if individual i is in group j, and the value 0 otherwise. Then the pattern of 1 and 0 values as they apply to the set of observations is called *the structural matrix* (or sometimes the *design matrix*) of the experiment being described.

For example, this time imagine a group of eight subjects, assigned by twos to four groups, each representing a qualitatively different treatment. Then the structure of the experiment would be shown as in Table 10.3.1, where the rows stand for the subjects, and the columns contain the x_{ij} values.

Table 10.3.1

The structure of an experiment involving eight subjects in four groups

	Subjects	x_0	x_{i1}	x_{i2}	x_{i3}	x_{i4}
	1	1	1	0	0	0
	2	1	1	0	0	0
groups	3	1	0	1	0	0
	4	1	0	1	0	0
	5	1	0	0	1	0
	6	1	0	0	1	0
	7	1	0	0	0	1
	8	1	0	0	0	1

The display of the x_{ij} values for the experiment, such as that shown in Table 10.3.1, is the structural, or design, matrix. Any designed experiment can be represented by such a matrix of 1 and 0 values.

Now applying the general linear model to this specific situation, we have, for $J = 4$ groups,

$$y_i = a_0 x_0 + a_1 x_{i1} + a_2 x_{i2} + a_3 x_{i3} + a_4 x_{i4} + e_i. \qquad [10.3.1]$$

Thus we find that the eight observation values y_i should be made up as follows:

$$y_1 = a_0 + a_1 \qquad\qquad + e_1$$

$$y_2 = a_0 + a_1 \qquad\qquad + e_2$$

$$y_3 = a_0 \quad + a_2 \qquad\qquad + e_3$$

$$y_4 = a_0 \quad + a_2 \qquad\qquad + e_4$$

$$y_5 = a_0 \qquad\quad + a_3 \qquad + e_5$$

$$y_6 = a_0 \qquad\quad + a_3 \qquad + e_6$$

$$y_7 = a_0 \qquad\qquad\quad + a_4 + e_7$$

$$y_8 = a_0 \qquad\qquad\quad + a_4 + e_8$$

All observations share the constant a_0. Only Observations 1 and 2 share the "effect" a_1, only Observations 3 and 4 the effect a_2, and so forth. An error term e_i is then associated with any subject or observation i.

Naturally, we could have constructed such a model for any number of subjects arranged into any number of groups. Regardless of how many treatment groups, however, the score of each subject is assumed to consist of a constant part common to all subjects, an effect of the group to which the subject belongs, and an error term unique to that subject in that group. (The very simple matrix expression for such a linear model is given in Appendix D, Section 6.19).

10.4 LEAST SQUARES AND THE IDEA OF AN EFFECT

In the linear model describing the relationship of the variables X to variable Y, the various constant values a_0, a_1, a_2, and so on were described as weights or effects. The term *effect* is used especially when an analysis of variance model is being discussed. What are these effects, and how are they estimated?

Obviously, the a_j weights in such a linear model could be any real values provided that they represent some reasonable theory of how the X variables relate to Y. However, we want to employ a principle that will provide the "best weights" or "best-fitting weights" to a given data situation. That is, if we wish to use our linear model to describe the Y values that occur on the basis of the X values that entered into the experiment, or to predict the Y value for an individual on the basis of information about the X values that individual shows, we want weights a_j that will provide the best such description or prediction. As mentioned above, the most usual way of achieving a good fit between data and such unknown constants in statistics is the so-called principle of least squares (or equivalently, the principle of minimum variance). This principle has already been used in Section 4.15, where it was shown that taking deviations about the sample mean gives the lowest value of the average squared deviation, or variance. We will now invoke this principle to find the best-fitting a_j values for our analysis of variance model.

Since e_i is a part of the linear model underlying any y_i observation, then it must be true that for any i:

$$e_i = y_i - a_0 x_0 - a_1 x_{i1} - \cdots - a_J x_{iJ} . \qquad [10.4.1]$$

Then we choose values for $a_0, a_1, \cdots , a_J$ so as to minimize the variance of these e_i values. That is, we minimize

$$\sum_i^N \frac{e_i^2}{N} - \bar{e}^2 \qquad \text{error mean} \qquad [10.4.2]$$

where $\bar{e}$ symbolizes the mean error over all N cases.

We will also place two more restrictions or "side conditions" on our model: on the average, across all observations, we would like the value of e to be zero. That is, we set $\bar{e} = 0$. We also require the average of the a_j values to be zero over the J groups. (The first of these restrictions makes the problem of minimizing the variance of the errors equivalent to minimizing the sum of their squares.) Then when minimization methods from the differential calculus are applied, it turns out that the sum of the squared errors is a minimum when

$$a_0 = \bar{y}$$

$$a_1 = \bar{y}_1 - \bar{y}$$

$$a_2 = \bar{y}_2 - \bar{y} \qquad\qquad [10.4.3]$$

$$\cdots$$

$$a_J = \bar{y}_J - \bar{y}$$

In other words, in an analysis of variance model, where N observations are arranged among J different groups, if the sum of the squared errors is to be minimized, the value of a_0 must be the mean of the Y values. Furthermore, any effect a_j must be the difference between the mean of the Y values in group j and the mean of all of the Y values.

Now we have a model for describing a set of data, which must fit the data in the least squares sense:

$$y_{ij} = \bar{y} + (\bar{y}_j - \bar{y}) + e_{ij} . \qquad [10.4.4^*]$$

Equivalently, we may write

$$y_{ij} = a_0 + a_j + e_{ij} . \qquad [10.4.5^*]$$

Note that Eq. 10.4.4 implies that the best predication one can make about the score of any individual i in group j is $\bar{y}_j$, the mean score for that group:

$$y_{ij} = \bar{y} + (\bar{y}_j - \bar{y}) + e_{ij} = \bar{y}_j + e_{ij} .$$

10.5 POPULATION EFFECTS

We have just seen the results of applying the least squares criterion to a linear analysis of variance model, for a particular set of data containing N cases arranged among J experimental or treatment groups. However, we are usually not as interested in describing a particular set of data as we are in describing large populations of potential observations in the different experimental treatments. Fortunately, the model and the least squares argument apply just as well as to vast populations of potential observations as they do to a small set of N cases.

Suppose that we now think of a large population of potential observations within a particular experimental treatment j. Each observation i in treatment population j will have a dependent variable score y_{ij}. For the population, the model equivalent to Eq. 10.4.5 is, for each i and j,

$$y_{ij} = \alpha_0 + \alpha_j + e_{ij} \qquad [10.5.1^*]$$

(Here, j subscripts have been added to both y and e to emphasize that we are speaking about individual i in population j; this will help us keep up with individuals and populations later on. Furthermore, the a values in Eqs. 10.3.1 and 10.4.5 have been replaced with Greek letters alpha to emphasize that this is a model of the population.)

Then by the criterion of least squares it must be true that

$$\alpha_0 = \mu$$

and

$$\alpha_j = \mu_j - \mu.$$

The constant α_j is called the *true* or *population effect* of treatment j. The effect α_j is simply the difference between the mean of the population j, or μ_j, and the general mean over all treatment populations, μ.

To get an idea of what these effects actually represent, consider the grand population, formed by pooling the J treatment populations. Now let the proportionate representation of any population j in the grand population be $p(x_j)$, so that the probability that any observation drawn at random from the grand population comes from population j is $p(x_j)$. If the J treatment populations are equally represented in the grand population, then for every j the value of $p(x_j) = 1/J$.

Now under these conditions, the mean μ of the grand population would be defined as the weighted average of the several population means:

$$\mu = \sum_j p(x_j)\mu_j.$$

In the special case where $p(x_j) = 1/J$ for all j, the grand mean would be

$$\mu = \frac{\sum_j \mu_j}{J}.$$

Then, as we have seen, the population effect of treatment j is defined as the deviation of the mean of population j, μ_j, from the grand population mean, μ:

population effect of treatment $j = (\mu_j - \mu)$ [10.5.2*]

or α_j $= (\mu_j - \mu).$

This symbol α_j (*not* to be confused with the alpha standing for the probability of Type I error) will stand for the effect of any single treatment j.

Since the grand population mean μ is also the weighted sum of all of the treatment population means, it follows that the weighted sum of all of the effects must be zero:

$$\sum_j p(x_j)\alpha_j = \sum_j p(x_j)(\mu_j - \mu) = \mu - \mu = 0.$$ [10.5.3*]

In the special case where $p(x_j) = 1/J$ for all treatment populations j, then the sum of the effects is zero:

$$\frac{\sum_j \alpha_j}{J} = 0 \text{ implies that } \sum_j \alpha_j = 0.$$ [10.5.4*]

Now suppose that there is absolutely no effect associated with any treatment. This means that

$$\alpha_j = 0$$

for each and every treatment population j. This is equivalent to the statement that

$$\mu_1 = \mu_2 = \cdot\cdot\cdot = \mu_J = \mu,$$

where the index numbers $1, 2, \cdot\cdot\cdot, J$ designate the various treatments. The *complete absence of effects is equivalent to the absolute equality of all of the population means.*

Notice also that when there are no treatment effects,

$$\sum_j p(x_j)\alpha_j^2 = 0;$$ [10.5.5*]

that is, the weighted sum of the α_j^2 must be zero, since each and every α_j is zero when no treatment effects exist.

The expected score value over all individuals i in population j is

$$E(y_{ij}) = \mu + \alpha_j + E(e_{ij}),$$

or

$$\mu_j = \mu_j + E(e_{ij}).$$

Thus, for any population j, the expectation of e_{ij} over all individuals is zero:

$$E(e_{ij}) = 0.$$ [10.5.6*]

10.6. MODEL I: FIXED EFFECTS

It is important to distinguish between two different sampling situations to which the analysis of variance applies. These differ both in the way that experimental treatments are selected and in the kinds of inferences one makes from the analysis. The formal statistical models applying to these two sampling situations have become known as Model I or the *fixed effects model,* and Model II, the *random effects model* (or, sometimes, *components of variance model*). The situations calling for these two different models will be compared briefly.

Imagine a situation where several experimental treatments are to be administered. Suppose that there are *J* different such treatments and that each treatment is to be administered to one and only one experimental group. Each of the *J* groups consists of individuals chosen at random and independently and assigned at random to the groups. For example, four different tranquilizing drugs are to be compared for the effect each has on driving skill. Some *N* subjects are chosen at random and independently and allotted at random to four, nonoverlapping groups. The individuals in the first group are given Drug 1, those in the second Drug 2, and so on. Then the groups are to be compared on the dependent variable *Y*, a score on driving skill. Or, perhaps, six different models of teaching second-grade arithmetic are known. School children from a specific population are sampled and allotted at random to six different groups, each group representing one of the six instruction methods. The groups are then compared on their average achievement after a year's instruction. In both examples, the elements of some small set of treatments are to be compared, and each treatment of interest is actually used in the experiment.

Experiments to which Model I applies are distinguished by the fact that inferences are to be made only about differences among the *J* treatments actually administered, and about no other treatments that might have been included. In advance of the actual experiment, the experimenter decides to see if differences in effect exist among some fairly small set of treatments or treatment combinations. There is interest in these treatments or combinations, and in *no* others. Each treatment of immediate interest to the experimenter is actually included in the experiment, and the set of treatments or treatment combinations applied exhausts the set of treatments about which the experimenter wants to make inferences. The effect of any treatment is "fixed," in the sense that it must appear in any complete repetition of the experiment on new subjects.

As an example to which the fixed effects model does not apply, consider an experiment involving a projective test. This test consists of 10 different stimulus cards given in turn to the subject, who gives verbal responses to each. Among the things recorded about the behavior of a subject is the mean time between the presentation of a card and the first response. The experiment is designed to study the effect of the order of presentation of the cards on average first response time. However, there are 10! or 3,628,800 different possible orders of presentation of the cards. The experimenter takes a random sample of, say, 20 such orders, and tests a different group of randomly selected subjects under each. Here, the experimenter is not really interested in the 20 orders actually administered so much as in the possible effects of order *in general*. Any complete repetition of the experiment would involve a freshly selected set of orders. Thus, the fixed effects model is inappropriate for this problem.

The methods to be discussed in this chapter, as well as those in Chapter 11 and in Chapter 12, will assume the fixed-effects model. Other models will apply in Chapter 13, however.

10.7 DATA GENERATED BY A FIXED-EFFECTS MODEL

To gain some feel for this linear model, imagine three samples consisting of three observations each. Suppose that these three samples represent identical population distributions, and that there is *no* variability (that is, no error) within any of the populations. If the mean of each of the populations were $\mu = 40$, then our sample results should look like this:

Sample 1	Sample 2	Sample 3
40	40	40
40	40	40
40	40	40

There should be no differences either between or within samples, if this were the true situation. When this is true, the linear model of Eq. 10.5.1 becomes simply

$$y_{ij} = \mu,$$

since $\alpha_j = 0$, $e_{ij} = 0$ for all i and j.

Now suppose that the three samples are given different treatments, and that treatment effects exist, but that there is once again no variability within a treatment population (again, no error). Our results might look like this:

Sample 1	Sample 2	Sample 3	
40 − 2 = 38	40 + 6 = 46	40 − 4 = 36	
40 − 2 = 38	40 + 6 = 46	40 − 4 = 36	
40 − 2 = 38	40 + 6 = 46	40 − 4 = 36	
$\bar{y}_1 = 38$	$\bar{y}_2 = 46$	$\bar{y}_3 = 36$	$\bar{y} = 40$

Here there are differences between observations in different treatments, but there are no differences within a treatment sample. The linear model here is

$$y_{ij} = \mu + \alpha_j,$$

since $\alpha_j \neq 0$ while $e_{ij} = 0$, for any i and j.

In actuality there is always variability in a population, so that there is sampling error. The actual data we might obtain would undoubtedly look something like this:

Sample 1	Sample 2	Sample 3	
40 − 2 + 5 = 43	40 + 6 − 5 = 41	40 − 4 + 3 = 39	
40 − 2 + 2 = 40	40 + 6 + 1 = 47	40 − 4 − 2 = 34	
40 − 2 − 3 = 35	40 + 6 + 8 = 54	40 − 4 + 1 = 37	
$\bar{y}_1 = 39.3$	$\bar{y}_2 = 47.3$	$\bar{y}_3 = 36.7$	$\bar{y} = 41.1$

Here, a random error component has been added to the value of μ and the value of α_j in the formation of each score. The linear model in this situation is

$$y_{ij} = \mu + \alpha_j + e_{ij}.$$

Notice that not only do differences exist between observations in different treatments, but also between observations in the same treatment.

If we estimate the effect of Treatment 1 by taking

est. $\alpha_1 = \bar{y}_1 - \bar{y} = 39.3 - 41.1 = -1.8$

[handwritten margin notes:]
$\alpha_2 = 47.3 - 41.1 = 6.2$
$\alpha_3 = 36.7 - 41.1 = -4.4$

it happens in this example that we are almost right since the data were fabricated so that $\alpha_1 = -2$. Likewise, our estimate of α_2 is in error by .2 and our estimate of α_3 in error by $-.4$. Although these errors may seen rather slight in this example, there is no guarantee in any given experiment that they will not be very large. Thus we need to evaluate how much of the apparent effect of any experimental treatment is, in fact, due to error before we can decide that something systematic is actually occurring.

This example should suggest that evidence for experimental effects has something to do with the differences *between* the different groups relative to the differences that exist *within* each group. Next, we will turn to the problem of separating the variability among observations into two parts: the part that should reflect both experimental effects and sampling error, and the part that should reflect sampling error alone.

10.8 THE PARTITION OF THE SUM OF SQUARES FOR ANY SET OF *J* DISTINCT SAMPLES

In this section we are going to leave the study of population effects for a while and show how the variability in any set of *J* experimentally different samples may be partitioned into two distinct parts. Actually, we will do this in terms of the sum of squared deviations about the grand mean for the samples, rather than in terms of the sample variance itself. Let n_j = number of cases in sample or group *j*. Then any score y_{ij} in sample *j* exhibits some deviation from the grand sample mean of all scores, $\bar{y}$:

deviation $= (y_{ij} - \bar{y})$

This deviation can be thought of as composed of two parts,

$$(y_{ij} - \bar{y}) = (y_{ij} - \bar{y}_j) + (\bar{y}_j - \bar{y}) \qquad [10.8.1]$$

[handwritten annotations under equation: Total ; WG ; BG]

the first part being the deviation of y_{ij} from the mean of group *j*, and the second being the deviation of the group mean from the grand mean.

Now suppose that we square the deviation from $\bar{y}$ for each score in the entire sample, and sum these squared deviations across all individuals *i* in all sample groups *j*:

$$\sum_j \sum_i (y_{ij} - \bar{y})^2 = \sum_j \sum_i [(y_{ij} - \bar{y}_j) + (\bar{y}_j - \bar{y})]^2$$

$$= \sum_j \sum_i (y_{ij} - \bar{y}_j)^2 + \sum_j \sum_i (\bar{y}_j - \bar{y})^2 \qquad [10.8.2]$$

$$+ 2 \sum_j \sum_i (y_{ij} - \bar{y}_j)(\bar{y}_j - \bar{y}) .$$

Now look at the last term on the right in Eq. 10.8.2 above:

$$2 \sum_j \sum_i (y_{ij} - \bar{y}_j)(\bar{y}_j - \bar{y}) = 2 \sum_j (\bar{y}_j - \bar{y}) \sum_i (y_{ij} - \bar{y}_j)$$

$$= 0$$

[10.8.3]

since the value represented by the term $(\bar{y}_j - \bar{y})$ is the same for all i in group j, and the sum of $(y_{ij} - \bar{y}_j)$ must be zero when taken over all i in any group j.

Furthermore,

$$\sum_j \sum_i (\bar{y}_j - \bar{y})^2 = \sum_j n_j(\bar{y}_j - \bar{y})^2$$

[10.8.4]

since, once again, $(\bar{y}_j - \bar{y})$ is a constant for each individual i figuring in the sum. Putting these results together, we have

$$\sum_j \sum_i (y_{ij} - \bar{y})^2 = \sum_j \sum_i (y_{ij} - \bar{y}_j)^2 + \sum_j n_j(\bar{y}_j - \bar{y})^2.$$

[10.8.5*]

This identity is usually called the *partition of the sum of squares,* and is true for any set of J distinct samples. Verbally, this fact can be stated as follows: the total sum of squared deviations from the grand mean can always be separated into two parts: the sum of squared deviations within groups, and the weighted sum of squared deviations of group means from the grand mean. It is convenient to call these two parts

$$SS \text{ within } = \sum_j \sum_i (y_{ij} - \bar{y}_j)^2$$

[10.8.6*]

for **sum of squares within groups,** and

$$SS \text{ between } = \sum_j n_j(\bar{y}_j - \bar{y})^2$$

[10.8.7*]

for **sum of squares between groups.** Thus, it is a true statement that

$$SS \text{ total } = SS \text{ within } + SS \text{ between.}$$

[10.8.8*]

The meaning of this partition of the sum of squares into two parts can easily be put into common sense terms: Individual observations in any sample will differ from each other, or show variability. These obtained differences among individuals can be due to two things. Some pairs of individuals are in different treatment groups, and their differences are due either to the different treatments, or to chance variation, or to both. The sum of squares between groups reflects the contribution of different treatments, as well as chance, to intergroup differences. On other hand, individuals in the *same* treatment groups can differ only because of chance variation since each individual within the group received exactly the same treatment. The sum of squares within groups reflects these intragroup differences due only to chance variation. Thus, in any sample two kinds of variability can be isolated: the sum of squares between groups, reflecting variability due to treatments *and* chance, and the sum of squares within groups, reflecting chance variation *alone*.

10.9 ASSUMPTIONS UNDERLYING INFERENCES ABOUT TREATMENT EFFECTS

The partition of the sum of squares is possible for any set of J distinct samples, and no special assumptions about populations or sampling are necessary in its derivation. However, before we can use sample data to make inferences about the existence of population effects, several assumptions must be made. These are as follows:

1. For each treatment population j, the distribution of e_{ij} is assumed normal.
2. For each population j, the distribution of e_{ij} has a variance σ_e^2, which is assumed to be the same for each treatment population.
3. The errors associated with any pair of observations are assumed to be independent. A consequence of this assumption is that if h and i stand for any pair of observations, and j and k for any pair of treatments, then

$$E(e_{ij}e_{hj}) = 0$$

and

$$E(e_{ij}e_{hk}) = 0.$$

In short, we are going to regard our observations as independently drawn from normal treatment populations, each having the same variance and with error components independent across all pairs of observations.

10.10 THE MEAN SQUARE BETWEEN GROUPS

The next question is how to use the partition of the sum of squares in making inferences about the existence of true treatment effects. First of all, we will examine the expectation of the sum of squares between groups. Let us define

$$\bar{e}_j = \sum_i \frac{e_{ij}}{n_j}$$

and

$$\sum_j \sum_i \frac{e_{ij}}{N} = \sum_j \frac{n_j \bar{e}_j}{N} = \bar{e}.$$

Then for any group j, a simple substitution from Eq. 10.5.1 above shows that

$$\bar{y}_j = \sum_i \frac{y_{ij}}{n_j} = \mu + \alpha_j + \bar{e}_j \qquad [10.10.1]$$

and

$$\bar{y} = \sum_j \sum_i \frac{y_{ij}}{N} = \mu + \bar{e} \qquad [10.10.2]$$

since

$$\sum_j \frac{n_j \alpha_j}{N} = \sum_j p(x_j)\alpha_j = 0.$$

Furthermore, let $\bar{y}_j$ be the mean of the sample given the particular treatment j, and let $\bar{y}$ stand for the mean over all individuals in all treatments for the experiment. Then over all possible samples of n_j individuals from population j,

$$E(\bar{y}_j) = \mu + \alpha_j + E(\bar{e}_j).$$

That is, the expected value of $\bar{y}_j$ depends on μ, on the effect α_j, and on the expected mean error over the n_j observations in group j. Since

$$E(\bar{y}_j) = \mu_j$$

it must be true that

$$E(\bar{e}_j) = 0 .$$

Thus, the deviation of any sample group mean from the grand sample mean is actually

$$(\bar{y}_j - \bar{y}) = \alpha_j + (\bar{e}_j - \bar{e}) .$$

From this it follows that

$$\text{SS between} = \sum_j n_j(\bar{y}_j - \bar{y})^2 = \sum_j n_j[\alpha_j + (\bar{e}_j - \bar{e})]^2. \qquad [10.10.3]$$

On taking the expectation of the SS between, we find

$$E(\text{SS between}) = E \sum_j n_j[\alpha_j + (\bar{e}_j - \bar{e})]^2$$

$$= \sum_j n_j\alpha_j^2 + E \sum_j n_j(\bar{e}_j - \bar{e})^2, \qquad [10.10.4]$$

since, within any j, $E(\bar{e}_j) = 0$, and $E(\bar{e}) = 0$. Then, since

$$E(\bar{e}_j^2) = \frac{\sigma_e^2}{n_j}$$

and

$$E(\bar{e}^2) = \frac{\sigma_e^2}{N}$$

$$E(\text{SS between}) = \sum_j n_j\alpha_j^2 + (J - 1)\sigma_e^2 . \qquad [10.10.5^*]$$

Ordinarily, we deal with the *mean square between,*

$$\text{MS between} = \frac{\text{SS between}}{J - 1}. \qquad [10.10.6^*]$$

Then

$$E(\text{MS between}) = \sigma_e^2 + \frac{\sum_j n_j \alpha_j^2}{J - 1}. \qquad [10.10.7^*]$$

The mean square between groups is an unbiased estimate of σ_e^2, the error variance, plus a term that can be zero only when there are no treatment effects at all. When the hypothesis of no treatment effects is absolutely true, then

$$E(\text{MS between}) = \sigma_e^2. \qquad [10.10.8^*]$$

If any true treatment effects at all exist, then

$$E(\text{MS between}) > \sigma_e^2. \qquad [10.10.9^*]$$

Accordingly, we can see that the mean square between groups gives one piece of the evidence needed to adjudge the existence of treatment effects. The sample value of MS between should be an unbiased estimate of error variance alone when no treatment effects exist. On the other hand, the value of MS between must be an estimate of σ_e^2 plus a positive quantity when any treatment effects exist.

Each sum of squares in the analysis of variance is associated with some number of degrees of freedom. What is the number of degrees of freedom for MS between? There are really only J different sample values that go into the computation of MS between: these are the J values of $\bar{y}_j - \bar{y}$. Thus there are *$J-1$ degrees of freedom for MS between* since these values must average to 0.

Naturally, MS between is always a sample quantity, and thus it must have a sampling distribution. However, it is easy to see what this sampling distribution must be: when there are no treatments effects, MS between is an unbiased estimate of σ_e^2. For this estimate, as for other estimates of σ^2 for normal populations (Section 9.3),

$$\frac{(\text{est. } \sigma_e^2)}{\sigma_e^2} = \frac{\chi_{(v)}^2}{v}.$$

The ratio of MS between to σ_e^2 must be a chi-square variable divided by degrees of freedom, *when* there are no treatment effects *and* the parent populations are normal (assumption 1, Section 10.9).

As yet we have no idea of the value of σ_e^2, so that the sampling distribution of MS between cannot be used directly to provide a test of the hypothesis of no treatment effects. Now, however, let us investigate the sampling distribution of MS within.

10.11 THE MEAN SQUARE WITHIN GROUPS

What population value is estimated by the mean square within groups? Under the fixed-effects model it is obvious that the treatments administered cannot be responsible for differences that occur among observations within any given group. This kind of within-groups variation should be a reflection of random error alone. Keeping this in mind, let us find the expectation of the sum of squares within groups.

$$E(\text{SS within}) = E \left[\sum_j \sum_i (y_{ij} - \bar{y}_j) \right]^2$$

For any given sample j,

$$E\left[\sum_i \frac{(y_{ij} - \bar{y}_j)^2}{(n_j - 1)}\right] = \sigma_e^2 \qquad [10.11.1]$$

since for any sample j this value is an unbiased estimate of the population error variance, σ_e^2. Thus,

$$\begin{aligned}
E(\text{SS within}) &= \sum_j E \sum_i (y_{ij} - \bar{y}_j)^2 \\
&= \sum_j (n_j - 1)\sigma_e^2 \\
&= (N - J)\sigma_e^2.
\end{aligned} \qquad [10.11.2^*]$$

We define

$$\text{Mean square within} = \text{MS within} = \frac{\text{SS within}}{N - J}.$$

Then it must be true that

$$E(\text{MS within}) = \sigma_e^2, \qquad [10.11.3^*]$$

the expected value of the mean square within is simply *error variance alone*. **The degrees of freedom for MS within is**

$$\sum_j (n_j - 1) = N - J. \qquad [10.11.4^*]$$

Surely you can anticipate the turn the argument takes now! We have MS between, which estimates σ_e^2 when there are no treatment effects, but a value greater than σ_e^2 when effects do exist. Moreover, we have another estimate of σ_e^2 given by MS within, which does *not* depend on the presence or absence of effects: Two variance estimates which *ought* to be the same under the null hypothesis suggest the F distribution (Section 9.8), and this is what we use to test the hypothesis.

10.12 THE *F* TEST IN THE ANALYSIS OF VARIANCE

The usual hypothesis tested using the analysis of variance is

$$H_0: \mu_1 = \cdots = \mu_j = \cdots \mu_J,$$

the hypothesis that all treatment population means are equal. The alternative is just

$$H_1: \text{not } H_0,$$

implying that some of the population means are different from others. As we have seen, these two hypotheses are equivalent to the hypothesis of no effects and its contrary:

$$H_0: \alpha_j = 0, \text{ for all } j$$

$$H_1: \alpha_j \neq 0, \text{ for some } j.$$

The argument in Sections 10.8 through 10.11 has shown that *when H_0 is true,*

$$E(\text{MS between}) = \sigma_e^2$$

and

$$E(\text{MS within}) = \sigma_e^2,$$

both the mean square between and the mean square within are unbiased estimates of the same value, σ_e^2. On the other hand, *when the null hypothesis is false,* then

$$E(\text{MS within}) < E(\text{MS between}).$$

Since these mean squares divided by σ_e^2 are each distributed as chi-square variables divided by their respective degrees of freedom when H_0 is true, it follows that their ratio should be distributed as F, provided that MS between and MS within are *independent* estimates of σ_e^2. From the principle of Section 6.6 the following can be proved: **For J samples of independent observations, each drawn from a normal population distribution, MS between and MS within are statistically independent.** For each sample, the mean $\bar{y}_j$ is independent of the variance estimate s_j^2, provided that the population distribution is normal. By an extension of the principle given in Section 6.6, MS between, based on the J values $\bar{y}_j$ must be independent of MS within, based on the several s_j^2 values; each piece of information making up MS between is independent of the information making up MS within, given normal parent distributions.

Finally, we have all the justification needed to say that the ratio

$$\frac{(\text{MS between}/\sigma_e^2)}{(\text{MS within}/\sigma_e^2)} = \frac{\text{MS between}}{\text{MS within}} \qquad [10.12.1^*]$$

is distributed as F with $J - 1$ and $N - J$ degrees of freedom, *when the null hypothesis is true*. This statistic is the ratio of two independent chi-square variables, each divided by its degree of freedom, and thus is exactly distributed as F when H_0 is true.

The F ratio used in the analysis of variance always provides a *one-tailed* test of H_0 in terms of the sampling distribution of F. Evidence for H_1 must show up as an F ratio greater than 1.00, and an F ratio less than 1.00 can signify nothing except sampling error (or perhaps nonrandomness of the samples or failure of the assumptions). Therefore, for the analysis of variance, the F ratio obtained can be compared directly with the one-tailed values given in Table V of Appendix E. An α level is chosen in advance, and this value determines the section of the table one uses to determine the significance of the obtained F ratio.

10.13 COMPUTATIONAL FORMS FOR THE ONE-WAY ANALYSIS OF VARIANCE

Although the argument given above dealt with sums of squares defined as follows:

$$\text{SS total} = \sum_j \sum_i (y_{ij} - \bar{y})^2$$

$$\text{SS within} = \sum_j \sum_i (y_{ij} - \bar{y}_j)^2$$

$$\text{SS between} = \sum_j n_j(\bar{y}_j - \bar{y})^2$$

most users of the analysis of variance find it more convenient to work with equivalent, but computationally simpler, versions of these sample values. These computational forms will be given below.

First, however, in order to save ourselves writing many summation signs, we can adopt the following useful convention in this and other discussions of analysis of variance: Let T_y represent the total of *all* of the y values for the set of N cases making up the data. That is,

$$T_y = \sum_j \sum_i y_{ij} .$$

Furthermore, let T_j stand for the total of the y values for the n_j cases making up any particular group j. That is,

$$T_j = \sum_i y_{ij} .$$

Now the total sum of squares can be shown to be equal to

$$\text{SS total} = \sum_j \sum_i y_{ij}^2 - \frac{T_y^2}{N}. \qquad\qquad [10.13.1^*]$$

It is easy to show that this is true:

$$\text{SS total} = \sum_j \sum_i (y_{ij} - \bar{y})^2 = \sum_j \sum_i (y_{ij}^2 - 2y_{ij}\bar{y} + \bar{y}^2)$$

$$= \sum_j \sum_i y_{ij}^2 - 2\bar{y} \sum_j \sum_i y_{ij} + \sum_j \sum_i \bar{y}^2$$

by Rules 3 and 4 in Appendix A. This last equation reduces further to

$$\sum_j \sum_i y_{ij}^2 - 2\bar{y}(N\bar{y}) + N\bar{y}^2$$

or

$$\sum_j \sum_i y_{ij}^2 - N\bar{y}^2 .$$

Then, by the definition of the sample grand mean, $\bar{y} = T_y/N$, we have Eq. 10.13.1.

The calculating formula for SS between can be worked out in a similar way:

$$\text{SS between} = \sum_j n_j(\bar{y}_j - \bar{y})^2 = \sum_j n_j(\bar{y}_j^2 - 2\,\bar{y}_j\bar{y} + \bar{y}^2)$$

$$= \sum_j n_j\left(\frac{T_j}{n_j}\right)^2 - N\bar{y}^2.$$

or

$$\text{SS between} = \sum_j \frac{T_j^2}{n_j} - \frac{T_y^2}{N} \qquad\qquad [10.13.2]$$

Finally, the computing formula for the sum of squares within groups is found by

SS within = SS total − SS between

$$= \sum_j \sum_i y_{ij}^2 - \left(\frac{T_y^2}{N}\right) - \sum_j \left(\frac{T_j^2}{n_j}\right) + \left(\frac{T_y^2}{N}\right) \qquad [10.13.3^*]$$

$$= \sum_j \sum_i y_{ij}^2 - \sum_j \left(\frac{T_j^2}{n_j}\right).$$

Ordinarily, the simplest computational procedure is to calculate both the sum of squares total and the sum of squares between directly, and then to subtract SS between from SS total in order to find the SS within.

10.14 A COMPUTATIONAL OUTLINE FOR THE ONE-WAY ANALYSIS OF VARIANCE

It is natural for the beginner in statistics to be a little staggered by all of the arithmetic that the analysis of variance involves. However, take heart! With a bit of organization and with the aid of a good calculator, simple analyses can be done quite quickly. The important thing is to form a clear mental picture of the different sample quantities you will need to compute and how they combine. Below is an outline of the steps to follow:

1. Start with a listing of the raw scores separated by columns into the treatment groups to which they belong.
2. Square each score (y_{ij}^2) and then add these squared scores over all individuals in all groups. The result is $\sum_j \sum_i y_{ij}^2$. Call this quantity A.
3. Now sum the *raw scores* over all individuals in all groups to find the total T_y.
4. Next, for any single group, say group j, sum all of the raw scores in that group to find T_j. Then square that sum and divide it by the number in that particular group to obtain

$$\frac{T_j^2}{n_j}.$$

5. Repeat Step 4 for each group, and then sum the results across the several groups to find

$$\sum_j \frac{T_j^2}{n_j}.$$

Call this quantity C.

6. The sum of squares total is found from $A - \left(\frac{T_y^2}{N}\right)$

7. The sum of squares between is $C - \left(\dfrac{T_y^2}{N}\right)$

8. The sum of squares within is

$$\text{SS total} - \text{SS between} = A - C.$$

9. Divide SS between by $J - 1$ to give **MS between.**
10. Divide SS within by $N - J$ to give **MS within.**
11. Divide MS between by MS within to find the F **ratio.**
12. Carry out the test by referring the F ratio to a table of the F distribution with $J - 1$ and $N - J$ degrees of freedom.

In actual research settings analyses of variance are now most often done by computer, by use of one of the statistical packages, such as SPSS, BMDP, and SAS, about which we will have more to say shortly (Section 10.21). Nevertheless, it is a good idea to have the computational routine in mind, at least for the simple analysis of variance, as an aid in understanding what this technique actually provides in the many statistical contexts in which it recurs.

10.15 THE ANALYSIS OF VARIANCE SUMMARY TABLE

The results of an anlysis of variance are often (though not invariably) displayed in a table similar to Table 10.15.1.

In practice, the column labeled SS in Table 10.15.1 contains the actual values of the sums of squares computed from the data. In the df column appear the numbers of degrees of freedom associated with each sum of squares; these numbers of degrees of freedom must sum to $N - 1$. The MS column contains the values of the mean squares, each formed by dividing the sum of squares by its degrees of freedom. Finally, the F statistic is formed from the ratio of the mean square between groups to the mean square within groups.

Table 10.15.1

Standard layout for the results of a simple analysis of variance.

Source	SS	df	MS	F
Treatments (between groups)	$\displaystyle\sum_j \left(\frac{T_j^2}{n_j}\right) - \left(\frac{T_y^2}{N}\right)$	$J - 1$	$\dfrac{\text{SS between}}{J - 1}$	$\dfrac{\text{MS between}}{\text{MS within}}$
Error (within groups)	$\displaystyle\sum_j \sum_i y_{ij}^2 - \sum_j \left(\frac{T_j^2}{n_j}\right)$	$N - J$	$\dfrac{\text{SS within}}{N - J}$	
Total	$\displaystyle\sum_j \sum_i y_{ij}^2 - \left(\frac{T_y^2}{N}\right)$	$N - 1$		

The student does well to form the habit of arranging the results of an analysis of variance in this way. Not only is it a good way to display the results for maximum clarity, but it also forms a convenient device for organizing and remembering the computational steps.

10.16 AN EXAMPLE OF SIMPLE, ONE-WAY, ANOVA

As an example of a set of data suitable for a simple analysis of variance, consider the following: A study was being made of the influence of certain visual and auditory stimuli on the ability of subjects to remember spoken materials. In particular, the investigator was interested in the effect of a lack of synchrony between visual and auditory cues upon the subjects' retention of a list of 50 spoken words.

Therefore, a sample of 30 college students was selected, and three groups were formed at random, with 10 in each group. In the first of these groups, labelled "fast sound," the subjects saw a film of a person repeating a list of 50 common words. However, the voice and lips were slightly out of synchrony, with the sound preceding the lip movements. In the third group, labelled "slow sound," the situation was just the opposite, and the lip movements were seen slightly before the sound was heard. In the second group the sound and movement were in regular synchrony, and this group was labelled "normal sound." After the film had been shown, each student individually recalled as many of the 50 words as he or she was able. The number of words recalled was then the dependent variable Y for the analysis. The data from this little three-group study are shown in Table 10.16.1.

It is very easy to apply the analysis of variance to these data. Following the steps outlined in Section 10.14, we first find the sum of all of the 30 squared values:

$$\sum_j \sum_i y_{ij}^2 = 23^2 + 22^2 + \cdots + 22^2 + 17^2 + 20^2 + 23^2$$

$$= 17,290$$

$$= A.$$

	Fast	Normal	Slow	
	23	27	23	
	22	28	24	
	18	33	21	
	15	19	25	
	29	25	19	
	30	29	24	
	23	36	22	
	16	30	17	
	19	26	20	
	17	21	23	
Totals	$T_1 = 212$	$T_2 = 274$	$T_3 = 218$	$T_y = 704$

Table 10.16.1

Data for the study on audiovisual synchrony

The grand sum of the raw y values is

$$T_y = 23 + 22 + \cdots + 22 + 17 + 20 + 23$$

$$= 704.$$

Then, it is a simple matter to find SS total:

$$\text{SS total} = A - \left(\frac{T_y^2}{N}\right)$$

$$= 17,290 - \left(\frac{704^2}{30}\right) = 769.47.$$

Next we need to determine T_j, the sum of each group j. Then we square each of these group sums, divide by the number in that group, and add the results together:

$$C = \frac{[212^2 + 274^2 + 218^2]}{10}$$

$$= 16,754.40.$$

Then the SS between is found to be

$$\text{SS between} = C - \left(\frac{T_y^2}{30}\right)$$

$$= 16,754.4 - \left(\frac{704^2}{30}\right)$$

$$= 233.87 .$$

Then, finally,

$$\text{SS within} = 769.47 - 233.87$$

$$= 535.60 .$$

When the summary table is completed with the F test, as shown in Table 10.16.2, we have significant evidence that lack of synchrony in the stimulus presentations did affect the extent to which subjects remembered the words presented.

Source	SS	df	MS	F	p
Between	233.87	2	116.93	5.89	<.01
Within	535.60	27	19.84		
Total	769.47	29			

Table 10.16.2

ANOVA summary table for the data in Table 10.16.1

10.17 THE DESCRIPTIVE STATISTICS OF ANOVA

Although an analysis of variance usually culminates in an F test of the hypothesis of no population effects, do not get the idea that the F test is the only information of value in ANOVA. Any analysis of variance also has utility as a description of a set of data. Even in circumstances where the sample is not random, or where one or more assumptions required for the F test may not be valid, the analysis of variance still shows the extent to which group membership tends to be associated with the dependent variable *in these data*.

To illustrate how this is true, let us look more closely at the connection between the sums of squares found in ANOVA and the sample effects present in the given set of data. Recall that the sample effect for any group j in a one-way analysis of variance was defined in Section 10.4 to be

$$a_j = (\bar{y}_j - \bar{y}) .$$

For the data in Table 10.16.1, which we have just subjected to an ANOVA, we can use the overall mean of 23.4667 and the different group means to find the sample effects, as follows:

$$a_1 = \bar{y}_1 - \bar{y}\ \ = 21.2 - 23.4667$$

$$= -2.2667$$

$$a_2 = \bar{y}_2 - \bar{y}\ \ = 27.4 - 23.4667$$

$$= 3.9333$$

$$a_3 = \bar{y}_3 - \bar{y}\ \ = 21.8 - 23.4667$$

$$= -1.6667$$

Recall also that the mean of these effects must be zero (within rounding error, which is why we are carrying four decimal places for each $\bar{y}$):

$$\bar{a} = \frac{\displaystyle\sum_j n_j a_j}{N}$$

$$= \frac{10\,(-2.2667 + 3.9333 - 1.6667)}{30}$$

$$= 0.$$

Then the SS between as defined by Eq. 10.8.7 can also be expressed in terms of the sample effects:

$$\text{SS between} = \sum_j n_j a_j^2 .$$

That is, the SS between found in ANOVA could have been found by squaring and summing the group effects, and then multiplying the total by n or 10. For our example:

$$\text{SS between} = 10[(2.2667)^2 + (3.9333)^2 + (1.6667)^2]$$

$$= 233.87 .$$

In addition you should observe that if SS between is divided by N, then we have the variance of the sample effects:

$$\frac{\text{SS between}}{N} = \sum_j \left(\frac{n_j \, a_j^2}{N} \right)$$

$$= \text{variance of the sample effects}$$

$$= S_A^2 \, .$$

(A method, for testing the significance of individual effects is given in Section 11.14.)

On the other hand, if the SS within is divided by N, we have the variance of the errors, e_{ij}, in the sample:

$$\frac{\text{SS within}}{N} = \sum_j \sum_i (y_{ij} - \bar{y}_j)^2 / N$$

$$= \sum_j \sum_i e_{ij}^2 / N$$

$$= S_e^2 \, .$$

Since we now know that

SS total = SS between + SS within

we also know that, in the sample, the total variance is divisible into two parts: the variance of the effects and the variance of the errors.

$$S^2 \text{ total} = S^2 \text{ effects} + S^2 \text{ errors}.$$

It is evident that the total variation of the Y scores has two sources: the effects of group membership, and the errors associated with individuals within the groups. What proportion of the total Y variability that we see among the N cases in a sample is actually due to X, or group membership effects? The answer is provided by a simple ratio of the effects variability to the total variability, as follows:

$$R_{y \cdot A}^2 = \text{proportion of } y \text{ variance due to group effects}$$

$$= \frac{S_A^2}{(S_A^2 + S_e^2)}$$

$$= \frac{\text{SS between}}{\text{SS total}} \, .$$

The symbol $R_{y \cdot x_1 x_2 \cdots x_K}^2$ is often used in statistics to denote the proportion of variance of Y that is accounted for by a set of variables $(X_1, X_2, \cdots, X_K)$. Such an expression is commonly encountered in multiple regression analysis, such as we will study in Chapter 15. For the sake of simplicity, this symbol is sometimes shortened to

$$R_{y \cdot \mathbf{x}}^2 ,$$

where $\mathbf{X}$ is here understood to mean the whole set of X variables. When we are dealing with the proportion of variance accounted for by some experimental factor A, with levels $A_1, A_2, \cdots, A_J$, it is thus convenient shorthand to write

$R^2_{y \cdot A}$ = proportion of Y variance accounted for by levels of factor A.

On other occasions we might want to relate y values to membership in one of a set of J groups (not necessarily levels of an experimental factor), and here we might wish to write

$R^2_{y \cdot G}$

to symbolize the proportion of variance in variable Y explained by group membership, where G here simply indicates groups.

Since we will shortly be emphasizing the close connection between analysis of variance and multiple regression (Chapter 16), it seems reasonable to point up this connection by using symbols such as $R^2_{y \cdot A}$ in analysis of variance. However, there is an old tradition of representing this proportion by the symbol η^2, or lower-case Greek eta squared, and referring to this proportion as the correlation ratio. If you prefer, you should feel perfectly free to use eta-square in place of R-square in this context, as the two are identical. In either event, however, you should bear in mind that, as they relate to analysis of variance, both eta-square and R-square are *purely descriptive statistics*. The ratio of SS between to SS total tells the strength of the relationship between group membership and the dependent variable Y *in this sample*.

In the analysis of variance just completed the value of R-square is given by

$$R^2_{y \cdot A} = \frac{\text{SS between}}{\text{SS total}}$$

$$= \frac{233.87}{769.47}$$

$$= \quad .30.$$

We are therefore justified in saying that for the cases in these data, three-tenths (or 30%) of the variability in their Y values can be explained by differences in the groups to which they belonged. There is, therefore, evidence for a substantial relationship between group membership and the variable Y in these data.

It should be obvious that since the ratio SS between/SS total shows the proportion of Y variance explained by group membership in the current sample, then the ratio of SS within to SS total gives the proportion of Y variance *unexplained* by group membership:

$$1 - R^2_{y \cdot A} = \frac{\text{SS within}}{\text{SS total}}$$

$$= \frac{535.6}{769.47}$$

$$= \quad .70.$$

Seventy percent of the variability in this set of data is unexplained by the groups to which individuals belonged.

10.18 AN EXAMPLE WITH UNEQUAL GROUPS

In the example of ANOVA just completed, the study showed the same number of cases in each of the experimental groups. However, it is equally possible to carry out ANOVA when the numbers of cases are different in the different groups. If it is possible to assume that the relative frequencies in the different experimental groups correspond to probabilities of membership in the corresponding populations, then the rationale for the ANOVA and F test is identical to that given earlier in this chapter. Alternatively, if the loss of subjects from otherwise equal groups can be viewed as a totally random event, unrelated to the true effects of the study, then the ANOVA and F test can still be justified. If neither of these assumptions is tenable, one can still apply the so-called *method of unweighted means,* which can be found explained in some detail in Keppel (1982) and in Edwards (1985).

Consider an experiment carried out to study the effect of a small lesion introduced into a particular structure in a rat's brain on the rat's ability to perform in a discrimination problem. The particular structure studied is bilaterally symmetric, so that the lesion could be introduced into the structure on the right side of the brain, the left side, both sides, or neither side (a control group). Four groups of randomly selected rats were formed, and given the various treatments. Originally the control group contained 7 rats and each of the experimental groups 14 rats, but due either to death or postoperative incapacity, only the numbers shown below were actually observed in the discrimination situation. The experimenter was prepared to assume that loss of rats during the study was a purely random event, which had no bearing on the results. The final data were as shown in Table 10.18.1.

Now the simple analysis of variance will be illustrated for these data. Given this listing of the raw scores according to treatment groups, we first square and sum over all individuals in all groups (Step 2 Section 10.14):

$$A = \sum_j \sum_i y_{ij}^2 = 20^2 + 18^2 + \cdots + 18^2 + 25^2 = 24{,}424.$$

Next, the raw scores over all observations are summed, and the result squared (Step 3):

$$T_y = \sum_j \sum_i y_{ij} = (20 + 18 + \cdots + 18 + 25) = 970$$

$$\frac{T_y^2}{N} = \frac{(970)^2}{40} = 23{,}522.50 .$$

Using Steps 4 and 5 we find

$$C = \sum_j \frac{T_j^2}{n_j} = \frac{(159)^2}{7} + \frac{(266)^2}{11} + \frac{(322)^2}{13} + \frac{(223)^2}{9} = 23{,}545.07.$$

Then (Steps 6, 7, and 8),

$$\text{SS total} = A - \left(\frac{T_y^2}{N}\right) = 24{,}424 - 23{,}522.50 = 901.50$$

$$\text{SS between} = C - \left(\frac{T_y^2}{N}\right) = 23{,}545.07 - 23{,}522.50 = 22.57$$

	Group		
I	II	III	IV
20	24	20	27
18	22	22	35
26	25	30	18
19	25	27	24
26	20	22	28
24	21	24	32
26	34	28	16
159	18	21	18
	32	23	25
	23	25	223
	22	18	
	266	30	
		32	
		322	

Table 10.18.1

Performance scores for four unequal groups of rats

$$\text{SS within} = A - C = \text{SS total} - \text{SS between}$$

$$= 901.50 - 22.57 = 878.93.$$

Steps 9, 10, and 11 are represented in the following summary table:

Source	SS	df	MS	F
Treatments (between groups)	22.57	3	7.5	$\dfrac{7.5}{24.4}$
Error (within groups)	878.93	36	24.4	
Totals	901.5	39		

Ordinarily, at this point, the obtained F ratio would be compared with the value shown in Table V of Appendix E for 3 and 36 degrees of freedom and the specified α level. For $\alpha = .05$, the required F is 2.84, with 40 degrees of freedom used as the value nearest to 36. However, this step is really not necessary for this example, since the obtained F value is less than one, and the null hypothesis cannot be rejected. There is not enough evidence to warrant the conclusion that mean differences, or effects, truly exist among these treatment populations.

10.19 THE *F* TEST AND THE *t* TEST

When only two independent groups are being compared in the experiment, and a non-directional alternative hypothesis is being considered, it makes no difference whether the analysis of variance or the *t* test shown in Section 8.9 is used. As noted in Section 9.11, the square of a variable distributed as *t* with $N - 2$ degrees of freedom will be distributed as F with 1 and $N - 2$ degrees of freedom. A simple analysis of variance for two groups always yields an F ratio that is the same as the *square* of the t ratio calculated as in Section 8.9 for the same data. If the obtained F value is significant for any α, the corresponding value of $t = \sqrt{F}$ will be significant at the same α level in a two-tailed test. However, if the alternative hypothesis is directional, the sign of the difference between the two means must be considered; in this situation if F is significant at the α level, the one-tailed t test will show significance at the $\alpha/2$ level, provided that the sign of the obtained difference is appropriate to the alternative hypothesis.

This direct parallel between the F test in the analysis of variance and the t test for a difference in means holds only for the case of two groups, with an important exception that will be discussed in Chapter 11. One is never really justified in carrying out all the $\binom{J}{2}$ different t tests for differences among J groups, and then regarding this as some kind of substitute for the analysis of variance. One reason is the inflation in the apparent Type I error rate, as discussed in Section 8.17. Another is that such t tests carried out on all pairs of means must necessarily extract redundant, or overlapping, information from the data, and as a result a complicated pattern of dependency must exist among the tests. In most instances the analysis of variance, testing a single meaningful hypothesis at a known rate of Type I error, is to be preferred. Then, as explained in Section 11.17, comparisons among pairs of means may be carried out on a post-hoc basis.

10.20 THE IMPORTANCE OF THE ASSUMPTIONS IN THE FIXED-EFFECTS MODEL

In the development of the fixed-effects model for the analysis of variance a number of assumptions were made. These assumptions help to provide the theoretical justification for the analysis and the F test. On the other hand, it is sometimes necessary to analyze data when these assumptions clearly are not met; indeed, it seldom stands to reason that they are exactly true. In this section we will examine the consequences of the application of the analysis of variance and the F test when these assumptions are not met.

In the first place, note that the inferences made in this chapter are about *means*. Other models to be described in a later chapter provide inferences about *variances*, and the remarks made in this section apply *only* to the present, *fixed-effects, model*.

The first assumption listed in Section 10.9 specifies a normal distribution of errors, e_{ij}, for any treatment population j. This is equivalent to the assumption that each population j has a normal distribution of scores, y_{ij}. What are the consequences for the conclusions reached from the analysis when this assumption is not true? It can be shown that, other things being equal, inferences made about means that are valid in the case of normal populations are *also* valid even when the forms of the population distributions depart considerably from normal, provided that the n in each sample is *relatively large*. There is no hard-and-fast rule about how large a sample needs to be in order to be safe in this respect.

There is, in fact, a steadily growing body of evidence suggesting that the probability of obtaining a significant F value will be quite close to the nominal alpha level even for small samples and for populations departing widely from the normal distribution form. A commonsense rule is not to worry unduly about the normal assumption, and to apply the analysis of variance and F test even with relatively small samples when you must. However, if you feel fairly sure that the population departs considerably from the normal distribution, then to be on the safe side you might make an effort to achieve a larger sample size than otherwise.

The second assumption listed in Section 10.9 states that the error variance, σ_e^2, must have the same value for all treatment populations. Ordinarily, other things being equal, this assumption of homogeneous variances can be violated *without serious risk,* provided that the *number of cases* in each sample is the *same.* On the other hand, when *very different* numbers of cases appear in the various samples, violation of the assumption of homogeneous variances can have *very serious* consequences for the validity of the final inference. Here, the best available evidence seems to be that unequal variances make little difference in the results of an ANOVA F test, so long as the ratio of the largest to the smallest group size is only about 1.5. The moral is again plain: whenever possible, an experiment should be planned so that the number of cases in each experimental group is the same, or nearly the same, unless the assumption of equal population variances is eminently reasonable in the experimental context.

Situations do arise in which it is known that the populations *cannot* have homogeneous variances. The experimenter knows this to be true for theoretical reasons, or because prior evidence suggests strongly that the variances are not homogeneous. This sort of problem often occurs when the dependent variable Y is a *proportion* or *percentage*. Because the variance of a proportion p is $p(1-p)/N$, populations with different true proportions will necessarily have different variances, thus violating the assumption underlying ANOVA. One solution to this and similar problems is to transform the Y variable into a new variable which will have the required properties. We will go no further into this topic here, but a good resource on transformations and their use exists in the book by Winer (1971), and this should be consulted if a difficulty of this sort is suspected.

Since the analysis of variance is based on the assumption of equal variances, it may seem quite sensible to carry out a test for homogeneous variances on the sample data and then use the result of that test to decide if the analysis of variance is legitimate. Such tests for the homogeneity of several variances exist, and some older statistics books advocate these procedures. However, as noted in Section 8.10 above, the standard tests for equality of several variances are extremely sensitive to any departure from normality in the populations, and modern opinion holds that the analysis of variance can and should be carried on without a preliminary test of variances, especially in situations where the number of cases in the various samples can be made equal or nearly so (see Box, 1953, 1954a, 1954b).

The third assumption in Section 10.9 requires statistical independence among the error components, e_{ij}. The assumption of independent errors is most important for the justification of the F test in the analysis of variance, and, unfortunately, violations of this assumption can have important consequences for the results of the analysis. If this assumption is *not* met, *very serious errors* in inference can be made. In general, great care should be taken to see that data treated by the fixed-effects analysis of variance are based on independent observations, both within and across groups (that is, each observation in no way relates to any of the other observations). This is most likely to present a problem in studies where repeated observations are made of the same experimental subjects, perhaps with each subject being observed under each of the experimental treatments. In some experiments of this sort there is good reason to believe that the performance of the subject on one occasion has a systematic effect on subsequent performances under the same or another experimental condition. In the fixed-effects model, such systematic connections or dependencies among observations amount to a lack of statistical independence among errors, in violation of the assumption (a more appropriate model for handling this situation will be examined in Chapter 13). For this reason some authors suggest that data based on repeated observations should never be treated under the fixed-effects model for analysis of variance. However, this seems to be a point of experimental technique on which the statistician must tread very lightly. The circumstance that observations were repeated does not, ipso facto, imply that observations must be regarded as statistically dependent; a shrewd experimenter can sometimes get subjects into a stable state very early by preexperimental warm-up techniques, and in such situations it may be quite reasonable to assume that repeated observations are statistically independent. In other kinds of experiments, still other grounds may exist for assuming independence of repeated observations. The point, as always, is that statistics is limited in its ability to legislate experimental practice. Statistical assumptions must not be turned into prohibitions against particular kinds of experiments, although these assumptions must be borne in mind by the experimenter exercising thought and care in matching the particular experimental situation with an appropriate form of analysis. More will be said about the problem of repeated observations in Chapter 13.

10.21 STRENGTH OF ASSOCIATION AND THE POWER OF F TESTS

Since it is true that

$$E(\bar{y}_j) = \mu_j \text{ and } E(\bar{y}) = \mu,$$

an unbiased estimate of the magnitude of any fixed population effect α_j can be obtained by taking

$$\text{est. } \alpha_j = \bar{y}_j - \bar{y}.$$

In this way the experimenter can gain an idea about the magnitudes of the various effects that appear to be present in the populations represented by the data.

Furthermore, following an analysis of variance it is often of interest to judge the extent to which the experimental treatments actually are accounting for variance in the dependent variable. As we have seen, a perfectly good *descriptive* statistic for this purpose is immediately at hand: the *ratio of the SS between to the SS total* is an index of the proportion of variance accounted for in this sample. This index can be symbolized by $R^2_{y \cdot A}$, as we noted previously.

However, it is also useful to have an estimate of the *population* strength of association between the independent variable X and the dependent variable Y. The population index ω^2 defined in Section 8.12 may be extended readily to the situation where there are J different populations:

$$\omega^2 = \frac{\sigma_Y^2 - \sigma_{Y|X}^2}{\sigma_Y^2} = \frac{\sum_j \alpha_j^2 p(x_j)}{\sigma_Y^2}. \qquad [10.21.1]$$

Then a rough sample estimate of this population value can be had by taking

$$\text{est. } \omega^2 = \frac{\text{SS between} - (J-1)\text{ MS within}}{\text{SS total} + \text{MS within}}. \qquad [10.21.2]$$

When F turns out to be less than 1.00, then this estimate will be negative. In that case, the estimated value of ω^2 is set at zero, of course.

Determining the power of an F test shares the difficulty we have already encountered in the case of t: When the null hypothesis of equal population means is true, then the ratio we compute and call F actually does have the F distribution given in Table V of Appendix E. On the other hand, when H_0 is not true, then the ratio we employ does not have an F distribution at all. Rather, it follows a sampling distribution known as *noncentral F*. Although *central F* has two parameters, ν_1 and ν_2, noncentral F has three parameters, ν_1, ν_2 and δ^2. Under our usual assumptions, when H_0 is not true the value of δ^2 can be expressed as

$$\delta^2 = \frac{\sum_j n_j \alpha_j^2}{\sigma_e^2}. \qquad [10.21.3]$$

Notice the connection between *the noncentrality parameter* δ^2 and the sum of the squares of the true effects, α_j^2. To the extent that the absolute values of the true effects tend to be large across the populations, so that the variability due to effects is large relative to error variability, then the value of δ^2 will be large. As noted in our earlier discussion of power in Section 7.9, the larger the true effects, the larger the power, other things being equal. On the other hand, if all effects are zero, then δ^2 is also zero, and the corresponding noncentral F distribution is just the ordinary central F distribution. It is also worth noting that the expected mean square between in ANOVA, as shown in Eq. 10.10.7, can also be expressed in terms of the noncentrality parameter δ^2, as follows:

$$E(\text{MS between}) = \sigma_e^2 + \left(\frac{\sigma_e^2 \, \delta^2}{J-1}\right) \qquad [10.21.4]$$

$$= \sigma_e^2 \left[1 + \frac{\delta^2}{(J-1)}\right].$$

In discussions of power and presentations of tables for power determination of F tests, some authors use δ, as taken from the positive square root of Eq. 10.21.3. Others use a function of δ, sometimes symbolized by φ (lowercase Greek phi):

$$\varphi = \frac{\delta}{\sqrt{J}}.$$

Power charts and tables based on δ or on φ are to be found in a number of texts, such as Kirk (1982) and Keppel (1982). An especially valuable treatment of the power of F and other tests may be found in Cohen (1977).

If you find it easier to think of the population situation in terms of true proportion of variance accounted for, or ω^2, rather than in terms of δ^2, the second index may be converted into the first by the relation

$$\omega^2 = \frac{\delta^2}{N + \delta^2}$$ [10.21.5]

when our assumptions of Section 10.5 are true.

Power may thus also be evaluated in terms of population ω^2, and Table XII in Appendix E can be useful in this connection. This table shows the approximate power of the ANOVA F tests for different true values of omega-square, for different numbers of groups J, for different numbers of cases n within groups, and for Type I alpha equal to .05 or to .01. (In using this table we will assume that each distinct group contains the same number of cases, n).

As an example of the use of this table, suppose that we are dealing with $J = 4$ groups, with $n = 8$ in each group. Our F test will be carried out with Type I error alpha = .05. Now suppose that the true proportion of variance accounted for (omega-square) is .50. What is the power of our F test? The answer, as given by the table is approximately .99. This says that if true omega-square is .50, then there is a probability of .99 of our data reflecting this situation by a significant F at the .05 level. On the other hand, it everything else is the same, but the true proportion of variance accounted for is only .20, then the power of our test is only .58 as shown by Table XII.

Like other power tables, Table XII can be used in deciding the appropriate size of a sample, provided that one can specify the power desired against a particular true value of omega-square. Thus, for example, suppose that we were planning an experiment with $J = 6$ groups. We will use an alpha level of .01 and we wish a power of at least .85 when the true proportion of variance due to group membership is .20 or more. How many cases should we have in each group in order to make all of these things true? Table XII shows that at least 15 cases per group will give the power desired, and this is the number we would choose. On the other hand, if the number required turns out to be too expensive or otherwise impracticable we might wish to lower our requirements about the power, to raise our requirements about the true omega-square to be detected, or to raise the probability of Type I error we intend to use. Thus, in the present example, if we are willing to use .05 instead of .01 as the value of Type I error alpha, and will settle for detecting a true omega-square of .40 as significant with a probability or power of .90, then we can get by with a sample size of about 5 per group.

10.22 ONE-WAY ANOVA IN THE COMPUTER PACKAGES

For simple problems in one-way analysis of variance, the SPSS and SPSS^x procedure ONEWAY is especially easy to use. After the specification of a variable list or an instruction to get a particular file, the procedure is put into operation by a control command of the form

ONEWAY   *dependent variable* BY *independent variable (min, max)/*

or

ONEWAY   *dependent variable list* BY *independent variable (min, max)/*

where *dependent variable* and *independent variable* are replaced by the names of the variables (or factor levels) desired, as they appear in the variable list. For ONEWAY there can be only a single independent variable, representing factor levels or groups, and it must take on only integer values, where *(min, max)* in the statements above refers to the lowest and highest of these values.

A single ONEWAY command may include dependent variable listings for up to 99 dependent variables in the same computer run, and the ouptut will contain a separate analysis of variance for each dependent variable. A ONEWAY command may be followed by an OPTIONS command, along with a numerical code for various optional features pertaining to the format of the input or the output. Similarly, the command STATISTICS may also be used along with a code number telling which of several optional statistics should be included with the output.

The BMDP program 1V is also designed for the one-way analysis of variance. Preliminary information is entered in the format of a paragraph /PROBLEM giving the job title, /INPUT, showing the number of variables and their entry format, and /VARIABLE, which lists the variable names and specifies the variable to be used for grouping (i.e., the independent variable). Then the key paragraph /GROUP lists codes or names for the groups to be examined, and the /DESIGN paragraph specifies one or more dependent variables in the format DEPEND = *variable list*. If no dependent variable is specified after /DESIGN, then a one-way ANOVA is carried out for each of the variables listed, except that designated as the grouping or independent variable. If only a selection of possible groups is to be used, then the /DESIGN paragraph may also include a statement such as GROUP = *group list*. In addition to the standard one-way ANOVA, a wide selection of output options can be obtained.

In the SAS statistical package, one-way ANOVA as well as certain of the more complicated versions of analysis of variance can be had from the PROC ANOVA command. Thus, if A represents a factor with J qualitatively different levels or groupings, and you wish to test for the presence of effects of this factor on dependent variable Y, the input to SAS would follow the basic format

```
PROC ANOVA;
     CLASS    A;
     MODEL    Y = A;
```

The result would be an ordinary one-way analysis of variance, based on data already on file in SAS memory or entered for the purpose. Various options are available, and, if desired, these are listed after the MODEL statement. Means and other statistics are available as well.

A one-way ANOVA can also be obtained from PROC GLM followed by CLASS and MODEL statements just as above. This procedure is advantageous if one wishes to examine particular combinations of the groups means, as outlined in Chapter 11 to follow.

EXERCISES

1. Following the pattern of Section 10.7, construct the scores for an experiment with 4 samples of 4 observations each where $\mu = 100$, $\mu_1 = 95$, $\mu_2 = 104$, $\mu_3 = 98$, and $\mu_4 = 103$. Let the error terms be as follows:

I	II	III	IV
-5	-2	0	3
1	5	-4	-6
0	6	-1	4
2	-2	3	-7

From these data, estimate μ and the separate effects, and compare them with the true values.

2. By use of the artifical data in Exercise 1, show it is indeed true that

$$\text{SS between} = \sum_j n(\alpha_j + \bar{e}_j - \bar{e})^2.$$

3. Three independent random samples were taken of U.S. males aged 13 years. Sample I was drawn from among those boys who are otherwise normal, but who have less than average motor coordination for their age level. Sample II consisted of boys with average motor coordination, and Sample III consisted of boys with greater than average motor coordination. There were 20 boys in each sample. Each boy then filled out a questionnaire that yielded a rating for him on interest in athletic participation. The data are shown below. Can we say that motor coordination is related to interest in athletic participation?

Group					
I		II		III	
7	8	15	19	6	20
3	3	17	4	24	17
9	4	15	4	9	7
12	7	6	12	11	19
3	2	4	19	8	24
10	3	19	20	12	12
13	10	20	5	15	15
14	5	20	7	8	7
8	2	4	4	25	16
11	7	6	8	18	23

4. In an effort to curb inflation, the U.S. government imposed certain restrictions on the use of credit cards. In order to check on the effects that this might be having on the use of a certain card to purchase gasoline, an investigator sampled 25 large service stations honoring this card in five different regions of the country. The average daily number of charges using this card at that station was taken for six months ago, and then the present daily average number of charges was subtracted for that station. The data for each station, arranged by region are shown below. Does decrease in the use of the credit card seem to vary across the different regions? Use $\alpha = .05$.

		Region		
A	B	C	D	E
6.3	9.3	1.5	−1.7	2.7
2.1	−2.1	2.7	−2.0	3.9
5.8	6.6	4.4	3.6	3.2
7.9	5.2	3.1	3.8	0.6
−4.2	−3.8	4.6	4.5	1.7

5. Carry out an analysis of variance on the data in Exercise 17, Chapter 8. How do the results of this analysis compare with those found using the t test for a difference between means?

6. Each football program in a college conference has a strength-training program for the players. There are seven teams in the conference. An investigator sampled eight players at random from each of these teams, and gave each a test of strength, producing the scores listed below. Can we say that the strength programs are actually different in the scores they produce?

			Teams			
1	2	3	4	5	6	7
65	76	61	70	66	65	75
64	69	72	66	67	70	74
60	69	67	65	68	73	69
64	74	64	53	65	65	76
61	79	76	69	65	65	77
63	68	69	64	66	63	70
64	69	73	68	62	71	72
61	72	70	62	72	66	71

7. For Exercise 3, estimate the effects associated with the three "coordination" groups.

8. In a study of the effects of sleep deprivation upon the ability of human subjects to solve intellectual problems, 60 subjects were distributed at random among six groups. Each person was first given a score on a test of reasoning. Then each individual was deprived of a certain number of hours of sleep, after which the reasoning test was administered once again. The dependent variable was the difference between the first and the second administration of the test. These difference scores are given below. Can one say that sleep loss affects the scores on this test?

		Hours deprived			
8	12	16	20	24	28
20	32	70	95	83	100
19	25	65	90	75	96
18	24	59	84	75	95
11	20	55	80	71	94
10	19	50	76	62	80
10	18	47	75	62	78
7	15	40	60	50	78
6	14	38	59	48	72
5	10	31	40	40	72
5	8	27	38	39	71

9. Estimate the effects associated with the sleep-deprivation groups in Exercise 8. What would you estimate the strength of association to be between sleep deprivation and loss in score?

10. Carry out an analysis of variance on the following data, corresponding to two randomly chosen and independent experimental groups of 10 cases each.

I		II	
19	16	18	21
20	20	19	19
24	20	19	23
20	18	24	17
22	24	18	18

11. Test the significance of the difference between the means of the two groups in Exercise 10 by using the t test. How does this compare with the results in Exercise 10?

12. Four groups of rats were observed running a complex maze. Each rat was given 10 trials in the same maze. For the first group, the maze was provided with visual cues as to correct path. In the second group, the maze contained olfactory cues. The third group received both visual and olfactory cues, whereas all cues were removed from the maze for the fourth group. Six rats were assigned at random to each of the different groups, and were tested individually. However, different numbers of rats provided usable data in the different groups by the end of the experiment. Their scores (showing average numbers of minutes used to run the maze) were as shown below. Do the groups appear to be significantly different?

Groups			
I	II	III	IV
1.5	1.7	1.4	1.5
2.3	2.1	1.1	1.3
1.9	1.8	1.7	1.9
1.8	1.9		1.8
	1.3		1.7
	1.6		

13. Carry out a one-way analysis of variance for the following data based on four independent groups:

A	B	C	D
29	19	31	33
41	13	37	47
27	21	23	33
17	23	21	25
33	17	31	37

14. For the data in Exercise 13 estimate the various treatment effects and the strength of association between the experimental and dependent variables.

15. The following experiment utilized three groups of differing numbers of cases. Test the hypothesis of the equality of all of the population means.

I	II	III
.5	.6	.8
.4	.6	.9
.6	.5	.8
1.0	1.0	1.1
.6	.9	1.2
.5	.4	1.1
1.1	.7	

16. Below are two hypothetical sets of $N = 16$ cases assigned to four experimental groups. The total sum of squares for each set of data is 59. Without carrying out any further calculations, complete and interpret the analysis of variance summary table for each set.

	Groups			
set A	1	2	3	4
	7	7	7	7
	9	9	9	9
	10	10	10	10
	5	5	5	5
set B	1	2	3	4
	7	9	10	5
	7	9	10	5
	7	9	10	5
	7	9	10	5

17. In the United States, engineering schools tend to be found in large state universities, medium-sized state universities, private universities, and technical institutes. A large proportion of the engineers graduating this past spring obtained very good jobs. You wish to see if the starting salaries tend to relate to type of school attended. A random sample of 28 newly graduated engineers is taken, 7 from each type of school. The results in terms of starting salary (in thousands of dollars) are as follows:

Medium state u.	Large state u.	Technical	Private u.
$\bar{y}_1 = 17.85$	$\bar{y}_2 = 22.29$	$\bar{y}_3 = 18.75$	$\bar{y}_4 = 24.00$
$S_1^2 = 35.84$	$S_2^2 = 18.20$	$S_3^2 = 16.24$	$S_4^2 = 35.42$

Do significant differences appear to exist among the four types of schools with reference to starting salaries? (**Hint:** Look at Eq. 10.8.5 in the text, and recall the basic definition of S^2. Then look at Eq. 10.8.6. How can you find the Ss within?)

18. Use the results from Exercise 17 to estimate the four fixed effects, and the strength of association.

19. Members of the clergy of the same large denomination from six large geographical areas of the United States were sampled at random. Each person sampled was given an attitude test, providing a score on the "liberalism" of his or her attitude toward modern life. The following data resulted:

S.E.	S.W.	N.E.	N.W	Midwest	Far West
27	29	34	44	32	45
43	49	43	36	28	50
40	27	30	30	54	30
30	46	44	28	50	33
42	26	32	42	46	35
29	48	42		36	47
30	28	41		41	
41	30	33			
28	47	31			
	50	40			

Estimate the effect associated with each region.

20. Test the hypothesis that the means of the populations in Exercise 19 are equal. What is the estimate of ω^2, the proportion of the variance in liberalism accounted for by geographical region?

21. Consider the following sets of data:

Group 1	Group 2	Group 3	Group 4
1.69	1.82	1.71	1.69
1.53	1.93	1.82	1.82
1.91	1.94	1.75	1.86
1.82	1.60	1.64	1.90
1.57	1.78	1.52	1.39
1.77	1.85	1.73	1.56
1.94	1.98	1.86	1.74
1.60	1.72	1.68	1.83
1.74	1.83	1.54	1.47
1.74	1.75	1.75	1.64

We wish to carry out an analysis of variance on these data and test for equality of means ($\alpha = .05$). However, we will simplify our computations by subtracting 1.00 from each number and multiplying by 100. Complete the analysis and carry out the F test. Should the transformation of the numbers ($X' = 100 (X - 1)$) affect the results of our F test? Does it affect the values of MS between and MS within? How?

22. Estimate ω^2 for the transformed data given in Exercise 21, as well as the effects associated with the four groups.

23. An experiment is being planned in which there are to be six experimental groups. The alpha level to be used is .01. It is desired to have a power of at least .80 if the true value of omega-square is .50 or more. How many subjects should be used in each group? How many should be used if it is desired to have a power of .80 when only 20% of the variance is explained by the experimental treatments?

24. Use the values of the sample effects in Exercise 12 to show that the SS between is actually the weighted sum of the squares of the group effects, where the weight for any group is the number of cases it contains.

Chapter *11*

Comparisons Among Means

Research is almost never conducted in the hermetically sealed atmosphere that statistical reports often suggest. Most emphatically, good research requires all the care and control that the experimenter can provide in the actual experiment. But research hypotheses are not entertained or dismissed in the cavalier way that a statistics book may imply. These books set down the formal rules of the game, a set of prearranged signals from one scientist to another that something worth looking into may have been found, but that the results contain sampling error. Research reports are written in this conventionalized language, and no doubt will continue to be until something better is devised. Nevertheless, the formal rules of the game should not be confused with the actual work of the scientist.

The novice in the use of statistics often feels that statistical tests somehow cut one off from the close examination of data. There is something that seems so final about an analysis of variance summary table, for example, that one is tempted to regard this as some ultimate distillation of all meaning in the experiment. Nothing could be further from the truth. Statistical summaries and tests are certainly useful, but they are not the *ends* of experimentation. The important part of data analysis often begins when the experimenter asks ''What accounts for the results I obtain?'' and turns to the detailed exploration of the data.

Considered by themselves, all that *t* tests or the *F* tests in an analysis of variance can tell you is that *something* seems to have happened. If the *F* is significant, then some effects presumably exist that can be expected to occur again under similar circumstances; if the test is not significant, something notable still may have happened, but if treatment effects exist they are at least partially obscured by other variation. Other than this, an *F* test alone tells almost nothing. If all experimenters were satisfied with the general statement that something did (did not) happen, then science would progress by slow stages indeed.

The material in this chapter deals with two devices for analyzing data in more detail than that provided by the ordinary analysis of variance. Both methods are designed for comparing means or groups of means in a variety of ways. First, the technique of **planned comparisons** or **contrasts** will be introduced. Here, instead of planning to analyze the data to see if any overall experimental effects exist, the researcher at the outset has a number of particular questions to be answered separately. This technique of planned comparisons is used *instead of* the ordinary analysis of variance and F test. Indeed, as will be shown, the usual analysis of variance actually summarizes the evidence for many possible such individual questions that might be asked of the data.

Next, the technique of **incidental** or **post hoc comparisons** (or **contrasts**) will be discussed. Here, means that are weighted and combined in any number of ways can be evaluated for signficance, *after* the overall F test has shown significance. This procedure of post hoc comparisons is an important supplement to the usual analysis of variance, and is very useful for the further exploration of data after the initial analysis has suggested the existence of real effects.

Both techniques, planned and post hoc comparisons, apply only to fixed-effects or mixed-model experiments (Section 10.6). The inferences to be made concern particular population means combined in particular ways, and these methods make most sense when applied to treatment groups representing *fixed-level, nonsampled* factors.

11.1 ASKING SPECIFIC QUESTIONS OF DATA

In Section 5.15 the idea of linear combinations of random variables was introduced, and principles for specifying the sampling distribution for a particular linear combination of random variables are given in Appendix C. Our concern with linear combinations is justified by the fact that the evidence pertaining to particular experimental questions can often be found from various ways of combining the data—specifically, by linear combinations of means. In this section we will see how special linear combinations are useful to the experimenter.

Consider a study of the influence of the manner of persuasion on the tendency of persons to change their attitude toward some institution or group. The particular attitude studied was that toward a minority group, and the experimental modes of persuasion used were:

1. a motion picture favorable to the minority group;
2. a lecture on the same topic, also favorable to the minority group;
3. a combination of the motion picture and the lecture.

Subjects were assigned at random to the different experimental groups, each having first been given a preliminary attitude test. Following the experimental treatment, each subject was given the test once again; the dependent variable here was the *change* in attitude score. It was recognized, however, that the mere repetition of a test may have some influence on an individual's changing score, and so a randomly selected control group was also used, in which the subjects were given no special experimental treatment.

However, during the design of the experiment the question arose, "Would the subjects perhaps change as a result of seeing *any* movie or hearing *any* lecture?" Thus it was decided to introduce two more control groups, the first to be shown a movie completely unrelated to the minority group, and the second given a lecture by the same person, but on a topic unrelated to minorities.

The final experimental design can be diagrammed as follows:

Experimental groups			Control groups		
I	II	III	IV	V	VI
Movie	Lecture	Movie and lecture	Nothing	Neutral movie	Neutral lecture

A total sample of 30 subjects was drawn at random, and subjects were assigned at random to the six conditions, with five subjects per condition.

In this study, the experimenter was not interested in the overall existence of treatment effects. Rather, from the outset interest lay in answering the following specific questions.

1. Do the experimental groups as a whole tend to differ from the control groups?
2. Is the effect of the experimental lecture–movie combination different from the average effect of either experimental movie or experimental lecture separately?
3. Is the effect of the experimental lecture different from the effect of the experimental movie?
4. Among the control groups, is there any effect of the neutral movie or lecture as compared with the group receiving no treatment?

In other words, the experimenter entertained a number of particular questions before the collection of the data, and wished to analyze the data to answer those questions. An overall analysis of variance and F test would give indication of the existence of *any* systematic effects. However, the main interest lay in the particular differences among population means corresponding to answers to these questions.

Notice that the evidence pertaining to each question comes from the various sample means combined in some special way. For example, the evidence for Question 3 involves only the difference between the means for Groups I and II. On the other hand, the evidence for Question 2 involves both the difference between Mean I and Mean III *and* the difference between Mean II and Mean III. Still other combinations of means pertain to the other questions asked. Can one, however, attach a significance level to each of these comparisons among means, permitting a statement about such differences among the *population* means?

This is a problem of planned comparisons among means, and the technique used will now be outlined. Let it be emphasized that this procedure applies only when the experimenter has specific questions to be asked *before* the data are analyzed, and this method may be used *instead* of the ordinary analysis of variance and F test.

11.2 PLANNED COMPARISONS

The basic theory underlying planned comparisons is briefly outlined in Section C.2 of Appendix C. There it is noted that given normally distributed variables, sampled independently and at random, values from any linear combination of those random variables will also be normally distributed (C.2.8). Furthermore, if the mean and variance for each variable are known, then the mean and variance of the sampling distribution of the linear combination are also known.

These principles were used in Section 8.7 to establish the rationale for the t test for the difference between means of independent groups. Now we will go far beyond the simple two-sample case and apply these principles for any linear combination of means.

First, we need to define a **population comparison or comparison among population means:** given the means of J distinct populations, $\mu_1, \cdot \cdot \cdot, \mu_J$, a comparison among those means is any linear combination or weighted sum, with weights c_j not all equal to zero:

$$\psi = c_1\mu_1 + c_2\mu_2 + \cdot \cdot \cdot + c_J\mu_J = \sum_{j=1}^{J} c_j\mu_j. \qquad [11.2.1^*]$$

We will use the symbol ψ (small Greek psi) to stand for the value of some particular *population* comparison. The weights for a comparison ψ are some set of *real numbers* $(c_1, \cdot \cdot \cdot, c_J)$ *not all zero.*

In addition, we impose the requirement that the *sum* of the weights c_j equals zero

$$\sum_{j=1}^{J} c_j = 0.$$

In the following, we will simply assume that this requirement is met and defer giving the reason for this stipulation until Section 11.7.

A **sample comparison** is defined exactly as for a population comparison, except that **sample means** are weighted and summed:

$$\text{est. } \psi = c_1\bar{y}_1 + c_2\bar{y}_2 + \cdot \cdot \cdot + c_J\bar{y}_J = \sum_{j} c_j\bar{y}_j. \qquad [11.2.2]$$

Once again, we will require that $\sum_{j} c_j = 0$. Be sure to notice that the symbol for a comparison, either ψ or est. ψ, stands for a single number, since it equals a *weighted sum of numbers*. Incidentally, although the use of est. ψ as a way of symbolizing a sample comparison is quite consistent with our previous use of *est.* to indicate an estimate of a population quantity made from a sample, this usage will prove a bit cumbersome in sections to come, where it will become necessary to use several est. indicators in the same expression. Therefore, we will adopt an alternative notation for a sample comparison in an effort to avoid such complexities, and the Greek psi symbol with a caret or "hat" over it will also be used to represent a sample comparison from time to time as in the following.

$$\hat{\psi} = \text{est. } \psi .$$

It is simple to illustrate the calculation of comparisons for a given set of data. For example, suppose that in an experiment there were four independent groups of equal size, labelled A_1, A_2, A_3 and A_4. Each group was given a different treatment. From the outset, three comparisons were planned: First, it was planned to compare the mean of the first two groups with the mean of the last two groups. Next the first two groups were to be compared with each other, and, finally, the last two groups were to be compared. These comparisons can be represented by the following sets of weights to be applied to the group means:

	A_1	A_2	A_3	A_4
comparison 1	1/2	1/2	−1/2	−1/2
comparison 2	1	−1	0	0
comparison 3	0	0	1	−1

Notice how the weights for Comparison 1 actually do involve averaging the means for Groups A_1 and A_2 (i.e., adding them up and dividing by 2) and then subtracting the average of the means for A_3 and A_4. On the other hand, the weights for the second and third comparisons lead to the simple difference between a pair of the groups. Notice also how each set of weights adds to zero across groups.

Now suppose that data were actually collected for the four groups, with four subjects assigned at random to each group. The data might look something like those in Table 11.2.1, where the raw scores are listed along with the group means. Then applying the three sets of weights listed in the previous table to the sample means, we have the three comparisons:

$$\text{comparison 1} = \hat{\psi}_1 = (1/2 \times 18) + (1/2 \times 25) - (1/2 \times 28) - (1/2 \times 22)$$

$$= -3.5$$

$$\text{comparison 2} = \hat{\psi}_2 = (1 \times 18) - (1 \times 25) + (0 \times 28) + (0 \times 22)$$

$$= -7$$

$$\text{comparison 3} = \hat{\psi}_3 = (0 \times 18) + (0 \times 25) + (1 \times 28) - (1 \times 22)$$

$$= 6$$

These values resulting from the actual weighing and summing of the means are the *estimated comparison values* themselves. The magnitude and sign of these comparison values provide the answers to the original questions that the comparisons represent. Thus, for example, here we now know that the first two groups average 3.5 units less than the last two groups (Comparison 1), that the second group is 7 units greater than the first (Comparison 2), and that the fourth group is less than the third by 6 units

	A_1	A_2	A_3	A_4	**Table 11.2.1**
	22	26	28	21	Scores for 16 subjects in four groups
	15	27	31	21	
	17	24	27	26	
	18	23	26	20	
means	$\overline{18}$	$\overline{25}$	$\overline{28}$	$\overline{22}$	

Source	SS	df	MS	F	p	**Table 11.2.2**
Groups	219	3	73	12.17	<.01	ANOVA for data of table
Error	72	12	6			
Total	291	15				

(Comparison 3). If we had planned to ask other questions, then other ways of weighting and summing the means would have been used, of course.

The analysis of variance for this same set of data is shown in Table 11.2.2. This analysis, like the comparious made above, also rests on differences among the group means, of course, and we will examine the connection between these two methods in the sections to follow.

11.3 STATISTICAL PROPERTIES OF COMPARISONS

Some of the statistical properties of comparisons will now be viewed in the light of the principles summarized in Appendix C. Since each sample comparison is a linear combination, it is true from C.2.2 that

$$E(\hat{\psi}) = E\left(\sum_j c_j \bar{y}_j\right) = \sum_j c_j E(\bar{y}_j) = \psi. \qquad [11.3.1]$$

Any sample comparison is an unbiased estimate of the population comparison involving the same weights c_j. If we wish to evaluate the weighted sum of population means, our best estimate is the weighted sum of sample means. In the examples above, each question asked by the experimenter can be answered by some particular comparison among population means, as reflected in the same weighted sum of the sample means.

Note especially that when all the population means are equal, so that $E(\bar{y}_j)$ is the same for all j, by C.2.3, $E(\hat{\psi}) = 0$.

Suppose that each of the sample means $\bar{y}_j$ is based on a different and independent sample of n_j cases drawn at random from a population with true mean μ_j. Furthermore, suppose that each population has the same variance, σ_e^2, so that the sampling distribution of $\bar{y}_j$ values has a variance σ_e^2/n_j.

By the principle of C.2.6 and C.2.7 in Appendix C, the sampling variance of an estimated comparison $\hat{\psi}$ based on independent means must be

$$\text{var}(\hat{\psi}) = \sum_j c_j^2 \, \text{var}(\bar{y}_j) = \sigma_e^2 \sum_j \frac{c_j^2}{n_j} .$$

[11.3.2]

Thus, for example, suppose that the weights in a sample comparison of five means were $(-2, -1, 3, 1, -1)$, so that

$$\hat{\psi} = -2\bar{y}_1 - 1\bar{y}_2 + 3\bar{y}_3 + 1\bar{y}_4 - 1\bar{y}_5 .$$

Furthermore, suppose that each mean is based on 10 cases, so that $n_j = 10$ for all j. Provided that σ_e^2 is the variance of each population, then the sampling distribution of values of $\hat{\psi}$ must have a variance given by

$$\text{var}(\hat{\psi}) = \sigma_e^2 \left[\frac{(-2)^2 + (-1)^2 + (3)^2 + (1)^2 + (-1)^2}{10} \right]$$

$$= \sigma_e^2 \left(\frac{16}{10} \right)$$

$$= \frac{8\sigma_e^2}{5} .$$

Obviously, in few practical situations will we know the exact value of σ_e^2. However, the value of σ_e^2 can be estimated from the data in exactly the same way that it is estimated in the analysis of variance:

est. $\sigma_e^2 = $ MS error.

(Here, σ_e^2 is the same as σ_e^2 in the expected mean squares for analysis of varaince, and mean square error is the same as mean square within groups in the one-way analysis.) Thus

$$\text{est. var}(\hat{\psi}) = (\text{MS error}) \sum_j \frac{c_j^2}{n_j} .$$

[11.3.3]

or

$$\text{estimated standard error of comparison} = \text{est. SE } (\hat{\psi})$$

[11.3.4]

$$= \sqrt{\text{est.var.}(\hat{\psi})} .$$

As an example, we can turn once again to the four groups shown in Table 11.2.1. For these data an analysis of variance gives the result presented in Table 11.2.2. We can see that for this example the mean square error, or MS error, is equal to 6. We can then estimate the variance and standard error for each of the three comparisons shown in the preceding section, as follows:

First, in order to estimate the sampling variance of comparison $\hat{\psi}_1$, we take

$$\text{est.var.}(\hat{\psi}_1) = 6 \left[(1/2)^2 + (1/2)^2 + (-1/2)^2 + (-1/2)^2 \right] / 4$$

$$= 3/2$$

so that the estimated standard error is

$$\text{est. SE } (\hat{\psi}_1) = \sqrt{3/2}$$

$$= 1.2247 .$$

We also find the sampling variance of comparisons $\hat{\psi}_2$ and $\hat{\psi}_3$:

$$\text{est. var.}(\hat{\psi}_2) = \frac{6 \, [(1)^2 + (-1)^2 + (0)^2 + (0)^2]}{4}$$

$$= 3$$

$$\text{est. SE}(\hat{\psi}_2) = \sqrt{3}$$

$$= 1.7321$$

and

$$\text{est. var.}(\hat{\psi}_3) = 3$$

$$\text{est. SE}(\hat{\psi}_3) = 1.7321 .$$

Now we know the important facts about a sample comparison $\hat{\psi}$ viewed as an estimator of the corresponding population comparison ψ: The sample comparison is an unbiased estimate of the population value, and the sampling variance is given by Eq. 11.3.2. Furthermore, if the underlying population distributions are normal, or if the size of our samples is relatively large, then the distribution of the sample values may also be regarded as normal. Since we are forced to estimate the value of the standard error, rather than knowing its value exactly, we are once again in the realm of the t distribution, where sample t may be defined by

$$t = \frac{\hat{\psi} - \psi}{\text{est. SE}(\hat{\psi})} .$$

[11.3.5]

This fact will later be used to test hypotheses and construct confidence intervals for the value of ψ.

Perhaps you noticed that the first comparison used as an example in this section involves weights that are whole numbers, whereas the second example employs fractional weights. The questions underlying a set of comparisons can often be seen most clearly when the weights are fractions reflecting different ways of averaging groups of cases, just as in Section 11.2. There is, however, nothing in the theory of comparisons restricting the weights to fractional values; given the means of J groups, a comparison may involve any J values (not all zero) so long as they sum to zero. Indeed, it is sometimes convenient to convert such comparison weights from fractions to whole numbers. This can be done quite simply by finding the lowest common denominator of the fractional weights, and then multiplying each weight by this common denominator, retaining only the numerator values of the results. Thus, for example, the comparison with weights of 1/3, 1/3, 1/3, −1/2, and −1/2 translates easily into a comparison with whole number weights of 2, 2, 2, −3, and −3 after multiplication by the common denominator of 6.

On the other hand, if you wish to calculate the value of the comparison that would have been found using the fractional weights, then you need to take the value found from the integer or whole number weights and divide it by the common denominator. Moreover, the estimated standard error (Eq 11.3.4) based on the integer weights must then also be divided by the common denominator value. It follows that the t value shown in Eq. 11.3.5 will be the same for the original comparison with fractional weights and for the equivalent comparison based on integer weights. That is, the significance test will not be affected by the decision to use the fractional or the corresponding integer weights. On the other hand, the confidence interval (Eq. 11.4.3) *is* different in these two situations. The point is that you may use either fractional weights or the equivalent integer weights in carrying out a comparison among means, provided that you are careful and consistent in doing so.

11.4 TESTS AND INTERVAL ESTIMATES FOR PLANNED COMPARISONS

In most instances, the only hypothesis of interest in planned comparisons is

H_0: $\psi = 0$.

Then a test for this hypothesis is given by

$$t = \frac{\hat{\psi}}{\sqrt{\text{est. var}(\hat{\psi})}},$$

[11.4.1]

distributed as t with degrees of freedom as for mean square error, or ($N - J$ in a single-factor experiment). If the hypothesis is to be tested against a directional alternative, then the sign of the t is considered, just as in the ordinary test for two means.

Notice that the t test for a difference between two independent means (Section 8.9) is merely a test of a comparison

$$\hat{\psi} = 1(\bar{y}_1) - 1(\bar{y}_2)$$

where the two weights are 1 and -1 respectively. However, the t test can be used to test any particular comparison, regardless of the number of groups involved.

When the hypothesis to be tested is nondirectional, the t test for any planned comparison may be replaced by the equivlent F test:

$$F = \frac{(\hat{\psi})^2}{\text{MS error} \left(\sum_j c_j^2/n_j \right)}$$

[11.4.2]

with 1 and $N - J$ degrees of freedom. (Recall, once again, that t^2 is distributed as an F variable with 1 and ν degrees of freedom.)

Alternatively, given the value of the planned sample comparison $\hat{\psi}$, a confidence interval for the population comparison value ψ can be found from

$$\hat{\psi} - t_{(\alpha/2;\nu)} \sqrt{\text{est. var}(\hat{\psi})} \leq \psi \leq \hat{\psi} + t_{(\alpha/2;\nu)} \sqrt{\text{est. var}(\hat{\psi})}.$$

[11.4.3]

For the $100(1 - \alpha)$ percent confidence interval, the value of $t_{(\alpha/2;\nu)}$ represents the value cutting off the *upper* $\alpha/2$ proportion of sample values in a distribution of t with ν degrees of freedom. The number of degrees of freedom is the same as the degrees of freedom for the mean square error used to estimate σ_e^2. This confidence interval can be used to test any hypothetical value of ψ at all. If some hypothetical value of ψ fails to be covered by the confidence interval, then that hypothesis may be rejected beyond the α level (two-tailed).

Our running example from Section 11.2 can be employed once again to illustrate these tests and confidence intervals.

The three t tests for these comparisons are

$$\text{for } \hat{\psi}_1, \; t = (-3.5)/(1.2247) = -2.86,$$

$$\text{for } \hat{\psi}_2, \; t = (-7)/(1.7321) = -4.04,$$

and

$$\text{for } \hat{\psi}_3, \; t = (6)/(1.7321) = 3.46.$$

Each of these t tests exceeds the value of 2.179 required for .05 significance (two-tailed) with 12 degrees of freedom.

In calculating the confidence intervals, let us look at comparison $\hat{\psi}_1$, previously found to have a value of -3.5, with a standard error of 1.2247. To set up the 95% confidence limits for this comparison we need first to find the required value of t at the .05 level (two-tailed) when there are 12 degrees of freedom, as in this example. As noted, this value turns out to be 2.179. Then, given all of this information we determine the lower and upper values of our confidence interval to be

$$-3.5 - 2.179(1.2247) = -6.17$$

and

$$-3.5 + 2.179(1.2247) = -.83$$

respectively. Obviously, the confidence interval defined by these limits does not include the value 0, and thus Comparison 1 can be considered significant at the .05 level for these data. In other words, there is a significant difference between the means of the first two and the last two groups, exactly as shown by the t test above.

The 95% confidence intervals for Comparisons 2 and 3 are found to have limits:

$$-7 - 2.179(1.7321) = -10.77$$

and

$$-7 + 2.179(1.7321) = -3.23$$

for $\hat{\psi}_2$ and, for $\hat{\psi}_3$,

$$6 - 2.179(1.7321) = 2.23$$

$$6 + 2.179(1.7321) = 9.77 .$$

Therefore, since none of the three confidence intervals covers 0, each comparison may be declared significantly different from zero. Such a t test or confidence interval for ψ

rests on these usual assumptions for a fixed-effects analysis of variance: normally distributed populations, each with variance σ_e^2, and independent errors. As in the overall F test for such an analysis, the assumption of normality is relatively innocuous even for fairly small samples.

On the other hand, violation of the assumption of homogeneity of error variances can have rather serious consequences in some instances, although such violations have their minimum effect when n is constant, or nearly so, over all groups (cf. Section 10.20), and when the comparison weights c_j are equal in absolute value. (Some evidence suggests that unequal variances can affect the validity of the t or F tests for comparisons when the weights differ considerably from each other across the groups. In the absence of more information it is hard to know what to do about this problem. Counseling that such comparisons should be avoided on the off chance of differing variances seems unnecessarily cautious under the circumstances.)

11.5 INDEPENDENCE OF PLANNED COMPARISONS

A single planned comparison among means is usually tested by a simple t ratio, or F ratio, as we have seen. However, seldom does an experimenter have interest in only one comparison on the data. Usually there are sets of questions to be answered, each corresponding to some comparison among means. This brings up the very critical problem of the independence of comparisons.

Just as the possible answers to some questions may depend logically on answers to others, so the values of some comparisons made on a given set of means may depend on the values of other comparisons. This fact has important consequences for estimates and tests of several comparisons, since the questions involved in the respective comparisons cannot be given truly separate and unrelated answers unless the comparisons are statistically independent of each other. Fortunately, a simple method exists for determining whether or not two comparisons are independent, given normal population distributions with equal variances σ_e^2. **The determination of the independence of two comparisons depends only on the weights each involve, and in no way on the means actually observed.** One can plan comparisons that will be independent *before* the data are collected.

The solution to the problem of independent comparisons rests on still another general principle having to do with linear combinations of variables, C.2.9 of Appendix C. This principle is another instance of the extraordinary utility of the assumption of normal distributions in statistical theory.

Suppose that there were J independent samples, each representing a normal population. Each population is assumed to have the same variance σ_e^2. Also, for the moment assume that within each of the samples there will be the same number n of observations. Any group j will, of course, have a mean $\bar{y}_j$. Now imagine *two* comparisons among these same means, the first, $\hat{\psi}_1$, with weights symbolized by c_{1j}, and the second, $\hat{\psi}_2$, with weights symbolized by c_{2j}. From C.2.9 of Appendix C, we know that the values $\hat{\psi}_1$ and $\hat{\psi}_2$ are themselves independent *provided* that

$$\sum_j c_{1j}c_{2j} = 0. \qquad\qquad [11.5.1]$$

In other words, **given the means of J independent samples, each of size n, one can decide if two different linear combinations of those means are independent simply by seeing if the products of the weights assigned to each mean sum to zero:**

$$c_{11}c_{21} + \cdots + c_{1J}c_{2J} = 0.$$

This principle is necessarily true only for samples from normal populations with equal variances.

Two comparisons satisfying this condition are said to be **orthogonal comparisons;** orthogonality of two comparisons is equivalent to the statistical independence of sample comparisons only when the populations are normal with equal σ_e^2, however. When two comparisons are statistically independent, the information each provides is actually nonredundant and unrelated to the information provided by the other. The estimate $\hat{\psi}_1$ is unrelated to the estimate $\hat{\psi}_2$. Thus, seeing if comparisons are orthogonal lets the experimenter judge whether or not they give unrelated, nonoverlapping pieces of information about the experiment.

For example, think once again of four groups, representing levels of some factor A. Below are shown the weights for two comparisons that we can test for orthogonality:

group	A_1	A_2	A_3	A_4
comparison 1	$-1/3$	1	$-1/3$	$-1/3$
comparison 2	$-1/2$	0	$-1/2$	1

Taking the product of the weights across the four groups we find

$$\sum_j c_{1j}c_{2j} = (-1/3 \times -1/2) + (1 \times 0) - (1/3 \times -1/2) - (1/3 \times 1) = 0$$

showing that these two comparisons actually are orthogonal.

Notice also that if the comparison weights were changed to the corresponding whole numbers of -1, 3, -1, -1 and -1, 0, -1, and 2, the comparisons would still be orthogonal:

$$(-1)(-1) + (3)(0) + (-1)(-1) + (-1)(2) = 0.$$

On the other other hand, suppose that we had another pair of comparisons, as represented by their weights:

	A_1	A_2	A_3	A_4
comparison 1	$-1/3$	1	$-1/3$	$-1/3$
comparison 2	$1/2$	$1/2$	$-1/2$	$-1/2$

Here, when we take the sum of the products of the weights, we find

$$(-1/3 \times 1/2) + (1 \times 1/2) - (1/3 \times -1/2) - (1/3 \times -1/2) = 4/6 = 2/3$$

showing that these two comparisons are *not* orthogonal. Each may be a perfectly fine comparison when used to answer some question of interest. The problem, however, is that, unlike the first pair examined, these comparisons give values that are, to some extent, related or redundant.

When the J distinct samples have different sizes, symbolized by n_j, the criterion for orthogonality for two comparisons among the means becomes

$$\sum_j \frac{c_{1j}c_{2j}}{n_j} = 0, \qquad [11.5.2]$$

so that the product of comparison weights for each sample is weighted inversely by n_j before the sum is taken. If this weighted sum of products is zero, then the comparisons may be regarded as orthogonal.

In the following discussion, the terms *independent* and *orthogonal* will be used interchangeably, but this is proper only because normal distributions with homogeneous variances are assumed throughout the discussion.

Speaking very strictly, however, the fact that two or more comparisons are independent over all possible samples of the populations in question *does not guarantee* that the t (or F) tests for these comparisons are independent. The problem comes about because we must estimate the error variance that figures in the denominator of t. To do this we use the same sample data on which the comparisons are made, and we use the same estimated quantity in each t test made on the same sample. This has the result of introducing a *dependency* between two or more such t tests. In practice, however, if the sample size is sufficiently large, the t tests may still be regarded as "approximately" independent. This probably should not be a serious concern except when there are very small numbers of error degrees of freedom. More will be said on a related problem in Section 12.17.

11.6 AN ILLUSTRATION OF INDEPENDENT AND NONINDEPENDENT PLANNED COMPARISONS

Suppose, once again, that four treatment groups were observed in an experiment, this time with a sample of six randomly assigned subjects in each. The means of the four groups were

I	*II*	*III*	*IV*
17	24	27	16.

The mean square error, found by the usual method for a one-way analysis of variance, was 5.6, with $24 - 4$, or 20, degrees of freedom. Before the data were seen, the experimenter expressed interest basically in the following questions:

1. Does Mean I differ from the average of Means II, III, and IV?
2. Does Mean II differ from the average of Means III and IV?
3. Does Mean III differ from Mean IV?
4. Does the average of I and II differ from the average of III and IV?

		Means		
Question	I	II	III	IV
1	1	$-\dfrac{1}{3}$	$-\dfrac{1}{3}$	$-\dfrac{1}{3}$
2	0	1	$-\dfrac{1}{2}$	$-\dfrac{1}{2}$
3	0	0	1	-1
4	$\dfrac{1}{2}$	$\dfrac{1}{2}$	$-\dfrac{1}{2}$	$-\dfrac{1}{2}$

Table 11.6.1

Weights corresponding to comparisons among four means.

It is assumed that each population is normally distributed, with the same variance in each. Now, each of the questions can be put into the form of comparisons among means by using the following sets of weights. In Table 11.6.1 the rows represent the various questions (comparisons) and the columns the various samples. In the cells the numbers are the weights c_j to be assigned to a given sample mean for a given comparison.

There is certainly nothing mysterious about the way these weights were chosen. The first comparison is designed to investigate a difference between Mean I and the *average* of Means II, III, and IV. This calls for a weighting of Mean I by unity, and subtracting the average of the three other means, which is tantamount to weighting each one by a *negative* 1/3 in the comparison with Mean I. Similarly, Mean I does not figure in the second question, and so gets weight 0, whereas Mean II is contrasted with the *average* for Means III and IV. The other weights are found in similar ways. Note that the sum of weights for each row in the table is zero, as it should be for comparisons among means.

Are these four comparisons orthogonal, and thus independent of each other? Consider the weights in rows 1 and 2; the sum of their products across columns is

$$\sum_j c_{1j}c_{2j} = (1 \times 0) + \left(\frac{-1}{3} \times 1\right) + \left(\frac{-1}{3} \times \frac{-1}{2}\right) + \left(\frac{-1}{3} \times \frac{-1}{2}\right) = 0,$$

so that Comparison 1 and Comparison 2 are orthogonal and hence independent. In the same way, by computing the sum of the products of weights we see that Comparison 2 and Comparison 3, and Comparison 1 and Comparison 3 are also independent.

Now look at Comparison 1 and Comparison 4. Here the sum of the products of weights is

$$\left(1 \times \frac{1}{2}\right) + \left(\frac{-1}{3} \times \frac{1}{2}\right) + \left(\frac{-1}{3} \times \frac{-1}{2}\right) + \left(\frac{-1}{3} \times \frac{-1}{2}\right) = \frac{2}{3}$$

so that Comparisons 1 and 4 are *not* orthogonal. Neither are Comparisons 2 and 4. On the other hand, Comparisons 3 and 4 are orthogonal.

The computation and tests for the first three comparisons will next be illustrated. Presumably, our experimenter would handle Comparison 4 separately because of its nonindependence of the others. The information gained from Comparison 4 is redundant, depending on the outcomes of the first three comparisons. However, please do not get the idea that just because a comparison turns out to be nonorthogonal, as referred to some other set of planned comparisons, that it is somehow inferior or "bad." The only problem with a nonorthogonal comparison is that the answer it gives is somewhat redundant with those given by one or more of the other comparisons. It is perfectly permissible to establish confidence limits for, or to test, such a comparison. The only thing this and other nonorthogonal comparisons will *not* do is to equate in particular ways to the sums of squares and degrees of freedom in ANOVA, as we will discuss in Section 11.9.

The value of Comparison 1 is

$$\hat{\psi}_1 = (1 \times 17) + \left(\frac{-1}{3}\right)(24 + 27 + 16) = -5.33.$$

The estimated variance of this comparison is, from Eq. 11.3.3,

$$\text{est. var.}(\hat{\psi}_1) = \frac{5.6}{6}\left[(1)^2 + \left(\frac{-1}{3}\right)^2 + \left(\frac{-1}{3}\right)^2 + \left(\frac{-1}{3}\right)^2\right].$$
$$= .93(1.33) = 1.24.$$

Under the hypothesis that $\psi_1 = 0$, the t ratio is, from Eq. 11.4.2,

$$t = \frac{-5.33}{\sqrt{1.24}} = -4.79.$$

For a nondirectional test, with 20 degrees of freedom, this result is significant beyond the 1% level, so that we reject the hypothesis that the true comparison value is zero. Our experimenter can assert confidently that Population Mean I does differ from the average of Means II, III, and IV.

In a similar fashion we find

$$\hat{\psi}_2 = (1 \times 24) + \left(\frac{-1}{2} \times 27\right) + \left(\frac{-1}{2} \times 16\right) = 2.50$$

with

$$\text{est. var}(\hat{\psi}_2) = \frac{5.6}{6}\left[(1)^2 + \left(\frac{-1}{2}\right)^2 + \left(\frac{-1}{2}\right)^2\right]$$
$$= 1.40.$$

$$t = \frac{2.5}{\sqrt{1.40}} = 2.11.$$

This is just significant at the 5% level for a nondirectional test; Mean II does differ significantly from Means III and IV. Finally,

$$\hat{\psi}_3 = 27 - 16 = 11,$$

with

$$\text{est. var}(\hat{\psi}_3) = \frac{5.6}{6}(1 + 1) = 1.87.$$

Here,

$$t = \frac{11}{\sqrt{1.87}} = 8.04,$$

which is significant far beyond the .05 level. We can say with confidence that Population Mean III is different from Population Mean IV. (See Section 11.9 for a method of testing the entire set of three comparisons on these four means.)

11.7 THE INDEPENDENCE OF SAMPLE COMPARISONS AND THE GRAND MEAN

Comparisons in which the sum of the weights is zero (that is, contrasts) represent weighted differences among sets of means. Such constrasts are usually of most interest to the experimenter. In principle, any set of weights (not all zero) could be used in carrying out a comparison among several sample means; however, as we have seen, one usually uses only sets of weights such that $\sum_j c_j = 0$ for each comparison. Now one reason for this requirement can be shown: by insisting that the weights applied in any comparison sum to zero, we make each comparison value be *independent* of the value of the *grand sample mean*.

The mean of any sample is a linear combination of random variables (Appendix C, comment following C.2.8). Given N individuals divided among J sample groups, the grand mean over all individuals in all groups is a linear combination

$$\bar{y} = \sum_j \sum_i \frac{y_{ij}}{N} \qquad [11.7.1]$$

where each score gets a weight of $1/N$. Furthermore, again in terms of the individual scores, any sample comparison can be written variously as

$$\hat{\psi} = \sum_j c_j \bar{y}_j = \sum_j c_j \sum_i \frac{y_{ij}}{n_j} = \sum_j \sum_i \left(\frac{c_j}{n_j}\right) y_{ij}. \qquad [11.7.2]$$

Each *individual* score y_{ij} actually gets a weight (c_j/n_j), depending on the group j to which it belongs.

In most instances, we want to ask questions about combinations of population means that will be unrelated to any consideration of what the overall mean of the combined populations is estimated to be. Thus, we choose weights such that any comparison to be made is orthogonal to the linear combination standing for the grand mean. In terms of the basic scores, it should be true that

$$\sum_j \sum_i \left(\frac{1}{N}\right)\left(\frac{c_j}{n_j}\right) = 0, \qquad [11.7.3]$$

which in turn implies that

$$\sum_j c_j = 0.$$

(This is the basis for the comments in Section C.4 of Appendix C about the unit vector, since the weights applied to the data to find the grand mean are simply $1/N$ times the unit vector. Then, by C.4.9 and the following, orthogonality with the unit vector implies orthogonality with the mean.)

The comparison for the mean can also be the basis of a test of a hypothesis about the population mean, such as H_0: $\mu = k$. One takes $\hat{\psi}_M = \bar{y}$, and then $F_{(1, N-J)} = (\hat{\psi}_M - k)^2/(\text{MSW}/N) = t^2_{(N-J)}$.

11.8 THE NUMBER OF POSSIBLE INDEPENDENT COMPARISONS

It stands to reason that given any finite amount of data, only a finite set of questions may be asked of those data if one is to get nonredundant, nonoverlapping answers. There is just so much information in any given set of data; once this information has been gained, asking further questions leads to answers that depend on the answers already learned.

This idea of the amount of information in a set of data has a statistical parallel in the number of possible independent (orthogonal) comparisons to be made among J means:

Given J independent sample means, there can be no more than $J - 1$ comparisons, each comparison being independent both of the grand mean and of each of the others.

In other words, the experimenter can frame no more than $J - 1$ different comparisons standing for questions about the sample means, if these comparisons are to be completely independent of each other and the grand mean. This does not say that many different sets of $J - 1$, mutually independent comparisons cannot be found for any set of data. It does say that once a set of $J - 1$ comparisons has been found, where the comparisons are independent of each other and the grand mean, it is impossible to find one more comparison which is also independent both of the grand mean and all of the rest. Thus the number of questions the experimenter may ask of the data as planned comparisons is limited, if the statistical answers are to be regarded as independent.

On the other hand, there are a great many *sets* of orthogonal comparisons that may be applied to any given set of data. In fact, there is no limit at all to the number of ways in which one might choose weights for the various comparisons, such that a set of $J - 1$ such comparisons could be formed. The point is that in any set of such comparisons, that set may consist of no more than $J - 1$ which are mutually orthogonal.

This was the difficulty in the example in Section 11.6. There were only four sample means, and four comparisons were planned; it is impossible to find four mutually orthogonal comparisons *whatever* the weights used, if they all are to be orthogonal to the grand mean in addition. This principle also illuminates the dependencies among multiple t tests; given J samples, there will be $(J)(J - 1)/2$ possible t tests between pairs of means. If J is greater than two, this will always be more than the number of possible independent comparisons. This, then, is another reason why making ordinary t tests on all the differences among J means is not a very good idea (Section 8.17). Not only is the nominal α value for such tests misleading; the pieces of information contributed by the different tests are partially redundant.

On the other hand, must the experimenter be limited only to inferences corresponding to orthogonal comparisons in the data? Fortunately, the answer is no, since there are proper ways of inspecting all differences or comparisons of interest, orthogonal or not, in a set of data, where the significance level attached to each comparison is quite meaningful. However, these methods are most easily understood under post hoc rather than planned comparisons, and will be discussed beginning in Section 11.14.

11.9 PLANNED COMPARISONS AND THE ANALYSIS OF VARIANCE

There is a very intimate connection between analysis of variance and the technique of planned comparisons. **Each and every degree of freedom associated with treatments in any fixed-effects analysis of variance corresponds to some possible comparison of means.** The number of degrees of freedom for the mean square between is the number of possible *independent* comparisons to be made on the means. Any analysis of variance is equivalent to a breakdown of the data into *sets* of orthogonal comparisons (an idea we will develop at length in Chapter 16).

To show how this is true, we need to define the **sum of squares for a comparison:** the sum of squares for any comparison $\hat{\psi}_g$ is

$$SS\,(\hat{\psi}_g) = \frac{(\hat{\psi}_g)^2}{w_g}$$ [11.9.1]

where

$$w_g = \sum_j \frac{c_j^2}{n_j}.$$ [11.9.2]

For any comparison $\hat{\psi}_g$ this sum of squares has *one* degree of freedom. It follows that

$$MS\,(\hat{\psi}_g) = SS\,(\hat{\psi}_g).$$ [11.9.3]

Notice that the F ratio for a comparison $\hat{\psi}_g$ could be written as

$$F_g = \frac{MS\,(\hat{\psi}_g)}{MS\ error},$$ [11.9.4]

with 1 and $N - J$ degrees of freedom, and where $t = \sqrt{F}$.

Extending this idea further, suppose that there were $J - 1$ independent comparisons on the data, and that the SS were calculated for each of them. Then

$$SS\,(all\ \hat{\psi}_g) = SS\,(\hat{\psi}_1) + \cdots + SS\,(\hat{\psi}_g) + \cdots + SS\,(\hat{\psi}_{J-1})$$ [11.9.5]

with $J - 1$ degrees of freedom.

It can be shown that **for any set of $J - 1$ independent sample comparisons on any set of J means,**

$$SS\,(all\ \hat{\psi}_g) = \sum_g SS\,(\hat{\psi}_g) = SS\ between\ groups.$$ [11.9.6]

The total of the sum of squares for any $J - 1$ independent comparisons on J means is always equal to the sum of squares between the J groups.

Furthermore, using Eqs. 11.9.4 and 11.9.6 above, it must then be true that the average F value over the $J - 1$ comparisons is the overall F value:

$$F_{overall} = \frac{\left(\sum_{g=1}^{J-1} F_g \right)}{(J - 1)} = \frac{MS \text{ between}}{MS \text{ within}}.$$ [11.9.7]

Complicated as they may seem on first reading, these principles can be illustrated quite simply in terms of a set of data. Turn once again to the four-group example first introduced in Section 11.2. These data are shown in Table 11.2.1, and the corresponding ANOVA is displayed in Table 11.2.2. Three orthogonal comparisons were carried out (you might want to go back and check the orthogonality for yourself) with the following results:

Comparison 1: -3.5
Comparison 2: -7
Comparison 3: 6

We convert each of these comparisons into its corresponding sum of squares, using the principle embodied in Eq. 11.9.1. Thus, for the first comparison, the denominator term w_1 is found to be

$$w_1 = [(1/2)^2 + (1/2)^2 + (-1/2)^2 + (-1/2)^2]/4$$

$$= 1/4$$

and the sum of squares for this comparison is therefore

$$SS(\hat{\psi}_1) = (-3.5)^2/(1/4)$$

$$= 49.$$

Similarly, we find that the denominator terms and sums of squares for Comparisons 2 and 3 are

$$w_2 = [1^2 + (-1)^2 + 0^2 + 0^2]/4 = 1/2$$

$$SS(\hat{\psi}_2) = (-7)^2/(1/2)$$

$$= 98$$

and

$$w_3 = 1/2$$

$$SS(\hat{\psi}_3) = (6)^2/(1/2)$$

$$= 72.$$

Now, when we add these three SS values representing three orthogonal comparisons, we should get the value of SS between in the ANOVA.

$$SS\ (\hat{\psi}_1) + SS\ (\hat{\psi}_2) + SS\ (\hat{\psi}_3) = 49 + 98 + 72$$
$$= 219.$$

This is, in fact, exactly what happens. Notice that this is precisely the value found for SS between in the ANOVA of Table 11.2.2.

We can then find an F value to test the significance of each comparison.

Thus, for Comparison 1,

$$F = \frac{SS\ (\hat{\psi}_2)}{MS\ within} = 49/6 = 8.17$$

For Comparison 2,

$$F = \frac{SS\ (\hat{\psi}_2)}{MS\ within} = 98/6 = 16.33$$

and for Comparison 3,

$$F = \frac{SS\ (\hat{\psi}_3)}{MS\ within} = 72/6 = 12.00.$$

The average of these F values is then $(8.17 + 16.33 + 12)/3 = 12.17$, which is exactly the overall F value found in the ANOVA of Table 11.2.

11.10 POOLING THE SUMS OF SQUARES FOR "OTHER COMPARISONS"

Quite often it happens that no set of prior questions exists that would use up all the possible $J - 1$ independent comparisons in the data. The experimenter may have only one or two questions that particularly need to be answered, and he is content to lump all other comparisons into a single test, corresponding to "are there any other differences?" Then a combination of planned comparisons and analysis of variance techniques is possible.

Suppose that out of $J - 1$ possible comparisons we decide that only two are of overriding interest. Let us call these comparisons $\hat{\psi}_1$ and $\hat{\psi}_2$. We calculate SS $(\hat{\psi}_1)$ and SS $(\hat{\psi}_2)$ and carry out the tests for each. Now what should we do about all the remaining $J - 1 - 2$ independent comparisons we might make? A very simple way exists for finding and testing the pooled SS value for those remaining. First of all, by the analysis of variance, SS between groups may be calculated in the ordinary way. Then

$$SS\ (\text{all } \hat{\psi} \text{ independent of } \hat{\psi}_1 \text{ and } \hat{\psi}_2) = SS\ \text{between} - SS\ (\hat{\psi}_1) - SS\ (\hat{\psi}_2).$$
$$[11.10.1]$$

In other words, one simply subtracts the sum of squares for Comparisons 1 and 2 from the sum of squares between groups to find the sum of squares for *all remaining* comparisons independent of the first two. Instead of only one F test, the experimenter now makes three tests: one for Comparison 1, one for Comparison 2, and one for all remaining comparisons, given by

$$F = \frac{\text{SS between} - \text{SS}(\hat{\psi}_1) - \text{SS}(\hat{\psi}_2)}{(J - 3)\,\text{MS error}} \qquad [11.10.2]$$

with $(J - 3)$ and $(N - J)$ degrees of freedom.

Any subset of some total possible set of independent comparisons may be planned for in this way: the comparisons of special interest are checked for independence and then tested. All remaining comparisons independent of those tested are embodied in the difference between the SS between and the sum of the SS for the special comparisons tested. The F test for "other comparisons" has degrees of freedom equal to $J - 1$ reduced by one for each comparison tested separately. If this F value is significant, then these comparisons of secondary interest can be examined individually by the post hoc methods discussed later.

11.11 A COMPLETE EXAMPLE USING PLANNED COMPARISONS

The experiment outlined in Section 11.1 was carried out with a total of 30 subjects, assigned at random into groups of 5. The weights for the comparisons representing the four basic questions are shown in Table 11.11.1. A check using the criterion for orthogonality (Section 11.5) shows that these comparisons can be regarded as independent.

It was decided to test each of these comparisons at the .05 level. The data turned out as shown in Table 11.11.2. The numbers in the table are changes in attitude score for each person.

Table 11.11.1

Comparison weights for the experiment of Section 11.1

| Comparison | Treatments | | | Controls | | |
	Movie	Lecture	Movie & lecture	Nothing	Neutral movie	Neutral lecture
1	$\frac{1}{3}$	$\frac{1}{3}$	$\frac{1}{3}$	$-\frac{1}{3}$	$-\frac{1}{3}$	$-\frac{1}{3}$
2	$\frac{1}{2}$	$\frac{1}{2}$	-1	0	0	0
3	1	-1	0	0	0	0
4	0	0	0	1	$-\frac{1}{2}$	$-\frac{1}{2}$

Table 11.11.2

Changes in attitude scores, following indicated treatments.

	Movie	Lecture	Movie & lecture	Nothing	Neutral movie	Neutral lecture	
	6	3	7	−6	5	−1	
	10	6	9	0	−5	3	
	1	−1	4	−5	3	2	
	6	5	9	2	−4	−1	
	4	2	3	2	5	−6	
Totals	27	15	32	−7	4	−3	68
Means	5.4	3	6.4	−1.4	.8	−.6	

The numerical computations for the ordinary one-way analysis of variance are first carried out; not only does this give the MS error needed for the comparisons tests, but also the SS between, which is useful in testing any remaining comparisons.

$$\text{SS total} = 720 - \frac{(68)^2}{30} = 565.87$$

$$\text{SS between} = \frac{(27)^2 + \cdots + (-3)^2}{5} - \frac{(68)}{30} = 256.27$$

$$\text{SS error} = 309.60$$

$$\text{MS error} = \frac{309.60}{24} = 12.90.$$

Now for Comparison 1:

$$\hat{\psi}_1 = \left(\frac{1}{3} \times 5.4\right) + \left(\frac{1}{3} \times 3\right) + \left(\frac{1}{3} \times 6.4\right) - \left(\frac{1}{3} \times -1.4\right) - \left(\frac{1}{3} \times .8\right) -$$

$$\left(\frac{1}{3} \times -.6\right) = 5.33$$

$$w_1 = \frac{1}{5}\left[\frac{1}{9} + \frac{1}{9} + \frac{1}{9} + \frac{1}{9} + \frac{1}{9} + \frac{1}{9}\right] = \frac{2}{15}$$

$$\text{SS}(\hat{\psi}_1) = 15(5.33)^2/2 = 213.07.$$

The test for Comparison 1 is given by

$$F = 213.07/12.90 = 16.52,$$

which is significant far beyond the 5% level for 1 and 24 degrees of freedom. There does seem to be a reliable difference between experimental and control groups, in general.

For Comparison 2,

$$\hat{\psi}_2 = \left(\frac{1}{2} \times 5.4\right) + \left(\frac{1}{2} \times 3\right) - (1 \times 6.4) = -2.20$$

$$w_2 = \frac{1}{5}\left[\left(\frac{1}{2}\right)^2 + \left(\frac{1}{2}\right)^2 + 1\right] = .30$$

$$SS(\hat{\psi}) = \frac{(2.2)^2}{.3} = 16.13$$

so that the F test for this comparison is

$$F = \frac{16.13}{12.90} = 1.25 .$$

This is not significant. There is not enough evidence to say that the combined movie–lecture effect is different from the average of their separate effects.

Comparison 3 gives

$$\hat{\psi}_3 = (1 \times 5.4) - (1 \times 3) = 2.40$$

with

$$w_3 = \frac{1}{5}(1 + 1) = .40$$

so that

$$SS(\hat{\psi}_3) = \frac{(2.4)^2}{.4} = 14.40.$$

The F test for Comparison 3 is then

$$F = \frac{14.4}{12.9} = 1.12,$$

again not significant. There is not enough evidence to say that a movie–lecture difference exists in the populations.

The value for Comparison 4 is

$$\hat{\psi}_4 = (1 \times -1.4) + \left(\frac{-1}{2} \times .8\right) + \left(\frac{-1}{2} \times -.6\right) = -1.50$$

with

$$w_4 = \frac{1}{5}\left[1 + \frac{1}{4} + \frac{1}{4}\right] = .30.$$

The sum of squares

$$SS(\hat{\psi}_4) = \frac{(-1.5)^2}{.3} = 7.50$$

is less than MS error, so that the F test is definitely not significant.

Source	SS	df	Ms	F
Between groups	256.27	5		
Comparison:				
1	213.07	1	213.07	16.52
2	16.13	1	16.13	1.25
3	14.40	1	14.40	1.12
4	7.50	1	7.50	—
Remainder	5.17	1	5.17	—
Error (within groups)	309.60	24	12.90	—
Totals	$\overline{565.87}$	$\overline{29}$		

Table 11.11.3

Summary table showing four planned comparisons.

Only four of the five possible independent comparisons have been made. The sum of squares for the fifth comparison can be found from

$$\text{SS between} - \text{SS} (\hat{\psi}_1) - \text{SS} (\hat{\psi}_2) - \text{SS} (\hat{\psi}_3) - \text{SS} (\hat{\psi}_4)$$

$$= 256.27 - 213.07 - 16.13 - 14.40 - 7.50 = 5.17$$

This last comparison might be tested, but we can see that since its sum of squares is less than MS error, it cannot be significant.

The results of this analysis can be put into tabular form as shown in Table 11.11.3.

11.12 THE CHOICE OF THE PLANNED COMPARISONS

Given any J independent means, there are any number of ways to choose the $J - 1$ independent comparisons to be made among these means. The important thing is that the experimenter have definite prior questions to be answered, and that, ideally, these questions can be framed as orthogonal, or independent, comparisons. The data that are actually collected must then be adequate to provide answers to these questions in terms of such comparisons. For example, in the experiment just analyzed, the experimenter might have had the following initial concerns:

1. Is there any effect of showing the experimental movie, as opposed to the initial lecture or nothing?
2. Is there any effect of the experimental lecture, as opposed to the experimental movie or nothing?
3. Is the effect of the experimental movie the same whether or not it is accompanied by a lecture?
4. Does the neutral movie have the same effect as the neutral lecture?

These four questions can be embodied in four comparisons quite different from those employed above (Table 11.12.1). A check shows that each of these comparisons is independent of each of the others. Notice, however, that some of these comparisons are *not* independent of some of the first set.

Table 11.12.1

Alternative comparisons for the experiment in Section 11.1

Comparisons	Movie	Lecture	Movie & lecture	Nothing	Neutral movie	Neutral lecture
1	$\frac{1}{2}$	$-\frac{1}{2}$	$\frac{1}{2}$	$-\frac{1}{2}$	0	0
2	$-\frac{1}{2}$	$\frac{1}{2}$	$\frac{1}{2}$	$-\frac{1}{2}$	0	0
3	$-\frac{1}{2}$	$-\frac{1}{2}$	$\frac{1}{2}$	$\frac{1}{2}$	0	0
4	0	0	0	0	1	-1

For any given experiment, there will undoubtedly be many interesting questions that could be framed as planned comparisons. Only the particular experimenter can decide, of course, which comparisons are of interest. As a rule, one should not try to think up a sensible question to correspond to each of the $J - 1$ degrees of freedom associated with J means just because it can, in principle, be done. Rather, the technique of planned comparisons should be used only when there are a few important specific questions to ask of the data that will clarify the whole experiment if answered. Once these questions have been decided on, they must be phrased so that the answer will be evident from some way of weighting and combining means. Usually, such questions resolve themselves into differences between groups of means, just as, in the last set of comparisons, the answer to Question 1 depends on the difference between the averages of two groups of means:

$$\frac{(\text{movie}) + (\text{movie \& lecture})}{2} - \frac{(\text{lecture}) + (\text{nothing})}{2}.$$

This difference weights the (movie) mean and the (movie & lecture) means each by 1/2, and the (lecture) and (nothing) group means each by $-1/2$, so that this difference gives us our weights for that comparison. The same is true for each of the other comparisons discussed here.

If a few comparisons have already been planned, and there is interest in finding other comparisons that would be orthogonal to these, then the Gram–Schmidt method outlined in C.5 of Appendix C may be used. Thus, as shown in Section C.5, the fifth comparison orthogonal to each of the comparisons outlined in Table 11.12.1 has weights given by

$$1/4, \quad 1/4, \quad 1/4, \quad 1/4, \quad -1/2, \quad -1/2$$

or, equivalently,

$$1, \quad 1, \quad 1, \quad 1, \quad -2, \quad -2.$$

Finally, if, for some reason, you need to find the weights for $J - 1$ orthogonal comparisons among J means, when the emphasis is on the orthogonality rather than on the meaning of specific comparisons, you can *always* do so using so-called *Helmert contrasts*. These are easily worked out as follows:

1. Choose a group, and give it the weight 1. Then give each of the remaining groups the weight $-1/(J - 1)$. These weights define the first comparison.
2. Assign 0 to the group chosen in Step 1. Choose a second group and give it the weight 1. Assign the weight $-1/(J - 2)$ to the remaining groups. This defines the second orthogonal comparison.
3. Assign 0 to both of the groups chosen in Steps 1 and 2. Choose a third group and assign the weight 1. Then give the weight $-1/(J - 3)$ to the remaining groups. These weights define the third comparison.
4. Keep going in this way until you reach comparison number $J - 1$, which will have one group with weight 1, one with weight -1, and the remainder with weights of 0. For example, if there are $J = 5$ groups then a set of orthogonal comparisons found by this method might look like this:

Group	1	2	3	4	5
comparison 1	1	$-1/4$	$-1/4$	$-1/4$	$-1/4$
comparison 2	0	1	$-1/3$	$-1/3$	$-1/3$
comparison 3	0	0	1	$-1/2$	$-1/2$
comparison 4	0	0	0	1	-1

On the other hand, if you had started with the difference between Groups 1 and 2 only, and then gone on to compare the mean of 1 and 2 with 3 only, and so forth, these would be called *reverse Helmert contrasts*.

Group	1	2	3	4	5
comparison 1	1	-1	0	0	0
comparison 2	1/2	1/2	-1	0	0
comparison 3	1/3	1/3	1/3	-1	0
comparison 4	1/4	1/4	1/4	1/4	-1

The Helmert contrasts do not necessarily have to follow the original ordering of the groups. If we decided to let Group 3 play the role of the first group, Group 1 the role of the second, Group 5 the role of the third, and so on, then the set of Helmert contrasts would look like this:

Group	1	2	3	4	5
comparison 1	$-1/4$	$-1/4$	1	$-1/4$	$-1/4$
comparison 2	1	$-1/3$	0	$-1/3$	$-1/3$
comparison 3	0	$-1/2$	0	$-1/2$	1
comparison 4	0	1	0	-1	0

Obviously, each of the comparisons found by this method focuses on one group, which is then compared with the average of the remaining groups, and where any group that has previously been used as the focus is omitted from consideration. Section 11.14 shows a variant of this method, suitable for testing the individual group effects.

11.13 ERROR RATES: PER COMPARISON AND FAMILYWISE

So far in our discussion of comparisons and their associated tests, we have treated each comparison as somehow "standing alone," and have been concerned only about the possibility of a Type I error on a single test. That is, the usual "nominal" alpha level such as .05 or .01 has been used for each test, just as though no other comparisons were being made and tested on the data. In modern statistical usage this simple, nominal alpha value is often called the *Type I error rate per comparison*, or the *PC error rate*. A convenient symbol for the PC error rate is α_{PC}.

Furthermore, up to this point we have been dealing with planned orthogonal comparisons and have, therefore, followed the usual practice of focussing only on the PC error rate. On the other hand, comparisons are almost always tested in families or sets based on the same set of data, and this introduces the possibility of the other kind of error rate mentioned in Section 8.17. That is, when each of a set of comparisons is to be tested, then we should also be concerned over the probability of making *at least one* Type I error in the entire set or family of comparisons. This probability of making one or more Type I errors in the set of comparison tests is known as the *familywise error rate*, or *FW error rate,* and may be symbolized by α_{FW}, or

α_{FW} = prob.(at least one Type I error in K tests of a family).

If we are dealing with K tests that actually are statistically independent of each other, where the same alpha level (or α_{PC}) is used for each test, then according to Eq. 8.17.1 the relationship of the familywise rate α_{FW} to the error rate per comparison α_{PC} can be stated precisely:

$$\alpha_{FW} = 1 - (1 - \alpha_{PC})^K. \qquad [11.13.1]$$

When the various tests are not independent, the relation between the familywise error rate α_{FW} and the error rate per comparison α_{PC} is much more difficult to specify, however. Nevertheless, as we also learned in Section 8.17, when any K tests are carried out, each employing the same value of α for Type I error, it must still be true that

probability of at least one Type I error $\leq K\alpha$.

(you may remember that in Section 8.17 this principle was identified as one of the Bonferroni inequalities.)

In the present context of testing a family of K comparisons we can identify the constant per-comparison error rate α_{PC} with α, so that we thus have

$$\alpha_{FW} \leq K\alpha_{PC} . \qquad [11.13.2]$$

More generally, the investigator might employ a different level for each of a set of K tests. One reason may be the desire to ensure more power for some, presumably more important, questions than for others. He or she does this by making the designated alpha level *larger* for these important tests than otherwise. Now assuming such a situation to exist, we can represent the K alpha levels associated with the tests by

$\alpha_1, \alpha_2, \cdots , \alpha_K.$

Then by another inequality due to Bonferroni, we have the relationship

$$\alpha_{FW} \le \alpha_1 + \alpha_2 + \cdots + \alpha_K \qquad [11.13.3]$$

The familywise error rate must always be less than or equal to the sum of the error rates over the individual tests. When all of the K tests are carried out at the same error rate, or α_{PC}, then Eq. 11.13.3 becomes identical to Eq. 11.13.2.

Clearly, the familywise error rate α_{FW} has an upper limit that depends on K, the number of tests to be made, and on the error rate per test, or α_{PC}. It follows that by making certain adjustments in the per-comparison error rate one can establish a satisfactory *maximum familywise error rate*, or *protection level* against the possibility of at least one Type I error. In particular, if we want to make the FW rate no larger than some value, say α_{FW*}, then we can always do so by setting the PC rate for *each test* at

$$\alpha_{PC} = (\alpha_{FW*})/K$$

so that

$$\alpha_{FW} \le \alpha_{FW*} . \qquad [11.13.4]$$

Obviously, the familywise error rate can be of concern whenever one makes more than a single test of significance on the same data. However, current practice often tends to ignore the familywise rate in carrying out a fairly small number of tests that can be regarded as (more or less) independent, as with a set of planned orthogonal comparisons. On the other hand, when the set of comparisons is not orthogonal, when there are many comparisons to be tested, or when comparisons are made on a post hoc basis, steps should be taken to control for excessive familywise error, or α_{FW}. This is sometimes done by setting a maximum familywise error rate using one of the Bonferroni principles just outlined. Alternatively, post hoc comparisons are often tested by special methods devised to achieve a relatively small value of α_{FW}.

We will now explore the use of the Bonferroni principle Eq. 11.13.2 to examine planned (though not necessarily orthogonal) sets of comparisons, and then we will turn to methods devised specifically for testing post hoc comparisons.

11.14 USING BONFERRONI TESTS FOR COMPARISONS

Should we so desire, the Bonferroni inequality (Eq. 11.13.2) mentioned in the previous section can be used directly for controlling the FW error rate in either orthogonal or nonorthogonal sets of planned comparisons. Thus, suppose that we plan to examine and test a set of K comparisons of interest, where K may or may not equal $J - 1$, the number of degrees of freedom between the groups in question. Then, you must decide on an acceptable familywise rate of error for the set. Let us again call this α_{FW*}.

Let us also assume that a two-tailed test is desired for each comparison. If the FW rate is to be fixed at α_{FW*}, then the effective PC rate according to the Bonferroni principle, or α_{BPC}, that will be used on any single test is

$$\alpha_{BPC} = \alpha_{FW*}/K. \qquad [11.14.1]$$

The symbol t_B will stand for the value of t that is required for two-tailed significance at the α_{BPC} level, with degrees of freedom equal to those for error, or $N - J$. We can then evaluate any of the required set of comparisons in the usual way (Eq. 11.2.2) and find the corresponding t value (Eq. 11.4.2). This value is checked against the value of t_B, as required by the value of α_{BPC} under this Bonferroni criterion. The result is declared significant if the comparison t equals or exceeds t_B in absolute value. (Although we will confine our attention here to situations where the investigator uses the same error rate α_{PC} for each of a set of comparisons, similar procedures are available based on the second Bonferroni inequality given above as Eq. 11.13.3. Such procedures are, therefore, applicable to the situation where different alpha levels are selected for different tests among a set of comparisons. Details of this alternative Bonferroni approach may be found in Kirk (1982).)

It will seldom be true, however, that the t_B corresponding to a desired α_{BPC} level will be one of those listed in a standard table such as Table III, Appendix E, and specialized tables giving t values for Bonferroni tests can be found in a number of texts such as Edwards (1985) and Kirk (1982).

On the other hand, for the most common situations where there are 10 or more *error* or *denominator* degrees of freedom, an adequate approximation to the t_B value required can be found as follows:

1. For a desired FW error rate α_{FW*} find the per-comparison rate α_{BPC}. Then let $z*$ be the standardized value cutting off the upper $(\alpha_{BPC})/2$ proportion in a normal distribution. This value can usually be found directly from Table I of Appendix E, or from special Table I-A.

2. Let ν symbolize the number of degrees of freedom in the t ratio for each comparison (that is, $N - J$, the usual number of degrees of freedom for MS error or MS within).

3. Then the required value of t_B, cutting off the upper $(\alpha_{BPC})/2$ proportion in a distribution with ν degrees of freedom, is given approximately by

$$t_B = z* + \{(z*)^3 + z*\}/4(\nu - 2) . \qquad [11.14.2]$$

4. If, for any comparison, the calculated t equals or exceeds the value of t_B found in Step 3 above, the comparison is declared to be significant. Otherwise, the comparison is considered nonsignificant.

A potentially useful modification of this procedure was suggested by Keppel (1982), for the situation where the number of comparisons exceeds $J - 1$, the number of degrees of freedom between groups. Keppel suggests setting $\alpha_{FW*} = (J - 1)(.05)$ or $\alpha_{FW*} = (J - 1)(.01)$ and then carrying out the procedure exactly as outlined above. This will make the familywise rate for the entire set of comparisons no larger than it would have been had the set of comparisons actually been orthogonal and the .05 or .01 level been used for each test.

For example, consider the means given below, representing 24 cases divided at random into four different treatment groups of 6 cases each. The analysis of variance on the data gave an MS error value of 7.82, with 20 degrees of freedom.

	Group			
	1	2	3	4
mean	21.83	18.33	11.17	23.33

Now suppose that our interest lies in four planned comparisons, equivalent to the four sample effects (Section 10.4). Since each effect is defined as the difference between a group mean and the overall mean, a little simple algebra shows that if n is constant over groups, each effect can be found by making a comparison of the form

(mean of a group − mean of the remaining groups)$(J − 1)/J$.

In other words, the basic information desired can be gained by contrasting each group with the mean of the remaining groups. These four comparisons would have weights as shown in Table 11.14.1, which also shows the comparison value for each.

Observe that these comparisons do not make up an orthogonal set, nor should they, since we are making four comparisons on data with only three degrees of freedom. Therefore, in making these tests we will use the Keppel modification as outlined above.

Following the steps recommended, if we let $\alpha_{FW*} = 3(.05) = .15$ then the value of $\alpha_{BPC} = .15/4 = .0375$, and $.0375/2 = .01875$. Table I in Appendix E shows that the proportion .01875 is cut off on the upper tail of a standardized normal distribution by a value of about 2.08. Hence $z* = 2.08$. Then when ν is set equal to 20, the degrees of freedom for MS error, the required value of t_B is estimated to be

$$t_B = z* + [(z*)^3 + z*]/4(\nu − 2)$$

$$= 2.08 + [(2.08)^3 + 2.08]/4(18)$$

$$= 2.23$$

This value of t_B tells us that if we calculate the t value for each of the four comparisons, we should call it significant only if the t equals or exceeds 2.23 in absolute value.

Table 11.14.1.

Four comparisons on the means of table as tested under the modified Bonferroni criterion

	Group					
	A_1	A_2	A_3	A_4	$\hat{\psi}$ value	t
mean	21.83	18.33	11.17	23.33		
comp. 1	1	−1/3	−1/3	−1/3	4.22	3.20 sig. $< .05$
comp. 2	−1/3	1	−1/3	−1/3	−0.45	−0.34
comp. 3	−1/3	−1/3	1	−1/3	−9.99	−7.58 sig. $< .05$
comp. 4	−1/3	−1/3	−1/3	1	6.22	4.72 sig. $< .05$

Therefore, on actually calculating the first t value we have

$$t = (4.22)/\sqrt{7.82(12/9)(1/6)}$$

$$= 3.20.$$

which is greater than 2.23, and hence judged to be significant.

The remainder of the t values and the conclusions about their significance are shown in Table 11.14.1. Note that the first, third, and fourth comparisons gave t values exceeding 2.23, and were thus declared significant. Over these four comparison t tests we know that the probability of at least one Type I error is at most .15, which is the same maximum familywise rate as if only three orthogonal comparisons had been tested.

11.15 INCIDENTAL OR POST HOC COMPARISONS IN DATA

Even though tests for planned comparisons form a useful technique in experimentation, it is far more common for the experimenter to have no special questions to begin with. Initial concern is to establish only that some real effects or comparison differences do exist in the data. Given a significant over-all test, the task is then to explore the data to find the source of these effects and to try to explain their meaning.

Post hoc comparisons are of three broad types, each handled in somewhat different ways. The first type of comparison may correspond to any question of interest about the J experimental groups, where the answer can be framed as a weighted sum of the means, and the weights themselves sum to zero. Such a comparison looks exactly like a planned comparison, and the actual comparison value, its sum of squares, and its t or F value are found as described in Sections 11.2 to 11.4 above. However, unlike a planned comparison, this way of weighting and summing the data is often suggested by the inspection of the results *after* an overall F test has been found to be significant. Furthermore, no requirement of orthogonality is placed on post hoc comparisons, and any number of such comparisons may be made and tested, once a significant overall F has been found.

The post hoc situation therefore calls for an approach to significance which makes allowance for the fact that comparisons can be suggested by inspection of the data, and that many tests are often made on the same set of data, leading to an inflation of the familywise error rate. The problem is complicated by the fact that such tests will not usually be independent. The Bonferroni method just outlined can be applied conveniently to post hoc comparisons, although it becomes much too conservative to be practicable when many comparisons are to be tested. Another standard and very useful approach to testing post hoc comparisons is the so-called Scheffé method, to be discussed in the next section.

A second strategy of post hoc comparisons involves examining all pairs of group means, and determining which of those pairs shows significant differences. Essentially, this approach uses $J(J - 1)/2$ comparisons, in which the weights are always 1 for one group's mean, and -1 for the mean of another group. Often, this is called the problem of *multiple comparisons* or *pair comparisons*. Clearly, such a set of comparisons cannot be orthogonal, and hence the comparison values and their tests must be interdependent to some extent. Several methods exist for handling such pair comparisons of means, and two of the most commonly used will be discussed in Section 11.17.

Still other specialized comparisons may be made for different purposes. One example is the comparison of each experimental group with a control group, which is handled by a method devised by Dunette. A description of this and related methods for specialized post hoc comparisons can be found in Kirk (1982). The limited space available here will be devoted to the frequently encountered types listed above.

11.16 SCHEFFÉ COMPARISONS

Perhaps the most versatile of the methods for testing post hoc comparisons is that due to Scheffé (1959). This method has the advantages of simplicity, applicability to groups of unequal size, and suitability for any comparison (with one exception to be noted below). The Scheffé method is known to be relatively insensitive to departures from normality and homogeneity of variance. The method is also conservative, meaning that in some situations (notably, comparisons of pairs of means) the Scheffé method tends to have low power compared with some other applicable methods.

The idea of a comparison here is exactly the same as defined in Section 11.2. A sample comparison $\hat{\psi}$ is a linear combination of sample means

$$\hat{\psi} = \sum_j c_j \bar{y}_j$$

where $\Sigma\, c_j = 0$.

After the overall F has been found significant, then any comparison $\hat{\psi}$ may be made and tested. Unlike planned comparisons, in the post hoc procedure, there is no advantage in requiring that these comparisons be orthogonal. Any and all comparisons of interest may be made and tested.

Given any comparison g made on the data after a significant F has been found for the relevant factor, the significance of the comparison value $\hat{\psi}_g$ may be found by use either of a t or an F ratio. If t is used, then one again takes

$$t = (\hat{\psi}_g)/\sqrt{\text{est. var}(\hat{\psi}_g)}$$

with $N - J$ degrees of freedom. Just as before, the estimated variance of the comparison is found from

$$\text{est. var}(\hat{\psi}_g) = (\text{MS error})w_g$$

with w_g defined by Eq. 11.9.2.

However, instead of referring this t value to the usual table showing probabilities in a t distribution, one uses the so-called Scheffé criterion, given by

$$\mathbf{S} = \sqrt{(J - 1)F_\alpha}. \qquad [11.16.1]$$

(Take care to note that boldface **S** as used here is *not* the sample standard deviation.) Here, F_α is the value required for significance at the α level, with $J - 1$ and $N - J$ degrees of freedom (that is, the F required for significance at the α level for an *overall* test of the J means).

If you are interested in a two-tailed test for your comparison, an equivalent procedure to t is an F test (by the relationship $t^2 = F$, with df numerator $= 1$). Hence we can also use

$$F = (\hat{\psi}_g)^2/\text{est. var}(\hat{\psi}_g)$$

or

$$F = \text{SS}(\hat{\psi}_g)/\text{MS error}$$

with 1 and $N - J$ degrees of freedom. For the F test, however, the Scheffé criterion becomes

$$\mathbf{S}^2 = (J - 1)F_\alpha$$

where F_α is defined just as above.

On the other hand, it may be preferable to construct confidence intervals for the value of $\hat{\psi}_g$, based on the sample value. This is done by taking the confidence limits

$$\hat{\psi}_g - \mathbf{S}\sqrt{\text{est.var}(\hat{\psi}_g)}$$

and [11.16.2]

$$\hat{\psi}_g + \mathbf{S}\sqrt{\text{est.var}(\hat{\psi}_g)}.$$

For any α, this gives the $100(1 - \alpha)$ percent confidence interval for ψ_g, the true value of the comparison. When the confidence interval fails to cover zero, the comparison is said to be significant, and identified as one possible contributor to the overall significance of F. (Actually, all one really needs to do to determine if the confidence interval covers the value of zero in the procedure given above is to calculate $\mathbf{S}\sqrt{\text{est.var}(\hat{\psi}_g)}$. If this is smaller than the absolute value of $\hat{\psi}_g$, then the confidence interval does *not* cover zero, and the comparison is significant.)

To illustrate the Scheffé method for post hoc comparisons, let us use the four means given in Section 11.14 once again, and pretend that the four comparisons carried out on these means (as shown in Table 11.14.1) actually followed the significant F in the analysis of variance. Then, in order to test the comparison values according to the Scheffé criterion, we need the value of F required for significance at the .05 level for 3 and 20 degrees of freedom, or $F_\alpha = 3.10$. The value of $\mathbf{S}$ is then

$$\mathbf{S} = \sqrt{3(3.10)} = 3.05 .$$

The estimated sampling variance for the first comparison is given by

$$\text{est. var}(\hat{\psi}_1) = \text{MS error}(1 + 1/9 + 1/9 + 1/9)/6$$

$$= 7.82(2/9) = 1.74.$$

The t ratios for each of these comparisons (also as shown in Table 11.14.1) are repeated below:

$$t_1 = 4.22/\sqrt{1.74} = 3.20$$

$$t_2 = -0.45/\sqrt{1.74} = -0.34$$

$$t_2 = -9.99/\sqrt{1.74} = -7.58$$

$$t_4 = 6.22/\sqrt{1.74} = 4.72$$

Now only the second t value fails to exceed the value of **S**, or 3.05. Hence the first, third, and fourth comparisons are significant beyond the .05 level according to the Scheffé criterion.

We could have reached these same conclusions by the use of confidence limits, of course. Thus, the confidence interval for Comparison 1 then has limits of

$$4.22 - 3.05\sqrt{1.74} \text{ or } 0.197$$

and

$$4.22 + 3.05\sqrt{1.74} \text{ or } 8.243 .$$

Obviously, these limits do not include 0, and hence this comparison would be judged significant. The other comparisons can be given confidence limits in exactly this same way. (To make these be confidence limits for the effects, α_j, we would simply multiply the limits by $(J - 1)/J$.)

The meaning of these tests and confidence intervals using the Scheffé criterion requires some special comment. In the Scheffé t and F tests, the FW error rate is actually the α value used to find the F_α that figures into the **S** criterion value. Thus, for our example the familywise error rate is .05. However, what is the *family* here? The family used as the basis for the Scheffé procedure is actually *all possible comparisons* that might have been carried out on this same set of data (orthogonal, nonorthogonal, or whatnot). This is a very large family indeed! Nevertheless, if we were able to discover, evaluate, and test every conceivable comparison among the four groups represented by these data, and if there really were no population differences, then the proportion of significant comparisons should be α_{FW}, the familywise rate.

The meaning of the confidence intervals for these post hoc comparisons is much the same: If we could work out all possible comparisons on the data, and for each comparison calculate a 95% confidence interval, then the chances are 95 in 100 that *every one* of these confidence intervals would contain the true value for that comparison. There is only a 5% chance that one or more confidence intervals will not cover the corresponding true comparison value.

Put in a different way, the *familywise* error rate for Scheffé tests is fixed when one chooses the value of α that determines F_α and **S**. However, the family in question consists of *all* comparisons that could be carried out on the given set of data. **If the overall F test is significant at the α level, then some possible comparison $\hat{\psi}_g$, must be significant at or beyond the same level.** Indeed, a significant F test can be interpreted as evidence that at least one true comparison value among all those possible is not zero. This does not mean that just because the overall F was significant you will necessarily find the significant comparisons, but only that they exist. Hence our interpretation of a significant F as a signal, "Something's here—start looking."

This statement is not to be interpreted to mean that post hoc comparisons are somehow illegal or immoral if the original F test is not significant at the required α level. It means only that the probability statements that one makes about such comparisons are not necessarily true when F does not reach that level, and that going through the procedure may be something of a waste of time if one really is interested in accurate probability statements. Nevertheless, you can investigate comparisons among means whenever and wherever you like, and such comparisons may be quite suggestive about what is going on in the data or about new questions calling for new experiments. What one cannot do is to attach an unequivocal probability statement to such post hoc comparisons, unless the conditions underlying the method have been met.

11.17 PAIR COMPARISONS AMONG MEANS

As we have seen, the Scheffé method can be applied to any comparisons among the means, and to as many comparisons as you wish to test, orthogonal or not. However, a very common problem is the investigation of all comparisons corresponding to differences between sample means, and, unfortunately, for tests of all differences between pairs of means, the Scheffé method may be unnecessarily conservative. There are, however, several methods available for the post hoc testing of all differences between means. Space limitations permit discussion of only two of these methods here, however: the Tukey HSD method, and the Newman-Keuls procedure. A good discussion of a number of the other available methods may be found in Kirk (1982).

Both the Tukey HSD and the Newman-Keuls methods rely on a statistic known as *the studentized range,* which may be defined as follows: Suppose that J independent samples are employed, each yielding a mean $\bar{y}_j$. These means are arranged in order of magnitude, and the largest and smallest mean noted. Let us symbolize these largest and smallest mean values by $\bar{y}_{max}$ and $\bar{y}_{min}$. The *range* of these J means is given by the difference between the largest and smallest, or $\bar{y}_{max} - \bar{y}_{min}$. Then the definition of *the studentized range statistic,* symbolized by q, is

$$q = \frac{\bar{y}_{max} - \bar{y}_{min}}{\sqrt{\text{MS error}/n}} \qquad [11.17.1]$$

where MS error is found from a preceding analysis of variance. The n here is the sample size, if all samples are of equal size. However, if the samples have different sizes, an approximate q statistic can be found by using $2n_1 n_2 / (n_1 + n_2)$ in the place of n, where n_1 is the size of the sample with the largest mean, or $\bar{y}_{max}$, and n_2 is the sample size for $\bar{y}_{min}$. Alternatively, n may be set equal to the harmonic mean of all the sample sizes. (See Section 4.2.) Other, somewhat better, procedures for use in this situation are outlined in Kirk (1982, pp. 118 f.f.).

The statistic q has a sampling distribution which depends upon two parameters: k, the number of means actually covered by the range making up the numerator of q, and v, the number of degrees of freedom associated with MS error. Table XI in Appendix E gives the .05 and the .01 points in a studentized range distribution, where k is shown by the column entry, and the degrees of freedom by the row. Thus, for $k = 3$ and 10 degrees of freedom, the .05 value is 3.15, meaning that in a studentized range distribution with these parameters, .05 of all q values fall at or above 3.15.

In the Tukey HSD test, we always take $k = J$, the entire number of groups. Having decided on the value of α to use, we next look under the correct degrees of freedom to find the q value that cuts off the upper required α proportion of the distribution. Let us call this value q_α. Then,

Source	SS	df	MS	F	**Table 11.7.1**
Between groups	2,942.4	4	735.6	4.13	
Error (within groups)	9,801.0	55	178.2		
Totals	12,743.4				

$$ \text{HSD} = \text{honestly significant difference} = q_\alpha \sqrt{\frac{\text{MS error}}{n}}. \qquad [11.17.2] $$

(Note the comment on n following Eq. 11.17.1) Then we examine all pairs of means and note their differences. If the absolute difference between any pair of means equals or exceeds the HSD, then we reject the hypothesis of the equality of means of the populations represented by the samples. Over the family of *all experiments* identical to the one being analyzed, the probability is α that a difference between means will exceed HSD, by chance alone. That is, for this method α_{FW} is equal to the value of α we use when finding the required value of q in Table XI. This method will now be illustrated.

In a psychological study subjects were assigned randomly to five different groups, representing five different experimental treatments. Twelve subjects were used in each group. The means of the various groups were

I	II	III	IV	V
63	82	80	77	70

and the analysis of variance is summarized in Table 11.17.1. For an α of .05, the required F for 4 and 55 degrees of freedom is approximately 2.53, so that this obtained value is significant. Hence post hoc comparisons may be tested for significance. The .05 level will be used in these tests as well.

First the Tukey HSD method will be illustrated for the differences between all pairs of means. For this purpose it is convenient to arrange the means in order, as in Table 11.17.2 so that the difference between any pair is immediately apparent. The numbers in the body of Table 11.17.2 represent the difference between the mean shown in the column and the mean shown in the row. Since both columns and rows list the means in order of magnitude, then all that need be shown are the differences above the diagonal; the same differences would appear with a negative sign below the diagonal.

		I	V	IV	III	II
		63	70	77	80	82
I	63	0	7	14	17	19
V	70		0	7	10	12
IV	77			0	3	5
III	80				0	2
II	82					0

Table 11.17.2

All pairwise differences for five means.

To apply the Tukey HSD criterion to these differences, we first note that $k = 5$, the total number of groups, and that MS error in Table 11.17.1 has 55 degrees of freedom. Using 60 as the closest degrees of freedom in Table XI and using $\alpha = .05$, we find the value of $q_{.05}$ to be 3.98. We already know that MS error $= 178.2$, and that $n = 12$. Thus, we take

$$\text{HSD} = 3.98\sqrt{178.2/12} = 15.34.$$

Now we look at Table 11.17.2 once again, and see that two differences exceed this HSD value: the difference between Groups I and II, and between Groups I and III. Hence, we can say that these two differences are significant beyond the .05 level.

The Newman-Keuls test procedure uses the same basic mechanics as the Tukey HSD procedure. That is, one starts with an ordered table of differences exactly like that in Table 11.17.2. However, rather than testing all the differences in the table in the same way, the table is regarded as divided into "layers," and differences occurring in the different layers are subjected to different test criteria.

The idea of a layered table is illustrated in Table 11.17.3. The top layer of the table, Layer 4 in this example, is shown in the upper right-hand corner. This layer contains the difference between means that are $K - 1 = 4$ steps apart in order. Obviously, for five means there is only one such difference, which is 19 in our example. The difference contained in this top layer is tested by use of the Tukey HSD, exactly as shown above:

$$\text{HSD, layer } K - 1 = q_{(\alpha,K,v)}\sqrt{\text{MS error}/n}$$

$$= 15.34 \ .$$

Therefore, our obtained difference of 19 in Layer 4 is significant. However, this is the only difference tested in exactly this way under the Newman-Keuls procedure.

The second highest layer of the table contains difference means that are $K - 2 = 3$ steps apart in order. In our example these are the differences 17 and 12. These differences are tested by applying the HSD criterion once again, with the same MS error and n values, but with q found for $K - 1 = 4$ rather than $K = 5$ groups. For this example, the q value at .05 for $K - 1 = 4$ groups and 55 degrees of freedom is about 3.74. Thus, the new HSD value for this layer is

$$\text{HSD, Layer 3} = q_{(\alpha,K-1,v)}\sqrt{\text{MS error}/n}$$

$$= 3.74 \ (3.85)$$

$$= 14.40$$

	I	*V*	*IV*	*III*	*II*	
I	0	7	14	17	19	← Layer 4
V		0	7	10	12	← Layer 3
IV			0	3	5	← Layer 2
III				0	2	← Layer 1
II					0	

Table 11.17.3

The differences of Table 11.17.2 shown in a "layered" table

Then, in Layer 3 only one difference, between Groups I and III, is found it be greater than 14.40, and thus significant.

We keep going in exactly this way, with the HSD criterion being modified for each layer by finding q for one fewer group than in the preceding layer, until finally we reach Layer 1. For this example, the required HSD values for Layers 2 and 1 turn out to be

HSD, Layer 2 $= 3.40(3.85) = 13.09$

so that the difference of 14 between Groups IV and I is significant, and

HSD, Layer 1 $= 2.83(3.85) = 10.90$.

Observe that by use of the Newman-Keuls procedure, as just illustrated, these data show one more significant difference than when the Tukey HSD method was used for the entire table. Naturally, this is the whole point of the Newman-Keuls and related methods: to improve the power of the post hoc pair comparison tests so as to detect more differences as significant. Unlike the Tukey HSD, however, the Newman-Keuls method does not control the familywise error rate at the nominal α used to find the value of q.

If, for illustrative purposes, we apply the Scheffé confidence interval method to this set of pairwise differences, we find that it takes a difference of 17.33 or more to be called significant at the .05 level. Thus, only the difference between Groups I and II is significant under the Scheffé method, whereas we found two significant differences using the Tukey HSD method, and three for Newman-Keuls. This illustrates the point made earlier about the conservatism of the Scheffé method when applied to pairwise comparisons.

By way of comparison, the Bonferroni inequality implies that we should divide $\alpha = .05$ by the number of pairs, or 10, if we wish the probability to be no more than .05 that at least one Type I error occurs, or $\alpha_{FW} \le .05$. Then, using a (two-tailed) significance level of .005, and approximating the required t_B value of about 2.92 by use of Eq. 11.14.1 we find that in order to be significant a difference must equal or exceed (in absolute value)

$$2.92 \sqrt{\frac{2MS \text{ error}}{12}} = 2.92 \sqrt{\frac{178.2}{6}} = 15.91.$$

Two differences actually do exceed this value, and are significant. In making this test for all 10 differences, we know that the probability of a Type I error in at least one test is no larger than .05. Notice that exactly the same pair of differences turn out to be significant in this method as in the Tukey HSD.

11.18 PLANNED OR POST HOC COMPARISONS?

It is obvious that in any given experiment it is always possible either to plan comparisons to be tested in lieu of the overall F test, or to perform post hoc comparisons should the overall F be significant. What are the arguments for and against these two ways of proceeding?

An important point in favor of planned comparisons is this: consider any true comparison ψ among J means $(J > 2)$, such that $\psi \neq 0$. **The probability of a test's detecting that ψ is not zero is greater with a planned than with an unplanned comparison on the same sample means.** Thus, for any particular comparison, the test is more powerful when planned than when post hoc. **Put differently, the confidence interval for any given comparison is shorter when that comparison is planned than when it is post hoc.**

The practical implication is that the *importance* of the comparison should dictate whether or not it is tested by the planned or the post hoc procedure. If the experimental question represented by the comparison is an important one for the interpretation of the experiment, and it is essential that a Type II error not be made in the accompanying test, then a planned comparison should be carried out. On the other hand, if the question is a minor one, and a considerable risk of overlooking a true nonzero value can be tolerated, then the post hoc method suffices.

However, we have also seen that if we wish to preserve the link to ANOVA, the number and variety of planned comparisons to be made and tested is limited by the independence requirement. No such requirement holds for the post hoc method. The data may be only partly explored by the planned method, but fully by post hoc methods.

In summary, then, let it be said that the planned comparisons method is best suited to situations where a few overriding concerns dictate the interpretation of the whole experiment. Here one must have the most powerful tests possible for resolving these issues. The post hoc method is suited to trying out hunches gained during the data analysis and for inferring the sources of the significant overall F test. We will return to the subject of planned and post hoc comparisons in Chapter 16, where their applicability to still another kind of experiment will be pointed out.

11.19 PLANNED AND POST HOC COMPARISONS IN COMPUTER PACKAGES

All of the popular statistical packages for computers include procedures for both planned and post hoc comparisons or contrasts. Although one-way ANOVA can be carried out on the subprogram ANOVA, the SPSS (and SPSS^x) subprogram ONEWAY, as introduced in the preceding chapter, also includes both kinds of comparisons as optional features. After the initial control statement

ONEWAY *dependent variable list* BY *independent variable (min,max)/*

planned comparisons (or, as the SPSS manual calls them, a priori contrasts) may be carried out by use of the keyword CONTRAST $=$, followed by the actual weights you wish to have applied to the groups, listed in order of the groups, *not* separated by commas, and ending with a slash (/). The user must check to see that the sum of the weights is zero, and that the comparisons are orthogonal (if such are desired). Up to 10 different contrasts are allowed for each ONEWAY statement, with each list of weights terminated by a slash.

The ONEWAY procedure also allows for post hoc comparisons (or, as the SPSS manual terms them, a posteriori contrasts). The keyword RANGES = is a call for post hoc contrasts on the set of means, and one may specify the Scheffé, Tukey, and Newman-Keuls tests, as well as several others not discussed here. (Other comparison options included in ONEWAY will be discussed following Chapter 16).

In both SPSS and SPSS[x] the MANOVA procedure also permits comparisons to be made, under the keyword CONTRAST. A choice is given among six different types of comparisons, or the user may select the comparison weights to fill special needs.

In the BMDP program 1V, planned comparisons or contrasts can also be requested as part of the DESIGN paragraph already referred to in Section 10.22. This takes the format

> / DESIGN   GROUP = group list (or all, if unspecified)
> DEPEND = variable list
> CONTRAST = list of weights
> (repeated for additional contrasts)

The comparisons specified are then subjected to t tests.

Within the SAS PROC ANOVA, the MEANS statement permits one to examine any of the commonly employed post hoc comparisons such as Scheffé, Tukey, Newman-Keuls, and Bonferroni criterion tests, as well as a number of newer and less familiar procedures. On the other hand, planned comparisons are most conveniently handled in the procedure PROC GLM, where CONTRAST statements are used to specifiy the weights to be used in specific planned comparisons. In addition, PROC GLM also provides for post hoc comparisons by use of a MEANS statement, just as in PROC ANOVA.

EXERCISES

1. A highway department was concerned with developing an effective reflective sign for use at certain critical locations. Originally they planned to look at all combinations of three background colors (red, blue, and black) with three possible letter colors (white, yellow, and green). However, three of these combinations proved to have insufficient contrast and they were dropped from the study. Visual stimuli were made up in the remaining six combinations, and flashed on a screen for a subject, who reacted to the stimuli as quickly as possible. Ten subjects saw all stimuli, and each subject was given a total reaction time score under each condition. The average scores across subjects for each color combination are shown below (in units of .1 s). The mean square used for error was 109.87, with 45 degrees of freedom. Set up the weights for comparisons representing the following questions:

(a) Do white letters produce different reaction times from yellow or green letters?
(b) Given white letters, does a red background give a different reaction time from that for a blue or black background?
(c) Given white letters, is a blue background different from a black?
(d) Given yellow letters, is a blue background different from black?
(e) Is Combination 6 different from Combinations 4 or 5?

Combination	Background	Letters	Mean
1	red	white	13
2	blue	white	15
3	black	white	10
4	blue	yellow	20
5	black	yellow	18
6	red	green	35

2. Check on the orthogonality of each pair of comparison weights found in Exercise 1. To which of these comparisons would a sixth comparison corresponding to the following also be orthogonal?

Comparison Question 6: Is the average of Combinations 2 and 3 different from the average of Combinations 4 and 5?

3. Test the significance of each of the planned comparisons outlined in Exercise 1. Use $\alpha = .05$.

4. Analyze the following data using Helmert comparisons, starting with Group A1 versus the remaining groups, proceeding to A2 against the other groups, and so on.

Groups			
A1	A2	A3	A4
2	4	15	10
8	7	20	12
6	8	18	15
4	6	14	17
5	10	10	10
3	12	11	8
2	7	9	8
1	9	12	13
4	8	14	8
3	12	15	9

5. Consider the data in Exercise 21, Chapter 10 once again. Design and carry out a set of orthogonal comparisons, beginning with the comparison of means for Group 1 with Groups 2, 3 and 4. (You should transform these scores as called for in the original exercise.)

6. Using Exercise 5 above, and the data in Exercise 21, Chapter 10, demonstrate the relation between the comparison sums of squares and the sum of squares between groups.

7. Suppose that the experimenters in Exercise 1 decided that instead of planned comparisons, they would carry out an overall F test for possible differences among the means of the six combinations. In view of the results for Exercise 3 above, find the value of this F statistic. Is this significant for $\alpha = .05$?

8. Below are data for four independent groups, with $n = 3$ cases per group. Use the data to do the following:

(a) Test the hypothesis of no group effects.
(b) If the preceding step shows significance, then carry out Tukey HSD tests for all pairs of these means. Use $\alpha = .05$.

A1	A2	A3	A4
5	8	16	7
7	10	15	8
9	9	12	14

9. In a study of the effects of reward on learning in small children, an experimenter used 20 children divided equally into four groups. Each child was given a puzzle which could be solved only if the steps were learned in order. For Group I, the child was rewarded for every correct move until the puzzle was learned. In Group II, 75% of a child's moves were rewarded on a random schedule. In Group III, only 25% of correct moves were rewarded, and in Group IV no moves were rewarded. The table below shows the number of trials it took each child to learn the correct sequence of moves. The experimenter entertained the following hypotheses:

(a) Constant reward will produce faster learning than the average of the other conditions.
(b) Frequent reward will produce faster learning than the average of infrequent or no reward.
(c) Infrequent reward will produce faster learning than no reward.

Test each of the experimenter's hypotheses, using $\alpha = .05$.

Group			
Constant Reward	Frequent	Infrequent	Never
12	9	15	17
13	10	16	18
11	9	17	12
12	13	16	17
12	14	16	19

10. Instead of treating the questions in Exercise 9 through planned comparisons, see if they can be answered in terms of post hoc comparisons, using the Scheffé criterion following a significant F test. Again use $\alpha = .05$.

11. If each of the comparisons outlined in Exercise 9 is tested for significance with $\alpha = .05$, what is the probability of a Type I error for at least one of these tests? If each were tested using, say, $\alpha = .017$, what is the probability that one or more will represent a Type I error?

12. Consider a set of seven groups, each containing 10 subjects. Planned comparisons are to be made with the following weights:

			Groups			
I	II	III	IV	V	VI	VII
4	4	4	−3	−3	−3	−3
1	1	−2	0	0	0	0
0	0	0	1	1	−1	−1

See if you can complete the set of orthogonal comparisons. (**Hint:** Look for sets of two or more groups which have, so far, received only the same signs or zeroes.)

13. Explain why there are only $J − 1$ rather than J orthogonal comparisons among J independent groups. What does this have to do with the requirement that the sum of each group of comparison weights must be zero?

14. Use the data in Exercise 4 to carry out Newman-Keuls comparisons on the means, for $\alpha = .05$.

15. For the data in Table 10.16.1 carry out Tukey HSD tests, using the .05 level.

16. Consider four groups, and let there be two orthogonal comparisons as follows. Now pick a number and multiply the weights for Comparison 1 by this number. Then pick another number and multiply the weights for Comparison 2 by that number. Are the comparisons based on the new weights still orthogonal? Why should this be?

	I	II	III	IV
ψ_1	1	1	−1	−1
ψ_2	1	−1	−1	1

17. Suppose that in Exercise 12 the first two comparisons shown were to be tested for significance as planned comparisons. The hypotheses to be tested are $H_0: \psi_1 \geq 0$, versus $H_1: \psi_1 < 0$, and $H_0: \psi_2 \leq 0$ versus $H_1: \psi_2 > 0$. An α level of .05 is to be used in each test. Explain the procedures you would use.

18. Consider the following set of means, each corresponding to one fixed experimental treatment on a separate group of 10 independent subjects:

I	II	III	IV	V
86	95	92	80	104

If the mean square error is 40, test the significance of the planned comparisons having the following weights:

I	II	III	IV	V
1	−1	0	0	0
0	0	0	1	−1
−1	−1	0	1	1
1	1	−4	1	1

19. Suppose that in Exercise 18 there had been prior interest only in the first two comparisons individually. Make an analysis of variance summary table showing the resulting mean squares and tests of significance.

20. Given the overall significance level established for Exercise 8, Chpater 10, carry out post hoc tests for the difference between all pairs of means, using the Tukey HSD.

21. The following data come from four groups with varying numbers of cases per group, for a total of $N = 22$. Below the group means are the weights for three comparisons. Show that these make up a set of orthogonal comparisons according to the criteria of Section 11.5. Then show that the sums of squares for these three comparisons add up to equal the sum of squares between groups.

	A_1	A_2	A_3	A_4
	10	12	9	11
	8	9	19	10
	7	7	16	14
	6	13	18	16
	11	12	11	
		13	10	
			13	
means	8.40	11.00	13.71	12.75
wts.	4.25	−1.50	−1.75	−1
	0	2.75	−1.75	−1
	0	0	1	−1

22. Carry out Tukey HSD tests for all pairs of means in Exercise 21. Use the harmonic mean of the sample sizes to calculate n, as suggested in Section 11.17.

23. Perform Newman-Keuls tests on all pairs of the six means in Table 11.11.2, using a nominal alpha value of .05 (that is, use .05 for looking up the value of q for each test).

24. Carry out tests on all fifteen pairs of means in Table 11.11.2 using the Bonferroni procedure with $\alpha_{FW*} = .15$.

Chapter *12*

Factorial Designs and Higher Order ANOVA

Our discussion of the analysis of variance in the two preceding chapters presupposed a set of data derived from the "classic" experimental situation, in which two or more different groups of cases are formed at random and then exposed to different treatments. In Chapter 8 we talked about the simplest instance where there are only two groups, one of which receives an experimental treatment, and the other, getting either no treatment or an imitation "placebo" treatment, serves as the control or contrast group. For two or more groups, these simple experimental situations are sometimes termed *randomized groups designs*.

The experimental attribute, consisting of the entire set of mutually exclusive and exhaustive experimental treatments, is usually called a *factor*. Then any given one of the treatments constituting a factor is called a *level* of that factor. Randomized groups designs can therefore be described as one-factor designs with J levels, where each distinct level is administered to a different randomly formed group.

It sometimes happens that the experimenter is interested in studying more than one factor in the same experiment, however. In that instance the study is often carried out in the form of a *factorial design*, which will be illustrated and discussed in detail in the next section.

12.1 FACTORIAL DESIGNS

Suppose that an experimenter is interested in "level of aspiration" as the dependent variable in an experiment. An experimental task has been developed consisting of a difficult game apparently involving motor skill, yielding a numerical score that can be attached to a person's performance. But this appearance is deceptive: unknown to the subject, the game is actually under the control of the experimenter, so that each subject is made to obtain exactly the same score. After a fixed number of trials, during which the subject unknowingly receives the preassigned score, the individual is asked to predict what the score will be on the next group of trials. However, before this prediction, the subject is given "information" about how the score compares with some fictitious norm group. In one experimental condition the subject is told that the first performance is *above average* for the norm group, in the second that it is *average,* and in the third that it is *below average.* There are thus three possible experimental "standings" that might be given to any subject. (Of course, after the experiment, each subject is fully informed of this little ruse by the experimenter.)

On the other hand, *two different norm groups* are used in the information given subjects. One-half of the subjects are told that they are being compared with college men, and the other half are told that they are being compared with professional athletes. Hence, there are two additional experimental treatments: "collge norms" and "professional athlete norms."

The dependent score value y is based on the report the subject makes about anticipated performance in the next group of trials. Since each subject has obtained the same score, this anticipated score on the next set of trials is treated as equivalent to a level of aspiration that the subject has set. Each subject is tested privately, and no communication is allowed between subjects until the entire experiment is completed.

In this example interest is focused on two distinct experimental factors: Factor B with $K = 3$ levels is the information about standing that the subject is given, and Factor A with $J = 2$ levels is the norm groups used for comparison. Either or both of these experimental factors might possibly influence the value that a subject anticipates as the next score (the level of aspiration). A random sample of 60 male college students is selected and assigned at random, 10 each to the six possible treatment *combinations.* This can be diagrammed as shown in Table 12.1.1.

This experiment represents an instance where two different sets of experimental treatments are *crossed,* or given in every combination. Here there are six distinct *cells,* or sample groups, each group being given a particular combination of two kinds of treatments. Three questions are of interest:

1. Are there systematic effects due to experimental set alone (averaged over the norm groups)?
2. Are there systematic effects due to norm information alone (averaged over the experimental sets)?
3. Are there systematic effects due neither to norm information alone nor to experimental set alone, but attributable only to the *combination* of a particular norm group with a particular experimental set?

| | | Standing | | |
		Above	Average	Below
Norm	Athlete	10	10	10
information	College	10	10	10

Table 12.1.1

Assignment of subjects to treatment combinations, by numbers in each combination.

Notice that this study could be viewed as two separate experiments carried out on the same set of subjects: There are three groups of 20 subjects each, differing only in experimental set, or standing; exactly the same set of norm-group conditions are represented in each experimental standing group. On the other hand, looking at Table 12.1.1 by rows rather than by columns, we see that there are also two samples of 30 subjects each, differing systematically by norm group. Each norm-group sample has exactly the same representation of the other experimental conditions within it.

Question 3 above cannot, however, be answered by the comparison of norm groups alone or by the comparison of experimental standing groups alone. This is a question of "interaction," the *unique effects of combinations of treatments*. This is the important new feature of the two-way (or higher) analysis of variance: we will be able to examine *main effects* of the separate experimental variables or factors just as in the one-way analysis, as well as *interaction effects,* differences apparently due only to the unique combinations of treatments. Before we go further into the topic of experiments where two or more experimental factors are to be studied, some other important terms must be introduced.

In the experiment described above, two experimental factors are under study, the first being the information given the subject about his standing, and the second being the norm information. Each subject is to get a combination of a category of standing information and a category of norm information. In experiments such as this, where two or more experimental factors are present and each category or level of one factor occurs with each level of every other, the factors are said to be **completely crossed.** All possible combinations of levels of two (or more) experimental factors occur in a completely crossed experiment. Furthermore, if all possible combinations of factor levels occur an equal number of times, the experiment is said to be **balanced.** Note that the experiment described above is thus completely crossed and balanced.

The remainder of this chapter will be devoted to experiments where there are two (or more) crossed and balanced factors, and then a different random sample of *n* subjects is assigned to each particular *combination* of the levels of the factors: that is, each combination of treatments is administered to a separate and distinct set of randomly selected subjects. This arrangement is called a **factorial design.** However, this is but one of the simplest kinds of experimental designs, and it is entirely possible to design and analyze a wide variety of experiments where different factors are not completely crossed or balanced in this way.

The discussion of the two-way analysis of variance will also be limited here to *replicated experiments*. For our purposes this means that **within each treatment combination there are at least two independent observations made under identical experimental circumstances.** The requirement that the experiment be replicated is introduced here so that an error sum of squares will be available, permitting the study of tests both for treatment effects and for interaction. For this reason, our discussion for the fixed-effect model will be confined to replicated experiments of the factorial type outlined above. Certain types of unreplicated experiments will be considered in Chapter 13, however.

An *orthogonal design* for an experiment can be defined as a way of collecting observations that will permit one to estimate and test for the various treatment effects and for interaction effects separately. The potential information in the experiment can be "pulled apart" for study in an orthogonal design. Any factorial layout can be regarded as an orthogonal design provided that: (1) the observations within a given treatment *combination* are sampled at random and independently from a normal population, and (2) the design is *balanced,* so that the number of observations in *each possible combination* of treatments is the same. Thus, the usual procedure in setting up an experiment to be analyzed by the two-way (or higher) analysis of variance is to assign subjects at random and independently to each combination of treatments so as to have an equal number in each combination. This means that in a table representing the experimental groups, such as Table 12.1.1, the cells of the table all contain the same number of observations. Let us continue to call the number of "row" treatments J and the number of "column" treatments K. For experiments of this sort where each cell in a $J \times K$ data table contains the same number n of cases, each *row* will contain Kn cases, and each *column Jn* cases. If at all possible, experiments should be set up in this way, not only to ensure orthogonality, but also to minimize the effect of nonhomogeneous population variances should they exist (see Section 10.20.)

It is also possible to design two-factor experiments that will be orthogonal even though the number of cases in the various cells differ, provided that **proportionality** holds within the cells of any given row or column. This means that for any treatment combination jk,

$$n_{jk} = \frac{n_j n_k}{N} .$$

This special situation calls for somewhat different computational forms and, as it is relatively rare, it will not be considered further here.

When a design is planned as a balanced factorial, but turns out to be nonorthogonal because of subject loss or for other reasons, there are remedies available. One of these is a set of methods called *unweighted means analysis;* these are described in various sources such as Edwards (1985). Other ways of handling the problem are given in some of the computer packages, and these will be mentioned in Chapter 16.

12.2 EFFECTS IN A FACTORIAL DESIGN

Suppose that the experiment described in the previous section were actually carried out. Now consider the y value produced by the individual subject i in the cell formed by row j and column k; we will symbolize this value by y_{ijk}. The set of y_{ijk} values for all of the $N = 60$ cases are displayed later on in this chapter in Table 12.7.1, and we will use the entire set of data in the analysis to be carried out in that section. However, for the time being our attention will be focused on the means of the rows, the means of the columns, and the cell means for this experiment.

Table 12.2.1 shows the six row and column combinations or cells for this experiment, along with the mean value $\bar{y}_{jk}$ for the 10 subjects in any cell formed by row j and column k. In addition, the Table 12.2.1 shows the mean for each row j, as well as the mean for each column k. This table also shows the overall mean across all of the $N = 60$ subjects in this experiment. (If you think that the number of decimal places being used in this

Table 12.2.1

Means actually obtained for each of the rows, columns, and cells of the experiment outlined in Table 12.1.1.

	B_1	B_2	B_3	Row means, $\bar{y}_{j.}$
A_1	46.4000	30.2000	17.8000	31.4667
A_2	36.8000	37.8000	21.4000	32.0000
column means $\bar{y}_{.k}$	41.6000	34.0000	19.6000	$31.7333 = \bar{y}_{..}$

example is rather excessive, you are quite right! However, in the discussion to follow it will be best to avoid rounding errors, if at all possible, and employing a fair number of decimal places is one way of doing so.)

Here and in later discussions of designs with two factors we will let the symbol $\bar{y}_{..}$ stand for the overall or grand mean of the sample values. The mean of any row level or Group A_j will be represented by $\bar{y}_{j.}$, and the mean of any column level or group B_k will be symbolized by $\bar{y}_{.k}$. The symbol $\bar{y}_{jk}$ will of course represent the mean of the cell in row j and column k. These symbols will be modified in obvious ways when we begin to talk about three or more factors. Futhermore, since we are assuming balanced designs, we will denote the number of cases in any cell jk of the experiment by the constant number $n = n_{jk}$.

To understand what is going on in a factorial design and how this guides our analysis, it is helpful to examine the different kinds of sample effects that underlie the data. These sample effects will be illustrated through use of the mean data in Table 12.2.1.

You will recall from Section 10.17 that in the ordinary one-way ANOVA, as applied to data from a randomized groups experiment, we found it convenient to discuss the sample effect of each of the groups:

$$a_j = \text{sample effect of Group } A_j = (\bar{y}_j - \bar{y}).$$

Remember that if the number n_j in each group is the same, then the sum of the effects a_j will be zero over the groups. (Even if the numbers in the groups are different, the weighted sum of the effects equals 0 as does the mean effect.)

We also saw in Section 10.17 that the sum of squares between groups is actually a weighted sum of the squared effects:

$$SS \text{ between } = \sum_j n_j a_j^2 \ .$$

In addition, we found that the sample effect a_j is an unbiased estimate of the population effect α_j, associated with the population j that Group A_j represents.

Now in discussions of factorial designs where two or more factors are utilized in the experiment, it is convenient to distinguish among different kinds of effects: *cell effects*, *main effects*, and *interaction effects*. We will deal with each of these types in turn.

The idea of a cell effect is best conveyed by remembering that each of the *JK* cells in a factorial experiment is actually a *distinct and independent* group of subjects. These subjects receive exactly one of the possible combinations of levels of the different factors. What is the effect on a subject of belonging to a particular cell, or, equivalently, of getting a certain combination of experimental treatments? This is called the *cell effect* associated with the combination of row j and column k. Such a cell effect is merely the difference between $\bar{y}_{jk}$, the mean of the cell formed by row j and column k, and $\bar{y}_{..}$, the overall mean of all N cases. That is

$$\text{cell effect } [ab]_{jk} = (\bar{y}_{jk} - \bar{y}_{..}) . \qquad [12.2.1]$$

Notice that there will be as many cell effects as there are combinations of row and column levels in our basic factorial design. For a balanced design, the sum of all of these cell effects must be zero:

$$\sum_{j}\sum_{k} [ab]_{jk} = 0 , \qquad [12.2.2]$$

and even for an unbalanced design the sum of the cell effects, each weighted by the number in that cell, must be zero.

You should bear in mind that the cells effects are fundamental, in the sense that they are derived from the basic groups of observations making up the data; a cell is such a basic "group" in a factorial design. In terms of their meaning, however, cell effects can represent relatively complex combination of other kinds of effects.

For example, Table 12.2.2 shows the cell effects for the data of Table 12.2.1. Each cell of the original design is associated with an effect, found as the difference between the mean of a cell and the grand mean of the sample. Thus, for the cell in Row 1 and Column 1 of Table 12.2.1, the effect is found to be

$$[ab]_{11} = \bar{y}_{11} - \bar{y}_{..}$$

$$= 46.4000 - 31.7333$$

$$= 14.6667$$

Once again, four decimal places are being carried here so that certain sums will agree closely when we compare them at a later stage. Similarly, the effect for the cell in, say, Row 2 and Column 3 is

$$[ab]_{23} = 21.4000 - 31.7333 = -10.3333,$$

and so on for all six cells in Table 12.2.2. You can check for yourself to see that, within rounding error, all of the cell effects for the sample sum to zero.

It stands to reason that *all of the systematic features of the experiment should be captured by the cell effects*. That is, cases in any cell are subjected to the possible

	B_1	B_2	B_3	Row effect
A_1	14.6667	−1.5333	−13.9333	−.2667
A_2	5.0667	6.0667	−10.3333	.2667
column effect	9.8667	2.2667	−12.1333	0

Table 12.2.2

Cell effects and main effects for data of Table 12.2.1.

effects of its row membership, of its column membership, and of the possible interaction between the row and column memberships. Thus, if the effects of the cells differ, this can be because of row main effects, column main effects, interaction effects, or the combination of all of these things. The *total systematic variation* in the experiment is therefore tied up in the weighted sum of the squared cell effects:

$$\text{SS AB cells} = N(\text{total systematic variance}) \qquad [12.2.3]$$
$$= \sum_j \sum_k n\, ([ab]_{jk})^2 \,.$$

For our example of Table 12.2.2, the weighted sum of the squares of the cell effects over the six cells turns out to be

$$\text{SS AB cells} = 10(14.6667)^2 + \cdots + 10(-10.3333)^2 = 5{,}808.53.$$

We will return to this SS value and its use shortly, after we discuss the other types of effects.

The effect of membership of a case in a given level of Factor A, or the row factor, shows how much the *y* value of that case tended to be "pushed" up or down relative to the average, simply as a result of that treatment, and without regard to Factor B. This push, which is attributable to membership in one level of a single factor, is known as a *main effect*.

The sample main effects for the rows (Factor A) are defined as follows:

$$a_j = \bar{y}_{j\cdot} - \bar{y}_{\cdot\cdot} \,. \qquad [12.2.4]$$

That is, the main effect of any level *j* of the row Factor A is just the difference between the mean of that row, or $\bar{y}_{j\cdot}$, and the grand mean $\bar{y}_{\cdot\cdot}$.

If the design is balanced, then the sum of the sample main effects for Factor A will be zero:

$$\sum_j a_j = 0 \,. \qquad [12.2.5]$$

The sample variability that is due to the fact that different groups got different levels of Factor A is then reflected by the weighted sum of the squares of the A main effects:

$$\text{SS Factor A} = N(\text{variance of A effects}) \qquad [12.2.6]$$
$$= \sum_j Kn(a_j)^2$$

Similarly, the sample main effects for the columns are:

$$b_k = \bar{y}_{\cdot k} - \bar{y}_{\cdot\cdot} \,. \qquad [12.2.7]$$

The sum (and hence the mean) of the sample effects b_k in a balanced design is then zero:

$$\sum_k b_k = 0 \,.$$

The column main effects are also shown in Table 12.2.2, for the data set of Table 12.2.1.

A main effect of Factor B also represents the push up or down an individual's y value receives from membership in a particular level of B. Then the systematic variability due to the different B levels is represented by the weighted sum of the squared b_k effects:

$$\text{SS Factor B} = N \text{ (variance of } B \text{ effects)} \qquad [12.2.8]$$
$$= \sum_k Jn(b_k)^2$$

For the A and the B main effects found as in Table 12.2.2, the respective sums of squares are

$$\text{SS Factor A} = 30(0.2667)^2 + 30(-0.2667)^2$$
$$= 4.27$$

$$\text{SS Factor B} = 20(9.8667)^2 + 20(2.2667)^2 + 20(-12.1333)^2$$
$$= 4,994.13$$

Notice that the sample row effects (the a_j) and the sample column effects (the b_k) are each found exactly as though we were dealing with a one-way design, with the experimental groups defined either by rows alone or by columns alone.

Finally, for each cell jk there is *a sample interaction effect,* symbolized by $(ab)_{jk}$. This interaction represents the unique effect of being in the combination of row j and the column k, quite over and above the effect of row j alone or the effect of column k alone. That is, the interaction effect for cell jk is the cell effect minus the main effects of the row and the column:

$(ab)_{jk}$ = interaction effect of cell jk

$\quad$ = effect for cell jk − effect row j − effect column k $\qquad [12.2.9]$

$\quad = [ab]_{jk} - a_j - b_k.$

This is equivalent to the statement that

$$\text{interaction } (ab)_{jk} = \bar{y}_{jk} - \bar{y}_{..} - (\bar{y}_{j.} - \bar{y}_{..}) - (\bar{y}_{.k} - \bar{y}_{..}) \qquad [12.2.10]$$
$$= \bar{y}_{jk} - \bar{y}_{j.} - \bar{y}_{.k} + \bar{y}_{...}$$

Thus, for example, the interaction effect for the cell in Row 1 and Column 1 of our example is given by

$$(ab)_{11} = 14.6667 - (9.8667) - (-0.2667) = 5.0667,$$

that for the cell in Row 2 and Column 1 by

$$(ab)_{21} = 5.0667 - (9.8667) - (0.2667) = -5.0667,$$

and so on for all the other cells of Table 12.2.2. Table 12.2.3 shows the set of all six interaction effects for this example. Notice that (within rounding error) the interaction effects add to zero within each row, within each column, and for the entire table.

If the interaction effects are each squared, multiplied by the number per cell, $n = n_{jk}$, and then summed, we have

	B_1	B_2	B_3	
A_1	5.0667	-3.5334	-1.5334	0
A_2	-5.0667	3.5334	1.5334	0
	0	0	0	0

Table 12.2.3

Interaction effects for the data of Table 12.2.1.

SS A×B interaction = N(variance of interaction effects)

$$= \sum_j \sum_k n(ab)_{jk}^2 .$$

Using the interaction effect values in Table 12.2.3 we can therefore find

SS A×B interaction = $10(5.0667)^2 + \cdots + 10(1.5334)^2$

$$= 810.15$$

The linear model that underlies ANOVA for a two-factor design can thus be written, in sample terms, as

$$y_{ijk} = \bar{y} + a_j + b_k + (ab)_{jk} + e_{ijk} \qquad [12.2.11]$$

where a_j is the sample effect of membership in row j, b_k is the sample effect of membership in B_k, and $(ab)_{jk}$ is the *interaction effect* associated with the cell in row j and column k.

It is well to bear in mind that the model for this two-factor experiment can *also* be expressed in terms of cell effects alone:

$$y_{ijk} = \bar{y}_{..} + [ab]_{jk} + e_{ijk} \qquad [12.2.12]$$

or

$$y_{ijk} = \bar{y}_{..} + (\bar{y}_{jk} - \bar{y}_{..}) + e_{ijk}$$

$$= \bar{y}_{jk} + e_{ijk} .$$

Equation 12.2.12 thus is equivalent to the model for a *one-way* analysis of variance in which there are JK independent groups, with n cases per group.

12.3 THE PARTITION OF THE SUMS OF SQUARES FOR A TWO-FACTOR DESIGN

In this section it will be shown that by use of the definitions of the different kinds of effects, as given in the previous section, the SS total for any set of data can be partitioned, first into

SS total = SS AB cells + SS within cells,

and then into

SS AB cells = SS Factor A + SS Factor B + SS A×B interaction, so that, altogether, we have

SS total = SS Factor A + SS Factor B + SS A×B interaction + SS within.

Let us begin with the model of our data given by Eq. 12.2.12:

$$y_{ijk} = \bar{y}.. + [ab]_{jk} + e_{ijk}$$

which says that the y score of individual i in cell jk is essentially the value of the grand mean, plus the effect of cell jk, plus an error term depending both on individual i and the cell jk into which the individual happens to fall.

A slight rearrangement changes this into a deviation:

$$(y_{ijk} - \bar{y}..) = [ab]_{jk} + e_{ijk}.$$

If such deviations are squared and summed over all individuals in all cells, to produce the SS total, we have

$$
\begin{aligned}
\text{SS total} &= \sum_k \sum_j \sum_i ([ab]_{jk} + e_{ijk})^2 \\
&= \sum_k \sum_j \sum_i ([ab]_{jk}^2 + e_{ijk}^2 + 2\,[ab]_{jk}e_{ijk}) \qquad [12.3.1] \\
&= \sum_k \sum_j n\,[ab]_{jk}^2 + \sum_k \sum_j \sum_i e_{ijk}^2
\end{aligned}
$$

since the cell effect $[ab]_{jk}$ is a constant for any case i in the cell, and since the errors within any given cell total to zero. Therefore, we can say that

SS total = SS AB cells + SS within cells   [12.3.2]

or

SS within cells = SS total − SS AB cells,   [12.3.3]

where, you recall, SS cells is just the weighted sum of the squared effects of the cells:

$$\text{SS AB cells} = \sum_k \sum_j n(\bar{y}_{jk} - \bar{y}..)^2.$$

Now we can also break up the SS AB cells into the SS for the main effects and the SS for interaction. Remember that the cell effect contains both main effects and an interaction effect.

$$[ab]_{jk} = a_j + b_k + (ab)_{jk}$$

Therefore,

$$
\begin{aligned}
\text{SS AB cells} &= \sum_j \sum_k (a_j + b_k + (ab)_{jk})^2 \\
&= \sum_j \sum_k [a_j^2 + b_k^2 + (ab)_{jk}^2 + 2\,a_j b_k \\
&\quad + 2\,a_j (ab)_{jk} + 2\,b_k (ab)_{jk}] \qquad\qquad [12.3.4] \\
&= \text{SS Factor A} + \text{SS Factor B} + \text{SS A×B inter.};
\end{aligned}
$$

since given any level A_j, the B effects and interaction effects sum to zero within that level, and given any level B_k, the same is true for the A and interaction effects. Hence, putting these facts together, we have the parittion of the SS total in a two-way analysis of variance:

$$\text{SS total} = \text{SS A} + \text{SS B} + \text{SS A} \times \text{B} + \text{SS within cells (or error).} \qquad [12.3.5]$$

Here, the different sums of squares have exactly the same definitions as in Section 12.2. That is,

$$\text{SS A} = \text{SS rows} = Kn \sum_j a_j^2 = Kn \sum_j (\bar{y}_{j\cdot} - \bar{y}_{\cdot\cdot})^2$$

$$\text{SS B} = \text{SS columns} = Jn \sum_k b_k^2 = Jn \sum_k (\bar{y}_{\cdot} - \bar{y}_{\cdot\cdot})^2$$

$$\text{SS A} \times \text{B} = \text{SS interaction} = n \sum_j \sum_k (ab)_{jk}^2 = n \sum_j \sum_k (\bar{y}_{jk} - \bar{y}_{j\cdot} - \bar{y}_{\cdot k} + \bar{y}_{\cdot\cdot})^2,$$

and where

$$\text{SS AB cells} = n \sum_j \sum_k [ab]_{jk}^2 = n \sum_j \sum_k (\bar{y}_{jk} - y_{\cdot\cdot})^2$$

with

$$\text{SS within} = \text{SS error} = \sum_j \sum_k \sum_i (y_{ijk} - \bar{y}_{jk})^2.$$

In short, the total variability of the y values in the data, as represented by SS total, is divisible into *four* components, representing variability due to A main effects, to B main effects, to $A \times B$ interaction effects, and to error, just as we showed in the previous section.

Next we are going to establish the connection between these sample quantities and the corresponding population values. Then we will deal with practical computational aspects of carrying out a two-way analysis of variance.

12.4 POPULATION EFFECTS FOR A TWO-WAY ANALYSIS

As we have just seen, sample data from a factorial design involving two fixed factors A and B can be represented by a model including the main effects for each factor, as well as $A \times B$ interaction effects and error. If, instead of a sample of data we wish to describe a set of *populations* in these same terms, we can conceive of the population main effect associated with the treatment A_j, or $\alpha_j = (\mu_{j\cdot} - \mu)$, where $\mu_{j\cdot}$ is the mean of population j and μ is the general mean of all of the populations.

In an identical way we can also discuss the population main effect associated with treatment B_k, or $\beta_k = (\mu_{\cdot k} - \mu)$. That is, the true or population effect of the kth level of Factor B is given by the difference between the mean of the population getting that treatment, and the mean of the grand population.

Finally, it is equally possible to think about the interaction effect $(\alpha\beta)_{jk}$, which is associated with the (hypothetical) population receiving both treatment A_j and B_k.

Then the model of the data, in terms of these population effects, becomes

$$y_{ijk} = \mu + \alpha_j + \beta_k + (\alpha\beta)_{jk} + e_{ijk}. \qquad [12.4.1]$$

The new features of Eq. 12.4.1 are the inclusion of a term representing the effect of row treatment k, or β_k, and a term standing for the interaction effect, $(\alpha\beta)_{jk}$. Remember that the interaction effect is the experimental effect created by the combination of treatments j and k *over and above* any effects associated with treatments j and k considered separately:

$$\begin{aligned}(\alpha\beta)_{jk} &= \mu_{jk} - \mu - \alpha_j - \beta_k \\ &= \mu_{jk} - \mu_{j\cdot} - \mu_{\cdot k} + \mu.\end{aligned} \qquad [12.4.2]$$

The interaction effect μ_{jk} is thus equal to the mean of the population given both of the treatments j and k, minus the mean of the treatment population j, minus the mean of the population given treatment k, plus the grand mean.

It is equally possible to conceive the presence of a cell effect in any one of the populations defined by a level of A combined with a level of B. That is, consider a population of cases receiving both the treatment A_j and the treatment B_k. Then, the cell effect for that population is

$$[\alpha\beta]_{jk} = \mu_{jk} - \mu. \qquad [12.4.3]$$

Just as in the simple one-way analysis of variance under the fixed-effects model, in the two-way situation it is assumed that the experimental treatments and treatment combinations are fixed, and that the only inferences to be made are about those treatments and treatment combinations actually represented in the experiment.

The following equalities are true of the effects:

$$\sum_j \alpha_j = 0 \qquad [12.4.4]$$

$$\sum_k \beta_k = 0 \qquad [12.4.5]$$

$$\sum_j (\alpha\beta)_{jk} = 0 \qquad [12.4.6]$$

$$\sum_k (\alpha\beta)_{jk} = 0. \qquad [12.4.7]$$

The effects, being deviations from a grand mean μ, sum to zero over all the different levels of a given kind of treatment. However,

$$\sum_j \alpha_j^2 > 0$$

unless $\alpha_j = 0$ for each j;

$$\sum_k \beta_k^2 > 0$$

unless $\beta_k = 0$ for each k;

$$\sum_j \sum_k (\alpha\beta)_{jk}^2 > 0$$

unless $(\alpha\beta)_{jk} = 0$ for each and every combination j,k.

12.5 THE MEAN SQUARES AND THEIR EXPECTATIONS

Before we turn to an examination of the sampling distribution of the various mean squares, the assumptions we must make to determine these sampling distributions will be stated:

1. The errors e_{ijk} are normally distributed with expectation of zero for each treatment-combination population jk.
2. The errors e_{ijk} have exactly the same variance σ_e^2 for each treatment-combination population.
3. The errors e_{ijk} are independent, both within each treatment combination and across treatment combinations.

You will note that these are essentially the same assumptions made for the one-way model, except that now we deal with treatment-combination populations, the entire set of potential observations to be made under any combination of treatments.

We begin by finding the expectation of the *error* sum of squares:

$$E(\text{SS within cells}) = E(\text{SS error})$$

$$= E\sum_k\sum_j\sum_i (y_{ijk} - \bar{y}_{jk})^2 \qquad [12.5.1]$$

$$= JK(n-1)\,\sigma_e^2$$

and

$$\text{MS error} = \frac{\text{SS error}}{JK(n-1)}. \qquad [12.5.2]$$

Then, it will be true that

$$E(\text{MS error}) = \frac{E(\text{SS error})}{JK(n-1)}$$

$$= \sigma_e^2. \qquad [12.5.3^*]$$

The expected value of the mean square error is simply the error variance σ_e^2.

The first hypothesis to be tested is that there are *no true row effects* among the populations represented by the rows of our data table:

$H_0 : \alpha_j = 0$, for every population j.

The alternative hypothesis is, of course, that *some true row effects are not equal to zero:*

$H_1 : \alpha_j \neq 0$, for some populations j.

Let us take another look at the mean square between rows, representing Factor A. Within any given row both the effects of B and the interaction effects sum to zero:

$$\sum_k \beta_k = 0 \text{ and } \sum_k (\alpha\beta)_{jk} = 0 .$$

Then, defining

$$\text{MS rows} = \text{MS A} = \frac{\text{SS rows}}{J-1} \qquad [12.5.4]$$

the expectation is

$$E(\text{MS A}) = \sigma_e^2 + \sum_j (Kn\,\alpha_j^2)/(J-1), \qquad [12.5.5]$$

the expectation of the MS A is error variance, plus the sum of the squared alpha effects multiplied by Kn, or the number in each A category, and divided by $J-1$.

The important point for our immediate purposes is that the expectation of the mean square for rows can be exactly σ_e^2 only when the hypothesis of zero row effects is true; otherwise,

$$E(\text{MS rows}) > \sigma_e^2. \qquad [12.5.6]$$

The mean square between rows (Factor A) is an unbiased estimate of σ_e^2 when the null hypothesis of no row effects is true, and it is independent of the mean square error. The hypothesis of no row effects is then tested by the ratio

$$F = \frac{\text{MS rows}}{\text{MS error}} = \frac{\text{MS A}}{\text{MS error}} \qquad [12.5.7]$$

with $J-1$ and $JK(n-1)$ (or $N - JK$) degrees of freedom.

Now we wish to test the hypothesis concerning the true column effects β_k in the populations represented in our data. The null hypothesis says that *these effects are all 0:*

$$H_0 : \beta_k = 0, \text{ for each population } k$$

as opposed to the alternative

$$H_1 : \beta_k \neq 0, \text{ for some populations } k.$$

Since the sample effects b_k estimate the population effects β_k our SS B contains the information relevant to testing this hypothesis.

Now look at the mean square between columns (Factor B). Since there are K columns, the mean square between columns is found from

$$\text{MS columns} = \frac{\text{SS columns}}{K-1} = \frac{\text{SS B}}{K-1} \qquad [12.5.8]$$

in exactly the same way as for the MS between in a one-way analysis of variance.

Since, within any column, $\sum_j \alpha_j = 0$, and $\sum_j (\alpha\beta)_{jk} = 0$ it follows that

$$E(\text{MS columns}) = E(\text{MS B}) = \sigma_e^2 + \frac{Jn \sum_k \beta_k^2}{(K-1)}. \qquad [12.5.9^*]$$

The mean square between columns (Factor B) and the mean square error are independent and unbiased estimates of the same variance σ_e^2 when the null hypothesis of no column or B effects is true. This hypothesis can be tested by the F ratio,

$$F = \frac{\text{MS columns}}{\text{MS error}} = \frac{\text{MS B}}{\text{MS error}} \qquad\qquad [12.5.10]$$

with $K - 1$ and $JK(n - 1)$ degrees of freedom. The rational for these tests is precisely the same as that given in Sections 10.10 through 10.12.

Finally, we wish to test the hypothesis that *no interaction effects at all* exist among the populations represented in our data. That is,

$H_0 : (\alpha\beta)_{jk} = 0$, for all combination populations jk

against the alternative

$H_1 : (\alpha\beta)_{jk} \neq 0$, for some combination populations jk.

The sum of squares for interaction, SS $A \times B$, is based on the sample interaction effects, and thus is relevant to the test of this null hypothesis.

If we define

$$\text{MS interaction} = \frac{\text{SS interaction}}{(J - 1)(K - 1)} = \frac{\text{SS A} \times \text{B}}{(J - 1)(K - 1)} \qquad\qquad [12.5.11]$$

then

$$E(\text{MS interaction}) = \frac{E(\text{SS interaction})}{(J - 1)(K - 1)}$$
$$= \sigma_e^2 + \frac{n \sum_k \sum_j (\alpha\beta)_{jk}^2}{(J-1)(K-1)}. \qquad\qquad [12.5.12^*]$$

When there are no interaction effects *at all*, then

$$E(\text{MS interaction}) = \sigma_e^2. \qquad\qquad [12.5.13]$$

The mean square for interaction is also an unbiased estimate of the error variance σ_e^2. Otherwise,

$$E(\text{MS interaction}) > \sigma_e^2. \qquad\qquad [12.5.14]$$

The mean square for interaction is independent of the mean square for error, and so the hypothesis of no interaction effects may be tested by

$$F = \frac{\text{MS interaction}}{\text{MS error}}, \qquad\qquad [12.5.15]$$

with $(J-1)(K-1)$ and $JK(n-1)$ degrees of freedom. (The independence of the MS interaction and of the MS error is a result of the principle, stated in Section 6.6, that for normally distributed variables, sample means are independent of sample variances. The interaction effects are, as we have seen, simply weighted sums of the cell, row, column, and general means. They can therefore always be written as weighted sums of all of the various cell means. If the basic populations are normal, these cell means should be independent of the within-cell variances. Weighted sums of the cell means will then also be independent of the within-cell variances. Thus, the MS $A \times B$ interaction, which involves the sum of squares of the interaction effects, is also independent of MS error, which is based on the within-cell variances. A similar rationale obtains for the independence of the MS for the main effects and MS error.)

We now see that it is possible to make separate tests of the hypothesis of no row effects, the hypothesis of no column effects, and the hypothesis of no interaction effects, all from the same data. Furthermore, under the fixed-effects model, and given a balanced experimental design, estimates of the three different kinds of effects are independent of each other.

12.6 COMPUTATIONAL FORMS FOR A TWO-WAY ANOVA

The meaning of the different SS values obtained in a two-way ANOVA is often most clearly seen in terms of the different kinds of sample effects, and, in principle, one can actually perform the two-way analysis of variance by use of the sample effects, as shown in Section 12.2. Nevertheless, it is much more efficient to use special computing forms for this purpose.

In the following, these symbols will be used to stand for the different sums that our computations will require:

$$T_y = \text{grand total of all } y \text{ values} = \sum_j \sum_k \sum_i y_{ijk}$$

$$T_{j\cdot} = \text{total of the row representing Level A} = \sum_k \sum_i y_{ijk}$$

$$T_{\cdot k} = \text{total of the column representing Level B} = \sum_j \sum_i y_{ijk}$$

$$T_{jk} = \text{total of the cell formed by levels A}_j \text{ and B}_k = \sum_i y_{ijk}.$$

In terms of this notation the computing formulas required by the two-way analysis of variance are:

$$\text{SS total} = \sum_j \sum_k \sum_i y_{ijk}^2 - \frac{T_y^2}{N} \qquad [12.6.1]$$

$$\text{SS AB cells} = \sum_j \sum_k \frac{T_{jk}^2}{n} - \frac{T_y^2}{N} \qquad [12.6.2]$$

$$\text{SS error} = \text{SS total} - \text{SS AB cells} \qquad [12.6.3]$$

$$\text{SS A} = \text{SS rows} = \sum_j \frac{T_{j\cdot}^2}{Kn} - \frac{T_y^2}{N} \qquad [12.6.4]$$

$$\text{SS B} = \text{SS columns} = \sum_k \frac{T_{\cdot k}^2}{Jn} - \frac{T_y^2}{N} \qquad [12.6.5]$$

SS A $\times$ B $=$ SS interaction

$$= \sum_j \sum_k \frac{T_{jk}^2}{n} - \sum_j \frac{T_{j\cdot}^2}{Kn} - \sum_k \frac{T_{\cdot k}^2}{Jn} + \frac{T_y^2}{N} \qquad [12.6.6]$$

$$= \text{SS AB cells} - \text{SS A} - \text{SS B}$$

or

SS A $\times$ B $=$ SS total $-$ SS A $-$ SS B $-$ SS error.

These computational formulas boil down to a few simple steps which apply to any two-way ANOVA under the fixed-effects model. We do the following:

1. for a Factor A with J levels and a Factor B with K levels, and n cases per cell, lay out the $J \times K$ table of data. Sum each cell jk to find the total T_{jk}.
2. Sum each row j to find $T_{j\cdot}$, each column k to get $T_{\cdot k}$, and sum the entire table to obtain T_y.
3. Take the total T_y and square it, and divide the result by N. Call this quantity G.

   $G = T_y^2/N$.

4. Now take each y score and square it, and add these values together. Subtract G from the result. This gives SS total.
5. Take each cell total T_{jk} and square it. Add these squared values together and divide by n, the number of cases per cell. Then subtract G from the result. This gives SS AB cells.
6. Subtract the result of Step 5 from the result of Step 4. This gives SS error:

   SS error $=$ SS total $-$ SS AB cells.

7. Take the total $T_{j\cdot}$ for each row (level of A) and square it. Add the results together and divide by (K times n), the number of cases in a row. Subtract G from the answer. This is SS A or SS rows.
8. Take the total $T_{\cdot k}$ for each column (level of B) square it, and add the results. Then divide the sum by (J times n), the number of cases in a column. Subtract G. This is SS B or SS columns.
9. Take the result of Step 5 and subtract the results of Steps 7 and 8. This is SS A $\times$ B interaction.
10. Arrange the results in an ANOVA summary table, as outlined in Table 12.6.1.
11. Divide the SS A (or SS rows) by $J - 1$ to find MS A.
12. Divide the SS B (or SS columns) by $K - 1$ to find MS B.
13. Divide the SS A $\times$ B interaction by $(J-1)(K-1)$ to find MS A $\times$ B.
14. Divide the SS error by $N - JK$ to find MS error.
15. Test the hypothesis of no A or row effects by taking

    $F = $ MS A/MS error

    with $J - 1$ and $N - JK$ degrees of freedom.
16. Test the hypothesis of no B or column effects by taking

    $F = $ MS B/MS error

Table 12.6.1

Model summary table for a two-way fixed-effects ANOVA

Source	SS	df	MS	F
A (rows)	$\sum_{j} T_{j.}^2/Kn - T_y^2/N$	$J-1$	SS A/$(J-1)$	MS A/MS err
B (cols.)	$\sum_{k} T_{.k}^2/Jn - T_y^2/N$	$K-1$	SS B/$(K-1)$	MS B/MS err
A×B (inter.)	$\sum_{j}\sum_{k} T_{jk}^2/n - \sum_{j} T_{j.}^2/Kn$ $- \sum_{k} T_{.k}^2/Jn + T_y^2/N$	$(J-1)(K-1)$	$\dfrac{\text{SS A×B}}{(J-1)(K-1)}$	$\dfrac{\text{MS A×B}}{\text{MS err}}$
Error (within cells)	$\sum_{j}\sum_{k}\sum_{i} y_{ijk}^2 - \sum_{j}\sum_{k} T_{jk}^2/n$	$N-JK$	SS err/$(N-JK)$	
Total	$\sum_{j}\sum_{k}\sum_{i} y_{ijk}^2 - T_y^2/N$	$N-1$		

with $K-1$ and $N-JK$ degrees of freedom.

17. The hypothesis of no interaction is tested by

$$F = \text{MS A×B/MS error}$$

with $(J-1)(K-1)$ and $N - JK$ degrees of freedom.

12.7 AN EXAMPLE

Suppose that the experiment on level of aspiration, outlined in Section 12.1 had actually been carried out, and the data shown in Table 12.7.1 were obtained.

We wish to examine three null hypotheses: (1) there is no effect of the standing given the subject, corresponding to the hypothesis of no column effects; (2) the actual norm group given the subjects has no effect, corresponding to the hypothesis of no row effects; and (3) the norm-group-standing combination has no unique effect, corresponding to the hypothesis of no row–column interaction. The α level chosen for each of these three tests will be .05.

The sum of the scores in each cell (Step 13) is given in Table 12.7.2. Taking the sum of the cell sums gives the total sum,

$$T_y = \sum_{j}\sum_{k}\sum_{i} y_{ijk} = 464 + 302 + \cdots + 214 = 1{,}904.$$

Following the computational outline given in Section 12.6 we next find the square of each of the scores, and sum:

$$\sum_{j}\sum_{k}\sum_{i} y_{ijk}^2 = (52)^2 + (48)^2 + \cdots + (22)^2 + (17)^2 = 66{,}872.$$

Norms	Standing		
	Above	Average	Below
	52	28	15
	48	35	14
	43	34	23
	50	32	21
	43	34	14
College men	44	27	20
	46	31	21
	46	27	16
	43	29	20
	49	25	14
	464	302	178
	38	43	23
	42	34	25
	42	33	18
	35	42	26
Professional athletes	33	41	18
	38	37	26
	39	37	20
	34	40	19
	33	36	22
	34	35	17
	368	378	214

Table 12.7.1

Results for the level of aspiration experiment.

Hence the total sum of squares is

$$\sum_j \sum_k \sum_i y_{ijk} - \frac{T_y^2}{N} = 66{,}872 - \frac{(1{,}904)^2}{60} = 6{,}451.73.$$

The next step (5) involves finding SS AB cells. This consists of squaring the sum for each of the cells of the table, adding these squared values, dividing by n, and then subtracting G-square (from Step 3):

$$\frac{(464)^2 + (368)^2 + (302)^2 + (378)^2 + (178)^2 + (214)^2}{10} - \frac{(1{,}904)^2}{60}$$

$$= \text{SS AB cells} = 5{,}808.53.$$

In Step 6 the difference between SS total and SS AB cells is taken, to yield the value of SS error (or SS within cells):

SS error = SS total − SS AB cells

$$= 6{,}451.73 - 5{,}808.53$$

$$= 643.20.$$

Next, following Step 7, the row totals T_j. are taken

$$T_{1\cdot} = \sum_k \sum_i y_{i1k} = 464 + 302 + 178 = 944$$

$$T_{2\cdot} = \sum_k \sum_i y_{i2k} = 368 + 378 + 214 = 960.$$

The sum of squares for rows is found from

$$\sum_j T_j^2/Kn - G = \frac{(944)^2 + (960)^2}{30} - \frac{(1,904)^2}{60}$$

$$= 4.27$$

In a similar way, we find the sum of squares for columns by first summing cell totals for each column

$$T_{\cdot 1} = \sum_j \sum_i y_{ij1} = 464 + 368 = 832$$

$$T_{\cdot 2} = \sum_j \sum_i y_{ij2} = 302 + 378 = 680$$

$$T_{\cdot 3} = \sum_j \sum_i y_{ij3} = 178 + 214 = 392.$$

The sum of squares for columns is found from

$$SS\ B = SS\ cols. = \frac{(832)^2 + (680)^2 + (392)^2}{20} - \frac{(1,904)^2}{60}$$

$$= 4,994.13.$$

The only remaining value to be calculated is the sum of squares for interaction; this is done by subtraction, as follows:

$$SS\ A \times B\ interaction = SS\ AB\ cells - SS\ A - SS\ B$$

$$= 5,808.53 - 4.27 - 4,994.13$$

$$= 810.13$$

or

$$SS\ total - SS\ rows - SS\ cols. - SS\ error$$

$$= 6,451.73 - 4.27 - 4,994.13 - 643.2$$

$$= SS\ interaction = 810.13.$$

(At this point you may wish to compare these sums of squares with those found directly from the effects in Section 12.2.)

Table 12.7.2 is the summary table for this analysis of variance.

The hypothesis of no row effects cannot be rejected, since the F value is less than unity. For the hypothesis of no column effects, an F of approximately 3.15 is required for rejection at the 5% level; the obtained F of 209.66 far exceeds this, and so we may conclude with considerable confidence that column effects exist. In the same way, the

Source	SS	df	MS	F
Rows (norm groups)	4.27	1	4.27	0.36
Columns (standings)	4,994.13	2	2,497.07	209.66
Interaction	810.13	2	405.07	34.01
Error (within cells)	643.20	54	11.91	
Totals	6,451.73	59		

Table 12.7.2

Summary of analysis for level of aspiration study.

F for interaction effects greatly exceeds that required for rejecting the null hypothesis, and so there seems to be reliable evidence for such interaction effects.

Our conclusions from this analysis of variance are:

1. There is apparently little or no effect of norm group alone on level of aspiration.
2. The experimental standing does seem to affect level of aspiration when considered over the different norm groups.
3. There is apparently an interaction between norm group and standing, meaning that the magnitude and direction of the effects of standing differ for different norm groups.

In short, one's standing makes a difference in the aspiration level, but the kind and extent of difference that it makes depends on the norm group to which the individual is being compared.

The different column effects can be estimated from the column means and the overall mean:

est. $\beta_1 = 41.60 - 31.73 = 9.87$

est. $\beta_2 = 34.00 - 31.73 = 2.27$

est. $\beta_3 = 19.60 - 31.73 = -12.13$

(Because of rounding error these do not quite total zero, as they should.) In a similar way, interaction effects $(\alpha\beta)_{jk}$ may be estimated from the means of cells, the rows, and the columns:

est. $(\alpha\beta)_{11} = 46.40 - 31.47 - 41.60 + 31.73 = 5.06$

est. $(\alpha\beta)_{12} = 30.20 - 31.47 - 34.00 + 31.73 = -3.54$

and so on. (Within rounding differences, these are the same as given in Table 12.2.3, of course.) The estimated total effect of the "above" treatment on a subject in the college norm group is thus

est. $(\beta_1 + (\alpha\beta)_{11}) = 9.87 + 5.06 = 14.93$

Note that for Group 1 ("above" standing) with unspecified norm group, the best guess we can make about the effect of this treatment on any individual is about 9.9 units. However, if we are told that this individual also belongs to the group given the "college men" norms, our best bet is about 14.9 as the amount of effect. In the same way, the effect of any column treatment k within a row-treatment population j is estimated to be

$$\beta_k + (\alpha\beta)_{jk}$$

since there is little evidence for row effects.

12.8 THE INTERPRETATION OF INTERACTION EFFECTS

The presence or absence of true interaction effects, as inferred from the F test for interaction, can have a very important bearing on how one interprets and uses the results of an experiment. When the presence of column effects is inferred, this implies that the populations represented by the columns have means that differ; the amount and direction of difference between sample means for any pair of columns provides an estimate of the corresponding difference between population means. When interaction effects are absent, differences among the means representing different column-treatment populations have the same size and sign, even though the populations are conceived as receiving still another treatment represented by one of the rows. This suggests that the difference between a pair of column means in the data is our best bet about the difference to be expected between a pair of individuals given different column treatments, quite irrespective of the particular row treatment that might have been administered. On the other hand, when interaction effects exist, *varying differences* exist between the means of populations representing different column treatments, depending on the particular row treatment that is applied. It is still true that differences between column means in the data provide an estimate of the difference we should expect between individuals given the particular treatments, but only on the average, over all of the different row treatments that might have been applied. When a particular row treatment is specified, it may be that quite another size and direction of difference should be expected between individuals given different column treatments. In short, interaction effects lead to a *qualification* on the estimate one makes of the differences attributable to different treatments; when interaction effects exist, the best estimate one can make of a difference attributable to one factor depends on the particular level of the other factor.

For example, suppose that an experimenter is comparing two methods of instruction in golf. Let us represent these two methods as the column treatments in the data table that follows. The other factor considered is the sex of the student; the study employs a group of 50 boys and a group of 50 girls, with 25 subjects in each group taught by Method I and the remainder by Method II. After a fixed period of instruction by one or the other method, each member of the sample is given a proficiency test. Suppose that the sample means for the four subgroups turn out as follows:

		Method		
		I	II	
	Girls	55	65	60
Sex	Boys	75	45	60
		65	55	

For a small enough estimated error variance, such data would lead to the conclusion that no difference exists between boys and girls in terms of performance on the proficiency test, but that *both column effects and interaction effects do exist.* Now suppose that you want to decide which method to use for the instruction of an individual student. If *low* scores indicate good performance, but you don't know or don't wish to specify the sex of the student, then Method II clearly is called for, since the best estimate is that Method I gives a higher mean than Method II over both sexes. However, suppose that you know that the individual to be instructed is a girl: in this case, you do *much* better to choose Method I, since you have evidence that *within the population of girls,* Mean II is higher than Mean I.

Significant interaction effects usually reflect a situation very like this: overall estimates of differences due to one factor are fine as predictors of average differences over *all possible levels of the other factor,* but it will not necessarily be true that these are good estimates of the differences to be expected when information about the category on the other factor is given. Significant interaction serves as a warning: treatment differences *do* exist, but to specify exactly *how* the treatments differ, and especially to make good individual predictions, one must look *within* levels of the *other* factor. The presence of interaction effects is a signal that in any predictive use of the experimental results, effects attributed to particular treatments representing one factor are best qualified by specifying the level of the *other* factor. This is extremely important if one is going to try to use estimated effects in forecasting the result of applying a treatment to an individual; when interaction effects are present, the best forecast can be made only if the individual's status on *both* factors is known.

For these reasons, the presence of interaction effects can be most important to the interpretation of the experiment. The estimated effects of any given treatment are not ''best bets'' about any randomly selected individual when interaction effects are present; the best prediction entails knowing the other treatment or treatments administered.

It is often helpful to display the interaction effects present in a set of data in the form of a graph. Such graphs can be laid out in many ways, but the simple type to be discussed is both convenient and popular.

To demonstrate that interaction effects are present or absent in a given set of data for a two-factor design, one can do the following:

1. Choose the factor with the larger number of levels. (This is Factor B in our current example.)
2. Lay out a graph with Factor B shown as the horizontal axis, with the K different levels of Factor B represented as equally spaced points on this axis. Let the vertical axis represent the different possible values of the cell means or $\bar{y}_{jk}$ values.
3. Now for each point on the horizontal axis, plot the cell mean values that correspond to that particular level of the factor defining that axis. For the current example, this involves plotting the two cell means within each column (or B level) of the design.
4. When all the cell mean points are plotted, then connect the points that stand for the same level of the *other* factor. (In our example, since Factor B or columns defines the horizontal axis, we would connect points that represent the same levels of Factor A.)
5. To the extent that the resulting lines on the graph are parallel, interaction is *not* present in the data. However, if the different lines have different slopes, and in particular if they cross, then there is evidence for interaction.

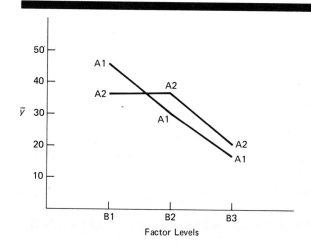

Figure 12.8.1

Representation of the AxB interaction in the data in Table 12.2.2.

For the cell means given in Table 12.2.1, and using the factor B or columns as the horizontal axis, we have Figure 12.8.1.

In this example, where there were no significant A effects, but both Factor B and $A \times B$ interaction proved significant, the graph shows how the presence of $A \times B$ interaction serves as a "qualifier" to statements about the results. Clearly, for Level B_1, the combination with A_1 creates a higher mean than the combination with A_2; on the other hand, just the reverse situation is true for Levels B_2 and B_3.

Conversely, suppose that the original data in Table 12.2.1 had turned out exactly as before, except that all of the interaction effects are now zero. This would have produced the set of mean values shown in Table 12.8.1. Then, when the graph is constructed just as before, with Factor B (or columns) defining the horizontal axis, and with cell means plotted and connected, we would have the result shown in Figure 12.8.2. Clearly, the

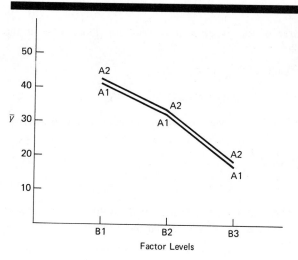

Figure 12.8.2

Representation of the AxB interaction in the data in Table 12.8.1.

	B_1	B_2	B_3	Row mean
A_1	41.33	33.73	19.33	31.46
A_2	41.87	34.27	19.87	32.00
col. mean	41.60	34.00	19.60	31.73

Table 12.8.1

Cell means for data of Table 12.2.1, but with sample interaction effects absent

total absence of interaction effects shows up as parallel lines in such a graph. Most real data situations are, of course, somewhere between these two extremes, but graphing the the interaction such as we have done here can help suggest the changes that do occur in the effects of one factor when the other factor is specified.

12.9 PROPORTION OF VARIANCE ACCOUNTED FOR IN TWO-FACTOR EXPERIMENTS

Just as in a one-way experimental situation, following a two-way analysis of variance one may wish a descriptive statistic for the proportion of variance accounted for by each separate factor, and by their interaction. Simply by taking each of these sums of squares and dividing them by the total sum of squares shows how much variance each type of effect accounts for in this sample. Thus, in the preceding example of Table 12.7.1, the sample proportion of y variance accounted for by A, the row factor, is given by

$$R^2_{y \cdot A} = \text{SS A/SS total} = 4.27/6,451.73 = .0007.$$

which is indeed negligible. The sample proportion of y variance explained by B, the column factor, is much larger, however:

$$R^2_{y \cdot B} = \text{SS B/SS total} = 4,994.13/6,451.73 = .77.$$

That is, Factor B accounts for about 77%. The $A \times B$ interaction then accounts for 13% of the y variability in the sample:

$$R^2_{y \cdot A \times B} = \text{SS } A \times B \text{ int./SS total} = 810.13/6,451,73 = .126.$$

Altogether, the systematic features of the experiment account for about 90% of the variability observed in this sample:

$$R^2_{y \cdot A, B, A \times B} = \text{SS AB cells/SS total} = R^2_{y \cdot A} + R^2_{y \cdot B} + R^2_{y \cdot A \times B}$$

$$= 5,808.53/6,451.73 = .90.$$

This implies that only 10% of the variability we observe in this sample is due to error, or is variance unaccounted for:

$$1 - R^2_{y \cdot A, B, A \times B} = \text{SS error/SS total}$$

$$= .10.$$

These are all *descriptive* statements, and are necessarily meaningful only in terms of *this* sample of data. Furthermore, we can be very sure that the strength of association between y and Factors A, B, and A$\times$B interaction is *less* than the sample suggests. Once again, just as in the case of one-way, randomized groups, we need a way to infer the strength of population associations that these findings suggest.

The strength of association measure ω^2 for a one-way experimental design (Section 10.21) may be extended in a straightforward way to the situation where there are two independent variables. When the experimental design has an equal number of observations in each combination of two independent factors, A with J levels and B with K levels, then the definition of $\omega^2_{y\cdot A}$, the proportion of variance accounted for by A alone in the population, is

$$\omega^2_{y\cdot A} = \frac{\left(\sum_j \alpha_j^2\right)\Big/ J}{\sigma_Y^2}. \tag{12.9.1}$$

Similarly, we can define

$$\omega^2_{y\cdot B} = \frac{\left(\sum_k \beta_k^2\right)\Big/ K}{\sigma_Y^2}, \tag{12.9.2}$$

and

$$\omega^2_{y\cdot A\times B} = \frac{\left(\sum_j \sum_k (\alpha\beta)_{jk}\right)\Big/ JK}{\sigma_Y^2}. \tag{12.9.3}$$

This last index is the proportion of variance accounted for uniquely by the interaction of *both* A and B.

Given these definitions, and our results about the expectations of mean squares for the two-way analysis of variance (Section 12.5) we can form rough estimates of these values of ω^2 by taking

$$\text{est. } \omega^2_{y\cdot A} = \frac{\text{SS A} - (J-1)\text{ MS error}}{\text{MS error} + \text{SS total}} \tag{12.9.4*}$$

$$\text{est. } \omega^2_{y\cdot B} = \frac{\text{SS B} - (K-1)\text{ MS error}}{\text{MS error} + \text{SS total}} \tag{12.9.5*}$$

$$\text{est. } \omega^2_{y\cdot A\times B} = \frac{\text{SS interaction} - (J-1)(K-1)\text{ MS error}}{\text{MS error} + \text{SS total}}. \tag{12.9.6*}$$

For the example in Section 12.7, these estimated values are

$$\text{est. } \omega^2_{y\cdot B} = \frac{4{,}994.13 - (2)(11.9)}{11.9 + 6{,}451.73} = .77$$

$$\text{est. } \omega^2_{y\cdot A\times B} = \frac{810.13 - (2)(11.9)}{11.9 + 6{,}451.73} = .12.$$

Since the F ratio shows a value less than 1.00 in the test for row differences, the estimate of $\omega_{y \cdot A}^2$ is set equal to zero. These estimates also suggest that a strong association exists between the treatments symbolized by B and the dependent variable Y. Knowing B alone tends to reduce our "uncertainty" about Y by about 77%. There is apparently a further accounting for around 12% of the variance of Y if one knows *both* of the categories represented by the A and B treatment combination.

12.10 PLANNED COMPARISONS AND TESTS IN FACTORIAL DESIGNS

Given a factorial design based on W different factors, then in principle it is possible to represent each degree of freedom for main effects and interactions by one of a set of orthogonal planned comparisons. Since in a factorial design the cells are the basic experimental "groups," we can think of all of the systematic degrees of freedom as *corresponding to different orthogonal comparisons among the cells.*

Thus, for example, if a two-factor design involves Factor A at J levels, and Factor B at K levels, then we know that there must exist a set of $JK - 1$ orthogonal comparisons among the cells of our design, just as there would be for any set of JK independent groups. Furthermore, if our design is balanced, then among this total set of comparisons among the cell means it is possible to find a set of $J - 1$ orthogonal comparisons that contain all the information about y variation due to the levels of Factor A. Another set of $K - 1$ orthogonal comparisons represents variation due to the differing levels of B, and a third set of $(J - 1)(K - 1)$ orthogonal comparisons must contain all of the information about variation due to $A \times B$ interaction. In other words, each of the $JK - 1$ degrees of freedom for AB cells can be identified with one of a set of orthogonal comparisons, and sets of these orthogonal comparisons provide information about Factor A, about Factor B, and about $A \times B$ interaction, exactly parallel to the breakdown provided in ANOVA.

We will first illustrate these facts for a two-factor design. However, the principle extends just as well to higher order designs.

If we consider a two-factor design, with cell means denoted by $\bar{y}_{jk}$, then it is possible to weight and sum the cell means so as to compare or contrast the levels of Factor A. That is, let a comparison among the J levels of Factor A be denoted by

$$\text{comparison among levels of Factor A} = \sum_{j=1}^{J} \sum_{k=1}^{K} c_{j \cdot} \bar{y}_{jk} \qquad [12.10.1^*]$$

where the symbol $c_{j \cdot}$ for a comparison weight will be used here for the time being to denote one of a set of $J - 1$ orthogonal comparisons among means of the different levels of Factor A. As usual with such comparisons, we want the sum of the weights to be 0:

$$\sum_{j} c_{j \cdot} = 0.$$

In addition, we can also find a set of $K - 1$ orthogonal comparisons among levels of the B factor. Any one of these orthogonal comparisons among the B levels may be expressed as

$$\text{comparison among levels of the B factor} = \sum_{k=1}^{K} \sum_{j=1}^{J} d_{\cdot k} \bar{y}_{jk} \qquad [12.10.2^*]$$

Again, just for the time being, we will let $d_{\cdot k}$ stand for the weight given to a cell in one of the comparisons among the B levels. Then as always for such weights

$$\sum_{k} d_{\cdot k} = 0.$$

If the factorial design in question is truly crossed and balanced, with each and every one of the possible combinations of the factor levels represented exactly n times, then any comparison made on the levels of A will be orthogonal to any comparison made on the levels of B. The whole point of the exercise of crossing and balancing your experiment is to make sure that the set of comparisons summarized under SS A will all be orthogonal to each of the comparisons summarized under SS B. You should check this out for yourself as we find the comparisons for A levels and B levels given below.

For example, suppose that the experimenter in Section 12.1 had, from the very outset, been interested in planned comparisons reflecting differences in the levels of Factor A, differences among the levels of Factor B, and certain interaction effects among the AB combinations. In particular, she wished to ask:

1. Does the mean of A_1 differ from the mean of A_2?
2. Does the mean of B_1 differ from the average of B_2 and B_3?
3. Does the mean of B_2 differ from the mean of B_3, irrespective of B_1?
4. Does the answer to Question 2 change if we first examine it within Level A_1 and then in Level A_2?
5. Does the answer to Question 3 change if we first answer it within Level A_1 and then within Level A_2?

It will be demonstrated that these planned comparisons each correspond to one of the five systematic degrees of freedom for this six-celled design. Furthermore, the sums of squares for A, for B, and for $A \times B$ are all immediately available from the values corresponding to these comparisons.

If we use the $JK = 6$ means of the cells shown in the data of Table 12.2.1, we can carry out one comparison on the means of the two levels of Factor A (i.e., two rows of the table), as follows:

	cell:	A_1B_1	A_2B_1	A_1B_2	A_2B_2	A_1B_3	A_2B_3	$\hat{\psi}$ value
	mean:	46.4	36.8	30.2	37.8	17.8	21.4	
A	$c_{j\cdot}$	1/3	−1/3	1/3	−1/3	1/3	−1/3	−.5333

Notice that this comparison encompasses the first question listed above, and provides the actual $\hat{\psi}$ value of about $-.5333$.

In terms of levels of Factor B, the two orthogonal comparisons that correspond to the two questions expressed above are:

cell:	A_1B_1	A_2B_1	A_1B_2	A_2B_2	A_1B_3	A_2B_3	$\hat{\psi}$ value
mean:	46.4	36.8	30.2	37.8	17.8	21.4	
1st B $d_{\cdot k}$	1/2	1/2	$-1/4$	$-1/4$	$-1/4$	$-1/4$	14.8
2nd B $d_{\cdot k}$	0	0	1/2	1/2	$-1/2$	$-1/2$	14.4

Taken together, these two sets of orthogonal comparisons, one set for Factor A and one set for Factor B, account for $J-1 + K-1$ or 3 degrees of freedom. Of course the sums of squares for these two sets of orthogonal comparisons, turn out to be identical to the corresponding ANOVA sums of squares in Table 12.7.2, by the principle we have already encountered in Section 11.9,

SS A comparison $= (-.5333)^2/(6/9(10)) = 4.27$
SS first B comparison $= (14.8)^2/(3/4(10)) = 2,920.53$
SS second B comparison $= (14.4)^2/(1/10) = 2,073.60$.

That is, the SS for the A comparison is exactly the same as SS A from ANOVA, and that the sum of the *two* comparison SS representing differences among the B levels is identical to SS B from ANOVA (Table 12.7.2):

SS first B comp. + SS second B comp. $= 2,920.53 + 2,073.60$

$$= 4,994.13$$

$$= \text{SS B}.$$

Clearly each of the SS values for main effects summarizes the sums of squares for a set of orthogonal comparisons equal in number to the listed degrees of freedom.

However, what of the $(J-1)(K-1)$ degrees of freedom associated with A×B interaction? Interaction comparisons are also to be found by weighting and summing the cell means, but where the weights can be found as shown below.

Given a set of orthogonal comparison weights for Factor A, and given another set of orthogonal comparison weights for Factor B, then a set of orthogonal comparisons containing the information in A×B interaction can be found by *multiplying each A set of weights by each set of the B weights*. That is, suppose that one of the comparisons among the A levels has weights $c_1, \ldots, c_J$, and a comparison on the B levels has weights $d_{\cdot 1}, \ldots, d_{\cdot K}$; then there exists *an interaction comparison* among the AB cells with weights given by

$$g_{jk} = c_j \cdot d_{\cdot k} \qquad \qquad [12.10.3^*]$$

In other words, given the weights c for a comparison among the A levels, and the weights d for a comparison among the B levels, then an interaction comparison can always be found by multiplying each c weight by its corresponding d weight, cell by cell. The new weights $g = cd$ will sum to zero just as they should, and the resulting comparison will be orthogonal both to the A comparison and to the B comparison (provided that the design is balanced, of course.)

Furthermore, if the weights for all $J-1$ of the orthogonal A comparisons are multiplied by all of the corresponding weights for the set of $K-1$ orthogonal B comparisons, the result will be the weights for a new set of $(J-1)(K-1)$ comparisons, which will also be pairwise orthogonal, and each of which will be identified with one degree of freedom for interaction.

This principle is probably easier to grasp from an example than from a lengthy explanation. Consider again our two-factor data. Let us now carry out two more comparisons, the first having weights that are products of the A comparison given above and the first B comparison. That is, for this new comparison we will take

$(1/3)(1/2) = 1/6$ as the weight for cell A_1B_1

$(-1/3)(1/2) = -1/6$ as the weight for cell A_2B_1

$(1/3)(-1/4) = -1/12$ as the weight for cell A_1B_2

$(-1/3)(-1/4) = 1/12$ as the weight for cell A_2B_2

$(1/3)(-1/4) = -1/12$ as the weight for cell A_1B_3

$(-1/3)(-1/4) = 1/12$ as the weight for cell A_2B_3.

Furthermore, to find a second interaction comparison, we can multiply the weights for the first A comparison by the weights for the second B comparison, to get

$(1/3)(0) = 0$ as the weight for cell A_1B_1

$(-1/3)(0) = 0$ as the weight for cell A_2B_1

$(1/3)(1/2) = 1/6$ as the weight for cell A_1B_2

$(-1/3)(1/2) = -1/6$ as the weight for cell A_2B_2

$(1/3)(-1/2) = -1/6$ as the weight for cell A_1B_3

$(-1/3)(-1/2) = 1/6$ as the weight for cell A_2B_3.

Once again let us display these two new sets of weights horizontally, and together. Then it is easy to confirm that the weights do add to zero over the cells, just as any respectable set of comparison weights should, and also that the two new sets of weights are themselves orthogonal (try it and see!).

cell:	A_1B_1	A_2B_1	A_1B_2	A_2B_2	A_1B_3	A_2B_3	$\hat{\psi}$ value
mean:	46.4	36.8	30.2	37.8	17.8	21.4	
A, 1st B	1/6	−1/6	−1/12	1/12	−1/12	1/12	2.5333
A, 2nd B	0	0	1/6	−1/6	−1/6	1/6	−0.6667

The sum of squares for this first comparison is found from

SS A and 1st B interaction $= (2.5333)^2/(1/12(10))$

$$= 770.11.$$

Similarly, for the second interaction comparison

$$\text{SS A and 2nd B interaction} = (-0.6667)^2/(1/9(10))$$

$$= 40.00.$$

The sum of these two sums of squares is then

$$770.11 + 40 = 810.11,$$

which agrees (within rounding) with

$$\text{SS interaction} = 810.13$$

from Table 12.7.2, also with two degrees of freedom.

Interaction comparisons found through multiplication as outlined above can be regarded as sets of comparisons made not on means, but on the values of other comparisons. Essentially, in an interaction comparison such as those made above, we are making comparisons of the levels of one factor within a single level of the second factor, and then looking for changes in the first comparison value as the second factor is varied.

This example demonstrates that two-way ANOVA corresponds to three sets of orthogonal comparisons among the cells in a two-factor design, just as one-way ANOVA corresponds to one set of orthogonal comparisons among independent, randomized groups, or the levels of a single factor. In principle for data qualifying under the fixed-effects model, it is always possible to carry out some or all of a set of orthogonal, planned comparisons corresponding to the sum of squares of one or both of the factors, or to the interaction sum of squares.

Each of the $JK - 1$ orthogonal comparisons that one has planned for a $J \times K$ factorial design may be given a confidence interval, or may be tested by a t or an F test, exactly as shown in Section 11.4. The MS error as found in the two-way ANOVA is used in the intervals and tests, and the F tests each have 1 and $N - JK$ degrees of freedom. Additional information on defining and testing interaction comparisons is given in Kirk (1982). An excellent discussion of both planned and post hoc comparisons for factorial designs may be found in Keppel (1982).

12.11 POST HOC TESTS IN A FACTORIAL DESIGN

It is quite possible to carry out posterior comparisons on the means of the rows, the columns, or of the cells in a factorial design with fixed effects. The issues of comparisonwise and familywise error rates occur in this context just as they do for comparisons in the randomized groups or one-factor situation (Section 11.13). However, in dealing with two-way or higher factorial designs, it is customary to define the "families" of comparisons according to the major breakdown of the ANOVA. That is, for two factors A and B, there will be a family of A-level comparisons, one of B-level comparisons, and a family of interaction comparisons. On occasion one may also wish to deal with the family of all cell comparisons. All of the methods discussed in Chapter 11 for dealing with post hoc comparisons apply directly to the families of comparisons in a factorial design.

If the F test for any one of the factors, or "main effects," proves to be significant at the α level, then the Scheffé criterion can be applied to judging the significance of any comparison made on the levels of that factor. Thus, if Factor A with J levels produces a significant F, then for a post hoct t test the Scheffé criterion is taken to be

$$\mathbf{S} = \sqrt{(J-1)\,F_\alpha}$$

where F_α is the F value required for significance at the α level with $J-1$ and $N-JK$ degrees of freedom. A confidence interval may then be found using MS error just as shown in Section 11.16. If the posterior comparisons are to be made on the K different levels of B, then the value of $\mathbf{S}$ is based on the multiplier $(K-1)$ and the F_α value for $K-1$ and $N-JK$ degrees of freedom. Interaction comparisons call for the multiplier $(J-1)\,(K-1)$ and the F_α required for $(J-1)\,(K-1)$ and $N-JK$ degrees of freedom.

The Tukey HSD and the Newman-Keuls techniques also apply to pair comparisons on the results of a factorial design. The only real modifications required are these: The definition of "the number of groups" as used in Section 11.17 changes to become "the number of levels" for the factor being examined. In addition, the value of n is replaced by Kn or Jn in the formula for the Tukey HSD. Thus, to test differences between all pairs of the J levels of Factor A, the Tukey HSD is found from

$$\mathrm{HSD}_A = q_{\alpha,J,N-JK}\ \sqrt{\mathrm{MS\ error}/nK}.$$

Similarly, for Factor B,

$$\mathrm{HSD}_B = q_{\alpha,K,N-JK}\ \sqrt{\mathrm{MS\ error}/nJ}.$$

For the pair comparisons among cells that might be of most interest when the ANOVA findings include significant interaction, the Tukey HSD is found from

$$\mathrm{HSD}_{AB} = q_{\alpha,JK,N-JK}\ \sqrt{\mathrm{MS\ error}/n}.$$

Given these modified versions of the Tukey HSD, then the Newman-Keuls procedure can be carried out exactly as described in Section 11.17.

12.12 THREE-FACTOR AND HIGHER ANALYSIS OF VARIANCE

The principles that we have discussed in connection with the two-factor ANOVA extend immediately to the situation where there are three or more factors. If one is solving such problems by hand, then the higher the order of the design, the more computational labor is involved. However, this is hardly a consideration when the analysis is done by computer, and, in principle, ANOVAs may be carried out for designs of almost any size.

The analysis of variance for a three-factor design is actually done as if for a set of 3 two-factor designs. Hence if you know how to do a two-way analysis of variance, then you also have the basic skills required for a three-way analysis. For a three-factor ANOVA, with the factors labeled as A with J levels, B with K levels, and C with H levels, the steps are these:

1. Find the SS total is the usual way, based on all N cases.
2. Now find the total T_{jkh} of each of the JKH basic *cells,* or ABC combinations in the data, and use these totals to find SS ABC cells:

3. Find the value of SS error by taking

$$\text{SS error} = \text{SS within cells} = \text{SS total} - \text{SS ABC cells}.$$

4. Ignore Factor C, and find the totals of the AB cells, T_{jk}. Carry out the complete two-way analysis on the AB table, to find SS A, SS B, and SS A×B.
5. Next ignore Factor B and carry out a two-way analysis on the AC totals, $T_{j \cdot h}$ to find SS C and SS A×C (recall that SS A has already been found). Then ignore Factor A and do the two-way analysis on the BC totals, $T_{\cdot kh}$. Use these to find SS B×C.
6. Find SS A×B×C by taking

$$\text{SS A} \times \text{B} \times \text{C} = \text{SS ABC cells} - \text{SS A} - \text{SS B}$$

$$- \text{SS C} - \text{SS A} \times \text{B} - \text{SS A} \times \text{C} - \text{SS B} \times \text{C}.$$

This procedure can be illustrated by use of the data in Table 12.12.1. Suppose that these data represent the outcomes of a study of the possible effects of caffeine consumption in soft drinks on retention of learned materials. Some 36 college women were chosen at random, and then $n = 2$ were assigned at random to each of 18 groups, representing combinations of levels of Factors A, B, and C. Each student was given a set of materials to study for a period of eight hours (interspersed with short rest breaks). Each hour the student was required to drink one soft drink supplied by the experimenter, although no other food or drink was permitted. Then, after another rest period, a test of recall was given over the material studied. The dependent variable value was the number of recall *errors* made by a student on this test.

The actual experimental factor in this study concerned the soft drinks consumed by the subjects. Factor A was the proportion of sugar used as the sweetener in each soft drink, with $A_1 =$ all sugar, $A_2 =$ half sugar and half artificial sweetener, and $A_3 =$ all artificial sweetner. Factor B concerned the amount of caffeine in the drinks, with $B_1 =$ none, $B_2 =$ moderate, and $B_3 =$ heavy. The third factor, C, had to do with the carbonation of the soft drinks, with $C_1 =$ noncarbonated, and $C_2 =$ carbonated. Other elements such as flavoring agents, etc. were exactly the same for all of the soft drinks consumed.

	B_1	B_2	B_3	B_1	B_2	B_3
A_1	12	7	19	13	11	13
	10	12	21	10	9	17
	22	19	40	23	20	30
A_2	6	10	21	4	16	25
	3	16	29	5	12	16
	9	26	50	9	28	41
A_3	15	8	24	10	5	13
	12	6	13	15	10	16
	27	14	37	25	15	29
		C_1			C_2	

$T_y = 464$

Table 12.12.1

Recall errors made by students under three levels of sugar consumption, three levels of caffeine, and two levels of carbonation of soft drinks consumed during study

The raw data for this experiment are shown in Table 12.12.1. To analyze these data, we begin by following Step 1 as outlined above, and finding the SS total:

$$\text{SS total} = (12)^2 + (10)^2 + \cdots + (13)^2 + (16)^2 - (464)^2/36$$

$$= 1{,}281.56 \; .$$

Next we find the totals for each of the 18 cells for this design. (Remember that for cell total T_{jkh}, the j stands for the A level, k for the B level, and h for the C level.) Then,

$$\text{SS ABC cells} = \frac{\sum_{j}\sum_{k}\sum_{h} T_{jkh}^2}{n} - \frac{T_y^2}{N}$$

$$= \frac{(22)^2 + (9)^2 + \cdots + (29)^2}{2} - \frac{(464)^2}{36}$$

$$= 1{,}050.56.$$

Immediately, then, we have the value of SS error

$$\text{SS error} = \text{SS total} - \text{SS ABC cells}$$

$$= 1{,}281.56 - 1{,}050.56$$

$$= 231.00 \; .$$

Our next step is to ignore one factor (say C) and put all the data together in an AB table such as this, showing the AB cell totals:

	B_1	B_2	B_3	tots.
A_1	45	39	70	154
A_2	18	54	91	163
A_3	52	29	66	147
tots.	115	122	227	464

Basing our calculations on this two-factor table, we find that

$$\text{SS AB cells} = \frac{\sum_{j}\sum_{k} T_{jk\cdot}^2}{Hn} - \frac{T_y^2}{N}$$

or

$$\text{SS AB cells} = \frac{(45)^2 + \cdots + (66)^2}{4} - \frac{(464)^2}{36}.$$

$$= 986.56.$$

Furthermore,

$$\text{SS A} = \frac{(154)^2 + (163)^2 + (147)^2}{12} - \frac{(464)^2}{36}$$

$$= 10.72$$

$$SS\ B = \frac{(115)^2 + (122)^2 + (227)^2}{12} - \frac{(464)^2}{36}$$

$$= 656.06$$

and

$$SS\ A \times B = SS\ AB\ cells - SS\ A - SS\ B$$

$$= 986.56 - 10.72 - 656.06$$

$$= 319.78.$$

Next we construct and use the table of AC cell totals:

	C_1	C_2	*tot.*
A_1	81	73	154
A_2	85	78	163
A_3	78	69	147
tots.	244	220	464

and find

$$SS\ AC\ cells = \frac{\sum_{j}\sum_{h} T_{j \cdot h}^2}{Kn} - \frac{T_y^2}{N},$$

or

$$SS\ AC\ cells = \frac{(81)^2 + \cdots + (69)^2}{6} - \frac{(464)^2}{36}$$

$$= 26.89,$$

and

$$SS\ A = 10.72$$

$$SS\ C = \frac{(244)^2 + (220)^2}{18} - \frac{(464)^2}{36}$$

$$= 16.00$$

$$SS\ A \times C = 26.89 - 10.72 - 16.00$$

$$= .17.$$

The last table required is that for Factors B and C, with the BC cell totals:

	B_1	B_2	B_3
C_1	58	59	127
C_2	57	63	100

This provides values for

$$\text{SS BC cells} = \frac{\displaystyle\sum_{k}\sum_{h} T_{\cdot kh}^{2}}{Jn} - \frac{T_{y}^{2}}{N},$$

or

$$\text{SS BC cells} = \frac{(58)^{2} + \cdots + (100)^{2}}{6} - \frac{(464)^{2}}{36}$$

$$= 718.22 .$$

In addition

$$\text{SS B} \quad = 656.06$$

$$\text{SS C} \quad = \quad 16.00$$

$$\text{SS B} \times \text{C} = 718.22 - 656.06 - 16$$

$$= \quad 46.16.$$

Lastly, we find the value of the SS $A \times B \times C$ interaction:

$$\text{SS } A \times B \times C = \text{SS ABC cells} - \text{SS } A \times B - \text{SS } A \times C$$

$$- \text{ SS } B \times C - \text{SS } A - \text{SS } B - \text{SS } C$$

$$= 1{,}050.56 - 319.78 - .17 - 46.16 - 10.72 - 656.06 - 16.00$$

$$= 1.67 .$$

The summary table (Table 12.12.2) for this or any other three-factor ANOVA is just an extension of the table used for a two-factor design.

The results of this three-way analysis of variance show that the Factor B, amount of caffeine, gave a highly significant F value, giving us confidence that there are different retention effects for differing amounts of caffeine. On the other hand, the evidence of

Table 12.12.2

Summary table for three-way ANOVA

Source	SS	df		MS	F	P
A	10.72	$J-1=$	2	5.36	0.42	NS
B	656.06	$K-1=$	2	328.03	25.57	<.01
C	16.00	$H-1=$	1	16.00	1.25	NS
A×B	319.78	$(J-1)(K-1)=$	4	79.95	6.23	<.01
A×C	0.17	$(J-1)(H-1)=$	2	0.085	0.007	NS
B×C	46.16	$(K-1)(H-1)=$	2	23.08	1.80	NS
A×B×C	1.67	$(J-1)(K-1)(H-1)=$	4	0.42	0.033	NS
Error (within cells)	231.00	$N-JKH=$	18	12.83		
Total	1,281.56	$N-1=35$				

true effects for differing amounts of sugar failed to be significant, although a significant caffeine-by-sugar interaction was found. Factor C as well as all of the remaining interactions failed to reach significance.

The calculation of the R-square values for this sample, showing how the factors and interactions account for variance in the y values, only reinforces the picture given by the F tests. We find that for Factor B, caffeine, the proportion of variance accounted for is

$$R_{y \cdot B}^2 = 656.06/1,281.56 = .51 .$$

Thus, in *these* data just over 50% of the y variance can be explained by the amount of caffeine the students consumed. Futhermore, about 25% of additional variance can be explained in terms of the caffeine–sugar interaction:

$$R_{y \cdot A \times B}^2 = 319.78/1,281.56 = .25 .$$

It is obvious that no other effects or interactions account for more than a very small proportion of the y variance. If desired, the R-square values for the sample as shown above can be modified into estimates of omega-square by taking

$$\text{est. } \omega_{y \cdot B}^2 = \frac{656.06 - (2 \times 12.83)}{1,281.56 + 12.83}$$

$$= .49$$

and

$$\text{est. } \omega_{y \cdot A \times B}^2 = \frac{319.17 - (4 \times 12.83)}{1,281.56 + 12.83}$$

$$= .21$$

just as suggested in Section 12.9. If estimates of proportions of variance accounted for are desired for other interactions in three-way or higher analyses they can be found in completely analogous ways.

In addition, planned or post hoc comparisons can be carried out for three-way or higher designs. Since the appropriate methods are obvious extensions of those we have already discussed for two-way designs, that ground will not be retraced here.

In short, our conclusions from this realtively elaborate experiment turn out to be pretty simple: The amount of caffeine consumed makes a significant difference in the errors made in a test of recall, with (as the data show) the most errors associated with the most caffeine, the second most with the next most caffeine, and the fewest errors with no caffeine. However, the significant interaction makes it necessary to qualify this statement: moderate caffeine leads to fewest errors in the presence of high sugar or no sugar, whereas no caffeine gives fewest errors when there is moderate sugar. These results can be seen in the AB table which is shown above. These can also be inferred from the $A \times B$ interaction graph based on the means of this AB table (Figure 12.12.1).

Figure 12.12.2 is a representation of the $A \times B \times C$ interaction in the three-factor experiment of Table 12.12.1. Here the graph of the $A \times B$ interaction is constructed much as in Figure 12.12.1, but this time a separate $A \times B$ graph is drawn for each level of the third factor, or C. The important thing here is to compare the $A \times B$ graphs across

Figure 12.12.1

Two representations of the A × B interaction in the data in Table 12.12.1.

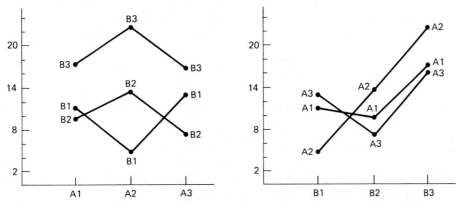

the levels of C; as you can see from the figure for these data the two graphs are virtually identical and each is very close to the A × B interaction graph of Figure 12.12.1. This is evidence for *no* A × B × C interaction in these data: the A × B interaction effects do not tend to vary across the levels of C. This supports the findings in the ANOVA for these data, in which the test for A × B × C interaction produced an *F* value less than 1.00. On the other hand, if the A × B × C interaction had turned out to be significant, we should have expected these graphs to show marked differences across the levels of C.

Figure 12.12.2

Representation of the A × B × C interaction in the data in Table 12.12.1.

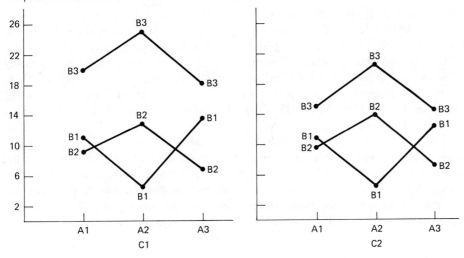

12.13 A GENERAL ALGORITHM FOR ANALYZING DESIGNS WITH MANY FACTORS

If a factorial design is based on several factors, say W in all, then it is possible to partition the total sum of squares into 2^W components. Of these sums of squares, exactly W will represent main effects, $\binom{W}{2}$ will be *first-order interactions,* of the form $A \times B$; an addtional $\binom{W}{3}$ will be *second-order interactions* of the form $A \times B \times C$; some $\binom{W}{4}$ will be *third-order interactions* of the type $A \times B \times C \times D$, and so on. Therefore, the ANOVA for a large design can be very complicated. Nevertheless, a few simple principles guide how such an analysis is carried out.

The analysis of data coming from a factorial design with any number W of factors A, B, C, $\cdots$, G, can always be carried out by following these steps:

1. Find the usual SS total.
2. Find the sum of each of the basic cells in the data table and calculate the SS ABC $\cdots$ G cells, based on these sums.
3. Find the SS error by taking SS total $-$ SS AB $\cdots$ G cells
4. Ignore one factor, and set up a new data table based on the sums of cells collapsed over that factor. Carry out the complete $(W-1)$ way analysis of variance for this table.
5. Carry out Step 4 for a total of W times, ignoring each factor in turn.
6. Find the highest order $A \times B \times C \times \cdots \times G$ interaction, by taking SS ABC $\cdots$ G cells $-$ (SS for all other main effects and interactions)

If the number of levels of the various factors are J for A, K for B, H for C, L for D, and so on until V levels for G, then the degrees of freedom for any single factor will always be (levels -1). Thus, the degrees of freedom for Factor A are $(J-1)$, for B they are $(K-1)$, and so on. The degrees of freedom for any interaction are then the *product of the degrees of freedom for the individual factors* involved in the interaction. Thus, the df for the first-order interaction $A \times B$ are $(J-1)(K-1)$. The df for the second-order interaction $A \times B \times C$ are $(J-1)(K-1)(H-1)$; for $A \times B \times C \times D$ they are found from $(J-1)(K-1)(H-1)(L-1)$, and so on.

This algorithm for finding the ANOVA sums of squares will work for any factorial design with W factors, where, of course, the factors are completely crossed, and the design is balanced, with at least $n = 2$ cases per cell. Obviously, the amount of work can quickly become prohibitive if there are more than three or four factors, and you will almost always want to handle such problems by computer anyway. Nevertheless, it is well to know the steps that have been followed to produce the output you will receive.

12.14 EXAMINING AND INTERPRETING SECOND- AND HIGHER ORDER INTERACTIONS

We have seen that first-order interactions such as $A \times B$ can play an important role in qualifying or otherwise limiting your conclusions about the factors that the interaction involves. This principle also extends in force to higher order interactions, where a significant $A \times B \times C$ interaction sum of squares forces you to qualify what is said about the $A \times B$ interaction (which depends on C), about $A \times C$ interaction (which depends on B) and $B \times C$ interaction (depending on A). In addition, the presence of such second-order interaction effects makes it necessary to qualify statements about main effects of A (in the light of BC), about B (in the light of AC), and about C (in the light of AB). Certainly, the presence even of a significant second-order interaction greatly complicates both the interpretation and the application of the results of any experiment, not to speak of the difficulties introduced by possible interactions of even higher order.

The meaning of second-order or higher order interactions can be clarified by the use of the notion of so-called simple effects (which perhaps might more appropriately be called "conditional" effects). A simple effect is the effect associated with some level of a given factor (or interaction) when the level of at least one other factor is specified. Thus, for example, consider an experiment with three factors, A, B, and C. Given Factor A with J levels, the ordinary sample main effects of this factor would be defined by

$$a_j = \bar{y}_{j\cdot\cdot} - \bar{y}_{\cdots}$$

However, now let us choose one level k of Factor B, and look only at the effects of Factor A *within* level B_k. Each of these A effects within a given level of B would be called *a simple main effect of* A, *given level* B_k, and would be defined by:

$$\text{simple main effect of } A_j \text{ given } B_k = (a|b)_{jk} \qquad [12.14.1]$$
$$= \bar{y}_{jk\cdot} - \bar{y}_{\cdot k\cdot},$$

the difference between the mean of the cells in Levels A_j and B_k, and the mean of Level B_k. The simple main effect of A_j within B_k can also be found from the effects

$$(a|b)_{jk} = [ab]_{jk} - b_k, \qquad [12.14.2]$$

showing that the simple main effect for A_j given B_k is the difference between the $A_j B_k$ cell effect and the main effect of B_k.

Observe that, for equal n per cell, within any B_k the simple main effects of A must add to zero:

$$\sum_j (a|b)_{jk} = 0. \qquad [12.14.3]$$

By an exactly parallel argument, we could also define the simple main effects of B_k given any level A_j of Factor A:

$$(b|a)_{jk} = \bar{y}_{jk\cdot} - \bar{y}_{j\cdot\cdot} \qquad [12.14.4]$$
$$= [ab]_{jk} - a_j.$$

Now, obviously, for any level A_j the simple main effect $(a|b)_{jk}$ given any B_k may differ from the regular main effect a_j for that level of A. The extent of that difference is then the *interaction effect* associated with Levels A_j and B_k:

$$\text{interaction effect for } A_j \text{ and } B_k = (ab)_{jk} \qquad [12.14.5]$$

$$= (a|b)_{jk} - a_j.$$

Equivalently, we may also write that

$$\text{interaction effect for } A_j \text{ and } B_k = (ab)_{jk} \qquad [12.14.6]$$

$$= (b|a)_{kj} - b_k.$$

If the simple main effect of one factor were the same regardless of the level of the other factor, then the simple main effect at *any* level would be equal to the regular main effect at that level, and *no interaction* would be present. In other words, a first-order interaction effect exists only when the simple main effect of one factor given the level of the other is *not* the same as the regular main effect, ignoring the other factor. This is the basis for our statement that the presence of an interaction acts to *qualify* the interpretation placed on an ordinary main effect: to specify exactly what the effect of a factor level will be we must also specify the level of the second factor.

The sum of the squares of the simple main effects for, say factor A_j given a specific level B_k, can be found as follows:

$$\text{SS simple effects } A|B_k = \sum_j (a|b)_{jk}^2$$

$$= \frac{\sum_j T_{jk}^2}{n} - \frac{T_{\cdot k}^2}{Jn}, \qquad [12.14.7]$$

with $J-1$ degrees of freedom.

Under certain circumstances it may be desirable to test the hypothesis of no simple effects of Factor A given some particular level of B, and this may be done on the basis of the MS formed from this sum of squares, as compared with the mean square for error found in the usual way for a fixed-effects model.

It will then also be true that

$$\text{SS A} + \text{SS A} \times \text{B} = \sum_k \text{SS simple effects } A|B_k$$

$$= \text{SS AB cells} - \text{SS B}, \qquad [12.14.8]$$

with $K(J-1)$ degrees of freedom.

It also follows that

$$\text{SS A} \times \text{B interaction} = \sum_k \text{SS simple effects } A|B_k - \text{SS A}. \qquad [12.14.9]$$

When there are three factors under consideration, the notion of a simple main effect can be generalized to that of *a simple interaction effect*. That is, the simple interaction for the combination $A_j B_k$, given some level m of Factor C is defined by

$$(ab|c)_{jkm} = \text{simple interaction effect of } A_j B_k \text{ given } C_m \qquad [12.14.10]$$

$$= \bar{y}_{jkm} - \bar{y}_{j\cdot m} - \bar{y}_{\cdot km} + \bar{y}_{\cdot\cdot m} .$$

Looking exclusively within Level C_m, we therefore examine the possible interaction between A_j and B_k, and this is defined as *the simple interaction effect* of A_j *and* B_k *given* C_m.

For any combination A_j and B_k the simple interaction within C_m can then be compared with the ordinary interaction $(ab)_{jk}$. If a difference exists between the simple interaction $(ab|c)_{jkm}$ and the regular interaction $(ab)_{jk}$ for that combination, then we may say that *a second-order interaction exists among A, B, and C.* That is, the simple interaction effect between A and B tends to vary with changes in C.

The same sort of statement can be made if the simple interaction of A_j and C_m within B_k differs from the regular interaction of A_j and C_m, or if the simple interaction of B_k and C_m given A_j differs from the regular interaction of B_k and C_m. In each of these circumstances one can say that the simple interaction of two factors varies with changes in a third factor, and this, in essence, is the meaning of a second-order interaction. When such interaction effects exist, then in order to specify how the interaction between two factors actually works, one must specify the level of the third factor.

Given four factors, in an analogous way one may discuss the simple interaction among Factors A, B, and C, given the level of Factor D:

$$(abc|d)_{jkmt} = \text{simple interaction of } A_jB_kC_m \text{ given } D_t$$

$$= \bar{y}_{jkmt} - \bar{y}_{jk\cdot t} - \bar{y}_{j\cdot mt} - \bar{y}_{\cdot kmt} \qquad [12.14.11]$$

$$+ \bar{y}_{j\cdot\cdot t} + \bar{y}_{\cdot k\cdot t} + \bar{y}_{\cdot\cdot mt} - \bar{y}_{\cdot\cdot\cdot t} .$$

On the other hand, the "regular" $A \times B \times C$ interaction effect, ignoring D, is given by

$$(abc)_{jkm\cdot} = \bar{y}_{jkm\cdot} - \bar{y}_{jk\cdot\cdot} - \bar{y}_{j\cdot m\cdot} - \bar{y}_{\cdot km\cdot} + \bar{y}_{j\cdot\cdot\cdot} + \bar{y}_{\cdot k\cdot\cdot} \qquad [12.14.12]$$

$$+ \bar{y}_{\cdot\cdot m\cdot} - \bar{y}_{\cdot\cdot\cdot\cdot} .$$

Then if there is a difference between the simple and the regular second-order interaction effect, we have evidence for a *third-order interaction effect* associated with this particular $A_jB_kC_mD_t$ combination.

In general, G-th order interaction effects exist if the simple interaction effects of order $(G-1)$, given one additional factor, tend to vary with variation in the additional factor. Hence, first-order interactions reflect changes in simple main effects; second-order interactions reflect changes in simple first-order interactions; third-order interactions show variations among simple second-order interactions, and so forth. (Although this discussion of simple effects has been restricted to sample effects, these definitions translate completely into population effects as well. This implies that interactions in population situations can be discussed in exactly the same terms we have just been using to describe sample interactions.)

The ideas of simple effects can be a very useful approach to examining a set of data to try to figure out the location and possible reasons for first-order or higher interaction effects. They can also be turned into sums of squares and used for making tests in certain kinds of experimental designs. Although we will not go further into this topic here, the reader is referred to Winer (1971) and to Keppel (1982) for extended discussions of simple effects and their various uses.

12.15 ASSUMPTIONS IN TWO-WAY (OR HIGHER) ANALYSIS OF VARIANCE WITH FIXED EFFECTS

The list of assumptions in Section 12.5 is an almost exact parallel to the list in Section 10.9. Similar assumptions are made for more complex experiments requiring a higher order analysis, provided that the fixed-effects model is appropriate.

As you may have anticipated, the same relaxation of assumptions is often possible in the two- or multiway analysis as in the one-way analysis. For experiments with a relatively large number of observations per cell, the requirement of a normal distribution of errors seems to be rather unimportant. In an experiment where it is suspected that the parent distributions of the values of dependent variables are very unlike a normal distribution, perhaps a correspondingly large number of observations per cell should be used.

When the data table represents an equal number of observations in each cell, the requirement of equal error variance in each treatment combination population may also be violated without serious risk, at least in terms of Type I error. However, in some circumstances the power of the F test may be affected. On the whole, there are two good reasons for planning experiments with equal n per cell: the experimental design will thus be orthogonal (Section 12.1) and the possible consequences of nonhomogeneous variances on the probabilty of Type I error will be minimized.

Regardless of the simplicity or complexity of the experiment, however, the error portions entering into the respective observations should be independent if the fixed-effects model is to apply.

12.16 THE ANALYSIS OF VARIANCE AS A SUMMARY OF DATA

It may appear that the main use of the analysis of variance, particularly for two-factor or multifactor experiments, is in generating a number of F tests on the same set of data, and that the partition of the sum of squares is only a means to this end. However, this is really a very narrow view of the role of this form of analysis in experimentation. The really important feature of the analysis of variance is that it permits the separation of all of the potential information in the data into distinct and nonoverlapping portions, each reflecting only certain aspects of the experiment.

Similarly, for a two-factor experiment we arrive at a mean square for one treatment factor and a separate mean square for the other. These two mean squares reflect quite nonredundant aspects of the experiment, even though they were each based on the same basic data: the first sum of squares reflects only the effects attributable to the first experimental factor (plus error), and the second those effects attributable to the other (plus error). Under the statistical assumptions we make, these two mean squares are independent of each other. Furthermore, the mean squares for interaction and for error are independent of each other and of the treatment mean squares. The analysis of variance lets the experimenter ''pull apart'' the factors that contribute to variation in the experiment, and identify them exclusively with particular summary statistics. For experiments of the orthogonal, balanced type considered here, the analysis of variance is a routine method for finding the statistics that reflect particular, meaningful aspects of the data. The real importance of the analysis of variance lies in the fact that it routinely provides such succinct overall "packaging" of the data.

For the moment, however, let us consider the several F tests obtained from a two- or multifactor experiment. The sums of squares and mean squares for columns, for rows, for interaction, and for error are all, under the assumptions made, independent of each other. However, are the three or more F tests themselves independent? Does the level of significance shown by any one of the tests in any way predicate the level of significance shown by the others? Unfortunately, it can be shown that although the mean squares are independent such F tests are *not* independent. Some connection exists among the various F values and significance levels. This is due to the fact that each of the F ratios involves the same mean square for error in the denominator; the presence of this same sample value in each of the ratios creates some statistical dependency among them.

However, according to the theory of estimation, the mean square error values should tend to stabilize toward a constant value over samples if the number of degrees of freedom for error is very large. We also know that F ratios with independent numerators and constant denominators will be independent, and so our F tests should approach independence for large samples. Empirical studies have also suggested that for other than quite small samples the nonindependence of the F tests need not be a particular concern. (However, for some recent and rather discouraging results on this question see Feingold and Korsog (1986))

For a factorial design with W factors there are also $2^W - 1$ potential F tests, and this can be a fairly large "family" of tests carried out on the same set of data, particularly when the number of factors W is greater than 2 or 3. The unadjusted familywise error rate can thus be considerable, even when the F tests are independent.

Although users of statistics should be aware that these problems exist, and that caution must be exercised against overinterpretation of a set of significant results from a high-order factorial design, the practical convention is to evaluate and report each of the F tests at the regular level for significance (like the comparisonwise rate of Chapter 11), and not to bother with the formal evaluation of the familywise rates for ANOVA F tests unless an unusually large number are required.

The experimenter should not pay too much attention to isolated results that happen to be significant. Rather, the pattern and interpretability of results, as well as the strength of association represented by the findings, form a more reasonable basis for the overall evaluation of the experiment. A great deal of thought must go into the interpretation of a complicated experiment, quite over and above the information provided by the significance tests. This caution is especially timely in the age of the computer package, when statistical analyses yielding many results are easy to apply to complicated data. The computer will give lots of F tests and other results, but it certainly will not interpret them for you!

For these reasons, very complicated experiments with many factors are somewhat uneconomical to perform, since they require a large number of observations as a rule, and a complete analysis of the data yields so many statistical results that the experiment as a whole is often very difficult to interpret in a statistical light. In planning an experiment, it is a temptation to throw in many experimental treatments, especially if the data are inexpensive and the experimenter is adventuresome. However, this is *not* always good policy if the investigator is interested in finding meaning in the results. Other things being equal, the simpler the experiment the better will be its execution, and the more likely will one be able to decide what actually happened and what the results actually mean.

The fixed-effects analysis of variance of data from a factorial design is one of the statistical techniques encountered most in current social and behavioral research. Its advantages are many: the general technique is extremely flexible and applies to a wide variety of experimental arrangements. Indeed, the availability of a statistical technique such as the analysis of variance has done much to stimulate inquiry into the logic and economics of the *planning* of experiments. Statistically, the F test in the fixed-effects analysis of variance is relatively *robust;* as we have seen, the failure of at least two of the underlying statistical assumptions does not necessarily disqualify the application of this method in practical situations. The computational analysis is relatively simple and routine and provides a condensation of the main statistical results of an experiment into an easily understood form.

However, the application of the analysis of variance never transforms a sloppy experiment into a good one, no matter how elegant the experimental design appears on paper, or how neat and informative the summary table appears to the reader. Furthermore, experiments should be planned so as to capture the phenomena under study in the clearest, most easily understood form, and this does not necessarily mean that one of the ''textbook'' experimental designs or a treatment by the analysis of variance will best clarify matters. The experimental *problem* must come first in planning, and not the requirements of some particular form of analysis, even though, ideally, both should be considered together from the outset. If it should come to a choice between preserving the essential character of the experimental problem, or using a relatively elegant technique such as the analysis of variance, then the problem should come first.

12.17 HIGHER ORDER ANOVA IN THE STATISTICAL PACKAGES

All of the popular statistical packages for the computer make provision for higher order analysis of variance based on data from a factorial design, as discussed in this chapter. Although some of the programs will yield a complete analysis for designs with no more than four or five factors, they are more than adequate to handle most experimental situations likely to occur in the social and behavioral sciences. In this section we will describe only those programs designed for the *factorial* setup, where each cell is an independent group and the cases within a cell are likewise independent. Procedures for matched groups or repeated observations of the same subjects will be described following Chapter 13. User manuals for the statistical packages should be consulted for ways to handle data collected in other nonfactorial designs.

In both the SPSS and SPSS^x packages, the subprogram ANOVA is designed specifically to handle factorial designs. Although one-way (i.e., one-factor) data may also be analyzed by ANOVA, these are usually handled more appropriately by the ONEWAY program mentioned in Chapter 10, especially if planned or post hoc comparisons among means are to be made. On the other hand, when the experiment is not carried out as a factorial design (as in Chapter 13 to follow), then the ANOVA procedure should be replaced by another such as MANOVA. Although ANOVA is set up to handle unbalanced factorial designs, we will defer consideration of these features until Chapter 16. Furthermore, to a limited extent the ANOVA procedure can also handle the so-called analysis of covariance, which we will be discussing in Chapter 17.

The basic command for this procedure is of the form

ANOVA *dependent variable list* BY *independent variable list (min,max)/.*

That is, the variables already specified through a VARIABLE LIST statement are listed either as a dependent variable or as one of the factors or independent variables. Any variable named as an independent variable or factor must, however, have *only* integer values, indicating the different levels of that factor. For a two-way or higher factorial design, one gives a list of the independent variables or factors, with the minimum and maximum values shown after each one. Thus, consider a two-factor design in which dependent variable Y, independent variable (Factor) A with three levels, and variable or Factor B with four levels have already been declared in the initial VARIABLE LIST. Then a two-way analysis of variance would be requested through the command.

ANOVA Y BY A (1,3) B (1,4)/.

In a single ANOVA command statement, a list of up to five dependent variables may be specified, and up to five variables may be named as independent variables (up to ten in SPSS[x]). This will produce a separate multiway analysis of variance for each dependent variable named. Furthermore, any ANOVA command can be followed by up to five different combinations of dependent and independent variable lists, where each set of lists is followed by a slash (/).

A number of OPTIONS and STATISTICS can then be specified following the ANOVA command. As these will be more relevant to the methods to be introduced in later chapters, their discussion will be deferred.

In addition to the usual analysis of variance results for a factorial design, SPSS ANOVA will, on request under OPTIONS, produce what is called a *multiple classification analysis*. This is, essentially, a set of sample effects and adjusted effects of various sorts, shown along with measures of association such as eta and R-square.

Factorial designs may be handled in the BMDP series by program 2V, which is also capable of analyzing a wide variety of other experimental designs. In order to specify ANOVA for a design with two or more factors and one dependent variable, the set of input commands would consist of the keyword /PROBLEM followed by the name of the analysis, /INPUT followed by the number of variables and their format, /VARIABLE NAMES, and a /DESIGN paragraph, indicating the dependent variable, and the grouping variables. The paragraph /GROUP can be used to assign codes and names to the levels of the grouping variables (i.e., the factor levels). The basic output of program 2V is the standard analysis of variance table, plus the mean, frequency, and standard deviation of each cell in the factorial design.

The SAS package will analyze factorial designs either through PROC ANOVA, applicable only to balanced designs (equal n per cell), or through PROC GLM, which will handle unbalanced designs as well. Following the PROC ANOVA command, one uses the statement CLASS followed by the names of variables (factors) that are categorical, and thus consist of discrete factor levels. Then a MODEL statement is required, naming the dependent variable(s) and independent effects. A MEANS statement may then be included to request the means of some or all of the factor levels and combinations. Thus, the statements for a complete three-way analysis of variance might look like this:

PROC ANOVA;

     CLASS   A,B,C;

     MODEL Y=A B C A*B A*C B*C A*B*C;

     MEANS   A*B*C;

This will produce the full three-way analysis of variance as well as the means of the ABC cells. The format for PROC GLM is very similar to that for PROC ANOVA, although many more options are possible.

EXERCISES

1. A study was set up as a 3×3 factorial design, with $n = 4$ cases assigned at random to each cell, for a total $N = 36$. Suppose that the sample effects of Factor A (rows) were

$a_1 = 4$, $a_2 = -1$, and $a_3 = -3$.

The sample effects of Factor B (columns) were

$b_1 = 2$, $b_2 = 3$, and $b_3 = -5$.

The sample interaction effects $(ab)_{jk}$ were:

	B_1	B_2	B_3
A_1	2	-1	-1
A_2	.5	1	-1.5
A_3	-2.5	0	2.5

The overall sample mean was 72. On the basis of this information, reconstruct the means of the cells, columns, and rows of the data table.

2. In the construction of a projective test, 40 more or less ambiguous pictures of two or more human figures were used. In each picture, the sex of at least one of the figures was only vaguely suggested. In a study of the influence of the introduction of extra cues into the pictures, one set of 40 was retouched so that the vague figure looked slightly more like a woman, in another set each was retouched to make the figure look slightly more like a man. A third set of the original pictures was used as a control. The 40 pictures were administered to a group of 18 male college students and an independent group of 18 female college students. Six members of each group saw the pictures with female cues, six the pictures with male cues, and six the original pictures. Each subject was scored according to the number of pictures in which the indistinct figure was interpreted as a female. The results follow:

	Female cues		Male cues		No cues	
Female subjects	29	36	14	5	22	25
	35	33	8	7	20	30
	28	38	10	16	23	32
Male subjects	25	35	3	5	18	7
	31	32	8	9	15	11
	26	34	4	6	8	10

Complete the analysis of variance.

3. For Exercise 2 above, estimate the column effects, the row effects, and the interaction effects. What must be true for each of these sets of effects?

4. Suppose that in the experiment in Exercise 2 everything had turned out as shown, *except* that the distinction between males and females (the row categories) was not made. Instead, imagine that these data were presented only for the three column groups, and a one-way analysis had been carried out. What would happen to the error sum of squares in such an analysis, relative to the error sum of squares in Exercise 2? What would this new error sum of squres actually include?

5. Given that SS total $= 2,030$ for the data described in Exercise 1 above, carry out and make a summary table for the two-way analysis of variance, including the appropriate F tests.

6. In a study of postmeningitic and postencephalitic brain damage, each of 36 subjects was given a battery of tests, providing a composite score for each. Low scores on this composite measure presumably represented a considerable degree of residual brain damage. The subjects were divided into three groups according to type of initial infection, and into three crossed groups according to time since apparent physical recovery from the illness. The data follow:

	1–2 years		3–5 years		7–10 years	
Postencephalitic	76	73	69	53	59	43
	75	62	72	55	41	57
Postmeningitic	81	89	82	70	68	50
	83	75	91	74	75	47
Postoperative control	75	84	85	79	98	100
	65	63	76	87	82	79

Do there seem to be significant differences in performance among the postencephalitic, postmeningitic, and control groups? Among the groups according to time since recovery? Is there apparent interaction between type of illness and time since recovery? (Use $\alpha = .05$ here.) State, verbally, your conclusions from these data.

7. Estimate the main effects and the interaction effects for Exercise 6.

8. An experiment was carried out on the relation between the size and the wall color of a room used for a standardized interview, and the measured anxiety level of the respondent. The following results were obtained:

Room color

		Red	Yellow	Green	Blue
	Small	160 155 170	134 139 144	104 175 96	86 71 112
Room size	Medium	175 152 167	150 156 159	83 89 79	110 87 100
	Large	180 154 141	170 133 128	84 86 83	105 93 85

Complete the analysis of variance on these data, and estimate the effects.

9. Suppose that α level of .05 is used for each of the tests carried out in Exercise 8. If each test is assumed independent of every other, what is the probability of a Type I error on at least one of these tests? Suppose that $\alpha = .01$. Now what is the probability of a Type I error on at least one such test? If a number of F tests are to be made following an analysis of variance, should a relatively large or a relatively small value of α be preferred for each? Why?

10. Suppose that, given the information about effects in Exercise 8, one was told that an interview of a subject was to be held in some unspecified room. What anxiety level should be predicted? If the room is specified as being green, what anxiety level should be predicted? If the information is that the room is small and red what should one predict?

11. The data below resulted from a 3×3 factorial design, where row Factor A had levels A_1, A_2, A_3, and column Factor B had levels B_1, B_2, and B_3. There were $n = 2$ cases per cell. However, here the cells are laid out as though this were a one-way design based on nine independent groups. Work out the weights and find the sums of squares for eight orthogonal comparisons that corrrespond to the eight systematic degrees of freedom that would be shown in the two-way ANOVA for these data. (Use the principle of Section 12.10 for finding interaction comparisons). Then check your results by actually doing the two-way ANOVA.

cell

A_1B_1	A_2B_1	A_3B_1	A_1B_2	A_2B_2	A_3B_2	A_1B_3	A_2B_3	A_3B_3
5	14	5	8	4	4	3	6	17
9	11	6	12	2	7	4	10	10

12. Draw a graph showing the $A \times B$ interaction for the data in Exercise 11 above.

13. In an experiment involving two factors, each at three levels, each combination of factors contained only one subject. Subjects had been assigned to the nine possible combinations totally at random, however. The data were as follows:

Factor A

		A_1	A_2	A_3
	B_1	37	36	40
Factor B	B_2	25	35	28
	B_3	20	22	30

Based on previous studies, the experimenter involved felt quite confident that no interaction effects should exist involving Factors A and B. Given this assumption, complete the analysis.

14. Suppose that the experimenter in Exercise 13 is wrong about the interaction between Factors A and B. How might this affect the results here, relative to a larger study in which interaction was also examined?

15. In a certain experiment, two sets of experimental factors were used. The first (columns) contained three levels, and the second (rows) four levels. Twelve independent subjects assigned at random to the treatment combinations provided the following data. Test for significant row and column effects. What must one assume in doing so?

	I	*II*	*III*
A	92	40	24
B	−13	98	16
C	12	−8	64
D	82	83	46

16. A factorial design involved Factor A at three levels, Factor B at four levels, and Factor C at five levels. There were $n = 8$ cases per cell. The total sum of squares for dependent variable Y was 362. Given the information below, construct the summary table for ANOVA, showing the sources of variance, the sums of squares, and the degrees of freedom.

SS_y ABC cells $= 167$ SS_y AB cells $= 66$
SS_y A $= 25$ SS_y AC cells $= 68$
SS_y B $= 36$ SS_y BC cells $= 90$
SS_y C $= 14$

17. Carry out Tukey HSD tests for all pairs of *cell means* of the data in Exercise 11, using alpha $= .05$.

18. Carry out Scheffé post hoc tests on the cell means in Exercise 11, using the weights for the four "interaction comparisons" found in that exercise. Once again use alpha $= .05$.

19. If estimated main effects are found by subtracting the grand mean from a marginal mean, as in

$$\text{est. } \alpha_j = (\bar{y}_{j.} - \bar{y})$$

and estimated $A \times B$ interaction effects are found by taking

$$\text{est}(\alpha\beta)_{jk} = (\bar{y}_{jk} - \bar{y}_j - \bar{y}_k + \bar{y}) \, ,$$

how do you estimate $A \times B \times C$ interaction effects in a three-factor design?

20. A factorial experiment was based on three factors: A with three levels, B with four levels, and C with two levels. Within each cell there were $n = 2$ cases assigned at random. The data are given below. Carry out the complete analysis of variance on these data. How would you interpret the results here?

| | \multicolumn{4}{c}{C_1} | | \multicolumn{4}{c}{C_2} |
	B_1	B_2	B_3	B_4	B_1	B_2	B_3	B_4
A_1	16	22	16	23	19	20	18	30
	18	18	17	28	17	24	17	25
A_2	15	19	23	19	24	14	19	26
	23	12	17	26	15	21	24	19
A_3	18	23	26	27	16	27	22	24
	14	25	21	23	20	24	27	28

21. Estimate all of the main and interaction effects for Exercise 20.

22. Find the sample proportions of variance accounted for by main effects and interactions in Exercise 20, and the corresponding estimated values for the populations.

Chapter *13*

The Analysis of Variance Models II and III: Random Effects and Mixed Models

13.1 RANDOMLY SELECTED TREATMENT LEVELS

The previous chapters were concerned exclusively with the variation of the general linear model appropriate for the fixed-effects analysis of variance. Now we will turn to still another model for the analysis of variance. This is known variously as the *random-effects model, components of variance model,* or Model II. This, too, is an instance of the general linear model, but where the effects associated with the treatments or treatment combinations are no longer conceived as fixed. Rather, different effects may appear over different repetitions of the same experiment. Such effects are thus unlike fixed effects, which are thought of as constants, irrespective of how many times the experiment is carried out.

The random-effects model is designed especially for experiments in which inferences are to be drawn about an entire set of distinct treatments or factor levels, including some not actually observed. For such experiments, many more treatment categories or factor levels are possible than can be actually represented in a given experiment. However, our interest lies in the whole range of treatment possibilities, even though what we use in a given experimental setup is only a random sample of the potential treatments or levels we might have observed. Before the experiment is carried out, a random sample of levels is drawn from among all possible such levels of a given factor, and then inferences are made about all such levels on the basis of this limited sample.

For example, suppose that in an experimental procedure we suspect that the personality of the experimenter may be having an effect on the results. In principle, there are a vast number of people, each presumably having a distinct personality, who might possibly serve as the person who conducts the experiment. Obviously, it is reasonable to narrow this population of potential experimenters down to people having the requisite technical skills to conduct the experiment (for example, English-speaking behavioral scientists), but even this leaves a great many people from which to choose. To put it mildly, trying out each such person in our experiment would complicate matters! So, instead, we draw a random sample from among such qualified persons. Suppose that five different persons are chosen; each of the five persons performs this procedure on a different sample of n subjects assigned at random. Each experimenter is actually an experimental treatment given to one group of subjects; in all other respects subjects are treated exactly the same. Since the experimental treatments employed are themselves a random sample, this experiment fits the random-effects, rather than the fixed-effects, model: an inference is to be made about experimenter effects in general from observation of only five experimenters sampled at random.

Parenthetically, the student is warned that designation of the fixed-effects model as Model I, of the random-effects model as Model II, and the later mixed model as Model III, though convenient for the organization adopted here, is not a standard accepted by all authors. It is probably better for the student to learn to think of these three models as *fixed effects, random effects* (or *components of variance*), and *mixed* from the very outset, rather than relying on any number designation alone.

It should also be kept in mind that although, for simplicity's sake the examples in this chapter will be restricted to one or, at most, two experimental factors, there is nothing inherent in the models to prevent their extension to any number of factors. Such higher order Model II or Model III designs will usually have more complex rules determining the appropriateness of F tests. However, these follow a definite pattern, discussed under the rubric ''expected mean squares'' in most experimental design texts (see Edwards (1985) as an example).

13.2 RANDOM EFFECTS AND MODEL II

Imagine a factor A representing an experimental manipulation that can be broken down into a very large number of qualitatively different categories or levels. Thus, in the preceding example, use of any given experimenter represents a qualitatively different category from the use of another experimenter. Here there are thousands of experimenters, who might be used, and hence thousands of levels of Factor A. Returning from the example to the general situation, let us designate one such level of A as Level g. Associated with any g will be a whole population of potential observations i, each having a y_{ig} value for the dependent variable. The linear model showing the makeup of any value y_{ig} in any such population g would be just as in the preceding chapter:

$$y_{ig} = \mu + \alpha_g + e_{ig}, \qquad\qquad [13.2.1]$$

where μ is the grand mean of all the populations, α_g is the effect of being in population g, and e_{ig} is the error component of y_{ig}. Over all of the different populations that can be symbolized by g, there will be a whole distribution of values α_g. This distribution of effects will have a mean μ_α, and a variance σ_α^2, over all possible populations g. Let us assume further that each population g is equally represented in the grand population. Then, since we are still dealing here with effects α, even when the number of groups is very large it will be true that $\mu_\alpha = 0$. However, only when $\alpha_g = 0$ for every single population g will we know the value of σ_α^2, since then it must be true that $\sigma_\alpha^2 = 0$.

Now in an actual experiment, a sample of N subjects will be assigned at random to a sample of J different treatment groups. Note that *only* J of the many possible treatments will be used. Before the experiment, a random sample of treatments is drawn. One of the selected treatments is designated randomly as Group 1, another as Group 2, and so on up to Group J. Let us again call any one of these treatment groups j.

Now consider an individual i who is a member of treatment group j. What is the model for the dependent variable value y_{ij}? It is given by the linear model

$$y_{ij} = \mu + a_j + e_{ij}. \tag{13.2.2*}$$

Here, however, the sample effect may be thought of in a different way than in the preceding chapters.

$$a_j = \sum_g \alpha_g v_{gj}$$

where v_{gj} is an indicator variable defined as follows:

$$v_{gj} = \begin{cases} 1 \text{ if treatment } g \text{ is selected and labeled as Group } j \\ 0 \text{ otherwise.} \end{cases}$$

The effect a_j of being in treatment group j is thus the same as the effect of being in exactly one of the treatment populations. However, the exact value of a_j depends on *which* treatment was selected and labeled j in the sample. For a group labeled j, the value of a_j will thus vary over repetitions of the experiment, depending on which treatment is selected and called j. Thus in this sampling scheme a_j is a *random variable,* and we call values such as a_j *random effects.* The a_j values have a distribution with a mean of zero, and a variance equal to σ_A^2, where $\sigma_A^2 = \sigma_\alpha^2$.

The assumptions to be made in deriving a test of the hypothesis of no treatment effects are as follows:

1. The possible effect values a_j represent a random variable having a distribution with a mean of zero, and a variance σ_A^2.
2. For any treatment j, the errors e_{ij} are normally distributed with a mean of zero and a variance σ_e^2, which is the same for each possible treatment j.
3. The J values of the random variable a_j occurring in the experiment are all independent of each other.
4. The values of the random variable e_{ij} are all independent.
5. Each pair of random variables a_j and e_{ij} are independent.

You will note that here we are making assumptions about two different kinds of distributions. First of all there is a distribution of possible values for the effects a_j that might appear in the experiment. For the experiment proper, a random sample is taken from among all possible such effects values when a random sample of treatments is drawn. For the time being, no special assumptions need be made about the distribution of effects from which the sample of effects is drawn, other than that the mean is zero and that the variance has some (finite) value σ_A^2. Later, when we discuss this model more generally, we will assume that the distribution of effects is normal.

The second kind of distribution is that of errors within particular treatments. Just as in Model I, we assume errors to be normally and independently distributed, with the same variance irrespective of the particular treatment under which the observation is made. Notice that we are, at present, making no assumptions about the number of potential treatments from which a sample of J is taken for the experiment, except that this number is greater than J. We are, however, assuming that within any given treatment population an infinite number of potential units for observation exist.

13.3 THE MEAN SQUARES FOR MODEL II

In this section the mean squares for the one-way analysis will be examined under Model II. The discussion will be restricted to the situation where *exactly the same number of observations* n *are made under each treatment.* Although the treatments may have different numbers of sample observations in the one-way case for the fixed-effects model, some difficulties arise in Model II unless the numbers of observations are equal.

The partition of the sum of squares for the random effects model is carried out in exactly the same way as for Model I. Furthermore, the mean squares are found just as before:

$$\text{MS between} = \frac{\text{SS between}}{J-1} = \frac{\sum_j n(\bar{y}_j - \bar{y})^2}{J-1}$$

and

$$\text{MS within} = \frac{\text{SS within}}{N-J} = \frac{\sum_j \sum_i (y_{ij} - \bar{y}_j)^2}{N-J}.$$

Bear in mind that for any random sample of J treatments or factor levels, there is a corresponding random sample of effects a_j. Let the sample mean of these effects be denoted by

$$\bar{a} = \frac{\sum_j a_j}{J}.$$

It is then true that over all samples $E(\bar{a}) = 0$. On the other hand, although the mean of the effects over all of the *possible* treatments must be zero, any sample value of $\bar{a}$ need not be zero however, since this is but the mean of a sample of effects actually occurring in the experiment.

In the same way, sample means for error may be defined:

$$\bar{e}_j = \frac{\sum_i^n e_{ij}}{n}$$

and

$$\bar{e} = \frac{\sum_j \sum_i e_{ij}}{N} = \frac{\sum_j \bar{e}_j}{J}.$$

Then the deviation of any mean $\bar{y}_j$ from the grand sample mean $\bar{y}$ can be written as

$$(\bar{y}_j - \bar{y}) = (a_j - \bar{a}) + (\bar{e}_j - \bar{e}) ,$$

the deviation of effect a_j from the mean of all of the effects in the sample, plus the deviation of the average error in Group j from the grand mean of all the errors in the sample. Because of the makeup of the deviation of $\bar{y}_j$ from $\bar{y}$, it can be shown that

$$E(\text{MS between}) = n\sigma_A^2 + \sigma_e^2. \qquad [13.3.1^*]$$

The expected mean square between is the weighted sum of two variances, that of the population of treatment effects and that of the error.

Now we turn to the mean square within treatments, where

$$E(\text{MS within}) = \sum_j \frac{\sigma_e^2}{J} = \sigma_e^2. \qquad [13.3.2^*]$$

Thus the mean square within is always an unbiased estimate of error variance alone.

In Model II, if there are no treatment effects at all, either for those represented in the experiment or for any other possible treatment in the set sampled, then it must be true that $\sigma_A^2 = 0$. Hence the null hypothesis of no treatment effects for Model II is usually written

$$H_0: \sigma_A^2 = 0, \text{ with } H_1: \sigma_A^2 > 0. \qquad [13.3.3^*]$$

Now suppose that this null hypothesis were true. In this case

$$E(\text{MS between}) = \sigma_e^2$$

$$E(\text{MS within}) = \sigma_e^2$$

so that *both* mean squares are unbiased estimates of error variance alone.

The mean square between can be shown to be independent of the mean square within under the assumptions of Section 13.2, so that when H_0 is true, the ratio

$$F = \frac{\text{MS between}/\sigma_e^2}{\text{MS within}/\sigma_e^2} = \frac{\text{MS between}}{\text{MS within}} \qquad [13.3.4^*]$$

can be referred to the F distribution with $J - 1$ and $N - J$ degrees of freedom, as a test of this null hypothesis. Significant values of F lead to a rejection of $H_0: \sigma_A^2 = 0$ *in favor of* $H_1: \sigma_A^2 > 0$.

13.4 AN EXAMPLE

Suppose that the experiment described in the introduction to this chapter was actually carried out. Five experimenters chosen at random carry out the same procedure, each on a different set of eight subjects randomly sampled and assigned at random among the experimental groups. The data are as shown in Table 13.4.1.

The hypothesis that $\sigma_A^2 = 0$ (that there are no experimenter effects) is to be tested using $\alpha = .01$.

Experimenter					**Table 13.4.1**
1	2	3	4	5	Results of using five different experimenters, chosen at random
5.8	6.0	6.3	6.4	5.7	
5.1	6.1	5.5	6.4	5.9	
5.7	6.6	5.7	6.5	6.5	
5.9	6.5	6.0	6.1	6.3	
5.6	5.9	6.1	6.6	6.2	
5.4	5.9	6.2	5.9	6.4	
5.3	6.4	5.8	6.7	6.0	
5.2	6.3	5.6	6.0	6.3	
44.0	49.7	47.2	50.6	49.3	

$$\sum_j \sum_i y_{ij} = 240.8$$

Source	SS	df	MS	E(MS)	F	**Table 13.4.2**
Between (experimenters)	3.48	4	.87	$8\sigma_A^2 + \sigma_e^2$	10.72	
Within	2.84	35	.08	σ_e^2		
Total	6.32	39				

The computations are:

$$\text{SS total} = 1{,}455.94 - \frac{(240.8)^2}{40} = 6.32$$

$$\text{SS between} = \frac{(44.0)^2 + \cdots + (49.3)^2}{8} - \frac{(240.8)^2}{40}$$

$$= 1{,}453.10 - 1{,}449.62 = 3.48$$

$$\text{SS within} = \text{total} - \text{between} = 6.32 - 3.48 = 2.84.$$

Table 13.4.2 is the summary table for this analysis. The F value required for rejection at the .01 level for 4 and 35 degrees of freedom is between 4.02 and 3.83 (that is, between the values for 30 and 40 degrees of freedom denominator). Accordingly, the hypothesis of no experimenter effects may be rejected. There is sufficient evidence to say that experimenter effects exist and contribute to the variance of Y.

13.5 ESTIMATION OF VARIANCE COMPONENTS IN A ONE-WAY ANALYSIS

Instead of estimating effects directly by taking differences of the treatment means from the grand mean, as in the fixed-effects model, in Model II we will estimate σ_A^2, the true variance due to treatments. For the foregoing experiment, this is the true variance attributable to experimenters.

Since the expectation of the mean square between is

$$E(\text{MS between}) = n\sigma_A^2 + \sigma_e^2,$$

an unbiased estimate of σ_A^2 may be found by taking

$$\frac{\text{MS between} - \text{MS within}}{n} = \text{est. } \sigma_A^2. \qquad [13.5.1]$$

(Note: est. $\sigma_A^2 = 0$ when MS within $\geq$ MS between.)

For the example, this estimate is

$$\text{est. } \sigma_A^2 = \frac{.87 - .08}{8} = .099.$$

The variance of y_{ij} over the population of all possible potential observations is

$$\sigma_Y^2 = E(y_{ij} - \mu)^2 = \sigma_A^2 + \sigma_e^2$$

so that the true variance consists of two independent parts or components: the variance due to treatments, and that due to error alone. The best estimate of the total variance σ_Y^2 is given by

$$\text{est.} \sigma_Y^2 = \text{est. } \sigma_A^2 + \sigma_e^2 = \frac{\text{MS between} + (n - 1) \text{ MS within}}{n}. \qquad [13.5.2]$$

For the example,

$$\text{est. } \sigma_Y^2 = .099 + .08$$

$$= .18.$$

This fact that the total variance must consist of two components allows one to make a somewhat more informative use of the estimates of σ_A^2 and σ_e^2. We can take the ratio of the estimated σ_A^2 to the estimated total variance to find the estimated proportion of variance accounted for by the treatments,

$$\text{est. proportion of variance accounted for} = \frac{\text{est. } \sigma_A^2}{\text{est. } \sigma_Y^2} \qquad [13.5.3]$$

$$\text{est. proportion of variance accounted for by experimenters} = \frac{(.099)}{.18} = .55.$$

Here we estimate that over one-half of the variance among observations seems to be due to experimenter differences. This would be a most important finding in a real experiment, as it would suggest rather strongly that different experimenters' repetitions of this experiment would not necessarily be comparable.

One way of expressing the idea that a factor accounts for a given amount of a variance is by the index known as the *population intraclass correlation coefficient*:

$$\rho_I = \frac{\sigma_A^2}{\sigma_A^2 + \sigma_e^2}. \qquad [13.5.4^*]$$

The intraclass correlation coefficient for the grand population will be zero only when σ_A^2 is zero, and will reach unity only when $\sigma_e^2 = 0$, given that $\sigma_Y^2 > 0$. Notice that this index is simply another way of expressing the proportion of variance attributable to the factor A. The population index ρ_I is identical to population ω^2 in its general form and meaning, although ρ_I applies to the random-effects and ω^2 to the fixed-effects model so that slightly different estimation methods apply in the two situations.

Quite often the intraclass correlation is used to express the fact that observations in the same category are related, or tend on the average to be more like each other than observations in different categories. The larger the value of ρ_I, the more similar do observations in the same treatment category tend to be, relative to observations in different categories. For example, in a study of the similarity in intelligence of twins, a random sample of sets of twins might be taken, each pair of twins constituting a natural "treatment" with $n = 2$. An estimated intraclass correlation greater than zero would indicate that some of the variability in intelligence is accounted for by variation among sets of twins, so that pairs of twins tend to be more alike in this respect than are pairs of nontwins. The value of ρ_I is thus a measure of the homogeneity of observations within classes, relative to such observations between classes.

Still another special application of this idea of an intraclass correlation occurs in the study of the reliability of repeated measurements of individuals. The reliability of a single measurement y_{ij} for an individual j can be defined in a way equivalent to ρ_I for a population of such individuals. The estimate of ρ_I given by Eq. 13.5.3 is then an estimate of the true reliability for such measurements. A modern discussion of these methods for estimating reliability in terms of the random-effects analysis of variance is given in Winer (1971), and in Guilford and Fruchter (1978).

Under Model II, it is also possible to estimate intervals for the proportion of variance accounted for by a factor. Here, however, we must assume that the basic distribution of effects is normal. Given this assumption, the required limits can be found from the distribution of the quantity θ, where

$$\theta = \frac{\sigma_A^2}{\sigma_e^2} \qquad [13.5.5]$$

and

$$\text{true proportion of variance accounted for} = \frac{\theta}{1 + \theta} = \rho_I. \qquad [13.5.6]$$

The required confidence interval is found as follows: Let F' be the value in an F distribution with $J - 1$ and $N - J$ degrees of freedom, cutting off the *upper* $\alpha/2$ proportion in the distribution, and let F'' be the value cutting off the *lower* $\alpha/2$. For this distribution, it must be true that

prob. $(F'' \leq F \leq F') = 1 - \alpha.$

Furthermore, regardless of the true value of σ_A^2, it will be true that the ratio

$$\frac{\text{MS between}/(\sigma_e^2 + n\sigma_A^2)}{\text{MS within}/(\sigma_e^2)} \qquad [13.5.7^*]$$

is distributed as an F variable with $J - 1$ and $N - J$ degrees of freedom, since numerator and denominator *are* independent chi-square variables divided by degrees of freedom when Model II is true, and the basic distribution of effects is normal. This ratio is equivalent to

$$\left(\frac{\text{MS between}}{\text{MS within}}\right)\left(\frac{1}{1 + n\theta}\right) = F,$$ [13.5.8]

so that

$$\text{prob}\left(F'' \le \frac{\text{MS between}}{\text{MS within}} \frac{1}{1 + n\theta} \le F'\right) = 1 - \alpha.$$

By algebraic operation on this inequality, it can be shown that the $100(1 - \alpha)$ percent confidence interval for θ is

$$\frac{1}{n}\left[\frac{\text{MS between}}{(\text{MS within}) F'} - 1\right] \le \theta \le \frac{1}{n}\left[\frac{\text{MS between}}{(\text{MS within}) F''} - 1\right].$$ [13.5.9*]

This may be turned into a confidence interval for ρ_I very easily by the relation given by Eq. 13.5.6.

The value of F' can be found directly from the F tables for $\alpha/2$, and the degrees of freedom $J - 1$, and $N - J$. However, the value of F'' must be found by the methods in Section 9.9 for values on the *lower* tail of an F distribution. These confidence intervals, like those for σ^2 given in Section 9.5, need not be optimal.

As an example let us find the 95% confidence interval for θ and ρ_I for the data in Section 13.4 Here,

$$\frac{\text{MS between}}{\text{MS within}} = 10.72.$$

The first thing we need is the value of F', that value cutting off the *upper* 2.5% in an F distribution with 4 and 35 degrees of freedom. Table V in Appendix E shows this to be about 3.2. Next, we must find F'', the value on the lower tail of this same distribution cutting off the lower 2.5%. From the same table we take the value cutting off the *upper* 2.5% in a distribution with 35 and 4 degrees of freedom: this is about 8.43. Thus,

$$F'' = \frac{1}{8.43} = .119.$$

The approximate 95% confidence interval for θ is then

$$\frac{1}{8}\left[\frac{10.72}{3.2} - 1\right] \le \theta \le \frac{1}{8}\left[\frac{10.72}{.119} - 1\right]$$

$$.29 \le \theta \le 11.14.$$

The corresponding interval for ρ_I is

$$\frac{.29}{1 + .29} \le \rho_I \le \frac{11.14}{1 + 11.14}$$

or

$$.22 \le \rho_I \le .92.$$

13.6 TESTING OTHER HYPOTHESES AND CALCULATING POWER UNDER MODEL II

The hypothesis tested in the example above was that there exist no effects of the individual treatments, so that $\sigma_A^2 = 0$. The theory is not limited to this situation, however. When the distribution of effects can be assumed to be normal, it is possible to test many other hypotheses about the ratio $\theta = \sigma_A^2/\sigma_e^2$ or, equivalently, about ρ_I.

It was noted in Eq. 13.5.8 that when there is a value of $\theta > 0$ which characterizes the experimental population, the random variable

$$\frac{\text{MS between}}{(\text{MS within})(1 + n\theta)} \qquad [13.6.1]$$

actually is distributed as F with $J - 1$ and $J(n - 1)$ degrees of freedom. On the other hand, the ordinary ratio (MS between/MS within) is *not* distributed as F in this instance.

In the general case, the hypothesis tested is

H_0: $\theta \le \theta_0$ versus H_1: $\theta_0 < \theta$

where θ_0 is any hypothetical value, $0 \le \theta_0$.

This general hypothesis is tested as follows: after α is chosen, we determine the F value cutting off the upper α proportion of cases in a distribution with $J - 1$ and $N - J$ degrees of freedom. Call this value F_α. Now we find the F ratio for the sample in the usual way (Section 13.3). The hypothesis is rejected in favor of H_1 when

$$\frac{\text{MS between}}{\text{MS within}} \ge (1 + n\theta_0)F_\alpha. \qquad [13.6.2^*]$$

In Chapter 10 we learned how power could be ascertained for Model I situations by the use of special tables. These power calculations, are, as we have seen, an aid in making decisions about sample size. Similarly, in Model II experiments we also wish to have power information for the same reasons as in Model I. The relationship given by Eq. 13.6.2 permits one to find the power of F in a much simpler way under Model II than under Model I. The problem in finding the power of the F test when a given value of θ is true is that of finding the probability

$$\text{prob.}\left(\frac{\text{MS between}}{\text{MS within}} > F_\alpha; \theta, v_1, v_2\right).$$

That is, the power of the test corresponds to the probability of rejecting the null hypothesis at the α level, when thre are v_1 and v_2 degrees of freedom, and the particular value of θ is indeed true.

Now since the ratio in Eq. 13.6.1 above is distributed as F, it follows that

$$\text{prob.}\left(\frac{\text{MS between}}{\text{MS within}} > F_\alpha\right) = \text{prob.}[F(1 + n\theta) > F_\alpha] \qquad [13.6.3^*]$$

$$= \text{prob.}[F > F_\alpha/(1 + n\theta)].$$

In other words, the power of the F test can be found from the probability that an F value exceeds the value of $F_\alpha/(1 + n\theta)$. This last probability can be calculated from the distribution of F, and thus the power can be determined. Bear in mind, however, that this method of determining power depends on the assumption that effects are sampled randomly from a normal distribution of such effects. This method may also require fairly extensive tables of the distribution of F, and such tables are not always easy to find. [One reasonably extensive set of tables for F is found in Dixon and Massey (1969)].

13.7 IMPORTANCE OF THE ASSUMPTIONS IN MODEL II

By now it should surely be clear that the actual *arithmetic* of the simple analysis of variance is the same regardless of the model adopted. The analysis of variance is, strictly speaking, only a way of arranging this arithmetic. However, the inferences made from the sample values are really quite different, depending on the model invoked. All the inferences made under Model I concern means (and differences between means). On the other hand, the inferences made using Model II deal with variances: that is, the basic inference has to do with the *variance of the population of effects* actually sampled by the experimenter. This distinction between the two models has an influence on the importance of the assumptions made in each.

In the first place, the *assumption of normality* can be quite important in Model II. Provided that one is concerned only with a test of the hypothesis $\sigma_A^2 = 0$, then slight departures from normality among the error distributions should have only minor consequences for the conclusions reached using a *reasonably large sample size*. Here, no assumption at all about the distribution of the α values need be made. On the other hand, for tests of the more general hypothesis given in Section 13.6 it is most important that the distributions *both* of effects and of errors be normal in form. In the same way, interval estimates for θ and for ρ_I depend heavily on the assumptions of normal distributions of the α and the e values for their validity.

The assumption of equality of variances has a somewhat different status in the random-effects than in the fixed-effects model. If the random-effects model applies, and if the effects α_j are independent of the errors e_{ij}, we need assume only one distribution of errors with variance σ_e^2 so that in a sense, the error variance for all possible observations given some treatment j must be the same as for any other treatment. However, the important assumptions are those involving independence: the errors *must* be independent both of the particular treatment effects and of each other. The simple random-effects model is not really applicable to data where the *errors* in observations must be related to the effect of the treatment applied, or to each other. [In another model (Section 13.19), we will modify the assumptions so as to permit one special kind of dependency to exist between errors.]

Although statistical dependency among the error components of scores creates a special problem in any analysis of variance, it is not true that the *observations themselves* must be completely independent for Model II to apply. We have just seen that the intraclass correlation coefficient expresses the degree of similarity or dependency among observations given the same treatment. When σ_A^2 is greater than zero, some statistical relation must exist between pairs of score values within the possible treatment groups. As we saw in Section 13.6 hypotheses about the degree of relatedness can be tested via

Model II. However, the form of relatedness that makes the simple Model II analysis inapplicable is most often a trend or serial relation, implying nonindependence of the error terms.

13.8 TWO-FACTOR EXPERIMENTS WITH SAMPLING OF LEVELS

Now we turn to the analysis of experiments involving two factors, each of which is represented in the experiment only by a sample of its levels.

For example, suppose that a projective test involves 10 cards administered individually to a subject. The subject must respond by giving as many free asociations to each card as possible. These responses are scored in a number of ways, but the total number of responses is an important index of overall performance. The developer of this test has some idea that the *order* in which cards are presented has a bearing on the total number of responses given, and so would like to see if this factor of order is an important one in accounting for variation in test performance. If it should turn out that order is important, an order of presentation for the cards will be sought that will be optimal in evoking responses. Furthermore, this investigator has worked out a standard set of instructions for test administrators, in the hope that administrator effects on performance are thereby made negligible. However, it is desirable to see if this is true.

Here are two factors, order of presentation and test administrator, that may account for variance in total response to the test. Obviously, neither factor can be represented at all levels in any experiment, since there are exactly 3,628,800 different ways of presenting the 10 cards, and a very large number of persons who might be trained to administer the test. Hence, the experimenter decides to conduct a study in which each of these factors will be sampled. From a single set of data the question of the relative contribution to variance of each of the factors can be answered, as well as the secondary question of possible interaction between test administrator and order of presentation.

This hypothetical experiment will be developed as an example in Section 13.12. For the time being we will turn our attention to extending Model II to cover such situations.

13.9 MODEL II FOR TWO-FACTOR EXPERIMENTS

The two different factors in the experiment will be designated A and B. A random sample of J different levels of A will be drawn, and shown as rows in the data table, and a random sample of K different levels of B will appear as the columns in the table. Within each combination shown by a cell of the data table, n observations are to be made at random.

The score of individual i in row j and column k of the table will be thought of as a sum:

$$y_{ijk} = \mu + a_j + b_k + (ab)_{jk} + e_{ijk}.$$

Here, a_j is the random variable standing for the effect of the sample treatment appearing in the data table as row j, b_k is the random variable indicating the effect of the sample treatment in column k, and $(ab)_{jk}$ is the random interaction effect associated with cell *jk*. The term e_{ijk} is a random variable standing for the error effect of the observation of individual *i* under the joint conditions indicated by row *j* and column *k*. Observe that in this linear model all components of a score except the grand mean μ are values of *random variables,* the sampled effects.

The assumptions made are:

1. The a_j are normally distributed random variables with mean zero and variance σ_A^2.
2. The b_k are normally distributed with mean zero and variance σ_B^2.
3. Each $(ab)_{jk}$ is a normal random variable with mean zero and variance σ_{AB}^2.
4. The e_{ijk} are normally distributed with mean zero and variance σ_e^2.
5. The a_j, the b_k, the $(ab)_{jk}$, and the e_{ijk} are pairwise independent.

13.10 THE MEAN SQUARES

The computations for the two-way analysis under Model II are exactly the same as for Model I (Section 12.6). Thus, the total sum of squares is partitioned into a sum of squares for rows, a sum of squares for columns, a sum of squares for interaction, and an error sum of squares. On examining the mean squares, however, we will find that the really important differences between Models I and II appear. You may recall that in Section 12.4 it was pointed out that for the fixed-effects model, both row and column effects sum to zero. Also the following things are true of the true interaction effects, $(\alpha\beta)_{jk}$:

$$\sum_j (\alpha\beta)_{jk} = 0, \sum_k (\alpha\beta)_{jk} = 0, \text{ and } \sum_j \sum_k (\alpha\beta)_{jk} = 0.$$

That is, in Model I, over the column treatments the interaction effects sum to zero, over the row treatments interaction effects sum to zero, and over both rows and columns these effects sum to zero. The net result is that in the sum of squares for, say, rows, only row effects and error can possibly be included since this sum of squares is based on squared deviations such as

$$(\bar{y}_{j\cdot} - \bar{y})^2 = (\alpha_{j\cdot} + \bar{e}_{j\cdot} - \bar{e})^2.$$

Any such row deviation will reflect only the effect $\alpha_{j\cdot}$ and mean errors, where the effects $\beta_{\cdot k}$ and $(\alpha\beta)_{jk}$ have all summed to 0. Notice especially that for the fixed-effects model, any interaction effect cannot possibly contribute to the sum of squares for rows because this sum of squares is itself based on the data *summed* over columns, automatically making the interaction effects sum to zero. For the same reason, the sum of squares for columns does not include any of the interaction effects.

In Model II the set of *J* row treatments represents a *sample* from a large set of possible such treatments, and the *K* column treatments another sample from a large set. Here it is no longer true that the a_j effects present in the sample must sum to zero over the sample rows, or that the b_k effects that are present must sum to zero over the sample columns. Nevertheless, the average column effects are removed when the grand mean is subtracted from the mean of any row. Moreover, the average row effects are removed when the grand mean is subtracted from the mean of any column.

However, interaction effects are defined only so that

$$E(ab)_{jk} = 0$$
$$\phantom{E(ab)_{jk} = 0}_j$$

$$E(ab)_{jk} = 0$$
$$\phantom{E(ab)_{jk} = 0}_k$$

$$E(ab)_{jk} = 0.$$
$$\phantom{E(ab)_{jk} = 0.}_{jk}$$

That is, the expected value of the interaction term $(ab)_{jk}$ is zero over *all possible* treatments that might have been selected for column k with a fixed row treatment j. Furthermore, the expected value of the interaction term $(ab)_{jk}$ is zero over all possible treatments that might have been selected for row j with a fixed column treatment k. Note that these requirements hold only when one considers *all possible* row or column treatments that might have been selected. Unlike the situation in the fixed-effects model, here there is *no requirement* at all that the interaction effects *sum to zero* over the particular set of J row treatments or over the particular set of K column treatments that just happened to appear in the sample. For this reason, no sum of interaction effects must be zero in any given set of data. The important consequence of this fact is that the sum of squares for rows reflects deviations due not only to row treatment effects, *but also to interaction effects*. In the same way the sum of squares for columns includes *both column and interaction* effects.

That is, the grand sample mean $\bar{y}_{..}$ consists of mean row effects, mean column effects, mean interaction effects, and mean error,

$$\bar{y} = \bar{a}_{..} + \bar{b}_{..} + \overline{(ab)}_{..} + \bar{e}_{...}$$

On the other hand, the mean $\bar{y}_{j.}$ of any row must consist of the effect of that row, the mean of the column effects, the mean of the interaction effects within that row, and the mean error in that row:

$$\bar{y}_{j.} = a_{j.} + \bar{b}_{..} + \overline{(ab)}_{j.} + \bar{e}_{j..}$$

Therefore, the difference between the row mean and the grand mean does not include any column effects, but it does include average interaction effects as well as row effects and error.

$$\bar{y}_{j.} - \bar{y}_{..} = a_{j.} + \bar{b}_{..} + \overline{(ab)}_{j.} + \bar{e}_{..} - (\bar{a}_{..} + \bar{b}_{..} + \overline{(ab)}_{..} + \bar{e}_{..})$$

$$= (\bar{a}_{j.} - \bar{a}_{..}) + (\overline{(ab)}_{j.} - \overline{(ab)}_{..}) + (\bar{e}_{j.} - \bar{e}_{..}).$$

Similarly, for the deviation of any column mean from the grand mean we have

$$\bar{y}_{.k} - \bar{y}_{..} = (b_{.k} - \bar{b}_{..}) + (\overline{(ab)}_{.k} - \overline{(ab)}_{..}) + (\bar{e}_{.k} - \bar{e}_{..}),$$

a column deviation contains a column effect, average interaction effects, and average errors, but no row effect. These facts have a most important bearing on estimates and tests made under Model II, as we shall see.

It then will always be true that under Model II for two factors,

$$E(\text{MS } A) = E(\text{MS between rows}) = \sigma_e^2 + n\sigma_{AB}^2 + Kn\sigma_A^2. \qquad [13.10.1^*]$$

Unlike the expectation for fixed effects, here the mean square for rows estimates a *weighted sum* of the error variance, the variance due to row treatments, and the interaction variance, σ_{AB}^2.

A parallel situation holds for Factor B, represented by the columns:

$$E(\text{MS columns}) = \sigma_e^2 + n\sigma_{AB}^2 + Jn\sigma_B^2 \qquad [13.10.2^*]$$

Turning to the expected value for the mean square interaction, we find that

$$E(\text{MS interaction}) = \sigma_e^2 + n\sigma_{AB}^2. \qquad [13.10.3^*]$$

The expectation of the mean square error is, as always,

$$E(\text{MS error}) = \sigma_e^2. \qquad [13.10.4^*]$$

13.11 HYPOTHESIS TESTING IN THE TWO-WAY ANALYSIS UNDER MODEL II

We have just seen that the expected values both for the mean square for rows and the mean square for columns are different in Model II from those in Model I. The practical implication of this fact is that the hypotheses of no row and no column effects are tested in a different way for this model than for Model I. Consider the test of the hypothesis

$$H_0: \sigma_A^2 = 0.$$

When this hypothesis is true, then

$$E(\text{MS A}) = \sigma_e^2 + n\sigma_{AB}^2 \qquad [13.11.1]$$

which is *not* the same as the expectation of mean square error, but rather that of *mean square interaction:*

$$E(\text{MS interaction}) = \sigma_e^2 + n\sigma_{AB}^2. \qquad [13.11.2]$$

These two mean squares (MS columns and MS interaction) can be shown to be independent, and when each is divided by $\sigma_e^2 + n\sigma_{AB}^2$, each is distributed as χ^2 divided by degrees of freedom. Thus, the hypothesis of no A, or row, effects is usually tested by

$$F = \frac{\text{MS A}}{\text{MS interaction}} \qquad [13.11.3^*]$$

with $J - 1$ and $(J - 1)(K - 1)$ degrees of freedom.

In the same way, the hypothesis

$$H_0: \sigma_B^2 = 0$$

(or no Factor B effects) is tested by a comparison of MS B, or columns, with the interaction MS, since if this hypothesis is true,

$$E(\text{MS B}) = \sigma_e^2 + n\sigma_{AB}^2, \qquad [13.11.4]$$

which is the same as the expected value of the mean square for interaction. The hypothesis is tested by the ratio

$$F = \frac{MS\ B}{MS\ \text{interaction}}$$

[13.11.5*]

with $K - 1$ and $(J - 1)(K - 1)$ degrees of freedom. The hypothesis of no interaction is tested just as for Model I:

$$H_0: \sigma_{AB}^2 = 0$$

tested by

$$F = \frac{MS\ \text{interaction}}{MS\ \text{error}}$$

[13.11.6*]

with $(J - 1)(K - 1)$ and $JK(n - 1)$ degrees of freedom.

It should be emphasized that the test of row and column effects against interaction is really appropriate *only* when the factors have been randomly sampled, as in Model II (and in some instances of the mixed models introduced in sections to follow). When the experiment qualifies for Model I, the ratio of the mean square for a factor to the interaction mean square is *not necessarily* distributed as F, even when the null hypothesis is true. This is one of the major practical distinctions between the fixed-effects and the random-effects models.

There is one situation, however, when the main effects are not tested against interaction in Model II, but rather against a pooled estimate formed from SS error plus SS interaction. This occurs when the experimenter has decided that the interaction variance σ_{AB}^2 is zero. There is either some theoretical or some empirical reason to believe that interaction effects should not occur, or there is *strong* evidence for no interaction in the data. Of course, there is a difficult problem involved in deciding when there really *is* strong evidence that interaction effects do not exist.

When this decision of no true interaction effects can be made, then there are often real advantages to be gained by pooling the SS interaction with the SS error, and using the results to form a new *pooled mean square for error*. That is, when the F value for interaction is so small that you feel safe in doing so, then the two sums of squares—error and interaction—are pooled as follows:

$$\text{pooled MS error} = \frac{SS\ \text{interaction} + SS\ \text{error}}{(J - 1)(K - 1) + JK(n - 1)}.$$

[13.11.7]

When $\sigma_{AB}^2 = 0$, the expectation of this pooled MS error is σ_e^2. The expectation of MS A is also σ_e^2 under the hypothesis of no A effects, so that the test may be carried out by use of the ratio

$$F = \frac{MS\ A}{\text{pooled MS error}}$$

[13.11.8*]

with $J - 1$ and $JKn - J - K + 1$ degrees of freedom. In the same way a test of no B effects is carried out by

$$F = \frac{MS\ B}{pooled\ MS\ error}$$

[13.11.9*]

with $K - 1$ and $JKn - J - K + 1$ degrees of freedom.

Notice that when one uses the data to make the decision that interaction variance is zero (or that interaction effects do not exist) this is actually a decision to *accept* a null hypothesis, $\sigma^2_{AB} = 0$. As we saw in Chapter 7, the danger such a decision holds is that we will make a Type II error and conclude interaction effects do not exist when, in reality, they do. You may also recall that one of the ways to buy power, or insurance against a Type II error, is to set a large probability value for a Type I error. That is, when faced with the decision about whether or not to pool the SS for interaction with the SS error, you will want to use an α value of at least .10, and perhaps even of .25 or larger. Then, if the F test for interaction does not reach the level required even for one of these relatively large values, you can feel fairly safe in concluding that interaction effects do not exist, and then proceeed to pool. (If you want to be *really* safe from Type II error, than choose α of about .50, which requires an F value ≤ 1.00 for the null hypothesis of no interaction to be accepted. However, this stringent criterion will not let you pool interaction and error very often.)

The advantages gained by pooling interaction with error are twofold: the increase in the number of degrees of freedom for error will tend to give you both a better and (often) smaller estimate of the true error variance; and, you will usually have a more powerful F test when the pooled MS error is used in place of MS interaction in the denominator of the other tests. Since Model II (and Model III) designs often have rather low numbers of degrees of freedom for interaction, and therefore give rather low-powered tests for main effects, it is wise to pool whenever the opportunity presents itself. For this reason, one ordinarily tests for interaction first, and then decides either to pool interaction with error, or to maintain a separate interaction mean square for use in the remaining tests. Remember: even though the interaction term may be nonsignificant by the usual .05 or .01 criteria, that is not enough! Rather, one chooses an alpha value of .10. .25, or even larger, and does not accept the null hypothesis unless the F value is small enough to fail this criterion of Type I error. Or, put in another way, one does not accept the null hypothesis of no interaction unless the F value *fails* to reach the criterion level for the relatively large α chosen.

13.12 AN EXAMPLE OF A TWO-FACTOR, MODEL II DESIGN

As an example, suppose that the developer of the projective test mentioned in Section 13.8 actually carried out the experiment. Recall that this study was designed to test both for the presence of effects associated with order of presentation and with particular administrators of the test. Four qualified persons were selected at random and trained to administer the test. Also, four orders of presentation of the test were selected at random. Each adminstrator gave the test to a different pair of randomly selected adults under each one of the selected order conditions, so that a total of 32 different test performances were observed in all. The dependent variable was the total number of responses a subject gave to the test cards. The experimenter was prepared to assume normal distributions of order effects, of administrator effects, of interaction effects, and of error. The data were as shown in Table 13.12.1. The α level chosen for the test of each hypothesis was .05.

The analysis proceeds in the usual manner for two-way designs (Section 12.6):

$$\text{SS total} = \sum_j \sum_k \sum_i y_{ijk}^2 - \frac{T_y^2}{N}$$

$$= 27,965 - \frac{(931)^2}{32}$$

$$= 878.72$$

$$\text{SS AB cells} = \sum_j \sum_k \frac{T_{jk}^2}{n} - \frac{T_y^2}{N}$$

$$= \frac{(51)^2 + (50)^2 + \cdots + (64)^2}{2} - \frac{(931)^2}{32}$$

$$= 568.22$$

$$\text{SS error} = \text{SS total} - \text{SS AB cells} = 310.50$$

$$\text{SS A} = \sum_j \frac{T_{j\cdot}^2}{Kn} - \frac{T_y^2}{N}$$

$$= \frac{(220)^2 + \cdots + (277)^2}{8} - \frac{(931)^2}{32} = 363.09$$

$$\text{SS B} = \sum_k \frac{T_{\cdot k}^2}{Jn} - \frac{T_y^2}{N}$$

Table 13.12.1

Results of the projective test study, with four orders and four administrators.

		Administrator (B)				
Order		1	2	3	4	Totals
	I	26	30	25	28	
		25	33	23	30	220
	II	26	25	27	27	
		24	33	17	26	205
(A)	III	33	26	30	31	
		27	32	24	26	229
	IV	36	37	37	39	
		28	42	33	25	277
Totals		225	258	216	232	931

$$= \frac{(225)^2 + \cdots + (232)^2}{8} - \frac{(931)^2}{32} = 122.34$$

$$\text{SS A} \times \text{B} = \text{SS AB cells} - \text{SS A} - \text{SS B}$$

$$= 568.22 - 363.09 - 122.34$$

$$= 82.79$$

Table 13.12.2 summarizes the analysis. Ordinarily, the tests both for row and for column effects would be carried out by dividing the mean square for rows or columns by the mean square for interaction. However, it is immediately apparent that here the mean square for error is larger than the mean square for interaction, so that the estimate of σ^2_{AB} is zero. Notice that the F is therefore smaller than required for $\alpha = .10, .25,$ or .50. There is little risk of Type II error in stating that no interaction effects exist. Thus, we find

$$\text{pooled MS error} = \frac{82.79 + 310.50}{9 + 16} = 15.7,$$

which has $9 + 16$ or 25 degrees of freedom.

We test for the A, or row, effects by finding

$$F = \frac{121.0}{15.7} = 7.7.$$

For 3 and 25 degrees of freedom, this value exceeds that required for the .05 level, and so the hypothesis of no row effects is rejected. The same procedure is carried out for Factor B, or columns, and here

$$F = \frac{\text{MS}}{\text{pooled MS error}} = 2.6$$

for 3 and 25 degrees of freedom. This fails to reach the value of 2.99 required to reject the null hypothesis. (These F values are enclosed in parenthesis in the summary table

Table 13.12.2

Source	SS	df	MS	E(MS)	F
Rows (orders)	363.09	3	121.0	$\sigma^2_e + 2\sigma^2_{AB} + 8\sigma^2_B$	$\left(\frac{121.0}{15.7} = 7.7\right)$
Columns (admin.)	122.34	3	40.8	$\sigma^2_e + 2\sigma^2_{AB} + 8\sigma^2_A$	$\left(\frac{40.8}{15.7} = 2.6\right)$
Interaction	82.79	9	9.2	$\sigma^2_e + 2\sigma^2_{AB}$	
Error	310.50	16	19.4	σ^2_e	
Totals	878.72	31			

to show that they are not obtained in the usual way. In reports of such data, the F tests would usually be accompanied by an explanatory footnote to the summary table.)

The conclusions from the experiment are, then, as follows:

1. Order *does* seem to have an effect on the total number of responses given in the test.
2. There is insufficient evidence to determine if true administrator differences exist.
3. There is virtually no evidence for administrator–order interaction.

13.13 POINT ESTIMATION OF VARIANCE COMPONENTS

The four expressions in the E(MS) column of the summary table allow us to estimate the components of variance, as follows:

$$\text{est. } \sigma_A^2 = \frac{\text{MS A} - \text{MS interaction}}{Kn} \qquad [13.13.1]$$

$$\text{est. } \sigma_B^2 = \frac{\text{MS B} - \text{MS interaction}}{Jn} \qquad [13.13.2]$$

$$\text{est. } \sigma_{AB}^2 = \frac{\text{MS interaction} - \text{MS error}}{n} \qquad [13.13.3]$$

$$\text{est. } \sigma_e^2 = \text{MS error}. \qquad [13.13.4]$$

If any of the first three estimates turn out to be negative in sign, then that component is estimated to be zero.

In the example above, the estimate of σ_{AB}^2 is zero, and the pooled MS error was taken to represent σ_e^2. Thus, to be consistent, one should estimate the other components as follows:

$$\text{est. } \sigma_A^2 = \frac{\text{MS rows} - \text{pooled MS error}}{Kn}$$

$$= \frac{121.0 - 15.7}{8} = 13.16$$

$$\text{est. } \sigma_B^2 = \frac{40.8 - 15.7}{8} = 3.14$$

$$\text{est. } \sigma_e^2 = 15.7.$$

The proportion of variance accounted for by Factor A (rows) is estimated from

$$\frac{\text{est. } \sigma_A^2}{\text{est. } (\sigma_e^2 + \sigma_{AB}^2 + \sigma_A^2 + \sigma_B^2)} = \frac{13.16}{15.7 + 0 + 13.16 + 3.14} = .41.$$

An estimated 41% of the variance in total response is attributable to order of presentation of this projective test.

For Factor B (columns) the estimated proportion of variance accounted for is

$$\frac{3.14}{15.7 + 0 + 13.16 + 3.14} = .098$$

so that if Factor B accounts for any variance at all, our best guess is that this is less than 10% of the total variance. Approximate confidence intervals for the variance components σ_A^2, σ_B^2, and so forth, can be found by methods outlined in Scheffé (1959, pp. 231–235).

13.14 MODEL III: A MIXED MODEL

Multifactor experiments involving Model II are relatively rare in social and behavioral research. It is far more common to encounter experiments where one or more factors have fixed levels and the remaining factors are sampled. This situation calls for a third model of data, in which each individual observation results in a score that is a sum of *both* fixed and random effects. Obviously, mixed models such as this apply only to experiments where two or more factors are under study.

As an example of an experiment fitting a mixed model, suppose that a study is concerned with the muscular tension induced in subjects by three different varieties of task. The subject performs using pencil and paper with the preferred hand, meanwhile holding the bulb of a sensitive pressure recording gauge in the other. The mean reading on this gauge during the performance provides the dependent variable score. Three separate kinds of tasks are administered: in one type the subject solves relatively complicated problems in arithmetic; in a second type a short, imaginative, composition is written; and in the third a careful tracing of a line drawing must be made.

The subjects in this experiment are to be children in the fifth grade in public schools in a large city. Since there may be sex differences in terms of the three tasks involved, for this study the experimenter decided to use only male subjects, on the supposition that males are more likely to exhibit strong emotional responses to the three tasks given. Furthermore, different school settings may differ in their emphasis on these sorts of skills. Thus, it was decided to sample from among the large number of fifth-grade classrooms in the city, and select at random six classrooms to be observed, with six boys then selected at random from each classroom. From a given classroom, two boys each are given one of the three tasks, for a total of 6 observations per classroom, or 36 observations in all. Thus "classrooms" represents a random-effects factor, and "tasks" is a fixed-effects factor.

This time, let the factor having fixed levels be labeled A, and represented by the J columns of the table, and let the randomly sampled factor be B, and shown by the K rows of the data table. Now it is assumed that

$$y_{ijk} = \mu + \alpha_j + b_k + (\alpha b)_{jk} + e_{ijk} \qquad [13.14.1]$$

where α_j is the fixed effect of the treatment indicated by the column j, b_k is the random variable associated with the kth row, $(\alpha b)_{jk}$ the random interaction effect operating in the cell jk, and e_{ijk} is the random error associated with observation i in the cell jk. We make the following assumptions:

1. The b_k and the $(\alpha b)_{jk}$ are jointly normal, each with mean of zero and with variances σ_B^2 and σ_{AB}^2 respectively.
2. The e_{ijk} are normally distributed, with a mean of zero and variance σ_e^2.
3. The e_{ijk} are independent of the b_k and the $(\alpha b)_{jk}$.
4. All error terms e_{ijk} are independent of each other.

13.15 THE EXPECTED MEAN SQUARES IN A MIXED MODEL

Within this mixed model, the sum of the α_j over all of the J fixed treatments must, as usual, be equal to zero:

$$\sum_j \alpha_j = 0.$$

Furthermore, within any row k the sum of the interaction effects, $(\alpha b)_{jk}$ will also be equal to zero. Thus,

$$\overline{(\alpha b)}_{\cdot k} = \sum_j (\alpha b)_{jk}/J = 0. \qquad [13.15.1]$$

From this it follows that the mean of the interaction effects for the entire experiment must be zero:

$$\overline{(\alpha b)}_{\cdot\cdot} = \sum_j \sum_k (\alpha b)_{jk}/JK = 0. \qquad [13.15.2]$$

In other words, within any B row, the fixed A effects and the interaction effects must both sum (and average) to zero. On the other hand, it need not be true that within any A column the b_k effects or the $(\alpha b)_{jk}$ effects will sum to zero. The $b_{\cdot k}$ are random effects arising from a sampled factor, and within any column there will be an average of these row effects:

$$\overline{b}_{\cdot\cdot} = \sum_k b_{\cdot k}/K .$$

However, the value for $\overline{b}_{\cdot\cdot}$ will be the same for each column.

The situation is different for the interaction effects, however. Within any column j there will be an average interaction effect

$$\overline{(\alpha b)}_{j\cdot} = \sum_k (\alpha b)_{jk}/K \qquad [13.15.3]$$

which is not necessarily equal to zero, and which will differ from column to column.

Now we will put these facts together as we examine the difference between a factor-level mean and the overall mean, in terms of effects. First let us look at *the random factor* (Factor B, or rows, in this example). Here the difference between the mean of any row B$_{\cdot k}$ and the grand mean is

$$\overline{y}_{\cdot k} - \overline{y}_{\cdot\cdot} = b_k - \overline{b}_{\cdot\cdot} + \overline{e}_{\cdot k} - \overline{e} . \qquad [13.15.4]$$

Notice how this deviation from the grand mean represents only the random effect $b_{\cdot k}$, the mean effect $b_{\cdot\cdot}$, and an average error component. Neither $\alpha_{j\cdot}$ effects nor interaction effects form a part of the B deviations.

Therefore, if one takes SS B, which is just the weighted sum of squares of such deviations as in Eq. 13.15.4, and then finds MS B for this random factor, the expectation is

$$E(\text{MS B}) = Jn\sigma_B^2 + \sigma_e^2 .$$ [13.15.5]

The expected mean square for Factor B, the random factor, consists of Jn times variability due to B effects, plus variability due to error. It follows that a test for the hypothesis of no Factor B effects, or $\sigma_B^2 = 0$, is given by the F ratio = (MS B/MS error), with the usual degrees of freedom equal to $K - 1$ and $N - JK$.

Now, however, let us take a look at the composition of the difference between the mean of one of the fixed A levels or columns, and the overall mean. In this instance we have

$$\bar{y}_{j.} - \bar{y}_{..} = \alpha_j + \bar{b}_{..} + \overline{(\alpha b)}_{j.} + \bar{e}_{j.} - \bar{b}_{..} - \bar{e}_{..}$$ [13.15.6]
$$= \alpha_j + \overline{(\alpha b)}_{j.} + \bar{e}_{j.} - \bar{e}_{..} .$$

Notice that **any deviation of a column or A mean from the grand mean includes not only the column effect α_j and a mean error deviation, but also an average interaction effect $\overline{(\alpha b)}_{j.}$.** This is because there is no requirement at all that interaction effects must sum to zero within columns when the row effects, and thus the interaction effects, are only a random sample of such effects. Thus column means, formed by summing observations across rows, include not only the fixed column effects, but also some of the random interaction effects as well.

For this reason, given the mean square for columns calculated in the usual way, it can be shown that the expectation for this mean square is

$$E(\text{MS A}) = \sigma_e^2 + n\sigma_{AB}^2 + \frac{Kn\left(\sum_j \alpha_j^2\right)}{J - 1}.$$ [13.15.7]

In Model III, the mean square for the fixed-effects factor is an estimate of a weighted sum of the error variance, the sum of the squared effects associated with the factor itself, *and* the interaction variance.

This often strikes students as a violation of intuition, but a little reflection should convince you that this is reasonable. Consider the example of Section 13.14 once again which produced the data arranged into the rows and columns of Table 13.16.1. Why should the means corresponding to the different tasks (columns) in the experiment tend to differ? In the first place there actually may be systematic differences between the population means, or effects, connected with the different tasks. Furthermore, since we deal only with sample means rather than with population means, a certain amount of random error may enter into the average differences we observe between the tasks. However, there is still another reason why the column means might differ in any sample. The subjects in any given class might find one or two of the tasks quite stressful and tension-producing and the others quite easy. Furthermore, the pattern of such hard and easy tasks might vary over classes, so that the effect that a task has *within* a particular class might be different across classes. Subjects in class 1 find arithmetic problems extremely challenging, although imaginative composition is a breeze for them. On the other hand, subjects in classroom 2 find arithmetic problems no challenge at all, but are brought to a standstill when faced with a writing task. Such systematic differences among the tasks *within* particular classrooms but differing from one classroom to another, is what we mean by "classroom by task interaction." Since classrooms were sampled for the experiment, it is entirely possible for such interaction effects to tend to "pile up" in one or another of the columns of the data table, and thus to contribute to the apparent differences between means for the tasks; whether or not this happens depends on the particular sample of classrooms we obtain. Such apparent differences among means need not be due to the main effects α_j of the tasks themselves, since interaction effects representing differences within *particular* classrooms need not have any special connection with the differences between means for the tasks across *all possible* classrooms. On the other hand, this is not the same as random error, since these differences do reflect systematic effects occurring in the experiment. The presence of interaction effects in the deviation of column mean $\bar{y}_{j.}$ from the general mean $\bar{y}_{..}$ is actually due to the circumstance that one set of factor levels is sampled, so that the interaction effects are under no necessity to "cancel out" in the particular sample drawn.

$$(\bar{y}_{jk} - \bar{y}_{..}) - (\bar{y}_{j.} - \bar{y}_{..}) - (\bar{y}_{.k} - \bar{y}_{..}) = (\overline{(\alpha b)}_{jk} - \overline{(\alpha b)}_{j.} + \bar{e}_{jk} - \bar{e}_{j.} - \bar{e}_{.k} + \bar{e}_{..}).$$

Finally, as shown above, the deviations underlying the interaction sum of squares, or SS $A \times B$, consist exclusively of average interaction effects and average error. Consequently, it can be shown that

$$E(\text{MS interaction}) = \sigma_e^2 + n\sigma_{AB}^2. \qquad [13.15.8]$$

That is, the interaction mean square is an estimate of a weighted sum of the error variance and the interaction variance.

When the hypothesis of no column or A effects is true, so that $\alpha_j = 0$ for each of the column treatments j, then

$$E(\text{MS columns}) = \sigma_e^2 + n\sigma_{AB}^2. \qquad [13.15.9]$$

Notice that this is exactly the same as the expectation for the MS interaction. Under our assumptions (particularly assumption 4) the ratio

$$F = \frac{\text{MS columns}}{\text{MS interaction}} \qquad [13.15.10]$$

with $J - 1$ and $(J - 1)(K - 1)$ degrees of freedom provides an appropriate test of the hypothesis of no column effects. Be careful to notice that it is the *fixed-effects factor* that is tested against interaction in this mixed model.

On the other hand, when the hypothesis of no row or B effects is true,

$$E(\text{MS rows}) = \sigma_e^2, \qquad [13.15.11]$$

which is the same as

$$E(\text{MS error}) = \sigma_e^2.$$

Then the test for the existence of row effects (that is a test of H_0: $\sigma_B^2 = 0$) is given by

$$F = \frac{\text{MS rows}}{\text{MS error}}$$

[13.15.12]

with $J - 1$ and $JK(n - 1)$ degrees of freedom, as noted above.

Thus, we see that there are two distinctly different kinds of F tests employed in the analysis of variance for this mixed model: the hypothesis that all the fixed effects are zero is tested by comparing the mean square for that factor against the mean square for interaction. On the other hand, the hypothesis associated with the random-effects factor is tested by comparing its associated mean square with the mean square for error. The important procedural differences among these three models thus have to do with the *way in which hypotheses are tested,* and especially with the *denominators* used in the various F ratios.

The test for interaction in a two-way analysis is the same for all three models:

$$F = \frac{\text{MS interaction}}{\text{MS error}}$$

[13.15.13]

with $(J - 1)(K - 1)$ and $JK(n - 1)$ degrees of freedom, as noted above.

When there is good reason to believe that interaction effects do not exist, there are advantages in pooling the interaction and the error sums of squares to get a new estimate of the error variance. This is done exactly as outlined in Section 13.11

13.16 AN EXAMPLE FITTING MODEL III

Let us suppose that the experiment outlined in Section 13.14 had actually been carried out, and that the data were as shown in Table 13.16.1. In this table the two values appearing under each combination of classroom and task represent the independent performances of the two subjects in that combination.

The .05 level is chosen for α, and the analysis proceeds as follows:

$$\text{SS total} = (7.8)^2 + \cdots + (10.5)^2 - \frac{(358.5)^2}{36} = 123.57$$

$$\text{SS AB cells} = \frac{(16.5)^2 + \cdots + (19.1)^2}{2} - \frac{(358.5)^2}{36} = 109.03.$$

$$\text{SS error} = \text{SS total} - \text{SS AB cells}$$
$$= 123.57 - 109.03 = 14.54$$

$$\text{SS columns} = \frac{(103.5)^2 + (136.0)^2 + (119.0)^2}{12} - \frac{(358.5)^2}{36} = 44.04$$

$$\text{SS rows} = \frac{(61.3)^2 + \cdots + (62.9)^2}{6} - \frac{(358.5)^2}{36} = 6.80$$

$$\text{SS interaction} = \text{SS AB cells} - \text{SS rows} - \text{SS columns}$$
$$= 109.03 - 6.80 - 44.04 = 58.19$$

Table 13.16.2 is the summary table for this analysis.

Table 13.16.1

Results of the experiment on handpressure.

Classrooms	Tasks I	II	III	Total
1	7.8	11.1	11.7	
	8.7	12.0	10.0	
	16.5	23.1	21.7	61.3
2	8.0	11.3	9.8	
	9.2	10.6	11.9	
	17.2	21.9	21.7	60.8
3	4.0	9.8	11.7	
	6.9	10.1	12.6	
	10.9	19.9	24.3	55.1
4	10.3	11.4	7.9	
	9.4	10.5	8.1	
	19.7	21.9	16.0	57.6
5	9.3	13.0	8.3	
	10.6	11.7	7.9	
	19.9	24.7	16.2	60.8
6	9.5	12.2	8.6	
	9.8	12.3	10.5	
	19.3	24.5	19.1	62.9
Totals	103.5	136.0	119.0	358.5

First of all, for 10 and 18 degrees of freedom and for $\alpha = .05$, the F test for interaction exceeds the required F of 2.41, and the hypothesis of no interaction *is* rejected.

For 2 and 10 degrees of freedom an F value of 4.10 is required for significance, so that the hypothesis of no column effects is *not* rejected. Notice that the F ratio for columns is formed by dividing the column mean square by the mean square for interaction. If this experiment had been incorrectly analyzed under Model I, quite a different

Table 13.16.2

Source	SS	df	MS	E(MS)	F
Columns (tasks)	44.04	2	22.02	$\sigma_e^2 + 2\sigma_{AB}^2 + \dfrac{12\sum_j \alpha_j^2}{2}$	3.78
Rows (class)	6.80	5	1.36	$\sigma_e^2 + 6\sigma_B^2$	1.68
Interaction (tasks by class)	58.19	10	5.82	$\sigma_e^2 + 2\sigma_{AB}^2$	7.19
Error	14.54	18	.81	σ_e^2	
Total	123.57	35			

conclusion would have been reached; this illustrates the importance of using the proper model in testing hypotheses.

For 5 and 18 degrees of freedom an F value of 2.77 is required in order to reject the null hypothesis at the 5% level. The F ratio formed by the mean square for rows over the mean square error gives a value smaller than this, so that the hypothesis of no classroom effects is also *not* rejected.

In summary, there is insufficient evidence to permit us to conclude either that there are task effects or that classroom effects exist in this experiment. There is, however, fairly strong evidence for the presence of interaction effects. There is something about the combination of a particular classroom with a particular task that accounts for variance in the data. Thus, within classrooms, task differences apparently exist, but these tend to be different for different classrooms. Perhaps the capacity of a task to produce tension in a given boy depends on his classroom background.

13.17 RANDOMIZATION AND CONTROLS

The basic statistical tools in experimentation are randomization and control. As distinct from informal observation, every experiment worthy of the name is conducted under specifc controls. All of the physical and situational factors that the experimenter can manage are kept as constant as possible in the course of the experiment. Efforts are made to have the units that are the basis for observations in the experiment, whether they be humans, animals, plants, or what not, be just as homogeneous as possible. Differences among the units might affect the outcome or the interpretability of the experiment. Often, the mere effect of giving a subject a treatment of any kind is checked by the use of control groups, participating in the experimental setting but receiving no treatment. Controls are exercised in hundreds of ways in any well-planned experiment. Nevertheless, there are countless factors that might affect the outcome of the experiment, but which the experimenter does not know to control, or know how to control.

Randomization enters into the experiment in at least two different ways. The actual individuals observed in any sample may differ in any number of ways from the "typical" individual in the population as a whole. It is entirely possible to obtain a sample of N individuals who are "peculiar" as compared with the population they represent, and the conclusions reached from this sample need not be indicative of what the population as a whole tends to be like. By sampling at random from the population, the experimenter is able to identify the peculiarities of the sample with random error, and to allow for the possibility of an atypical sample in the conclusions.

Furthermore, given some sample it will always be true that factors other than the ones manipulated by the experimenter will contribute to the observed differences between subjects in the particular situation. If it should happen that some extraneous factor operates unevenly over several treatment groups or over different subjects, this can create spurious differences, or mask true effects in the data. Such a factor is a "nuisance factor," playing a role analogous to "noise" in communication. Randomization of subjects over treatments is one device for "scattering" the effects of these nuisance factors through the data. Often, when particular nuisance factors are known, levels of these factors are scattered at random throughout the design. In order to randomize an experiment, the researcher uses some scheme such as a table of random numbers to allot individual subjects to experimental groups or nuisance levels to subgroups in a purely random, unsystematic, "chancy" way. By randomization, the possibility of pileups of nuisance effects in particular treatment groups is identified with random error, and the experimenter can rest assured that over all repetitions of the experiment under the same conditions, true effects will eventually emerge if they exist.

Every experiment is randomized to some extent. Ordinarily, a study having one or more experimental factors and certain constant controls where every other factor contribution to variance is randomized is called a **fully randomized** design: usually a random sample is drawn, and then subjects are allotted to treatments or treatment combinations purely at random. In addition, levels of one or more nuisance factors may be assigned at random. The experiments used as examples in Chapter 10 were of this type. Furthermore, many other experiments fitting either the fixed-effects or the random-effects models can also be thought of in this way if nature or some other agency is conceived as carrying out the random assignment of individuals to groups.

On the other hand, sometimes it is advantageous to the experimenter *not* to randomize the effects of particular nuisance factors, but rather to represent them systematically in the experiment. These nuisance factors are treated as though they were experimental factors, when actually the purpose of representing them in the experiment in the first place is to control them and thereby reduce error variance. It should be obvious from our discussion of the indices ω^2 and ρ_I that deliberate introduction of any systematic effects that account for a portion of the variance σ_Y^2 must also *reduce* the error variance σ_e^2. That is, the true variance of any dependent variable is thought of as a sum of components:

$$\sigma_Y^2 = \text{(the variance attributable to systematic features of the experiment)}$$

$$+ \text{(the variance attributable to unsystematically represented factors).}$$

The more *relevant* factors that can be introduced into the experiment in a systematic way, the smaller will be the variance considered as error. Furthermore, the smaller the error variance, then the more precise will be the experiment, in that confidence intervals will be smaller and true effects more likely to be detected if present.

A large part of experimental design deals with this strategy of deliberately introducing nuisance factors into experiments so as remove them from our estimates of error variance. Two of the most common and simple of these strategies will be examined next.

13.18 RANDOMIZED BLOCKS DESIGNS

In the randomized groups and factorial designs emphasized so far, each level or cell in the experiment has received a separate and distinct sample of cases, assigned purely at random to that group or cell. Clearly, however, there will be all sorts of differences among the individuals who happen to be grouped together. One reason for assigning subjects to these groups or cells purely at random is to try to ensure that such individual differences, if they affect the dependent variable values, will do so in a way that can be identified only with *error,* rather than being mistaken for part of the systematic variance due to our experimental treatments. We hope to reduce the risk that error due to individual differences will tend to pile up in certain groups or cells. The mean square for error in our analysis then reflects these individual differences among subjects, which we have scattered "at random" throughout the data, but otherwise made no attempt to control.

It may well be that some of these individual difference variables are themselves appreciably related to the dependent variable used in the experiment. Such variables, which are not of primary interest in our experiment, but which could well influence how our experiment comes out, are the nuisance variables of the preceding section. A very common example is the intelligence level of a subject. It may be that we wish, for example, to compare four tranquilizing drugs for their influence on cognitive performance of subjects who take them. Inevitably, the subjects assigned to the different tranquilizer groups will differ somewhat in intelligence, and this variable will in turn affect the measure of cognitive performance. To some extent, then, our results are going to depend on exactly where the cases "fall" in our experiment as the result of randomization. An unfortunate pattern of how cases fell among groups or cells could make us think we have results due to our experimental treatments, when, actually, all we are seeing are the effects of the nuisance variables. Conversely, the wrong pattern could obscure real differences among the drug groups.

One way of keeping this sort of thing from happening, so that we can feel confident that our results are not due to nuisance variables, is to hold these variables constant. Hence, if we suspect that intelligence makes a difference in how a subject performs on a cognitive task, we use only subjects with the same IQ scores in the experiment. This is certainly a highly effective way to rule out the influence of nuisance variables, but it also has the very unfortunate result of limiting our conclusions only to those levels of the nuisance variable that we happened to use. What we need is an experimental design that will give us the advantages of constant control of certain nuisance variables without sacrificing the generalizability of our results.

As we have seen, randomization leads us to expect that possible pile-ups of high or low values of nuisance variables will tend to balance out in the long run across the experimental groups or levels. Nevertheless, the nuisance variables are still very much present in our data, in the form of error variability. Randomization helps us to identify nuisance variation or individual differences with random error, but it alone does nothing to eliminate such variability. However, if we are able to eliminate some of this type of known but uncontrolled variability from our estimate of error variance (MS error), we can often achieve a considerable reduction in the size of this estimate. Remember that the *power* of our statistical tests tends to *increase* with any *decrease* in the error variability. Therefore, removing such identifiable but unwanted variability from MS error helps to increase the power of our tests, other things being equal. In short, what we want is an experimental strategy that will not only lead to control over nuisance variables, but will also deduct their contribution from our estimates of error variance.

One strategy for eliminating such nuisance variability from MS error and thus achieving relatively more power, is called *blocking.* In simple language, blocking is just the creation (or the selection) of matched groups of subjects.

A block is a matched group of subjects who are as similar as possible on one or more nuisance variables.

Then, the entire set of experimental treatments or combinations of treatments is carried out *within* this block of subjects. By so doing we make sure that within this block there can be no differences among subjects on the nuisance variables used for matching, In other words, variability within blocks *cannot* be due to the nuisance variables on which the matching took place.

On the other hand, from experimental results found within any *single* block we can safely draw conclusions only about a population matched in exactly the same way as the block in question. The solution to this problem of generalizability is pretty straightforward, however: simply take *a number* of blocks, *differing widely* in the levels of the nuisance variables used to create the matching, and carry out all of the experimental treatments *within each block*. Essentially, given K blocks, then the entire "experimental" part of the experiment is carried out K times, once within each block.

In short, we wish to control for the nuisance variables by matching subjects into blocks, but not at the risk of losing generality in our conclusions. Therefore, to make this blocking strategy most effective, we need to create sets of blocks that are just as homogeneous as possible within each block, but as heterogeneous between the blocks as we can make them with respect to the basis for matching. In addition we wish to block on the basis of variables that are appreciably related to Y.

A so-called randomized blocks experiment or RB design thus employs several blocks of subjects, with the subjects within each block matched on the basis of their values on one or more nuisance variables. Then within each block of subjects the entire set of treatments (factor levels or combinations) is carried out. The subjects within each block are assigned to the different treatments or treatment combinations completely at random (hence the name, *randomized* blocks).

The diagram for a randomized blocks experiment with one experimental factor looks something like a factorial design with two factors, since the J treatment levels are completely crossed with the K blocks, and there are consequently JK cells, with the same $n \geq 1$ subjects per cell. However, this appearance is somewhat deceiving, as the two designs differ in terms of what one actually *does* to carry them out. In a factorial design all of the JK combinations of two factors A and B are first worked out, and then $N = nJK$ subjects are assigned at random and independently to the JK cells representing those combinations. As we emphasized in Chapter 12, the most basic way to look at the systematic variability in the data in a factorial design is through the variability among cells.

The procedure for a randomized blocks design is, however, quite different from that for a factorial design. In the simplest RB situation, the following steps are followed:

1. First the experimental factor of interest is chosen with, let us say, J levels.
2. Next the nuisance variables to be used in blocking must be decided on. As indicated above, these must be variables related to the dependent variable Y. Sometimes matched groups of subjects already exist and need only to be sampled (e.g., litters of animals, matched on heredity variables.). Otherwise blocks of subjects

may have to be formed by selecting individuals who have (or who have been made to have) equal (or very nearly equal) values on the nuisance variables. Subjects within blocks *must* be quite homogeneous on the nuisance variables in question.

3. If the blocking is to be effective, then there *must* be heterogeneity on the nuisance variables across the blocks. That is, every effort is made to create or find matched groups that differ widely on the values of the nuisance variables used for matching.

4. The size of any matched group or block must be at least equal to J, the number of factor levels. All blocks used in the experiment must be of the same size.

5. If the number of blocks you created or selected is much greater than the number you will finally need, you will probably randomly select blocks for use in the experiment.

6. For each block, the subjects are allocated at random among the J treatment levels. If there are only J subjects per block, then each combination of a block and a treatment level will contain *one* subject. If, however, there are nJ subjects per block, then each combination of a block and a treatment level will contain n subjects.

For example, suppose that a corporation hires people for the job of microscopic inspection of circuits on computer chips. The employee training division of the company had devised two new methods (A_2, and A_3) of training for this job, which they wished to compare with the old method (A_1) through an experiment using new hires for the job. This amounted to a factor of "method" with $J =$ three fixed levels, with the dependent variable Y consisting of a score evaluating the individual's performance on the job six weeks after training. On the other hand, it is known that an individual's performance on this job is related both to his or her age and education. It was therefore desired to block for these two nuisance variables, so that the conclusions about training method would be uninfluenced by those considerations.

From the large number of age–education combinations available, a set of blocks of three subjects each was identified. Then, from this pool of possible blocks, 10 blocks were selected at random. Each block or matched group contained $J = 3$ subjects, who had exactly the same age and the same level of education (non-HS grad, HS grad, some college, college grad, graduate study).

Now within each of these 10 blocks of three subjects each, one subject was assigned at random to Method A_1, a second subject to the Method A_2, and the third subject to the Method A_3. The same instructor was responsible for each method, and the subjects were all treated as alike as possible in every other respect. Then six weeks after training and on the job, each subject received a performance evaluation score y.

The results of this experiment are shown in Table 13.18.1. In that table the blocks of subjects matched on age and educational level are shown as the *rows* of the table. Then the three levels of Factor A, or methods, are shown as the *columns* of the table. Notice how each level of A is represented in each block, with $n = 1$ per cell in this example. We will discuss the analysis of this set of data later in this section.

The schematic layout for a summary table produced by the analysis of variance on data from a randomized blocks design is given in Table 13.18.2.

A randomized blocks experiment such as this, where blocks are sampled, is usually

	A_1	A_2	A_3	block totals
block 1	41	44	42	127
2	39	44	47	130
3	28	33	37	98
4	26	41	30	97
5	41	42	41	124
6	28	36	36	100
7	26	33	35	94
8	24	33	34	91
9	27	49	30	106
10	31	41	28	100
treat. totals	311	396	360	$1{,}067 = T_y$

Table 13.18.1

Data for a randomized blocks experiment, with three treatments in 10 blocks

analyzed as a **mixed model design without replication.** The analysis of variance proceeds just as for any two-way analysis, except that the partition of the sum of squares contains only three terms:

$$\text{SS total} = \text{SS blocks} + \text{SS treatments} + \text{SS interaction.}$$

There are no degrees of freedom for error, and hence no available information about error variability. Since there is only one individual per cell in this instance ($n = 1$) the only F test to be made is that for treatments, which involves the MS treatments divided by the MS interaction. The blocks (rows or matched groups effects) cannot be tested in the usual way unless at least two observations are made in each block and treatment combination. Note also that this table is arranged in two main parts, blocks and within blocks, as is conventional for such designs. The information about the treatments is subsumed under the "within-blocks" category, as is the information about the interac-

Table 13.18.2

Summary table for a Model III randomized blocks design, $n = 1$.

Source	SS	df	MS	E(MS)	F
Blocks		$K - 1$	$\dfrac{\text{SS rows}}{K - 1}$	$\sigma_e^2 + J\sigma_B^2$	
Within blocks				$\sigma_e^2 + \sigma_{AB}^2$	
Treatments (Factor A)		$J - 1$	$\dfrac{\text{SS columns}}{J - 1}$	$+ \dfrac{K\sum\limits_{j} \sigma_j^2}{J - 1}$	$\dfrac{\text{MS treat.}}{\text{MS blocks by treat.}}$
Blocks by treatments		$(J - 1)(K - 1)$	$\dfrac{\text{SS int.}}{(J - 1)(K - 1)}$	$\sigma_e^2 + \sigma_{AB}^2$	
Total		$JK - 1$			

tion between treatments and blocks. This is as it should be, as the experimental treatments are literally carried out on subjects assigned at random *within* each block.

The actual computations for a simple randomized blocks design set up such as shown in Table 13.18.1 are identical to those for a two-way factorial design. That is, the sum of squares total for the data in Table 13.18.1 is

$$SS \text{ total} = 41^2 + \cdots + 28^2 - (1,067)^2/30 = 1,365.37.$$

That for blocks is based on the block totals (rows here), as follows:

$$SS \text{ blocks} = \frac{127^2 + \cdots 100^2}{3} - (1,067)^2/30 = 640.70.$$

for the treatments SS we have

$$SS \text{ treatments} = \frac{311^2 + 396^2 + 360^2}{10} - (1,067)^2/30 = 364.07.$$

Finally,

$$SS \text{ treatments} \times \text{blocks} = SS \text{ total} - SS \text{ blocks} - SS \text{ treatments}$$

$$= 360.6.$$

The analysis is then displayed in a summary table following the format of Table 13.18.3.

For this example, where we are assuming that the blocks actually are a random sample of a much larger set that might have been used, it is perfectly appropriate to test the null hypothesis of no true treatment effects by taking the ratio of MS treatments to the MS for block $\times$ treatment interaction. When we do so, we find that the null hypothesis can be rejected even for $\alpha < .01$. Clearly, in this study the three training methods are producing significantly different average job performances among the subjects. Our estimates of these fixed treatment effects are

$$a_1 = (\bar{y}_{\cdot 1} - \bar{y}_{\cdot \cdot}) = -4.4667$$

$$a_2 = (\bar{y}_{\cdot 2} - \bar{y}_{\cdot \cdot}) = 4.0333$$

$$a_3 = (\bar{y}_{\cdot 3} - \bar{y}_{\cdot \cdot}) = 0.4333.$$

Source	SS	df	MS	F	p
Blocks	640.70	9	71.19		
Within blocks					
Treatment	364.07	2	182.03	9.09	<.01
Block × treat.					
interaction	360.60	18	20.03		
Total	1,365.37	29			

Table 13.18.3

ANOVA summary table for the randomized blocks data of Table 13.18.1

Obviously, the largest effect is produced by method A_2, and the smallest by A_1. Method A_2 seems quite superior to the old method of training.

Observe that there is no test of block effects; since there is no way to find MS error, which is required for the test of the random factor in a mixed model design. This probably would not concern the experimenter unduly, as the nuisance variables of age and education would not have been used to form blocks in the first place unless there were already pretty strong evidence that these variables are associated with the y values. The strategy of using blocks is effective only when the nuisance variables actually *do* relate fairly strongly to the dependent variable, and here is one of the places where a familiarity with the research literature on such variables can be extremely valuable.

13.19 ADDITIVITY AND GENERALIZED RANDOMIZED BLOCKS

As we already noted and illustrated above, the classic randomized blocks situation has exactly as many observations in each block as there are treatments, leading to $n = 1$ per cell. As we have also seen this introduces no problem for the F test of the main treatment effects *provided* that the blocks can be regarded as randomly sampled.

Serious practical difficulties tend to arise, however, when the randomized blocks design is classic, or unreplicated, thus providing no MS for error, *and* both blocks and treatments are fixed. In this situation, one typically assumes that there is *no true block by treatment interaction*. Indeed, even in the ANOVA summary table for this situation it is customary to take the SS and MS line usually called "block $\times$ treatment interaction" and to label it *residual* (even though the computations and the degrees of freedom remain exactly the same as if this were called interaction).

Here, where there are fixed-treatment and fixed-block effects and *no interaction* is assumed, the underlying model simplifies to

$$y_{ijk} = \mu + \alpha_j + \beta_k + e_{ijk}$$

[13.19.1]

where the combined effects of treatments (α_j) and of the blocks (β_k) simply add to the general mean and the error to produce the observed y value. Such a model, lacking interaction, is known as *additive*, and the property of no interaction is called *additivity*. Then, when the additive model is true, both blocks and treatments can be tested against MS residual, which becomes the estimate of error variance alone. (Actually, this is true for the additive model regardless of whether the blocks are considered random or fixed, as the former complications involving interaction are no longer present.)

Moreover, even when Model III provides an F test, there is still a problem of interpretation when we assume that the full model shown as Eq. 13.14.1 underlies the data for such a randomized blocks design. This model contains interaction between blocks and treatments. According to our interpretation of such interaction, its presence forces us to add a qualification to any statement we may wish to make about the effects of the treatments. That is, if interaction is actually present, then a detailed description of the effects of the three treatments in the example above really must be set in terms of the particular age–education blocks to which they are applied. On the other hand, it would be ideal if we could make statements about the three treatments (or training methods) in this example without having to worry about the nuisance factors of age and education.

However, it sometimes happens that blocks can be formed in which there are nJ subjects per block, where $n > 1$. This means that there are enough cases per block to permit allocation of n cases to each block-treatment combination. Here, in addition to the SS for blocks, the SS for treatments, and the SS interaction within blocks, we can find a within-blocks error term, based on the variation within the block-treatment cells.

This sort of situation is sometimes referred to as a *randomized blocks design with replication,* or a *generalized randomized blocks design.* Then, if the blocks happen to be randomly sampled, one can go directly to the tests outlined for a mixed model in Section 13.14. That is, MS blocks may be tested against MS error if desired, and MS treatments against MS blocks by treatments. The advantage to be gained here, however, is that blocks $\times$ treatments interaction may itself be tested *first* against MS error. If that F test fails to reach the value required even for a large α level, such as .10 or .25, as outlined in Section 13.11, the SS interaction may be pooled with the SS error to form a new estimate of error variability, or pooled MS error. This also provides a new number of degrees of freedom for pooled error, consisting of the original df for interaction plus the original df for error. Then, once the experimenter has decided that no interaction exists, the tests under Model III are effectively reduced to those of Model I. This implies that *both* blocks and treatments can be tested against MS pooled error.

On the other hand, if the randomized design is replicated and *both* blocks and treatments are fixed, then there is again no big problem. Model I applies here, and all the tests are carried out against MS error. Even so, significant interaction introduces the usual complications for interpretation.

Let us return to the unreplicated situation once more. How do you know when additivity truly holds? When is it safe simply to use MS residual in F tests for a nonreplicated design with fixed blocks? One way you might know is from previous experiments in which little or no evidence for interaction between blocks and treatments has been found. Conceivably, theory and your review of the literature may be enough to convince you that the underlying model is additive. Nevertheless in the classic $n = 1$ situation it is handy to have a way of testing for the presence of nonadditivity (interaction) before you proceed to test for main effects. Such a test exists, called *the Tukey test for non-additivity* (Tukey, 1949).

The null hypothesis to be tested is that interaction effects equal 0. Essentially, the Tukey test for nonadditivity is an application of the principle we discussed in Section 12.10, in which comparisons representing interaction can be found by multiplying comparison weights for main effects. Tukey's test simply uses the sample main effects for treatments, and the sample effects for blocks, as though these were the weights for two comparisons. (Notice that for equal n per cell, both the treatment and block effects do have the property of summing to zero, and thus are legitimate comparison weights.) Then these effect-weights are multiplied together, creating a new "interaction comparison" weight for each cell, just as we did in Section 12.10. These new weights are then used to multiply the value in each cell, and a sum of squares for the resulting comparison value found in the usual way. That is, if we let "SS nonadd" stand for the SS for the comparison found in the way just described, it will be true that

$$\text{SS nonadd} = \frac{JK\left[\sum_j \sum_k a_j b_k y_{jk}\right]^2}{(\text{SS treat})(\text{SS blocks})},$$

[13.19.2]

where a_j is the sample effect for the treatment j in Factor A, and b_k is the sample effect for block k. Since this actually *is* a comparison among the JK cell means, then this SS will have 1 degree of freedom.

Now if SS residual (the alias, remember, for the interaction SS value) is taken, and SS nonadd is subtracted from it, we have an SS remainder:

$$\text{SS remainder} = \text{SS residual} - \text{SS nonadd} \qquad [13.19.3]$$

with degrees of freedom given by

$$\text{df remainder} = \text{df residual} - \text{df nonadd} = (J - 1)(K - 1) - 1. \qquad [13.19.4]$$

Then the F test for nonadditivity is provided by

$$F = \text{MS nonadd/MS remainder} \qquad [13.19.5]$$

with 1 and $(J - 1)(K - 1) - 1$ degrees of freedom.

For example, suppose that the blocks in the example of Table 13.18.1 had not been sampled, but rather represented a set of 10 matched groups specifically chosen by the experimenter to vary widely on both age and educational level. Each group, however, contained three individuals identical in both age and education. We will assume that the data turned out just as in Table 13.18.1, however.

The Tukey test for nonadditivity will be carried out for these data, permitting us to decide if the treatments and blocks may safely be tested against MS residual. As in Section 13.11 where a similar sort of decision must be made, we face a danger of making a Type II error by deciding that there *is* additivity or no interaction. Therefore, we want to use a value of α such as .10 or .25 that is larger than the usual .05 or .01. The Tukey test for nonadditivity is then carried out using Table 13.19.1, which shows the original data of Table 13.18.1, with the sample effects a_j of the treatments (columns) and the sample effects b_k of the blocks shown by the rows.

Table 13.19.1

The original data of Table 13.18.1 shown in the format for the Tukey Test for Nonadditivity.

	Treatments			Sample Effects b_k
	A_1	A_2	A_3	
blocks				
1	41	44	43	6.7667
2	39	44	47	7.7667
3	28	33	37	-2.9000
4	26	41	30	-3.2333
5	41	42	41	5.7667
6	28	36	36	-2.2333
7	26	33	35	-4.2333
8	24	33	34	-5.2333
9	27	49	30	-0.2333
10	31	41	28	-2.2333
sample effects a_j	-4.4667	4.0333	0.4333	

Then, according to Eq. 13.19.2, the value of SS nonadd is found from

$$SS\ nonadd = \frac{JK\left[\sum_j\sum_k a_j b_k y_{jk}\right]^2}{(SS\ treatments)(SS\ blocks)}$$

$$= \frac{30[-4.4667(6.7667)(41) + \cdots + 0.4333(-2.333)(28)]^2}{(364.07)(640.7)}$$

$$= 33.86.$$

Then since SS residual $= 360.6$

SS remainder $= 360.6 - 33.86 = 326.74,$

with degrees of freedom of $(9)(2) - 1 = 17$. The F ratio is thus

$$F = 33.86/(326.74/17) = 1.76.$$

For 1 and 17 degrees of freedom this is not significant even at the $\alpha = .10$ level. It is probably safe here to conclude that the model *is* additive, and that both blocks and treatments can be tested against MS residual (previously listed as interaction).

13.20 SOME PROS AND CONS OF RANDOMIZED BLOCKS DESIGNS

Randomized blocks designs can be quite effective in controlling for nuisance variables and in reducing error variability in an experiment. How much the blocks will help to reduce error depends primarily on the extent of the association between the nuisance variable(s) and the dependent variable. Unless the experimenter is sure that there is a substantial relationship between the nuisance variable values and the values on the dependent variable, blocking may not be worth the time and effort. In fact, if the experimenter guesses wrong, and actually blocks on a nuisance variable that has little to do with y values, she may even suffer a loss of power relative to the unblocked situation.

Finding or creating blocks of subjects can be a laborious, time-consuming, and expensive exercise. The standards of homogeneity within and heterogenity between the blocks must be taken seriously; blocks composed of individuals who are just "sort of alike" won't do. Insofar as possible, individuals in blocks must be *exactly* alike on the variables in question. Other difficult considerations include the number of blocks to be used, and whether they should be fixed or a sample from a larger set. Issues of the number of blocks are considered in a number of places, such as Feldt (1958).

Finally, other methods of reducing error variability, such as the analysis of covariance to be discussed in Chapter 17, may be much easier and less expensive to carry out than blocking. Nevertheless, use of randomized blocks is a good, standard approach to many experimental situations, especially when combined in sophisticated ways with other design features such as those of a factorial design. Any modern book on experimental design such as Kirk (1982), Keppel (1982), or Edwards (1985) will illustrate these further possibilities.

13.21 REPEATED MEASURES (OR WITHIN-SUBJECTS) EXPERIMENTS

An individual subject used in an experiment brings along a regular "package" of nuisance variables that may affect his or her performance on dependent variable y. Some of these things are "organismic," and have to do with biological characteristics and features (age, sex, health factors, and all the rest). Other nuisance variables have to do with societal factors and experiences the subject may or may not have had. Still other variables may be situational with highly temporary values (did you have a fight with your spouse this morning?) Despite the different origins and nature of these nuisance variables, they all share the ability to affect the score an individual will produce on dependent variable y in a certain experimental situation. These are the nuisance variables that underlie the individual differences that we see in a set of experimental data.

However, just as we argued in dealing with randomized blocks, there are advantages in identifying and measuring variability due to individual differences, and removing it from our estimate of error variability. Since many nuisance variables are tied to a single subject, and since on some (though not all) of the nuisance variables the subject is matched with himself or herself throughout the experiment, it is quite possible to extend the ideas underlying the randomized blocks experiment to the situation where the same subject performs over several experimental situations or trials, providing a measured value of Y on each trial. Since each subject is thus measured several times during the course of the experiment, such experiments are often called *repeated measures* experiments. Alternatively, such experimental layouts are sometimes described as *within-subjects designs*.

The logical extension of the idea of randomized blocks designs is, therefore, to make an individual subject stand for a block, and to give each of J different treatments to each subject. Ordinarily, the treatments are assigned in some random order to a given subject. However, special "counterbalancing" techniques are used when possible order effects are *themselves* of interest to the experimenter (see Keppel, 1982, pp 372–377, for an excellent discussion of this possibility). Exactly K subjects are used in all. Every subject is observed under J different treatments, and every treatment is applied to K different subjects, for a total of JK observations in all. (When the emphasis is on the differences among J treatment levels, this layout is sometimes called a *treatment by subjects design*. However, when the emphasis is on the possible effects of the order of the trials that a subject receives, the design may be called a *trials by subjects* design.) In such a design the subjects are almost invariably regarded as having been chosen at random, and thus Model III ordinarily applies.

The expected mean squares then become as follows:

$$E(\text{MS between subjects}) = \sigma_e^2 + J\sigma_{sub.}^2 \qquad [13.21.1]$$

$$E(\text{MS between treatments}) = \sigma_e^2 + \sum_j \frac{K\alpha_j^2}{J-1} + \sigma_{sub.\ by\ A}^2 \qquad [13.21.2]$$

$$E(\text{MS subjects by treatments}) = \sigma_e^2 + \sigma_{sub.\ by\ A}^2 \qquad [13.21.3]$$

Here, $\sigma_{sub.\ by\ A}^2$ represents subject-treatment interaction variance. In such an analysis, MS between subjects is calculated just as for MS between for any Factor B in a two-way analysis, except that this time the Factor B is "subjects." The MS between treat-

Table 13.21.1

Summary table for the Model III analysis of variance for a repeated measures design.

Source	SS	df	MS	E(MS)	F
Between subjects (rows)	$K - 1$		$\dfrac{SS\ \text{sub.}}{K - 1}$	$\sigma_e^2 + J\sigma_{sub.}^2$	
Within subjects					
Between treatments (columns)	$J - 1$		$\dfrac{SS\ \text{treat.}}{J - 1}$	$\sigma_e^2 + \sum_j \dfrac{K\alpha_j^2}{J - 1} + \sigma_{sub.\ by\ A}^2$	$\dfrac{MS\ \text{treat.}}{MS\ \text{sub. by treat.}}$
Sub. by treat.	$(J - 1)(K - 1)$		$\dfrac{SS\ \text{sub.} \times \text{treat.}}{(J - 1)(K - 1)}$	$\sigma_e^2 + \sigma_{sub.\ by\ A}^2$	

ments is found as for any fixed factor A, and MS subjects by treatments is computed as for interaction. However, in the summary table for a subjects-by-treatments, or repeated measures design, it is customary first to break the total sum of squares into two major divisions, SS between subjects, and SS within subjects, where

SS within subjects = SS between treatments + SS subjects by treatments

The summary table for such an analysis takes the form shown in Table 13.21.1. As you can see from Table. 13.21.1, the main breakdown of the results is into variability attributable to difference among subjects, and variability within subjects. As in any randomized blocks set up, the entire set of experimental treatments (or combinations) is applied within each block, which is now a subject. It is for this reason that repeated measures designs are sometimes referred to as within-subjects designs, since a complete experiment occurs within each subject employed.

This way of summarizing the arithmetic (Table 13.21.1) really has no bearing on the numerical results, but it does help in keeping up with which F tests are appropriate in such a setup, particularly in the more complex designs.

Given the expected mean squares above, it can be seen that under Model III the only meaningful F test here must be

$$F = \frac{MS\ \text{between treatments}}{MS\ \text{subjects by treatments}}$$

[13.21.4]

with $J - 1$ and $(K - 1)(J - 1)$ degrees of freedom. This is the only F ratio where numerator and denominator have the same expected values when the null hypothesis, $H_0: \alpha_j = 0$ for all j, is true. However, note that in this simple subjects-by-treatments design, there is no way to test for subject effects. This creates no big problem, since "subjects" is ordinarily introduced as a nuisance factor anyway.

For example, a study was carried out on the role of certain stimulus factors on the ability of subjects to carry out fairly complex perceptual problem-solving tasks, such as solving jigsaw puzzles. Four puzzles were constructed which differed on two factors:

puzzles were either round (A_1) or square (A_2), and they were either monochromatic (B_1) or contained color variations (B_2). In every other respect an attempt was made to have the puzzles of equivalent difficulty.

Twelve randomly selected adult subjects were used, and each subject was given a fixed amount of training and experience in solving jigsaw puzzles. Then, each subject individually solved the four experimental puzzles. The order in which the puzzles were given was varied systematically so that each puzzle appeared in the first-order position for exactly three subjects, although subjects were assigned to particular orders at random. The dependent variable was the amount of time, in minutes, required for each subject to solve each puzzle. The data are shown in Table 13.21.2.

The analysis of variance for this repeated measures design is given in Table 13.21.3. Since the subjects are viewed as randomly sampled, and the factors defining the puzzles can be considered as fixed, this mixed model calls for a test of "puzzle" effects by taking the MS for puzzles against the MS for subject-by-puzzle interaction, as shown in the table.

For this example, the MS for puzzles divided by the MS for subject-by-puzzle interaction gives an F ratio of 4.60. If we are using an alpha level of .05 for this test, this result would be called significant. The four puzzle types appear to be having different effects with respect to solution time required by a subject. However, if interaction does indeed exist, then one would need to know the individual subject to be able to predict exactly how the puzzle affects the time. A test for nonadditivity could also be used here if we wished.

It is, by the way, also worth pointing out that the data in Table 13.21.2 actually represent a repeated measures design in which there are two "within-subjects" factors: Factor A, or puzzle shape, and Factor B, puzzle color, with all four AB combinations given to each subject. Therefore, the total sum of squares for this example could be broken down even further, as shown in Table 13.21.4. In this further breakdown of within-subjects variability, the test for each factor and interaction is carried out by use of the corresponding interaction involving subjects: that is, Factor A is tested against

Table 13.21.2

Time in minutes required for 12 subjects to solve four puzzles

| subject | Puzzle Type | | | |
	A_1B_1	A_2B_1	A_1B_2	A_2B_2
1	49	48	49	45
2	47	46	46	43
3	46	47	47	44
4	47	45	45	45
5	48	49	49	48
6	47	44	45	46
7	41	44	41	40
8	46	45	43	45
9	43	42	44	40
10	47	45	46	45
11	46	45	45	47
12	45	40	40	40

Source	SS	df	MS	F	p	
Subjects	226.5	11				**Table 13.21.3**
Within subjects						ANOVA for the repeated measures data of Table 13.21.2
Puzzles	24.0	3	8	4.60	(see text)	
Subs. × Puzzles	57.5	33	1.74			
Total	308.0	47				

subjects by A interaction, Factor B against subjects by B interaction, and A × B against subjects by A × B interaction. Unfortunately, we can pursue this idea no further here, but the example does illustrate how repeated measures designs lend themselves to combinations with other arrangements such as factorial designs. Any good text such as Edwards (1985) may be consulted for further details.

In any repeated measures experiment such as this, in which there is a fixed, within-subjects factor with J levels, our interest might focus on planned comparisons among the factor levels themselves rather than in an overall test for the presence of treatment effects. So long as the factor of interest represents fixed effects, planned comparisons may be carried out just as for any fixed-effects design, except that in the mixed model situation as illustrated here, the comparisons are each tested by use of *MS subject × treatment interaction,* just as we did above for the overall F test. Furthermore, as we shall see in Section 13.25, subject to certain assumptions the Scheffé and Tukey procedures for post hoc comparisons may be carried out as before, but using MS subject × treatment interaction in lieu of MS error.

On the other hand, a repeated measures analysis such as we have been discussing requires a quite specialized assumption in order to justify the tests for main effects or for interactions. Before going further we must examine this assumption and our options for dealing with its possible violation.

Source	SS	df	MS	F	p	
Subjects	226.5	11				**Table 13.21.4**
Within subjects						A further ANOVA on the data of Table 13.21.2
A (shape)	12	1	12	7.55	<.05	
Sub. × A	17.5	11	1.59			
B (color)	12	1	12	13.95	<.01	
Sub. × B	9.5	11	.86			
A × B	0	1	0			
Sub. × A × B	30.5	11	2.77			
Total	308.0	47				

13.22 SPECIAL ASSUMPTIONS IN RANDOMIZED BLOCKS AND REPEATED MEASURES DESIGNS

Randomized blocks and repeated measures designs rest on the usual assumptions that we have been making in connection with Model I or Model III situations. On the other hand, both randomized blocks and repeated measures have a feature that is usually absent in the randomized groups or factorial designs we have discussed. In a randomized blocks or repeated measures design the experimenter (or perhaps nature or society) has intervened to make individuals in some parts of the experiment (the blocks) *more alike* than those in other parts. This introduces an element of dependency or "correlation" across the scores of individuals given repeated measurements, or among members of the some matched group.

Although the following argument actually applies both to randomized blocks and to repeated measures designs, it is probably more important to understand the issues as they affect repeated measures experiments. Therefore, the discussion here will be framed in terms of the same subject given repeated treatments or measurements, with the tacit understanding that at any point the word *subject* could be replaced with *block of subjects*.

To show how the fact of dependency or correlation enters into the mean square values we get for repeated measures data, let us consider the very simplest of the mixed models that might apply:

$$y_{ij} = \mu + \alpha_j + b_i + e_{ij}$$

where here, α_j is the fixed effect of treatment j, b_i is the random effect associated with an individual subject i, and e_{ij} is the error term associated with subject i receiving treatment j. Notice also that in this very simplest model we are assuming additivity, or no interaction between subjects and treatments.

The true mean or expectation of the subject effects b_i can be assumed to be 0. However, if any of the subject effects b_i are nonzero, there will be nonzero true variance of the subject effects:

$$\sigma_B^2 = \text{true variance of the subject effects}$$

$$= E(b_i^2).$$

In addition, there will be a true variance of the random errors e_{ij}:

$$\sigma_e^2 = \text{true variance of the errors.}$$

Now let us single out two of the J possible treatments in our design, and call these the treatment j and the treatment m. Clearly, any subject i will have a value y_{ij} on treatment j and a value y_{im} on treatment m. Over all possible subjects who *might* be given those treatments, there will be a variance of the y_{ij} values, or σ_j^2 and a variance of the y_{im}, or σ_m^2. In addition, over all subjects i there will be a *covariance*, or $\text{cov}(y_{ij}, y_{im}) = \rho_{jm}\sigma_j\sigma_m$, where ρ_{jm} is the true subject-by-subject *correlation* between the treatment population j and the treatment population m. (See Appendix B.3 if you need refreshing on the notions of covariance and correlation). The key point is that the identity or the matching of subjects not only introduces a dependency among the resulting scores under different treatments, but also creates dependencies among the treatment populations, as reflected in the covariances.

The set of variance values for the J populations, along with the value of the covariance for each *pair* of such populations is called *the variance-covariance matrix*. Such a variance-covariance matrix can always be shown as a $J \times J$ array of values, where, in principle, each variance could have a different value, and the covariances might also differ from each other. One such array or matrix of population variances and covariances is symbolized below:

$$
\begin{bmatrix}
\sigma_1^2 & \rho_{12}\sigma_1\sigma_2 & \rho_{13}\sigma_1\sigma_3 & \cdots & \rho_{1J}\sigma_1\sigma_J \\
\rho_{21}\sigma_2\sigma_1 & \sigma_2^2 & \rho_{23}\sigma_2\sigma_3 & \cdots & \rho_{2J}\sigma_2\sigma_J \\
\cdots & \cdots & \cdots & \cdots & \cdots \\
\rho_{J1}\sigma_J\sigma_1 & \rho_{J2}\sigma_J\sigma_2 & \rho_{J3}\sigma_J\sigma_3 & \cdots & \sigma_J^2
\end{bmatrix}
$$

(You may wish to look at Sections C.3 and D.6 for more about such variance-covariance matrices.)

The original rationale for the F test in randomized blocks designs (including repeated measures designs) rested on a very specialized assumption concerning the true variance-covariance matrix of the treatments across the blocks. Originally it was assumed that the condition termed *compound symmetry* exists across the population of blocks or of subjects. Essentially, the compound symmetry assumption requires that each treatment have exactly the same true variance over all possible blocks, and that the *covariance* for each pair of treatments also have a constant value. Thus, it is assumed that for any pair of treatment populations the variances are equal:

$$
\sigma_j^2 = \sigma_m^2 = \sigma_T^2 \tag{13.22.1}
$$

where σ_T^2 will be used here to symbolize the value of the variance for *any* of these treatment populations. Furthermore, it is assumed that the *covariance* for each pair of populations j and m is the same. In the light of the foregoing assumption this is equivalent to the assumption that

$$
\text{cov}(y_{ij}, y_{im}) = \rho\sigma_T^2 \tag{13.22.2}
$$

for all pairs j and m, where ρ stands for a constant value of the correlation.

In short, under compound symmetry the entire array or matrix of the variances and covariances for the J populations would look like this:

Treatment populations

	A_1	A_2	A_3	$\cdots$	A_J
A_1	σ_T^2	$\rho\sigma_T^2$	$\rho\sigma_T^2$	$\cdots$	$\rho\sigma_T^2$
A_2		σ_T^2	$\rho\sigma_T^2$	$\cdots$	$\rho\sigma_T^2$
$\cdots$					
A_J				$\cdots$	σ_T^2

(For simplicity of presentation only the upper portion of the variance-covariance matrix is shown here; the lower half is exactly like the upper.)

Given the truth of these assumptions, then for any of the treatment populations the variance σ_T^2 must consist only of error variance and subjects (or blocks) variance:

$$\sigma_T^2 = \sigma_e^2 + \sigma_B^2 \qquad\qquad [13.22.3]$$

In addition, the value of the correlation ρ is the same as the intraclass correlation (Section 13.5) for subjects:

$$\rho = \sigma_B^2/\sigma_T^2 \qquad\qquad [13.22.4]$$

so that

$$\rho\sigma_T^2 = \sigma_B^2 \qquad\qquad [13.22.5]$$

and

$$\sigma_e^2 = \sigma_T^2 (1 - \rho) . \qquad\qquad [13.22.6]$$

Finally, a few algebraic substitutions yield the following new versions (and new interpretations) of the expected mean squares. Given these assumptions, then for a sample of K subjects the expected mean squares have the following compositions:

$$E(\text{MS subjects}) = \sigma_e^2 + J\sigma_B^2 \qquad\qquad [13.22.7]$$

$$= \sigma_T^2 [1 + (J - 1)\rho]$$

$$E(\text{MS treatments}) = \sigma_e^2 + K \sum_j \alpha_j^2/(J - 1) \qquad\qquad [13.22.8]$$

$$= \sigma_T^2 (1 - \rho) + K \sum_j \alpha_j^2/(J - 1)$$

and, using *residual*, or *error*, to designate the only remaining term,

$$E(\text{MS residual}) = \sigma_e^2 = \sigma_T^2(1 - \rho) . \qquad\qquad [13.22.9]$$

By use of these expected mean squares it is easy to show how the dependencies between the matched observations play a role in determining the outcome of the ANOVA. In the mean squares for between subjects, or MS B, a large correlation implies a large MS value, and hence a large proportion of variance accounted for by subject differences. Bear in mind that the point of matching or repeated measures is to *remove* large amounts of variance due to nuisance variables in the experiment, so that such variability will not show up as error variance. A large correlation *within* subjects or blocks accomplishes that goal. On the other hand, a large correlation ρ actually lowers the proportion of MS treatments or (MS A) that is due to error variability. Similarly, the size of the residual MS, standing for error variability is smaller for larger values of the correlation ρ, other things being equal. This demonstrates one of the reasons for matching or using repeated measures in the first place: to achieve control and a consequent reduction in the error variability in the data.

A similar argument can be used for the model which contains treatment by interaction effects as well. Although we will not go further into this model here, the implications are the same: some assumption must be made about the set of true variances and co-variances (the true variance-covariance matrix) for the J treatments populations represented in the data, if the usual F test for treatments is to be justified.

The compound symmetry assumption just discussed is *sufficient* to give a mathematical justification for the use of the F test in randomized blocks and repeated measures problems. However, it is *not* a *necessary condition* in the mathematical sense, meaning that other assumptions can also be used to justify these procedures. Furthermore, the compound symmetry assumption appears to be very stringent and unlikely to be satisfied in practice, especially in the context of repeated measures experiments.

In recent years it has been shown that another assumption can be made which is mathematically *both necessary and sufficient* to establish the validity of the use of the F test in these procedures. Known as the *circularity assumption,* (or *sphericity assumption*), this also appears to be somewhat less objectionable than the original assumption of compound symmetry, as applied to actual experimental situations. The details of this rather sophisticated argument, due to a number of authors, including Huynh and Feldt (1970) and Rouanet and Lepine (1970), need not detain us at this point. If you are interested, you can find the circularity assumption outlined in matrix terms in Section D.7 of Appendix D. A good discussion of these assumptions may also be found in Kirk (1982). (Incidentally, the assumption of compound symmetry implies that the assumption of circularity will be true; the reverse implication does not hold, however.)

The important point here is that if the assumption of circularity is not true, one runs a considerable risk of a misleading F test. In particular, the failure of this assumption tends to make the *probability of a Type I error greater than it appears to be when F is evaluated in the usual way*. A test of the circularity assumption is possible (Mauchley, 1940). However, this test seems to be unduly sensitive to things such as nonnormal distributions, and is therefore not usually performed.

Fortunately, it is also possible to adjust the degrees of freedom in the F test so as to compensate for the possible failure of the circularity assumption. This procedure, which we will refer to here as the *Box adjustment,* is often used as part of a three-step procedure for carrying out the F test in repeated measures designs, as discussed in the following sections. Strictly speaking, the Box adjustment applies both to randomized blocks and to repeated measures designs, when the circularity assumption cannot be made. Although the circularity assumption often appears rather innocuous in the case of randomized blocks designs, it may be quite unreasonable for repeated measures. The Box adjustment is therefore usually recommended when one is dealing with a repeated measures design, but has reservations about making the circularity assumption. It is also convenient that the Box adjustment procedure to be discussed next is now a routine feature of some computer programs for analyzing repeated measures data, such as PROC GLM in SAS., where the Box adjustment is labelled H-F for Huynh and Feldt (1970).

Finally, before we go further it should be mentioned that other ways exist for avoiding objectionable assumptions such as circularity in the analysis of repeated measures. One of these options is the use of *multivariate analysis of variance,* or MANOVA, to analyze such data, where the different values within each subject are treated as though they were values on different variables. This strategy produces results comparable to repeated measures ANOVA but without the necessity for the assumptions that ANOVA F tests require. The statistical packages often include repeated measures analysis as a part of their overall MANOVA subprograms for this, among other, reasons. Unfortunately, MANOVA is beyond the scope of this text. However, any of a number of good texts in multivariate methods, such as Timm (1975), Harris (1985), and Cliff (1987), will shed light on how this is done.

13.23 THE BOX ADJUSTMENT TO DEGREES OF FREEDOM

Here we are going to work through a method for adjusting the degrees of freedom for F in a randomized blocks or repeated measures experiment so as to avoid reaching erroneous conclusions due to the failure to satisfy the circularity assumption. We start with a matrix of sums of squares and sums of products (see Appendix C.3). (In the following, we can cut down on notational complexity just a bit if we use SP_{jj} instead of our usual SS symbol to stand for the sum of squares within treatment j, and also use SP_{jk} to stand for the sum of products for the two treatments j and k).

That is for J different experimental treatments or trials the following computations will yield a $J \times J$ matrix, in which the principal diagonal contains SP_{jj} or sums of squares values calculated *within* the various treatments j, and across the K various blocks or subjects:

$$SP_{jj} = \sum_{k=1}^{K} y_{jk}^2 - T_j^2/K. \qquad [13.23.1]$$

The off-diagonals of the matrix will contain SP_{jk} or sums of products calculated for each *pair* of treatments j and g, and across the different blocks or subjects:

$$SP_{jg} = \sum_{k=1}^{K} y_{jk}y_{gk} - (T_j)(T_g)/K \qquad [13.23.2]$$

(At this point, some descriptions of this method convert the SP values into variances and covariances by dividing by N or $N - 1$. However, this makes no difference at all in the results, and so we will save a little effort by sticking with the SP values.)

Now, given the completed $J \times J$ matrix of SP values, we first need to take the mean of the diagonal (or SP_{jj}) values:

$$\overline{D} = \text{mean of diagonal values} \qquad [13.23.3]$$

$$= \sum_{j} SP_{jj}/J$$

Next we take the mean of the entire matrix, and call this $\overline{SP}_{..}$:

$$\overline{SP}_{..} = \sum_{j=1}^{J} \sum_{g=1}^{J} SP_{jg}/(J^2) \qquad [13.23.4]$$

In addition, it is necessary to find the mean of each row of the matrix:

$$\overline{SP}_{j.} = \sum_{g=1}^{J} SP_{jg}/J . \qquad [13.23.5]$$

Then, the estimated correction factor symbolized by $\hat{\epsilon}$ (epsilon) is found by taking

$$\hat{\epsilon} = \frac{J^2 \, (\overline{D} - \overline{SP}..)^2}{(J-1) \left(\sum_j \sum_g SP_{jg}^2 - 2J \sum_j \overline{SP}_{j.}^2 + J^2 \overline{SP}_{..}^2 \right)} \qquad [13.23.6]$$

This estimated correction factor $\hat{\epsilon}$ will always lie between $1/(J - 1)$ and 1.00. Then in the "adjusted" test for differences among the treatments, the original value of F is unchanged. However, new degrees of freedom for numerator and denominator are found:

   v_1^* = adjusted numerator degrees of freedom

      = $\hat{\epsilon}$ times original numerator degrees of freedom

   v_2^* = adjusted denominator degrees of freedom

      = $\hat{\epsilon}$ times original denominator degrees of freedom

The F value is looked up in the table for these adjusted degrees of freedom, and a final conclusion reached.

Geisser and Greenhouse (1958) pointed out that the lowest possible value of $\hat{\epsilon}$ must be $1/(J - 1)$ when there are J different treatments, and that a conservative test of the hypothesis of no treatment differences is therefore provided by using this $\hat{\epsilon}$ value. Such a test, though conservative, does not depend on the actual variance-covariance matrix that is true for the population. Hence, if both the regular F test and the Geisser-Greenhouse procedure give significant results, then there is no question that the outcome is genuinely significant, regardless of the circularity assumption. In that circumstance, where the Geisser-Greenhouse procedure leads to a significant F, there is no need to calculate the Box correction. Only when the original F test is significant and the Geisser-Greenhouse procedure gives a nonsignificant F is the Box calculation really necessary. (A question for you to think about: What happens when $J = 2$?)

Obviously, the calculation of the Box correction factor $\hat{\epsilon}$ is a bit of bother when one is working by hand, and it is desirable not to have to go through this procedure unless it is truly necessary. Therefore, in the interest of avoiding unnecessary calculations, the following three-step procedure is often used for randomized blocks and (especially) repeated measures designs:

1. Analyze the data by ANOVA, as shown in Section 13.18 or Section 13.21. Find the significance of F in the usual way. If this F is not significant at the chosen alpha level then *stop*. However, if F is significant go on to Step 2.
2. Carry out the Geisser-Greenhouse procedure by multiplying both numerator and denominator degrees of freedom by $1/(J - 1)$. Look up F using these new degrees of freedom. If significant, stop . . . your result is *really* significant. However, if the F is not significant under the Geisser-Greenhouse degrees of freedom, go on to Step 3.
3. Carry out the Box adjustment procedure for degrees of freedom. Look up the F once again, using the adjusted degrees of freedom. Accept the result as *the* significance of the obtained F ratio.

13.24 AN EXAMPLE OF THE THREE-STEP PROCEDURE AND BOX ADJUSTMENT IN REPEATED MEASURES

This three-step procedure, including calculation and use of the Box adjustment to degrees of freedom can be illustrated readily by use of the repeated measures data of Table 13.21.2. Applying Step 1 to these data, we note immediately from the ANOVA summary table the F value of 4.60, which, for 3 and 33 degrees of freedom, is clearly significant beyond the .05 level. In Step 2 we apply the Geisser-Greenhouse criterion, in which the numerator degrees of freedom from F are reduced to $3/3 = 1$ and the denominator degrees of freedom to $33/3 = 11$. For 1 and 11 degrees of freedom, our obtained F of 4.60 is no longer significant at the .05 level. Then since there is a disagreement in the conclusions between Steps 1 and 2 of this procedure, we now turn to the estimated Box correction to degrees of freedom as a way to solve the problem.

To calculate the Box correction for the degrees of freedom in these data, we must first find the sample matrix of sums of squares and sums of products as outlined in the previous section. Thus, for example, the sum of squares for the first treatment (i.e., puzzle A_1B_1) is found to be

$$SP_{11} = SS_1 = (49)^2 + \cdots + (45)^2 - (552)^2/12$$

$$= 52 .$$

On the other hand, the sum of products for Treatments 1 and 2 (i.e., puzzles A_1B_1 and A_2B_1) is found from

$$SP_{12} = (49)(48) + \ldots + (45)(40) - (552)(540)/12$$

$$= 36, \text{ and so on for other pairs of puzzles.}$$

For these data, the entire SS and SP matrix thus turns out to be as follows, in terms of the different puzzles, or treatments:

	A_1B_1	A_2B_1	A_1B_2	A_2B_2	row mean
A_1B_1	52	36	50	50	47
A_2B_1	36	66	65	52	54.75
A_1B_2	50	65	84	58	64.25
A_2B_2	50	52	58	82	60.5

It follows that from this matrix we can easily find

$$\overline{D} = \text{diagonal mean} = (52 + 66 + 84 + 82)/4 = 71$$

$$\overline{SP}_{..} = \text{table mean} = (52 + 36 + \ldots + 58 + 82)/16 = 56.625$$

$$\sum_j \sum_g SP_{jg}^2 = \text{sum of all squared entries} = 54{,}018$$

$$\sum_j \overline{SP}_j^2 = \text{sum of squared row means} = 12{,}994.875$$

Then from Eq. 13.23.6 the value of $\hat{\epsilon}$ is immediately found to be

$$\hat{\epsilon} = \frac{16(71 - 56.625)^2}{(3)[54,018 - 2(4)(12,994.875) + 16(56.625)^2]}$$

$$= 0.81$$

The most accurate F test available under these circumstances, allowing for the degree of departure from strict circularity suggested by the data, is given by multiplying both numerator and denominator degrees of freedom by $\hat{\epsilon}$, or 0.81, to give

adjusted numerator degrees of freedom $= 0.81(3) = 2.43$

adjusted denominator degrees of freedom $= 0.81(33) = 26.73.$

As a rough approximation, we can find the F required for significance at the .05 level by taking 27 denominator degrees of freedom and then estimating a value about 43% of the way between that for 2 degrees of freedom or (3.35) and that for 3 degrees of freedom (or 2.96).

estimated value of F required $= 3.35 - .43(3.35 - 2.96)$

$$= 3.18.$$

If this criterion value is applied, in this example the result is clearly significant beyond the .05 level. The shape and color characteristics of the puzzles apparently do make a difference in the time it takes the subjects to put them together.

In summary, the circularity assumption is generally not too offensive when it is applied to a study in which blocks or matched groups have been created or chosen at random. Thus, it is probably safe to omit the three-step procedure outlined above for such randomized blocks designs. However, the circularity assumption seldom is appealing, on the face of it, when one is dealing with repeated measures of the same randomly selected subjects. Here is where the three-step strategy, culminating in the Box procedure, provides a good "insurance policy" against possibly misleading results.

13.25 POST HOC TESTS OF MEANS OF A WITHIN-BLOCKS OR WITHIN-SUBJECTS FACTOR, IN A RANDOMIZED BLOCKS OR REPEATED MEASURES DESIGN

With some modifications both the Scheffé and the Tukey HSD procedures discussed in Sections 11.16 and 11.17 can be applied to the sample means found in a randomized blocks or repeated measures design. Such post hoc tests apply to the fixed within-blocks or within-subjects factor (or factors) in such a design.

Let us use MS residual to stand for the mean square term that is used in the denominator of the F test for the within-blocks (or within-subjects) factor. Thus, if additivity is *not* assumed, MS residual is the same as MS treatments $\times$ blocks, as used to test for treatment effects in a randomized blocks design with randomly sampled blocks. Equivalently, MS residual is equal to MS treatments $\times$ subjects used to test for treatment effects when the subjects are sampled at random. On the other hand, when additivity is assumed, MS residual is the equivalent term used to test for treatment effects, but taken to represent error alone.

Now, suppose that we are dealing with K blocks (or subjects) and a single "within" factor A with J levels. If we are confident that the circularity assumption is actually true, then following a significant F test for Factor A, any comparison among the means of levels of A can be made, and tested by use of MS residual and the Scheffé criterion

$$S^2 = (J - 1) F_{\alpha, \nu_1, \nu_2} \qquad [13.25.1]$$

where F_{α, ν_1, ν_2} here represents the value required for significance at the α level with $\nu_1 = J - 1$ and $\nu_2 = (J - 1)(K - 1)$ degrees of freedom. As before, the Scheffé criterion for the t test is simply $\sqrt{S^2}$.

The Tukey HSD test may be applied to any *pair* of the means of the within factor A, by use of the expression

$$\text{HSD} = q_{\alpha, J, \nu_2} \sqrt{(\text{MS res})/K} \qquad [13.25.2]$$

where q is once again the value of the studentized range statistic for $\alpha = .05$ or $.01$, as found from Table XI (Appendix E), for J groups and $\nu_2 = (J - 1)(K - 1)$ degrees of freedom.

Conversely, if the circularity assumption is *not* tenable, then a different procedure is used, where the MS residual (i.e., the term calculated for MS block $\times$ treatment or MS subject $\times$ treatment interaction) is replaced by *a separate MS residual term for each comparison to be tested*. That is, consider an experiment with J treatments given to each of $K = n$ subjects in a repeated measures design, resulting in a total of $N = Jn$ values of Y. If the F test made by taking MS treatments divided by MS residual is significant, and we wish to follow up with Scheffé, Bonferroni, or Tukey comparisons, then we do the following:

1. Take the weights c_{jg} for any within-subject factor comparison $\hat{\psi}_g$ and apply these weights directly to the J values found *within each individual i*. The result will be a comparison value for subject i, which we can symbolize as $\hat{\psi}_{gi}$ (to distinguish this value from the usual comparison made on the means, or $\hat{\psi}_g$).

2. Over all n of the subjects take the sum of squares based on these $\hat{\psi}_{gi}$ values, and call this SS_g:

$$\text{SS}_g = \sum_{i=1}^{n} \hat{\psi}_{gi}^2 - n(\hat{\psi}_g)^2. \qquad [13.25.3]$$

3. In addition, as in Eq. 11.9.2, let us once again use the symbol w_g to stand for the sum of the squared weights for a particular comparison g, with each square divided by the number of cases n:

$$w_g = \sum_j c_{jg}^2/n . \qquad [13.25.4]$$

4. Then, we can define MS res_g, the mean square residual for a particular comparison, as

$$\text{MS res}_g = \frac{\text{SS}_g}{w_g (n - 1)n} . \qquad [13.25.5]$$

Such a value of MS res_g is found for each comparison g that you wish to test, and then an F test can be carried out, using the usual format:

$$F = \frac{SS\,(\hat{\psi}_g)}{MS\,res_g}\,.$$ [13.25.6]

This is compared with the Scheffé criterion found exactly as in Section 11.16, *except* that the value of F in Eq. 11.16.1 is found for $J - 1$ and $n - 1$ degrees of freedom and the desired alpha level:

$$S^2 = (J - 1)\,F_{(\alpha, J-1, n-1)}\,.$$

Similarly, if it is your intention to do pair comparisons, where each comparison g has one weight of 1, one weight of -1, and the rest of the weights are 0, then the equivalent Tukey HSD procedure is

$$HSD = q\,\sqrt{MS\,res_g/n}\,.$$ [13.25.7]

Here the studentized range statistic q is looked up in Table XI of Appendix E exactly as in Section 11.17, but for $n - 1$ degrees of freedom and J groups.

For example, suppose that a repeated measures experiment involved $J = 4$ treatments given to each of $n = 10$ subjects. The raw data are given in Table 13.25.1, and the ANOVA summary is shown as Table 13.25.2.

Table 13.25.1

Repeated measures data for four treatments and 10 subjects

	A_1	A_2	A_3	A_4
S_1	11	9	5	17
S_2	14	12	10	18
S_3	15	7	6	21
S_4	17	10	13	22
S_5	15	7	6	15
S_6	14	8	13	22
S_7	9	6	7	15
S_8	17	11	10	19
S_9	10	13	14	23
S_{10}	12	8	11	20
means	13.4	9.1	9.5	19.2	$\bar{y} = 12.8$

Table 13.25.2

ANOVA summary table for the data in Table 13.25.1

Source	SS	df	MS	F	p
Subjects	156.9	9	17.43		
Within Subjects					
Treatments	659.0	3	219.67	43.41	$<<.01$
Treat. × Sub. (residual)	136.5	27	5.06		
Total	952.4				

Here it is assumed that the subjects were sampled at random, so that the F test for treatments can be carried out using the MS treatments × subjects. However, even if the subjects had not been a random sample, assuming additivity makes the corresponding mean square (or MS residual) still a proper denominator for our F test.

As it happens, with these data the three-step procedure outlined in Section 13.23 gives significance beyond the .01 level both for the original F test and for the conservative Geisser-Greenhouse test. Although in this situation it is not strictly necessary to calculate the Box adjustment to degrees of freedom, doing so anyway does suggest that the assumption of circularity is probably not justified here. That is, the estimated value of the Box $\hat{\epsilon}$ is here found to be .739, or a good deal below the value of 1.00 we would expect if circularity were strictly true. Therefore, assuming that circularity does not hold for this example, we will proceed to illustrate the calculation of separate MS residual terms for each of a set of post hoc comparisons.

Table 13.25.3 repeats the data in Table 13.25.1, but this time with the weights for four comparisons shown across the top. The first three of these are comparisons of certain pairs of means, which thus can be tested through the Tukey HSD procedure (or, if we wish, by the Bonferroni t method of Section 11.14.) The fourth comparison involves all four means, and thus will be tested by use of the Scheffé criterion. At the right side of the table are shown the comparison values $\hat{\psi}_{gi}$ found within each subject i on each one of these comparisons, $g = 1, 2, 3,$ and 4. Then at the bottom right of the table are shown the SS_g values calculated over subjects' comparison values $\hat{\psi}_{gi}$ and the corresponding value of the MS res_g found for the particular comparisons using Eq. 13.25.5.

Table 13.25.3

Calculation of the MS res_g values for post hoc comparisons among the treatments

Comparison	Weights							
1	1	−1	0	0				
2	1	0	−1	0				
3	1	0	0	−1				
4	1	−1	−1	1				
	A_1	A_2	A_3	A_4	$\hat{\psi}_{1i}$	$\hat{\psi}_{2i}$	$\hat{\psi}_{3i}$	$\hat{\psi}_{4i}$
S_1	11	9	5	17	2	6	−6	14
S_2	14	12	10	18	2	4	−4	10
S_3	15	7	6	21	8	9	−6	23
S_4	17	10	13	22	7	4	−5	16
S_5	15	7	6	15	8	9	0	17
S_6	14	8	13	22	6	1	−8	15
S_7	9	6	7	15	3	2	−6	11
S_8	17	11	10	19	6	7	−2	15
S_9	10	13	14	23	−3	−4	−13	6
S_{10}	12	8	11	20	4	1	−8	13
means	13.4	9.1	9.5	19.2				
			SS_g =		106.1	148.9	113.6	186.0
			MS res_g =		5.89	8.27	6.31	5.17

Now using the first set of comparison weights shown in Table 13.25.3, we have the mean of the A_1 treatment minus that for the A_2 treatment, or $\hat{\psi}_1 = 13.4 - 9.1 = 4.3$. Then, the relevant Tukey HSD value at the .05 level for this difference is found by taking $q = 4.42$, for four groups and $n - 1 = 9$ degrees of freedom, from Table XI. Then using the MS $res_1 = 106.1/(18) = 5.89$ as found in Table 13.25.3, we have

$$\text{HSD} = 4.42 \sqrt{5.89/10} = 3.39.$$

Since the pair comparison value of 4.3 exceeds this, we can say that it is significant for alpha $= .05$. Similarly the second comparison gives $13.4 - 9.5 = 3.9$, and an HSD value of

$$\text{HSD} = 4.42 \sqrt{8.27/10} = 4.02,$$

implying a nonsignificant difference between this pair of means. The third comparison produces a difference of -5.8, which, when compared *in absolute terms* to an HSD value of 3.51, is significant. Although we have illustrated the Tukey procedure for only three of the six possible differences among the four treatment means, the remaining three can be tested in exactly this same way.

To test the fourth comparison, which involves more than two means, we will use the Scheffé criterion. Thus, the comparison value is found to be $13.4 - 9.1 - 9.5 - 19.2 = 14$. This is first turned into a sum of squares in the usual way for any comparison among means:

$$\text{SS} (\hat{\psi}_g) = (14)^2/[(1 + 1 + 1 + 1)/10] = 490.$$

That is then divided by the MS $res_4 = 5.17$, as found above for this comparison. The resulting F of $490/5.17$, or 94.78, is evaluated by the Scheffé criterion value for alpha $= .05$, or

$$S^2 = 3 \times 3.86 = 11.58,$$

where 3.86 is the F value required for significance for $J - 1 = 3$ and $n - 1 = 9$ degrees of freedom. In this instance the comparison F is far larger than required for significance, of course. Any other comparison of interest could be tested in exactly this same way, using its own value of MS res_g as illustrated here.

This appears to be a relatively good way to approach post hoc tests in repeated measure designs using the Scheffé method for general post hoc comparisons. As we have seen, the same principle applies to the Tukey HSD method (which seems to work best when the number of pairs of means being compared is relatively small), and even to the Bonferroni t tests, as in Section 11.14 and 11.17.

On the other hand, a little thought may convince you that these are fairly low-powered methods, as compared to the methods you would employ given that the circularity assumption happened to be true. This is because the degrees of freedom for "residual" has been reduced here from the usual $(J - 1)(n - 1)$ degrees of freedom to only $n - 1$. This is the price one pays for avoiding the circularity assumption. The methods discussed here can also be applied without modification to unreplicated randomized blocks experiments, simply by substituting "blocks" for "subjects" throughout.

13.26 THE GENERAL PROBLEM OF EXPERIMENTAL DESIGN

The factorial layout mentioned in the last chapter, as well as the randomized blocks and repeated measures designs just discussed are but some of the simplest and commonly used ways to design a study. The literature of experimental design is full of ways for collecting and analyzing data in response to particular sorts of research questions. Many of these designs in common use may be found in references such as Edwards (1985), Kirk (1982), Myers (1979), Keppel (1982), and particularly in Winer (1971). A more advanced treatment of the subject may be found in Cochran and Cox (1957), whereas an informative, nonmathematical introduction to design is given in D. R. Cox (1958).

From a broad point of view, the problem of choosing a design for an experiment is a problem in economics. The experimenter has some question or questions to answer, and a budget of time, effort, and money wherewith to achieve answers. The following points must be assured:

1. The actual data collected will contain all of the information needed to make inferences about the population of interest, and this information can be extracted from the data.
2. The important hypotheses can be tested validly, and separately.
3. The level of precision reached in estimation and the power of statistical tests will be satisfactory for the purposes of the experiment.

Consideration number 1 involves the actual selection of treatments and their combinations that the experimenter will employ. What are the factors of interest? How many levels of each will be observed? Will these levels be sampled or regarded as fixed? Which factors will be crossed in the design, and which may, perhaps, be "nested" within levels of other factors? Consideration 2 is closely related to the first, of course: must all combinations of factors be observed, or is there actual interest only in particular treatment combinations? If only a part of the possible set of treatment combinations is to be observed, is it still possible to make logically and statistically separate inferences about the factors and interactions of interest? Consideration 3 involves the choice of sample size and of experimental controls, both of which must influence power. Is the sample size contemplated large enough to give the precision of estimation or the power the experimenter feels necessary? Does the research literature suggest that certain types of controls are especially important as ways to reduce the size of error variance, as represented by MS error in the sample?

However, attendant to each of these considerations are parallel considerations of cost.

4. The more different treatments administered, necessitated by the more questions asked of the data, the more the experiment will cost in time, subjects, effort, and other expenses.
5. The more kinds of information the experimenter wants to gain, the larger the set of assumptions that may be required to obtain valid inferences.
6. The more hypotheses that can be tested validly and separately, the greater the

number of treatment combinations necessary, and the larger may be the required number of subjects. Furthermore, the clarity of the statistical findings may be lessened, and the experiment as a whole may be harder to interpret.

7. The experimenter can increase precision and power by larger samples, or by exercising additional controls in the experiment, either as constant control, or by a matching procedure. Each possibility has its real costs in time, effort, and perhaps, money.

Other things being equal, the experimenter would like to get by with as few subjects as possible. Even "approximately" random samples are extremely difficult to obtain, and often the sheer cost of the experiment in time and effort goes up with each slight increase in sample size. Furthermore, the experiment may not be carried out as carefully with a large number as with only a few subjects. Throwing in lots of ill-considered treatments just to see what will happen can be an expensive pastime for the serious experimenter. The number of treatment combinations can increase very quickly when many factors are added, and some of these combinations may be of no interest at all to the experimenter or add little or nothing to precision, so that a high price may be paid for discovering "garbage" effects in the data. If a large sample is out of the question, power and precision can be bought at the price of constant or matching controls. However, remember that constant control reduces the generality of the conclusions the experimenter can reach from a set of data, and matching may be extremely hard to carry out.

All in all, there are several things the experimenter wants from the experiment and several ways to get them. Each choice has its price, however, and the experimenter must somehow decide if the gain in designing the experiment in a particular way is offset by the loss that may be incurred in so doing. This is why the problem of experimental design has strong economic overtones.

Texts in experimental design present ways of laying out the experiment so as to get "the most for the least" in a given situation. Various designs emphasize one aspect or another of the considerations and costs involved in getting and analyzing experimental data. Texts in design can give only a few standard types or layouts that study and experience have shown optimal in one or more ways, with, nevertheless, some price paid for using each design. Obviously, the best design for every conceivable experiment does not exist "canned" somewhere in a book, and experienced researchers and statisticians often come up with novel ways of designing an experiment for a special purpose. On the other hand, a study of the standard experimental designs is very instructive for any experimenter, if only to appreciate the strengths and weaknesses of different ways of laying out an experiment.

In any event, a principle for the beginner to remember is, *keep it simple!* Concentrate on how well and carefully you can carry out a few *meaningful* manipulations on a few subjects, chosen, *if possible,* randomly from some well-defined population and *randomized* among treatments. The experiment and its meaning is the thing to keep in mind, and not some fancy way of setting up the experiment that has no real connection with the basic problem or the economics of the situation. Then, when the novice finally knows his way around in his experimental area, the refinements of design are open for exploration.

13.27 MIXED MODEL, RANDOMIZED BLOCKS, AND REPEATED MEASURES ANALYSES IN THE COMPUTER PACKAGES

The latest available versions of the popular computer packages tend to be much more sophisticated than the earlier versions in handling mixed model, within-subject, and the more complicated designs for experimental data. Although the ONEWAY and ANOVA subprograms in SPSS and SPSS[x] are not readily adaptable to the sorts of designs discussed in this chapter, this limitation is more than compensated for by the newer MANOVA subprogram, with its wealth of user options (which, unfortunately, make it more difficult to learn to use.)

The fundamental MANOVA command applicable to analyzing data from a designed experiment takes the following form when there are two or more factors A, B, C, etc.:

MANOVA *dependent variable* BY *A (min,max) B (min, max), . . . /*

Here, of course, the words *dependent variable* and the letters A, B, and so forth are replaced by the names used in the initial VARIABLE LIST in the set of commands. The *(min, max)* indicates the index numbers $(1, J)$ when you are dealing with Factor A with J levels, and similarly for other factors such as B and C.

Then under the general MANOVA command will be listed a subcommand of the form

DESIGN = *effect 1, effect 2, effect 3, . . . /*

which specifies the effects to be assumed present in the model underlying the data. Thus, if there are three factors A, B, and C where the model includes the main effects and the interactions $A \times B$ and $A \times C$, but no $B \times C$, or $A \times B \times C$, this is indicated by

DESIGN = *A, B, A* BY *B, A* BY *C / .*

When no DESIGN statement is included after MANOVA then all effects are assumed present.

A randomized blocks design like the example used in Section 13.18, with three treatment levels A, and 10 blocks could be specified by

MANOVA *Y* BY *BLKS* (1,10) *A* (1,3)/

ERROR = *R* /

DESIGN = *BLKS, A* /

A DESIGN statement including an interaction provides the Tukey test. Although the R following the ERROR = command specifies that residual is used to test for the main effects (which also happens to be the correct test even if blocks are assumed to be sampled with A fixed), this will happen by default in this design with $n = 1$. However, within, and combined within + residual terms, can be specified to fit other situations by use of the subcommand ERROR = followed by W, R, or WR.

Repeated measures designs are also handled under the general MANOVA subprogram. The very simplest situation, where there is only a single factor, say A with J levels and these are applied within each subject, is indicated by a WSFACTORS subcommand. Then the fundamental command structure is just

MANOVA WSFACTORS $= A(J)$ /

      PRINT $=$ SIGNIF(BRIEF)/

      ANALYSIS(REPEATED)/

This says that Factor A only is to be analyzed within subjects, following the requirements for a repeated measures design. The PRINT $=$ command included here is just a way to keep from getting a lot of irrelevant output from what the computer is treating as a problem in multivariate analysis. A DESIGN statement would be used only if there were also one or more *between-subject* factors, as there are in some designs.

In a situation like that discussed in Section 13.21, where two factors, A and B, are employed in the form of a factorial design *within* each subject, the necessary modifications would be

MANOVA WSFACTORS $= A(2)$ BY $B(2)$/

      WSDESIGN $= A, B, A$ BY B/

      PRINT $=$ SIGNIF(BRIEF)/

      ANALYSIS(REPEATED)/

This general outline is capable of extension to all sorts of experimental designs involving within-subjects and between-subjects factors.

In the BMDP series of programs, program 2V appears to be the simplest procedure that permits considerable flexibility for handling repeated measures designs. To carry out analysis of a simple repeated measures situation, in which Factor A, with four levels, is administered to each subject, one would first need an /INPUT and /VARIABLE statement, followed by /DESIGN paragraph something like this:

/ PROBLEM   TITLE IS ' . . . ' .

/ INPUT   VARIABLES ARE 4.

               FORMAT IS '(4F3.0)'.   (for example)

/ VARIABLE   NAMES ARE A1, A2, A3, A4.

/ DESIGN   DEPENDENT ARE 1 TO 4.

               LEVEL IS 4.

               NAME IS FACTA.

/END

(data list follows)

This program also permits more than one within-subjects factor, as illustrated in Section 13.21, and between-subjects factors may also be specified for more complex designs. A test of circularity is provided through the statement

  SYM yes.

under DESIGN. Program P4V goes even further and provides adjusted degrees of freedom by the Box method.

Randomized blocks can be analyzed in SAS either in PROC ANOVA or PROC GLM. In the simplest situation the latter uses statements such as

PROC GLM;

    CLASS LEVEL BLOCK;

    MODEL Y = LEVEL BLOCK;

Here, the dependent variable Y is stated by the model to depend only on the factor level and the block; therefore, the model is assumed to be additive, and the test of levels is carried out against residual.

The SAS program PROC ANOVA also contains a REPEATED statement that is introduced following the MODEL statement. Thus, for the simple situation with one within-subjects (repeated measures) factor A with four levels, we might have

PROC ANOVA;

    CLASS A;

    MODEL Y = A;

    REPEATED A 4;

Much the same setup is also used for REPEATED in PROC GLM. The PROC GLM also permits a test of the circularity assumption and will provide estimates of the Box correction to degrees of freedom.

EXERCISES

1. A large chain of retail stores was used in a study of the effects of management style on employee satisfaction. A random sample of eight stores having different managers was taken. Then, within each store each of a random sample of 10 employees was given a questionnaire, which provided a score on "employee satisfaction." Do store managers appear to account for variation in employee satisfaction within this chain of stores? The data were as follows:

			Manager				
A	B	C	D	E	F	G	H
20	16	19	26	23	27	20	19
21	17	18	25	21	23	20	18
19	16	18	20	22	23	20	21
18	15	20	25	21	25	21	21
20	17	16	26	23	28	23	19
23	19	20	21	22	29	22	19
20	18	17	21	22	30	22	18
21	16	17	25	24	26	21	20
19	14	18	25	20	25	21	18
20	16	18	27	21	28	21	19

2. Estimate the proportion of variance that seems to be attributable to managers in Exercise 1. Does it seem reasonable to suppose that managers alone could account for this much variance in employee satisfaction? Do stores differ in other ways besides managers, and could these other things be accounting for some of the differences in satisfaction?

3. An interstate highway runs across a large state. The highway patrol of the state monitors this highway and, among other things, gives tickets to speeders. However, does the patrol tend to cover some short sections more thoroughly than others? To find evidence for this question, the highway was divided into five-mile stretches, and a random sample of seven of these stretches was taken. Then, for each stretch, 10 days were chosen at random, and the number of speeding tickets given on each day recorded. These data are given below.

Highway stretches

A	B	C	D	E	F	G
2	13	7	17	9	4	12
0	5	8	4	3	5	4
5	10	6	9	4	6	5
5	4	6	5	6	16	4
15	7	16	5	7	10	6
9	5	2	6	4	4	5
3	6	7	8	13	5	7
21	8	4	5	5	6	5
4	7	5	6	5	8	6
2	10	2	3	10	1	9

Would you say that the highway patorl's coverage is spread evenly along this highway, or does it tend to vary over different stretches? Estimate the proportion of variance in tickets given that is accounted for by highway stretch. What does this result mean?

4. As part of an experiment on psychogalvanic reactions of dogs to unexpected visual stimuli, "artificial" human faces were constructed by combining elements of actual human features. There were six positions in which such elements might appear in a given face, and six possible elements in each position. Thus there were 6^6 or 46,656 different faces possible. As part of a pretest of the main experiment, eight of these possible stimulus faces were chosen at random, and each presented to a different group of two dogs. The results were as follows:

Stimuli

1	2	3	4	5	6	7	8
18	12	20	15	16	19	17	22
19	14	21	13	18	20	19	25

Test the hypothesis of no true between-stimulus variance.

5. From the results of Exercise 4 what proportion of the total variance in dogs' responses would you judge to be due to stimulus differences?

6. A random sample of five supermarkets was selected from a very large chain, and within each supermarket the number of items purchased on a single day by each of 10 customers selected at random was noted. The entries in the table below are the numbers of items purchased, arranged by store.

Test the hypothesis of no variance in number of items purchased attributable to stores. (Use $\alpha = .05$.)

		Stores			
1	2	3	4	5	6
13	11	18	45	26	18
16	14	27	48	29	36
19	17	51	27	32	14
16	17	57	33	32	14
19	25	9	18	38	28
22	26	18	27	41	31
19	24	19	36	18	14
19	26	28	40	21	25
13	17	30	28	12	28
6	23	34	10	25	30

7. In Exercise 6 test the hypothesis that variance between stores accounts for 50% or more of the total variance in number of items purchased.

8. An investigator was interested in the possible effects of the measured body-build or *somatotype* of a man, together with the ethnic background of that person, on his tendency to be overweight. Thus, a sample of five different somatotypes was chosen, along with a sample of four different ethnic-geographical groupings. Two men were found in each somatotype and ethnic combination, and then a typical week's calorie consumption of each was found. The table shows the average daily consumption in thousands of calories for each man. Does there seem to be a relation between somatotype, ethnic-geographic background, or their combination, and average daily calorie consumption? (Use $\alpha = .05$.)

		Ethnic-geographic group			
		A	B	C	D
Somatotype	I	4.21 3.38	4.37 3.29	2.91 3.35	3.22 2.46
	II	3.58 1.82	3.58 1.37	2.31 2.96	3.02 2.11
	III	3.30 1.89	3.01 2.34	2.01 1.85	2.17 1.22
	IV	3.37 2.21	2.49 3.86	3.36 3.10	2.78 2.79
	V	3.56 2.58	3.70 3.48	3.44 2.78	3.74 2.35

9. An experiment was carried out on the ability of children to solve puzzles of different types. Groups of six children each were formed on the basis of age, sex, and IQ. From a large number of such groups, a random selection of 10 such groups were actually used in the experiment described here. The dependent variable Y was the number of puzzles of a certain type, of 20 possible, actually solved by a child within a fixed time period. Within each matched group, each child got only one type of puzzle, but all types were presented within a given group. Children worked alone. The data are shown below, with *block* indicating a matched group of children. Carry out the appropriate analysis.

			Puzzle type			
	A_1	A_2	A_3	A_4	A_5	A_6
1	5	14	8	10	11	6
2	7	10	7	9	12	5
3	11	9	10	11	14	6
4	9	10	6	13	15	7
5	13	12	7	14	16	11
Block 6	7	9	8	6	11	5
7	10	11	8	12	13	8
8	4	8	5	7	9	4
9	14	13	11	15	17	12
10	9	9	8	11	14	9

10. An automobile company was interested in the comparative efficiency of three different sales approaches to one of their products. They selected a random sample of 10 different large cities, and then assigned the various selling approaches at random to three agencies within the same city. The results in terms of sales volume over a fixed period for each agency were as follows:

		Approach	
	A	B	C
1	38	27	28
2	47	45	48
3	40	24	29
4	32	23	33
5	41	34	26
City 6	39	23	31
7	38	29	34
8	42	30	25
9	45	31	26
10	41	27	34

Does there seem to be a significant difference ($\alpha = .05$) among these three sales approaches? Would you test for city differences here? Explain.

11. Suppose that the data given in Exercise 8, Chapter 12 had represented two factors, each with randomly selected levels. Carry out the appropriate analysis of variance and F tests.

12. In a pilot study preliminary to a large experiment there was interest in the proportion of variance attributable to a particular experimental factor. Four levels of this factor were selected at random, and three subjects were assigned at random to each. The data turned out as follows:

Level 1	Level 2	Level 3	Level 4
3.9	1.9	5.3	4.5
4.7	2.8	3.0	5.4
2.4	3.2	4.5	4.7

Establish 95% confidence limits for the true proportion of variance of the dependent variable accounted for by the experimental factor.

13. In Exercise 9, carry out Tukey HSD tests for the pairs of means of puzzle types (Factor A). (For this purpose assume that the assumption of circularity is true.)

14. In a given experiment, 12 randomly selected pairs of children, each pair matched with respect to age, sex, and intelligence, were used. The members of each pair were randomly assigned to two fixed experimental conditions. The data in terms of the dependent variable are given below:

Pair	Condition I	Condition II
1	18	23
2	19	34
3	27	26
4	25	23
5	28	29
6	15	30
7	17	14
8	29	41
9	36	37
10	25	24
11	46	45
12	31	32

Carry out an analysis of variance and F test on these data. Then analyze the data via a t test for matched pairs, and compare the result with that found in the preceding exercise. Do the two methods agree as they should?

15. In an experiment on paired-associate learning, eight randomly chosen subjects were presented with three different lists of 35 pairs of words to learn. Each subject was successively given the three lists in some randomly chosen order. The score for a subject was the number of pairs correctly recalled on the first trial. Do these three lists seem to be significantly different ($\alpha = .05$) in their difficulty for subjects?

		List A	List B	List C
	1	22	15	18
	2	15	9	12
	3	16	13	10
	4	19	9	10
Subject	5	20	12	13
	6	17	14	12
	7	14	13	10
	8	17	19	18

16. In an experiment on attention, eight subjects chosen at random were utilized. Each subject was given five pages of printed matter, where the pages were in a randomly assigned order. Each page was of a different level of reading difficulty, but each page contained the same number of typographical errors. The task of any subject was to locate the typographical errors present on each page, and the dependent variable Y was the number of correct identifications of errors. The data were as shown below. Carry out the appropriate analysis, including the three-step procedure of Section 13.23, as required.

Pages

	A₁	A₂	A₃	A₄	A₅
1	76	70	73	64	67
2	79	65	75	68	63
3	74	67	72	63	65
Subjects 4	68	69	74	70	72
4	64	68	70	71	68
5	65	66	71	72	68
6	66	60	74	73	67
7	78	73	72	70	65
8					

17. A one-way, random-effects analysis of variance with 6 and 21 degrees of freedom gave an F value just significant at the .01 level for the hypothesis that $\sigma_A^2 = 0$. Test the hypothesis that $\rho_I \le .10$ against the alternative that $\rho_I > .10$.

18. Suppose that in Exercise 13, Chapter 12 both Factor A and Factor B had been sampled to produce the three actual levels used for each. How would this fact change the analysis? How would it change the assumptions that need to be made to carry out the tests?

19. Below is an analysis of variance summary table. Carry out F tests ($\alpha = .05$ throughout) and interpret the findings under

   (a) A fixed-effects model
   (b) A random-effects model (both variables).
   (c) Factor A fixed, Factor B random.

Source	SS	df
Factor A	386.0	5
Factor B	1,427.2	9
Interaction	2,699.6	45
Error	3,109.3	120
Total	7,622.1	179

20. Subjects were given a driving simulation task in which traffic situations were shown on a screen, and the subject steered, applied the brake, etc. exactly as though truly driving. Twelve subjects were selected at random for this experiment, where each was an experienced driver. Amid normal driving situations, seven emergency situations requiring a full stop were introduced, in a different randomly chosen order for each driver. The dependent variable was the number of elapsed seconds between the beginning of the emergency and the full stop by the driver. The data are show below. Analyze these data as appropriate.

Emergency situations

	A₁	A₂	A₃	A₄	A₅	A₆	A₇
1	.53	.50	.52	.41	.51	.38	.49
2	.55	.43	.51	.42	.52	.42	.41
3	.59	.47	.56	.45	.51	.42	.41
4	.50	.49	.47	.44	.60	.47	.55
5	.52	.54	.55	.48	.39	.56	.53
Subjects 6	.51	.56	.48	.39	.56	.40	.43
7	.57	.47	.59	.36	.57	.39	.45
8	.51	.48	.60	.45	.49	.47	.47
9	.53	.58	.65	.47	.58	.44	.49
10	.54	.52	.55	.49	.51	.43	.53
11	.50	.55	.53	.50	.52	.55	.51
12	.47	.49	.50	.48	.54	.47	.50

21. For the data in Exercise 8, see if you can estimate the proportion of variance accounted for by each of the two main factors, and by their interaction.

22. In a certain one-way random-effects design with $J = 11$ and $n = 4$, the mean square between treatments was 28.9 and the mean square within was 3.7. The mean square between had 10 degrees of freedom, and the mean square within had 33 degrees of freedom. Find the 95% confidence limits for ρ_I, variance accounted for.

23. What is the power of the test described in Exercise 22 if the alternative hypothesis that $\rho_I = .50$ is true? Use $\alpha = .05$.

24. In a study of human memory, subjects were given lists of 50 words to memorize. A priori, however, the experimenter felt that the four lists chosen (regarded as fixed treatments in the experiment) might differ considerably in difficulty. Therefore each of 10 subjects chosen at random was given all four lists in a randomly assigned order. The data, in terms of words correctly recalled, turned out as follows. Are the lists significantly different from each other?

Lists

	I	II	III	IV
1	15	21	24	20
2	27	35	32	34
3	26	28	30	30
4	38	41	40	39
5	14	22	19	23
Subjects 6	29	33	30	31
7	32	37	34	35
8	45	45	40	42
9	28	38	35	32
10	33	34	33	29

25. For the data shown in Table 13.16.1, how might you estimate the average squared effects of the tasks? How could you estimate the interaction variance? The variance for classes? How might these estimates be used to find the proportion of total variance each apparently accounts for in the experiment?

26. Repeat Exercise 5 of Chapter 12, but this time assume that the first factor (A) is fixed and the second factor (B) is random. What would be used for error in the test for Factor A? Factor B?

27. Use the data in Exercise 9 above to apply the Tukey test for nonadditivity. What implication does this test have for our ability to test for "blocks"?

Chapter 14

Problems in Regression and Correlation

So far in applying the different variations of the general linear model we have confined our attention to situations where the independent variable X represents a set of essentially *qualitative* distinctions. These distinctions have sometimes represented preformed groups of subjects, and at other times they have stood for different levels of some treatment actually administered by the investigator. The model adopted in these instances has treated the variable X strictly as an indicator, taking on only the values 0 and 1 depending on which of a set of groups is being specified.

Now we are going to consider situations in which the independent variable X takes on *any real number value*. Such a variable may, for example, represent the amount of some treatment that a subject received in an experiment, or it may be the score that the subject earned on some test. In short, we are now going to deal with quantitative independent, or predictor, variables as well as a quantitative dependent variable Y.

First, we are going to deal with situations where the variable Y is clearly *the dependent variable* (DV) or the score to be predicted from information about X, *the independent* or *predictor variable* (IV). This will be the model when we refer a little later to such asymmetric situations as "problems in regression." There we will find that the sampling scheme which a problem in regression assumes is similar to that which we have used for fixed-effects analysis of variance. That is, in a problem in regression the values of the IV X are treated as though they were deliberately preselected, whereas the accompanying values of the DV Y are free to vary, given X for any individual.

In a problem in regression, the focus tends to be on the *rule* for predicting Y values from known values of X, and, in particular on the weight given to values of X. That is, how does one weight the known IV value of X for an individual in order to arrive at the best possible prediction of the DV value of Y for that same individual? Furthermore, we may want to test the hypothesis that no such prediction is possible in the population of cases from which our sample was drawn.

Next, we will be concerned with situations where the focus is on the *extent* of the linear relationship, if any, between two variables X and Y, where neither is viewed as *the* dependent variable to be predicted. Rather, in such symmetric "problems in correlation" the role of X and Y can be reversed, and we can still arrive at meaningful or useful prediction rules and inferences about population relations. Furthermore, we will see that such problems in correlation generally assume a sampling model somewhat like that for the random-effects analysis of variance. That is, each individual included in the sample simply "brings along" a value of X and a value of Y, where each is free to vary given the value of the other variable.

Nevertheless, the basic mechanics are very much the same in both problems of regression and problems of correlation, and so we will begin by examining the descriptive statistics of regression and correlation which are common to both. Then we will turn to inferential methods, where, for the first time, the distinction between these two possible approaches will become important.

14.1 SIMPLE LINEAR RELATIONS

Problems in correlation and in regression are both concerned with these main questions:

1. Does a statistical relation affording some predictability appear between the variables X and Y?
2. How strong is the apparent statistical relation, in the sense of the possible predictive ability the relation affords?
3. Can an simple linear rule be formulated for predicting Y and X, and if so, how good is this rule?

The ordinary techniques we have studied heretofore apply to the first two of these questions, but the third is a new feature. In this chapter we are going to study the possibility of applying a simple bivariate *linear model as a rule for the prediction of Y from X*. We are going to act *as though* the true relation actually were a linear function, and, using a linear function rule, make predictions or bets about Y values from knowledge of X values. Then we are going to evaluate the *goodness* of this prediction rule in terms of how well one actually would do by predicting according to the rule. If the statistical relation actually is a function, then some rule exists that affords perfect prediction; for the usual statistical relation, no rule permits perfect prediction, but some function rule may nevertheless provide a good "fit" to the relation under study. Proceeding in this fashion gives us two important advantages: quite often we are able to achieve a fair degree of predictive ability by adopting a particular linear function rule, even though the true relation itself is not really a precise function. Second, by studying our errors using this rule, we gain information about how the rule might be made better and how the general form of the relation might be specified more adequately.

The model we will use to describe the relation between X and Y will be linear, just as in Chapters 10 through 13. However, unlike the model of ANOVA, the model we will employ here will relate two quantitative variables. That is, we assume that

$$y_i = a_0 + bx_i + e_i,$$

[14.1.1]

where, as before, a_0 is a constant that enters into the value of y_i for any indivudal i, b is a constant weight that applies to the value x_i, and e_i stands for error. Alternatively, we may write

$$y_i = y_i' + e_i$$

where y_i' stands for the prediction we make about the y score of individual i based on the x score for that individual, and e_i stands for the amount by which that prediction is in error. Note that if there were no errors, then

$$y_i' = a_0 + bx_i, \qquad\qquad\qquad\qquad\qquad\qquad [14.1.2]$$

which is strictly linear in the sense that a plot of all of the (x_i, y_i') pairs of values would fall along a straight line. Naturally, there is no law that says the relationship between values such as x_i and values such as y_i must be linear. The best description of how the x_i are related to y_i values in a given set of data might call for a very different mathematical function. Why, then, do we emphasize such linear rules or models?

The reasons for starting with linear rules for prediction are several: linear functions are the simplest to discuss and understand; such rules are often good approximations to other, much more complicated, rules; and we will find that in certain circumstances the only prediction rule that *can* apply is linear. However, do not jump to the conclusion that just because we deal first with linear prediction this is the only important way to predict, or that all real relationships must be more or less like linear functions. In succeeding chapters we will find that there are many other, nonlinear, function rules that might also be applied to a given problem.

Before we can turn to inferential methods appropriate to problems in regression or correlation, we need some terminology for describing a linear relation between variables X and Y in any given set of N observations, or sample of N cases. Thus, the succeeding few sections will deal with the descriptive statistics of regression.

14.2 THE REGRESSION EQUATION FOR PREDICTING Y FROM X

Just as we discussed how one could find and interpret the mean and variance as descriptions of particular aspects of a set of data, we will now turn to the problem of finding a linear rule that fits a given set of data as well as possible. For the moment, our interest is only in a specific set of data, the scores for some particular set of N observed individuals.

Imagine this kind of situation; a teacher of a large introductory college course is interested in the possible relationship between the high-school preparation in mathematics that a student has and success in the course. In a particular semester the teacher has a class of 91 students, and at the outset each student is asked the number of mathematics courses taken in high school (four years). The teacher weights these course in a routine way and assigns scores running from 2 through 8 to the students. Let us call these "mathematics scores" X.

The teacher, however, files these reports away and does not look at them until after the final examination in the course has been given. The actual raw scores on this examination will be called the variable Y. After both scores for each student are known, the teacher asks this question: "To what extent is there a linear relation between the *X* and the *Y* scores?" In other words, how well does a simple linear rule allow one to predict the *Y* score of a student drawn at random from this group, given the information about the *X* score? The problem is to find the best possible linear rule for predicting from these data, and then to evaluate the goodness of such a rule.

As noted in the previous section, since a linear rule is to be used for our predictions, this means a rule of the general form

$$y_i' = a + bx_i \qquad\qquad [14.2.1]$$

where *a* and *b* are constant values. (Here, and in the following, we will use the lower-case letters *a* and *b* when we are discussing a rule for predicting raw values, y', of the dependent variable *Y*. However, shortly we will also develop a rule for predicting standardized or *z* values on variable *Y*, and then we will use capital letters such as *A* and *B* to stand for the constants that apply in such a standardized prediction rule.)

The problem of prediction using a linear rule is shown graphically in Figure 14.2.1. Here the horizontal axis represents possible values of *x*, the vertical axis the possible values for *y*, and any point within the plane defined by these two axes is a pair of scores (x,y) that might be associated with any individual observation. Points that are in the functional relation $y' = a + bx$ lie along the straight line in the figure. The point at

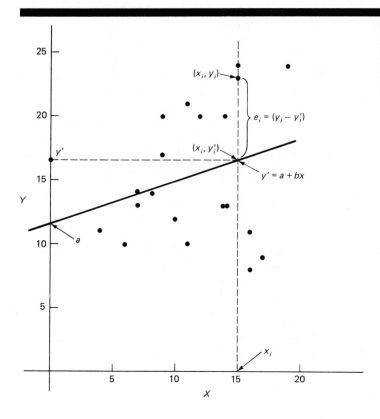

Figure 14.2.1.

Scatterplot of 20 *X* and *Y* values, and raw score regression line for predicting *Y* from *X*.

which the line intercepts the vertical axis indicates the value of a, and the slope of the line relative to the x axis reflects the value of b.

For the particular value of x_i indicated in the figure, note the two different points that are shown, (x_i, y_i') and (x_i, y_i). That is, the linear rule affords a *predicted value* y_i'. However, this need not correspond to the actual value y_i shown by individual i. The extent of the "miss" or "error" between the predicted and the actual value is represented by the vertical distance between the two points (x_i, y_i') and (x_i, y_i). We would like our prediction rule to be such that the fit between the predicted and actual scores is as close as possible.

You may be puzzled by the use of the term *prediction* in this context: the teacher actually *has* the two scores for each of the 91 students. Why not merely look at the Y score for any student? Actually the methods to be developed here will apply to situations where the user wants to go beyond the immediate data, and to forecast the Y or z_Y score for an individual for which this information is not already available. However, the basis for these methods is best seen if one deals only with one intact group of N cases, each having two scores, X and Y. For the moment, prediction consists of drawing one case at random from this particular group, noting the value of X and then finding a predicted value of Y by use of the linear rule.

The choice of an optimum linear prediction rule of the form $y_i' = a + bx_i$ boils down to the question of the best possible values for a and b, in the sense of providing accurate predictions. Any individual i is going to have an actual value y_i as well as the predicted value y_i' as found by our rule. A *deviation* or *error* occurs when these two values are not identical, and the magnitude of the error is given by

$$e_i = \text{amount of deviation or error} = (y_i - y_i') \ . \qquad [14.2.2]$$

Now there are two things that are usually held to be desirable in the selection of the constants a and b for a linear prediction rule. In the first place, we would like for the predictions to be *unbiased*, so that *on the average our predictions are right*. This implies that the mean predicted value $\bar{y}'$ should be equal to the mean of the actual values, or $\bar{y}$. Second, we would like the average of the squared errors in prediction to be as *small as possible*. (This is another instance of the criterion of least squares, which we have already encountered in Chapters 4 and 10.) That is, we want to make

$$S_e^2 = \sum_i (y_i - y_i')^2/N \qquad [14.2.3^*]$$

to take on its minimum value.

We will show that if the predictions are to be both unbiased and "best" from the least squares point of view, the constants a and b should be found from

$$a = \bar{y} - b(\bar{x}) \qquad [14.2.4^*]$$
$$\text{and } b = \text{cov}(x,y)/S_x^2 \ .$$

(Please understand that in the following we are able to use elementary algebra to solve this problem only because the problem situation is very simple here. In actual practice, the methods of the differential calculus would be used to solve this or any other least squares problem.) The required value of the intercept a is found by taking the mean of the predicted values, or $\bar{y}'$, and simply setting it equal to $\bar{y}$.

$$\bar{y}' = \sum_i (a + bx_i)/N = a + b\bar{x} \qquad [14.2.5]$$

Hence, since we wish $\bar{y}' = \bar{y}$, the required value of a is

$$a = \bar{y} - b\bar{x} , \qquad [14.2.6^*]$$

the mean of the y values minus b times the mean of the x values.

Now we will substitute the value for a we have just found as we show that $b = \text{cov}(x,y)/S_x^2$ is the least squares solution for b. This can be done as follows: First take

$$S_e^2 = \sum_i (y_i - y_i')^2/N$$

$$= \sum_i [(y_i - \bar{y}) - b(x_i - \bar{x})]^2/N \qquad [14.2.7]$$

$$= \sum_i [(y_i - \bar{y})^2 + b^2(x_i - \bar{x})^2 - 2b(x_i - \bar{x})(y_i - \bar{y})]/N$$

$$= S_y^2 + b^2 S_x^2 - 2b\, \text{cov}(x,y) .$$

Here $\text{cov}(x,y)$ is just as defined in Appendix B and used in Chapter 8. Let us define

$$b = d + c$$

where

$$d = \text{cov}(x,y)/S_x^2,$$

so that

$$dS_x^2 = \text{cov}(x,y),$$

and where c stands for *any* real number. Then, on substituting and rearranging we have

$$S_e^2 = S_y^2 + (c + d)^2 S_x^2 - 2(c + d)\, \text{cov}(x,y) \qquad [14.2.8]$$
$$= S_y^2 - d\, \text{cov}(x,y) + c^2 S_x^2 .$$

Now, for what value of c is this expression at its minimum? Notice that c is in the expression *only* as a squared term, so that, other things being constant, the entire expression is at its smallest when $c = 0$. However, this is equivalent to saying that the average squared error is at its minimum when

$$b = d = \text{cov}(x,y)/S_x^2 . \qquad [14.2.9]$$

Thus, we have established what we wished to show in the first place: The raw score regression equation for predicting y from x is given by

$$y_i' = \bar{y} + b_{y \cdot x}(x_i - \bar{x})$$

or $\qquad\qquad\qquad\qquad\qquad\qquad\qquad\qquad\qquad\qquad\qquad [14.2.10^*]$

$$y_i' = (\bar{y} - b_{y \cdot x}\bar{x}) + b_{y \cdot x}x_i$$

where the *raw score regression coefficient* or *slope* is given by

$$b_{y \cdot x} = \text{cov}(x,y)/S_x^2$$

(you can read the subscript symbol $y \cdot x$ as meaning "for y predicted from x.")

The regression coefficient is fairly simple to find directly from a set of x and y values for each of a set of N cases. Essentially these quantities are needed:

$$T_y = \text{the total of the } y \text{ values} = \sum_i y_i$$

$$T_x = \text{the total of the } x \text{ values} = \sum_i x_i$$

$$\sum_i x_i y_i = \text{the sum of each } x \text{ multiplied by its corresponding } y$$

$$\sum_i x_i^2 = \text{the } x \text{ values squared and summed.}$$

We define

$$SP_{xy} = \sum_i x_i y_i - T_x T_y/N \qquad\qquad [14.2.11^*]$$

and

$$SS_x = \sum_i x_i^2 - T_x^2/N \;, \qquad\qquad [14.2.12^*]$$

with

$$SS_y = \sum_i y_i^2 - T_y^2/N$$

as before. Notice that in ANOVA terminology

$$SS_x = SS_x \text{ total, and } SS_y = SS_y \text{ total.}$$

Then

$$\text{sample cov}(x,y) = SP_{xy}/N \qquad\qquad [14.2.13^*]$$

and

$$\text{sample } S_x^2 = SS_x/N. \qquad\qquad [14.2.14^*]$$

Therefore the sample regression coefficient is

$$b_{y \cdot x} = \text{cov}(x,y)/S_x^2 \qquad\qquad [14.2.15^*]$$

Or, for a little simpler version, take

$$b_{y \cdot x} = SP_{xy}/SS_x \;. \qquad\qquad [14.2.16^*]$$

(Incidentally, when two or more variables are under discussion each variable will have a sums of squares or SS value, and each pair of variables will have a sum of products or SP value. When we wish to refer to the entire set of these values it is useful to employ the symbol **SSCP**, standing for "the matrix of sums of squares and sums of (cross-) products."

When, as here, there are only two variables, X and Y, the **SSCP** matrix consists only of the following:

$$\mathbf{SSCP} = \begin{bmatrix} SS_x & SP_{xy} \\ SP_{yx} & SS_y \end{bmatrix}$$

or, since $SP_{yx} = SP_{xy}$,

$$\mathbf{SSCP} = \begin{bmatrix} SS_x & SP_{xy} \\ & SS_y \end{bmatrix}$$

We will have many occasions to use these terms in sections to come. Also, look at Sections C3 and D6.)

As we have just seen, the regression coefficient for predicting Y from X can always be found from the ratio of the sum of products, SP_{xy}, to the sum of squares for X, or SS_x.

For example, suppose that the teacher mentioned in Section 14.2 finds that the X and Y values in the data for 91 students are as given in Table 14.2.1. Let us use these values as the basis for constructing a linear regression equation for predicting Y, the scores on a college course examination, on the basis of the mathematics scores background X.

We begin by finding the two means, which turn out to be

$$\bar{x} = T_x/N = 381.5/91 = 4.19$$

and

$$\bar{y} = T_y/N = 2{,}169/91 = 23.84.$$

x	y	x	y	x	y	x	y
4	36	4	25	3.5	25	7.5	41
3.5	19	4	19	3.5	22	6.5	44
6	38	3	24	3	22	7.5	35
7	52	3	9	2.5	6	5	32
4	20	2	7	2	5	2	3
3.5	12	2	2	3	29	2.5	12
2	10	3.5	34	2.5	26	4.5	16
8	53	3	23	2	17	4	27
3	16	3.5	26	5	41	3.5	17
3	26	6	33	4	24	5.5	25
4.5	27	3.5	17	2.5	7	3.5	17
3	8	4	18	5	19	2.5	16
3	24	2.5	13	2	9	8	40
2.5	23	2.5	10	4.5	28	7.5	38
5.5	32	5	27	6.5	28	6	27
6.5	37	7.5	42	6	34	3.5	23
8	40	8	48	4	18		
4	19	3	26	3.5	23		
3.5	22	2.5	10	3.5	8		
4	35	2	22	6	46		
2.5	18	2	16	6.5	32		
4	25	4	30	7	37		
4	12	4	3	2	14		
2.5	18	6	20	5.5	25		
3.5	18	7	46	4	21		

Table 14.2.1

High-school mathematics scores (x) and final examination scores (y).

The SS_x value is then found from

$$SS_x \text{ total} = (4^2 + 3.5^2 + \cdots + 3.5^2) - (381.5^2)/91$$

$$= 275.51$$

It follows that the variance of variable X is

$$S_x^2 = 275.51/91 = 3.03.$$

Next the sum of products, SP_{xy}, is found by taking

$$SP_{xy} \text{ total} = [(4 \times 36) + (3.5 \times 19) + \cdots + (3.5 \times 23)]$$

$$- (381.5 \times 2{,}169)/91 = 1{,}489.71$$

so that

Then the value of the regression coefficient, $b_{y \cdot x}$, is found to be

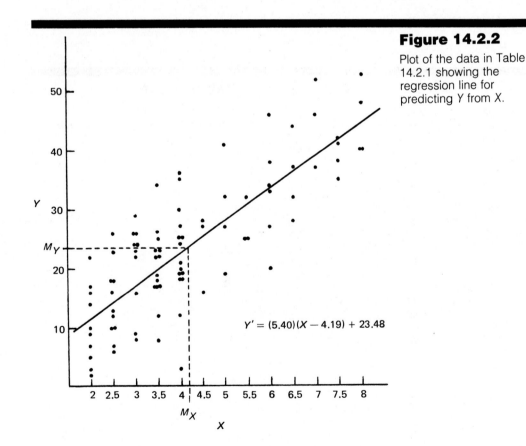

Figure 14.2.2

Plot of the data in Table 14.2.1 showing the regression line for predicting Y from X.

$$Y' = (5.40)(X - 4.19) + 23.48$$

$$b_{y \cdot x} = SP_{xy} \text{ total}/SS_x \text{ total}$$

$$= \text{cov}(x,y)/S_x^2$$

$$= 5.40$$

Using this regression coefficient we find that the raw score regression equation for predicting Y from X is

$$y_i' = 23.84 + (5.4)(x_i - 4.19).$$

For instance, given that an individual has a high-school mathematics score of 5, the teacher can predict that the score on the course examination is

$$y_i' = 23.84 + (5.4)(5 - 4.19)$$

$$= 28.21.$$

Figure 14.2.2 shows a plot of the regression equation, together with the actual (x, y) pairs for these data. Notice that although the actual pairs of scores do tend to cluster about the predicted (x, y') pairs, there is nevertheless some "scatter" of the actual Y scores about the predicted value for each X.

14.3 THE STANDARD ERROR OF ESTIMATE FOR RAW SCORES

We have just seen how a regression equation can be found for predicting the value of an individual's y score from the known value of x for that individual. On the other hand, we also saw that we stand a considerable chance of being wrong in any such prediction, and that the amount of the error that will be made for any individual i can be expressed as

$$e_i = (y_i' - y_i) . \qquad [14.3.1]$$

Our method for finding the values of the intercept a and the slope b required by the regression equation was to minimize the average of the squares of such errors.

However, the *average squared error in prediction* is an important statistic in its own right, and sometimes goes under the name *the variance of estimate:*

$$\text{sample variance of estimate} = S_{y \cdot x}^2 \qquad [14.3.2^*]$$

$$= S_y^2 - b_{y \cdot x}^2 S_x^2.$$

Thus, for example, the data in Table 14.2.1, which gave us values

$$S_x^2 = 3.03 , S_y^2 = 135.65$$

and

$$b_{y \cdot x} = 5.40$$

lead to a sample variance of estimate value of

$$S_{y \cdot x}^2 = 135.65 - (5.40^2 \times 3.03)$$

$$= 47.30 .$$

That is, if the y value for each of the 91 individuals in the sample were predicted, using the x score and the regression equation we have just found, then the mean of the errors would be zero, and the variance of the errors would be equal to 47.30, the sample variance of estimate.

The square root of the sample variance of estimate is usually called *the sample standard error of estimate,* and should be reported for any regression equation. For our example, the sample standard error of estimate is:

$$S_{y \cdot x} = \sqrt{47.30}$$

$$= 6.88 \; .$$

Speaking fairly loosely, one can say that the sample standard error of estimate shows about how "off" one can reasonably expect to be in a prediction made for any case in a given set of data. We will refine this definition a bit more in the sections to come.

14.4. THE REGRESSION EQUATION FOR STANDARDIZED SCORES

It might be that the teacher in our example is not especially interested in predicting the raw Y score of a student so much as in the *relative* performance of the student in terms of Y. That is, the teacher would like to be able to predict the standard score z_Y, given by

$$z_Y = \frac{y - \bar{y}}{S_Y}.$$

This prediction is to be based on the standard score z_X, where

$$z_X = \frac{x - \bar{x}}{S_X}.$$

Since a linear rule is to be used for this prediction, this means a rule of the form

$$z_Y' = A + B z_X \qquad\qquad [14.4.1]$$

where B and A are constants. This situation is portrayed in Figure 14.4.1. The least squares criterion requires that we choose A and B in such a way that the average squared error over individual predictions be as small as possible. Thus, given N individuals i, we want to choose A and B so as to make

$$\frac{\sum_i (z_{Yi}' - z_{Yi})^2}{N} \qquad\qquad [14.4.2]$$

have its minimum possible value.

Exactly the same least squares argument can be applied to the standardized as to the raw score regression equation, with the result that for standardized values

$$A = \text{intercept} = 0 \qquad\qquad [14.4.3]$$

and

$$B = \text{regression coefficient or slope}$$

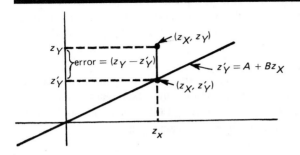

Figure 14.4.1

Plot of a linear regression equation for the prediction of the standard score on Y from the standard score on X.

$$= \frac{\sum_i z_{Xi} z_{Yi}}{N} = r_{xy}. \tag{14.4.4}$$

This value of B for standardized scores is actually the **correlation coefficient** (or Pearson product-moment correlation coefficient) r_{xy}, about which we will have much to say. By definition,

$$r_{xy} = \frac{\sum_i z_{Xi} z_{Yi}}{N} \tag{14.4.5}$$

$$= \text{sample covariance of } z_X \text{ and } z_Y \text{ values}.$$

Notice that the correlation coefficient r_{xy} is actually the *standardized covariance* between two variables x and y. Thus, following the definition of the regression coefficient that we developed in Section 14.2, we have

$$b_{y \cdot x} = \text{cov}(x,y)/S_x^2$$

becoming, for standardized variables,

$$B_{y \cdot x} = r_{xy}/1$$

since the variance of the standardized X variable is 1.00. Thus, when both variable X and variable Y are in standardized form, the regression equation is very simple:

$$z'_{yi} = r_{xy} z_{xi}. \tag{14.4.6}$$

This shows that the correlation coefficient acts as a rate of exchange between a standardized value of x and a standardized predicted value of y.

It is a simple matter to find the value of the correlation coefficient if the value of the regression coefficient for raw scores has already been found. We can simply take

$$r_{xy} = b_{y \cdot x} S_x / S_y. \tag{14.4.7}$$

For the data of Section 14.2, where we have already found the regression coefficient to be 5.40, the standard deviation of X to be 1.74, and that for Y to be 11.65, then

$$r_{xy} = 5.40(1.74)/11.65$$

$$= .81 ,$$

the correlation coefficient is .81. This means that to predict the standardized value on Y, or z_y, for any individual, we take the standardized value on X, or z_x, and multiply by 0.81. That is,

$$z'_y = (.81)z_x .$$

For our teacher's class of 91 students, one standard deviation's worth of mathematics preparation is "worth" 81% of a standard deviation of course examination score.

Once again, the notion of the mean squared error (Eq. 14.4.2) is an important one in its own right, and we will give it a special symbol and name. Let

$$S^2_{z_Y \cdot z_X} = \frac{\sum_i (z'_{Yi} - z_{Yi})^2}{N}$$
$$= 1 - r^2_{xy}$$

[14.4.8]

be called the **sample variance of estimate for standard scores.** This variance of estimate reflects the *poorness* of the linear rule for prediction of standard scores, the extent to which squared error is, on the average, large. Most often, however, this index is discussed in terms of its positive square root

$$S_{z_Y \cdot z_X} = \sqrt{1 - r^2_{xy}},$$

[14.4.9]

which is called the **sample standard error of estimate for predicting standard scores.**

For our example, once again, the standard error of estimate for the prediction of standardized Y values is

$$S_{z_Y \cdot z_X} = \sqrt{1 - r^2_{xy}}$$
$$= \sqrt{1 - (.81)^2}$$
$$= .586 .$$

Therefore, speaking rather loosely once again, if we actually predict the standardized value on Y for an individual by use of the standardized value of X, then we stand a very good chance of being "off" by about .586 in either direction.

Obviously, there is a close connection between the size of the standard error of estimate in a sample and the value of r in the regression equation: **the larger the absolute value of r_{xy}, the smaller is the standard error of estimate.** Now we turn to some interpretations of the index r_{xy}.

14.5 SOME PROPERTIES OF THE CORRELATION COEFFICIENT IN A SAMPLE

The connection between the variance of estimate and the correlation coefficient shows us at once that r_{xy} can take on values *only* between -1 and 1. Notice that the variance of estimate, being an averaged sum of squares, can be only a positive number or zero. If r_{xy} were less than -1, or greater than $+1$, then the variance of estimate could not be positive. Hence, $-1 \leq r_{xy} \leq 1$.

What does it mean when r_{xy} is exactly zero? When this is true, one predicts $z'_Y = 0$, corresponding to the mean of Y, *regardless* of the value of X; for any X, the mean of Y is the best linear prediction when the correlation is zero. Furthermore,

$$S^2_{z_Y \cdot z_X} = 1$$

for $r_{xy} = 0$. This means that when the correlation coefficient is zero, the variance of estimate for standard scores is exactly the same as the variance of the standard scores z_Y with X *unspecified*. Thus, when r_{xy} is zero, predicting by the linear rule *does not* reduce the variability of z_Y below the variability present when z_X is unknown. In short, the fact that r_{xy} is zero implies that if a predictive statistical relation exists for the set of data, it is not linear, and the linear rule gives no predictive power.

On the other hand, when r_{xy} is either $+1$ or -1, the variance of error in prediction is zero, so that each prediction is exactly right. These ultimate limits for r_{xy} can occur only when X and Y *are* functionally related, *and* follow a linear rule.

All intermediate values of r_{xy} indicate that some prediction is possible using the linear rule, but that this prediction is not perfect, and some error in prediction exists. Any value of r_{xy} between 0 and 1 in absolute magnitude indicates either that the relationship is not functional or that if it is a function, the rule is not exactly linear, although a linear rule does afford some predictability.

In the regression equation for standard scores the correlation coefficient plays the role of converting a standard score in X into a predicted standard score in Y. The correlation coefficient can be said to be "the rate of exchange," the value of a standard deviation of X in terms of *predicted* standard deviation units of Y.

This is easily seen from the plot of a standardized regression equation, such as that in Figure 14.5.1. Regardless of the "true" relationship between variables X and Y, z_x and z'_y will plot as a straight line. Given a particular z_X, as shown in the figure, then the ratio of the predicted value to the known standard score value for X,

$$\frac{z'_Y}{z_X} = \tan \theta, \qquad [14.5.1]$$

is the trigonometric tangent of the angle θ. However, this ratio is also r_{xy}, so that

$$r_{xy} = \tan \theta, \qquad \begin{cases} 0° \le \theta \le 45° \\ 135° \le \theta \le 180°. \end{cases} \qquad [14.5.2]$$

The tangent of the angle θ gives the *slope* of the regression line for predicting standard scores, and the value of r_{xy} corresponds to the slope of this line. When r_{xy} is between 0 and 1, the angle θ in such a plot must be between 0 and 45°. For an r_{xy} between 0 and -1, the angle θ must be between 135 and 180°, so that in this instance a high standing on X yields a prediction of low standing on Y, and vice versa.

This point can be reinforced by considering the raw score regression coefficient once again, but this time written in terms of the correlation, r_{xy}, and the two standard deviations, S_x and S_y. From Eq. 14.4.7 we see that the raw score regression coefficient or regression weight, $b_{y \cdot x}$, can also be written as

$$b_{y \cdot x} = \frac{r_{xy} S_y}{S_x}. \qquad [14.5.3]$$

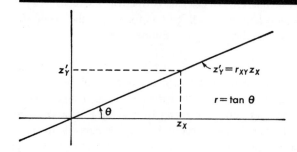

Figure 14.5.1

The correlation coefficient as the slope of the regression line for predicting the value of z_Y from z_X.

This is also true because

$$r_{xy} = \text{cov}(x,y)/(S_x S_y)$$

$$r_{xy} S_y = \text{cov}(x,y)/S_x.$$

On dividing both sides of this second equation by S_x, we have

$$r_{xy} S_y/S_x = \text{cov}(x,y)/S_x^2 = b_{y \cdot x} .$$

Thus, the raw score regression coefficient also acts as a *rate of exchange* in the regression equation, but in two ways: not only does the $b_{y \cdot x}$ value translate the standardized standing on X into a standing on Y, it also uses the ratio of S_y to S_x to convert a unit of measurement in terms of X into the unit of measurement in terms of Y, as it must when we are using raw X values to predict raw values on Y.

A closely related point is that the correlation coefficient is a dimensionless or "pure" measure of the linear association between two variables X and Y. This is true because the definition of the correlation coefficient is based on standardized values for each of the variables. Thus, X can be altered by adding (or subtracting) a constant to each value, and by multiplication (or division) by a positive number, and the standardized values, and hence the correlation coefficient, will be unaffected. Variable Y may also be transformed by similar operations without affecting the the correlation coefficient. On the other hand, multiplying either or both variables by different constant numbers *will* affect the raw score regression coefficient (but *not* the correlation coefficient.)

There is, incidentally, a large literature devoted to special-purpose correlations of various kinds. Among these are the *point-biserial correlation,* designed for the situation in which there are only two independent variable or X values, each accompanied by an array of Y values as mentioned in Section 8.12. Another historically important formulation of the correlation coefficient is the *tetrachoric correlation,* computed when both X and Y are arranged into widespread dichotomous classes. Still another, the *phi coefficient,* is mentioned in Chapter 18. In addition, various corrections to the correlation coefficient may be made to allow for various kinds of groupings or curtailments of the X and Y values (One example is given in Section 14.10). We will not devote further space to these matters here. A modern summary of some of these techniques may be found in Guilford and Fruchter (1978), however.

14.6 THE PROPORTION OF VARIANCE ACCOUNTED FOR BY A LINEAR RELATIONSHIP

Just how good is a linear rule for predicting values of Y from values of X in a given sample? In order to answer this question, we need an index of the *strength of linear relationship* between X and Y in the data. We can approach this problem as follows:

We already know that the variance of any set of standard values z_Y has to be 1.00. However, what is the variance of the *predicted* values z'_{Yi}? We can think of this as variance accounted for, variability among the z_{Yi} values for different observations i *directly attributable to the fact that they have different z_X scores*. When we take the variance of the predicted z'_Y values we have

$$S_{z'_Y}^2 = \frac{\sum_i (z'_{Yi})^2}{N} - \left(\sum_i \frac{z'_{Yi}}{N} \right)^2. \qquad [14.6.1]$$

First of all notice that

$$\sum_i \frac{z'_{Yi}}{N} = \sum_i \frac{r_{xy}\, z_{Xi}}{N} = 0,$$

since the sum of all of the z_{Xi} values must be zero. Thus, the second term on the right, the squared mean of all the predicted z'_Y values, is itself zero. Then we find

$$\sum_i \frac{(z'_{Yi})^2}{N} = \sum_i \frac{r_{xy}^2\, (z_{Xi})^2}{N} = r_{xy}^2. \qquad [14.6.2]$$

The variance of the predicted z'_Y values, or the *variance explained by the z_{Xi} values, is thus r_{xy}^2.*

Since the variance of the original z_Y values has to be 1.00, then if we take the ratio of the variance of the predicted values to the total variance of the z_Y values, we find that

$$\left(\begin{array}{l} \text{proportion of variance explained} \\ \text{by linear regression of } Y \text{ on } X \end{array} \right) = \frac{S_{zY'}^2}{S_{zY}^2} = r_{xy}^2, \qquad [14.6.3]$$

the proportion of variance in Y accounted for, or explained, by X is given by r_{xy}^2. Thus, an index of the "goodness" of the linear rule for predicting X from Y is given by r_{xy}^2, the proportion of variance in X accounted for by Y under the linear rule. The index r_{xy}^2 is usually termed the **coefficient of determination.** You can always think of r_{xy}^2 as representing the strength of *linear* relationship in a given set of data. Furthermore, although it was convenient to discuss the proportion of variance accounted for initially in terms of z values, r_{xy} is the same either for z scores or for raw values so that these interpretations are valid for r_{xy}^2 whether we are discussing standardized or raw scores.

Thus, if the value of a correlation coefficient is .50 (positive or negative in sign) then some .25 of the variability in Y is accounted for by specifying the linear rule and X. If the correlation is .80, then 64% of the variance in Y is accounted for in this way. A correlation of positive or negative 1.00 means that 100% of variability in Y can be accounted for by the linear rule and X, but if $r_{xy} = 0$, none of the variability is thereby accounted for. All in all, not the correlation coefficient per se but the *square* of the correlation coefficient informs us of the "goodness" of the linear rule for prediction.

14.7 THE IDEA OF REGRESSION TOWARD THE MEAN

The term *regression* has come to be applied to the general problem of prediction by use of a wide variety of rules, although the original application of this term had a very specific meaning. The term *regression* is a shortened form of **regression toward the mean in prediction.** The general idea is that **given any standard score** z_X, **the best linear prediction of the standard score** z_Y **is one relatively nearer the mean of zero than is** z_X.

This can be illustrated quite simply from our standardized regression equation 14.4.6. Suppose that an individual has a standard score z_X of 2. Also suppose that the regression equation we have found for the group to which the individual belongs is

$$z_Y' = .5z_X.$$

Then we predict this individual to have a z_Y score of 1, since

$$z_Y' = .5(2) = 1.$$

Notice that we predict the individual to fall relatively *nearer* the mean on Y than on X. That is, we predict in accordance with *regression toward the mean*. For another set of data, the regression equation might be

$$z_Y' = -.75z_X.$$

Now in this instance, suppose that the z_X for some randomly selected individual were 1.5. Then

$$z_Y' = (-.75)(1.5) = -1.125.$$

Since the correlation coefficient is negative, the prediction is that this individual falls *below* the mean of Y, given that the X value is above the mean.

However, in absolute terms, we again predict a standing relatively *closer* to the mean on Y than on X.

This principle of predicting relatively closer to the mean, or regression toward the mean, is a feature of any linear prediction rule that is best in the "least squares" sense. The idea is that if we are going to use such a linear rule for prediction, then it is always a good bet that an individual will fall *relatively closer to the group mean on the thing predicted than on the thing actually known*. This does *not* imply that the actual Y value *must* fall relatively closer to the mean than does the value of X, however, but only that our *best bet* is that it will do so. Although regression toward the mean does occur in many contexts, it is not some immutable law of nature. Rather it is, at least in part, a statistical consequence of our choosing to predict in this linear way, using the criterion of least squares in the choice of a rule.

14.8 THE REGRESSION OF z_X ON z_Y

In a true correlation problem, nothing makes it necessary to think of X as the independent variable, or the value somehow known first or predicted from. It is entirely possible to consider a situation where one might want to predict z_X from knowing z_Y.

What does this do to the linear prediction rule, to the correlation coefficient, and so on?

In the first place, the same argument used in Section 14.4 shows that for predicting z_X from z_Y,

$$z'_{Xi} = r_{xy} z_{Yi} \qquad [14.8.1]$$

where the correlation coefficient is, just as before, given by Eq. 14.4.4.

It is tempting to ask why we do not just solve the original regression equation for z'_{Yi} in terms of z_{Xi} in order to get

$$z_{Xi} = z'_{Yi}/r_{xy}.$$

However, recall what the symbols z'_Y and z'_X actually represent. These are *predicted* values and do not necessarily symbolize the actual values of z_Y and z_X at all. Solving Eq. 14.4.6 for z_X *might* be useful if one wanted to know the value of z_X *known*, given that z'_Y were the predicted value. (This is sometimes called the problem of ''inverse regression,'' and can arise in situations where calibration of x values is desired in terms of predicted y values.)

The form of the regression equation that is used (Eq. 14.4.6 or 14.8.1) depends strictly on which variable, X or Y, is designated as the independent variable, or the thing assumed to be known first in a prediction situation.

In prediction of X from Y, the sample variance of estimate for standard scores is

$$S^2_{z_X \cdot z_Y} = 1 - r^2_{xy} \qquad [14.8.2]$$

(notice the reversal in subscripts from when z_Y is predicted from z_X). The proportional variance in X accounted for by Y is, once again, r^2_{xy}.

This brings up the point that the correlation coefficient is a *symmetric* measure of linear relationship. So long as we are talking about the *correlation coefficient alone,* it is *immaterial* which we designate as the independent and which the dependent variable; the measure of possible linear prediction is the same. However, when we deal with the actual regression equations themselves, this symmetry is not usually present. As we have already seen, it does make a difference whether you are predicting Y from X or X from Y when it comes to finding the regression equations and errors of estimate for *raw* scores.

Notice that when no specification is put on which is to be regarded as the independent or predictor variable, there are two possible raw score regression coefficients, $b_{Y \cdot X}$ and $b_{X \cdot Y}$, and that

$$\sqrt{b_{Y \cdot X} b_{X \cdot Y}} = r_{xy}, \qquad [14.8.3]$$

the square root of the product of the two regression coefficients (i.e., their geometric mean) is the correlation coefficient.

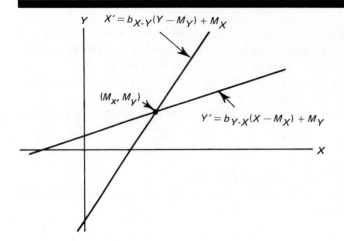

Figure 14.8.1

Plot of the two regression lines for predicting Y from X and X from Y.

Once again this points up the fact that r_{xy} is a symmetric measure of relationship. This is why we can use a single symbol, such as r_{xy} to stand both for the correlation of X with Y and the correlation of Y with X, and $r_{yx} = r_{xy}$.

Figure 14.8.1 shows the two raw score regression lines that might apply to a given set of data.

When prediction of raw scores is to be carried out, the **sample variance of estimate** for predicting Y from X is

$$S^2_{Y \cdot X} = S^2_Y(1 - r^2_{xy})$$
[14.8.4*]

and the **sample standard error of estimate** is

$$S_{Y \cdot X} = S_Y \sqrt{1 - r^2_{xy}}.$$
[14.8.5*]

Similarly, the sample variance of estimate for predicting X from Y is

$$S^2_{X \cdot Y} = S^2_X(1 - r^2_{xy})$$
[14.8.6*]

and the sample standard error of estimate is

$$S_{X \cdot Y} = S_X \sqrt{1 - r^2_{xy}}.$$
[14.8.7*]

Finally, remember that the proportion of variance accounted for by linear relationship (either in predicting Y from X or X from Y) is r^2_{xy}, just as for standardized scores.

14.9 COMPUTATIONAL FORMS FOR r_{xy} AND $b_{Y \cdot X}$

Although the sample correlation coefficient was actually defined in Section 14.4 as a summed product of standard scores (Eq. 14.4.5), in practice the index is seldom calculated in this way. An equivalent raw score computational form will now be found. We will show that the raw score form of the correlation coefficient is

$$r_{xy} = \frac{\left(\sum_i x_i y_i / N \right) - \overline{xy}}{S_x S_Y}. \qquad [14.9.1^*]$$

Starting with the definition of r_{xy} given by Eq. 14.4.5, and substituting the raw score equivalents of z_{Xi} and z_{Yi}, we have

$$r_{xy} = \frac{\sum_i (x_i - \bar{x})(y_i - \bar{y})}{NS_X S_Y}$$

$$= \frac{1}{NS_X S_Y} \left[\sum_i x_i y_i - \sum_i x_i \bar{y} - \sum_i y_i \bar{x} + N\overline{xy} \right]$$

$$= \frac{1}{NS_X S_Y} \left[\sum_i x_i y_i - 2N\overline{xy} + N\overline{xy} \right] \qquad [14.9.2^*]$$

$$= \frac{\left(\sum_i x_i y_i / N \right) - \overline{xy}}{S_X S_Y}.$$

Still another computing form for r_{xy} that is often useful when work is being done on a pocket or desk calculator is

$$r_{xy} = \frac{N \sum_i x_i y_i - \left(\sum_i x_i \right) \left(\sum_i y_i \right)}{\sqrt{\left[N \sum_i x_i^2 - \left(\sum_i x_i \right)^2 \right] \left[N \sum_i y_i^2 - \left(\sum_i y_i \right)^2 \right]}}. \qquad [14.9.3^*]$$

$$= \frac{SP_{xy}}{\sqrt{(SS_x)(SS_y)}}.$$

Given the correlation coefficient, it is then possible to find the sample raw score regression coefficient $b_{Y \cdot X}$ directly from

$$b_{Y \cdot X} = r_{xy} \left(\frac{S_Y}{S_X} \right). \qquad [14.9.4^*]$$

However, for many problems it is desirable to calculate $b_{Y \cdot X}$ without first finding r_{xy}. This may be done most simply by taking

$$b_{Y \cdot X} = \frac{\sum_i x_i y_i - N\overline{xy}}{\sum_i x_i^2 - N\bar{x}^2} \qquad [14.9.5^*]$$

or

$$b_{Y \cdot X} = \frac{N \sum_i x_i y_i - (T_x)(T_y)}{N \sum_i x_i^2 - (T_x)^2} = \frac{SP_{xy}}{SS_x}. \qquad [14.9.6^*]$$

A great many hand-held or pocket calculators of the "scientific" type are preprogrammed to provide the value of the correlation coefficient, along with b and other
values. All one has to do is to enter each of the pairs of (x_i, y_i) values, and the calculator
produces r_{xy}, in addition to the two means and variances, a and $b_{Y\cdot X}$ if desired. Many
also include the feature of giving the predicted y' value in terms of some entered x
value, and x' from y. The ease of use and wide availability of these inexpensive calculators has taken much of the former labor out of doing correlation and regression computations. Of course, problems of any size will now be done by computer.

Finally, it is also useful to remember that the proportion of variance accounted for,
or r_{xy}^2, can be found directly from the regression coefficient and the X and Y variance
values:

$$r_{xy}^2 = b_{y\cdot x}^2 S_X^2 / S_Y^2$$

$$= \frac{\left(\sum_i x_i y_i - T_x T_y / N \right)^2}{\left(\sum_i x_i^2 - T_x^2 / N \right) \left(\sum_i y_i^2 - T_y^2 / N \right)}. \qquad [14.9.7^*]$$

14.10 PROBLEMS ENCOUNTERED IN CALCULATING REGRESSION AND CORRELATION COEFFICIENTS FOR SAMPLE DATA

A few comments are in order about the propriety of computing correlations and regression equations for sample data. In some of the older literature in social and behavioral
research several misleading ideas appear about when it is proper to compute these
indices and equations as *descriptive* statistics. We need to be clear about these matters.
It is *not* necessary to make any assumptions at all about the form of the distribution,
the variability of Y scores within X columns or "arrays," or the true level of measurement represented by the scores in order to employ linear regression and correlation
indices to *describe* a given set of data. So long as there are N distinct cases, each having
two numerical scores, X and Y, then the *descriptive* statistics of correlation and regression may be used. In so doing, we describe the data *as though* a linear rule were to be
used for prediction, and this is a perfectly adequate way to talk about the tendency for
these numerical scores to associate or "go together" in a linear way *in these data*.

The confusion has arisen because in inference about true linear relationships in populations, and in some applications of regression equations to predictions beyond the
sample, assumptions do become necessary, as we shall see presently. However, one
may apply correlation techniques to any set of paired-score data, and the results are
valid descriptions of two things: the particular linear rule that best applies, and the
goodness of the linear prediction rule as a summary of the tendency of Y scores to differ
systematically with differences in X *in these data*.

It is true, however, that the possible values that r_{xy} may assume depend to some
extent on the forms of *marginal* distributions of both X and Y in the joint data table.
Unless the distributions for X and Y are similar in form, it is not necessarily true that
the obtained value of r_{xy} can range between -1 and $+1$. In fact, it is possible to

produce examples where the forms of the distributions of X and Y are very different, and the maximum possible absolute value of r_{xy} is only .3 or less. The fact that the value of the correlation coefficient can, in principle, range from -1.00 to $+1.00$ does not imply that the opportunity for a linear predictive relation to appear in a sample has nothing to do with the marginal distributions of the X and Y scores. In the same way, the actual possible range of the correlation in a population depends on the marginal distributions. This fact has very important implications for those who study the patterns of correlations in multivariate studies, and particularly for those who must employ some variant on, or approximation to, the correlation coefficient in such studies. An informative discussion of these issues is given in Carroll (1961).

A commonly encountered and troublesome point in this connection with a correlation or regression situation occurs where there is a restriction on the range of either variable X or Y. (The problem is usually discussed in terms of restrictions on the IV or X, and we will do so here.) That is, relative to the settings when one would really like to use the regression equation or the correlation coefficient, the range of X values that can actually appear in the data is limited.

Thus, for example, an employer may be interested in predicting the performance of individuals hired to be computer data-entry operators. Among other things, a test of general educational achievement X may be used as a predictor of job performance Y. It stands to reason that there should be at least a moderate correlation between X and Y. However, the chances are very good that this correlation will be less than you would expect it to be. The reason is that people with very low educational achievement will not be hired (or perhaps even apply) in the first place. Thus the distribution of X values will be truncated, meaning that the X scores that can be paired with Y performance scores in these data will be much more limited than in the general population. This will tend to produce a relatively lowered correlation (in absolute terms) as compared with the unconstrained or untruncated situation.

In those instances where the standard deviation of the unrestricted X distribution is actually known, then a correction formula may be used to estimate the value of the correlation that should have resulted if the full range of X values had been present. That is, let

$_t r_{xy}$ = the correlation obtained with truncated X

$_t S_X$ = standard deviation of X in the truncated distribution

S_X = standard deviation of X in the unrestricted distribution

and

est. r_{xy}^* = an estimate of the correlation to be obtained without truncation .

Then

$$\text{est. } r_{xy}^* = \frac{_t r_{xy} \, (S_X/_t S_X)}{\sqrt{1 + _t r_{xy}^2 \, [(S_X^2/_t S_X^2) - 1]}} . \qquad [14.10.1]$$

Guilford and Fruchter (1978) is a good source for further information on this and closely related matters.

A related point is that a correlation can be shown to exist only where there is variability. If you seek to establish the existence of a correlation between variables X and Y, but either one or the other exhibits little or no variability, then you are very likely doomed to failure. Indeed, if one of the variables is actually a constant, with no variability, then the correlation coefficient is not even defined. The moral is that if you are looking for relationships among variables, you must be sure that the variables in question actually vary over the cases for which this relationship is being sought.

Finally, a sample correlation may suggest that two variables fail to have a linear relationship, when, in fact, part of the problem lies with the scale of measurement of one or both of the variables. It may be that if one or both of the X or Y variables were transformed, by the creation of new variables such as, for example, $W = \log(X)$ or $V = 1/X$, a larger absolute correlation would be found. It is up to the investigator to assure himself that the X and Y variables are indeed measured in the appropriate units and on the appropriate scale for the purpose at hand. Therefore an initial inspection of the sample distributions of X and Y as well as of the joint distribution of their values (i.e., the scatterplot) can be worthwhile in forestalling this sort of difficulty. A very sophisticated yet readable and informative survey of such problems and their solutions is given in Mosteller and Tukey (1977), and this should be required reading for anyone beginning to get seriously involved in the applications of regression theory.

14.11 AN EXAMPLE OF A "CLASSIC" REGRESSION PROBLEM

A social psychologist was interested in problem solving carried out cooperatively by small groups of individuals. The theory upon which her experiments were based suggested that within a particular range of possible group sizes, the relationship between group size and average time to solution for a particular kind of problem should be linear and negative: the larger the group, the less time on the average should it take for the problem to be solved. In order to check on this theory, the psychologist decided to form a set of experimental groups, ranging in size from groups consisting of a single individual to groups consisting of six individuals. Thus, 105 individuals were chosen at random from some specific population, and then formed at random into five groups consisting of 1 individual each, five groups each consisting of 2 individuals, five groups of 3, and so on until there were six different and nonoverlapping sets of five groups each, the last consisting of five groups each of size 6. Each individual subject participated in one and only one group in the study.

Each experimental group was given the same set of problems to solve, and given a score Y based on the average time to solve the problems. (The normal distribution of such scores might be open to some question, but we'll simply ignore this problem.) Thus, the experimental unit of observation was a group, and the experimental design can be represented in terms of the following six samples of 5 groups each:

	5 groups	5 groups	5 groups	5 groups	5 groups	5 groups
Group size	1	2	3	4	5	6

Notice that in this example $N = 30$, the number of distinct experimental *groups*.

Now the independent variable X in this experiment is the group size; X can take on only the whole values 1 through 6 in this experiment, and each value occurs with frequency 5. The dependent variable Y is the score attached to each distinct group. Does a linear relation appear to exist between X and Y in this range of group sizes?

Although the units for observation happen to be groups rather than individuals in this study, in all other respects it is a classic example of a problem in regression. That is, the different values of X were preselected, and then cases (groups) within a certain X level were sampled and their Y values noted. Problems in regression do not necessarily correspond to formal experiments like this example, and a little later we will consider a somewhat different sort of example. Nevertheless, the clearest examples of regression problems are usually to be found in single-factor experiments where the factor levels are made to correspond to different values of some quantitative variable X.

The raw data were as shown in Table 14.11.1. Preliminary computations show that

$$T_x = \sum_j \sum_i x_j = (1 \times 5) + (2 \times 5) + (3 \times 5) + (4 \times 5) + (5 \times 5)$$
$$+ (6 \times 5) = 105$$

$$\sum_j \sum_i x_j^2 = (1 \times 5) + (4 \times 5) + \cdots + (36 \times 5) = 455$$

$$\sum_j \sum_i y_{ij}^2 = (32)^2 + \cdots + (16)^2 = 16{,}813$$

$$T_y = \sum_j \sum_i y_{ij} = 697$$

$$\bar{x} = 105/30 = 3.5$$

$$\bar{y} = 697/30 = 23.23$$

$$S_x^2 = 455/30 - (105/30)^2$$
$$= 2.92$$

$$S_y^2 = 16{,}813/30 - (697/30)^2$$
$$= 20.65$$

$$\sum_j \sum_i y_{ij}x_j = (32 \times 1) + (29 \times 1) + \cdots + (18 \times 6) + (16 \times 6)$$
$$= 2{,}231.$$

Table 14.11.1

Group scores arranged according to group size

			x		
1	2	3	4	5	6
32	30	27	19	23	20
29	30	26	21	19	19
30	26	24	20	20	16
28	27	24	20	22	18
26	26	23	18	18	16
145	139	124	98	102	89

Given these values it is a simple matter to find the regression coefficient $b_{y \cdot x}$ for these data. Substituting into Eq. 14.9.6 we have

$$b_{y \cdot x} = \frac{N \left(\sum_i x_i y_i \right) - T_x T_y}{N \left(\sum_i x_i^2 \right) - T_x^2}$$

$$= \frac{(30 \times 2{,}231) - (105 \times 697)}{(30 \times 455) - (105)^2} = -2.383.$$

Notice the negative sign of $b_{y \cdot x}$, indicating that the apparent linear relationship *is* negative. The regression equation is then given by

$$y_i' = 23.23 - 2.383 \, (x_i - 3.5)$$

or

$$y_i' = 31.57 - 2.383 x_i \, .$$

The sample variance of estimate is, from Eq. 14.3.2,

$$S_{y \cdot x}^2 = S_y^2 - b_{y \cdot x}^2 S_x^2$$

$$= 20.65 - (2.383^2 \times 2.92)$$

$$= 4.07.$$

This can then be turned into the sample standard error of estimate:

$$S_{y \cdot x} = \sqrt{4.07}$$

$$= 2.02$$

We can also find the proportion of Y variance accounted for by X in these data:

$$r_{xy}^2 = b_{y \cdot x}^2 \, S_Y^2 / S_X^2$$

$$= (2.383^2 \times 2.92)/20.65$$

$$= .80.$$

Therefore, for these sample data, some 80% of the variance of Y is accounted for by differences in the X values. For actual data this would indicate a relationship of extraordinary strength. Now we are going to use these sample values as we begin to make inferences about the population situation that these data represent.

14.12 POPULATION REGRESSION

Imagine a population where each distinct elementary event qualifies for one and only one joint (x,y) event, where X and Y are random variables having some known bivariate distribution. (The general features of bivariate distributions are outlined in Appendix

C.) That is, each individual i is assumed to be associated with a pair of values, (x_i, y_i), and, under random sampling of individuals, there is a probability density to be assigned to each conceivable (x_i, y_i) pair. Let us for the moment adopt the stance of a problem in regression, and consider the variable Y as the dependent variable, to be predicted from information about variable X for any individual. For each possible value x of variable X there will then be a distribution of possible Y values within that subpopulation or "level" x.

Moreover, let us suppose that we are interested in some finite number J of values of X, any one of which can be represented by the value x_j. Each value or level x_j then defines a whole population of potential observations each having the same value of X, but with an entire distribution or "array" of possible values of Y.

Now using these populations as our frame of reference, we may state the linear relation between y_{ij}, the Y value for individual i in the population j, and x_j, the value of X shared by all members of that population, in this form:

$$y_{ij} = \alpha_0 + \beta_{Y \cdot X} x_j + e_{ij} .$$ [14.12.1]

Here α_0 is the intercept, equivalent to a in Eq. 14.2.1, but representing the population situation, and $\beta_{Y \cdot X}$ is the regression coefficient equivalent to b in Eq. 14.2.1, but again representing the population relationship.

If, given this population situation, we apply the same least squares argument we have already used for sample data, we find that

$$\beta_{Y \cdot X} = \text{cov}(X, Y)/\sigma_X^2$$

where, as defined in Section B.3,

$$E[(X - \mu_X)(Y - \mu_Y)] = \text{cov}(X, Y).$$

That is, the population regression coefficient $\beta_{Y \cdot X}$ is the covariance of the random variables X and Y, divided by the variance of variable X. The required value of the constant α_0, or intercept, for predicting raw values is

$$\alpha_0 = \mu_Y - \beta_{Y \cdot X} \mu_X.$$ [14.12.2]

Then

$$y'_{ij} = \mu_Y + \beta_{Y \cdot X}(x_j - \mu_X),$$ [14.12.3]

which is the population regression equation.

Notice that for an example such as that in Section 14.11, within each category j, each and every individual exhibits the same value of X or x_j. It will be assumed that the J treatment categories are fixed, just as in fixed-effects analysis of variance. This means that the *same sample distribution of x_j values* will appear in any *exact repetition* of the experiment for a different sample of N individuals, although the distribution of Y values within a category *may differ* from sample to sample.

Given this linear model, and the fixed values assumption for the categories j, then the true mean of X for each category j must be x_j, or

$$\mu_{x_j} = x_j.$$

This implies that the model may also be written

$$y_{ij} = \mu_Y + \beta_{Y \cdot X} (\mu_{x_j} - \mu_X) + e_{ij}.$$ [14.12.4]

Be sure to notice the close similarity to the fixed-effects analysis of variance model; if we identify

$$\beta_{Y \cdot X} (\mu_{x_j} - \mu_X) = \alpha_j$$ [14.12.5]

we see this to be only a special version of Model I for the one-way analysis of variance. It should therefore come as no surprise that problems in regression can be, and often are, handled by ANOVA methods. This will be demonstrated in the next section.

Given random sampling of the Y values, then the sample value of the regression coefficient $b_{y \cdot x}$ is our best available estimate of the population regression coefficient $\beta_{Y \cdot X}$. In fact,

$$E(b_{y \cdot x}) = \beta_{Y \cdot X}.$$ [14.12.6]

It is also true that our best estimate of $\beta_{Y \cdot X} (\mu_x)$ is given by $b_{y \cdot x} (\bar{x})$, and, as always, the best estimate of μ_Y is $\bar{y}$.

Since each of these estimates corresponds to a term in the sample regression equation, we can use the sample equation itself as our best estimate of the population equation. Thus, for example, the regression equation found in Section 14.11,

$$y_i' = 23.23 - 2.383(x_i - 3.5)$$

$$= 31.57 - 2.383x_i$$

is our best estimate of the regression equation that applies to the population(s) represented in this study. This can be used in place of the unknown population equation to predict the y value for any case whose x value is known.

In addition, an unbiased estimate of the *true* variance of estimate for predicting the value of Y from the value of X is given by

$$\text{est. } \sigma_{Y \cdot X}^2 = \left(\frac{N}{N - 2} \right) S_{Y \cdot X}^2$$

$$= \left(\frac{N}{N - 2} \right) (S_y^2 - b_{y \cdot x}^2 S_x^2)$$ [14.12.7*]

so that our best available estimate of the true standard error of estimate is

$$\text{est. } \sigma_{Y \cdot X} = \sqrt{\text{est. } \sigma_{Y \cdot X}^2}.$$ [14.12.8*]

For our example of Section 14.11 the estimate of the true standard error of estimate is

$$\text{est. } \sigma_{Y \cdot X} = \sqrt{\left(\frac{30}{28} \right) (4.07)}.$$

$$= 2.09.$$

The true standard error of the regression coefficient $b_{y \cdot x}$ is then given by

$$\sigma_b = \sigma_{Y \cdot X} / \sigma_X \sqrt{N}.$$ [14.12.9]

However, since in a regression problem we generally assume that the same distribution of X values appears in every sample, the sample value S_X can be substituted for σ_X in the expression above, giving

$$\sigma_b = \sigma_{Y \cdot X}/S_X \sqrt{N}.$$
[14.12.10]

Our best available estimate of the standard error of the regression coefficient is then

$$\text{est.}\sigma_b = \text{est.}\sigma_{Y \cdot X}/(S_X \sqrt{N}).$$
[14.12.11*]

14.13 ASSUMPTIONS IN A REGRESSION PROBLEM

For a genuine regression problem, where the various x_j values are selected beforehand for observation, the following assumptions are made, in addition to the assumption of the general model:

$$y_{ij} = \mu_Y + \beta_{Y \cdot X}(x_j - \mu_X) + e_{ij}$$

1. within each population j, the distribution of y_{ij} values is normal;
2. within each population j, the variance σ_e^2 is the same;
3. the errors e_{ij} are completely independent.

In other words, here we must make the assumption of a normal distribution for the Y values within each population having the same X value. We must also assume equal variances for each of these populations (sometimes referred to as the assumption of *homoscedasticity*).

Notice the parallel between these assumptions and those made in a simple, fixed-effects analysis of variance. Notice also that no explicit assumption is made about the distribution of the X variable in a problem in regression, where the different X values to be represented in the data are chosen in advance. In an example such as that in Section 14.11, we implicitly assume that the true distribution of X values among the populations is like that found in the sample. In a sense, the inferences made in a problem in regression are conditional upon the distribution of X as obtained or created in the sample.

14.14 INTERVAL ESTIMATION IN A REGRESSION PROBLEM

The sample regression coefficient $b_{y \cdot x}$ is really just a weighted sum of Y means over the J different groups representing X values. Thus, given the assumption that each population defined by an X value actually has a normal distribution of Y values, it is easy enough to see that the coefficient $b_{y \cdot x}$ must itself have a normal distribution over samples.

Given the normal sampling distribution of the regression coefficient $b_{y \cdot x}$, with expectation $\beta_{Y \cdot X}$, it should be possible to create a confidence interval for $\beta_{Y \cdot X}$ by use of the normal distribution, *provided* that we had the true value of the standard error rather than only an estimate. Instead we do the usual thing for this circumstance, and turn to the t distribution, which allows for the estimation of the error term, as in Eq. 14.12.11.

Under the regression model, it is therefore quite possible to form confidence intervals for $\beta_{Y \cdot X}$, the true regression coefficient. The $100(1 - \alpha)$ percent confidence interval is found from

$$b_{Y \cdot X} - \frac{\text{est. } \sigma_{Y \cdot X} t_{(\alpha/2)}}{S_X \sqrt{N}} \leq \beta_{Y \cdot X} \leq b_{Y \cdot X} + \frac{\text{est. } \sigma_{Y \cdot X} t_{(\alpha/2)}}{S_X \sqrt{N}} \qquad [14.14.1^*]$$

where here

$$\text{est. } \sigma_{Y \cdot X} = \sqrt{\frac{N S_Y^2 - N b_{Y \cdot X}^2 S_X^2}{N - 2}} \qquad [14.14.2^*]$$

as shown in Eq. 14.12.7. The $t_{(\alpha/2)}$ value is found for $N - 2$ degrees of freedom. For our example,

$$\text{est. } \sigma_{Y \cdot X} = \sqrt{[(30 \times 20.65) - (30 \times 2.383^2 \times 2.92)]/28}$$

$$= 2.09 .$$

For $30 - 2 = 28$ degrees of freedom, the t value cutting off the uper .025 proportion of the distribution is 2.048. Hence, the lower bound to the 95% confidence interval is

$$-2.383 - [(2.09)(2.048)/\sqrt{30(2.92)}] = -2.84$$

and the upper bound is

$$-2.383 + [(2.09)(2.048)/\sqrt{30(2.92)}] = -1.92 .$$

Notice that this confidence interval does *not* include the value 0. Therefore, we could reject the hypothesis that the true value of $\beta_{Y \cdot X} = 0$, and therefore we say that the sample $b_{y \cdot x}$ value is significant beyond the .05 level (two-tailed).

Occasionally, interval estimates are desired for the predicted values y' using the *population* regression rule, and some specific value x_j. Remember that the predicted y' value found using a *sample* regression equation does not necessarily agree with that to be found using the population equation. In a regression problem, there are two possible sources of disagreement between the sample-based y' and the value based on the population equation: the sample mean may be in error to some extent as an estimate of μ_Y, and the sample regression coefficient may be wrong to some extent as a representation of true $\beta_{Y \cdot X}$. Considering both of these sources of error, we have for a given score x the following $100(1 - \alpha)$ percent confidence limits for the mean predicted y' given x_j according to the population regression equation:

$$y' \pm t_{(\alpha/2)} \text{ est. } \sigma_{Y \cdot X} \sqrt{(1/N) + \frac{(x_j - \bar{x})^2}{N S_X^2}}. \qquad [14.14.3^*]$$

The number of degrees of freedom is again $N - 2$. Be sure to notice the interesting fact that the regression equation found for a sample is *not equally good* as an approximation to the population equation over all possible values of x. The sample rule is at its best in substituting for the population rule when $x_j = \bar{x}$, the mean of all of the x values, since the confidence interval is smallest at this point. However, as x values are increasingly deviant from the mean $\bar{x}$ in either direction, the confidence intervals grow wider. This indicates that for really extreme values of x we really cannot be sure that the predicted y' from the sample comes anywhere near the value we would have predicted using the population regression equation.

However, you will also notice that three things operate to make the predicted y' values agree better with their population counterparts: large sample size; a large value of S_X^2, such as would be achieved with a wide range of x values each with equal n; and a small value of est. $\sigma_{Y \cdot X}$, such as might result from controls operating in the experiment as a whole.

Finally, if we wish, we can establish a confidence interval for the *actual score* (not the predicted score) of a *single* individual given that the overall relationship is linear. In this instance, the predicted score of an individual based on a sample will differ from the prediction based on population values, and even the "best" prediction, or true y'_j, will still differ from the individual's actual score, y_{ij}. Hence in the distribution of predictions of actual scores there are three kinds of variability: of $\bar{y}$, over the various samples, of $b_{Y \cdot X}$, also over samples, and of the differences between y_{ij} and true y'_j within the population. These three kinds of variability are reflected in the confidence interval found from

$$y'_j - t_{(\alpha/2)} \text{ est. } \sigma_{Y \cdot X} S_j \leq y_{ij} \leq y'_j + t_{(\alpha/2)} \text{ est. } \sigma_{Y \cdot X} S_j \qquad [14.14.4]$$

where

$$S_j = \sqrt{1 + \frac{1}{N} + \frac{(x_j - \bar{x})^2}{NS_X^2}}.$$

For small samples, this may be a very wide interval indeed, particularly for extreme values of x_j.

Obviously, for the case in which N approaches an infinite value, the confidence interval for the prediction of an individual score becomes simply

$$y'_j - z_{(\alpha/2)} \sigma_{Y \cdot X} \leq y_{ij} \leq y'_j + z_{(\alpha/2)} \sigma_{Y \cdot X} \qquad [14.14.5]$$

since the variability of $\bar{y}$ and $b_{Y \cdot X}$ over samples approaches zero, and since we have assumed that each of the populations j is normally distributed with the same variance $\sigma_{Y \cdot X}^2$.

The assumptions of normal distributions within each X population and equality of variance within populations are quite important when one is using these interval estimation methods. It is possible to make an estimate of the regression equation, and to use this equation for prediction without assuming anything except random sampling, but *interval estimates depend heavily on these normal assumptions for their validity.*

On the other hand, the *test* of the hypothesis that $\beta_{Y \cdot X}$ is zero apparently shares the features of the fixed-effects ANOVA model, in that the assumptions of normality and of homogeneous variances are somewhat less important than in other situations. In particular, arranging to have an equal number of cases per X grouping is good policy in a regression study if one suspects that the variances are unequal.

14.15 A TEST FOR THE REGRESSION COEFFICIENT

The same principle that permits the establishment of a confidence interval for the value of the population regression coefficient also yields a test of an exact hypothesis about $\beta_{Y \cdot X}$, based on the sample value $b_{y \cdot x}$.

Thus, in order to test the exact hypothesis

$$H_0: \beta_{Y \cdot X} = k \text{ (a constant)}$$

we can form the t ratio

$$t = (b_{y \cdot x} - k)(S_x \sqrt{N})/\text{est. } \sigma_{Y \cdot X} \qquad [14.15.1^*]$$

where the degrees of freedom are $N - 2$, and the value of est. $\sigma_{Y \cdot X}$ is found exactly as in Eq. 14.14.2. This test may be either one- or two-tailed, depending, as always, on the alternative hypothesis of interest. Most commonly the value to be tested is $\beta_{Y \cdot X} = 0$, in which instance the significance test becomes

$$t = b_{y \cdot x} (S_x \sqrt{N})/\text{est. } \sigma_{Y \cdot X} \qquad [14.15.2^*]$$

For example, the data of Table 14.11.1 provide this t ratio:

$$t = (-2.383 \times 1.71 \times 5.48)/2.09$$

$$= -10.68$$

where the value of est. $\sigma_{Y \cdot X}$ is that found in Eq. 14.12.7. This t value is, of course, extremely significant, two-tailed, for $N - 2 = 28$ degrees of freedom.

On the other hand, if we had wished to test another hypothesis, say

$$H_0 : \beta_{Y \cdot X} = -2.5$$

against the alternative

$$H_1 : \beta_{Y \cdot X} > -2.5$$

then the t value would have been

$$t = [(-2.383 + 2.5) \times 1.71 \times 5.48]/2.09$$

$$= 0.52$$

which, for 28 degrees of freedom, is definitely not significant. This hypothesis would certainly not be rejected.

In the next section we will see how the t test for the regression coefficient may also be carried out as an F test in the format of ANOVA.

14.16 REGRESSION PROBLEMS IN ANOVA FORMAT

In a regression problem think of the score y_{ij}, which belongs to case i in X level j. Then the deviation of this score from the grand mean $\bar{y}$ can be partitioned into two parts, as follows:

$$(y_{ij} - \bar{y}) = (y_{ij} - y'_{ij}) + (y'_{ij} - \bar{y}) \qquad [14.16.1^*]$$

Then by an argument exactly like that for ANOVA (Section 10.8), it can be shown that the total sum of squares for Y can be partitioned into

$$SS_y \text{ total} = SS_y \text{ regression on } X + SS_y \text{ residual}.$$

Just as in ANOVA

$$SS_y \text{ total} = \sum_j \sum_i y_{ij}^2 - (T_y^2/N). \qquad [14.16.2^*]$$

In addition, there is a total SS to be found for the x values as well:

$$SS_x \text{ total} = \sum_j n_j x_j^2 - (T_x^2/N). \qquad [14.16.3^*]$$

Once again letting

$$SP_{xy} = \sum_j \sum_i x_j y_{ij} - (T_x T_y/N) \qquad [14.16.4^*]$$

we may find

$$b_{y \cdot x} = SP_{xy}/SS_x \text{ total} \qquad [14.16.5^*]$$

and thus

$$SS_y \text{ regression on } X = b_{y \cdot x}^2 \, SS_x \text{ total}. \qquad [14.16.6^*]$$

Then the remaining term, SS_y residual, will be

$$SS_y \text{ residual} = SS_y \text{ total} - SS_y \text{ regression on } X$$

$$= SS_y \text{ total} - b_{y \cdot x}^2 \, SS_x \text{ total} .$$

These sums of squares may then be summarized in an ANOVA-style table, on the order of Table 14.16.1. Thus, for our example of Section 14.11, we have the following:

$$SS_x \text{ total} = (1 \times 5) + (4 \times 5) + \cdots + (36 \times 5) - [(105)^2/30]$$

$$= 87.5$$

$$SS_y \text{ total} = 32^2 + \cdots + 16^2 - [(697)^2/30]$$

$$= 619.37$$

and

$$SP_{xy} = (1 \times 145) + (2 \times 139) + \cdots + (6 \times 89) - [(105 \times 697)/30]$$

$$= -208.5$$

Source	SS	df	MS	F	p	**Table 14.16.1**
Regression on X	496.89	1	496.89	113.70	<.01	ANOVA table for the data in Section 14.11
Residual	122.48	28	4.37			
Total	619.37	29				

so that

$$b_{y \cdot x} = -208.5/87.5$$

$$= -2.383.$$

On substituting into Eq. 14.16.6 we find

$$\text{SS}_y \text{ regression on } X = 2.383^2 \times 87.5$$

$$= 496.89$$

which leaves

$$\text{SS}_y \text{ residual} = 619.37 - 496.89$$

$$= 122.48$$

The SS_y regression on X always has one degree of freedom, for a reason that can be explained as follows. Since

$$\text{SS}_y \text{ regression on } X = b_{y \cdot x}^2 \text{ SS}_x \qquad [14.16.7]$$

$$= N \, [\text{cov}(x,y)]^2/S_x^2$$

so that the only things that determine this sum of squares are the sample N, the sample variance of the x values, and the sample covariance, $\text{cov}(x,y)$. However, N and the variance of the x values are actually assumed to be the same over replications of this experiment. Hence the SS regression really depends only on the sample covariance.

Look for a moment at the sample covariance,

$$\text{sample cov}(x,y) = \frac{\sum\limits_j \sum\limits_i x_j y_{ij}}{N} - \overline{xy} \qquad [14.16.8]$$

Since x_j is the same for each individual i in group j, the sample covariance in this instance is

$$\text{sample cov}(x,y) = \frac{\sum\limits_j n_j(\bar{y}_j)(x_j)}{N} - \frac{\sum\limits_j n_j(\bar{y}_j)\,\bar{x}}{N} \qquad [14.16.9]$$

$$= \sum\limits_i \frac{n_j(x_j - \bar{x})(\bar{y}_j)}{N}.$$

You are probably unaware of it, but we have just shown a most interesting fact: the sample covariance in a regression problem is *a linear combination* or comparison among the means $\bar{y}_j$ for J samples, each mean being weighted by a value

$$c_j = \frac{n_j(x_j - \bar{x})}{N}. \qquad [14.16.10]$$

Furthermore,

$$\sum\limits_j c_j = \sum\limits_j \frac{n_j(x_j - \bar{x})}{N} = 0, \qquad [14.16.11]$$

since the average of deviations of individual x_j values about the mean $\bar{x}$ must be zero for any sample. Thus, formally, **the sample covariance in a regression problem is simply a comparison among sample means.**

Now the information about comparisons in Section 11.9 can be used to find what the expectation of SS linear regression must be. For any sample comparison $\hat{\psi}$, we found that

$$\text{SS}(\hat{\psi}) = \frac{\hat{\psi}^2}{w}$$

is a sum of squares with 1 degree of freedom. For the covariance comparison

$$w = \sum_j \frac{c_j^2}{n_j} = \sum_j \frac{n_j^2 (x_j - \bar{x})^2}{n_j N^2}$$

$$= \frac{S_X^2}{N}. \tag{14.16.12}$$

The sum of squares for the covariance comparison is thus

$$\frac{[\text{sample cov}(x,y)]^2}{(S_x^2/N)} = \frac{N[\text{sample cov}(x,y)]^2}{S_x^2}$$

$$= \text{SS linear regression}. \tag{14.16.13*}$$

The sum of squares for linear regression has one degree of freedom, since it corresponds to the sum of squares for a single comparison among sample means. Then

$$E(\text{MS linear regression}) = E(\hat{\psi}^2/w)$$

$$= \frac{N[\text{cov}(x,y)]^2}{S_X^2} + \sigma_e^2 \tag{14.16.14}$$

$$= N\beta_{Y \cdot X}^2 S_X^2 + \sigma_e^2.$$

When $\beta_{Y \cdot X}$ is truly zero, so that the true $\text{cov}(x,y)$ is zero, then

$$E(\text{MS linear regression}) = \sigma_e^2. \tag{14.16.15}$$

Now under our assumption that the linear model is a true description of the relationship between the X and the Y variables, then any variability over and above that due to regression can reflect only error. Furthermore, since there can be only $N - 1$ degrees of freedom in all, the degrees of freedom associated with residual should be $N - 1 - 1 = N - 2$. Therefore, in this model it should be true that

$$\text{MS}_y \text{ residual} = \text{SS}_y \text{ residual}/(N - 2) = \text{est. } \sigma_{Y \cdot X}^2 \tag{14.16.16}$$

and

$$E(\text{MS}_y \text{ residual}) = \sigma_e^2 = \sigma_{Y \cdot X}^2. \tag{14.16.17}$$

Obviously, the appropriate F test for the hypothesis of zero regression will then be

$$F = \frac{\text{MS regression on } X}{\text{MS residual}} \qquad\qquad [14.16.18^*]$$

with 1 and $N - 2$ degrees of freedom.

For our sample regression-problem summarized in Table 14.16.1 the F value turns out to be 113.70, which exceeds most of the common criteria for significance with 1 and 28 degrees of freedom.

A little bit of algebra is sufficient to show that the F test arrived at in this way is simply the square of the t test for a regression coefficient as given by Eq. 14.15.2.

$$\sqrt{F} = t = \frac{b_{Y \cdot X} S_x \sqrt{N - 2}}{S_{Y \cdot X}}. \qquad\qquad [14.16.19]$$

Thus, the two approaches are equivalent. It is, however, somewhat simpler to use the t-test approach if one is interested in a nonzero hypothesis about the regression coefficient, or in a one-tailed test. (In Chapter 16 we are going to have occasion to test not only the significance of the regression coefficient, but also to ask if the linear model is itself appropriate. In so doing we will use the fact illustrated here, that the linear regression is itself representable as a comparison.)

Just as in ANOVA for qualitatively different groups, it is possible to express the sample proportion of Y variance accounted for by X. This is done by taking

$$R^2_{Y \cdot X} = r^2_{xy} = (SS_y \text{ regression on } X)/SS_y \text{ total}, \qquad\qquad [14.16.20]$$

exactly as in the usual ANOVA situation. The sample proportion of variance accounted for by X in a regression situation is, of course, the square of the correlation between X and Y.

Once again, the sample proportion of variance accounted for by X tends to be larger than the true proportion of variance due to X in the population. An adjusted version along the lines of the estimated omega-square indices used earlier is

$$\text{est. population } R^2_{Y \cdot X} = \frac{\text{MS}_y \text{ regression} - \text{MS}_y \text{ residual}}{(\text{MS}_y \text{ regression}) + (N - 1)(\text{MS}_y \text{ residual})}$$

$$= \frac{(N - 1)r^2_{xy} - 1}{(N - 1) - r^2_{xy}} \qquad\qquad [14.16.21]$$

With very large N this is simply r^2_{xy}. When MS regression is less than MS residual (as when the F value is less than 1.00), then the estimate is set at 0.

For our example of Table 14.16.1 the sample proportion of Y variance explained by X is

$$R^2_{Y \cdot X} = r^2_{xy} = 496.89/619.37$$

$$= .80.$$

Then the estimated proportion of variance accounted for in the population is

$$\text{est. pop. } R^2_{Y \cdot X} = \frac{(29 \times .80) - 1}{29 - .80}$$

$$= .787$$

so that the estimated proportion of variance accounted for is reduced from the sample value of .80 to about .79.

Finally, given the results of a regression analysis, where F (or t) has turned out to be significant, how sure can you be that the regression equation will have any *practical* utility, beyond any theoretical interest that you may have in the extent of linear relationship between X and Y? In other words, are the predictions made by the regression equation different enough from each other to be "visible to the naked eye," so to speak? A generic answer to this question is hard to give, since ideally it would depend on the context and the utility or cost that is attached to errors of different sizes.

On the other hand, the problem has been studied in general terms, and especially in terms of the F ratio as an indicator of probable practical utility. Work by Box and Wetz which is reported in Draper and Smith (1981) defines such practical utility as a situation where the predicted Y difference between individuals with different X values tends to be quite large as compared with error. Their work suggests that you can be fairly sure of "practically useful" regression when the F ratio is not only significant, but exceeds the value required for significance by a factor of at least 4 or 5.

In terms of our example, the F value of 113.70 is 27.0 times the F required for significance at the .05 level, and 14.9 times the F needed for significance at the .01. In this example, then, there would be no problem in forecasting that this regression equation would produce predictions different enough from each other to make a practical difference in a real application.

14.17 ANOTHER PROBLEM IN REGRESSION

The problem in regression given in Section 14.11 described an experimental situation in which the independent variable X was represented by quantitatively different levels or groupings, as defined by the experimenter. This sort of situation is a "pure" or "classic" example of a problem in regression, where the assumptions outlined in Section 14.13 are usually most reasonable.

However, the methods developed for problems in regression can also be applied to situations that are not experimental, in the sense that the experimenter neither creates or manipulates the conditions represented by the X values in the data. Rather, the investigator chooses some set of possible X values to examine, and then compares the different X "arrays" or groupings in terms of their Y values. Then, in order to treat this as a problem in regression, it is necessary only that the distribution of Y values within each X array or grouping can be assumed to represent an independent normal population, and that the variances of these populations be equal. In addition, one must be prepared to treat the X values in the data as though they were fixed, so that inferences drawn will have a conditional character, "Given a true distribution of X values like that found in this sample, then. . . ." Even with these drawbacks, it is often worthwhile to use the methods of a problem in regression to examine the relation between independent variable X and dependent variable Y in such nonexperimental data.

Consider this study: A psychologist interested in attitudes toward music of different sorts was curious about the possible relationship between the number of years of formal musical training that a person had as a child or adolescent, and the depth of that person's interest in classical music as an adult. It was decided to look at whole numbers of years of musical study, ranging from 0 to 10. The investigator took a sample of $N = 50$ university seniors, and, through use of a questionnaire, grouped them into 11 categories according to the number of years of formal musical training each had received up to and including high school. Music majors were excluded from the study, and this eliminated three cases, bringing the final N to 47. Then, for each group, each individual was asked to respond to an attitude questionnaire on classical music, where a large positive score showed a strongly favorable attitude, and a low score a negative attitude.

The data for this study are shown in Table 14.17.1. This can be viewed as a regression problem because the experimenter actually chose the levels of musical training as defining the independent variable X in advance of collecting the data, and the variable Y, attitude toward classical music, is clearly the dependent variable of interest in this study. A regression equation will be formulated and used to predict the attitude score for an individual, based on the number of years of musical training he or she has received.

First we find the total of the x values to be

$$T_x = (0 \times 9) + (1 \times 8) + \cdots + (10 \times 2) = 176$$

so that the mean

$$\bar{x} = 176/47 = 3.74.$$

The sum of squares for the x values is then

$$\text{SS}_x \text{ total} = (0^2 \times 9) + (1^2 \times 8) + \cdots + (10^2 \times 2) - (176)^2/47$$

$$= 460.94$$

so that the variance of the x values is

$$S_x^2 = 460.94/47 = 9.81.$$

Similarly, we can find the total and mean of the y values:

Table 14.17.1

Attitude scores toward classical music (Y), according to years of childhood music study (X)

					Y values					
22										
16	25									
15	22									
17	16									
31	25					27				
20	17	11		18		25	20			
16	24	29	24	20	25	20	22	12		
15	20	22	26	25	22	20	26	28	22	24
18	14	16	24	22	12	21	26	22	33	27
$X =$ 0	1	2	3	4	5	6	7	8	9	10
$T_{yj} =$ 170	163	78	74	85	59	113	94	62	55	51

$$T_y = 22 + 16 + \cdots + 24 + 27 = 1,004$$

and

$$\bar{y} = 1,004/47 = 21.36.$$

Next the total sum of squares for the y values is taken:

$$SS_y \text{ total} = 22^2 + 16^2 + \cdots + 27^2 - 1,004^2/47$$

$$= 1,160.85$$

and the variance of the y values is

$$S_y^2 = 1,160.85/47 = 24.70.$$

The sum of products of the x and y values is next found from

$$SP_{xy} = (0 \times 22) + (0 \times 16) + \cdots + (10 \times 24) + (10 \times 27) - (176 \times 1,004)/47$$

$$= (0 \times 170) + (1 \times 163) + \cdots + (9 \times 55) + (10 \times 51) - (176 \times 1,004)/47$$

$$= 253.34$$

The value of the sample covariance, $\text{cov}(x,y)$, is then

$$\text{cov}(x,y) = SP_{xy}/N = 5.39$$

and the sample regression coefficient is

$$b_{y \cdot x} = \text{cov}(x,y)/S_x^2 = .55 \ .$$

The regression equation for predicting Y, attitude toward classical music, from X, years of formal study as a child, is given by

$$y_i' = 21.36 + .55(x_i - 3.74)$$

or

$$y_i' = 19.30 + .55 \, x_i.$$

The sample value of the variance of estimate turns out here to be

$$S_{y \cdot x}^2 = S_y^2 - b_{y \cdot x}^2 S_x^2$$

$$= 24.70 - (.55^2 \times 9.81) = 21.73$$

which converts into a standard error of estimate for this sample of

$$S_{y \cdot x} = \sqrt{21.73}$$

$$= 4.66$$

The proportion of Y variance accounted for by the variable X is here found to be

$$r_{xy}^2 = b_{y \cdot x}^2 \, S_X^2/S_Y^2$$

$$= (.55^2 \times 9.81)/24.70$$

$$= .12 \ .$$

Thus, in this example only about 12% of the variability among attitudes toward classical music appears to be explained by the amount of formal musical training an individual received as a child. From Eq. 14.16.21 we also have

est. $R_{y \cdot x}^2 = [(46 \times .12) - 1]/(46 - .12) = .10$

Given the results above, we can find the 95% confidence limits for the true regression coefficient by taking

$t = 2.01$ (approximately) for $N - 2 = 45$ degrees of freedom

est. $\sigma_{Y \cdot X} = \sqrt{47 [24.70 - (.55)^2 \times (9.81)]/45} = 4.76$

so that the limits are defined by

$b_{y \cdot x} \pm$ est. $\sigma_{Y \cdot X} (t_{\alpha/2})/(S_X \sqrt{N})$

or

$.55 \pm (4.76 \times 2.01)/(3.13 \sqrt{47})$.

Therefore the confidence interval is

$.104 \leq \beta_{Y \cdot X} \leq .996$.

Since this interval does not include 0, we can reject the hypothesis of no linear relationship between X and Y at or beyond the .05 level, two-tailed.

The equivalent t test is

$t = b_{y \cdot x} (S_X \sqrt{N})/$est. $\sigma_{Y \cdot X}$

$= .55 (3.13 \sqrt{47})/4.76$

$= 2.48$

which is, of course, larger than the 2.01 required for significance at the .05 level.

If we had preferred to approach this problem through ANOVA, then the summary table would have been as follows:

Source	SS	df	MS	F	p
Regression on X	139.24	1	139.24	6.13	<.05
Residual	1,021.61	45	22.70		
Total	1,160.85	46			

Observe that the square root of the F value found here is 2.48, which agrees with the t found just above. In addition, if we take the proportion of Y variance accounted for by linear regression on X, then

$$R_{Y \cdot X}^2 = \frac{SS_y \text{ regression on } X}{SS_y \text{ total}}$$

$= 139.24/1,160.85$

$= .12$.

14.18 ERRORS IN THE PREDICTOR VARIABLE X

An assumption that is implicit in our use of the model of linear regression is that the X values are measured *without error*. In consequence, if a case in a sample has a value $X = x$, then that X value is assumed to be *right*, and any replication of the study in which that case appears should show exactly that same value of X. In practice, of course, this assumption can be pretty unrealistic. Particularly in the social sciences we take a fairly tolerant attitude toward some of our measures, and understand that errors of measurement will affect the variable values that we use. When errors are present, however, it can make a difference in our conclusions from a regression analysis.

To show how this problem affects the regression coefficient, think of a variable X^* which is the *true* or *error-free* measure of the independent variable. Then the value of the X variable we actually observe is

$$x_i = x_i^* + \epsilon_i \qquad\qquad [14.18.1]$$

where ϵ is *an error of measurement*. Presumably the value of x_i^* would be fixed and unchanging for a given individual i, but the value of ϵ_i, the error of measurement, will be different from measurement occasion to occasion, even though the long-run average values of both x and x^* will be the same, or μ_X. For an individual i on a given occasion we will know the value of Y and the value of X, but not the true value of X^*. Yet, what we would really like to have for our regression equation is

$$y_i' = \mu_Y + \beta_{Y \cdot X}^* (x_i^* - \mu_X) . \qquad\qquad [14.18.2]$$

Here the symbol $\beta_{Y \cdot X}^*$ stands for the regression coefficient based on the true, error-free, values of variable X^*. The symbol $\beta_{Y \cdot X}$ will be used to represent what we actually get, the regression coefficient between Y and the fallible measures X.

Under certain circumstances it is then possible to determine the effect of these errors on the regression coefficient we obtain. That is, if the errors of measurement ϵ, the residual errors e, and the X^* are all independent, and each follows a normal distribution, then it can be shown that

$$\beta_{Y \cdot X} = \beta_{Y \cdot X}^* / (1 + \lambda) \qquad\qquad [14.18.3]$$

where

$$\lambda = \sigma_\epsilon^2 / \sigma_X^2$$

with

$$\sigma_X^2 = \text{variance of the true } X^* \text{ values}$$

and

$$\sigma_\epsilon^2 = \text{variance of the measurement errors, } \epsilon.$$

The obvious consequence is that if there are measurement errors at all, so that σ_ϵ^2 is not equal to zero, then the regression coefficient we calculate (or estimate) is *biased* relative to the true, error-free, regression coefficient. In particular, the obtained regression coefficient tends to be *too small* in absolute value.

In addition, where there are errors in the X variable values, the standard error of estimate is relatively too large. That is, if we let $\sigma_{Y \cdot X^*}$ stand for the standard error of estimate free of measurement error, then the standard error of estimate we actually calculate (or estimate), represented by $\sigma_{Y \cdot X}$, is given by

$$\sigma_{Y \cdot X} = \sqrt{\sigma_{Y \cdot X^*}^2 + [\lambda/(1 + \lambda)][\sigma_Y^2 - \sigma_{Y \cdot X^*}^2]} \qquad [14.18.4]$$

This says that so long as errors of measurement exist in X, then the calculated standard error of estimate will be too large, and, interestingly enough, the amount of the bias will tend to be larger the stronger the relationship between Y and true X^*.

Methods exist for estimating the variance of the errors of measurement, or σ_ϵ^2 (i.e., the squared standard error of measurement), and such values can be substituted into Eq. 14.18.4 to correct for the bias in the regression coefficient. Thus, for example, suppose that in the data in Section 14.17.1 it is known that the reliability coefficient of the measure X (here symbolized by r_{XX}) is only .60. Given that the unbiased estimate of the variance of this measure X is

$$s_X^2 = (47/46)(9.81) = 10.02$$

the estimated variance of the errors of measurement is then

$$\text{est. } \sigma_\epsilon^2 = s_X^2 (1 - r_{XX})$$
$$= 10.02 (1 - .60) = 4.01$$

so that the constant λ is

$$\lambda = 4.01/10.02 = .40.$$

(Notice, by the way, that in this sort of situation the constant λ turns out to be the estimated *unreliability* of the X measures, or $1 - r_{XX}$.)

Then we are able to estimate the amount of bias that exists in the regression coefficient of .55 found from the sample. Since

$$\beta_{Y \cdot X} = \beta_{Y \cdot X}^*/(1 + \lambda),$$

then the true regression coefficient $\beta_{Y \cdot X}^*$ is estimated to be

$$\beta_{Y \cdot X}^* = \beta_{Y \cdot X} (1 + \lambda)$$
$$= .55 (1 + .4) = .77 .$$

The unbiased estimate of the variance of the Y values is found to be

$$\text{est. } \sigma_Y^2 = (NS_Y^2)/(N - 1)$$
$$= 47(24.70)/46$$
$$= 25.24 .$$

In addition, we find the unbiased estimate of $\sigma_{Y \cdot X}^2$ from our data to be

$$\text{est. } \sigma_{Y \cdot X}^2 = NS_{Y \cdot X}^2/(N - 2)$$
$$= 47(21.73)/45$$
$$= 22.70 ,$$

or

$$\text{est. } \sigma_{Y \cdot X} = 4.76 \ .$$

On rearranging Eq. 14.18.4, we find that

$$\text{est. } \sigma_{Y \cdot X*}^2 = \text{est. } \sigma_{(Y \cdot X')}^2 (1 + \lambda) - \lambda \text{ est. } \sigma_Y^2$$

$$= 22.70 (1 + 0.4) - (0.4 \times 25.24)$$

$$= 21.68,$$

or

$$\text{est. } \sigma_{Y \cdot X*} = 4.66 \ .$$

Thus, our estimate of the true standard error of estimate is reduced slightly, from 4.76 to 4.66, when the presence of errors of measurement is figured in.

The good news is this: Even though it may not be possible to assume that the X values are basically normally distributed, it is still possible to use the method outlined above to correct for possible bias in the regression coefficient *provided* that N is large. Even with relatively small N the method can be used if the ratio λ turns out to be fairly small.

If your interests are principally in examining the relation that exists between two variables X and Y, and you are not especially concerned with prediction and with errors of estimate, the problem of errors of measurement can often be ignored. On the other hand, if you wish to establish a useful prediction rule based on a fallible predictor measure, this problem really needs to be faced.

Obviously, this little exposition and example has only scratched the surface of an important area of concern. The reader is referred to Snedecor and Cochran (1980) and to Draper and Smith (1981) for further details.

14.19 POPULATION CORRELATION

For the past several sections we have been emphasizing problems in regression, where the main focus of attention is on the sample regression coefficient, $b_{y \cdot x}$, as an estimate of the population regression coefficient, $\beta_{Y \cdot X}$. Now we are going to shift our interest back to the correlation coefficient, once again, as we prepare to talk about inference in so called "problems in correlation." Just as we can define the regression coefficient and the regression equation in population terms, so also is it possible to think about the correlation coefficient in a population.

Think once more of a population of potential observations, where each member of the population has both a value on variable X and a value on variable Y. Then, we can assume a *bivariate density function* which associates a probability density with each *possible* (x,y) pair. There will be a true variance σ_X^2 of variable X, a true variance σ_Y^2 for variable Y, and a true covariance $\text{cov}(X,Y)$ defined on all such pairs of values. The population correlation coefficient ρ_{XY} (lowercase Greek "rho") is then defined by

$$\rho_{XY} = \frac{\text{cov}(X,Y)}{\sigma_X \sigma_Y}, \qquad\qquad [14.19.1]$$

In this conception, the relation between X and Y is a symmetric one, and we really do not think either of X or of Y as *the* dependent variable. Thus, in this population we can easily think of there being two regression equations, just as we discussed in Section 14.8.

It follows that

$$\beta_{Y \cdot X} = \rho_{XY} \frac{\sigma_Y}{\sigma_X} \qquad\qquad [14.19.2]$$

and

$$\beta_{X \cdot Y} = \rho_{XY} \frac{\sigma_X}{\sigma_Y}. \qquad\qquad [14.19.3]$$

In fact, we could define the true correlation coefficient in terms of these two regression coefficients, as follows:

$$\rho_{XY} = \sqrt{\beta_{Y \cdot X} \beta_{X \cdot Y}} , \qquad\qquad [14.19.4]$$

the correlation coefficient is the geometric mean of the two regression coefficients, for predicting Y from X and X from Y.

For any such population, we can discuss the *true* variance of estimate for Y predicted from X,

$$\sigma_{Y \cdot X}^2 = \sigma_Y^2 (1 - \rho_{XY}^2), \qquad\qquad [14.19.5]$$

as well as the true variance of estimate for X predicted from Y,

$$\sigma_{X \cdot Y}^2 = \sigma_X^2 (1 - \rho_{XY}^2). \qquad\qquad [14.19.6]$$

The two **true standard errors of estimate** are thus

$$\sigma_{Y \cdot X} = \sigma_Y \sqrt{1 - \rho_{XY}^2} \qquad\qquad [14.19.7]$$

and

$$\sigma_{X \cdot Y} = \sigma_X \sqrt{1 - \rho_{XY}^2}. \qquad\qquad [14.19.8]$$

Notice that just as for a sample, one can interpret the square of the correlation coefficient as the **proportion of variance accounted for by linear regression.** That is, $\sigma_{Y \cdot X}^2$ is "error" variance in the use of the linear rule, the variability *not* accounted for by linear regression. Thus,

$$\sigma_Y^2 - \sigma_{Y \cdot X}^2$$

is *the true variance accounted for by linear regression,* or the reduction in variance accomplished by using a linear prediction rule. It follows that

$$\rho_{XY}^2 = \frac{\sigma_Y^2 - \sigma_{Y \cdot X}^2}{\sigma_Y^2}, \qquad\qquad [14.19.9]$$

the square of the population correlation coefficient is the *true proportion* of variance accounted for by the use of a linear prediction rule either for Y predicted from X, or X predicted from Y.

14.20 CORRELATION IN BIVARIATE AND MULTIVARIATE NORMAL POPULATIONS

In a problem in correlation our main interest often focuses on the value of r_{xy} itself, especially as an estimate of ρ_{XY} for the population. Given a particular assumption about the population distribution of joint (x,y) events we can test not only if X and Y are linearly related, but also if any systematic relationship at all exists between the two variables.

As shown in Section C.1 of Appendix C, in a joint distribution of discrete random variables, a probability is associated with each possible X and Y pair (x,y). A similar conception holds when X and Y are continuous variables, and a probability is associated with any joint interval of values. Such joint distributions of two random variables are known as bivariate distributions. Although any number of theoretical bivariate distributions are possible in principle, by far the most studied is the **bivariate normal distribution.** The density function for this joint distribution has a rather elaborate-looking rule. However, for standardized variables, this can be condensed to

$$f(z_X, z_Y) = \frac{e^{-G}}{K} \qquad\qquad [14.20.1]$$

where

$$G = \frac{(z_X^2 + z_Y^2 - 2\rho_{XY}z_X z_Y)}{2(1 - \rho_{XY}^2)}$$

and

$$K = 2\pi\sqrt{(1 - \rho_{XY}^2)}.$$

Notice that in a bivariate normal distribution, the population correlation coefficient ρ_{XY} appears as a parameter in the rule for the density function. Thus, even though z_X and z_Y are both standardized variables, the particular bivariate distribution cannot be specified unless the value of the correlation ρ_{XY} is known. (Another way of defining a bivariate normal distribution is given in Section C.3.)

In a bivariate normal distribution, the marginal distribution of X over all observations is itself a normal distribution, and the marginal distribution of Y is also normal. Furthermore, given any X value, the *conditional* distribution of Y is normal; given any Y, the conditional distribution of X is normal. In other words, if a bivariate normal distribution is conceived in terms of a table of joint events where the number of possible X values, of Y values, and of possible joint (x,y) events is infinite, then within any possible row of the table one would find a normal distribution; a normal distribution also exists within any possible column, and the marginals of the table also exhibit normal distributions.

For our purposes, however, the feature of most importance in a bivariate normal distribution is this: **given that densities for joint (x,y) events follow the bivariate normal rule, then X and Y are independent if and only if $\rho_{XY} = 0$.** For *any* joint distribution of (x,y) the independence of X and Y implies that $\rho_{XY} = 0$, but it may happen that $\rho_{XY} = 0$ even though X and Y are *not* independent. However, for this special joint distribution, the bivariate normal, $\rho_{XY} = 0$ both implies and is implied by the *statistical independence of X* and *Y*. The only predictability possible in a bivariate normal distribution is that based on a *linear rule*.

On the other hand, just because the distribution of X and the distribution of Y both happen to be normal, when considered as marginal distributions, this does *not* necessarily mean that the joint distribution of (x,y) values is bivariate normal. Hence, it is entirely possible for a nonlinear statistical relation to exist even though both X and Y are normally distributed when considered separately. It is not, however, possible for any but a linear relation to exist when X and Y jointly follow the bivariate normal law.

Most of the classical theory of inference about correlation and regression was developed in terms of the bivariate normal distribution. **If one can assume such a joint population distribution, inferences about correlation are equivalent to inferences about independence or dependence between two random variables.** For the kinds of problems here called **correlation problems,** the assumption of a bivariate normal distribution is usually made. When this assumption is valid, any inference about the value of ρ_{XY} is equivalent to an inference about the independence, or degree of dependence, between two variables; this is not, however, a feature of regression problems where the bivariate normal assumption need not be made. As always, by adopting more stringent assumptions about the form of the population distribution, one is able to make much more definite statements from sample results.

The generalization of the idea of a bivariate normal distribution to more than two variables is the *multivariate normal distribution*. Here, too, the parameters of the multivariate normal distribution are the mean and variance of each variable, as well as the correlation between each pair of variables. The conditional distribution of any variable holding the others constant is normal, the conditional distribution of any pair of variables holding the others constant is bivariate normal, and so on. As in the case of the bivariate distribution, the only predictability possible in a multivariate normal distribution is linear: $\rho_{XY} = 0$ for any pair of variables X and Y implies and is implied by the independence of X and Y. Further discussion of multivariate normal distributions will be found in any text on multivariate statistics, such as that by Timm (1975). We will have more to say about multivariate normal distributions in Chapters 15 and 17. Also, see Section C.3 and Section D.6 for more on these concepts.

14.21 INFERENCES ABOUT CORRELATIONS

For the typical problem in correlation, the investigator simply draws a random sample of N cases, and notes the X and the Y value for each individual. Often there is no compelling reason to view one of the variables as the dependent or criterion variable and the other as the predictor or independent variable. Rather, interest focuses on the extent of the linear relation, if any, between X and Y. This is the context in which the tests to be discussed below most often apply, and each of these tests presupposes a bivariate normal distribution in the population from which the sample was drawn.

To illustrate these tests and intervals, the data in Section 14.2 will be used once again, and we will assume that the sampling approach corresponds to a problem in correlation, where the population of X and Y values is bivariate normal.

For many correlation problems the only hypothesis of interest is

$$H_0 : \rho_{XY} = 0,$$

where the alternative hypothesis may be either directional or nondirectional. When this hypothesis is true and the population can be assumed to be bivariate normal in form, the distribution of the sample correlation coefficient tends, rather slowly, toward a normal distribution for increasing N. The standard error of this distribution of sample correlations r_{xy} is approximately

$$\sigma_r = 1/\sqrt{N}. \qquad [14.21.1]$$

When sample size is reasonably large (say, $N \geq 50$), then it is possible to test the significance of r_{xy} in this way, by forming the usual z statistic and referring it to the normal distribution.

On the other hand, it is also important to realize that a situation where the bivariate normal distribution applies also satisfies the three assumptions listed in Section 14.13 for a problem in regression. Therefore, it is possible to convert the t test for regression outlined in Section 14.15 into a t test for correlation:

$$t = \frac{b_{y \cdot x} S_x \sqrt{N - 2}}{S_{y \cdot x}} \qquad [14.21.2]$$

$$= \frac{r_{xy} \sqrt{N - 2}}{\sqrt{(1 - r_{xy}^2)}}.$$

This test has $N - 2$ degrees of freedom, just as before.

Hence, for our example of Section 14.2, where the correlation coefficient was equal to .81 and $N = 91$, we find that the t test of significance is

$$t = \frac{.81 \sqrt{89}}{\sqrt{1 - (.81)^2}}$$

$$= 13.03 .$$

This, for $N - 2 = 89$ degrees of freedom, is significant beyond both the .05 and the .01 levels.

Unlike the t test for a regression coefficient, only the hypothesis that the true correlation is 0 may be tested by this method. Confidence intervals and hypotheses other than zero call for a method outlined below.

Under the assumptions made in a problem in correlation, the value of r_{xy} may be used directly as an estimator of ρ_{XY} for the population. Although it is a sufficient and consistent estimator for ρ_{XY}, the sample correlation is slightly biased; however, the amount of bias involves terms of the order of $1/N$, and for most practical purposes can be ignored.

As mentioned earlier, for very large samples the distribution of the sample correlation coefficient may be regarded as approximately normal when $\rho_{XY} = 0$. Even for relatively small samples ($N > 4$) this sampling distribution is unimodal and symmetric. However, when ρ_{XY} is other than zero, the distribution of r_{xy} tends to be very skewed. The more that ρ differs from zero, the greater is the skewness. When ρ_{XY} is greater than zero, the skewness tends to be toward the left, with intervals of high values of r_{xy} relatively more probable than similar intervals of negative values. When ρ_{XY} is negative, this situation is just reversed, and the distribution is skewed in the opposite direction. The fact that the particular form of the sampling distribution depends on the value of ρ_{XY} makes it impossible to use the t test for other hypotheses about the value of the population correlation, or to set up confidence intervals for this value in some direct elementary way. Although the sampling distribution of r_{xy} for $\rho_{XY} \neq 0$ has been fairly extensively tabled, it is much simpler to employ the following method.

R. A. Fisher showed that tests of hypotheses about ρ_{XY}, as well as confidence intervals, can be made from moderately large samples (about $N \geq 10$) from a bivariate normal population if one uses a particular *function* of r_{xy}, rather than the sample correlation coefficient itself. The function used is known as the Fisher r-to-Z transformation, given by the rule

$$Z = \frac{1}{2} \log_e \left(\frac{1 + r_{xy}}{1 - r_{xy}} \right).$$

[14.21.3]

Fisher showed that for virtually any value of ρ_{XY}, for samples of moderate size the sampling distribution of Z values is *approximately normal*, with an expectation given approximately by

$$E(Z) = \zeta = \frac{1}{2} \log_e \left(\frac{1 + \rho_{XY}}{1 - \rho_{XY}} \right).$$

[14.21.4]

(The population value of Z, corresponding to ρ, is denoted by ζ, small Greek zeta.) The sampling variance of Z is approximately

$$\text{var}(Z) = \frac{1}{N - 3}.$$

[14.21.5]

These approximations tend to improve with *smaller* absolute values of ρ_{XY} and *larger* sample sizes. For moderately large samples the hypothesis that ρ_{XY} is equal to any value ρ_0 (not too close to 1 or -1) can be tested. This is done in terms of the test statistic

$$\frac{Z - \zeta}{\sqrt{1/(N - 3)}}$$

[14.21.6*]

referred to a *normal* distribution. The value taken for $E(Z)$ or ζ depends on the value given for ρ_0 by the null hypothesis:

$$\zeta = E(Z) = \frac{1}{2} \log_e \left(\frac{1 + \rho_0}{1 - \rho_0} \right)$$

[14.21.7*]

and the sample value of Z is taken from the sample correlation,

$$Z = \frac{1}{2} \log_e \left(\frac{1 + r_{xy}}{1 - r_{xy}} \right).$$

[14.21.8*]

It should be emphasized that the use of this r-to-Z transformation *does* require the assumption that the (x,y) events have a bivariate normal distribution in the population.

On its face, this assumption seems to be a very stringent one, which may not be reasonable in some situations, though there is some evidence that the assumption may be relatively innocuous in others. However, the consequences of this assumption's not being met seem largely to be unknown. Perhaps the safest course is to require rather larger samples in uses of this test when the assumption of a bivariate normal population is very questionable.

Table VI in Appendix E gives the Z values corresponding to various values of r. This table is quite easy to use, and makes carrying out the test itself extremely simple. Only positive r and Z values are shown, since if r is negative, the sign of the Z value is taken as negative also.

For example, suppose that we wanted to test the hypothesis that $\rho_{XY} = .50$ in some bivariate normal population. A sample of 100 cases drawn at random gives a correlation r_{xy} of .35. The hypothesis is to be tested with $\alpha = .05$, two-tailed. Then, from the Table VI, we find that for $r_{xy} = .35$,

$$Z = .3654.$$

For $\rho_{XY} = .50$, we find

$$\zeta = E(Z) = .5493.$$

The test statistic is then

$$\frac{.3654 - .5493}{\sqrt{1/97}} = -1.81.$$

In a normal sampling distribution, a standard score of 1.96 in absolute value is required for rejecting the hypothesis at the .05 level, two-tailed. Thus, we do not reject the hypothesis that $\rho_{XY} = .50$ on the basis of this evidence. Observe that the test made in terms of Z leads to an inference in terms of ρ.

Occasionally one has two *independent* samples of N_1 and N_2 cases respectively, where each is regarded as drawn from a bivariate normal distribution, and one computes a correlation coefficient for each. The question to be asked is, "Do both of these correlation coefficients represent populations having the *same* true value of ρ_{XY}?" Then a test of the hypothesis that the two populations show equal correlation is provided by the ratio

$$\frac{Z_1 - Z_2}{\sigma_{(Z_1 - Z_2)}} \qquad [14.21.9]$$

where Z_1 represents the transformed value of the correlation coefficient for the first sample, Z_2 the transformed value for the second, and

$$\sigma_{(Z_1 - Z_2)} = \sqrt{\frac{1}{N_1 - 3} + \frac{1}{N_2 - 3}}. \qquad [14.21.10]$$

For reasonably large samples (say, 10 in each) this ratio can be referred to the normal distribution. Remember, however, that the two samples must be independent (in particular, not involving the same or matched subjects) and the population represented by each must be bivariate normal in form.

More generally, suppose that there are J independent samples, each drawn from a bivariate normal distribution of (x, y) pairs. Each sample j yields a sample correlation r_j between X and Y. Then the hypothesis that the true value ρ_{XY} is the same for all of the populations can be tested by the statistic

$$V = \sum_j (n_j - 3)(Z_j - U)^2 \qquad [14.21.11]$$

which is distributed as chi-square with $J - 1$ degrees of freedom when the null hypothesis that $\rho_1 = \rho_2 = \cdots = \rho_J$ is true. Here, n_j is the number of observations in the sample j, and

$$U = \frac{\sum_j (n_j - 3)Z_j}{\sum_j (n_j - 3)}. \qquad [14.21.12]$$

14.22 CONFIDENCE INTERVALS FOR ρ_{XY}

If the population has a bivariate normal distribution of (x, y) events, then the r-to-Z transformation can be used to find confidence intervals, very much as for a mean of a large sample. It is approximately true that for random samples of size N, an interval such as

$$Z - z_{(\alpha/2)}\sqrt{\frac{1}{N - 3}} \leq \zeta \leq Z + z_{(\alpha/2)}\sqrt{\frac{1}{N - 3}} \qquad [14.22.1^*]$$

will cover the true value of ζ with probability $1 - \alpha$. Here, Z is the sample value corresponding to r_{xy}, ζ is the Z value corresponding to ρ_{XY}, and $z_{(\alpha/2)}$ (*definitely* to be distinguished from Z) is the value cutting of the upper $\alpha/2$ proportion in a normal distribution. Thus, Eq. 14.22.1 gives the $100(1 - \alpha)$ percent confidence interval for ζ. On changing the limiting values of Z back into correlation values, we have a confidence interval for ρ_{XY}.

In the example given in the preceding section,

$$Z = .3654$$

$$N = 100,$$

so that

$$\sqrt{\frac{1}{N - 3}} = .1.$$

For $\alpha = .05$, $z_{(\alpha/2)} = 1.96$, so that the 95% confidence interval for ζ is given approximately by

$$.3654 - (1.96 \times .1) \leq \zeta \leq .3654 + (1.96 \times .1)$$

or

$$.1694 \leq \zeta \leq .5614.$$

The corresponding interval for ρ_{XY} is then approximately

$.168 \le \rho_{XY} \le .510$

(the correlation values here are taken to correspond to the nearest tabled Z values). We can assert that the probability is about .95 that sample intervals such as this cover the true value of ρ_{XY}.

14.23 OTHER INFERENCES IN CORRELATION PROBLEMS

In Section 14.14 we saw that confidence intervals for a regression coefficient and for the predicted value found from a regression equation can be established in a problem in regression. Happily, if one is interested in using the results of a correlation study for prediction, then exactly these same methods can be applied there as well, provided that a bivariate normal distribution can be assumed and given that the population mean and variance of X is like that in the sample.

Each of the methods of Section 14.14 demands that within each level or array of the predictor variable X there be a normal distribution of Y values, and that the true variance of the Y values is the same for each X. These requirements are necessarily satisfied *whenever one can assume a bivariate normal distribution*, as we routinely do in problems of correlation. Hence, almost all of the machinery of regression analysis that we have discussed in previous sections is available when a problem in correlation is undertaken. Furthermore, this machinery is available for predicting Y from X, or X from Y.

14.24 EXAMINING RESIDUALS

The correlation coefficient found for two variables X and Y is a good index of the extent of linear relationship that appears to exist between them in a given sample. Thus, a correlation coefficient with an absolute value close to 1.00 shows that the linear rule is an excellent way to relate the value to X to the value of Y, or vice-versa. However, the correlation coefficient represents only the adequacy of fit of a linear rule as a description of how the two variables are related. A correlation at or near zero does not imply that no relationship exists between X and Y, but only that the relationship, whatever it may be, is not well described as a linear function. If two variables are indeed unrelated to each other, then we should expect a correlation of 0, but a zero correlation alone does not establish that X and Y actually are unrelated.

This principle is especially important to remember in discussing a population correlation: If random variables X and Y are independent, then they must have a true correlation ρ_{XY} equal to 0. However, the fact that $\rho_{XY} = 0$ is not ordinarily sufficient to establish the independence of X and Y. The exception is in a bivariate normal population, where $\rho_{XY} = 0$ both implies and is implied by the independence of variables X and Y.

It can happen that two variables X and Y show little or no correlation in a given set of data, and are therefore viewed (incorrectly) as only slightly related, when a closer inspection may show some other, perhaps nonlinear, form of relationship, and perhaps a stronger overall relationship than was at first indicated. That is, it may be that the usual linear model just does not fit your particular situation. When this happens, the obtained value of r_{xy}^2 may tend to underestimate the relationship between X and Y. Or it may be that one or both of the variables need to be transformed into new units before a linear function can correctly reflect their relationship.

Other problems may also exist, such as an error in the analysis, so that the apparent fit of the data to the model is deceptive. Or, it may be that one or more of the assumptions required for an inference about the population are clearly not supported by the data. Finally, there may be "outliers," which are atypical cases in the data which may be exerting a disproportionate influence on the regression or correlation results.

All of these issues can arise in a regression or a correlation problem, and the presence of these complications can often be detected by an examination of the residuals or "errors," which, you may recall, are simply the differences between the predicted values y_i' and the actual values y_i for each of the individuals i making up the data.

It is easy to show that the residuals e_i must be uncorrelated with the predicted values y_i'. That is, the covariance of the residuals and the predicted values is

$$\text{cov}(e,y') = \sum_i e_i y_i'/N - (0)(\bar{y})$$

$$= \sum_i (y_i - y_i')y_i'/N$$

$$= r_{xy}^2 S_y^2 + \bar{y}^2 - r_{xy}^2 S_y^2 - \bar{y}^2$$ [14.24.1]

$$= 0$$

so that the residuals e are uncorrelated with the y' values. In much the same way it can be shown that the residuals e are uncorrelated with the x values as well. This implies that if a plot is made of residual value (vertical axis) against the predicted y' values (horizontal axis), the results ought to look approximately like Figure 14.24.1, and the same thing should be true for residuals against the x values. However, when the plots depart from this form, this is a pretty good clue that something is amiss.

Examination of residuals is extremely easy to carry out, and is well worth doing in any problem concerned with the extent of linear relation between two variables. Modern

Figure 14.24.1

"Normal" pattern of residuals plotted against predicted Y values.

e

y'

authorities (cf. Draper and Smith, 1981) recommend that the residuals routinely be calculated for the N data cases, and that plots of these values be made

1. for the simple distribution of the residuals, and
2. either against the predicted values y', or against the predictor variable values x.

Furthermore, if *time* is a factor in the experiment, where some of the data have been collected at an early point in time and others at later points, then it is recommended that a plot of residuals also be made against time order.

   A plot of the distribution of the residuals should give the appearance *not* of a normal distribution but of a sample that might reasonably have come from a normal distribution. The mean of the e values should be zero, of course, but one needs to be especially alert to very skewed or otherwise highly irregular arrangements that are unlikely to occur in sampling from a normal distribution. Much more accurate judgments on this issue can be arrived at through plotting the residuals on normal probability paper (again, see Draper and Smith (1981) for details.).

   Now, as noted, if the linear model is appropriate and all requirements are supported by the data, the plot of residuals against y' or x variables should give an appearance somewhat similar to that of Figure 14.24.1, reflecting the lack of correlation between the residuals and these features of the experiment. However, if the plot of residuals against y' values should have an appearance similar to Figure 14.24.2A, then there is evidence that the variance of the Y variable is not constant within levels or arrays of X,

Figure 14.24.2

"Abnormal" patterns of residuals plotted against predicted Y values.

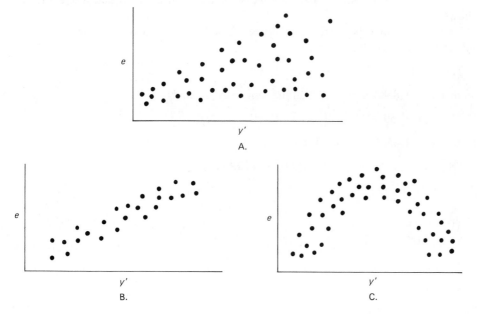

and that the assumption of homogeneous variances appears to be violated. Other methods, such as so-called weighted least squares (Draper and Smith, 1981) or a transformation of the Y values should probably be used in place of an ordinary regression analysis.

If the plot against y' values turns out approximately like Figure 14.24.2B, this may be evidence for a systematic error made in the analysis or the calculation of the residuals. A thorough recheck is in order here. A result like Figure 14.24.2C (or some other clearly curved plot) shows that simple linear model is probably not fully appropriate to describe the relationship between X and Y, as represented in the data. Very similar conclusions are reached when the plots of residuals against x values show the forms of Figures 14.24.2A, B, or C.

An outlier is a point on the residual plot that falls far from the bulk of the points and therefore reflects an extreme position among the data cases. Outliers should be identified and studied, not so much with the aim of discarding them as ''wrong,'' as in an attempt to understand the particular combination of circumstances that made this case (or set of cases) different from the others. Of course, some cases that appear to be outliers can result from simple errors in coding or recording the data, and these, when discovered, *should* be either corrected or discarded.

14.25 AN EXAMPLE

As an example to illustrate a problem in correlation with an examination of residuals, consider the following. As a portion of a larger study of factors affecting the muscular coordination of male college athletes, two measures were used. The first, called X here, was a measure of the large-muscle coordination exhibited by a subject. The second measure, Y, represented the athlete's small-muscle coordination. A random sample of 25 such athletes was drawn, and both measures taken on each subject. Table 14.25.1 shows the results.

First we will find the correlation between X and Y, and test this correlation for significance. From the data we have

$$T_x = 73 + 76 + \ldots + 78 = 1,868$$

so that $\bar{x} = 1,868/25 = 74.72$. In the same way we find

$$T_y = 64 + 59 + \ldots + 55 = 1,450$$

so that $\bar{y} = 1,450/25 = 58$. Then

$$SS_x \text{ total} = (73)^2 + \ldots + (78)^2 - (1,868)^2/25$$

$$= 127$$

giving

$$S_x^2 = 5.08$$

and

$$S_x = 2.25 \ .$$

Furthermore,

$$SS_y \text{ total} = (64)^2 + \ldots + (55)^2 - (1,450)^2/25$$

$$= 212$$

so that

$$S_y^2 = 212/25$$

$$= 8.48$$

and

$$S_y = 2.91 \; .$$

The correlation coefficient may then be found by taking

$$SP_{xy} = (73 \times 64) + \ldots + (78 \times 55) - (1,868 \times 1,450)/25$$

$$= -71$$

or

$$\text{cov}(x,y) = SP_{xy}/N = -2.84 \; .$$

Then,

$$r_{xy} = \frac{SP_{xy}}{\sqrt{(SS_x \text{ total})(SS_y \text{ total})}}$$

$$= \text{cov}(x,y)/(S_X S_Y)$$

$$= -.43 \; .$$

It appears that in this sample $(.43)^2 = .1849$, or a little over 18%, of the variance in Y can be explained by X variation. On the face of it, this seems a little low for two measures each representing a type of muscle coordination. In addition, the negative sign of the correlation might seem a little surprising for two such measures.

Then, if we wish to predict Y from X, we can calculate the regression coefficient

$$b_{y \cdot x} = r_{xy}(S_y/S_x)$$

or, a bit more accurately, perhaps,

$$b_{y \cdot x} = (SP_{xy})/(SS_x \text{ total})$$

$$= -71/127 = -.559 \; ,$$

and the raw score regression equation would be

$$y_i' = \bar{y} + b_{y \cdot x}(x_i - \bar{x})$$

$$= 58 - (.559 \, (x_i - 74.72))$$

$$= 99.77 - .559 x_i \; .$$

On the other hand, if we wish to predict X from Y, then the regression coefficient is

$$b_{x \cdot y} = r_{xy}(S_x/S_y)$$

$$= -.43(2.25/2.91) = -.332.$$

Thus, the raw score regression equation is

$$x_i' = 74.72 - .332 \, (y_i - 58)$$

or

$$x_i' = 93.98 - .332 \, y_i.$$

For predictions of Y from X, the sample standard error of estimate is then

$$S_{y \cdot x} = S_y \sqrt{1 - r_{xy}^2}$$

$$= 2.91 \sqrt{(1 - 0.1849)}$$

$$= 2.63 \, .$$

The significance of the correlation can be tested by use of t, as follows:

$$t = \frac{r_{xy} \sqrt{N - 2}}{\sqrt{1 - r_{xy}^2}}$$

or

$$t = \frac{-.43 \sqrt{23}}{\sqrt{1 - (-.43)^2}}$$

$$= -2.28.$$

For $N - 2 = 23$ degrees of freedom this is significant beyond the .05 level, two-tailed.

In the remainder of this discussion we are going to assume that the problem is to predict Y from X. If the problem is to predict X from Y then the same procedures can be used, with obvious adjustments, of course.

Now, let us examine the residuals from predicting Y from X, or

$$e_i = (y_i - y')$$

The last two columns in Table 14.25.1 show, for each of the 25 cases, the predicted value y_i' and the residual e_i. The values of x_i and of y_i are shown in the first two columns, of course.

Please observe that two checks are available on the accuracy of our computation of the residuals: In the first place, the mean of the values should turn out to be 0, and the standard deviation of the residuals should equal the sample standard error of estimate (both within rounding error, of course). Thus, for our set of residuals as given in Table 14.25.1, we have $\bar{e} = 0.0004$, and $S_e = 2.63$ just as we should expect from the value of $S_{y \cdot x}$ found above.

The plot of the distribution of the residuals is shown as Figure 14.25.1. As you can see, there is no apparent abnormality here, and this plot might well represent a sample from a normal population distribution. Thus, no further analysis seems to be required of this aspect of the residual information.

Subject	x value	y value	y' value	Residual
1	73	64	58.96	5.04
2	76	59	57.28	1.72
3	72	60	59.52	0.48
4	77	53	56.73	−3.73
5	73	58	58.96	−0.96
6	75	57	57.84	−0.84
7	78	53	56.17	−3.17
8	71	58	60.08	−2.08
9	76	55	57.28	−2.28
10	78	56	56.17	−0.17
11	72	57	59.52	−2.52
12	74	64	58.40	5.60
13	77	60	56.73	3.27
14	73	61	58.96	2.04
15	76	57	57.28	−0.28
16	71	56	60.08	−4.08
17	74	60	58.40	1.60
18	75	60	57.84	2.16
19	72	57	59.52	−2.52
20	77	55	56.73	−1.73
21	78	56	56.17	−0.17
22	73	57	58.96	−1.96
23	75	60	57.84	2.16
24	74	62	58.40	3.60
25	78	55	56.17	−1.17

Table 14.25.1

Large-muscle (X) and small-muscle (Y) coordination measures for 25 male college athletes

Next, Figure 14.25.2 shows the plot of the residuals against the predicted values y'. It is obvious that this plot does not resemble Figure 14.24.1, as we expected it to do. Rather, a curved relationship resembling Figure 14.24.2C emerges from this plot. This becomes even clearer if we use the mean Y value for each of the different X values, as shown in Figure 14.25.2.

This sort of result, where the residuals suggest a curved relationship when plotted against the predicted y' values, gives evidence that there is more to the relationship between X and Y than the simple linear model can accommodate. In fact, in this example there is a good deal more Y variance to be accounted for by X than the 18% or so suggested by the squared correlation coefficient. We will consider such situations further in Chapter 16, and illustrate how a larger proportion of variance may sometimes be accounted for with an alternative model.

Figure 14.25.1

Distribution of residuals in example of large- and small-muscle measures.

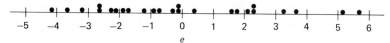

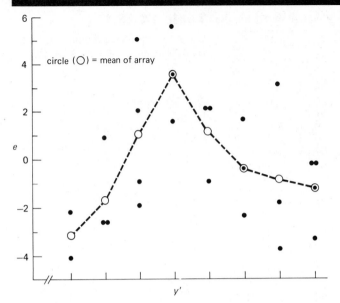

Figure 14.25.2

Plot of residuals against predicted Y values for large- and small-muscle measures.

14.26 COMPUTER PACKAGE PROGRAMS FOR REGRESSION AND CORRELATION

Although the commonly employed computer packages tend to be very rich in regression methods, most of these will be more appropriately discussed following Chapters 15 and 16. For the moment, then, we will confine our attention to those programs where the emphasis tends to be on bivariate regression, or on the correlations between pairs of variables.

In both SPSS and SPSS[x] the subprogram PEARSON CORR will generate correlations among up to 500 variables, through a command of the form

PEARSON CORR *variable list* WITH *variable list/*

If the variable lists contain the same set of K variables, then the output will be the entire intercorrelation matrix among these variables. The outcome of a one-tailed t test of significance is reported for each correlation, unless OPTION 3 is specified, which leads to two-tailed tests. User choices for STATISTICS include means and standard deviations and covariances for pairs of variables.

An alternative subprogram, more suited to exploring regression coefficients for pairs of variables, is SCATTERGRAM. This control command usually takes the form

SCATTERGRAM *variable list* WITH *variable list/*.

If you wish only a part of the full range for either variable, you can specify a lowest and highest value for either (or both). The output for each pair of variables specified to left and right of the keyword WITH is a two-dimensional *plot* of the data points. A number of OPTIONS may be requested, and STATISTICS available include Pearson's r for any pair, the square of r, the significance of the correlation coefficient, the standard error of estimate, the value of the intercept a, and the slope b. Although PEARSON CORR is also available in SPSS[x], in the latest version the SCATTERGRAM subprogram has been replaced by the subprogram PLOT.

In the BMDP series of programs, the simplest approach to bivariate regression is the program P6D. This takes the general paragraph form

/PLOT

    XVAR $=$ *variable list*

    YVAR $=$ *variable list*

    STAT. *yes*

This produces a plot of each pair from the XVAR and the YVAR lists. Also given are the means and standard deviations of both variables, and the two regression lines with slope and intercept. Program P8D also produces correlation matrices with a significance test for each correlation, along with means and standard deviations for the variables.

In SAS, the basic regression procedure is PROC REG. Specification of the very simplest bivariate regression model, relating variable Y to variable X, would look like this:

PROC REG;

    MODEL $Y = X$;

On this basis one receives output for the intercept, the slope, the significance test, and a variety of other pieces of information about the bivariate regression of Y on X. Other information such as predicted values and residuals may also be obtained. PROC PLOT permits bivariate graphs to be constructed either for the original values or for residuals. However, the "big guns" of PROC REG are brought into play in the multiple regression problems to be discussed in the next chapter.

EXERCISES

1. List the assumptions made in a test of significance in a linear regression problem; compare them with the assumptions made in fixed-effects analysis of variance. Compare the assumptions made in significance tests for correlation problems with the assumptions made in random-effects analysis of variance.

2. An experimenter was interested in the possible linear relation between the time spent per day in practicing a foreign language and the ability of the person to speak the language at the end of a six-week period. Some 50 students were assigned at random among five experimental conditions ranged from 15 min practice daily to 3 h practice per day. Then at the end of six weeks, each student was scored for proficiency in the language. The data follow:

Proficiency scores, by daily practice time x = practice, in hours				
.25	.50	1	2	3
117	106	86	140	105
85	81	98	128	149
112	74	125	108	110
81	79	123	104	144
105	118	118	132	137
109	110	94	133	151
80	82	93	96	117
73	86	91	101	113
110	111	122	103	142
78	113	130	135	112

Find the linear regression equation for predicting Y, the proficiency of a student, from X, the practice time per day.

3. For Exercise 2, plot the obtained data and the straight line representing the regression equation.

4. Construct the ANOVA summary table for the regression analysis of the data in Exercise 2.

5. For the data in Exercise 8 Chapter 10, test the hypothesis of no linear regression of performance on the hours of sleep deprivation. Use $\alpha = .01$ in this instance.

6. A study dealt with the ability of teachers to judge a child's age from drawings. In order to gather data on this, the investigator selected at random 12 drawings by children aged 4, 12 by children aged 5, 12 from age 6, and 12 from age 7. Then each of a panel of teachers assigned a "guessed age" to each drawing. The data below show the four age groups together with the average age guessed for each child from drawings.

Average ages guessed for children, arranged according to actual age. x = actual age			
4	5	6	7
5.87	6.46	6.01	6.09
4.49	4.80	5.03	5.87
5.83	4.12	6.10	7.25
4.83	4.13	5.13	5.77
5.60	5.25	6.75	6.51
5.18	5.78	5.27	6.69
4.85	5.59	5.99	6.40
4.77	4.64	6.19	6.10
5.79	4.56	5.90	7.40
4.41	4.66	5.60	5.59
4.23	6.09	6.90	6.09
5.62	5.87	5.09	5.87

Test the hypothesis of no linear regression, by use of fixed-effects analysis of variance methods. Use $\alpha = .05$.

7. Find the 95% confidence interval for the true value of the linear regression coefficient for the data in Exercise 6.

8. A college mathematics teacher was studying the relationship between the number of years of mathematics taken by a student in high school and his or her score in mathematics on a college entrance examination. A random sample of 85 freshmen students was taken, and the test scores (Y) were grouped according to the number of years of high-school mathematics taken (X). Use the results as shown below to find the linear regression equation for predicting the value of Y from X.

	X								
	0	**.5**	**1.0**	**1.5**	**2.0**	**2.5**	**3.0**	**3.5**	**4.0**
y	480	443	462	497	502	502	500	500	524
	470	435	464	486	498	513	521	523	526
		439	447	493	492	493	502	520	560
			461	483	496	507	515	505	552
				500	490	501	501	506	535
				487	497	492	516	515	531
					490	482	499	506	553
						509	508	502	551
						508	490	513	532
						494	507	517	546
						507	488	502	
						501	503	505	
						490	494	518	
						493	505	510	
						490	499		
						491	508		
						501	498		
						490	505		
							496		
							506		
							494		

9. Show the linear regression analysis of the data in Exercise 8 in analysis of variance format, and carry out the F test of significance. What must one assume here?

10. Suppose that a student in Exercise 8 had exactly two years of high-school mathematics. Find the 95% confidence limits for his or her predicted y score.

11. Fifteen political scientists ranked each of three countries in terms of the perceived "political responsibility" each manifests. The data follow. Calculate the correlation between the ratings given to countries A and B. Then calculate the correlations between the ratings for B and C, and for A and C.

Countries			
Raters	A	B	C
1	5	5	6
2	11	11	9
3	6	8	5
4	12	14	11
5	8	6	6
6	5	8	6
7	7	7	5
8	5	8	4
9	12	14	9
10	8	8	4
11	5	8	4
12	6	8	5
13	7	7	8
14	6	7	7
15	7	9	13

12. Using the data in Exercise 11, from the regression equations for predicting

(a) The z score on A from the z score on B.
(b) The z score on B from the z score on C.
(c) The z score on C from the z score on A.

13. Calculate the sample standard error of estimate for each of the regression equations found in Exercise 12.

14. Using the data in Exercise 11 once again, find the raw score regression equations for

(a) Predicting the rating on A (or x_A) from the rating on B (or x_B).
(b) Predicting the rating on C from the rating on A.
(c) Predicting the rating on B from the rating on C.

Then find the sample standard error of estimate for each of the regression equations.

15. In Exercise 11 what percentage of the variance in ratings in B is accounted for by linear relationship with A? What percentage of the variance in C is accounted for by linear relationship with A? What percentage is accounted for in C by linear relationship with B?

16. An experimenter was interested in the possible linear relationship between the measure of finger dexterity X, and another measure representing general muscular coordination Y. A random sample of 25 persons showed the following scores:

Person	X value	Y value
1	75	84
2	77	94
3	75	90
4	76	90
5	75	91
6	76	86
7	73	87

8	75	95
9	74	83
10	75	85
11	76	88
12	74	91
13	72	80
14	75	85
15	73	87
16	75	82
17	78	86
18	76	83
19	74	85
20	74	88
21	77	100
22	75	98
23	76	89
24	74	91
25	75	99

Compute the correlation coefficient, and test its significance ($\alpha = .05$).

17. Based on the data in Exercise 16, find the regression equation for predicting X from Y. Plot this regression equation along with the raw data. What is the appropriate measure of the "scatter" or horizontal deviations of the obtained points in this plot about the regression line?

18. A developmental psychologist believes that the age at which a normal child begins to speak words clearly is closely related to the age at which the child first begins to use complete sentences. A random sample of of 33 normal children was taken, and very careful records kept for each. Let X be the age at which words are first clearly used, and let Y be the age at which complete sentences are used. The data below give the values of X and Y in months. Find the correlation between X and Y.

Child	X	Y	Child	X	Y	Child	X	Y
1	15.1	25.2	12	14.3	25.7	23	13.6	24.3
2	12.7	24.3	13	11.5	23.4	24	15.2	26.3
3	11.7	22.1	14	13.4	25.7	25	12.1	23.4
4	13.1	23.3	15	13.7	24.5	26	12.6	24.5
5	13.0	24.1	16	13.5	26.0	27	14.1	26.2
6	11.2	23.6	17	12.8	24.6	28	11.2	23.0
7	13.3	25.5	18	13.2	25.4	29	14.0	24.3
8	12.3	24.3	19	14.7	26.3	30	13.1	25.3
9	13.7	25.5	20	12.2	25.2	31	11.5	24.2
10	12.2	23.2	21	14.7	26.4	32	14.9	27.2
11	13.3	27.1	22	14.6	25.8	33	13.8	26.3

19. An investigator in early childhood education conceived the notion that experienced elementary school teachers are able to identify the actual age of a child fairly accurately from inspection of drawings that the child had produced. To test this idea, a sample of 48 children of various ages between 4 and 7 years was drawn at random. Each produced four free drawings of subjects suggested by the investigator. These drawings were then studied by a panel of 10 teachers working independently, and an age estimated for each child. The data below show the actual age X and the average estimated age Y for each child. Find the correlation coefficient between actual and average estimated age.

X	Y	X	Y	X	Y	X	Y
4.1	5.7	6.5	6.5	5.5	5.7	6.8	5.9
6.5	6.9	5.3	5.3	4.0	5.2	6.9	6.3
5.1	5.4	6.8	7.2	7.2	7.0	5.2	4.8
7.6	7.6	5.9	6.4	6.4	6.5	7.4	6.2
5.6	6.0	6.3	6.8	5.2	5.3	5.8	6.1
4.2	4.0	6.0	4.5	7.1	7.0	6.7	5.8
7.1	5.2	7.9	7.5	6.7	6.5	5.6	5.8
6.3	6.1	5.3	5.6	5.7	6.1	4.8	6.1
6.1	5.5	7.3	7.5	6.5	6.7	6.7	7.0
5.7	4.4	6.3	6.8	6.1	5.6	5.8	7.2
7.4	7.1	7.1	5.9	5.6	4.9	6.6	6.5
6.4	6.7	5.4	5.5	6.8	7.2	5.7	5.6

Is there a significant linear relation between a child's actual age X, and the average age Y guessed by the panel of teachers?

20. In a study of the relation between the age at which study of the piano was begun and the eventual proficiency of the student after five years practice, a random sample of 100 children and adults each just completing five years of piano study, was selected. Each student was given the same piece to play, and a panel of judges rated performance. Summary statistics emerging from the final data are given below:

$$\sum X = 1{,}475 \qquad \sum Y = 12{,}890$$

$$\sum X^2 = 24{,}459 \qquad \sum Y^2 = 1{,}714{,}421$$

$$\sum XY = 186{,}659$$

Establish the 95% confidence limits for the true correlation between age at beginning piano lessons X, and fifth-year proficiency Y. Test the hypothesis that the true correlation is $-.50$. (**Hint:** Use the r-to-Z transformation.)

21. Suppose that a person began piano lessons at age 35. On the basis of the data in Exercise 20 predict proficiency at age 40. Given that proficiency after five years of study is normally distributed, establish the 95% confidence limits for this person's actual proficiency.

22. Under what circumstances must uncorrelated variables also be independent variables? Under what circumstances might two independent variables be correlated?

23. In an investigation of the relation between the height of an abolescent boy and a measure of his physical stamina, three random samples of 40 boys each were used. In the first sample, all boys were 15 years old; in the second, all were 16 years old, and in the third, the boys were 17 years of age. The following correlations were obtained:

Sample 1	Sample 2	Sample 3
$r = .11$	$r = .23$	$r = .19$
$N_1 = 40$	$N_2 = 40$	$N_3 = 40$

Test the hypothesis that the correlation between height and stamina is the same for the populations of boys represented by the first two samples. (**Hint:** Use the r-to-Z transformation.)

24. Test the hypothesis that all the sample correlations shown in Exercise 23 represent equal population correlations. Use $\alpha = .05$.

25. In a study of the origins of gender stereotyping of young girls, a random sample of 35 intact families was taken, in which there was an oldest (or only) girl in the ninth grade. The father answered a questionnaire about his interest in sports and received a score X_1. The mother answered a similar questionnaire, and received a score X_2. The physical education instructor of each girl rated her on general athletic ability, and this was used as variable X_3. The dependent variable Y was the girl's own score on a questionnaire on interest in sports. Use the data given below to form the correlation matrix relating the three X variables to Y for this sample. Also find the mean and standard deviation for each variable. Test the significance of each correlation.

Family	X_1	X_2	X_3	Y
1	23	25	8	24
2	32	30	13	30
3	25	25	15	25
4	36	31	10	35
5	37	36	9	39
6	31	33	10	30
7	28	22	8	27
8	26	18	5	14
9	20	14	13	39
10	35	30	15	42
11	42	36	11	41
12	40	30	11	38
13	32	27	12	39
14	33	38	10	37
15	45	40	18	47
16	30	31	13	30
17	29	25	12	36
18	28	24	4	12
19	24	27	10	26
20	18	20	8	20
21	37	35	16	43
22	27	29	15	31
23	39	32	14	40
24	34	30	7	31
25	48	40	12	36
26	37	31	6	21
27	44	41	19	44
28	23	26	5	11
29	29	24	8	27
30	38	30	9	24
31	41	20	15	40
32	39	31	7	32
33	27	29	6	10
34	48	43	14	37
35	33	30	7	19

26. For the data in Exercise 25, find the raw score regression equation for predicting Y from X_1. Actually carry out these predictions, and then plot the residuals against the values of X_1. Does this plot seem to suggest anything unusual about the relationship between X_1 and Y?

Chapter 15

Partial and Multiple Regressioin

15.1 QUESTIONS INVOLVING MORE THAN TWO VARIABLES

We now turn to a group of methods which extend the basic ideas of bivariate correlation and regression to situations where more than two variables are involved. Certainly, in practical situations it is seldom that only two items of information about an individual will be known. Thus, in the employment setting several ratings of background, education, and experience may be used, and several preemployment tests may be administered. Each of these things will be involved to some extent in the decision to hire or not to hire. The admissions officer in a university may have not only college entrance examination scores, but also other information such as the high-school record, the applicant's out-of-class record, and the educational level of the parents, each of which may be useful in judging the potential for academic success shown by an applicant.

However, measures such as these are not only related to something that must be predicted, such as job performance or success in college; they are also related to each other. The extent to which measures are correlated with each other is an index of the overlap or redundancy of the information that they provide. It may well be that the reason that college entrance scores appear to predict success in university work is because such scores supply some of the same information that is given by another excellent predictor, high-school record. The question then might well be asked, "Do college entrance examination scores still predict success in university work if we look only at students with the *same* high-school record?" Even in a purely research setting, where a number of variables may be investigated at the same time, one often needs to trace the extent to which one variable or set of variables accounts for or "explains" the variability in others. The question often arises, "If I were able to hold one or more of these variables constant across all of the subjects in my study, what would happen to the relationships that I observe among the remaining variables?'' These are the sorts of questions that the *coefficient of partial correlation,* to be discussed in the next section, was designed to answer.

The partial correlation coefficient lets one examine a linear relationship between two variables as though values on at least one other variable had been held constant. Next the closely related idea of a *part correlation* (or *semipartial correlation*) will be introduced. Unlike a partial correlation, the part correlation between two variables presumes that one, but not both, of the two variables has been adjusted for the linear relationship with a third variable. Thus, one might have a question such as this: ''Granted that a correlation exists between a student's high-school record and college entrance examination scores, suppose that the entrance exam scores were actually adjusted to remove the influence of high-school performance. How well would the adjusted entrance scores then correlate with university grades?'' Such a question would be answered by the part correlation between university scores and adjusted entrance exam scores, where the linear regression on high-school grades has been removed from the exam scores, but not from the university grades. The square of such a part correlation would reflect the proportion of variance in university grades due to entrance exam scores, *over and above* any variance that might be due to high-school grades.

Finally, the bulk of this chaper will be devoted to concepts of *multiple regression* and *multiple correlation* as ways of representing the relationship of a set of two or more predictor variables and a single dependent variable. Our first attack on the problem of multiple regression will be through the ''classical'' or ''simultaneous'' approach. Later however, we will deal with multiple regression through use of a series of part correlations, in a method that is often termed *stepwise* regression.

15.2 PARTIAL CORRELATIONS

The partial correlation coefficient is a valuable device for examining the interrelationships that exist among *three or more variables*. Specifically, the coefficient of partial correlation between two variables X and Y is an index of the linear relationship that would still exist between these variables if all linear influences of one or more other variables were removed.

In order to deal with the simplest version of a partial correlation, we will need to consider three variables, which, for the time being, we will refer to as X_1, X_2, and X_3. Each individual i out of N cases will then have three values (x_{i1}, x_{i2}, and x_{i3}). First we may calculate the ordinary Pearson correlation (or zero-order correlation) for each pair of these variables; this would yield the three correlations that we can symbolize as r_{12}, r_{13}, and r_{23}.

Now we wish to find the partial correlation $r_{12 \cdot 3}$, for variables X_1 and X_2, with both adjusted for variable X_3, or with X_3 held constant.

In simplest terms, the partial correlation $r_{12 \cdot 3}$ is merely the correlation between the residual from predicting X_1 from X_3, and the residual from predicting X_2 from X_3.

In neither of these residuals can the linear regression on X_3 make a difference, and so variable X_3 is effectively held constant. That is,

$r_{12 \cdot 3}$ = correlation between X_1 and X_2 where X_3 is held constant.

It is possible to calculate this partial correlation directly, by literally adjusting X_1 and X_2 to remove the regression of X_3 from each. This process will be illustrated next. Then we will discuss how this same result can be achieved with a good deal less work, through use of the intercorrelations among the variables.

To find the partial correlation $r_{12 \cdot 3}$ directly from the raw data, we would first find new values of X_1, adjusted to remove the part that is predictable from the linear relationship between X_1 and X_3. We already know that for any individual i the predicted value on variable X_1, or x'_{i1}, can be found from x_{i3} by the regression equation

$$x'_{i1} = \bar{x}_1 + [b_{1 \cdot 3}(x_{i3} - \bar{x}_3)] \qquad [15.2.1]$$

where

$$b_{1 \cdot 3} = r_{13} (S_1/S_3).$$

The adjusted value of x_{i1}, after elimination of the part that can be predicted from the value of x_{i3}, is just the difference between the actual value and this predicted value. If we use $x_{i(1 \cdot 3)}$ to symbolize the adjusted value, then it is true that

$$x_{i(1 \cdot 3)} = x_{i1} - x'_{i1}$$
$$= x_{i1} - \bar{x}_1 - [b_{1 \cdot 3}(x_{i3} - \bar{x}_3)]. \qquad [15.2.2]$$

Note that $x_{i(1 \cdot 3)}$ is the part of the x_{i1} that *cannot* be predicted linearly from x_3 and is thus the same as the error or residual in predicting x_1 from x_3. Such an adjusted x_{i1} value, or $x_{i(1 \cdot 3)}$, can be found for each individual i in the data, based on the x_{i1} and x_{i3} values.

Note also that unlike the original x_{i1} values, the adjusted values $x_{i(1.3)}$ are *not correlated* with the x_{i3} values since, as indicated in Section 14.24, residuals are not correlated with the predictor values.

In exactly this same way, the x_{i2} values may also be adjusted for regression on the x_{i3} scores. Thus, for any individual i we may take

$$x_{i(2.3)} = x_{i2} \text{ value adjusted for } x_{i3}$$
$$= x_{i2} - \bar{x}_2 - [b_{2 \cdot 3}(x_{i3} - \bar{x}_3)] \qquad [15.2.3]$$

where

$$b_{2 \cdot 3} = r_{23} (S_2/S_3).$$

Once all of the x_{i1} scores have been adjusted to form $x_{i(1 \cdot 3)}$ and all of the x_{i2} scores adjusted to form $x_{i(2 \cdot 3)}$ values, then the partial correlation $r_{12 \cdot 3}$ is just the ordinary correlation between the new adjusted $x_{i(1 \cdot 3)}$ and the $x_{i(2 \cdot 3)}$ values:

$$r_{12 \cdot 3} = \text{correlation of } x_{(1 \cdot 3)} \text{ and } x_{(2 \cdot 3)} . \qquad [15.2.4]$$

The partial correlation cannot, therefore, depend in any linear way on the x_3 values, since this regression has literally been removed from both the adjusted $x_{(1 \cdot 3)}$ and the $x_{(2 \cdot 3)}$ values. The result is equivalent to holding variable X_3 at a *constant* value for each individual i making up the data, and then assessing the resulting correlation between variable X_1 and X_2.

Table 15.2.1

Direct calculation of the partial correlation

case	x_1	x_2	x_3	x_1'	$x_{(1 \cdot 3)}$	x_2'	$x_{(2 \cdot 3)}$
1	15	6	12	10.975	4.025	8.763	−2.763
2	8	7	7	7.585	0.415	10.458	−3.458
3	9	12	6	6.907	2.093	10.797	1.203
4	5	11	8	8.263	−3.263	10.119	0.881
5	8	13	10	9.619	−1.619	9.441	3.559
6	10	9	13	11.653	−1.653	8.424	0.576

	x_1	x_2	x_3	x_1'	$x_{(1 \cdot 3)}$	x_2'	$x_{(2 \cdot 3)}$
mean	9.17	9.67	9.33	9.167	0.000	9.667	0.000
SD	3.02	2.56	2.56	1.736	2.475	0.868	2.409

$r_{13} = .5742,$ $r_{23} = -.3390$ $r_{12} = -.6173$

$b_{1 \cdot 3} = .6780,$ $b_{2 \cdot 3} = -.3390$ $r_{12 \cdot 3} = -.5487$

As a very simple example of the direct calculation of a partial correlation through adjustment of each value of two variables for regression on a third, look at the six cases shown in the rows of Table 15.2.1. Each has values on three variables, shown in the columns headed x_1, x_2, and x_3.

Now suppose that we wish to adjust both x_1 and x_2 values in order to remove the linear regression of each on x_3. Then the correlation of these adjusted values gives the partial correlation $r_{12 \cdot 3}$, showing the linear relation that remains between x_1 and x_2 when the value of x_3, is, in effect, held constant. Table 15.2.1 shows correlation r_{13} as .5742 with the regression coefficient $b_{1 \cdot 3} = .6780$. Furthermore, the correlation $r_{23} = -.3390$, with $b_{2 \cdot 3} = -.3390$. Then the values of x_1 as predicted from x_3 are shown in the column labeled x_1', and the residual $(x_1 - x_1')$ for each individual is shown in the column labeled $x_{(1 \cdot 3)}$. Similarly, the x_2 values as predicted from the x_3 are shown as x_2', and the residuals $(x_2 - x_2')$ in the column headed $x_{(2 \cdot 3)}$. Then the ordinary Pearson correlation between the residuals (or adjusted values) $x_{(1 \cdot 3)}$ and $x_{(2 \cdot 3)}$ yields the *partial correlation* $r_{12 \cdot 3} = -.5487$.

This direct way of calculating a partial correlation coefficient is not too difficult to do for small sample N, as in Table 15.2.1, although it can be laborious for large N. It is usually easier to employ a formula relating the value of the partial correlation to the original correlations among the variables. That is, it is always possible to find the partial correlation between X_1 and X_2 with X_3 held constant by taking

$$r_{12 \cdot 3} = \frac{r_{12} - (r_{13})(r_{23})}{\sqrt{(1 - r_{13}^2)(1 - r_{23}^2)}}.$$
[15.2.5*]

Since $r_{12} = -.6173$ in our example, then

$$r_{12.3} = \frac{-.6173 - (.5742)(-.3390)}{\sqrt{(1 - (.5742)^2)(1 - (.3390)^2)}}$$

$$= -.5487$$

This form for calculating the partial correlation can be justified quite easily if one remembers that the adjusted variables $X_{(1 \cdot 3)}$ and $X_{(2 \cdot 3)}$ are just the *residuals* or *errors* from predicting X_1 from X_3, and X_2 from X_3. Therefore, the the mean of the $X_{(1 \cdot 3)}$ values must be zero, and so must the mean of the $X_{(2 \cdot 3)}$. Furthermore, the standard deviations of the adjusted $X_{(1 \cdot 3)}$ and $X_{(2 \cdot 3)}$ variables must be the standard errors of estimate:

$$S_{1 \cdot 3} = S_1 \sqrt{1 - r_{13}^2}, \qquad\qquad\qquad [15.2.6^*]$$

and

$$S_{2 \cdot 3} = S_2 \sqrt{1 - r_{23}^2}$$

respectively.

As another example, consider a sample of $N = 75$ male executives in a corporation, who participated in an intensive fitness program sponsored by the company. Each man was put on a program of exercise, and was given a dieting regimen where caloric intake was recorded for each meal. The weight loss for each man was measured after two months on this routine. The relevant variables were then

X_1 = weight loss in pounds.

X_2 = average daily amount of exercise in hours.

X_3 = average daily calorie intake.

X_4 = age of the man.

For the time being we will deal only with the first three of these variables, and reserve variable X_4, or age, for future consideration.

The sample correlations among these three variables are displayed in Table 15.2.2. (Here, as in all the tables of intercorrelations, it is really necessary to write down only the half of the table that is either above or below the main diagonal. Since the Pearson correlation is a symmetric index of linear association, the correlation r_{jk} between the two variables X_j and X_k must be the same as r_{kj} between the variables X_k and X_j.)

Suppose now that initially we are interested in the correlation between weight loss (X_1) and exercise (X_2). On the face of it, a moderate correlation (.43) seems to exist between these two variables. However, weight loss is also correlated moderately, though negatively, ($-.51$) with calorie intake. A moderate positive correlation also exists for exercise and calorie consumption. A reasonable question is, then, "What would the relationship between exercise and weight loss have been if each man had consumed the same daily average number of calories?" In other words, what is the value of the partial correlation, $r_{12 \cdot 3}$ for these data?

	X_1	X_2	X_3	X_4
X_1	1.00	.43	$-.51$	$-.54$
X_2		1.00	.41	$-.12$
X_3			1.00	.23
X_4				1.00

Table 15.2.2

Correlations for exercise example

Thus,

$$r_{12 \cdot 3} = \frac{r_{12} - (r_{13})(r_{23})}{\sqrt{(1 - r_{13}^2)(1 - r_{23}^2)}}$$

$$= \frac{.43 - (-.51)(.41)}{\sqrt{(1 - (-.51)^2)(1 - (.41)^2)}}$$

$$= .8146.$$

The correlation between exercise and weight loss is raised dramatically, from .43 to .81, when calorie consumption is controlled.

Obviously, for three variables X_1, X_2, and X_3, there are *three* partial correlations that can be calculated: $r_{12 \cdot 3}$, $r_{13 \cdot 2}$, and $r_{23 \cdot 1}$. (The value of $r_{21 \cdot 3} = r_{12 \cdot 3}$, $r_{31 \cdot 2} = r_{13 \cdot 2}$, and $r_{23 \cdot 1} = r_{32 \cdot 1}$, of course).

For the present example, these values are

$$r_{12 \cdot 3} = .8146$$

as just calculated, and

$$r_{13 \cdot 2} = \frac{-.51 - (.41)(.43)}{\sqrt{(1 - (.41)^2)(1 - (.43)^2)}}$$

$$= -.8334$$

and

$$r_{23 \cdot 1} = \frac{.41 - (.43)(-.51)}{\sqrt{(1 - (.43)^2)(1 - (-.51)^2)}}$$

$$= .8103.$$

In short, the partial correlations here show that there is a much larger positive correlation between weight loss and exercise when calorie consumption is controlled than is apparent from the original correlation between these two variables. In addition, the partial correlation with exercise held constant reveals a much stronger negative relationship between calorie consumption and weight loss than we might otherwise suspect. Finally, the positive connection between exercise and calorie consumption is strengthened when the variable of weight loss is held constant. Obviously, converting a set of Pearson correlations into partial correlations can affect the impressions one receives about the interrelationships among three or more variables.

15.3 HIGHER ORDER PARTIAL CORRELATIONS

We have so far considered only the simplest case of partial correlation, in which the correlation for two variables is adjusted for a third. This is called a *first-order partial correlation*. These ideas extend quite directly to the situation where there are four or more variables to be considered. Thus, consider four variables, X_1, X_2, X_3, and X_4, where each of N cases has a value on each of these variables. The set of all correlations between pairs of these variables can be calculated, and there will be *six* of these inter-correlations. Then the second-order partial correlation between any two of these variables, with the other two variables held constant may be determined. Thus, the partial correlation of X_1 and X_2, with X_3 and X_4 held constant can be found as follows:

$$r_{12\cdot 34} = \frac{r_{12\cdot 3} - (r_{14\cdot 3})(r_{24\cdot 3})}{\sqrt{(1 - (r_{14\cdot 3})^2)(1 - (r_{24\cdot 3})^2)}} \qquad [15.3.1*]$$

As you can see, the second-order partial correlation is found by use of three of the first-order partial correlations. One of the two variables to be held constant (X_3 in this case) is held constant throughout the expression, and then the remaining variable to be held constant is treated just like this were a first-order partial, in which X_1 and X_2 were to be adjusted for X_4.

Such a second-order partial correlation $r_{12\cdot 34}$ represents the correlation between variables X_1 and X_2 after each has been adjusted to remove linear regression on *both* variables X_3 and X_4. In essence, such a second-order partial correlation gives the correlation between X_1 and X_2 that would have resulted if both X_3 and X_4 had been held constant. (Incidentally, the *order* of a partial correlation is just the number of variables held constant.)

For the data given in the previous section let us find the second-order partial correlation $r_{12\cdot 34}$. By use of the correlations given in Table 15.2.2, one can calculate the first-order partial correlations involving variable X_4 as follows:

$$r_{14\cdot 3} = -.5049$$

$$r_{24\cdot 3} = -.2414$$

Then using the value of $r_{12\cdot 3} = .8146$ already found above, we determine the value of the second-order partial as follows:

$$r_{12\cdot 34} = \frac{.8146 - (-.5049)(-.2414)}{\sqrt{(1 - (-.5049)^2)(1 - (-.2414)^2)}}$$

$$= .8270 \ .$$

Note that for this example, holding both calorie consumption and age constant yields a correlation between weight loss and exercise that is just a little larger than the correlation holding only calorie count constant, as found in the previous section. In a similar way we could find the the second-order partial correlation for any other pair of variables holding the remaining two constant.

Given more variables, even higher order partial correlations can be found. In general, let the subscript a stand for one variable X_a, b for another variable X_b, c for a third variable X_c, and also let the symbol G stand for a set of $K - 1$ other variables to be held constant. Then the Kth-order partial correlation between X_a and X_b, with the variable X_c as well as those in set G held constant, is

$$r_{ab\cdot cG} = \frac{r_{ab\cdot G} - r_{ac\cdot G}\, r_{bc\cdot G}}{\sqrt{(1 - r_{ac\cdot G}^2)\,(1 - r_{bc\cdot G}^2)}}. \qquad [15.3.2]$$

Obviously, partial correlations find their main use in the study of the interrelations that exist in a set of three or more variables, especially in terms of how the apparent linear relationship between a pair of variables can be influenced by some other variable or set of variables. In addition, we will see that the concept of partial correlation is an important building block in the theory of multiple regression analysis, to be introduced later in this chapter.

15.4 PART OR SEMIPARTIAL CORRELATIONS

Recall that in finding a partial correlation such as $r_{12 \cdot 3}$ in Section 15.2 above, both the variable X_1 and the variable X_2 were adjusted to remove the linear regression on variable X_3. That is, the partial correlation $r_{12 \cdot 3}$ is actually the correlation between the two adjusted variables that can be symbolized by $X_{(1 \cdot 3)}$ and $X_{(2 \cdot 3)}$.

Now, however, imagine the situation where we want to remove the part of X_2 due to X_3, to form the new variable $X_{(2 \cdot 3)}$, but we *do not* wish to adjust variable X_1 at all. What is desired is the correlation between the unadjusted variable X_1 and the adjusted variable $X_{(2 \cdot 3)}$, as symbolized by $r_{1(2 \cdot 3)}$. Such a correlation between an adjusted and an unadjusted variable is known as a *part correlation* (or sometimes a *semipartial correlation*). One speaks of ''the part correlation between variable X_1 and variable X_2, where X_2 has been adjusted for X_3.''

Bear in mind that, unlike partial correlations, where both of the variables to be correlated have been adjusted for one or more other variables, in part correlations only *one* of the variables being correlated has been adjusted. For this reason partial and part correlations do not usually agree in value. Although they do agree in sign, part correlations have smaller absolute values than partial correlations.

Thus, for example, for the data on men in the fitness program, given in Section 15.2, we might be interested in the correlation between weight loss, variable X_1, and exercise, X_2, but where exercise has been adjusted for calorie consumption, or X_3, to yield variable $X_{(2 \cdot 3)}$. This is *not* the same as the partial correlation of weight loss and exercise where calorie consumption has been held constant, and thus removed as a source of variation from both exercise and weight loss. Here, with a part correlation, we wish to remove calorie consumption as a source of variation from the exercise scores, but not from the weight-loss numbers.

Like partial correlations, part correlations can also be found directly, by adjusting one variable and then correlating the results with the unadjusted values of the other variable. However, again as is the case with partial correlations, it is usually much easier to find the desired value by use of the original correlations and the following equations than by adjusting one of the variables to remove the linear effects of a third.

The equation for the part correlation between X_1 and X_2, where only variable X_2 has been adjusted for a third variable X_3, is:

$$r_{1(2 \cdot 3)} = \frac{r_{12} - r_{13} r_{23}}{\sqrt{(1 - r_{23}^2)}}. \qquad [15.4.1^*]$$

Hence, for our example from Table 15.2.2,

$$r_{1(2 \cdot 3)} = \frac{.43 - (-.50)(.41)}{\sqrt{(1 - (.41)^2)}}$$

$$= .70.$$

Note that, for this example, the correlation between weight loss and exercise is raised from .43 to .70 if the relationship to calorie consumption is removed only from exercise, whereas the partial correlation, with both variables adjusted for calorie consumption was found to be about .81.

Given any three variables such as X_1, X_2, and X_3, it is possible to find *six* part correlations, including $r_{1(2 \cdot 3)}$, as found above, as well as:

$$r_{1(3 \cdot 2)} = \frac{r_{13} - r_{12} \, r_{32}}{\sqrt{(1 - r_{32}^2)}} \qquad [15.4.2]$$

$$r_{2(1 \cdot 3)} = \frac{r_{21} - r_{23} \, r_{13}}{\sqrt{(1 - r_{13}^2)}} \qquad [15.4.3]$$

and so on, *mutatis mutandis*, for $r_{2(3 \cdot 1)}$, $r_{3(1 \cdot 2)}$, and $r_{3(2 \cdot 1)}$.

Like a partial correlation, a part correlation can also be found for two variables where one of the variables has been adjusted for more than one other variable. Such a K-order part correlation, where there is adjustment for K other variables, can be found from the *partial correlations* of order $K - 1$. Thus, the second-order part correlation symbolized by $r_{1(2 \cdot 34)}$ is found from the first-order *partial* correlations as follows:

$$r_{1(2 \cdot 34)} = [r_{12 \cdot 3} - (r_{14 \cdot 3})(r_{24 \cdot 3})] \sqrt{(1 - r_{13}^2)/(1 - r_{24 \cdot 3}^2)}. \qquad [15.4.4]$$

Thus, for example, the correlations in Table 15.2.2 give the following partial correlations:

$$r_{12 \cdot 3} = .8146, \; r_{14 \cdot 3} = -.5049, \; r_{24 \cdot 3} = -.2414$$

so that the value of the part correlation $r_{1(2 \cdot 34)}$ is

$$r_{1(2 \cdot 34)} = [.8146 - (-.5049)(-.2414)] \sqrt{[1 - (.51)^2]/[1 - (.2414)^2]}$$

$$= .6140.$$

This second-order part correlation shows the relationship between weight loss and exercise, where the measure of exercise has been adjusted both for regression on calorie intake and on age. Alternatively, such a part correlation could be found from

$$r_{1(2 \cdot 34)} = r_{12 \cdot 34} \sqrt{(1 - r_{13}^2)(1 - r_{14 \cdot 3}^2)}. \qquad [15.4.5]$$

Similar expressions can be written for finding even higher order part correlations in terms of partial correlations. One seldom needs to calculate such high-order part correlations, however, as the information they contain can usually be gained more efficiently through multiple regression methods, as we shall see.

15.5 EXPLAINING VARIANCE THROUGH PART AND PARTIAL CORRELATIONS

In Chapter 14 we learned that one important feature of the correlation coefficient is the interpretation that can be given to its square. That is, given the correlation coefficient r_{xy} then, the value of r^2 can be interpreted as the proportion of Y variance that can be accounted for by linear regression on X, or, alternatively, as the proportion of X variance that can be accounted for by regression on Y. A simple diagram such as that given in Figure 15.5.1 is sometimes useful in helping one visualize this interpretation of r-square. The circle in Figure 15.5.1 can be taken to represent the total variance of variable Y, or S_Y^2. Then the shaded area of the circle can be thought of as the variance in Y that is actually due to regression on X. As we already know, the Y variance due to regression on X is

$$\text{variance due to regression on } X = r_{xy}^2 S_Y^2 \qquad [15.5.1^*]$$

and

$$\text{variance not due to regression on } X = S_Y^2 (1 - r_{xy}^2). \qquad [15.5.2^*]$$

Therefore, in the figure the ratio of the shaded area to the total area is, obviously,

$$\frac{r_{xy}^2 S_Y^2}{S_Y^2} = \text{proportion of } Y \text{ variance explained by regression on } X$$

$$= r_{xy}^2. \qquad [15.5.3^*]$$

Now consider once again three variables labeled X_1, X_2, and X_3. The first circle shown in Figure 15.5.2 can then be taken to represent the total variance of variable X_1, with the shaded sector standing for the variance that can be explained by linear regression on X_3. The ratio of the shaded sector to the whole circle then represents the value of the squared correlation r_{13}^2.

Now take a look at the second circle shown in Figure 15.5.2. This time the area representing the variance accounted for by X_3 is still shown, just as before. However,

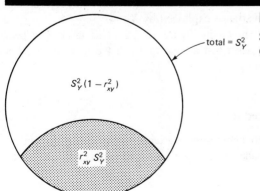

Figure 15.5.1

Schematic representation of Y variance accounted for by variable X.

total = S_Y^2

$S_Y^2 (1 - r_{xy}^2)$

$r_{xy}^2 S_Y^2$

Figure 15.5.2

Schematic representation of X_1 variance accounted for by X_3, the squared part correlation $r^2_{1 \cdot (23)}$, and the squared partial correlation $r^2_{12 \cdot 3}$.

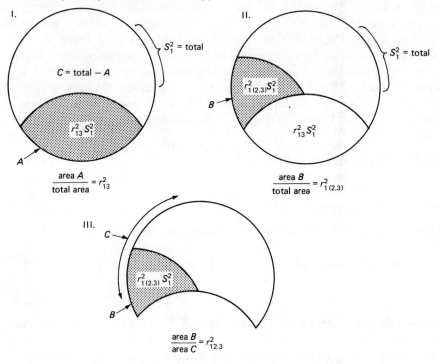

the shaded sector now represents X_1 variance that is explained by variable X_2 but *not* by X_3. The ratio of this shaded sector to the total circle shows the squared part correlation $r^2_{1(2 \cdot 3)}$, so that $r^2_{1(2 \cdot 3)}$ = proportion of total X_1 variance due to X_2 but *not* to X_3. That is, as the variable $X_{(2 \cdot 3)}$ entering into this part correlation has been adjusted to remove regression on X_3, then any X_1 variance it accounts for *cannot* be due to X_3. A squared part correlation such as $r^2_{1(2 \cdot 3)}$ then can be interpreted as the proportion of X_1 variance explained by X_2, over and beyond any variance explained by X_3.

Now take a look at the third circle shown in Figure 15.5.2. This circle has been altered to remove the portion that was described as due to X_3 in the first circle. Now the remaining area of the circle is equal to

$$S^2_{1 \cdot 3} = S^2_1 (1 - r^2_{13})$$

> = residual variance of X_1 adjusted for X_3 .

Again in this third figure the shaded area represents variance explained by X_2 but not by X_3. The relative proportion of this shaded area to the unshaded area is now, however,

$$r^2_{12 \cdot 3} = r^2_{1(2 \cdot 3)} S^2_Y / (S^2_Y (1 - r^2_{13}))$$

> = proportion of X_1 variance *not* due to X_3 that *is* explained by regression on X_2.

In other words, the squared part correlation $r^2_{1(2\cdot3)}$ represents a proportion relative to the *total* variance of X_1, *not* adjusted for X_3, whereas the squared partial correlation $r^2_{12\cdot3}$ represents a proportion relative to the *residual variance* of X_1, after extraction of the variance that is due to linear regression on X_3. Since the residual variance of X_1 after removal of regression on X_3 will be less than or equal to the total variance of unadjusted X_1, this explains why the squared partial correlation $r^2_{12\cdot3}$ may never be less than the squared part correlation $r^2_{1(2\cdot3)}$.

In future sections we are going to use these ideas of squared part and partial correlation, and drawings such as Figures 15.5.1–2 to help us understand certain aspects of multiple regression. For the moment, however, we need to discuss inferences concerning partial and part correlations.

15.6 INFERENCES ABOUT PARTIAL AND PART CORRELATIONS

A partial correlation coefficient may be tested for significance in much the same way as a regular (zero-order) Pearson correlation. That is, a sample of N cases is assumed to come from a multivariate normal distribution. Then it is permissible to use the Fisher r to Z transformation, just as in Section 14.21 to test hypotheses and to find confidence intervals for $\rho_{12\cdot3}$, the population partial correlation. However, the standard error of Z here becomes

$$\sigma_z = \frac{1}{\sqrt{N - 3 - (K - 2)}},$$ [15.6.1]

where K is the total number of variables considered. Thus, for three variables ($K = 3$), when the partial corrlation $r_{12\cdot3}$ is used to test the hypothesis that $\rho_{12\cdot3}$ is, say 0, the value of $r_{12\cdot3}$ is changed to a corresponding Z value via Table VII (Appendix E); and the test is carried out using the normal distribution of the statistic

$$\frac{Z - 0}{[1/\sqrt{N - 3 - (3 - 2)}]} = Z[\sqrt{(N - 4)}].$$ [15.6.2]

If this statistic shows a value significant at the α level, then the hypothesis that $\rho_{12\cdot3} = 0$ can be rejected at the α level. All in all, when the problem is one of *correlation*, and the assumption of a multivariate normal population is tenable, any procedure using the r to Z transformation applicable to a simple correlation also applies to a partial correlation, provided that the standard error of Z is adjusted. However, the assumption of a multivariate normal distribution (all variables normally distributed, all pairwise joint events such as (x_1, x_2) bivariate normal, and so on) is very unreasonable in many instances, and such tests should not be undertaken lightly.

If a partial correlation such as $r_{12\cdot3}$ is 0, then the associated part correlations, $r_{1(2\cdot3)}$ and $r_{2(1\cdot3)}$ must also be 0, as all three share the same numerator. Hence, a significance test for a partial correlation can also be regarded as a test for the part correlations that share this feature.

There are occasions, nevertheless, when we will want to test the significance of a part correlation, without having to worry about the related partial correlations. These occasions will arise in connection with multiple regression methods, however, and so we will defer their consideration until later in this chapter.

15.7 TESTING SIGNIFICANCE FOR INTERCORRELATIONS

Before we leave the topic of partial correlation, a word must be said about significance tests for intercorrelations. It is quite common to find research in psychology and other fields where a number of different variables are studied in the same sample, and all sample intercorrelations are found among these variables. For example, a study may concern three variables, X_1, X_2, and X_3, and values are found for r_{12}, r_{13}, and r_{23}. This in itself is fine as a description of linear relations in the data, and is the first step in virtually any multivariate analysis, such as finding a multiple-regression equation by the methods in the remainder of this chapter.

However, one often finds the experimenter testing the significance of each one of these $\binom{K}{2}$ intercorrelations by the method of Section 14.21 as though each one were based on a different sample. The resulting significance levels are largely meaningless, for reasons much like those making t tests for all differences among a set of means a dubious procedure. In the first place, even for independent tests of significance, when so many tests are carried out the probability that some Type I errors are being made may be very high. Even worse, the t tests for correlations are quite redundant and are far from statistically independent when carried out on a table of intercorrelations. Consequently, the set of results can be grossly misleading. In particular, one should ordinarily expect *more* than $\binom{K}{2} \alpha$ such tests to show significance by chance alone.

It is simple to illustrate that dependencies must exist among intercorrelations. Imagine a sample of N cases, each of which provides three score values X_1, X_2, and X_3. Imagine that r_{12} turns out to be $-.80$ and that r_{13} is also equal to $-.80$. What is the smallest value that r_{23} may be? We can determine this by examining the partial correlation coefficient for $r_{23.1}$. Since we know that this partial correlation coefficient must be greater than or equal to -1, we take

$$r_{23.1} = \frac{r_{23} - r_{12}r_{13}}{\sqrt{(1 - r_{12}^2)(1 - r_{13}^2)}}$$

$$= \frac{r_{23} - (-.80)^2}{[1 - (-.80)^2]} \geq -1.00,$$

and then

$$r_{23} - .64 \geq -1 + .64,$$

so that

$$r_{23} \geq .28.$$

Th smallest value that r_{23} can possibly have, and still be consistent with r_{12} and r_{13} is thus .28. On the other hand, if $r_{12} = -.5$ and $r_{13} = -.5$, then r_{23} can also be as low as $-.5$. This value would make the partial correlation $r_{23.1}$ be equal to exactly -1. Fixing the value of two of the correlations determines the necessary *lower* limit of the third. Higher order patterns of dependency exist among any set of intercorrelations, and these patterns may be used to fix a lower bound for any subset of the intercorrelations. Such lower bounds may be calculated in a variety of ways. However, the following rule gives an absolute lower bound for the average of the intercorrelations.

In general, for K variables, the average of the $\binom{K}{2}$ intercorrelations among these variables must be greater than (or equal to) $- 1/(K - 1)$.

It follows that given the values of some of the intercorrelations, the average lower limit for all the other correlations is not -1, but some number greater than -1. The larger K is, the closer this lower limit comes to 0. Hence, it is somewhat pointless to treat each of the correlations in turn as though the sampling distribution of values could extend from -1 through $+1$, when with each successive value of r known from the sample the possible lower limit to the next set of values is raised. One should either not test for significance in the ordinary way in dealing with intercorrelations found for a single sample, or should interpret the significance levels with *considerable* latitude.

(Note that the Bonferroni approach of dividing α by the number of tests, or $\binom{K}{2}$, is not very practical when K is large.)

15.8 MULTIPLE REGRESSION

As we saw in the preceding chapter, bivariate regression theory is concerned with the linear association between two variables: a predictor variable X and a dependent variable Y. If we desire to use X to find the best-fitting linear prediction of the value of Y, or y', then, according to the principle of least squares, the required regression equation is of the form

$$y' = a + bx$$

where

$$a = (\bar{y} - b\bar{x})$$

and

$$b = b_{y \cdot x} = \text{cov}(x,y)/S_X^2$$

or, equivalently,

$$b_{y \cdot x} = (r_{xy}S_Y)/S_X$$

(The matrix version of these concepts may be found in Appendix D. Eq. 6.31 and the following.)

This same basic idea extends directly to the situation where there is a single dependent variable Y, but two or more distinct predictor variables, $X_1, X_2, \ldots, X_K$. Our goal is to form a prediction equation, or multiple regression equation, by weighting and summing the values of the X variables in such a way as to get the best possible prediction of the value of Y for any individual.

You will recognize that this is but another application of the general linear model (GLM) we first discussed in Section 10.2. At first we applied the GLM to the situation where each X stands for the group membership of a particular case, and this led to the model for ANOVA, as applied in Chapters 10 and 12. Moreover, in the preceding chapter, the GLM was seen in another version, where there was one quantitative predictor variable X to go with one quantitative dependent variable Y: This led to the model of bivariate linear regression.

Now we are ready to use the more extended version of the GLM, as in Eq. 10.2.1, where there are two or more X variables, each of which represents a quantitative value or score on some measured property of an individual. These values are to be weighted and summed to provide a prediction of the quantitative variable Y. This version of the GLM, the *multiple linear regression model,* will be developed and used in the sections to follow.

15.9 MULTIPLE REGRESSION EQUATIONS

Think of a sample of N distinct individuals, each of whom has values on each of K variables, X_1 through X_K. We can think of the set of these X variables as the *independent variables,* or the *predictor variables*. Any individual i will have a value x_{ij} on each of these predictor variables, X_j.

In addition, within a given set of data each individual i has a value y_i on the dependent variable Y. The point is then to use the predictor values x to find a good bet or estimate of the value of y. The linear model relating the value of y to the values of x shown by an individual then takes the form of a weighted sum:

$$y_i = a + b_1 x_{i1} + b_2 x_{i2} + b_3 x_{i3} + \cdots + b_K x_{iK} + e_i. \qquad [15.9.1^*]$$

Here each individual i out of a set of N individual cases has not only a value of y, but also values on each of the K predictor variables, $x_{i1}, \ldots, x_{iK}$. The model that we will assume to be true states that the observed value of y for individual i consists of a constant a, plus the weighted sum of all of the x values. Finally, the score y also contains a component e_i, the "error," or the part of the y value for individual i that cannot be predicted in a linear way from the information contained in one or more of the x values.

Now suppose that we do not know the exact y value for some individual. Nevertheless, we would still like to be able to make a guess or prediction about what that value might be, based on the set of x values that we do know. Given the value of the constant a and all of the different weights b, then we could make a prediction of the value of y on the basis of the x values. Let us symbolize this predicted y value for individual i by

y'_i once again. Then the prediction equation or regression equation for the value of y'_i is given by

$$y'_i = a + b_1 x_{i1} + b_2 x_{i2} + \cdots + b_K x_{iK}. \qquad [15.9.2^*]$$

Notice that for any individual i and for any set of X variables used as predictors, we would not expect the prediction of the actual y_i value to be exactly right. On the other hand, it stands to reason that the accuracy of our prediction should tend to improve as we increase the number of items of relevant X information that we have available, and this is indeed the case. Nevertheless, many other things besides the factors represented by the x values may be influencing and helping to determine the value of y for any given individual. Thus, it follows that our prediction will always be in error to some extent. The amount of the error is e_i, which is the difference between the predicted value based on the X variables alone, and the actual value of variable Y for individual i.

That is

$$e_i = \text{error for individual } i \qquad [15.9.3]$$
$$= (y_i - y'_i).$$

Furthermore, if such predictions were made for each of the N cases then there would be an average error

$$\bar{e} = \sum_i e_i, \qquad [15.9.4]$$

and, like any set of values, these errors will also have a variance:

$$S_e^2 = \sum_i (e_i - \bar{e})^2/N. \qquad [15.9.5]$$

The variance of the errors in prediction in a given sample is called the *sample variance of multiple estimate*. Furthermore, the standard deviation of these errors of prediction, or the square root of the variance of multiple estimate, is called the *sample standard error of multiple estimate*. These are important concepts, and we will put them to good use in the sections to follow.

15.10 STANDARDIZED MULTIPLE REGRESSION EQUATIONS

The multiple regression equation shown above as Eq. 15.9.2 is based on the raw values of variable X, and is designed for the prediction of raw Y values. However, it is equally possible to define a multiple regression equation for standardized value z_{yi} based on the predictor variable values $z_{i1}, z_{i2}, \ldots, z_{iK}$. Then the multiple regression equation becomes

$$z'_{yi} = A + B_1 z_{i1} + B_2 z_{i2} + \cdots + B_K z_{iK}. \qquad [15.10.1^*]$$

(As in the preceding chapter, the capital letters A and B will be used to stand for the constant and the various regression weights when we are discussing standardized values, whereas lowercase letters a and b will be used for a raw-score or unstandardized multiple regression equation.)

Bear in mind that each standardized variable z_y and each z_1, z_2, and so on, has a mean of 0 and a standard deviation of 1.00. Once again, when the z'_{yi} value is predicted for an individual i, there will be a certain amount of error:

$$e_{zi} = (z'_{yi} - z_{yi}).$$

For the entire set of N cases there will be an average error

$$\bar{e}_z = \sum_i (z'_{yi} - z_{yi})/N.$$

The variance of these errors is called the *sample variance of multiple estimate for standardized values,* and is defined by

$$S^2_{ez} = \sum_i e^2_{zi}/N - \bar{e}^2_z. \qquad [15.10.2]$$

The *sample standard error of multiple estimate for standardized values* is then the square root of the variance of multiple estimate.

15.11 FINDING THE STANDARDIZED REGRESSION WEIGHTS

To find and use a multiple regression equation, it is, of course, necessary to determine the value of the constant (A or a) and the weights (B or b) to be applied to the X variables. In this section we will discuss how these values are found. For the sake of simplicity the multiple regression equation for standardized variables will be considered first. (The correct term for the B weights that enter into a regression equation for standardized values is *standardized partial regression weights* or *standardized partial regression coefficients*. In the same way the b values should be called the *unstandardized partial regression weights or coefficients*. However, in this chapter where we will be discussing only multiple regression situations, the simpler and less awkward terms *standardized regression weights* and *unstandardized regression weights* will be used. Furthermore, some authors use rather elaborate symbols such as $B_{yj \cdot 12 \cdots (j) \cdots K}$ and $b_{yj \cdot 12 \cdots (j) \cdots K}$ to stand for these partial regression weights, and to emphasize how these values depend on the whole set of X variables. In the interest of simplifying an overly complex notation, we will symbolize the standardized partial regression coefficient for variable X_j by B_j, with b_j standing for the unstandardized version. Even so, it must always be remembered that these values are relative to the particular set of predictor variables X to which the variable X_j belongs. The value of B_j or b_j relative to one set of predictors may well be different from the values these symbols represent when an even slightly different set of predictors is being used.)

Clearly, at least two things are desirable in prediction: First, we would like for our predictions to be *unbiased,* so that we are exactly right on the average. This implies that the mean of the errors, symbolized by $\bar{e}_z$, should be 0 over all N cases. However, for standardized values, imposing this requirement determines the value of the constant a, as follows:

If

$$\bar{e}_z = \sum_i e_{zi}/N = 0$$

then

$$\sum_i (z_{yi} - z'_{yi})/N = 0$$

so that

$$\sum_i (z_{yi} - A - B_1 z_{i1} - \cdots - B_K z_{iK})/N = 0$$

It follows that

$$\bar{z}_y - A - B_1\bar{z}_1 - \cdots - B_K\bar{z}_K = 0.$$

However, since we are dealing with standardized values, *all* of the means in the equation above are equal to zero. Therefore, it must be true that for standardized scores,

$$A = 0. \qquad [15.11.1]$$

What we have shown is that if the predictions z'_y are to be unbiased, with zero error on the average, then the constant A must be equal to 0 in the multiple regression equation for standardized scores.

The second thing we want to be true in prediction is that the errors tend to be as small as possible. We have just seen that by our choice of the constant A we can make the mean error equal to zero, which implies that the sum of the errors over all N individuals must also be zero. Hence, instead of dealing with the raw errors, we consider the squared errors, e_{zi}^2. Applying the principle of least squares once again allows one to choose values of the B weights that will make the average squared errors as small as possible (i.e., minimizing the variance of multiple estimate for the standardized scores.)

This involves minimizing the expression

$$S_{ez}^2 = \sum_i (z_{yi} - z'_{yi})^2/N$$

$$= \sum_i (z_{yi} - B_1 z_{i1} - B_2 z_{i2} - \cdots - B_K z_{iK})^2/N$$

$$= \left[\sum_i z_{yi}^2 + \sum_i \sum_j B_j^2 z_{ij}^2 + 2 \sum_i \sum_{j<k} B_j B_k z_{ij} z_{ik} - 2 \sum_i \sum_j B_j z_{ij} z_{yi} \right]/N$$

$$= 1 + \sum_j B_j^2 + 2 \sum_{j<k} B_j B_k r_{jk} - 2 \sum_j B_j r_{yj}. \qquad [15.11.2]$$

Now this expression, which involves the B weights, the correlations r_{jk} between pairs of predictors, X_j and X_k, and the correlations r_{yj} between Y and each predictor X_j, is treated by methods from the differential calculus to determine the B values which make the average squared errors a minimum. (The details of how this is done need not detain us here; however, if you have some background in the calculus you may recognize that what is required is the partial derivative taken relative to each of the B_j, with each derivative then set equal to zero.)

The result of applying the least squares criterion then is the following set of equations, with one equation for each of the K predictor variables z_j: That is, we have

$$B_1 \quad + B_2 r_{12} + \cdots + B_K r_{1K} = r_{1y}$$

$$B_1 r_{21} + B_2 \quad + \cdots + B_K r_{2K} = r_{2y}$$

$$\cdots\cdots\cdots\cdots\cdots\cdots\cdots\cdots\cdots\cdots\cdots\cdots$$

$$B_1 r_{j1} + B_2 r_{j2} + \cdots + B_K r_{jK} = r_{jy}$$

$$\cdots\cdots\cdots\cdots\cdots\cdots\cdots\cdots\cdots\cdots\cdots\cdots$$

$$B_1 r_{K1} + B_2 r_{K2} + \cdots + B_K \quad = r_{Ky}.$$

[15.11.3*]

These equations, one for each X variable, are known as the set of *normal equations* for standardized scores. Notice that the normal equations for standardized variables each express the correlation between Y and one of the X variables as a weighted sum of the correlations among the X variables.

The problem is still not really solved until we determine the K different values of B required by the multiple regression equation, and in order to do so we have to solve the normal equations. In fact, the main computational problem in multiple regression is usually the simultaneous solution of the K normal equations to determine the required values of the B weights. Pencil and paper solution of simultaneous equations such as these can be very laborious unless the number of predictor variables is quite small, and we can all be thankful that these days computer methods are employed to solve almost all multiple regression problems.

However, it is still instructive to see the steps that these computations require, even in problems of very limited size, and therefore in the next section we will use one of the standard "hand" methods to solve a problem with two predictors. Later we will be able to deal with somewhat larger problems employing a simple method that is practicable for use with up to six or eight variables. This will be introduced in Section 15.27. For the moment we will merely illustrate the normal equations and their solution for the simplest possible situation, involving only two normal equations. (This classic method of finding the weights required in a multiple regression equation is sometimes called *simultaneous regression* because it requires solving simultaneous equations, and to distinguish it for the "hierarchical" and "stepwise" methods of solution to be discussed toward the end of this chapter. There we will consider the X variables one at a time rather than all together.)

15.12 A SIMPLE EXAMPLE OF MULTIPLE REGRESSION

Suppose that an employer has a measure of job productivity for the person he or she hires. Let us call this productivity measure Y. In addition, there are two other variables to be used as predictors:

X_1 = score on a test of manual dexterity

X_2 = score on a test of attention to detail.

These two measures are believed to be potentially useful in predicting productivity on the job. To test this notion, the employer gives both the manual dexterity test and the attention test to $N = 50$ new employees. Then, after six weeks, the employer records the score for each of these employees on the productivity measure. The question is, "How well can you predict the relative productivity of employees based on their relative standing on the test scores?"

We interpret this as an attempt to find the multiple regression equation for predicting z_y, the standardized value on Y, from z_1, the standardized value on X_1, and z_2, the standardized value on X_2.

In such a problem the first step is to calculate the intercorrelations among all of the variables. Suppose that for these 50 employees the correlations turned out to be those shown in the table below:

	X_1	X_2	Y
X_1	1	$-.32$	.41
X_2	$-.32$	1	.56
Y	.41	.56	1

Given these correlations, then to find the standardized multiple regression equation we must solve two normal equations of the form shown by Eq. 15.11.3 above:

$$B_1 - .32 B_2 = .41$$

$$-.32 B_1 + B_2 = .56$$

Equations such as these can be solved in any of several ways. However, here we will use the *method of determinants,* as shown in Section D.2 of Appendix D.

First we find the determinant of the matrix involving only the correlations among the X variables:

$$\mathbf{D}(X) = \textbf{deter.} \begin{bmatrix} 1 & -.32 \\ -.32 & 1 \end{bmatrix}$$

$$= 1 - (-.32)(-.32)$$

$$= .8976.$$

Next, we substitute the correlations with Y into the column for the X_1 and take the determinant once again:

$$\mathbf{D}(1) = \textbf{deter.} \begin{bmatrix} .41 & -.32 \\ .56 & 1 \end{bmatrix}$$

$$= (.41)(1) - (.56)(-.32)$$

$$= .5892.$$

Then the value of B_1 is easily found by taking

$$B_1 = \mathbf{D}(1)/\mathbf{D}(X)$$

$$= .5892/.8976$$

$$= .6564.$$

Next we substitute the correlations with Y into the column for X_2 and take the determinant:

$$\mathbf{D}(2) = \mathbf{deter.} \begin{bmatrix} 1 & .41 \\ -.32 & .56 \end{bmatrix}$$

$$= (1 \times .56) - (-.32 \times .41)$$

$$= .6912.$$

The value of B_2 is found from

$$B_2 = D(2)/D(X)$$

$$= .6912/.8976$$

$$= .7701.$$

The problem of the B weights to be applied in the multiple regression equation is therefore solved. We now know that to predict the z'_y for an individual, on the basis of the z_1 and the z_2 scores, we would use the standardized regression equation

$$z'_y = .6564(z_1) + .7701(z_2).$$

Notice that the weight B_2 given to the standardized value z_2 is somewhat greater than the weight B_1 given to the standardized value z_1. This attests to the fact that the standardized value on variable X_2 carries somewhat more predictive power relative to z_y than does the standardized score on variable X_1.

15.13 MULTIPLE REGRESSION EQUATIONS FOR RAW SCORES

The rationale for finding the b weights in the multiple regression equation for raw scores is identical to that for finding the B weights applicable to standardized values, except that the normal equations for raw scores which once again demand solution are formed in terms of variances and covariances rather than correlations. Even so, their rationale is really the same and the raw score b weights can be found directly from the standardized B weights (and vice-versa), without the necessity for solving another set of normal equations. All that is needed in addition to the correlations among all the variables is the mean and variance (or the standard deviation) for each one.

The transformation of standardized B weights to unstandardized b values can be shown easily enough by converting the z values in the standardized regression equation back into the raw scores they represent:

$$z'_{yi} = B_1 z_{i1} + B_2 z_{i2} + \cdots + B_K z_{iK}$$

or

$$\frac{(y'_i - \bar{y})}{S_y} = \frac{B_1(x_{i1} - \bar{x}_1)}{S_1} + \frac{B_2(x_{i2} - \bar{x}_2)}{S_2} + \cdots + \frac{B_K(x_{iK} - \bar{x}_K)}{S_K}$$

Rearranging terms we get

$$y_i' = \bar{y} + B_1(S_y/S_1)(x_{i1} - \bar{x}_1) + \cdots + B_K(S_y/S_K)(x_{iK} - \bar{x}_K)$$

$$= (\bar{y} - b_1\bar{x}_1 - b_2\bar{x}_2 - \cdots - b_K\bar{x}_K) + b_1x_{i1} + \cdots + b_Kx_{iK}$$

$$= a + b_1x_{i1} + b_2x_{i2} + \cdots + b_Kx_{ik}$$

Thus, the weight b_j to be applied to any predictor variable value x_j in the raw-score multiple regression equation is given by

$$b_j = B_j(S_y/S_j) \qquad [15.13.1^*]$$

where S_j represents the standard deviation of the x_j values in the sample, and S_y the standard deviation of the dependent variable values y. Note, however, that for the raw-score regression equation, the constant a is not 0, but rather is found from

$$a = (\bar{y} - b_1\bar{x}_1 - b_2\bar{x}_2 - \cdots - b_K\bar{x}_K)$$

$$= \bar{y} - \sum_{j}^{K} b_j\bar{x}_j. \qquad [15.13.2^*]$$

That is, in the raw-score multiple regression equation, the required constant a consists of the Y mean, minus the sum of the X means each weighted by the corresponding b value.

For the current example, suppose that the means and standard deviations of the variables were as follows:

variable	mean	standard deviation
X_1	52	5.8
X_2	46	7.3
Y	63	11.0

Then the unstandardized b weights, as found from the standardized B weights of Section 15.12 turn out to be

$$b_1 = (0.6564 \times 11.0)/(5.8)$$

$$= 1.24$$

$$b_2 = (0.7701 \times 11.0)/(7.3)$$

$$= 1.16.$$

The constant a for the regression equation is then

$$a = 63 - (1.24 \times 52) - (1.16 \times 46)$$

$$= -54.84.$$

The multiple regression equation for predicting raw y values, given the raw values, x_1 and x_2 for any individual is thus

$$y' = -54.84 + 1.24x_1 + 1.16x_2.$$

For instance, if an individual had a value $x_1 = 58$, and a value $x_2 = 43$, then the predicted value y' would be found from

$$y' = -54.84 + (1.24 \times 58) + (1.16 \times 43)$$

$$= 66.96$$

and so on for any other pair of X variable values.

15.14 THE COEFFICIENT OF MULTIPLE CORRELATION

Now that we know how to form a multiple regression equation for predicting either standardized or raw scores, the next question to come to mind is, "Just how good are the predictions made by use of this equation?" This is the information provided by the *coefficient of multiple correlation,* usually symbolized by R, and by R^2, sometimes called the *coefficient of multiple determination* (but already familiar to us as the *proportion of variance accounted for*).

Suppose that we have a set of N cases, and have constructed the multiple regression equation,

$$z'_{yi} = B_1 z_{i1} + B_2 z_{i2} + \cdots + B_K z_{iK}$$

for predicting the standardized value of Y from the standardized values of X_1, X_2, and so on up to X_K.

There will be a predicted value z'_{yi} possible for each case i in our sample, and an actual value z_{yi}. If our equation is doing a good job of predicting, then the predicted and the actual values should agree closely with each other, and there should be a high correlation between the predicted and the actual values. This correlation between actual values and predicted values is the *multiple correlation coefficient, R:*

R = Pearson correlation between z'_y and z_y values.

Therefore, it must also be true that

R = Pearson correlation between y' and y values.   [15.14.1]

Indeed one can think of B_j weights for standardized variables as being those values that will *maximize* the Pearson correlation between the original z_y and the predicted z'_y values. In the same way the unstandardized values b_j are found so as to maximize the correlation between the original y values, and the predicted y' values for a given set of data.

Fortunately, it is not really necessary to predict for each case and then correlate these predictions with the actual values in order to determine the R value. Rather, the value of the multiple correlation R can be found quite simply by taking

$$R_{y \cdot 12 \cdots K} = \sqrt{B_1 r_{y1} + B_2 r_{y2} + \cdots + B_K r_{yK}}.$$ [15.14.2*]

That is, the multiple correlation between Y and the set of K predictor variables X_1 through X_K is given by the square root of the weighted sum of the correlations of each X variable with Y, where the weights are the regression weights B_j that apply to each variable X_j in the standardized regression equation. (The subscript $y \cdot 12 \cdots K$ is the conventional way to symbolize, "for Y predicted on the basis of K variables, X_1, X_2, etc. However, this subscript format can prove awkward, and from time to time we will use other symbols for this purpose.)

Thus, in our example of Section 15.12, the correlation of Y with X_1 is .41 and the standardized regression weight B_1 is .6564. We also have a correlation between Y and X_2 of .56 with a B_2 value of .7701. Then the value of the multiple correlation coefficient is

$$
\begin{aligned}
R_{y \cdot 12} &= \sqrt{B_1 r_{y1} + B_2 r_{y2}} \\
&= \sqrt{(.6564 \times .41) + (.7701 \times .56)} \\
&= .8369, \text{ or about } .84.
\end{aligned}
$$

Clearly, there is excellent agreement here between the predicted and actual values, as shown by the value of R.

The derivation of this formula for finding the value of R in terms of the B values and the correlations with Y is not difficult to show. What we want is the correlation between the standardized values and the predictions, or

$$
r \text{ between } z_y \text{ and } z_y' = \sum_i z_{yi} z_{yi}' / (N S_{z_y} S_{z_y'}) \tag{15.14.3}
$$

Now for the moment let us consider only the numerator of this expression, along with the divisor N:

$$
\begin{aligned}
\sum_i z_{yi} z_{yi}'/N &= \sum_i z_{yi}(B_1 z_{i1} + B_2 z_{i2} + \cdots + B_K z_{iK})/N \\
&= B_1 r_{y1} + B_2 r_{y2} + \cdots + B_K r_{yK}.
\end{aligned} \tag{15.14.4}
$$

On turning back to the denominator terms above, we know that the standard deviation of the z_{yi} values must be 1. However, we must see what the standard deviation of the predicted z_{yi}' values equals:

$$
\begin{aligned}
S_{z_y'} &= \sqrt{\sum_i z_{yi}'^2 / N} \\
&= \sqrt{\sum_i (B_1 z_{i1} + \cdots + B_K z_{iK})^2 / N} \tag{15.14.5} \\
&= \sqrt{\sum_j B_j^2 + \sum_{j \neq k} \sum B_j B_k r_{jk}}.
\end{aligned}
$$

By use of the normal equations (15.11.3) we can convert this expression into

$$
\begin{aligned}
S_{z_y'} &= \sqrt{\sum_j B_j \left(\sum_k B_k r_{jk} \right)} = \sqrt{\sum_j B_j r_{yj}} \\
&= \sqrt{B_1 r_{y1} + B_2 r_{y2} + \cdots + B_K r_{yK}}.
\end{aligned} \tag{15.14.6}
$$

Then on substituting back into Eq. 15.11.3, we have

$$R_{y \cdot 12 \cdots K} = r_{z_y z_y'} = \sqrt{B_1 r_{y1} + B_2 r_{y2} + \cdots + B_K r_{yK}}.$$

We have, incidentally, also shown that the standard deviation of predicted z_y' values is equal to R, so that the variance of the predicted z' values is equal to R^2:

$$R_{y \cdot 12 \cdots K}^2 = \sum_i (z_{yi}')^2 / N. \qquad [15.14.7]$$

In terms of the unstandardized b values this expression for the multiple correlation coefficient becomes

$$R_{y \cdot 12 \cdots K} = \sqrt{b_1 (S_1/S_Y) r_{y1} + b_2 (S_2/S_Y) r_{y2} + \cdots + b_K (S_K/S_Y) r_{yK}}. \qquad [15.14.8]$$

It should be noted that, unlike the ordinary correlation, the *value of R can never be negative*. Thus, there is always a positive relationship between the predicted and the actual scores in multiple regression. In addition, it is important to realize that the value of R^2 based on two or more predictors will always be *at least as large as* the square of any of the correlations betweeen a single X variable and Y. One can never account for *less* of the Y variance by adding more predictor variables.

15.15. THE PROPORTION OF VARIANCE ACCOUNTED FOR AND THE STANDARD ERROR OF MULTIPLE ESTIMATE

Although the goodness of a multiple regression equation for prediciting Y from K variables X can be discussed in terms of the multiple correlation coefficient R, it is much more common for the measure of predictive ''goodness'' to be stated in terms of R^2:

R^2 is the proportion of Y variance that can be accounted for by multiple regression on the K predictor variables.

The easiest way to see that R^2 has this interpretation is to recall that R is an ordinary Pearson correlation coefficient between the actual y scores and the predicted y' values (Eq. 15.14.1). Therefore, as with any other correlation r, the squared value shows the proportion of Y variance that is accounted for by the multiple regression predictions, Y'.

Since we know how to find R from the standardized or unstandardized regression weights and the correlations, it is obvious that the proportion of variance accounted for can be found from the square of either Eq. 15.14.2 or Eq. 15.14.8.

In our two-predictor example, we have

$$R_{y \cdot 12}^2 = B_1 r_{y1} + B_2 r_{y2}$$

$$= (.6564 \times .41) + (.7701 \times .56) \qquad [15.15.1]$$

$$= .70.$$

Therefore, we can see that 70% of the variance in the Y values is attributable to variations in the X_1 and the X_2 scores.

The *sample standard error of multiple estimate* is the standard deviation of the errors or "misses" in the prediction of a Y value from the set of X predictors. There are two versions of the standard error of multiple estimate, depending on whether standardized or raw values are being predicted.

The *sample standard error of multiple estimate for standardized values* is found from

$$S_{z_y \cdot 12 \cdots K} = \sqrt{1 - R^2_{y \cdot 12 \cdots K}}. \qquad [15.15.2^*]$$

The fact that this is so can be shown as follows: The variance of the errors in prediction for standardized scores is

$$S^2_{z_y \cdot 12 \cdots K} = \sum_i (z_{yi} - z'_{yi})^2 / N$$

$$= \sum_i z^2_{yi}/N + \sum_i z'^2_{yi}/N - 2 \sum_i z_{yi} z'_{yi}/N.$$

By virtue of the fact that R is defined as the correlation of the actual and predicted z values, and that the variance of the predicted z values is R^2, as shown in the preceding section, we have

$$S^2_{z_y \cdot 12 \cdots K} = 1 + R^2_{y \cdot 12 \cdots K} - 2R^2_{y \cdot 12 \cdots K}$$

$$= 1 - R^2_{y \cdot 12 \cdots K}$$

or, on taking the square root,

$$S_{z_y \cdot 12 \cdots K} = \sqrt{1 - R^2_{y \cdot 12 \cdots K}}.$$

The sample standard error of multiple estimate for standardized values is the square root of 1 minus the squared multiple correlation.

For example, for the two-predictor example of Section 15.14, the value of the multiple correlation coefficient was found to be about .84. Therefore, the sample standard error of multiple estimate for standardized scores is equal to

$$S_{z_y \cdot 12} = \sqrt{1 - (.84)^2}$$

$$= .54.$$

Thus, if you are going to predict the standardized value on a normally distributed variable Y, based on the standardized values of variables X_1 and X_2, then only 5% of the time should you expect to be in error by more than 1.96(.54) or 1.06; only about 1% of the time should your error exceed 2.58(.54) or 1.39, and so on.

An argument very much like that above, but carried out in terms of raw rather than standardized y scores, gives the *sample standard error of multiple estimate for raw scores:*

$$S_{y \cdot 12 \cdots K} = S_Y \sqrt{1 - R^2_{y \cdot 12 \cdots K}}. \qquad [15.15.3^*]$$

This has the usual meaning of a standard error of estimate: a "typical" amount of error to be anticipated in the prediction of a raw Y' value, based on the K different predictor variables.

15.16 AN EXAMPLE WITH THREE PREDICTORS

Once again let us use the example of the employer of Section 15.12. So far in this example, the employer was thought of as using two predictor variables, X_1 = a measure of manual dexterity, and X_2 = a test of attention to detail, to predict Y = job productivity. It was found in Section 15.15 that R^2 = .70 for these two variables and Y, so that about 70% of the variance can be accounted for in terms of manual dexterity and attention to detail.

However, the employer also has other information available. Among this additional information may be a rating on the stability of the previous employment (i.e., how long at one place) for the subject. This too should have some bearing on the job productivity, and so this variable was also included as X_3 in the multiple regression equation for predicting Y, or productivity. Is it possible to account for appreciably more variance in productivity by adding this third variable to the equation, or is its contribution to prediction only a neglible one?

Table 15.16.1 shows the correlations for all four of variables in this new version of the problem, along with the mean and standard deviation of each variable. Once again, the complete table of intercorrelations is shown, since we will be using determinants to solve for the regression weights.

First, we will solve for the B values that enter into the standardized score regression equation

$$z'_{yi} = B_1 z_{i1} + B_2 z_{i2} + B_3 z_{i3}.$$

Employing the method of determinants (Section D.2) again to solve for the B values we have

$\mathbf{D}(X)$ = determinant of matrix of correlations among X variables

$$= \begin{bmatrix} 1.00 & -.32 & .06 \\ -.32 & 1.00 & .15 \\ .06 & .15 & 1.00 \end{bmatrix} \begin{bmatrix} 1.00 & -.32 & .06 \\ -.32 & 1.00 & .15 \\ .06 & .15 & 1.00 \end{bmatrix}$$

$$= (1 \times 1 \times 1) + (-.32 \times .15 \times .06) + (.06 \times -.32 \times .15)$$
$$- (.06 \times 1 \times .06) - (.15 \times .15 \times 1) - (1 \times -.32 \times -.32) = .86574$$

Next the value of D(1) is found from the determinant of the X matrix but with the column for X_1 replaced by that for Y:

	X_1	X_2	X_3	Y	Mean	SD
X_1	1	-.32	.06	.41	52	5.8
X_2	-.32	1	.15	.56	46	7.3
X_3	.06	.15	1	.38	14.7	3.3
Y	.41	.56	.38	1	63	11.0

Table 15.16.1

Correlations for the three predictor example

$$D(1) = \begin{bmatrix} .41 & -.32 & .06 \\ .56 & 1.00 & .15 \\ .38 & .15 & 1.00 \end{bmatrix} \begin{bmatrix} .41 & -.32 & .06 \\ .56 & 1.00 & .15 \\ .38 & .15 & 1.00 \end{bmatrix}$$

$$= (.41 \times 1 \times 1) + (-.32 \times .15 \times .38) + (.06 \times .56 \times .15)$$
$$- (.38 \times 1 \times .06) - (.15 \times .15 \times .41) - (1 \times .56 \times -.32) = .54398$$

so that

$$B_1 = D(1)/D(X) = .54398/.86574 = .6283.$$

(You really should carry a good many decimals for determinants such as these, and then round to three or four decimals after each B value is found.)

Next we find the value of B_2, using the determinant $D(2)$ divided by $D(X)$:

$$D(2) = \begin{bmatrix} 1.00 & .41 & .06 \\ -.32 & .56 & .15 \\ .06 & .38 & 1.00 \end{bmatrix} \begin{bmatrix} 1.00 & .41 & .06 \\ -.32 & .56 & .15 \\ .06 & .38 & 1.00 \end{bmatrix}$$

$$= (1 \times .56 \times 1) + (.41 \times .15 \times .06) + (.06 \times -.32 \times .38)$$
$$- (.06 \times .56 \times .06) - (.38 \times .15 \times 1) - (1 \times -.32 \times .41) = .62858$$

so that

$$B_2 = D(2)/D(X) = .62858/.86574 = .7261.$$

Finally, we take

$$D(3) = \begin{bmatrix} 1.00 & -.32 & .41 \\ -.32 & 1.00 & .56 \\ .06 & .15 & .38 \end{bmatrix} \begin{bmatrix} 1.00 & -.32 & .41 \\ .32 & 1.00 & .56 \\ .06 & .15 & .38 \end{bmatrix}$$

$$= (1 \times 1 \times .38) + (-.32 \times .56 \times .06) + (.41 \times -.32 \times .15)$$
$$- (.06 \times 1 \times .41) - (.15 \times .56 \times 1) - (.38 \times -.32 \times -.32)$$
$$= .20206.$$

Then the value of the standardized regression weight B_3 is

$$B_3 = D(3)/D(X) = .20206/.86574 = .2334.$$

The completed regression equation for standardized scores is then

$$z'_{yi} = .6283\ z_{i1} + .7261\ z_{i2} + .2334\ z_{i3}.$$

Suppose now that an individual subject has these standardized values on the three predictors: $-.2$ for variable X_1, 1.3 for variable X_2, and 3.1 for variable X_3. What should we predict as the standardized value on Y, or job performance? The answer is given by

$$z'_{yi} = .6283(-.2) + .7261(1.3) + .2334(3.1)$$

$$= 1.54.$$

Even though this individual is slightly below average on manual dexterity, and a little over one standard deviation above average on attention to detail, the exceptional standard score on job stability makes him or her a good bet to be about 1.5 standard deviations above the mean on job productivity.

The proportion of Y variance accounted for by the three variables is then

$$R^2_{y \cdot 123} = (.6283 \times .41) + (.7261 \times .56) + (.2334 \times .38)$$

$$= .7529, \text{ or about } .75 .$$

About 75% of the variance in the job performance scores can be explained by differences among individuals in manual dexterity, attention to detail, and previous employment stability.

When we used only the first two of these variables, as in Section 15.12, we were able to account for about 70% of the Y variance. Now with the addition of the third variable, it appears that we have gained $100(.75 - .70)$ or 5% in our ability to predict Y.

The multiple correlation between the predicted and the actual Y values is then the square root of the R-square value, or about .868, and the sample standard error of multiple estimate for standardized scores is

$$S_{z_y \cdot 123} = \sqrt{1 - .75} = .50, \text{ approximately.}$$

If it is our intention to predict raw Y scores by use of the three raw X scores, we can do so in terms of the raw score regression equation, where the required b values are

$$b_1 = B_1(S_Y/S_1) = .6283(11.0/5.8) = 1.1916$$

$$b_2 = B_2(S_Y/S_2) = .7261(11.0/7.3) = 1.0941$$

$$b_3 = B_3(S_Y/S_3) = .2334(11.0/3.3) = 0.7780$$

Consequently, the raw score version of the multiple regression equation for predicting Y from the three X variables is

$$y_i' = 63 + 1.1916(x_{i1} - 52) + 1.0941(x_{i2} - 46) + 0.7780(x_{i3} - 14.7)$$

or, if we prefer,

$$y_i' = -60.73 + 1.1916 \, x_{i1} + 1.0941 \, x_{i2} + 0.7780 \, x_{i3} .$$

If an employee had a score x_1 of 59, and x_2 of 51, and an x_3 value of 10.2, then we would predict a job performance score of

$$y_i' = -60.73 + (1.1916 \times 59) + (1.0941 \times 51) + (0.7788 \times 10.2)$$

$$= 73.32$$

or about 10 points above the mean.

The sample value of the standard error of multiple estimate for these raw score predictions is

$$S_{y \cdot 123} = S_Y \sqrt{1 - .75} = 5.5 .$$

15.17 MULTIPLE REGRESSION SUMMARIZED IN ANOVA FORMAT

In Section 14.16 it was shown that bivariate regression analysis can be put into the format of ANOVA, with a sum of squares representing the regression of dependent variable Y on predictor variable X, and a sum of squares for residual, adding up to a sum of squares total.

In very much the same way the results of a multiple regression analysis can also be displayed in the form of an analysis of variance. Here, the basic partition of the sum of squares for variable Y is

$$SS_y \text{ total} = SS \text{ multiple regression on the } X \text{ variables} + SS \text{ residual} \qquad [15.17.1^*]$$

where

$$SS_y \text{ multiple regression on the } X \text{ variables} = \sum_i (y_i' - \bar{y})^2$$

$$= SS_y \text{ total} (R^2_{y \cdot 12 \cdots K}) \qquad [15.17.2^*]$$

$$= N\, S_Y^2\, (R^2_{y \cdot 12 \cdots K})$$

with K degrees of freedom, and

$$SS_y \text{ residual} = \sum_i (y_i - y_i')^2$$

$$= SS_y \text{ total} (1 - R^2_{y \cdot 12 \cdots K}) \qquad [15.17.3^*]$$

$$= N\, S_Y^2\, (1 - R^2_{y \cdot 12 \cdots K})$$

with $N - K - 1$ degrees of freedom.

The typical ANOVA summary table that displays the results of a multiple regression analysis with K predictors then looks like this:

Source	SS	df	MS	F	p
Multiple reg. on X var.	SS_y tot. (R^2)	K	(SS mult. reg.)/K	$\dfrac{MS \text{ reg.}}{MS \text{ resid.}}$	
Residual	SS_y tot.$(1 - R^2)$	$N - K - 1$	$\dfrac{SS \text{ resid.}}{N - K - 1}$		
Total	SS_y total	$N - 1$			

In other words, the sum of squares based on the y' values for each of the N individuals is the SS_y multiple regression, and it always equals the SS_y total for the y values multiplied by the squared multiple correlation $R^2_{y \cdot 12 \cdots K}$. The other sum of squares, SS_y residual, is based on the squared residuals or errors, and must always equal SS_y total times $(1 - R^2_{y \cdot 12 \cdots K})$, which is just N times the sample variance of multiple estimate.

As an illustration, once again employing our three-predictor variable example, we have

$$SS_y \text{ total} = 50\,(121) = 6{,}050$$

and

$$SS_y \text{ regression on the } X \text{ variables } = SS_y \text{ total } (R^2_{y \cdot 12 \cdots K})$$

$$= 6,050 \times .75 = 4,537.5$$

so that

$$SS_y \text{ residual } = SS_y \text{ total } - SS_y \text{ multiple regression}$$

$$= 6,050 - 4,537.5 = 1,512.5 \, .$$

The analysis of variance summary table is then as shown below.

Source	SS	df	MS	F	p
Multiple regression	4,537.5	3	1,512.5	46	<.01
Residual	1,512.5	46	32.88		
Total	6,050	49			

The proportion of Y variance accounted for by the three X variables is, of course,

$$R^2_{y \cdot 12 \cdots K} = SS_y \text{ multiple regression}/SS_y \text{ total } = .75$$

The rationale for the F test included in this table will be discussed in Section 15.18.

15.18 INFERENCES IN MULTIPLE REGRESSION PROBLEMS

Everything we have said so far about methods of multiple regression analysis has been confined to the description of a single sample. Clearly, however, in most uses of these methods our interests would extend far beyond the immediate sample. In some instances we would like to try to arrive at some general relationships that exist among a set of variables and a dependent measure of practical or theoretical interest; in other instances we may be, like the employer in our example, concerned with predicting a future value of some variable having special meaning. In both kinds of situations we want to be able to make statements that extend well beyond the evidence given by our immediate sample.

For any well-defined population of potential subjects, it is possible to conceive of the linear relationships that exist between some variable Y and some number K of well-defined predictor variables X. Each of these variables will have a true correlation ρ_{yj} between Y and variable X_j, and any pair of variables X_j and X_k will will have a true correlation ρ_{jk}.

Then by the same logic that applies to the sample situation, the criterion of least squares produces a set of normal equations for the population. Thus, for a population of cases, each of which has a set of values $(y_i, x_{i1}, x_{i2}, \cdots, x_{iK})$, we may think of the true standardized regression coefficients, or B^* values, as solutions to the K normal equations, from

$$B^*_1 + \rho_{12}B^*_2 + \cdots + \rho_{1K}B^*_K = \rho_{y1}$$

to   [15.18.1]

$$\rho_{1K}B_1^* + \rho_{2K}B_2^* + \cdots + B_K^* = \rho_{yK}.$$

The corresponding unstandardized regression coefficients are then defined by

$$\beta_k = B_k^* \left(\frac{\sigma_Y}{\sigma_k}\right).$$ [15.18.2]

The squared multiple correlation coefficient for the population may be denoted by $P_{y\cdot12\cdots K}^2$, where

$$P_{y\cdot12\cdots K}^2 = B_1^* \rho_{y1} + \cdots + B_K^* \rho_{yK}.$$ [15.18.3]

The true variance of multiple estimate is then

$$\sigma_{y\cdot12\cdots K}^2 = \sigma_Y^2 (1 - P_{y\cdot12\cdots K}^2).$$ [15.18.4]

An unbiased estimate of this variance can be made from a sample, by taking

$$\text{est. } \sigma_{y\cdot12\cdots K}^2 = S_Y^2 \left(\frac{N}{N - K - 1}\right) (1 - R_{y\cdot12\cdots K}^2).$$ [15.18.5*]

The sample value of $R_{y\cdot12\cdots K}^2$ also tends to be too large as an estimate of $P_{y\cdot12\cdots K}^2$. One commonly used correction for this bias is to take

$$\text{est. } P_{y\cdot12\cdots K}^2 = 1 - \frac{(1 - R_{y\cdot12\cdots K}^2)(N - 1)}{(N - K - 1)}.$$ [15.18.6*]

By use of this formula for correcting the bias in the squared multiple correlation, one always arrives at smaller estimated value for the population than that achieved in the sample. This bit of conservatism is appropriate, since a regression equation applied to a new sample will almost always result in a smaller proportion of variance accounted for than in the sample for which the equation was derived. This is sometimes called the "shrunken" value of R^2.

Much as we discussed in Chapter 14, there are two sampling models or schemes that are commonly encountered in multiple regression studies:

Model I. Fixed X values, random Y values
Model II. Both X and Y values are random variables.

Under the fixed model, which we can here call Model I, the following assumptions are made:

1. Within every X value or combination of X values appearing in the study there is an independent normal distribution of Y values in the population.
2. The variance of each of the distributions of Y values has the same true value, or σ_e^2.

Since we are also implicitly assuming that the linear model of multiple regression is *correct* as a description of the relationships among the variables in the population, it will also be true that the variance within any population will be the same as the true variance of multiple estimate, or

$$\sigma_e^2 = \sigma_{y\cdot12\cdots K}^2. \tag{15.18.7}$$

Furthermore, because of our assumption of the correctness of the linear multiple regression model, then within each population defined by a combination of X values $(x_{i1}, x_{i2}, \ldots, x_{iK})$ the population mean symbolized by $\mu_{(12\cdots K)}$ is the same as the value y' predicted for each member of that population.

The alternative model, Model II, requires the assumption of a multivariate normal distribution. Since the methods to be talked about in the next few sections apply under both of these models, we will simply assume that one of these two sets of assumptions is true. However, a little later we will deal with one method appropriate only under the assumptions of Model II.

15.19 THE F TEST FOR MULTIPLE REGRESSION

In Section 15.17 we saw how the total sum of squares for variable Y can be divided into two parts consisting of a sum of squares for multiple regression, and a sum of squares for residual.

Recall once more that the squared multiple correlation coefficient for the population may be denoted by

$$P_{y\cdot12\cdots K}^2 = B_1^* \rho_{y1} + \cdots + B_K^* \rho_{yK}.$$

Now if we take a look at the expected mean squares under the assumptions for either model as outlined above, we find that

$$E(\text{MS regression}) = \sigma_e^2 + NP_{y\cdot12\cdots K}^2 \sigma_Y^2 \tag{15.19.1}$$

and

$$E(\text{MS residual}) = \sigma_e^2 . \tag{15.19.2}$$

Then the null hypothesis states that $P_{y\cdot12\cdots K}^2 = 0$. Therefore, by the usual argument familiar from ANOVA in Chapter 10, we have

$$F = \text{MS multiple regression/MS residual} \tag{15.19.3*}$$
$$= \frac{(\text{SS multiple regression})}{K} \Big/ \frac{(\text{SS residual})}{N - K - 1}$$

with K and $N - K - 1$ degrees of freedom.

In our example of Section 15.17, the ANOVA table gives

$$F = 1,512.5/32.88 = 46$$

which is significant well beyond the .01 level with 3 and 46 degrees of freedom.

It is sometimes convenient to carry out the F test for a value of R-square without going through all the detail of ANOVA, and this can be done as follows:

$$F = \frac{R_{y\cdot12\cdots K}^2 (N - K - 1)}{(1 - R_{y\cdot12\cdots K}^2)(K)}$$

Just as in the other version, this F test has K and $N - K - 1$ degrees of freedom.

For example, if we take the $R^2_{y \cdot 123}$ value of .75, found in Section 15.17, the significance of this squared multiple correlation can be found from

$$F = .75 (50 - 3 - 1)/.25(3)$$

$$= 46$$

and this, for 3 and 46 degrees of freedom, is significant far beyond the .01 level.

15.20 SOME USEFUL NOTATION

The standard notation we have been using, where $R^2_{y \cdot 12 \cdots K}$ represents the squared multiple correlation for dependent variable Y and predictor variables X_1 through X_K, is probably satisfactory for many purposes. On the other hand, when one wishes to discuss situations where the set of predictors may change, making these distinctions using the ellipsis becomes very awkward. Therefore, we are going to adopt the following conventions, using just a little bit of the language of set theory.

Let $H = \{X_1, \cdots, X_K\}$ stand for the full set of K predictor variables which will be under discussion in a given problem. Furthermore, let $\{X_j\}$ represent a set consisting only of the variable X_j. The set $G = H - \{X_j\}$ then represents all of the variables *except* X_j. Although later we will wish to expand even this convention a bit, it will prove useful and save a lot of subscript changing in this and the following sections. Thus

$$R^2_{y \cdot H} = \text{proportion of variance in } Y \text{ accounted for by } \{X_1, \cdots, X_K\} \text{ or } H$$

$$R^2_{y \cdot G} = \text{proportion of variance accounted for by } G, \text{ where } H - \{X_j\} = G.$$

15.21 TESTS AND INTERVAL ESTIMATES FOR REGRESSION COEFFICIENTS

Under sampling Model I outlined in Section 15.18, or under Model II, conditional on the distribution of X values in the sample truly representing the population distribution, we can make a variety of inferences about the the regression coefficients that apply in the population. In the first place, the sample value of the standardized regression coefficient B_j for variable X_j is an unbiased estimate of the corresponding value B_j^* in the population,

$$E(B_j) = B_j^* . \qquad [15.21.1]$$

Hence, the sample regression equation for standardized values is the best available estimate of the regression equation that applies to the population that has been sampled.

In a similar way, the sample value b_j of the unstandardized regression coefficient for each predictor can be used as an unbiased estimate of its population counterpart, and the sample constant a gives an estimate of the corresponding value in the population.

Among the more common tests associated with multiple regression analysis is that for the significance of a single regression coefficient, B_j. The hypothesis tested is, of course, that the corresponding value in the population, B_j^*, is truly zero. However, a one-tailed alternative hypothesis is possible, and may be of interest. Hence we will give this test in its "t" form with $N - K - 1$ degrees of freedom.

$$t = \frac{B_j}{\text{est. } \sigma_{B_j}}$$ 15.21.2*]

where

$$\text{est. } \sigma_{B_j} = \sqrt{\frac{1 - R_{y \cdot H}^2}{(1 - R_{j \cdot G}^2)(N - K - 1)}}.$$ [15.21.3*]

Note that H contains all X variables, whereas G contains all but X_j.

A confidence interval may also be formed for the population value B_j^* corresponding to variable X_j. When the value of est σ_{B_j} is found from Eq. 15.21.3 above, then the $100 (1 - \alpha)$ percent confidence interval is given by

$$B_j - t_{(\alpha/2)} \text{ est. } \sigma_{B_j} \leq B_j^* \leq B_j + t_{(\alpha/2)} \text{ est. } \sigma_{B_j}$$ [15.21.4*]

Here, as usual, $t_{(\alpha/2)}$ is that t value cutting off the upper $\alpha/2$ proportion of a distribution with $N - K - 1$ degrees of freedom.

In the example in Section 15.17, the variable X_1 was found to have a standardized regression weight B_1 equal to .6564. If we want to test the significance of this B value, then we need the information that $N = 50$, $K = 2$, and that $R_{y \cdot 12}^2 = .70$. In addition, we need the value of $R_{j \cdot G}^2 = R_{1 \cdot 2}^2$, the proportion of variance in X_1 that is accounted for by X_2. However, we know from the original correlation matrix that this is just $r_{12}^2 = (-.32)^2 = .1024$. Therefore, from Eq. 15.21.3, we can find

$$\text{est. } \sigma_{B_1} = \sqrt{(1 - .70)/[(1 - .1024)(47)]} = .0843$$

To test the null hypothesis that the true B^* value is 0, we can form the t ratio

$$t = .6564/.0843 = 7.79$$

which greatly exceeds the value of 2.02 required for significance (two-tailed) at the .05 level for $N - K - 1 = 47$ degrees of freedom.

Alternatively, we can form the 95% confidence limits by taking

$$.6564 \pm (2.02 \times .0843)$$

or

$$.4861 \text{ and } .8267.$$

As the value of zero does not fall between these limits, we know that the sample value of B_1 is significant beyond the .05 level, two-tailed.

You will note that just as in the case of bivariate regression in Chapter 14, hypotheses other than zero may also be tested for these standardized regression weights, either through use of a confidence interval or a t test under Model I assumptions.

Matters are a bit more complicated, however, when we wish to check the significance of a B value when there are three or more predictor variables. Thus, suppose that we take the three-variable example of Section 15.16, and test the significance of the weight B_3, for variable X_3 in predicting Y. Now in addition to the overall proportion of variance accounted for, $R^2_{y \cdot H} = R^2_{y \cdot 123} = .75$, we also need to know the proportion of variance in X_3 that is due to X_1 and X_2. That is, we have to find the value of $R^2_{3 \cdot G} = R^2_{3 \cdot 12}$.

Solving the multiple regression problem for X_3 predicted from X_1 and X_2, using the method of determinants once again, gives

$$\mathbf{D(X)} = \begin{bmatrix} 1 & -.32 \\ -.32 & 1 \end{bmatrix}$$

$$= 1 - (-.32)^2 = .8976$$

$$\mathbf{D(1)} = \begin{bmatrix} .06 & -.32 \\ .15 & 1 \end{bmatrix}$$

$$= (.06 \times 1) - (-.32 \times .15) = .1080$$

so that

$$B_{3 \cdot 1} \text{ (for predicting } X_3 \text{ from } X_1) = .1080/.8976 = .1203,$$

and

$$\mathbf{D(2)} = \begin{bmatrix} 1 & .06 \\ -.32 & .15 \end{bmatrix}$$

$$= (1 \times .15) - (-.32 \times .06) = .1692$$

with

$$B_{3 \cdot 2} \text{ (for predicting } X_3 \text{ from } X_2) = .1692/.8976 = .1885 .$$

Therefore, $R^2_{3.12} = .1203(.06) + .1885(.15) = .0355$.
Now on substituting into Eq. 15.21.3, we have

$$\text{est. } \sigma_{B_3} = \sqrt{(1 - .75)/[(1 - .0355)(50 - 3 - 1)]} = .0751.$$

Since $B_3 = .2334$, and $t_{.025}$ is about 2.02 for 46 degrees of freedom, the 95% confidence limits are therefore

$$.2334 \pm (2.02)(.0751)$$

or

$$.0817 \text{ and } .3851.$$

Once again, zero in not included, and so we may say that B_3 is significant beyond the .05 level.

The theory of hypothesis testing and interval estimation extends directly from the methods just shown for standardized coefficients B, to methods for the unstandardized weights b. For the unstandardized or raw-score coefficient

$$b_j = (S_y/S_j)B_j$$

the estimated standard error is given by

$$\text{est. } \sigma_{b_j} = (S_y/S_j) \text{ est. } \sigma_{B_j} \qquad [15.21.5]$$

$$= (S_y/S_j) \sqrt{\frac{(1 - R_{y \cdot H}^2)}{(1 - R_{j \cdot G}^2)(N - K - 1)}}$$

where, as usual, H is the entire set of K predictor variables, and G is the set *excluding* variable X_j.

It is also possible to establish confidence limits for the y' or predicted values from a given regression equation, as referred to a population of cases from which the sample was drawn. We will not devote space to these methods here, but they can be found in sources such as Snedecor and Cochran (1980).

15.22 THE FISHER TRANSFORMATION FOR A MULTIPLE CORRELATION

Another useful test for the multiple correlation coefficient is an extension of the Fisher r-to-Z transformation, as discussed in Section 14.21. This method, devised by Konishi (1981), is especially appropriate for data collected under the "correlation" or Model II sampling scheme of Section 15.18, as one must assume that the joint distribution of the dependent variable Y and the K different X variables is multivariate normal, with a true value of multiple R denoted by

$$\text{true multiple correlation} = P = P_{y \cdot 12 \cdots K}$$

Then it is possible to test any hypothesis one desires (not too close to 1.00) about the true value $P \geq 0$ by letting

Z_R = Fisher Z value corresponding to the sample R value

ζ_P = Fisher Z value corresponding to the population P value

where each of these values of Z or ζ can be found directly from Table VI in Appendix E. We then proceed by taking

$$C_R = [Z_R - \zeta_P - (K - 1 + P^2)/(2P(N - 1))] \sqrt{(N - 1)} \qquad [15.22.1]$$

This statistic C_R is referred to the normal distribution, exactly as in Section 14.21, and a decision reached about the hypothesis.

For example, suppose that in a problem with $K = 5$ predictor variables X we wish to test the hypothesis that the multiple correlation in the population is .50. That is, our null hypothesis is

$$H_0 : P = .50$$

against the alternative

$$H_1 : P < .50.$$

We draw a sample of $N = 40$ cases, and find the value of sample $R^2 = .11$, so that sample $R = .33$ approximately. Now to test the hypothesis we must first find the Fisher

Z value for $R = .33$, which is determined from Table VI to be .3428. Next we find the corresponding population ζ for $P = .50$. This is .5493. Then our test statistic is given by

$$C_R = [.3428 - .5493 - (4 + .25)/(2 \times .5 \times 39)] \sqrt{39}$$

$$= -1.97 .$$

In a standard normal distribution, a value of -1.97 is significant beyond the .05 level, and hence we may reject the hypothesis that the true value of the multiple correlation is equal to .50 (or greater) in the light of the sample evidence.

   This method is available for testing any exact hypothesis (except a value very close to 1.00) for the true or population value of R, where the value must, of course, be nonnegative, and less than 1.00. However, just as the Fisher r-to-Z transformation in Section 14.21 demands a bivariate normal distribution of X and Y in the population, this method assumes a multivariate normal distribution of the entire set of $K + 1$ variables consisting of the K predictors X and the dependent variable Y.

15.23 MULTIPLE REGRESSION IN TERMS OF PART AND PARTIAL CORRELATIONS

As we have just seen, in the classic approach to multiple regression it is possible to find the standardized regression weights such as B_j, or the unstandardized weights such as b_j for each of a set of K predictor variables relative to a dependent variable Y. Then the proportion of Y variance explained by the set of predictor variables can be found from

$$R^2_{y \cdot 12 \cdots K} = B_1 r_{y1} + B_2 r_{y2} + \cdots + B_{yK} r_{yK} \qquad [15.23.1]$$

in terms of the the standardized B weights, or

$$R^2_{y \cdot 12 \cdots K} = b_1 (S_1/S_y) r_{y1} + \cdots + b_K (S_K/S_y) r_{yK}$$

in terms of the unstandardized b weights.

   However, there are useful alternative approaches to determining the strength of relationship, R^2, between Y and a set of predictor X variables. One approach uses the partial correlations between Y and each X variable taken in turn, with other X variables held constant. Thus, for a simple example, consider Y and the two predictors X_1 and X_2 once again. Then it will be true that the value of $R^2_{y \cdot 12}$ will depend on the partial correlations as follows:

$$1 - R^2_{y \cdot 12} = (1 - r^2_{y1})(1 - r^2_{y2 \cdot 1}). \qquad [15.23.2]$$

That is, the proportion of Y variance *not* explained by the set X_1 and X_2 is the proportion unexplained by X_1, times the proportion unexplained by X_2 when X_1 is held constant. It is thus true that the proportion of Y variance explained by X_1 or X_2 is

$$R^2_{y \cdot 12} = 1 - (1 - r^2_{y1})(1 - r^2_{y2 \cdot 1}). \qquad [15.23.3]$$

In a similar way, if we are dealing with three X variables, then

$$R^2_{y \cdot 123} = 1 - (1 - r^2_{y1})(1 - r^2_{y2 \cdot 1})(1 - r^2_{y3 \cdot 12})$$

For four predictor variables we have

$$R^2_{y\cdot 1234} = 1 - (1 - r^2_{y1})(1 - r^2_{y2\cdot 1})(1 - r^2_{y3\cdot 12})(1 - r^2_{y4\cdot 123}) \qquad [15.23.4]$$

and so on for any number of predictors X. In principle, one could always find the strength of the relationship between a set of X variables and a dependent variable Y by dealing only with a set of partial correlations.

Thus, for example, consider once again the correlations presented in Table 15.2.2. This time let us suppose that the place of the dependent variable Y is taken by X_1. In Section 15.2 we found that the partial correlation of variables X_1 and X_3 with X_2 held constant is $-.8334$. We also know from Table 15.2.1 that the correlation between X_1 and X_2 is .43. Therefore, it must be true that

$$(1 - R^2_{1\cdot 23}) = (1 - r^2_{12})(1 - r^2_{13\cdot 2})$$

$$= (1 - .43^2)(1 - .8334^2)$$

$$= .2490$$

Therefore,

$$R^2_{1\cdot 23} = 1 - .2490$$

$$= \text{about } .75.$$

In addition, if you already know $R^2_{Y\cdot G}$ the proportion of Y variance accounted for by a set G consisting of K predictor variables, and you wish to find the variance that would be accounted for if *one more* predictor variable X_j were added to the set, you can do so by finding the Kth-order partial correlation $r_{yj\cdot G}$ between Y and X_j, holding the variables in the set G constant. Then, if H represents the set of $K + 1$ variables including both the set G and the variable X_j,

$$(1 - R^2_{y\cdot H}) = (1 - R^2_{y\cdot G})(1 - r^2_{yj\cdot G}) \qquad [15.23.5]$$

Notice that this also gives a way to determine the squared value of a high-order partial correlation:

$$(1 - r^2_{yj\cdot G}) = (1 - R^2_{y\cdot H})/(1 - R^2_{y\cdot G}) \qquad [15.23.6]$$

For example, we saw above that for the "exercise" data of Section 15.2. the R-square value for variable $Y = X_1$ as predicted from variables X_2 and X_3 is about .75. What would happen if we also included variable X_4, or age, as a predictor? From the partial correlations already found in Section 15.3 we know that

$$r_{12\cdot 3} = .8146, \; r_{14\cdot 3} = -.5049, \text{ and } r_{24\cdot 3} = -.2414$$

so that

$$r_{14.23} = \frac{r_{14\cdot 3} - (r_{12\cdot 3})(r_{24\cdot 3})}{\sqrt{(1 - r^2_{12\cdot 3})(1 - r^2_{24\cdot 3})}}$$

$$= -.55$$

Therefore,

$$(1 - R^2_{1 \cdot 234}) = (1 - R^2_{1 \cdot 23})(1 - r^2_{14 \cdot 23}) = .1744$$

so that

$$R^2_{1 \cdot 234} = 1 - .1744 = .8256 \text{ or about } .83.$$

The variable X_4, or age, adds about 8% to our ability to predict weight loss, over and above the predictive ability given by the variables of exercise and calorie consumption.

On the other hand, an even simpler connection exists between the value of R^2 and a set of part or semipartial correlations relating Y with the various X variables. That is, again with only two predictors X_1 and X_2, we have

$$R^2_{y \cdot 12} = r^2_{y1} + r^2_{y(2 \cdot 1)} .$$ [15.23.7]

This says that the total proportion of variance of Y explained by the X variables consists of the portion explained by X_1 alone, plus the portion that is explained by X_2 *over and above that explained by X_1*.

The order of the variables really makes no difference, however, Equivalently, we could write

$$R^2_{y \cdot 12} = r^2_{y2} + r^2_{y(1 \cdot 2)} .$$ [15.23.8]

This says that the total Y variance explained by the two X predictor variables consists of the portion explained by X_2 plus that portion explained by X_1, *over and above that already explained by X_2*.

This way of describing a squared muliple correlation as a sum of squared part correlations can be applied to any number of predictor variables X. For three predictors we can thus write

$$R^2_{y \cdot 123} = r^2_{y1} + r^2_{y(2 \cdot 1)} + r^2_{y(3 \cdot 12)}$$ [15.23.9]

(For an illustration of this point see Figure 15.23.1.) You can also see that this expression could equally well be written in five other ways, as the order in which we adjust and correlate the X variables with Y is irrelevant to the final outcome. Thus,

Figure 15.23.1

Two ways of showing the proportion of Y variance accounted for by X_1, X_2, and X_3.

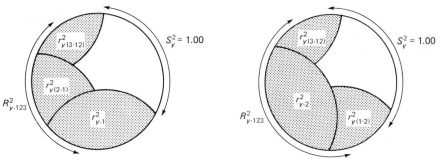

$$R^2_{y\cdot123} = r^2_{y1} + r^2_{y(3\cdot1)} + r^2_{y(2\cdot13)} \qquad\qquad [15.23.10]$$

$$= r^2_{y2} + r^2_{y(1\cdot2)} + r^2_{y(3\cdot12)}$$

$$= r^2_{y2} + r^2_{y(3\cdot2)} + r^2_{y(1\cdot23)}$$

$$= r^2_{y3} + r^2_{y(1\cdot3)} + r^2_{y(2\cdot13)}$$

$$= r^2_{y3} + r^2_{y(2\cdot3)} + r^2_{y(1\cdot23)}$$

In general, if G represents a set of K predictor variables, and H represents a set of $K + 1$ such variables, consisting of the set G plus a new variable X_j, then the proportion of Y variance accounted by the total set H may be found from:

$$R^2_{y\cdot H} = R^2_{y\cdot G} + r^2_{y(j\cdot G)} \qquad\qquad [15.23.11^*]$$

Notice how this expression also provides a definition of the square of a K-order part (or semipartial) correlation. That is

$$r^2_{y(j\cdot G)} = R^2_{y\cdot H} - R^2_{y\cdot G} \qquad\qquad [15.23.12^*]$$

The squared part correlation for Y with X_j adjusted for the K variables in the set G is the difference in the proportion of variance of Y accounted for by the total set of variables including X_j, and the proportion of variance accounted for by the original set G alone.

15.24 AN IMPORTANT SPECIAL CASE: UNCORRELATED PREDICTORS

The relations between multiple correlation and part correlations just illustrated can be used to demonstrate the special situation where the predictor variables X are themselves uncorrelated. When this is true, then the value of each B weight in the standardized multiple regression equation is simply the correlation of the variable in question with Y. That is,

$$B_j = r_j$$

when the correlation $r_{jk} = 0$ for variable X_j and with all other predictor variables X_k. It follows that for this situation

$$R^2_{y\cdot12\cdots K} = r^2_{y1} + r^2_{y2} + \cdots + r^2_{yK} . \qquad\qquad [15.24.1^*]$$

This is really just a special case of Eq. 15.23.11 since if all of the X variables are uncorrelated, then the part correlation of any Y with any variable X_j, adjusted for one or more of the other X variables, is just the original correlation r_{yj}.

Since we know from Eq. 15.13.2 that

$$R^2_{y\cdot12\cdots K} = B_1 r_{y1} + B_2 r_{y2} + \cdots + B_K r_{yK} ,$$

then when the X variables are uncorrelated it must also be true that

$$R^2_{y\cdot12\cdots K} = r^2_{y1} + r^2_{y2} + \cdots + r^2_{yK} \qquad\qquad [15.24.2]$$

This principle will be used in Chapter 16, where we will show how orthogonal (i.e., uncorrelated) comparisons in ANOVA are equivalent to uncorrelated predictors in multiple regression.

15.25 INCREMENTS IN PREDICTIVE ABILITY THROUGH ADDITION OF MORE VARIABLES

In this section interest will focus on the changes that occur in a multiple regression situation when predictor variables are added to or subtracted from the set originally used. In doing this, we will rely on the connections between part and multiple regression outlined in Section 15.23.

Let us start off with the simplest situation, once again, with two predictor variables. Then the proportion of variance accounted for by both variables can be thought of as divisible into two parts: that which is due to X_1 alone, and that which is due to X_2 once the influence of X_1 has been removed from X_2. That is,

$$R^2_{y \cdot 12} = r^2_{y1} + r^2_{y(2 \cdot 1)} \qquad [15.25.1]$$

(or, in our new notation, where $H - \{X_2\} = G = \{X_1\}$)

$$R^2_{y \cdot H} = r^2_{yG} + r^2_{y(2 \cdot G)}. \qquad [15.25.2]$$

The first of these two parts shown in Eq. 15.25.1 is simply the squared correlation between Y and X_1, reflecting the proportion of variance accounted for by X_1. Next is the square of the *part correlation* for Y and X_2; here all the predictability of X_2 from X_1 has been removed. Notice that Eq. 15.25.1 also gives us another definition of a squared part correlation:

$$r^2_{y(2 \cdot 1)} = R^2_{y \cdot 12} - r^2_{y1}.$$

Thus the squared part correlation reflects the absolute increase in proportion of variance explained when variable X_2 is added to variable X_1 in a multiple regression equation for predicting Y.

We could also write

$$R^2_{y \cdot 12} = r^2_{y2} + r^2_{y(1 \cdot 2)} \qquad [15.25.3]$$

so that

$$r^2_{y(1 \cdot 2)} = R^2_{y \cdot 12} - r^2_{y2}, \qquad [15.25.4]$$

the absolute increase in proportion of variance of Y accounted for by X_1 is the squared part correlation between Y and X_1 with the variance due to X_2 removed from X_1.

In general, given any complete set H of K predictors, including X_j,

$$R^2_{y \cdot H} = R^2_{y \cdot G} + r^2_{y(j \cdot G)}, \qquad H - \{X_j\} = G. \qquad [15.25.5^*]$$

Then the absolute increase in proportion of variance accounted for by the additon of X_j to a set of predictors G is the squared part correlation $r^2_{y(j \cdot G)}$:

$$r^2_{y(j \cdot G)} = R^2_{y \cdot H} - R^2_{y \cdot G}. \qquad [15.25.6^*]$$

For example, a multiple regression problem was worked having first two and then three predictors. What was the absolute increase in predictability afforded by the addition of X_3? The original squared multiple correlation was .187, and the squared multiple correlation was .256 with all three variables included. Thus,

$$r^2_{y(3\cdot12)} = .256 - .187 = .069.$$

We thus also know that the square of the part correlation between Y and adjusted X_3 in this example is about .07.

This same idea can be turned around, of course, to find the absolute increase in predictability afforded by adding all of the other variables in G to an original variable X_j:

$$R^2_{y(G\cdot j)} = R^2_{y\cdot H} - r^2_{yj}, \qquad H - G = \{X_j\}. \qquad [15.25.7]$$

Thus, in the example, how much more predictive power do we gain by addition of variables X_1 and X_3 to the variable that already correlates best with Y, or X_2 where $r^2_{y2} = .16$?

$$R^2_{y(13\cdot2)} = R^2_{y\cdot123} - r^2_{y2}$$

$$= .256 - .160$$

$$= .096.$$

The answer is that about 10% more variance is accounted for using three variables rather than X_2 alone.

Sometimes we are interested not so much in the absolute gain in predictive ability as the *relative gain* in view of all the variance that remains to be accounted for. Thus, for two predictor variables once again, we may take

$$r^2_{y2\cdot1} = \frac{R^2_{y\cdot12} - r^2_{y1}}{1 - r^2_{y1}}$$

$$= \frac{r^2_{y(2\cdot1)}}{1 - r^2_{y1}}. \qquad [15.25.8]$$

In other words, once variable X_1 has been used to predict Y, there is still variance left to be accounted for. This proportion of unexplained variance is $1 - r^2_{y1}$. Then finding the additional variance accounted for by X_2, or $r^2_{y(2\cdot1)}$, and dividing by the amount left over from X_1, gives the relative increment due to X_2. Notice, however, that *this is also the square of the partial correlation*, $r^2_{y2\cdot1}$. (c.f. Figure 15.5.2, number III.)

In general, the relative gain in proportion of variance accounted for by adding variable X_j to an existing set of variables G is given by

$$r^2_{yj\cdot G} = \frac{R^2_{y\cdot H} - R^2_{y\cdot G}}{1 - R^2_{y\cdot G}} \qquad [15.25.9]$$

where, again, $H - \{X_j\} = G$. In other words, when the relative increment in found, resulting from the addition of variable X_j to the existing set of $K - 1$ predictors G, this is the same as the squared partial correlation between Y and X_j, holding all the other variables constant. The partial correlation is then of order $K - 1$.

For example, in our three-predictor problem above, the relative proportion of increase in variance accounted for in Y due to the addition of X_3 is given by

$$r^2_{y3 \cdot 12} = \frac{.256 - .187}{1 - .187} = .085.$$

The absolute value of the partial correlation $r_{y3 \cdot 12}$ is thus the square root of this value, or .29. Notice that this is a second-order partial, since two variables are held constant.

We may also ask about the relative contribution to explained variance made by the addition of the $K - 1$ variables in the set G to the single variable X_j:

$$R^2_{yG \cdot j} = \frac{R^2_{y \cdot H} - r^2_{yj}}{1 - r^2_{yj}}. \qquad [15.25.10]$$

Although this is not, strictly speaking a squared partial correlation, it does have much the same interpretation: given that variable X_j accounts for a certain amount of variance, relatively how much of what is left is accounted for by the remaining variables making up the set G?

The standardized partial regression weight B_j for variable X_j is related to the absolute increment $r^2_{y(j \cdot G)}$ as follows:

$$B^2_j = \frac{r^2_{y(j \cdot G)}}{1 - R^2_{j \cdot G}}. \qquad [15.25.11]$$

That is, B^2_j is the squared part correlation between Y and X_j, with the effects of G removed from X_j, but relative to the variance left over in X_j from the set G. The value of B^2_j may also be found from the squared partial correlation $r^2_{yj \cdot G}$:

$$B^2_j = \frac{r^2_{yj \cdot G} (1 - R^2_{y \cdot G})}{(1 - R^2_{j \cdot G})}. \qquad [15.25.12]$$

Thus, for two X variables we have, from Eq. 15.25.11 and Eq. 15.25.12

$$B^2_1 = \frac{r^2_{y(1 \cdot 2)}}{1 - r^2_{12}} = \frac{r^2_{y1 \cdot 2} (1 - r^2_{y2})}{1 - r^2_{12}}$$

$$B_1 = \frac{r_{y1} - r_{y2} r_{12}}{1 - r^2_{12}}$$

and

$$B^2_2 = \frac{r^2_{y(2 \cdot 1)}}{1 - r^2_{12}} = \frac{r^2_{y2 \cdot 1} (1 - r^2_{y1})}{1 - r^2_{12}}$$

$$B_2 = \frac{r_{y2} - r_{y1} r_{12}}{1 - r^2_{12}}.$$

An important thing to notice in Eqs. 15.25.11 and 15.25.12 is that if all the variance in X_j is accounted for by the set G, the value of B_j is indeterminate. Even when true $R^2_{j \cdot G}$ is quite large, estimates of the true value of B_j will tend to be very unstable. We will refer to this as the problem of *multicollinearity* in Section 15.26.

Perhaps the significance tests of most utility in multiple regression problems are those that have to do with the proportions of variance explained when new variables are added to the set of predictors. Thus, a test of significance that goes with $r^2_{y(j\cdot G)}$, the absolute increment in proportion of variance accounted for by X_j when it is added to the set G, where $H - \{X_j\} = G$, is given by

$$F = \frac{r^2_{y(j\cdot G)}}{(1 - R^2_{y\cdot H})/(N - K - 1)} \, .$$

[15.25.13.*]

The degrees of freedom here are 1 and $N - K - 1$. This is, of course, also a test of significance for the part correlation $r_{y(j\cdot G)}$, representing the relation between Y and X_j when all the influence of the variables in G have been removed from X_j. Since a part correlation of zero in the population implies a partial correlation of zero as well, this may also be regarded as a significance test for a partial correlation represented by $r_{yj\cdot G}$.

Thus, for example suppose that in a study with $N = 60$, when three predictor variables X_1, X_2, and X_3, were used, the value of R-square was .27. However, when variable X_4 was added to the regression equation, then the R-square value increased to .30. Is this gain in predictive ability significant? We take

$$F = \frac{(R^2_{y\cdot 1234} - R^2_{y\cdot 123})(N - K - 1)}{(1 - R^2_{y\cdot 1234})}$$

[15.25.14]

$$= 2.36 \, .$$

This, for 1 and $N - 4 - 1 = 55$ degrees of freedom, fails to reach significance at the .05 level, and hence we can feel safe in saying that variable X_4 does not add significantly to the predictive power of variables X_1, X_2, and X_3.

It is a very common practice in multiple regression studies to examine the contribution to variance made by each variable X_j, beyond any variance attributable to all the rest of the variables in the total set H, as given by Eq. 15.25.6. However, an alternative strategy is to determine the contribution to variance made by the variables as they are added to the set in some predetermined, *hierarchical order*. That is, for convenience, suppose that the numbers appearing as subscripts also stand for the order in which the variables are to be added to compose the total set H. Then the tests might start with

$$F = \frac{r^2_{y1} \, (N - K - 1)}{1 - R^2_{y\cdot H}}, \qquad H = \{X_1, \cdots, X_K\}$$

[15.25.15]

with 1 and $N - K - 1$ degrees of freedom, which reflects the contribution of variable X_1 alone. Then the contribution of variable X_2, removing the linear regression on X_1, could be tested by taking

$$F = \frac{r^2_{y(2\cdot 1)} \, (N - K - 1)}{1 - R^2_{y\cdot H}}$$

[15.25.16]

with degrees of freedom 1 and $N - K - 1$.

Next the contribution of X_3, holding X_1 and X_2 constant, is tested by using the ratio

$$F = \frac{r^2_{y(3\cdot 12)} \, (N - K - 1)}{1 - R^2_{y\cdot H}},$$

[15.25.17]

again with the same numbers of degrees of freedom as before. This procedure thus is carried out until all the variables to be included in the set H have been examined.

Finally, suppose that we have a total set of variables H that can be divided into two subsets, the variables in set I and the variables in set G, so that $H - I = G$. There are K variables in the total set H, and L variables in the set I. The absolute increment in proportion of variance accounted for by variables in the set I, above that accounted for in the set G, is given by

$$R^2_{y(I \cdot G)} = R_{y \cdot H} - R^2_{y \cdot G}, \qquad (H - I = G). \qquad \qquad [15.25.18]$$

A test of the hypothesis that set G accounts for all the explainable variance is then given by

$$F = \frac{R^2_{y(I \cdot G)} (N - K - 1)}{(1 - R^2_{y \cdot H})(L)} \qquad \qquad [15.25.19]$$

with L and $N - K - 1$ degrees of freedom.

15.26 SOME HAZARDS OF MULTIPLE REGRESSION ANALYSIS

Multiple regression analysis is such a versatile and attractive technique, especially as run through a sophisticated computer package, that it is easy to lose sight of the problems that can arise in its application. There are, however, a few things that must be kept in mind if multiple linear regression is chosen as the way to handle a set of data.

In the first place, multiple regression methods are only as good as the data to which they are applied. If data collection methods were sloppy and there is a lot of experimental error hiding among the numbers given to the computer, the results can contain unexpected and quite meaningless relationships. These serve as "static" in a scientific investigation, and they can be fatal to any attempt to use the regression equation in some practical way. To be meaningful, a set of data need to be as clean as we can make them before multiple regression, or any other sophisticated statistical method, is applied. The data should be inspected for unusually skewed or artificially restricted distributions, missing data, unusually deviant cases or outliers, and a host of other signs of potential trouble even before the original correlation matrix is found. Multiple regression is, after all, based on the analysis of the linear relationships that exist among a set of data. If there is a good bit of evidence for relationships that are nonlinear, then multiple regression may afford a misleading picture of the genuine structure of the data. Fortunately, even messy data can often be cleaned up enough to be used, but doing so requires many choices and, generally, a good bit of help, so that the least one can do is to familiarize oneself with the details of the data as they come to be analyzed. Perhaps the best source for methods to help with this aspect of multiple regression analysis, and data analysis more generally, is Mosteller and Tukey (1977).

For many of the same reasons, examination of residuals (Section 14.24) is important for multiple regression analysis just as it is in bivariate regression. Generally, it is most useful to examine plots of residuals, or $(y - y')$ values for each case, against y' values, or against time if the data were collected in some known sequence. What one looks for is patterns similar to those of Section 14.24 in the plot, other systematic patterning among the residuals, or noticeable outliers or deviant cases. Such patterns or outliers are, of course, only symptoms that some features of the data need to be examined in more detail before the results of the analysis are used or given serious interpretation.

In almost all "real" research settings, subjects have already been preselected on a number of variables, some perhaps unknown to the investigator. Thus, for example, think of a multiple regression study carried out on college students, and think of the many intellectual, motivational, economic, social, and even physical variables on which these subjects have been preselected, often repeatedly. Each such preselection can alter the predictive configuration of the set of remaining variables, and thus can introduce a very large problem into the interpretation and use of multiple regression. The book by Nunnally (1978) provides an excellent discussion of this problem.

Among the problems that can afflict a set of data being analyzed is that called *singularity* or *multicollinearity*. As we saw in Sections 15.11 and 15.15 the actual calculation of the standardized regression coefficients involves taking the determinant of the matrix of the X correlations and using this as a divisor. However, it can work out that the determinant of this matrix is zero, a mathematical condition known as singularity. Since the solution for the regression weights requires division by this determinant value, the consequence of a zero value is the inability to solve the problem for the unknown B or b values. The form of singularity known as multicollinearity comes about because two or more of the X variables are highly correlated, or because almost all of the variance in one X variable can be accounted for by a set of other predictor variables. If multicollinearity is to be avoided, then X variables should *not* tend to be highly related to each other. As Eqs. 15.25.11 and 15.25.12 demonstrate, even if the R-square for one X variable as predicted by a set of others is very close to 1.00, the estimates of the B weights and their standard errors can fluctuate a great deal due to division by small, rounded numbers. Although there may not be a lot of recourse when multicollinearity is found, other than to throw out one or more X variables that are highly correlated with the others, it is still worth paying attention to possible advance symptoms of this problem. One naive, though common, way in which multicollinearity is introduced into data is through inclusion of X variables that are actually the same measures under different names. Height in inches and height in centimeters are guaranteed to blow your computer program a mile high if they are both included in the same set of predictors. Variables that *must* correlate highly because of the influence of a third variable can also be a problem, as are those linked in a functional relationship. This will be illustrated in the next chapter.

Many computer programs for multiple regression help guard against multicollinearity by reporting a "tolerance" figure for each of the variables entering into a regression equation. This tolerance is simply the proportion of the variance for the variable in question that is *not* due to other X variables. A tolerance value close to 1 means that you are very safe, whereas a value close to 0 shows that you run the risk of multicollinearity, and possibly no solution, by including this variable.

Even sets of variables that are close to collinearity, through large intercorrelations, can be a problem, as the regression coefficients found in this situation can be very unstable and, therefore, unrepresentative of what one might find with other samples. A method that is sometimes recommended in situations such as this is called *ridge regres-*

sion. Many modern treatments of multiple regression, such as Draper and Smith (1981), contain descriptions of ridge regression techniques.

Another problem that can bring one to grief is the use of samples that are much too small relative to the numbers of variables that are under study. In the extreme situation where there are more variables than cases, no unique solution would be achieved. A look at the degrees of freedom for the ANOVA table for multiple regression shows that *at the very least* the number of cases N must be greater than $K + 1$, or else the number of residual degrees of freedom turns out to be 0 or less. In addition, even with adequate, though smallish, samples the B and b weights can turn out to be poor estimates of the true (and perhaps useful) values that would allow good predictability. There is no hard and fast rule about this, but some authors recommend as many as 20 times as many cases as predictor variables. A sophisticated scheme for arriving at sample size through power analysis is given in Cohen and Cohen (1983).

One precaution extends not only to multiple regression, but also to the simple regression situation of Chapter 14 as well. This is the phenomenon of "shrinkage" already referred to in Section 15.18. Essentially, shrinkage means that if a regression equation is developed on a sample, and then actually used to make predictions about another sample or about cases in the population as a whole, the chances are very good that the predictions will not be as accurate as the original data suggested. That is, the apparent predictive power in a sample tends to shrink when one goes beyond that sample. The method gives the best way of combining the variables for predictions *in that sample*, even though some of the relations between the X values and Y, and between the X values themselves may simply be chance occurrences that we have no reason to expect in future samples. For this reason, when people try to work out regression equations for future application to subjects not in the original group, as in the employment or college admissions examples, then they are very careful to cross-validate the original results on one or more additional samples. This lets them evaluate how the predictions are likely to hold up in the larger group in which they will actually be used.

A problem that is often overlooked is that the weights that the predictors carry in the multiple regression equation are *always relative to the entire set of predictors employed*. The predictive strength that some variable X seems to show for some variable Y may be very different if the set of remaining variables is altered. It is risky to talk about the regression weights or the proportion of variance that seems to be due to one or another variable unless the entire set of variables is specified. An even more difficult problem occurs when one tries to find optimal weights for a population situation. Perhaps there are variables X that *should* have been, but were not, included in the sample. The optimal regression weight of X in the population should take account of those additional X variables. The regression weights found in the sample can be very misleading as inferences about the predictive power of those variables in the population unless the missing, very relevant X variables are also included. When one is truly interested in trying to establish the best predictive weights for a set of variables, as referred to a population, then care should go into seeing that every variable known to be truly relevant is taken into consideration. This is, of course, no small task.

With respect to the consequences of violations of the assumptions that underlie the tests and other inferential methods for multiple regression, not a great deal seems to be known. Even so, it seems a safe bet that for relatively large samples, the methods of multiple regression are fairly robust, much like ANOVA. However, it stands to reason that violations of the assumption of equal variances could lead to special difficulties, especially if these violations are extreme.

15.27 THE SWEEP-OUT METHOD FOR MULTIPLE REGRESSION

A very handy method exists for carrying out multiple regression for problems where only a few variables are under consideration. This is generally known as "the sweep-out method," and will be presented here in a version slightly simplified from that given in Snedecor and Cochran (1980). This is an especially useful way to approach small problems in hierarchical or stepwise regression, to be discussed in the next section. Granted that these days almost all such problems will be handled by computer, it is still valuable to have a method that illustrates what is really going on inside the computer program, and permits one to get a "quick fix" on what the results will be for a relatively small problem. Given a reasonably sophisticated hand calculator, of the scientific type, the sweep-out method can be carried out fairly easily, as illustrated below.

Although the sweep-out method will be illustrated here for only three predictor variables and for problems in which the standardized regression weights are desired, Appendix D shows how the method can be extended to more variables and to the solution of other common problems in statistics, such as finding the determinant of a square matrix, and finding what is known as the "inverse" of such a matrix.

The so-called sweep operator works as follows: We start with a square matrix of intercorrelations. Then we choose one variable X to "sweep." Let us call the X variable that is being swept from the matrix variable X_v. Essentially, in the sweep operation we adjust all of the other variables by removing the linear regression for each of them on variable X_v. Thus, starting with a correlation matrix, if we wish to find the standardized regression coefficients, the steps in applying the sweep-out operator are these:

1. Write down the basic correlation matrix, showing the correlation of each pair of X variables, and of each X variable with Y. Arrange the correlations so that dependent variable Y appears as the last row and last column in the matrix. Also, make sure that each diagonal entry, showing the correlation of a variable with itself, contains the value 1.00.
2. Now suppose that we wish to sweep the correlation matrix of some variable, say variable X_a. Then we take $v = a$, symbolizing the variable we will sweep. Locate the diagonal entry corresponding to variable X_v. Call this entry r_{vv}.
3. Now look at any row j. If $j = v$ then divide each entry r_{vk} in that row and any column k by r_{vv}, to find

$$r'_{vk} = r_{vk}/r_{vv}.$$ [15.27.1]

Write these r'_{vk} values down as row v of a new matrix symbolized by SWP(a).
4. Now take any row j where j is *not* equal to v. For each column k not equal to v in that row take

$$r'_{jk} = r_{jk} - (r_{jv} \times r_{vk})/r_{vv}. \qquad\qquad [15.27.2]$$

Enter these r'_{jk} values in row j of the new matrix SWP(a).

5. Omit the column for X_a from the new "swept" matrix SWP(a), but include rows for all of the X variables and for Y.

(The application of the sweep operator to any cell r_{jk}, as outlined in step 4, is easy to visualize if you will simply ignore all of the rows and columns in your matrix *except* those symbolized by row j, column k, and row and column v. Then think of a quadrangle inside the matrix, with corners given by r_{jv}, r_{jk}, r_{vv}, and r_{vk}, like this:

	col. *v.*	col. *k*			col. *v*	col. *k*
row *v*	4	3	or	row *j*	2	1
row *j*	2	1		row *v*	4	3

Then through use of the sweep operator, the new cell value r'_{jk} for cell jk is always just

$$r'_{jk} = \text{cell } 1 - (\text{cell } 2 \times \text{cell } 3)/(\text{cell } 4).$$

However, if row j happens to be row v, then

$$r'_{vk} = r_{vk}/r_{vv} = (\text{cell } 3)/(\text{cell } 4)$$

as shown in step 3. Also notice that this sweep operation is symmetric, so that $r'_{kj} = r'_{jk}$.)

It is simple to illustrate the sweep operator, and in order to do so let us consider the following matrix of correlations for $K = 3$ predictors X and one dependent variable Y:

	X_1	X_2	X_3	Y
X_1	1	.18	$-.11$	.30
X_2	.18	1	.21	.42
X_3	$-.11$	.21	1	$-.21$
Y	.30	.42	$-.21$	1

Now following the steps outlined above we begin to construct the new matrix SWP(1) by sweeping out variable 1. First we divide each entry in row 1 by the value in the diagonal entry for row and column 1, or r_{11}, and enter this row in our new matrix, SWP(1). Thus, we have

$$r'_{12} = .18/1, \; r'_{13} = -.11/1, \text{ and } r'_{1y} = .30/1.$$

Next we take any cell not in row 1 or column 1 and carry out step 2. Thus, for the cell in row 2 and column 2 we take

$$r'_{22} = r_{22} - r_{21} \times r_{12}/r_{11}$$

$$= 1 - (.18) \times (.18)/1$$

$$= .9676.$$

Doing the same thing for row 2 and column 3 we have

$$r'_{23} = r_{23} - r_{21} \times r_{13}/r_{11}$$

$$= .21 - (.18 \times -.11)/1$$

$$= .2298.$$

In addition, since $r'_{jk} = r'_{kj}$, we also know that $r'_{32} = .2298$. (Incidentally, it is a good idea to carry four or more decimal places in each sweep-out operation, in the interest of accuracy in the final results.)

Proceeding in exactly this same way for the other cells we find

$$r'_{2Y} = .42 - (.18 \times .30)/1 = .3660$$

$$r'_{33} = 1 - (-.11 \times -.11)/1 = .9879$$

$$r'_{3Y} = -.21 - (-.11 \times .30)/1 = -.1770$$

$$r'_{YY} = 1 - (.30 \times 30)/1 = .9100.$$

This completes the table SWP(1), which appears below. Notice that this table omits column 1, standing for the swept X_1.

SWP(1) =

	X_2	X_3	Y	part r^2	B
X_1	.18	$-.11$	.30	—	.30
X_2	.9676	.2298	.3660	.1384	—
X_3	.2298	.9879	$-.1770$	.0317	—
Y	.3660	$-.1770$	.9100		

$$R^2_{y \cdot 1} = 1 - .9100 = .09$$

Even though we have so far swept out only variable X_1, the resulting SWP(1) table contains a number of interesting items of information. If we want to predict Y from variable X_1 alone, then the standardized B weight is the entry in row 1 and column Y, or .30. Second, the squared part correlation between Y and X_2 adjusted for X_1 is

$$r^2_{y(2.1)} = (r'_{2y})^2/r'_{22}$$

$$= (.3660)^2/.9676$$

$$= .1384$$

and the square part correlation for Y with variable X_3 adjusted for X_1 is

$$r^2_{y(3.1)} = (r'_{3y})^2/r'_{33}$$

$$= (-.1770)^2/.9879$$

$$= .0317.$$

If you also wish to know the squared partial correlation, where both the predictor variable and variable Y have been adjusted for the swept variable, this can be found by dividing the squared part correlation value, as found above, by the value in the YY cell of the swept table. Thus, for example,

$$r_{y2.1}^2 = r_{y(2.1)}^2/(r_{yy}') = .1384/.9100 = .1521$$

and

$$r_{y3.1}^2 = r_{y(3.1)}^2/(r_{yy}') = .0317/.9100 = .0348.$$

Finally, the proportion of Y variance *unaccounted for* by variable X_1 is shown in row Y and column Y. This is .9100, so that the proportion of variance accounted for by this one variable is $1 - .91 = .09$.

In general, once the new matrix SWP(a) has been determined by sweeping out variable X_a, then these things will be true:

1. The standardized regression weight B_a for predicting Y from variable X_a *alone* is found from the entry for row a and column Y, or

$$B_a = r_{ay}'. \qquad [15.27.3]$$

2. For any other variable X_j, where j does not equal a, the squared part correlation between Y and X_j after adjustment for X_a is given by the square of r_{jy}', divided by the diagonal entry r_{jj}':

$$r_{y(j \cdot a)}^2 = (r_{jy}')^2/r_{jj}' \qquad [15.27.4]$$

3. The squared partial correlation of variable X_j with Y, with X_a held constant, equals the square of r_{jy}', divided by the diagonal entries for variable X_j and for Y:

$$r_{yj \cdot a}^2 = (r_{jy}')^2/r_{jj}'r_{yy}' \qquad [15.27.5]$$

4. The proportion of Y variance *not* accounted for by the swept variable X_a is given by the diagonal entry r_{yy}', so that the proportion of variance that *is* explained by linear regression on X_a equals.

$$R_{y \cdot a}^2 = 1 - r_{yy}'. \qquad [15.27.6]$$

Since typically we will be interested in predicting Y on the basis of more than one X variable, we need to learn to sweep the matrix for other variables as well. To sweep the matrix of another variable, say X_b, we forget all about the original correlation matrix and deal only with the last matrix SWP(a). That is, we treat the values in SWP(a) as though they were the original correlations r_{jk}. We then let $v = b$, whereas before v stands for the variable to be swept. Then the five steps given above are carried out, to give a new $J \times (J - 2)$ matrix SWP(ab), which omits both the columns that corresponded to X_a and X_b.

Returning to our example, we can now forget all about our original correlation matrix, and deal with the entries in SWP(1) as though they were the original r values. (We will continue to symbolize the new, swept, values by r'.) Then we repeat the sweep operation just as before, but this time using variable $V = X_2$ in order to find a new matrix SWP(12).

First, we divide each entry in row 2 of SWP(1) by the value in row 2 and column 2, and place these results in row 2 of our new matrix, as shown below. Then we apply the sweep operator to the remaining cells, just as before. Thus, for the cell in row 1 and column 3 we have

$$r'_{13} = r_{13} - (r_{12} \, r_{23})/r_{22}$$

$$= -.11 - (.18 \times .2298)/.9676$$

$$= -.1527.$$

For the cell in row 1 and column Y we have
$$r'_{1y} = .30 - (.18 \times .3660)/.9676$$

$$= .2319.$$

For row 3 and column 3 we find

$$r'_{33} = .9878 - (.2298 \times .2298)/.9676$$

$$= .9333$$

and so on until we have completed the table SWP(12), which omits both columns 1 and 2:

SWP(12) =

	X_3	Y	part r^2	B
X_1	$-.1527$	.2319	—	.2319
X_2	.2375	.3783	—	.3783
X_3	.9333	$-.2639$	.0746	—
Y	$-.2639$	.7716	$(R^2_{y \cdot 12} = .2284)$	

From the table SWP(12) we see that in order to predict Y from X_1 and X_2 alone, the standardized regression weights needed are

$$B_1 = r'_{1y} = .2319 \text{ and } B_2 = r'_{2y} = .3783.$$

We also see immediately that the proportion of variance not accounted for either by X_1 or X_2 is $r'_{yy} = .7716$. Therefore, the proportion of variance that *is* accounted for by these two variables must be

$$R^2_{y \cdot 12} = 1 - .7716 = .2284.$$

Finally, we see that the squared part correlation of X_3 with Y, after X_3 has been adjusted for both X_1 and X_2, is

$$r^2_{y(3.12)} = (r'_{3y})^2/r'_{33}$$

$$= (-.2639)^2/.9333 = .0746.$$

(From the relationship given in Eq. 15.23.11, we also know that the proportion of variance that will be accounted for when all three variables are used as predictors is .2284 + .0746 = .3030.)

Finally, we carry out the sweep operation for variable X_3, proceeding exactly as before, but using the SWP(12) matrix. Then we find

SWP(123) =

$$Y = B$$

	$Y = B$
X_1	.1887
X_2	.4455
X_3	$-.2828$
Y	.6970 $(R^2_{y \cdot 123} = .3030)$

As a check we may note that, by Eq. 15.14.2,

$$R^2_{y \cdot 123} = .1887(.30) + .4455(.42) + (-.2828)(-.21)$$

$$= .3031, \text{ a close agreement.}$$

In short, when we have achieved the last table, from which all of the X variables have been swept, the remaining Y column contains the standardized regression weights that apply for predicting standardized Y from all of the standardized X values. Furthermore, the last cell in this column gives the proportion of variance *not* explained by these X variables, and 1 minus the entry in this cells gives R-square, the proportion of variance accounted for.

With a little practice and a good hand calculator, one can do small multiple regression problems surprisingly quickly and easily by the sweep-out method, even for problems with six or eight X variables. However, as mentioned above, one would want to use the computer for problems of any real magnitude, and the sweep-out procedure is principally useful where one would like a "quick fix" on how a few variables relate to some dependent variable of interest.

Incidentally, the sweep-out method can also be used in this same way to find the unstandardized, rather than the standardized, regression weights for predicting raw Y from raw X values. However, if unstandardized weights are desired, one starts out with a matrix containing variances (rather than ones) in the diagonal, and covariances (rather than correlations) in the remaining cells. Then the final result is a set of b or unstandardized weights, and the value of R-square is determined by dividing the final value in the Y column by the *original* variance of Y, and then subtracting the result from 1.00. Be warned, however, that when variances and covariances are used, the numbers that result from the calculations can sometimes become very large or very small, and therefore hard to handle accurately on a calculator. Therefore, if this method is to be used to find unstandardized weights, it is probably best to start with a matrix of correlations, as illustrated above, and then to find the unstandardized b weights by applying the standard deviations to the standardized B weights, as shown in Section 15.13.

You should also be aware that multicollinearity, as mentioned in Section 15.26, is a problem for the sweep-out method just as in the other solutions to multiple regression problems. However, in the sweep-out method you can at least "see it coming" in the way the diagonal cells behave when multicollinearity is a problem. Whenever the diagonal cells of a swept matrix become near zero in value, multicollinearity is present. The result is that the obtained B or b values are often very unreliable, and, if the diagonal entries actually become zero then the method cannot be continued. When this begins to happen with the sweep-out method, one would usually drop the offending X variables from further consideration and proceed to solve the problem with the X variables that are left.

15.28 HIERARCHICAL AND STEPWISE REGRESSION

The method involving the solution of the normal equations to find the regression weights is sometimes called *simultaneous regression,* ostensibly because it demands the simultaneous solution of a set of K normal equations in K unknowns. However, other approaches to multiple regression are often useful.

The sweep-out method we have been discussing assumes that the investigator is interested in building a multiple regression equation by adding one predictor variable at a time. Here interest usually focuses in the amount of predictive power that each additional variable contributes, and on the search for a small but effective set of predictor variables. In the *hierarchical* approach to multiple regression, the investigator specifies from the outset the *order* or *hierarchy* in which the predictor X variables are to be introduced into the regression equation. Then, just as in our example of the sweep-out method, the variable designated as first in order is used as the first variable v to be swept from the matrix. Then the second variable in order is used as the basis for the second sweep until, finally, all of the variables of interest have been swept, and thus included in the final regression equation.

It must be kept in mind, however, that so long as the investigator includes exactly the same set of predictor variables, the regression weights and the R-square value will be exactly the same at the end *regardless* of the order in which the variables are included in the regression equation. This is true not only for hierarchical regression but also for the forward-selection, stepwise methods we will talk about next.

In the example just given, the hierarchy can be considered to be the "natural" order X_1, X_2, and X_3, and the problem solved by sweeping out each of these variables in turn. The hierarchical approach is particularly appropriate when there are good theoretical or practical reasons for believing that one variable, or a small set of variables, is quite important for predicting Y, and thus is to be examined first. This single variable, or one of the "important set," is then chosen for the initial sweep operation. Another variable which is judged to be second in importance is subsequently swept-out next, and so on for the remaining variables in their predetermined order. Or, the hierarchical approach can be useful when one is really interested in a set of high-order part or partial correlations, and several variables are first to be "adjusted out" or "held constant" so that the predictive efficacy of a second set of variables may be examined.

An even more popular approach to multiple regression is known as *stepwise regression*. For the time being we will confine our attention to the strategy usually known as *forward selection*. Unlike the hierarchical approach, in stepwise regression with forward selection there is no preset order in which the X variables will be introduced to the regression equation. Rather, the point is to include the variables one at a time, generally with the goal of accounting for the largest additional amount of Y variance with each new X variable added to the equation. Thus, the selection is "forward" in the sense that we will always be looking ahead to add the best possible predictor variable at each step.

A word of caution: By use of hierarchical or stepwise regression methods one actually gets the same sample information as if the problem had been approached by use of the "simultaneous" strategy, with the addition of some values of part or partial correlations. However, statistical inference becomes more complicated with hierarchical, and especially with stepwise, regression because a whole series of statistical tests can be involved at each stage of the analysis. Some of the problems outlined in Section 15.26 can also be exacerbated in a hierarchical or stepwise approach to regression. For these and a variety of other reasons, some users of statistics have come to feel that these methods should be restricted to situations where the interests of the investigator are primarily descriptive and exploratory, and that stepwise methods should be avoided or used with special caution when the aim is primarily the development of an optimal regression equation for actual predictive purposes.

You will recall from Section 15.23 that with any set of V different X variables, X_a, X_b, . . . , X_V, we can always represent the value of R-square, or the proportion of variance accounted for, as the sum of the set of squared part correlations:

$$R^2_{y'ab\cdots V} = r^2_{ya} + r^2_{y(b'a)} + r^2_{y(c\cdot ab)} + \cdots + r^2_{y(V\cdot ab\cdots V - 1)} \qquad [15.28.1]$$

(Here, the subscripts a, b, c etc. stand for "first variable included," "second variable included," "third variable included," and so on, as distinct from the usual subscripts 1, 2, . . . , K, which are in effect the "names" of the variables.)

Now suppose we are going to choose a set of X variables to predict Y, selecting one variable at a time with the aim of maximizing the proportion of Y variance accounted for at every step. Then obviously the first variable to select (symbolized here by X_a) should have the largest squared correlation with Y, or maximum r^2_{ya}.

Now, after selecting the first X variable, we want the next one selected to have the *largest squared part correlation* with Y, after adjustment for the first, or X_a. Let us use X_b to symbolize the second selected, or the one having the largest squared part correlation after adjustment on X_a. Then the third variable selected, symbolized here by X_c, should be the one having the largest squared part correlation with Y after adjustment for both X_a and X_b, and so on. Indeed, proceeding in this way we can either keep going until all of the available X variables are selected, or until the squared part correlations for the remaining variables are too small to add substantially to the value of R-square.

Such a *stepwise procedure with forward selection* is easy to illustrate by use of the sweep-out method introduced above. Once again we can use the same three X variables employed in the previous example, yielding just as before the correlation matrix:

	X_1	X_2	X_3	Y
X_1	1	.18	−.11	.30
X_2	.18	1	.21	.42
X_3	−.11	.21	1	−.21
Y	.30	.42	−.21	1

Now, in order to proceed in stepwise fashion, we first inspect the correlation matrix above to find the X variable having the largest squared correlation with Y. Since the variable labeled X_2 has a correlation of .42 with Y, this gives the largest squared correlation with Y, or .1764. This then is the variable chosen to be swept-out first (that is, X_2 is the first variable v).

Sweeping out variable X_2 and omitting the second column, which corresponds to this variable, we find the matrix SWP(2):

SWP(2) =

	X_1	X_3	Y	Part r^2	B
X_1	.9676	−.1478	.2244	.0520	—
X_2	.18	.21	.42	—	.42
X_3	−.1478	.9559	−.2982	.0930	—
Y	.2244	−.2982	.8236		

$(R_{y \cdot 2}^2 = .1764)$

This SWP(2) table shows that if we wish to predict the standardized Y value from standardized X_2 alone, the required standardized regression weight B_2 is .42. The proportion of variance accounted for by X_2 alone is then $1 - .8236 = .1764$.

Next we want to pick a second variable to sweep out, or include in our regression equation along with X_2. By looking at the squared part correlations of the remaining (unswept) X variables, one can see that the largest squared part correlation (.0930) belongs to variable X_3. Hence, we proceed to sweep variable X_3 to find the matrix SWP(23):

SWP(23) =

	X_1	Y	Part r^2	B
X_1	.9447	.1783	.0336	—
X_2	.2125	.4855	—	.4855
X_3	−.1546	−.3119	—	.3119
Y	.1783	.7306		

$(R_{y \cdot 23}^2 = .2694)$

From table SWP(23) we see that the regression weights for predicting standardized Y from standardized X_2 and standardized X_3 are then .4855 and .3119. The multiple regression equation involving these variables yield predictions that account for about 27% of the variance of Y.

The largest (indeed the only) squared part correlation after adjustment for X_2 and X_3 then belongs to X_1. Hence, variable X_1 will be the basis for our third and final sweep.

After sweeping for variable X_1, we find the matrix SWP(231):

SWP (231) =

	$Y = B$
X_1	.1887
X_2	.4454
X_3	−.2828
Y	.6970

$(R_{y \cdot 231}^2 = .3030)$

Notice for this example how the final value of R-square is actually the sum of the squared part correlations of the X variables in the order in which they were introduced to the regression equation, and adjusted for the variables previously entered. That is

$$R^2_{y\cdot231} = R^2_{y\cdot123} = r^2_{y\cdot2} + r^2_{y(3\cdot2)} + r^2_{y(1\cdot23)}$$

$$= .1764 + .0930 + .0336$$

$$= .3030.$$

This, of course, illustrates the principle in Eq. 15.23.11.

Also observe that this stepwise solution of the multiple regression problem gives the same final results as the solution found in Section 15.27.

So long as the same X variables or predictor variables are included in the regression equation, the order of inclusion has no bearing on the final regression weights nor on the final value of R-square obtained. In addition, the final B and R-square values will be the same, whether found through hierarchical, stepwise, or simultaneous procedures, provided that exactly the same set of data involving the same set of variables is used in each procedure, and that the same set of these variables is included in the final regression equation.

Just as for other regression analyses, it is customary to show the results of a hierarchical or stepwise analysis in the form of an ANOVA table. Thus, suppose that for the $N = 50$ subjects in this study, the total sum of squares for dependent variable Y was 1,289. Then the stepwise analysis we have just completed might be represented in an ANOVA table something like Table 15.28.1.

Note that the F values in this table are exactly the same as one would obtain by using Eqs. 15.25.15 through 15.25.17. In addition, a table of gains in proportion of variance accounted for (that is, the squared part correlations) is often included. Although this sort of summary is fairly standard for a stepwise multiple regression problem, computer printouts vary widely both in the ANOVA format and in the amount of information provided, depending on the package employed.

Source	SS	df	MS	F	p
Regression on X_2	227.38	1	227.38	11.64	<.01
Regression on X_3, beyond X_2	119.88	1	119.88	6.14	<.05
Regression on X_1, beyond X_2, X_3	43.31	1	43.31	2.22	NS
Residual	898.43	46	19.53		
Total	1,289	49			

Table 15.28.1

ANOVA summary for a stepwise regression analysis

15.29 SIGNIFICANCE AS A CRITERION FOR INCLUSION

When stepwise regression is carried out by computer, it is usual to apply still another criterion to the choice of the variable to be selected at each step: This is the *significance of the part correlation*, which is considered along with the magnitude of the squared part correlation for the variable in question. As we saw in Section 15.25 when significance is tested for $r^2_{y(j \cdot G)}$, the squared part correlation between Y and some variable X adjusted for some other set of variables G, this is the same as testing the significance of the *gain* in R-square due to the addition of variable X_j to a regression equation already containing the variables in set G.

There is, by the way, some disagreement among the experts about the proper denominator for the F test in this context. Some would say that this should correspond to the mean square residual based on the regression equation including all of the variables added to the regression equation up to and including the new variable being tested for significance. Others would say that the denominator of F should be an MS residual based on the entire set of K variables eventually to be included in the regression equation just as in Section 15.25. Although there is something to be said for both points of view, here we will adopt the position that the appropriate denominator in the F test for gain is based on *the entire set* of variables eventually to be included. That is, the F test we will use will be based on these conventions: For the moment let G stand for all of the variables included in the regression equation *up to* but not including variable X_j. Then let H stand for the entire set of K variables included in the final regression equation. In practice this means that an overall simultaneous analysis for the entire set of K variables is needed to establish the value of $R^2_{y \cdot H}$ prior to the analysis by stepwise regression. This is a bit of a nuisance for hand calculation but no strain for a computer. Our F test for the significance of the gain in predictability due to the added variable X_j is then

$$F = v^2_{y(j \cdot G)} (N - K - 1)/(1 - R^2_{y \cdot H}) \qquad [15.29.1^*]$$

or

$$F = (R^2_{y \cdot jG} - R^2_{y \cdot G})(N - K - 1)/(1 - R^2_{y \cdot H})$$

with 1 and $N - K - 1$ degrees of freedom. Notice that here $R^2_{y \cdot jG}$ symbolizes the variance accounted for by the set G plus variable X_j. This is different from $R^2_{y \cdot H}$, which is based on the set of all predictor variables to be considered in the problem. This procedure then has the advantage of making all of the F tests in a forward-selection process comparable in terms of the MS residual each involves and the degrees of freedom.

When the criterion of significance is added to forward selection in multiple regression, only those variables are chosen that yield the *largest significant gain in variance accounted for at each step*. The F ratio that is used as the basis for this decision is sometimes called the "F to enter."

In the example of stepwise regression just given, suppose that we had added the requirement that the squared part correlation (or gain in Y variance accounted for) had to be significant at or beyond the .05 level for any new variable to be selected for inclusion in our regression equation. Furthermore, suppose that the original correlation matrix is based on a sample of $N = 50$ cases. We already know that the entire set of three predictor variables gives an R-square value of .3030.

Then, when we undertake the stepwise analysis, the first variable selected, or X_2 shows a squared correlation with Y of .1764, giving an F ratio of

$$F = r_{y2}^2 (N - 3 - 1)/(1 - R_{y \cdot H}^2)$$

$$= (.1764 \times 46)/(1 - .3030)$$

$$= 11.64$$

for 1 and $N - 3 - 1 = 46$ degrees of freedom, this is significant beyond the .05 level.

Next, we selected variable X_3 by noticing that its squared part correlation of .0930, as shown in table SWP(2), was larger than that for variable X_1, the other remaining variable. Does the square of this part correlation represent a significant gain in our ability to predict Y, over and above what we can do with X_2 alone? This question is answered by the F ratio

$$F = r_{y(3 \cdot 2)}^2 (N - 3 - 1)/(1 - R_{y \cdot H}^2)$$

$$= (.0930 \times 46)/(1 - .3030)$$

$$= 6.14 \ .$$

For 1 and $N - 3 - 1 = 46$ degrees of freedom, this too is significant beyond the .05 level, and thus eligible for inclusion.

In table SWP(23) the squared part correlation $r_{y(1 \cdot 23)}^2$ is found to be .0336. The F test for this part correlation is thus

$$F = r_{y(1 \cdot 23)}^2 (N - 3 - 1)/(1 - R_{y \cdot H}^2)$$

$$= (.0336 \times 46)/(1 - .3030)$$

$$= 2.22.$$

For 1 and 46 degrees of freedom, this F value fails to reach the .05 level. In other words, if the .05 level of significance were used as a criterion in this forward-selection procedure, we would stop after the inclusion of only variables X_2 and X_3.

It should be obvious that the procedure of stepwise regression with forward selection can be applied to any number of X variables. The limitations noted in Section 15.26 all apply, just as they do in simultaneous and in hierarchical regression.

15.30 BACKWARD ELIMINATION OF VARIABLES

Sometimes, however, the problem is not to add variables to a regression equation so as to increase predictive power at every step, as in forward selection. Rather, we may start off with a fairly large number of variables, and then ask, "Which of these variables might be eliminated as predictors with least damage to the predictive ability of the entire equation?" This is the problem of *backward elimination*, and it is handled as follows:

First the simultaneous or classical solution to the multiple regression equation is found, using all of the predictor variables. Then the variable is sought which has the *smallest* (and nonsignificant) squared part correlation with Y after adjustment for all of the other variables. This variable is then eliminated from the equation. Next the R-square for the remaining variables is calculated, and if this is still a satisfactory value, still another variable is eliminated by the same process as the first.

These steps are repeated until a small set of variables remains, sufficient to provide an acceptable proportion of variance accounted for, or R-square. Tests of significance often accompany these decisions to eliminate or not eliminate variables, with the small and nonsignificant contributors to R-square being the likely candidates for elimination.

When the problem is to find the *best K* out of a larger set of predictor variables, it is common practice to have the computer produce the solutions for all possible sets of K variables, and then to examine the R-square and significance values to decide on the best of these possibilities. Needless to say, this method has become popular only with the availability of computer packages for multiple regression.

In short, in both the forward-selection and the backward-elimination approaches one is generally looking for the most predictive power as given by the fewest variables. However, in the former method this goal is accomplished by finding the most predictive variables and adding them one at a time to the equation, whereas in the backward method one finds and throws out those variables that do not seem to carry their share of the load.

15.31 MULTIPLE REGRESSION IN THE STATISTICAL PACKAGES

The regression programs of the major statistical packages tend to offer elaborate sets of input and output features and options. The format and the information provided for multiple regression analysis vary considerably from package to package, and one should study the manual for the particular package with special care before these programs are attempted.

For a multiple regression problem with K predictor variables in which a stepwise analysis is desired, the SPSS subprogram NEW REGRESSION follows a simple format such as

NEW REGRESSION   VARIABLES = $X1$ TO Y/

DEPENDENT = Y/

STEPWISE/

Notice that the variables line includes all the variables, one of which is then singled out as the dependent variable. If methods other than stepwise are desired, the alternative keywords FORWARD (for forward inclusion), BACKWARD (for backward elimination), or ENTER = *variable list* (showing that the variables listed are to be entered hierarchically in the order shown). If ENTER is shown without a list, then all predictor variables are entered at once, and a simultaneous regression analysis is provided. A great many options exist for descriptive statistics, inclusion and exclusion criteria, and residuals analysis. Correlations or variances = covariances may be entered initially.

The REGRESSION subprogram in SPSS[x] is rather similar to the SPSS NEW REGRESSION, except that it offers somewhat more choices and options for the user.

Although they differ slightly in details, both SPSS and SPSS[x] offer a subprogram PARTIAL CORR in which first-order or higher order partial correlations may be found for specified sets of variables. Analysis of part correlations would, however, usually be carried out most easily through a series of hierarchical or stepwise multiple regression analyses.

The simplest and most straightforward of the BMDP programs for handling multiple regression is P1R, which can handle up to about 145 predictors. For an ordinary problem in simultaneous multiple regression, the setup is

/ PROBLEM	TITLE IS . . .
/ INPUT	VARIABLES ARE (total number of variables)
	FORMAT (entry format for variables in data)
/ REGRESS	DEPENDENT IS (variable name)
	INDEPENDENT ARE (variable names)
/ END	

Output for P1R includes means and standard deviations for each variable, along with minimum and maximum values. Multiple R, R-square, and the standard error of estimate are then given, accompanied by an ANOVA table for the regression of Y on the predictor variables. Both unstandardized and standardized regression weights are reported, with standard errors and t tests. The tolerance for each predictor variable is also listed.

Program P2R has the same general input format, although the analysis is stepwise. Both forward and backward selection are permitted, as well as hierarchical inclusion. A good many options for input and output, and for selection are provided in this program. Program P9R performs regression analysis on all possible sets of predictor variables. This is done in an economical and efficient manner, saving computer time for large sets of variables, and permitting choice of the "best" subset according to user-defined criteria.

As usual, in its programs for multiple regression SAS provides some of the most sophisticated options, but probably requires more knowledge on the part of the user than some of the other packages. The easiest SAS procedure to use for multiple regression seems to be PROC REG, where for K predictors and simultaneous regression the control statements are simply

PROC REG;

    MODEL Y = X1 X2 X3 $\cdots$ XK;

The principal output then includes an ANOVA for the multiple regression, including R-square, and the unstandardized weights and intercepts, with standard errors and t tests. Other options for the method of analysis are afforded by the keywords FORWARD, BACKWARD, STEPWISE, and RSQUARE (the last of which provides for selection among all possible subsets of predictors). Many "diagnostics" and residual procedures are also available. The PROC GLM also offers multiple regression, geared toward hierarchical and stepwise methods, and open to a wide variety of options.

Incidentally, the output from these and other multivariate programs can be enormous, especially if many subanalyses and a large number of options are requested. In the interests of economy, it is well to form the habit of not requesting additional output beyond the essentials unless you have thought about *how* you intend to use it, and if it is worth the additional expense. Computer paper may be relatively cheap, but it is not usually free, and the expense can mount up more rapidly than you might think.

EXERCISES

1. In a study of 50 cases sampled at random, the correlations among three variables, X_1, X_2, and X_3, were as follows:

$r_{12} = .38$

$r_{13} = .45$

$r_{23} = -.17$

Find the partial correlation of variables X_1 and X_2, holding X_3 constant. Find the partial correlation of variable X_2 and X_3, holding X_1 constant.

2. Calculate all of the possible part (or semipartial) correlations involving the variables in Exercise 1.

3. Suppose that the intercorrelation among three variables could be $-.75$, $-.75$, and $-.75$. What would this imply about the partial correlation between the first and second variables, holding the third constant? Can this situation actually exist? What are the *smallest* intercorrelations that can exist among three variables, if all three intercorrelations are the same?

4. Use the data in Exercise 11, Chapter 14, to find the raw-score regression equation for predicting the value of B from A. Actually make this prediction for each of the cases in the data. Then find the residual, or error, for each such prediction. Finally, correlate these residuals (or B values adjusted for A) with the variable C. Show that this gives the same result as taking the part correlation, $r_{C(B \cdot A)}$.

5. Use the data in Exercise 11 of Chapter 14 once again, but this time find the C value for each individual after adjustment for variable A. Correlate these adjusted C values with the adjusted B values found in Exercise 4 above, and show that this correlation is equal to the partial correlation, $r_{BC \cdot A}$.

6. For Exercise 1 above, find the multiple regression equation for predicting the standardized value z_1 from z_2 and z_3.

7. From Exercise 6, what would predicted z_1 be if $z_2 = 1.9$ and $z_3 = -1.2$?

8. Use the results of Exercise 25, Chapter 14, to find the following partial and part correlation values:

(a) $r_{y2 \cdot 3}$
(b) $r_{y(2 \cdot 1)}$

(c) $r_{y3 \cdot 1}$
(d) $r_{y(3 \cdot 1)}$
(e) $r_{y(1 \cdot 2)}$

9. Take the correlations found in Exercise 25, Chapter 14 and calculate the second-order partial correlation symbolized by $r_{y3 \cdot 12}$.

10. On the assumption that the data in the problem just above represent a random sample from a multivariate normal distribution, test the significance of the obtained partial correlation.

11. For the date in Exercise 11, Chapter 14, find the multiple regression equation for the prediction of the raw rating for country A from the ratings of countries B and C.

12. Calculate the value of the coefficient of multiple correlation R for the data in Exercise 11, Chapter 14. For these data, what is the proportion of variance in the ratings of A accounted for by the ratings of B and C?

13. If the ratings in Exercise 12 above represent a random sample with $N = 15$ from a multivariate normal distribution, what can we say about the significance level of the multiple correlation obtained?

14. Carry out a multiple regression analysis based on the data in Exercise 25, Chapter 14, by setting up and solving the normal equations for the standardized regression weights. Calculate and test the significance of the R^2 value.

15. Use the results of Exercise 14 above, along with the means and standard deviations found in Exercise 25 of Chapter 14 to construct the multiple regression equation for predicting the raw value on variable Y from the raw values of X_1, X_2, and X_3.

16. Set up the ANOVA table summarizing the results of the multiple regression analysis in Exercises 14 and 15.

17. By use of the part and partial correlation values found in Exercises 8 and 9 above, find and test the significance of the value of $R^2_{y \cdot 123}$ for the prediction of Y from X_1, X_2, and X_3.

18. A test of scholastic aptitude was composed of four subtests, with scores symbolized here by X_1, X_2, X_3, and X_4. This test was so designed that for the population of recent graduates of U.S. high schools, the four subtest scores would be uncorrelated with each other. However, if we let Y symbolize the final high-school grade point average of a graduate, the correlations were

	X_1	X_2	X_3	X_4	Y
X_1	1	0	0	0	.36
X_2		1	0	0	−.42
X_3			1	0	.19
X_4				1	.50
Y					1

Find the multiple regression equation for predicting standardized grade point values from standardized X values. What is the proportion of variance in final grade point averages that is accounted for by the scholastic aptitude test?

19. Below is the intercorrelation matrix for three predictor variables X, and dependent variable Y. There were $N = 86$ cases.

	X_1	X_2	X_3	Y
X_1	1	$-.31$	.27	.22
X_2	$-.31$	1	.15	.36
X_3	.27	.15	1	$-.23$
Y	.22	.36	$-.23$	1

Use the sweep-out method to carry out a hierarchical regression analysis, taking the variables in the order X_1, X_2, and X_3. Show the squared part correlations for each of these variables, adjusted for any previously swept variables.

20. Use the correlation matrix in Exercise 19 once again, but this time use the sweep-out method to carry out a stepwise analysis, with forward selection based on maximizing gain in proportion of variance accounted for. Test for the significance of the gain in proportion of variance accounted for after the inclusion of each variable.

21. A multiple regression problem with $N = 152$ involved four predictors X, and a dependent variable Y. The matrix of intercorrelations is shown below. Use the sweep-out method and the arbitrary hierarchical order of X_1, X_2, X_3, X_4 to find the standardized regression equation and the proportion of variance accounted for by the X variables.

	X_1	X_2	X_3	X_4	Y
X_1	1	.15	.11	$-.18$	.26
X_2	.15	1	$-.13$	.21	$-.15$
X_3	.11	$-.13$	1	.07	.42
X_4	$-.18$	.21	.07	1	$-.20$
Y	.26	$-.15$	.42	$-.20$	1

22. In Exercise 21, what is the proportion of Y variance accounted for by variables X_1 and X_2? Is a significant amount of additional varaince accounted for by the further inclusion of variables X_3 and X_4? By the inclusion of X_3 alone without X_4?

Chapter *16*

Further Topics in Regression

In this chapter we are going to look at further applications of the general linear model and the concepts of multiple regression to various problems of data analysis. As we shall see, these methods apply most directly to experimental situations where the levels of the experimental factor are chosen in advance, or in situations where only certain values of the predictor variables are examined or permitted to occur. In other words, here the sampling strategy will be that of a problem in regression.

We will find that in this model questions both of the linear and possible nonlinear relationships may be examined. We will start by looking at the intimate connection between the analysis of variance and multiple regression methods. Indeed, we will show that ANOVA corresponds directly to multiple regression with dummy variables used as predictors, as suggested in Chapter 10.

16.1 MULTIPLE REGRESSION WITH DUMMY VARIABLES

It is perfectly possible for one or more of the predictor, or X, variables used in multiple regression analysis to represent qualitative or categorical distinctions in the data. Thus, for example, the qualitative distinction "male" versus "female" might serve as a very good predictor of some dependent variable Y, and so might other distinctions such as "married" versus "unmarried," "college graduate" versus "non college graduate," and so on. Nor must the qualitative distinctions consist only of two categories or a dichotomy; whole attributes consisting of J mutually exclusive and exhaustive classes, just like the "factors" used in analysis of variance, can also be employed in the role of predictors in multiple regression.

Just as with the analysis of variance discussed in Section 10.2, these qualitative distinctions can be handled through the device of "dummy variables." That is, membership in a particular qualitative grouping j is shown by use of a dummy variable, where the value of variable $X_j = 1$ when a case in question qualifies for the category j, and $X_j = 0$ otherwise.

However, one must be alert for two things when such qualitative distinctions are introduced into multiple regression problems. In the first place, such 1 and 0 dummy variables standing for a set of J mutually exclusive and exhaustive categories are constrained to sum to 1.00 for each individual in the data (since everyone belongs somewhere). This implies that in the data the dummy variables X_1 standing for membership in Group 1, X_2 for membership in Group 2, and so on, *must* fall into the relationship

$$X_1 + X_2 + \cdots + X_J = 1 .$$

The result of this dependency among the dummy variables must be multicollinearity among this set of X variables and, thus, an inability to yield a multiple regression solution unless some corrective steps are taken.

It is very simple to illustrate that this is true. Take the example of Table 10.16.1 once again. Subjects in these data were divided into three groups, and thus we might classify each of the subjects into his or her group by the use of three dummy variables, as shown in Table 16.1.1 below. For our immediate purposes we need only the values of the dummy variables X, and not the values of the dependent variable Y. Now treating these dummy variables as though they were ordinary predictor variables in multiple regression, we can find the intercorrelation matrix among these X variables, shown below as Table 16.1.2 (remember that only the correlations among the X variables need to be shown at this time).

Now, suppose that we attempt to solve for the standardized regression weights, by use, let us say, of the method of determinants as illustrated in Section 15.16. Here, as in that section, we start by taking the value of the determinant based on the correlations among the three X variables:

$$D(X) = \begin{bmatrix} 1.0 & -.5 & -.5 \\ -.5 & 1.0 & -.5 \\ -.5 & -.5 & 1.0 \end{bmatrix} \begin{bmatrix} 1.0 & -.5 & -.5 \\ -.5 & 1.0 & -.5 \\ -.5 & -.5 & 1.0 \end{bmatrix} \begin{bmatrix} 1.0 & -.5 & -.5 \\ -.5 & 1.0 & -.5 \\ -.5 & -.5 & 1.0 \end{bmatrix}$$

$$= 1 - .125 - .125 - .25 - .25 - .25 = 0.$$

Table 16.1.1

Data of Table 10.16.1 coded by use of three dummy variables

Case	X_1	X_2	X_3	Case	X_1	X_2	X_3	Case	X_1	X_2	X_3
1	1	0	0	11	0	1	0	21	0	0	1
2	1	0	0	12	0	1	0	22	0	0	1
3	1	0	0	13	0	1	0	23	0	0	1
4	1	0	0	14	0	1	0	24	0	0	1
5	1	0	0	15	0	1	0	25	0	0	1
6	1	0	0	16	0	1	0	26	0	0	1
7	1	0	0	17	0	1	0	27	0	0	1
8	1	0	0	18	0	1	0	28	0	0	1
9	1	0	0	19	0	1	0	29	0	0	1
10	1	0	0	20	0	1	0	30	0	0	1

	X_1	X_2	X_3
X_1	1.0	$-.5$	$-.5$
X_2	$-.5$	1.0	$-.5$
X_3	$-.5$	$-.5$	1.0

Table 16.1.2

Correlatations among the dummy variable predictors of Table 16.1.1

The value of this determinant based on the X values is 0, and yet this is the determinant that must serve in the denominator of the ratios used to find the B values. Obviously, the B values cannot be determined here, and this is an instance of multicollinearity among the X variables. The dependency among the dummy X values shows up as multicollinearity and an inability to apply ordinary multiple regression methods to these data.

On the other hand, if we eliminate one of these dummy variables, then there is no problem with multicollinearity, and the multiple regression analysis may be completed, as illustrated in the next section.

The solution to this problem is thus surprisingly simple: Instead of using J dummy variables to stand for the distinctions among J qualitatively different groups, we use only $J - 1$ such variables for each set of J groupings. That is, we use the number of dummy variables which is equivalent to the number of degrees of freedom for J groups in ANOVA, and this avoids the multicollinearity problem. Therefore, for three qualitatively different categories, we employ two dummy variables; for four categories, we use three dummy variables, and so forth.

In the second place, dummy variables fit easily into the role of predictor variables *only* when the Model I sampling scheme for regression is employed and, ideally, with fixed categories as in Model I ANOVA. It is just not reasonable to think of normal distributions of $1-0$ dummy variables, and thus the sampling models that depend on the multivariate normal distribution of both X and Y variables are unrealistic when dummy variables are involved.

In short, it is perfectly legitimate to introduce qualitative distinctions as part of the set of predictor variables in multiple regression problems. However, in doing so, be sure that you use *only* $J - 1$ such dummy variables to represent each attribute, or set of J mutually exclusive and exhaustive categories. Also, dummy variables are most appropriate when the sampling model supposes fixed X and random Y values.

16.2 AN EXAMPLE OF SIMPLE ANOVA IN MULTIPLE REGRESSION FORMAT

Now that we have some of the mechanical techniques needed to deal with multiple regression, we can see how a typical ANOVA problem translates into a problem in multiple regression, through use of the device of dummy variables, just as suggested in Section 10.2.

Like Table 16.1.1, Table 16.2.1 is actually a reworking of Table 10.16.1, in which 30 subjects are divided into three qualitatively different groups. Each subject has not only a value on dummy variable X_1, showing if he or she was a member of Group 1,

Table 16.2.1

The data of Table 10.16.1 translated into dummy variable form

Case	X_1	X_2	Y	Case	X_1	X_2	Y	Case	X_1	X_2	Y
1	1	0	23	11	0	1	27	21	0	0	23
2	1	0	22	12	0	1	28	22	0	0	24
3	1	0	18	13	0	1	33	23	0	0	21
4	1	0	15	14	0	1	19	24	0	0	25
5	1	0	29	15	0	1	25	25	0	0	19
6	1	0	30	16	0	1	29	26	0	0	24
7	1	0	23	17	0	1	36	27	0	0	22
8	1	0	16	18	0	1	30	28	0	0	17
9	1	0	19	19	0	1	26	29	0	0	20
10	1	0	17	20	0	1	21	30	0	0	23

but also a value on X_2, indicating membership in Group 2. For the reasons outlined in the previous section, we do not use dummy variable X_3, however, Finally, Table 16.2.1 shows the value for each case on variable Y.

The data in Table 16.2.1 translate immediately into a set of correlations, shown here with means and standard deviations.

	X_1	X_2	Y	mean	SD
X_1	1	−.50	−.3165	.3333	.4714
X_2		1	.5492	.3333	.4714
Y			1	23.4667	5.0645

When the simultaneous multiple regression is carried out in the usual way we have $R^2_{y \cdot 12} = .3040$. The F test for this $R^2_{y \cdot 12}$ value is

$$F = .3040 \, (30 - 2 - 1)/(1 - .3040)(2) = 5.90 \, ,$$

which is significant beyond the .01 level for 2 and 27 degrees of freedom.

The SS_y regression on X_1 and $X_2 = NS^2_y \, (.3040) = 233.92$.

Then the summary table that accompanies this multiple regression analysis is the following:

Source	SS	df	MS	F	p
Multiple regression on X_1 and X_2	233.92	2	116.96	5.90	<.01
Residual	535.55	27	19.84		
Total	769.47	29			

Now compare the results of this summary table with the summary table for the ANOVA of Section 10.16. It is easy to see that, within small rounding error, these two tables are identical, with SS multiple regression having the same value as SS between, and with SS residual agreeing with SS within. Furthermore, in both tables, the ratio of SS between or SS multiple regression to SS total is equal to about .304, the value of $R^2_{y \cdot 12}$. As advertised, ANOVA is multiple regression carried out on dummy variables *denoting group membership*.

Although we have demonstrated this correspondence with dummy variables, the same principle holds if we let the X variables be defined in any of several other ways. Thus, this same sort of demonstration of the connection between ANOVA and multiple regression can be done by use of so-called effects coding, where, for example, variable $X_j = 1$ if i is in A_J; $X_j = -1$ if i is in A_J and $X_j = 0$ otherwise (see Pedhazer, 1982). In the next section we will also show the relationship of sets of orthogonal comparisons in ANOVA to analysis through multiple regression methods.

16.3 ORTHOGONAL COMPARISONS AND MULTIPLE REGRESSION

The illustration of the parallel between ANOVA and multiple regression for J qualitatively different groups of cases is but one of the many linkages between ordinary ANOVA and the methods of multiple regression. Another involves a set of $J - 1$ orthogonal comparisons made on J qualitatively distinct experimental groups. It is easy to show that applying orthogonal comparisons is equivalent to transforming the set of $J - 1$ dummy variables X, as used in the previous section, into a set of $J - 1$ new variables V, which have the property of being uncorrelated with each other. Then the relationships between orthogonal comparisons and ANOVA, which we have already examined in Chapter 11, have exact parallels in multiple regression.

Take, for example, the data of Table 16.2.1, once again. However, instead of coding each case according to the $1 - 0$ dummy variable scheme showing group membership as in that table, let us now apply two sets of orthogonal comparison weights to the three groups as follows, but here letting the symbol v_j instead of c_j stand for the weight to be applied to a group j in comparison number k, where $k = 1, \ldots, J - 1$. Thus, for three groups we might have:

	Group I	Group II	Group III
v_{1j}	2	-1	-1
v_{2j}	0	1	-1

Notice that these two sets of v values are simply two sets of comparison weights which are orthogonal.

Now in terms of the variables V_k with values v_{kj}, the data of Table 16.2.1 can be tranformed into that of Table 16.3.1, where each subject now has a value on variable V_1, a value on variable V_2, and a value on dependent variable Y.

As shown in Table 16.3.1, in this experimental situation all n individuals in the same group j receive exactly the same v_{kj} value on the variable V_k. Therefore, we really do

Table 16.3.1

Data of Table 16.2.1 coded according to orthogonal comparison weights

Case	V_1	V_2	Y	Case	V_1	V_2	Y	Case	V_1	V_2	Y
1	2	0	23	11	−1	1	27	21	−1	−1	23
2	2	0	22	12	−1	1	28	22	−1	−1	24
3	2	0	18	13	−1	1	33	23	−1	−1	21
4	2	0	15	14	−1	1	19	24	−1	−1	25
5	2	0	29	15	−1	1	25	25	−1	−1	19
6	2	0	30	16	−1	1	29	26	−1	−1	24
7	2	0	23	17	−1	1	36	27	−1	−1	22
8	2	0	16	18	−1	1	30	28	−1	−1	17
9	2	0	19	19	−1	1	26	29	−1	−1	20
10	2	0	17	20	−1	1	21	30	−1	−1	23

not need to distinguish among individuals in the same group according to their V_k values.

The correlation between dependent variable Y and variable V_k is thus

$$r_{yk} = \frac{\sum_j v_{kj} \sum_i y_{ij}}{\sqrt{\sum_j nv_{kj}^2}\,\sqrt{NS_Y^2}} = \frac{\sum_j v_{kj}\bar{y}_j}{\left(\sqrt{\sum_j v_{kj}^2/n}\right)\left(\sqrt{NS_Y^2}\right)} \qquad [16.3.1]$$

since the V_k values depend only on groups, not on individuals, and the mean of V_k is zero. On the other hand, since the V variables are orthogonal to each other, the correlation between any pair of these variables, say r_{kg} for the pair V_k and V_g, *must* be zero. In consequence, the normal equations for multiple regression become simply

$$B_1 = r_{y1}$$

$$B_2 = r_{y2}$$

$$\cdot \cdot \cdot$$

$$B_{J-1} = r_{y(J-1)},$$

and by Eq. 15.24.1 the square of the multiple correlation coefficient is just

$$R_{y \cdot H}^2 = r_{y1}^2 + r_{y2}^2 + r_{y3}^2 + \cdots + r_{y(J-1)}^2 = \sum_{k=1}^{J-1} r_{yk}^2. \qquad [16.3.2]$$

Here again the symbol H represents the entire set of $J - 1$ predictor variables, $H = \{V_1, \cdots, V_{J-1}\}$. The square of the multiple correlation, or $R_{y \cdot H}^2$ is thus the sum of the squared correlations between Y and each of the variables V_k.

Thus, for our example of Table 16.3.1, it turns out that the correlation matrix and the means and standard deviations are:

	V_1	V_2	Y	Mean	SD
V_1	1	0	$-.3165$	0	1.4142
V_2		1	.4514	0	0.8165
Y			1	23.4667	5.0645

Then, since there is a zero correlation between the two orthogonal predictor variables V_1 and V_2, we have

$$R_{y \cdot 12}^2 = R_{y \cdot H}^2 = (-.3165)^2 + (.4514)^2$$

$$= .3039 \text{ or } .304 ,$$

exactly as found in the previous section using dummy variables.
 In this analysis,

$$\text{SS regression on } V_1 \text{ and } V_2 = 233.84$$

$$= \text{SS between} ,$$

the same, within rounding, as found in the original analysis of variance in Section 10.16, and

$$\text{SS residual} = 535.63$$

$$= \text{SS within} ,$$

which also agrees with the previous results, as does the F value

$$F = .3039 (30 - 2 - 1)/(1 - .3039)(2) = 5.89$$

with 2 and 27 degrees of freedom.
 Furthermore, if we let

$$NS_k^2 = \sum_j \sum_i v_{kj}^2 = \sum_j n v_{kj}^2 ,$$

and recall that

$$b_k = B_k \left(\frac{S_Y}{S_k} \right) = r_{yk} \left(\frac{S_Y}{S_k} \right),$$

then

$$b_k S_k = r_{yk} S_Y .$$

However, we can rewrite r_{yk} as

$$r_{yk} = \frac{n \hat{\psi}_k}{\sqrt{\sum_j n \, v_{kj}^2} \, \sqrt{NS_Y^2}} = \frac{\hat{\psi}_k}{\sqrt{w_k} \, \sqrt{NS_Y^2}} . \tag{16.3.3*}$$

Then

$$b_k^2 (NS_k^2) = r_{yk}^2 (NS_Y^2) = SS(\hat{\psi}_k) ,$$

so that

$$r_{yk}^2 = \frac{SS(\hat{\psi}_k)}{SS_Y \text{ total}} \qquad\qquad [16.3.4^*]$$

and

$$b_k^2 = \frac{SS(\hat{\psi}_k)}{n^2 w_k} . \qquad\qquad [16.3.5^*]$$

Thus

$$b_k = \frac{\hat{\psi}_k}{nw_k} = \frac{\hat{\psi}_k}{\sum_k v_k^2} . \qquad\qquad [16.3.6^*]$$

An orthogonal comparison formed with weights v_{kj} and divided by the sum of v_{kj}^2, is actually the unstandardized regression weight for predicting Y from the values of V_k.

Suppose that we have available a set of comparison weights (the v_{kj} values), and we wish to construct the regression equation for predicting the value y_{ij}' for any individual in group j. Any comparison is associated with a regression weight by Eq. 16.3.6. Then the required regression equation is

$$y_{ij}' = \bar{y} + \sum_{k=1}^{J-1} b_k v_{kj} \qquad\qquad [16.3.7]$$

or

$$y_j' = \bar{y} + \sum_{k=1}^{J-1} \frac{\hat{\psi}_k}{nw_k} v_{kj}. \qquad\qquad [16.3.8]$$

In the population, this is

$$y_j' = \mu_Y + \sum_k \frac{\psi_k}{nw_k} v_{kj}. \qquad\qquad [16.3.9]$$

However, this regression equation reduces to just

$$y_{ij}' = \mu_Y + \alpha_j, \qquad\qquad [16.3.10]$$

which is the population model for one-way analysis of variance as found with the method of least squares (Section 10.5).

In summary, the one-way analysis of variance under the fixed-effects model is also identical to a multiple regression analysis when the indicator or dummy variables for group membership are transformed into a corresponding set of $J - 1$ orthogonal variables. Such orthogonal variables are equivalent to a set of $J - 1$ orthogonal vectors of comparison weights.

These basic connections are summarized in Table 16.3.2

Analysis of variance	Multiple regression	**Table 16.3.2**
SS between/SS total	$R^2_{y \cdot H}$, where $H =$ set of $J - 1$ dummy variables or $H =$ set of $J - 1$ orthogonal comparisons	Some connections between a fixed-effects one-way analysis of variance and multiple regression.
SS within/SS total	$1 - R^2_{y \cdot H}$	
Comparison $\hat{\psi}_k / n w_k$	b_k	
SS($\hat{\psi}_k$)/SS total	r^2_{yk}	
MS between/MS within	$\dfrac{R^2_{y \cdot H}(N - J)}{(1 - R^2_{y \cdot H})(J - 1)}$	

16.4 AN EXAMPLE OF A FACTORIAL EXPERIMENT ANALYZED THROUGH MULTIPLE REGRESSION

One of the main reasons for exploring the connections between analysis of variance and multiple regression analysis is the ability we gain to handle certain situations that are difficult to manage through ANOVA, but much easier if approached by way of multiple regression. One of these situations is the analysis of an unbalanced design with two or more factors. This creates problems for simple ANOVA approaches, but can be solved fairly neatly through multiple regression.

Before we can investigate how multiple regression can help with the problem of an unbalanced design, we need to have a little better understanding of how multiple regression applies to a two-factor (or higher) design, with membership in the factor levels coded by the use of dummy variables. Thus, in preparation for the next section, we will work through a simple example of a new two-factor design which is crossed and balanced.

Below are the data representing a two-factor design, in which Factor A (rows) has three levels, and Factor B (columns) has two levels. Then in each of the six possible cells representing a combination of an A level with a B level, three subjects were assigned at random, for a total of $N = 18$ cases. The data are given in Table 16.4.1

	B_1	B_2	**Table 16.4.1**
			A 3 × 2 factorial design with three per cell
A_1	18	15	
	12	19	
	15	21	
A_2	7	16	
	11	10	
	6	11	
A_3	11	5	
	19	7	
	13	4	

Cell	Case	X_1	X_2	X_3	X_4	X_5	Y
A_1B_1	1	1	0	1	1	0	18
	2	1	0	1	1	0	12
	3	1	0	1	1	0	15
A_1B_2	4	1	0	0	0	0	15
	5	1	0	0	0	0	19
	6	1	0	0	0	0	21
A_2B_1	7	0	1	1	0	1	7
	8	0	1	1	0	1	11
	9	0	1	1	0	1	6
A_2B_2	10	0	1	0	0	0	16
	11	0	1	0	0	0	10
	12	0	1	0	0	0	11
A_3B_1	13	0	0	1	0	0	11
	14	0	0	1	0	0	19
	15	0	0	1	0	0	13
A_3B_2	16	0	0	0	0	0	5
	17	0	0	0	0	0	7
	18	0	0	0	0	0	4

Table 16.4.2

Data for 3×2 factorial design coded by dummy variables

arranged in the usual way by cells. Then Table 16.4.2 shows the same set of 18 subjects coded for factor-level membership by the use of dummy variables.

For purposes of later checking, let us first take a look at the ordinary two-way AN-OVA results for these data:

Source	SS	df	MS	F	p
Factor A	178.11	2	89.06	9.66	<.01
Factor B	0.89	1	0.89	—	—
A × B Inter.	165.44	2	82.72	8.97	<.01
Error	110.67	12	9.22		
Total	455.11	17			

In Table 16.4.2, we have the same basic data as in Table 16.4.1. except that now five dummy variables have been used to represent the location of any individual in the design and possible interaction. That is, the dummy variables are

X_1 = 1 if a case is in Level A_1 of Factor A
X_1 = 0 otherwise

X_2 = 1 if a case is in Level A_2 of Factor A
X_2 = 0 otherwise

X_3 = 1 if a case in Level B_1 of Factor B
X_3 = 0 otherwise

$$X_4 = X_1 \text{ times } X_3$$

$$X_5 = X_2 \text{ times } X_3$$

The last two dummy variables, found by multiplication, will represent the interaction information in the data.

Bear in mind that we *are* here once again dealing with dummy variables, just as we did in Section 16.2. Later in the present section we will repeat the analysis using orthogonal comparisons just as we did in Section 16.3. Now we will treat this as a problem in multiple regression, in which the five X variables will be used as predictors relative to dependent variable Y. The simplest way to achieve the output we need will be through hierarchical regression, in which we will first include variables X_1 and X_2 representing Factor A, then variable X_3, representing Factor B, and then finally variables X_4 and X_5, which represent interaction.

For this problem, the correlation matrix based on Table 16.4.2 is as follows:

	X_1	X_2	X_3	X_4	X_5	Y	Mean	SD
X_1	1	−.5000	0	.6325	−.3162	.6250	.33	.47
X_2		1	0	−.3162	.6325	−.2891	.33	.47
X_3			1	.4472	.4472	.0442	.50	.50
X_4				1	−.2000	.2471	.17	.37
X_5					1	−.3755	.17	.37
Y						1	12.22	5.03

$$SS_y \text{ tot} = 455.11$$

If the sweep-out method of Section 15.26 is employed, then the R-square value after sweeping variables X_1 and X_2 (i.e., Factor A) is

$$R^2_{y \cdot 12} = .3914.$$

Then, after variable X_3 is swept the R-square value is

$$R^2_{y \cdot 123} = .3933.$$

The proportion of variance accounted for solely by Factor B (or variable X_3) over and beyond any variance due to Factor A is then

$$R^2_{y(3 \cdot 12)} = R^2_{y \cdot 123} - R^2_{y \cdot 12}$$

$$= .3933 - .3914 = .0019$$

by Eq. 15.22.12 Factor B is obviously accounting for almost no Y variance at all.

Next in our hierarchical progression variables X_4 and X_5 are added to the regression equation (or swept-out). This results in an R-square value of

$$R^2_{y \cdot 12345} = .7568.$$

Therefore, the variance that is accounted for by dummy variables X_4 and X_5 (i.e., the A $\times$ B interaction) over and above variance accounted for by any other variables is

$$R^2_{y(45 \cdot 123)} = R^2_{y \cdot 12345} - R^2_{y \cdot 123}$$

$$= .7568 - .3933$$

$$= .3635.$$

Since the SS_y total $= 455.11$ the SS_y for regression on X_1 and X_2 is then $.3914(455.11) = 178.13$. Then SS_y regression on X_3 beyond X_1 and $X_2 = .0019(455.11) = .86$. Next SS_y for X_3 and X_4 beyond X_1, X_2, and X_3 is $.3635(455.11) = 165.43$, and SS_y residual is then $455.11 - 178.13 - .86 - 165.43 = 110.69$.

This regression analysis can be reported in a summary table like this:

Source	SS	df	MS	F	p
Regression on X_1 and X_2	178.13	2	89.07	9.66	<.01
on X_3 beyond X_1 and X_2	0.86	1	0.86	—	—
on X_4 and X_5 beyond X_1, X_2, X_3	165.43	2	82.72	8.97	<.01
Residual	110.69	12	9.22		
Total	455.11	17			

If allowances are made for small rounding differences, we once again have the original ANOVA table for these data. That is, if the first line, "Regression on X_1 and X_2" is renamed "Factor A," the second line "on X_3 beyond X_1 and X_2" is relabelled "Factor B," the third line "on X_4 and X_5 beyond X_1, X_2, X_3" is renamed "A $\times$ B interaction," and "Residual" is called "Error," we have the ANOVA summary table for the two-way analysis on these data. Finally, the three F tests for gain in proportion of variance accounted for, as outlined in Section 15.24, are identical to the three F tests for Factor A, Factor B, and for A $\times$ B interaction.

Alternatively, the dummy variables shown in Table 16.4.2 could have been turned into new variables corresponding to orthogonal comparisons for A, B, and A $\times$ B interaction, by use of the Gram-Schmidt method of Section C.5 (Appendix C). This would result in a set of comparisons with weights as shown in the Table 16.4.3 (notice that each case in a given cell gets the same weights as the mean of that cell). We will call each of these sets of comparison weights a V variable, just as we did in Section 16.3.

Then the correlation matrix of these five comparison variables and Y is as follows:

	V_1	V_2	V_3	V_4	V_5	Y
V_1	1	0	0	0	0	.6250
V_2		1	0	0	0	.0271
V_3			1	0	0	.0442
V_4				1	0	$-$.2656
V_5					1	$-$.5413
Y						1

Cell	Case	V_1	V_2	V_3	V_4	V_5	Y
A_1B_1	1	2	0	1	2	0	18
	2	2	0	1	2	0	12
	3	2	0	1	2	0	15
A_1B_2	4	2	0	-1	-2	0	15
	5	2	0	-1	-2	0	19
	6	2	0	-1	-2	0	21
A_2B_1	7	-1	1	1	-1	1	7
	8	-1	1	1	-1	1	11
	9	-1	1	1	-1	1	6
A_2B_2	10	-1	1	-1	1	-1	16
	11	-1	1	-1	1	-1	10
	12	-1	1	-1	1	-1	11
A_3B_1	13	-1	-1	1	-1	-1	11
	14	-1	-1	1	-1	-1	19
	15	-1	-1	1	-1	-1	13
A_3B_2	16	-1	-1	-1	1	1	5
	17	-1	-1	-1	1	1	7
	18	-1	-1	-1	1	1	4

Table 16.4.3

Five orthogonal comparisons corresponding to the dummy variables of Table 16.4.2

For this table, since the intercorrelations among the V variables are all 0, the R-square values are easy to find from the sum of the squres of the correlations of the V variables with Y. Thus

$$R^2_{y\cdot12} = (.6250)^2 + (.0271)^2 = .3914$$

This implies that the variance accounted for by comparisons 1 and 2 (Factor A) is

$$SS_y \text{ regression on } V_1 \text{ and } V_2 = SS_y \text{ total } (.3914)$$

$$= 178.13$$

exactly as shown for Factor A in the ANOVA table for these data.
    Similarly,

$$R^2_{y\cdot123} - R^2_{y\cdot12} = (.0442)^2 = .0019,$$

so that

$$SS \text{ factor } B = SS_y \text{ regression on } V_3, \text{ beyond } V_1 \text{ and } V_2$$

$$= SS_y \text{ total } (.0019) = .86.$$

Then finally,

$$R^2_{y\cdot12345} - R^2_{y\cdot123} = (.2656)^2 + (.5413)^2$$

$$= .3635,$$

implying that

SS$_y$ regression on V_4 and V_5, above V_1, V_2, V_3 = SS$_y$ total (.3635)

$$= 165.44$$

$$= SS_y \, A \times B,$$

agreeing with the value found in the ANOVA for these data.

16.5 HANDLING UNBALANCED DATA THROUGH MULTIPLE REGRESSION METHODS: METHOD H, THE HIERARCHICAL APPROACH

As noted above, one of the main reasons to study the connection between multiple regression and multifactor ANOVA is the ability that we gain to solve the troublesome problem of unbalanced data. As given in Chapter 12, a balanced factorial design either has equal numbers of cases in the various possible cells, or number of cases in the cells that are proportional to the numbers in the various levels of each factor. Any set of data otherwise fitting the format of a factorial design, but lacking equality or proportionality of cases among the cells, is called *unbalanced*. If one tries to apply the usual methods of ANOVA to a set of unbalanced data the results can be, to say the least, misleading. In a two-factor situation, for example, the SS for Factor A may also contain information about Factor B and the A × B interaction; the SS Factor B may contain information about Factor A and interaction; and the SS A × B interaction may be contaminated to some extent by main effects of Factors A and B. In other words, the clean separation into variability due to different types of effects, which is routinely possible for a balanced factorial design, is no longer guaranteed when the data are unbalanced.

Rather, when the data are unbalanced, the accepted procedure is to regard the problem in terms of multiple regression rather than the usual ANOVA, and to follow one of several available approaches for completing the analysis, and finding SS values for main effects and for interactions. Here we will go into detail about only two of these approaches, which we will refer to as the *hierarchical approach* or Method H and the *regression approach* or Method R. This section will be devoted to Method H, and we will look at Method R in the next section. Still other methods are possible, as will be mentioned later.

In Method H, the unbalanced data are treated by *hierarchical regression,* exactly as illustrated for the 3 × 2 balanced design of Section 16.4. That is, the investigator decides in advance that one factor will be investigated for its relation to Y, without adjustment for any of the other factors. Thus, suppose that in a two-factor design Factor A has some theoretical or practical priority for the investigator, and so it is selected as the first in the hierarchy of factors to be analyzed.

Here, however, there are advantages to using the effects coding based on values of 1, 0, and -1, rather than the usual 1 and 0 values for our unbalanced data. Then as we have already shown, this matrix of correlations is swept for the $J - 1$ variables X representing levels of Factor A. The result gives

$R^2_{y \cdot A}$ = proportion of Y variance due to Factor A,

so that

SS$_y$ regression on Factor A = SS$_y$ total [$R^2_{y \cdot A}$).

Next the matrix SWP(A), which has been adjusted for Factor A, is swept-out for the $K - 1$ variables X representing Factor B. This produces

$R^2_{y \cdot A,B}$ = proportion of Y variance accounted for by Factors A and B.

However, the item of most importance is the proportion of Y variance due to Factor B *after adjustment* to remove possible effects of Factor A:

$$R^2_{y(B \cdot A)} = R^2_{y \cdot A,B} - R^2_{y \cdot A}. \qquad [16.5.1^*]$$

This in turn provides

$$\text{SS}_y \text{ regression on B, beyond A} = \text{SS}_y \text{ total } (R^2_{y(B \cdot A)}). \qquad [16.5.2^*]$$

That is, at this stage we have a sum of squares for unadjusted A, and for B adjusted for A.

Finally, for a two-factor design, the $(J - 1)(K - 1)$ variables standing for the interaction A × B are swept out, providing the value of

$$R^2_{y \cdot A,B,A \times B} = \text{proportion of } Y \text{ variance due to A, B, and A × B.} \qquad [16.5.3^*]$$

This, then, is the basis for finding the sum of squares for A × B interaction, after adjustment to remove possible effects of A or B:

$$R^2_{y(A \times B \cdot A,B)} = R^2_{y \cdot A,B,A \times B} - R^2_{y \cdot A,B}, \qquad [16.5.4^*]$$

so that

SS$_y$ regression on A × B beyond A and B = SS$_y$ total [$R^2_{y(A \times B \cdot A,B)}$].

Finally, the sum of squares for residual is found from

$$\text{SS}_y \text{ residual} = \text{SS}_y \text{ total } [1 - R^2_{y \cdot A,B,A \times B}]. \qquad [16.5.5^*]$$

The sums of squares found through this hierarchical regression procedure, or Method H, will add to SS$_y$ total, just as they do for any ANOVA. However, these sums of squares have a somewhat different interpretation than in an ordinary analysis of variance: Through the sweep-out procedures, representing adjustments for certain variables, we are artificially "pulling apart" the information on factors and on interactions that has been mixed up in the unbalanced data. The first factor in the hierarchy (Factor A here) can be looked at essentially as though this were data from a one-factor experiment, with possible Factor B and interaction effects not controlled. The evaluation of Factor B, however, is conditional on adjustment to separate out possible Factor A effects. The sum of squares for A × B interaction is, finally, conditional on the adjustment process for both Factors A and B.

It is easy to illustrate the Method H analysis for a set of unbalanced data. To do so, let us use the factorial data of Section 16.4 once more, but this time we will suppose that two of the cases in cell A_1B_2 were not observed, and that one of the cases in cell A_2B_1 was also missing. This, would of course produce an unbalanced design with a total of $N = 15$ rather than $N = 18$ cases. The new, unbalanced, data are shown in Table 16.5.1.

	B_1	B_2	**Table 16.5.1**
A_1	18 12 15	19	Data from an unbalanced 3×2 factorial design
A_2	7 6	16 10 11	
A_3	11 19 13	5 7 4	

These same data, after effects coding into 1, 0, and -1, are given in Table 16.5.2. Once again, $X_4 = X_1 X_3$ and $X_5 = X_2 X_3$. The correlation matrix that results from the data of Table 16.5.2 is shown in Table 16.5.3.

Using Method H, a hierarchical regression analysis will be carried out. The order of the factors in the regression analysis should be decided on at this point, of course. However, for our demonstration purposes let us simply take the factors in the arbitrary order A, B, and A $\times$ B (which is equivalent to taking the dummy X variables in the order X_1 and X_2, followed by X_3, and ending with X_4 and X_5.) On sweeping the matrix for variables X_1 and X_2 (or Factor A), we have

$$R_{y \cdot A}^2 = R_{y \cdot 12}^2 = .3011.$$

(Note that this is the same of the two squared part correlations,

$$r_{y \cdot 1}^2 + r_{y(2 \cdot 1)}^2 = R_{y \cdot 12}^2$$

just as described in Section 15.23.)

Now given that the SS_y total $= 361.73$, we have

$$SS_y \text{ Factor A } = SS_y \text{ total } (R_{y \cdot A}^2) = 108.92.$$

case	X_1	X_2	X_3	X_4	X_5	Y	**Table 16.5.2**
1	1	0	1	1	0	18	The unbalanced data of Table 16.5.1 shown with effects coding
2	1	0	1	1	0	12	
3	1	0	1	1	0	15	
4	1	0	-1	-1	0	19	
5	0	1	1	0	1	7	
6	0	1	1	0	1	6	
7	0	1	-1	0	-1	16	
8	0	1	-1	0	-1	10	
9	0	1	-1	0	-1	11	
10	-1	-1	1	-1	-1	11	
11	-1	-1	1	-1	-1	19	
12	-1	-1	1	-1	-1	13	
13	-1	-1	-1	1	1	5	
14	-1	-1	-1	1	1	7	
15	-1	-1	-1	1	1	4	

	X_1	X_2	X_3	X_4	X_5	Y
X_1	1	.5687	.1769	.2329	−.0129	.4730
X_2		1	−.0730	.0129	−.0976	.0403
X_3			1	−.1769	−.0730	.2376
X_4				1	.5946	−.4056
X_5					1	−.6276
Y						1

Table 16.5.3

Correlation matrix for data of Table 16.5.2

mean Y = 11.53
SD Y = 4.91
SS total = 361.73

Next, the remaining matrix SWP(12) or SWP(A) is swept out for variable X_3, standing for Factor B. This gives

$$R^2_{y \cdot A,B} = R^2_{y \cdot 123} = .3109.$$

Therefore, the proportion of Y variance due to Factor B, over and above that due to Factor A, is

$$R^2_{y(B \cdot A)} = R^2_{y \cdot A,B} - R^2_{y \cdot A} = .3109 - .3011 = .0098.$$

(This is also the squared part correlation $r^2_{y(3 \cdot 12)}$.) Hence,

$$\text{SS}_y \text{ Factor B, beyond Factor A} = \text{SS}_y \text{ total } [R^2_{y(B \cdot A)}]$$

$$= 361.73(.0098) = 3.55.$$

Finally, when dummy variables X_4 and X_5 are swept from the SWP(123) matrix, we find

$$R^2_{y \cdot A,B,A \times B} = R^2_{y \cdot 12345} = .7830.$$

The proportion of Y variance that is due to A × B interaction, as adjusted both for Factor A and Factor B, is thus

$$R^2_{y(A \times B \cdot A,B)} = R^2_{y \cdot A,B,A \times B} - R^2_{y \cdot A,B}$$

$$= .7830 - .3109 = .4721.$$

(This value is equivalent to the sum of squared part correlations

$$r^2_{y(4 \cdot 123)} + r^2_{y(5 \cdot 1234)}.$$

This implies that

$$\text{SS}_y \text{ A} \times \text{B, beyond Factors A and B} = 361.73(.4721) = 170.78.$$

Finally, the residual (or error) sum of squares is found to be

$$\text{SS}_y \text{ residual} = \text{SS}_y \text{ total } (1 - R^2_{y \cdot A,B,A \times B})$$

$$= 361.73(1 - .7830) = 78.50.$$

The results can be displayed in an ANOVA summary table like Table 16.5.4.

Source	SS	df	MS	F	p
Factor A	108.92	2	54.46	6.25	<.05
Factor B, beyond A	3.54	1	3.54	—	—
A × B, beyond Factors A, B	170.77	2	85.39	9.79	<.01
Residual	78.50	9	8.72		
Total	361.73	14			

Table 16.5.4

Results of first hierarchical analysis

Clearly, Factor A, and especially A × B interaction, account for a large proportion (about 77%) of the variance in these data. Furthermore, each is clearly significant beyond the .05 level.

An example of Method H such as this could equally well have begun with Factor B, and then proceeded to Factor A and A × B interaction, where each new term is adjusted for the ones that have gone before. Although in actual situations one would choose a hierarchical order and stick with it, we will look at a second order B, A, and A × B, simply for purposes of comparison.

When the analysis is done with this changed hierarchical ordering, a slight difference in the results is seen (as shown in Table 16.5.5), where Factor B considered alone accounts for about 6% of the variance, and Factor A adjusted for Factor B then accounts for an additional 25% of the Y variance. The A × B interaction still accounts for 47% of the variance, however, just as before. This second analysis demonstrates that the results of using Method H depend on the order in which the factors of the experiment are introduced into the hierarchical analysis.

Furthermore, just as we have illustrated here, it is important to adjust for each of the factors (main effects) before introducing the interactions for any of these factors into the hierarchical analysis. There is a form of logical dependency between an interaction

Source	SS	df	MS	F	p
Factor B	20.44	1	20.44	2.34	NS
Factor A, beyond B	92.02	2	46.01	5.28	<.05
A × B Int. above A, B	170.77	2	85.39	9.79	<.01
Residual	78.50	9	8.72		
Total	361.73				

Table 16.5.5

Results of second hierarchical analysis

effect and the factors that define it, and this can render the adjustment of factors for interactions a meaningless procedure in some contexts, whereas there is no problem in the adjustment of interactions for main effects. This comment holds in force for inter-actions of different orders, and a fair amount of caution is appropriate in using this method to look at higher order interactions adjusted for lower order interactions by use of this method, in complex factorial design. In such instances it is probably wise to stop after first-order interactions, and simply take a "remainder" term to indicate all the remaining variance to be accounted for.

What if you make a mistake, and subject a set of data that are actually balanced to a Method H analysis? Fortunately, there is no harm done, as this analysis will turn out to be identical to the results of an ordinary ANOVA when the data are actually balanced.

16.6 HANDLING UNBALANCED DESIGNS THROUGH MULTIPLE REGRESSION: METHOD R, THE REGRESSION APPROACH

In the previous section we saw how unbalanced data can be analyzed through hierarchical analysis according to a predecided order of precedence among the factors, with each successive factor examined after adjustment for those that have gone before. Now we will consider an alternative approach, Method R, which is often called simply the *regression approach*. In Method R the unbalanced data are once again subjected to multiple regression analysis, but this time to one in which each factor or interaction is examined *only after* adjustment for all of the remaining factors and interactions.

Once more consider our example of a two-factor unbalanced design. Suppose that we start off by examining Factor A once again. Then the sweep-out procedure is first applied to the variables standing for Factor B, then to those for A × B, and finally to those for Factor A. The results provided are, first of all

$$R^2_{y \cdot B, A \times B} = \text{proportion of } Y \text{ variance due to B and A} \times \text{B} \qquad [16.6.1]$$

and

$$R^2_{y \cdot A, B, A \times B} = \text{proportion of } Y \text{ variance due to A and B and A} \times \text{B}. \qquad [16.6.2]$$

Then

$$R^2_{y(A \cdot B, A \times B)} = R^2_{y \cdot A, B, A \times B} - R^2_{y \cdot B, A \times B} \qquad [16.6.3^*]$$

(or, as part correlations, $r^2_{y(1 \cdot 3,4,5)} + r^2_{y(2 \cdot 1,3,4,5)}$). Thus,

$$\text{SS}_y \text{ Factor A, beyond B, A} \times \text{B} = \text{SS}_y \text{ total } [R^2_{y(A \cdot B, A \times B)}]. \qquad [16.6.4^*]$$

Next, the procedure is altered, with the variables representing Factor A first swept out of the matrix, and then followed by those for A × B. Last, the variables representing Factor B are swept out, providing

$$R^2_{y(B \cdot A, A \times B)} = R^2_{y \cdot A, B, A \times B} - R^2_{y \cdot A, A \times B} \qquad [16.6.5^*]$$

(that is, $r^2_{y(3 \cdot 1,2,4,5)}$ for this example). Hence,

$$\text{SS}_y \text{ Factor B, beyond A, A} \times \text{B} = \text{SS}_y \text{ total } [R^2_{y(B \cdot A, A \times B)}]. \qquad [16.6.6^*]$$

Last, the proportion of Y variance due to A $\times$ B interaction, over and beyond variance due either to A or to B is given by first finding

$$R^2_{y \cdot A,B} = \text{proportion of variance due to A or B,} \qquad [16.6.7]$$

as well as

$$R^2_{y \cdot A,B,A \times B} = \text{proportion of variance due to A, or B, or} \qquad [16.6.8]$$
$$\text{A} \times \text{B,}$$

and then taking

$$R^2_{y(A \times B \cdot A,B)} = R^2_{y \cdot A,B,A \times B} - R^2_{y \cdot A,B}, \text{ (or, } r^2_{y(4 \cdot 1,2,3)} + r^2_{y(5 \cdot 1,2,3,4)}). \qquad [16.6.9^*]$$

The equivalent sum of squares value is:

$$\text{SS}_y \text{ A} \times \text{B interaction, beyond A and B} = \text{SS}_y \text{ total } [R^2_{y(A \times B \cdot A,B)}] \qquad [16.6.10^*]$$

As usual, the sum of squares for residual is found from

$$\text{SS}_y \text{ residual} = \text{SS}_y \text{ total } [1 - R^2_{y \cdot A,B,A \times B}] \qquad [16.6.11^*]$$

The F tests of significance then each employ MS residual as the denominator, with $N - JK$ denominator degrees of freedom, where JK is the number of cells in the two-factor design.

Now, think once again of the unbalanced data used in Section 16.5, and shown in dummy variable form in Table 16.5.2, and in terms of correlations in Table 16.5.3. Applying the regression Method R to these data, we have first

$$R^2_{y \cdot B,A \times B} = .4309,$$

and then

$$R^2_{y \cdot A,B,A \times B} = .7830,$$

giving

$$R^2_{y(A \cdot B,A \times B)} = .7830 - .4309 = .3521.$$

This implies that

$$\text{SS}_y \text{ Factor A, beyond B and A} \times \text{B} = 361.73(.3521) = 127.37.$$

Next we look at Factor B adjusted for Factor A and A $\times$ B by taking

$$R^2_{y \cdot A,A \times B} = .7823,$$

so that

$$R^2_{y(B \cdot A,A \times B)} = R^2_{y \cdot A,B,A \times B} - R^2_{y \cdot A,A \times B}$$
$$= .7830 - .7823$$
$$= .0007.$$

Consequently, we have

$$\text{SS}_y \text{ Factor B, beyond A, A} \times \text{B} = 361.73(.0007) = .25.$$

As for A × B adjusted for both Factors A and B we have

$$R^2_{y \cdot A,B,A \times B} = .7830$$

and

$$R^2_{y(A \times B \cdot A,B)} = .7830 - .3109 = .4721.$$

Therefore,

SS$_y$ A × B, beyond A and B = 361.73(.4721) = 170.77

and

SS$_y$ residual = 361.73(1 − .7830) = 78.50.

These results may also be put into a summary table like 16.6.1. It is obvious that for these data the results of the Method R analysis differ from those obtained from the two analyses done under Method H. The main difference here is that Factor A seems to account for a shade more variance under this model, whereas Factor B also accounts for less variance than was suggested in either of the Method H analyses.

You observe, however, that whereas the sum of squares add up to the total in the usual way under a Method H, or hierarchical, analysis, under Method R the sums of squares *no longer* have to add up to the original total. On the other hand, in Method H we had to decide in advance on the hierarchy, or order in which the factors would be examined, whereas Method R does not require any particular order. Both of these approaches should yield the usual ANOVA results for balanced data, however.

Both the hierarchical, Method H, and the regression, Method R, approaches are quite feasible for use in most circumstances, but the user must realize that they are *different in the results* they provide, as just illustrated. Moreover, they differ in the interpretation of their results as "adjusted" values.

It is not easy to say that one approach is uniformly better than the other. However, in many ways the results of the Method R regression approach seem to be closer in intuitive meaning to those of an ordinary ANOVA based on balanced data, and this may be reason enough to prefer the regression method for most applications. Certainly this seems to be the most popular method, at this writing.

On the other hand, when one or more of the factors really have been introduced into the design in an effort to increase controls, and there is genuine interest in looking at

Source	SS	df	MS	F	p	
Factor A, beyond B, A × B	127.37	2	63.68	7.30	<.05	**Table 16.6.1**
Factor B, beyond A, A × B	.25	1	.25	–	–	Results of regression analysis (method R)
A × B Int., beyond A, B	170.77	2	85.39	9.79	<.01	
Residual	78.50	9	8.72			

the remaining factors after successive adjustments for variables in this original set, then the hierarchical procedure of Method H may be just what the doctor ordered.

A fairly extensive literature exists on this topic, and those with interests in pursuing it further may wish to look into sources such as Overall and Spiegel (1969); Overall and Klett (1973); Overall, Spiegel, and Cohen (1975); and Speed, Hocking, and Hackney (1978).

There is, by the way, no universal agreement on what these methods for handling unbalanced designs should be called. As noted below in Section 16.21, the SAS computer package refers to our hierarchical method (Method H) as TYPE I SS, but to the regression strategy as Type II or Type III. On the other hand, manuals for SPSS and SPSS^x refer not only to our hierarchical and our regression approaches but also to the default option of a "classic regression approach," wherein each main effect is adjusted for other main effects, each interaction by others of its order, and lower orders. The BMDP program P2V for unbalanced designs offers only the regression approach, whereas the user is given options in the SPSS programs. In SAS the default GLM procedure offers both TYPE I and TYPE III SS analyses. Obviously, it is wise to read the manual very carefully before analyzing unbalanced data using one of these packages.

16.7 TWO-WAY (OR HIGHER) FIXED-EFFECTS REGRESSION ANALYSIS

A fixed-effects linear regression problem is capable of extension in a variety of ways. For example, imagine a factorial experiment with *two* quantitative factors, X_1 and X_2. The J levels of X_1 and the K levels of X_2 are fixed ahead of the study. Furthermore, for the sake of simplicity, let us suppose that each combination of level j from X_1 and level k from X_2 contains the same number of observations n. In this instance, the simplest linear two-variable model is written, in population terms, as

$$y_{ijk} = \mu_y + \beta_1(x_{1j} - \mu_1) + \beta_2(x_{2k} - \mu_2) + e_{ijk}.$$

Here x_{1j} stands for the value of X_1 received by members of group j, and x_{2k} the value on X_2 received by members of group k. Note that no interaction is assumed: the only influences on the variance of Y are the linear regression on X and the linear regression on X_2.

This is a straightforward application of multiple regression to the prediction of Y from X_1 and X_2. However, since we are assuming an orthogonal factorial design with equal n per cell, $r_{12} = 0$, the two variables are themselves uncorrelated. Thus, we can take

$$b_1 = \frac{\sum_j (x_{1j} - \bar{x}_1)\bar{y}_j}{\sum_j (x_{ij} - \bar{x}_1)^2} \qquad [16.7.1]$$

and

$$b_2 = \frac{\sum_k (x_{2k} - \bar{x}_2)\bar{y}_k}{\sum_k (x_{2k} - \bar{x}_2)^2}. \qquad [16.7.2]$$

Then, we have

$$\text{SS linear regression on } X_1 = b_1^2 \, \text{SS}_{X_1} = r_{y1}^2 \, \text{SS}_Y \text{ total}$$

$$\text{SS linear regression on } X_2 = b_2^2 \, \text{SS}_{X_2} = r_{y2}^2 \, \text{SS}_Y \text{ total}$$

$$\text{SS deviations from regression} = \text{SS}_Y \text{ total} - \text{SS regression on } X_1$$
$$- \text{SS linear regression on } X_2,$$

so that

$$\text{MS linear regression on } X_1 = \text{SS linear regression on } X_1$$

$$\text{MS linear regression on } X_2 = \text{SS linear regression on } X_2$$

$$\text{MS deviations} = \frac{\text{SS deviations from regression}}{N - 3}.$$

Thus, both regressions may be tested for significance by comparison with MS deviations, with 1 and $N - 3$ degrees of freedom.

Obviously, such a design can be analyzed directly by multiple regression. However, the actual hand computations for such a problem go as follows:

$$\text{SS}_{X_1} = \sum_j n_j x_{1j}^2 - \frac{\left(\sum_j n_j x_{1j} \right)^2}{N}$$

$$\text{SS}_{X_2} = \sum_k n_k x_{2k}^2 - \frac{\left(\sum_k n_k x_{2k} \right)^2}{N}. \qquad [16.7.3]$$

We also calculate *sums of products,* as follows:

$$\text{SP}_{YX_1} = \sum_j x_{1j} \sum_k \sum_i y_{ijk} - \frac{\left(\sum_j n_j x_{1j} \right)\left(\sum_j \sum_k \sum_i y_{ijk} \right)}{N}$$

$$\text{SP}_{XY_2} = \sum_k x_{2k} \sum_j \sum_i y_{ijk} - \frac{\left(\sum_k n_k x_{2k} \right)\left(\sum_k \sum_j \sum_i y_{ijk} \right)}{N} \qquad [16.7.4]$$

with SS_Y total figured as usual. Here $n_j = Kn$, $n_k = Jn$, of course.
Then

$$\text{SS linear regression on } X_1 = \frac{(\text{SP}_{YX_1})^2}{\text{SS}_{X_1}}$$

$$\text{SS linear regression on } X_2 = \frac{(\text{SP}_{XY_2})^2}{\text{SS}_{X_2}}$$

and

$$\text{SS deviations from regression} = \text{SS}_Y \text{ total} - \text{SS linear on } X_1 - \text{SS linear on } X_2.$$

The same general procedures apply to factorial design with any number of quantitative factors, provided that no interactions are assumed to exist. Models with interactions may be handled by multiplication as suggested in Section 12.10. This setup will be illustrated for two factors in the next section.

16.8 AN EXPERIMENT WITH TWO QUANTITATIVE FACTORS

In Section 14.11 we considered an example in which a social psychologist was studying the behavior of small groups in problem-solving situations. There, some 30 groups of varying size were used to establish the regression equation for predicting dependent variable Y, or time taken by the group, working cooperatively, to solve an assigned problem. The theory upon which these experiments were based suggested that within a particular range of possible group sizes, the relationship between group size and average time to solution for a particular kind of problem should be such that the larger the group, the less time on the average it should take for the problem to be solved. Therefore, the psychologist decided to form a set of experimental groups, ranging in size from a single individual to six individuals. Some 105 individuals were chosen at random, and then formed at random into five groups consisting of 1 individual each, five groups each consisting of 2 individuals, five groups of 3, and so on until there were six different and nonoverlapping sets of five groups each, the last consisting of five groups each of size 6. Each individual subject participated in one and only one group in the study.

Now, let us suppose that five different problems were available for use with these groups. These problems had been scaled in difficulty, from 5 for the most difficult to 1 for the least difficult. In order to see if this scaling correlated with the time it took the groups to solve these problems, and also as a way to reduce error variance, the experimenter decided to give all five problems within each set of groups of the same size, with each group receiving one problem assigned at random. Thus, every combination of problem difficulty and size was represented by exactly one group.

Each experimental group then solved its assigned problem, and the time taken for them to do so was noted; this time to completion was used as the dependent variable. Table 16.8.1 gives the results, with group size shown as the columns, and problem difficulty by the rows.

		$X_1 = $ Group size					
		1	2	3	4	5	6
$X_2 = $ difficulty	5	32	22	27	19	25	20
	4	29	22	26	21	22	18
	3	30	24	24	20	23	16
	2	28	23	24	20	25	18
	1	26	22	23	18	21	16

Table 16.8.1

Results for groups of different sizes and problems of varying levels of difficulty.

Then, we take

$$SS_Y \text{ total} = 16{,}058 - \frac{684^2}{30} = 462.8$$

$$SS_{X_1} = 455 - \frac{105^2}{30} = 87.50$$

$$SS_{X_2} = 330 - \frac{90^2}{30} = 60$$

$$SP_{YX_1} = 2{,}243 - \frac{105 \times 684}{30} = -151$$

$$SP_{YX_2} = 2{,}090 - \frac{90 \times 684}{30} = 38.$$

Then,

$$SS \text{ linear regression on } X_1 = \frac{151^2}{87.5} = 260.58$$

$$SS \text{ linear regression on } X_2 = \frac{38^2}{60} = 24.07$$

and

$$SS \text{ deviations} = SS_Y \text{ total} - SS \text{ linear } X_1 - SS \text{ linear } X_2$$
$$= 178.15.$$

We also find that

$$b_1 = \frac{SP_{YX_1}}{SS_{X_1}} = \frac{-151}{87.5} = -1.73$$

and

$$r_{y1}^2 = \frac{SS \text{ linear regression on } X_1}{SS_Y \text{ total}} = \frac{260.58}{462.80} = .56$$

$$r_{y1} = -.75,$$

as well as

$$b_2 = \frac{SP_{YX_2}}{SS_{Y_2}} = \frac{38}{60} = .63$$

$$r_{2y}^2 = \frac{SS \text{ linear regression on } X_2}{SS_Y \text{ total}} = .05$$

$$r_{2y} = .23.$$

Source	SS	df	MS	F
Linear regression on X_1	260.58	1	260.58	$\dfrac{260.18}{6.60} = 39.4$
Linear regression on X_2	24.07	1	24.07	$\dfrac{24.07}{6.60} = 3.65$
Deviations (error)	178.15	27	6.60	
Total	462.80	29		

Table 16.8.2

Summary table for data in Table 16.8.1.

The total proportion of variance accounted for is

$$R^2_{y \cdot 12} = \frac{\text{SS linear } X_1 + \text{SS linear } X_2}{\text{SS}_Y \text{ total}}$$

$$= .62.$$

Note that the correlation between group size and time is negative, as the experimenter expected. The results are displayed in summary Table 16.8.2.

The first of these F tests is significant well beyond the .01 level. The other fails to reach the .05 level. There is thus very strong evidence that group size is negatively related to the time required for the solution of a problem, but little evidence that the problems scaled as more difficult do tend to require more time. However, note that by including the different problems in this experiment, the researcher accounted for 5% of the total variance (i.e., $24.07/462.80 = .05$ for linear regression on X_2), which otherwise would have gone into deviations from regression, and thus would have tended to lower the power of the test of the other regression.

16.9 CURVILINEAR REGRESSION

All the discussion so far in this chapter has centered on linear regression, the use of a linear function rule for the prediction of Y from X. However, the theory of regression is much more extensive than the preceding discussion of linear regression might suggest. Indeed, the linear rule for prediction is only the simplest of a large number of such rules that might apply to a given statistical relation. Linear regression equations may serve quite well to describe many statistical relations that are roughly like linear functions, or that may be treated as linear as a first approximation. Nevertheless, there is no law of nature requiring all important relationships between variables to have a linear form. It thus becomes important to extend the idea of regression equations to the situation where the relation is *not necessarily* best described by a linear rule. Now we are going to consider problems of curvilinear regression, problems where the best rule for prediction need not specify a simple linear function.

Our study of nonlinear prediction rules will be confined to *problems of regression,* as defined previously. An independent variable X is identified, and various values of X are chosen in advance to be represented in the experiment. Ordinarily, each X value corresponds to an experimental treatment administered in some specific quantity.

A regression problem where a curvilinear rule for prediction might make sense is the following: there is an investigation of the effect of environmental noise on the human subject's ability to perform a complex task. Several experimental treatments are planned, each treatment differing from the others only in the intensity of background noise present while the subject performs the task. Each treatment represents a one-step interval in the intensity of noise within a particular range. A group of some N subjects chosen at random are assigned at random to the various groups, with n subjects per group. In the experiment proper, each subject works on the same problem individually in the presence of the assigned intensity of background noise. The dependent variable is thus a subject's score on the task, and the independent variable is the noise intensity. The experimenter is interested not only in the possible existence of a linear relation between noise intensity and performance, but also a possible curvilinear relationship between X and Y.

In the next section, a general model for dealing with curvilinear, as well as linear, regression will be explored. Using this model we will see first how a test may be made for the existence of a nonlinear relationship. Then we will deal with the problem of inferring the form of the statistical relation by use of planned comparisons among means. Finally, inferences about form of relation made from post hoc comparisons will also be considered.

16.10 THE MODEL FOR LINEAR AND CURVILINEAR REGRESSION

So far, in this chapter inferences about the existence and extent of linear regression were made in terms of the population model

$$y_{ij} = \mu_Y + \beta_{Y \cdot X}(x_j - \mu_X) + e_{ij},$$

which we saw to be an instance of the fixed-effects model for analysis of variance when the X values are fixed. In adopting this model, we assume that the only systematic tendency of Y to vary with X is due to the linear regression of Y on X.

Now we will extend this model to allow for the existence of other kinds of systematic dependence of Y on X. The model we will use can be stated as follows:

$$y_{ij} = \beta_0 + \beta_1(x_j - \mu_X) + \beta_2(x_j - \mu_X)^2 + \cdots \qquad [16.10.1]$$
$$+ \beta_{J-1}(x_j - \mu_X)^{J-1} + e_{ij}.$$

That is, we assume that the value of Y might depend on the value of $(x_j - \mu_X)$, on $(x_j - \mu_X)^2$, and, indeed, on any power of $(x_j - \mu_X)$ up to and including the $J - 1$ power, each term weighted by a constant β_1, β_2, and so on.

Some intuitive feel for this model may be gained as follows. Suppose that it were really true that the best possible prediction of Y_{ij} were afforded by

$$y'_{ij} = \beta_0 + \beta_1(x_j - \mu_X) + \beta_2(x_j - \mu_X)^2. \qquad [16.10.2]$$

However, the experimenter does not know this, and applies the simple linear rule:

$$y'_{ij} = \beta_1(x_j - \mu_X) + \mu_Y.$$

Does this describe all the predictability possible in the relationship? No, since if β_2 is not zero, the residuals, or *deviations*, of the predicted from the obtained values of Y should be predictable using the rule

$$d'_{ij} = (y_{ij} - y'_{ij}) = \beta_2(x_j - \mu_X)^2 + e_{ij} + (\beta_0 - \mu_Y). \qquad [16.10.3]$$

In short, the deviations from linear regression here represent not only error, but also some further predictability depending on the *squared* value of $(x_j - \mu_X)$. We should expect that, when plotted, the deviations from linear regression for the various sample groups should tend to lie about a parabola, much as in Figure 16.10.1.

The constant β_2 is a coefficient of second-degree or **quadratic regression.** We need a way to decide if β_2 really is zero if we are going to tell whether or not the best prediction rule possible is at least of the second degree.

Extending this idea, suppose that the true relation between X and Y is given by the rule

$$y_{ij} = \beta_0 + \beta_1(x_j - \mu_X) + \beta_2(x_j - \mu_X)^2 + \beta_3(x_j - \mu_X)^3 + e_{ij}. \qquad [16.10.4]$$

However, the experimenter actually predicts using either a first- or a second-degree regression equation. Suppose that an equation of the second degree is used. Then, nevertheless, the deviations from the predicted y_{ij} values will tend to be systematically related to x values, since

$$(y_{ij} - y'_{ij}) = \beta_3(x_j - \mu_X)^3 + e_{ij} + \beta_0 - \mu_Y. \qquad [16.10.5]$$

Unless β_3 is zero, cubic trends exist between X and Y, and the plot of deviations from prediction for the various populations should tend to fall into a cubic function, something like that shown in Figure 16.10.2.

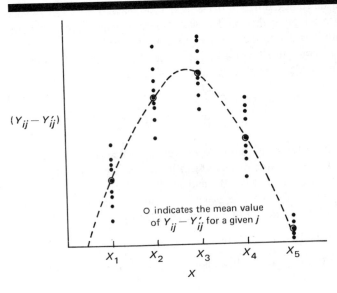

Figure 16.10.1

Plot of deviations from linear regression, suggesting the presence of second-degree, or quadratic, regression.

$(Y_{ij} - Y'_{ij})$

O indicates the mean value of $Y_{ij} - Y'_{ij}$ for a given j

$X_1 \qquad X_2 \qquad X_3 \qquad X_4 \qquad X_5$

X

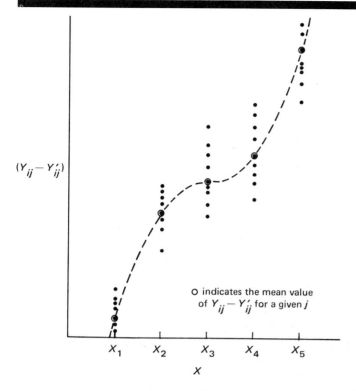

Figure 16.10.2

Deviations from linear or quadratic regression, suggesting the presence of third-degree, or cubic, regression.

o indicates the mean value of $Y_{ij} - Y'_{ij}$ for a given j

The same general ideas hold for any number of experimental groups each associated with a fixed value x_j. A regression equation of the first, second, third, and so on up to the degree $J - 1$ is assumed possible. It can be shown, however, that for any finite number J of groupings, each associated with some fixed value of x_j, the best prediction equation possible in this model is of *no higher degree than $J - 1$*. Thus, a first-degree equation always suffices to describe the relation between X values and mean Y values for *two* groups, a second-degree equation suffices to describe the relation between X and mean Y for *three* groups, a third-degree equation for *four* groups, and so on.

16.11 ANOTHER LOOK AT THE PARTITION OF THE SUM OF SQUARES FOR REGRESSION PROBLEMS

Under the model used in Section 14.12, nothing could contribute to deviations from linear regression except random errors e_{ij}. Consequently, the mean square for deviations from linear regression was called MS residual and used to estimate σ_e^2.

Now, however, we have a different model, which permits other, nonlinear, kinds of relations to exist. Nevertheless, MS within groups still reflects only error variance in the data. Furthermore, the SS linear regression depends only on the true regression coefficient β_1 and error, just as before. On the other hand, the status of the SS deviations from linear regression changes in this new model; this sum of squares reflects all the possible *nonlinear* relations as well as random error.

We already know, of course, that all predictability in the data is embodied in the sum of squares between groups. That is, in the terminology of Section 16.3, we know that

$$SS_Y \text{ total } (R^2_{y \cdot H}) = SS \text{ between groups} \qquad [16.11.1]$$

where H represents either $J - 1$ dummy variables or, equivalently, a set of $J - 1$ orthogonal variables based on the J fixed levels of X. We also saw in Section 16.3 that the sum of squares for linear regression is equivalent to the sum of squares for one variable, V_1, out of this set. Let us then relabel r^2_{yx} as $r^2_{y \cdot 1}$. Then, SS linear regression = $r^2_{y \cdot 1}$ (SS_Y total). Now in order to assess possible curvilinearity, we take

$$SS \text{ deviations from linear regression} = SS \text{ between} - SS \text{ linear} \qquad [16.11.2]$$

so that we have

$$SS \text{ deviations from linear regression} = (R^2_{y \cdot H} - r^2_{y \cdot 1})(SS_Y \text{ total}) \qquad [16.11.3]$$

or

$$SS \text{ deviations from linear regression} \qquad [16.11.4]$$
$$= R^2_{y(G \cdot 1)}(SS_Y \text{ total}), \ H - \{V_1\} = G.$$

In the symbolism adopted in Section 15.19, this index $R^2_{y(G \cdot 1)}$ represents the variance in Y explained by all other aspects of X, *after* the contribution made by linear regression is removed. Thus, $R^2_{y(G \cdot 1)}$ must represent *any possible curvilinear relationship* that still exists between X and Y, over and above the linear relationship represented by $r^2_{y \cdot 1}$.

The partition of the total sum of squares is thus

SS total = SS linear regression + SS deviations from linear + SS error

where

$$SS \text{ deviations from linear} = SS \text{ between} - SS \text{ linear regression} \qquad [16.11.5]$$

The computations thus needed are those for an ordinary one-way analysis of variance, and the SS linear regression, as found from Eq. 14.16.6 and Eq. 14.16.7.

The analysis ordinarily is summarized in a table such as that shown below.

Source	SS	df	MS	F
Between groups		$(J - 1)$	—	—
Linear regression		1	$\dfrac{SS \text{ lin.}}{1}$	$\dfrac{MS \text{ lin.}}{MS \text{ error}}$
Deviations from linear regression		$J - 2$	$\dfrac{SS \text{ dev.}}{J - 2}$	$\dfrac{MS \text{ dev.}}{MS \text{ error}}$
Error		$N - J$	$\dfrac{SS \text{ error}}{N - J}$	
Total		$N - 1$		

We know already that the degrees of freedom associated with SS linear regression must be 1. The degrees of freedom for SS between groups is, of course, $J - 1$. Hence, the degrees of freedom for SS deviations from linear regression (representing the number of remaining orthogonal variables after the first, linear, component is removed) must be $J - 2$.

In this instance, the test for linear regression is given by

$$F = \frac{\text{MS linear regression}}{\text{MS error}} \qquad\qquad [16.11.6]$$

with 1 and $N - J$ degrees of freedom. This, of course, tests the hypothesis H_0: $\rho_{y\cdot1}^2 = 0$, against the alternative H_1: $\rho_{y\cdot1}^2 > 0$.

To test the hypothesis of no curvilinear regression, we employ the F ratio

$$F = \frac{\text{MS deviations from linear}}{\text{MS error}} \qquad\qquad [16.11.7]$$

with $J - 2$ and $N - J$ degrees of freedom. Note that this is exactly the same as the test

$$F = \frac{R_{y(G\cdot1)}^2(N - J)}{(1 - R_{y\cdot H}^2)(J - 2)}, \qquad \text{where } H - \{V_1\} = G, \qquad [16.11.8]$$

given by Eq. 15.25.19 when the appropriate substitutions are made.

A significant value for the first of these F tests indicates that it is safe to conclude that some predictability is afforded by a linear rule. In the same way, a significant value for F for the second test lets one conclude that some further prediction is possible using a curvilinear rule. However, neither of these tests guarantees that there exists either a strong linear or a strong curvilinear relationship in the population sampled. The evidence for strength of linear relationship is given by (SS linear/SS$_Y$ total), which is, of course the sample value of $r_{y\cdot1}^2$. The evidence for strength of curvilinear relation is then given by (SS deviations from linear/SS$_Y$ total), which is the sample value of $R_{y(G\cdot1)}^2$. The assumptions underlying these tests are exactly the same as those made in the test for linear regression in Section 14.13.

Although the example to follow shows an experiment in which an equal number of subjects are assigned to each of J groups, this need not be true in general. Furthermore, although X values in equal steps are chosen for the groups in the example, this is not a general requirement in the simple tests for linear and curvilinear trend as illustrated here. However, methods such as those to be introduced in Section 16.13 must be modified if unequal numbers are used in the groups, or if the X values are chosen in unequal steps.

16.12 AN EXAMPLE OF TESTS FOR LINEAR AND CURVILINEAR REGRESSION

In Section 16.9 an example was suggested in which a curvilinear relation between experimental and dependent variables might possibly occur. Let us now suppose that this experiment is carried out. Six groups are formed, each of which is exposed to a different

Noise intensity levels, X_j						**Table 16.12.1**
1	2	3	4	5	6	Performance of subjects under differing levels of noise intensity.
18	34	39	37	15	14	
24	36	41	32	18	19	
20	39	35	25	27	5	
26	43	48	28	22	25	
23	48	44	29	28	7	
29	28	38	31	24	13	
27	30	42	34	21	10	
33	33	47	38	19	16	
32	37	53	43	13	20	
38	42	33	23	33	11	
270	370	420	320	220	140	

level of noise intensity. Sixty subjects chosen at random are assigned at random to these groups, with 10 per group. Each group represents a one-step increment in a scale of noise intensity. The exact units of noise intensity need not bother us here, and we can represent the levels of X as 1, 2, 3, and so on, since the different groupings represent equal intervals in noise intensity. The dependent variable is the score y_{ij} of a subject in a complex performance given the noise intensity level x_j. It is desired to test both for linear and for curvilinear regression, using $\alpha = .01$. The scores are shown in Table 16.12.1.

The usual computations for a one-way analysis of variance are carried out first:

$$\text{SS total} = 18^2 + 24^2 + \cdots + 11^2 - \frac{1,740^2}{60}$$

$$= 7,252$$

$$\text{SS between} = \frac{270^2 + \cdots + 140^2}{10} - \frac{1,740^2}{60}$$

$$= 5,200$$

$$\text{SS error} = \text{SS total} - \text{SS between} = 7,252 - 5,200 = 2,052.$$

Next the SS for linear regression is found from Eqs. 14.16.5 and 14.16.6. Since $n_j = 10$ for each group,

$$b_{Y \cdot X} = \frac{\sum_j \sum_i x_j y_{ij} - \left[\left(\sum_j 10 x_j \right) \left(\sum_j \sum_i y_{ij} \right) / 60 \right]}{\left(\sum_j 10 x_j^2 \right) - \left[\left(\sum_j 10 x_j \right)^2 / 60 \right]}.$$

Here

$$\sum_j \sum_i x_j y_{ij} = \sum_j x_j \left(\sum_i y_{ij} \right) = 1(270) + 2(370) + \cdots + 6(140) = 5,490$$

$$\sum_j 10x_j = 10(1) + \cdots + (10)(6) = 210$$

$$\sum_j 10x_j^2 = 10(1 + 4 + \cdots + 36) = 910$$

so that

$$b_{Y \cdot X}^2 \, SS_X = SS \text{ lin. reg.} = \frac{\{5{,}490 - [(210)(1{,}740)/60]\}^2}{910 - (210^2/60)}$$

$$= 2{,}057.14.$$

Then

$$SS \text{ dev. from lin. reg.} = SS \text{ between} - SS \text{ lin. reg.}$$

$$= 5{,}200 - 2{,}057.14$$

$$= 3{,}142.86$$

Thus,

$$r_{y \cdot 1}^2 = \frac{2{,}057.14}{7{,}252} = .28.$$

and

$$R_{y(G \cdot 1)}^2 = \frac{3{,}142.86}{7{,}252} = .43.$$

The completed summary table is thus

Source	SS	df	MS	F
Between groups	(5,200)	(5)	—	—
Linear reg.	2,057.14	1	2,057.1	54.1
Dev. from lin.	3,142.86	4	785.7	20.7
Error	2,052	54	38	
Totals	7,252	59		

On evaluating these two F tests, we find that each is significant far beyond the .01 level. In short, we can reject both the hypothesis of no linear regression *and* the hypothesis of no curvilinear regression, although we should recall that the probability of error in at least one of these conclusions is not necessarily .01. Nevertheless, here we may say with some confidence that both linear and curvilinear regression exist. Actually, in this sample, curvilinear regression accounts for more variance ($3{,}142.9/7{,}252 = .43$ or 43%) than does linear regression ($2{,}057.1/7{,}252 = .28$, or 28%). The best rule for population prediction apparently involves both x_j and powers of x_j, in terms of this model.

A plot of the Y means for these data supports our findings (Figure 16.15.1). The sample means tend, roughly, to fall along a straight line with negative slope (the dotted line in the figure). However, they would cluster even more closely along a curved line suggesting a parabolic arc. The experimenter would be likely to conclude that, in the range studied, relatively low levels of noise intensity may actually facilitate the performance of subjects in this task, but that beyond a particular point (about Level 3) average performance falls off rapidly with further increases in noise intensity.

16.13 PLANNED COMPARISONS FOR TREND: ORTHOGONAL POLYNOMIALS

The separation of sums of squares into components representing linear and nonlinear regression is but the simplest of the procedures that may be applied to problems of linear and curvilinear regression. Many times an experiment is set up to answer certain specific questions about the *form* of the relation existing between two variables. In general when the experimenter wants to look into the question of trend, or form of relationship, he should have some prior ideas about what the population relation should be like. These hunches about trend often come directly from theory, or they may come from the extrapolation of established findings into new areas. At any rate, specific questions are to be asked about the *degree* of the regression equation sufficient to permit prediction.

The technique of planned comparisons studied in Chapter 11 lends itself directly to this problem. Given J groups, each of which is associated with a value x_j of the experimental variable, each of the $J - 1$ degrees of freedom for comparisons can be allotted to the study of *one question* about the form or degree of the predictive relation. In this discussion, we will restrict our attention to the situation where the following conditions are satisfied.

1. The various x_j values represent *equally spaced unit intervals* on the X dimension or continuum. The units may be anything, but the difference between each x_j and x_{j+1} is precisely one of these units.
2. The number of observations made, n_j, is the same for each value of x_j.
3. Inferences are to be made about a hypothetical population in which the *only possible* values of x_j are those actually represented in the experiment.

The first two of these requirements are not especially severe limitations, since this technique is ordinarily applied to experimental studies where one is free to choose the x_j values and the n in a group quite arbitrarily. In such studies it is usually more convenient than otherwise to form equal spacings of x_j values, and to assign individuals at random and equally among the groups. The third restriction is quite serious, however. In uses of this method, really **nothing can be said about values of** x_j **that are not directly represented in the experiment,** insofar as statistical inference is concerned. The form of the relation may be quite different for X outside of the range represented, or even for potential x_j values intermediate to two actually represented. It is up to the experimenter to exercise caution and good scientific judgment in extrapolating beyond the x_j values actually observed. The statistician cannot really help here, since this is a question of scientific, not statistical, inference.

We have already seen (Section 14.16) that in the regression model the sum of squares for *linear* regression corresponds to the SS for one possible comparison among the sample means. Extending this idea, an orthogonal comparison can also be made reflecting second-degree, or quadratic, relationship. Still another orthogonal comparison can be carried out reflecting the third-degree, or cubic, relationship, and so on, until the experimenter has either investigated all the specific questions of interest, or $J - 1$ orthogonal comparisons have been made (whichever happens first). As for any set of *planned* comparisons, particular values from among the $J - 1$ possible comparisons are tested separately instead of the overall F test of between-group means. If any one of these comparisons is significant, then the experimenter has relatively strong evidence that the corresponding β value in the population is not zero, and that the best prediction rule is at least of the degree represented by that comparison.

For equally spaced x values, the weights which each trend comparison involves are completely dictated by J, the number of different x_j groups, and by the degree of the particular trend investigated. That is, given any number of groups J, one standard set of weights exists for investigating linear, or first-degree, trends, an orthogonal set exists for second-degree, or quadratic, trends, another for cubic, or third-degree, trends, and so on, up to and including trends of degree $J - 1$. These standard sets of comparison weights are called **coefficients of orthogonal polynomials,** and the method itself is sometimes called the **method of orthogonal polynomials.**

(The final example of the use of the Gram-Schmidt procedure in Section C.5 deals with a set of vectors where every vector represents a power of the entries in vector $\mathbf{x}_1$, so that $\mathbf{x}_2$ gives the squares, $\mathbf{x}_3$ the cubes, and so on, of the elements of $\mathbf{x}_1$. Then when an orthogonal basis is found, the vectors yield the orthogonal polynomial weights we will be using below. In other words, these sets of orthogonal polynomial weights are just an orthogonal basis for X, X^2, X^3, and so on. All of the points raised in Section 16.3 are thus relevant here as well.)

The orthogonal polynomial coefficients are so derived that the particular comparisons among means each reflect one kind of possible trend or form of relationship in the data.

This sounds much more complicated than it actually is. Perhaps a simple example may clarify this point. For four groups, equally spaced with regard to the x_j values, the three sets of standard orthogonal polynomial weights are

Comparison	$\bar{y}_1$	$\bar{y}_2$	$\bar{y}_3$	$\bar{y}_4$
linear	−3	−1	1	3
quadratic	1	−1	−1	1
cubic	−1	3	−3	1

The symbol $\bar{y}_1$ denotes the mean Y value for the group with the *lowest* X value, or x_1, $\bar{y}_2$ the mean Y for the next lowest value of X, or x_2, and so on. Applying the weights given in the first row to the means, we get the first, or *linear* comparison:

$$\hat{\psi}_1 = -3(\bar{y}_1) - 1(\bar{y}_2) + 1(\bar{y}_3) + 3(\bar{y}_4).$$

The sample value of $\hat{\psi}_1$ is an unbiased estimate of the *population* comparison

$$\psi_1 = -3(\mu_1) - 1(\mu_2) + 1(\mu_3) + 3(\mu_4).$$

If this population comparison is not equal to zero, then there is at least a linear trend in the relationship. Higher order trends may be present, however.

By applying the weights of the second comparison to the sample means, we get

$$\hat{\psi}_2 = 1(\bar{y}_1) - 1(\bar{y}_2) - 1(\bar{y}_3) + 1(\bar{y}_4)$$

which estimates ψ_2, the same linear combination applied to the population means. This comparison value in the population can be zero only if $\beta_2 = 0$, so that no quadratic trend exists in the statistical relationship. If this comparison value in the population is other than zero, then at least a quadratic trend exists in the statistical relationship.

Finally, the comparison for cubic trends is

$$\hat{\psi}_3 = -1(\bar{y}_1) + 3(\bar{y}_2) - 3(\bar{y}_3) + 1(\bar{y}_4).$$

This comparison estimates ψ_3. If the population value is other than zero, a cubic trend exists in the relationship.

Given only four x values, then if all three comparisons, ψ_1, ψ_2, ψ_3, are zero when applied to population values, no predicative relationship at all exists. On the other hand, large and significant sample comparison values in data provide us with evidence that trends of at least those particular degrees exist in the population relation.

One more thing should be pointed out: these weights used in the example are bona fide comparison weights, since the weights for each comparison sum to zero. Furthermore, the comparisons are all orthogonal, since the summed products of weights for any two comparisons is also zero. Thus the theory we have studied for any set of planned orthogonal comparisons applies perfectly well to trend comparisons of this type, as do the connections with ANOVA and multiple regression outlined earlier.

Table VII in Appendix E gives the required weights for trend comparisons for $J = 3$ through $J = 15$, and for trends through the sixth degree, which certainly should cover most social science research on such problems. One simply finds the appropriate section of the table, and reads off the required weights. Within each section, the rows represent the x_j groupings; here the lowest value of x_j is shown as the first or topmost row. The columns represent the *degree* of trend to be examined. Then the entry in any row j and column h gives the weight applied to the mean of the jth group in the comparison for a trend of degree h. A larger table of required weights is given in *Biometrika Tables for Statisticians* (1966).

This general procedure is followed until as many of the comparisons as have a priori interest for the experimenter have been made. These comparisons should actually be *planned before* the data are seen, and should really correspond to questions that are germane to the experimental problem and will clarify the interpretation of the experiment by their separate examination. For more than four or five experimental groups, it is seldom practical or useful to work out all the possible $J - 1$ trend comparisons. Usually, there is little understanding to be gained by finding, for example, that a sixth-degree or higher trend is significant. Ordinarily one stops after the cubic or perhaps quartic (fourth degree) comparison, and then relegates all other trend effects to the pooled sum of squares:

$$\text{SS other trends} = \text{SS between} - \sum_{h=1}^{L} \text{SS}(\hat{\psi}_h) \qquad [16.13.1^*]$$

where h symbolizes any one from the L comparisons made separately. For, say, five groups, we might be interested only in linear and quadratic, and then find the pooled SS for other trends from

$$\text{SS (cubic and quartic)} = \text{SS between} - \text{SS lin.} - \text{SS quad.} \qquad [16.13.2]$$

which has $4 - 2$ or 2 degrees of freedom.

The proportion of variance accounted for in the data by any trend h is, of course,

$$r_{y \cdot h}^2 = \frac{\text{SS}(\hat{\psi}_h)}{\text{SS}_Y \text{ total}}. \qquad [16.13.3^*]$$

Then the proportion of variance attributable to any set G of trends, any one of which is symbolized by g, is

$$R_{Y \cdot G}^2 = \sum_{g \text{ in } G} \text{SS}(\hat{\psi}_g)/\text{SS}_Y \text{ total}. \qquad [16.13.4^*]$$

16.14 AN EXAMPLE OF PLANNED COMPARISONS FOR TREND

Suppose that the experiment on noise intensity level and performance had actually been planned as a study of trend. The experimenter is especially interested in judging if linear and quadratic trends exist in the relation between X and Y. We will pretend that the data in Section 16.12 had been collected specifically for this purpose.

The SS total, the SS between, and the SS error are found just as before, of course. However, let us list the group means and find the linear and quadratic comparison values. Table 16.14.1 gives the six means and the weights that apply to each for linear and quadratic comparisons, as found from Table VII (Appendix E). The comparison values themselves are

$$\hat{\psi}_1 = (-5 \times 27) - (3 \times 37) - (1 \times 42) + (1 \times 32) + (3 \times 22) + (5 \times 14)$$
$$= -120,$$

and

$$\hat{\psi}_2 = (5 \times 27) - (1 \times 37) - (4 \times 42) - (4 \times 32) - (1 \times 22) + (5 \times 14)$$
$$= -150.$$

The w values for the corresponding sums of squares are found from

$$w_1 = \frac{25 + 9 + 1 + 1 + 9 + 25}{10} = 7$$

and

$$w_2 = \frac{25 + 1 + 16 + 16 + 1 + 25}{10} = 8.4,$$

	Group						**Table 16.14.1**
	1	2	3	4	5	6	Linear and quadratic orthogonal polynomial weights for the noise intensity data.
Means	27	37	42	32	22	14	
Weights							
linear	−5	−3	−1	1	3	5	
quadratic	5	−1	−4	−4	−1	5	

so that

$$\text{SS linear} = \frac{(-120)^2}{7} = 2{,}057.14$$

$$\text{SS quad.} = \frac{(-150)^2}{8.4} = 2{,}678.57$$

The proportions of variance accounted for are then, respectively,

$$r_{y\cdot1}^2 = \frac{2{,}057.14}{7{,}252} = .28,$$

just as found in Section 16.12, and

$$r_{y\cdot2}^2 = \frac{2{,}678.57}{7{,}252} = .37.$$

Then

$$\text{SS other trends} = 5{,}200 - 2{,}057.14 - 2{,}678.57 = 464.29$$

with $J - 1 - 2 = 3$ degrees of freedom. The proportion of variance accounted for by cubic and higher trends is thus

$$R_{y\cdot345}^2 = \frac{464.29}{7{,}252} = .06.$$

The summary table for this analysis is

Source	SS	df	MS	F
Between	(5,200)	(5)	—	—
Linear	2,057.14	1	2,057.1	54.1
Quadratic	2,678.57	1	2,678.6	70.5
Other trends	464.29	3	154.8	4.1
Error	2,052	54	38	
Totals	7,252	59		

The F tests for both linear and quadratic trends greatly exceed the value required for the .01 level of significance, but the overall test for other trends does not reach the .01

level. Thus, we can be fairly sure that in the population, predictive association does exist between X and Y, and that the general rule for prediction includes both linear and quadratic components. We cannot, however, really be sure that the population prediction rule does not include higher order components as well. Even so, we infer from these data that a curvilinear regression equation affording at least some prediction has the general form

$$y'_{ij} = \beta_0 + \beta_1(x_j - \mu_x) + \beta_2(x_j - \mu_x)^2,$$

where the β_1 and β_2 are the linear and quadratic regression coefficients in the population.

16.15 ESTIMATION OF A CURVILINEAR PREDICTION FUNCTION

Sometimes it is convenient to find the actual pairings of x_j and predicted y'_j values given by the curvilinear regression equation estimated from a sample. The experimenter may be interested in examining the departures the data tend to show from this estimated function, or perhaps in displaying in graphic form the relation suggested by the significant trends in the data. This is rather tedious to do in terms of the original x_j values, but when the x_j values are equally spaced a very simple substitute exists, involving the orthogonal polynomial coefficients once again. Instead of finding the predicted y directly as a function of x_j, we may represent y' as a function of the *polynomial weights* given to groups, and then use these predicted y values to find and plot all (x_j, y'_j) pairs.

A way to find such a prediction rule has already been given in Section 16.3. There it was pointed out that when $J - 1$ variables are orthogonal, both to each other and to the general mean, each such variable V_k has a regression coefficient given by Eq. 16.3.6. The regression equation in terms of v values is then given by Eq. 16.3.8.

This idea applies directly to orthogonal polynomial comparisons. Thus, the linear trend comparison $\hat{\psi}_1$ can be turned into a regression weight b_1 by taking

$$b_1 = \frac{\hat{\psi}_1}{nw_1},$$ [16.15.1]

where nw_1 is equal to the sum of the squared weights. In exactly this same way the regression weight b_2 can be found for the quadratic comparison, b_3 for the cubic comparison, and so on. The regression equation employing all $J - 1$ comparisons would then be (Eq. 14.1.13):

$$y'_j = \bar{y} + \sum_{k=1}^{J-1} b_k c_{kj},$$ [16.15.2]

where c_{kj} is the weight given to group j in the comparison for trend k. However, it is not very useful to employ the b values for *all* of the $J - 1$ trends in setting up this equation. The reason is that, if we do so, the predicted value for any individual in any group j will simply be $\bar{y}_j$, the mean for that group.

Instead, what we do is to take only those trends that have proved significant in the original analysis, and then use these comparison values to build the regression equation.

Thus, if we let g stand for one of the set G of *significant* trends, then the prediction rule of interest will be

$$y'_{j \cdot G} = \bar{y} + \sum_{g \in G} \frac{\hat{\psi}_g c_{gj}}{nw_g}.$$ [16.15.3*]

The set G could, of course, be defined in any other appropriate way, although generally it is of most interest to use only those trends which we have some confidence do exist in the population.

After the original comparisons have been carried out using the orthogonal polynomial weights, it is a fairly simple matter to construct this multiple regression equation, as we shall now see.

Recall first of all that for this example we concluded that *both* linear and quadratic trends exist. From the evidence at hand, these trends should be relatively strong in the population, and so we will construct a function relating y'_j to x_j which reflects *only* linear and quadratic trends. First of all, we take the value already found for the linear comparison,

$$\hat{\psi}_1 = -120,$$

and divide by

$$nw_1 = 10 \times 7$$

to find

$$b_1 = \frac{-120}{70} = -1.71.$$

Next we take the value for the quadratic comparison,

$$\hat{\psi}_2 = -150$$

and

$$nw_2 = 10 \times 8.4 = 84,$$

to find

$$b_2 = \frac{-150}{84} = -1.79.$$

Then our prediction rule for y'_j, *in terms of the polynomial weights* c_{gj} is

$$y'_j = \bar{y} + b_1 c_{1j} + b_2 c_{2j}$$

$$= 29 + (-1.71)c_{1j} + (-1.79)c_{2j}.$$

Now suppose that we want to predict for the group with the *lowest* value of x_1. A look at Table VII (Appendix E) for $J = 6$ shows that in the comparison for linear trends the mean of this group is weighted by -5, and in the comparison for quadratic by $+5$, so that

$$c_{11} = -5$$

$$c_{21} = 5.$$

For this group, the predicted value of y_j' in terms of linear and quadratic trends is thus

$$y_1' = 29 - (1.71 \times -5) - (1.79 \times 5)$$

$$= 28.60.$$

Our best guess is that an observation in the lowest x-value group will have a y score of about 28.60.

Now we predict for x_2, the second lowest value of x. Here, in the comparison for linear trends the weight assigned is -3, and for quadratic, -1. Thus

$$y_2' = 29 - (1.71 \times -3) - (1.79 \times -1)$$

$$= 35.92.$$

Therefore, we predict the mean of the second group to be 35.92.

Going on in the same way for the other groups, we find the following functional pairings of x_j and *predicted* y_j' values:

for x_1, $y_1' = 28.60$

for x_2, $y_2' = 35.92$

for x_3, $y_3' = 37.87$

for x_4, $y_4' = 34.45$

for x_5, $y_5' = 25.66$

for x_6, $y_6' = 11.50$.

These (x_j, y_j') are plotted in Figure 16.15.1 together with the actual $(x_j, \bar{y}_j)$ pairings for the sample. The actual data points are shown as the heavy dots, and the fitted curve is drawn *as though* this function obtained over the whole range of X values. Notice that although the fit between the predicted values and the sample means is not perfect (since each mean point departs somewhat from its predicted position on the regression curve), the general form of the curve does more or less parallel the systematic differences among the sample means. By way of contrast, the predictions using the linear regression equation alone correspond to points falling on the dotted line; it is apparent that adding a quadratic component to the prediction equation greatly improves our description of the relationship between X and Y exhibited in this sample, since the various means depart considerably from the regression line. On the other hand, if we added cubic, quartic, and quintic terms to our prediction equation, then we would find a curve on which the mean points for each group would fall exactly. There is not much profit in this, however, as the best description of a set of data is a curve fitting the points most nearly with a *simple* function rule (one of the lowest possible degree). Any set of means from J sampled groups can be fitted with a curve of at most degree $J - 1$.

The goodness of fit of this curve to the y_{ij} values is of course reflected in the proportion of the total variance accounted for by these trends:

$$R_{y \cdot \mathbf{G}}^2 = R_{y \cdot 12}^2 = \frac{\text{SS linear} + \text{SS quadratic}}{\text{SS total}} = .65$$

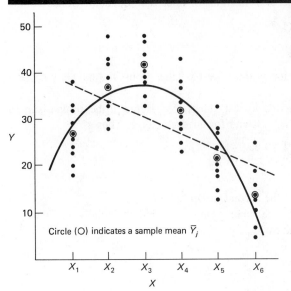

Figure 16.15.1

Linear and quadratic functions fitted to the data of Section 16.12

Circle (O) indicates a sample mean $\bar{Y}_j$

Thus, the multiple correlation $R_{y \cdot 12}$ is about .81. The curve consisting only of linear and quadratic trends appears to fit the data very well indeed.

This device for actually obtaining predicted points using a curvilinear regression equation is quite easy to apply after a trend analysis, since most of the computations have already been done. It is useful as a way to visualize how the kind of trend suggested by the significant comparisons might actually appear in a plot, as well as a device to dramatize the predictive strength, or the weakness, of the relation inferred from the data. Remember, however, that such an obtained regression curve is actually only a description of *these* data. It can be regarded as an estimate of the population regression function only for hypothetical populations showing *exactly the same representation* of X values as used in the sample. Free and easy extrapolation outside these values is done at the user's peril!

An additional word of caution: curve fitting by use of orthogonal polynomials is but the simplest approach to the problem of fitting a function to a set of data. A number of other approaches to this problem exist. Merely because we can reproduce the general form of relation in the data by, say, a function of the third degree, it does not follow that the true relation is best expressed in these terms. It may well be that another, and perhaps simpler, form of function rule best describes the relation, with parameters that can be linked to the social or psychological factors involved. In particular, when one has definite ideas about the mathematical form of the relation, and wishes to estimate parameters for this function, other methods are called for. The book by Lewis (1960) gives a useful introduction to the general subject of curve fitting for various purposes. Some important modern ideas on this whole subject may be found in Mosteller and Tukey (1977) and in Draper and Smith (1981).

16.16 LOOKING AT CURVILINEAR REGRESSION IN OTHER WAYS

At this point the question may have occurred to you, ''Why not investigate trends for some variable X by simply including different powers or other functions of X as separate variables in a multiple regression analysis?'' This can certainly be done when, for example, one suspects that a simple second degree, or quadratic, function of variable X is an important predictor of dependent variable Y, as in

$$y' = a + b_1 x + b_2 x^2.$$

Then X may be introduced twice into the regression equation, once as

$$X_1 = X.$$

and a second time as

$$X_2 = X^2$$

This will certainly introduce a quadratic term into the regression equation, which may be all that is desired. However, this may introduce multicollinearity as well, since X and powers of X tend to be very highly correlated.

Furthermore, for expressions involving higher powers of X, the interpretation of the results may not be as clear as with orthogonal polynomials, where the quadratic trend clearly represents that variation in Y values that is unrelated to any linear regression on X, the cubic trend represents variation that has nothing to do either with linear or quadratic regression, and so on for the higher order polynomials. This is not always easy to see clearly in the results of multiple regression simply based on powers of variables, especially when powers greater than two are used, or when more than one basic X variable is involved.

In short, powers of variables may be included as new variables in multiple regression analyses under certain conditions in order to represent nonlinear relationships. However, if carried beyond the very simplest examples, risks such as multicollinearity are introduced, and the results may lack the clarity given by orthogonal polynomials.

We have also confined our attention here to curvilinear relations that are expressed as polynomials in some variable X. However, there are many other curvilinear relationships that for good theoretical or empirical reasons are known to have forms such as

1. $y = (b/x) + c,$
2. $y = ae^x$

to list but two of the simplest possibilities.

Such curvilinear expressions can often be changed into linear relationships by simple transformations of X and Y into new variables U and V, where U and V are linearly related. Thus, for example, the relation given by the first possibility above can easily be changed into

$$u = bv + c,$$

where $u = y$ and $v = 1/x$, where the new variables shown on the left- and right-hand sides of this equation are in a linear relationship (though no longer necessarily in units with the same spacing, of course.). Similarly, possibility 2 can be coverted to

$$u = \log_e a + v$$

where $u = \log_e y$ and $v = x$.

When the apparent curvilinear relationships in the data can be traced back to simple forms such as these, then appropriate transformations of the data should be considered before linear regression methods are employed. Once again, good sources of further information on such transformations are Mosteller and Tukey (1977) and Draper and Smith (1981). Modern computer programs make it easy to try out different transformations in order to find those that may convert apparent nonlinear relations into forms that at least approach linearity.

16.17 TREND ANALYSIS IN POST HOC COMPARISONS

It should be mentioned that the trend comparisons can perfectly well be made *after* the overall analysis of variance has shown significant evidence for effects of the experimental variable. In particular, it may be informative to investigate trends of various degrees after significant evidence for overall curvilinear regression has been found. The Scheffé (1959) method discussed in Chapter 11 applies to any post hoc comparison among means, and one can use the orthogonal polynomial weights in the usual way to make post hoc trend comparisons. However, just as for any set of post hoc comparisons, the probability of our detecting a true trend as a significant result is ordinarily less in the test for a post hoc than in the test for a planned comparison for that trend. Even so, post hoc trend comparisons provide a most useful device for exploring data after the overall F for between groups has shown significance, when the independent variable groupings are equally spaced quantitative distinctions. Unless definite questions exist about particular trends prior to accumulating the data, tests for trend are usually made by the post hoc method. Estimation of the function can then be carried out as in Section 16.15.

16.18 TREND ANALYSIS IN MULTIFACTOR DESIGNS

As we have seen, factorial designs sometimes include one or more factors that are quantitative in character, so that the successive levels of one of these factors represent successively greater amounts of some treatment or condition. Such factors may be accompanied by other factors having the usual qualitative interpretation of the levels that constitute them. Alternatively, all of the factors may be quantitative. In either circumstance, the machinery of trend analysis through orthogonal polynomials translates directly into the analysis of multifactor data.

First let us consider the simple situation where a design involves two factors, one of which (say Factor A with J Levels) is *quantitative*. Although the general theory applies to levels representing values that are not necessarily equally spaced, for convenience we will restrict our discussion once again to equally spaced levels of A. Let us then also think of another Factor B, which has K *qualitatively* distinct levels. The balanced factorial design defined by these two factors then consists of the JK cells formed from each level of A combined with each level of B, and with n cases assigned at random to each such cell. Each case in the experiment then produces a value of the dependent variable Y.

If the quantitative nature of Factor A is ignored, the results of such an experiment look exactly like those from any other $J \times K$ design. That is, the essential breakdown of the total sum of squares is

   SS A with $J - 1$ degrees of freedom
   SS B with $K - 1$ degrees of freedom
   SS A $\times$ B interaction, with $(J - 1)(K - 1)$ degrees of freedom
   SS error, or within cells.

However, Factor A, being quantitative, may be broken down into any or all of the $J - 1$ possible trends in relation to Y. Generally, if one goes to the trouble to build a quantitative factor into a design, especially where the different levels represent equal increments in value, then the experimenter usually has some special questions he or she wants answered about trends. There may also be planned comparisons bearing on qualitative Factor B as well, although these may not be as characteristic of such a design as those asked about the quantitative factor.

An interesting new feature of such a design lies in the ability to break the interaction sum of squares into parts that reflect the *consistency of a trend* over the levels of the other, qualitative, factor. This can be done in any of several ways, but one of the simplest is as follows:

1. Work out $K - 1$ orthogonal comparisons among the levels of Factor B. These can be comparisons of genuine interest to the investigator, or they can be quite arbitrary, just as long as they are orthogonal (the Helmert comparisons of Section 11.12 can be useful here). Call any comparison out of this set of $K - 1$ the comparison v. For the moment we will symbolize the weight for level B_k in comparison v by $d_{(v)k}$.

2. Now find the orthogonal polynomial weights for all trends of interest on Factor A. This will very likely include the linear trend, and the second- and third- order trend. (On occasion you may wish to include even higher order trends). Here the weight that gets assigned to level A_j in any trend comparison u is symbolized by $c_{j(u)}$.

3. Follow the procedure of Section 12.10 to calculate the interaction comparisons and sum of squares, based on the products of the weights in each trend comparison u of Factor A and each comparison v of Factor B. That is, for each pair of comparisons u and v, and for each cell $A_j B_k$ of the table find the weight $g_{j(uv)k}$ by taking the product

$$g_{j(uv)k} = c_{j(u)} \times d_{(v)k} \qquad\qquad [16.18.1]$$

Then the interaction or product comparison (uv) is calculated over all the cell means, by taking

$$\hat{\psi}_{uv} = \sum_j \sum_k g_{j(uv)k} \times \bar{y}_{jk} \qquad\qquad [16.18.2^*]$$

and

$$SS\,(\hat{\psi}_{uv}) = \frac{[(\hat{\psi}_{uv})^2]}{\left[\sum_j \sum_k g^2_{j\,(uv)k}/n\right]}.$$ [16.18.3*]

There will, of course, be $(J - 1)\,(K - 1)$ possible product, or interaction, comparisons, and each will have one degree of freedom. Adding the sums of squares for all of these comparisons yields, of course, the SS_y A × B interaction.

However, in this sort of problem, with quantitative A and qualitative B, it is usually more interesting to take a given trend, such as linear, or $u = 1$, and then to add the sums of squares over all the B comparisons v to get

$$SS_y\,B \times \text{trend } u = \sum_v SS_y(\hat{\psi}_{uv})$$ [16.18.4*]

Accordingly, you could have SS_y B by linear trend, SS_y B by quadratic trend, and so on, each with $K - 1$ degrees of freedom. Each one of these interaction sums of squares shows how the trend in question *varies* over the different levels of B. A small, nonsignificant, interaction implies that the trend, if any, is stable over the B groupings; large and significant interactions say that the regression coefficient for the trend changes over the different groups.

As an illustration, think of a factorial design in which Factor A (or rows) stands for different numbers of hours of practice by subjects on a learning task. Let's say that the numbers of hours go in even steps from 1 to 4, with independent groups receiving the different amounts of practice. However, the subjects are made to practice in the presence of distractions. The other factor might then be the type of distraction encounterd during practice, with three qualitatively different levels:

B_1 = auditory only, B_2 = visual only, B_3 = auditory and visual.

The dependent variable would be a measure of performance based on the practice. There are n subjects per cell. One question of interest might, then, be the extent to which the relation between A and Y tends to differ under different sorts of distraction (B). This would call for an examination of B by trend interactions, perhaps as follows:

Source	SS	df	MS	F	p
(Factor A)		(3)			
linear		1			
quad.		1			
other		1			
Factor B		2			
(A × B)		(6)			
B × linear		2			
B × quad		2			
B × other		2			
Error		$12(n - 1)$			
Total		$12n - 1$			

Then large and significant trends for Factor A would show the form of relation that apparently exists between hours of practice and performance, where performance is averaged over the three types of distraction. The Factor B row in the ANOVA table simply reflects the tendency of performance to be different over the three distractions, as averaged over the hours of practice. However, the various A × B interaction components show the tendency of the trends, linking practice to performance, to be different over the different B levels. Thus, the B × linear row in the ANOVA table compares the different B levels for their linear trend linking A with performance, the B × quad compares the B levels for the quadratic trend of A with performance, and so on. Since a fixed-effects model is assumed to be true here, each of these mean squares may be tested by dividing by MS error and referring the result to an F distribution with the appropriate number of degrees of freedom.

The problem is handled in much the same way when both of the factors in the design are quantitative, with equal spacing of levels. Now, the J levels of Factor A are examined with up to $J - 1$ orthogonal polynomial comparisons, and the K levels of Factor B are also investigated for trends of up to degree $K - 1$. Then, much as discussed above, the actual comparison weights for A and B are multiplied to find interaction comparisons. Thus, with $J = 4$ and $K = 3$ levels in the two quantitative factors, for Factor A we can find

SS   A linear
SS   A quadratic
SS   A cubic

in the usual way, through orthogonal polynomials. In addition, we can find

SS   B linear
SS   B quadratic

since there are only three levels of B. Finally, there are six degrees of freedom associated with *trend interactions* as follows:

SS   A linear × B linear
SS   A linear × B quadratic
SS   A quadratic × B linear
SS   A quadratic × B quadratic
SS   A cubic × B linear
SS   A cubic × B quadratic

These interaction comparisons and their 1 degree of freedom sums of squares are found exactly as indicated above and in Section 12.10.

The meaning of a large and significant "A-linear by B-linear" interaction goes something like this: The tendency of Factor A to relate in a linear way to Y *changes* across the levels of B, and the pattern of these changes is itself linear. Thus, if you look at the linear regression coefficient for Y predicted from Factor A, these regression coefficients are linearly related to the B values. In geometric terms these relationships can be pictured as a *plane surface* in three or more dimensions, so that a linear by linear interac-

Figure 16.18.1 (1)

Schematic representation of a linear by linear interaction.

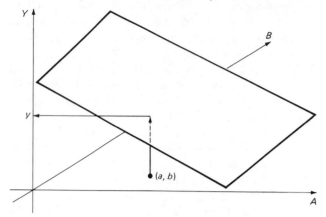

tion implies a plane surface something like Figure 16.18.1(1). Here if Y values were plotted against (a, b) values for Factors A and B, the result would tend to be a regular plane surface, as illustrated in the figure.

On the other hand, a large and significant "A-linear by B-quadratic" interaction implies that the linear relation between A values and Y changes across changing values of Factor B. Geometrically, this suggests a ridge or "hump" on an otherwise plane surface, something like Figure 16.18.1(2), or, perhaps a trough on such a plane as in 16.18.1(3). A quadratic by quadratic trend would be analogous to a "peak" or to a "crater" in a plane surface, or perhaps even a "saddle" such as illustrated in Figure

Figure 16.18.1 (2)

Schematic representation of linear by quadratic trend interaction.

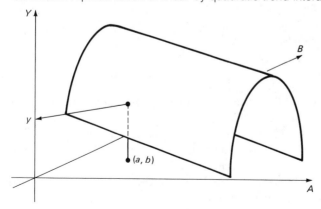

Figure 16.18.1 (3)

Another possibility for linear by quadratic interaction.

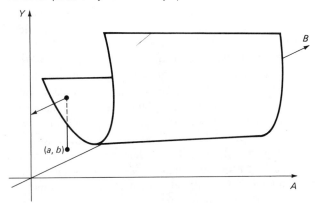

16.18.1.(4). Other, more complex patterns of interaction can also be interpreted in to-pological or geometric terms that yield insights of theoretical or practical value. One reason to investigate such interactions and the surfaces they represent is to help in locating optimal combinations of values of factor levels so as to produce certain desired results in terms of Y. Unfortunately, we will be unable to consider these possibilities more fully here. Suffice it to say that a well-developed and very useful branch of statistics called *response surface analysis* takes its point of departure from just such concepts as we have just been discussing. An introduction to some of these ideas is contained in Snedecor and Cochran (1980).

Figure 16.18.1 (4)

Possible schematic form of quadratic by quadratic interaction.

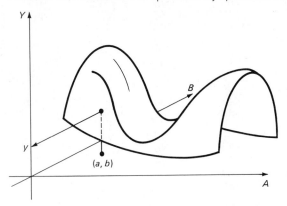

16.19 AN EXAMPLE OF A TWO-FACTOR DESIGN WITH TREND ANALYSIS

A study was carried out in a program designed to help people stop smoking cigarettes. This two-week program had two aspects: an intensive series of group and individual therapy sessions designed to break the smoking habit, and a course of medication to reduce the craving for nicotine. The study was carried out on five groups of subjects: B_1 consisted of 16 men who regularly smoked 20 cigarettes a day, group B_2 smoked 30 cigarettes, B_3 smoked 40, B_4 smoked 50, and B_5 smoked 60 cigarettes a day. Within each group, four men received 1 unit of medication (A_1), four received 2 units (A_2), four received 3 units (A_3), and a final four received 4 units (A_4). The medications were assigned in a completely "blind" fashion. After the two-weeks' program was concluded, during which none of the men had been permitted to smoke, and after the termination of the medication, each man filled out a rating scale reporting the intensity of his current desire to resume smoking. The men's self-ratings on this 100-point scale were treated as the dependent variable Y. The raw data for this study are shown in Table 16.19.1

Let us assume that from the outset it was the investigator's purpose to examine the trends that relate Factor A to Y and Factor B to Y, and that, in addition, it was the intention to look at trend interactions in these data. This implies that the following sets of orthogonal polynomials are to be used for columns (A), for rows (B), and for inter-

Table 16.19.1

Ratings Given by 80 Men in the Antismoking Study (Means in Parenthesis)

	$B_1 = 20$ (42)	$B_2 = 30$ (61)	$B_3 = 40$ (52.06)	$B_4 = 50$ (46)	$B_5 = 60$ (63.75)
$A_1 = 1$ (56.85)	58	47	39	44	73
	40	89	45	47	76
	31	83	69	47	70
	40	65	75	38	61
	(42.25)	(71)	(57)	(44)	(70)
$A_2 = 2$ (55.15)	49	70	76	43	76
	37	55	76	43	70
	46	67	55	40	55
	40	76	4	58	67
	(43)	(67)	(52.75)	(46)	(67)
$A_3 = 3$ (50.85)	47	56	57	55	55
	35	95	42	52	61
	32	56	45	34	64
	53	17	54	43	64
	(41.75)	(56)	(49.5)	(46)	(61)
$A_4 = 4$ (49)	44	50	40	57	63
	38	50	85	42	48
	32	65	16	33	54
	50	35	55	60	63
	(41)	(50)	(49)	(48)	(57)

(grand mean $\bar{y} = 52.96$)

actions. (In Table 16.19.2 the symbol R_1 stands for the linear comparison for rows, R_2 is the quadratic comparison for rows, etc. In the same way C_1 is the linear trend comparison for columns, C_2 the quadratic comparison, etc. Then R_1C_1 stands for the row-linear by column-linear interaction, R_1, C_2 is the row-linear by column-quadratic comparison, and so forth. Although 19 such comparisons are possible, only R_1 and R_2, along with C_1, C_2, and C_3 and their interactions will be listed.)

Thus, for instance, the linear trend for the A rows is evaluated by using the R_1 orthogonal polynomial weights on the cell means, as follows:

$$\hat{\psi}_{A\ lin.} = (-3 \times 42.25) + (-1 \times 43) + (1 \times 41.75) + (3 \times 41) + \ldots$$

$$+ (-3 \times 70) + (-1 \times 67) + (1 \times 61) + (3 \times 57) = -139.25$$

so that

$$SS_y\ A\ lin. = (139.25)^2/(100/4) = 775.62,$$

where 100 is the sum of the squared weights over all the cells. In the same way, the other comparisons for trends and interaction may be carried out. Thus for R_1C_1 we have

$$\hat{\psi}_{lin. \times lin.} = (6 \times 42.25) + (2 \times 43) + (-2 \times 41.75) + (-6 \times 41) + \ldots$$

$$+ (-6 \times 70) + (-2 \times 67) + (2 \times 61) + (6 \times 57) = -6$$

Table 16.19.2

Comparison weights for data of Table 16.19.1

cell	R_1	R_2	C_1	C_2	C_3	R_1C_1	R_1C_2	R_1C_3	R_2C_1	R_2C_2	R_2C_3
A_1B_1	−3	1	−2	2	−1	6	−6	3	−2	2	−1
A_2B_1	−1	−1	−2	2	−1	2	−2	1	2	−2	1
A_3B_1	1	−1	−2	2	−1	−2	2	−1	2	−2	1
A_4B_1	3	1	−2	2	−1	−6	6	−3	−2	2	−1
A_1B_2	−3	1	−1	−1	2	3	3	−6	−1	−1	2
A_2B_2	−1	−1	−1	−1	2	1	1	−2	1	1	−2
A_3B_2	1	−1	−1	−1	2	−1	−1	2	1	1	−2
A_4B_2	3	1	−1	−1	2	−3	−3	6	−1	−1	2
A_1B_3	−3	1	0	−2	0	0	6	0	0	−2	0
A_2B_3	−1	−1	0	−2	0	0	2	0	0	2	0
A_3B_3	1	−1	0	−2	0	0	−2	0	0	2	0
A_4B_3	3	1	0	−2	0	0	−6	0	0	−2	0
A_1B_4	−3	1	1	−1	−2	−3	3	6	1	−1	−2
A_2B_4	−1	−1	1	−1	−2	−1	1	2	−1	1	2
A_3B_4	1	−1	1	−1	−2	1	−1	−2	−1	1	2
A_4B_4	3	1	1	−1	−2	3	−3	−6	1	−1	−2
A_1B_5	−3	1	2	2	1	−6	−6	−3	2	2	1
A_2B_5	−1	−1	2	2	1	−2	−2	−1	−2	−2	−1
A_3B_5	1	−1	2	2	1	2	2	1	−2	−2	−1
A_4B_5	3	1	2	2	1	6	6	3	2	2	1

so that

$$\text{SS lin.} \times \text{lin.} = 6^2/(200/4) = 0.72,$$

with $200 =$ the sum of the squared weights, and so on for all other comparisons.

Each of these trend comparisons is then applied to the means of the cells of the table, and the SS value found just as shown. The summary for this analysis is shown in Table 16.19.3.

Here it can be seen that a pronounced linear trend exists for Factor A. The larger the dosage of the nicotine-curbing medication that a subject received the less tended to be his preceived need to continue smoking. No other significant trend emerged for Factor A.

On the other hand, not only was there a significant linear trend for Factor B (the more cigarettes ordinarily smoked, the stronger the desire to keep smoking), there was an even more significant curvilinear trend of order 3, or cubic form, in the data. A curious reversal of the apparent overall linear trend seems to have occurred for groups B_3 and B_4, where both exhibit means that are lower than those for groups B_2 and B_5.

Finally, in the interaction trend comparisons the linear by cubic interaction of trends was found to be significant. This indicates that the tendency for there to be a linear relation between Factor A levels and Y *changes* in a cubic way across changing levels of B. This interaction finding is illustrated in Figure 16.19.1, showing, for each of the B levels, the value of the linear comparison for the A means within that B. Thus, within level B_1, the four means were 42.25 for A_1B_1, 43 for A_2B_1, 41.75 for A_3B_1, and 41 for A_4B_1. The value of the linear comparison among those four means is then

$$\hat{\psi} \text{ linear A in } B_1 = (-3 \times 42.25) - (1 \times 43)$$

$$+ (1 \times 41.75) + (3 \times 41) = -5.$$

This is the first value plotted, for B_1, in Figure 16.19.1, and the remaining values are found in this same manner. You can see from this figure that the linear trend values for

Source	SS	df	MS	F	P	
(Factor A)	(801.24)	(3)				**Table 16.19.3**
linear	775.62	1	775.62	3.23	<.05	Summary table for data in Table 16.19.2
other	25.62	2	12.81	—	—	
(Factor B)	(5,606.95)	(4)				
linear	1,299.60	1	1,299.60	5.42	<.025	
quadratic	0.16	1	0.16	—	—	
cubic	4,284.90	1	4,284.90	17.86	<.01	
other	22.29	1	22.29	—	—	
(A × B int.)	(941.45)	(12)				
lin. × lin.	0.72	1	0.72	—	—	
lin. × quad.	3.89	1	3.89	—	—	
lin. × cubic	898.88	1	898.88	3.75	<.05	
other	37.96	(9)	4.22	—	—	
Error	14,393.25	60	239.89			
Total	21,742.89	79				

Figure 16.19.1

Plot of the linear comparison values for the A means within the different B levels, and showing linear by cubic interaction.

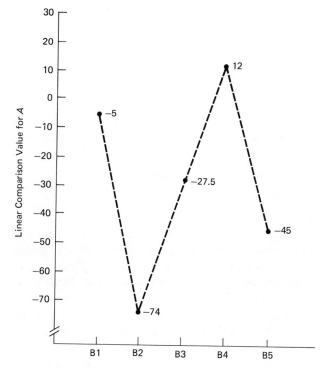

the A means tend to change across the B levels, and the way in which these linear trends change itself suggests a cubic trend.

Indeed, the SS for linear by cubic interaction can itself be found from these five linear comparison values, weighted and summed using the orthogonal weights for a cubic trend. That is

cubic comparison among linear values for As within Bs =
$(-1 \times -5) + (2 \times -74) + (0 \times -27.5)$

$$- (2 \times 12) + (1 \times -45) = -212$$

SS cubic among linear =

$$\frac{212^2}{[(3^2 + 1^2 + 1^2 + 3^2)(1^2 + 2^2 + 2^2 + 1^2)]/4}$$

$$= 898.88$$
$$= \text{SS linear} \times \text{cubic}$$

Even though the linear trend was significant for Factor A, or levels of medication, the proportion of Y variance accounted for by linear regression on Factor A is only

$$r^2_{\text{linear on } A} = 775.62/21{,}742.89 = .036$$

or just under 4%. The variance accounted for by linear regression on the levels of B is then

$$r^2_{\text{linear on } B} = 1{,}299.6/21{,}742.89 = .060$$

However, the largest proportion of variance is accounted for by the cubic trends linking Y with the levels of B:

$$r^2_{\text{cubic on } B} = 4{,}284.9/21{,}742.89 = .197$$

or right at 20%. We also know that the variance accounted for in this sample by linear-cubic interaction is

$$r^2_{\text{lin} \times \text{cubic}} = 898.88/21{,}742.89 = .04$$

Taken together, the significant linear and cubic trends in the data then account for a total of

$$R^2_{y \cdot \text{sign} \cdot \text{trends}} = \frac{(775.62 + 1{,}299.6 + 4{,}284.9 + 898.88)}{21{,}742.89} = .33$$

or about a third of the total variance of the Y values.

Such a finding would doubtless send the investigator back to take another hard look at the data. In circumstances such as these there are no very compelling reasons leading one to expect such a curvilinear relationship, and particularly a strong cubic trend, between intensity of smoking behavior and desire to continue smoking. Instead, one could, perhaps, interpret this as a signal that, by itself, the number of cigarettes smoked per day may not be a very good indicator of the degree of the person's nicotine addiction. It may be, for example, that in this study the men in the groups smoking relatively few cigarettes tended to smoke higher nicotine brands than some of the other groups who, ostensibly, were heavier smokers. Therefore, the real factor B of interest, degree of nicotine addiction is itself not necessarily shown by the number of cigarettes smoked by a man. This is just a simple illustration of one of the possible by-products of a trend analysis: unexpected and unexplained trends should send you back to the data looking for possible other circumstances and factors that may have been operating in the study, and thereby producing the apparently complex results.

16.20 ANALYZING TRENDS IN REPEATED MEASURES DESIGNS

An important application of trend analysis through orthogonal polynomials occurs in repeated measures or within-subjects designs, such as those discussed in Section 13.21. Thus, for example, studies of learning with practice, or of change in performance in time, are often oriented toward possible curvilinear relationships in the data. A factor of interest in such studies is usually "trials," which we will here suppose to be J experimental occasions equally spaced in time. When defined in this way the factor "trials" can usually be treated just like any other fixed-effects factor.

Furthermore, as pointed out in Section 13.21, when each of n randomly selected subjects receives each level of a fixed factor, and when the assumptions of the model are met, comparisons may be planned and carried out in the usual way, *except* that the mean square for subject by treatment interaction is used in the test of each of the planned comparisons among the means of the factor levels. Therefore, planned comparisons for trend over the J trials may be carried out using the orthogonal polynomials as the source of the comparison weights, and the mean square for "subject by trial" rather than "subject by treatment" interaction as the denominator term in the F tests.

On the other hand, the procedure just outlined *does* depend on validity of the circularity assumption, which we discussed in Section 13.22. In instances in which the circularity assumption does not hold, then special methods should be employed for obtaining the mean square error term used in the F tests for the orthogonal polynomial comparisons. The same considerations hold, of course, for post hoc comparisons for trend, as noted in Section 13.25. The texts by Kirk (1982) and by Edwards (1985) are good resources in these circumstances.

It should be apparent that use of a repeated measures design for the study of trends over trials generally rests on the assumption that the same experimental conditions exist for each subject on each trial. That is, of course, not always the situation, and often different subjects get different treatments on different trials, or the same subject may even receive different treatments across the trials. There are a number of ways that such experimental arrangements may be worked out, of course, but the main interest of the experimenter will usually focus on the question, "Do the different experimental treatments produce the same trends over the trials, or do significant treatment by trend interactions exist?"

The analysis of experimental data such as these combines features of the repeated measures analysis discussed in Chapter 13 and the mixed quantitative–qualitative setup outlined in Section 16.18. Exposition of this type of experimental setup, along with the appropriate ANOVA breakdown and tests may be found in most standard texts on experimental design, and we will devote no further space to it here. An unusually good treatment of this problem is given in Edwards (1985).

16.21 COMPUTER PROGRAMS FOR UNBALANCED DESIGNS AND FOR TREND ANALYSIS

As indicated previously in this chapter, it is possible to employ regression programs to carry out the analysis of variance, and this possibility underlies several of the modern approaches to the analysis of unbalanced experimental designs. However, in practice it is usually far more convenient to rely on the options offered in the computer statistical packages for handling this sort of problem.

In SPSS Releases 7–9 (Hull & Nie, 1981), the ANOVA program permits the analysis of unbalanced designs by the use of OPTION 9, which is the regression approach discussed in Section 16.6. On the other hand, one may instead choose OPTION 10, which yields the hierarchical approach of Section 16.5. In SPSS^x the same options exist. The default is called the "classic experimental approach." Under this approach, each factor is adjusted for all other factors, and each interaction is adjusted for all main effects and interactions of the same or lower order. (See Herr, 1986, for a critique of these approaches in the statistical packages). The MANOVA subprogram in each package also permits either a regression or a hierarchical approach. The first is the default option, and the second is tied to the order of factors and interactions specified in a DESIGN = subcommand.

In the BMDP series, unbalanced designs may be handled through the program P2V, using a format similar to that already discussed in connection with Chapter 12. The regression approach as outlined here is routinely applied in P2V (The considerably more sophisticated program P4V may also be applied to unbalanced designs.)

The SAS program PROC GLM provides four options for treating unbalanced designs in ANOVA, of which two are routinely utilized. These are designated as

TYPE I SS, which is equivalent to our hierarchical method, (Method H), in which the order of factors and interactions as listed in the MODEL statement dictates the hierarchy (any interaction should be listed in order following the main effects for the factors involved in that interaction; thus, A * B must follow both A and B, and so forth.)

TYPE II SS, which is equivalent to the classic regression approach mentioned above.

The TYPE III SS is similar to TYPE II SS, but with special restrictions on the factors or interactions for which adjustment may be made at any step. The TYPE IV SS can be thought of as a variation on TYPE III SS. (See the chapter on "estimable functions" in the SAS manuals for a further explanation of these distinctions.)

Trend analysis is also very easy to apply in the computer packages. The simple SPSS subprogram ONEWAY for J independent groups uses the keyword POLYNOMIAL to call for orthogonal polynomial comparisons. The highest order trend that can be investigated is the *smaller* of 5 or $J - 1$. The subprogram MANOVA, however, uses the instruction TRANSFORM = POLYNOMIAL to transform one or more of the variables, and thus can be employed for trend analysis in factorial designs as outlined in this chapter, as well as in other types of designs. Either equal or unequal spacing of categories may be specified. The P2V program, like other programs in the BMD "P" series, also allows full use of trend analysis, with either equal or unequal spacing of values.

The SAS package allows the widest latitude in the treatment of nonlinear relations, both through the use of polynomial regression as touched on here, and through a special program for nonlinear regression. The novice user will almost surely need help in making his or her way through the many possibilities available, however.

EXERCISES

1. The data given below show $N = 16$ cases arranged equally in four independent groups. The value on dependent variable Y is shown for each case. Show these same 16 cases in terms of three dummy variables, X_1, X_2, and X_3, and the dependent variable Y. Calculate all of the correlations among the X and the Y variables for these data.

	Group		
A_1	A_2	A_3	A_4
26	22	26	25
19	29	13	32
24	30	20	33
27	23	25	29

2. Take the correlations among the X and the Y variables found in Exercise 1, and perform a multiple regression analysis. Show that the sum of squares for regression on the X variables is equal to the SS between in ANOVA for these data, and that SS residual or SS deviations from regression equals the SS within or SS error.

3. Carry out a hierarchical analysis on the correlations from Exercise 1, and for the hierarchical order of X_1, X_2, and X_3 find the gain in proportion of Y variance accounted for at each step.

4. Take the data in Exercise 1 and carry out the following Helmert comparisons on the means:

	A_1	A_2	A_3	A_4
comp. 1	3	-1	-1	-1
comp. 2	0	2	-1	-1
comp. 3	0	0	1	-1 .

Then show that if the sum of squares for the first comparison is divided by SS total, the result will correspond to the gain due to X_1, as found in Exercise 3; similalry, the sum of squares for Comparison 2 divided by SS total will equal the gain due to X_2, and so on.

5. The data below come from Exercise 21 of Chapter 11, except that here they have been arranged according to the three dummy variables representing membership among the four groups, and the dependent variable Y. Find the matrix of correlations for this set of four variables as shown in these data. Then carry out a multiple regression analysis and show that

$$SS_y \text{ between} = SS_y \text{ total } (R^2_{y \cdot A})$$

just as explained in this chapter.

Subjects	Variables			
	A_1	A_2	A_3	Y
1	1	0	0	10
2	1	0	0	8
3	1	0	0	7
4	1	0	0	6
5	1	0	0	11
6	0	1	0	12
7	0	1	0	9
8	0	1	0	7
9	0	1	0	13
10	0	1	0	12
11	0	1	0	13
12	0	0	1	9
13	0	0	1	19
14	0	0	1	16
15	0	0	1	18
16	0	0	1	11
17	0	0	1	10
18	0	0	1	13
19	0	0	0	11
20	0	0	0	10
21	0	0	0	14
22	0	0	0	16

6. Test the significance of the squared multiple correlation found in Exercise 5. Show that this is exactly the same F test value as would be found in ANOVA for these same data.

7. In Exercise 21 of Chapter 11, three comparisons were shown that also apply to the data in Exercise 5 above. Carry out a hierarchical multiple regression analysis on the correlations found from these data, and show that if only variable A_1 is included in the regression equation, the proportion of Y variance accounted for is the same as that accounted for by Comparison 1 in Exercise 21 of Chapter 11. Furthermore, show that if both variables A_1 and A_2 are included, the proportion of Y variance explained is the same as for the first *and* second comparisons in Exercise 21 of Chapter 11.

8. Below is a 2 by 2 factorial design, with $n = 4$ cases per cell, for a total $N = 16$ independent cases. The value on dependent variable Y is shown for each case. Translate this set of data into a dummy variable format, using variable X_1 for the rows or Factor A, X_2 for the columns or Factor B, and X_3 for the interaction variable X_1 times X_2. Find the correlations among these variables, and use hierarchical multiple regression to show the gain in proportion of Y variance accounted for by the introduction of variable X_1, of X_2 and by X_3.

	B_1	B_2
A_1	68	79
	63	71
	69	75
	60	78
A_2	45	29
	47	42
	61	36
	51	37

9. Use the results of Exercise 8 to demonstrate that the proportion of variance accounted for by introduction of X_1 is identical to SS A divided by SS total; that the proportion of variance accounted for by X_2 is identical to SS B divided by SS total; and that the proportion of additional variance accounted for by X_3 is the same as SS A × B/SS total.

10. Below are data from a two-factor design with $N = 15$, but with different numbers of cases in the four cells. Analyze this design by use of the regression approach outlined in this chapter.

	B_1	B_2
A_1	10	23
	15	24
	19	32
	16	
A_2	14	17
	24	19
	28	18
	20	20

11. Analyze the data in Exercise 10 by use of the hierarchical approach outlined in this chapter, where the effects are examined in the hierarchical order of A, B, and A × B interaction. Compare the obtained results with those from Exercise 10.

12. Suppose that the four cells of the 2 by 2 factorial design in Exercise 10 had been treated simply as four independent groups of cases and the orthogonal comparison weights shown below the dotted line were found. Actually evaluate these comparisons, and find the sum of squares for each. Then show that the sum of squares for Comparison 1 is equal to the SS A found in Exercise 11, that the sum of squares for the second comparison is equal to SS B in that exercise, and that the third sum of squares equals to SS A × B as found in Exercise 11. In other words, show that the hierarchical method is equivalent to a set of orthogonal comparisons for unequal groups as found in Appendix C.

	A_1B_1	A_2B_1	A_1B_2	A_2B_2
$n_j =$	5	3	3	4
mean:	14.80	24.00	26.33	18.50
	. . . .			
A_1 vs. A_2	1.667	−1.143	1.000	−1.524
B_1 vs. B_2	1.094	1.000	−1.094	−1.000
A × B	1.000	−1.000	−1.000	1.000

13. A study used small rodents to investigate the possible influence of infantile deprivation on the tendency of the adult animal to be physically active. The total of 72 newborn experimental animals used in the study were assigned at random and independently to 24 cells, with three animals per cell. Each cell in this 4 by 6 factorial represented a combination of a level of Factor A (rows), consisting of four ages (in weeks) at which the animals began to be deprived of their normal food allocation. Factor B (columns) consisted of six levels of deprivation, ranging from -5, or five food units below normal, to 0, indicating a normal diet. As adults the animals were given activity scores, and this served as the dependent variable. The mean activity score for each cell is shown below. Assuming that the SS total for this set of data was 3,097.43, analyze these results along the lines of the example in Section 16.8.

	B_1	B_2	B_3	B_4	B_5	B_6
	-5	-4	-3	-2	-1	0
A_1 4	20	19	15	14	12	10
A_2 3	19	23	22	29	29	17
A_3 2	24	27	23	21	18	16
A_4 1	26	20	22	15	13	11

14. Look at the sleep deprivation data in Exercise 8, Chapter 10. Carry out an analysis using orthogonal polynomials to investigate possible trends in these data.

15. Form the regression equation for predicting group means on the basis of the significant trends in the data in Exercise 14. How close do your predictions come to the actual group means?

16. In a study of factors in alcoholism, samples of U.S. cities were taken, and the rate of verified alcoholism over the past 10 years in each found. The sizes of the cities ranged from 10,000 through 109,000, and the cities were classified into five class intervals according to population size, as shown below. Five cities were selected at random from each population grouping. The average alcoholism rate per 1,000 population for each city over the past 10 years is shown below:

Alcoholism rate, per thousand

	$x =$ population, in thousands			
10–29	30–49	50–69	70–89	90–109
17.57	19.08	18.75	20.07	24.74
16.35	18.02	17.99	19.01	23.30
17.37	18.89	18.78	20.16	24.09
16.26	18.03	18.08	19.29	22.86
14.19	17.26	18.59	18.48	23.90

Assuming that each city in a group falls at the midpoint of its interval, test for linear, quadratic, cubic, and quartic trends. What would you conclude about the relation of alcoholism rate and size of city, based on these samples?

17. Based on the significant trends found in Exercise 16, find the curvilinear regression equation. Use the method of orthogonal polynomials to find the required equation. Plot the regression equation.

18. Make a freehand extrapolation of the regression curve found in Exercise 17 in order to guess at the alcoholism rate for a city with 1,000,000 population. Does this rate seem particularly reasonable? What does this illustrate about extrapolating findings beyond the range actually sampled in such studies?

19. Why is the discussion of the statistical theory of curvilinear regression limited to "regression problems" (as defined in Chapter 14) rather than "problems of correlation"?

20. Show that the polynomial weights for linear, quadratic, and cubic comparisons are orthogonal when, say, the number of experimental groups is 7. Attempt to explain why the highest possible trend is of the sixth degree when there are only seven sample means.

21. Suppose that in an experiment such as that in Exercise 14, the investigator is specifically interested in the presence or absence or a particular trend (say, the quadratic). The others are of secondary interest and can be saved for post hoc examination. How might the analysis of variance summary table then appear?

22. For the data in Exercise 2, Chapter 14, carry out tests for linear and for curvilinear trends. (**Hint:** see Section 14.16.)

23. For Exercise 2, Chapter 14, estimate the proportion of variance accounted for by linear trends. Estimate the proportion of variance accounted for by curvilinear trends. Which sort of trend appears to be more pronounced in these data?

24. Take the data in Exercise 13, and carry out a two-way analysis using orthogonal polynomials. Investigate linear and quadratic trends both for the A and B factors, as well as the four interactions involving linear and quadratic trends.

25. For the data in Exercise 13 as analyzed in Exercise 24 above, find the sample proportion of variance accounted for by each trend and interaction of trends.

26. Reanalyze the data in Exercise 13, but this time investigate linear and curvilinear trends for Factor A, and linear and curvilinear trends for Factor B, without going into the specific nonlinear trends. Similarly look only at linear by linear interaction, and relegate the other interactions to a term called simply "other." Now, complete the analysis on this basis. Is there much difference from the results of Exercise 24?

Chapter **17**

The Analysis of Covariance

In this chapter we will discuss an important technique that represents an extension of the fundamental ideas of the analysis of variance. This technique is *the analysis of covariance,* or ANCOVA, which provides a type of ''after-the-fact,'' statistical control for one or more variables that may have affected the data resulting from an experiment. The object is to find out what the analysis of variance results might have been like if these other variables actually had been held constant.

Uses of ANCOVA have become more common in recent years due to the wide availability of the statistical packages for the computer, which make this method extremely easy to apply. On the other hand, ANCOVA is a very sophisticated statistical technique, which needs to be applied carefully and under the right circumstances. Even though the examples we will treat are very simple ones as compared with those one can handle on a computer, the principles at work here are the same as in the more complicated situations. It is, therefore, hoped that even the fairly brief and low-level coverage that will be provided here will be of assistance to you in knowing what to expect and in understanding what is going on within the rather mysterious-seeming computer programs.

17.1 THE ANALYSIS OF COVARIANCE AS STATISTICAL CONTROL

Whereas the usual analysis of variance applies to experimental factors treated qualitatively, and regression analysis typically deals with quantitative variables, the analysis of covariance represents a link between these two approaches. In a simple analysis of covariance, there is at least one experimental factor, A, divided into J levels, and there is also a quantitative dependent variable Y, just as in the analysis-of-variance situation. In addition, however, there is yet another quantitative variable X which represents a source of variation in Y, over and above that due to the experimental factor A. The different levels of Factor A may also differ in terms of the average values of X that they display. When this is true, the differences in mean Y found among the levels of A may be due not only to genuine A effects, but also to regression of Y on X.

Ideally, the problem of such a "nuisance variable" X would be avoided by arranging the experiment so that the groups standing for differing levels of A have exactly the same mean values on X. Indeed, as we saw in Sections 13.18 through 13.21, the main reasons for doing experiments in the form of randomized blocks or repeated measures designs is to achieve control over such nuisance variables and to lower our estimates of error variability by eliminating the influence of such factors.

On the other hand, in many experiments use of constant controls, or control of nuisance variables through blocking or repeated measures, is simply not feasible, experimentally, or economically. What is needed instead is a way to adjust the levels of the experimental factor A *after the fact,* so that the results may be interpreted *as though* nuisance variable X really had been held constant. In the following it will be useful to distinguish between the qualitative experimental factor (the indpendent variable) and the quantitative nuisance factor, X. The first we will call simply the experimental factor; the second we will call the *concomitant variable* or the *covariate*. The dependent variable Y is of course quantitative.

A typical use of the analysis of covariance is exemplified in the following situation: Suppose that an experiment were planned on the relationship between different methods of teaching students how to program in FORTRAN for the computer and their performance at the end of a year's instruction. Although originally none of the students knows how to program, it is clear that they vary in general intelligence. It might be possible to match the students on intelligence prior to the experiment, and then analyze the results accordingly. An alternative strategy is not to match the students, but rather to measure the general intelligence of each and to adjust for this factor by use of the analysis of covariance. This technique permits the experimenter to adjust the results after the fact, in such a way that performance differences among the different treatment groups, due to the linear relationship of performance and intelligence, are effectively removed from consideration.

Generally speaking, the analysis of covariance permits a post hoc, *statistical control* for one or more concomitant variables, removing their influence from the comparison of groups on the experimental factor(s). Naturally there are some important qualifications to the use of this procedure, and it is not so universally useful as it might at first appear. In particular, this method is effective only where the relationship between the concomitant and the dependent variable is linear, and where the degree of this relationship does not itself depend on the experimental variable. Nevertheless, in situations in which its requirements are satisfied, the use of the analysis of covariance can be quite effective in helping to clarify statistical relationships.

The model for simple analysis of covariance is yet another version of the general linear model, first introduced in Section 10.2. Let A be the main, qualitative experimental factor, represented by J distinct levels. Each observation i within any group j yields two score values. The first is a value y_{ij} on the dependent variable; the second is a value x_{ij} on the concomitant variable. Then the linear model that is assumed to hold is

$$y_{ij} = \mu + \alpha_j + \beta_{Y \cdot X}(x_{ij} - \mu_X) + e_{ij}. \qquad [17.1.1^*]$$

That is, y_{ij} depends on the general mean μ, plus an effect due to treatment j; it also depends on the value of x_{ij}, as weighted by the linear regression coefficient $\beta_{Y \cdot X}$. There is also a random error component e_{ij}. We assume further that only a linear relationship exists between X and Y, and that the value of $\beta_{Y \cdot X}$ does not depend on j. In other words, within each treatment population, the relationship between X and Y is linear and has the same regression coefficient $\beta_{Y \cdot X}$. The other, usual, analysis-of-variance assumptions are also made.

17.2 PARTITIONING SUMS OF SQUARES AND SUMS OF PRODUCTS

Like the analysis of variance, the analysis of covariance rests on the partition of the sum of squares total for the Y values, defined in the usual way:

$$SS_y \text{ total} = \sum_j \sum_i (y_{ij} - \bar{y})^2$$

$$SS_y \text{ between} = \sum_j n_j (\bar{y}_j - \bar{y})^2$$

and

$$SS_y \text{ within} = \sum_j \sum_i (y_{ij} - \bar{y}_j)^2$$

$$= SS_y \text{ total} - SS_y \text{ between} .$$

In addition, the analyis of covariance demands the partition of the X total sum of squares in exactly the same way:

$$SS_x \text{ total} = \sum_j \sum_i (x_{ij} - \bar{x})^2$$

$$SS_x \text{ between} = \sum_j n_j(\bar{x}_j - \bar{x})^2$$

and

$$SS_x \text{ within} = \sum_j \sum_i (x_{ij} - \bar{x}_j)^2$$

$$= SS_x \text{ total} - SS_x \text{ between}.$$

Finally, the name *analysis of covariance* comes about from the fact that we actually partition *sums of cross-product* terms representing covariances (just as we partition SS terms representing variances in analysis of variance). This is done by methods already used in Sections 12.23 and 14.2, as follows: We define the *total sum of products*, or SP_{xy} total, by

$$SP_{xy} \text{ total} = \sum_j \sum_i (x_{ij} - \bar{x})(y_{ij} - \bar{y}) . \qquad [17.2.1^*]$$

Notice that this is simply N times the sample covariance, $\text{cov}(x,y)$, based on all of the cases, without regard to groups.

The *sum of cross-products between groups* or SP_{xy} between is next defined by

$$SP_{xy} \text{ between} = \sum_j n_j(\bar{x}_j - \bar{x})(\bar{y}_j - \bar{y}) .$$ [17.2.2.*]

If every case were given x and y values equal to the X and Y means of the group to which that case belongs, then N times the covariance would be equal to SP_{xy} between. Then, just as for the sums of squares, a SP_{xy} within can be found:

$$SP_{xy} \text{ within} = \sum_j \sum_i (x_{ij} - \bar{x}_j)(y_{ij} - \bar{y}_j)$$

$$= SP_{yx} \text{ total} - SP_{xy} \text{ between} .$$ [17.2.3*]

In short, the basic computations for ANCOVA are actually threefold:

1. the SS values for an analysis of variance carried out on X
2. the SS values for an analysis of variance carried out on Y
3. the SP values for an analysis of covariance carried out on XY products.

A little later we will find it convenient to discuss the SS and SP values in the form of a rectangular array, or *matrix*, of numbers. Such a matrix will be called an *SSCP matrix*, since it will contain both SS (sum of squares) and SP (sum of cross products) values. Thus, for the simple analysis of covariance, we will have three SSCP matrices:

$$\textbf{SSCP total} = \begin{bmatrix} SS_x \text{ total} & SP_{xy} \text{ total} \\ SP_{yx} \text{ total} & SS_y \text{ total} \end{bmatrix}$$

(Notice that the two SS values, for X and for Y, are placed in the main left-to-right diagonal of the matrix, whereas the SP value is written down twice, once in each of the off-diagonal cells, since $SP_{xy} = SP_{yx}$.)

There will also be an *SSCP between* matrix of values:

$$\textbf{SSCP between} = \begin{bmatrix} SS_x \text{ between} & SP_{xy} \text{ between} \\ SP_{yx} \text{ between} & SS_y \text{ between} \end{bmatrix}$$

and a matrix *SSCP within:*

$$\textbf{SSCP within} = \begin{bmatrix} SS_x \text{ within} & SP_{xy} \text{ within} \\ SP_{yx} \text{ within} & SS_y \text{ within} \end{bmatrix}$$

Then if we symbolize the addition of equivalent cells of two or more matrices using a plus $(+)$ sign, it is true that

$$\text{SSCP total} = \text{SSCP between} + \text{SSCP within} .$$ [17.2.4]

Now we will use these concepts in arriving at "adjusted" sums of squares for Y, as though the X values had actually been held constant.

17.3 FINDING THE ADJUSTED SUMS OF SQUARES

One commonly used rationale for the analysis of covariance goes like this: If we look at all of the data, the values of Y will contain variation that is due to variable X as well as to linear regression on Factor A. The proportion of this total Y variance that is due to variable X must then be reflected in the ordinary squared Pearson correlation r_{xy}^2, based on the total set of cases without regard to the A groups. The variance that is *not* accounted for by regression on variable X will be

variance of Y not accounted for by regression on $X = S_Y^2(1 - r_{xy}^2)$

or, in terms of the sum of squares,

$$SS_y \text{ adjusted total} = NS_y^2 (1 - r_{xy}^2) = SS_y \text{ total} - \frac{(SP_{xy} \text{ total})^2}{SS_x \text{ total}}. \qquad [17.3.1]$$

Bear in mind that this Y variance that is not accounted for by regression on variable X *does* include variance attributable to Factor A.

On the other hand, it is possible to hold the effects of Factor A quite literally constant by looking at variance only *within* the A groups. Then, within the groups, we can find the sum of squares for X, the SS for Y, and the sum of XY products, or SP within. Observe that a correlation based strictly on these "within" quantities does indeed hold the effects of Factor A constant, and is, therefore, under our assumptions, equivalent to a partial correlation, holding A effects constant:

$$r_{yx \cdot A} = \text{correlation within the groups}$$

$$= \frac{SP_{xy} \text{ within}}{\sqrt{(SS_x \text{ within})(SS_y \text{ within})}}. \qquad [17.3.2]$$

Therefore, it is true that

$$r_{yx \cdot A}^2 (SS_y \text{ within}) = (SP_{xy} \text{ within})^2/(SS_x \text{ within}). \qquad [17.3.3]$$

The variation within the groups, adjusted to remove that portion due to X as well as that due to Factor A, is thus reflected in

$$SS_y \text{ adjusted error} = SS_y \text{ within } (1 - r_{yx \cdot A}^2). \qquad [17.3.4]$$

Finally, the variation that is due to A effects, but adjusted for variable X, must then be shown by

$$SS_y \text{ adjusted means} = SS_y \text{ adjusted total} - SS_y \text{ adjusted error}. \qquad [17.3.5]$$

Achieving this last sum of squares, showing how Factor A relates to Y after the levels of Factor A have been equated for variable X, is the main point of the analysis of covariance. However, this translates into practical computational terms as follows:

$$SS_y \text{ adjusted means} = SS_y \text{ adjusted total} - SS_y \text{ adjusted error}$$

$$= SS_y \text{ total} - SS_y \text{ total } (r_{yx}^2) - SS_y \text{ within} + SS_y \text{ within } (r_{y(x \cdot A)}^2)$$

$$= SS_y \text{ between groups} + \frac{(SP_{xy} \text{ within})^2}{SS_x \text{ within}} - \frac{(SP_{xy} \text{ total})^2}{SS_x \text{ total}}.$$

$$[17.3.6^*]$$

In addition, we already have

$$SS_y \text{ adjusted error} = SS_y \text{ within} - \frac{(SP_{xy} \text{ within})^2}{SS_x \text{ within}}.$$

$$[17.3.7^*]$$

These three sums of squares, for adjusted total (Eq. 17.3.1), for adjusted means (Eq. 17.3.6), and for adjusted error (Eq. 17.3.7), are actually the main results to come out of a one-way analysis of covariance, with a single covariate X.

The reasoning behind this method may perhaps be seen a little more clearly in the terminology of multiple regression. We know already from the discussion in Section 16.1 that

$$R^2_{y \cdot A} = SS_y \text{ between groups}/SS_y \text{ total}.$$

$$[17.3.8]$$

We also know that by working exclusively within groups one can find the equivalent of the part correlation of Y with variable X, holding possible group effects constant:

$$r^2_{y(x \cdot A)} = \text{squared part correlation of } Y \text{ with } X \text{ adjusted for A.} \quad [17.3.9]$$

Then, according to the principle of Eq. 15.23.11, the total proportion of variance accounted for either by X or by Factor A must be given by

$$R^2_{y \cdot xA} = R^2_{y \cdot A} + r^2_{y(x \cdot A)}.$$

$$[17.3.10]$$

However, what we *really* want is the proportion of variance due to Factor A adjusted for X, or

$$R^2_{y(A \cdot x)} = \text{proporation of variance due to A, adjusted for } X.$$

We note that we can also write the proportion of variance due either to A or X (again by Eq. 15.23.11) as

$$R^2_{y \cdot xA} = r^2_{yx} + R^2_{y(A \cdot x)},$$

$$[17.3.11]$$

where, you recall, r^2_{yx} is simply the proportion of Y variance due to X in the total sample without regard to A. Through a simple substitution and a rearrangement of terms we arrive at

$$R^2_{y(A \cdot x)} = R^2_{y \cdot A} + r^2_{y(x \cdot A)} - r^2_{yx},$$

$$[17.3.12]$$

so that

$$SS_y \text{ adjusted means} = SS_y \text{ total } (R^2_{y(A \cdot x)})$$

$$= SS_y \text{ total } (R^2_{y \cdot A}) + SS_y \text{ total } (r^2_{y(x \cdot A)}) - SS_y \text{ total } (r^2_{xy})$$

$$= SS_y \text{ between} + \frac{(SP_{xy} \text{ within})^2}{SS_y \text{ within}} - \frac{(SP_{xy} \text{ total})^2}{SS_y \text{ total}} \quad [17.3.13]$$

exactly as given in Eq. 17.3.6.

This little exercise in algebra shows once again the very close connection between the theory of partial and multiple regression and other highly useful techniques such as analysis of covariance.

17.4 COMPUTATIONS IN THE ANALYSIS OF COVARIANCE

A simple one-way analysis of covariance requires three sets of calculations. First, for the concomitant variable X, one finds the usual sums of squares:

$$SS_X \text{ between } = \sum_j \frac{\left(\sum_i x_{ij}\right)^2}{n_j} - \frac{(T_x)^2}{N}, \qquad [17.4.1^*]$$

where, as before

$$T_x = \sum_j \sum_i x_{ij}, \qquad [17.4.2^*]$$

and

$$SS_X \text{ total } = \sum_j \sum_i x_{ij}^2 - \frac{(T_x)^2}{N},$$

$$SS_X \text{ within } = SS_X \text{ total } - SS_X \text{ between.} \qquad [17.4.3^*]$$

The same sorts of computations are then carried out for the Y variable:

$$SS_Y \text{ between } = \sum_j \frac{\left(\sum_i y_{ij}\right)^2}{n_j} - \frac{(T_y)^2}{N}, \qquad [17.4.4^*]$$

with

$$T_y = \sum_j \sum_i y_{ij}, \qquad [17.4.5^*]$$

and

$$SS_Y \text{ total } = \sum_j \sum_i y_{ij}^2 - \frac{(T_y)^2}{N},$$

$$SS_Y \text{ within } = SS_Y \text{ total } - SS_Y \text{ between.} \qquad [17.4.6^*]$$

Finally, we must calculate sums of products of X and Y, both within and between groups. These sums of products form the basis for estimates of the regression coefficient. The calculations are

$$SP_{XY} \text{ total } = \sum_j \sum_i x_{ij} y_{ij} - \frac{(T_x)(T_y)}{N}, \qquad [17.4.7^*]$$

$$SP_{XY} \text{ between} = \sum_j \frac{(T_{xj})(T_{yj})}{n_j} - \frac{(T_x)(T_y)}{N}, \qquad [17.4.8^*]$$

$$SP_{XY} \text{ within} = SP_{XY} \text{ total} - SP_{XY} \text{ between.}$$

Then, by Eqs. 17.3.6 and 17.3.7,

$$SS_y \text{ adjusted means} = SS_Y \text{ between} + \frac{(SP_{XY} \text{ within})^2}{SS_X \text{ within}} - \frac{(SP_{XY} \text{ total})^2}{SS_X \text{ total}}, \qquad [17.4.9^*]$$

$$SS_y \text{ adjusted error} = SS_Y \text{ within} - \frac{(SP_{XY} \text{ within})^2}{SS_X \text{ within}}. \qquad [17.4.10^*]$$

The analysis of adjusted means can be summarized thus:

Source	SS	df	MS	F
Adjusted means	SS$_Y$ ad. means	$J - 1$	$\dfrac{SS_Y \text{ ad. means}}{J - 1}$	$\dfrac{MS_Y \text{ ad. means}}{MS_Y \text{ ad. err.}}$
Adjusted error	SS$_Y$ ad. err.	$N - J - 1$	$\dfrac{SS_Y \text{ ad. err.}}{N - J - 1}$	
Adjusted total		$N - 2$		

Some interesting and potentially useful by-products of an analysis of covariance are also available. Thus, for example, the Pearson correlation between the comcomitant variable X and the dependent variable Y over the entire sample is found from

$$r_{yx} = \frac{(SP_{xy} \text{ total})}{\sqrt{(SS_x \text{ total})(SS_y \text{ total})}}. \qquad [17.4.11]$$

The sum of squares that is due to the regression of Y on the variable X is then given by

$$SS_y \text{ regression on } X = SS_y \text{ total } (r_{xy}^2)$$

$$= \frac{(SP_{xy} \text{ total})^2}{SS_x \text{ total}}. \qquad [17.4.12]$$

On the other hand, the sum of squares that is due to residuals, or errors, in predicting Y from X, is equal to

$$SS_y \text{ residuals} = SS_y \text{ total } (1 - r_{xy}^2) \qquad [17.4.13]$$

$$= SS \text{ adjusted total.}$$

Therefore, a test of significance for the regression of Y on the concomitant variable X is given by

$$F = \frac{[(SS_y \text{ regression on } X)(N - 2)]}{SS_y \text{ residuals}} \qquad [17.4.14]$$

with 1 and $N - 2$ degrees of freedom. (A little study of this F ratio will reveal that it is just the square of the t test for a regression coefficient discussed in Section 14.15).

In addition, it is possible to define and calculate the average correlation *within* the various groups or factor levels. That is, we can take

$$r_w = \frac{(SP_{xy} \text{ within})}{\sqrt{(SS_x \text{ within})(SS_y \text{ within})}}$$

[17.4.15]

Since the effects of Factor A are neutralized in this correlation based strictly within the groups, and since all relationships in the data are assumed to be truly linear, then, as shown in Eq. 17.3.2, this correlation is logically equivalent to the partial correlation between X and Y, with the effects of Factor A held constant, or $r_{yx \cdot A}$.

It is the squared value of this correlation coefficient that shows how effective our adjustment of Y for X actually proves to be, since the square of this correlation represents not just variance accounted for, but, more importantly, the relative reduction in *error* variance that can be attributed to our introduction of variable X in the analysis.

Finally, it is possible to find the regression coefficient that corresponds to the correlation r_w by taking:

$$b_w = \frac{(SP_{xy} \text{ within})}{(SS_x \text{ within})}$$

[17.4.16]

This is the regression coefficient that we would wish to use to predict the Y value from X in a manner that is free of any influence of Factor A. Actually, we will be using this regression coefficient as we calculate the adjusted means in Section 17.7.

17.5 A SIMPLE EXAMPLE OF COVARIANCE ANALYSIS

In Section 17.1 a situation was described in which several different methods of teaching the FORTRAN programming language were being compared. However, the experimenter had reason to believe that general ability of the students might also affect their achievement under any given method. It was then desired to remove the possible linear effects

Table 17.5.1

Ability and performance scores for 25 subjects in five learning groups.

Groups									
1		2		3		4		5	
X	y	X	y	X	y	X	y	X	y
10	18	22	40	30	38	35	25	11	15
20	17	31	22	31	40	37	45	16	17
15	23	16	28	18	41	41	50	19	20
12	19	17	31	22	40	30	51	25	23
57	77	86	121	101	159	143	171	71	75

of general ability on programming achievement through use of the analysis of covariance.

The subjects were 20 female college freshmen who were assigned at random to five groups, each representing a different method of teaching this material. Prior to the experimental sessions, each subject was given a general ability test (score x_{ij}) and, after the training, achievement was measured (score y_{ij}). No student had prior experience with programming languages. The results are shown in Table 17.5.1 (each x and y value has been reduced by 100 for computational convenience; this has no effect whatsoever on the final analysis). The desire of the experimenter was to treat the general ability score as the concomitant variable, and to ask if differences due to method exist even after adjustment for general ability.

Then for the analysis of covariance, the following are found for the X variable:

$$SS_X \text{ total} = 12{,}066 - \frac{(458)^2}{20} = 1{,}577.80$$

$$SS_X \text{ between} = \frac{46{,}336}{4} - \frac{(458)^2}{20} = 1{,}095.80$$

$$SS_X \text{ within} = 1{,}577.8 - 1{,}095.8 = 482.00.$$

Next, for the Y,

$$SS_Y \text{ total} = 20{,}851 - \frac{(603)^2}{20} = 2{,}670.55$$

$$SS_Y \text{ between} = \frac{80{,}717}{4} - \frac{(603)^2}{20} = 1{,}998.80$$

$$SS_Y \text{ within} = 2{,}670.6 - 1{,}998.8 = 671.75$$

Finally, the sums of products are

$$SP_{XY} \text{ total} = 15{,}140 - \frac{458 \times 603}{20} = 1{,}331.30$$

$$SP_{XY} \text{ between} = \frac{60{,}632}{4} - \frac{458 \times 603}{20} = 1{,}349.30$$

$$SP_{XY} \text{ within} = 1{,}331.3 - 1{,}349.3 = -18.00$$

(note that the SP_{XY} terms *can* be negative.) The adjusted sums of squares are then

$$SS_y \text{ adjusted means} = 1{,}998.80 + \frac{(-18)^2}{482} - \frac{(1{,}331.3)^2}{1{,}577.8} = 876.16$$

$$SS_y \text{ adjusted error} = 671.75 - \frac{(-18)^2}{482} = 671.08$$

The summary table for this analysis of covariance is given on the following page.

Source	SS	df	MS	F
Adjusted means	876.16	4	219.1	4.6
Adjusted error	671.08	14	47.9	
Adjusted total	1,547.24	18		

The F value necessary for significance at the .05 level, for 4 and 14 degrees of freedom is 3.11. Thus, the hypothesis of no difference between the *adjusted* means may be rejected. Even removing the effects of general ability, the groups do seem to differ in performance.

On the other hand, had the experimenter analyzed the Y values through an ordinary one-way analysis of variance, without adjusting for general ability, then an F value of 11.16 would have been calculated for 4 and 15 degrees of freedom. Adjusting for general ability did reduce the apparent discrepancies between instructional methods.

Nevertheless, we have evidence that the inclusion of the concomitant variable X in this analysis was really quite ineffective, if the aim was to reduce error variance. Here the value of the average correlation within, r_w, was found to be only

$$r_w = \frac{(SS_{xy} \text{ within})}{\sqrt{(SS_x \text{ within})(SS_y \text{ within})}}$$

$$= \frac{-18}{\sqrt{(482)(671.75)}}$$

$$= -.032$$

so that the proportional reduction in error variance was only

$$(-.032)^2 = .001 \ ,$$

or about one tenth of one percent! Fortunately, this situation will not always exist, and considerable percentage reductions in error variance are possible.

17.6 TESTING FOR HOMOGENEITY OF REGRESSION

Perhaps the biggest hitch in the application of the analysis of covariance is the assumption that the regression coefficient is constant within each of the treatment populations. Furthermore, it is necessary to assume that the regression of X on Y is linear within each such population. Fortunately, one can examine the data to see if the first of these assumptions is at least reasonable. Let β_j be the true regression coefficient within population j. Then we wish to test the hypothesis H_0: $\beta_j = \beta$ for every j.

The test for homogeneity of regression ("parallelism") rests on a partition of the SS_y adjusted error, as found in the routine ANCOVA calculations. First we can find a new sum of squares which is associated with variation among the regression coefficients across the different groupings or levels of A:

$$SS_y \text{ between regressions} = \sum_j \frac{SP_{xyj}^2}{SS_{xj}} - \frac{(SP_{xy} \text{ within})^2}{SS_x \text{ within}} \qquad [17.6.1^*]$$

with $J - 1$ degress of freedom. Here, for any group j

$$SS_{xj} = \sum_i x_{ij}^2 - \frac{T_{xj}^2}{n_j} \qquad [17.6.2^*]$$

(just an ordinary SS for x carried out in group j), and

$$SP_{xyj} = \sum_i x_{ij}y_{ij} - \frac{T_{xj}T_{yj}}{n_j} \qquad [17.6.3^*]$$

(just the ordinary sum of products, but done within group j).
Then

$$SS_y \text{ remainder} = SS_y \text{ adjusted error} - SS_y \text{ between regressions}$$

$$= SS_y \text{ within} - \sum_j \frac{SP_{xyj}^2}{SS_{xj}} \qquad [17.6.4^*]$$

with $N - 2J$ degrees of freedom.

(If, as is really proper, you are carrying out the test for homogeneity of regression prior to doing the ANCOVA calculations, you can make good use of the fact that

$$SS_x \text{ within} = \sum_j SS_{xj} \qquad [17.6.5]$$

and that

$$SP_{xy} \text{ within} = \sum_j SP_{xyj} \qquad [17.6.6]$$

so that the SS within for x and the SP within do not have to be recalculated once you have found the SS for x and the SP for each group.)

These results for a homogeneity of regression test can be displayed in a special ANOVA table like the following:

Source	SS	df	MS	F	p
Between regressions		$J - 1$	$\dfrac{SS \text{ bet.}}{N - 1}$	$\dfrac{MS \text{ bet.}}{MS \text{ remain.}}$	
Remainder		$N - 2J$	$\dfrac{SS \text{ remain.}}{N - 2J}$		
Adjusted error		$N - J - 1$			

The F test for homogeneity of regression is carried out by forming the ratio of MS between regressions and MS remainder. Ideally, we would like this F ratio to be *non-significant* even for a large value of α, with perhaps $F = 1.00$ or less. This would mean that it is safe to assume homogeneous regression coefficients for the J different populations, and that our F test on the adjusted means is justified. In using this test, it is probably a good idea to proceed with the ANCOVA when the F test for homogeneity of regression *fails* to reach significance even at the .25 level, However, if the homogeneity of regression test reaches significance even for α of .10 or less, there may be good reason to doubt the validity of the F test in ANCOVA. (No one is going to throw you in jail if you should carry out ANCOVA when there is some small evidence for nonhomogeneous regressions. However, you are taking a bit of an extra chance of being wrong when you accept the results of the ANOVA F test and related procedures.)

As an example, we will test the homogeneity of regression for the example analyzed by ANCOVA in Section 17.5. For each group we find the SS for x and the SP values as shown below.

group j	SS_{xj}	SP_{xyj}	$(SP_{xyj})^2/SS_{xj}$
1	56.75	− 4.25	0.3183
2	141.00	−64.50	29.5053
3	118.75	−16.75	2.3626
4	62.75	6.75	0.7261
5	102.75	60.75	35.9179
totals	482.00	− 18.00	68.8302

Then

$$SS_y \text{ between regressions} = 68.8302 - \frac{(-18)^2}{482}$$

$$= 68.16$$

$$SS_y \text{ remainder} = 671.75 - 68.83 = 602.92$$

$$SS_y \text{ adjusted error} = 68.16 + 602.92 = 671.08 \ .$$

The completed table for the homogeneity of regression test is shown below.

Source	SS	df	MS	F	p
Between reg.	68.16	4	17.04	0.28	—
Remainder	602.92	10	60.29		
Adj. error	671.08	14			

The F test here represents the highly desirable situation where F fails to reach significance even for large α, meaning that we can have considerable confidence in making the assumption of no differences among the regression coefficients for the populations in question.

You will notice that the presence of nonhomogeneous regressions among a set of populations is tantamount to interaction between the covariate X and the Factor A. That is, when the regressions are not homogeneous the linear relationship between X and Y

is dependent on the level of A that is being discussed. Clearly, if Y is to be adjusted for X across the different levels of A, we do not want such an interaction to exist.

The test for homogeneity of regression is therefore framed as a test for the presence of X-by-A interaction in some of the computer packages providing ANCOVA. (See the comment at the end of Section 17.17).

17.7 CALCULATING ADJUSTED MEANS FOR THE GROUPS

Often it is of interest to calculate the values of the means of the various experimental groups after adjustment for the average X value in that group. In a sense, the adjusted means show what would have happened, "on the average," if the concomitant variable X really had been held constant.

To see how this may be done, recall the model for simple ANCOVA given by Eq. 17.1.1, or

$$y_{ij} = \mu_Y + \alpha_j + \beta_{Y \cdot X}(x_{ij} - \mu_X) + e_{ij} .$$

The part of y_{ij} that can be predicted linearly from X is given by

$$\beta_{Y \cdot X}(x_{ij} - \mu_X) . \qquad [17.7.1]$$

Hence, if the true regression due to X could be removed from each case, the mean of these values for any group A_j should represent the mean that *would have* occurred if X had been held constant.

In order to make such an adjustment we use the fact that $\bar{x}_j$ and $\bar{x}$ are unbiased estimates of population values μ_{Xj} and μ_X, and the fact that when the least squares criterion is applied to the model given above, subject to the restriction that the sum of the α_j effects must be zero, the estimate of $\beta_{Y \cdot X}$ is provided by

$$\text{est. } \beta_{Y \cdot X} = b_w = \frac{SP_{XY} \text{ within groups}}{SS_X \text{ within groups}} , \qquad [17.7.2^*]$$

exactly as given by Eq. 17.4.15 as well.

Therefore, the adjusted mean value for level j of Factor A is found from

$$\text{adjusted } \bar{y}_j = \bar{y}_j - b_W(\bar{x}_j - \bar{x}) . \qquad [17.7.3^*]$$

Such adjusted mean values will have a mean equal to $\bar{y}$, the same as the unadjusted values. However, it will *not ordinarily be true* that N times the variance of these adjusted values will equal the SS adjusted means term in the ANCOVA table, as one might be led to expect.

If we find the adjusted means for the example of Section 17.5 we come up with the following:

$$b_W = (-18)/482 = -.0373$$

Then the means can be adjusted as follows:

group y mean $-$ b_W (group x mean $-$ grand x mean) $=$ adjusted mean

group 1 $19.25 - (-.0373)(14.25 - 22.9) = 18.93$

group 2 $30.25 - (-.0373)(21.5 - 22.9) = 30.20$

group 3 $39.75 - (-.0373)(25.25 - 22.9) = 39.84$

group 4 $42.75 - (-.0373)(35.75 - 22.9) = 43.23$

group 5 $18.75 - (-.0373)(17.75 - 22.9) = \underline{18.56}$

mean $= 30.15 = \bar{y}$

In this example, the adjusted mean values do not differ radically from the original means, although minor adjustments up and down do occur. In some problems, however, rather large differences may exist between the adjusted and the original means.

17.8 POST HOC COMPARISONS IN ANCOVA

Naturally, following an analysis of covariance, principal interest will usually focus on the adjusted means. These adjusted means may be subjected to any of the usual variety of post hoc tests such as the Scheffé and the Tukey HSD.

In applications of the Scheffé method to ANCOVA situations, comparisons are first worked out for the adjusted means, corresponding to the investigator's interests in the usual way. That is, a sample comparison can be represented by

$$\hat{\psi} = c_1(\text{adj. } \bar{y}_1) + c_2(\text{adj. } \bar{y}_2) + \cdots + c_J(\text{adj. } \bar{y}_J) . \qquad [17.8.1]$$

However, since these are comparisons on *adjusted* means, the sampling variance of any comparison is different from that discussed in Section 11.2. If, in the experimental setup, the J different treatments have been assigned at random to the experimental units so that any differences among the X means over the experimental groups can also be viewed as random, then the variance of the comparison may be estimated by taking

$$\text{est. } \sigma_{\hat{\psi}}^2 = (\text{MS}_y \text{ adj. error}) \left[1 + \left(\frac{\text{MS}_x \text{ between}}{\text{SS}_x \text{ within}} \right) \right] \left(\sum_j \frac{c_j^2}{n_j} \right) . \qquad [17.8.2]$$

Hence, a t test may be carried out by taking

$$t = \hat{\psi}/\text{est. } \sigma_{\hat{\psi}} \qquad [17.8.3]$$

with $N - J - 1$ degrees of freedom, or an F test by using the square of the t ratio. However, in order to be declared significant, the absolute value of the t ratio must exceed the Scheffé criterion,

$$\mathbf{S} = \sqrt{(J - 1)F_{(\alpha, J-1, N-J-1)}} \qquad [17.8.4]$$

or, the calculated F must exceed the square of this value.

On the other hand, suppose that the different groups being compared in the data are actually preformed or otherwise intact groups chosen by the experimenter, which fact alone may be responsible for the differences of the X means among the groups. In this instance the standard error of the comparison has a different format.

Let the comparison among the adjusted Y means as shown in Eq. 17.8.1 above now be labeled as $\hat{\psi}_Y$, and let us use the symbol $\hat{\psi}_X$ to stand for the same comparison (i.e., same weights) done on the X means rather than the adjusted Y means. Then the sampling variance of the comparison among the adjusted Y means may be estimated from

$$\text{est. } \sigma_{\hat{\psi}}^2 = (\text{MS adjusted error})[\sum_j \frac{c_j^2}{n_j} + \frac{(\hat{\psi}_X)^2}{\text{SS}_x \text{ within}}] \qquad [17.8.5]$$

with this modification, then exactly the same procedure as outlined above is used for the Scheffé test.

The Tukey HSD procedure as outlined in Section 11.17 is no longer appropriate when one is dealing with pair comparisons among adjusted means. Rather, a generalization of Tukey's HSD procedure, devised by Bryant and Paulson (1976) should be used as the basis for all such pair comparisons. This requires a special set of tables, which may be found in Huitema (1980), Kirk (1982), and a variety of other sources.

17.9 USE OF ANCOVA IN PLACE OF DIFFERENCE SCORES

There are many research situations where it seems most natural to define the dependent variable as the difference between Y scores measured before and after some experimental treatment. Thus, for example, we may measure the arithmetic acheivement of a class of students, then give a special course of instruction, followed by a repetition of the same achievement test. Because we wish to see how much the students' scores have improved, presumably because of the instruction, we might think first of simply taking the difference for each student of the second or postscore, minus the first or prescore, and then using these difference scores as the dependent variable. This method of difference scores appeals to our intuition for many reasons, including its simplicity and the fact that studying differences between physical measures has been standard practice in the natural sciences since time immemorial. On the other hand, difference scores may not be the best way to equate for original standing in some areas of the social and behavioral sciences, as it generally is for physical measurements.

An alternative to using difference scores is to let the first measure or prescore be the concomitant variable X, and the second measure or postscore be the dependent variable Y. Then, through the analysis of covariance the experimental groups can be equated after the fact, *as though* each had had the same average standing at the outset. However, this is equivalent to taking difference scores as the dependent variable only under the special circumstance where the *only* sort of systematic change that can occur from the first time to the second time is due to the experimental treatment, implying that the true regression coefficient $\beta_{y \cdot x}$ is actually 1.00. Such an assumption is generally not particularly reasonable for social or behavioral measures, and therefore it is considered good practice to treat change through ANCOVA rather than through the use of difference scores in these areas. The next example, to be discussed in Section 17.12, deals with such a study of before-and-after change through use of ANCOVA.

17.10 ANCOVA FOR A FACTORIAL DESIGN

The general ideas of analysis of covariance or ANCOVA apply directly to the situation where the data have been collected in the format of a factorial design. As you will recall from Section 12.1, in a factorial design there are two or more factors that are crossed, so that each of the possible combinations of the factor levels is represented in the data, and balanced, meaning that each combination contains the same number of cases, n. In the ANCOVA situation there is also a concomitant variable, or covariate, X in addition to the dependent variable Y. The point, as usual, is to examine the results following statistical adjustments to remove variablility due to X, or to control for X after the fact.

Suppose, as the simplest case, that we are dealing with a two-factor design, where one of the factors is denoted by A, and has J levels, and the other factor is B, with K levels. Each of the JK possible combinations of treatments constitutes a cell of the design, and each cell contains exactly n cases. Each case shows both a value of X, the covariate, and a value of Y, the dependent variable.

To carry out an analysis of covariance for such data, we begin with the same basic steps as in Section 17.2, and divide the SS_x total, the SP_{xy} total, and the SS_y total into components. However, since here we are dealing with a two-factor design, each of these totals is broken down into the usual portions corresponding to main effects, interaction, and within cells, or error:

$$SS_x \text{ total } = SS_x \text{ A } + SS_x \text{ B } + SS_x \text{ A} \times \text{B } + SS_x \text{ within cells}$$

$$SP_{xy} \text{ total } = SP_{xy} \text{ A } + SP_{xy} \text{ B } + SP_{xy} \text{ A} \times \text{B } + SP_{xy} \text{ within cells}$$

$$SS_y \text{ total } = SS_y \text{ A } + SS_y \text{ B } + SS_y \text{ A} \times \text{B } + SS_y \text{ within cells .}$$

That is, a two-way ANOVA is carried out on the X values, on the Y values, and SP total is also broken down in the analogous way. Thus,

$$SP_{xy} \text{ total } = \sum_j \sum_k \sum_i x_{ijk} y_{ijk} - T_x T_y / N \qquad [17.10.1]$$

$$SP_{xy} \text{ AB cells } = \sum_j \sum_k T_{xjk} T_{yjk} / n - T_x T_y / N \qquad [17.10.2]$$

$$SP_{xy} \text{ within cells } = SP_{xy} \text{ total } - SP_{xy} \text{ AB cells} \qquad [17.10.3]$$

$$SP_{xy} \text{ A } = \sum_j T_{xj} T_{yj} / Kn - T_x T_y / N \qquad [17.10.4]$$

$$SP_{xy} \text{ B } = \sum_k T_{xk} T_{yk} / Jn - T_x T_y / N \qquad [17.10.5]$$

$$SP_{xy} \text{ A} \times \text{B } = SP_{xy} \text{ AB cells } - SP_{xy} \text{ A } - SP_{xy} \text{ B .} \qquad [17.10.6]$$

Then the adjusted SS_y within can be found directly by taking

$$\text{adj. } SS_y \text{ error } = SS_y \text{ within cells } - \frac{(SP_{xy} \text{ within cells})^2}{SS_x \text{ within cells}} . \qquad [17.10.7]$$

The degrees of freedom for the adjusted within cells or df adjusted error $= (N - JK - 1.)$ Notice that this is one less than the error degrees of freedom for a regular two-way ANOVA.

Next the adjusted sum of squares for Factor A is found:

$$\text{adj. } SS_y \text{ A } = SS_y \text{ A } + \frac{(SP_{xy} \text{ within cells})^2}{SS_x \text{ within cells}}$$

$$- \frac{(SP_{xy} \text{ A } + SP_{xy} \text{ within cells})^2}{SS_x \text{ A } + SS_x \text{ within cells}} . \qquad [17.10.8]$$

As you can see, here the role played by SP_{xy} total and SS_x total in the adjustment for a one-factor design (Eq. 17.3.6) is now played by the sum $(SP_{xy} A + SP_{xy}$ within) and the sum $(SS_x A + SS_x$ within). Otherwise, we run the risk of overadjusting by taking away variability due to Factor B and A $\times$ B interaction, as well as the unwanted X variability.

In a similar way the adjusted SS for Factor B is found from

$$\text{adj. } SS_y B = SS_y B + \frac{(SP_{xy} \text{ within cells})^2}{SS_x \text{ within cells}}$$

$$- \frac{(SP_{xy} B + SP_{xy} \text{ within cells})^2}{SS_x B + SS_x \text{ within cells}} . \qquad [17.10.9]$$

Finally, for the adjusted interaction sum of squares we take

$$\text{adj. } SS_y A \times B = SS_y A \times B + \frac{(SP_{xy} \text{ within cells})^2}{SS_x \text{ within cells}}$$

$$- \frac{(SP_{xy} A \times B + SP_{xy} \text{ within cells})^2}{SS_x A \times B + SS_x \text{ within cells}} . \qquad [17.10.10]$$

The degrees of freedom for adjusted A, adjusted B, and adjusted A $\times$ B sums of squares are all exactly the same as the degrees of freedom in the usual analysis of variance for two factors.

The general rule for adjustment, which holds for a factorial design of any size is this: For any type of effect (or interaction),

$$\text{adj. } SS_y \text{ effect} = SS_y \text{ effect} + \frac{(SP_{xy} \text{ within cells})^2}{SS_x \text{ within cells}}$$

$$- \frac{(SP_{xy} \text{ effect} + SP_{xy} \text{ within cells})^2}{SS_x \text{ effect} + SS_x \text{ within cells}} . \qquad [17.10.11^*]$$

Furthermore,

$$SS_y \text{ adjusted error} = SS_y \text{ within cells} - \frac{(SP_{xy} \text{ within cells})^2}{SS_x \text{ within cells}} . \qquad [17.10.12^*]$$

The degrees of freedom for the adjusted SS for any effect or interaction is always identical to the corresponding degrees of freedom used in ANOVA. The adjusted SS_y within cells, or adjusted error has degrees of freedom given by $N - $ cells $- 1$.

The summary table for such an ANCOVA follows the same pattern as the ANOVA summary table for a factorial design (Section 12.6), except that adjusted sums of squares are presented. This will be illustrated in Section 17.12.

17.11 FINDING ADJUSTED MEANS FOLLOWING FACTORIAL ANCOVA

If one desires, adjusted means can be found following ANCOVA for a factorial design, much as adjusted means may be determined following a one-way ANCOVA. Generally, one will be interested in the *adjusted cell means* in such a design, and these can be found as follows.

Let $\bar{y}_{jk}$ stand for the mean in cell A_jB_k for a two-factor design. In addition let $\bar{x}_{jk}$ symbolize the cell mean on the covariate X, and let $\bar{x}$ and $\bar{y}$ stand for the grand means on X and Y respectively. Then let

$$b_w = SP_{xy} \text{ within cells}/SS_x \text{ within cells} .$$ [17.11.1*]

The adjusted mean for cell A_jB_k is then

$$\text{adj. } \bar{y}_{jk} = \bar{y}_{jk} - b_w (\bar{x}_{jk} - \bar{x})$$ [17.11.2*]

Once the adjusted means for the cells are determined, then any other adjusted means (row, columns, etc.) can be found from them, and they may also be used as the basis for post hoc comparisons, as suggested in Section 17.8 above.

17.12 AN EXAMPLE OF ANCOVA IN A TWO-FACTOR DESIGN

Suppose that a study was carried out on the attitudes of junior-high school students toward careers in science. Some $N = 30$ junior-high school students were selected at random. Of this total group, once-half were girls (Group A_1) and the other half boys (Group A_2). Each child was given one of two parallel forms of a test of attitude toward a career in science, in which a high score reflected a favorable attitude. Then three groups of girls were formed at random, and similarly three groups of boys were created. One group of boys and one group of girls saw a series of professionally prepared videotapes on science careers. This was Treatment B_1. A second group of boys and a similar group of girls heard a series of live lectures on careers in science, given by junior-high science teachers. This was treatment Group B_2. A third group among the girls and among the boys heard a series of live presentations by active scientists in different fields, and this constituted Treatment B_3. Following each series of presentations, each child filled out the second form of the attitude questionnaire.

The main question in this study concerned the extent to which the children tended to change their attitudes as a result of the treatments they received. In particular, there was interest in any differential tendency of girls and boys generally to change, in differences among types of presentation, and in the possible interaction between the treatments and the gender groupings. The score on the pretest of attitude was treated as the covariate X, and the attitude score after the presentations was the dependent variable Y. The data that emerged from this study are shown in Table 17.12.1.

The sums of squares and the sums of products for these data were then analyzed as follows:

SS_x total = 1,124.17	SP_{xy} total = −354.00	SS_y total = 1,727.20
SS_x AB cells = 1,022.17	SP_{xy} AB cells = −378.40	SS_y AB cells = 1,540.00
SS_x within = 102.00	SP_{xy} within = 24.40	SS_y within = 187.20
SS_x A = 472.03	SP_{xy} A = 150.73	SS_y A = 48.13
SS_x B = 388.27	SP_{xy} B = −560.80	SS_y B = 1,298.60
SS_x A × B = 161.87	SP_{xy} A × B = 31.67	SS_y A × B = 193.27

	B_1		B_2		B_3	
	x	y	x	y	x	y
	18	34	21	17	16	23
	15	33	20	21	17	26
A_1	15	31	19	18	18	23
	19	34	18	24	21	26
	15	30	21	17	20	23
	16	36	30	23	31	30
	19	37	34	17	32	36
A_2	15	32	29	14	31	35
	19	33	32	12	30	35
	20	34	27	14	27	30

Table 17.12.1.

Pretest and posttest attitude scores for 30 junior-high school students

These SS and SP values are the raw materials for the adjusted SS values to be calculated next:

$$\text{adj. } SS_y \text{ within cells} = 187.20 - (24.40^2/102)$$

$$= 181.36$$

$$\text{adj. } SS_y \text{ A} = 48.13 + (24.4^2/102) - [(24.40 + 150.73)^2/(102 + 472.03)]$$

$$= .54$$

$$\text{adj. } SS_y \text{ B} = 1{,}298.60 + (24.4^2/102) - [(24.40 - 560.80)^2/(388.27 + 102)]$$

$$= 717.56$$

$$\text{adj. } SS_y \text{ A} \times \text{B} = 193.27 + (24.4^2/102) - [(31.67 + 24.4)^2/(161.87 + 102)]$$

$$= 187.19$$

The ANCOVA summary table for this example is as shown below. Notice that the ANCOVA summary table does not contain an entry for adj. SS total, as the separate adjusted components do not add up to adjusted total in this analysis.

Source	SS	df	MS	F	p
Adj. A	0.54	1	0.54	—	—
Adj. B	717.56	2	358.78	45.50	<.01
Adj. A × B	187.19	2	93.60	11.87	<.01
Adj. error	181.36	23	7.89		

Here we would conclude that after adjustment for the pretest attitude, the posttest measures do not appear to differ by the sex of the student. The different methods of presentation do seem to be having significantly different effects on attitude change, although the significant F for interaction suggests that the direction and magnitude of the effects must be qualified in terms of the sex of the student.

For this example the adjusted cell means are found as follows: First we find the regression coefficient within, or b_w

$$b_w = (SP_{xy} \text{ within cells})/SS_x \text{ within cells}$$

$$= 24.4/102$$

$$= .2392.$$

Then for cell A_1B_1, where $\bar{x}_{11} = 16.4$, $\bar{y}_{11} = 32.4$, and with the grand means $\bar{x} = 22.17$, $\bar{y} = 26.60$, the adjusted mean value for this cell is

$$\text{adj. } \bar{y}_{11} = \bar{y}_{11} - b_w (\bar{x}_{11} - \bar{x})$$

$$= 32.40 - 0.2392(16.40 - 22.17)$$

$$= 33.78.$$

Proceeding in this same way we can find the adjusted means for the other cells to be

$$\text{adj. } \bar{y}_{21} = 34.40 - 0.2392 (17.80 - 22.17)$$

$$= 35.45$$

$$\text{adj. } \bar{y}_{12} = 19.40 - 0.2392 (19.80 - 22.17)$$

$$= 19.97$$

$$\text{adj. } \bar{y}_{22} = 16.00 - 0.2392 (30.40 - 22.17)$$

$$= 14.03$$

$$\text{adj. } \bar{y}_{13} = 24.20 - 0.2392 (18.40 - 22.17)$$

$$= 25.10$$

$$\text{adj. } \bar{y}_{23} = 33.20 - 0.2392 (30.20 - 22.17)$$

$$= 31.28$$

mean of adjusted means $= 26.60 = \bar{y}$.

Since through our adjustment each cell has been placed on an equal footing with respect to initial attitude, differences in adjusted Y means can be interpreted directly as differences in attitude change. On this basis, it is apparent that in these data the greatest attitude changes occurred in connection with Method B_1, where girls changed a bit more than boys. Method B_2 was not so effective, although it seemed to be better for boys than girls. Just the reverse was true for Method B_3, where girls showed a much larger change than did boys.

As desired, on the basis of these adjusted cell means, estimated adjusted effects may be calculated, or post hoc comparisons among the adjusted cell means may be carried out, along the lines suggested in Section 17.8.

Analysis of covariance on a factorial design demands the assumption of homogeneity of regression among the populations represented by the *cells*. This assumption may be tested exactly as shown in Section 17.6, except that a *group* in this situation is a cell. Furthermore, if, as is usually the case, the x values are produced at random rather than being fixed in advance, then a bivariate normal distribution of x and y values is assumed for each population represented by a cell in the data. Such bivariate normal distributions are further assumed to have the same variance–covariance matrix throughout.

Although the application of ANCOVA has been discussed here only for factorial designs, the principle extends readily to other types of designs. The book by Kirk (1982) should be consulted for applications to other types.

17.13 ANALYSIS OF COVARIANCE WITH TWO OR MORE COVARIATES

The ideas of analysis of covariance extend in a fairly straightforward way to the situation where not just one but several nuisance variables have been measured for each case in a simple, one-factor design. Then it is desired to carry out an after-the-fact statistical adjustment on the data to equate the experimental groups on all of these concomitant variables, or covariates.

The rationale for such a covariance adjustment flows directly from the reasoning underlying simple ANCOVA. You may recall from Section 17.3 that in simple ANCOVA one way to write the sum of squares for adjusted means is

$$SS_y \text{ adjusted means} = SS_y \text{ adjusted total} - SS_y \text{ adjusted error}$$

$$= SS_y \text{ total}(1 - r_{yx}^2) - SS_y \text{ within}(1 - r_{y(x \cdot A)}^2)$$

where r_{yx} is the XY correlation over the total set of data, and $r_{y(x \cdot A)}$ stands for the part correlation of Y with X calculated within the A levels or groups.

Exactly this same principle is used when there are two or more covariates or X values involved. However, now the role of the simple correlation between X and Y is assumed by the multiple correlation $R_{y \cdot G}$, where G here stands for the entire set of K different covariates X. Furthermore, the role of the part correlation $r_{y(x \cdot A)}$ calculated within the groups is taken by another multiple correlation, symbolized by $R_{y(G \cdot A)}$, and also calculated within the A levels or groups.

That is,

$$SS_y \text{ adjusted means}$$

$$= SS_y \text{ total}(1 - R_{y \cdot G}^2) - SS_y \text{ within}(1 - R_{y(G \cdot A)}^2). \qquad [17.13.1]$$

In addition, we have

$$SS_y \text{ adjusted error} = SS_y \text{ within}(1 - R_{y(G \cdot A)}^2) \qquad [17.13.2]$$

This sounds a good bit harder than it turns out to be in practice, especially for a small number of covariates and if the sweep-out procedure of Section 15.26. is used. The actual procedure that can be employed is this:

1. Just as in simple ANCOVA, calculate the SS total, SS between, and SS within values for each X variable and for Y. Moreover, this time you will need to calculate SP total, SP between, and SP within not only for the products of X and Y values for each variable X, but also for the products of each pair of the X variables.

2. Form the SSCP matrices as discussed in Section 17.2. If there are K different covariates, then each matrix will have $K + 1$ rows and $K + 1$ columns. If there is only one Factor A then there will be three of these SSCP matrices: SSCP total, containing the SS total and SP total values for all of the X variables and Y; SSCP between containing the SS and SP values between the A groups; and SSCP within containing the SS and SP values within the groups.

3. Now take the SSCP total matrix. Carry out the sweep-out procedures (Section 14.26 and Appendix D.3) on this matrix, sweeping out all of the X variables. Then the last entry in the Y column of the final SWP matrix is equal to

$$\text{SS}_y \text{ total}(1 - R^2_{y \cdot G}) \qquad\qquad [17.13.3]$$

the adjusted total.

4. Next take the SSCP-within matrix. Perform the sweep operation repeatedly on this matrix, sweeping out the X variables until only the Y column remains. Then the bottom entry in the Y column is

$$\text{SS}_y \text{ within}(1 - R^2_{y(G \cdot A)}) \qquad\qquad [17.13.4]$$

or SS_y adjusted error.

5. Subtract the result of step 4 from the result of step 3. This gives the SS_y adjusted means.

For K covariates and J groups, the sum of squares for adjusted error has $N - J - K$ degrees of freedom. On the other hand, the degrees of freedom for the adjusted SS between remains $J - 1$, regardless of the number of covariates. The usual ANCOVA table is constructed to display the results of such an analysis, as will be shown in Section 17.14. (If you try out these steps using the SSCP-total and SSCP-within matrices for a single covariate and Y, you can easily see that you get the same adjustment procedures as described in Section 17.3).

When ANCOVA with two or more covariates is applied to data from a factorial design, the procedures remain much the same as shown above, except that instead of only three SSCP matrices you will have an SSCP matrix for each factor or interaction, as well as an SSCP matrix for within cells. Then, for each factor or interaction sum of squares to be adjusted, the role of the SSCP-total matrix is taken by a matrix formed from the sum of the SSCP matrix for that factor or interaction and the SSCP-within-cells matrix.

17.14 AN EXAMPLE WITH TWO COVARIATES

Consider a study of the effects of very small, and apparently nonintoxicating, amounts of alcohol on the retention of learned material. A sample of $N = 40$ male college freshmen was drawn, and subjects were assigned at random to four equal groups. Each man had to study a list of 100 uncommon technical terms and their definitions, and then, after a given period of time, take a test on the material studied. Motivation was maintained by an arrangement to pay the subjects at varying rates, in accordance with

Table 17.14.1

Covariate and error scores of 40 men in four experimental conditions.

A_1			A_2			A_3			A_4		
X_1	X_2	Y	X_1	X_2	Y	X_1	X_2	Y	X_1	X_2	Y
11	13	8	−3	31	17	2	36	21	7	17	19
5	15	11	7	29	12	6	34	17	4	26	31
−4	19	9	21	12	19	8	18	18	4	35	15
6	25	7	−15	25	13	12	28	22	6	43	24
18	26	10	6	17	29	−4	26	33	−8	36	27
−2	23	6	8	16	24	2	51	14	10	22	23
5	33	14	11	42	14	−6	30	26	14	19	21
−3	18	12	−2	30	23	5	23	25	−6	21	20
4	25	13	−8	26	16	11	20	23	4	20	25
9	38	17	13	35	20	12	15	21	5	11	27

their scores on the test. In Group A_1 men studied individually under fixed conditions and without ingesting any alcohol. In Condition A_2 the men studied under the same conditions, but after ingesting one small unit of alcohol. In Condition A_3 the men studied after ingesting two such units, and in A_4 each man consumed three units of alcohol before study. All subjects were deprived of food for 3 hours prior to the study period, and until after the test had been taken.

In this study two covariates were used, IQ and body weight. However, to simplify the calculations these covariates were transformed into the following: covariate $X_1 =$ (IQ − 120), and $X_2 =$ (weight in pounds − 125). The variable Y was errors made in recalling the definitions of the list of words. The resulting data are shown in Table 17.14.1.

Now given these data the first step in the ANCOVA is to find the SSCP matrices, for the total, the between groups, and the within groups. These matrices are shown below:

$$\text{SSCP total} = \begin{array}{c} \\ X_1 \\ X_2 \\ Y \end{array} \begin{bmatrix} \begin{array}{ccc} X_1 & X_2 & Y \\ 2{,}221.375 & -451.875 & -58.750 \\ -451.875 & 3{,}309.975 & -73.850 \\ -58.750 & -73.850 & 1{,}767.100 \end{array} \end{bmatrix}$$

$$\text{SSCP between} = \begin{array}{c} \\ X_1 \\ X_2 \\ Y \end{array} \begin{bmatrix} \begin{array}{ccc} X_1 & X_2 & Y \\ 9.275 & -2.175 & -44.850 \\ -2.175 & 114.475 & 223.750 \\ -44.850 & 223.750 & 951.300 \end{array} \end{bmatrix}$$

$$\text{SSCP within} = \begin{array}{c} \\ X_1 \\ X_2 \\ Y \end{array} \begin{bmatrix} \begin{array}{ccc} X_1 & X_2 & Y \\ 2{,}212.100 & -449.700 & -13.900 \\ -449.700 & 3{,}195.500 & -297.600 \\ -13.900 & -297.600 & 815.800 \end{array} \end{bmatrix}$$

Now following the steps given above, for Step 3 we carry out the sweep-out operation on the SSCP-total matrix:

$$\textbf{total} = \begin{bmatrix} 2{,}221.375 & -451.875 & -58.750 \\ -451.875 & 3{,}309.975 & -73.850 \\ -58.750 & -73.850 & 1{,}767.100 \end{bmatrix}$$

$$\textbf{SWP(1)} = \begin{bmatrix} -.2033 & -.0264 \\ 3{,}218.054 & -85.8010 \\ -85.8010 & 1{,}765.5462 \end{bmatrix}$$

$$\textbf{SWP(12)} = \begin{bmatrix} -.0318 \\ -.0266 \\ 1{,}763.2585 \end{bmatrix}$$

Hence,

SS_y adjusted total $= 1{,}763.2585$

Next we carry out the same process using the SSCP-within matrix:

$$\textbf{Within} = \begin{bmatrix} 2{,}212.100 & -449.700 & -13.900 \\ -449.700 & 3{,}195.500 & -297.600 \\ -13.900 & -297.600 & 815.800 \end{bmatrix}$$

$$\textbf{SWP(1)} = \begin{bmatrix} -0.2033 & -.0063 \\ 3{,}104.0800 & -300.4257 \\ -300.4257 & 815.7127 \end{bmatrix}$$

$$\textbf{SWP(12)} = \begin{bmatrix} -.0260 \\ -.0968 \\ 786.6363 \end{bmatrix}$$

Thus,

SS_y adjusted error $= 786.6363$

Then immediately we can find the adjusted SS between:

SS_y adjusted means $= \mathrm{SS}_y$ adjusted total $-$ SS_y adjusted error

$$= 1{,}763.2585 - 786.6363$$

$$= 976.62$$

The summary table for this completed analysis is as shown below.

Source	SS	df	MS	F	p
adj. means	976.62	3	325.54	14.07	<.01
adj. error	786.64	34	23.14		
adj. total	1,763.26	37			

Clearly, the amount of alcohol consumed seems to make a difference in the recall of the subjects, even controlling for IQ and body weight. We could, if we wished, go on to find adjusted means for the groups by setting up an unstandardized multiple regression equation of the form

$$\text{adj. } \overline{y}_j = \overline{y}_j - b_{w1}(\overline{x}_{j1} - \overline{x}_1) - b_{w2}(\overline{x}_{j2} - \overline{x}_2) \tag{17.14.1}$$

where the b_{wj} values are given by the Y column in the within SWP(12) matrix (these values are $-.0260$ and $-.0968$ for b_{w1} and b_{w2} respectively.) Tests for such unstandardized weights are also possible along lines suggested in Snedecor and Cochran (1980).

17.15 ASSUMPTIONS AND PROBLEMS IN ANCOVA

As pointed out above, the analysis of covariance demands the same set of assumptions that apply to any fixed-effects analysis of variance. In addition, as we have seen, one must assume homogeneity of regression among the various population levels represented by the groups or treatment levels, and that the residuals or deviations from the regression of Y on X are normally distributed with a mean of 0 and the same variance within each population. Tacitly we then must assume that the model of Eq. 17.1.1 is indeed correct and that the relationship between X and Y is linear within each of the populations in question. (Consult an authoritative text such a Huitema, 1980, for a discussion of homogeneity of regressions, and possible options to employ when this fails to be true for one or more covariates.)

The "classic" theory of analysis of covariance we have described here was developed from a regression model, in which the X values are viewed as fixed, whereas the Y values are random. On the other hand, it is much more likely that the data will represent a sampling situation in which both X and Y are random. This seems to present no problem as long as one is able and willing to assume a multivariate normal distribution of the X and the Y values within the populations in question (an assumption that is not always realistic in practice) or to restrict inferences to the conditional form, "Given the distribution of X values such as those in the sample, then. . . ."

Another important assumption, often overlooked, is that the X values are measured without error. The presence of errors of measurement can produce misleading conclusions, either by inflating or by obscuring the true differences among the treatment levels. The best insurance against such misleading conclusions is probably to make sure of the reliability of the measurements of X, whatever they may be. An excellent discussion of these issues will also be found in Huitema (1980).

A related type of error, common in social and behavioral sciences, is sometimes called *error of specification*. An error of specification occurs when one measures a trait or concept that is actually different in some degree from what one hopes to measure. Thus, for example, if a test that purports to measure academic achievement in some area mainly reflects "test-wiseness" on the part of subjects, then an error of specification is made.

Still another problem, similar to an issue that often arises in multiple regression, is the failure of the study to include a possible covariate that relates to Y and that exhibits important differences among the populations under study. Probably the best insurance against errors of this sort is to know your material thoroughly, and especially the research literature showing the importance of different variables in accounting for variance in the phenomena under study.

Several additional precautions apply when one is dealing with more than one covariate X. Thus, for any group of n cases, the SSCP matrix calculated within that group and divided cell-by-cell by n defines the *variance–covariance* matrix within that group (c.f., Section 13.19). Furthermore, if we were dealing with a set of populations instead of a set of samples, within each population there would be a true variance–covariance matrix.

When more than one covariate is used, then the assumption of homogeneity of regression becomes equivalent to the assumption that there is the same true variance–covariance matrix within each population.

This is, of course, a much stiffer assumption of homogeneity of regression than for a single covariate, and the assumption certainly appears harder to satisfy the larger the number of covariates. As it happens, there is a test for homogeneity of variance–covariance matrices, which is listed in many computer programs as the *Box M* test. Some authorities recommend carrying out the Box M test for within-groups variance–covariance matrices any time that two or more covariates are to be used. However, this test is overly sensitive in some contexts and must be used with caution.

In addition, most common applications of ANCOVA with two or more variates will probably fit the sampling pattern where the covariates of X values are random. Once again, such models generally demand the assumption of the multivariate normal distribution in order to justify the inferences made in the ANCOVA. This too may be a far from reasonable assumption in many contexts, and one should keep it in mind in choosing to use a fairly elaborate ANCOVA.

For reasons that should be obvious from the method that we used to find the adjusted sums of squares, the problem of multicollinearity can rear its ugly head when two or more covariates are used, and it can happen that one or more of these variables have to be eliminated before the analysis can be completed. In addition, as in any multiple regression analysis, the number of covariates must not be allowed to become excessive in the light of the total number of cases. Although a concrete rule is hard to give, the value of N should certainly be many times the number of covariates employed.

It may seem that this litany of assumptions and potential problems is presented here in an effort to discourage you from using the analysis of covariance in most practical situations. This is far from the case, however. The point is that the analysis of covariance is a very potent and sophisticated method that should not be used only as a blunt instrument to pound the data. Rather, the present easy availability of demanding methods such as this should serve as a challenge to apply the research knowledge, care, and skill that a truly competent researcher brings to a problem. The statistical method will do the job it was designed for, but only if the researcher has the background judgment to apply it in an appropriate way.

17.16 ALTERNATIVES TO ANCOVA

Analysis of covariance is not the only available technique for introducing controls into your study, of course. The method of controlling certain nuisance variables by holding them at a constant value is a feature of every truly experimental study; when all is said and done this is probably the most effective way to control for any nuisance variable. On the other hand, constant controls are often difficult and expensive for the experimenter to bring about, and each constant control does introduce a possible limitation on the generality of the conclusions that flow from any study.

Another strategy for introducing controls into an experiment is "matching", or the formation of randomized blocks, as discussed in Section 13.18. In many contexts this is a more effective method than ANCOVA for removing unwanted nuisance variation from the estimates of error variance, and thus of improving the power of tests of experimental effects. Certainly, the use of randomized blocks may involve less restrictive assumptions than ANCOVA, and the relationships between the matching (i.e., concomitant variables) and the dependent variable Y do not even have to be linear for randomized blocks to be applied. The drawback is that the formation of randomized blocks may be difficult or even impossible in some research situations where ANCOVA is still possible. The point is that both strategies are possible, in principle, and both should probably be considered in the early stages of planning an experiment, when issues of control must be faced and decided on. Even a combination of randomized blocks and analysis of covariance can be useful in some circumstances. A text such as Snedecor and Cochran (1980) can be consulted for various possibilities of combining ANCOVA with other designs.

Finally, the objectives of ANCOVA can sometimes be met through an entirely different approach, such as an analysis based on ranks rather than scores. See Huitema (1980) for details.

17.17 ANCOVA IN THE COMPUTER PACKAGES

The analysis of covariance appears as a routine option in the most commonly used programs for ANOVA in the statistical packages. Thus, in SPSS, the subprogram ANOVA permits adjustment for up to five concomitant variables. The format of the control command includes the usual ANOVA specification of *dependent variable* BY *independent variable(s)* along with their minimum and maximum values. However, for covariance analysis the keyword WITH also appears, followed by the names of one or more concomitant variables. Thus, in a situation where a single dependent variable named Y is to be analyzed for effects of Factor A with three levels and Factor B with four levels, after adjustment for numerical variables named X_1 and X_2, the ANOVA subcommand would be entered as:

ANOVA   *Y* BY *A* (1,3) *B* (1,4) WITH *X1,X2*/

The output will then give a two-way analysis of covariance, after adjustment for the two concomitant variables (or covariates) X_1 and X_2. The ANOVA procedure in SPSS[x] follows this same format, but will allow up to 10 covariates to be specified.

In unbalanced designs the same options apply as discussed in Section 16.21, in both SPSS and SPSS^x analysis of covariance procedures. Under the regression OPTION 9, each type of effect is assessed following adjustment for the covariates as well as for all other types of effects. In the hierarchical OPTION 10, effects are first adjusted for covariates and then assessed in the order specified by the factor list following the ANCOVA command.

In neither SPSS nor SPSS^x does the ANOVA program give a test of homogeneity of regression (or as the manuals describe it, "factor-by-covariate interactions"), nor does this procedure handle repeated measures or nonfactorial designs. For these purposes the subprogram MANOVA should be used. This will also provide adjusted means through use of the keyword PMEANS after the PRINT= subcommand.

Although the BMDP series provides one-way analysis of covariance in the program P1V, much more versatility is given by programs P2V and P4V, which will handle repeated measures and complicated experimental designs. Thus, in P2V if A and B stand for the names of two factors, Y is the name of a single dependent variable, and there are two covariates with names symbolized by X_1 and X_2, ANCOVA is requested through a DESIGN paragraph entered among the control statements, as follows:

/DESIGN DEPENDENT IS Y.

GROUPING ARE A, B.

COVARIATES ARE X1, X2.

This will produce the two factor ANCOVA with adjustment for two covariates. In addition, the adjusted cell means will be provided.

In the SAS package the GLM procedure permits many different types of covariance analysis. The basic command structure, for, say, dependent variable Y, one factor A, and two covariates X_1 and, X_2 would look like this

PROC GLM;

CLASS A;

MODEL Y = A X1 X2/ SOLUTION;

LSMEANS A;

This set of commands produces the analysis of variance for variable Y in terms of the effects of Factor A (this is listed under TYPE I SS). Then the analysis of covariance is given for variable Y in terms of Factor A, after adjustment for covariates X_1 and X_2 (this is listed under TYPE III SS). Finally, the means of the A levels are given, after adjustment for the two covariates. Balanced and unbalanced designs may both be subjected to ANCOVA by this procedure.

For a test of homogeneity of regression, the MODEL statement may be replaced by a statement of the form

MODEL Y = A X1 X2 X1*A X2*A X1*X2*A/SOLUTION;

This model permits examination of the homogeneity of regression for each covariate considered separately as well as jointly.

EXERCISES

1. In a study of factors related to success in medical school, the dependent variable Y was an overall numerical rating of success during the four years of study. The independent variable was a measure of the "authoritarian personality" characteristics of the individual student. Students were arranged into four groups, from very low to very high, on this personality trait. Finally a concomitant variable X was used, representing the average anxiety level of the student. Carry out an analysis of covariance on the resulting data, and draw conclusions about the relation between the independent and dependent variable when anxiety is controlled as a factor.

Authoritarianism							
Very low		Low		High		Very high	
X	Y	X	Y	X	Y	X	Y
30	23	22	17	27	21	19	23
30	20	26	20	26	20	24	28
24	20	24	19	30	24	28	32
27	19	28	23	27	19	30	36
23	15	26	18	28	20	29	33
28	20	25	22	27	19	24	30

2. Evaluate the adjusted means for the data in Exercise 1. How do the conclusions from the analysis of covariance differ from those that would be reached if anxiety were not controlled in this way?

3. Does the assumption of homogeneity of regression in Exercise 1 appear to be justified? Use $\alpha = .05$ for the test.

4. Suppose that the experiment on "level of aspiration" described in Section 12.7 had actually been carried out as a one-factor experiment, involving the three groups labeled "above average," "average," and "below average," but without the second factor. Each group contained 20 randomly chosen and assigned subjects. However, prior to the experiment, each subject had been tested on a game very similar to that used in the experiment proper, and a "skill score" X obtained for each. This was used as the concomitant variable in the experiment. The data are given below. Do differences appear to exist among the groups when the effects of the prior skill factor X are removed? Use $\alpha = .05$.

Groups					
Above average		Average		Below average	
X	Y	X	Y	X	Y
44	52	38	28	23	15
47	48	26	35	17	14
30	43	36	34	31	23
38	50	30	32	25	21
40	43	36	34	27	14
45	44	23	27	35	20
36	46	45	31	25	21
41	46	28	27	28	16
40	43	34	29	30	20
43	49	37	25	37	14
48	38	40	43	32	23
24	42	36	34	32	25
39	42	41	33	34	18
36	35	29	42	48	26
46	33	39	41	39	18
33	38	37	37	38	26
38	39	47	37	30	20
26	34	34	40	24	19
41	33	47	36	31	22
36	34	31	35	19	17

5. Complete the analysis of variance for the data in Exercise 4 ignoring the variable X. Then compare these results with those obtained in Exercise 4. How do you account for this difference in results? What would this imply about the original experiment?

6. An experiment involved different treatments given to three randomly constituted groups of subjects. In addition to the dependent variable Y, there was also a concomitant variable X. Carry out an ANOVA on the data listed below, and then follow this with an ANCOVA on the same data. How do the conclusions differ? When does this sort of difference between ANOVA and ANCOVA tend to occur, and what does it imply?

Experimental group					
A_1		A_2		A_3	
X	Y	X	Y	X	Y
31	31	37	26	44	30
27	30	33	29	47	26
33	25	34	30	49	24
30	27	39	27	45	31
33	22	30	34	39	35
26	27	37	22	46	28

7. Find the adjusted means for the data in Exercise 6. Then carry out the Scheffé tests comparing the adjusted means for the first two groups with that for the last group, and the adjusted mean for the first group to those for the last two.

8. The data shown in Exercise 4 are actually divided into six cells by the broken line running through the table; these six cells correspond to those in Table 12.7.1. Perform the analysis of covariance for this 2-by-3 factorial design, using, as in Exercise 4, the concomitant variable X.

9. Use the data in Exercise 8 to test the hypothesis of homogeneous regression among the populations represented by the six *cells* of the experimental design. (**Hint:** regard each cell as a "group" and run the test for homogeneity of regression among these groups, just as for a one-factor ANCOVA situation.)

10. An experiment compared three methods for teaching a course in statistics to upper-division college students. One group of students (A_1) was given only lectures by the professor in charge; Group A_2 attended the lectures, and also recitation sessions conducted by the teaching assistants; in Group A_3 the students studied on their own and then attended recitation sessions. Because of scheduling conflicts, the three groups had different numbers of students. The concomitant variable X here is the grade point average of the student on entering the course, and the dependent variable Y is the score on the final examination in the course. Do the methods seem to be having different effects on the final examination scores, after adjustment for initial grade point average?

A_1		A_2		A_3	
X	Y	X	Y	X	Y
2.67	71	2.67	68	2.16	62
3.29	92	2.13	62	3.15	85
3.51	84	2.03	70	2.04	84
2.73	78	2.14	63	4.00	97
3.11	70	2.77	75	2.44	78
3.22	73	3.19	61	3.20	82
4.00	96	3.73	94	3.11	75
3.18	83	2.15	65	3.16	73
2.50	75	3.70	95	3.11	85
2.14	72	3.72	86	2.17	50
2.39	78	3.51	84	3.21	81
3.15	90	3.39	70	3.59	87
		3.81	93	2.53	47
		2.71	84	3.60	87
		2.23	80	2.68	77
				3.18	82
				3.25	86

11. A study was carried out to investigate possible psychological differences between people who suffer from two different types of arthritis. Group A_1 consisted of individuals with well-established cases of rheumatoid arthritis, and Group A_2 consisted of patients with diffuse arthritic symptoms. Measures Y_1, Y_2, and Y_3 were three well-known tests of personality characteristics. Can one say that the two groups appear to be significantly different on the mean Y_3 values they tend to exhibit after adjustment for Y_1? After adjustment for Y_2?

		Measures		
		Y_1	Y_2	Y_3
Group A_1				
Patient	1	2	1	9
	2	1	7	5
	3	3	6	10
	4	3	4	15
	5	5	2	11
	6	5	6	7
	7	5	9	14
	8	7	3	14
	9	7	7	16
	10	10	10	12
Group A_2				
Patient	1	8	5	5
	2	9	8	8
	3	11	6	8
	4	13	1	9
	5	13	6	9
	6	13	10	11
	7	11	10	13
	8	15	8	13
	9	14	12	15
	10	15	8	15

12. Do the two A groups in Exercise 11 appear to be significantly different on variable Y_3 after it is adjusted for both variables Y_1 and Y_2?

13. In a study of children's ability to solve puzzles requiring a certain amount of insight for solution, a sample of 54 fourth-grade children was taken, and divided at random into 18 groups of three subjects each. Each child was given two puzzles to solve. Under Condition C_1, after the first puzzle the child was given training in how such puzzles should be approached. However, in Condition C_2 this training was not given. In experimental Condition B_1 the two puzzles were very similar to each other, in B_2 they were somewhat similar, but in B_3 the puzzles were quite differ- ent. Finally, in Condition A_1 the experiment was conducted under highly competetive conditions, in A_2 the atmosphere was moderately competetive, and in A_3 the situation was noncompetetive. The time (in seconds) required to solve the first puzzle is shown here as X and the time required for the second puzzle is shown as the dependent variable Y. Using X as the concomitant variable, carry out an analysis of covariance on these data.

		B_1		B_2		B_3	
		X	Y	X	Y	X	Y
		82	66	79	65	74	82
	A_1	75	60	81	68	70	80
		73	68	80	69	73	71
		74	63	72	60	84	74
C_1	A_2	79	65	76	64	79	80
		81	72	69	65	83	81

		87	73	70	67	80	75
	A_3	76	70	75	70	81	79
		72	71	73	73	85	80
		81	77	76	75	78	72
	A_1	83	74	75	73	77	70
		87	76	70	70	69	73
C_2		80	78	73	71	64	70
	A_2	74	84	72	74	85	75
		83	75	86	75	83	87
		81	71	74	76	82	81
	A_3	80	70	83	70	84	80
		87	72	75	74	80	78

14. For the data in Exercise 13 how would you find the adjusted means for Factor B or columns, and for Factor C or "tables"?

15. For Exercise 13 above, state the assumptions that one must make in order to carry out this analysis.

16. A sample of 40 college men kept diaries of their activities, hour-by-hour for a week. The sample was equally divided among those men classified as freshmen, sophomores, juniors, and seniors. The data below show the hours per week spent by each man on the following activities: Y_1 = classroom or lab attendance; Y_2 = study or library work; Y_3 = recreation. Would you say that the four groups are alike in average Y_3 after adjustment for Y_1 and Y_2?

Freshmen			Sophomores		
Y_1	Y_2	Y_3	Y_1	Y_2	Y_3
14	39	12	15	53	10
15	42	15	15	49	12
14	50	13	18	60	7
17	48	13	16	50	9
18	53	15	15	51	11
19	51	11	19	62	8
14	40	17	17	57	8
15	47	10	20	65	7
15	25	14	15	53	9
12	30	16	21	68	5

Juniors			Seniors		
19	73	3	20	61	3
16	60	7	15	42	8
18	72	4	17	34	10
14	53	13	18	45	7
18	68	5	15	51	5
20	71	4	16	42	15
15	50	8	17	44	12
17	65	2	23	63	2
12	41	12	17	62	3
18	67	1	18	52	9

Chapter *18*

Analyzing Qualitative Data: Chi-Square Tests

There are many research problems in which one wants to make direct inferences about two or more distributions, either by asking if a population distribution has some particular specifiable form, or by asking if two or more population distributions are identical. These questions occur most often when *both* variables in some experiment are qualitative in character, making it impossible to carry out the usual inferences in terms of means or variances. In these instances we need methods for studying independence or association from categorical data. Other situations exist, however, when we wish to ask if a population distribution of a random variable has some precise theoretical form, such as the normal distribution, without having any special interest in summary properties such as the mean.

The first topic of this chapter is the comparison of a sample with a hypothetical population distribution. We would like to infer whether or not the sample result actually does represent some particular population distribution. We will deal only with discrete or grouped population distributions, and so inferences will be made through an approximation to the exact multinomial probabilities. Such problems are said to involve ''goodness of fit'' between a single sample and a single population distribution.

Next, we will extend this idea to the simultaneous comparison of several discrete distributions. Ordinarily, the reason for comparing such distributions in the first place is to find evidence for association between two qualitative attributes. In short, we are going to employ a test for independence between attributes, which can be regarded as based on the comparison of *sample* distributions.

Also, we will deal with the problem of measuring the strength of association between two attributes from sample data. Tests and measures of association for qualitative data are very important for the social sciences, where many of the most important distinctions to be made are, essentially, categorical or qualitative in character. The methods in this chapter are widely used, both because of the kinds of data social scientists collect, and because of the simplicity of their application. However, the theory underlying these tests is not simple, and misapplication of these tests is very common. For this reason a good deal of space will be devoted to discussing some of the basic ideas underlying these methods, and some pains will be taken to emphasize their inherent limitations.

18.1 COMPARING SAMPLE AND POPULATION DISTRIBUTIONS: GOODNESS OF FIT

A study was carried out surveying the cigarette-smoking habits of American men, 40 to 50 years of age. A random sample of 863 men in this age group was taken, and the individuals were classified into the following categories according to average numbers of packages of cigarettes smoked per day (rounded up to the next whole number). The obtained frequencies in each group are shown below:

Category	Obtained frequency
A_0 = none	406
A_1 = one	164
A_2 = two	189
A_3 = three	78
A_4 = four or more	26
	$\overline{863}$ = N

The investigator also knew that 10 years ago the distribution of average numbers of packages of cigarettes smoked a day in this age group of men was as follows:

category	relative frequency
A_0	.43
A_1	.17
A_2	.24
A_3	.10
A_4	.06
	$\overline{1.00}$

Can we say on the basis of this sample that the smoking habits of American men in the 40–50 age group are the same now as they were 10 years ago? In Table 18.1.1 below the obtained frequencies are laid out once again, alongside the frequencies to be expected in each category if the present and the former population distributions are actually the same.

Category	Obtained	Expected	**Table 18.1.1**
A_0 = none	406	371.09	Obtained and expected frequencies of 863 U.S. men
A_1 = one	164	146.71	in the 40–50 age group, according to average
A_2 = two	189	207.12	numbers of packs of cigarettes smoked each day.
A_3 = three	78	86.30	
A_4 = four or	26	51.78	
more			
	863	863.00	

The last column on the right gives the *expected* frequencies under the hypothesis that the population has the same distribution as 10 years ago. For each category, the expected frequency is

$$Np_j = m_j = \text{expected frequency} \qquad [18.1.1]$$

where p_j is the relative frequency for category j dictated by the hypothesis. Notice that unlike the obtained frequencies, the theoretical expected frequencies need not be whole numbers.

How well do these two distributions, the obtained and the expected, agree? At first blush, you might think that the mean difference in frequency obtained and expected across the categories, or

$$\sum_j (f_j - m_j),$$

would describe the difference in the two distributions. However, it must be true that

$$\sum_j (f_j - m_j) = \sum_j f_j - \sum_j m_j$$
$$= N - N$$
$$= 0,$$

so that this is definitely not a satisfactory index of disagreement.

On the other hand, the sum of the *squared* differences in observed and expected frequencies does begin to reflect the extent of disagreement:

$$\sum_j (f_j - m_j)^2.$$

This quantity can be zero only when the fit between the obtained and expected distributions is perfect, and must be large when the two distributions are quite different.

An even better index might be

$$\sum_j \frac{(f_j - m_j)^2}{m_j},$$

where each squared difference in frequency is weighted inversely by the frequency expected in that category. This weighting makes sense if we consider that a departure from expectation should get relatively more weight if we expect rather few individuals in that category than if we expect a great many. Somehow, we are more "surprised" to get many individuals where we expected to get few or none, than when we get few or none where we expected many; thus, the squared departure from expectation is appropriately weighted inversely by the frequency expected in the first place, when an index of overall departure from expectation is desired.

Remember, however, that our real purpose in this example is to test the hypothesis that the true current distribution is the same as it was 10 years ago. The exact probability of obtaining any particular result for a sample of N cases, given the distribution stated in the hypothesis, can be found from the multinomial distribution rule of Section 3.19. In principle, then, one might find the probability of a result as much or more deviant from expectation as we actually obtained, and this would correspond to the significance level for rejecting the null hypothesis. On the other hand, a staggering amount of computation would be involved in doing this for N and J this large.

Rather, in situations such as this the statistican looks around for an approximation device. In this particular instance, it turns out that the multivariate normal distribution provides a way of approximating the multinomial probabilities, and this permits a solution. We will not attempt to go into this involved derivation here; suffice it to say that when the probability of any sample distribution follows the multinomial rule, and when N is very large, the following goodness-of-fit test procedure is justified:

We form the statistic

$$\chi^2 = \sum_j \frac{(f_j - m_j)^2}{m_j}$$

[18.1.2*]

where, as before, the symbol m_j stands for the *expected* frequency in category j, or $m_j = Np_j$. This is known as the Pearson χ^2 (chi-square) statistic (after its inventor, Karl Pearson). Given that the exact probabilities for samples follow a multinomial distribution, and given a very large N, **when H_0 is true, this statistic χ^2 is distributed approximately as chi-square with $J - 1$ degrees of freedom** (see Section 9.1). Probabilities arrived at using this statistic are *approximately* the same as the exact multinomial probabilities we would like to be able to find for samples as much or more deviant from expectation as the sample obtained. The larger the sample N, the better should this approximation be.

Note that the number of degrees of freedom here is $J - 1$, the number of distinct categories in the same distribution, or J, minus 1. You may have anticipated this from the fact that the sum of the differences between observed and expected frequencies is zero; given any $J - 1$ such differences, the remaining difference is determined. This is very similar to the situation for deviations from a sample mean, and the mathematical argument for degrees of freedom here would be much the same as for degrees of freedom in a variance estimate.

To return to our example, the value of the goodness-of-fit chi-square statistic is

$$\chi^2 = \frac{(406 - 371.09)^2}{371.09} + \frac{(164 - 146.71)^2}{146.71} + \frac{(189 - 207.12)^2}{207.12}$$

$$+ \frac{(78 - 86.30)^2}{86.30} + \frac{(26 - 51.78)^2}{51.78}$$

$$= 20.54.$$

For $J - 1 = 5 - 1 = 4$ degrees of freedom this is significant beyond the .01 level. Thus, rather strong evidence exists that the distribution of cigarette-smoking behavior among this group of men differs now from the situation of 10 years ago.

Chi-square tests of goodness of fit may be carried out for any hypothetical population distribution we might specify, provided that the population distribution is discrete, or is thought of as grouped into some relatively small set of class intervals. However, in the use of the Pearson χ^2 statistic to approximate multinomial probabilities, it *must* be true that:

1. each and every sample observation falls into one and only one category or class interval;
2. the outcomes for the N respective observations in the sample are independent;
3. sample N is large.

The first two requirements stem from the multinomial sampling distribution itself: the multinomial rule for probability holds only for mutually exclusive and exhaustive categories, and for independent observations in a sample (random sampling with replacement). The third requirement comes from the use of the chi-square distribution to approximate these exact multinomial probabilities: this approximation is good only for large sample size. Furthermore, unless N is infinitely large, the Pearson χ^2 itself is not distributed exactly as the chi-square variable.

The fact that when H_0 is true, the Pearson χ^2 statistic for goodness of fit is not distributed exactly as the random chi-square variable, can be seen from the expected value and the variance of the Pearson χ^2, where v symbolizes the degrees of freedom, $J - 1$:

$$E(\chi^2) = v$$

$$\text{var}(\chi^2) = 2v + \frac{1}{N}\left(\sum_j \frac{1}{p_j} - v^2 - 4v - 1\right).$$

Recall that we learned in Chapter 9 that the expected value of a chi-square variable with v degrees of freedom is v, and that the variance is $2v$. Although the expected value of the Pearson χ^2 statistic is also v, the variance of this statistic need not equal the variance of the random variable chi-square, unless N is infinitely large. This implies that the Pearson χ^2 is not ordinarily distributed as the chi-square variable for samples of finite size. Note that the expression for the variance of the Pearson χ^2 indicates that the goodness of the chi-square distribution as an approximation to the distribution of the Pearson statistic depends on several things, including not only the size of N but also the true probabilities p_j associated with the various categories and the number of degrees of freedom.

How large should sample size be to permit the use of the Pearson χ^2 goodness-of-fit tests? Opinions vary on this question, and some fairly sharp debate has been raised by the issue over the years. Many rules of thumb exist, but as a conservative rule one is usually safe in using this chi-square test for goodness of fit if each *expected* frequency, m_j, is 10 or more when the number of degrees of freedom is 1 (that is, two categories), or if the expected frequencies are each 5 or more where the number of degrees of freedom is greater than 1 (more than two categories). We will have more to say about sample size and Pearson χ^2 tests in Section 18.5.

Be sure to notice, however, that these rules of thumb apply to *expected,* not observed, frequencies per category.

One of the common uses of a goodness-of-fit test is to compare a sample distribution to the expected frequencies, given a population distribution with some specified theoretical form. Thus, for example, we might wish to ask if a sample distribution fits the frequencies to be expected given a *normal* population distribution.

To test the hypothesis of a normal distribution, we must first decide on a number of class intervals of some given size, and then think of both the sample and the population as divided into these class intervals. Then, we can determine what proportion of cases should fall into each such interval in a normal distribution, and use these proportions to find our expected frequencies, just as shown above. Table 19.1.1 gives an example of a set of intervals having equal expected frequencies in a normal distribution. The chi-square test is based on the comparisons of the obtained frequencies f_j with the expected frequencies m_j over the J different class intervals j, exactly as in Eq. 18.1.2. When neither the mean nor the variance of the population distribution is known, and must be estimated from the sample in order to find the probabilities of the various intervals, then there are $J - 3$ degrees of freedom for this test.

(Incidentally, other methods exist that are often superior to the Pearson chi-square test of goodness of fit when the aim is to compare obtained and theoretical distributions of random variables. Among these are the Kolmogorov-Smirnov tests, which can be found in Section 19.1 where the test of a normal hypothesis is illustrated.)

18.2 THE RELATION TO LARGE-SAMPLE TESTS OF A SINGLE PROPORTION

The goodness-of-fit test with 1 degree of freedom is formally equivalent to the large-sample test of a proportion, based on the normal approximation to the binomial (Section 6.4). That is, imagine a distribution with only two categories.

Category j	Expected frequency	Obtained frequency	Expected proportion	Obtained proportion
1	m_1	f_1	p	P
2	m_2	f_2	q	Q
	N	N	1.00	1.00

Suppose that the normal approximation to the binomial is to be used to test the hypothesis that the true population proportion in category 1 is p. Then we would form the test statistic

$$z = \frac{NP - Np}{\sqrt{Npq}} = \frac{-(NQ - Nq)}{\sqrt{Npq}} \qquad [18.2.1]$$

or

$$z = \frac{f_1 - m_1}{\sqrt{m_1 (N - m_1)/N}} .$$ [18.2.2]

For very large N, this can be regarded as a standardized normal variable. Now consider the *square* of this standardized variable z:

$$z^2 = \frac{N(f_1 - m_1)^2}{m_1 (N - m_1)} = \frac{(m_1)(f_1 - m_1)^2 + (N - m_1)(f_1 - m_1)^2}{m_1 (N - m_1)}$$ [18.2.3]

$$= \frac{(f_1 - m_1)^2}{m_1} + \frac{(f_1 - m_1)^2}{(N - m_1)} .$$

Since

$$(f_1 - m_1) = -(f_2 - m_2), \text{ and } m_1 + m_2 = N,$$

then we may write

$$z^2 = \frac{(f_1 - m_1)^2}{m_1} + \frac{(f_2 - m_2)^2}{m_2}$$ [18.2.4]

$$= \chi^2 .$$

When the frequency distribution has only two categories, the Pearson χ^2 statistic has exactly the same value as the square of the standardized variable used in testing for a single proportion, using the normal approximation to the binomial.

From the definition of a chi-square variable with 1 degree of freedom (Section 9.1), it can be seen that if $E(P) = p$, and if N is very large, then sample values of χ^2 should be distributed approximately as a chi-square variable with 1 degree of freedom. For a large sample, and so long as a two-tailed test is desired, it is immaterial whether we use the normal-distribution test for a single proportion or the Pearson χ^2 test for a two-category problem. Furthermore, taking the square root of this χ^2 value and adding the appropriate sign gives the equivalent z value if one does desire a one-tailed test. This direct equivalence between z and $\sqrt{\chi^2}$ holds only for 1 degree of freedom, however.

Again, only for 1 degree of freedom, the Pearson χ^2 test may be improved somewhat if the test statistic is found by taking

$$\chi^2 = \frac{(|f_1 - m_1| - .5)^2}{m_1} + \frac{(|f_2 - m_2| - .5)^2}{m_2}$$ [18.2.5*]

so that the *absolute value* of the difference between observed and expected frequencies is reduced by .5 for each category before the squaring is carried out. This is known as **Yates' correction** and depends on the fact that the binomial is a discrete, and the normal a continuous, distribution (cf. Eqs. 6.4.4 and 6.4.5.). However, Yates' correction applies only when there is 1 degree of freedom.

18.3 PEARSON CHI-SQUARE TESTS OF ASSOCIATION

In this section we are going to deal with data arranged into a two-dimensional array, just as we used to represent joint events in Section 1.15. In such a two-dimensional array, the rows of the table stand for one attribute, divided into J mutually exclusive and exhaustive categories, and the columns represent a second attribute consisting of K such categories. Then each of the $J \times K$ cells of the table contains the frequency f_{jk}, showing how often that particular combination of a row category and a column category occurred in the data. Clearly, the sum of all the possible cell frequencies must equal to the total N of the sample. A sample of N cases laid out in such a two-dimensional array is usually called a *contingency table*.

In discussing a $J \times K$ contingency table and its analysis, it will be convenient to adopt a notational scheme like that outlined below:

	B_1	B_2	...	B_k	...	B_K	row totals
A_1	f_{11}	f_{12}	...	f_{1k}	...	f_{1K}	f_{1+}
A_2	f_{21}	f_{22}	...	f_{2k}	...	f_{2K}	f_{2+}
A_j	f_{j1}	f_{j2}	...	f_{jk}	...	f_{jK}	f_{j+}
A_J	f_{J1}	f_{J2}	...	f_{Jk}	...	f_{JK}	f_{J+}
column totals	f_{+1}	f_{+2}	...	f_{+k}	...	f_{+K}	N

Here, the symbol f_{jk} stands for the frequency actually obtained in the cell formed by the rows A_j and the column B_k. Then the total frequency in any row A_j is symblized by f_{j+}, and that within any column by f_{+k}.

Parallel to the contingency table showing the actual data, one can think of a similar table for the population of potential observations, where the frequency for any cell is replaced by the probability of a case occurring in that cell, or p_{jk}. Then we can think of the probability of the row category A_j as represented by the the symbol p_{j+}, and that of the column category B_k by the symbol p_{+k}.

Given a situation that is true of the population represented by our sample, where the probability of any cell jk is actually some value p_{jk}, then we should *expect* the sample frequency in cell jk to be equal to

$$E(f_{jk}) = m_{jk} = Np_{jk}$$

[18.3.1]

for each of the possible $J \times K$ cells.

In particular, suppose that in the population the events represented by the row categories are actually independent of the events represented by the columns. That is, there is truly *no association* between the two kinds of events. Then, as we know from Section 1.15, it must be true that, in the population,

$$p_{jk} = p_{j+}p_{+k} \qquad \text{for every cell } jk.$$

[18.3.2]

Now we would like to see how well this model of independence in the population actually fits the data in our table. However, we do not, as a rule, actually know the values of the population probabilities such as p_{j+} and p_{+k} for the rows and columns respectively. Rather, we estimate these unknown probabilities by using the sample row and column frequencies relative to sample N:

$$\text{estimate of } p_{j+} = f_{j+}/N$$

and

$$\text{estimate of } p_{+k} = f_{+k}/N \ .$$

If the population model of independence is really true, then for any cell

$$\hat{m}_{jk} = \text{estimated } m_{jk} = N (\text{estimated } p_{j+})(\text{estimated } p_{+k}) \qquad [18.3.3^*]$$
$$= (f_{j+})(f_{+k})/N.$$

In other words, if the row and column attributes are actually independent in the population, then the frequency we should expect in any cell jk may be estimated by taking the row frequency times the column frequency and dividing by N.

Then by the same logic used in the Pearson chi-square test of goodness of fit, as discussed in Section 18.1, we may employ the *Pearson chi-square test of association*:

$$\chi^2 = \sum_j \sum_k \frac{(f_{jk} - \hat{m}_{jk})^2}{\hat{m}_{jk}} \ , \qquad [18.3.4^*]$$

where

$$\hat{m}_{jk} = f_{j+} f_{+k}/N \text{ for each cell.}$$

When the null hypothesis

$$H_0: p_{jk} = p_{j+}p_{+k}, \text{ for all } j \text{ and } k,$$

is true, then this statistic behaves like a chi-square variable with $(J - 1)(K - 1)$ degrees of freedom, *provided* that

1. each and every observation is independent of each other observation;
2. each observation qualifies for one and only one cell in the table:
3. sample size N is large.

If we actually knew all the different probabilities p_{jk} for the population, or if we knew the exact values of the p_{j+} and p_{+k} probabilities, this statistic could be referred to a theoretical chi-square distribution with $JK - 1$ degrees of freedom. However, in order to estimate the expected frequencies m_{jk} it is necessary to form estimates of the p_{j+} and the p_{+k} by the use of the sample row and column frequencies. The result is that the number of degrees of freedom is here reduced to $JK - 1 - (J - 1) - (K - 1)$ or $(J - 1)(K - 1)$.

Although the Pearson chi-square test for association is usually discussed in terms like Eq. 18.3.4, other ways of calculating the chi-square test of association can be much easier to use, especially when the work is being done on a good hand calculator. For example, we may also use the equation below, which is algebraically identical to Eq. 18.3.4:

$$\chi^2 = N \left[\sum_j \sum_k \frac{f_{jk}^2}{f_{j+}f_{+k}} - 1 \right].$$ [18.3.5*]

Here, for each cell one need take only the frequency and square it, divide by the row total and the column total, and sum the results. Then when this has been done over all the cells, 1.00 is subtracted from the sum, and the ensuing value is multipled by N to find the chi-square value.

The following example is fairly typical of the chi-square analysis of a contingency table. The juvenile justice system in a large metropolitan area cooperates with the school system to offer education and rehabilitation programs for teenagers who have a first-offense conviction of drug possession or use. Four of these programs are in operation, as denoted in Table 18.3.1 by the symbols A_1, A_2, A_3, and A_4. This table also shows the conviction history of these individuals for three years after completion of one of the programs. Thus, the columns of the table show B_1 (no further convictions), B_2 (one more conviction), and B_3 (two or more further convictions).

Note that in Table 18.3.1 the figure in parentheses within each cell is the *estimated expected frequency*,

$$\hat{m}_{jk} = \frac{f_{j+}\, f_{+k}}{N}.$$

Thus, for the cell in row 1 and column 1 the estimated expected frequency is

$$\hat{m}_{11} = (200 \times 461)/797 = 115.68$$

that for the cell in row 2 column 1 is

$$\hat{m}_{21} = (195 \times 461)/797 = 112.79$$

and so on for all the other cells.

program	B_1 (no more)	B_2 (one more)	B_3 (more than one)	f_{j+}
A_1	122 (115.68)	70 (73.27)	8 (11.04)	200
A_2	141 (112.79)	39 (71.44)	15 (10.77)	195
A_3	106 (117.42)	79 (74.37)	18 (11.21)	203
A_4	92 (115.11)	104 (72.91)	3 (10.99)	199
f_{+k}	461	292	44	797

Table 18.3.1

Educational program attended (A) and later conviction record (B) of 797 teenaged students with drug-related first offenses, showing expected frequencies in parentheses.

For this table, we can find the value of the Pearson chi-square test of association by taking

$$\chi^2 = \frac{(122 - 115.68)^2}{115.68} + \frac{(70 - 73.27)^2}{73.27} + \cdots + \frac{(3 - 10.99)^2}{10.99}$$

$$= 54.00$$

for $(4 - 1)(3 - 1) = 6$ degrees of freedom. This is significant beyond the .01 level.

Alternatively, we could have skipped the step of finding the expected frequencies directly, and calculated chi-square as follows:

$$\chi^2 = N \left[\sum\sum \frac{f_{jk}^2}{f_{j+} f_{+k}} - 1 \right]$$

$$= 797 \left[\frac{(122)^2}{(200 \times 461)} + \frac{(70)^2}{(200 \times 292)} + \cdots \right.$$

$$\left. + \frac{(3)^2}{(199 \times 44)} - 1 \right]$$

$$= 797 [1.067754 - 1] = 54.00 .$$

Clearly, this table shows a very significant association between the program that the teenaged offender attended, and the number of subsequent offences recorded for him or her.

Following such an analysis of a two-dimensional contingency table, it is often informative to examine the individual cells to locate the large departures from expectation, both in the positive and the negative directions. In fact, if N is large, then for any cell we may take the *standardized residual*

$$z_{jk} = (f_{jk} - \hat{m}_{jk})/\sqrt{\hat{m}_{jk}} \qquad [18.3.6^*]$$

and regard this as an approximately normally distributed random variable under the hypothesis of no association. The magnitudes of these standardized residuals show the cells that tend to bear the heaviest share in the association reflected in the overall chi-square value, and the sign of the standardized residual shows whether the association is positive (more than expectation) or negative (fewer than expected) in that particular combination of row and column categories.

(You will note that the chi-square test for association is equal to the sum of the squares of these standardized residual values:

$$\chi^2 = \sum_j \sum_k z_{jk}^2 .) \qquad [18.3.7^*]$$

If you recall from Section 9.1 the definition of a chi-square variable, it is easy to see that if the standardized residuals are approximately normally distributed for large N, then the distribution of the Pearson chi-square statistic for association should approximate a theoretical chi-square distribution. The numerators of the standardized residuals must sum to zero across rows, across columns, and across the entire table, reflecting the fact that there are $(J - 1)(K - 1)$ degrees of freedom.

Standardized residuals also provide a device for judging the post hoc significance of the departure from independence in one or more of the cells. In making these evaluations across the entire table, however, it is wise to adopt a conservative approach, such as use of the Bonferroni procedure of Section 11.14. That is, after deciding on a familywise error rate for the set of tests, such as $\alpha_{FW}* = .05$ or $.01$, then the Bonferroni rate per test (i.e., per comparison) might be set at

$$\alpha_{BPC} = \alpha_{FW*}/JK.$$

Therefore, for this example, if we wished to test the significance for all the residuals, or departures from independence in the cells, then we could set the familywise rate $\alpha_{FW}* = .05$ and then take $\alpha_{BPC} = .05/12 = .004$ as the standard for each test. For two-tailed tests this would imply that a standardized residual be declared significant only if its z value equals or exceeds 2.86 in absolute value (since Table IA of Appendix E shows that this value cuts off approximately the upper .002 proportion of a normal distribution).

In Table 18.3.1 the standardized residual is significant, for example, in the cell for row A_2 and column B_2. Here, the standardized residual is found to be

$$z_{22} = (39 - 71.44)/\sqrt{71.44} = -3.84,$$

which exceeds in absolute value our criterion of 2.86 for significance at the .05 level. The standardized residuals for each of the other cells may be calculated and evaluated in exactly this same way.

Overall, it appears from Table 18.3.1 that Program A_2 was the most successful in keeping these people from further offenses, since the number of cases in cell A_2B_2 was significantly lower than expectation, and that in cell A_2B_1 turned out a good bit (though not quite significantly) higher than expectation. Just the opposite situation seems to exist for Program A_4, however. On the other hand, the frequencies in rows A_1 and A_3 accord fairly well with expectation under the hypothesis of independence of rows and columns.

18.4 THE SPECIAL CASE OF A FOURFOLD TABLE

Two-by-two contingency, or joint frequency, tables are especially common in social and behavioral research. For such fourfold tables, computations for the Pearson χ^2 test can be put into a very simple form. Consider the following table:

a	b	$a + b$
c	d	$c + d$
$a + c$	$b + d$	N

Here, the small letters a, b, c, and d represent the frequencies in the four cells, respectively. Then, the value of χ^2 can be found quite easily from

$$\chi^2 = \frac{N(ad - bc)^2}{(a + b)(c + d)(a + c)(b + d)} \qquad [18.4.1*]$$

with, of course, one degree of freedom.

This value of χ^2 is usually corrected to give a somewhat better approximation to the exact probability. With the correction, the value is found from

$$\chi^2 = \frac{N(|ad - bc| - N/2)^2}{(a + b)(c + d)(a + c)(b + d)}$$ [18.4.2*]

which is compared with Table IV (1 degree of freedom) in the usual way. This is another instance of **Yates' correction for continutiy,** which we encountered first in Section 18.2. This correction should be applied only when the number of degrees of freedom is one, however.

18.5 THE ASSUMPTIONS IN CHI-SQUARE TESTS OF ASSOCIATION

Chi-square tests are among the easiest for the novice in statistics to carry out, and they lend themselves to a wide variety of social and behavioral data. This computational simplicity is deceptive, however, as the use of the chi-square approximation to find multinomial probabilities is based on a fairly elaborate mathematical rationale, requiring a number of very important assumptions. This rationale and the importance of these assumptions has not always been understood, even by experienced researchers, and there is probably no other statistical method that has been so widely misapplied.

In the first place, since the exact probabilities to be approximated are assumed to follow the multinomial rule, each and every observation categorized should be independent of each other observation. In particular, this means that caution may be required in the application of χ^2 tests to data where dependency among observations may be present, as is sometimes the case in repeated observations of the same individuals. As always, however, it is not the mere fact that observations were repeated, but rather the nature of the experiment and the type of data, that let one judge the credibility of the assumption of independent observations. Nevertheless, the novice user of statistical methods does well to avoid the application of Pearson χ^2 tests to data where each individual observed contributes more than a single entry to the joint frequency table.

In the second place, the joint frequency table must be complete, in the sense that each and every observation made must represent one and only one joint-event possibility. This means that each distinct observation made must qualify for one and only one row, one and only one column, and one and only one cell in the contingency table.

The stickiest question of all concerns sample size and the minimum expected frequency in each cell. Probabilities found from the chi-square tables for such tests are always approximate. Only when the sample size is infinite must these probabilities be exact. The larger the sample size, the better this approximation generally is, but the goodness of the approximation also depends on such things as true marginal distributions of events, the number of cells in the contingency table, and the significance level employed.

Without going further into the complexities of the matter, we will simply state a rule that is at least current, fairly widely endorsed, and generally conservative.

For tables with more than a single degree of freedom, a minimum expected frequency of 5 can be regarded as adequate, although when there is only a single degree of freedom, a minimum expected frequency of 10 is much safer.

This rule of thumb is ordinarily conservative, however, and circumstances may arise where smaller expected frequencies can be tolerated. In particular, if the number of degrees of freedom is large, then it is fairly safe to use the χ^2 test for association even if the minimum expected frequency is as small as 1, provided that there are only a few cells with small expected frequencies (such as one out of five or fewer).

A word must also be said about the practice of pooling categories to attain large expected frequencies after the data are seen. This has been done routinely for many years, and many statistics texts advise this as a way out of the problem. However, this may amount to jumping from the frying pan into the fire! The whole rationale for the chi-square approximation rests on the randomness of the sample, and that the categories into which observations may fall are chosen in advance. When we start pooling categories after the data are seen, we are doing something to the randomness of the sample, with unknown consequences for our inferences. The manner in which categories are pooled can have an important effect on the inferences drawn. This practice is to be avoided if at all possible; nevertheless, when expected frequencies of less than 1 are encountered, pooling may be the only recourse.

18.6 THE POSSIBILITY OF EXACT TESTS FOR ASSOCIATION

As we have mentioned, the Pearson χ^2 tests of association and of goodness of fit give approximations to exact probabilities, which, in principle, may be found using the multinomial (or in some instances, the hypergeometric) rule. The basic reason for using the chi-square approximation is that actual computation of these exact probabilities is extremely laborious or downright impossible. However, in some situations where the sample size is so small that the use of the χ^2 tests is ruled out, it may be practicable to compute probabilities exactly. This will be illustrated only for a simple test of association, although other possibilities exist.

The test we will discuss is commonly known as **Fisher's exact test** for a 2×2 contingency table. This is not appropriately called a χ^2 test, since it does not use the chi-square approximation at all. Instead, the exact probability is computed for a sample showing as much or more evidence for association than that obtained, given only the operation of chance.

Suppose that some N subjects are categorized into the following 2×2 table:

	A_1	A_1	
B_1	a	b	$a + b$
B_2	c	d	$c + d$
	$a + c$	$b + d$	N

Now suppose for a moment that the N subjects actually make up the population and that the distribution in the population shows $a + c$ individuals in the first *column* category, A_1, and $b + d$ individuals in the second column category, A_2. Some $n = a + b$ individuals are sampled at random and *without* replacement. What is the probability that in *this sample* exactly a individuals fall into A_1 and b into A_2? In Section 3.20 it was mentioned that the probability of a sample drawn *without* replacement from a finite population can be found by the *hypergeometric rule*. Applying the rule, we find that the probability of a in A_1 and b in A_2 *within the sample* represented by row B_1 is just

$$\frac{\binom{a + c}{a}\binom{b + d}{b}}{\binom{N}{a + b}}$$

which is the same as

$$\frac{(a + b)! \, (c + d)! \, (a + c)! \, (b + d)!}{N! \, a! \, b! \, c! \, d!}.$$

Now consider the marginals of the sample table as fixed, so that regardless of the arrangement within the table we know the totals in rows and columns. Imagine that occurrences of the events represented by the rows and columns have absolutely nothing to do with each other, so that the two attributes are independent. Any sample result in the cells of the table occurs as though individuals in the columns were assigned to the rows at random. Then the probability of any particular random arrangement is given by the use of the hypergeometric rule, just as for the probability found above.

If one finds the probability of the arrangement actually obtained, as well as every other arrangement giving as much or more evidence for association, then one can test the hypothesis that the obtained result is *purely a product of chance* by taking this probability as the significance level.

This amounts to finding both the probability of the obtained table, and every other table (with the same marginals) showing more disproportion between cells a and c than in the table obtained.

Unlike the χ^2 test, the Fisher exact test is essentially one-tailed. The probabilities are calculated for all possible results departing as much or more in a specific direction from the marginal distribution of A as does our sample. When the rows (or columns) have equal frequencies (as in the example) the final probability can be doubled to arrive at the two-tailed significance level. However, when both the two rows and the two columns have unequal marginal frequencies, the probability should be found for each possible table where the absolute difference.

$$\left| \frac{a}{a + b} - \frac{c}{c + d} \right|$$

is as great or greater than in the table actually obtained.

Convenient tables are available for the Fisher exact test, so that it is not usually necessary to carry out all the computations given here in order to perform this test. These tables are available in the *Biometrika Tables for Statisticians* (1966) and in Siegel (1956), among other places.

18.7 A TEST FOR CORRELATED PROPORTIONS IN A TWO-BY-TWO TABLE

A problem that often arises in research is somewhat different from the usual problem of association between attributes. As an illustration of this problem, suppose that some N individual subjects are each observed by two independent judges. Each judge places each subject into one of two mutually exclusive and exhaustive categories, such as "high leadership potential" versus "low leadership potential." It is assumed that a judge's ratings of different individuals are independent. Let us call these categories simply H and L for the moment. We would like to ask if these two judges, given all possible subjects in the population, would show the same true proportion of individuals rated in Category H. In other words, in the population of all subjects to be rated, does $p_1(H), = p_2(H)$, where $p_1(H)$ is the proportion rated in Category H by Judge 1, and $p_2(H)$ is the proportion rated in that category by Judge 2?

This is a problem of **correlated proportions,** since each of the two sample proportions will be based in part on the same individuals. A test devised by McNemar (1975) applies to this situation. Suppose that the sample of N individuals were arranged into the following 2 × 2 table:

Judge 1

		H	L
Judge 2	H	a	b
	L	c	d

An *exact* test uses twice the binomial prob. $(x \le h|p = .5, N = a + c)$, where h is the lesser of b or c. When N is fairly large, the exact probability may be approximated by use of χ^2, where

$$\chi^2 = \frac{(|b - c| - 1)^2}{b + c} \tag{18.7.1*}$$

with one degree of freedom.

For our example, a significant result would let one conclude that the *true* distributions of judgments for the two judges differ. Be sure to notice that this is not an ordinary test of association for a contingency table, but rather a test of the equality of two proportions where each sample proportion involves some of the same observations, making the two sample proportions dependent. This test has been extended to more complicated situations by Cochran (1950); the procedure is discussed in Section 19.3.

18.8 MEASURES OF ASSOCIATION IN CONTINGENCY TABLES

One of the oldest problems in descriptive statistics is that of indexing the strength of statistical association between qualitative attributes. Although a number of simple and meaningful indices exist to describe association in a fourfold table, this problem grows more complex for larger tables, and has perhaps never been solved to everyone's real satisfaction.

Before we go into the problem of describing statistical association in a sample, a general way of viewing statistical association in a population will be introduced. This is the **index of mean square contingency,** originally suggested by Karl Pearson, the originator of the χ^2 test for association. Imagine a *discrete joint probability distribution* represented in a table with K columns and J rows. The columns represent the qualitative attribute B, and the rows A. Then the mean square contingency is defined to be

$$\Phi^2 = \sum_j \sum_k \frac{p(A_j, B_k)^2}{p(A_j)p(B_k)} - 1 \qquad [18.8.1]$$

This population index Φ^2 (upper case Greek phi, squared) can be zero only when there is complete independence, so that

$$p(A_j, B_k) = p(A_j)p(B_k)$$

for each joint event (A_j, B_k). However, when there is *maximum association* in the table, the value of Φ^2 is given by

$$\text{max. } \Phi^2 = L - 1, \qquad [18.8.2]$$

where L is the *smaller* of the two numbers J or K (number of rows or columns in the table).

The sample version of the index of mean square contingency, symbolized by φ^2 (lower case phi, squared) can be found from

$$\varphi^2 = \sum_j \sum_k \frac{f_{jk}^2}{f_{j+}f_{+k}} - 1 \qquad [18.8.3]$$

which, by use of Eq. 18.3.5, can be turned into

$$\varphi^2 = \chi^2/N. \qquad [18.8.4]$$

We will have more to say about this relationship in Section 18.11.

Now consider a sample of data arranged into a fourfold contingency table, as in Section 18.4. The *sample* value φ (lower case phi) is given by

$$\varphi = \frac{(ad - bc)}{\sqrt{(a + b)(c + d)(a + c)(b + d)}}. \qquad [18.8.5]$$

Notice that this is almost exactly the square root of the expression for χ^2 in a 2×2 table, given by Eq. 18.4.1. In fact,

$$\chi^2 = N\varphi^2$$

and

$$\phi = \sqrt{\frac{\chi^2}{N}}.$$ [18.8.6]

Since both χ^2 and population Φ^2 reflect the degree to which there is nonindependence between A and B, a test for the hypothesis

H_0: true $\Phi^2 = 0$

is provided by the ordinary χ^2 test for association in a 2 × 2 table.

In a 2 × 2 table there is an important link between sample ϕ and the correlation coefficient r. Let the categories within each attribute A and B be thought of as ordered, and suppose that each individual i among the N cases making up the sample is assigned dummy variable values as follows:

$x_i = 1$ if i falls in the higher category of A

$x_i = 0$ otherwise

and

$y_i = 1$ if i falls in the higher category of B

$y_i = 0$ otherwise.

The data in the table would then be of this form

	y	
	0	1
0	a	b
1	c	d

Suppose that the correlation between these scores were computed across the N individuals i. We would find that

$$r_{xy} = \frac{\sum_i x_i y_i / N - \overline{xy}}{S_X S_Y}$$

$$= \frac{Nd - (c + d)(b + d)}{\sqrt{(a + b)(c + d)(a + c)(b + d)}}$$ [18.8.7]

$$= \frac{ad + bd + cd + d^2 - bc - cd - bd - d^2}{\sqrt{(a + b)(c + d)(a + c)(b + d)}}$$

$$= \phi$$

What we have shown is that the coefficient ϕ may be regarded as the correlation between the attributes A and B when the categories are associated with ''scores,'' or dummy-variable values, of 0 and 1. Because of this connection with r, φ is often called the **fourfold point correlation.**

As you can see, the *sign* of the correlation coefficient (and thus of φ) is meaningful only if there is a meaningful order for the two row categories, and for the two columns. That is, assigning 1 and 0 differently to rows or to columns will only affect the sign of the correlation. However, even if the row categories are unordered, and the same is true for the column categories, the *squared* value of r or of φ is meaningful as an expression of the strength of association between row and column membership.

The idea of φ^2 extends to samples in larger contingency tables as well. As we have already seen, for a set of data arranged into a $J \times K$ table, the *sample* value of φ^2 is simply

$$\varphi^2 = \frac{\chi^2}{N}.$$

However, the value of φ^2 needs to be corrected if it is to be an index ranging between zero and one. A convenient way to describe the apparent strength of association in a sample is *Cramér's (1946) statistic,*

$$V_C = \sqrt{\frac{\varphi^2}{L - 1}} = \sqrt{\frac{\chi^2}{N(L - 1)}} \qquad [18.8.8^*]$$

which must lie between 0, reflecting complete independence, and 1, showing complete dependence, of the attributes. For Cramér's statistic, the Pearson χ^2 is computed in the ordinary way as for a test of association, and L is the lesser of J, the number of rows, or K, the number of columns.

This index V_C (Cramér's statistic) is not to be confused with the ordinary *coefficient of contingency,* sometimes used for the same purpose. The coefficient of contingency is defined by

$$C_{AB} = \sqrt{\frac{\chi^2}{N + \chi^2}}.$$

This last index has the disadvantage that it cannot attain an upper bound of 1.00 unless the number of categories for A and B is infinite. Obviously, this limits the usefulness of C_{AB} as a descriptive statistic, and the index given by V_c is superior. A pair of slightly different indices for a fourfold table were provided by G. U. Yule:

$$Q = \frac{ad - bc}{ad + bc} \qquad [18.8.9]$$

and

$$Y = \frac{\sqrt{ad} - \sqrt{bc}}{\sqrt{ad} + \sqrt{bc}}. \qquad [18.8.10]$$

The Q index is very similar to φ, except that its denominator does not have the same form. The Y index is constructed on the supposition that the marginals of the table all have equal frequencies, and thus this index is less sensitive to marginal inequalities than are φ and Q.

18.9 A MEASURE OF PREDICTIVE ASSOCIATION FOR CATEGORICAL DATA

Now we will discuss a different approach to describing the structure of a contingency table, based upon an **index of predictive association.** This index, which was developed by Goodman and Kruskal (1979), will be called λ_B, (small Greek lambda, sub B):

$$\lambda_B = \frac{p(\text{error}|A_j \text{ unknown}) - p(\text{error}|A_j \text{ known})}{p(\text{error}|A_j \text{ unknown})}. \qquad [18.9.1]$$

This index shows the proportional reduction in the *probability* of error in predicting B afforded by specifying A_j. If the information about the A category does not reduce the probability of error in predicting B at all, the index is zero, and one can say that there is no predictive association. On the other hand, if the index is 1.00, no error is made in predicting B given the A_j classification, and there is complete predictive association.

It must be emphasized that this idea is not completely equivalent to independence and association as reflected in χ^2 and V_c. It is quite possible for some statistical association to exist even though the value of λ_B is zero. In this situation, A and B are not independent, but the relationship is not such that giving A_j causes one to change one's bet about B_k; the index λ_B is other than 0 only when *different* B_k categories would be predicted for different A_j information.

On the other hand, if there is complete proportionality throughout the table, so that φ' is zero, then λ_B must be zero. Furthermore, when there is complete association, so that perfect prediction is possible, both λ_B and V_c must be 1.00.

Sample values of λ_B can be calculated quite easily from a contingency table. Here, the sample is regarded as though it were the population, and probabilities are taken from the relative frequencies in the sample. Thus we interpret λ_B as the proportional reduction in the probability of error in prediction for cases drawn at random from *this* sample, or, if you will, a population exactly like this sample in its joint distribution.

In terms of the frequencies in the sample, we find

$$\lambda_B = \frac{\sum_j \max_k f_{jk} - \max_k f_{\cdot k}}{N - \max_k f_{\cdot k}} \qquad [18.9.2]$$

where

f_{jk} is the frequency observed in cell (A_j, B_k)

$\max_k f_{jk}$ is the *largest* frequency in row A_j

$\max_k f_{\cdot k}$ is the largest *marginal* frequency among the columns B_k.

When two or more cells in any row have the same frequency, larger than any others in that row, then the frequency belonging to any single one of those cells is used as the maximum value for the row. Similarly, if several column marginals each exhibit the same frequency, which is largest among the columns, that frequency is used.

The index λ_B is an *assymetric* measure, much like ω^2 and ρ_I. It applies when A is *the* independent variable, or the thing ordinarily known first, and B is the thing predicted. However, for the same set of data, it is entirely possible to reverse the roles of A and B, and obtain the index

$$\lambda_A = \frac{p(\text{error}|B_k \text{ unknown}) - p(\text{error}|B_k \text{ known})}{p(\text{error}|B_k \text{ unknown})}$$

which is suitable for predictions of A from B. In terms of frequencies:

$$\lambda_A = \frac{\sum_k \max_j f_{jk} - \max_j f_{j\cdot}}{N - \max_j f_{j\cdot}}. \qquad [18.9.3]$$

In general, the two indices λ_B and λ_A will not be identical; it is entirely possible to have situations where B may be quite predictable from A, but not A from B.

Finally, in some contexts it may be desirable to have a *symmetric* measure of the power to predict, where neither A nor B is specially designated as the thing predicted from or known first. Rather, we act as though sometimes the A and sometimes the *B* information is given beforehand. In this circumstance the index λ_{AB} can be computed from

$$\lambda_{AB} = \frac{\sum_j \max_k f_{jk} + \sum_k \max_j f_{jk} - \max_k f_{\cdot k} - \max_j f_{j\cdot}}{2N - \max_k f_{\cdot k} - \max_j f_{j\cdot}}. \qquad [18.9.4]$$

Furthermore, Goodman and Kurskal (1979) showed that when the true λ_B value in the population is neither exactly 0 nor exactly 1.00, the sampling distribution of λ_B is approximately normal for large N, with a sampling variance which can be estimated from

$$\sigma_{\lambda_B}^2 = \frac{\left(N - \sum_j \max_k f_{jk}\right)\left(\sum_j \max_k f_{jk} + \max_k f_{\cdot k} - 2\sum_j{}^* \max_k f_{jk}\right)}{(N - \max_k f_k)^3} \qquad [18.9.5]$$

where $\sum_j{}^* \max f_{jk}$ is the summation of the maximum frequencies in the columns, taken only over those columns where the maximum falls in the same row as $\max_k f_k$.

These measures of *predictive* association form a valuable adjunct to the tests given by χ^2 methods. When the value of χ^2 turns out to be significant, one can say with confidence that the attributes A and B are not independent. Nevertheless, the significance level alone tells almost nothing about the strength of association. If we want to say something about the predictive strength of the relation, or have the remotest interest in actual predictions using the relation studied, then the λ measures are worthwhile.

18.10 CHI-SQUARE METHODS AND MULTIPLE REGRESSION

Although the methods appear very different on the surface, there are intimate connections between the Pearson chi-square statistic as calculated for $J \times K$ contingency tables, and the multiple regression methods surveyed in Chapter 15. In particular, if either rows or columns are a dichotomy, so that $J = 2$ or $K = 2$, then the connection between chi-square methods and multiple regression is especially simple.

The obtained Pearson chi-square value is equal to R-square times N, where R-square refers to the multiple regression of the single dummy variable representing the two-category attribute on the set of $J - 1$ or $K - 1$ dummy variables representing the other attribute.

Thus, for example, suppose that there are only $K = 2$ column categories in a contingency table, but $J \geq 2$ categories for the rows. Then the squared multiple correlation showing the relation of the column dummy variable to the $J - 1$ row dummy variables is equal to chi-square divided by sample N: $R^2_{\text{col. row}} = \chi^2/N$.

Before we go further, a word of warning is in order, operative both in this and the next two sections. The connections between multiple regression and chi-square methods outlined here are *purely algebraic,* and have no implications for the inferential aspects of these techniques, which apply to very different situations and require very different assumptions. Significance tests and the other inferential aspects of multiple regression usually presuppose quantitative variables following normal distribution laws, and, as we have seen, chi-square tests provide large-sample approximations to probabilities that are, essentially, multinomial. In terms of assumptions and the situations to which their inferential aspects apply, these two types of methods *are* very different, even though, as we shall see, the same logico-mathematical machinery drives each type of method.

To illustrate this connection between chi-square and multiple regression, consider a set of hypothetical data representing a group of 216 students admitted into a large PhD program in psychology during a five-year period. The row classifications in the resulting data show the types of undergraduate majors presented by the students. This attribute, A, consists of the five categories

    A_1 = psychology
    A_2 = other nonpsychology social or behavioral science
    A_3 = humanities or fine arts
    A_4 = physical or biological science, or mathematics
    A_5 = other.

The other attribute, B, consisted only of the two categories:

    B_1 = completed PhD program within five years
    B_2 = did not complete the program within five years (including dropped from
        the program).

Our question: ''Is the undergraduate major of a PhD student related to finishing within five years?'' The data are in Table 18.10.1.

On the assumption that this set of 216 cases represents a random sample of all graduate admissions in psychology during this same period, we can carry out a significance test. As it happens, the chi-square test of association gives a value of

$$\chi^2 = 10.8336$$

for these data, with $(5 - 1)(2 - 1) = 4$ degrees of freedom. This is significant beyond the .05 level. In addition, this implies that the value of the Cramér V_C statistic is

$$V_C = \sqrt{10.8336/216} = .2240 .$$

Now we can also approach this problem through multiple regression, using $J - 1 = 4$ dummy variables X to stand for row membership, and one dummy variable Y to stand for column category membership. That is, we let

$x_1 = 1$ if a case is a member of row A_1

$x_1 = 0$ otherwise,

$x_2 = 1$ if a case is a member of row A_2

$x_2 = 0$ otherwise,

and so on for the two other row and one column dummy variables.

Then it is possible to find the correlations among all of these dummy variables, using the format of the phi coefficient discussed in Section 18.8. Thus, for variables X_1 and Y we have

	$y = 1$	$y = 0$	
$x_1 = 1$	40	36	76
$x_1 = 0$	63	77	140
	103	113	

$$r_{11} = \varphi_{11} = \frac{(40 \times 77) - (63 \times 36)}{\sqrt{(103)(113)(140)(76)}}$$

$$= .0730$$

	$B_1 = 5$ years or less	$B_2 = $ more than 5 (or dropped)	totals	**Table 18.10.1**
$A_1 = $ psychology	40	36	76	Undergraduate major and time to complete PhD degree shown for 216 students admitted to a graduate psychology program.
$A_2 = $ other social science	13	19	32	
$A_3 = $ humanities	5	19	24	
$A_4 = $ physical sciences	23	15	38	
$A_5 = $ other	22	24	46	
totals	103	113	216	

	X_1	X_2	X_3	X_4	Y
X_1	1	−.3073	−.2605	−.3404	.0730
X_2		1	−.1474	−.1927	−.0589
X_3			1	−.1634	−.1901
X_4				1	.1188
Y					1

Table 18.10.2

Intercorrelations among dummy variables representing rows and columns of Table 18.10.1.

and similarly for all of the other X and Y correlations, as shown in Table 18.10.2. (Bear in mind that the order of the row and column categories, as reflected in the assignment of the dummy variable values, is arbitrary. Even though this set of choices will determine the signs of the correlations obtained, if one is consistent, the final R-square value found on the basis of these correlations will be unaffected.)

The correlations among the X variables can be found in much the same way. Thus for variables X_1 and X_2 we take

	$x_2 = 1$	$x_2 = 0$	
$x_1 = 1$	0	76	76
$x_1 = 0$	32	108	140
	32	184	

$$r_{x_1 x_2} = \frac{(0 \times 108) - (32 \times 76)}{\sqrt{(32)(184)(76)(140)}}$$

$$= -.3073,$$

and so on for the other pairs of X variables. These too are shown in Table 18.10.2. When the table of correlations shown as 18.2.2 is subjected to multiple regression analysis by one of the methods discussed in Chapter 15, it is found that

$$R^2_{y \cdot x_1 x_2 x_3 x_4} = .0502$$

$$= 10.8336/216$$

$$= \chi^2/N.$$

It is not all that difficult to show that this must be true for a $J \times 2$ (or a $2 \times K$) table. We can start out by remembering that in the one-way ANOVA for J independent groups, the ratio of SS between to SS total is equal to the R-square value associating the dependent variable Y to the $J - 1$ dummy variables standing for groups (Section 10.17). Therefore let the J rows of our contingency be the groups and let Y be the dummy variable standing for membership in column B_1. Then it will be true that

$$\text{SS}_y \text{ between} = \sum_j \frac{f_{j1}^2}{f_{j+}} - \frac{f_{+1}^2}{N} \qquad [18.10.1]$$

where

f_{j1} = the frequency in row A_j and column B_1

f_{j+} = the marginal frequency in row A_j

and

f_{+1} = the marginal frequency in column B_1.

Note also that $f_{+1} + f_{+2} = N$, and that $f_{j1} + f_{j2} = f_{j+}$.
The ANOVA SS_y total can also be written as

$$SS_y \text{ total} = f_{+1} - f_{+1}^2/N$$

$$= f_{+1}(N - f_{+1})/N \hspace{3cm} [18.10.2]$$

$$= f_{+1}f_{+2}/N.$$

Hence the squared multiple correlation is

$$R_{y \cdot \text{rows}}^2 = SS_y \text{ between rows} / SS_y \text{ total}$$

$$= \frac{\left[\sum_j (f_{j1}^2/f_{j+}) - (f_{+1}^2/N) \right]}{(f_{+1}f_{+2}/N)}$$

$$= \sum_j \frac{N f_{j1}^2}{f_{j+}f_{+1}f_{+2}} - \frac{f_{+1}}{f_{+2}} \hspace{2cm} [18.10.3]$$

$$= \sum_j \frac{f_{+2}f_{j1}^2 + f_{+1}f_{j1}^2}{f_{j+}f_{+1}f_{+2}} - \frac{f_{+1}}{f_{+2}}$$

$$= \sum_j \frac{f_{j1}^2}{f_{j+}f_{+1}} + \sum_j \frac{f_{j1}^2}{f_{j+}f_{+2}} - \frac{f_{+1}}{f_{+2}}.$$

However, if we substitute $(f_{j+} - f_{j2})$ for f_{j1} in the second term, we have

$$R_{y \cdot \text{rows}}^2 = \sum_j \left[f_{j1}^2/(f_{j+}f_{+1}) \right] + \sum_j \left[f_{j2}^2/(f_{j+}f_{+2}) \right] - 2 + [N/f_{+2}] - [f_{+1}/f_{+2}]$$

$$= \sum_j \sum_k \left[f_{jk}^2/(f_{j+}f_{+k}) \right] - 1 \hspace{2cm} [18.10.4]$$

$$= \chi^2/N,$$

by Eq. 18.3.5. Rows and columns are simply reversed to show that this argument is true for a $2 \times K$ table as well. Incidentally, this also demonstrates the relation between chi-square and a one-way ANOVA done on frequencies, excluding the F test, of course.

18.11 CHI-SQUARE AND MULTIPLE REGRESSION IN A $J \times K$ TABLE

Consider a $J \times K$ table in which, let us say, there are equal numbers of rows and columns, or there are *fewer* columns, so that $J \geq K$. Then if the dummy variables representing column membership are transformed into $L = K - 1$ orthogonal (and thus uncorrelated) variables d_h, where $h = 1, 2, \cdots, L$, it will be true that the chi-square value divided by N will be equal to the sum of $K - 1$ values of R^2, each corresponding to the multiple regression of a single column variable d_h on the dummy variables representing rows:

$$\chi^2/N = \sum_{h=1}^{L} R^2_{d_h \cdot \text{rows}}.$$

[18.11.1]

That is, the value of chi-square for the total table, divided by N, is equal to the sum of $K - 1$ squared multiple correlations, one for each of $L = K - 1$ *orthogonal variables* d, as predicted from $J - 1$ dummy variables representing rows.

The rationale of this statement actually goes back about a half-century to the work of Sir Ronald Fisher (1940), who argued as follows: For any $J \times K$ contingency table, let L be the *smaller* of $J - 1$ or $K - 1$. Essentially, Fisher then showed that for any such contingency table (containing no zero cells) there must exist L pairs of standardized *latent* or *canonical* variables (x_{jg}, y_{kg}) where $g = 1, 2, \cdots, L$, and where x values are functions of the rows and the y values are functions of the columns. Across different sets the x values are uncorrelated, and so are the y values across sets. However, for a given set g, and the cell frequencies f_{jk}, there will be a correlation between the x_{jg} and y_{kg} values, as symnbolized by ρ_g, or "rho sub g," the *canonical correlation* for that set of values. The canonical correlation ρ_g is just the ordinary Pearson correlation between the x and y values in a given set g of such pairs of values, as taken over the frequencies f_{jk} in cells of the contingency table, or

$$\rho_g = \sum_{j}^{J} \sum_{k}^{K} \frac{x_{jg} y_{kg} f_{jk}}{N}.$$

(Remember that the canonical x and y variables are thought of as standardized, with mean of 0 and variance of 1.00.) However, the canonical x and y variables are, by definition, such that the first canonical correlation will be the largest, the second will be next largest, and so on:

$$1.00 \geq \rho_1 \geq \rho_2 \geq \cdots \geq \rho_L.$$

Fisher then pointed out that for *any* cell with observed frequency f_{jk} and corresponding expected frequency $f_{j+}f_{+k}/N$, it must be true that

$$(\text{observed frequency/expected frequency}) = N f_{jk}/(f_{j+}f_{+k})$$

[18.11.2]

$$= 1 + \sum_{g} \rho_g x_{jg} y_{kg}.$$

In other words, for any cell the obtained frequency divided by the expected frequency is equal to 1 plus the sum of the products of the canonical values for that cell, where each product is weighted by the corresponding canonical correlation. Thus, the ratios of the observed and expected frequencies give information about the canonical x and y values that (potentially at least) underlie the contingency table.

However, it is possible to take Fisher's results a step further and to establish the link between the value of chi-square and the squares of the underlying canonical correlations: First, we rewrite Eq. 18.11.2 slightly, and then take the average across all the cells, by multiplying Eq. 18.11.2 by the frequency of cell jk, summing across cells, and then dividing by N. That is, on rearranging,

$$N f_{jk}/(f_{j+}f_{+k}) - 1 = \sum_g \rho_g x_{jg} y_{kg}. \qquad [18.11.3]$$

On multiplying by frequency f_{jk} we have

$$N f_{jk}^2/(f_{j+}f_{+k}) - f_{jk} = \sum_g \rho_g x_{jg} y_{kg} f_{jk},$$

so that summing over all cells we find

$$N \sum_j \sum_k f_{jk}^2/(f_{j+}f_{+k}) - N = \sum_g \rho_g \sum_j \sum_k x_{jg} y_{kg} f_{jk}$$

$$= \sum_g \rho_g N \rho_g \qquad [18.11.4]$$

$$= N \sum_g \rho_g^2.$$

However, the expression on the left is just Eq. 18.3.5, one of our ways of writing chi-square, whereas the right-hand term is N times the sum of the squared canonical correlations. Dividing both sides by N we have

$$\chi^2/N = \sum_g \rho_g^2. \qquad [18.11.5]$$

Therefore,

---

the value of chi-square divided by N is the sum of the squared canonical correlations for the L sets of latent variables underlying the contingency table.

Chi-square divided by N is equal to a sum of squared correlations between (potential) variables representing the rows and columns.

Since L is the smaller of $J - 1$ or $K - 1$, and since there will be exactly L canonical correlations, we then have

$$\chi^2/NL = \text{mean } \rho_g^2 \qquad [18.11.6]$$

$$= V_C^2.$$

This says that the squared value of Cramér's statistic, V_C^2, is just the *average squared canonical correlation*.

Think again of our $J \times K$ table as having fewer columns than rows, so that $L = K - 1$. We can continue to show row category membership by dummy variables, of course. However, let us transform the $K - 1$ dummy variables standing for column category membership into $K - 1$ orthogonal (that is, uncorrelated) variables d_h, where $h = 1, \cdots, K - 1$. These are *not* the canonical variables alluded to above, although it will be convenient for these new column variables also to have means of zero and variances of 1.00, as well as being uncorrelated with each other.

Now we may think of the squared multiple correlation indicating the multiple regression of each of the column variables d_h on the dummy variables standing for rows, and there will be $K - 1$ of these R-square values, or one for each column variable d_h. It can then be shown that under these circumstances, the sum of the R-square values is equal to the sum of the squared canonical correlations:

$$\sum_{h}^{K-1} R^2_{d_h \cdot \text{rows}} = \sum_{g}^{L} \rho^2_g. \tag{18.11.7}$$

(For those of you with some background in matrix algebra, the basis for this relationship may be clarified by a couple of well-known principles that arise in the theory of canonical correlations and of multiple regression. These appear in Section D.8 of Appendix D).

Recall that the value of chi-square divided by N must also equal the sum of the canonical correlations, as shown in Eq. 18.11.10. Therefore,

$$\chi^2/N = \sum_{h=1}^{K-1} R^2_{d_h \cdot \text{rows}} \tag{18.11.8}$$

where, remember, we are thinking of the columns in terms of $K - 1$ orthogonal variables, each symbolized by d_h. Naturally, if we had a table where the number of rows was less than the number of columns, then we could just reverse the argument and talk about a set of orthogonal row variables to be predicted from the columns.

Suppose further that we have a set of uncorrelated row variables c_g, where $g = 1, 2, \cdots, J - 1$, and another set of uncorrelated column variables d_h, where $h = 1, 2, \cdots, K - 1$. (As we shall see, such variables can actually be obtained by the methods outlined in Section C.7 in Appendix C). Then suppose that we found all of the $(J - 1)(K - 1)$ correlations between the c and d variables across the tabled frequencies, f_{jk}. Then Eq. 18.11.8 implies that the sum of all of these squared correlations would equal to chi-square divided by N:

$$\chi^2/N = \sum_{g}\sum_{h} r^2_{gh}. \tag{18.11.9}$$

Moreover, each squared correlation could be turned into a *partial chi-square*, symbolized by ch^2_{gh}, by taking

$$ch^2_{gh} = Nr^2_{gh}. \tag{18.11.10}$$

These partial chi-squares will sum to form the chi-square value for the entire original table

$$\sum_{g}\sum_{h} ch^2_{gh} = \chi^2. \tag{18.11.11}$$

The point of this discussion is that a connection does exist between the ordinary Pearson chi-square test of association and multiple regression, in that chi-square divided by total N equals a sum of R-square values, where there are $L =$ minimum $(J - 1, K - 1)$ R-square values in the sum.

Finally, given the above, we can easily see that it must be true that

$$V^2_C = \chi^2/N(L) \tag{18.11.12}$$

$$= \text{average of } R^2_{d \cdot \text{rows}}.$$

The squared value of Cramér's V is equal to the average squared multiple correlation between columns and rows, where columns are represented by a set of $K - 1$ orthogonal variables d.

Actually, it is also true that if the rows are associated with $J - 1$ orthogonal variables or columns identified with $K - 1$ orthogonal variables, regardless of whether $J > K$ or $K > J$, then the value of chi-square divided by N is still equal to the sum of a set of R-square values, where the number of squared multiple correlations is the smaller of $J - 1$ or $K - 1$.

The next section will show how the chi-square value calculated on a $J \times K$ table can be divided into partial chi-square values, such that the sum of the $(J - 1)(K - 1)$ separate chi-square values is equal to the value of chi-square for the total table. This principle is equivalent to the creation of $(J - 1)(K - 1)$ "shadow tables" or 2×2 partial subtables of the total.

18.12 PARTITIONING A $J \times K$ TABLE INTO $(J - 1)(K - 1)$ DISTINCT "SHADOW TABLES"

As stated in the preceding section, it is possible to take a two-dimensional contingency table and to transform the dummy variables standing for row membership into new, uncorrelated, variables, each with mean of zero and variance of 1.00. (That is, the dummy variables are transformed into *orthonormal* form.) Similarly, dummy variables showing column membership can also be transformed in this way. Let c_{jg} (where $g = 1, 2, \ldots, J - 1$) stand for the value assigned to row j on one of these new, uncorrelated row variables, and let d_{kh} (where $h = 1, 2, \ldots, K - 1$) stand for the value for column k on one of the new column variables. Then, the correlation between the variables over all N cases in the contingency table is given by

$$r_{gh} = \sum_j \sum_k c_{jg} d_{kh} f_{jk} / N.$$

[18.12.1]

(In the vector and matrix terminology of Appendices C and D, Eq. 18.12.1 may be written as

$$r_{gh} = \mathbf{c}'_g \, \mathbf{F} \mathbf{d}_h N^{-1},$$

where $\mathbf{c}'_g$ is the transpose of the $J \times 1$ orthonormal vector number g associated with the rows, $\mathbf{d}_h$ is the $K \times 1$ orthonormal vector number h associated with the columns, and $\mathbf{F}$ here represents the $J \times K$ matrix of frequencies, f_{jk}.)

As we saw in the previous section, the sum of the squares of all of these $(J - 1)(K - 1)$ correlations is equal to the value of chi-square divided by N.

For example, consider once again the contingency table shown as Table 18.3.1. Section C.7 of Appendix C shows how the dummy variables standing for rows and for columns of this particular table can be turned into standardized, uncorrelated row variables c and column variables d. For this example the three sets of row variable values can be shown as the three orthonormal vectors

$$\mathbf{c}_1 = \begin{bmatrix} 1.7276 \\ -.5786 \\ -.5786 \\ -.5786 \end{bmatrix}, \quad \mathbf{c}_2 = \begin{bmatrix} 0 \\ 1.6590 \\ -.8046 \\ -.8046 \end{bmatrix}, \quad \mathbf{c}_3 = \begin{bmatrix} 0 \\ 0 \\ 1.3943 \\ -1.4219 \end{bmatrix}$$

The corresponding values for the three columns are given by the two orthonormal $\mathbf{d}'$ vectors:

$$\mathbf{d}'_1 = [.8538, -1.1713, -1.1713] \qquad \mathbf{d}'_2 = [0, .5982, -3.9680]$$

Table 18.12.1 shows the original contingency table, but with these new vectors of variable values listed for rows and for columns.

The correlation between c_1 and d'_1 values is then found from Eq. 18.12.1 to be:

$$\begin{aligned} r_{11} = \ & (1/797) \left[(1.7276 \times .8538 \times 122) + (1.7276 \times -1.1713 \times 70) \right. \\ & + (1.7276 \times -1.1713 \times 8) - (.5786 \times .8538 \times 141) - (.5786 \times -1.1713 \times 39) \\ & - (.5786 \times -1.1713 \times 15) - (.5786 \times .8538 \times 106) - (.5786 \times -1.1713 \times 79) \\ & - (.5786 \times -1.1713 \times 18) - (.5786 \times .8538 \times 92) - (.5786 \times -1.1713 \times 104) \\ & \left. - (.5786 \times -1.1713 \times 3) \right] \\ = \ & .037 \end{aligned}$$

Since each c or d' vector contains only two different, nonzero, values, whereas zero values effectively eliminate rows or columns from the computation, this is equivalent to finding the correlation for a collapsed 2×2 table, such as this:

	.8538	-1.1713
1.7276	122	78
-.5786	339	258

Table 18.12.1

The frequencies of Table 18.3.1 shown along with the new orthonormal vectors representing rows and columns.

c_1	c_2	c_3	d'_1: .8538	-1.1713	-1.1713
			d'_2: 0	.5982	-3.9680
1.7276	0	0	122	70	8
-.5786	1.6590	0	141	39	15
-.5786	-.8046	1.3943	106	79	18
-.5786	-.8046	-1.4219	92	104	3

or

$$r_{11} = (1/797)[(1.7276 \times .8538 \times 122) + (1.7276 \times -1.1713 \times 78)$$

$$- (.5786 \times .8538 \times 339) - (.5786 \times -1.1713 \times 258)]$$

$$= .0370.$$

Note, however, that the sample N remains the same, or 797, for each table. Such a 2×2 table defined by a pair of row and column variables that have been made orthogonal to the other rows and columns, is what is here referred to as a *shadow table*. For such a table, the square of this correlation, multiplied by N, can be thought of as a partial chi-square, contributing to the overall chi-square value. For this first shadow table the partial chi-square value is found from

$$ch_{11}^2 = Nr_{11}^2 = 797(.037)^2 = 1.09.$$

The remaining correlations, shown below for the relevant 2×2 tables, are:

	.8538	−1.1713
1.6590	141	54
−.8046	198	204

$$r_{21} = (1/797) [(1.6590 \times .8538 \times 141) + (1.6590 \times -1.1713 \times 54)$$

$$-(.8046 \times .8538 \times 198) - (.8046 \times -1.1713 \times 204)]$$

$$= .1895$$

so that

$$ch_{21}^2 = Nr_{21}^2 = 797(.1895)^2 = 28.62.$$

Next,

	.8538	−1.1713
1.3943	106	97
−1.4219	92	107

so that

$$r_{31} = (1/797)[(1.3943 \times .8538 \times 106) + (1.3943 \times -1.1713 \times 97)$$

$$- (1.4219 \times .8538 \times 92) - (1.4219 \times -1.1713 \times 107)].$$

$$= .0430$$

Here, $ch_{31}^2 = 797(.0430)^2 = 1.47.$

The fourth table and correlation is

	.5982	−3.9680
1.7276	70	8
−.5786	222	36

$$r_{12} = (1/797)[(1.7276 \times .5982 \times 70) + (1.7276 \times -3.9680 \times 8)$$
$$- (.5786 \times .5982 \times 222) - (.5786 \times -3.9680 \times 36)]$$
$$= .0293$$

so that

$$ch_{12}^2 = 797(.0293)^2 = .68.$$

For the fifth,

	.5982	−3.9680
1.6590	39	15
−.8046	183	21

$$r_{22} = (1/797)[(1.6590 \times .5982 \times 39) + (1.6590 \times -3.9680 \times 15)$$
$$- (.8046 \times .5982 \times 183) - (.8046 \times -3.9680 \times 21)]$$
$$= -.1017.$$

and

$$ch_{22}^2 = 797(.1017)^2 = 8.24.$$

Finally, we have

	.5982	−3.9680
1.3943	79	18
−1.4219	104	3

so that

$$r_{32} = (1/797)[(1.3943 \times .5982 \times 79) + (1.3943 \times -3.9680 \times 18)$$
$$-(1.4219 \times .5982 \times 104) - (1.4219 \times -3.9680 \times 3)]$$
$$= -.1320.$$

and

$$ch_{32}^2 = 797(.1320)^2 = 13.89.$$

The intercorrelation matrix, then, formed by the c variables standing for the rows, and the d variables for the columns is

	c_1	c_2	c_3	d_1	d_2
c^1	1	0	0	.0370	.0293
c_2		1	0	.1895	$-.1017$
c_3			1	.0430	$-.1320$
d_1				1	0
d_2					1

From this table it is easy to see that the squared multiple correlation for predicting d_1 from the set of c variables must be

$$R^2_{d_1 \cdot c_1 c_2 c_3} = (.0370)^2 + (.1895)^2 + (.0430)^2 = .0391$$

and that for predicting d_2 from the set of c variables is

$$R^2_{d_2 \cdot c_1 c_2 c_3} = (.0293)^2 + (.1017)^2 + (.1320)^2 = .0286.$$

The value of chi-square divided by N is then the sum of these two values of R-square, predicting column variables made orthogonal from row variables that are orthogonal:

$$\chi^2/N = 54.00/797 = (.0391 + .0286) = .0677.$$

Thus, the average of these two R^2 values turns out to be

$$(.0391 + .0286)/2 = .0339 = V^2_C$$

$$= \chi^2/2N$$

exactly as stated in Eq. 18.11.12, and agreeing with Section 18.3.

In this same way, the value of χ^2 divided by N can be regarded as the sum of three squared multiple correlations for predicting rows from columns. That is,

$$R^2_{c_1 \cdot d_1 d_2} = (.0370)^2 + (.0293)^2 = .0022$$

$$R^2_{c_2 \cdot d_1 d_2} = (.1895)^2 + (.1017)^2 = .0463$$

$$R^2_{c_3 \cdot d_1 d_2} = (.0430)^2 + (.1320)^2 = .0192.$$

Hence,

$$.0022 + .0463 + .0192 = .0677 = \chi^2/N.$$

Furthermore, the partial chi-squares that result from each of the 2×2 tables listed above will also total to form the overall chi-square value found in 18.3:

$$ch^2_{11} + ch^2_{21} + ch^2_{31} + ch^2_{12} + ch^2_{22} + ch^2_{32}$$

$$= 1.09 + 28.62 + 1.47 + .68 + 8.24 + 13.89$$

$$= 54.00$$

$$= \chi^2.$$

Unfortunately, space considerations forbid our going further along these lines. Suffice it here to say that each of the 2×2 tables, or shadow tables, on which the correlation and partial chi-square value is calculated, represents a distinct part of the total association present in the entire table, as represented in the total chi-square. By examining these partial chi-squares, one can begin to form a picture about where the association between row and column membership may be especially strong, and where it is relatively weak. Thus, examination of such shadow tables can form an important aid in the interpretation of the total association to be found in a contingency table.

18.13 A MEASURE OF INTERJUDGE AGREEMENT

A troublesome problem in many social and behavioral studies is that of assessing the agreement between two raters or judges, viewing the same set of people or objects. A useful index for such agreement, over and above the agreements that should occur even if the judgments were strictly independent, is given by Cohen (1960). Suppose that the ratings for two judges were arranged in an $I \times I$ table. Then Cohen's index "kappa"

$$K = \frac{N \sum_i x_{ii} - \sum_i x_{i+} x_{+i}}{N^2 - \sum_i x_{i+} x_{+i}} \qquad [18.13.1]$$

is a measure of their agreement, over and above the agreement to be expected for independent ratings.

Here, the judges are presumed to be rating on the same set of categories; x_{ii} symbolizes the number of agreements about category i, x_{i+} stands for the number of times Judge 1 used category i altogether, and x_{+i} symbolizes the number of times that the other judge used category i. Then N is the number of ratings given, or number of things rated.

If the agreements between the two raters are exactly what one would expect under independence, then K will be zero. If there is perfect agreement between the two raters or judges, then K will be 1.00.

Any exact hypothesis about the value of K may be tested, for large N, by use of the normal distribution, with an estimated value of the standard error of K given by the square root of

$$\text{est } \sigma_K^2 = \frac{1}{N} \left(\frac{\theta_1(1 - \theta_1)}{(1 - \theta_2)^2} + \frac{2(1 - \theta_1)(2\theta_1\theta_2 - \theta_3)}{(1 - \theta_2)^3} + \frac{(1 - \theta_1)^2(\theta_4 - 4\theta_2^2)}{(1 - \theta_2)^4} \right)$$

where

$$\theta_1 = \sum_i x_{ii}/N$$

$$\theta_2 = \sum_i (x_{i+} x_{+i})/N^2$$

$$\theta_3 = \sum_i x_{ii} (x_{i+} + x_{+i})/N^2 \qquad [18.13.2]$$

$$\theta_4 = \sum_i \sum_j (x_{ij})(x_{i+} + x_{+j})^2/N^3.$$

Furthermore, this standard error may be used to establish confidence limits for the true value of K when N is large. Thus, the hypothesis that long-run K is equal to zero may be tested by setting up the $100(1 - \alpha)$ percent confidence limits, and checking to see if this interval includes zero.

The hypothesis of *total* independence between the judges may be tested using K divided by the square root of

$$\text{est. } \sigma_K^2 = \frac{\theta_2 + \theta_2^2 - \sum_i x_{i+} x_{+i} (x_{i+} + x_{+i})/N^3}{N(1 - \theta_2)^2} \qquad [18.13.3]$$

once again with the normal distribution.

This index has been extended for application to more than two judges by Fleiss (1971).

18.14 ANALYZING LARGER TABLES THROUGH LOG-LINEAR MODELS

So far, the discussion of analyzing contingency tables has been restricted to the situation where the table is two-dimensional. That is, only two qualitative attributes, one represented by the rows and one represented by the columns, constitute the data table. However, in the social sciences it is quite common for data to be collected in terms of more than two attributes at a time. Thus, for example, for some population we might be interested in three attributes of a voter: age, represented by the groupings "under 40," "between 40 and 60," and "over 60"; political philosophy, as represented by the three groupings, "conservative," "moderate," and "liberal"; and party preference, represented by "Democratic," "Republican," "Independent," and "other." Each voter sampled would then be categorized in all three of these ways. The entire set of N cases would then be shown as frequencies in a three-dimensional table having $3 \times 3 \times 4$, or 36, cells. If even more categories or qualitative distinctions are drawn, we might have a four-way, or five-way, or twenty-way table.

Within such a complex table we may still be interested in the statistical association of the attributes represented. In the simple case of a two-way table, we can ask only about the relationship between the two attributes represented by the rows and columns. However, just as in the analysis of variance, higher order contingency tables generate higher order questions about the relationships in the table. How do you examine such complex relationships among qualitative attributes, and test for their existence in the population which the data represent?

This is an important problem for social scientists, and only in fairly recent years has it been given a manageable solution, largely due to the work of Leo Goodman (1978) and his colleagues. Although, for reasons of space, we cannot give an adequate treatment of this subject here, the methods are becoming sufficiently important in the social science literature that the student should have at least some limited acquaintance with what they are and how they work.

Modern methods for analyzing a large contingency table are based on a *log-linear model*. Although the model and method extend to tables of any dimensions, in the following, only the model for a three-dimensional table will be described briefly.

Imagine a three-dimensional table, representing some attribute A with I rows, Attribute B with J columns, and Attribute C with K layers. Let the frequency of observations in any cell (i, j, k) be denoted by x_{ijk}. Now the population from which the random sample of N was drawn has a probability associated with any cell (i, j, k) and we will call this probability p_{ijk}. Furthermore, for a sample of N independent observations, there will be an *expected frequency* $m_{ijk} = Np_{ijk}$ for any cell. Now let the *natural logarithm* (that is, the log to the base e) of the expected frequency m_{ijk} be symbolized by L_{ijk},

$$L_{ijk} = \log_e m_{ijk}.$$

Furthermore, let L (without a subscript) stand for the *average* L_{ijk} over all the IJK cells, let L_i stand for the average L_{ijk} value in column i, let L_{ij} stand for the average L_{ijk} value in combination ij, and so on for all of the other combinations of subscripts i, j, and k taken alone or in combination. Then the log-linear model is

$$L_{ijk} = u + u_i + u_j + u_k + u_{ij} + u_{ik} + u_{jk} + u_{ijk}. \qquad [18.14.1]$$

That is, the logarithm of the expected frequency in cell (i,j,k) is the sum of a general effect u; the effect u_i of being in row i; the effect of being in column j, or u_j; the effect u_k of being in layer k; plus an interaction effect u_{ij}; an interaction effect u_{ik}; an interaction effect u_{jk}; and finally a second-order interaction effect u_{ijk}. The u_i, u_j, and u_k are the *marginal* effects, the interaction effects u_{ij}, u_{ik}, u_{jk} represent the influence of *combinations* of two categories on frequencies in the table, and u_{ijk} represents the effect of *all three* categories in combination.

These population effects are defined very much as in the fixed-effects analysis of variance, and each is associated with a number of degrees of freedom. Thus given $u = L$, with df $= 1$,

$$u_i = L_i - L, \qquad\qquad \text{df} = (I - 1)$$

$$u_j = L_j - L, \qquad\qquad \text{df} = (J - 1)$$

$$u_k = L_k - L, \qquad\qquad \text{df} = (K - 1)$$

$$u_{ij} = L_{ij} - L_i - L_j + L, \qquad \text{df} = (I - 1)(J - 1)$$

$$u_{ik} = L_{ik} - L_i - L_k + L, \qquad \text{df} = (I - 1)(K - 1)$$

$$u_{jk} = L_{jk} - L_j - L_k + L, \qquad \text{df} = (J - 1)(K - 1)$$

$$u_{ijk} = L_{ijk} - L_{ij} - L_{ik} - L_{jk}$$
$$+ L_i + L_j + L_k - L, \ \text{df} = (I - 1)(J - 1)(K - 1)$$

Given this general model, it is possible to test various hypotheses in the form of specialized models including or excluding the various effects. The main restriction is that any model to be tested is regarded as *hierarchical*, which means simply that if, in a given model, a lower order effect, such as u_{ij} is assumed to be zero, then any higher order effect involving the same subscripts plus others must also be assumed to be zero. Thus, in a given hypothesis, $u_{ij} = 0$ implies $u_{ijk} = 0$. This is not an especially harsh restriction, however. (For example, as might at first appear, it does not necessarily rule out equal marginal frequencies. See Bishop, Fienberg, and Holland (1975) for details.)

Now, given the hypothesis that any effect or group of effects is zero (subject to the hierarchical principle just stated), the model permits one to derive estimates of the expected frequencies in the data. Once we have these estimated expected frequencies, it is possible to employ either chi-square tests or the so-called likelihood ratio test to assess the goodness of fit of the model to the data. In the three-dimensional situation, the likelihood ratio test, often called G^2, is made by taking

$$G^2 = -2 \sum_i \sum_j \sum_k x_{ijk} \log_e(\text{est. } m_{ijk}/x_{ijk}).$$ [18.14.2]

(Here, $0 \log_e 0$ is defined to be 0.) Under the null hypothesis generating the particular values of est. m_{ijk}, G^2 is distributed as chi-square with degrees of freedom equal to the *sum of the degrees of freedom* associated with *each term set equal to zero* in the model being tested. A significant result means, of course, that one can reject the hypothesis that these particular effects are all zero, implying that the particular model tested does not fit the data.

For example, suppose that the hypothesis to be tested for a three-way table were

H_0: $u_{ij} = u_{ik} = u_{jk} = u_{ijk} = 0$.

That is, we are testing the hypothesis that the three attributes are all independent of each other. Then the log-linear model implies that $L_{ijk} = L_i + L_j + L_k - 2L$. Sample estimates are then formed by taking, for any cell (i,j,k),

est. $L_{ijk} = \log_e(\text{est. } m_{ijk})$

$$= \log_e x_{i++} + \log_e x_{+j+} + \log_e x_{++k} - 2 \log_e N$$ [18.14.3]

where x_{i++} is the sum of row i, x_{+j+} is the sum of column j, and x_{++k} is the sum of layer k.

Then on substituting these $\log_e(\text{est. } m_{ijk})$ values into G^2 as given above, we have

$$G^2 = 2\left[\sum_i \sum_j \sum_k x_{ijk}\log_e x_{ijk} - \sum_i x_{i++}\log_e x_{i++} - \sum_j x_{+j+}\log_e x_{+j+} \right.$$

$$\left. - \sum_k x_{++k}\log_e x_{++k} + 2N \log_e N \right],$$ [18.14.4]

with $(I - 1)(J - 1) + (I - 1)(K - 1) + (J - 1)(K - 1) + (I - 1)(J - 1)(K - 1) = IJK - I - J - K + 2$ degrees of freedom.

If we wish to test a second model in which only one pair of attributes, say A and B, are related, this can be represented by the hypothesis

H_0: $u_{ik} = u_{jk} = u_{ijk} = 0$

Here, $L_{ijk} = L_{ij} + L_k - L$ so that

$$\log_e(\text{est. } m_{ijk}) = \log_e x_{ik+} + \log_e x_{++k} - \log_e N,$$

where x_{ij+} is the sum of the row-column combination i and j, and so on. Once again, a likelihood ratio test can be carried out using these values as the expected frequencies, and this time with $(I - 1)(K - 1) + (J - 1)(K - 1) + (I - 1)(J - 1)(K - 1)$ or $(K - 1)(IJ - 1)$ degrees of freedom:

$$G^2 = 2 \left[\sum_i \sum_j \sum_k x_{ijk} \log_e x_{ijk} - \sum_i \sum_j x_{ij+} \log_e x_{ij+} \right. $$

$$\left. - \sum_k x_{++k} \log_e x_{++k} + N \log_e N \right]. \qquad [18.14.5]$$

To test the hypothesis that a single pair of attributes, say A and B, are *not* related, while allowing for the possibility that A and C, and B and C *might* be related, we could frame the hypothesis

$$H_0: u_{ij} = u_{ijk} = 0.$$

Then, the model provides the logs of the estimated expected frequencies to be

$$\log_e (\text{est. } m_{ijk}) = \log_e x_{i+k} + \log_e x_{+jk} - \log_e x_{++k}$$

and the likelihood ratio test has $(I - 1)(J - 1) + (I - 1)(J - 1)(K - 1)$ degrees of freedom:

$$G^2 = 2 \left[\sum_i \sum_j \sum_k x_{ijk} \log_e x_{ijk} - \sum_i \sum_k x_{i+k} \log_e x_{i+k} \right. $$

$$\left. - \sum_j \sum_k x_{+jk} \log_e x_{+jk} + \sum_k x_{++k} \log_e x_{++k} \right]. \qquad [18.14.6]$$

Finally, suppose that we wish to examine the model which includes all possible effects *except* second-order interaction among A, B, and C.

That is, we wish to test the hypothesis

$$H_0: u_{ijk} = 0.$$

In this instance, the model does not provide a direct way to estimate the expected frequencies. Rather, these values are found by a process of *iteration*, in which the expected frequencies are approximated by successive "fits" to the table's marginals. This is laborious to do by hand but extremely simple to do on a computer. We will not describe this process here, however. A good description of the iterative process may be found in Bishop, Fienberg, and Holland (1975, pp. 84–89). Suffice it to say that once these estimated expected frequencies are found, they, too, are used in a likelihood ratio test, this time with $(I - 1)(J - 1)(K - 1)$ degrees of freedom.

The application of the log-linear model to a three-dimensional table may be illustrated as follows: Consider this 24-cell table, consisting of all combinations of an attribute A with two categories, an attribute B with three categories, and an attribute C with four categories. The frequencies shown in the cells of this table add up to 946 independent observations.

	A_1					A_2			
	B_1	B_2	B_3			B_1	B_2	B_3	
C_1	11	6	18	35	C_1	24	13	92	129
C_2	10	2	6	18	C_2	40	18	134	192
C_3	0	1	13	14	C_3	13	6	92	111
C_4	13	4	21	38	C_4	130	37	242	409
	34	13	58	105		207	74	560	841

Then preliminary calculations give

$$\sum_i \sum_j \sum_k x_{ijk} \log_e x_{ijk} = 11 \log_e 11 + \ldots + 242 \log_e 242 = 4{,}196.81$$

$$\sum_i \sum_j x_{ij+} \log_e x_{ij+} = 34 \log_e 34 + \ldots + 560 \log_e 560 = 5{,}354.76$$

$$\sum_i \sum_k x_{i+k} \log_e x_{i+k} = 35 \log_e 35 + \ldots + 409 \log_e 409 = 4{,}970.36$$

$$\sum_j \sum_k x_{+jk} \log_e x_{+jk} = (11 + 24)\log_e(11 + 24) + \ldots$$

$$+ (21 + 242)\log_e(21 + 242) = 4{,}507.83$$

$$\sum_i x_{i++} \log_e x_{i++} = 105 \log_e 105 + 841 \log_e 841 = 6{,}152.46$$

$$\sum_j x_{+j+} \log_e x_{+j+} = (34 + 207)\log_e(34 + 207) + \ldots$$

$$+ (58 + 560)\log_e(58 + 560) = 5{,}681.94$$

$$\sum_k x_{++k} \log_e x_{++k} = (35 + 129)\log_e(35 + 129) + \ldots$$

$$+ (38 + 409)\log_e(38 + 409) = 5{,}290.65.$$

$$N \log_e N = 946 \log_e 946 = 6{,}482.22.$$

Now the first hypothesis stated above (complete independence of A, B, and C) is tested by taking Eq. 18.14.4:

$$G^2 = 2(4{,}196.81 - 6{,}152.46 - 5{,}681.94 - 5{,}290.65 + 2(6{,}482.22))$$

$$= 72.40.$$

For 17 degrees of freedom, this is clearly significant, and so we can reject the model of total independence among A, B, and C on the basis of these data.

The second hypothesis corresponds to a model having no interactions involving variable C, but having an A × B interaction. Then using Eq. 18.14.5 we take

$$G^2 = 2(4{,}196.81 - 5{,}354.76 - 5{,}290.65 + 6{,}482.22) = 67.24.$$

For 15 degrees of freedom, this too is significant beyond .05, so that we can reject the model containing only an A × B interaction.

Next, the model containing A × C and B × C interactions, but no A × B or A × B × C interaction is examined (hypothesis 3 above) by using Eq. 18.14.6

$$G^2 = 2(4{,}196.81 - 4{,}970.36 - 4{,}507.83 + 5{,}290.65) = 18.54.$$

For 8 degrees of freedom, this too is significant beyond the .05 level.

Finally, if we test the hypothesis that there is no second-order interaction, but that all first-order interactions exist (finding the expected frequencies by iteration) we have a G^2 value of 11.96 with 6 degrees of freedom, which is not significant at the .05 level. Thus, it appears that the best fit to our data is a model with interactions A × B, A × C, B × C present, but with little or no A × B × C interaction effects.

In short, the log-linear model for a three-dimensional table permits one to test hypotheses about any, or any hierarchical set, of the seven effects (other than general u) entering into the model. The model then dictates what the expected frequencies should be, and how these should be estimated from the data. A likelihood ratio test, comparing the obtained and the expected frequencies, can then be used to test the hypothesis in question.

This method extends to tables of almost any number of attributes or dimensions. Naturally, for larger tables the number of interaction effects that can enter into the model becomes very much larger. Even so, in principle each effect and each combination of effects (under the hierarchical principle) may be examined. The assumptions are the same as for any ordinary chi-square test: independent observations representing multinomial sampling from some population. For very large tables, large samples are required in order to have a reasonable chance of some expected frequency (1 or 2) in every cell.

Unfortunately, space does not permit our going further into log-linear models here. Several good books exist, however, showing many of the methods flowing from the log-linear model. Perhaps the most complete discussion is to be found in Bishop, Fienberg, and Holland (1975). A good simple introduction to log-linear models is provided by Kennedy (1983). Original papers on these topics are found in Goodman (1978). Extensions to tables with ordered categories are considered in Goodman (1984).

Specialized computer programs also exist for analyzing large contingency tables by these methods, and these are now to be found in major research and university computing centers. The best known of these programs, which goes by the acronym ECTA (for Everyman's Contingency Table Analysis) was developed by Goodman and his colleagues at the University of Chicago (Fay and Goodman, 1975), and is widely available. It provides all required iterations, expected values, G^2 values, and a variety of statistics for contingency tables of virtually any size. Another, more versatile, program for this type of analysis is MULTIQUAL (Bock & Yates, 1973). Some packaged programs for these analyses are mentioned in the next section.

18.15 QUALITATIVE DATA ANALYSIS IN THE COMPUTER PACKAGES

All of the popular statistical packages for the computer include procedures for handling qualitative data and contingency tables. In SPSS the subprogram CROSSTABS arranges raw data into contingency tables, ranging in size from two-way to ten-way tables. Each such table will show the cell frequencies along with the marginal frequencies, and, if desired, the percentage of the total that each marginal represents. Statistics that may be requested for any table include the Pearson chi-square for association, (or Fisher's exact test when there are fewer than 21 cases), phi or Cramér's V, as appropriate, the Goodman and Kruskal lambda measures, and various other measures of association. In addition, the CROSSTABS subprogram in the SPSS[x] package permits one to request the expected frequency for each cell, the residuals, and standardized residuals. SPSS[x] also includes the McNemar test for correlated proportions, and the one-sample goodness-of-fit chi-square procedure as a part of the NPAR subprogram.

In addition to CROSSTABS, the SPSS[x] package also includes the analysis of log-linear models for multiway tables. These procedures are divided into two distinct sub-programs: LOGLINEAR, which includes the ordinary log-linear models mentioned in this chapter, as well as a variety of closely related techniques; and HILOGLINEAR, which gives a somewhat more efficient approach to the estimation of log-linear parameters and the selection of the best fitting model.

In the BMDP packages, the program P4F covers almost exactly the same ground as discussed above in connection with SPSS and SPSS[x]. There are, however, a number of additional features, such as, for example, collapsing of adjacent categories to reach a minimum cell frequency specified by the user. Log-linear models may also be examined in P4F, and there are many available options available for doing so.

The SAS package of programs includes contingency table construction and analysis as part of the FREQ procedure. Pearson chi-square tests and a wider than usual range of measures are available as options. Log-linear model analysis is only one of many approaches available in the CATMOD procedure of SAS, some of which apparently are to be found nowhere else. Use of this SAS procedure probably requires a somewhat more experienced or sophisticated user than do the log-linear routines referred to above, however.

EXERCISES

1. In the Midwest, a large number of sportscasters make predictions each Thursday about the outcomes of Saturday's football games during the season. On a certain Thursday, each of 50 sportscasters made predictions about the same eight games. The numbers of correct predictions of the winner are as follows:

Number correct	Number of sportscasters
8	1
7	3
6	5
5	11
4	15
3	8
2	5
1	1
0	1

Test the hypothesis that the correct predictions of any given sportscaster are outcomes of a stable and independent binomial process with $p = .5$. (**Hint:** Recall Chapter 3 and the binomial distribution for $p = .5$ and $N = 8$.)

2. Suppose that a theory predicts that if the average daily calorie consumption of healthy U.S. adolescent boys were measured, the relative frequency distribution would be like that below. Use the sample data in Exercise 9, Chapter 4, to test this theory.

Calories (in hundreds)	Relative frequency
50.00–54.99	.01
45.00–49.99	.04
40.00–44.99	.11
35.00–39.99	.20
30.00–34.99	.28
25.00–29.99	.20
20.00–24.99	.11
15.00–19.99	.04
10.00–14.99	.01

3. Again use the theoretical distribution in Exercise 2, but this time test the hypothesis that the population of adolescent girls follows this distribution, using Exercise 9 of Chapter 4 once again.

4. Take the two sample distributions in Exercise 9, Chapter 4, and test the hypothesis that they represent exactly the same population distribution. (**Hint:** If both samples come from the same population, what is your best estimate of the population relative frequencies?)

5. Suppose that the 1,000 words mentioned in Exercise 13, Chapter 2, were actually a random sample of all such foreign words. Test the hypothesis that if a word entered English after 1451 it was equally likely to have appeared in any half-century between that date and 1900.

6. Let us assume that the 270 neighborhoods in Exercise 28, Chapter 1, actually represent a random sample of a large population of such neighborhoods. Test the hypothesis that income and ethnic balance are independent properties of such neighborhoods.

7. Describe the degree of relationship that seems to exist between income and ethnic balance in Exercise 5 above by use of Cramér's statistic (V_C). How do you test the significance of Cramér's statistic?

8. An experimenter has developed an "excitability score" for rats, as a composite of a number of behavior indices. The experimenter is looking for evidence of association between the excitability score and the strain the rat belongs to. Some 25 rats are selected at random from each of the four pure strains, giving a total of 100 individuals in all. Within each strain used in the experiment a grouped frequency distribution of "excitability scores" is found, the same class intervals being used for each distribution. Test the hypothesis that all four strains have the same distribution, on the basis of the data below:

		Strain			
Scores	1	2	3	4	Total
9 and over	5	8	6	4	23
6–8	10	7	8	6	31
3–5	5	4	7	8	24
0–2	5	6	4	7	22
	25	25	25	25	100

9. Repeat Exercise 6, but this time describe the extent to which ethnic balance predicts income, by means of the λ_A measure. How well does income predict ethnic balance?

10. Consider the table in Exercise 32, Chapter 1. Test the hypothesis that interest pattern is independent of occupation, given that this is a random sample of independent observations.

11. To what extent is there predictive association from interest pattern to occupation in Exercise 10? How well can one predict from occupation to interest pattern?

12. For Exercise 31, Chapter 1, test the two 3×3 tables separately for association, and calculate Cramér's statistic for each. Do these two tables, each holding a third factor constant, seem to be different in the degree of association each represents? How might such an analysis of subtables shed light on partial association (analogous to partial correlation) in contingency tables?

13. In a study of the possible relationship between the number of years of nursery school and kindergarten training experienced by a child, and the rated deportment in the first grade, a random sample of 150 first-graders was obtained, and each was rated in terms of behavior:

	Deportment in 1st grade		
Prior experience	Poorly behaved	Moderately well behaved	Very well behaved
2 yrs. + kindergarten	6	12	0
1 yr. + kindergarten	12	25	6
Kindergarten only	14	31	12
No kindergarten	2	23	7

Is there significant association ($\alpha = .05$) between amount of nursery school and kindergarten and rated deportment in the first grade, according to these data? As judged from the data given above, to what extent does knowing the number of years in preschool and kindergarten permit us to predict the behavior rating of a child in first grade? (**Hint:** Compute the index of predictive association.) Comment on this result in the light of the first finding above.

14. Four random samples of 44 subjects each were drawn, and each sample assigned to a different experimental condition. Each subject was given the same problem-solving test, with a possible score of 0 through 12. The results yielded four frequency distributions, as follows. Test the hypothesis that these four samples represent identical population distributions ($\alpha = .05$):

Scores	Sample 1	Sample 2	Sample 3	Sample 4
12	1	5	3	2
11	1	2	3	6
10	1	2	3	5
9	2	6	3	6
8	3	2	3	1
7	8	4	4	1
6	12	2	6	2
5	8	4	4	1
4	3	2	3	1
3	2	6	3	6
2	1	2	3	5
1	1	2	3	6
0	1	5	3	2
	44	44	44	44

15. By inspecting the data table in Exercise 14, see if you can conclude what the F value would have been if an analysis of variance had been used. Would this be significant? Why? How do these sample distributions differ? What does this show about a comparison of samples via a chi-square test as opposed to a comparison via the F test in analysis of variance?

16. In a comparison of child-rearing practices within two cultures, a researcher drew a random sample of 100 families representing Culture I, and another sample of 100 families representing Culture II. Each family was classified according to whether the family was father-dominant or mother-dominant, in terms of administration of discipline. The results follow:

	Culture I	Culture II
Father-dominant	53	37
Mother-dominant	47	63

At approximately what α level can one reject the hypothesis of no association between culture and the dominant parent in a family?

17. For the data in Exercise 16 find the coefficient of predictive association (λ) for predicting parent dominance of a family from the culture. Do these data suggest the presence of a very strong predictive association here? Explain.

18. Four large Midwestern universities were compared with respect to the fields in which graduate degrees are given. The graduation rolls for last year from each university were taken, and the results put into the following contingency table:

University	Law	Medicine	Science	Humanities	Other
A	29	32	81	87	73
B	31	59	128	100	87
C	35	51	167	112	252
D	30	49	152	98	215

Is there significant association ($\alpha = .05$) between the university and the fields in which it awards graduate degrees? What are we assuming when we carry out this test?

19. In a study of the effect of a particular kind of cortical lesion on the ability of a monkey to learn a discrimination problem, two groups of six monkeys each were used. The following data were found:

	Solved problem	Did not solve problem
Experimental group	1	5
Control group	5	1

Is there a significant difference ($\alpha = .05$) between these two groups of monkeys, so that one may say that the experimental group is less likely to solve the problem?

20. A researcher was interested in the stability of political preference among American women voters. A random sample of 80 women who voted in the elections of 1976 and 1972 showed the following results:

1976 vote

		Republican	Democrat	
	Republican	34	11	45
1972 Vote				
	Democrat	5	30	35
		39	41	

Do these data afford significant evidence ($\alpha = .05$) that the true proportion of women who voted Republican in 1976 was different from the true proportion who voted Republican in 1972? (**Hint:** This problem involves *overlapping* or *correlated* proportions.)

21. In a study of the possible relationship between a male military officer's own confidential judgment of his effectiveness as a leader, and the judgment of him by his immediate superior, a sample of 112 officers was drawn at random. Each officer rated himself with respect to his leadership ability, and then his immediate superior was asked to rate him. These ratings were made independently. The data turned out as follows:

		Rating by superior			
		Low	Mod low	Mod high	Very high
	Very high	1	9	7	6
Self-rating	High	2	5	8	12
	Mod low	4	12	15	3
	Low	5	10	8	5

Does there seem to be significant ($\alpha = .05$) association between an officer's own judged leadership ability and the judgment of his immediate superior? Compute the Cramér coefficient showing the relative degree of association beween self and superior's ratings.

22. Use Cohen's kappa to assess the agreement in Exercise 19 between superior's ratings and self-rating. (**Note:** Be sure to observe the "High-Low" direction for each marginal of the table.)

23. In a study of the possibility of a sex linkage with the occurrence of identical twins, a random sample was taken of records of normal deliveries from a particular set of hospitals over a one-year period. These records revealed the following:

	Male	Female
Single births	658	688
Identical twins	39	34

Is there significant evidence that identical twins are more likely to be a given sex than are infants born singly? What does one assume about the respective deliveries recorded in this table?

24. Take the contingency table 18.3.1 once again, and carry out procedures for finding six shadow tables and associated partial chi-squares for this table, as illustrated in Section 18.12. However, this time, use the second sets of row and column variables given in Section C.7 of Appendix C. Demonstrate that the sum of the squared correlations for the six shadow tables equals chi-square divided by N, and that the sum of the partial chi-squares equals the overall chi-square for the original table.

Chapter **19**

Some Order Methods

In this chapter we are going to discuss methods based primarily on the **order relations** among observations in a set of data. The reasoning behind each of these tests involves relatively simple applications of probability theory; it sometimes happens that discrete sampling distributions often can be found in particularly simple ways if only the order features of the data are considered.

There are at least two reasons why we should be interested in analyzing data in terms of ordinal properties: In the first place, the only relevant information in a set of numerical scores may actually be ordinal. That is, it may well be true that the operation used to measure some psychological or social characteristic is valid only at the ordinal level, so that the numerical scores obtained actually give information only about relative magnitudes of the underlying property, and arithmetic differences between scores have no particular meaning in terms of this property. This is a common situation in measurements employed by social scientists.

An equally common reason for using order methods is that one or more assumptions about the population distributions, strictly necessary for one of the standard or parametric methods to apply, may be quite unreasonable. Rather than use the parametric method anyway and wonder about the validity of the conclusion, the experimenter prefers to change the question in such a way that another, *nonparametric* method does apply. Ordinarily, such a method will require that only certain features of the raw numerical data be considered if the objectionable assumptions are to be avoided, so that all of the numerical information in the scores is not used. Consequently, the experimenter may lose something by deciding to use the nonparametric method: the question answered by the nonparametric method may not be the same as that answered by the corresponding parametric method, and for a given sample size the nonparametric test may represent a considerably weaker use of the evidence.

Because of their direct dependency on elementary probability theory, and their comparative freedom from assumptions about population distributions and parameters, order techniques are usually classified among the nonparametric methods. However, not all nonparametric techniques involve considerations of order, nor are all order methods completely free of assumptions about distributions and parameters. Therefore, in this chapter, the author prefers to limit discussion to a few methods based on order, and more or less beg the complicated question of parametric versus nonparametric. For this, and virtually every other question concerned with such statistical methods, the reader is referred to the monumental *Handbook* by Walsh (1962, 1965). This chapter simply shows some techniques that are often found useful, and that happen to involve order.

Even so, a word of warning is called for in the use of order methods as stand-ins for the classical methods in situations where both kinds of methods are appropriate. At least two things must be borne in mind:

First, the actual hypothesis tested by a given order method is seldom exactly equivalent to the hypothesis tested by a parametric technique. These order tests can be regarded essentially as testing the hypothesis of identical population distributions. Regardless of how well the actual test statistic agrees with expectation, however, the population distributions still *might* be different in particular ways. In most order tests the sampling distribution implied by a true H_0 can also obtain when H_0 is not true in various ways. Departures from strict identity among the population distributions may exist to which the order methods are very insensitive. If one is willing to make only minimal assumptions about the population distributions, then the kinds of true differences among populations that the test fails to detect may be quite unknown. Of course, various additional assumptions can be made about the population distributions, and then the sensitivity of these tests to various alternatives can be studied. Nevertheless, these gratuitous assumptions may be fully as ill-considered as the assumptions the test was designed to avoid in the first place. It is sad but true: the specificity of our final conclusion is more or less bought in terms of what we already know or can at least assume to be true. If we do not know or assume anything, we cannot conclude very much.

Second, when both the order method and a parametric method actually do apply (that is, when the parametric assumptions are true), the power of the two kinds of tests may be compared, given α, the sample N, and the true situation. Order techniques, along with many of the nonparametric methods, will usually be relatively low-powered as compared with parametric tests. This means that, other things such as α and N being equal, one is taking a greater risk of a Type II error in using the order method. If Type II errors are to be avoided, then a relatively larger sample size (or a larger α value) may be required to use the order technique than for a parametric method.

The decision to use or not to use order methods in a given problem cannot be given a simple prescription. This is but another place where the experimenter has to think about what is to be accomplished and how. It is wrong to conclude that statistical assumptions are bad, that only bona fide interval scale data can be subjected to the classical statistical treatment, that tests with relatively low power efficiency are useless, and so on. None of these statements is necessarily true in all situations. Through practice and using all the help available, the experimenter must learn to pick and choose among all of the various methods available, finding the one that most clearly, economically, and reasonably sheds light on the particular question to be answered. This is not a simple task, and a brief discussion such as this must of necessity be only a cursory look at these issues.

In the sections to follow, several types of order statistics and tests will be discussed. First of all, tests will be mentioned that are appropriate to experimental data where the experimental factor is categorical and the dependent variable is treated at the ordinal level. Then some correlation-like methods for ordinal data will be surveyed, and finally an index of association for ordered classes will be outlined.

19.1 THE KOLMOGOROV-SMIRNOV TEST FOR GOODNESS OF FIT

An important though simple technique for comparing distributions is known as the Kolmogorov-Smirnov test (Kolmogorov, 1941). Problems can call for the comparison of a sample and a theoretical distribution, or in the comparison of two sample distributions. When the distributions each refer to a set of qualitative categories, then the chi-square methods of Sections 18.1 and 18.3 apply. However, when the distributions refer to categories that are at least ordered, so that the distributions can be put into cumulative form (Section 2.19) then the Kolmogorov-Smirnov test applies.

Suppose that we have a sample of N cases, producing a frequency distribution for some random variable X. There is also a theoretical distribution of the same random variable, and our task is to ask how well the theory fits the results of the sample. Let the sample cumulative relative frequency at or below any value x be represented by $F_S(x)$, and let the corresponding cumulative relative frequency in the theoretical distribution be symbolized $F_T(x)$. Then the statistic in the Kolmogorov-Smirnov test is

$$D_K = \underset{x}{\text{maximum}} \left| F_S(x) - F_T(x) \right|. \qquad [19.1.1]$$

That is, the statistic D_K is the *largest absolute difference* between the obtained and the theoretical *cumulative relative frequencies* that occurs over all the possible values of x. Table XIII in Appendix E gives the upper percentage points of this test.

For example, a researcher had the theory that a random variable should follow a normal distribution law within a certain population of cases. If this were truly the case, then in the population exactly 1/8 of the cases should fall into the class limits (in z-score terms) listed in Table 19.1.1 below. The researcher then drew a random sample of $N = 400$ cases from that population, and found the observed frequencies of z values listed under column f_j in the table. On the other hand, the expected frequencies, as shown in the column headed m_j, were 50 per class interval. How well does the theory fit the data?

Table 19.1.2 shows the cumulative relative frequency distribution for the data, under the column headed $F_S(x)$, and the cumulative distribution of the expected relative frequencies under the column headed $F_T(x)$. It is easy to see that the largest absolute difference between cumulative relative frequencies occurs for upper limit value .68, where we have

$$D_K = .9225 - .75 = .1725.$$

Class limits in terms of standardized variable	f_j	Observed relative frequencies	m_j	Expected relative frequencies
1.15 and above	14	.035	50	.125
0.68 to 1.15	17	.0425	50	.125
0.32 to 0.68	76	.190	50	.125
0.00 to 0.32	105	.2625	50	.125
−0.32 to 0.00	71	.1775	50	.125
−0.68 to −0.32	76	.190	50	.125
−1.15 to −0.68	31	.0775	50	.125
below −1.15	10	.025	50	.125

Table 19.1.1

Observed and expected frequencies from a normal distribution.

Value of standardized variable, x	$F_S(x)$	$F_T(x)$	$\lvert F_S(x) - F_T(x)\rvert$
1.15	.965	.875	.09
0.68	.9225	.75	.1725
0.32	.7325	.625	.11
0.00	.47	.50	.03
−0.32	.2925	.375	.0825
−0.68	.1025	.25	.1475
−1.15	.025	.125	.10

Table 19.1.2

Kolmogorov-Smirnov computations for the example.

The value of D_K is illustrated in Figure 19.1.1. On looking at Table XIII (Appendix E) we find that for $N = 400$ a value of

$$D_K = 1.63/\sqrt{400} = .0815$$

is required for significance at the .01 level. Our value is larger than this, and the researcher could reject the hypothesis of a normal population, beyond the .01 level for alpha.

The value of D_K obtained in a test such as this *does* depend on the choice of the class limits when one is dealing with a presumably continuous variable, as here. One should give a good bit of thought to the class intervals to be employed, and special considerations may dictate intervals of a certain size, or even unequal intervals. However, other things being equal, division into equal intervals provides some advantages. Actually, one really doesn't have to find class intervals at all as long as cumulative frequencies can be found for the values of X in each distribution. However, more complex procedures may have to be used to find D_K in that circumstance.

In the situation where two independent sample distributions are to be compared, the cumulative frequency from the second sample takes the role of $F_T(x)$, and D_K may be found in exactly the same way as above:

$$D_K = \underset{x}{\text{maximum}} \lvert F_{S_1}(x) - F_{S_2}(x)\rvert. \tag{19.1.2}$$

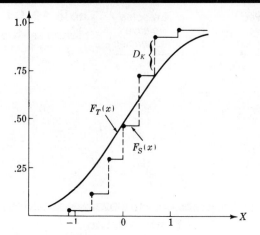

Figure 19.1.1

Curve illustrating value of D_K.

If the variable x is grouped, then the same class intervals should be employed in the distribution for each sample.

For small samples a special table is required for the Kolmogorov-Smirnov test in the two-sample case (Siegel, 1956). For larger samples, however, the test statistic may be converted to a chi-square variable by taking

$$\chi^2 = 4 D_K^2 \left(\frac{N_1 N_2}{N_1 + N_2} \right) \qquad [19.1.3]$$

with 2 degrees of freedom. This test should be adequate when the *smaller* of N_1 or N_2 is at least 20.

19.2 THE SIGN TEST FOR MATCHED PAIRS

In Section 3.16 we encountered a test applicable when the number of experimental treatments is only two, and *pairs* of observations are matched, the so-called sign test, which is based on the binomial distribution. We will now examine this test somewhat more closely.

Let N be the number of pairs of observations, where one member of each pair belongs to experimental (or natural) Treatment 1 and the other to Treatment 2. Here, the only relevant information given by the two scores for the pair is taken to be the **sign of the difference** between them. If the two treatments actually represent identical populations, and chance is the only determiner of which member of a pair falls into which treatment, we should expect an equal number of differences of plus and of minus sign. The theoretical probability of a plus sign is .5, and so the probability of a particular number of plus (or minus) signs can be found by the binomial rule with $p = .5$ and N. Notice, however, that we must assume either that the population distributions are continuous so that exact quality between scores has probability zero, or that ties are otherwise impossible. Ordinarily, pairs showing zero differences are simply dropped from the sample, although this makes the final conclusion have a conditional character, "In a population of untied pairs . . . ," and so on. The test is carried out as follows:

First, the direction of the difference (that is, the sign of the difference) between the two observations in each pair is noted, with the same order of subtraction always maintained. Thus each *pair* is given a classification of Plus or Minus, according to the sign of the difference between scores. If the null hypothesis of no difference between the two matched populations were true, one would expect half the nonzero differences to show a positive sign, and half to show a negative sign. Thus, one may simply take the *proportion* of plus differences, and test the hypothesis that the sample proportion arose from a true proportion of .50.

The normal approximation to the binomial can be used if the number of sample pairs is large (20 or more):

$$z = \frac{|P - p| - 1/(2N)}{\sqrt{pq/N}} = \frac{|P - .50| - 1/(2N)}{\sqrt{(.50)(.50)/N}}.$$ [19.2.1]

Here, $1/(2N)$ is a correction for continuity, as in Section 6.4. Although this form holds for a two-tailed test, either a one- or two-tailed test may be carried out, depending on the alternative hypothesis appropriate to the particular problem. Naturally, in the one-tailed test the sign of the difference between P and .5 is considered.

If the sample size is fewer than 20 pairs, the binomial distribution should be used to give an exact probability. That is, the binomial table (Table II, Appendix E) with $p = .5$ should be used to find the probability that the *obtained frequency* of the *more frequent sign* should be equaled or exceeded by chance, given a true p of .50. These tables of exact probabilities are given in Siegel (1956) as well as many other places. For a one-tailed test, one also checks that the more frequent sign accords with the alternative hypothesis before carrying out the test, of course. For a two-tailed test, this one-tailed binomial probability is doubled.

As an example of the sign test, consider this situation: An experimenter wants to see the effect that a frustrating experience might have upon the "social age" of a child. Each child could be age-rated from his play by a trained observer. Each of a random sample of preschool children was first rated by the observer according to the social-age level of his play during a free-play period. Pairs of children were then formed having the same rated social age. One randomly chosen member of each pair was frustrated by being allowed to play with a desirable toy that was then taken away, whereas the pair-mates were not frustrated. The observer did not know which children were frustrated, and the children were given these experimental conditions separately. Finally, in a post-experimental session the children were again rated at free play. Suppose that 20 pairs were used, and that the postexperimental ratings were as shown in Table 19.2.1.

The final rating of the frustrated child was subtracted from that of the pair-mate in order to find the sign of the difference between them. Eleven pairs showed a positive sign, seven showed a negative sign, and two showed a zero difference. Did presence or absence of frustration seem to be related to the difference in rated social age? The null and alternative hypotheses are

H_0: $p = .5$

H_1: $p \neq .5$,

Pair	Frustrated	Not frustrated	Sign of diff.
1	32	36	+
2	35	34	−
3	33	34	+
4	36	40	+
5	44	42	−
6	41	40	−
7	32	35	+
8	38	40	+
9	37	38	+
10	35	35	0
11	29	35	+
12	34	32	−
13	50	51	+
14	40	38	−
15	39	42	+
16	31	33	+
17	47	46	−
18	41	42	+
19	30	29	−
20	35	35	0

Table 19.2.1

Differences in observed "social age" of pairs of children.

where p is the population proportion of plus changes among pairs (plus being the more frequent sign of change in the sample). The statistic for this test is thus

$$z = \frac{|11/18 - .50| - 1/36}{\sqrt{(.50 \times .50)/18}} = \frac{.611 - .500 - .028}{.118} = .70.$$

Note that the two pairs showing no difference were excluded from the number of pairs figuring in the test, so that N is reduced to 18. When referred to a normal distribution, this z value leads to the conclusion of no significant difference. There is not enough evidence to conclude that frustration did lead to a difference in social-age rating. Table II of Appendix E gives an exact probability of .24 for 11 or more "successes" when $p = .50$ and $N = 18$, supporting our conclusion.

19.3 COCHRAN'S Q TEST

A test that seems to stand midway between methods designed for contingency table data and methods based on order was devised by Cochran (1950). This test can be viewed as a generalization of the McNemar two-sample test mentioned in Section 18.7, and it is appropriate in an experiment involving repeated observations (or matched groups) where the dependent variable can take on only two values:

$y_{jk} = 1$ (for "success," "pass," and so on, recorded for individual k in treatment j).

$y_{jk} = 0$ (for "fail," and so on, recorded for individual k in treatment j).

For example, suppose that in some experiment K subjects were observed in a standard situation where each subject performed individually under each of J different experimental conditions. Each subject was assigned the conditions in some random order. In each condition, the task of the subject was to solve one of a set of simple reasoning problems. If the problem was solved correctly within one minute, the performance was recorded as a "success," and as a "failure" otherwise. Suppose that the interest of the experimenter was in seeing if the problems had equal difficulty for the subjects—that is, if the true proportion of successes was constant over the problems. In this situation the Cochran test would apply.

As usual, let the experimental treatments (problems, in the example) be shown as columns in the data table. The subjects are shown as the rows. Then the entry in the cell formed by column j and row k contains y, which is 1 for a success and 0 for a failure. Let

$$y_k = \sum_j y_{jk}$$

be the marginal total for row k and let

$$y_j = \sum_k y_{jk}$$

be the marginal total for column j. Finally, let

$$\overline{T} = \frac{\sum_j y_j}{J}.$$

Then the statistic for Cochran's test is given by

$$Q = \frac{J(J - 1) \sum_{j=1}^{J} (y_j - \overline{T})^2}{J\left(\sum_k y_k\right) - \left(\sum_k y_k^2\right)}. \qquad [19.3.1]$$

For relatively large K, this is distributed approximately as chi-square with $J - 1$ degrees of freedom, when the hypothesis is true that the probability of a success is constant over all treatments J.

To continue the example, suppose that the experiment outlined above had been carried out. Twenty randomly selected subjects were given each of four problems, and each subject got the problems in some different, randomly chosen, order. A 1 was recorded for a successful solution, and a 0 for a failure. The data turned out to be as shown in Table 19.3.1. For 3 degrees of freedom, the χ^2 table shows this value not significant at either the .05 or .01 level. Thus, the hypothesis of no difference between problems is not rejected.

So far, all the tests discussed in this chapter have actually depended on some way of arranging the observations into only two ordered classes: above or below the median in the median tests, plus or minus differences in the sign test, 1 and 0 categories in the Cochran test. The only role of the numerical scores has been to assign individuals to

| | Problem | | | | |
Subject	1	2	3	4	Y_k
1	1	1	1	0	3
2	0	1	1	1	3
3	0	0	1	0	1
4	1	1	1	1	4
5	0	1	0	0	1
6	0	0	1	0	1
7	1	0	0	0	1
8	0	0	1	1	2
9	0	0	0	0	0
10	1	0	0	0	1
11	1	0	1	0	2
12	0	0	1	1	2
13	0	1	0	1	2
14	1	0	0	0	1
15	0	1	0	0	1
16	1	0	1	1	3
17	0	1	0	0	1
18	0	0	1	0	1
19	0	1	1	0	2
20	0	0	1	1	2
Y_j	7	8	12	7	34

$$\bar{T} = \frac{34}{4} = 8.5$$

$$\sum_k y_k^2 = 76$$

Table 19.3.1

Successes of 20 subjects on four problems.

$$Q = \frac{(4 \times 3)[(7 - 8.5)^2 + \cdots + (7 - 8.5)^2]}{(4 \times 34) - 76}$$

$$= 3.40$$

one of these two categories. When numerical data are reduced to only two categories, it is obvious that much of the possible information in the data is sacrificed. However, other techniques exist which use somewhat more of the information in the scores themselves; that is, observations are rank-ordered in terms of their scores in these methods. Next we turn to one of the simplest of the tests where the numerical score serves to give each observation a place in order.

19.4 THE WALD-WOLFOWITZ "RUNS" TEST FOR TWO SAMPLES

The Wald-Wolfowitz (1940) test applies to the situation where two unmatched samples are to be compared, and each observation is paired with a numerical score. The underlying variable that these scores represent is assumed to be continuously distributed.

Suppose that the numbers of observations in the two experimental groups are N_1 and N_2 respectively. All of the $N_1 + N_2$ observations in these samples are drawn independently and at random. For convenience, we will call any observation appearing in Sample 1 an A and any observation in Sample 2 a B. Now suppose that all these sample observations, irrespective of group, are arranged in order according to the magnitude of the scores shown. Then there will be some *arrangement* or pattern of A's and B's in order. In particular there will be **runs** or "clusterings" of the A's and B's. This is easily illustrated by an example.

Suppose that in some two-sample experiment the data turned out as shown in Table 19.4.1. When these scores are combined into a single set and arranged in order of magnitude, we get the following:

```
 B  B  B   A  A  A   B  B   A  A   B   A  A   B  B   A  A   B  B   A
 1  1  2   3  4  4   5  5   6  6   7   8  8  12 13  14 15  18 19  20
 _____/_____/_____/_____/\___/_____/_____/_____/_____/\___/
     1          2         3       4      5      6        7        8       9     10
```

Above each observation's score is an A or a B, denoting the group to which that observation belongs. Now notice that there are runs of A's and B's. That is, the ordering starts off with a run of three B's . . . this run is underlined and numbered 1. Then there is a run of three A's, which is run number 2. Proceeding in this way, and counting the beginning of a new run whenever an A is succeeded by a B or vice versa, we find that there are 10 runs.

It should be obvious that there must be at least two runs in any ordering of scores from two groups. If the groups are of equal size N there can be no more than $2N$ runs in all. In general, the number of runs cannot exceed $N_1 + N_2$.

Now suppose that the two groups were random samples from *absolutely identical* population distributions. In this instance we should expect many runs, since the values for the two samples should be well "mixed up" when put in order. On the other hand, if the populations differ, particularly in central tendency, we should expect there to be less tendency for runs to occur in the sample ordering. This principle provides the basis for a test based on fairly simple probability considerations.

Sample 1(A)	Sample 2(B)
8	12
6	13
8	19
4	18
14	7
4	2
15	1
20	1
3	5
6	5

Table 19.4.1

Data from a two-sample experiment.

For moment, let R symbolize the total number of runs appearing for the samples. Then it can be shown that if R is any *odd* number, $2g + 1$, the probability for that number of runs is

$$\text{prob}(R = 2g + 1) = \frac{\binom{N_1 - 1}{g - 1}\binom{N_2 - 1}{g} + \binom{N_1 - 1}{g}\binom{N_2 - 1}{g - 1}}{\binom{N_1 + N_2}{N_1}},$$

when all arrangements in order are equally likely.

If R is an *even* number, $2g$, then

$$\text{prob}(R = 2g) = \frac{2\binom{N_1 - 1}{g - 1}\binom{N_2 - 1}{g - 1}}{\binom{N_1 + N_2}{N_1}}.$$

On this basis, the exact sampling distribution of R can be worked out, given equal probability for all possible arrangements of A and B observations. It turns out that for fairly large samples the distribution of R can be approximated by a normal distribution with

$$E(R) = \frac{2N_1N_2}{N_1 + N_2} + 1 \qquad\qquad [19.4.1]$$

and

$$\sigma_R^2 = \frac{2N_1N_2(2N_1N_2 - N_1 - N_2)}{(N_1 + N_2)^2(N_1 + N_2 - 1)}. \qquad\qquad [19.4.2]$$

Thus, an approximate large-sample test is given by

$$z = \frac{R - E(R)}{\sigma_R} \qquad\qquad [19.4.3]$$

referred to a normal distribution. Since the usual alternative hypothesis entails "too few" runs, the test is ordinarily *one-tailed* with only negative values of z leading to rejection of the hypothesis of identical distributions.

For sample size less than or equal to 20 in *either* sample, exact values of R required for significance are given in Siegel (1956).

There is some reason to believe that the runs test generally has rather low power when it is compared either with a t test for means or with other order tests for identical populations. However, it is mentioned here because of its utility in various problems where other methods may not apply. Actually, A and B may designate any dichotomy within a *single* sample, and any principle at all that gives an ordering to A's and B's may be used. For example, it may be of interest to see if there is a time-related trend such as learning in a *single* set of data. In this instance, we might find the overall median for the set of scores, calling above the median, A, and below, B. Here time is used as the ordering principle, and the occurrence of few runs is treated as evidence that time trends do exist. In any problem where the data may be given a dichotomous classification in one respect and then simply ordered in another respect, the runs test gives a way to answer the question of possible association between the basis for the ordering and the categorization. This makes the runs test useful in a variety of problems where other tests do not apply directly, and especially to problems where the *experimental* variable is ordinal and the *dependent* variable categorical.

A major technical problem with this test is the treatment of ties. In principle, ties should not occur if the scores themselves represent a continuous random variable. But, of course, ties do occur in actual practice, since we seldom represent the underlying variable directly or precisely in the data. If tied scores all occur among observations in the *same* group (as in the example above), then there is no problem: the value of R is unchanged by any method of breaking these ties. However, if members of *different* groups show tied scores, the number of runs depends on how these ties are resolved in the final ordering. One way of meeting this problem is to break all ties in a way *least* conducive to rejecting the null hypothesis (so that the number of runs is made as large as possible). This at least makes for a conservative test of H_0. However, if cross-group ties are very numerous, this test is really inapplicable.

19.5 THE MANN-WHITNEY TEST FOR TWO INDEPENDENT SAMPLES

Unlike the runs test, the Mann-Whitney (1947) test employs the actual *ranks* of the various observations directly as a device for testing hypotheses about the identity of two population distributions. It is apparently a good and relatively powerful alternative to the usual t test for equality of means. We assume that the underlying variable on which two groups are to be compared is continuously distributed. The null hypothesis to be tested is that the two population distributions are identical. Then we proceed as follows:

The scores from the combined samples are arranged in order (much as in the runs test). However, now we assign a *rank* to each of the observations, in terms of the magnitude of the original score. That is, the lowest score gets rank 1, the next lowest 2, and so on. Now choose one of the samples, say Sample 1, and find the *sum* of the ranks associated with observations in that sample. Call this T_1. Then find

$$U = N_1N_2 + \frac{N_1(N_1 + 1)}{2} - T_1. \qquad [19.5.1]$$

If the resulting value of U is *larger* than $N_1N_2/2$, take

$$U' = N_1N_2 - U.$$

The statistic used is the *smaller* of U or U'. (Incidentally, this choice of the smaller of the two values for U is important in using tables to find significance for small samples, but is really immaterial in the large-sample test to be described later.)

As an example, consider the following data:

Sample 1(A)	Sample 2(B)
8	1
3	7
4	9
6	10
	12

Arranged in order and ranked, these data become

	B	A	A	A	B	A	B	B	B
Score	1	3	4	6	7	8	9	10	12
Rank	1	2	3	4	5	6	7	8	9

The sum of the ranks for the A observations (Group 1) is

$$T_1 = 2 + 3 + 4 + 6 = 15.$$

This is turned into a value of U by taking

$$U = 4(5) + \frac{4(5)}{2} - 15$$

$$= 15.$$

This larger than $\frac{(4)(5)}{2}$ or 10, and so we take

$$U' = 20 - 15 = 5$$

as the value we will use.

Now notice that given these 9 *scores,* the value of U depends only on how the A's and B's happen to be arranged over the rank order. The number of possible random arrangements is just $\binom{N_1 + N_2}{N_1}$, and if the hypothesis of completely identical popula-tions is true, the random assignment of individuals to groups should be the only factor entering into variation among obtained U values. Under the null hypothesis, all arrange-ments should be equally likely, and this gives the way for finding the probability asso-ciated with various values of U. For large samples, this sampling distribution of U is approximately normal, with

$$E(U) = \frac{N_1 N_2}{2} \qquad\qquad [19.5.2]$$

and

$$\sigma_U^2 = \frac{N_1 N_2 (N_1 + N_2 + 1)}{12}.$$ [19.5.3]

Thus, for large samples, the hypothesis of no difference in the population distributions is tested by

$$z = \frac{U - E(U)}{\sigma_U}.$$ [19.5.4]

For a two-tailed test, either U or U' may be used, since the absolute value of z will be the same for either. However, if the alternative hypothesis is such that one of the populations should tend to have a lower average than the other (assuming distributions of similar *form*), then a one-tailed test is appropriate, and the sign of the z should be considered.

For situations where the larger of the two samples is 20 or more and the samples are not too different in size, the normal approximation given above should suffice. However, when the larger sample contains fewer than 20 observations, tables given in Siegel (1956) should be used to evaluate the significance of U.

This test is one of the best of the nonparametric techniques with respect to power. It seems to compare quite well with t when assumptions for both tests are met. For some special situations, it is even superior to t. This makes it an extremely useful device for the comparison of two independent groups.

Ordinarily, ties are treated in the Mann-Whitney test by giving each of a set of tied scores the *average* rank for that set. Thus, if three scores are tied for fourth, fifth, and sixth place in order, each of the scores gets rank 5. If two scores are tied for ninth and tenth place in order, each gets rank 9.5, and so on. This introduces no particular problem for larger sample size when the normal approximation is used and ties are relatively infrequent. However, when ties exist σ_U^2 becomes

$$\sigma_U^2 = \frac{N_1 N_2}{12} \left[N_1 + N_2 + 1 - \frac{\sum_{i=1}^{G} (b_i^3 - b_i)}{(N_1 + N_2)(N_1 + N_2 - 1)} \right],$$ [19.5.5]

where there are some G distinct *sets* of tied observations, i represents any *one* such set, and b_i is the number of observations tied in set i. For a small number of ties and for large $N_1 + N_2$, this correction to σ_U^2 can safely be ignored.

19.6 THE WILCOXON TEST FOR TWO MATCHED SAMPLES

As we saw in Section 19.2, two matched samples can be compared by the sign test if the only feature of the data considered is the sign of the difference between each pair. However, this still overlooks one other important property of any pair of scores: not only does a difference have a direction, but also a size that can be ranked in order among the set of all such differences. The Wilcoxon (1949) test takes account of both features in the data, and thus uses somewhat more of the available information in paired scores than does the sign test.

The mechanics of the test are very simple: the signed difference between each pair of observations is found, just as for the t test for matched groups (Section 8.16). Then these differences are rank-ordered in terms of their absolute size. Finally, the sign of the difference is attached to the rank for that difference. The test statistic is T, the *sum of the ranks with the less frequent sign.*

Suppose that in some experiment involving a single treatment and one control group, subjects were first matched pairwise, and then one member of each pair was assigned to the experimental group at random. In the experiment proper, each subject received some Y score. Perhaps the data turned out as shown in Table 19.6.1. Here, the differences are found, their absolute size ranked, and then the sign of the difference attached to the rank. The less frequent sign is minus, and so

$$T = 3.5 + 1 + 3.5 + 7 = 15.$$

The hypothesis tested by the Wilcoxon test is that the two populations represented by the respective members of matched pairs are identical. When this hypothesis is true, then each of the 2^N possible sets of *signed* ranks obtained by arbitrarily assigning $+$ or $-$ signs to the ranks 1 through N is *equally* likely. The random assignment of subjects to experimental versus control group in the example is tantamount to such a random assignment of signs to ranks when the null hypothesis is true. On this basis, the exact distribution of T over all possible randomizations can be worked out. For large N (number of pairs), the sampling distribution is approximately normal with

$$E(T) = \frac{N(N + 1)}{4} \qquad [19.6.1]$$

and

$$\sigma_T^2 = \frac{N(N + 1)(2N + 1)}{24} \qquad [19.6.2]$$

Pair	Treatment	Control	Difference	Rank	Signed rank
1	83	75	8	8	8
2	80	78	2	2	2
3	81	66	15	10	10
4	74	77	−3	3.5	−3.5
5	79	80	−1	1	−1
6	78	68	10	9	9
7	72	75	−3	3.5	−3.5
8	84	90	−6	7	−7
9	85	81	4	5	5
10	88	83	5	6	6

Table 19.6.1

Differences and signed ranks for 10 matched pairs.

so that a large-sample test is given by

$$z = \frac{T - E(T)}{\sigma_T}.$$

[19.6.3]

This test can be either directional or nondirectional, depending on the alternative hypothesis. However, a directional test usually makes sense only if one is prepared to assume that the distributions have the same form, and that a signed deviation of T from $E(T)$ is equivalent to a particular difference in central tendency between the two populations. For samples larger than about $N = 8$, this normal approximation is adequate. For very small samples, a table given in Siegel (1956) should be used. Since only one set of differences is ranked, ties present no special problem unless they occur for zero differences. If an *even* number of zero differences occur, each zero difference is assigned the average rank for the set (zero differences, of course, rank lowest in absolute size), and then half are arbitrarily given positive and half negative sign. If an odd number of zeros occur, one randomly chosen zero difference is discarded from the data, and the procedure for an even number of zeros followed, except that N is reduced by 1, of course. For other kinds of tied differences, the method used in the example may be followed. Be sure to notice that even when several pairs are tied in absolute size so that they all receive the midrank for that set, the sign given to that midrank for different pairs may be different. For fairly large samples with relatively few ties, this procedure of assigning average ranks introduces negligible error.

All in all, the Mann-Whitney and the Wilcoxon tests are generally regarded as the best of the order tests for two samples. They both compare well with t in the appropriate circumstances, and when the assumptions for t are not met they may even be superior to this classical method. However, each is fully equivalent to a classical test of the hypothesis that the *means* of two groups are equal only when the assumptions appropriate to t are true. Unless additional assumptions are made, these tests refer to the hypothesis that two population distributions of unspecified form are *exactly* alike. In many instances, this is the hypothesis that the experimenter wishes to test, especially if interest lies only in the possibility of statistical independence or of association between experimental and dependent variables. However, to make particular kinds of inferences, particularly about population *means,* other assumptions become necessary. Without these assumptions, the rejection of H_0 implies only that the populations differ in *some* way, but the test need not be equally sensitive to all ways that population distributions might differ.

19.7 THE KRUSKAL-WALLIS "ANALYSIS OF VARIANCE" BY RANKS

The same general argument for the Mann-Whitney test may be extended to the situation where J independent groups are being compared. The version of a J-sample rank test given here was devised by Kruskal and Wallis (1952). This test has very close ties to the Mann-Whitney and Wilcoxon tests just discussed, and can properly be regarded as a generalized version of the Mann-Whitney method.

Imagine some J experimental groups in which each observation is associated with a numerical score. As usual, we assume that the underlying variable is continuously distributed. Now, just as in the Mann-Whitney test, the scores from all groups are pooled, arranged in order of size, and ranked. Then the rank sum attached to *each separate group* is found. Let us denote this sum of ranks for group j by the symbol T_j.

T_j = sum of ranks for group j.

For example, suppose that three groups of small children were given the task of learning to discriminate between pairs of stimuli. Each child was given a series of pairs of stimuli, in which each pair differed in a variety of ways. However, attached to the choice of one member of a pair was a reward, and within an experimental condition, the cue for the rewarded stimulus was always the same. On the other hand, the experimental treatments themselves differed in the *relevant* cue for discrimination: in Treatment I, the cue was form, in Treatment II, color, and in Treatment III, size. Some 36 children of the same sex and age were chosen at random and assigned at random to the three groups, with 12 children per group. The dependent variable was the number of trials to a fixed criterion of learning. Suppose that the data turned out to be as shown in Table 19.7.1. Here the numbers in parentheses are the ranks assigned to the various score values in the entire set of 36 cases. Then the sum of the ranks for each particular group j is found, and designated T_j. The value of T is the sum of these rank sums; if the ranking has been done correctly, it will be true that

$$T = \frac{N(N + 1)}{2}.$$

Note here that

$$T = \frac{36 \times 37}{2} = 666.$$

	Treatment		**Table 19.7.1**
I	II	III	Scores and ranks for three independent groups.
6 (1)	31 (34.5)	13 (10)	
11 (7)	7 (2)	32 (36)	
12 (9)	9 (4)	31 (34.5)	
20 (19)	11 (7)	30 (33)	
24 (23)	16 (14)	28 (31)	
21 (20)	19 (17.5)	29 (32)	
18 (16)	17 (15)	25 (24)	
15 (13)	11 (7)	26 (26.5)	
14 (11.5)	22 (21)	26 (26.5)	
10 (5)	23 (22)	27 (29.5)	
8 (3)	27 (29.5)	26 (26.5)	
14 (11.5)	26 (26.5)	19 (17.5)	
T_j 139.0	200.0	327.0	

$T = 666$

For large samples, a fairly good approximate test for identical populations is given by

$$H = \frac{12}{N(N + 1)} \left[\sum_j \frac{T_j^2}{n_j} \right] - 3(N + 1). \tag{19.7.1}$$

This value of H can be referred to the chi-square distribution with $J - 1$ degrees of freedom for a test of the hypothesis that all J population distributions are identical.

For the example,

$$H = \frac{12}{36 \times 37} \left(\frac{139^2 + 200^2 + 327^2}{12} \right) - 3(37)$$

$$= 13.81.$$

However, since there were ties involved in the ranking, this value of H really should be corrected by dividing through by a value found from

$$C = 1 - \left[\frac{\sum_i^G (t_i^3 - t_i)}{N^3 - N} \right] \tag{19.7.2}$$

where G is the number of sets of tied observations, and t_i is the number tied in any set i. For example, there are 4 sets of two tied observations, 1 set of three ties, and 1 set of four tied observations. Thus,

$$C = 1 - \left[\frac{4(2^3 - 2) + (3^3 - 3) + (4^3 - 4)}{36^3 - 36} \right]$$

so that

$$C = .997$$

Finally, the corrected value of H is

$$H' = \frac{H}{C} = \frac{13.81}{.997} = 13.85.$$

Unless N is small, or unless the number of tied observations is very large, relative to N, this correction will make very little difference in the value of H. Certainly, this was true here. Furthermore, when each set of tied observations lies within the same experimental group, the correction becomes unnecessary.

In terms of Table IV (Appendix E), for 2 degrees of freedom the value of H exceeds that required for the .01 level, and so the experimenter can be quite confident in saying that the population distributions are not identical. Apparently there is some association between the type of cue given in the discrimination problem and the number of trials to criterion.

It is somewhat difficult to specify the class of alternative hypotheses appropriate to the J-sample rank test, and so the question of power is somewhat more obscure than for the Mann-Whitney test. However, there is reason to believe that this test is about the best of the *J*-sample order methods. In comparisons with *F* from the analysis of variance, the Kruskal-Wallis test shows up extremely well.

When sample size is relatively small within the groups, the tables given by Siegel (1956) should be consulted.

19.8 THE FRIEDMAN TEST FOR *J* MATCHED GROUPS

Much as the Kruskal-Wallis test represents an extension of the Mann-Whitney test, so the Friedman (1937) test is related to the Wilcoxon matched-pairs procedure. The Friedman test is appropriate when some *K* sets of matched individuals are used, where each set contains *J* individuals assigned at random to the *J* experimental treatments. It also applies when each of *K* individuals is observed under each of *J* treatments in random order. Thus, it is useful in situations where data are collected much as in the randomized blocks or repeated measures experiments mentioned in Chapter 13.

The data are set up in a table as for a two-way analysis of variance with one observation per cell. The experimental treatments are shown by the respective columns, and the matched sets of individuals by the rows. Within each row (matched group) a rank order of the *J* scores is found. Then the resulting ranks are summed by columns, to give values of T_j.

For example, in an experiment with four experimental treatments ($J = 4$), 11 groups of four matched subjects apiece were used. Within each matched group the four subjects were assigned at random to the four treatments, one subject per treatment. The data can be represented as in Table 19.8.1

It is important to remember that the ranks are given to the scores *within* rows. Then the T_j values are simply the sums of those ranks within columns.

Groups	I	II	III	IV
1	1 (2)	4 (3)	8 (4)	0 (1)
2	2 (2)	3 (3)	13 (4)	1 (1)
3	10 (3)	0 (1)	11 (4)	3 (2)
4	12 (3)	11 (2)	13 (4)	10 (1)
5	1 (2)	3 (3)	10 (4)	0 (1)
6	10 (3)	3 (1)	11 (4)	9 (2)
7	4 (1)	12 (4)	10 (2)	11 (3)
8	10 (4)	4 (2)	5 (3)	3 (1)
9	10 (4)	4 (2)	9 (3)	3 (1)
10	14 (4)	4 (2)	7 (3)	2 (1)
11	3 (2)	2 (1)	4 (3)	13 (4)
T_j	30	24	38	18

Table 19.8.1

Scores and ranks within matched groups.

$$T = \frac{K(J)(J + 1)}{2}$$

$$= 110$$

The rationale for this test is really very simple. Suppose that within the population represented by a row, the distribution of values for all the *treatment* populations were identical. Then, under random sampling and randomization of observations, the probability for any given permutation of the ranks 1 through J within a given row should be the same as for any other permutation. Furthermore, across rows, each and every one of the $J!$ possible permutations of ranks across columns should be equally probable. This implies that we should expect the column sums of ranks to be identical under the null hypothesis. However, if there tend to be pile-ups of high or low ranks in particular columns, this is evidence against equal probability for the various permutations, and thus against the null hypothesis.

The test statistic for large samples is given by

$$\chi_r^2 = \frac{12}{KJ(J + 1)} \left[\sum_j T_j^2 \right] - 3K(J + 1) \qquad [19.8.1]$$

distributed approximately as chi-square with $J - 1$ degrees of freedom.

For the example, we would take

$$\chi_r^2 = \frac{12}{11(4 \times 5)} [(30)^2 + (24)^2 + (38)^2 + (18)^2] - 3(11 \times 5)$$

$$= 11.95.$$

For 3 degrees of freedom, this is just significant at the .01 level, and so the experimenter may say with some assurance that the treatment populations differ.

If ties in ranks within rows should occur, a conservative procedure is to break the ties so that the T_j values are as close together as possible.

This test may well be the best alternative to the ordinary two-way analysis of variance for matched groups or repeated measures. Once again, there is every reason to believe this test, much like the Kruskal-Wallis test for one-factor experiments, should compare well with F when both the classical and order methods apply.

The chi-square approximation given above is good only for fairly large K. However, the test should be satisfactory when $J \geq 4$ and $K \geq 10$. As usual, tables giving significance levels for small samples can be found in Siegel (1956).

19.9 RANK-ORDER CORRELATION METHODS

The tests discussed so far in this chapter are designed for situations where the experimental variable is categorical and the dependent variable is in ordinal terms, as given either by the original measurement procedure or by a transformation of the scores into ordered classes or ranks. In this section, measures and tests of association will be described for situations where *both* variables are represented in ordinal terms. First of all, two somewhat different measures of "agreement" or association between rank orders will be introduced. Next, a measure of association will be given for the special situation where both variables take the form of ordered classes. Finally, the problem of indexing simultaneous agreement among several rank orders will be treated.

It is customary to call some of these rank-order statistics "correlations," but this usage deserves some qualification. The Spearman rank correlation to be introduced next actually *is* a correlation coefficient computed for numerical values that happen to be ranks. However, the next index to be considered, Kendall's tau, is not a correlation coefficient at all. Neither of these indices is closely connected with the classical theory of *linear* regression. Note especially that the *square* of a correlation-like index on ranks is not to be interpreted in the usual way as a proportion of variance accounted for in the underlying variables. Instead, it is somewhat better to think of both the Spearman and the Kendall indices only as showing "concordance" or "agreement," the tendency of two rank orders to be similar. As descriptive statistics, both indices serve this purpose very well, although the definition of *disagree* is somewhat different for these two statistics.

19.10 THE SPEARMAN RANK CORRELATION COEFFICIENT

Imagine a group of N cases drawn at random as for a problem in correlation. However, instead of having an X score for each individual, we have only the *rank* in the group on variable X (say for low to high, the ranks 1 through N). In the same way, for each individual we have the rank in the group on variable Y. The question is "How much does the ranking on variable X tend to agree with the ranking on variable Y?" and a measure is desired to show the extent of the agreement. Or, perhaps, we have two judges who each rank the same set of N objects. We wish to ask, "How much does Judge A agree with Judge B?" In either instance, there are two distinct rank orders of the same N things, and these rank orders are to be compared for their agreement with each other. Furthermore, when these objects or individuals constitute a random sample from some population, we may wish to test the hypothesis that the true agreement in ranks is zero.

Two simple ways for comparison of two rank orders for agreement have already been mentioned: the first, and older, method is the Spearman rank correlation (Kendall, 1963) commonly symbolized r_S (although ρ is sometimes used, this symbol r_S will be used here to avoid confusion with the population correlation); the second method is the Kendall (1963) "tau" statistic, which will be discussed in the following section. Regardless of whether the data are two rank orders representing scores shown by individuals in a sample, or ranks of objects given by two judges, we can apply r_S. Suppose that in either circumstance we call the things ranked "individuals," and the two bases for ranking, the "variables." We can take the point of view that if rank orders agree, the ranks assigned to individuals should *correlate* positively with each other, whereas disagreement should be reflected by a negative correlation. A zero correlation represents an intermediate condition: no particular connection between the rank of an individual on one variable and the rank on the other.

For a descriptive index of agreement between ranks, the ordinary correlation coefficient can be computed on the *ranks* just as for any numerical scores, and this is how the Spearman rank correlation for a sample is defined:

r_S = correlation between ranks over individuals.

However, since the numerical values entering into the computation of the correlation coefficient actually are ranks in this instance, r_S can be given a very simple computational form when no ties in rank exist:

$$r_S = 1 - \left[\frac{6\left(\sum_i D_i^2\right)}{N(N^2 - 1)} \right].$$

[19.10.1]

Here D_i is the *difference* between ranks associated with the particular individual i, and N is the number of individuals observed.

The Spearman rank correlation is thus very easy to compute when ranks are untied. All one needs to know is N, the number of individuals ranked, and D_i, the difference in ranking for each individual. In spite of the apparently different computations involved, r_S is only an ordinary correlation coefficient calculated on ranks.

The computation of r_S will be illustrated in a problem dealing with agreement between judges' rankings of objects. In a test of fine weight discrimination, two judges each ranked 10 small objects in order to their judged heaviness. The results are shown in Table 19.10.1. Did the judges tend to agree? The Spearman rank correlation is only .22, so that agreement between the two judges was not very high, although there was some slight tendency for similar ranks to be given to the same objects by the judges.

Incidentally, notice that if the relative true weights of the objects are known, we could also find the agreement of each judge with the true ranking. When some criterion ranking is known, it may be useful to compare a judged ranking with this criterion to evaluate the accuracy or "goodness" of the judgments—this is not the same as the agreement of judges with each other, of course.

Equation 19.10.1 may not be used if there are ties in either or both rankings, since the means and variances of the ranks then no longer have the simple relationship to N present in the no-tie case. When ties exist, perhaps the simplest procedure is to assign

Object	Judge I	Judge II	D_i	D_i^2
1	6	4	2	4
2	4	1	3	9
3	3	6	3	9
4	1	7	6	36
5	2	5	3	9
6	7	8	1	1
7	9	10	1	1
8	8	9	1	1
9	10	3	7	49
10	5	2	3	9
				128

Table 19.10.1

Rank orders of "heaviness" given by two judges.

$$r_s = 1 - \frac{6(128)}{10(10^2 - 1)} = .22.$$

mean ranks to sets of tied individuals; that is, when two or more individuals are tied in order, each is assigned the mean of the ranks they would otherwise occupy. Next an ordinary correlation coefficient is computed, using the ranks as though they were simply numerical scores. The result is a Spearman rank correlation that can be regarded as corrected for ties. On the other hand, if a test of the significance of r_S is the main object of the analysis, a conservative course of action is to find a way to break the ties that will make the absolute value of r_S as small as possible.

When no ties in rank exist, the exact sampling distribution of r_S can be worked out for small samples. Exact tests of significance for r_S are based on the idea that if one of the two rank orders is known, and the two underlying variables are independent, then each and every permutation in order of the individuals is equally likely for the other ranking. On this basis, the exact distribution of $\sum_i D_i^2$ can be found, and this can be converted into a distribution of r_S. Exact probability tables for $\sum_i D_i^2$ and r_S based on small N are to be found in Kendall (1963), Siegel (1956), and elsewhere.

The hypothesis of the independence of the two variables represented by rankings can also be given an approximate, large-sample test in terms of r_S. This test has a form very similar to that for r_{XY}:

$$t = \frac{r_S \sqrt{N - 2}}{\sqrt{1 - r_S^2}}$$ [19.10.2]

with $N - 2$ degrees of freedom. This test is really satisfactory, however, only when N is fairly large; N should be at least greater than or equal to 10.

Under the assumption of a bivariate normal distribution, the value of r_S from a large sample can be treated as an estimate of the value of ρ for the variables underlying the ranks. When the population is bivariate normal with $\rho = 0$, values of r_S and r_{XY} agree highly over samples. On the other hand, this assumption is rather special, and the status of r_S as an estimator of ρ *in general* is open to considerable question.

19.11 THE KENDALL TAU COEFFICIENT

A somewhat different approach to the problem of agreement between two rankings is given by the τ coefficient (small Greek tau) devised by M. G. Kendall (1963). Instead of treating the ranks themselves as though they were scores and finding a corrrelation coefficient, as in r_S, in the computation of τ we depend only on the number of *inversions* in order for pairs of individuals in the two rankings. A single inversion in order exists between *any pair* of individuals b and c when $b > c$ in one ranking and $c > b$ in the other. When two rankings are *identical,* no inversions in order exist. On the other hand, when one ranking is exactly the reverse of the other, an inversion exists for *each* pair of individuals; this means that complete disagreement corresponds to $\binom{N}{2}$ inversions. If the two rankings *agree* (show noninversion) for as many pairs as they disagree about (show inversion), the tendency for the two rank orders to agree or disagree should be exactly zero.

This leads to the following definition of the τ statistic:

$$\tau = 1 - \left[\frac{2(\text{number of inversions})}{\text{number of pairs of objects}} \right].$$

[19.11.1]

This is equivalent to

$$\tau = \frac{(\text{number of times rankings agree about a pair}) - (\text{number of times rankings disagree})}{\text{total number of pairs}}.$$

[19.11.2]

It follows that the τ statistic is essentially a difference between two proportions: the proportion of pairs having the *same* relative order in both rankings minus the proportion of pairs showing *different* relative order in the two rankings.

Viewed as coefficients of agreement, r_S and τ thus rest on somewhat different conceptions of "disagree." In the computation of r_S, a disagreement in ranking appears as the *squared* difference between the ranks themselves over the individuals. In τ, an inversion in order for any *pair* of objects is treated in the same way as evidence for disagreement. Although these two conceptions are related, they are not identical: the process of squaring differences between rank values in r_S places somewhat different weight on *particular* inversions in order, whereas in τ all inversions are weighted equally by a simple frequency count. Values of the statistics r_S and τ are correlated over successive random samples from the same population, but the extent of the correlation depends on a number of things, including sample size and the character of the relation between the underlying variables in the population. Nevertheless, the two statistics are closely connected, and a number of mathematical inequalities must be satisfied by the values of the two statistics. For example,

$$-1 \le 3\tau - 2r_S \le 1.$$

It will be convenient to discuss the numerator term in the τ coefficient separately, and thus we will define

$$S_T = (\text{number of agreements in order}) - (\text{number of disagreements in order})$$

$$= \binom{N}{2} - 2(\text{number of inversions}).$$

Various methods exist for the computation of S_T and τ, but the simplest is a graphic method. In this method, all one does is to list the individuals or objects ranked, once in the order given by the first ranking, and again in the order given by the second. For example, suppose that in some problem there were seven objects {a,b,c,d,e,f,g} ranked by each of two judges, and that the rankings came out like this:

				Rank			
	1	2	3	4	5	6	7
Judge 1	c	a	b	e	d	g	f
Judge 2	a	c	e	b	f	d	g

Now straight lines are drawn connecting the same objects in the two parallel rankings, thus:

```
1    2    3    4    5    6    7

c    a    b    e    d    g    f
  ><      ><      ><
a    c    e    b    f    d    g
```

Then the number of times that pairs of lines *cross* is the number of inversions in order. Here, the number of crossings is 4, and so

$$S_T = \binom{7}{2} - (2)(4)$$

$$= 21 - 8$$

$$= 13.$$

The sample value of ρ is

$$\tau = \frac{13}{21}$$

$$= .62.$$

Although r_S from a sample has a rather artificial interpretation as a correlation coefficient, the interpretation of the obtained value of τ is quite straightforward: if a *pair* of objects is drawn at random from among those ranked, the probability that these two objects show the *same* relative order in both rankings is .62 *more* than the probability that they would show a different order. In other words, from the evidence at hand it is a considerably better bet that the two judges will tend to order a randomly selected pair in the *same* way than in a different way.

This graphic method of computing τ is satisfactory only when no ties in ranking exist. For nontied rankings, however, it is very simple to carry out, even when moderately large numbers of individuals have been ranked. Notice that although both the examples of computations for r_S and τ featured judges' rankings of objects, exactly the same methods apply when the ranking principle is provided by scores shown by individuals on each of two variables.

Still another way of computing S_T will be mentioned, since we will find an extension of this method very convenient when ties exist in one or both rankings. The data are organized as in Table 19.11.1. Notice that this is really just a joint-frequency table based on ranks. Each distinct observation (object) has a pair of rank numbers, and this determines the cell in the table in which that observation falls. As a convention for the procedure described below, the ranks along the margins of the table are listed with the lowest ranks in the upper lefthand corner, as shown here.

Now S_T is found as follows: we compute a value S_+ by first taking any cell with nonzero frequency and, ignoring its row and column, counting the number of entries *to the right and below that cell.* That, for the cell containing the object c, there are five entries to the right and below. For the cell with object b, there are three entries to the right and below, and so on. Then S_+ is the sum of these numbers over cells. Over all nonzero cells, the value of S_+ is found to be:

		Judge 1					
	1	2	3	4	5	6	7
1		1(a)					
2	1(c)						
3				1(e)			
Judge 2 4			1(b)				
5							1(f)
6					1(d)		
7						1(g)	

Table 19.11.1

Rankings of seven objects by two judges.

$$
\begin{array}{ll}
5 & \text{(for cell a)} \\
5 & \text{(for cell c)} \\
3 & \text{(for cell e)} \\
3 & \text{(for cell b)} \\
1 & \text{(for cell d)} \\
0 & \text{(for cell f)} \\
\underline{0} & \text{(for cell g)}
\end{array}
$$

$S_+ = 17.$

Then

$$ S_T = 2(S_+) - \binom{N}{2} = 34 - 21 = 13 $$

and

$$ \tau = .62. $$

As we shall see, this general method is advantageous when ties exist in the data or when the data are put into ordered classes rather than ranks.

Notice that we might also compute a value S_- by taking the sum of the frequencies to the left and below the various nonzero cells. Then

$$ S_T = \binom{N}{2} - 2(S_-) $$

since

$$ (S_+) + (S_-) = \binom{N}{2}, $$

where there are no tied ranks.

An exact test of significance for τ may be had, based on the assumption that the variables underlying the ranks are continuously distributed, so that ties are impossible.

However, for our purposes, the fact of importance is that the exact sampling distribution of τ approaches a normal distribution very quickly with successive increases in the size of N. Even for fairly small values of N, the distribution of τ is approximated relatively well by the normal distribution. Of course, this is true only when H_0 is true, so that the rankings are each equally likely to show any of the $N!$ permutations in order, and $E(\tau) = 0$. The distribution of τ is not simple to discuss when other conditions hold, and so we will test only the hypothesis of independence between rankings (implying equal probability of occurrence for each and every possible ordering of N observations on the second variable, given their ordering on the first).

For N of about 10 or more, the test is given by

$$z = \frac{\tau}{\sigma_\tau}$$

referred to a normal distribution, where

$$\sigma_\tau^2 = \frac{2(2N + 5)}{9N(N - 1)}. \qquad [19.11.3]$$

Equivalently, in terms of S_T,

$$z = \frac{S_T}{\sigma_S} \qquad [19.11.4]$$

where

$$\sigma_S^2 = \frac{N(N - 1)(2N + 5)}{18}.$$

For small N, the exact tables in terms of S_T given by Kendall (1963) or Siegel (1956) can be used.

This approximate test can be improved if a correction for continuity is made. The correction for continuity involves subtraction of $1 \Big/ \binom{N}{2}$ from the absolute value of τ, or 1 from the absolute value for S_T, before the z statistic is formed.

19.12 KENDALL'S TAU WITH TIES

The determination of the exact sampling distribution of S_T (or τ) depends on the assumption of *no* ties. However, if the underlying model is altered in a particular way, τ can be computed and tested even though ties occur in one or both rankings. First of all, we will consider the method for finding S_T when ties occur, and then turn to the problem of a significance test.

The simplest method for finding S_T and τ when ties occur is to use the joint-frequency-table method suggested in Section 19.11. However, here it will occur that more than one entry may fall into a particular row, column, or cell. This method is best shown by example.

Subject	X ranking	Y ranking
1	5	8.5
2	7	11
3	5	12
4	1	8.5
5	2	8.5
6	3	6
7	5	5
8	11.5	8.5
9	11.5	2
10	9.5	2
11	9.5	2
12	8	4

Table 19.12.1

Two rankings with ties.

Suppose that some 12 subjects were observed, and ranked on each of two variables, so that the ranked data were as shown in Table 19.12.1. A table is formed in which the first ranking is represented by the columns and the second ranking by the rows. (Actually, the score values themselves could have been used to form this table, since the actual rank values are not used at all after the data are tabled.) Table 19.12.2 shows these data.

Now we proceed to find S_+ just as above, except that the number of cases to right and below a given cell is weighted by the number of cases in that cell. For example, for the cell given by the row labeled 2 and the column labeled 9.5, there is only 1 case to the right and below; however, since there are 2 cases *in* the cell, in the sum for S_+

Table 19.12.2

Rankings with ties in a frequency table.

Y ranking	X ranking								Total
	1	2	3	5	7	8	9.5	11.5	
2						2	1		3
4					1				1
5			1						1
6		1							1
8.5	1	1		1			1		4
11				1					1
12			1						1
Total	1	1	1	3	1	1	2	2	12

we enter 2(1) or 2 for this cell. Proceeding in this way over all of the cells with nonzero frequencies, we have

$$S_+ = 2(1) + 1 + 1(2) + 1(4) + 1(2) + 1(2) + 1(1)$$

$$= 14.$$

Now we find a value S_-, computed exactly as for S_+ except that we find the number of cases to the *left and below* a given cell, and weight this number by the cell frequency. Here

$$S_- = 2(8) + 1(8) + 1(7) + 1(3) + 1(2) + 1(2) + 1(1)$$

$$= 39.$$

Then,

$$S_T = S_+ - S_-,$$

so that for this example

$$S_T = 14 - 39 = -25.$$

Kendall suggests calculating the denominator of τ by taking

$$\sqrt{\left(\frac{N(N-1)}{2} - T_1\right)\left(\frac{N(N-1)}{2} - T_2\right)},$$ [19.12.1]

where

$$t_1 = \frac{\sum\limits_{j} n_j(n_j - 1)}{2}$$

n_j = the marginal total of column j

$$T_2 = \frac{\sum\limits_{k} n_k(n_k - 1)}{2}$$

n_k = the marginal total for row k.

For this example,

$$T_1 = \frac{1}{2}[3(2) + 2(1) + 2(1)]$$

$$= 5$$

and

$$T_2 = \frac{1}{2}[3(2) + 4(3)]$$

$$= 9.$$

Then

$$\tau = \frac{S_T}{\sqrt{\left(\frac{N(N-1)}{2} - T_1\right)\left(\frac{N(N-1)}{2} - T_2\right)}}$$

$$= \frac{-25}{\sqrt{(66 - 5)(66 - 9)}}$$

$$= \frac{-25}{59}$$

$$= -.42.$$

The apparent degree of agreement between these two rankings is a *negative* .42.

A large-sample test of significance for tau with ties can be constructed if one forms the ratio

$$z = \frac{S_T}{\sigma_S}$$

and refers this to a normal distribution. However, it must be emphasized that when ties exist in either or both rankings, this test is *conditional* to the distribution of ties, and the significance level refers to a probability of occurrence among samples each showing *exactly* the same distribution of ties as appeared in the sample actually obtained. When ties exist, the sampling variance of S_T or σ_S^2, is found from

$$\sigma_S^2 = \frac{N(N-1)(2N+5) - \sum_j n_j(n_j - 1)(2n_j + 5) - \sum_k n_k(n_k - 1)(2n_k + 5)}{18}$$

$$+ \frac{\left[\sum_j n_j(n_j - 1)(n_j - 2)\right]\left[\sum_k n_k(n_k - 1)(n_k - 2)\right]}{9(N)(N-1)(N-2)} \qquad [19.12.2]$$

$$+ \frac{\left[\sum_j n_j(n_j - 1)\right]\left[\sum_k (n_k)(n_k - 1)\right]}{2(N)(N-1)}.$$

After this rather forbidding-looking calculation for σ_S^2 has been carried out, the value of z may be found and referred to a normal table, provided that N is rather sizable. Unless N is quite large and ties are infrequent, a correction for continuity is required; this is given in Kendall (1963).

It is interesting to observe that even when ties exist, one really does not have to assign ranks at all to compute τ; we are really dealing with ordered categories of observations, and when we assign the midrank to a set of observations we are, in effect, allotting them to the same ordered category. Thus, in principle this procedure may be carried out on a joint *grouped* frequency distribution; the actual computations, however, really involve only the *ordinal* properties of the scores on which the distributions were based, and the class intervals are treated as ordered classes.

Furthermore, even when data are grouped in this way, the τ index can be regarded as reflecting monotonicity in the relationship between the underlying scores. Large values of τ lead to the conclusion that the relationship tends to be monotone (i.e. regularly increasing or decreasing), and small absolute values may indicate either that there is no statistical association, or that the form of the relationship tends to be nonmonotone.

19.13 A MEASURE OF ASSOCIATION IN ORDERED CLASSES

It has just been noted that the method for computing S_T where the rankings contain ties is tantamount to arranging the data into ordered classes. Unfortunately, the value of the τ statistic itself does not seem to have a very simple interpretation when ties are present in either ranking. This difficulty is removed if one uses the γ (Greek gamma) statistic suggested by Goodman and Kruskal (1979) specifically for data arranged into ordered classes. Actually, this γ statistic has the same numerator term as τ; the S_+ and S_- values are found exactly the same way. However, its denominator differs from that of τ, and permits γ to have a simpler interpretation.

In terms of the quantities S_+ and S_- computed above, the statistic γ is just

$$\gamma = \frac{S_+ - S_-}{S_+ + S_-}.$$ [19.13.1]

For the example above, this is

$$\gamma = \frac{-25}{14 + 39}$$

$$= -.47.$$

This value can be interpreted as follows: suppose that a pair of subjects were drawn at random from the 12 actually observed. Given that these subjects were tied in neither of the rankings, is it a better bet that they show the same or a different ordering on X and Y? The value of γ shows that it is a much better bet that an untied pair has different ordering on the two variables, since the probability of finding a pair with a different ordering is .47 *more* than the probability of a pair with the same ordering, among all possible untied pairs we might draw. Put more formally,

$$\gamma = p(\text{same ordering|untied pairs}) - p(\text{different ordering|untied pairs})$$

for a pair chosen at random with replacement from among these 12 subjects.

If there are no tied ranks, then

$$\gamma = \tau.$$

The index γ has the same basic interpretation as τ: a difference in probability for same versus different ordering on the underlying variables for a randomly selected pair. The only difference is that γ is conditional to the set of *untied* such pairs. The index γ

is much like the λ measures for categorical data (Section 18.9), since it has a simple interpretation in a predictive sense, although γ refers to *ordered* categories and thus reflects the general tendency toward monotonicity in the relationship, whereas λ reflects *any* form of predictive relationship.

19.14 KENDALL'S COEFFICIENT OF CONCORDANCE

Sometimes we want to know the extent to which members of a set of m distinct rank orderings of N things tend to be similar. For example, in a beauty contest each of 7 judges ($m = 7$) gives a simple rank order of the 10 contestants ($N = 10$). How much do these rank orders tend to agree, or show "concordance"?

This problem is usually handled by application of Kendall's (1963) statistic W, the *coefficient of concordance*. As we shall see, the coefficient W is closely related to the average r_S among the m rank orders.

The coefficient W is computed by putting the data into a table with m rows and N columns. In the cell for column j and row k appears the rank number assigned to individual object j by judge k. Table 19.14.1 might show the data for the judges and the beauty contestants. It is quite clear that the judges did not agree perfectly in their rankings of these contestants. However, what should the column totals of ranks, T_j, have been if the judges had agreed exactly? If each judge had given exactly the same rank to the same girl, then one column should total to 7(1), another to 7(2), and so on, until the largest sum should be 7(10). On the other hand, suppose that there were complete disagreement among the judges, so that there was no tendency for high or low rankings to pile up in particular columns. Then we should expect each column sum to be about

Table 19.14.1

Rankings given to 10 beauty contestants by seven judges.

Judges	Contestants									
	1	2	3	4	5	6	7	8	9	10
1	8	7	5	6	1	3	2	4	10	9
2	7	6	8	3	2	1	5	4	9	10
3	5	4	7	6	3	2	1	8	10	9
4	8	6	7	4	1	3	5	2	10	9
5	5	4	3	2	6	1	9	10	7	8
6	4	5	6	3	2	1	9	10	8	7
7	8	6	7	5	1	2	3	4	10	9
T_j	45	38	43	29	16	13	34	42	64	61

$$T = \frac{m(N)(N + 1)}{2}$$

$$= 385$$

the same. In this example, the column sums of ranks are not identical, so that apparently some agreement exists, but neither are the sums as different as they should be when absolutely perfect agreement exists.

This idea of the extent of variability among the respective sums of ranks is the basis for Kendall's W statistic. Basically,

$$W = \frac{\text{variance of rank sums}}{\text{maximum possible variance of rank sums}}.$$

Because the mean rank and the variance of the ranks each depend only on N and m, this reduces to

$$W = \left(\frac{12 \sum_j T_j^2}{m^2 N(N^2 - 1)}\right) - \frac{3(N + 1)}{N - 1}. \qquad [19.14.1]$$

For the example, we find

$$W = \left(\frac{12[(45)^2 + \cdots + (61)^2]}{(49 \times 10 \times 99)}\right) - \frac{3(11)}{9}$$

$$= 4.28 - 3.66$$

$$= .62.$$

There is apparently a moderately high degree of concordance among the judges, since the variance of the rank sums is 62% of the maximum possible. Note that by its definition, W cannot be negative, and its maximum value is 1.

The value of the concordance coefficient is somewhat hard to interpret directly in terms of the tendency for the rankings to agree, but an interpretation can be given in terms of the average value of r_S over all possible *pairs* of rank orders. That is,

$$\text{average } r_S = \frac{mW - 1}{m - 1}. \qquad [19.14.2]$$

For the example,

$$\text{average } r_S = \frac{7(.62) - 1}{7 - 1}$$

$$= .56.$$

If we took all of the possible $\binom{7}{2}$ or 21 *pairs* of judges, and found r_S for each such pair, the average rank correlation would be about .56. Thus, on the average, judge-pairs do tend to give relatively similar rankings. The advantage of reporting this finding in terms of W rather than average r_S is that

$$\frac{-1}{m - 1} \leq \text{average } r_S \leq 1$$

whereas, regardless of the values for N or m,

$0 \leq W \leq 1$.

This makes W values more immediately comparable across different sets of data. Nevertheless, the clearest interpretation of W seems to be in terms of average r_S.

Recall that this is essentially the idea employed in the Friedman test for matched groups. In the Friedman procedure a matched group takes the place of a judge, and the rank order of scores for different treatments within a group is like an ordering of objects of judgment. Then, if the scores for different treatments tend to show up in substantially the same order for the various groups, a true difference in treatments is inferred.

An exact test is possible for the hypothesis that there is no actual agreement among judges (see Kendall, 1963; and Siegel, 1956, for tables). For m of at least 8, an approximate test is given by

$$\chi^2 = m(N - 1)W \qquad\qquad [19.14.3]$$

referred to the chi-square distribution with $N - 1$ degrees of freedom. This test is really a good one only for fairly sizable m and N, however.

19.15 ORDER METHODS IN THE STATISTICAL PACKAGES

Statistics packages for the computer without exception provide order or nonparametric methods, including those outlined here, as well as some others. The SPSS package (as well as SPSS^x) divides these methods between two subprograms. The subprogram NPAR TESTS gives the one- and two-sample Kolmogorov-Smirnov tests, the sign test, Cochran's test, the Wilcoxon, the Friedman, and the Mann-Whitney tests, along with the Wald-Wolfowitz and Kruskal-Wallis tests. In addition, a few methods discussed in Chapter 18, such as the McNemar test and the chi-square test for goodness of fit are included in this group of methods. Tests not currently included in this Chapter but available on SPSS also include the two-sample and many-sample median tests (these tests were dropped from this edition of this text because of the very low power they appear to show in many situations.)

The subprogram NONPAR CORR calculates Spearman and Kendall rank-order correlations, as discussed in this Chapter. Other indices of ordinal relationship discussed in this Chapter, such as tau and gamma, may be obtained from the subprogram CROSS-TABS applied to data in tables with ordered categories. One- or two-tailed tests are available for each type of measure.

The BMDP program 3S (nonparametric statistics) combines the options of the two SPSS subprograms into one convenient grouping. This program offers the sign, Wilcoxon, Friedman, Kendall concordance, Mann-Whitney, and Kruskal-Wallis analyses, along with the Spearman and Kendall (tau) measures of association. As in the SPSS programs, the data read into the computer do not have to be ranks, or anything other than the raw scores, measured at least at the ordinal level.

The SAS procedure PROC NPAR1WAY concentrates on those methods that are analogous to ANOVA carried out on ranks. These include the Wilcoxon, Mann-Whitney, and Kruskal-Wallis tests, the median tests mentioned above, and two other procedures (Van der Waerden and Savage tests) not discussed in the preceding chapter. On the other hand, association measures based on order, such as the Spearman correlation, Kendall's tau, and gamma, are to be found as part of the FREQ procedure, along with some of the methods covered in Chapter 18.

EXERCISES

1. Each of a randomly selected group of 214 preschool children, age 5, was given a reading readiness test. The resulting scores are shown below, arranged into a distribution with a class interval size of 7. The mean of this distribution is 37.78, and the standard deviation, based on the unbiased estimate of the variance, is 15.57. Use the Kolmogorov-Smirnov procedure to test the null hypothesis that the true distribution of scores among children of this age is normal, with this same mean and the unbiased estimate of the standard deviation.

Class limits	Frequency
66–72	5
59–65	17
52–58	20
45–51	29
38–44	35
31–37	39
24–30	27
17–23	20
10–16	12
3– 9	10
	214

2. Use the Kolmogorov-Smirnov procedure to reanalyze the data in Exercise 1, Chapter 18. Did your conclusion change with the change in procedure?

3. In addition to the random sample of 214 children age 5 as shown in Exercise 1, reading readiness scores were also obtained from a sample of 183 children age 6. Both sets of scores are shown below, arranged into distributions with the same class intervals. Use the Kolmogorov-Smirnov procedure to test the hypothesis that the two population distributions are identical.

Class interval	Age 5	Age 6
73–79	0	3
66–72	5	12
59–65	17	29
52–58	20	27
45–51	29	31
38–44	35	28
31–37	39	23
24–30	27	12
17–23	20	9
10–16	12	7
3– 9	10	2
	214	183

4. Take the data in Exercise 19, Chapter 14, and test the hypothesis that the teachers were equally likely to judge a child's age too high or too low. Use the sign test for this purpose.

5. In a study of engaged couples' preferences with respect to size of family, some 26 engaged couples were selected at random and were asked to state, independently, the ideal number of children they would like to have. Responses of men and women from each couple are listed below:

Couple	Men	Women
1	3	2
2	0	1
3	1	0
4	2	2
5	0	3
6	2	3
7	1	2
8	2	3
9	2	3
10	1	3
11	2	4
12	0	1
13	3	4
14	5	2
15	7	2
16	1	2
17	0	3
18	2	4
19	0	3
20	2	3
21	2	4
22	0	2
23	1	3
24	3	2
25	5	2
26	0	1

Use the sign test to decide if men and women differ significantly on this issue.

6. A psychologist was interested in the verb-adjective ratio as an index of the habitual pattern of verbal expression for an individual. In a comparison of science majors and English majors in terms of verb-adjective ratio, 10 science majors and 12 English majors were sampled at random, and a selection of the free-writing of each subject taken. Each such selection was scored according to the ratio of number of verbs used relative to the number of adjectives. In terms of the resulting data, use the Mann-Whitney test to describe if science majors and English majors differ significantly in their relative usage of verbs and adjectives.

Verb-adjective ratios	
Science majors	English majors
1.32	1.04
2.30	.93
1.98	.75
.59	.33
1.02	1.62
.88	.76
.92	.97
1.39	1.21
1.95	.80
1.25	1.16
	.71
	.96

7. Eighteen subjects were asked to respond to very faint tones by pushing a key specific to each tone. The task was carried out under five different conditions of noise. In each condition 100 tones were presented, and the number of correct responses recorded. The data follow:

		Condition				
		1	2	3	4	5
	1	73	75	71	74	64
	2	91	89	83	80	82
	3	92	94	90	93	83
	4	84	82	76	73	75
	5	56	58	54	57	47
	6	60	58	52	49	51
	7	73	75	69	72	62
	8	70	68	64	61	63
	9	87	89	83	86	76
Subject	10	75	73	69	66	68
	11	77	75	69	72	62
	12	68	70	64	61	63
	13	73	75	69	72	62
	14	75	73	68	65	67
	15	93	95	89	92	82
	16	90	89	85	82	84
	17	84	86	80	83	73
	18	69	67	63	60	62

Are there significant differences among conditions ($\alpha = .05$)? Analyze these data using the Friedman matched-groups analysis by ranks.

8. Use the Kruskal-Wallis analysis of variance by ranks to compare the groups represented by the columns in Table 12.7.1.

9. In a study of conformity to social norms, and how this might vary with age, four samples of eight subjects each were selected at random. However, each sample came from a different age group. Each subject was observed driving a specially equipped car, and the average amount of time spent at a stop sign (at an otherwise clear intersection) was recorded for each in units of .1 seconds. The data follow:

Age groups			
18–25	35–40	45–50	over 65
11.7	41.2	59.0	61.6
28.9	39.5	62.1	70.3
26.2	52.2	57.5	59.2
31.8	47.6	58.3	25.8
59.4	38.9	42.6	74.1
27.6	37.7	44.5	60.9
50.5	46.3	40.9	65.4
29.2	57.7	52.2	53.7

Since it does not seem reasonable to suppose that such observations should represent a normally distributed population, it was decided to employ the Kruskal-Wallis method of analysis. Do these age groups differ in this form of behavior?

10. For the data in Exercise 11, Chapter 14, carry out the Wilcoxon test for the ratings given to countries A and B. Do these countries tend to be rated differently over political scientists? Use $\alpha = .05$.

11. Once again using the data in Exercise 11, Chapter 14, carry out the Wilcoxon test using countries B and C, and countries A and C, with $\alpha = .05$ for each. Are these tests independent of each other and of the test in the previous exercise? What can one say about the probability of making a Type I error in at least one of these three tests?

12. Suppose that the rows of Table 13.16.1 actually represented a random sample of 12 different classrooms, with all three tasks then given to each classroom. Under this assumption, what does the Friedman test reveal about the differences among tasks?

13. Analyze the data in Exercise 4, Chapter 10, in terms of the Kruskal-Wallis analysis of variance by ranks.

14. Analyze the data in Exercise 5 above by means of the Wilcoxon test for matched samples.

15. An experiment involved two randomly selected groups of 10 subjects each. The scores on a reading comprehension test given each person are shown below. Group I read with a background of music of a loud "heavy metal" character. Group II read with a background of standard popular music of the 1930s. Compare the average reading comprehension scores for the two groups by way of the Mann-Whitney test.

I	II
128	136
131	141
146	139
157	157
129	136
151	157
142	147
136	135
140	131

16. How does a two-tailed t test for the data in Exercise 15 compare with the Mann-Whitney results?

17. Some 24 subjects were selected at random, and each was given a series of six puzzles requiring a certain amount of "insight" to solve. After a fixed allotment of time on each, the subject was scored 1 if the puzzle was solved, and 0 if the puzzle was still unsolved. Then each subject moved on to the next puzzle until all six had been attempted. Give the data table below, does it appear that the puzzles differed significantly in difficulty?

			Puzzles			
Subjects	I	II	III	IV	V	VI
1	1	0	1	1	1	0
2	0	0	1	0	1	0
3	1	1	0	1	1	1
4	1	1	1	0	0	1
5	1	1	0	1	1	0
6	1	1	0	1	1	1
7	1	1	1	1	1	1
8	1	0	0	1	1	1
9	1	1	1	1	0	0
10	1	1	0	1	0	1
11	0	1	1	1	1	1
12	1	0	0	1	1	1
13	1	1	0	1	1	0
14	1	1	1	1	1	1
15	1	0	0	0	1	1
16	1	1	0	1	1	0
17	1	0	0	1	1	1
18	0	1	1	1	1	1
19	1	0	0	0	1	0
20	1	1	0	0	0	1
21	1	0	0	1	1	1
22	1	1	1	1	1	1
23	1	0	0	1	1	1
24	1	1	1	1	1	1

18. Calculate the Spearman rank correlation for the data in Exercise 18, Chapter 14. You may assign average ranks to any group of tied values of X or of Y.

19. Test the significance of the Spearman correlation found in Exercise 18 above.

20. In a psychological study, 20 subjects were used, each subject being tested on a different day. Although each subject was cautioned not to tell anyone about the experiment, the experimenter felt that the later subjects were performing systematically better than the earlier, and suspected that information about the experiment was being gradually "leaked." Test the hypothesis that there is no correlation between the day on which a subject was tested and the score. (**Hint:** Compute a Spearman rank correlation between days and scores, and test its significance.)

Day	Score	Day	Score
1	120	11	132
2	124	12	136
3	123	13	133
4	128	14	119
5	125	15	140
6	127	16	138
7	134	17	139
8	129	18	161
9	130	19	142
10	137	20	145

21. Use the runs test to answer the question in Exercise 20. (**Hint:** Divide the data according to position above and below the median score, and then look for runs above and below the median in terms of days.)

22. Analyze the data in Exercise 9, Chapter 11, by use of the Kruskal-Wallis analysis of variance by ranks for independent groups.

23. Suppose that the data of Table 13.4.1 were actually the results of five repeated observations on each of eight subjects. Furthermore, let a score value greater than or equal to 6.00 be counted a success, and any value which falls below 6.00 a failure. Compare the five sets of repeated observations by use of Cochran's test.

24. Apply the Kruskal-Wallis analysis of variance by ranks to the data in Exercise 21, Chapter 10.

25. In a study of upward mobility trends in American society, a random sample of 107 U.S. families with at least one adult son was taken. The occupational level of the father and that of his oldest son was taken, and the data put into the following table:

Father

	Unskilled labor	Skilled labor	White collar	Professional-executive
Unskilled labor	2	3	7	3
Skilled labor	3	4	20	9
White collar	1	2	14	19
Professional-executive	0	4	2	14

(Son, rows)

Index the extent to which an ordinal relation seems to exist between the occupational level of the father and the son.

26. Test the hypothesis that no ordinal relationship exists in the data in Exercise 25.

27. Compare the two groups in Exercise 6 above in terms of the *runs* test.

28. In Exercise 5, Chapter 2, responses are given for 30 subjects on each of three questionnaire items. The data for Items I and II can be put into the following table. Index the extent to which there seems to be an ordinal association between I and II. Does a high response to Item I tend to go with a high response to Item II? Use Kendall's tau with ties.

Item II

Item I	1	2	3	4	5
1	0	2	1	1	1
2	1	1	2	5	0
3	1	1	5	0	1
4	0	3	2	0	0
5	0	0	2	1	0

29. Make a table like that in Exercise 28 for Items II and III and for Items I and III in Exercise 5, Chapter 2. Then examine the γ index of predictive relationship for all three tables. Which pair of items predict order best? Which have the least relationship of this sort?

30. In a study of attitudes toward international affairs, each of a group of 15 subjects ranked 10 countries according to their reputation for democracy. The data follow:

					Country					
		A	B	C	D	E	F	G	H	I	J
	1	2	1	3	5	4	6	7	8	9	10
	2	8	7	6	4	9	1	2	10	5	3
	3	3	4	2	1	6	5	7	8	9	10
	4	9	10	8	6	7	1	3	4	5	2
	5	5	2	3	4	6	1	7	8	9	10
	6	2	4	3	1	5	8	9	10	7	6
	7	4	3	2	5	1	6	9	10	8	7
Subject	8	2	4	1	6	5	3	8	7	9	10
	9	9	10	6	7	8	1	4	3	5	2
	10	5	4	1	10	2	3	7	8	6	9
	11	2	4	1	5	3	6	7	9	10	8
	12	3	4	2	5	1	10	9	6	8	7
	13	4	3	5	2	10	1	8	6	9	7
	14	5	3	4	2	1	8	9	6	10	7
	15	4	5	10	1	3	2	8	9	6	7

Find the coefficient of concordance and the average rank correlation for this group.

31. Assuming that the 15 subjects in Exericse 30 constitute a random sample, test the hypothesis of zero true concordance among subjects in terms of their rankings of countries.

32. Two judges each ranked 10 pictures (labeled A, B, C, etc.) in order of beauty. The rankings are given below. Use Kendall's tau coefficient to describe the agreement between the judges.

Judge 1	Judge 2
C	A
A	C
B	B
D	G
F	D
E	F
G	E
J	H
I	J
H	I

33. Use the data in Exercise 32 to test the hypothesis of zero true agreement between the judges.

Appendix *A*

Rules of Summation

A.1 SUBSCRIPTS

In the discussion to follow, let X and Y stand for numerical variables. Ordinarily, the numerical values for these variables are scores in some given set of data, where each individual observed in a group of N cases is paired either with a single score, such as x, or with a pair of scores (x, y).

The *subscript* notation is useful in keeping track of the various score values across the different individuals or observations. For the moment, we will consider N distinct individuals, arbitrarily indexed by the set of numbers $(1, \cdots, i, \cdots, N)$. The "running subscript" i denotes any one of these individuals. Then, X_i denotes the particular value of X associated with any particular individual i. That is, the variable X_i ranges only over the set of particular values

$$(x_1, x_2, \cdots, x_N).$$

Now we need a way to discuss the *sum* of these particular values of X over the entire set of N different individuals i. The symbol Σ (capital Greek sigma) stands for the operation "the sum of," referring to the values symbolized immediately after the summation sign. Thus,

$$\sum_{i=1}^{N} x_i$$

is read as "the sum of the N values x_i, beginning with $i = 1$ and ending with $i = N$."

That is,

$$\sum_{i=1}^{N} x_i = x_1 + x_2 + \cdots + x_N.$$

Suppose, for example, that $N = 4$, and the actual values associated with the four different individuals were

6, 8, 3, 11.

Then,

$$\sum_{i=1}^{N} x_i = 6 + 8 + 3 + 11 = 28.$$

In many simple contexts where it is clear that the sum extends across *all* of a set of N individuals, the subscripts and superscripts on the summation sign are simplified or omitted. Thus, we often write:

$$\sum_{i} x_i$$

to indicate the sum of values across all N individuals i, or even

$$\sum x_i$$

to stand for this sum. However, the subscript is usually omitted only in very simple expressions where the set of values to be included in the sum is perfectly clear.

Quite frequently, symbols for values corresponding to individual observations are given two (or more) subscripts. That is, it happens that sometimes we want to discuss individual observations arranged into several groupings or cross-classifications. Then two (or more) sets of index numbers are used. For example, suppose that there were some J different *groups* of observations; these different groups are identified by being paired arbitrarily with the numbers

$$(1, \cdots, j, \cdots, J).$$

Furthermore, within any particular group j, there is some number n_j of distinct individuals or observations, each of which is associated with an X score. Then, the particular value of the score for the ith individual in the group j is symbolized by x_{ij}. If we wanted to symbolize the sum of the scores for the particular group j, then we would write

$$\sum_{i=1}^{n_j} x_{ij}$$

which stands for "the sum of values for all of the n_j different individuals in the particular group labeled j." Where it is clear that all individual scores in a particular group j are to be summed, this can be abbreviated to

$$\sum_{i} x_{ij}.$$

Notice that this quite literally is the symbol for "sum over i" for a set of scores x_{ij}; the fact that j appears as a subscript for x but that i alone appears below the sigma is a cue that all the scores summed in this particular operation *must* have the same subscript j (that is, must be in the same group of observations).

However, often one wants to indicate the sum of X_{ij} over *all* individuals in *all* groups. Written out in full, this would be symbolized by

$$\sum_{j=1}^{J} \sum_{i=1}^{n_j} x_{ij} = (x_{11} + x_{21} + \cdots + x_{n_j-1,J} + x_{n_j,J}).$$

This *double* summation is interpreted as "Take a particular group j, and sum scores over all the n_j different observations in that group; then, having done this for each of the J different groups, sum the results over all of these groups." Once again, this can be simplified to

$$\sum_{j} \sum_{i} x_{ij}$$

when the meaning of the symbol is perfectly clear in terms of which values are to be summed.

This idea can be extended, although in our discussions a *triple* summation is the largest such symbol used. In some problems, each individual i belongs simultaneously to *two* groupings. That is, suppose that all the scores in a set of data were arranged into a table with K rows, J columns, and JK cells, and that each individual observation belongs to one and only one row, one and only one column, and one and only one cell in the data table. Now let the various *column* groupings be symbolized by the numbers

$$(1, \cdots, j \cdots, J).$$

The *rows* of the data table are indexed by the set of numbers

$$(1, \cdots, k, \cdots, K).$$

Then any cell of the table is indexed by a *pair* of numbers jk. The score of the ith individual in the cell jk of the data table is thus x_{ijk}. The number of individuals in the cell jk is shown by n_{jk}.

Now, when there are three subscripts like this,

$$\sum_{j=1}^{J} \sum_{k=1}^{K} \sum_{i=1}^{n_{jk}} x_{ijk},$$

the following is symbolized: "Take a particular cell jk and sum the scores of all individuals i in that cell; then, still within a particular column j, sum the results of doing this for each row k over the K respective rows; finally, sum the results for each column j over the J respective columns." A little thought should convince you that this is simply the sum of all individual scores in all cells, or the sum of all scores in the table.

In this instance, it is immaterial whether we write the sum for the rows or the sum for the columns first, since the total number symbolized is the same. However, this is not universally true, and the order of the summation signs can be important.

Sometimes we want to symbolize only the sum for a particular cell, row, or column: First of all, letting n stand for the number in cell jk, we write

$$\sum_{i=1}^{n} x_{ijk} = (x_{1jk} + \cdot \cdot \cdot + x_{njk}).$$

This denotes the sum of all individual scores in the *single* cell jk. If we write

$$\sum_{j=1}^{J} \sum_{i=1}^{n} x_{ijk} = (x_{11k} + x_{21k} + \cdot \cdot \cdot + x_{njk}),$$

we intend that first the sum of individuals in a particular cell in column j and row k shall be taken, and then, still within the particular row k, the result for all the different columns j shall be summed. This amounts to the sum of all scores within the particular row k. On the other hand, if we want the sum of all the scores in a particular column j, we could write

$$\sum_{k=1}^{K} \sum_{j=1}^{n} x_{ijk} = (x_{1j1} + \cdot \cdot \cdot + x_{njK})$$

indicating the sum for a particular cell jk, and then the sum over the results for all rows k within the particular column j. In short, unless a subscript appears both in the symbol for the x values and under the summation sign, the total value symbolized pertains to some particular set of observations, represented by that subscript *not appearing* under the summation symbol. Notice that when the sum of all observations in a particular row k of the data table was intended, the sum was indicated over i and over j, but *not* over k. On the other hand, when the sum of observations in some column j was desired, the sum was indicated over i and over k, but not over j.

In learning to work with summation signs, it is well to start the habit of reading the summation symbols from *right to left,* and from *inside the punctuation outward.* The *innermost* summation indicated in a mathematical expression is ordinarily to be carried out first, then the next summation symbolized to the left, and then finally the summation appearing closest to the margin on the left. Furthermore, although we have symbolized a row by k and a column by j in this discussion, this symbolism is arbitrary and, often, we will use j for rows and k for columns.

A.2 RULES OF SUMMATION

A few simple rules can help one learn to interpret and carry out the computations symbolized by statistical expressions involving summations. Furthermore, elementary statistical derivations almost always involve one or more of these summation rules used as an algebraic equivalence. First of all, some simple rules will be given that apply when values for only one variable are being summed, and then we will consider situations involving two or more variables.

RULE 1. If a is some constant value over the N different observations i, then

$$\sum_{i=1}^{N} a = Na.$$

For example, suppose that 5 observations are considered, and paired with each observation is the constant number 10. Then

$$\sum_{i} a = 10 + 10 + 10 + 10 + 10 = 5 \times 10.$$

Anytime that, in the particular sum indicated, each and every individual observation entering into the sum is paired with exactly the same constant value a, then the sum is equal to the number of individuals considered times the value of a.

Extending this rule to double or triple summations, we have

$$\sum_{j=1}^{J} \sum_{i=1}^{nj} a = \sum_{j=1}^{J} (n_j)a$$

$$\sum_{k=1}^{K} \sum_{j=1}^{J} \sum_{i=1}^{njk} a = \sum_{k=1}^{K} \sum_{j=1}^{J} (n_{jk})a$$

and so on.

RULE 2. Given the value a, which is constant over all individuals entering into the summation, then

$$\sum_{i=1}^{N} ax_i = a \sum_{i=1}^{N} x_i.$$

For example, consider the five individual scores

6, 7, 3, 9, 20.

Suppose that each of these scores is multiplied by the constant 2. Then

$$\sum_{i} 2x_i = 2(6) + 2(7) + 2(3) + 2(9) + 2(20)$$

$$= 2(6 + 7 + 3 + 9 + 20) = 2 \sum_{i} x_i.$$

Once again, this rule can be applied to several sums:

$$\sum_{j=1}^{J} \sum_{i=1}^{nj} ax_{ij} = a \sum_{j=1}^{J} \sum_{i=1}^{nj} x_{ij}.$$

Notice here that the a symbol has no subscript, and thus may be regarded as a constant over the sum. However, it might happen that the sum is such that a_j appears instead of a, meaning that a_j is a constant for all observations i in the *same* group j, but is different for different groups. Then,

$$\sum_{j=1}^{J} \sum_{i=1}^{nj} a_j x_{ij} = \sum_{j=1}^{J} a_j \left(\sum_{i=1}^{nj} x_{ij} \right).$$

One must be sure to notice the set of observations for which some value *is* constant in multiple summations.

RULE 3. If some operation is to be carried out on the individual values of X before the summation, this is indicated by mathematical punctuation, such as parentheses, or equivalent symbols having the force of punctuation. Unless the summation symbol is included within this punctuation, the summation is to be done after the other operation: for example,

$$\sum_{i}^{N} (x_i)^2 = \sum_{i}^{N} x_i^2 = (x_1^2 + x_2^2 + \cdots + x_N^2)$$

$$\sum_{i} \sqrt{x_i} = \sqrt{x_i} + \sqrt{x_2} + \cdots + \sqrt{x_N}.$$

However,

$$\left(\sum_{i} x_i \right)^2 = (x_1 + \cdots + x_N)^2$$

$$\sqrt{\sum_{i} x_i} = \sqrt{x_1 + \cdots + x_N}.$$

Be sure to notice that where the parentheses (or other symbols such as exponent and radical signs having the same force as parentheses) happen to appear makes a big difference in the sum symbolized. In general,

$$\sum_{i} x_i^2 \neq \left(\sum_{i} x_i \right)^2$$

$$\sum_{i} \sqrt{x_i} \neq \sqrt{\sum_{i} x_i}$$

$$\log \left(\sum_{i} x_i \right) \neq \sum_{i} \log x_i,$$

and so on.

This is especially important when there is multiple summation. For example,

$$\sum_{j=1}^{n} \left(\sum_{i} x_{ij} \right)^2 = \sum_{j} (x_{1j} + \cdots + x_{nj})^2.$$

The notation here tells us *first* to take the sum of all the individual values in a single group j; then, having done this for each group j, we square the *sum* for each group; finally, the sum of the *squared* sums for the various groups is found. Squaring at the wrong place in this sequence of steps will give quite a different result.

RULE 4. If the only operation to be carried out before a sum is taken is itself a sum (or difference), then the summation may be distributed. Thus,

$$\sum_{i} (x_i - 4) = \sum_{i} x_i - \sum_{i} 4$$

$$\sum_{i} (x_i^2 + 3x_i + 10) = \sum_{i} x_i^2 + \sum_{i} 3x_i + \sum_{i} 10.$$

This rule is very important in algebraic manipulations on expressions involving sums. For example, using this rule we can make the following algebraic changes:

$$\sum_{j=1}^{J} \sum_{i=1}^{nj} (x_{ij} - m)^2 = \sum_{j=1}^{J} \sum_{i=1}^{nj} (x_{ij}^2 - 2mx_{ij} + m^2)$$

$$= \sum_{j=1}^{J} \sum_{i=1}^{nj} x_{ij}^2 - \sum_{j=1}^{J} \sum_{i=1}^{nj} 2mx_{ij} + \sum_{j=1}^{J} \sum_{i=1}^{nj} m^2.$$

Invoking Rules 1 and 2, we can reduce this expression to

$$\sum_{j} \sum_{i} x_{ij}^2 - 2m \sum_{j} \sum_{i} x_{ij} + m^2 \sum_{j} n_j.$$

The value represented by this last expression is algebraically equivalent to the value of the first expression, but it may be far more convenient actually to compute this number in the second way, or to use this second expression in a mathematical argument.

A.3 SUMMATION RULES FOR TWO OR MORE VARIABLES

So far we have acted as though each distinct individual or observation were associated with one and only one numerical score. However, it may be that two distinct values x_i and y_i are paired with each individual or observation i. Then the rules given above may be extended:

RULE 5. If each of N distinct individuals i has two scores, x_i and y_i, then

$$\sum_{i=1}^{N} x_i y_i = x_1 y_1 + x_2 y_2 + \cdots + x_N y_N.$$

Notice that this symbol means that the *product of the pair of scores* belonging to any given individual i is to be found *first,* and then when this has been done for the entire set of individuals, the *sum* of these products is to be found. Be sure to notice that *in general*

$$\sum_i x_i y_i \neq \left(\sum_i x_i \right) \left(\sum_i y_i \right).$$

The other rules extend quite easily to the situation where two (or more) variables are considered:

RULE 6. $\sum_i a x_i y_i = a \sum_i x_i y_i.$

RULE 7. Given the constants a and b,

$$\sum_i (a x_i + b y_i) = a \sum_i x_i + b \sum_i y_i.$$

Sometimes it happens that when observations are arranged into groups, all observations i in the same group j will have the same value x_j on one variable, but different values y_{ij} on the other. Then, the following rule applies:

RULE 8. $\sum_{j=1}^{J} \sum_{i=1}^{n_j} x_j y_{ij} = \sum_j x_j \left(\sum_i^{n_j} y_{ij} \right).$

All of these rules may be extended by obvious analogy when there are more than three variables involved, or when there is multiple summation.

Like most of the mechanical skills of mathematics, real facility in algebraic manipulation of summations comes only with practice. However, when puzzled or in doubt about the equivalence to two sums, you can usually check this equivalence by actually writing out the terms in the sums symbolized.

EXERCISES—RULES OF SUMMATION

Write out in extended form the sums represented by each of the following expressions:

1. $\sum_{i=1}^{5} x_i$

2. $\sum_{i=1}^{6} x_i^2$

3. $\sum_{i=1}^{3} x_i^3 + \sum_{i=5}^{8} x_i^2$

4. $\sum_{i=2}^{4} 10(x_i + y_i)$

5. $\sum_{i=1}^{3} 4 \left(\sum_{j=1}^{2} x_{ij} \right)$

6. $\sum_{j=1}^{4} \left(\sum_{i=1}^{3} x_{ij} \right)^2$

7. $\left(\sum_{i=1}^{4} x_i^2 g_i \right) \Big/ \left(\sum_{i=1}^{4} g_i \right)$

8. $\sum_{j=2}^{5} n_j \left(\sum_{i=1}^{3} x_{ij} \right)^2$

9. $\sum_{i=1}^{5} \sum_{j=2}^{4} (x_{ij}^2 + y_i)$

10. $\sum_{i=4}^{6} \sqrt{x_i + 10}$

Express each of the following extended sums in the appropriate summation notation:

11. $(x_1 + x_2 + x_3 + x_4)/N$

12. $x_1 y_1 + x_2 y_2 + x_3 y_3 + \cdots + x_N y_N$

13. $x_2 f(x_2) + x_3 f(x_3) + x_4 f(x_4)$

14. $p(x_1) + p(x_2) + p(x_3) + p(x_{10}) + p(x_{11}) + p(x_{12})$

15. $3(x_1 + y_1)^2 + 3(x_1 + y_2)^2 + \cdots + 3(x_1 + y_N)^2$

16. $12(x_1^2 - 2x_1 + 3) + 12(x_2^2 - 2x_2 + 3) + 12(x_3^2 - 2x_3 + 3)$

17. $[3x_1 + 3x_2 + 3x_3 + 3x_4][5x_1 + 5x_2 + 5x_3 + 5x_4]$

18. $\sqrt{(x_1 - 2)(y_1 + 5) + (x_2 - 2)(y_2 + 5) + \cdots + (x_N - 2)(y_N + 5)}$

19. $[4(x_2 - 5)^2/y_2] + [4(x_3 - 5)^2/y_3] + [4(x_4 - 5)^2/y_4]$

20. $(x_{11} - M_1)^2 + (x_{12} - M_1)^2 + (x_{13} - M_1)^2 + (x_{21} - M_2)^2 + (x_{22} - M_2)^2 + (x_{23} - M_2)^2$

21. $(x_1 + x_2 + x_3 + x_4)^2 (y_1^2 + y_2^2 + y_3^2 + y_4^2)$

Reduce each of the following expressions into simplest form by application of the rules of summation:

22. $\sum_{i=1}^{10} (5x_i - 3)$

23. $\left(\sum_{i=1}^{5} (x_i + 2c) \right) \left(\sum_{j=1}^{3} 7 \right)$

24. $\sum_{i=4}^{6} (x_i - 2a)^2 + \sum_{i=1}^{3} (x_i + 2a)^2$

25. $\sum_{i=1}^{N} \left[x_i - \left(\sum_{i=1}^{N} x_1/N \right) \right]$

26. $\sum\limits_{i=1}^{N} (x_i - 3)(x_i + 3)/N$

27. $\sqrt{\sum\limits_{i=1}^{N} (a + b)} \sqrt{\sum\limits_{i=1}^{N} c}$

28. $\left[\left(\sum\limits_{i=1}^{N} x_i \right)^2 - \left(\sum\limits_{i=1}^{N} y_i \right)^2 \right] \Big/ \sum\limits_{i=1}^{N} (x_i + y_i)$

29. $\dfrac{\sum\limits_{i=1}^{N} \left[x_i^2 - \left(\sum\limits_{i=1}^{N} x_i/N \right)^2 \right]}{N}$

30. $(1/N) \sum\limits_{i=1}^{N} \left[x_i - \left(\sum\limits_{i=1}^{N} x_i/N \right) \right] \left[y_i - \left(\sum\limits_{i=1}^{N} y_i/N \right) \right]$

Consider the following values:

$$x_1 = 7 \qquad x_4 = 4 \qquad x_7 = 2 \qquad x_{10} = 5$$
$$x_2 = 3 \qquad x_5 = 0 \qquad x_8 = 8$$
$$x_3 = 9 \qquad x_6 = 1 \qquad x_9 = 8$$

In terms of these values for x_i, evaluate the following sums:

31. $\sum\limits_{i=1}^{10} (x_i - 5)$

32. $\sum\limits_{i=1}^{10} (x_i^2 - 3x_i + 1)$

33. $\sum\limits_{i=1}^{5} x_i - \sum\limits_{i=6}^{10} (x_i + 5)$

34. $\sum\limits_{i=1}^{10} \left[x_i - \left(\sum\limits_{i=1}^{10} x_i/10 \right) \right] \Big/ 10$

35. $\left(\sum\limits_{i=1}^{10} x_i \right) \Big/ \left(\sum\limits_{i=1}^{5} x_i \right)$

36. $\left(\sum\limits_{i=1}^{10} x^2/10 \right) - \left(\sum\limits_{i=1}^{10} x_i/10 \right)^2$

Appendix *B*

The Algebra of Expectations

B.1 EXPECTATIONS OF RANDOM VARIABLES

A very prominent place in theoretical statistics is occupied by the concept of the mathematical expectation of a random variable X. If the distribution of X is discrete, then the expectation (or expected value) of X is defined to be

$$E(X) = \sum_x xp(x)$$

where the sum is taken over all of the different values that the variable X can assume, and

$$\sum_x p(x) = 1.00.$$

For a continuous random variable X ranging over all the real numbers, the expectation is defined by

$$E(X) = \int_{-\infty}^{\infty} xf(x)\, d(x)$$

where

$$\int_{-\infty}^{\infty} f(x)\, d(x) = 1.00.$$

In essence, the expectation defined in either of these ways is a kind of weighted sum of values, and thus the rules of summation have very close parallels in the rules for the algebraic treatment of expectations. These rules apply either to discrete or to continuous random variables if particular boundary conditions exist; for our purposes these rules can be used without our going further into these special qualifications, however.

RULE 1. If a is some constant number, then

$E(a) = a.$

That is, if the same constant value a were associated with each and every elementary event in some sample space, the expectation or mean of the values would most certainly be a.

RULE 2. If a is some constant real number and X is a random variable with expectation $E(X)$, then

$E(aX) = aE(X).$

Suppose that a new random variable is formed by multiplying each value of X by the constant number a. Then the expectation of the new random variable is just a times the expectation of X. This is very simple to show for a discrete random variable X: By definition,

$$E(aX) = \sum_x axp(ax).$$

However, the probability of any value aX must be exactly equal to the probability of the corresponding X value, and so

$$E(aX) = \sum_x axp(x) = a \sum_x xp(x) = aE(X).$$

RULE 3. If a is a constant real number and X is a random variable, then

$E(X + a) = E(X) + a.$

This can be shown very simply for a discrete variable. Here,

$$E(X + a) = \sum_x (x + a)p(x + a)$$

$$= \sum_x xp(x + a) + a \sum_x p(x + a).$$

However, $p(X + a) = p(X)$ for each value of X, so that

$$E(X + a) = E(X) + a \sum_x p(x) = E(X) + a.$$

The expectations of functions of random variables, such as

$E[(X + 2)^2]$

$E(\sqrt{X + b})$

$E(b^X),$

to give only a few examples, are subject to the same algebraic rules as summations. That is, the operation indicated within the punctuation is to be carried out *before* the expectation is taken. It is most important that this be kept in mind during any algebraic argument involving expectations. In general,

$E[(X + 2)^2] \neq [E(X) + E(2)]^2$

$E(\sqrt{X}) \neq (\sqrt{E(X)})$

$E(b^X) \neq b^{E(X)}$

and so forth.

The next few rules concern two (or more) random variables, symbolized by X and Y.

RULE 4. If X is a random variable with expectation $E(X)$, and Y is a random variable with expectation $E(Y)$, then

$E(X + Y) = E(X) + E(Y).$

Verbally, this rule says that the expectation of a sum of two random variables is the sum of their expectations. Once again, the proof is simple for two discrete variables X and Y. Consider the new random variable $(X + Y)$. The probability of a value of $(X + Y)$ involving a *particular* X value and a *particular* Y value is the joint probability $p(x,y)$. Thus,

$$E(X + Y) = \sum_x \sum_y (x + y)p(x,y).$$

Notice that here the expectation involves the sum over all possible *joint* events (x,y). This could be written as

$$E(X + Y) = \sum_x \sum_y (x + y)p(x,y)$$

$$= \sum_x \sum_y xp(x,y) + \sum_x \sum_y yp(x,y).$$

However, for any fixed x.

$$\sum_y p(x,y) = p(x)$$

and for any fixed y,

$$\sum_x p(x,y) = p(y).$$

Thus,

$$E(x + y) = \sum_x xp(x) + \sum_y yp(y) = E(X) + E(Y).$$

In particular, one of the random variables may be in a functional relation to the other. For example, let $Y = 3X^2$. Then

$$E(X + Y) = E(X + 3X^2)$$
$$= E(X) + E(3X^2)$$
$$= E(X) + 3E(X^2).$$

This principle lets one *distribute* the expectation over an expression which itself has the form of a sum. We will make a great deal of use of this principle.

This rule may also be extended to any finite number of random variables:

RULE 5. Given some finite number of random variables, the expectation of the sum of those variables is the sum of their individual expectations. Thus,

$$E(X + Y + Z) = E(X) + E(Y) + E(Z)$$

and so on.

In particular, some of these random variables may also be in functional relations to others. Let $Y = 6X^4$, and let $Z = \sqrt{2X}$. Then

$$E(X + Y + Z) = E(X) + 6E(X^4) + E(\sqrt{2X}).$$

B.2 THE VARIANCE OF A RANDOM VARIABLE.

More useful rules involve the variance of a random variable. The variance is defined by

$$\text{var}(X) = \sigma_X^2 = E[X - E(X)]^2$$

or

$$\sigma_X^2 = E(X^2) - [E(X)]^2.$$

The standard deviation of the random variable X is then $\sigma_X = \sqrt{\text{var}(X)}$.

The following rules give the effect of a transformation of X on the variance.

RULE 6. If a is some constant real number, and if X is a random variable with expectation $E(X)$ and variance σ_X^2, then the random variable $(X + a)$ has variance σ_X^2.

This can be shown as follows:

$$\text{var}(X + a) = E[(X + a)^2] - [E(X + a)]^2.$$

By Rule 3 above, and expanding the squares, we have

$$E[(X + a)^2] - [E(X + a)]^2 = E[(X^2 + 2Xa + a^2)] - [E(X) + a]^2$$
$$= E(X^2 + 2aX + a^2) - [E(X)]^2 - 2aE(X) - a^2.$$

Then by Rules 5 and 1, we have

$$\text{var}(X + a) = E(X^2) + 2aE(X) + a^2 - [E(X)]^2 - 2aE(X) - a^2$$
$$= E(X^2) - [E(X)]^2$$
$$= \sigma_X^2.$$

RULE 7. If a is some constant real number, and if X is a random variable with variance σ_X^2, the variance of the random variable aX is

$$\text{var}(aX) = a^2\sigma^2{}_X.$$

In order to show this, we take

$$\text{var}(aX) = E[(aX)^2] - [E(aX)]^2$$
$$= a^2E(X^2) - a^2[E(X)]^2$$
$$= a^2(E(X^2) - [E(X)]^2)$$
$$= a^2\sigma_X^2.$$

In short, adding a constant value to each value of a random variable leaves the variance unchanged, but multiplying each value by a constant multiplies the variance by the square of the constant.

RULE 8. If X and Y are independent random variables, with variances σ_X^2 and σ_Y^2 respectively, then the variance of the sum $X + Y$ is

$$\sigma_{(X+Y)}^2 = \sigma_X^2 + \sigma_Y^2.$$

Similarly, the variance of $X - Y$ is

$$\sigma_{(X-Y)}^2 = \sigma_X^2 + \sigma_Y^2.$$

This principle also extends to any number of independent random variables (see Eq. C.2.6 in Appendix C).

B.3 INDEPENDENCE AND COVARIANCE

The next rule is a most important one that applies only to *independent* random variables (see Section C.1 in Appendix C).

RULE 9. Given random variable X with expectation $E(X)$ and the random variable Y with expectation $E(Y)$, then if X and Y are independent,

$$E(XY) = E(X)E(Y).$$

This rule states that if random variables are *statistically independent,* the expectation of the product of these variables is the product of their separate expectations. An important corollary to this principle is:

If $E(XY) \neq E(X)E(Y)$, the variables X and Y are not independent.

The basis for Rule 9 can also be shown fairly simply for discrete variables. Since X and Y are independent, $p(x,y) = p(x)p(y)$. Then,

$$E(XY) = \sum_x \sum_y (xy)p(x)p(y) = \sum_x \sum_y xp(x)yp(y).$$

However, for any fixed x, $yp(y)$ is perfectly free to be any value, so that

$$E(XY) = \sum_x xp(x) \sum_y yp(y) = E(X)E(Y).$$

Definition: Given the random variable X with expectation $E(X)$ and the random variable Y with expectation $E(Y)$, then the covariance of X and Y is

$$\text{cov}(X,Y) = E(XY) - E(X)E(Y),$$

the expected value of the product of X and Y minus the product of the expected values.

The covariance is thus a reflection of the departure from independence of X and Y. When X and Y are independent,

$$\text{cov}(X,Y) = 0,$$

by the rule given above. When random variables are *independent,* their covariance is *zero.* However, it is not necessarily true that zero covariance implies that the variables are independent. On the other hand, $\text{cov}(X,Y) \neq 0$ always implies that the variables X and Y are not independent.

Definition: Given two random variables X and Y, then the covariance X and Y divided by the standard deviation of each variable,

$$\rho_{XY} = \frac{\text{cov}(X,Y)}{\sigma_X \sigma_Y},$$

is known as the coefficient of correlation between X and Y.

The correlation coefficient ρ_{XY} may be any value between -1 and 1. However, note that if X and Y are independent, $\rho_{XY} = 0$. The converse is not true, however, since $\rho_{XY} = 0$ does not necessarily imply the independence of X and Y.

By extension of Rule 9 to any finite number of random variables, we have:

RULE 10. Given any finite number of random variables, if all the variables are independent of each other, the expectation of their product is the product of their separate expectations: thus,

$$E(XYZ) = E(X)E(Y)E(Z),$$

and so on.

EXERCISES—THE ALGEBRA OF EXPECTATIONS

Consider the following probability distributions of discrete random variables. Find the expectation $E(X)$, for each.

1.

x	$p(x)$
1	.75
0	.25
	1.00

2.

x	$p(x)$
5	4/18
0	12/18
-5	2/18
	18/18

3.

x	$p(x)$
5	.1
4	.4
3	.1
2	.3
1	.1
	1.0

4.

x	$p(x)$
399	.20
154	.20
125	.20
100	.20
-200	.20
	1.00

5.

x	$p(x)$
36	.12
30	.18
24	.20
18	.32
12	.09
6	.05
0	.04
	1.00

Consider a discrete random variable taking on only the values $x_1, x_2, \cdots, x_{10}$. Symbolize the following as expectations:

6. $x_1 p(x_1) + x_2 p(x_2) + \cdots + x_{10} p(x_{10})$

7. $x_1^2 p(x_1) + x_2^2 p(x_2) + \cdots + x_{10}^2 p(x_{10})$

8. $[x_1^2 p(x_1) + x_2^2 p(x_2) + \cdots + x_{10}^2 p(x_{10})] - [x_1 p(x_1) + x_2 p(x_2) + \cdots + x_{10} p(x_{10})]^2$

9. $[x_1 - E(X)]^2 p(x_1) + [x_2 - E(X)]^2 p(x_2) + \cdots + [x_{10} - E(X)]^2 p(x_{10})$

10. $(x_1^2 - 4x_1 + 5)p(x_1) + (x_2^2 - 4x_2 + 5)p(x_2) + \cdots + (x_{10}^2 - 4x_{10} + 5)p(x_{10})$

Consider two random variables, X and Y. Simplify the following:

11. $E[(X + 35)/10]$

12. $E[X - 14Y + E(Y) + 7] - E(X + Y - 5)$

13. $E[X^2 - 2XE(X) + E^2(X)]$

14. $E[X^2 + Y^2 - 2(X + Y)^2]$

15. $E(17)$

16. $E[(X - E(X))(Y - E(Y))]$

See if you can prove the following for discrete random variables: (**Hint:** Turn each expectation into the equivalent weighted sum.)

17. $E(aX) = aE(X)$

18. $E(aX + b) = aE(X) + b$

19. $E[X - E(X)] = 0$

For the variables defined in Exercises 1, 2, and 3, find

20. var(X), or σ_X^2

21. $E(5X - 12)$

22. $E[(X^2 + 2)/10]$

23. Suppose that the random variables given in Exercises 3 and 4 were independent. Let us call the variable in Exercise 3, X, and that in exercise 4, Y. Then, find the value of $E(XY)$.

Appendix C

Joint Random Variables and Linear Combinations

C.1 JOINT RANDOM VARIABLES AND INDEPENDENCE

Think of two discrete random variables X and Y. If all possible joint events ($X = a$, $Y = b$) are considered, each such joint event will have a *joint probability $p(a,b)$*. Then the function relating each possible such joint event to its probability is the *joint distribution* or *bivariate distribution* of X and Y.

When the two variables X and Y are continuous, so that the event $X = a$ is associated with a probability density $g(a)$, and the event $Y = b$ is associated with the probability density $h(b)$, then the joint event ($X = a$, $Y = b$) is associated with a *joint density* $f(a,b)$. Once again, the pairing of joint events involving the random variables with the probability density of each joint event can be thought of as the joint distribution of X and Y, or the bivariate distribution of X and Y.

Once again considering two discrete random variables X and Y, we know from Section 1.14 that the conditional probability that $X = a$, given that $Y = b$ is

$$p(X = a | Y = b) = \frac{p(X = a, Y = b)}{p(Y = b)}.$$

In the same way, the conditional probability of $Y = b$ given $X = a$ is found from

$$p(Y = b | X = a) = \frac{p(X = a, Y = b)}{p(X = a)}.$$

The corresponding *conditional densities* for continuous random variables are

$$r(X = a | Y = b) = \frac{f(a,b)}{h(b)} \qquad \text{[C.1.1]}$$

and

$$s(Y = b|X = a) = \frac{f(a,b)}{g(a)}.$$

Associated with any joint distribution of X and Y there is also a *marginal distribution* of X which gives a probability value $p(X = a)$ or a density value $g(X = a)$, to the event $(X = a)$. Then, in the case of discrete random variables,

$$p(X = a) = \sum_{y} p \ (X = a, Y = y),$$ [C.1.2]

or, in the case of continuous X and Y,

$$g(a) = \int_{\text{all } y} f(a,y)dy.$$ [C.1.3]

Similar expressions hold for the marginal probability or density of the event $(Y = b)$ of course.

For a joint distribution of discrete random variables X and Y, one may also define a *conditional expectation*, as follows

$$E(Y|X = a) = \sum_{\text{all } y} y \, p(Y = y|X = a).$$ [C.1.4]

If the joint distribution is pictured as a two-way table, this may be thought of as the expectation of the distribution formed by the Y distribution *within the array* (or column of the table) *standing for* $X = a$. The conditional expectation of X given $Y = b$ is defined similarly for rows of the table. Furthermore, in an analogous way one may define the *conditional variance* $\sigma_{Y|a}^2$ given $X = a$, and the conditional variance $\sigma_{X|b}^2$ given $Y = b$. Precisely these same ideas hold for continuous random variables, if the necessary adjustments in notation are made as in Eq. C.1.7.

We know from Section 1.14 that independence of events C and D exists if and only if

$$p(C \cap D) = p(C)p(D),$$

so that

$$p(C|D) = p(C) \text{ and } p(D|C) = p(D).$$

It follows that **two discrete random variables are independent if and only if**

$$p(X = a, Y = b) = p(X = a)p(Y = b)$$ [C.1.5*]

for all pairs of values a **and** b. Furthermore, $p(X = a|Y = b) = p(X = a)$ and $p(Y = b|X = a) = p(Y = b)$, when X and Y are independent.

The principle extends immediately to continuous random variables:

A random variable X **with density** $g(a)$ **at value** a, **and a random variable** Y **with density** $h(b)$ **at value** b **are independent if and only if for all** (a,b)

$$f(a,b) = g(a)h(b),$$ [C.1.6*]

where $f(a,b,)$ **is the** *joint* **density for the event** (a,b).

Just as for conditional probability, it is also possible to define **conditional density:**

$$w(b|a) = \frac{f(a,b)}{g(a)},$$

[C.1.7]

where $w(b|a)$ symbolizes the density for $Y = b$ given some value a of X. The *conditional distribution* of Y given X will be *exactly the same as* the distribution of Y with X *left unspecified when X and Y are independent:* for all (a,b),

$$w(b|a) = h(b), \quad X \text{ and } Y \text{ independent.}$$

In the same way,

$$w(a|b) = g(a)$$

when the variables are independent.

The idea of independence of random variables can be extended to functions of random variables as well. That is, suppose we have a random variable X with a density $g(x)$, and an *independent* random variable Y with density $h(y)$, as before.

Now imagine some function of X: associated with each possible x value is a new number $v = t(x)$. For example, it might be that

$$v = ax + b$$

so that V is some linear function of X. Or perhaps,

$$v = 3x^2 + 2x,$$

and so on for any other function rule. Furthermore, let there be some function of Y:

$$w = s(y),$$

[C.1.8]

so that a new number w is associated with each possible value of Y. Then V and W are *independent random variables,* provided that X and Y are *independent*. This principle applies both to continuous and to discrete random variables, and to all the ordinary functions studied in elementary mathematical analysis.

Although we have dealt so far only with the bivariate distribution of two variables such as X and Y, the ideas extend readily to the situation where an event may consist of values of more than two random variables. Thus, for example, perhaps each individual i in a sample is associated with three scores, the first representing a value of the random variable X, the second a value of the random variable Y, and the third a value of a random variable W. Then the event (x_i, y_i, w_i) is a joint occurrence of the event in which individual i receives the value x_i on variable X, y_i on variable Y, and w_i on variable W. Furthermore, associated with each such joint event is a probability or probability density. The function giving the relation between each possible such joint event and its probability (density) is the *multivariate distribution function* of X, Y, and W.

The ideas of conditional probabilities or conditional densities, conditional expectations, and conditional variances extend to multivariate distributions as well. We may also consider the marginal distribution of any single variable, such as X, summed (or integrated) over the remaining variables, or the bivariate distribution of two variables summed or integrated over the others, et cetera.

Independence among three variables, or any number of variables, is also defined in a way completely analogous to the concept of independence of any two variables. Thus, if the joint density of X, Y, and W is represented by $t(x,y,w)$ for any particular set of values, $X = x$, $Y = y$, $W = w$, and the marginal densities of the individual variables for these values are $g(x)$, $h(y)$, and $f(w)$, then

$$t(x,y,w) = g(x)h(y)f(w) \qquad \text{[C.1.9*]}$$

if and only if X, Y, and W are *pairwise independent*.

(See also Rule 10 of Appendix B for the relation among the expectations of such independent random variables.)

C.2 LINEAR COMBINATIONS OF RANDOM VARIABLES

Consider a set of J random variables, $(X_1, X_2, \ldots, X_J)$, having some multivariate distribution function. Now also consider a set of J weights, not all zero, $(c_1, c_2, \cdots, c_J)$. Then a *linear combination* of the random variables is a weighted sum

$$Y = c_1 X_1 + c_2 X_2 + \cdots + c_J X_J. \qquad \text{[C.2.1*]}$$

The linear combination Y is then a random variable, with a probability (density) which depends on the joint probability (density) of the set of random variables X_1 and so on. (When at least some of the weights are not zero, then the linear combination is said to be "nontrivial.")

The random variable Y (the linear combination) will have an expectation and a variance, of course:

Given J random variables, and some linear combination,
$Y = c_1 X_1 + \cdots + c_J X_J,$

the expected value of Y is

$$E(Y) = c_1 E(X_1) + c_2 E(X_2) + \cdots + c_J E(X_J). \qquad \text{[C.2.2.*]}$$

The expectation of any linear combination is the same linear combination of the expectations.

A special case of this principle gives us the following:

Given n sample values from the same distribution with mean μ, then the expectation of any linear combination of those sample values is

$$E(Y) = \mu(c_1 + c_2 + \cdots + c_n). \qquad \text{[C.2.3*]}$$

If, under these circumstances,

$$c_1 + c_2 + \cdots + c_k = 0, \, E(Y) = 0. \qquad \text{[C.2.4]}$$

In addition, we know the variance of a linear combination:

For any set of random variables $(X_1, \cdots, X_J)$ and the linear combination Y with weights $(c_1, \cdots, c_J)$, not all zero, then the variance of Y is

$$\sigma_Y^2 = \sum_j c_j^2 \sigma_j^2 + 2 \sum\sum_{j>k} c_j c_k \, \text{cov}(X_j, X_k), \qquad \text{[C.2.5]}$$

where σ_j^2 symbolizes the variance of variable j (See Appendix B for a definition of $\text{cov}(X_j, X_k)$). This says that the variance of linear combination Y is a sum of the variances of the X variables, each weighted by the squared value of the original combination weight, plus a weighted sum of covariances between all pairs of X variables. However, by Appendix B, Rule 9, $\text{cov}(X_j, X_k) = 0$ if X_j and X_k are independent. Thus,

Give J independent random variables, with variances $\sigma_1^2, \sigma_2^2, \cdots, \sigma_J^2$ respectively, and the linear combination

$$Y = c_1 X_1 + c_2 X_2 + \cdots + c_J X_J, \qquad \text{[C.2.6*]}$$

then the distribution of Y has variance given by

$$\sigma_Y^2 = c_1^2 \sigma_1^2 + \cdots + c_J^2 \sigma_J^2.$$

The variance of a linear combination of independent random variables is a weighted sum of their separate variances, each weight being the *square* of the original weight given to the variable.

It immediately follows from the above that in the special case where the variance of each variable X_1 through X_J is equal to the *same* value, σ^2, then

$$\sigma_Y^2 = \sigma^2 \sum_j c_j^2, \qquad \text{[C.2.7*]}$$

the common value of σ^2 times the sum of the squared weights c_j.

Now for an important principle involving the normal distribution:

PRINCIPLE 1: Given some J independent random variables $X_1, \cdots, X_J$, each normally distributed, then any nontrivial linear combination of these variables is also a normally distributed random variable.

For example, the sample mean of N cases is a linear combination, where the score of Case 1, representing random variable 1, is given the weight $1/N$, and so on for all other scores. Hence, if the scores of the N individuals are independent of each other and are drawn from a normal population, the sampling distribution of the mean is exactly normal, regardless of how large or small N may be.

Suppose that each of J variables $X_1, X_2, \cdots, X_J$ represents an independent observation from a normal distribution, with variance σ^2. Now two linear combinations are formed,

$$Y_1 = c_{11} X_1 + \cdots + c_{1j} X_j + \cdots + c_{1J} X_J$$

and

$$Y_2 = c_{21} X_1 + \cdots + c_{2j} X_j + \cdots + c_{2J} X_J,$$

where in both Y_1 and Y_2 some of the c weights are other than zero. Then, the two variables Y_1 and Y_2 are *independent* if and only if

$$\sum_j c_{1j} c_{2j} = 0. \qquad \text{[C.2.8*]}$$

That is, the independence of Y_1 and Y_2 implies and is implied by the fact that the sum of the products of weights used in each is zero, when each X_j is a normally distributed variable.

Finally, a very general principle can be stated about linear combinations of sample means when the size of each sample is large. This is a direct consequence of the central limit theorem (Section 6.7) and can be stated as follows:

PRINCIPLE 2: Given J independent samples, containing N_1, N_2, $\cdots$, N_J independent observations, respectively, then the sampling distribution of any nontrivial linear combination of means of those samples approaches a normal distribution as the size of each sample grows large.

C.3 BIVARIATE NORMAL AND MULTIVARIATE NORMAL DISTRIBUTIONS

In using methods such as those in Chapter 14 onward it often becomes necessary to assume a particular form of joint distribution for two random variables. This is known as a *bivariate normal distribution*. In Section 14.20 this notion is discussed briefly along with its generalization, *the multivariate normal distribution*. Here, we will attempt to tie these two important theoretical distributions into the ideas of linear combinations of random variables, such as we have just been discussing.

That is, consider two random variables, say X_1 and X_2. These variables have means, μ_1, and μ_2, and variances, σ_1^2 and σ_2^2. In addition, the two variables have a covariance, $\text{cov}(X_1, X_2)$. Note that since the variables may be correlated, the covariance need not be zero. Often the two variances and the covariance of two such random variables are referred to in the form of a 2×2 *array* or *matrix,* with the variances in the main diagonal, and the covariance in the off-diagonal cells. Such a population variance–covariance matrix is often symbolized by a boldface capital sigma:

population variance-covariance matrix =

$$\mathbf{\Sigma} = \begin{bmatrix} \sigma_1^2 & \text{cov}(X_1,X_2) \\ \text{cov}(X_2,X_1) & \sigma_2^2 \end{bmatrix} \qquad [\text{C.3.1}]$$

PRINCIPLE 3: The variables X_1 and X_2 follow a bivariate normal distribution if and only if for every possible linear combination y,

$Y = c_1 X_1 + c_2 X_2,$

(with the c values not both zero) the distribution of Y is normal.

That is, any nontrivial weighted sum of the two variables will also yield a normally distributed variable.

The particular bivariate normal distribution that holds for two such variables depends on the variance–covariance matrix Σ, as defined above.

This general idea extends directly to the situation where there are K distinct random variables, $X_1, X_2, \cdots, X_K$. Here, there are K population means $\mu_1, \mu_2, \cdots, \mu_K$, and a $K \times K$ variance–covariance matrix Σ such as this:

$$\Sigma = \begin{bmatrix} \sigma_1^2 & \text{cov}(X_1,X_2) & \cdots & \text{cov}(X_1,X_K) \\ \text{cov}(X_2,X_1) & \sigma_2^2 & \cdots & \text{cov}(X_2,X_K) \\ \cdots\cdots\cdots\cdots\cdots\cdots\cdots\cdots\cdots\cdots\cdots\cdots\cdots\cdots\cdots \\ \text{cov}(X_K,X_1) & \text{cov}(X_K,X_2) & \cdots & \sigma_K^2 \end{bmatrix}$$

[C.3.2]

PRINCIPLE 4: The K variables follow a joint multivariate normal distribution if and only if for every linear combination Y, where

$$Y = c_1X_1 + c_2X_2 + \cdots + c_KX_K$$

(and where the c values are not all zero) the distribution of Y is normal.

In other words, if a set of K variables follow a joint multivariate normal distribution, then any nontrivial way of weighting and summing those variables yields another normally distributed variable. Once again, the particular multivariate normal distribution is characterized by the variance–covariance matrix Σ. More will be said about such variance–covariance matrices in Section D.6.

It is worth noting that the situation covered by Principle 1 above is a special case of a multivariate normal distribution, where the covariance between any pair of variables is 0. However, in general the covariances, and hence the correlations, for pairs of variables will tend to be other than zero.

It is frequently necessary to estimate the population matrix of variances and covariances of a set of variables by use of a sample of N cases. Suppose that each case i in the sample has a value x_{ij} on each of J variables $X_1, X_2, \ldots, X_J$. We would then start by finding the *sum of squares total* for each variable, X_j:

$$SS_j \text{ total} = \sum_{i=1}^{N} x_{ij}^2 - T_j^2/N$$

where

$$T_j = \sum_{i=1}^{N} x_{ij}.$$

Then

$$\text{est. } \sigma_j^2 = SS_j \text{ total}/(N-1).$$

That is, the unbiased estimate of the variance of any variable X_j is the sum of squares total for that variable divided by $N - 1$, so that

$$E(\text{est. } \sigma_j^2) = \sigma_j^2, \text{ for each } X_j.$$

Similarly, if we want to estimate the population covariance $\text{cov}(X_j,X_k)$ for any two of the variables X_j and X_k, we first find the total sum of products, as follows:

$$\text{SP}_{jk}\text{total} = \sum_{i=1}^{N} x_{ij}x_{ik} - (T_j)(T_k)/N.$$

The unbiased estimate of the population $\text{cov}(X_j,X_k)$ is then given by

$$\text{est. cov}(X_j,X_k) = (\text{SP}_{jk}\text{ total}/(N - 1)).$$

In this way the (usually) unknown matrix of variances and covariances Σ can be estimated by a matrix $\hat{\Sigma}$, based on the sample SS and SP values divided by $N - 1$.

C.4 VECTOR NOTATION

In working with statistics, quite often one needs to speak of a particular set of values taken on by a variable, as, for example, when a sample of N observations is taken on some variable X. In other contexts, we need to represent the joint event consisting of the values taken on by several variables. For such situations, the use of vector and matrix notation is a great convenience, and becomes an absolute necessity in the study of multivariate statistical methods. We will need some rudiments of vector algebra in order to explain and use the methods in the following sections.

A vector is an ordered set of values. Any number appearing as a part of a vector is referred to as an *element* of that vector. The position of any element of a vector has meaning, as it often tells to whom or to what that value belongs, or which variable that value represents. Any vector has a *dimension,* which indicates how many elements that vector contains, or how many "places" in order it has. Thus the vector $[10, 25, -2, 4]$ is of dimension 4. A vector of dimension n contains n elements. The vectors of variables and of weights in Section C.2 have dimension J.

In the following a boldface lowercase letter will represent a vector. Thus, if the letter **a** is used for a 5-dimensional vector, this may be thought of as an ordered set of 5 values

$$\mathbf{a} = [a_1, a_2, a_3, a_4, a_5]$$

where the ordinary italic letters represent the various elements of **a**. Here, **a** is shown as a *row vector,* so that the values are written horizontally, or along one row. However, it is equally possible to have a column vector,

$$\mathbf{a} = \begin{bmatrix} a_2 \\ a_2 \\ a_3 \\ a_4 \\ a_5 \end{bmatrix}.$$

It doesn't make a lot of difference whether we think in terms of row or column vectors; however, once we have decided that a vector will be a row or a column vector in a given problem, we stick with that same definition throughout. As it happens, in the following it will be most convenient to work with column vectors. In general, for dimension n, any column vector such as $\mathbf{a}$ will thus be representable as

$$\mathbf{a} = \begin{bmatrix} a_1 \\ a_2 \\ \cdot \\ \cdot \\ \cdot \\ a_j \\ \cdot \\ \cdot \\ \cdot \\ a_n \end{bmatrix}. \qquad \text{[C.4.1]}$$

If we wish to speak of an individual element in $\mathbf{a}$, we will use a_j to stand for such an element.

The *transpose* of a column vector $\mathbf{a}$, written as $\mathbf{a}'$, consists of exactly the same elements as those in $\mathbf{a}$, but written as a row vector: thus for the column vector shown in Expression 4, Section C.3.1, we have

$$\mathbf{a}' = [a_1, a_2, \cdots, a_j, \cdots, a_n].$$

In the same way, the transpose of any row vector is the corresponding column vector.

A special vector of interest is the *unit vector* **1**, which has entries that are all "ones"; that is

$$\mathbf{1} = \begin{bmatrix} 1 \\ 1 \\ \cdot \\ \cdot \\ \cdot \\ 1 \end{bmatrix}. \qquad \text{[C.4.2]}$$

On the other hand, the *zero vector*, or the "null vector," has entries which are all zeros.

When a vector is multiplied by a constant, this is called *scalar multiplication*. Each entry in the vector is then multiplied by that constant. Thus, if d is a constant,

$$d\mathbf{a} = \begin{bmatrix} da_1 \\ da_2 \\ \cdot \\ \cdot \\ \cdot \\ da_n \end{bmatrix}. \qquad \text{[C.4.3]}$$

Now consider two vectors, **x** and **y**, each of the same dimension n:

$$\mathbf{x} = \begin{bmatrix} x_1 \\ x_2 \\ \cdot \\ \cdot \\ \cdot \\ x_n \end{bmatrix}, \; \mathbf{y} = \begin{bmatrix} y_1 \\ y_2 \\ \cdot \\ \cdot \\ \cdot \\ y_n \end{bmatrix}. \qquad [\text{C.4.4}]$$

Then the *sum* of **x** and **y** is given by their element-by-element or place-by-place sum:

$$\mathbf{x} + \mathbf{y} = \begin{bmatrix} x_1 + y_1 \\ x_2 + y_2 \\ \cdot \\ \cdot \\ \cdot \\ x_n + y_n \end{bmatrix}. \qquad [\text{C.4.5}]$$

In a similar way we can define the difference, $\mathbf{x} - \mathbf{y}$.

Furthermore, suppose that c and d are two constant numbers. Then a *linear combination* of the vectors **x** and **y** is

$$\mathbf{w} = c\mathbf{x} + d\mathbf{y} = \begin{bmatrix} cx_1 + dy_1 \\ cx_2 + dy_2 \\ \cdot \\ \cdot \\ \cdot \\ cx_n + dy_n \end{bmatrix}. \qquad [\text{C.4.6}^\star]$$

In other words, each weighted value for **x** is summed with the corresponding weighted value of **y.** Then, the elements of vector **w** are composed of these weighted sums. The same idea applies to weighted sums of any number of vectors.

If two vectors have exactly the same dimension, then their *inner product* may be found. That is, let **x** and **y** be two row vectors, each of the same dimension n. Alternatively, **x** and **y** may both be column vectors of the same dimension n.

The *inner product* of two vectors **x** and **y**, represented by (**x,y**), is simply the product of each element x_j with its corresponding element y_j, and then summed over all j:

$$(\mathbf{x,y}) = \sum_{j=1}^{n} x_j y_j. \qquad [\text{C.4.7}^\star]$$

On the other hand, when **x** and **y** are both row vectors of the same dimension, the inner product can also be symbolized by

$$\mathbf{xy}' = \text{inner product} = (\mathbf{x,y}).$$

Or, when **x** and **y** are both column vectors, the inner product may be written as

$$\mathbf{x}'\mathbf{y} = \text{inner product} = (\mathbf{x,y}).$$

Regardless of which is used to represent the inner product, however, what is being symbolized is a *single value* found by multiplying the corresponding elements of the two vectors and then summing.

The concept of the inner product of two vectors is an especially useful and important way of discussing linear combinations of random variables, as in Section C.2. That is, suppose that we wish to discuss a set of K different random variables. These random variables can be displayed in terms of the vector.

$$\mathbf{x} = [X_1, X_2, \ldots, X_k, \ldots, X_K].$$

In addition, we can also consider the vector $\mathbf{a}$, consisting of coefficients or "weights" a, or

$$\mathbf{a} = [a_1, a_2, \ldots, a_k, \ldots, a_K].$$

Then the linear combination or weighted sum

$$Y = a_1X_1 + a_2X_2 + \cdots + a_KX_K$$

can be written very economically as the product

$$Y = \mathbf{ax'} \quad \text{or} \quad Y = (\mathbf{a,x}).$$

Another use of the notion of the inner product of vectors occurs when we have a set of scores for N cases on some variable X, and a set of scores for the same N cases on another variable Y. The first set of scores can be symbolized by the vector $\mathbf{x}$, and the second set by the vector $\mathbf{y}$. Often in statistical calculations we need the sum of the products of these scores, and this sum of products can be symbolized very neatly by the inner product $(\mathbf{x,y})$.

The inner product of any vector $\mathbf{x}$ with itself, represented by $\|\mathbf{x}\|^2$ is called the *squared length* of the vector:

$$\|\mathbf{x}\|^2 = \sum_{j=1}^{n} x_j^2 \qquad\qquad \text{[C.4.8*]}$$

The *length* of vector $\mathbf{x}$ is then the *positive square root*, $\sqrt{\|\mathbf{x}\|^2} = \|\mathbf{x}\|$.

When two vectors have an inner product which is equal to zero,

$$(\mathbf{x,y}) = 0, \qquad\qquad \text{[C.4.9*]}$$

the two vectors are said to be *orthogonal*. An *orthogonal set* of vectors is one in which each pair of vectors is orthogonal. An orthogonal set of vectors in which each vector also has a length of 1.00 is called an *orthonormal set* or an *orthonormal basis*.

Suppose that we have two vectors which are orthogonal. Now if we form two new vectors by taking some constant c times $\mathbf{x}$, and some constant d times $\mathbf{y}$, so that $\mathbf{w} = c\mathbf{x}$ and $\mathbf{z} = c\mathbf{y}$, then $(\mathbf{w,z}) = 0$, and the vectors $\mathbf{w}$ and $\mathbf{z}$ are also orthogonal.

One vector $\mathbf{x}$ is said to be *linearly dependent* on another vector $\mathbf{y}$ if it is possible to write $\mathbf{x}$ as some constant c times $\mathbf{y}$:

$$\mathbf{x} = c\mathbf{y}. \qquad\qquad \text{[C.4.10]}$$

A vector $\mathbf{x}$ is also said to be linearly dependent on some set of other vectors if it is possible to write **x** as a linear combination of two or more of these other vectors. Thus, if **x** is linearly dependent on **y** and **z**, then there exist constants c and d such that

$$\mathbf{x} = c\mathbf{y} + d\mathbf{z}. \tag{C.4.11}$$

On the other hand, *two vectors which are orthogonal are linearly independent.* A set of vectors is said to be linearly independent if it is impossible to find sets of constants permitting any vector to be written as a linear combination of some subset of the others. Any orthogonal set of vectors (in which every vector is orthogonal to every other) is a *linearly independent set.*

Now suppose that we have a whole set consisting of m vectors, $\mathbf{X} = (\mathbf{x}_1, \cdots, \mathbf{x}_k, \cdots, \mathbf{x}_m)$, each member of which, such as $\mathbf{x}_k$, is a column vector of dimension n. Such a set of vectors is called a *matrix of dimension $n \times m$,* and has elements in the rows and columns as follows:

$$\mathbf{X} = \begin{bmatrix} x_{11} \cdots x_{1k} \cdots x_{1m} \\ x_{j1} \cdots x_{jk} \cdots x_{jm} \\ x_{n1} \cdots x_{nk} \cdots x_{nm} \end{bmatrix}$$

(We will have a great deal more to say about matrices and their manipulation in Appendix D.) The vectors making up the columns of the matrix **X** may or may not be linearly dependent. However, it is *always possible* to find a matrix **V** made up of q column vectors $v_1, \cdots, v_i, \cdots, v_q$ each of dimension n, and with $q \leq n$, with **V** as an orthogonal set of vectors. Then every vector in **X** can be shown as a linear combination of the vectors in **V**, and we say that **V** is an *orthogonal basis* for the set of vectors **X**. As the text shows, particularly in Chapter 11, there are distinct advantages to be gained by replacing a set of linearly dependent vectors **X** with an orthogonal basis **V**.

Given any small set of vectors **X** it is fairly simple and straightforward to find an orthogonal basis. This is done through the *Gram-Schmidt orthogonalization process,* which will be outlined next. This well-known mathematical method is described fully in sources such as Korn and Korn (1968, p. 491) and Noble and Daniel (1977, pp 138–143).

C.5 FINDING AN ORTHOGONAL BASIS FOR A SET OF VECTORS

The Gram-Schmidt process requires that we already have an initial set of vectors **X**. The process is begun by choosing one vector out of **X** to serve as the first vector, $\mathbf{v}_0$, in the set **V**. In most statistical work, the set **X** will include the unit vector **1**. Although this is not formally necessary, the vector **1** is thus almost always chosen to be the vector $\mathbf{v}_0$.

The process continues when we choose another vector from **X**. Let us call this $\mathbf{x}_1$. Then we take

$$\mathbf{v}_1 = \mathbf{x}_1 - b_{10}\mathbf{v}_0, \text{ where } b_{10} = \frac{(\mathbf{x}_1, \mathbf{v}_0)}{\|\mathbf{v}_0\|^2} \tag{C.5.1}$$

This gives a new vector $\mathbf{v}_1$ which is orthogonal to $\mathbf{v}_0$, or **1**. Now we take the next vector from **X**, or $\mathbf{x}_2$. We find a new vector $\mathbf{v}_2$ by taking

$$\mathbf{v}_2 = \mathbf{x}_2 - b_{20}\mathbf{v}_0 - b_{21}\mathbf{v}_1, \text{ where } b_{20} = \frac{(\mathbf{x}_2, \mathbf{v}_0)}{\|\mathbf{v}_0\|^2}, b_{21} = \frac{(\mathbf{x}_2, \mathbf{v}_0)}{\|\mathbf{v}_1\|^2}. \tag{C.5.2}$$

This new vector $\mathbf{v}_2$ is then orthogonal to both $\mathbf{v}_0$ and $\mathbf{v}_1$. Proceeding in this way, we find the vector $\mathbf{v}_3$ by taking

$$\mathbf{v}_3 = \mathbf{x}_3 - b_{30}\mathbf{v}_0 - b_{31}\mathbf{v}_1 - b_{32}\mathbf{v}_2, \text{ where } b_{30} = \frac{(\mathbf{x}_3, \mathbf{v}_0)}{\|\mathbf{v}_0\|^2}, \quad b_{31} = \frac{(\mathbf{x}_3, \mathbf{v}_1)}{\|\mathbf{v}_1\|^2}, \qquad \text{[C.5.3]}$$

$$b_{32} = \frac{(\mathbf{x}_3, \mathbf{v}_2)}{\|\mathbf{v}_2\|^2},$$

which yields a vector $\mathbf{v}_3$, orthogonal to $\mathbf{v}_0$, to $\mathbf{v}_1$, and to $\mathbf{v}_2$.

This process is repeated until all m vectors in $\mathbf{X}$ have been used or until each additional vector yields a $\mathbf{v}$ vector consisting only of zero entries.

The Gram-Schmidt process will be illustrated by use of a set of five vectors making up $\mathbf{X}$ as follows. Here the dimensionality n is 4.

$$\mathbf{x}_0 = \mathbf{1} = \begin{bmatrix} 1 \\ 1 \\ 1 \\ 1 \end{bmatrix}, \ \mathbf{x}_1 = \begin{bmatrix} 1 \\ 0 \\ 0 \\ 0 \end{bmatrix}, \ \mathbf{x}_2 = \begin{bmatrix} 0 \\ 1 \\ 0 \\ 0 \end{bmatrix}, \ \mathbf{x}_3 = \begin{bmatrix} 0 \\ 0 \\ 1 \\ 0 \end{bmatrix}, \ \mathbf{x}_4 = \begin{bmatrix} 0 \\ 0 \\ 0 \\ 1 \end{bmatrix}.$$

The first step in this process is to choose one vector from $\mathbf{X}$ to be the initial vector in $\mathbf{V}$. As is often the case in statistical problems, here set $\mathbf{X}$ includes the unit vector $\mathbf{1}$, so that this will be taken as the initial vector in the basis, or $\mathbf{v}_0$. Then by Eq. C.5.1 take

$$b_{10} = \frac{(\mathbf{x}_1, \mathbf{v}_0)}{\|\mathbf{v}_0\|^2} = \frac{(1)\,(1) + (0)\,(1) + (0)\,(1) + (0)\,(1)}{1 + 1 + 1 + 1} = \frac{1}{4}.$$

On substituting into Eq. C.5.1, we have

$$\begin{bmatrix} 1 \\ 0 \\ 0 \\ 0 \end{bmatrix} - \left(\frac{1}{4}\right) \begin{bmatrix} 1 \\ 1 \\ 1 \\ 1 \end{bmatrix} = \begin{bmatrix} \dfrac{3}{4} \\ -\dfrac{1}{4} \\ -\dfrac{1}{4} \\ -\dfrac{1}{4} \end{bmatrix} = \mathbf{v}_1,$$

or, since multiplying through by the common denominator still yields an orthogonal vector relative to $\mathbf{v}_0$, we can simplify matters by taking

$$\mathbf{v}_1 = \begin{bmatrix} 3 \\ -1 \\ -1 \\ -1 \end{bmatrix}.$$

Next we take vector $\mathbf{X}_2$ and find that

$$b_{20} = \frac{1}{4}, \ b_{21} = \frac{0(3) - (1)(1) + (0)(-1) + (0)(-1)}{9 + 1 + 1 + 1} = \frac{-1}{12}$$

so that

$$\begin{bmatrix} 0 \\ 1 \\ 0 \\ 0 \end{bmatrix} - \left(\frac{1}{4}\right) \begin{bmatrix} 1 \\ 1 \\ 1 \\ 1 \end{bmatrix} - \left(-\frac{1}{12}\right) \begin{bmatrix} 3 \\ -1 \\ -1 \\ -1 \end{bmatrix} = \begin{bmatrix} 0 \\ \frac{2}{3} \\ -\frac{1}{3} \\ -\frac{1}{3} \end{bmatrix} \text{ or } \begin{bmatrix} 0 \\ 2 \\ -1 \\ -1 \end{bmatrix} = \mathbf{v}_2,$$

where the last vector is once again cleared of fractions.

As the next step, $\mathbf{x}_3$ is taken, and we find

$$b_{30} = \frac{1}{4}, \ b_{31} = -\frac{1}{12}, \ b_{32} = -\frac{1}{6}$$

so that

$$\begin{bmatrix} 0 \\ 0 \\ 1 \\ 0 \end{bmatrix} - \left(\frac{1}{4}\right) \begin{bmatrix} 1 \\ 1 \\ 1 \\ 1 \end{bmatrix} - \left(-\frac{1}{12}\right) \begin{bmatrix} 3 \\ -1 \\ -1 \\ -1 \end{bmatrix} - \left(-\frac{1}{6}\right) \begin{bmatrix} 0 \\ 2 \\ -1 \\ -1 \end{bmatrix} = \begin{bmatrix} 0 \\ 0 \\ \frac{1}{2} \\ -\frac{1}{2} \end{bmatrix} \text{ or } \begin{bmatrix} 0 \\ 0 \\ 1 \\ -1 \end{bmatrix} = \mathbf{v}_3$$

Finally, if we take $\mathbf{x}_4$ and find $b_{40} = 1/4$, $b_{41} = -1/12$, $b_{42} = -1/6$, and $c_{43} = -1/2$, we have

$$\begin{bmatrix} 0 \\ 0 \\ 0 \\ 1 \end{bmatrix} - \left(\frac{1}{4}\right) \begin{bmatrix} 1 \\ 1 \\ 1 \\ 1 \end{bmatrix} - \left(\frac{-1}{12}\right) \begin{bmatrix} 3 \\ -1 \\ -1 \\ -1 \end{bmatrix} - \left(-\frac{1}{6}\right) \begin{bmatrix} 0 \\ 2 \\ -1 \\ -1 \end{bmatrix} - \left(-\frac{1}{2}\right) \begin{bmatrix} 0 \\ 0 \\ 1 \\ -1 \end{bmatrix} = \begin{bmatrix} 0 \\ 0 \\ 0 \\ 0 \end{bmatrix}.$$

In other words, no other vector (not all zero) exists which is orthogonal to $\mathbf{v}_0$, $\mathbf{v}_1$, $\mathbf{v}_2$, and $\mathbf{v}_3$. Thus an orthogonal basis for the set of vectors $\mathbf{X}$ is provided by the vectors

$$\mathbf{x}_0 = \mathbf{v}_0 = \begin{bmatrix} 1 \\ 1 \\ 1 \\ 1 \end{bmatrix}, \ \mathbf{v}_1 = \begin{bmatrix} 3 \\ -1 \\ -1 \\ -1 \end{bmatrix}, \ \mathbf{v}_2 = \begin{bmatrix} 0 \\ 2 \\ -1 \\ -1 \end{bmatrix}, \ \mathbf{v}_3 = \begin{bmatrix} 0 \\ 0 \\ 1 \\ -1 \end{bmatrix}.$$

Each vector in $\mathbf{X}$ is some linear combination of the vectors in $\mathbf{V}$ (or, like $\mathbf{x}_0$, is identical to a vector in $\mathbf{V}$). Thus, for example, we can form $\mathbf{x}_3$ by the linear combination

$$\mathbf{x}_3 = b_{30}\mathbf{v}_0 + b_{31}\mathbf{v}_1 + b_{32}\mathbf{v}_2 + \frac{1}{2}(\mathbf{v}_3)$$

and so on for the other vectors in **X**. Furthermore, note that the number of vectors q required in **V** is $\leq m$, the number of original vectors in **X**, and also that $q \leq n$.

Such sets of orthogonal vectors are often converted into an *orthonormal basis*, which means that each vector $\mathbf{v}_1$ is multiplied by $1/\|\mathbf{v}_i\|$. (This step is actually a built-in part of many statistical procedures, such as the orthogonal comparisons discussed in Chapter 11. This is why we can be fairly free and easy about clearing away fractions, leaving only whole-number vectors to work with, since this will be corrected later.) Values making up the **V** vectors in the preceding example correspond to the weights used in a set of comparisons, such as those made in Chapter 11.

Occasionally, the problem is to start with a set of fewer than $J - 1$ comparisons among J groups, and to complete the set by finding other comparisons which will be orthogonal not only to each other, but to those in the original set. Thus, suppose that two orthogonal comparisons are planned, out of $J - 1$ that are possible. Then, the weights already being used form three of the vectors in the orthogonal basis **V**: the two sets of weights for the planned comparison plus the unit vector **1**. Using any arbitrary vector **x** to stand for a member of **X** and carrying out the Gram-Schmidt procedure will give either a new orthogonal vector of weights, or will give the zero vector. If the zero vector is found, a new selection for **x** is made and the process repeated. Any nonzero vector **v** found from this process will be orthogonal to the vectors corresponding to the original comparison weights and to the unit vector. Going on in this way, one can find a complete set of $J - 1$ orthogonal comparison weights.

For example, let us find the weights for the fifth orthogonal comparison for the example shown in Table 11.12.1. Arbitrarily, we choose an initial vector, say $\mathbf{x}' = [1,0,0,0,0,0]$. Then, going through the steps of Eq. C.5.1 through Eq. C.5.3, and eliminating denominators, we have

$$
\begin{bmatrix} 1 \\ 0 \\ 0 \\ 0 \\ 0 \\ 0 \end{bmatrix} - \frac{1}{6}\begin{bmatrix} 1 \\ 1 \\ 1 \\ 1 \\ 1 \\ 1 \end{bmatrix} - \frac{1}{4}\begin{bmatrix} 1 \\ -1 \\ 1 \\ -1 \\ 0 \\ 0 \end{bmatrix} - \frac{(-1)}{4}\begin{bmatrix} -1 \\ 1 \\ 1 \\ -1 \\ 0 \\ 0 \end{bmatrix} - \frac{(-1)}{4}\begin{bmatrix} -1 \\ -1 \\ 1 \\ 1 \\ 0 \\ 0 \end{bmatrix}
$$

$$
- \frac{(0)}{2}\begin{bmatrix} 0 \\ 0 \\ 0 \\ 0 \\ 1 \\ -1 \end{bmatrix} = \begin{bmatrix} 1/12 \\ 1/12 \\ 1/12 \\ 1/12 \\ -2/12 \\ -2/12 \end{bmatrix} \quad \text{or} \quad \begin{bmatrix} 1 \\ 1 \\ 1 \\ 1 \\ -2 \\ -2 \end{bmatrix}
$$

Thus, the set of weights $(1, 1, 1, 1, -2, -2)$ provides a fifth orthogonal comparison among the six means. If this first trial vector of weights, **x**, had given us only zero weights, we would have chosen another trial vector **x** and kept on going until a set of nonzero weights was found.

Another example of the Gram-Schmidt process will yield the orthogonal polynomial values such as those employed in Section 16.13. Suppose that four experimental groups are being used, each containing the same number of cases. The groups are administered quantitatively different treatments, equally spaced in amount. Thus, the independent variable X could be represented by the vector

$$\mathbf{x}' = [1, 2, 3, 4].$$

These values will be taken as the trial vector for the first or linear, comparison among the groups. The second trial vector will use the squares of these values (for the quadratic comparison), the third vector the cubes of these values, and finally, the fourth trial vector (if required) would use the values to the fourth power.

Here, the vectors in the original set $\mathbf{X}$ are then:

$$\mathbf{x}_0 = \begin{bmatrix} 1 \\ 1 \\ 1 \\ 1 \end{bmatrix}, \ \mathbf{x}_1 = \begin{bmatrix} 1 \\ 2 \\ 3 \\ 4 \end{bmatrix}, \ \mathbf{x}_2 = \begin{bmatrix} 1 \\ 4 \\ 9 \\ 16 \end{bmatrix}, \ \mathbf{x}_3 = \begin{bmatrix} 1 \\ 8 \\ 27 \\ 64 \end{bmatrix}, \ \mathbf{x}_4 = \begin{bmatrix} 1 \\ 16 \\ 81 \\ 256 \end{bmatrix}.$$

Notice here that $\mathbf{x}_2$ is composed of the squared entries in $\mathbf{x}_1$, $\mathbf{x}_3$ of the cubes of $\mathbf{x}_1$ values and the entries in $\mathbf{x}_4$ are the $\mathbf{x}_1$ entries raised to the fourth power.

Then applying the Gram-Schmidt process and reducing the results to their simplest forms we have

$$\begin{bmatrix} 1 \\ 2 \\ 3 \\ 4 \end{bmatrix} - \frac{10}{4} \begin{bmatrix} 1 \\ 1 \\ 1 \\ 1 \end{bmatrix} = \begin{bmatrix} -\dfrac{6}{4} \\ -\dfrac{2}{4} \\ \dfrac{2}{4} \\ \dfrac{6}{4} \end{bmatrix} \text{ or } \begin{bmatrix} -3 \\ -1 \\ 1 \\ 3 \end{bmatrix} = \mathbf{v}_1,$$

$$\begin{bmatrix} 1 \\ 4 \\ 9 \\ 16 \end{bmatrix} - \left(\frac{30}{4}\right) \begin{bmatrix} 1 \\ 1 \\ 1 \\ 1 \end{bmatrix} - \left(\frac{50}{20}\right) \begin{bmatrix} -3 \\ -1 \\ 1 \\ 3 \end{bmatrix} = \begin{bmatrix} 1 \\ -1 \\ -1 \\ 1 \end{bmatrix} = \mathbf{v}_2,$$

$$\begin{bmatrix} 1 \\ 8 \\ 27 \\ 64 \end{bmatrix} - \left(\frac{100}{4}\right) \begin{bmatrix} 1 \\ 1 \\ 1 \\ 1 \end{bmatrix} - \left(\frac{208}{20}\right) \begin{bmatrix} -3 \\ -1 \\ 1 \\ 3 \end{bmatrix} - \left(\frac{30}{4}\right) \begin{bmatrix} 1 \\ -1 \\ -1 \\ 1 \end{bmatrix}$$

$$= \begin{bmatrix} -\dfrac{6}{20} \\ \dfrac{18}{20} \\ -\dfrac{18}{20} \\ \dfrac{6}{20} \end{bmatrix} \text{ or } \begin{bmatrix} -1 \\ 3 \\ -3 \\ 1 \end{bmatrix} = \mathbf{v}_3,$$

and

$$\begin{bmatrix} 1 \\ 16 \\ 81 \\ 256 \end{bmatrix} - \left(\frac{354}{4}\right)\begin{bmatrix} 1 \\ 1 \\ 1 \\ 1 \end{bmatrix} - \left(\frac{830}{20}\right)\begin{bmatrix} -3 \\ -1 \\ 1 \\ 3 \end{bmatrix} - \left(\frac{160}{4}\right)\begin{bmatrix} 1 \\ -1 \\ -1 \\ 1 \end{bmatrix} - \left(\frac{60}{20}\right)\begin{bmatrix} -1 \\ 3 \\ -3 \\ 1 \end{bmatrix} = \begin{bmatrix} 0 \\ 0 \\ 0 \\ 0 \end{bmatrix}.$$

An orthogonal basis for **X** is thus provided by the four vectors

$$\mathbf{x}_0 = \mathbf{v}_0 = \begin{bmatrix} 1 \\ 1 \\ 1 \\ 1 \end{bmatrix}, \ \mathbf{v}_1 = \begin{bmatrix} -3 \\ -1 \\ 1 \\ 3 \end{bmatrix}, \ \mathbf{v}_2 = \begin{bmatrix} 1 \\ -1 \\ -1 \\ 1 \end{bmatrix}, \ \mathbf{v}_3 = \begin{bmatrix} -1 \\ 3 \\ -3 \\ 1 \end{bmatrix}.$$

Such vectors are the *orthogonal polynomials* employed in Section 16.13.

When the original X values representing quantitative treatments happen *not* to be equally spaced, then one simply substitutes these values for the original **x** trial vector used above, along with their squares, cubes, etc. for the remaining trial vectors. In this way orthogonal polynomials can be found fairly easily for nonequally spaced situations.

C.6 FINDING ORTHOGONAL COMPARISONS AMONG MEANS OF UNEQUAL GROUPS

The Gram-Schmidt method, as described in Section C.5 above can be used without modification to find orthogonal comparisons among means where the J groups being compared all have equal numbers of cases. With very slight modifications, these same methods can also be used to find orthogonal comparisons among groups with unequal numbers of cases.

Thus, suppose that we are dealing with J groups, where the number of cases in any group j is symbolized by n_j, and the total number of cases is N. Then, for any trial vector $\mathbf{x}_t$ and for any previously determined vector $\mathbf{v}_s$, we may define

$$(\mathbf{x}_t, \mathbf{v}_s) = \sum_j n_j x_{jt} v_{js} \qquad \text{[C.6.1]}$$

and

$$\|\mathbf{v}_s\|^2 = \sum_j n_j v_{js}^2. \qquad \text{[C.6.2]}$$

Then, proceeding exactly as in the previous section, we find **V**, the set of $J - 1$ vectors of values v_j. However, we also need to make one adjustment, so that the sum of any set of comparison weights will be 0, and to try to ensure that the weights obtained are of a convenient magnitude. Thus, we transform each of the vectors $\mathbf{v}_t$ in the set **V** into a new vector $\mathbf{c}_t$ by taking

$$c_{jt} = n_j v_{jt}/L_t \qquad \text{[C.6.3]}$$

where L_t is equal to $|$minimum $n_j v_{jt}|$, the absolute value of the minimum product of n_j times the value of obtained weight v_{jt}. Then, for these new vectors $\mathbf{c}_t$ consisting of comparison weights c_{jt}, the sum of the weights will be equal to zero, as should be true for any comparison among means. It will also be true that for any comparisons represented by the vectors $\mathbf{c}_{jt}$ and $\mathbf{c}_{js}$, the criterion of orthogonality given in Eq. 11.5.2 will be satisfied. That is, the sum of the weighted products of the c values corresponding to the same group j, with each product divided by the number of cases in the group, or n_j, will equal to zero.

For example, consider four groups of unequal size in which the numbers of cases were, respectively

$$n_1 = 5, n_2 = 2, n_3 = 3, n_4 = 4.$$

Now we will find a set of three orthogonal vectors representing comparison weights to be applied to the means of these four groups. We will do so by using the same set of four trial vectors $\mathbf{X}$ as used in the first example of the section above. That is, letting

$$\mathbf{x}_0 = \begin{bmatrix} 1 \\ 1 \\ 1 \\ 1 \end{bmatrix} = \mathbf{v}_0, \text{ and then taking } \mathbf{x}_1 = \begin{bmatrix} 1 \\ 0 \\ 0 \\ 0 \end{bmatrix}$$

by use of Eq. C.6.1 we have

$$(\mathbf{x}_1, \mathbf{v}_0) = (5 \times 1 \times 1) + (2 \times 0 \times 1) + (3 \times 0 \times 1) + (4 \times 0 \times 1) = 5$$

and from Eq. C.6.2

$$\|\mathbf{v}\|^2 = (5 \times 1 \times 1) + (2 \times 1 \times 1) + (3 \times 1 \times 1) + (4 \times 1 \times 1) = 14$$

so that

$$\begin{bmatrix} 1 \\ 0 \\ 0 \\ 0 \end{bmatrix} - (5/14) \begin{bmatrix} 1 \\ 1 \\ 1 \\ 1 \end{bmatrix} = \begin{bmatrix} 9/14 \\ -5/14 \\ -5/14 \\ -5/14 \end{bmatrix} \text{ or } \begin{bmatrix} 9 \\ -5 \\ -5 \\ -5 \end{bmatrix} = \mathbf{v}_1.$$

Then, taking $\mathbf{X}_2 = \begin{bmatrix} 0 \\ 1 \\ 0 \\ 0 \end{bmatrix}$ we have

$$(\mathbf{x}_2, \mathbf{v}_0) = (5 \times 0 \times 1) + (2 \times 1 \times 1) + (3 \times 0 \times 1) + (4 \times 0 \times 1) = 2$$

$$(\mathbf{x}_2, \mathbf{v}_1) = (5 \times 0 \times 9) + (2 \times 1 \times -5) + (3 \times 0 \times -5) + (4 \times 0 \times -5)$$

$$= -10$$

$$\|\mathbf{v}_1\|^2 = 5(9)^2 + 2(-5)^2 + 3(-5)^2 + 4(-5)^2 = 630,$$

so that

$$\begin{bmatrix} 0 \\ 1 \\ 0 \\ 0 \end{bmatrix} - (2/14) \begin{bmatrix} 1 \\ 1 \\ 1 \\ 1 \end{bmatrix} - (-10/630) \begin{bmatrix} 9 \\ -5 \\ -5 \\ -5 \end{bmatrix} = \begin{bmatrix} 0 \\ (490/630) \\ (-140/630) \\ (-140/630) \end{bmatrix} \text{ or } \begin{bmatrix} 0 \\ 7 \\ -2 \\ -2 \end{bmatrix} = \mathbf{v}_2.$$

Finally, for $\mathbf{X}_3 = \begin{bmatrix} 0 \\ 0 \\ 1 \\ 0 \end{bmatrix}$ we have

$(\mathbf{x}_3, \mathbf{v}_0) = (5 \times 0 \times 1) + (2 \times 0 \times 1) + (3 \times 1 \times 1) + (4 \times 0 \times 1) = 3$

$(\mathbf{x}_3, \mathbf{v}_1) = (5 \times 0 \times 9) + (2 \times 0 \times -5) + (3 \times 1 \times -5) + (4 \times 0 \times -5)$

$\qquad = -15$

$(\mathbf{x}_3, \mathbf{v}_2) = (5 \times 0 \times 0) + (2 \times 0 \times 7) + (3 \times 1 \times -2) + (4 \times 0 \times -2)$

$\qquad = -6$

$\|\mathbf{v}_2\|^2 = 126.$

Then

$$\begin{bmatrix} 0 \\ 0 \\ 1 \\ 0 \end{bmatrix} - (3/14) \begin{bmatrix} 1 \\ 1 \\ 1 \\ 1 \end{bmatrix} - (-15/630) \begin{bmatrix} 9 \\ -5 \\ -5 \\ -5 \end{bmatrix} - (-6/126) \begin{bmatrix} 0 \\ 7 \\ -2 \\ -2 \end{bmatrix}$$

$$= 1/630 \begin{bmatrix} 0 \\ 0 \\ 360 \\ -270 \end{bmatrix} \text{ or } \begin{bmatrix} 0 \\ 0 \\ 4 \\ -3 \end{bmatrix} = \mathbf{v}_3.$$

If we tried a fourth trial vector $\mathbf{x}_4$, we would obtain a vector of zeroes for $\mathbf{v}_4$, since there can be only three orthogonal comparisons among four means.

Now we transform these vectors $\mathbf{v}_t$ into vectors $\mathbf{c}_t$ as follows: For vector $\mathbf{v}_1$, $L_1 = |\text{minimum } n_j v_{j1}| = |4(-5)| = 20$. Then

$c_{11} = 5(9)/20 = 2.25$

$c_{21} = 2(-5)/20 = -0.5$

$c_{31} = 3(-5)/20 = -0.75$

$c_{41} = 4(-5)/20 = -1.00.$

These are the weights for the first comparison among the means.

In finding the weights for the second comparison, we use $L_2 = |\text{minimum } n_j v_{j2}| = 8$, so that

$c_{12} = 0/8 = 0$

$c_{22} = 2(7)/8 = 1.75$

$c_{32} = 3(-2)/8 = -0.75$

$c_{33} = 4(-2)/8 = -1.00$

The weights for the third comparison are then

$$c_{13} = 0/12 = 0$$

$$c_{23} = 0/12 = 0$$

$$c_{33} = 3(4)/12 = 1.00$$

$$c_{43} = 4(-3)/12 = -1.00.$$

Finally, the mutual orthogonality of these comparisons may be checked by application of Eq. 11.5.2 to each pair of weighted vectors **c**. Thus for the first and second comparison we have

$$(2.25)(0)/5 + (-0.50)(1.75)/2 + (-0.75)(-0.75)/3 + (-1.00)(-1.00)/4 = 0.$$

Similarly for the first and third:

$$(2.25)(0)/5 + (-0.50)(0)/2 + (-0.75)(1.00)/3 + (-1.00)(-1.00)/4 = 0,$$

and for the second and third,

$$(0)(0)/5 + (1.75)(0)/2 + (-0.75)(1.00)/3 + (-1.00)(-1.00)/4 = 0.$$

Thus, these three comparisons constitute an orthogonal set for the three group means.

C.7 CONVERTING SETS OF DUMMY VARIABLES INTO ORTHOGONAL SETS

In Sections 18.12 and 18.13 there is occasion to deal with a set of J or of K dummy variables, indicating category membership, and to convert these into a set of $J - 1$ or $K - 1$ orthogonal variables, each with mean 0 and variance 1.00. Actually, this can be done by the same procedure outlined in Section C.6 above to find orthogonal comparison weights for means of unequal groups, *except* that the step represented by Eq. C.6.3 is replaced by a different step when we are dealing with dummy variables. That is, in place of Eq. C.6.3 for each vector **v** we calculate the standard deviation

$$S_v = \sqrt{\sum_j v_j^2 n_j}$$

and then take the values

$$c_j = v_j/S_v$$

The purpose of this step is, of course, to standardize the **c** vectors so that the variance of the **c** values will be 1.00. We also note that since the **v** vectors will have means of 0, and since any pair of **v** vectors will have 0 correlation, the same things will be true of the new **c** vectors.

Take, for example, the contingency table shown as Table 18.3.1., which has row frequencies as follows:

Category	Frequency	Dummy 1	Dummy 2	Dummy 3	Dummy 4
A_1	200	1	0	0	0
A_2	195	0	1	0	0
A_3	203	0	0	1	0
A_4	199	0	0	0	1

Applying the procedures of Section C.6 to the rows of this table we obtain

$$
\mathbf{v}_1 =
\begin{matrix} 200 \\ 195 \\ 203 \\ 199 \end{matrix}
\begin{bmatrix} 1 \\ 0 \\ 0 \\ 0 \end{bmatrix}
- \frac{200}{797}
\begin{bmatrix} 1 \\ 1 \\ 1 \\ 1 \end{bmatrix}
=
\begin{bmatrix} 597/797 \\ -200/797 \\ -200/797 \\ -200/797 \end{bmatrix}
=
\begin{bmatrix} 0.7491 \\ -0.2509 \\ -0.2509 \\ -0.2509 \end{bmatrix}
$$

$$
\mathbf{v}_2 =
\begin{matrix} 200 \\ 195 \\ 203 \\ 199 \end{matrix}
\begin{bmatrix} 0 \\ 1 \\ 0 \\ 0 \end{bmatrix}
- \frac{195}{797}
\begin{bmatrix} 1 \\ 1 \\ 1 \\ 1 \end{bmatrix}
- \frac{(-48.93)}{149.81}
\begin{bmatrix} 0.7491 \\ -0.2509 \\ -0.2509 \\ -0.2509 \end{bmatrix}
=
\begin{bmatrix} 0 \\ 0.6734 \\ -0.3266 \\ -0.3266 \end{bmatrix}
$$

and

$$
\mathbf{v}_3 =
\begin{matrix} 200 \\ 195 \\ 203 \\ 199 \end{matrix}
\begin{bmatrix} 0 \\ 0 \\ 1 \\ 0 \end{bmatrix}
- \frac{203}{797}
\begin{bmatrix} 1 \\ 1 \\ 1 \\ 1 \end{bmatrix}
- \frac{(-50.93)}{149.81}
\begin{bmatrix} 0.7491 \\ -0.2509 \\ -0.2509 \\ -0.2509 \end{bmatrix}
- \frac{(-66.30)}{131.31}
\begin{bmatrix} 0 \\ 0.6734 \\ -0.3266 \\ -0.3266 \end{bmatrix}
$$

$$
=
\begin{bmatrix} 0 \\ 0 \\ 0.4951 \\ -0.5049 \end{bmatrix}.
$$

The standard deviation of $\mathbf{v}_1 = 0.4336$, so that we take

$$
\mathbf{c}_1 = \frac{1}{0.4336}\mathbf{v}_1 =
\begin{bmatrix} 1.7276 \\ -0.5786 \\ -0.5786 \\ -0.5786 \end{bmatrix}.
$$

Similarly, the standard deviation of $\mathbf{v}_2 = 0.4059$, and that for $\mathbf{v}_3$ is 0.3551. Thus

$$
\mathbf{c}_2 = \frac{1}{0.4059}\mathbf{v}_2 =
\begin{bmatrix} 0 \\ 1.6590 \\ -0.8046 \\ -0.8046 \end{bmatrix}
$$

and

$$\mathbf{c}_3 = \frac{1}{0.3551} \mathbf{v}_3 = \begin{bmatrix} 0 \\ 0 \\ 1.3943 \\ -1.4219 \end{bmatrix}$$

Each variable represented by these three $\mathbf{c}$ vectors has (within rounding error) a mean of zero and a variance of 1.00, and they are pairwise uncorrelated.

Similarly, there are three categories shown as columns in Table 18.3.1 (but more conveniently shown as rows here). These categories have the following frequencies:

Category	Frequency	Dummy 1	Dummy 2	Dummy 3
B_1	461	1	0	0
B_2	292	0	1	0
B_3	44	0	0	1

Once again following our procedure of Section C.5, as modified in Section C.6, and letting each dummy variable in turn stand as the "trial vector," we have

$$\mathbf{v}_1 = \frac{461}{292} \begin{bmatrix} 1 \\ 0 \\ 0 \end{bmatrix} - \frac{461}{797} \begin{bmatrix} 1 \\ 1 \\ 1 \end{bmatrix} = \begin{bmatrix} 336/797 \\ -461/797 \\ -461/797 \end{bmatrix} = \begin{bmatrix} 0.4216 \\ -0.5784 \\ -0.5784 \end{bmatrix}$$

and

$$\mathbf{v}_2 = \frac{461}{292} \begin{bmatrix} 0 \\ 1 \\ 0 \end{bmatrix} - \frac{292}{797} \begin{bmatrix} 1 \\ 1 \\ 1 \end{bmatrix} - \frac{(-168.89)}{194.35} \begin{bmatrix} 0.4216 \\ -0.5784 \\ -0.5784 \end{bmatrix} = \begin{bmatrix} 0 \\ 0.1310 \\ -0.8690 \end{bmatrix}.$$

After division by the standard deviations of 0.4938 and 0.2190 respectively, then new vectors are

$$\mathbf{d}_1 = \frac{1}{0.4938} \mathbf{v}_1 = \begin{bmatrix} 0.8538 \\ -1.1713 \\ -1.1713 \end{bmatrix}$$

$$\mathbf{d}_2 = \frac{1}{0.2190} \mathbf{v}_2 = \begin{bmatrix} 0 \\ 0.5982 \\ -3.9680 \end{bmatrix}.$$

Each new variable for columns, with values shown by the $\mathbf{d}$ vectors, has a mean of 0, a variance of 1.00, and is uncorrelated with the other variable in its set. These are the vectors of values actually applied in Section 18.12.

Still other orthogonal sets for rows or columns of a contingency table may be found by different choices for the trial vectors, of course. In particular, placing 1 in two or more categories, -1 in other categories, and 0 in the remainder is equivalent to combining categories with the same signs, and omitting categories with 0 entries. Thus, for the first example of row categories above, suppose we were interested in Rows 1 and 2 combined, as contrasted with the combination of Rows 3 and 4. This would give the trial vector and consequent vector of c as follows:

$$\text{trial } \mathbf{x}_1 = \begin{bmatrix} 1 \\ 1 \\ -1 \\ -1 \end{bmatrix} \quad \text{final } \mathbf{c}_1 = \begin{bmatrix} 1.0088 \\ 1.0088 \\ -0.9912 \\ -0.9912 \end{bmatrix}.$$

Then, we might follow this by ignoring Rows 3 and 4 and contrasting only Rows 1 and 2

$$\text{trial } \mathbf{x}_2 = \begin{bmatrix} 1 \\ -1 \\ 0 \\ 0 \end{bmatrix} \quad \text{final } \mathbf{c}_2 = \begin{bmatrix} 1.4026 \\ -1.4387 \\ 0 \\ 0 \end{bmatrix}.$$

Finally, we can ignore Rows 1 and 2 and compare Rows 3 and 4:

$$\text{trial } \mathbf{x}_3 = \begin{bmatrix} 0 \\ 0 \\ 1 \\ -1 \end{bmatrix} \quad \text{final } \mathbf{c}_3 = \begin{bmatrix} 0 \\ 0 \\ 1.3945 \\ -1.4218 \end{bmatrix}.$$

This also produces a set of three uncorrelated variables, each with mean 0 and variance 1.00 (within rounding).

If we decided to do the same sort of thing to the three columns of this same table in the example, we might start by combining Columns 1 and 2, and comparing them to Column 3:

$$\text{trial } \mathbf{y}_1 = \begin{bmatrix} 1 \\ 1 \\ -1 \end{bmatrix} \quad \text{final } \mathbf{d}_1 = \begin{bmatrix} 0.2417 \\ 0.2417 \\ -4.1369 \end{bmatrix}.$$

Finally, to complete the set of column variables we can take

$$\text{trial } \mathbf{y}_2 = \begin{bmatrix} 1 \\ -1 \\ 0 \end{bmatrix} \quad \text{final } \mathbf{d}_2 = \begin{bmatrix} 0.8188 \\ -1.2927 \\ 0 \end{bmatrix}.$$

These new row and column variables can be used in place of the ones found above to find a new set of shadow tables, as illustrated in Section 18.12.

Appendix **D**

Some Principles and Applications of Matrix Algebra

This appendix contains some of the more elementary aspects of matrix algebra, and a few applications of matrix principles to some of the concerns of this text. In order to understand the very brief introduction to matrix algebra given in the first section below, it may be helpful first to read Section C.4 (Appendix C) in which a few basic ideas of vector theory are presented. It will then be apparent how matrix theory builds on some of these principles that apply to vectors.

For a more extended treatment of vectors and matrices, a text on linear algebra such as Noble and Daniel (1977) should be consulted.

D.1 SOME BASIC MATRIX ALGEBRA

If, as suggested above, you have looked over the discussion of vectors in Section C.4, you know that a row vector of dimension C is simply a horizontal array or ordered display of C elements. Similarly, a column vector of dimension R is just a vertical array of R elements.

Now picture a set of $R \times C$ elements in a rectangular arrangement, where there are R rows and C columns. Such a rectangular array of elements is called a *matrix*. Each element in a matrix is either a value or a symbol, as in these examples of different matrices:

$$\mathbf{X} = \begin{bmatrix} 2 & -1 \\ 0 & 4 \end{bmatrix}, \qquad \mathbf{Y} = \begin{bmatrix} 5 & 3 & 6 \\ 4 & 1 & 12 \\ 5 & 0 & 9 \\ 7 & -6 & 2 \end{bmatrix}, \qquad \mathbf{W} = \begin{bmatrix} -4 & -5 & 3 & 7 \\ 1 & 1 & 2 & -4 \\ 2 & 7 & 10 & 11 \\ 5 & 9 & 3 & -2 \end{bmatrix}$$

or

$$\mathbf{A} = \begin{bmatrix} a_{11} & a_{12} & a_{13} \\ a_{21} & a_{22} & a_{23} \\ a_{31} & a_{32} & a_{33} \end{bmatrix}, \qquad \mathbf{B} = \begin{bmatrix} b_{11} & b_{12} & b_{13} & b_{14} \\ b_{21} & b_{22} & b_{23} & b_{24} \end{bmatrix}$$

and so forth. When a matrix has R rows and C columns then one says it is a *matrix of order* $R \times C$. Thus, the first example is of order 2×2, and the second of order 3×4, whereas the third example shows a matrix of order 4×4, and so on. A matrix is said to be *square* when $R = C$; otherwise, the matrix is rectangular.

(In Section C.4 we followed the practice of symbolizing a vector by a lowercase letter printed in boldface, to distinguish these sets of values or variables from single values or variables, which are generally represented by letters printed in italics. Now, we will extend this principle by representing any $R \times C$ matrix by a capital letter printed in boldface. Thus, a symbol such as X will ordinarily represent a variable, whereas a symbol such as $\mathbf{X}$ will stand for a matrix of dimension $R \times C$ and consisting of elements symbolized by x.)

Notice that *each row* of any $R \times C$ matrix can be thought of as a *row vector* with C elements, and the matrix then shows a set of R such vectors. On the other hand, *any column* of an $R \times C$ matrix can be viewed as a *column vector* with R elements, with the whole matrix considered a set of such vectors. Indeed, a $1 \times C$ matrix is just a single row vector, and any $R \times 1$ matrix is just a single column vector.

Just as *scalar multiplication* of a vector by a constant means that each element is multiplied by that constant (Eq. C.4.10), exactly the same thing is true for a matrix. That is, if d is some constant number or *scalar,* and $\mathbf{A}$ is a matrix of order $R \times C$ consisting of elements such as a_{jk}, then scalar multiplication of $\mathbf{A}$ by d is symbolized by $d\mathbf{A}$, and results in a new $R \times C$ matrix in which every element a_{jk} of $\mathbf{A}$ has been replaced by the element da_{jk}.

Now consider two matrices $\mathbf{A}$ and $\mathbf{B}$, which are each of the same order, $R \times C$. An element a_{jk} of $\mathbf{A}$ is said to correspond to an element b_{jk} of $\mathbf{B}$ if each occupies the same row-column location, jk.

Two matrices, $\mathbf{A}$ and $\mathbf{B}$, are equal, $\mathbf{A} = \mathbf{B}$, if and only if they are of the same order and their corresponding elements are equal. That is, for every row $j = 1, \ldots, R$ and every column $k = 1, \ldots, C$ it must be true that $a_{jk} = b_{jk}$.

The *transpose* of an $R \times C$ matrix $\mathbf{A}$, symbolized by $\mathbf{A}'$, is a $C \times R$ matrix whose rows are the columns of $\mathbf{A}$, and vice-versa. Thus, given the 3×2 matrix

$$\mathbf{A} = \begin{bmatrix} 1 & -10 \\ -5 & 6 \\ 3 & 7 \end{bmatrix}$$

the transpose $\mathbf{A}'$ is

$$\mathbf{A}' = \begin{bmatrix} 1 & -5 & 3 \\ -10 & 6 & 7 \end{bmatrix}. \tag{D.1.1}$$

One can also take the transpose of a transposed matrix, or $(\mathbf{A}')'$. Notice that the transpose of a transpose is simply the original matrix:

$$(\mathbf{A}')' = \mathbf{A}. \tag{D.1.2}$$

A special matrix of great importance is the *identity matrix,* which is symbolized by **I**. The identity matrix is always square, or of order $R \times R$, and has 1 in each diagonal cell, but 0 in each nondiagonal cell. Thus:

$$\mathbf{I} = \begin{bmatrix} 1 & 0 & 0 \\ 0 & 1 & 0 \\ 0 & 0 & 1 \end{bmatrix}$$

[D.1.3]

is the 3×3 identity matrix.

When two matrices **A** and **B** are of the same order, they can be added or subtracted in a cell-by-cell manner. Thus, if

$$\mathbf{A} = \begin{bmatrix} 2 & 3 & 5 \\ -6 & 0 & 1 \\ 4 & 3 & -2 \\ 1 & 1 & 5 \end{bmatrix} \text{ and } \mathbf{B} = \begin{bmatrix} 4 & 1 & 3 \\ 1 & 2 & 3 \\ -1 & 1 & 0 \\ -2 & 3 & -5 \end{bmatrix}$$

then

$$\mathbf{A} + \mathbf{B} = \begin{bmatrix} 2+4 & 3+1 & 5+3 \\ -6+1 & 0+2 & 1+3 \\ 4-1 & 3+1 & -2+0 \\ 1-2 & 1+3 & 5-5 \end{bmatrix} = \begin{bmatrix} 6 & 4 & 8 \\ -5 & 2 & 4 \\ 3 & 4 & -2 \\ -1 & 4 & 0 \end{bmatrix}$$

and

$$\mathbf{A} - \mathbf{B} = \begin{bmatrix} 2-4 & 3-1 & 5-3 \\ -6-1 & 0-2 & 1-3 \\ 4+1 & 3-1 & -2-0 \\ 1+2 & 1-3 & 5+5 \end{bmatrix} = \begin{bmatrix} -2 & 2 & 2 \\ -7 & -2 & -2 \\ 5 & 2 & -2 \\ 3 & -2 & 10 \end{bmatrix}.$$

[D.1.4]

Two matrices **A** and **B** can also be multiplied, although matrix multiplication differs from ordinary arithmetic multiplication in that it is not necessarily symmetric. That is, the matrix product **A** times **B** is not necessarily the same as matrix product **B** times **A**. One can find the matrix product **A** times **B**, or **AB**, only if the *number of columns of **A** is equal to the number of rows of **B**.* This means that the new product matrix **AB** will always have as many rows as **A** and as many columns as **B**.

Matrix multiplication is carried out through a process similar to finding the inner product of a pair of vectors, shown by Eq. C.4.7. However, in finding the product of two matrices, **A** and **B**, one must find the *inner product of each row vector in **A** with each column vector in **B**.* That is, given that the number of columns of matrix **A** is equal to the number of rows of matrix **B**, as in this example,

$$\mathbf{A} = \begin{bmatrix} a_{11} & a_{12} & a_{13} \\ a_{21} & a_{22} & a_{23} \end{bmatrix} \qquad \mathbf{B} = \begin{bmatrix} b_{11} & b_{12} \\ b_{21} & b_{22} \\ b_{31} & b_{32} \end{bmatrix},$$

then the matrix product **AB** is found as follows:

$$\mathbf{AB} = \begin{bmatrix} a_{11}b_{11} + a_{12}b_{21} + a_{13}b_{31} & a_{11}b_{12} + a_{12}b_{22} + a_{13}b_{32} \\ a_{21}b_{11} + a_{22}b_{21} + a_{23}b_{31} & a_{21}b_{12} + a_{22}b_{22} + a_{23}b_{32} \end{bmatrix}.$$

[D.1.5]

Thus, for example, given the 4×3 matrix **A** and the 3×2 matrix **B** as follows,

$$\mathbf{A} = \begin{bmatrix} 1 & 2 & 4 \\ 3 & 0 & 1 \\ 2 & 2 & 3 \\ 0 & 1 & 5 \end{bmatrix} \qquad \mathbf{B} = \begin{bmatrix} 2 & 3 \\ 1 & 2 \\ 3 & 1 \end{bmatrix}$$

the product **AB** turns out to be

$$\mathbf{AB} = \begin{bmatrix} 1(2)+2(1)+4(3) & 1(3)+2(2)+4(1) \\ 3(2)+0(1)+1(3) & 3(3)+0(2)+1(1) \\ 2(2)+2(1)+3(3) & 2(3)+2(2)+3(1) \\ 0(2)+1(1)+5(3) & 0(3)+1(2)+5(1) \end{bmatrix} = \begin{bmatrix} 16 & 11 \\ 9 & 10 \\ 15 & 13 \\ 16 & 7 \end{bmatrix}. \qquad [\text{D.1.6}]$$

Notice how multiplying the 4×3 and the 3×2 matrix produces a 4×2 product matrix. The order of multiplication of two matrices is *very* important, and ordianary matrix multiplication of **A** times **B** is not possible unless the number of columns of **A** matches the number of rows of **B**. When the order from left to right is **AB** one says that **B** is *premultiplied* by **A**. However, we can also say that **A** is *postmultiplied* by **B**. The product of any $R \times R$ matrix **A** and the corresponding identity matrix **I** is simply **A**. That is,

$$\mathbf{AI} = \mathbf{A} \text{ and } \mathbf{IA} = \mathbf{A}. \qquad [\text{D.1.7}]$$

Furthermore, if the $R \times C$ matrix **A** is premultiplied by an $R \times R$ matrix **I**,

$$\mathbf{IA} = \mathbf{A},$$

then once again the product is simply **A**. The same thing is true if the $R \times C$ matrix is postmultiplied by the identity matrix of dimension $C \times C$.

Given a square, $R \times R$ matrix **A**, if another matrix $\mathbf{A}^{-1}$ exists such that

$$\mathbf{A}^{-1}\mathbf{A} = \mathbf{I}, \text{ and } \mathbf{AA}^{-1} = \mathbf{I} \qquad [\text{D.1.8}]$$

then $\mathbf{A}^{-1}$ is called the *inverse* of matrix **A**. In matrix algebra, the inverse of a matrix **A** is analogous to finding the reciprocal of a number or of an expression in ordinary algebra, and thus inverses permit one to carry out operations similar to division in matrix terms. Exactly as some numbers such as zero do not have reciprocals, the inverse of a square matrix does not always exist; such a matrix which has no inverse is called *singular*.

There are a number of ways that may be employed to find the inverse of a square matrix (if such exists), and we will describe one of these ways of obtaining the inverse in Section D.5.

D.2 DETERMINANTS OF SQUARE MATRICES

Still another important operation that may be carried out on a square matrix is called finding the *determinant* of the matrix.

A determinant is a single value that can be found for any square matrix. A determinant "determines" whether or not a square matrix will have an inverse. If the value of the determinant is zero, then no inverse exists.

Among other uses, determinants are important in the solution of simultaneous equations through matrix methods, as well as in certain multivariate methods. This use of determinants is illustrated later in this section, and in Chapter 15 of this text. Determinants are extremely easy to find for 2×2 or 3×3 matrices, although they can be much more complicated to work out for larger matrices. For any 2×2 matrix the determinant can be found as follows:

1. Multiply the two main, or "left-right," diagonal elements together, and write down the result.
2. Multiply the two off-diagonal elements together, and subtract the result from that found in Step 1. The result is the determinant of the matrix.

An easy way to visualize the determinant for a 2×2 matrix is as follows:

$$\text{deter.}\mathbf{A} = \begin{bmatrix} a_{11} & a_{12} \\ a_{21} & a_{22} \end{bmatrix}.$$

Give a positive sign to the results of multiplying if the arrow points downward, and a negative sign when the arrow points upward. Then sum the two products to find the determinant:

$$\text{deter.}\mathbf{A} = a_{11}a_{22} - a_{21}a_{12}. \qquad\qquad \text{[D.2.1]}$$

Thus, for example, if

$$\mathbf{A} = \begin{bmatrix} 5 & -2 \\ 4 & 1 \end{bmatrix}$$

then

$$\text{deter. } \mathbf{A} = (5 \times 1) - (4 \times -2) = 13.$$

A similar scheme helps one remember how to find the determinant of a 3×3 matrix. This time, the steps are these:

1. Augment the square matrix $\mathbf{A}$ by writing the entire matrix once again in an adjacent way, like this

$$\begin{bmatrix} a_{11} & a_{12} & a_{13} & a_{11} & a_{12} & a_{13} \\ a_{21} & a_{22} & a_{23} & a_{21} & a_{22} & a_{23} \\ a_{31} & a_{32} & a_{33} & a_{31} & a_{32} & a_{33} \end{bmatrix}$$

2. Now multiply along each of three diagonal lines extending downward like this, and sum the results:

$$\begin{bmatrix} a_{11} & a_{12} & a_{13} & a_{11} & a_{12} & a_{13} \\ a_{21} & a_{22} & a_{23} & a_{21} & a_{22} & a_{23} \\ a_{31} & a_{32} & a_{33} & a_{31} & a_{32} & a_{33} \end{bmatrix}$$

That is, take

$$a_{11}a_{22}a_{33} + a_{12}a_{23}a_{31} + a_{13}a_{21}a_{32} \qquad [\text{D.2.2}]$$

3. Now multiply along each of the digaonal lines extending upward like this, and sum the results:

$$\begin{bmatrix} a_{11} & a_{12} & a_{13} & a_{11} & a_{12} & a_{13} \\ a_{21} & a_{22} & a_{23} & a_{21} & a_{22} & a_{23} \\ a_{31} & a_{32} & a_{33} & a_{31} & a_{32} & a_{33} \end{bmatrix}$$

That is, take

$$a_{31}a_{22}a_{13} + a_{32}a_{23}a_{11} + a_{33}a_{21}a_{12} . \qquad [\text{D.2.3}]$$

4. Subtract the results of Step 3 from those of Step 2. The difference is the determinant of matrix **A.**

For example, consider the matrix **A** as follows:

$$\mathbf{A} = \begin{bmatrix} 8 & 6 & -2 \\ -3 & 1 & 3 \\ 2 & 1 & 4 \end{bmatrix}.$$

We then find the determinant of **A** as follows:

$$\begin{bmatrix} 8 & 6 & -2 & 8 & 6 & -2 \\ -3 & 1 & 3 & 3 & 1 & 3 \\ 2 & 1 & 4 & 2 & 1 & 4 \end{bmatrix}$$

$$(8 \times 1 \times 4) + (6 \times 3 \times 2) + (-2 \times -3 \times 1) = 74$$

$$\begin{bmatrix} 8 & 6 & -2 & 8 & 6 & -2 \\ -3 & 1 & 3 & 3 & 1 & 3 \\ 2 & 1 & 4 & 2 & 1 & 4 \end{bmatrix}$$

$$(2 \times 1 \times -2) + (1 \times 3 \times 8) + (4 \times -3 \times 6) = -52.$$

Then the determinant of matrix **A** $= 74 - (-52) = 126.$

Similar though somewhat more complicated methods exist for finding the determinant of a 4 × 4 or larger matrix. One such method will be outlined in Section D.5.

As noted above, one of the common mathematical uses of determinants is in the solution of a set of K linear equations in K unknowns. That is, consider a set of equations like those shown below:

$$a_{11}x_1 + a_{12}x_2 + \cdots + a_{1K}x_K = b_1$$
$$a_{21}x_1 + a_{22}x_2 + \cdots + a_{2K}x_K = b_2$$
$$\cdots\cdots\cdots\cdots\cdots\cdots\cdots\cdots\cdots\cdots\cdots\cdots\cdots\cdots$$
$$a_{K1}x_1 + a_{K2}x_2 + \cdots + a_{KK}x_K = b_K$$

Here there are K different variable values $x_1, x_2, \ldots, x_K$. These variable values are weighted and summed, with symbols such as a_{jk} standing for the coefficient or weight given to variable x_k in equation number j. Furthermore, each weighted sum of the x_k variables is equal to some value represented by b_j. We will assume that the number of equations is equal to the number of x variables, or K. The weights a_{jk} are assumed to be known, as are the various values of b. The problem is then to find the K values of x that will make these K equations simultaneously hold true.

We can also write the set of linear equations in matrix terms if we wish. Thus if **A** is the $K \times K$ matrix of coefficients or weights a_{jk}, **x** is the column vector of the different values of x, and **b** is the column vector of values of b, then the set of equations can be represented by

$$\mathbf{Ax} = \mathbf{b}. \qquad [\text{D.2.4}]$$

The theory of determinants can be used to solve such a set of K linear equations in K unknowns as follows. Let $D(\mathbf{A})$ stand for the determinant of the matrix of coefficients **A**:

$$D(\mathbf{A}) = \text{deter. } \mathbf{A}.$$

Then, let $D(1)$ stand for the determinant of the matrix **A**, but where the first column has been replaced by the vector **b**. Similarly, let $D(2)$ stand for the determinant of the matrix **A** but with the second column replaced by the vector **b**, and so on. In general, $D(k)$ will stand for the determinant of **A** after column k has been replaced by vector **b**. Then it will be true that any desired solution value x can be found by taking

$$x_k = D(k)/D(\mathbf{A}). \qquad [\text{D.2.5}]$$

All K of the values of x can be found in this way, of course.

For example, let us solve the following set of three linear equations in three unknowns:

$$2x_1 + 3x_2 - x_3 = 5$$

$$4x_1 - 2x_2 + 2x_3 = 6$$

$$x_1 + x_2 - 2x_3 = -3$$

To solve for the three x values, we first take the determinant of the matrix of coefficients, **A**:

$$D(\mathbf{A}) = \text{deter. } \begin{bmatrix} 2 & 3 & -1 \\ 4 & -2 & 2 \\ 1 & 1 & -2 \end{bmatrix} = 28.$$

Then we substitute the vector of b values for the first column in $\mathbf{A}$ and take the determinant:

$$D(1) = \text{deter.} \begin{bmatrix} 5 & 3 & -1 \\ 6 & -2 & 2 \\ -3 & 1 & -2 \end{bmatrix} = 28.$$

Hence, $x_1 = D(1)/D(\mathbf{A}) = 28/28 = 1.00.$

Next, to find the value of x_2 we take

$$D(2) = \text{deter.} \begin{bmatrix} 2 & 5 & -1 \\ 4 & 6 & 2 \\ 1 & -3 & -2 \end{bmatrix} = 56$$

so that $x_2 = 56/28 = 2.00.$

Finally, we have

$$D(3) = \text{deter.} \begin{bmatrix} 2 & 3 & 5 \\ 4 & -2 & 6 \\ 1 & 1 & -3 \end{bmatrix} = 84$$

so that $x_3 = 84/28 = 3.00.$ The vector $\mathbf{x}$ of solutions is thus $\mathbf{x} = \begin{bmatrix} 1 \\ 2 \\ 3 \end{bmatrix}$ for this example.

If it should turn out that $D(\mathbf{A})$, the determinant of the matrix of coefficients $\mathbf{A}$, has a value of 0, then the values making up the vector $\mathbf{x}$ of solutions cannot be determined. That is, the determinant of 0 implies that the solution does not exist, as well as the related fact that the matrix $\mathbf{A}$ has no inverse $\mathbf{A}^{-1}$.

While we are on the subject of simultaneous linear equations and their solution, it must also be pointed out that other ways exist as well for solving such equations. Thus, since we know from Eq. D.2.4 that

$$\mathbf{Ax} = \mathbf{b}$$

then, provided that matrix $\mathbf{A}$ actually has an inverse, $\mathbf{A}^{-1}$, the required vector $\mathbf{x}$ also may be found by taking

$$\mathbf{A}^{-1}\mathbf{Ax} = \mathbf{A}^{-1}\mathbf{b}$$

so that

$$\mathbf{Ix} = \mathbf{A}^{-1}\mathbf{b}$$

or, more simply,

$$\mathbf{x} = \mathbf{A}^{-1}\mathbf{b}. \tag{D.2.6}$$

Thus, for the example above, the inverse of the matrix $\mathbf{A}$ can be found by standard methods to be

$$\mathbf{A}^{-1} = \begin{bmatrix} .0714 & .1786 & .1429 \\ .3571 & -.1071 & -.2857 \\ .2143 & .0357 & -.5714 \end{bmatrix}$$

so that $\mathbf{A}^{-1}\mathbf{A} = \mathbf{I}$. Then we find that

$$\mathbf{A}^{-1}\mathbf{b} = \begin{bmatrix} .0714 & .1786 & .1429 \\ .3571 & -.1071 & -.2857 \\ .2143 & .0357 & -.5714 \end{bmatrix} \begin{bmatrix} 5 \\ 6 \\ -3 \end{bmatrix} = \begin{bmatrix} 1 \\ 2 \\ 3 \end{bmatrix} = \mathbf{x}$$

which is the same set of solution values we found before.

D.3 MORE ON THE SWEEP-OUT OPERATOR

In Section 15.27 the so called sweep-out operator was introduced as a device for solving multiple regression problems in a hierarchical or stepwise fashion. The examples in the text are all shown in terms of matrices of correlations, yielding the standardized regression weights. However, this same method may also be used with matrices of variances and covariances, to obtain *unstandardized* regression weights, or b values. If variances and covariances are used, the numbers involved can sometimes be very awkward in hand computations, becoming very large or very small. Therefore, it is often wise to apply the method to correlations when hand calculations are attempted. These calculations, however, will yield standardized regression weights, which are easily converted to unstandardized weights given the means and standard deviations of all the variables.

Nevertheless, it may be useful to see the method illustrated on a variance–covariance matrix, and for a few more predictors than we employed in Chapter 15. In this section, therefore, an example will be provided of hierarchical regression using the sweep-out method, where there are five predictor or X variables, and where the method is applied directly to a matrix of variances and covariances. The raw data for our example are given in Table D.3.1, where scores are shown for 30 subjects on each of five personality

Table D.3.1

Scores for 30 subjects on five personality subtests (X_1 to X_5) and a happiness measure (Y)

Sub.	X_1	X_2	X_3	X_4	X_5	Y	Sub.	X_1	X_2	X_3	X_4	X_5	Y
1	16	16	18	11	11	13	16	18	16	13	9	15	18
2	14	14	18	10	16	19	17	14	15	15	7	16	8
3	15	16	5	8	15	16	18	12	17	6	9	16	13
4	12	14	14	9	17	13	19	16	14	9	8	18	11
5	13	15	7	12	18	16	20	14	14	13	9	16	15
6	16	17	18	8	16	12	21	11	18	7	11	15	12
7	11	13	9	10	16	17	22	14	15	17	8	18	14
8	12	15	13	9	10	13	23	12	14	11	12	17	9
9	14	18	5	8	17	11	24	16	16	13	9	16	10
10	8	15	8	9	16	14	25	15	17	5	9	9	15
11	15	13	17	7	12	10	26	12	13	10	10	16	13
12	14	15	4	8	16	13	27	14	15	12	9	11	13
13	8	12	13	12	17	14	28	13	14	13	10	17	15
14	14	17	18	8	16	9	29	15	17	18	9	10	12
15	7	13	11	11	18	10	30	13	12	12	10	16	14

measures, as well as on a "happiness" measure, Y. Then Table D.3.2 exhibits the variances and covariances for these data. Our purpose is to find the unstandardized weight b_j applying to each variable, X_j in order to predict the raw value of Y.

In the general situation, let the $J \times J$ matrix to be swept contain entries w_{jk} in row j and column k. (These entries may or may not be correlations, so that here we will use the neutral symbol w in place of the symbol r used in Section 15.27.) Furthermore, in this regression situation, let the last row and column in the original matrix stand for the dependent variable Y.

One starts by choosing a column variable to be swept out; we will denote this variable by the subscript v. Next, we will form a new matrix **SWP**(v) by applying the sweep operator as follows:

1. For any entry w_{vk} in row v and column k, then the new value w_{vk}^* is found from

$$w_{vk}^* = w_{vk}/w_{vv} \qquad\qquad [D.3.1]$$

2. For any other cell w_{jk}, where neither j nor k is equal to v, then the new entry for **SWP**(v) is found by taking

$$w_{jk}^* = w_{jk} - (w_{jv})(w_{vk})/w_{vv} . \qquad\qquad [D.3.2]$$

The entire column v is deleted in the new matrix **SWP**(v), although the row v is maintained.

Suppose now for this example we first sweep out variable X_1, so that $v = 1$ here. The matrix **SWP**(1) shown below is the result of sweeping the variable X_1 from the variance–covariance matrix of Table D.3.2. (Notice that only the entries above the dotted line actually have to be calculated directly, as $w_{jk}^* = w_{kj}^*$ when neither j nor k equals v.)

Table D.3.2

Variance–covariance matrix for the data in Table D.3.1

	X_1	X_2	X_3	X_4	X_5	Y	*mean*	*SD*
X_1	6.0622	1.6667	2.6711	-1.6800	-1.8956	0.4156	13.2667	2.4622
X_2	1.6667	2.7333	-1.0667	-0.6667	-1.2333	-0.6000	15.0000	1.6533
X_3	2.6711	-1.0667	19.4622	-0.6200	-1.6711	-1.7156	11.7333	4.4116
X_4	-1.6800	-0.6667	-0.6200	1.8767	0.5300	0.8467	9.3000	1.3699
X_5	-1.8956	-1.2333	-1.6711	0.5300	6.4456	-0.2156	15.2333	2.5388
Y	0.4156	-0.6000	-1.7156	0.8467	-0.2156	6.8622	13.0667	2.6196

SWP (1):

	X_2	X_3	X_4	X_5	Y	b_j	$S_y^2 r_{y(j\cdot 1)}^2$
X_1	0.2749	0.4406	−0.2771	−0.3127	0.0686	.0686	
X_2	2.2751	−1.8011	−0.2048	−0.7121	−0.7142		$(0.7142)^2/2.2751$
	....						$= .2242$
X_3	−1.8011	18.2853	0.1202	−0.8359	−1.8987		.1972
		....					
X_4	−0.2048	0.1202	1.4111	0.0047	0.9619		.6557
			....				
X_5	− .7121	−0.8359	0.0047	5.8529	−0.0856		.0013
				....			
Y	−0.7142	−1.8987	0.9619	−0.0856	6.8337		$= S_y^2(1 - R_{y\cdot 1}^2)$
					....		

$$R_{y\cdot 1}^2 = 1 - (6.8337/6.8622) = .0042 .$$

Note that in **SWP**(1) the entry for Row 1 and Column 2 is found by taking

$$w_{12}^* = w_{12}/w_{11} = 1.6667/6.0622 = 0.2749,$$

and that for Row 1 and Column 3 from

$$w_{13}^* = w_{13}/w_{11} = 2.6711/6.0622 = 0.4406.$$

On the other hand, the new entry w^* in Row 2 Column 2 is found from

$$w_{22}^* = w_{22} - (w_{21}w_{12}/w_{11})$$

$$= 2.7333 - ((1.6667 \times 1.6667)/6.0622) = 2.2751 ;$$

the entry in Row 2 column 3 is found by taking

$$w_{23}^* = w_{23} - w_{21}w_{13}/w_{11}$$

$$= -1.0667 - ((1.6667 \times 2.6711)/6.0622) = -1.8011,$$

and so on for the other entries in **SWP**(1).

Since the the original w values here are variances and covariances, the entry $w_{vy}^* = w_{1y}^* = 0.0686$ in the last column of the new matrix **SWP**(1) is the unstandardized regression weight for the prediction of Y using *only* variable X_v, or X_1. On the other hand, if the variable symbolized by any row j has already been swept-out in a previous step, then the entry w_{jy}^* is the unstandardized regression weight for variable X_j in an equation for predicting raw Y from all X variables up to and including variable X_v. (If the original w values are correlations, these same results represent *standardized* regression weights, B_j.)

In addition, as the original w values are variances and covariances in this example, then in any row j not equal to v the ratio

$$(w_{jy}^{*2})/w_{jj}^* = S_j^2 r_{y(j\cdot v)}^2 \qquad \text{[D.3.3]}$$

is the Y variance accounted for by variable X_j, after X_j has been adjusted for variable X_v, as well as for any other previously swept variables. Thus, the variance accounted for by X_2 after adjustment for X_1 is $(.7142)^2/(2.2751)$ or .2242. Similarly, the variance accounted for by X_3 after adjustment for X_1 is .1972, and so on. (If the original w values are correlations, then the Y variance S_y^2 can here be considered to equal 1.00, and the ratios shown above are simply the squared part correlations.)

Finally, since our original matrix consisits of variances and covariances, the entry w_{yy}^* in the last row and column of matrix **SWP**(1) is equal to the Y variance *unaccounted for* or

$$w_{yy}^* = S_y^2(1 - R_{y\cdot 1}^2) = 6.8337$$

or, more generally,

$$w_{yy}^* = S_y^2(1 - R_{y\cdot G}^2) , \qquad\qquad [D.3.4]$$

where $R_{y\cdot G}^2$ symbolizes the squared multiple correlation between y and the set G consisting of all variables swept from the matrix, up to and including the most recent variable, X_v. Hence, the proportion of Y variance accounted for by the swept variable(s) is equal to

$$R_{y\cdot G}^2 = 1 - (w_{yy}^*)/S_y^2 . \qquad\qquad [D.3.5]$$

(If, however, the original matrix consists of correlations, then

$$R_{y\cdot G}^2 = 1 - w_{yy}^* .) \qquad\qquad [D.3.6]$$

This same method is next applied to each cell of the matrix **SWP**(1) to produce the new matrix **SWP**(12). That is, treating **SWP**(1) as though it were the original matrix, we next find **SWP**(12) by sweeping out variable X_2, as follows:

SWP(12):

	X_3	X_4	X_5	Y	b	$S_y^2 r_{y(j\cdot 12)}^2$
X_1	0.6582	−0.2524	−0.2267	0.1549	.1549	
X_2	−0.7917	−0.0900	−0.3130	−0.3139	−.3139	
X_3	16.8594	−0.0419	−1.3996	−2.4641		.3601
X_4	−0.0419	1.3927	−0.0594	0.8976		.5785
X_5	−1.3996	−0.0594	5.6300	−0.3091		.0170
Y	−2.4641	0.8976	−0.3091	$6.6095 = S_y^2(1 - R_{y\cdot 12}^2)$		

$$R_{y\cdot 12}^2 = 1 - (6.6095/6.8622) = .0368 .$$

Proceeding in exactly this same way to sweep variable X_3, we have
SWP(123):

	X_4	X_5	Y	b_j	$S_y^2 r_{y(j\cdot 123)}^2$
X_1	−0.2508	−0.1721	0.2511	.2511	
X_2	−0.0920	−0.3787	−0.4296	−.4296	
X_3	−0.0025	−0.0830	−0.1462	−.1462	
X_4	1.3926	−0.0629	0.8915		.5707
X_5	−0.0629	5.5138	−0.5137		.0479
Y	0.8915	−0.5137	$6.2494 = S_y^2 (1 - R_{y\cdot 123}^2)$		

$$R_{y\cdot 123}^2 = 1 - (6.2494/6.8622) = .0893 .$$

On sweeping out variable X_4 we have
SWP(1234):

	X_5	Y	b_j	$S_y^2 r_{y(j \cdot 1234)}^2$
X_1	-0.1834	0.4117	$.4117$	
X_2	-0.3829	-0.3707	$-.3707$	
X_3	-0.0831	-0.1446	$-.1446$	
X_4	-0.0452	0.6402	$.6402$	
X_5	5.5110	-0.4734		$.0407$
Y	-0.4734	$5.6787 = S_y^2 (1 - R_{y \cdot 1234}^2)$		

$$R_{y \cdot 1234}^2 = 1 - (5.6787/6.8622) = .1725 .$$

Finally, on sweeping out variable X_5, we have
SWP(12345):

	Y	b_j
X_1	0.3959	$.3959$
X_2	-0.4036	$-.4036$
X_3	-0.1517	$-.1517$
X_4	0.6363	$.6363$
X_5	-0.0859	$-.0859$
Y	$5.6380 = S_y^2(1 - R_{y \cdot 12345}^2)$	

$$R_{y \cdot 12345}^2 = 1 - (5.6380/6.8622) = .1784 .$$

The final, unstandardized regression equation can then be written as

$$y_i' = \bar{y} + .3959 (x_{i1} - \bar{x}_1) - .4036 (x_{i2} - \bar{x}_2) - .1517 (x_{i3} - \bar{x}_3)$$

$$+ .6363 (x_{i4} - \bar{x}_4) - .0859 (x_{i5} - \bar{x}_5) .$$

D.4 FINDING THE DETERMINANT OF A SQUARE MATRIX WITH THE SWEEP OPERATOR

Another handy use of the sweep operator is to find the determinant of a square matrix. This is especially useful for a matrix of the type (**AA′** or **A′A** positive definite) usually found in statistics. This method works as follows:

As we saw in Section 15.27 and in Section D.3 just above, when a variable X_v, represented by column v, is being swept out of a matrix, one identifies and uses the diagonal element w_{vv} as the basis for the sweep operation. Suppose that we are sweeping each column in turn from a $J \times J$ matrix. Let us denote the value of the element w_{vv} in the sweep operation for column j by

$$x^{(j)} = \text{value of } w_{vv} \text{ in the sweep operation for column } j . \qquad \text{[D.4.1]}$$

The determinant may be found by carrying out the sweep operation for $J - 1$ of the columns. If at any stage a column v shows a diagonal cell $w_{vv} = 0$, then you go on to another column, and don't sweep the original until w_{vv} is no longer 0. The determinant value is then

$$\text{deter. } \mathbf{A} = x^{(1)} x^{(2)} x^{(3)} \cdots x^{(J-1)} x^{(J)} , \qquad \text{[D.4.2]}$$

where $x^{(J)}$ is the value left in the single cell in column J after the first $J - 1$ columns of matrix **A** have been swept out.

For example, consider the following square matrix:

$$\mathbf{A} = \begin{bmatrix} 8 & 9 & 4 & 3 \\ 5 & 7 & 6 & 2 \\ 3 & 1 & 4 & 1 \\ 2 & 9 & 5 & 6 \end{bmatrix}.$$

Now, to find the determinant of matrix **A**, we first sweep out Column 1, where w_{11} has the value 8. That is

$$x^{(1)} = w_{vv} \text{ for first sweep} = 8.$$

Then, leaving out Row 1 as well, we have the new swept table

$$\mathbf{SWP}(1) = \begin{bmatrix} 1.375 & 3.5 & 0.125 \\ -2.375 & 2.5 & -0.125 \\ 6.750 & 4.0 & 5.250 \end{bmatrix}.$$

Now, when we next sweep for Column 2, the value of w_{22} is 1.375, or

$$x^{(2)} = 1.375.$$

On omitting both Column 2 and Row 2 our swept matrix becomes

$$\mathbf{SWP}(12) = \begin{bmatrix} 8.545455 & 0.090909 \\ -13.181818 & 4.636364 \end{bmatrix}.$$

For sweeping Column 3, the value for w_{33} is used, so that

$$x^{(3)} = 8.545455 ,$$

and the results of sweeping that column (also eliminating the row) is

$$\mathbf{SWP}(123) = 4.776596.$$

Therefore,

$$x^{(4)} = 4.776596.$$

Then, by Eq. D.4.2 above, the value of the determinant of this matrix is given by

$$\text{deter. } \mathbf{A} = 8 \times 1.375 \times 8.545455 \times 4.776596 = 449.$$

In the interest of getting an accurate final result with this method it is very important to *carry a good many decimal places* (say, six or more) for all intermediate steps, and *then to round the final result* as necessary. With a good hand calculator this should prove easy enough to do for most small matrices. For big problems you will, of course, wish to use the computer.

Notice that if any of the values represented here by $x^{(j)}$ can only be zero, then the determinant is also zero. As noted above, this implies that no inverse of the matrix exists.

D.5 FINDING THE INVERSE OF A SQUARE MATRIX BY USE OF THE SWEEP OPERATOR

Although there are several standard mathematical procedures for doing so (see Winer, 1962, pp 120–126), the sweep-out operator discussed above gives a relatively simple and routine way of finding the *inverse* of a square matrix, especially of the sort (**A′A** or **AA′** positive definite) often found in statistics. The inverse of a matrix **A** (if it exists), is another $J \times J$ matrix, symbolized by $\mathbf{A}^{-1}$, such that

$$\mathbf{A}^{-1}\mathbf{A} = \mathbf{A}\,\mathbf{A}^{-1} = \mathbf{I}. \qquad\qquad [\text{D.5.1}]$$

Here, as before, **I** symbolizes the identity matrix, having 1 in each main diagonal cell, and 0 elsewhere.

To find the inverse of a square matrix using the sweep-out operator, we can do the following:

1. Write down the $J \times J$ square matrix **A**, but augment it by writing down J additional columns corresponding to the matrix **I**. This creates the new $J \times (2J)$ matrix,

   $$\text{augmented matrix} = [\mathbf{A} \mid \mathbf{I}]. \qquad\qquad [\text{D.5.2}]$$

2. Now start with the first column, and carry out the sweep operation on each entry in the augmented matrix.
3. Eliminate a column of **A** after each sweep, but *retain all J rows throughout*. (To ensure accuracy in the final result, carry six or more decimal places in using this method.)
4. When the first J columns have been eliminated, and only the last J columns remain, then the resulting $J \times J$ matrix is the inverse $\mathbf{A}^{-1}$ of the matrix **A**. (To check your result, simply take the product $\mathbf{A}^{-1}$ times **A**, or **A** times $\mathbf{A}^{-1}$. Either of these products should equal **I**, the identity matrix.)

For example, consider the 4×4 square matrix we used to demonstrate the determinant in Section D.4. We will now use the sweep-out operator to find the inverse. First, augmenting the matrix we have

$$[\mathbf{A} \mid \mathbf{I}] = \begin{bmatrix} 8 & 9 & 4 & 3 & \vdots & 1 & 0 & 0 & 0 \\ 5 & 7 & 6 & 2 & \vdots & 0 & 1 & 0 & 0 \\ 3 & 1 & 4 & 1 & \vdots & 0 & 0 & 1 & 0 \\ 2 & 9 & 5 & 6 & \vdots & 0 & 0 & 0 & 1 \end{bmatrix}.$$

Now we sweep out the Column 1, obtaining

$$\mathbf{SWP}(1) = \begin{bmatrix} 1.125 & 0.5 & 0.375 & \vdots & 0.125 & 0 & 0 & 0 \\ 1.375 & 3.5 & 0.125 & \vdots & -0.625 & 1 & 0 & 0 \\ -2.375 & 2.5 & -0.125 & \vdots & -0.375 & 0 & 1 & 0 \\ 6.750 & 4.0 & 5.250 & \vdots & -0.250 & 0 & 0 & 1 \end{bmatrix}.$$

Then, eliminating Column 2, we have

$$
\textbf{SWP}(1,2) = \begin{bmatrix}
-2.363636 & 0.272727 & 0.636364 & -0.818182 & 0 & 0 \\
2.545454 & 0.090909 & -0.454545 & 0.727273 & 0 & 0 \\
8.545454 & 0.090909 & -1.454545 & 1.727273 & 1 & 0 \\
-13.181818 & 4.636363 & 2.818182 & -4.909091 & 0 & 1
\end{bmatrix},
$$

and after elimination of Column 3, the remaining matrix is

$$
\textbf{SWP}(1,2,3) = \begin{bmatrix}
0.297872 & 0.234043 & -0.340425 & 0.276596 & 0 \\
0.063830 & -0.021276 & 0.212766 & -0.297872 & 0 \\
0.010638 & -0.170213 & 0.202130 & 0.117021 & 0 \\
4.776596 & 0.574469 & -2.244681 & 1.542552 & 1
\end{bmatrix}.
$$

Finally, after elimination of Column 4, the remaining matrix of values is the inverse (shown rounded to four decimals):

$$
\textbf{A}^{-1} = \textbf{SWP}(1,2,3,4) = \begin{bmatrix}
0.1982 & -0.2004 & 0.1804 & -0.0624 \\
-0.0290 & 0.2428 & -0.3185 & -0.0134 \\
-0.1715 & 0.2071 & 0.1136 & -0.0022 \\
0.1203 & -0.4699 & 0.3229 & 0.2094
\end{bmatrix}.
$$

This is easily checked by multiplying the original matrix by the inverse (or inverse by the original matrix) to see if the result is the identity matrix $\textbf{I}$. For this example, the results do indeed check out within a very small rounding error:

$$
\begin{bmatrix}
8 & 9 & 4 & 3 \\
5 & 7 & 6 & 2 \\
3 & 1 & 4 & 1 \\
2 & 9 & 5 & 6
\end{bmatrix}
\begin{bmatrix}
0.1982 & -0.2004 & 0.1804 & -0.0624 \\
-0.0290 & 0.2428 & -0.3185 & -0.0134 \\
-0.1715 & 0.2071 & 0.1136 & -0.0022 \\
0.1203 & -0.4699 & 0.3229 & 0.2094
\end{bmatrix}
$$

$$
= \begin{bmatrix}
1 & 0 & 0 & 0 \\
0 & 1 & 0 & 0 \\
0 & 0 & 1 & 0 \\
0 & 0 & 0 & 1
\end{bmatrix}.
$$

If you reach a stage where some column is being eliminated and the diagonal entry w_{vv} is zero, then that column is skipped until it no longer has such a zero value. However, if all remaining non-swept columns have zero diagonals, then this is evidence that no inverse exists to be found.

D.6 MATRIX TERMINOLOGY FOR SOME OF THE KEY STATISTICAL TERMS USED IN THIS TEXT

In the following, consider N cases, each with a score value y_i on a variable symbolized by Y. Then the entire set of values can be shown as a matrix **Y**, of order $1 \times N$:

$$\mathbf{Y} = [y_1, y_2, \cdots, y_N] . \qquad [\text{D.6.1}]$$

Furthermore, consider a matrix of order $1 \times N$ (a row vector in other words), composed exclusively of 1's, as represented by the symbol **1**:

$$\mathbf{1} = [1, 1, \cdots, 1].$$

Also, let **0** stand for a row vector of zeroes, such as

$$\mathbf{0} = [0, 0, \cdots, 0].$$

Postmultiplying the matrix **Y** by the transpose of matrix **1**, or **1'**, yields the sum of the all of the individual y values. The mean of variable Y, symbolized by $\bar{y}$, is then the matrix product

$$\bar{y} = \mathbf{Y}\,\mathbf{1'}N^{-1}, \text{ where } N^{-1} \text{ is here just the scalar } 1/N.$$

The variance of variable Y can therefore be expressed by

$$\text{var}(Y) = \mathbf{YY'}N^{-1} - \bar{y}^2 . \qquad [\text{D.6.2}]$$

The sum of squares total for variable Y is then

$$\text{SS}_y \text{ total} = N\,\text{var}(Y) = \mathbf{Y}\,\mathbf{Y'} - \bar{y}^2 N . \qquad [\text{D.6.3}]$$

In other words, the mean and variance of a variable Y can be expressed quite simply in matrix terms. However, the great utility of matrix terminology becomes especially apparent when we wish to deal with several variables simultaneously. To show how this is done in matrix language, suppose that there are $K \geq 2$ different variables Y. Each of N cases has a value y_{ik} on each variable Y_k. Then the $K \times N$ data matrix can be represented by

$$\mathbf{Y} = \begin{bmatrix} y_{11} & y_{12} & \cdots & y_{1N} \\ \cdots\cdots\cdots\cdots\cdots\cdots\cdots\cdots \\ y_{K1} & y_{K2} & \cdots & y_{KN} \end{bmatrix} .$$

If we again let **1** symbolize a $1 \times N$ matrix of 1's, we can find a $K \times 1$ matrix of means of all of the variables by taking

$$\overline{\mathbf{Y}} = \mathbf{Y}\mathbf{1'}N^{-1} = \begin{bmatrix} \bar{y}_1 \\ \bar{y}_2 \\ \cdot \\ \cdot \\ \cdot \\ \bar{y}_K \end{bmatrix} . \qquad [\text{D.6.4}]$$

Notice that applying the principles of matrix multiplication to the product of the $K \times 1$ matrix $\overline{\mathbf{Y}}$ and its $1 \times K$ transpose, or $\overline{\mathbf{Y}}'$, gives the following:

$$\overline{\mathbf{Y}\mathbf{Y}}' = \begin{bmatrix} \bar{y}_1\bar{y}_1 & \bar{y}_1\bar{y}_2 & \cdots & \bar{y}_1\bar{y}_K \\ \cdots\cdots\cdots\cdots\cdots\cdots\cdots\cdots \\ \bar{y}_k\bar{y}_1 & \bar{y}_k\bar{y}_2 & \cdots & \bar{y}_k\bar{y}_K \\ \cdots\cdots\cdots\cdots\cdots\cdots\cdots\cdots \\ \bar{y}_K\bar{y}_1 & \bar{y}_K\bar{y}_2 & \cdots & \bar{y}_K\bar{y}_K \end{bmatrix}$$

which is the $K \times K$ matrix of products of all pairs of the mean values. The values in the diagonal cells of this matrix are, of course, the squares of the mean values for all of the different variables. The way of multiplying a $J \times 1$ vector by a $1 \times K$ vector to form a new $J \times K$ matrix is sometimes called the *outer product,* or the *major product,* of two vectors.

Remember the sum of squares for any variable Y may be found by taking the sum of the squared values minus the mean squared and multiplied by N:

$$SS_y \text{ total } = \sum_i y_i^2 - \bar{y}^2 N = NS_y^2 . \qquad [D.6.5]$$

Furthermore, for two different variables Y_j and Y_k, the sum of products or SP_{jk} value is found by taking

$$SP_{jk} = \sum_i y_{ij}y_{ik} - \bar{y}_j\bar{y}_k N = N \text{ cov}(y_j, y_k). \qquad [D.6.6]$$

All of the sums of squares for the K variables, and the sums of products for the $K(K - 1)/2$ pairs of variables are then contained in the matrix $\mathbf{SSCP}_{yy}$, found as follows:

$$\mathbf{SSCP}_{yy} = \mathbf{Y}\,\mathbf{Y}' - \overline{\mathbf{Y}}\,\overline{\mathbf{Y}}'N = \begin{bmatrix} SS_1 & SP_{12} & \cdots & SP_{1K} \\ \cdots\cdots\cdots\cdots\cdots\cdots\cdots\cdots \\ SP_{K1} & SP_{K2} & & SS_{KK} \end{bmatrix} \qquad [D.6.7]$$

Note that this matrix is symmetric, with $SP_{gh} = SP_{hg}$ for any two variables Y_g and Y_h.

Now if we wish to transform the $\mathbf{SSCP}_{yy}$ matrix into a matrix of *sample* variances and covariances, symbolized here by $\mathbf{VCV}_{yy}$ we may do so by taking

$$\mathbf{VCV}_{yy} = \mathbf{SSCP}_{yy}\, N^{-1} . \qquad [D.6.8]$$

On the other hand, we may estimate the matrix of population variance and covariance values by taking

$$\text{est. of population } \mathbf{VCV}_{yy} = \text{est. } \mathbf{\Sigma}_{yy} = \mathbf{SSCP}_{yy}\,(N - 1)^{-1}. \qquad [D.6.9]$$

(Once again, as in Section C.3, notice that this use of a capital Greek sigma written in boldface denotes the matrix of true, or population, variances and covariances among a set of variables.)

In Section C.3 we were able to define the multivariate normal distribution of J variables in terms of the normal distribution of each possible linear combination of those variables (where the weights applied are not all zero, of course). Now that we have some matrix terminology available, we can give an explicit formula for the *multivariate normal density function*. Let $\mathbf{x}$ stand for a J-dimensional vector of values on the different X variables. In addition, let $\boldsymbol{\mu}$ stand for the vector of expected values over the different variables, and let $\boldsymbol{\Sigma}$ be the $J \times J$ variance–covariance matrix we have just been discussing. Then if the joint distribution of the variables is multivariate normal, the probability density associated with the vector $\mathbf{x}$ is given by

$$f(\mathbf{x}) = \exp[-(1/2)(\mathbf{x} - \boldsymbol{\mu})' \, \boldsymbol{\Sigma}^{-1} \, (\mathbf{x} - \boldsymbol{\mu})]/\sqrt{(2\pi)^J |\boldsymbol{\Sigma}|} \qquad [\text{D.6.10}]$$

where the notation exp followed by an expression is equivalent to the constant e raised to the power indicated and $|\boldsymbol{\Sigma}|$ symbolizes the determinant of the matrix $\boldsymbol{\Sigma}$.

When there are only two variables, Eq. D.6.10 converts to the expression for the *bivariate normal* distribution:

$$f(\mathbf{x}) = (e^{-G})/K, \text{ where } K = 2\pi\sigma_1\sigma_2\sqrt{(1 - \rho_{12}^2)}$$

and

$$G = \frac{\left[\dfrac{(x_1 - \mu_1)^2}{\sigma_1^2} - 2\,\dfrac{\rho_{12}(x_1 - \mu_1)(x_2 - \mu_2)}{\sigma_1\sigma_2} + \dfrac{(x_2 - \mu_2)^2}{\sigma_2^2} \right]}{2(1 - \rho_{12}^2)}.$$

Returning to our Y variables, we now define a matrix $\mathbf{S}_Y$ composed of the standard deviations of the K different Y variables in the diagonals, and zeroes elsewhere:

$$\mathbf{S}_y = \begin{bmatrix} S_1 & 0 & 0 & \cdots & 0 \\ 0 & S_2 & 0 & \cdots & 0 \\ \cdots & \cdots & \cdots & \cdots & \cdots \\ 0 & 0 & 0 & \cdots & S_K \end{bmatrix}.$$

The original matrix $\mathbf{Y}$ can then be converted into a matrix of standardized values, $\mathbf{Z}_y$, by taking

$$\mathbf{Z}_y = (\mathbf{Y} - \overline{\mathbf{Y}}\mathbf{1})\mathbf{S}_y^{-1}. \qquad [\text{D.6.11}]$$

If the matrix $\mathbf{Y}$ is transformed into the corresponding matrix $\mathbf{Z}_Y$ in the way just shown, and if the variance–covariance matrix $\mathbf{VCV}$ found for this standardized matrix, the result is the correlation matrix $\mathbf{R}_{yy}$;

$$\mathbf{R}_y = \mathbf{Z}_y\mathbf{Z}'_y N^{-1} = \begin{bmatrix} 1 & r_{12} & r_{13} & \cdots & r_{1K} \\ r_{21} & 1 & r_{23} & \cdots & r_{2K} \\ \cdots & \cdots & \cdots & \cdots & \cdots \\ r_{K1} & r_{K2} & r_{K3} & \cdots & 1 \end{bmatrix}. \qquad [\text{D.6.12}]$$

Notice that the matrices $\mathbf{SSCP}_{yy}$ and $\mathbf{VCV}_{yy}$ can both be turned into the sample correlation matrix $\mathbf{R}_{yy}$ by taking

$$\mathbf{S}_y^{-1}\mathbf{SSCP}_{yy}\mathbf{S}_y^{-1}N^{-1} = \mathbf{S}_y^{-1}\mathbf{VCV}_{yy}\mathbf{S}_y^{-1} = \mathbf{R}_{yy}. \qquad [\text{D.6.13}]$$

Now consider a situation in which N cases are sampled, and each case produces a value of a single dependent variable Y, corresponding as before to the $1 \times N$ matrix or row vector

$$\mathbf{Y} = [y_1, y_2, \cdots, y_N]. \qquad [\text{D.6.14}]$$

In addition, each case exhibits values on some K different predictors or "independent variables" X, so that all of the values on the X variables can be exhibited in a matrix $\mathbf{X}$ of order $K \times N$ (actually, it is often more convenient to regard of the matrix $\mathbf{X}$ as being of order $(K + 1) \times N$, with a first row, standing for X_0, consiting of 1's.)

$$\mathbf{X} = \begin{bmatrix} 1 & 1 & 1 & \cdots & 1 \\ x_{11} & x_{12} & x_{13} & & x_{1N} \\ \cdots\cdots\cdots\cdots\cdots\cdots\cdots\cdots\cdots\cdots \\ x_{K1} & x_{K2} & x_{K3} & \cdots & x_{KN} \end{bmatrix}. \qquad \text{[D.6.15]}$$

Furthermore, let the $1 \times (K + 1)$ matrix $\mathbf{A}$ consist of weight values such as

$$\mathbf{A} = [a_0, a_1, a_2, \cdots, a_K]. \qquad \text{[D.6.16]}$$

Finally, let the $1 \times N$ matrix $\mathbf{E}$ consist of error terms e_i for each individual i among the N cases, or

$$\mathbf{E} = [e_1, e_2, \cdots, e_N]. \qquad \text{[D.6.17]}$$

As shown in Section 10.2, the *general linear model* is used to state the relationship between the Y variable and the $K + 1$ different variables X. In matrix terms, the general linear model may be written as

$$\mathbf{Y} = \mathbf{AX} + \mathbf{E}, \qquad \text{[D.6.18]}$$

with $\mathbf{Y}, \mathbf{A}, \mathbf{X}$, and $\mathbf{E}$ all defined as stated above. This simple matrix expression says that the values of variable Y are each composed of a weighted sum of the values of a set of $K + 1$ different X variables, plus values associated with error. This matrix equation is equivalent to Eq. 10.2.1 in the text.

In the experimental design or ANOVA model of Section 10.3, the N cases have all been allocated at random to J experimental groups or treatment levels. The X variables are then all indicators or dummy variables showing group membership: Thus, the $(J + 1) \times N$ "structural" matrix $\mathbf{X}$ becomes

$$\mathbf{X} = \begin{bmatrix} 1 & 1 & 1 & 1 & \cdots & 1 \\ 1 & 1 & 0 & 0 & & 0 \\ 0 & 0 & 1 & 1 & & 0 \\ 0 & 0 & 0 & 0 & & 0 \\ \cdots\cdots\cdots\cdots\cdots\cdots\cdots \\ 0 & 0 & 0 & 0 & \cdots & 1 \end{bmatrix} \qquad \text{[D.6.19]}$$

The model is then still

$$\mathbf{Y} = \mathbf{AX} + \mathbf{E} \qquad \text{[D.6.20]}$$

(or, as shown in Table 10.3.1 and the following,

$$\mathbf{Y}' = \mathbf{X}'\mathbf{A}' + \mathbf{E}'). \qquad \text{[D.6.21]}$$

However, as explained in Section 10.4, the criterion of least squares requires that the values in $\mathbf{A}$ be the *sample effects,* or

$$\mathbf{A} = [a_0, a_1, \cdots, a_j, \cdots, a_J]. \qquad \text{[D.6.22]}$$

where a_0 is now equal to the mean $\bar{y}$, and any effect a_j is equal to the difference between group mean $\bar{y}_j$ and $\bar{y}$. Moreover, in this situation, where there are n_j cases in any group j, it will be true that

$$
\mathbf{XX'} = \begin{bmatrix}
N & n_1 & n_2 & n_3 & \cdots & n_J \\
n_1 & n_1 & 0 & 0 & \cdots & 0 \\
n_2 & 0 & n_2 & 0 & \cdots & 0 \\
n_3 & 0 & 0 & n_3 & \cdots & 0 \\
\cdots\cdots\cdots\cdots\cdots\cdots\cdots\cdots\cdots \\
n_J & 0 & 0 & 0 & \cdots & n_J
\end{bmatrix}
\qquad \text{[D.6.23]}
$$

Furthermore, it will be true that

$$
\mathbf{XE'} = \mathbf{0} \quad \text{and} \quad \mathbf{EX'} = \mathbf{0}
\qquad \text{[D.6.24]}
$$

since the sum of deviations about the mean must be zero within any group. It will also be true that $\mathbf{AXX'1} = 2N\bar{y}$ where 1 is a $J \times 1$ matrix of 1's.

Therefore, it can be shown that

$$
\begin{aligned}
\text{SS}_y \text{ total} &= \mathbf{YY'} - N\bar{y}^2 \\
&= (\mathbf{AX} + \mathbf{E})(\mathbf{AX} + \mathbf{E})' - N\bar{y}^2 \\
&= (\mathbf{AX} + \mathbf{E})(\mathbf{E'} + \mathbf{X'A'}) - N\bar{y}^2 \\
&= \mathbf{AXE'} + \mathbf{EX'A'} + \mathbf{AXX'A'} + \mathbf{EE'} - N\bar{y}^2 \\
&= 0 + 0 + N\bar{y}^2 + \text{SS}_y \text{ between} + \text{SS}_y \text{ within} - N\bar{y}^2 \\
&= \text{SS}_y \text{ between} + \text{SS}_y \text{ within}.
\end{aligned}
\qquad \text{[D.6.25]}
$$

This is of course, the matrix version of the partition of the sum of squares in the analysis of variance, as shown in Section 10.8.

In problems of regression and correlation, we deal with one or more predictor variables X and one (or more) dependent variables Y. Here it is convenient to think of the data matrix as partitioned into two parts, consisting of values on the X variables, and those on the Y variable(s). When there are J different X variables, and K distinct Y variables, such a partitioned data matrix may be represented in a $(J + K) \times N$ matrix as follows:

$$
\begin{bmatrix} \mathbf{X} \\ \text{---} \\ \mathbf{Y} \end{bmatrix} = \begin{bmatrix}
x_{11} & x_{12} & x_{13} & \cdots & x_{1N} \\
x_{21} & x_{22} & x_{23} & \cdots & x_{2N} \\
\vdots & & & & \vdots \\
x_{J1} & x_{J2} & x_{J3} & \cdots & x_{JN} \\
\text{-----} & \text{-----} & \text{-----} & \text{-----} & \text{-----} \\
y_{11} & y_{12} & y_{13} & \cdots & y_{1N} \\
\cdot & \cdot & \cdot & \cdots & \cdot \\
y_{K1} & y_{K2} & y_{K3} & \cdots & y_{KN}
\end{bmatrix} .
\qquad \text{[D.6.26]}
$$

If this matrix is postmultiplied by a $N \times 1$ matrix **1,** consisting only of 1's, and the result multiplied by the scalar $(1/N)$, the result is a $(J + K) \times 1$ matrix of means:

$$\begin{bmatrix} \mathbf{X} \\ \hline \mathbf{Y} \end{bmatrix} \mathbf{1} N^{-1} = \begin{bmatrix} \overline{\mathbf{X}} \\ \hline \overline{\mathbf{Y}} \end{bmatrix}. \qquad [D.6.27]$$

It follows that the **SSCP** matrix, consisting of the sums of squares and the sums of products for all of the variables, both X and Y, can itself be partitioned into four parts, as follows

$$\mathbf{SSCP} = \begin{bmatrix} \mathbf{X} \\ \hline \mathbf{Y} \end{bmatrix} \begin{bmatrix} \mathbf{X} \\ \hline \mathbf{Y} \end{bmatrix}' - \begin{bmatrix} \overline{\mathbf{X}} \\ \hline \overline{\mathbf{Y}} \end{bmatrix} \begin{bmatrix} \overline{\mathbf{X}} \\ \hline \overline{\mathbf{Y}} \end{bmatrix}' N$$

$$= \begin{bmatrix} \mathbf{SSCP}_{xx} & \mathbf{SSCP}_{xy} \\ \mathbf{SSCP}_{yx} & \mathbf{SSCP}_{yy} \end{bmatrix}. \qquad [D.6.28]$$

Here, $\mathbf{SSCP}_{xx}$ represents a $J \times J$ matrix consisting of the SS and SP values for all of the X variables. Similarly, the $K \times K$ matrix $\mathbf{SSCP}_{yy}$ consists of the SS and SP values based on the Y variables. The $J \times K$ matrix $\mathbf{SSCP}_{xy}$ then includes the SP values for each X with each Y, and $\mathbf{SSCP}_{yx}$ is just the transpose of $\mathbf{SSCP}_{xy}$.

In the special case of bivariate regression, in which there is only one X and one Y variable, the **SSCP** matrix can be written as

$$\mathbf{SSCP} = \begin{bmatrix} SS_x & SP_{xy} \\ SP_{yx} & SS_y \end{bmatrix}. \qquad [D.6.29]$$

In this sort of problem the point is to find a single regression weight b and a constant a so that $\mathbf{Y^*}$, the $(1 \times N)$ matrix of predicted Y values, can be found from the $(1 \times N)$ matrix $\mathbf{X}$ by the linear rule

$$\mathbf{Y^*} = a + b\mathbf{X}. \qquad [D.6.30]$$

The least squares solution for a and b then dictates that

$$b = (SP_{xy})(SS_x)^{-1}, \qquad [D.6.31]$$

and

$$a = \overline{y} - b\overline{x}. \qquad [D.6.32]$$

The actual sum of squares due to regression of Y on X is then equal to

$$SS_y \text{ regression on } X = (SP_{yx})(SS_x)^{-1}(SP_{xy})$$

$$= (SP_{xy})^2/(SS_x). \qquad [D.6.33]$$

The sum of squares for deviations from regression (or error) is consequently

$$SS_y \text{ deviations} = SS_y - (SP_{yx})(SS_x)^{-1}(SP_{xy}). \qquad [D.6.34]$$

In addition, the proportion of Y variance accounted for by X is given by

$$r_{xy}^2 = (SS_y)^{-1}(SP_{yx})(SS_x)^{-1}(SP_{xy}) \qquad [D.6.35]$$

When the regression equation is stated in standardized form for both variables Y and X, or

$$Z_y = A + BZ_x \qquad\qquad [D.6.36]$$

the least squares solution is

$$B = R_{xy} = r_{xy} \qquad\qquad [D.6.37]$$

and $A = 0$.

In multiple regression problems, where there is a single Y variable, but two or more variables X, the solution follows exactly the same matrix form as for bivariate regression. That is, in the unstandardized case we now want a vector of constant values $\mathbf{a}$, and a $(1 \times K)$ matrix of weights $\mathbf{b}$ such that when the $(1 \times N)$ matrix $\mathbf{Y^*}$ is found from

$$\mathbf{Y^*} = \mathbf{a} + \mathbf{bX} \qquad\qquad [D.6.38]$$

average squared error will be minimized. This requires that the $(1 \times K)$ matrix $\mathbf{b}$ of weights be found from

$$\mathbf{b} = (\mathbf{SSCP}_{yx})(\mathbf{SSCP}_{xx})^{-1} \qquad\qquad [D.6.39]$$

(Notice the similarity of this expression to Eq. D.6.31 above for bivariate regression. However, the matrix $\mathbf{b}$ in Eq. D.6.39 gives the regression weights from predicting values of the Y variable from values on each of the X variables.)

Furthermore, the sum of squares for any Y variable due to regression on all of the X predictor variables is found from

$$SS_y \text{ regression on the } X \text{ variables} = (\mathbf{SSCP}_{yx})(\mathbf{SSCP}_{xx})^{-1}(\mathbf{SSCP}_{xy}) \qquad [D.6.40]$$

and

$$SS_y \text{ deviations from regression} = SS_y - (\mathbf{SSCP}_{yx})(\mathbf{SSCP}_{xx})^{-1}(\mathbf{SSCP}_{xy}). \qquad [D.6.41]$$

The proportion of Y variance due to regression on the set of X variables is then

$$R^2_{y \cdot X} = (SS_y)^{-1}(\mathbf{SSCP}_{yx})(\mathbf{SSCP}_{xx})^{-1}(\mathbf{SSCP}_{xy}). \qquad\qquad [D.6.42]$$

When one is dealing with standardized values for Y and X, $\mathbf{B}$, the $(1 \times K)$ vector of standardized B weights, is found from the correlation matrix $\mathbf{R}$ as follows:

The criterion of least squares dictates that to find $\mathbf{B}$ the following normal equations must be solved.

$$\mathbf{BR}_{xx} = \mathbf{R}_{yx}. \qquad\qquad [D.6.43]$$

This may be done by taking

$$\mathbf{BR}_{xx}\mathbf{R}_{xx}^{-1} = \mathbf{R}_{yx}\mathbf{R}_{xx}^{-1} \qquad\qquad [D.6.44]$$

so that

$$\mathbf{B} = \mathbf{R}_{yx}\mathbf{R}_{xx}^{-1}. \qquad\qquad [D.6.45]$$

Then

$$R^2_{y \cdot X} = \mathbf{BR}_{xy} = R_{yx}R_{xx}^{-1}R_{xy}. \qquad\qquad [D.6.46]$$

If, in addition to J variables X and K variables Y, some $L \geq 1$ concomitant variables U can be distinguished, the basic $(J + K + L) \times N$ data matrix, say $\mathbf{D}$, can be partitioned further into

$$\mathbf{D} = \begin{bmatrix} \mathbf{U} \\ \cdots \\ \mathbf{X} \\ \cdots \\ \mathbf{Y} \end{bmatrix}. \text{ Then } \overline{\mathbf{D}} = \mathbf{D1}'N^{-1} = \begin{bmatrix} \overline{\mathbf{U}} \\ \cdots \\ \overline{\mathbf{X}} \\ \cdots \\ \overline{\mathbf{Y}} \end{bmatrix}$$

so that

$$\mathbf{DD}' - \overline{\mathbf{DD}'}N = \begin{bmatrix} \mathbf{SSCP}_{uu} & \mathbf{SSCP}_{ux} & \mathbf{SSCP}_{uy} \\ \mathbf{SSCP}_{xu} & \mathbf{SSCP}_{xx} & \mathbf{SSCP}_{xy} \\ \mathbf{SSCP}_{yu} & \mathbf{SSCP}_{yx} & \mathbf{SSCP}_{yy} \end{bmatrix} \qquad \text{[D.6.47]}$$

Now if it is desired to remove all linear influence of the set of variables U from the remaining X and Y variables, this can be done by taking

$$\mathbf{SSCP}_{.U} = \begin{bmatrix} (\mathbf{SSCP}_{xx} - \mathbf{SSCP}_{xu}(\mathbf{SSCP}_{uu})^{-1}\mathbf{SSCP}_{ux}) & (\mathbf{SSCP}_{xy} - \mathbf{SSCP}_{xu}(\mathbf{SSCP}_{uu})^{-1}\mathbf{SSCP}_{uy}) \\ (\mathbf{SSCP}_{yx} - \mathbf{SSCP}_{yu}(\mathbf{SSCP}_{uu})^{-1}\mathbf{SSCP}_{ux}) & (\mathbf{SSCP}_{yy} - \mathbf{SSCP}_{yu}(\mathbf{SSCP}_{uu})^{-1}\mathbf{SSCP}_{uy}) \end{bmatrix}$$
$$\text{[D.6.48]}$$

The sweep-out operator we have been using is actually an application of the principle shown in Eq. D.6.48. That is, suppose that we are dealing with the matrix showing the correlations among some variable U, and X and Y variables. We wish to sweep the single variable U out of the matrix. Then the sweep-out operator, discussed in Section D.3, or

$$r_{jk}^* = r_{jk} - r_{jv}r_{vk}/r_{vv} \qquad \text{(for } v = u, j \neq v, \text{ and } k \neq v\text{)}$$

corresponds exactly to the matrix expression

$$\mathbf{R}_{.U} = \begin{bmatrix} (\mathbf{R}_{xx} - \mathbf{R}_{xu}\mathbf{R}_{uu}^{-1}\mathbf{R}_{ux}) & (\mathbf{R}_{xy} - \mathbf{R}_{xu}\mathbf{R}_{uu}^{-1}\mathbf{R}_{uy}) \\ (\mathbf{R}_{yx} - \mathbf{R}_{yu}\mathbf{R}_{uu}^{-1}\mathbf{R}_{ux}) & (\mathbf{R}_{yy} - \mathbf{R}_{yu}\mathbf{R}_{uu}^{-1}\mathbf{R}_{uy}) \end{bmatrix} \qquad \text{[D.6.49]}$$
$$= \mathbf{SWP}(U)$$

These same matrix operations form the basis for ANCOVA and for the multivariate analysis of covariance, or MANCOVA, as well.

D.7 THE CIRCULARITY ASSUMPTION

The *circularity assumption* discussed in Section 13.22 can be described in terms of the population matrix of variances and covariances, which we symbolized as $\boldsymbol{\Sigma}$ in Sections C.3 and D.6. That is, suppose there are J different treatments in an experiment, with any two of these treatments indexed by j and k. Furthermore, think of a population of subjects (or of blocks) each with values of Y for all J treatments. Let us symbolize the diagonal element in row j and column j of the matrix Σ by σ_{jj}. Furthermore, we will symbolize any off-diagonal element, such as that in row j and column k, by σ_{jk}. Then it will be true that σ_{jj} is the population variance σ_j^2 within treatment population A_j, and σ_{jk} will be the population covariance between treatments A_j and A_k. Thus, in this new notation we can write the population variance–covariance matrix Σ as

$$\Sigma = \begin{bmatrix} \sigma_{11} & \sigma_{12} & \cdots & \sigma_{1J} \\ \sigma_{21} & \sigma_{22} & \cdots & \sigma_{2J} \\ \cdots\cdots\cdots\cdots\sigma_{jk}\cdots\cdots \\ \sigma_{J1} & \sigma_{J2} & \cdots & \sigma_{JJ} \end{bmatrix}$$

Now think of the possibility of doing *orthogonal comparisons* among the σ_{jk} values in any column of Σ, just as we could do comparisons among the means of the J treatment populations. If they are to be orthogonal, then there will be $J - 1$ such comparisons possible, of course. Let us index these different possible comparisons by using $g = 1, 2, \ldots, J - 1$ and when we need to talk about *two* of these comparisons, let us call them g and h. The weights for any of the comparisons, say g, can be shown as a vector of values, such as

$$\begin{bmatrix} c_{1g} \\ c_{2g} \\ \cdot \\ \cdot \\ \cdot \\ c_{Jg} \end{bmatrix} = \mathbf{c}_g.$$

Just as for any comparison among means, we will require that the comparison values c_{jg} sum to zero over the treatments j.

In addition, we will impose the requirement that the sum of the squared weights for any comparison must be 1.00. That is, we require that

$$\mathbf{c}_g'\mathbf{c}_h = 1.00 \quad \text{if } g = h.$$

Since we already are dealing with a set of orthogonal comparisons, where

$$\mathbf{c}_g'\mathbf{c}_h = 0 \quad \text{if } g \neq h,$$

this new requirement leads to an *orthonormal* set (see Section C.4).

Now we can use the matrix $\mathbf{C}$ to represent the entire set of of weights applied to the J treatment groups (rows) in each of the $J - 1$ orthogonal comparisons (the columns). Since there are J treatments and $J - 1$ comparisons, this will be a $J \times (J - 1)$ matrix. Notice that our requirement that the set of comparisons be orthonormal makes it necessarily true that

$$\mathbf{C}'\mathbf{C} = \mathbf{I}$$

where $\mathbf{I}$ is the $(J - 1) \times (J - 1)$ identity matrix.

Then the *assumption of circularity* may be stated as

$$\mathbf{C}'\Sigma\mathbf{C} = \lambda\mathbf{I} \qquad\qquad\qquad \text{[D.7.1]}$$

where λ is simply a constant value.

Notice that the first part of Eq. D.7.1, or $\mathbf{C'\Sigma}$, represents a set of $J - 1$ comparisons made on the σ_{jk} values within each of the columns of $\mathbf{\Sigma}$. That is, a row g of matrix $\mathbf{C'}$ multiplied by a column k of matrix $\mathbf{\Sigma}$ corresponds to a comparison among the σ_{jk} values. We can represent this comparsion value by the symbol φ_{gk}

$$\varphi_{gk} = \sum_j c_{gj}\sigma_{jk}.$$

These comparison values will tend to be different from zero to the extent that the σ_{jk} values within that column are different from each other. Then the second part of Eq. D.7.1 amounts to a second set of comparisons among the columns of φ values based on the first sets of comparisons. This second type of comparison is symbolized by γ_{gh}:

$$\gamma_{gh} = \sum_k \varphi_{gk}c_{kh}.$$

Thus, Eq. D.7.1 represents a $(J - 1)(J - 1)$ matrix of comparison values, γ_{gh}, themselves found from comparisons φ_{gk} among variances and covariances in $\mathbf{\Sigma}$.

In essence, this assumption says that if one makes all $J - 1$ orthogonal comparisons among the σ_{jk} values within any column k of the matrix $\mathbf{\Sigma}$, and then compares these comparison values in the same ways across the columns, no systematic differences will be found: this implies that the *same basic structure* of variances and covariances obtains *within each of the treatment populations* relative to the remaining populations. Such differences as may exist among the variances and covariances making up matrix $\mathbf{\Sigma}$ have minimal influence on the outcome when this assumption of circularity is true.

D.8 A JUSTIFICATION FOR THE RELATIONSHIP SHOWN IN SECTION 18.11, LINKING CANONICAL CORRELATION AND MULTIPLE REGRESSION IN A CONTINGENCY TABLE

In Section 18.11 it is pointed out that when the overall chi-square value for a contingency table is divided by N, the result may be interpreted as the sum of $L - 1$ values of R-square, where L is the smaller of J (rows) or K (columns). This statement may be justified by the argument leading up to Eq. 18.11.7, which comes directly from the theory of canonical correlation.

Thus, let R_{aa} symbolize the $(J \times J)$ correlation matrix of rows (A) with rows. Similarly, let R_{bb} stand for the $(K \times K)$ correlation matrix of columns (B) with columns. In addition, let R_{ab} symbolize the $J \times K$ matrix correlations of rows with columns, and R_{ba} the $K \times J$ correlations of columns with rows. Obviously, $R'_{ab} = R_{ba}$.

In the theory of canonical correlation it is shown that

$$\sum_g^L \rho_g^2 = \text{trace } (R_{bb}^{-1}R_{ba}R_{aa}^{-1}R_{ab}),$$ [D.8.1]

where the *trace* of a square matrix is simply the sum of the elements in the main diagonal. In the theory of multiple regression it is also true that

$$\sum_h^{K-1} R_{d_h \cdot \text{rows}}^2 = \text{trace } (R_{ba}\, R_{aa}^{-1}\, R_{ab}).$$ [D.8.2]

See, for example, Eq. D.6.46.

On the other hand, if the column variables are orthonormal as in Section 18.11,

$$R_{bb}^{-1} = I \qquad\qquad\qquad\qquad\qquad [D.8.3]$$

and so the two sums Eqs. D.8.1 and D.8.2 are identical, and Eq. 18.11.7 is true.

Incidentally, the results outlined in Section 18.11 and the above are equivalent to the relationship between chi-square divided by N and the Pillai "trace criterion" as shown by Marascuilo and Levin (1983). However, the present rationale utilizing Fisher's work (1940) strikes me as a good deal more direct.

Tables

Table I

Cumulative normal probabilities

z	F(z)	z	F(z)	z	F(z)	z	F(z)
.00	.5000000	.21	.5831662	.42	.6627573	.63	.7356527
.01	.5039894	.22	.5870604	.43	.6664022	.64	.7389137
.02	.5079783	.23	.5909541	.44	.6700314	.65	.7421539
.03	.5119665	.24	.5948349	.45	.6736448	.66	.7453731
.04	.5159534	.25	.5987063	.46	.6772419	.67	.7485711
.05	.5199388	.26	.6025681	.47	.6808225	.68	.7517478
.06	.5239222	.27	.6064199	.48	.6843863	.69	.7549029
.07	.5279032	.28	.6102612	.49	.6879331	.70	.7580363
.08	.5318814	.29	.6140919	.50	.6914625	.71	.7611479
.09	.5358564	.30	.6179114	.51	.6949743	.72	.7642375
.10	.5398278	.31	.6217195	.52	.6984682	.73	.7673049
.11	.5437953	.32	.6255158	.53	.7019440	.74	.7703500
.12	.5477584	.33	.6293000	.54	.7054015	.75	.7733726
.13	.5517168	.34	.6330717	.55	.7088403	.76	.7763727
.14	.5556700	.35	.6368307	.56	.7122603	.77	.7793501
.15	.5596177	.36	.6405764	.57	.7156612	.78	.7823046
.16	.5635595	.37	.6443088	.58	.7190427	.79	.7852361
.17	.5674949	.38	.6480273	.59	.7224047	.80	.7881446
.18	.5714237	.39	.6517317	.60	.7257469	.81	.7910299
.19	.5753454	.40	.6554217	.61	.7290691	.82	.7938919
.20	.5792597	.41	.6590970	.62	.7323711	.83	.7967306

Note. From *Biometrika Tables for Statisticians*, Vol. 1, 3rd ed. (Table 1, pp. 104–110) by E. S. Pearson and H. O. Hartley (Eds.) 1966, Cambridge: Cambridge Univ. Press. Copyright 1954 by the Biometrika Trustees. Adapted by permission.

Table I   925

Table I (continued)

z	F(z)	z	F(z)	z	F(z)	z	F(z)
.84	.7995458	1.32	.9065825	1.79	.9632730	2.26	.9880894
.85	.8023375	1.33	.9082409	1.80	.9640697	2.27	.9883962
.86	.8051055	1.34	.9098773	1.81	.9648521	2.28	.9886962
.87	.8078498	1.35	.9114920	1.82	.9656205	2.29	.9889893
.88	.8105703	1.36	.9130850	1.83	.9663750	2.30	.9892759
.89	.8132671	1.37	.9146565	1.84	.9671159	2.31	.9895559
.90	.8159399	1.38	.9162067	1.85	.9678432	2.32	.9898296
.91	.8185887	1.39	.9177356	1.86	.9685572	2.33	.9900969
.92	.8212136	1.40	.9192433	1.87	.9692581	2.34	.9903581
.93	.8238145	1.41	.9207302	1.88	.9699460	2.35	.9906133
.94	.8263912	1.42	.9221962	1.89	.9706210	2.36	.9908625
.95	.8289439	1.43	.9236415	1.90	.9712834	2.37	.9911060
.96	.8314724	1.44	.9250663	1.91	.9719334	2.38	.9913437
.97	.8339768	1.45	.9264707	1.92	.9725711	2.39	.9915758
.98	.8364569	1.46	.9278550	1.93	.9731966	2.40	.9918025
.99	.8389129	1.47	.9292191	1.94	.9738102	2.41	.9920237
1.00	.8413447	1.48	.9305634	1.95	.9744119	2.42	.9922397
1.01	.8437524	1.49	.9318879	1.96	.9750021	2.43	.9924506
1.02	.8461358	1.50	.9331928	1.97	.9755808	2.44	.9926564
1.03	.8484950	1.51	.9344783	1.98	.9761482	2.45	.9928572
1.04	.8508300	1.52	.9357445	1.99	.9767045	2.46	.9930531
1.05	.8531409	1.53	.9369916	2.00	.9772499	2.47	.9932443
1.06	.8554277	1.54	.9382198	2.01	.9777844	2.48	.9934309
1.07	.8576903	1.55	.9394292	2.02	.9783083	2.49	.9936128
1.08	.8599289	1.56	.9406201	2.03	.9788217	2.50	.9937903
1.09	.8621434	1.57	.9417924	2.04	.9793248	2.51	.9939634
1.10	.8643339	1.58	.9429466	2.05	.9798178	2.52	.9941323
1.11	.8665005	1.59	.9440826	2.06	.9803007	2.53	.9942969
1.12	.8686431	1.60	.9452007	2.07	.9807738	2.54	.9944574
1.13	.8707619	1.61	.9463011	2.08	.9812372	2.55	.9946139
1.14	.8728568	1.62	.9473839	2.09	.9816911	2.56	.9947664
1.15	.8749281	1.63	.9484493	2.10	.9821356	2.57	.9949151
1.16	.8769756	1.64	.9494974	2.11	.9825708	2.58	.9950600
1.17	.8789995	1.65	.9505285	2.12	.9829970	2.59	.9952012
1.18	.8809999	1.66	.9515428	2.13	.9834142	2.60	.9953388
1.19	.8829768	1.67	.9525403	2.14	.9838226	2.70	.9965330
1.20	.8849303	1.68	.9535213	2.15	.9842224	2.80	.9974449
1.21	.8868606	1.69	.9544860	2.16	.9846137	2.90	.9981342
1.22	.8887676	1.70	.9554345	2.17	.9849966	3.00	.9986501
1.23	.8906514	1.71	.9563671	2.18	.9853713	3.20	.9993129
1.24	.8925123	1.72	.9572838	2.19	.9857379	3.40	.9996631
1.25	.8943502	1.73	.9581849	2.20	.9860966	3.60	.9998409
1.26	.8961653	1.74	.9590705	2.21	.9864474	3.80	.9999277
1.27	.8979577	1.75	.9599408	2.22	.9867906	4.00	.9999683
1.28	.8997274	1.76	.9607961	2.23	.9871263	4.50	.9999966
1.29	.9014747	1.77	.9616364	2.24	.9874545	5.00	.9999997
1.30	.9031995	1.78	.9624620	2.25	.9877755	5.50	.9999999
1.31	.9049021						

	Desired familywise error rate	
Number of tests	.05	.01
2	2.2414	2.8070
3	2.3954	2.9290
4	2.4949	3.0236
5	2.5758	3.0902
6	2.6350	3.1559
7	2.6874	3.1947
8	2.7370	3.2389
9	2.7703	3.2647
10	2.8070	3.2905
11	2.8338	3.3217
12	2.8627	3.3528
13	2.8943	3.3720
14	2.9112	3.3847
15	2.9290	3.4005
21	3.0351	3.4915
28	3.1214	3.5729
36	3.1947	3.6410
45	3.2647	3.6912

Table I-A

Some approximate normal deviate values (z^*) for use in finding Bonferroni t_B values from Table 11.14.1

(Note that if each of the tests corresponds to a normal deviate rather than a t ratio, these values may be used directly to establish the two-tailed significance for each test, given the desired familywise rate.)

Table II   **927**

Table II

Binomial probabilities $\binom{N}{r}p^r q^{N-r}$

N	r	.05	.10	.15	.20	.25	.30	.35	.40	.45	.50
							p				
1	0	.9500	.9000	.8500	.8000	.7500	.7000	.6500	.6000	.5500	.5000
	1	.0500	.1000	.1500	.2000	.2500	.3000	.3500	.4000	.4500	.5000
2	0	.9025	.8100	.7225	.6400	.5625	.4900	.4225	.3600	.3025	.2500
	1	.0950	.1800	.2550	.3200	.3750	.4200	.4550	.4800	.4950	.5000
	2	.0025	.0100	.0225	.0400	.0625	.0900	.1225	.1600	.2025	.2500
3	0	.8574	.7290	.6141	.5120	.4219	.3430	.2746	.2160	.1664	.1250
	1	.1354	.2430	.3251	.3840	.4219	.4410	.4436	.4320	.4084	.3750
	2	.0071	.0270	.0574	.0960	.1406	.1890	.2389	.2880	.3341	.3750
	3	.0001	.0010	.0034	.0080	.0156	.0270	.0429	.0640	.0911	.1250
4	0	.8145	.6561	.5220	.4096	.3164	.2401	.1785	.1296	.0915	.0625
	1	.1715	.2916	.3685	.4096	.4219	.4116	.3845	.3456	.2995	.2500
	2	.0135	.0486	.0975	.1536	.2109	.2646	.3105	.3456	.3675	.3750
	3	.0005	.0036	.0115	.0256	.0469	.0756	.1115	.1536	.2005	.2500
	4	.0000	.0001	.0005	.0016	.0039	.0081	.0150	.0256	.0410	.0625
5	0	.7738	.5905	.4437	.3277	.2373	.1681	.1160	.0778	.0503	.0312
	1	.2036	.3280	.3915	.4096	.3955	.3602	.3124	.2592	.2059	.1562
	2	.0214	.0729	.1382	.2048	.2637	.3087	.3364	.3456	.3369	.3125
	3	.0011	.0081	.0244	.0512	.0879	.1323	.1811	.2304	.2757	.3125
	4	.0000	.0004	.0022	.0064	.0146	.0284	.0488	.0768	.1128	.1562
	5	.0000	.0000	.0001	.0003	.0010	.0024	.0053	.0102	.0185	.0312
6	0	.7351	.5314	.3771	.2621	.1780	.1176	.0754	.0467	.0277	.0156
	1	.2321	.3543	.3993	.3932	.3560	.3025	.2437	.1866	.1359	.0938
	2	.0305	.0984	.1762	.2458	.2966	.3241	.3280	.3110	.2780	.2344
	3	.0021	.0146	.0415	.0819	.1318	.1852	.2355	.2765	.3032	.3125
	4	.0001	.0012	.0055	.0154	.0330	.0595	.0951	.1382	.1861	.2344
	5	.0000	.0001	.0004	.0015	.0044	.0102	.0205	.0369	.0609	.0938
	6	.0000	.0000	.0000	.0001	.0002	.0007	.0018	.0041	.0083	.0156
7	0	.6983	.4783	.3206	.2097	.1335	.0824	.0490	.0280	.0152	.0078
	1	.2573	.3720	.3960	.3670	.3115	.2471	.1848	.1306	.0872	.0547
	2	.0406	.1240	.2097	.2753	.3115	.3177	.2985	.2613	.2140	.1641
	3	.0036	.0230	.0617	.1147	.1730	.2269	.2679	.2903	.2918	.2734
	4	.0002	.0026	.0109	.0287	.0577	.0972	.1442	.1935	.2388	.2734
	5	.0000	.0002	.0012	.0043	.0115	.0250	.0466	.0774	.1172	.1641
	6	.0000	.0000	.0001	.0004	.0013	.0036	.0084	.0172	.0320	.0547
	7	.0000	.0000	.0000	.0000	.0001	.0002	.0006	.0016	.0037	.0078
8	0	.6634	.4305	.2725	.1678	.1001	.0576	.0319	.0168	.0084	.0039
	1	.2793	.3826	.3847	.3355	.2760	.1977	.1373	.0896	.0548	.0312
	2	.0515	.1488	.2376	.2936	.3115	.2965	.2587	.2090	.1569	.1094
	3	.0054	.0331	.0839	.1468	.2076	.2541	.2786	.2787	.2568	.2188
	4	.0004	.0046	.0185	.0459	.0865	.1361	.1875	.2322	.2627	.2734
	5	.0000	.0004	.0026	.0092	.0231	.0467	.0808	.1239	.1719	.2188
	6	.0000	.0000	.0002	.0011	.0038	.0100	.0217	.0413	.0703	.1094
	7	.0000	.0000	.0000	.0001	.0004	.0012	.0033	.0079	.0164	.0312
	8	.0000	.0000	.0000	.0000	.0000	.0001	.0002	.0007	.0017	.0039

Note. From *Handbook of Probability and Statistics with Tables,* 2nd ed., by R. C. Burington and D. C. May, 1970, New York: McGraw-Hill. Copyright 1953 by McGraw-Hill Co. Adapted by permission.

Table II (continued)

N	r	.05	.10	.15	.20	.25	.30	.35	.40	.45	.50
9	0	.6302	.3874	.2316	.1342	.0751	.0404	.0277	.0101	.0046	.0020
	1	.2985	.3874	.3679	.3020	.2253	.1556	.1004	.0605	.0339	.0176
	2	.0629	.1722	.2597	.3020	.3003	.2668	.2162	.1612	.1110	.0703
	3	.0077	.0446	.1069	.1762	.2336	.2668	.2716	.2508	.2119	.1641
	4	.0006	.0074	.0283	.0661	.1168	.1715	.2194	.2508	.2600	.2461
	5	.0000	.0008	.0050	.0165	.0389	.0735	.1181	.1672	.2128	.2461
	6	.0000	.0001	.0006	.0028	.0087	.0210	.0424	.0743	.1160	.1641
	7	.0000	.0000	.0000	.0003	.0012	.0039	.0098	.0212	.0407	.0703
	8	.0000	.0000	.0000	.0000	.0001	.0004	.0013	.0035	.0083	.0176
	9	.0000	.0000	.0000	.0000	.0000	.0000	.0001	.0003	.0008	.0020
10	0	.5987	.3487	.1969	.1074	.0563	.0282	.0135	.0060	.0025	.0010
	1	.3151	.3874	.3474	.2684	.1877	.1211	.0725	.0403	.0207	.0098
	2	.0746	.1937	.2759	.3020	.2816	.2335	.1757	.1209	.0763	.0439
	3	.0105	.0574	.1298	.2013	.2503	.2668	.2522	.2150	.1665	.1172
	4	.0010	.0112	.0401	.0881	.1460	.2001	.2377	.2508	.2384	.2051
	5	.0001	.0015	.0085	.0264	.0584	.1029	.1536	.2007	.2340	.2461
	6	.0000	.0001	.0012	.0055	.0162	.0368	.0689	.1115	.1596	.2051
	7	.0000	.0000	.0001	.0008	.0031	.0090	.0212	.0425	.0746	.1172
	8	.0000	.0000	.0000	.0001	.0004	.0014	.0043	.0106	.0229	.0439
	9	.0000	.0000	.0000	.0000	.0000	.0001	.0005	.0016	.0042	.0098
	10	.0000	.0000	.0000	.0000	.0000	.0000	.0000	.0001	.0003	.0016
11	0	.5688	.3138	.1673	.0859	.0422	.0198	.0088	.0036	.0014	.0005
	1	.3293	.3835	.3248	.2362	.1549	.0932	.0518	.0266	.0125	.0054
	2	.0867	.2131	.2866	.2953	.2581	.1998	.1395	.0887	.0513	.0269
	3	.0137	.0710	.1517	.2215	.2581	.2568	.2254	.1774	.1259	.0806
	4	.0014	.0158	.0536	.1107	.1721	.2201	.2428	.2365	.2060	.1611
	5	.0001	.0025	.0132	.0388	.0803	.1231	.1830	.2207	.2360	.2256
	6	.0000	.0003	.0023	.0097	.0268	.0566	.0985	.1471	.1931	.2256
	7	.0000	.0000	.0003	.0017	.0064	.0173	.0379	.0701	.1128	.1611
	8	.0000	.0000	.0000	.0002	.0011	.0037	.0102	.0234	.0462	.0806
	9	.0000	.0000	.0000	.0000	.0001	.0005	.0018	.0052	.0126	.0269
	10	.0000	.0000	.0000	.0000	.0000	.0000	.0002	.0007	.0021	.0054
	11	.0000	.0000	.0000	.0000	.0000	.0000	.0000	.0000	.0002	.0005
12	0	.5404	.2824	.1422	.0687	.0317	.0138	.0057	.0022	.0008	.0002
	1	.3413	.3766	.3012	.2062	.1267	.0712	.0368	.0174	.0075	.0029
	2	.0988	.2301	.2924	.2835	.2323	.1678	.1088	.0639	.0339	.0161
	3	.0173	.0852	.1720	.2362	.2581	.2397	.1954	.1419	.0923	.0537
	4	.0021	.0213	.0683	.1329	.1936	.2311	.2367	.2128	.1700	.1208
	5	.0002	.0038	.0193	.0532	.1032	.1585	.2039	.2270	.2225	.1934
	6	.0000	.0005	.0040	.0155	.0401	.0792	.1281	.1766	.2124	.2256
	7	.0000	.0000	.0006	.0033	.0115	.0291	.0591	.1009	.1489	.1934
	8	.0000	.0000	.0001	.0005	.0024	.0078	.0199	.0420	.0762	.1208
	9	.0000	.0000	.0000	.0001	.0004	.0015	.0048	.0125	.0277	.0537
	10	.0000	.0000	.0000	.0000	.0000	.0002	.0008	.0025	.0068	.0161
	11	.0000	.0000	.0000	.0000	.0000	.0000	.0001	.0003	.0010	.0029
	12	.0000	.0000	.0000	.0000	.0000	.0000	.0000	.0000	.0001	.0002

Table II   929

Table II (continued)

N	r	.05	.10	.15	.20	.25	.30	.35	.40	.45	.50
							p				
13	0	.5133	.2542	.1209	.0550	.0238	.0097	.0037	.0013	.0004	.0001
	1	.3512	.3672	.2774	.1787	.1029	.0540	.0259	.0113	.0045	.0016
	2	.1109	.2448	.2937	.2680	.2059	.1388	.0836	.0453	.0220	.0095
	3	.0214	.0997	.1900	.2457	.2517	.2181	.1651	.1107	.0660	.0349
	4	.0028	.0277	.0838	.1535	.2097	.2337	.2222	.1845	.1350	.0873
	5	.0003	.0055	.0266	.0691	.1258	.1803	.2154	.2214	.1989	.1571
	6	.0000	.0008	.0063	.0230	.0559	.1030	.1546	.1968	.2169	.2095
	7	.0000	.0001	.0011	.0058	.0186	.0442	.0833	.1312	.1775	.2095
	8	.0000	.0000	.0001	.0011	.0047	.0142	.0336	.0656	.1089	.1571
	9	.0000	.0000	.0000	.0001	.0009	.0034	.0101	.0243	.0495	.0873
	10	.0000	.0000	.0000	.0000	.0001	.0006	.0022	.0065	.0162	.0349
	11	.0000	.0000	.9000	.0000	.0000	.0001	.0003	.0012	.0036	.0095
	12	.0000	.0000	.0000	.0000	.0000	.0000	.0000	.0001	.0005	.0016
	13	.0000	.0000	.0000	.0000	.0000	.0000	.0000	.0000	.0000	.0001
14	0	.4877	.2288	.1028	.0440	.0178	.0068	.0024	.0008	.0002	.0001
	1	.3593	.3559	.2539	.1539	.0832	.0407	.0181	.0073	.0027	.0009
	2	.1229	.2570	.2912	.2501	.1802	.1134	.0634	.0317	.0141	.0056
	3	.0259	.1142	.2056	.2501	.2402	.1943	.1366	.0845	.0462	.0222
	4	.0037	.0349	.0998	.1720	.2202	.2290	.2022	.1549	.1040	.0611
	5	.0004	.0078	.0352	.0860	.1468	.1963	.2178	.2066	.1701	.1222
	6	.0000	.0013	.0093	.0322	.0734	.1262	.1759	.2066	.2088	.1833
	7	.0000	.0002	.0019	.0092	.0280	.0618	.1082	.1574	.1952	.2095
	8	.0000	.0000	.0003	.0020	.0082	.0232	.0510	.0918	.1398	.1833
	9	.0000	.0000	.0000	.0003	.0018	.0066	.0183	.0408	.0762	.1222
	10	.0000	.0000	.0000	.0000	.0003	.0014	.0049	.0136	.0312	.0611
	11	.0000	.0000	.0000	.0000	.0000	.0002	.0010	.0033	.0093	.0222
	12	.0000	.0000	.0000	.0000	.0000	.0000	.0001	.0005	.0019	.0056
	13	.0000	.0000	.0000	.0000	.0000	.0000	.0000	.0001	.0002	.0009
	14	.0000	.0000	.0000	.0000	.0000	.0000	.0000	.0000	.0000	.0001
15	0	.4633	.2059	.0874	.0352	.0134	.0047	.0016	.0005	.0001	.0000
	1	.3658	.3432	.2312	.1319	.0668	.0305	.0126	.0047	.0016	.0005
	2	.1348	.2669	.2856	.2309	.1559	.0916	.0476	.0219	.0090	.0032
	3	.0307	.1285	.2184	.2501	.2252	.1700	.1110	.0634	.0318	.0139
	4	.0049	.0428	.1156	.1876	.2252	.2186	.1792	.1268	.0780	.0417
	5	.0006	.0105	.0449	.1032	.1651	.2061	.2123	.1859	.1404	.0916
	6	.0000	.0019	.0132	.0430	.0917	.1472	.1906	.2066	.1914	.1527
	7	.0000	.0003	.0030	.0138	.0393	.0811	.1319	.1771	.2013	.1964
	8	.0000	.0000	.0005	.0035	.0131	.0348	.0710	.1181	.1647	.1964
	9	.0000	.0000	.0001	.0007	.0034	.0116	.0298	.0612	.1048	.1527
	10	.0000	.0000	.0000	.0001	.0007	.0030	.0096	.0245	.0515	.0916
	11	.0000	.0000	.0000	.0000	.0001	.0006	.0024	.0074	.0191	.0417
	12	.0000	.0000	.0000	.0000	.0000	.0001	.0004	.0016	.0052	.0139
	13	.0000	.0000	.0000	.0000	.0000	.0000	.0001	.0003	.0010	.0032
	14	.0000	.0000	.0000	.0000	.0000	.0000	.0000	.0000	.0001	.0005
	15	.0000	.0000	.0000	.0000	.0000	.0000	.0000	.0000	.0000	.0000
16	0	.4401	.1853	.0743	.0281	.0100	.0033	.0010	.0003	.0001	.0000
	1	.3706	.3294	.2097	.1126	.0535	.0228	.0087	.0030	.0009	.0002
	2	.1463	.2745	.2775	.2111	.1336	.0732	.0353	.0150	.0056	.0018

Table II (continued)

N	r	.05	.10	.15	.20	.25	.30	.35	.40	.45	.50
						p					
16	3	.0359	.1423	.2285	.2463	.2079	.1465	.0888	.0468	.0215	.0085
	4	.0061	.0514	.1311	.2001	.2252	.2040	.1553	.1014	.0572	.0278
	5	.0008	.0137	.0555	.1201	.1802	.2099	.2008	.1623	.1123	.0667
	6	.0001	.0028	.0180	.0550	.1101	.1649	.1982	.1983	.1684	.1222
	7	.0000	.0004	.0045	.0197	.0524	.1010	.1524	.1889	.1969	.1746
	8	.0000	.0001	.0009	.0055	.0197	.0487	.0923	.1417	.1812	.1964
	9	.0000	.0000	.0001	.0012	.0058	.0185	.0442	.0840	.1318	.1746
	10	.0000	.0000	.0000	.0002	.0014	.0056	.0167	.0392	.0755	.1222
	11	.0000	.0000	.0000	.0000	.0002	.0013	.0049	.0142	.0337	.0667
	12	.0000	.0000	.0000	.0000	.0000	.0002	.0011	.0040	.0115	.0278
	13	.0000	.0000	.0000	.0000	.0000	.0000	.0002	.0008	.0029	.0085
	14	.0000	.0000	.0000	.0000	.0000	.0000	.0000	.0001	.0005	.0018
	15	.0000	.0000	.0000	.0000	.0000	.0000	.0000	.0000	.0001	.0002
	16	.0000	.0000	.0000	.0000	.0000	.0000	.0000	.0000	.0000	.0000
17	0	.4181	.1668	.0631	.0225	.0075	.0023	.0007	.0002	.0000	.0000
	1	.3741	.3150	.1893	.0957	.0426	.0169	.0060	.0019	.0005	.0001
	2	.1575	.2800	.2673	.1914	.1136	.0581	.0260	.0102	.0035	.0010
	3	.0415	.1556	.2359	.2393	.1893	.1245	.0701	.0341	.0144	.0052
	4	.0076	.0605	.1457	.2093	.2209	.1868	.1320	.0796	.0411	.0182
	5	.0010	.0175	.0668	.1361	.1914	.2081	.1849	.1379	.0875	.0472
	6	.0001	.0039	.0236	.0680	.1276	.1784	.1991	.1839	.1432	.0944
	7	.0000	.0007	.0065	.0267	.1668	.1201	.1685	.1927	.1841	.1484
	8	.0000	.0001	.0014	.0084	.0279	.0644	.1143	.1606	.1883	.1855
	9	.0000	.0000	.0003	.0021	.0093	.0276	.0611	.1070	.1540	.1855
	10	.0000	.0000	.0000	.0004	.0025	.0095	.0263	.0571	.1008	.1484
	11	.0000	.0000	.0000	.0001	.0005	.0026	.0090	.0242	.0525	.0944
	12	.0000	.0000	.0000	.0000	.0001	.0006	.0024	.0081	.0215	.0472
	13	.0000	.0000	.0000	.0000	.0000	.0001	.0005	.0021	.0068	.0182
	14	.0000	.0000	.0000	.0000	.0000	.0000	.0001	.0004	.0016	.0052
	15	.0000	.0000	.0000	.0000	.0000	.0000	.0000	.0001	.0003	.0010
	16	.0000	.0000	.0000	.0000	.0000	.0000	.0000	.0000	.0000	.0001
	17	.0000	.0000	.0000	.0000	.0000	.0000	.0000	.0000	.0000	.0000
18	0	.3972	.1501	.0536	.0180	.0056	.0016	.0004	.0001	.0000	.0000
	1	.3763	.3002	.1704	.0811	.0338	.0126	.0042	.0012	.0003	.0001
	2	.1683	.2835	.2556	.1723	.0958	.0458	.0190	.0069	.0022	.0006
	3	.0473	.1680	.2406	.2297	.1704	.1046	.0547	.0246	.0095	.0031
	4	.0093	.0700	.1592	.2153	.2130	.1681	.1104	.0614	.0291	.0117
	5	.0014	.0218	.0787	.1507	.1988	.2017	.1664	.1146	.0666	.0327
	6	.0002	.0052	.0310	.0816	.1436	.1873	.1941	.1655	.1181	.0708
	7	.0000	.0010	.0091	.0350	.0820	.1376	.1792	.1892	.1657	.1214
	8	.0000	.0002	.0022	.0120	.0376	.0811	.1327	.1734	.1864	.1669
	9	.0000	.0000	.0004	.0033	.0139	.0386	.0794	.1284	.1694	.1855
	10	.0000	.0000	.0001	.0008	.0042	.0149	.0385	.0771	.1248	.1669
	11	.0000	.0000	.0000	.0001	.0010	.0046	.0151	.0374	.0742	.1214
	12	.0000	.0000	.0000	.0000	.0002	.0012	.0047	.0145	.0354	.0708
	13	.0000	.0000	.0000	.0000	.0000	.0002	.0012	.0045	.0134	.0327
	14	.0000	.0000	.0000	.0000	.0000	.0000	.0002	.0011	.0039	.0117

Table II   931

Table II (continued)

N	r	.05	.10	.15	.20	.25	.30	.35	.40	.45	.50
	15	.0000	.0000	.0000	.0000	.0000	.0000	.0000	.0002	.0009	.0031
	16	.0000	.0000	.0000	.0000	.0000	.0000	.0000	.0000	.0001	.0006
	17	.0000	.0000	.0000	.0000	.0000	.0000	.0000	.0000	.0000	.0001
	18	.0000	.0000	.0000	.0000	.0000	.0000	.0000	.0000	.0000	.0000
19	0	.3774	.1351	.0456	.0144	.0042	.0011	.0003	.0001	.0000	.0000
	1	.3774	.2852	.1529	.0685	.0268	.0093	.0029	.0008	.0002	.0000
	2	.1787	.2852	.2428	.1540	.0803	.0358	.0138	.0046	.0013	.0003
	3	.0533	.1796	.2428	.2182	.1517	.0869	.0422	.0175	.0062	.0018
	4	.0112	.0798	.1714	.2182	.2023	.1491	.0909	.0467	.0203	.0074
	5	.0018	.0266	.0907	.1636	.2023	.1916	.1468	.0933	.0497	.0222
	6	.0002	.0069	.0374	.0955	.1574	.1916	.1844	.1451	.0949	.0518
	7	.0000	.0014	.0122	.0443	.0974	.1525	.1844	.1797	.1443	.0961
	8	.0000	.0002	.0032	.0166	.0487	.0981	.1489	.1797	.1771	.1442
	9	.0000	.0000	.0007	.0051	.0198	.0514	.0980	.1464	.1771	.1762
	10	.0000	.0000	.0001	.0013	.0066	.0220	.0528	.0976	.1449	.1762
	11	.0000	.0000	.0000	.0003	.0018	.0077	.0233	.0532	.0970	.1442
	12	.0000	.0000	.0000	.0000	.0004	.0022	.0083	.0237	.0529	.0961
	13	.0000	.0000	.0000	.0000	.0001	.0005	.0024	.0085	.0233	.0518
	14	.0000	.0000	.0000	.0000	.0000	.0001	.0006	.0024	.0082	.0222
	15	.0000	.0000	.0000	.0000	.0000	.0000	.0001	.0005	.0022	.0074
	16	.0000	.0000	.0000	.0000	.0000	.0000	.0000	.0001	.0005	.0018
	17	.0000	.0000	.0000	.0000	.0000	.0000	.0000	.0000	.0001	.0003
	18	.0000	.0000	.0000	.0000	.0000	.0000	.0000	.0000	.0000	.0000
	19	.0000	.0000	.0000	.0000	.0000	.0000	.0000	.0000	.0000	.0000
20	0	.3585	.1216	.0388	.0115	.0032	.0008	.0002	.0000	.0000	.0000
	1	.3774	.2702	.1368	.0576	.0211	.0068	.0020	.0005	.0001	.0000
	2	.1887	.2852	.2293	.1369	.0669	.0278	.0100	.0031	.0008	.0002
	3	.0596	.1901	.2428	.2054	.1339	.0716	.0323	.0123	.0040	.0011
	4	.0133	.0898	.1821	.2182	.1897	.1304	.0738	.0350	.0139	.0046
	5	.0022	.0319	.1028	.1746	.2023	.1789	.1272	.0746	.0365	.0148
	6	.0003	.0089	.0454	.1091	.1686	.1916	.1712	.1244	.0746	.0370
	7	.0000	.0020	.0160	.0545	.1124	.1643	.1844	.1659	.1221	.0739
	8	.0000	.0004	.0046	.0222	.0609	.1144	.1614	.1797	.1623	.1201
	9	.0000	.0001	.0011	.0074	.0271	.0654	.1158	.1597	.1771	.1602
	10	.0000	.0000	.0002	.0020	.0099	.0308	.0686	.1171	.1593	.1762
	11	.0000	.0000	.0000	.0005	.0030	.0120	.0336	.0710	.1185	.1602
	12	.0000	.0000	.0000	.0001	.0008	.0039	.0136	.0355	.0727	.1201
	13	.0000	.0000	.0000	.0000	.0002	.0010	.0045	.0146	.0366	.0739
	14	.0000	.0000	.0000	.0000	.0000	.0002	.0012	.0049	.0150	.0370
	15	.0000	.0000	.0000	.0000	.0000	.0000	.0003	.0013	.0049	.0148
	16	.0000	.0000	.0000	.0000	.0000	.0000	.0000	.0003	.0013	.0046
	17	.0000	.0000	.0000	.0000	.0000	.0000	.0000	.0000	.0002	.0011
	18	.0000	.0000	.0000	.0000	.0000	.0000	.0000	.0000	.0000	.0002
	19	.0000	.0000	.0000	.0000	.0000	.0000	.0000	.0000	.0000	.0000
	20	.0000	.0000	.0000	.0000	.0000	.0000	.0000	.0000	.0000	.0000

Column header: *p*

Table III

Upper percentage points of the t distribution

ν	$Q = 0.4$ $2Q = 0.8$	0.25 0.5	0.1 0.2	0.05 0.1	0.025 0.05	0.01 0.02	0.005 0.01	0.001 0.002
1	0.325	1.000	3.078	6.314	12.706	31.821	63.657	318.31
2	.289	0.816	1.886	2.920	4.303	6.965	9.925	22.326
3	.277	.765	1.638	2.353	3.182	4.541	5.841	10.213
4	.271	.741	1.533	2.132	2.776	3.747	4.604	7.173
5	0.267	0.727	1.476	2.015	2.571	3.365	4.032	5.893
6	.265	.718	1.440	1.943	2.447	3.143	3.707	5.208
7	.263	.711	1.415	1.895	2.365	2.998	3.499	4.785
8	.262	.706	1.397	1.860	2.306	2.896	3.355	4.501
9	.261	.703	1.383	1.833	2.262	2.821	3.250	4.297
10	0.260	0.700	1.372	1.812	2.228	2.764	3.169	4.144
11	.260	.697	1.363	1.796	2.201	2.718	3.106	4.025
12	.259	.695	1.356	1.782	2.179	2.681	3.055	3.930
13	.259	.694	1.350	1.771	2.160	2.650	3.012	3.852
14	.258	.692	1.345	1.761	2.145	2.624	2.977	3.787
15	0.258	0.691	1.341	1.753	2.131	2.602	2.947	3.733
16	.258	.690	1.337	1.746	2.120	2.583	2.921	3.686
17	.257	.689	1.333	1.740	2.110	2.567	2.898	3.646
18	.257	.688	1.330	1.734	2.101	2.552	2.878	3.610
19	.257	.688	1.328	1.729	2.093	2.539	2.861	3.579
20	0.257	0.687	1.325	1.725	2.086	2.528	2.845	3.552
21	.257	.686	1.323	1.721	2.080	2.518	2.831	3.527
22	.256	.686	1.321	1.717	2.074	2.508	2.819	3.505
23	.256	.685	1.319	1.714	2.069	2.500	2.807	3.485
24	.256	.685	1.318	1.711	2.064	2.492	2.797	3.467
25	0.256	0.684	1.316	1.708	2.060	2.485	2.787	3.450
26	.256	.684	1.315	1.706	2.056	2.479	2.779	3.435
27	.256	.684	1.314	1.703	2.052	2.473	2.771	3.421
28	.256	.683	1.313	1.701	2.048	2.467	2.763	3.408
29	.256	.683	1.311	1.699	2.045	2.462	2.756	3.396
30	0.256	0.683	1.310	1.697	2.042	2.457	2.750	3.385
40	.255	.681	1.303	1.684	2.021	2.423	2.704	3.307
60	.254	.679	1.296	1.671	2.000	2.390	2.660	3.232
120	.254	.677	1.289	1.658	1.980	2.358	2.617	3.160
∞	.253	.674	1.282	1.645	1.960	2.326	2.576	3.090

Note. From *Biometrika Tables for Statisticians,* Vol. 1, 3rd ed. (Table 12, p. 138) by E. S. Pearson and H. O. Hartley (Eds.) 1966, Cambridge: Cambridge Univ. Press. Copyright 1954 by the Biometrika Trustees. Adapted by permission.

Table IV   933

Table IV

Upper percentage points of the χ^2 distribution

ν \ Q	0.995	0.990	0.975	0.950	0.900	0.750	0.500
1	$392704 \cdot 10^{-10}$	$157088 \cdot 10^{-9}$	$982069 \cdot 10^{-9}$	$393214 \cdot 10^{-8}$	0.0157908	0.1015308	0.454937
2	0.0100251	0.0201007	0.0506356	0.102587	0.210720	0.575364	1.38629
3	0.0717212	0.114832	0.215795	0.351846	0.584375	1.212534	2.36597
4	0.206990	0.297110	0.484419	0.710721	1.063623	1.92255	3.35670
5	0.411740	0.554300	0.831211	1.145476	1.61031	2.67460	4.35146
6	0.675727	0.872085	1.237347	1.63539	2.20413	3.45460	5.34812
7	0.989265	1.239043	1.68987	2.16735	2.83311	4.25485	6.34581
8	1.344419	1.646482	2.17973	2.73264	3.48954	5.07064	7.34412
9	1.734926	2.087912	2.70039	3.32511	4.16816	5.89883	8.34283
10	2.15585	2.55821	3.24697	3.94030	4.86518	6.73720	9.34182
11	2.60321	3.05347	3.81575	4.57481	5.57779	7.58412	10.3410
12	3.07382	3.57056	4.40379	5.22603	6.30380	8.43842	11.3403
13	3.56503	4.10691	5.00874	5.89186	7.04150	9.29906	12.3398
14	4.07468	4.66043	5.62872	6.57063	7.78953	10.1653	13.3393
15	4.60094	5.22935	6.26214	7.26094	8.54675	11.0365	14.3389
16	5.14224	5.81221	6.90766	7.96164	9.31223	11.9122	15.3385
17	5.69724	6.40776	7.56418	8.67176	10.0852	12.7919	16.3381
18	6.26481	7.01491	8.23075	9.39046	10.8649	13.6753	17.3379
19	6.84398	7.63273	8.90655	10.1170	11.6509	14.5620	18.3376
20	7.43386	8.26040	9.59083	10.8508	12.4426	15.4518	19.3374
21	8.03366	8.89720	10.28293	11.5913	13.2396	16.3444	20.3372
22	8.64272	9.54249	10.9823	12.3380	14.0415	17.2396	21.3370
23	9.26042	10.19567	11.6885	13.0905	14.8479	18.1373	22.3369
24	9.88623	10.8564	12.4011	13.8484	15.6587	19.0372	23.3367
25	10.5197	11.5240	13.1197	14.6114	16.4734	19.9393	24.3366
26	11.1603	12.1981	13.8439	15.3791	17.2919	20.8434	25.3364
27	11.8076	12.8786	14.5733	16.1513	18.1138	21.7494	26.3363
28	12.4613	13.5648	15.3079	16.9279	18.9392	22.6572	27.3363
29	13.1211	14.2565	16.0471	17.7083	19.7677	23.5666	28.3362
30	13.7867	14.9535	16.7908	18.4926	20.5992	24.4776	29.3360
40	20.7065	22.1643	24.4331	26.5093	29.0505	33.6603	39.3354
50	27.9907	29.7067	32.3574	34.7642	37.6886	42.9421	49.3349
60	35.5346	37.4848	40.4817	43.1879	46.4589	52.2938	59.3347
70	43.2752	45.4418	48.7576	51.7393	55.3290	61.6983	69.3344
80	51.1720	53.5400	57.1532	60.3915	64.2778	71.1445	79.3343
90	59.1963	61.7541	65.6466	69.1260	73.2912	80.6247	89.3342
100	67.3276	70.0648	74.2219	77.9295	82.3581	90.1332	99.3341
z_Q	-2.5758	-2.3263	-1.9600	-1.6449	-1.2816	-0.6745	0.0000

Note. From *Biometrika Tables for Statisticians,* Vol. 1, 3rd ed. (Table 8, pp. 130–131) by E. S. Pearson and H. O. Hartley (Eds.) 1966, Cambridge: Cambridge Univ. Press. Copyright 1954 by the Biometrika Trustees. Adapted by permission.

Table IV (continued)

ν \ Q	0.250	0.100	0.050	0.025	0.010	0.005	0.001
1	1.32330	2.70554	3.84146	5.02389	6.63490	7.87944	10.828
2	2.77259	4.60517	5.99147	7.37776	9.21034	10.5966	13.816
3	4.10835	6.25139	7.81473	9.34840	11.3449	12.8381	16.266
4	5.38527	7.77944	9.48773	11.1433	13.2767	14.8602	18.467
5	6.62568	9.23635	11.0705	12.8325	15.0863	16.7496	20.515
6	7.84080	10.6446	12.5916	14.4494	16.8119	18.5476	22.458
7	9.03715	12.0170	14.0671	16.0128	18.4753	20.2777	24.322
8	10.2188	13.3616	15.5073	17.5346	20.0902	21.9550	26.125
9	11.3887	14.6837	16.9190	19.0228	21.6660	23.5893	27.877
10	12.5489	15.9871	18.3070	20.4831	23.2093	25.1882	29.588
11	13.7007	17.2750	19.6751	21.9200	24.7250	26.7569	31.264
12	14.8454	18.5494	21.0261	23.3367	26.2170	28.2995	32.909
13	15.9839	19.8119	22.3621	24.7356	27.6883	29.8194	34.528
14	17.1170	21.0642	23.6848	26.1190	29.1413	31.3193	36.123
15	18.2451	22.3072	24.9958	27.4884	30.5779	32.8013	37.697
16	19.3688	23.5418	26.2962	28.8454	31.9999	34.2672	39.252
17	20.4887	24.7690	27.5871	30.1910	33.4087	35.7185	40.790
18	21.6049	25.9894	28.8693	31.5264	34.8053	37.1564	42.312
19	22.7178	27.2036	30.1435	32.8523	36.1908	38.5822	43.820
20	23.8277	28.4120	31.4104	34.1696	37.5662	39.9968	45.315
21	24.9348	29.6151	32.6705	35.4789	38.9321	41.4010	46.797
22	26.0393	30.8133	33.9244	36.7807	40.2894	42.7956	48.268
23	27.1413	32.0069	35.1725	38.0757	41.6384	44.1813	49.728
24	28.2412	33.1963	36.4151	39.3641	42.9798	45.5585	51.179
25	29.3389	34.3816	37.6525	40.6465	44.3141	46.9278	52.620
26	30.4345	35.5631	38.8852	41.9232	45.6417	48.2899	54.052
27	31.5284	36.7412	40.1133	43.1944	46.9630	49.6449	55.476
28	32.6205	37.9159	41.3372	44.4607	48.2782	50.9933	56.892
29	33.7109	39.0875	42.5569	45.7222	49.5879	52.3356	58.302
30	34.7998	40.2560	43.7729	46.9792	50.8922	53.6720	59.703
40	45.6160	51.8050	55.7585	59.3417	63.6907	66.7659	73.402
50	56.3336	63.1671	67.5048	71.4202	76.1539	79.4900	86.661
60	66.9814	74.3970	79.0819	83.2976	88.3794	91.9517	99.607
70	77.5766	85.5271	90.5312	95.0231	100.425	104.215	112.317
80	88.1303	96.5782	101.879	106.629	112.329	116.321	124.839
90	98.6499	107.565	113.145	118.136	124.116	128.299	137.208
100	109.141	118.498	124.342	129.561	135.807	140.169	149.449
z_Q	+0.6745	+1.2816	+1.6449	+1.9600	+2.3263	+2.5758	+3.0902

Table V   **935**

Table V

Percentage points of the F distribution,
UPPER 25% POINTS

v_2 \ v_1	1	2	3	4	5	6	7	8	9	10	12	15	20	24	30	40	60	120	∞
1	5·83	7·50	8·20	8·58	8·82	8·98	9·10	9·19	9·26	9·32	9·41	9·49	9·58	9·63	9·67	9·71	9·76	9·80	9·85
2	2·57	3·00	3·15	3·23	3·28	3·31	3·34	3·35	3·37	3·38	3·39	3·41	3·43	3·43	3·44	3·45	3·46	3·47	3·48
3	2·02	2·28	2·36	2·39	2·41	2·42	2·43	2·44	2·44	2·44	2·45	2·46	2·46	2·46	2·47	2·47	2·47	2·47	2·47
4	1·81	2·00	2·05	2·06	2·07	2·08	2·08	2·08	2·08	2·08	2·08	2·08	2·08	2·08	2·08	2·08	2·08	2·08	2·08
5	1·69	1·85	1·88	1·89	1·89	1·89	1·89	1·89	1·89	1·89	1·89	1·89	1·88	1·88	1·88	1·88	1·87	1·87	1·87
6	1·62	1·76	1·78	1·79	1·79	1·78	1·78	1·78	1·77	1·77	1·77	1·76	1·76	1·75	1·75	1·75	1·74	1·74	1·74
7	1·57	1·70	1·72	1·72	1·71	1·71	1·70	1·70	1·69	1·69	1·68	1·68	1·67	1·67	1·66	1·66	1·65	1·65	1·65
8	1·54	1·66	1·67	1·66	1·66	1·65	1·64	1·64	1·63	1·63	1·62	1·62	1·61	1·60	1·60	1·59	1·59	1·58	1·58
9	1·51	1·62	1·63	1·63	1·62	1·61	1·60	1·60	1·59	1·59	1·58	1·57	1·56	1·56	1·55	1·54	1·54	1·53	1·53
10	1·49	1·60	1·60	1·59	1·59	1·58	1·57	1·56	1·56	1·55	1·54	1·53	1·52	1·52	1·51	1·51	1·50	1·49	1·48
11	1·47	1·58	1·58	1·57	1·56	1·55	1·54	1·53	1·53	1·52	1·51	1·50	1·49	1·49	1·48	1·47	1·47	1·46	1·45
12	1·46	1·56	1·56	1·55	1·54	1·53	1·52	1·51	1·51	1·50	1·49	1·48	1·47	1·46	1·45	1·45	1·44	1·43	1·42
13	1·45	1·55	1·55	1·53	1·52	1·51	1·50	1·49	1·49	1·48	1·47	1·46	1·45	1·44	1·43	1·42	1·42	1·41	1·40
14	1·44	1·53	1·53	1·52	1·51	1·50	1·49	1·48	1·47	1·46	1·45	1·44	1·43	1·42	1·41	1·41	1·40	1·39	1·38
15	1·43	1·52	1·52	1·51	1·49	1·48	1·47	1·46	1·46	1·45	1·44	1·43	1·41	1·41	1·40	1·39	1·38	1·37	1·36
16	1·42	1·51	1·51	1·50	1·48	1·47	1·46	1·45	1·44	1·44	1·43	1·41	1·40	1·39	1·38	1·37	1·36	1·35	1·34
17	1·42	1·51	1·50	1·49	1·47	1·46	1·45	1·44	1·43	1·43	1·41	1·40	1·39	1·38	1·37	1·36	1·35	1·34	1·33
18	1·41	1·50	1·49	1·48	1·46	1·45	1·44	1·43	1·42	1·42	1·40	1·39	1·38	1·37	1·36	1·35	1·34	1·33	1·32
19	1·41	1·49	1·49	1·47	1·46	1·44	1·43	1·42	1·41	1·41	1·40	1·38	1·37	1·36	1·35	1·34	1·33	1·32	1·30
20	1·40	1·49	1·48	1·47	1·45	1·44	1·43	1·42	1·41	1·40	1·39	1·37	1·36	1·35	1·34	1·33	1·32	1·31	1·29
21	1·40	1·48	1·48	1·46	1·44	1·43	1·42	1·41	1·40	1·39	1·38	1·37	1·35	1·34	1·33	1·32	1·31	1·30	1·28
22	1·40	1·48	1·47	1·45	1·44	1·42	1·41	1·40	1·39	1·39	1·37	1·36	1·34	1·33	1·32	1·31	1·30	1·29	1·28
23	1·39	1·47	1·47	1·45	1·43	1·42	1·41	1·40	1·39	1·38	1·37	1·35	1·34	1·33	1·32	1·31	1·30	1·28	1·27
24	1·39	1·47	1·46	1·44	1·43	1·41	1·40	1·39	1·38	1·38	1·36	1·35	1·33	1·32	1·31	1·30	1·29	1·28	1·26
25	1·39	1·47	1·46	1·44	1·42	1·41	1·40	1·39	1·38	1·37	1·36	1·34	1·33	1·32	1·31	1·29	1·28	1·27	1·25
26	1·38	1·46	1·45	1·44	1·42	1·41	1·39	1·38	1·37	1·37	1·35	1·34	1·32	1·31	1·30	1·29	1·28	1·26	1·25
27	1·38	1·46	1·45	1·43	1·42	1·40	1·39	1·38	1·37	1·36	1·35	1·33	1·32	1·31	1·30	1·28	1·27	1·26	1·24
28	1·38	1·46	1·45	1·43	1·41	1·40	1·39	1·38	1·37	1·36	1·34	1·33	1·31	1·30	1·29	1·28	1·27	1·25	1·24
29	1·38	1·45	1·45	1·43	1·41	1·40	1·38	1·37	1·36	1·35	1·34	1·32	1·31	1·30	1·29	1·27	1·26	1·25	1·23
30	1·38	1·45	1·44	1·42	1·41	1·39	1·38	1·37	1·36	1·35	1·34	1·32	1·30	1·29	1·28	1·27	1·26	1·24	1·23
40	1·36	1·44	1·42	1·40	1·39	1·37	1·36	1·35	1·34	1·33	1·31	1·30	1·28	1·26	1·25	1·24	1·22	1·21	1·19
60	1·35	1·42	1·41	1·38	1·37	1·35	1·33	1·32	1·31	1·30	1·29	1·27	1·25	1·24	1·22	1·21	1·19	1·17	1·15
120	1·34	1·40	1·39	1·37	1·35	1·33	1·31	1·30	1·29	1·28	1·26	1·24	1·22	1·21	1·19	1·18	1·16	1·13	1·10
∞	1·32	1·39	1·37	1·35	1·33	1·31	1·29	1·28	1·27	1·25	1·24	1·22	1·19	1·18	1·16	1·14	1·12	1·08	1·00

Note. From *Biometrika Tables for Statisticians*, Vol. 1, 3rd ed. (Table 18, pp. 157–162) by E. S. Pearson and H. O. Hartley (Eds.) 1966, Cambridge: Cambridge Univ. Press. Copyright 1954 by the Biometrika Trustees. Adapted by permission.

Table V (continued)

UPPER 10% POINTS

ν_2 \ ν_1	1	2	3	4	5	6	7	8	9	10	12	15	20	24	30	40	60	120	∞
1	39·86	49·50	53·59	55·83	57·24	58·20	58·91	59·44	59·86	60·19	60·71	61·22	61·74	62·00	62·26	62·53	62·79	63·06	63·33
2	8·53	9·00	9·16	9·24	9·29	9·33	9·35	9·37	9·38	9·39	9·41	9·42	9·44	9·45	9·46	9·47	9·47	9·48	9·49
3	5·54	5·46	5·39	5·34	5·31	5·28	5·27	5·25	5·24	5·23	5·22	5·20	5·18	5·18	5·17	5·16	5·15	5·14	5·13
4	4·54	4·32	4·19	4·11	4·05	4·01	3·98	3·95	3·94	3·92	3·90	3·87	3·84	3·83	3·82	3·80	3·79	3·78	3·76
5	4·06	3·78	3·62	3·52	3·45	3·40	3·37	3·34	3·32	3·30	3·27	3·24	3·21	3·19	3·17	3·16	3·14	3·12	3·10
6	3·78	3·46	3·29	3·18	3·11	3·05	3·01	2·98	2·96	2·94	2·90	2·87	2·84	2·82	2·80	2·78	2·76	2·74	2·72
7	3·59	3·26	3·07	2·96	2·88	2·83	2·78	2·75	2·72	2·70	2·67	2·63	2·59	2·58	2·56	2·54	2·51	2·49	2·47
8	3·46	3·11	2·92	2·81	2·73	2·67	2·62	2·59	2·56	2·54	2·50	2·46	2·42	2·40	2·38	2·36	2·34	2·32	2·29
9	3·36	3·01	2·81	2·69	2·61	2·55	2·51	2·47	2·44	2·42	2·38	2·34	2·30	2·28	2·25	2·23	2·21	2·18	2·16
10	3·29	2·92	2·73	2·61	2·52	2·46	2·41	2·38	2·35	2·32	2·28	2·24	2·20	2·18	2·16	2·13	2·11	2·08	2·06
11	3·23	2·86	2·66	2·54	2·45	2·39	2·34	2·30	2·27	2·25	2·21	2·17	2·12	2·10	2·08	2·05	2·03	2·00	1·97
12	3·18	2·81	2·61	2·48	2·39	2·33	2·28	2·24	2·21	2·19	2·15	2·10	2·06	2·04	2·01	1·99	1·96	1·93	1·90
13	3·14	2·76	2·56	2·43	2·35	2·28	2·23	2·20	2·16	2·14	2·10	2·05	2·01	1·98	1·96	1·93	1·90	1·88	1·85
14	3·10	2·73	2·52	2·39	2·31	2·24	2·19	2·15	2·12	2·10	2·05	2·01	1·96	1·94	1·91	1·89	1·86	1·83	1·80
15	3·07	2·70	2·49	2·36	2·27	2·21	2·16	2·12	2·09	2·06	2·02	1·97	1·92	1·90	1·87	1·85	1·82	1·79	1·76
16	3·05	2·67	2·46	2·33	2·24	2·18	2·13	2·09	2·06	2·03	1·99	1·94	1·89	1·87	1·84	1·81	1·78	1·75	1·72
17	3·03	2·64	2·44	2·31	2·22	2·15	2·10	2·06	2·03	2·00	1·96	1·91	1·86	1·84	1·81	1·78	1·75	1·72	1·69
18	3·01	2·62	2·42	2·29	2·20	2·13	2·08	2·04	2·00	1·98	1·93	1·89	1·84	1·81	1·78	1·75	1·72	1·69	1·66
19	2·99	2·61	2·40	2·27	2·18	2·11	2·06	2·02	1·98	1·96	1·91	1·86	1·81	1·79	1·76	1·73	1·70	1·67	1·63
20	2·97	2·59	2·38	2·25	2·16	2·09	2·04	2·00	1·96	1·94	1·89	1·84	1·79	1·77	1·74	1·71	1·68	1·64	1·61
21	2·96	2·57	2·36	2·23	2·14	2·08	2·02	1·98	1·95	1·92	1·87	1·83	1·78	1·75	1·72	1·69	1·66	1·62	1·59
22	2·95	2·56	2·35	2·22	2·13	2·06	2·01	1·97	1·93	1·90	1·86	1·81	1·76	1·73	1·70	1·67	1·64	1·60	1·57
23	2·94	2·55	2·34	2·21	2·11	2·05	1·99	1·95	1·92	1·89	1·84	1·80	1·74	1·72	1·69	1·66	1·62	1·59	1·55
24	2·93	2·54	2·33	2·19	2·10	2·04	1·98	1·94	1·91	1·88	1·83	1·78	1·73	1·70	1·67	1·64	1·61	1·57	1·53
25	2·92	2·53	2·32	2·18	2·09	2·02	1·97	1·93	1·89	1·87	1·82	1·77	1·72	1·69	1·66	1·63	1·59	1·56	1·52
26	2·91	2·52	2·31	2·17	2·08	2·01	1·96	1·92	1·88	1·86	1·81	1·76	1·71	1·68	1·65	1·61	1·58	1·54	1·50
27	2·90	2·51	2·30	2·17	2·07	2·00	1·95	1·91	1·87	1·85	1·80	1·75	1·70	1·67	1·64	1·60	1·57	1·53	1·49
28	2·89	2·50	2·29	2·16	2·06	2·00	1·94	1·90	1·87	1·84	1·79	1·74	1·69	1·66	1·63	1·59	1·56	1·52	1·48
29	2·89	2·50	2·28	2·15	2·06	1·99	1·93	1·89	1·86	1·83	1·78	1·73	1·68	1·65	1·62	1·58	1·55	1·51	1·47
30	2·88	2·49	2·28	2·14	2·05	1·98	1·93	1·88	1·85	1·82	1·77	1·72	1·67	1·64	1·61	1·57	1·54	1·50	1·46
40	2·84	2·44	2·23	2·09	2·00	1·93	1·87	1·83	1·79	1·76	1·71	1·66	1·61	1·57	1·54	1·51	1·47	1·42	1·38
60	2·79	2·39	2·18	2·04	1·95	1·87	1·82	1·77	1·74	1·71	1·66	1·60	1·54	1·51	1·48	1·44	1·40	1·35	1·29
120	2·75	2·35	2·13	1·99	1·90	1·82	1·77	1·72	1·68	1·65	1·60	1·55	1·48	1·45	1·41	1·37	1·32	1·26	1·19
∞	2·71	2·30	2·08	1·94	1·85	1·77	1·72	1·67	1·63	1·60	1·55	1·49	1·42	1·38	1·34	1·30	1·24	1·17	1·00

Table V 937

Table V (continued)

UPPER 5% POINTS

$\nu_2 \backslash \nu_1$	1	2	3	4	5	6	7	8	9	10	12	15	20	24	30	40	60	120	∞
1	161.4	199.5	215.7	224.6	230.2	234.0	236.8	238.9	240.5	241.9	243.9	245.9	248.0	249.1	250.1	251.1	252.2	253.3	254.3
2	18.51	19.00	19.16	19.25	19.30	19.33	19.35	19.37	19.38	19.40	19.41	19.43	19.45	19.45	19.46	19.47	19.48	19.49	19.50
3	10.13	9.55	9.28	9.12	9.01	8.94	8.89	8.85	8.81	8.79	8.74	8.70	8.66	8.64	8.62	8.59	8.57	8.55	8.53
4	7.71	6.94	6.59	6.39	6.26	6.16	6.09	6.04	6.00	5.96	5.91	5.86	5.80	5.77	5.75	5.72	5.69	5.66	5.63
5	6.61	5.79	5.41	5.19	5.05	4.95	4.88	4.82	4.77	4.74	4.68	4.62	4.56	4.53	4.50	4.46	4.43	4.40	4.36
6	5.99	5.14	4.76	4.53	4.39	4.28	4.21	4.15	4.10	4.06	4.00	3.94	3.87	3.84	3.81	3.77	3.74	3.70	3.67
7	5.59	4.74	4.35	4.12	3.97	3.87	3.79	3.73	3.68	3.64	3.57	3.51	3.44	3.41	3.38	3.34	3.30	3.27	3.23
8	5.32	4.46	4.07	3.84	3.69	3.58	3.50	3.44	3.39	3.35	3.28	3.22	3.15	3.12	3.08	3.04	3.01	2.97	2.93
9	5.12	4.26	3.86	3.63	3.48	3.37	3.29	3.23	3.18	3.14	3.07	3.01	2.94	2.90	2.86	2.83	2.79	2.75	2.71
10	4.96	4.10	3.71	3.48	3.33	3.22	3.14	3.07	3.02	2.98	2.91	2.85	2.77	2.74	2.70	2.66	2.62	2.58	2.54
11	4.84	3.98	3.59	3.36	3.20	3.09	3.01	2.95	2.90	2.85	2.79	2.72	2.65	2.61	2.57	2.53	2.49	2.45	2.40
12	4.75	3.89	3.49	3.26	3.11	3.00	2.91	2.85	2.80	2.75	2.69	2.62	2.54	2.51	2.47	2.43	2.38	2.34	2.30
13	4.67	3.81	3.41	3.18	3.03	2.92	2.83	2.77	2.71	2.67	2.60	2.53	2.46	2.42	2.38	2.34	2.30	2.25	2.21
14	4.60	3.74	3.34	3.11	2.96	2.85	2.76	2.70	2.65	2.60	2.53	2.46	2.39	2.35	2.31	2.27	2.22	2.18	2.13
15	4.54	3.68	3.29	3.06	2.90	2.79	2.71	2.64	2.59	2.54	2.48	2.40	2.33	2.29	2.25	2.20	2.16	2.11	2.07
16	4.49	3.63	3.24	3.01	2.85	2.74	2.66	2.59	2.54	2.49	2.42	2.35	2.28	2.24	2.19	2.15	2.11	2.06	2.01
17	4.45	3.59	3.20	2.96	2.81	2.70	2.61	2.55	2.49	2.45	2.38	2.31	2.23	2.19	2.15	2.10	2.06	2.01	1.96
18	4.41	3.55	3.16	2.93	2.77	2.66	2.58	2.51	2.46	2.41	2.34	2.27	2.19	2.15	2.11	2.06	2.02	1.97	1.92
19	4.38	3.52	3.13	2.90	2.74	2.63	2.54	2.48	2.42	2.38	2.31	2.23	2.16	2.11	2.07	2.03	1.98	1.93	1.88
20	4.35	3.49	3.10	2.87	2.71	2.60	2.51	2.45	2.39	2.35	2.28	2.20	2.12	2.08	2.04	1.99	1.95	1.90	1.84
21	4.32	3.47	3.07	2.84	2.68	2.57	2.49	2.42	2.37	2.32	2.25	2.18	2.10	2.05	2.01	1.96	1.92	1.87	1.81
22	4.30	3.44	3.05	2.82	2.66	2.55	2.46	2.40	2.34	2.30	2.23	2.15	2.07	2.03	1.98	1.94	1.89	1.84	1.78
23	4.28	3.42	3.03	2.80	2.64	2.53	2.44	2.37	2.32	2.27	2.20	2.13	2.05	2.01	1.96	1.91	1.86	1.81	1.76
24	4.26	3.40	3.01	2.78	2.62	2.51	2.42	2.36	2.30	2.25	2.18	2.11	2.03	1.98	1.94	1.89	1.84	1.79	1.73
25	4.24	3.39	2.99	2.76	2.60	2.49	2.40	2.34	2.28	2.24	2.16	2.09	2.01	1.96	1.92	1.87	1.82	1.77	1.71
26	4.23	3.37	2.98	2.74	2.59	2.47	2.39	2.32	2.27	2.22	2.15	2.07	1.99	1.95	1.90	1.85	1.80	1.75	1.69
27	4.21	3.35	2.96	2.73	2.57	2.46	2.37	2.31	2.25	2.20	2.13	2.06	1.97	1.93	1.88	1.84	1.79	1.73	1.67
28	4.20	3.34	2.95	2.71	2.56	2.45	2.36	2.29	2.24	2.19	2.12	2.04	1.96	1.91	1.87	1.82	1.77	1.71	1.65
29	4.18	3.33	2.93	2.70	2.55	2.43	2.35	2.28	2.22	2.18	2.10	2.03	1.94	1.90	1.85	1.81	1.75	1.70	1.64
30	4.17	3.32	2.92	2.69	2.53	2.42	2.33	2.27	2.21	2.16	2.09	2.01	1.93	1.89	1.84	1.79	1.74	1.68	1.62
40	4.08	3.23	2.84	2.61	2.45	2.34	2.25	2.18	2.12	2.08	2.00	1.92	1.84	1.79	1.74	1.69	1.64	1.58	1.51
60	4.00	3.15	2.76	2.53	2.37	2.25	2.17	2.10	2.04	1.99	1.92	1.84	1.75	1.70	1.65	1.59	1.53	1.47	1.39
120	3.92	3.07	2.68	2.45	2.29	2.17	2.09	2.02	1.96	1.91	1.83	1.75	1.66	1.61	1.55	1.50	1.43	1.35	1.25
∞	3.84	3.00	2.60	2.37	2.21	2.10	2.01	1.94	1.88	1.83	1.75	1.67	1.57	1.52	1.46	1.39	1.32	1.22	1.00

Table V (continued)

UPPER 2.5% POINTS

$\nu_2 \backslash \nu_1$	1	2	3	4	5	6	7	8	9	10	12	15	20	24	30	40	60	120	∞
1	647.8	799.5	864.2	899.6	921.8	937.1	948.2	956.7	963.3	968.6	976.7	984.9	993.1	997.2	1001	1006	1010	1014	1018
2	38.51	39.00	39.17	39.25	39.30	39.33	39.36	39.37	39.39	39.40	39.41	39.43	39.45	39.46	39.46	39.47	39.48	39.49	39.50
3	17.44	16.04	15.44	15.10	14.88	14.73	14.62	14.54	14.47	14.42	14.34	14.25	14.17	14.12	14.08	14.04	13.99	13.95	13.90
4	12.22	10.65	9.98	9.60	9.36	9.20	9.07	8.98	8.90	8.84	8.75	8.66	8.56	8.51	8.46	8.41	8.36	8.31	8.26
5	10.01	8.43	7.76	7.39	7.15	6.98	6.85	6.76	6.68	6.62	6.52	6.43	6.33	6.28	6.23	6.18	6.12	6.07	6.02
6	8.81	7.26	6.60	6.23	5.99	5.82	5.70	5.60	5.52	5.46	5.37	5.27	5.17	5.12	5.07	5.01	4.96	4.90	4.85
7	8.07	6.54	5.89	5.52	5.29	5.12	4.99	4.90	4.82	4.76	4.67	4.57	4.47	4.42	4.36	4.31	4.25	4.20	4.14
8	7.57	6.06	5.42	5.05	4.82	4.65	4.53	4.43	4.36	4.30	4.20	4.10	4.00	3.95	3.89	3.84	3.78	3.73	3.67
9	7.21	5.71	5.08	4.72	4.48	4.32	4.20	4.10	4.03	3.96	3.87	3.77	3.67	3.61	3.56	3.51	3.45	3.39	3.33
10	6.94	5.46	4.83	4.47	4.24	4.07	3.95	3.85	3.78	3.72	3.62	3.52	3.42	3.37	3.31	3.26	3.20	3.14	3.08
11	6.72	5.26	4.63	4.28	4.04	3.88	3.76	3.66	3.59	3.53	3.43	3.33	3.23	3.17	3.12	3.06	3.00	2.94	2.88
12	6.55	5.10	4.47	4.12	3.89	3.73	3.61	3.51	3.44	3.37	3.28	3.18	3.07	3.02	2.96	2.91	2.85	2.79	2.72
13	6.41	4.97	4.35	4.00	3.77	3.60	3.48	3.39	3.31	3.25	3.15	3.05	2.95	2.89	2.84	2.78	2.72	2.66	2.60
14	6.30	4.86	4.24	3.89	3.66	3.50	3.38	3.29	3.21	3.15	3.05	2.95	2.84	2.79	2.73	2.67	2.61	2.55	2.49
15	6.20	4.77	4.15	3.80	3.58	3.41	3.29	3.20	3.12	3.06	2.96	2.86	2.76	2.70	2.64	2.59	2.52	2.46	2.40
16	6.12	4.69	4.08	3.73	3.50	3.34	3.22	3.12	3.05	2.99	2.89	2.79	2.68	2.63	2.57	2.51	2.45	2.38	2.32
17	6.04	4.62	4.01	3.66	3.44	3.28	3.16	3.06	2.98	2.92	2.82	2.72	2.62	2.56	2.50	2.44	2.38	2.32	2.25
18	5.98	4.56	3.95	3.61	3.38	3.22	3.10	3.01	2.93	2.87	2.77	2.67	2.56	2.50	2.44	2.38	2.32	2.26	2.19
19	5.92	4.51	3.90	3.56	3.33	3.17	3.05	2.96	2.88	2.82	2.72	2.62	2.51	2.45	2.39	2.33	2.27	2.20	2.13
20	5.87	4.46	3.86	3.51	3.29	3.13	3.01	2.91	2.84	2.77	2.68	2.57	2.46	2.41	2.35	2.29	2.22	2.16	2.09
21	5.83	4.42	3.82	3.48	3.25	3.09	2.97	2.87	2.80	2.73	2.64	2.53	2.42	2.37	2.31	2.25	2.18	2.11	2.04
22	5.79	4.38	3.78	3.44	3.22	3.05	2.93	2.84	2.76	2.70	2.60	2.50	2.39	2.33	2.27	2.21	2.14	2.08	2.00
23	5.75	4.35	3.75	3.41	3.18	3.02	2.90	2.81	2.73	2.67	2.57	2.47	2.36	2.30	2.24	2.18	2.11	2.04	1.97
24	5.72	4.32	3.72	3.38	3.15	2.99	2.87	2.78	2.70	2.64	2.54	2.44	2.33	2.27	2.21	2.15	2.08	2.01	1.94
25	5.69	4.29	3.69	3.35	3.13	2.97	2.85	2.75	2.68	2.61	2.51	2.41	2.30	2.24	2.18	2.12	2.05	1.98	1.91
26	5.66	4.27	3.67	3.33	3.10	2.94	2.82	2.73	2.65	2.59	2.49	2.39	2.28	2.22	2.16	2.09	2.03	1.95	1.88
27	5.63	4.24	3.65	3.31	3.08	2.92	2.80	2.71	2.63	2.57	2.47	2.36	2.25	2.19	2.13	2.07	2.00	1.93	1.85
28	5.61	4.22	3.63	3.29	3.06	2.90	2.78	2.69	2.61	2.55	2.45	2.34	2.23	2.17	2.11	2.05	1.98	1.91	1.83
29	5.59	4.20	3.61	3.27	3.04	2.88	2.76	2.67	2.59	2.53	2.43	2.32	2.21	2.15	2.09	2.03	1.96	1.89	1.81
30	5.57	4.18	3.59	3.25	3.03	2.87	2.75	2.65	2.57	2.51	2.41	2.31	2.20	2.14	2.07	2.01	1.94	1.87	1.79
40	5.42	4.05	3.46	3.13	2.90	2.74	2.62	2.53	2.45	2.39	2.29	2.18	2.07	2.01	1.94	1.88	1.80	1.72	1.64
60	5.29	3.93	3.34	3.01	2.79	2.63	2.51	2.41	2.33	2.27	2.17	2.06	1.94	1.88	1.82	1.74	1.67	1.58	1.48
120	5.15	3.80	3.23	2.89	2.67	2.52	2.39	2.30	2.22	2.16	2.05	1.94	1.82	1.76	1.69	1.61	1.53	1.43	1.31
∞	5.02	3.69	3.12	2.79	2.57	2.41	2.29	2.19	2.11	2.05	1.94	1.83	1.71	1.64	1.57	1.48	1.39	1.27	1.00

Table V   939

Table V (continued)

UPPER 1% POINTS

ν_2 \ ν_1	1	2	3	4	5	6	7	8	9	10	12	15	20	24	30	40	60	120	∞
1	4052	4999.5	5403	5625	5764	5859	5928	5982	6022	6056	6106	6157	6209	6235	6261	6287	6313	6339	6366
2	98.50	99.00	99.17	99.25	99.30	99.33	99.36	99.37	99.39	99.40	99.42	99.43	99.45	99.46	99.47	99.47	99.48	99.49	99.50
3	34.12	30.82	29.46	28.71	28.24	27.91	27.67	27.49	27.35	27.23	27.05	26.87	26.69	26.60	26.50	26.41	26.32	26.22	26.13
4	21.20	18.00	16.69	15.98	15.52	15.21	14.98	14.80	14.66	14.55	14.37	14.20	14.02	13.93	13.84	13.75	13.65	13.56	13.46
5	16.26	13.27	12.06	11.39	10.97	10.67	10.46	10.29	10.16	10.05	9.89	9.72	9.55	9.47	9.38	9.29	9.20	9.11	9.02
6	13.75	10.92	9.78	9.15	8.75	8.47	8.26	8.10	7.98	7.87	7.72	7.56	7.40	7.31	7.23	7.14	7.06	6.97	6.88
7	12.25	9.55	8.45	7.85	7.46	7.19	6.99	6.84	6.72	6.62	6.47	6.31	6.16	6.07	5.99	5.91	5.82	5.74	5.65
8	11.26	8.65	7.59	7.01	6.63	6.37	6.18	6.03	5.91	5.81	5.67	5.52	5.36	5.28	5.20	5.12	5.03	4.95	4.86
9	10.56	8.02	6.99	6.42	6.06	5.80	5.61	5.47	5.35	5.26	5.11	4.96	4.81	4.73	4.65	4.57	4.48	4.40	4.31
10	10.04	7.56	6.55	5.99	5.64	5.39	5.20	5.06	4.94	4.85	4.71	4.56	4.41	4.33	4.25	4.17	4.08	4.00	3.91
11	9.65	7.21	6.22	5.67	5.32	5.07	4.89	4.74	4.63	4.54	4.40	4.25	4.10	4.02	3.94	3.86	3.78	3.69	3.60
12	9.33	6.93	5.95	5.41	5.06	4.82	4.64	4.50	4.39	4.30	4.16	4.01	3.86	3.78	3.70	3.62	3.54	3.45	3.36
13	9.07	6.70	5.74	5.21	4.86	4.62	4.44	4.30	4.19	4.10	3.96	3.82	3.66	3.59	3.51	3.43	3.34	3.25	3.17
14	8.86	6.51	5.56	5.04	4.69	4.46	4.28	4.14	4.03	3.94	3.80	3.66	3.51	3.43	3.35	3.27	3.18	3.09	3.00
15	8.68	6.36	5.42	4.89	4.56	4.32	4.14	4.00	3.89	3.80	3.67	3.52	3.37	3.29	3.21	3.13	3.05	2.96	2.87
16	8.53	6.23	5.29	4.77	4.44	4.20	4.03	3.89	3.78	3.69	3.55	3.41	3.26	3.18	3.10	3.02	2.93	2.84	2.75
17	8.40	6.11	5.18	4.67	4.34	4.10	3.93	3.79	3.68	3.59	3.46	3.31	3.16	3.08	3.00	2.92	2.83	2.75	2.65
18	8.29	6.01	5.09	4.58	4.25	4.01	3.84	3.71	3.60	3.51	3.37	3.23	3.08	3.00	2.92	2.84	2.75	2.66	2.57
19	8.18	5.93	5.01	4.50	4.17	3.94	3.77	3.63	3.52	3.43	3.30	3.15	3.00	2.92	2.84	2.76	2.67	2.58	2.49
20	8.10	5.85	4.94	4.43	4.10	3.87	3.70	3.56	3.46	3.37	3.23	3.09	2.94	2.86	2.78	2.69	2.61	2.52	2.42
21	8.02	5.78	4.87	4.37	4.04	3.81	3.64	3.51	3.40	3.31	3.17	3.03	2.88	2.80	2.72	2.64	2.55	2.46	2.36
22	7.95	5.72	4.82	4.31	3.99	3.76	3.59	3.45	3.35	3.26	3.12	2.98	2.83	2.75	2.67	2.58	2.50	2.40	2.31
23	7.88	5.66	4.76	4.26	3.94	3.71	3.54	3.41	3.30	3.21	3.07	2.93	2.78	2.70	2.62	2.54	2.45	2.35	2.26
24	7.82	5.61	4.72	4.22	3.90	3.67	3.50	3.36	3.26	3.17	3.03	2.89	2.74	2.66	2.58	2.49	2.40	2.31	2.21
25	7.77	5.57	4.68	4.18	3.85	3.63	3.46	3.32	3.22	3.13	2.99	2.85	2.70	2.62	2.54	2.45	2.36	2.27	2.17
26	7.72	5.53	4.64	4.14	3.82	3.59	3.42	3.29	3.18	3.09	2.96	2.81	2.66	2.58	2.50	2.42	2.33	2.23	2.13
27	7.68	5.49	4.60	4.11	3.78	3.56	3.39	3.26	3.15	3.06	2.93	2.78	2.63	2.55	2.47	2.38	2.29	2.20	2.10
28	7.64	5.45	4.57	4.07	3.75	3.53	3.36	3.23	3.12	3.03	2.90	2.75	2.60	2.52	2.44	2.35	2.26	2.17	2.06
29	7.60	5.42	4.54	4.04	3.73	3.50	3.33	3.20	3.09	3.00	2.87	2.73	2.57	2.49	2.41	2.33	2.23	2.14	2.03
30	7.56	5.39	4.51	4.02	3.70	3.47	3.30	3.17	3.07	2.98	2.84	2.70	2.55	2.47	2.39	2.30	2.21	2.11	2.01
40	7.31	5.18	4.31	3.83	3.51	3.29	3.12	2.99	2.89	2.80	2.66	2.52	2.37	2.29	2.20	2.11	2.02	1.92	1.80
60	7.08	4.98	4.13	3.65	3.34	3.12	2.95	2.82	2.72	2.63	2.50	2.35	2.20	2.12	2.03	1.94	1.84	1.73	1.60
120	6.85	4.79	3.95	3.48	3.17	2.96	2.79	2.66	2.56	2.47	2.34	2.19	2.03	1.95	1.86	1.76	1.66	1.53	1.38
∞	6.63	4.61	3.78	3.32	3.02	2.80	2.64	2.51	2.41	2.32	2.18	2.04	1.88	1.79	1.70	1.59	1.47	1.32	1.00

Table V (continued)

UPPER .5% POINTS

v_2\\v_1	1	2	3	4	5	6	7	8	9
1	16211	20000	21615	22500	23056	23437	23715	23925	24091
2	198	199	199	199	199	199	199	199	199
3	55.55	49.80	47.47	46.20	45.39	44.84	44.43	44.13	43.88
4	31.33	26.28	24.26	23.16	22.46	21.98	21.62	21.35	21.14
5	22.78	18.31	16.53	15.56	14.94	14.51	14.20	13.96	13.77
6	18.64	14.54	12.92	12.03	11.46	11.07	10.79	10.57	10.39
7	16.24	12.40	10.88	10.05	9.52	9.16	8.89	8.68	8.51
8	14.69	11.04	9.60	8.81	8.30	7.95	7.69	7.50	7.34
9	13.61	10.11	8.72	7.96	7.47	7.13	6.88	6.69	6.54
10	12.83	9.43	8.08	7.34	6.87	6.54	6.30	6.12	5.97
11	12.23	8.91	7.60	6.88	6.42	6.10	5.86	5.68	5.54
12	11.75	8.51	7.23	6.52	6.07	5.76	5.52	5.35	5.20
13	11.37	8.19	6.93	6.23	5.79	5.48	5.25	5.08	4.94
14	11.06	7.92	6.68	6.00	5.56	5.26	5.03	4.86	4.72
15	10.80	7.70	6.48	5.80	5.37	5.07	4.85	4.67	4.54
16	10.58	7.51	6.30	5.64	5.21	4.91	4.69	4.52	4.38
17	10.38	7.35	6.16	5.50	5.07	4.78	4.56	4.39	4.25
18	10.22	7.21	6.03	5.37	4.96	4.66	4.44	4.28	4.14
19	10.07	7.09	5.92	5.27	4.85	4.56	4.34	4.18	4.04
20	9.94	6.99	5.82	5.17	4.76	4.47	4.26	4.09	3.96
21	9.83	6.89	5.73	5.09	4.68	4.39	4.18	4.01	3.88
22	9.73	6.81	5.65	5.02	4.61	4.32	4.11	3.94	3.81
23	9.63	6.73	5.58	4.95	4.54	4.26	4.05	3.88	3.75
24	9.55	6.66	5.52	4.89	4.49	4.20	3.99	3.83	3.69
25	9.48	6.60	5.46	4.84	4.43	4.15	3.94	3.78	3.64
26	9.41	6.54	5.41	4.79	4.38	4.10	3.89	3.73	3.60
27	9.34	6.49	5.36	4.74	4.34	4.06	3.85	3.69	3.56
28	9.28	6.44	5.32	4.70	4.30	4.02	3.81	3.65	3.52
29	9.23	6.40	5.28	4.66	4.26	3.98	3.77	3.61	3.48
30	9.18	6.35	5.24	4.62	4.23	3.95	3.74	3.58	3.45
40	8.83	6.07	4.98	4.37	3.99	3.71	3.51	3.35	3.22
60	8.49	5.80	4.73	4.14	3.76	3.49	3.29	3.13	3.01
120	8.18	5.54	4.50	3.92	3.55	3.28	3.09	2.93	2.81
∞	7.88	5.30	4.28	3.72	3.35	3.09	2.90	2.74	2.62

Table V   **941**

Table V (continued)

UPPER .5% POINTS

10	12	15	20	24	30	40	60	120	∞
24224	24426	24630	24836	24940	25044	25148	25253	25359	25465
199	199	199	199	199	199	199	199	199	200
43.69	43.39	43.08	42.78	42.62	42.47	42.31	42.15	41.99	41.83
20.97	20.70	20.44	20.17	20.03	19.89	19.75	19.61	19.47	19.32
13.62	13.38	13.15	12.90	12.78	12.66	12.53	12.40	12.27	12.14
10.25	10.03	9.81	9.59	9.47	9.36	9.24	9.12	9.00	8.88
8.38	8.18	7.97	7.75	7.64	7.53	7.42	7.31	7.19	7.08
7.21	7.01	6.81	6.61	6.50	6.40	6.29	6.18	6.06	5.95
6.42	6.23	6.03	5.83	5.73	5.62	5.52	5.41	5.30	5.19
5.85	5.66	5.47	5.27	5.17	5.07	4.97	4.86	4.75	4.64
5.42	5.24	5.05	4.86	4.76	4.65	4.55	4.44	4.34	4.23
5.09	4.91	4.72	4.53	4.43	4.33	4.23	4.12	4.01	3.90
4.82	4.64	4.46	4.27	4.17	4.07	3.97	3.87	3.76	3.65
4.60	4.43	4.25	4.06	3.96	3.86	3.76	3.66	3.55	3.44
4.42	4.25	4.07	3.88	3.79	3.69	3.58	3.48	3.37	3.26
4.27	4.10	3.92	3.73	3.64	3.54	3.44	3.33	3.22	3.11
4.14	3.97	3.79	3.61	3.51	3.41	3.31	3.21	3.10	2.98
4.03	3.86	3.68	3.50	3.40	3.30	3.20	3.10	2.99	2.87
3.93	3.76	3.59	3.40	3.31	3.21	3.11	3.00	2.89	2.78
3.85	3.68	3.50	3.32	3.22	3.12	3.02	2.92	2.81	2.69
3.77	3.60	3.43	3.24	3.15	3.05	2.95	2.84	2.73	2.61
3.70	3.54	3.36	3.18	3.08	2.98	2.88	2.77	2.66	2.55
3.64	3.47	3.30	3.12	3.02	2.92	2.82	2.71	2.60	2.48
3.59	3.42	3.25	3.06	2.97	2.87	2.77	2.66	2.55	2.43
3.54	3.37	3.20	3.01	2.92	2.82	2.72	2.61	2.50	2.38
3.49	3.33	3.15	2.97	2.87	2.77	2.67	2.56	2.45	2.33
3.45	3.28	3.11	2.93	2.83	2.73	2.63	2.52	2.41	2.29
3.41	3.25	3.07	2.89	2.79	2.69	2.59	2.48	2.37	2.25
3.38	3.21	3.04	2.86	2.76	2.66	2.56	2.45	2.33	2.21
3.34	3.18	3.01	2.82	2.73	2.63	2.52	2.42	2.30	2.18
3.12	2.95	2.78	2.60	2.50	2.40	2.30	2.18	2.06	1.93
2.90	2.74	2.57	2.39	2.29	2.19	2.08	1.96	1.83	1.69
2.71	2.54	2.37	2.19	2.09	1.98	1.87	1.75	1.61	1.43
2.52	2.36	2.19	2.00	1.90	1.79	1.67	1.53	1.36	1.00

Table VI

The transformation of r to Z

r	r (3rd decimal)					r	r (3rd decimal)				
	.000	.002	.004	.006	.008		.000	.002	.004	.006	.008
.00	.0000	.0020	.0040	.0060	.0080	.35	.3654	.3677	.3700	.3723	.3746
1	.0100	.0120	.0140	.0160	.0180	6	.3769	.3792	.3815	.3838	.3861
2	.0200	.0220	.0240	.0260	.0280	7	.3884	.3907	.3931	.3954	.3977
3	.0300	.0320	.0340	.0360	.0380	8	.4001	.4024	.4047	.4071	.4094
4	.0400	.0420	.0440	.0460	.0480	9	.4118	.4142	.4165	.4189	.4213
.05	.0500	.0520	.0541	.0561	.0581	.40	.4236	.4260	.4284	.4308	.4332
6	.0601	.0621	.0641	.0661	.0681	1	.4356	.4380	.4404	.4428	.4453
7	.0701	.0721	.0741	.0761	.0782	2	.4477	.4501	.4526	.4550	.4574
8	.0802	.0822	.0842	.0862	.0882	3	.4599	.4624	.4648	.4673	.4698
9	.0902	.0923	.0943	.0963	.0983	4	.4722	.4747	.4772	.4797	.4822
.10	.1003	.1024	.1044	.1064	.1084	.45	.4847	.4872	.4897	.4922	.4948
1	.1104	.1125	.1145	.1165	.1186	6	.4973	.4999	.5024	.5049	.5075
2	.1206	.1226	.1246	.1267	.1287	7	.5101	.5126	.5152	.5178	.5204
3	.1307	.1328	.1348	.1368	.1389	8	.5230	.5256	.5282	.5308	.5334
4	.1409	.1430	.1450	.1471	.1491	9	.5361	.5387	.5413	.5440	.5466
.15	.1511	.1532	.1552	.1573	.1593	.50	.5493	.5520	.5547	.5573	.5600
6	.1614	.1634	.1655	.1676	.1696	1	.5627	.5654	.5682	.5709	.5736
7	.1717	.1737	.1758	.1779	.1799	2	.5763	.5791	.5818	.5846	.5874
8	.1820	.1841	.1861	.1882	.1903	3	.5901	.5929	.5957	.5985	.6013
9	.1923	.1944	.1965	.1986	.2007	4	.6042	.6070	.6098	.6127	.6155
.20	.2027	.2048	.2069	.2090	.2111	.55	.6184	.6213	.6241	.6270	.6299
1	.2132	.2153	.2174	.2195	.2216	6	.6328	.6358	.6387	.6416	.6446
2	.2237	.2258	.2279	.2300	.2321	7	.6475	.6505	.6535	.6565	.6595
3	.2342	.2363	.2384	.2405	.2427	8	.6625	.6655	.6685	.6716	.6746
4	.2448	.2469	.2490	.2512	.2533	9	.6777	.6807	.6838	.6869	.6900
.25	.2554	.2575	.2597	.2618	.2640	.60	.6931	.6963	.6994	.7026	.7057
6	.2661	.2683	.2704	.2726	.2747	1	.7089	.7121	.7153	.7185	.7218
7	.2769	.2790	.2812	.2833	.2855	2	.7250	.7283	.7315	.7348	.7381
8	.2877	.2899	.2920	.2942	.2964	3	.7414	.7447	.7481	.7514	.7548
9	.2986	.3008	.3029	.3051	.3073	4	.7582	.7616	.7650	.7684	.7718
.30	.3095	.3117	.3139	.3161	.3183	.65	.7753	.7788	.7823	.7858	.7893
1	.3205	.3228	.3250	.3272	.3294	6	.7928	.7964	.7999	.8035	.8071
2	.3316	.3339	.3361	.3383	.3406	7	.8107	.8144	.8180	.8217	.8254
3	.3428	.3451	.3473	.3496	.3518	8	.8291	.8328	.8366	.8404	.8441
4	.3541	.3564	.3586	.3609	.3632	9	.8480	.8518	.8556	.8595	.8634

Note. From Biometrika Tables for Statisticians, Vol. 1, 3rd ed. (Table 14, p. 139) by E. S. Pearson and H. O. Hartley (Eds.) 1966, Cambridge: Cambridge Univ. Press. Copyright 1954 by the Biometrika Trustees. Adapted by permission.

Table VI **943**

TABLE VI (continued)

r	r (3rd decimal)					r	r (3rd decimal)				
	.000	.002	.004	.006	.008		.000	.002	.004	.006	.008
.70	.8673	.8712	.8752	.8792	.8832	.85	1.256	1.263	1.271	1.278	1.286
1	.8872	.8912	.8953	.8994	.9035	6	1.293	1.301	1.309	1.317	1.325
2	.9076	.9118	.9160	.9202	.9245	7	1.333	1.341	1.350	1.358	1.367
3	.9287	.9330	.9373	.9417	.9461	8	1.376	1.385	1.394	1.403	1.412
4	.9505	.9549	.9594	.9639	.9684	9	1.422	1.432	1.442	1.452	1.462
.75	0.973	0.978	0.982	0.987	0.991	.90	1.472	1.483	1.494	1.505	1.516
6	0.996	1.001	1.006	1.011	1.015	1	1.528	1.539	1.551	1.564	1.576
7	1.020	1.025	1.030	1.035	1.040	2	1.589	1.602	1.616	1.630	1.644
8	1.045	1.050	1.056	1.061	1.066	3	1.658	1.673	1.689	1.705	1.721
9	1.071	1.077	1.082	1.088	1.093	4	1.738	1.756	1.774	1.792	1.812
.80	1.099	1.104	1.110	1.116	1.121	.95	1.832	1.853	1.874	1.897	1.921
1	1.127	1.133	1.139	1.145	1.151	6	1.946	1.972	2.000	2.029	2.060
2	1.157	1.163	1.169	1.175	1.182	7	2.092	2.127	2.165	2.205	2.249
3	1.188	1.195	1.201	1.208	1.214	8	2.298	2.351	2.410	2.477	2.555
4	1.221	1.228	1.235	1.242	1.249	9	2.647	2.759	2.903	3.106	3.453

Table VII

Coefficients of orthogonal polynomials

	$J = 3$		$J = 4$			$J = 5$			
	1	2	1	2	3	1	2	3	4
X_1	-1	1	-3	1	-1	-2	2	-1	1
X_2	0	-2	-1	-1	3	-1	-1	2	-4
X_3	1	1	1	-1	-3	0	-2	0	6
X_4			3	1	1	1	-1	-2	-4
X_5						2	2	1	1
Σc_j^2	2	6	20	4	20	10	14	10	70

	$J = 6$					$J = 7$					
	1	2	3	4	5	1	2	3	4	5	6
X_1	-5	5	-5	1	-1	-3	5	-1	3	-1	1
X_2	-3	-1	7	-3	5	-2	0	1	-7	4	-6
X_3	-1	-4	4	2	-10	-1	-3	1	1	-5	15
X_4	1	-4	-4	2	10	0	-4	0	6	0	-20
X_5	3	-1	-7	-3	-5	1	-3	-1	1	5	15
X_6	5	5	5	1	1	2	0	-1	-7	-4	-6
X_7						3	5	1	3	1	1
Σc_j^2	70	84	180	28	252	28	84	6	154	84	924

	$J = 8$						$J = 9$					
	1	2	3	4	5	6	1	2	3	4	5	6
X_1	-7	7	-7	7	-7	1	-4	28	-14	14	-4	4
X_2	-5	1	5	-13	23	-5	-3	7	7	-21	11	-17
X_3	-3	-3	7	-3	-17	9	-2	-8	13	-11	-4	22
X_4	-1	-5	3	9	-15	-5	-1	-17	9	9	-9	1
X_5	1	-5	-3	9	15	-5	0	-20	0	18	0	-20
X_6	3	-3	-7	-3	17	9	1	-17	-9	9	9	1
X_7	5	1	-5	-13	-23	-5	2	-8	-13	-11	4	22
X_8	7	7	7	7	7	1	3	7	-7	-21	-11	-17
X_9							4	28	14	14	4	4
Σc_j^2	168	168	264	616	2184	264	60	2772	990	2002	468	1980

Table VII   **945**

Table VII (continued)

			$J = 10$							$J = 11$			
	1	2	3	4	5	6	1	2	3	4	5	6
X_1	-9	6	-42	18	-6	3	-5	15	-30	6	-3	15
X_2	-7	2	14	-22	14	-11	-4	6	6	-6	6	-48
X_3	-5	-1	35	-17	-1	10	-3	-1	22	-6	1	29
X_4	-3	-3	31	3	-11	6	-2	-6	23	-1	-4	36
X_5	-1	-4	12	18	-6	-8	-1	-9	14	4	-4	-12
X_6	1	-4	-12	18	6	-8	0	-10	0	6	0	-40
X_7	3	-3	-31	3	11	6	1	-9	-14	4	4	-12
X_8	5	-1	-35	-17	1	10	2	-6	-23	-1	4	36
X_9	7	2	-14	-22	-14	-11	3	-1	-22	-6	-1	29
X_{10}	9	6	42	18	6	3	4	6	-6	-6	-6	-48
X_{11}							5	15	30	6	3	15
Σc_j^2	330	132	8580	2860	780	660	110	858	4290	286	156	11220

			$J = 12$							$J = 13$			
	1	2	3	4	5	6	1	2	3	4	5	6
X_1	-11	55	-33	33	-33	11	-6	22	-11	99	-22	22
X_2	-9	25	3	-27	57	-31	-5	11	0	-66	33	-55
X_3	-7	1	21	-33	21	11	-4	2	6	-96	18	8
X_4	-5	-17	25	-13	-29	25	-3	-5	8	-54	-11	43
X_5	-3	-29	19	12	-44	4	-2	-10	7	11	-26	22
X_6	-1	-35	7	28	-20	-20	-1	-13	4	64	-20	-20
X_7	1	-35	-7	28	20	-20	0	-14	0	84	0	-40
X_8	3	-29	-19	12	44	4	1	-13	-4	64	20	-20
X_9	5	-17	-25	-13	29	25	2	-10	-7	11	26	22
X_{10}	7	1	-21	-33	-21	11	3	-5	-8	-54	11	43
X_{11}	9	25	-3	-27	-57	-31	4	2	-6	-96	-18	8
X_{12}	11	55	33	33	33	11	5	11	0	-66	-33	-55
X_{13}							6	22	11	99	22	22
Σc_j^2	572	12012	5148	8008	15912	4488	182	2002	572	68068	6188	14212

Table VII (continued)

			$J = 14$							$J = 15$		
	1	2	3	4	5	6	1	2	3	4	5	6
X_1	-13	13	-143	143	-143	143	-7	91	-91	1001	-1001	143
X_2	-11	7	-11	-77	187	-319	-6	52	-13	-429	1144	-286
X_3	-9	2	66	-132	132	-11	-5	19	35	-869	979	-55
X_4	-7	-2	98	-92	-28	227	-4	-8	58	-704	44	176
X_5	-5	-5	95	-13	-139	185	-3	-29	61	-249	-751	197
X_6	-3	-7	67	63	-145	-25	-2	-44	49	251	-1000	50
X_7	-1	-8	24	108	-60	-200	-1	-53	27	621	-675	-125
X_8	1	-8	-24	108	60	-200	0	-56	0	756	0	-200
X_9	3	-7	-67	63	145	-25	1	-53	-27	621	675	-125
X_{10}	5	-5	-95	-13	139	185	2	-44	-49	251	1000	50
X_{11}	7	-2	-98	-92	28	227	3	-29	-61	-249	751	197
X_{12}	9	2	-66	-132	-132	-11	4	-8	-58	-704	-44	176
X_{13}	11	7	11	-77	-187	-319	5	19	-35	-869	-979	-55
X_{14}	13	13	143	143	143	143	6	52	13	-429	-1144	-286
X_{15}							7	91	91	1001	1001	143
Σc_j^2	910	728	97240	136136	235144	497420	280	37128	39780	6446460	10581480	426360

Table VIII   **947**

Table VIII

Factorials of integers

n	$n!$	n	$n!$
1	1	26	4.03291×10^{26}
2	2	27	1.08889×10^{28}
3	6	28	3.04888×10^{29}
4	24	29	8.84176×10^{30}
5	120	30	2.65253×10^{32}
6	720	31	8.22284×10^{33}
7	5040	32	2.63131×10^{35}
8	40320	33	8.68332×10^{36}
9	362880	34	2.95233×10^{38}
10	3.62880×10^{6}	35	1.03331×10^{40}
11	3.99168×10^{7}	36	3.71993×10^{41}
12	4.79002×10^{8}	37	1.37638×10^{43}
13	6.22702×10^{9}	38	5.23023×10^{44}
14	8.71783×10^{10}	39	2.03979×10^{46}
15	1.30767×10^{12}	40	8.15915×10^{47}
16	2.09228×10^{13}	41	3.34525×10^{49}
17	3.55687×10^{14}	42	1.40501×10^{51}
18	6.40327×10^{15}	43	6.04153×10^{52}
19	1.21645×10^{17}	44	2.65827×10^{54}
20	2.43290×10^{18}	45	1.19622×10^{56}
21	5.10909×10^{19}	46	5.50262×10^{57}
22	1.12400×10^{21}	47	2.58623×10^{59}
23	2.58520×10^{22}	48	1.24139×10^{61}
24	6.20448×10^{23}	49	6.08282×10^{62}
25	1.55112×10^{25}	50	3.04141×10^{64}

Table IX

Binomial coefficients, $\binom{N}{r}$

N \ r	0	1	2	3	4	5	6	7	8	9	10
1	1	1									
2	1	2	1								
3	1	3	3	1							
4	1	4	6	4	1						
5	1	5	10	10	5	1					
6	1	6	15	20	15	6	1				
7	1	7	21	35	35	21	7	1			
8	1	8	28	56	70	56	28	8	1		
9	1	9	36	84	126	126	84	36	9	1	
10	1	10	45	120	210	252	210	120	45	10	1
11	1	11	55	165	330	462	462	330	165	55	11
12	1	12	66	220	495	792	924	792	495	220	66
13	1	13	78	286	715	1287	1716	1716	1287	715	286
14	1	14	91	364	1001	2002	3003	3432	3003	2002	1001
15	1	15	105	455	1365	3003	5005	6435	6435	5005	3003
16	1	16	120	560	1820	4368	8008	11440	12870	11440	8008
17	1	17	136	680	2380	6188	12376	19448	24310	24310	19448
18	1	18	153	816	3060	8568	18564	31824	43758	48620	43758
19	1	19	171	969	3876	11628	27132	50388	75582	92378	92378
20	1	20	190	1140	4845	15504	38760	77520	125970	167960	184756

Table X

Selected values of e^{-m}

	.0	.1	.2	.3	.4	.5	.6	.7	.8	.9
0.	1.00000	.90484	.81873	.74082	.67032	.60653	.54881	.49659	.44933	.40657
1.0	.36788	.33287	.30119	.27253	.24660	.22313	.20190	.18268	.16530	.14957
2.0	.13534	.12246	.11080	.10026	.09072	.08209	.07427	.06721	.06081	.05502
3.0	.04979	.04505	.04076	.03688	.03337	.03020	.02732	.02472	.02237	.02024
4.0	.01832	.01657	.01500	.01357	.01228	.01111	.01005	.00910	.00823	.00745
5.0	.00674	.00610	.00552	.00499	.00452	.00409	.00370	.00335	.00303	.00274
6.0	.00248	.00224	.00203	.00184	.00166	.00150	.00136	.00123	.00111	.00101
7.0	.00091	.00083	.00075	.00068	.00061	.00055	.00050	.00045	.00041	.00037
8.0	.00034	.00030	.00027	.00025	.00022	.00020	.00018	.00017	.00015	.00014
9.0	.00012	.00011	.00010	.00009	.00008	.00007	.00007	.00006	.00006	.00005

For higher or more precise values of m, note that $e^{-a-b} = e^{-a}e^{-b}$, and that $e^{-ab} = (e^{-a})^b$. Hence, within each row, the entry in any column is .9048 times the entry in the preceding column.

Table XI   949

Table XI

Studentized range statistic, q

UPPER 5% POINTS

ν \ k	2	3	4	5	6	7	8	9	10	11	12	13	14	15	16	17	18	19	20
1	18.0	27.0	32.8	37.1	40.4	43.1	45.4	47.4	49.1	50.6	52.0	53.2	54.3	55.4	56.3	57.2	58.0	58.8	59.6
2	6.09	8.3	9.8	10.9	11.7	12.4	13.0	13.5	14.0	14.4	14.7	15.1	15.4	15.7	15.9	16.1	16.4	16.6	16.8
3	4.50	5.91	6.82	7.50	8.04	8.48	8.85	9.18	9.46	9.72	9.95	10.15	10.35	10.52	10.69	10.84	10.98	11.11	11.24
4	3.93	5.04	5.76	6.29	6.71	7.05	7.35	7.60	7.83	8.03	8.21	8.37	8.52	8.66	8.79	8.91	9.03	9.13	9.23
5	3.64	4.60	5.22	5.67	6.03	6.33	6.58	6.80	6.99	7.17	7.32	7.47	7.60	7.72	7.83	7.93	8.03	8.12	8.21
6	3.46	4.34	4.90	5.31	5.63	5.89	6.12	6.32	6.49	6.65	6.79	6.92	7.03	7.14	7.24	7.34	7.43	7.51	7.59
7	3.34	4.16	4.68	5.06	5.36	5.61	5.82	6.00	6.16	6.30	6.43	6.55	6.66	6.76	6.85	6.94	7.02	7.09	7.17
8	3.26	4.04	4.53	4.89	5.17	5.40	5.60	5.77	5.92	6.05	6.18	6.29	6.39	6.48	6.57	6.65	6.73	6.80	6.87
9	3.20	3.95	4.42	4.76	5.02	5.24	5.43	5.60	5.74	5.87	5.98	6.09	6.19	6.28	6.36	6.44	6.51	6.58	6.64
10	3.15	3.88	4.33	4.65	4.91	5.12	5.30	5.46	5.60	5.72	5.83	5.93	6.03	6.11	6.20	6.27	6.34	6.40	6.47
11	3.11	3.82	4.26	4.57	4.82	5.03	5.20	5.35	5.49	5.61	5.71	5.81	5.90	5.99	6.06	6.14	6.20	6.26	6.33
12	3.08	3.77	4.20	4.51	4.75	4.95	5.12	5.27	5.40	5.51	5.62	5.71	5.80	5.88	5.95	6.03	6.09	6.15	6.21
13	3.06	3.73	4.15	4.45	4.69	4.88	5.05	5.19	5.32	5.43	5.53	5.63	5.71	5.79	5.86	5.93	6.00	6.05	6.11
14	3.03	3.70	4.11	4.41	4.64	4.83	4.99	5.13	5.25	5.36	5.46	5.55	5.64	5.72	5.79	5.85	5.92	5.97	6.03
15	3.01	3.67	4.08	4.37	4.60	4.78	4.94	5.08	5.20	5.31	5.40	5.49	5.58	5.65	5.72	5.79	5.85	5.90	5.96
16	3.00	3.65	4.05	4.33	4.56	4.74	4.90	5.03	5.15	5.26	5.35	5.44	5.52	5.59	5.66	5.72	5.79	5.84	5.90
17	2.98	3.63	4.02	4.30	4.52	4.71	4.86	4.99	5.11	5.21	5.31	5.39	5.47	5.55	5.61	5.68	5.74	5.79	5.84
18	2.97	3.61	4.00	4.28	4.49	4.67	4.82	4.96	5.07	5.17	5.27	5.35	5.43	5.50	5.57	5.63	5.69	5.74	5.79
19	2.96	3.59	3.98	4.25	4.47	4.65	4.79	4.92	5.04	5.14	5.23	5.32	5.39	5.46	5.53	5.59	5.65	5.70	5.75
20	2.95	3.58	3.96	4.23	4.45	4.62	4.77	4.90	5.01	5.11	5.20	5.28	5.36	5.43	5.49	5.55	5.61	5.66	5.71
24	2.92	3.53	3.90	4.17	4.37	4.54	4.68	4.81	4.92	5.01	5.10	5.18	5.25	5.32	5.38	5.44	5.50	5.54	5.59
30	2.89	3.49	3.84	4.10	4.30	4.46	4.60	4.72	4.83	4.92	5.00	5.08	5.15	5.21	5.27	5.33	5.38	5.43	5.48
40	2.86	3.44	3.79	4.04	4.23	4.39	4.52	4.63	4.74	4.82	4.91	4.98	5.05	5.11	5.16	5.22	5.27	5.31	5.36
60	2.83	3.40	3.74	3.98	4.16	4.31	4.44	4.55	4.65	4.73	4.81	4.88	4.94	5.00	5.06	5.11	5.16	5.20	5.24
120	2.80	3.36	3.69	3.92	4.10	4.24	4.36	4.48	4.56	4.64	4.72	4.78	4.84	4.90	4.95	5.00	5.05	5.09	5.13
∞	2.77	3.31	3.63	3.86	4.03	4.17	4.29	4.39	4.47	4.55	4.62	4.68	4.74	4.80	4.85	4.89	4.93	4.97	5.01

Note. From *Biometrika Tables for Statisticians*, Vol. 1, 3rd ed. (Table 29, pp. 176–177) by E. S. Pearson and H. O. Hartley (Eds.) 1966, Cambridge: Cambridge Univ. Press. Copyright 1954 by the Biometrika Trustees. Adapted by permission.

Table XI (continued)

UPPER 1% POINTS

ν \ k	2	3	4	5	6	7	8	9	10	11	12	13	14	15	16	17	18	19	20
1	90.0	135	164	186	202	216	227	237	246	253	260	266	272	277	282	286	290	294	298
2	14.0	19.0	22.3	24.7	26.6	28.2	29.5	30.7	31.7	32.6	33.4	34.1	34.8	35.4	36.0	36.5	37.0	37.5	37.9
3	8.26	10.6	12.2	13.3	14.2	15.0	15.6	16.2	16.7	17.1	17.5	17.9	18.2	18.5	18.8	19.1	19.3	19.5	19.8
4	6.51	8.12	9.17	9.96	10.6	11.1	11.5	11.9	12.3	12.6	12.8	13.1	13.3	13.5	13.7	13.9	14.1	14.2	14.4
5	5.70	6.97	7.80	8.42	8.91	9.32	9.67	9.97	10.24	10.48	10.70	10.89	11.08	11.24	11.40	11.55	11.68	11.81	11.93
6	5.24	6.33	7.03	7.56	7.97	8.32	8.61	8.87	9.10	9.30	9.49	9.65	9.81	9.95	10.08	10.21	10.32	10.43	10.54
7	4.95	5.92	6.54	7.01	7.37	7.68	7.94	8.17	8.37	8.55	8.71	8.86	9.00	9.12	9.24	9.35	9.46	9.55	9.65
8	4.74	5.63	6.20	6.63	6.96	7.24	7.47	7.68	7.87	8.03	8.18	8.31	8.44	8.55	8.66	8.76	8.85	8.94	9.03
9	4.60	5.43	5.96	6.35	6.66	6.91	7.13	7.32	7.49	7.65	7.78	7.91	8.03	8.13	8.23	8.32	8.41	8.49	8.57
10	4.48	5.27	5.77	6.14	6.43	6.67	6.87	7.05	7.21	7.36	7.48	7.60	7.71	7.81	7.91	7.99	8.07	8.15	8.22
11	4.39	5.14	5.62	5.97	6.25	6.48	6.67	6.84	6.99	7.13	7.25	7.36	7.46	7.56	7.65	7.73	7.81	7.88	7.95
12	4.32	5.04	5.50	5.84	6.10	6.32	6.51	6.67	6.81	6.94	7.06	7.17	7.26	7.36	7.44	7.52	7.59	7.66	7.73
13	4.26	4.96	5.40	5.73	5.98	6.19	6.37	6.53	6.67	6.79	6.90	7.01	7.10	7.19	7.27	7.34	7.42	7.48	7.55
14	4.21	4.89	5.32	5.63	5.88	6.08	6.26	6.41	6.54	6.66	6.77	6.87	6.96	7.05	7.12	7.20	7.27	7.33	7.39
15	4.17	4.83	5.25	5.56	5.80	5.99	6.16	6.31	6.44	6.55	6.66	6.76	6.84	6.93	7.00	7.07	7.14	7.20	7.26
16	4.13	4.78	5.19	5.49	5.72	5.92	6.08	6.22	6.35	6.46	6.56	6.66	6.74	6.82	6.90	6.97	7.03	7.09	7.15
17	4.10	4.74	5.14	5.43	5.66	5.85	6.01	6.15	6.27	6.38	6.48	6.57	6.66	6.73	6.80	6.87	6.94	7.00	7.05
18	4.07	4.70	5.09	5.38	5.60	5.79	5.94	6.08	6.20	6.31	6.41	6.50	6.58	6.65	6.72	6.79	6.85	6.91	6.96
19	4.05	4.67	5.05	5.33	5.55	5.73	5.89	6.02	6.14	6.25	6.34	6.43	6.51	6.58	6.65	6.72	6.78	6.84	6.89
20	4.02	4.64	5.02	5.29	5.51	5.69	5.84	5.97	6.09	6.19	6.29	6.37	6.45	6.52	6.59	6.65	6.71	6.76	6.82
24	3.96	4.54	4.91	5.17	5.37	5.54	5.69	5.81	5.92	6.02	6.11	6.19	6.26	6.33	6.39	6.45	6.51	6.56	6.61
30	3.89	4.45	4.80	5.05	5.24	5.40	5.54	5.65	5.76	5.85	5.93	6.01	6.08	6.14	6.20	6.26	6.31	6.36	6.41
40	3.82	4.37	4.70	4.93	5.11	5.27	5.39	5.50	5.60	5.69	5.77	5.84	5.90	5.96	6.02	6.07	6.12	6.17	6.21
60	3.76	4.28	4.60	4.82	4.99	5.13	5.25	5.36	5.45	5.53	5.60	5.67	5.73	5.79	5.84	5.89	5.93	5.98	6.02
120	3.70	4.20	4.50	4.71	4.87	5.01	5.12	5.21	5.30	5.38	5.44	5.51	5.56	5.61	5.66	5.71	5.75	5.79	5.83
∞	3.64	4.12	4.40	4.60	4.76	4.88	4.99	5.08	5.16	5.23	5.29	5.35	5.40	5.45	5.49	5.54	5.57	5.61	5.65

Table XII   951

Table XII

Approximate power of the ANOVA *F* test, for different true values of omega-square, for different numbers of groups (*J* = 2 to 8) and cases within groups (*n* = 2 to 15 and 20, 30).

$\omega^2 =$	0	.1	.2	.3	.4	.5	.6	.7	.8	.9
					J = 2 Groups					
n = 2	.05	.07	.09	.13	.17	.22	.29	.39	.56	.84
	.01	.01	.02	.03	.04	.05	.07	.10	.16	.30
n = 3	.05	.09	.15	.23	.33	.46	.62	.80	.95	.99*
	.01	.02	.04	.06	.10	.15	.24	.39	.63	.94
n = 4	.05	.12	.21	.34	.49	.66	.83	.96	.99	.99
	.01	.03	.06	.11	.19	.31	.48	.71	.93	.99
n = 5	.05	.15	.28	.44	.62	.80	.93	.99	.99	.99
	.01	.04	.09	.17	.30	.47	.69	.89	.99	.99
n = 6	.05	.17	.34	.53	.73	.89	.98	.99	.99	.99
	.01	.05	.12	.24	.41	.62	.83	.97	.99	.99
n = 7	.05	.20	.39	.61	.81	.94	.99	.99	.99	.99
	.01	.06	.16	.31	.52	.74	.92	.99	.99	.99
n = 8	.05	.22	.45	.69	.87	.97	.99	.99	.99	.99
	.01	.08	.20	.39	.62	.83	.97	.99	.99	.99
n = 9	.05	.25	.51	.75	.92	.99	.99	.99	.99	.99
	.01	.09	.24	.46	.70	.90	.99	.99	.99	.99
n = 10	.05	.28	.56	.80	.95	.99	.99	.99	.99	.99
	.01	.10	.28	.52	.77	.94	.99	.99	.99	.99
n = 11	.05	.30	.61	.85	.97	.99	.99	.99	.99	.99
	.01	.12	.32	.59	.83	.97	.99	.99	.99	.99
n = 12	.05	.33	.65	.88	.98	.99	.99	.99	.99	.99
	.01	.13	.36	.65	.88	.98	.99	.99	.99	.99
n = 13	.05	.36	.69	.91	.99	.99	.99	.99	.99	.99
	.01	.15	.41	.70	.91	.99	.99	.99	.99	.99
n = 14	.05	.38	.73	.93	.99	.99	.99	.99	.99	.99
	.01	.17	.45	.75	.94	.99	.99	.99	.99	.99
n = 15	.05	.41	.76	.95	.99	.99	.99	.99	.99	.99
	.01	.18	.49	.79	.96	.99	.99	.99	.99	.99
n = 20	.05	.53	.89	.99	.99	.99	.99	.99	.99	.99
	.01	.27	.67	.93	.99	.99	.99	.99	.99	.99
n = 30	.05	.73	.98	.99	.99	.99	.99	.99	.99	.99
	.01	.46	.90	.99	.99	.99	.99	.99	.99	.99

*Note. Each value ".99" is a lower bound to the actual power.

Table XII (continued)

$\omega^2 =$	0	.1	.2	.3	.4	.5	.6	.7	.8	.9
					J = 3 Groups					
n = 2	.05	.07	.10	.14	.18	.25	.35	.49	.70	.95
	.01	.02	.02	.03	.04	.06	.09	.15	.26	.55
n = 3	.05	.10	.16	.25	.38	.53	.72	.89	.99	.99
	.01	.02	.04	.08	.13	.21	.35	.56	.84	.99
n = 4	.05	.13	.23	.38	.56	.75	.91	.99	.99	.99
	.01	.03	.07	.14	.25	.42	.64	.87	.99	.99
n = 5	.05	.15	.31	.50	.70	.88	.98	.99	.99	.99
	.01	.04	.11	.22	.39	.61	.84	.97	.99	.99
n = 6	.05	.18	.38	.60	.81	.95	.99	.99	.99	.99
	.01	.06	.15	.31	.53	.77	.94	.99	.99	.99
n = 7	.05	.22	.45	.70	.89	.98	.99	.99	.99	.99
	.01	.07	.20	.40	.66	.87	.98	.99	.99	.99
n = 8	.05	.25	.51	.77	.94	.99	.99	.99	.99	.99
	.01	.09	.25	.49	.76	.94	.99	.99	.99	.99
n = 9	.05	.28	.58	.83	.97	.99	.99	.99	.99	.99
	.01	.10	.30	.58	.84	.97	.99	.99	.99	.99
n = 10	.05	.31	.63	.88	.98	.99	.99	.99	.99	.99
	.01	.12	.36	.66	.89	.99	.99	.99	.99	.99
n = 11	.05	.34	.69	.92	.99	.99	.99	.99	.99	.99
	.01	.14	.41	.73	.93	.99	.99	.99	.99	.99
n = 12	.05	.37	.73	.94	.99	.99	.99	.99	.99	.99
	.01	.16	.46	.78	.96	.99	.99	.99	.99	.99
n = 13	.05	.40	.77	.96	.99	.99	.99	.99	.99	.99
	.01	.18	.51	.83	.98	.99	.99	.99	.99	.99
n = 14	.05	.44	.81	.97	.99	.99	.99	.99	.99	.99
	.01	.20	.57	.87	.99	.99	.99	.99	.99	.99
n = 15	.05	.47	.85	.98	.99	.99	.99	.99	.99	.99
	.01	.22	.61	.91	.99	.99	.99	.99	.99	.99
n = 20	.05	.60	.95	.99	.99	.99	.99	.99	.99	.99
	.01	.34	.80	.98	.99	.99	.99	.99	.99	.99
n = 30	.05	.81	.99	.99	.99	.99	.99	.99	.99	.99
	.01	.56	.96	.99	.99	.99	.99	.99	.99	.99

Table XII   953

Table XII (continued)

$\omega^2 =$		0	.1	.2	.3	.4	.5	.6	.7	.8	.9
					$J = 4$ Groups						
$n = 2$	.05	.07	.11	.15	.21	.30	.42	.60	.82	.99	
	.01	.02	.02	.04	.05	.08	.13	.22	.40	.78	
$n = 3$	.05	.10	.18	.29	.44	.61	.80	.95	.99	.99	
	.01	.02	.05	.09	.17	.28	.47	.72	.94	.99	
$n = 4$	.05	.13	.26	.43	.64	.83	.96	.99	.99	.99	
	.01	.04	.09	.18	.33	.53	.78	.95	.99	.99	
$n = 5$	.05	.17	.35	.57	.78	.94	.99	.99	.99	.99	
	.01	.05	.13	.28	.50	.74	.93	.99	.99	.99	
$n = 6$	.05	.20	.43	.68	.88	.98	.99	.99	.99	.99	
	.01	.07	.19	.39	.65	.88	.98	.99	.99	.99	
$n = 7$	.05	.24	.51	.77	.94	.99	.99	.99	.99	.99	
	.01	.08	.25	.50	.77	.95	.99	.99	.99	.99	
$n = 8$	.05	.27	.58	.85	.97	.99	.99	.99	.99	.99	
	.01	.10	.31	.61	.86	.98	.99	.99	.99	.99	
$n = 9$	.05	.31	.65	.90	.99	.99	.99	.99	.99	.99	
	.01	.12	.37	.69	.92	.99	.99	.99	.99	.99	
$n = 10$	.05	.35	.71	.93	.99	.99	.99	.99	.99	.99	
	.01	.15	.45	.78	.96	.99	.99	.99	.99	.99	
$n = 11$	.05	.39	.76	.96	.99	.99	.99	.99	.99	.99	
	.01	.17	.51	.83	.98	.99	.99	.99	.99	.99	
$n = 12$	.05	.43	.81	.98	.99	.99	.99	.99	.99	.99	
	.01	.19	.56	.88	.99	.99	.99	.99	.99	.99	
$n = 13$	.05	.46	.85	.99	.99	.99	.99	.99	.99	.99	
	.01	.22	.62	.92	.99	.99	.99	.99	.99	.99	
$n = 14$	.05	.49	.88	.99	.99	.99	.99	.99	.99	.99	
	.01	.24	.67	.94	.99	.99	.99	.99	.99	.99	
$n = 15$	.05	.53	.91	.99	.99	.99	.99	.99	.99	.99	
	.01	.27	.72	.96	.99	.99	.99	.99	.99	.99	
$n = 20$	.05	.68	.98	.99	.99	.99	.99	.99	.99	.99	
	.01	.42	.89	.99	.99	.99	.99	.99	.99	.99	
$n = 30$	.05	.88	.99	.99	.99	.99	.99	.99	.99	.99	
	.01	.68	.99	.99	.99	.99	.99	.99	.99	.99	

Table XII (continued)

$J = 5$ Groups

$\omega^2 =$	0	.1	.2	.3	.4	.5	.6	.7	.8	.9
$n = 2$	.05	.08	.11	.17	.24	.34	.49	.69	.90	.99
	.01	.02	.03	.04	.06	.10	.17	.30	.54	.92
$n = 3$	.05	.11	.20	.33	.49	.69	.87	.98	.99	.99
	.01	.03	.06	.11	.21	.36	.58	.83	.98	.99
$n = 4$	.05	.14	.29	.49	.71	.89	.98	.99	.99	.99
	.01	.04	.10	.22	.40	.64	.87	.99	.99	.99
$n = 5$	.05	.18	.39	.63	.85	.97	.99	.99	.99	.99
	.01	.06	.16	.35	.59	.84	.97	.99	.99	.99
$n = 6$	.05	.22	.48	.75	.93	.99	.99	.99	.99	.99
	.01	.07	.23	.47	.75	.94	.99	.99	.99	.99
$n = 7$	.05	.26	.57	.84	.97	.99	.99	.99	.99	.99
	.01	.10	.30	.60	.86	.98	.99	.99	.99	.99
$n = 8$	.05	.30	.65	.90	.99	.99	.99	.99	.99	.99
	.01	.12	.37	.70	.93	.99	.99	.99	.99	.99
$n = 9$	.05	.35	.72	.94	.99	.99	.99	.99	.99	.99
	.01	.15	.45	.79	.97	.99	.99	.99	.99	.99
$n = 10$	.05	.39	.78	.97	.99	.99	.99	.99	.99	.99
	.01	.17	.52	.85	.98	.99	.99	.99	.99	.99
$n = 11$	.05	.43	.82	.98	.99	.99	.99	.99	.99	.99
	.01	.20	.59	.90	.99	.99	.99	.99	.99	.99
$n = 12$	.05	.47	.87	.99	.99	.99	.99	.99	.99	.99
	.01	.23	.66	.94	.99	.99	.99	.99	.99	.99
$n = 13$	.05	.51	.90	.99	.99	.99	.99	.99	.99	.99
	.01	.27	.72	.96	.99	.99	.99	.99	.99	.99
$n = 14$	.05	.55	.93	.99	.99	.99	.99	.99	.99	.99
	.01	.29	.77	.98	.99	.99	.99	.99	.99	.99
$n = 15$	.05	.59	.95	.99	.99	.99	.99	.99	.99	.99
	.01	.33	.81	.99	.99	.99	.99	.99	.99	.99
$n = 20$	.05	.74	.99	.99	.99	.99	.99	.99	.99	.99
	.01	.49	.95	.99	.99	.99	.99	.99	.99	.99
$n = 30$	.05	.92	.99	.99	.99	.99	.99	.99	.99	.99
	.01	.77	.99	.99	.99	.99	.99	.99	.99	.99

Table XII   955

Table XII (continued)

				$J = 6$ Groups						
$\omega^2 =$	0	.1	.2	.3	.4	.5	.6	.7	.8	.9
$n = 2$	.05	.08	.12	.18	.27	.39	.56	.77	.95	.99
	.01	.02	.03	.05	.08	.13	.22	.38	.67	.97
$n = 3$	.05	.11	.22	.36	.55	.75	.92	.99	.99	.99
	.01	.03	.07	.14	.25	.44	.68	.91	.99	.99
$n = 4$	.05	.15	.32	.54	.77	.93	.99	.99	.99	.99
	.01	.04	.12	.26	.48	.73	.93	.99	.99	.99
$n = 5$	.05	.20	.43	.69	.90	.98	.99	.99	.99	.99
	.01	.06	.19	.41	.68	.90	.99	.99	.99	.99
$n = 6$	.05	.24	.53	.81	.96	.99	.99	.99	.99	.99
	.01	.08	.27	.55	.83	.97	.99	.99	.99	.99
$n = 7$	.05	.29	.62	.88	.99	.99	.99	.99	.99	.99
	.01	.11	.35	.68	.92	.99	.99	.99	.99	.99
$n = 8$	.05	.33	.70	.94	.99	.99	.99	.99	.99	.99
	.01	.14	.44	.78	.96	.99	.99	.99	.99	.99
$n = 9$	.05	.38	.77	.97	.99	.99	.99	.99	.99	.99
	.01	.17	.52	.86	.99	.99	.99	.99	.99	.99
$n = 10$	.05	.43	.83	.98	.99	.99	.99	.99	.99	.99
	.01	.20	.60	.91	.99	.99	.99	.99	.99	.99
$n = 11$	.05	.47	.88	.99	.99	.99	.99	.99	.99	.99
	.01	.24	.67	.95	.99	.99	.99	.99	.99	.99
$n = 12$	.05	.52	.91	.99	.99	.99	.99	.99	.99	.99
	.01	.27	.74	.97	.99	.99	.99	.99	.99	.99
$n = 13$	.05	.56	.94	.99	.99	.99	.99	.99	.99	.99
	.01	.30	.79	.98	.99	.99	.99	.99	.99	.99
$n = 14$	.05	.60	.95	.99	.99	.99	.99	.99	.99	.99
	.01	.34	.84	.99	.99	.99	.99	.99	.99	.99
$n = 15$	.05	.64	.97	.99	.99	.99	.99	.99	.99	.99
	.01	.38	.87	.99	.99	.99	.99	.99	.99	.99
$n = 20$	.05	.80	.99	.99	.99	.99	.99	.99	.99	.99
	.01	.57	.97	.99	.99	.99	.99	.99	.99	.99
$n = 30$	.05	.96	.99	.99	.99	.99	.99	.99	.99	.99
	.01	.84	.99	.99	.99	.99	.99	.99	.99	.99

Table XII (continued)

$\omega^2 =$		0	.1	.2	.3	.4	.5	.6	.7	.8	.9
						J = 7 Groups					
$n = 2$	.05	.05	.08	.13	.20	.30	.44	.62	.83	.97	.99
	.01		.02	.03	.05	.09	.16	.27	.47	.78	.99
$n = 3$	.05	.05	.12	.24	.40	.60	.81	.95	.99	.99	.99
	.01		.03	.07	.16	.30	.51	.76	.95	.99	.99
$n = 4$	.05	.05	.16	.35	.59	.82	.96	.99	.99	.99	.99
	.01		.05	.14	.31	.55	.81	.96	.99	.99	.99
$n = 5$	.05	.05	.21	.47	.74	.93	.99	.99	.99	.99	.99
	.01		.07	.22	.47	.75	.94	.99	.99	.99	.99
$n = 6$	.05	.05	.25	.57	.85	.98	.99	.99	.99	.99	.99
	.01		.09	.31	.62	.88	.99	.99	.99	.99	.99
$n = 7$	.05	.05	.31	.67	.92	.99	.99	.99	.99	.99	.99
	.01		.12	.40	.75	.95	.99	.99	.99	.99	.99
$n = 8$	.05	.05	.36	.75	.96	.99	.99	.99	.99	.99	.99
	.01		.15	.50	.84	.98	.99	.99	.99	.99	.99
$n = 9$	.05	.05	.41	.82	.98	.99	.99	.99	.99	.99	.99
	.01		.19	.59	.91	.99	.99	.99	.99	.99	.99
$n = 10$	.05	.05	.47	.87	.99	.99	.99	.99	.99	.99	.99
	.01		.23	.67	.95	.99	.99	.99	.99	.99	.99
$n = 11$	.05	.05	.51	.91	.99	.99	.99	.99	.99	.99	.99
	.01		.26	.74	.97	.99	.99	.99	.99	.99	.99
$n = 12$	.05	.05	.56	.94	.99	.99	.99	.99	.99	.99	.99
	.01		.30	.80	.99	.99	.99	.99	.99	.99	.99
$n = 13$	.05	.05	.60	.96	.99	.99	.99	.99	.99	.99	.99
	.01		.34	.85	.99	.99	.99	.99	.99	.99	.99
$n = 14$	.05	.05	.65	.97	.99	.99	.99	.99	.99	.99	.99
	.01		.39	.89	.99	.99	.99	.99	.99	.99	.99
$n = 15$	.05	.05	.69	.98	.99	.99	.99	.99	.99	.99	.99
	.01		.43	.92	.99	.99	.99	.99	.99	.99	.99
$n = 20$	.05	.05	.85	.99	.99	.99	.99	.99	.99	.99	.99
	.01		.64	.99	.99	.99	.99	.99	.99	.99	.99
$n = 30$	.05	.05	.95	.99	.99	.99	.99	.99	.99	.99	.99
	.01		.89	.99	.99	.99	.99	.99	.99	.99	.99

Table XII   957

Table XII (continued)

				$J = 8$ Groups						
$\omega^2 =$	0	.1	.2	.3	.4	.5	.6	.7	.8	.9
$n = 2$	.05	.08	.14	.22	.33	.48	.68	.88	.99	.99
	.01	.02	.03	.06	.11	.18	.32	.55	.86	.99
$n = 3$	.05	.13	.25	.43	.65	.85	.97	.99	.99	.99
	.01	.03	.08	.18	.34	.58	.83	.98	.99	.99
$n = 4$	.05	.17	.38	.64	.86	.97	.99	.99	.99	.99
	.01	.05	.16	.35	.61	.86	.98	.99	.99	.99
$n = 5$	.05	.23	.51	.79	.95	.99	.99	.99	.99	.99
	.01	.08	.25	.53	.81	.97	.99	.99	.99	.99
$n = 6$	.05	.28	.62	.89	.99	.99	.99	.99	.99	.99
	.01	.11	.35	.68	.93	.99	.99	.99	.99	.99
$n = 7$	.05	.33	.71	.94	.99	.99	.99	.99	.99	.99
	.01	.14	.45	.80	.97	.99	.99	.99	.99	.99
$n = 8$	.05	.39	.79	.97	.99	.99	.99	.99	.99	.99
	.01	.17	.55	.89	.99	.99	.99	.99	.99	.99
$n = 9$	.05	.45	.86	.99	.99	.99	.99	.99	.99	.99
	.01	.21	.65	.94	.99	.99	.99	.99	.99	.99
$n = 10$	.05	.50	.90	.99	.99	.99	.99	.99	.99	.99
	.01	.25	.72	.97	.99	.99	.99	.99	.99	.99
$n = 11$	.05	.55	.94	.99	.99	.99	.99	.99	.99	.99
	.01	.29	.79	.99	.99	.99	.99	.99	.99	.99
$n = 12$	.05	.60	.96	.99	.99	.99	.99	.99	.99	.99
	.01	.34	.85	.99	.99	.99	.99	.99	.99	.99
$n = 13$	.05	.65	.97	.99	.99	.99	.99	.99	.99	.99
	.01	.39	.89	.99	.99	.99	.99	.99	.99	.99
$n = 14$	.05	.69	.98	.99	.99	.99	.99	.99	.99	.99
	.01	.44	.93	.99	.99	.99	.99	.99	.99	.99
$n = 15$	.05	.73	.99	.99	.99	.99	.99	.99	.99	.99
	.01	.49	.95	.99	.99	.99	.99	.99	.99	.99
$n = 20$	.05	.88	.99	.99	.99	.99	.99	.99	.99	.99
	.01	.71	.99	.99	.99	.99	.99	.99	.99	.99
$n = 30$	.05	.98	.99	.99	.99	.99	.99	.99	.99	.99
	.01	.93	.99	.99	.99	.99	.99	.99	.99	.99

Table XIII

Critical values of *D* in the Kolmogorov-Smirnov one-sample test

n	F(D)				
	.80	.85	.90	.95	.99
1	.900	.925	.950	.975	.995
2	.684	.726	.776	.842	.929
3	.565	.597	.642	.708	.828
4	.494	.525	.564	.624	.733
5	.446	.474	.510	.565	.669
6	.410	.436	.470	.521	.618
7	.381	.405	.438	.486	.577
8	.358	.381	.411	.457	.543
9	.339	.360	.388	.432	.514
10	.322	.342	.368	.410	.490
11	.307	.326	.352	.391	.468
12	.295	.313	.338	.375	.450
13	.284	.302	.325	.361	.433
14	.274	.292	.314	.349	.418
15	.266	.283	.304	.338	.404
16	.258	.274	.295	.328	.392
17	.250	.266	.280	.318	.381
18	.244	.259	.278	.309	.371
19	.237	.252	.272	.301	.363
20	.231	.246	.264	.294	.356
25	.21	.22	.24	.27	.32
30	.19	.20	.22	.24	.29
35	.18	.19	.21	.23	.27
Over 35	$1.07/\sqrt{n}$	$1.14/\sqrt{n}$	$1.22/\sqrt{n}$	$1.36/\sqrt{n}$	$1.63/\sqrt{n}$

Note. From "The Kolmogorov-Smirnov Test for Goodness of Fit," by F. J. Amssey, 1951, *Journal of the American Statistical Association, 46,* 70. Copyright 1951 by F. J. Amssey and the American Statistical Association. Adapted by permission.

Solutions to Selected Exercises

CHAPTER 1

1. *(a)* A particular 7-digit number appearing in that directory
  (b) A particular student currently enrolled in a college or university in the United States
  (c) A positive real number (including fractional numbers, of course)
  (d) Any current U.S. senator
  (e) Any whole positive number, including zero
  (f) Any positive or negative real number
  (g) Any positive real number between 0 and 100

3. *(a)* $A \cup B$, "A or B or both"
  (b) $A \cap \overline{B}$ "A and not B"
  (c) $\overline{(A \cap B)}$, "not both A and B"
  (d) $(A \cap \overline{B}) \cup (\overline{A} \cap B)$, "A and not B, or B and not A"
  (e) $(A \cap \overline{B} \cap \overline{C}) \cup (\overline{A} \cap B \cap \overline{C}) \cup (\overline{A} \cap \overline{B} \cap C)$, "only A, or only B, or only C"
  (f) $(A \cap B \cap \overline{C}) \cup (A \cap \overline{B} \cap C) \cup (\overline{A} \cap B \cap C)$, "only A and B, or only A and C or only B and C"
  (g) $\overline{(A \cup B)}$, "neither A nor B" (notice that this may or may not include C)

5. *(a)* $p(\mathcal{S}) = p(A) + p(B) + p(C) = (1/4) + (3/5) + (3/20) = 1.00$
  (b) $p(\varnothing) = 0$
  (c) $p(A \cup B) = p(A) + p(B) = (1/4) + (3/5) = 17/20$
  (d) $p(B \cup C) = p(B) + p(C) = (3/5) + (3/20) = 3/4$
  (e) $p(A \cap B) = p(\varnothing) = 0$
  (f) $p(\overline{B}) = 1 - (3/5) = 2/5$
  (g) $p(A \cap \overline{B}) = p(A) = 1/4$
  (h) $p(C \cap \overline{B}) = p(C) = 3/20$
  (i) $p(C \cup \overline{B}) = p(C) + p(A) = 2/5$

7. *(a)* $p(\text{ace}) = 1/13$
  (b) $p(\text{red}) = 1/2$
  (c) $p(\text{face card}) = 3/13$
  (d) $p(\text{even value}) = 5/13$
  (e) $p(\text{spade or diamond}) = (1/4) + (1/4) = 1/2$
  (f) $p(10 \text{ in red suit}) = 1/26$

9. *(a)* $p(\text{female}) = 2/3$
  (b) $p(\text{brunette}) = 1/3$
  (c) $p(\text{red-haired}) = 2/9$
  (d) $p(\text{red-haired male}) = 1/9$
  (e) $p(68 \text{ inches or more}) = 4/9$
  (f) $p(\text{blonde female under 65 inches}) = 2/9$

11. The 36 elementary events are shown as points in the following table:

White die

		1	2	3	4	5	6
black	1	.	.	.	.	.	.
die	2	.	.	.	.	.	.
	3	.	.	.	.	.	.
	4	.	.	.	.	.	.
	5	.	.	.	.	.	.
	6	.	.	.	.	.	.

(a) 6   *(e)* 9
(b) 15   *(f)* 27
(c) 6   *(g)* 22
(d) 30

13. *(a)* $p(\text{vowel}) = 3/13$
(b) $p(\text{consonant other than } s \text{ or } t) = 9/13$
(c) $p(\text{consonant, first half}) = 5/13$
(d) $p(\text{letter in } exodus) = 3/13$
(e) $p(\text{letter in } born \text{ or in } film) = (4/26) + (4/26) = 4/13$
(f) $p(\text{letter in } lead \text{ or in } load) = (4/26) + (4/26) - (3/26) = 5/26$

15. *(a)* $p(3) = 1/12$
(b) $p(\text{even red}) = 1/6$
(c) $p(5 \text{ or more blue}) = 2/9$
(d) $p(\text{red or } 8) = (12/36) + (3/36) - (1/36) = 7/18$
(e) $p(\text{white and odd}) = 1/6$
(f) $p(\text{white or odd}) = (12/36) + (18/36) - (6/36) = 2/3$

17. *(a)* $p(\text{less than } 19) = p(\text{boy})p(\text{less than } 19, \text{ given boy})$
$$+ \ p(\text{girl})p(\text{less than } 19, \text{ given girl}) = .75$$
(b) If independent, $p(19 \text{ or over, given boy}) = p(19 \text{ or over})$. However, this is not true here, so that these two events are not independent in this sample space.
(c) By Bayes's theorem, $p(\text{boy}|19 \text{ or over}) = \dfrac{.37 \times .52}{(.37 \times .52) + (.12 \times .48)}$
$$= .77$$
(d) $p(\text{girl}|\text{under } 19) = .56$

19. *(a)* marginal probabilities: $p(A_1) = .33, p(A_2) = .34, p(A_3) = .33$
(b) $p(0) = .2, p(1) = .35, p(2) = .45$
(c) $p(0|A_1) = .15$
(d) $p(2 \text{ or more}|A_2) = .44$
(e) $p(A_3|0) = .65$

21. *(a)* $p(\text{black or white}) = 4/9$
(b) $p(\text{yellow}) = 2/9$
(c) $p(\text{neither white nor yellow}) = 5/9$
(d) $p(\text{green or blue}) = 2/9$
(e) $p(\text{blue or not black}) = 7/9$

23. *(a)* $p(\text{one culture}) = 11/21$
(b) $p(\text{two cultures}) = 1/3$
(c) $p(\text{three cultures}) = 1/7$
(d) $p(\text{more than } 1) = 1 - (11/21) = 10/21$
(e) $p(\text{less than three}) = 1 - (1/7) = 6/7$

25. *(a)* $p(\text{productive}|A) = 5/7$

(b) $p(\text{productive}|C) = 2/3$

(c) $p(\text{productive}|\overline{A} \cap B \cap \overline{C}) = 1/3$

(d) $p(\text{productive}|A \cap \overline{B} \cap \overline{C}) = 2/3$

(e) $p(\text{productive}|A \cap B \cap C) = 2/3$

27. *(a)* $p = 5/6$ *(c)* $p = .5$

(b) $p = .3$

29. *(a)* .87   *(d)* .38

(b) .03   *(e)* .62

(c) .78

31. *(a)* .36   *(e)* .06

(b) .46   *(f)* .81

(c) .27   *(g)* .76

(d) .21   *(h)* .93

33. *(a)* .448   *(e)* .670

(b) .521   *(f)* .349

(c) .631   *(g)* .050

(d) .137   *(h)* .287

35. *(a)* 187:813, or about 23:100

(b) 264:736, or about 9:25

(c) 294:706, or about 21:50

(d) 255:745, or about 17:50

(e) 558:442, or about 63:50

37. *(a)* .078   *(e)* .003

(b) .065   *(f)* .019

(c) .035   *(g)* .542

(d) .063

CHAPTER 2

1. *(a)* Nominal

(b) Ratio

(c) Ordinal (often treated as interval)

(d) Ordinal

(e) Interval

(f) Ratio

(g) Interval

(h) Interval

(i) Ordinal

(j) Nominal

3.

Rating	f
very good	2
good	3
acceptable	7
poor	4
	16 = N

5.

Item I	
Response	f
5	3
4	5
3	7
2	9
1	6
	30 = N

Item II	
Response	f
5	2
4	7
3	12
2	7
1	2
	30 = N

Item III	
Response	f
5	6
4	9
3	7
2	5
1	3
	30 = N

7.

Class interval	Midpoint	f
3,793–3,913	3,853	3
3,672–3,792	3,732	1
3,551–3,671	3,611	4
3,430–3,550	3,490	5
3,309–3,429	3,369	3
3,188–3,308	3,248	2
3,067–3,187	3,127	4
2,946–3,066	3,006	2
2,825–2,945	2,885	0
2,704–2,824	2,764	$\underline{1}$
		$\overline{25} = N$

9. For $i = 5.1$:

Class interval	Midpoint	f
90.0–95.0	92.5	9
84.9–89.9	87.4	5
79.8–84.8	82.3	5
74.7–79.7	77.2	2
69.6–74.6	72.1	2
64.5–69.5	67.0	1
59.4–64.4	61.9	7
54.3–59.3	56.8	13
49.2–54.2	51.7	5
44.1–49.1	46.6	2
39.0–44.0	41.5	$\underline{1}$
		$\overline{52} = N$

11.

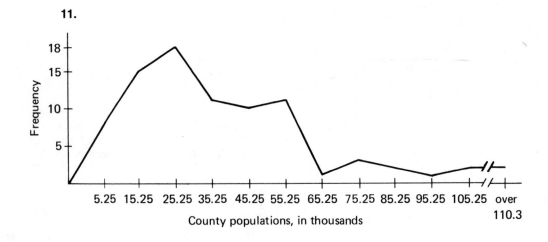

County populations, in thousands

13. (Figure not given.)

15. *(a)* .093   *(c)* .130
(b) .333   *(d)* .722

17. Frequency distribution of weights, using $i = 5$ pounds:

Class interval	Midpoint	f
149–153	151	2
144–148	146	5
139–143	141	1
134–138	136	8
129–133	131	3
124–128	126	13
119–123	121	12
114–118	116	15
109–113	111	9
104–108	106	7
99–103	101	3
		78 = N

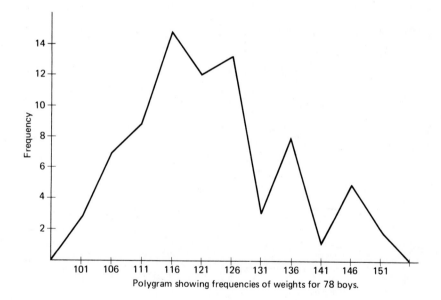

Polygram showing frequencies of weights for 78 boys.

19. *(a)* Probability mass function:

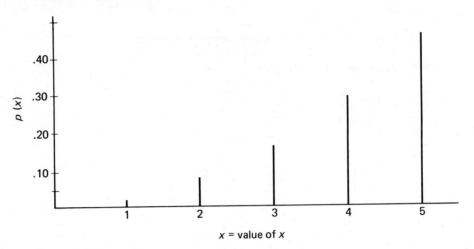

$p(1) = .018,\ p(2) = .073,\ p(3) = .164,\ p(4) = .291,\ p(5) = .454.$

(b) Step function:

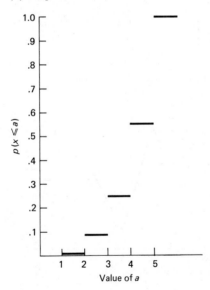

(c) $p(2, 3, \text{or } 4) = .073 + .164 + .291 = .528$

(d) $p(\text{some value}) = 1.00$

21. *(a)* Probability density function:

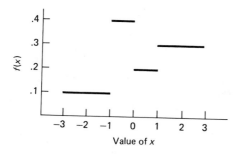

(b) If can be seen that this is not a strictly continuous distribution by the breaks or "jumps" in its probability function.

23.

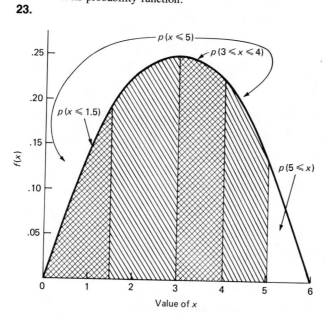

25.

Value of X = a	Expected frequency
20	.96
18	7.20
16	4.32
14	5.28
12	5.28
10	7.20
8	7.68
6	5.28
4	3.36
2	1.44
0	0
	48 = N

CHAPTER 3

1. $6! = 720$

3. There are $(12)^4 = 20,736$ sequences of four notes. There are only 12 sequences, how-ever, where the same note is struck four times. Hence, $p(\text{same note four times}) = 12/20,736 = .0006$. In addition, $p(\text{at least two different notes}) = 1 - p(\text{same note four times}) = .9994$. For 12 notes, there are nine possible four-note ascending scales, so that $p(\text{four-note scale}) = 9/20,736 = .0004$.

5. $7 \times 6 \times 5 \times 4 \times 3 \times 2 \times 1 = 7! = 5,040$ different ways

7. $\binom{20}{10} = 184,756$ ways for the first group

$\binom{10}{10} = 1$ way for the second group

probability of any given assignment $= \dfrac{1}{\binom{20}{10}} = \dfrac{1}{184,756}$

9. probability(one all-black group) $= \dfrac{6\binom{20}{5}\binom{30}{5}\binom{25}{5}\binom{20}{5}\binom{15}{5}\binom{10}{5}}{\binom{50}{5}\binom{45}{5}\binom{40}{5}\binom{35}{5}\binom{30}{5}\binom{25}{5}}$

probability(two black groups) $= \dfrac{\binom{6}{2}\binom{20}{5}\binom{15}{5}\binom{30}{5}\binom{25}{5}\binom{20}{5}\binom{15}{5}}{\binom{50}{5}\binom{45}{5}\binom{40}{5}\binom{35}{5}\binom{30}{5}\binom{25}{5}}$

11. number of possible committees $= \binom{10}{2}\binom{15}{3}\binom{20}{4} = 99,201,375$

13. probability that all four fall in same month $= \dfrac{\binom{12}{1}\left(\frac{1}{12}\right)^4}{} = \left(\dfrac{1}{12}\right)^3$

probability that no two fall in the same month $= \dfrac{\binom{12}{4}}{(12)^4}$

probability of two or more in the same month $= 1 - \dfrac{\binom{12}{4}}{(12)^4} = .976$

15. The probability of three repeated letters and three repeated digits is

$\dfrac{\binom{26}{1}\binom{10}{1}}{(26)^3(10)^3}$, or about $.000015$

The probability of three repeated letters and three different digits is

$\dfrac{\binom{26}{1}\binom{10}{3}3!}{(26)^3(10)^3} = .001065$

For three different letters and three different digits, the probability is

$$\frac{\binom{26}{3}\binom{10}{3}3!3!}{(26)^3(10)^3} = .6391$$

17. $p(0 \text{ or } 1: N = 15, .30) = \binom{15}{0}(.3)^0(.7)^{15} + \binom{15}{1}(.3)^1(.7)^{14}$

$$= .035$$

$p(2 \text{ or more}; N = 15, .30) = 1 - .035 = .965$

19. (a) $p(3; N = 18, .20) = .23$

(b) $p(7; 18, .20) = .035$

(c) $p(14; 18, .20) = $ almost zero

(d) $p(\text{number correct} > 5; 18, .20) = .13$

(e) $p(\text{number correct} < 2; 18, .2) = .10$

(f) $p(2 \leq \text{number correct} \leq 7; 18, .20) = .88$

21. $p(9 \text{ or more undernourished}; N = 20, .25) = .04$. The evidence supports the conconclusion that more than 25% are undernourished.

23. Here there are six pairs favoring Group II, and one tied pair, out of the 10 differences observed. Then

$$p(6 \text{ or more positive differences}; N = 9, p = .5) = .25$$

This is not an unlikely result to obtain when the groups are actually equal.

25. $p(0 \text{ errors}; m = .2) = \dfrac{e^{-.2}(.2)^0}{0!} = .82$

$p(1 \text{ or more errors}; m = .2) = 1 - .82 = .18$

27. $p(N = 4; 1, .60) = (.6)(.4)^3 = .038$

$p(N = 1; 1, .6) = .6$

29. The multinomial rule gives the probability of this distribution of preferences to be

$$\frac{20!}{5!3!4!0!8!}(.32)^5(.14)^3(.19)^4(.04)^0(.31)^8 = .0036$$

31. (a) $p(10 \text{ sophomores}) = \dfrac{\binom{18}{10}}{\binom{50}{10}}$

(b) $p(7 \text{ seniors and } 3 \text{ juniors}) = \dfrac{\binom{7}{7}\binom{10}{3}}{\binom{50}{10}}$

(c) $p(3 \text{ freshmen, } 4 \text{ sophomores, } 2 \text{ juniors, } 1 \text{ senior}) = \dfrac{\binom{15}{3}\binom{18}{4}\binom{10}{2}\binom{7}{1}}{\binom{50}{10}}$

CHAPTER 4

1. Mode is at classification B. This is chosen because the data are nominally scaled. Thus, $p(\text{wrong}) = 1 - (18/54) = 2/3$.
3. Mean = 24.44, median = 24.50
5. Mean = 3,404.52
7. $\bar{x} = .308$. Yes, since the mean is positive.

9.	*Boys*	*Girls*
	median = 34.16	median = 28.25
	mode = 37.495	mode = 27.495

11. $E(X) = 101.72$. The most likely value can be taken as the midpoint of the most frequent interval, or 103.
13. Median = 15.925, mean = 16.2, mode = 16. The differences among these indices reflect the distribution's positive skewness.
15. *(1)* $E(X) = (1/6) + 2(1/6) + 3(1/6) + 4(1/6) + 5(1/6) + 6(1/6) = 3.5$
   (2) $E(X) = (1/21) + 2(2/21) + 3(3/21) + 4(4/21) + 5(5/21) + 6(6/21) = 4.33$
   (3) $E(X) = (1/12) + 2(2/12) + 3(3/12) + 4(3/12) + 5(2/12) + 6(1/12) = 3.5$
17. $E(X) = -3(1/16) - 2(2/16) - 1(3/16) + 0(4/16) + (3/16) + 2(2/16) + 3(1/16) = 0$.
19. $S^2 = 21.43$, $S = 4.63$
21. $S^2 = 1.49$, $S = 1.22$
23. The boys' average is about 33.2 and the girls' average is 28.8. Since the boys have a standard deviation of 7.94 as compared with 7.20, the girls are more homogeneous than the boys in this respect.
25. $E(X) = 7$, $\text{var}(X) = 5.83$, standard deviation = 2.42
27. *(1)* $E(X) = 3.5$, $\text{var}(X) = 2.92$, standard deviation = 1.71
   (2) $E(X) = 4.33$, $\text{var}(X) = 2.25$, standard deviation = 1.50
   (3) $E(X) = 3.5$, $\text{var}(X) = 1.92$, standard deviation = 1.38
29. $E(X) = 0$, $\text{var}(X) = 2.5$, standard deviation = 1.58
31. $E(X) = 6$, $\text{var}(X) = 2.4$ standard deviation = 1.55

33.	*Divorced*	*Married*
(a)	.12	−.24
(b)	1.20	.96
(c)	−1.60	−2.16
(d)	2.71	2.64
(e)	−.96	−1.44

35. *(a)* Since $p(|z| \geq 4.318) \leq .054$, the maximum probability is .054.
   (b) $p(|z|) \geq 1.7) \leq (1/1.7^2)$, so the probability is less than or equal to .346.
   (c) Since $p(|z| \geq 1.3) \leq .59$, the proportion must be greater than or equal to .41.

CHAPTER 5

1. $E(X) = 1.5$, $\text{var}(X) = .25$
3. $E(G) = 4(.25) + 3(.50) + 2(.25) = 3 = 2E(X)$

5. The sampling distribution has a larger range of values than the population distribution on which it is based. There is also a distinct mode at $G = 3$, whereas the population distribution has no mode. Here, the expected value of G is twice the expected value of X and the variance of G is twice the variance of X.

7. $E(G) = 2(.125) + 1.667(.375) + 1.333(.375) + 1(.125) = 1.5 = E(X)$

$\sigma_G^2 = 4(.125) + (1.667)^2(.375) + (1.333)^2(.375) + (.125) - (1.5)^2$

$\qquad = .083 = .25/3 = \sigma_X^2/3$

9. The population distribution is

x	$p(x)$
1	.25
0	.50
-1	.25
	1.00

$E(X) = 0,\ \sigma_X^2 = .50,\ \sigma_X = .707$

11. $E(\bar{x}) = E(x) = 0,\ \sigma_M^2 = .25 = \sigma_X^2/2$

13. (Here a number of decimal places are carried so that the probabilities will sum to 1.00.) The distribution for the mean, given $N = 3$, is

$\bar{x}$	$p(\bar{x})$
1.0	.015625
.67	.093750
.33	.234375
.00	.312500
$-.33$	.234375
$-.67$	.093750
-1.00	.015625
	1.000000

We note that distribution is much more "bell-shaped" than the population distribution from which the samples are taken.

15. (a) $p(\bar{x} = 1) = .016$ (d) $p(\bar{x} \le 0) = .344$

(b) $p(\bar{x} = 0) = .313$ (e) $p(\bar{x} \le 2/3) = .984$

(c) $p(\bar{x} \le 1/3) = .891$ (f) $p(\bar{x} < -1/3) = .109$

17. $p(8$ females: $N = 15,\ p = .6) = .1387$

$p(8$ females; $N = 15,\ p = .4) = .0925$

Conclude that the cards belong to the group with 60% females, since this gives the highest prior likelihood to the sample results.

19. $p(3;\ N = 20,\ .05) = .06$

$p(3;\ N = 20,\ .20) = .21$

By the principle of maximum likelihood, we should choose .2 as the better estimate of p. However, the best estimate is .15.

21. For Typist I, the likelihood (by the Poisson rule, and letting $m = 5$ per 5 letters) is

$$\frac{e^{-5}(5)^2}{2!} = .084.$$

For Typist II, letting $m = 3$ per five letters, the likelihood is

$$\frac{e^{-3}(3)^2}{2!} = .224.$$

For Typist III, where $m = .5$ per five letters, we have

$$\frac{e^{-.5}(.5)^2}{2!} = .076.$$

Hence, by the principle of maximum likelihood, Typist II is the best bet.

23. Interviewer A: $p(N = 5; r = 3, p = .6) = .207$
Interviewer B: $p(N = 5; r = 3, p = .4) = .138$
Interviewer C: $p(N = 5; r = 3, p = .3) = .079$
Interviewer D: $p(N = 5; r = 3, p = .1) = .005$
Interviewer A is the best bet.

25. If the sample P value is .4, then, in the sampling distribution for $p = .52$, this amounts to a z score of $(.40 - .52)/.032 = -3.75$. Then, according to the Tchebycheff inequality, for a symmetric distribution,

$$p(|z| \geq 3.75) \leq \frac{4}{9(3.75)^2} \text{ or } .03.$$

Hence the probability of a deviation as large or larger than this, if $p = .52$ is true, is only .03 or less.

27. $E(\bar{x}) = \mu = 1.5$, $\sigma_M^2 = .25/50 = .005$, $\sigma_M = .071$, $z = .71$. The probability of a value this much or more deviant from expectation is something less than .88 according to the Tchebycheff inequality. Hence the result appears to be relatively likely.

29. Since $\sigma_M = 4.9/\sqrt{150} = .40$, the sample mean of 12.3 represents a z value of -12.75 if the true value of $\mu = 17.4$. Thus, by the Tchebycheff inequality, the chances are less than .003 for a sample result this much or more deviant, given $\mu = 17.4$. It appears to be quite safe to conclude that the new population is different on the average.

31. $\bar{x} = 30.8$ is the best available estimate of μ. The value $s^2 = 47.96$ is the best available estimate of σ^2.

33. This represents a finite population (208 students) from which this sample of 49 students was drawn. Thus, the estimate of the population mean is $\bar{x} = 684.85$. The estimate of the population variance is

$$\text{est. } \sigma^2 = \frac{N(T - 1)}{(N - 1)T} S^2 = \frac{49(207)}{48(208)} (14,219.19) = 14,445.64$$

Then the estimated standard deviation for the population is 120.19. The estimated total amount spent is $(208)(684.45) = 142,365.6$.

35. pooled estimate of $\mu = \dfrac{(10)(96) + (20)(105) + (15)(103)}{45} = 102.33$

pooled estimate of $\sigma^2 = \dfrac{(10)(22) + (20)(29) + (15)(31)}{45 - 3} = 30.12$

$$\text{est. } \sigma_M = \frac{5.49}{\sqrt{45}} = .82$$

37. $\mu = 422.25$, $\sigma^2 = 3,564$

39. $\bar{x}_1 - \bar{x}_2 = 4.12$ units approximately

CHAPTER 6

1. (a) .115   (d) .038
   (b) .831   (e) .334
   (c) .970   (f) .984
3. (a) .964   (d) .309
   (b) .067   (e) .773
   (c) .008

5. About .21 is the probability of this sample or one more deviant from expectation if the true mean is still 23.6. Thus, there is not strong reason to say that the mean has changed. Since the sample size is fairly large, the central limit theorem may be relied on to counter the effect of the nonnormal population.

7.

	$p = .5, N = 8$		$p = .4, N = 8$	
	Binomial	Approx. normal*	Binomial	Approx. normal*
8	.0039	.007	.0007	.001
7	.0312	.032	.0079	.008
6	.1094	.106	.0413	.040
5	.2188	.218	.1239	.123
4	.2734	.274	.2322	.241
3	.2188	.218	.2787	.278
2	.1094	.106	.2090	.198
1	.0312	.032	.0896	.084
0	.0039	.007	.0168	.027

*Values read from the normal table will vary depending on rounding methods used. Note that the fit between the normal and binomial probabilities is very nearly as close for $p = .4$ as for $p = .5$, even for N as low as 8.

9. Since for $\mu = .1$ and $\sigma = .03$, so that $\sigma_M = .0095$, the obtained z value is -31.58. We can say that this sample result is *most* unlikely.

11. The limits of the 99% confidence interval are approximately 3.73 and 3.83.

13. $p(303 \le \bar{x} \le 304) = .015; p(304 \le \bar{x} \le 305) = .0023$. The odds are about 6.5 to 1.

15. $p(\bar{x} \le 294) = .0014; p(306 \le \bar{x}) = .0014$

17. If $p(\text{cover}) = .90$, then $p(\text{not cover}) = .10$. Hence, $E(\text{number that do not cover}) = 200(.1) = 20$.

19. The limits for the interval are about .126 and .303.

21. The confidence limits may be found by taking

$$\frac{1,000}{1,000 + (2.58)^2} \left[.51 + \frac{(2.58)^2}{2,000} \pm 2.58 \sqrt{\frac{(.51)(.49)}{1,000} + \frac{(2.58)^2}{4(1,000)^2}} \right]$$

giving the limits as .47 and .55. This interval does not include the value of .46 favoring (.54 against). It does include the value of .50, however. Thus, there is some reason to believe that .50 or more might favor the issue.

23. Let $1.96 = .01/(.07/\sqrt{N})$. Then $N = 188.24$, or 189.

25. The probability is less than .01. Hence the risk in rejecting the possibility that the true proportion is .18 or more is also less than .01.

27. The limits are about 11.64 and 12.96. These limits define a narrower interval than before since the exact normal probabilities were used rather than the inexact Tchebycheff limits.

29. Based on the relation

$$\frac{K\sigma}{\sigma/\sqrt{N}} = 1.96,$$

the table is as follows:

N	K
1	1.96
5	.88
10	.62
20	.44
30	.36
40	.31
50	.28
100	.20

31. The 99% confidence interval has limits 645.62 to 723.28.

33. Pooled estimate of $\mu = 102.33$. Pooled estimate of $\sigma = \sqrt{30.12} = 5.49$. Since $z = .40$, the probability of a positive deviation this large or larger is about .345.

35. The sampling distribution of means drawn from theoretical distribution 3 would probably be most nearly normal, since the population itself is unimodal and symmetric. The third distribution would also have the smallest standard error, or .195.

CHAPTER 7

1. H_0: $p = .39$; H_1: $p \neq .39$. The two rejection regions correspond to the following:

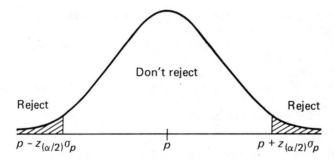

3. H_0: $\mu \geq 172$; H_1: $\mu < 172$. The value $\sigma = 16$ is assumed in either case. The rejection region is shown as follows:

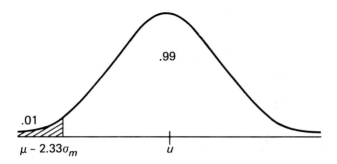

5.

	True situation	
	Child belongs in 7th grade	Child belongs in 6th grade
place in 7th grade decide	correct	error
place in 6th grade	error	correct

7. *(a)* Rule 6(c), under the minimax criterion
   (b) Rule 6(b) by minimax
   (c) Rule 6(a) by minimax
   (d) Rule 6(c) by minimax

9. *(a)* The table of expected losses is as follows:

		True situation	
		$p = .4$	$p = .2$
Rule	8(a)	833.70	6.40
	8(b)	46.30	642.20
	8(c)	382.20	120.90
	8(d)	167.20	322.20

The minimax criterion would indicate Rule 8(d).
   (b) For this set of loss values,

		True situation	
		$p = .4$	$p = .2$
Rule	8(a)	833.70	9.60
	8(b)	46.30	936.30
	8(c)	382.20	181.35
	8(d)	167.20	483.30

This suggests Rule 8(c).

(c) For this set,

	True situation	
	$p = .4$	$p = .2$
8(a)	4,168.50	6.40
Rule 8(b)	231.50	624.20
8(c)	1,911.00	120.90
8(d)	836.00	322.20

Rule 8(b) would fit the minimax criterion.

(d) For this set,

	True situation	
	$p = .4$	$p = .2$
8(a)	833.70	64.00
Rule 8(b)	46.30	6,242.00
8(c)	383.20	1,209.00
8(d)	167.20	3,222.00

This suggests Rule 8(a).

11. The value of z is 2.91. Since this exceeds 1.96, reject H_0.

13. H_0: $p \le .25$; H_1: $p > .25$. Then

$$z = \frac{.45 - .25 - (.5/20)}{\sqrt{(.25)(.75)/20}} = 1.81$$

This is significant beyond the .05 level, one-tailed; thus the null hypothesis is rejected.

15. In place of the unknown value of σ^2, since N is large we can here use est. $\sigma^2 = s^2 = 1.626$. Then $z = 2$, permitting us to reject H_0 for $\alpha = .05$ (two-tailed).

17. Assuming $\sigma = 2$, then the z value is 4.14, which exceeds the value for $\alpha = .01$, one-tailed. The H_0 that $\mu = 15$ may safely be rejected.

19. N should be about 27; $\bar{x} = 102.5$.

21. Since the critical value of $\bar{x}$ under H_0: $\mu = 28.6$ is 30.214 for the upper tail and 26.98 for the lower tail, the z for 30.214 is -2.17, and the z for the other critical value is -6.09 when $\mu = 32$. The power is .985. When $\mu = 25$, the equivalent z values are 6.33 and 2.40. The power is then about .99 for $\mu = 25$.

23. In this one-tailed test, the power is .9975 against the alternative $\mu = 16$, and over .9999 against $\mu = 20$.

25.

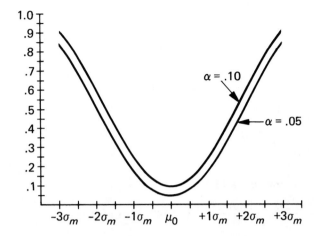

27. The probability that a Type I error is made on both tests is $(0.5)^2 = .0025$. The probability that neither involves a Type I error is $(.95)^2 = .9025$. Then the probability of at least one Type I error is $1 - .9025 = .0975$.

29.
$$\mathscr{P}(200|\text{data}) = \frac{(.15)(.5)}{(.15)(.50) + (.23)(.25) + (.20)(.25)} = .41$$

$$\mathscr{P}(210|\text{data}) = \frac{(.23)(.25)}{(.15)(.50) + (.23)(.25) + (.20)(.25)} = .32$$

$$\mathscr{P}(220|\text{data}) = \frac{(.20)(.25)}{(.15)(.50) + (.25)(.23) + (.20)(.25)} = .27$$

31. For Decision Rule B the probabilities are

		True		
		H_0	H_1	H_2
	H_0	.9599	.2266	.0006
accept	H_1	.0399	.6147	.0662
	H_2	.0002	.1587	.9332

CHAPTER 8

1. The t value of -1.26 with 185 degrees of freedom is not significant.

3. For H_0: $\mu \geq 17.34$ and H_1: $\mu < 17.34$, the obtained t value of -2.5 lets one reject the null hypothesis beyond the .01 level, one-tailed.

5. Employing the t value for $\alpha = .01$ (two-tailed) at ∞ degrees of freedom, we find the confidence limits to be 98.51 and 98.89

7. The required confidence interval has limits of 32.04 and 32.76.

9. For 79 degrees of freedom, the t value of -2.48 lets us reject H_0: $\mu \geq 115.2$ beyond the .01 level (one-tailed).

11. The limits for the 95% confidence interval are 34.74 and 37.46.

13. The H_0: $\mu_1 - \mu_2 = 0$ would be rejected in favor of the H_1: $\mu_1 - \mu_2 \neq 0$, since $t = 4.81$. This is significant beyond the .01 level (two-tailed).

15. The t value of 2.81 is significant beyond the .005 level, two-tailed.

17. The t value of -1.33 is not significant. We assume independent random samples from normal populations with the same variance.

19. On taking

$$\sigma_Y^2 = \sigma_{Y|X}^2 + \frac{(\mu_1 - \mu_2)^2}{4} = 16 + \left(\frac{9}{4}\right) = 18.25$$

so that

$$\omega^2 = \frac{9}{4(18.25)} = .123;$$

we have, from Eq. 8.13.2,

$$n \geq \frac{(1.282 + 1.96)^2(.877)}{2(.123)} = 37.47, \text{ or 38 per group.}$$

Therefore, $n = 39$ for each sample definitely should be sufficient.

21. For 38 degrees of freedom the t value of 2.02 is significant right at the .05 level, two-tailed.

23. For the hypotheses H_0: $\mu_1 - \mu_2 \leq 1,000$ and H_1: $\mu_1 - \mu_2 > 1,000$, although the t value was -3.66, this falls into the region of nonrejection, and thus H_0 is not rejected.

25. The t value of -1.17 does not permit us to reject the null hypothesis H_0: $\mu_D \geq 0$.

27. The confidence interval has limits of $-.35$ and 3.35.

29. The limits of the 95% confidence interval are 1.75 and 5.91.

CHAPTER 9

1. $p(|z| \geq 1.15) = .25$, so $p[\chi_{(1)}^2 \geq (1.15)^2] = .25$

3. $E(\chi_{(5)}^2) = 5$; $p(15 < \chi_{(5)}^2)$ is about .01, $p(11 < \chi_{(5)}^2$ is about .05, $p(\chi_{(5)}^2 < 1.6)$ is about .10.

5. The confidence limits are approximately 74.19 and 866.60.

7. The confidence limits for the mean are 46.57 and 55.43. For the variance the confidence limits are about 59.19 and 204.22

9. Since $N(S^2)/\sigma^2 = \chi^2$, if S is 1.3 times σ, the value of chi-square will be $(30) \times (1.3)^2 = 50.7$. For 29 degrees of freedom a value this large or larger has a probability of less than .01.

11. The limits of the confidence interval are about 5.25 and 81.01.

13. For 268 degrees of freedom, the chi-square values required for significance (two-tailed) may be found from

$$\chi_{(268;\ .025)}^2 = \frac{(1.96 + \sqrt{2(268) - 1})^2}{2} = 314.76$$

and

$$\chi_{(268;\ .975)}^2 = \frac{(-1.96 + \sqrt{2(268) - 1})^2}{2} = 224.09.$$

Hence the obtained value of 237.69 is not significant.

15. The null hypothesis H_0: $\sigma = 24$ may be rejected in favor of H_1: $\sigma \neq 24$ since the chi-square value of 41.02 is significant beyond the .05 level for 24 degrees of freedom.

17. Here, $s_1^2 = 984.38$, and $s_2^2 = 957.29$; thus $F = 1.03$ with 24 and 24 degrees of freedom. This is not significant.

19. Here, H_0: $\sigma_1^2 = \sigma_2^2$ against H_1: $\sigma_1^2 > \sigma_2^2$. The $F = 2.14$ is significant at the .05 level for 19 degrees of freedom. H_0 is thus rejected in favor of H_1.

21. The F value of 1.12 is not significant for 12 and 8 degrees of freedom.

23. Yes, when H_0 is true the ratio s_1^2/ks_2^2 should be distributed as F.

25. $E(\bar{y}) = 2.5$ minutes; $\sigma_M = .079$.

CHAPTER 10

1. $\bar{x} = 99.81$, est. $\alpha_1 = -5.31$ (actual $= -5$), est. $\alpha_2 = 5.941$ (actual $= 4$), est. $\alpha_3 = -2.31$ (actual $= -2$) est. $\alpha_4 = 1.69$ (actual $= 3$). The ''data'' are

I	II	III	IV
90	102	98	106
96	109	94	97
95	110	97	107
97	102	101	96

3.

Source	SS	df	MS	F	p
Between groups	603.63	2	301.82	9.1	<.01
Within groups	1,890.95	57	33.17		
Total	2,494.58	59			

Yes, motor coordination is apparently related to athletic interests.

5.

Source	SS	df	MS	F	p
Between conditions	160.00	1	160	1.78	not signif.
Within conditons	719.60	8	89.95		
Total	879.60	9			

Here, $t = -\sqrt{1.78}$, as before.

7. Estimated effect of Group 1 $= -4.0333$
Estimated effect of Group 2 $= 0.3167$
Estimated effect of Group 3 $= 3.7167$

9. Estimated effect of Group 1 $= -37.5$
Estimated effect of Group 2 $= -30.1$
Estimated effect of Group 3 $= -0.4$
Estimated effect of Group 4 $= 21.1$
Estimated effect of Group 5 $= 11.9$
Estimated effect of Group 6 $= 35.0$

In the sample the proportion of variance accounted for by the amount of sleep deprivation is .81. The value of est. $\omega^2 = .79$.

11. Here $t = .648$, so that $t^2 = .42 = F$.

13.

Source	SS	df	MS	F	p
Between groups	698.20	3	232.7	4.69	<.05
Within groups	793.60	16	49.6		
Total	1,491.80	19			

15.

Source	SS	df	MS	F	p
Between groups	0.4086	2	0.2045	4.06	<.05
Within groups	0.8569	17	0.0504		
Total	1.2655	19			

17.

Source	SS	df	MS	F	p
Between groups	177.39	3	59.13	1.92	not signif.
Within groups	739.90	24	30.77		
Total	917.29	27			

19. Effect for Group 1 = -3.151
Effect for Group 2 = 0.404
Effect for Group 3 = -0.596
Effect for Group 4 = -1.596
Effect for Group 5 = 3.404
Effect for Group 6 = -2.405

Here the sum of the effects weighted by group size equals to zero.

21. For the original data the results are:

Source	SS	df	MS	F	p
Between groups	0.1049	3	0.0350	1.89	not signif.
Within groups	0.6663	36	0.0185		
Total	0.7712	39			

Subtracting 1 makes no difference in MS between and MS within. Multiplying each value by 100 multiplies the mean squares by $(100)^2$. However, in either instance, F is unaffected.

23. Look at Table XII of Appendix E, for $J = 6$, and omega-square $= .50$, and then trace down the values of n until you locate one that gives a power at or just over .80 when the value of alpha $= .01$. Assigning 5 cases per group should be adequate here.

CHAPTER 11

1.

	Comparisons						
Means	1	2	3	4	5	6	(for Exercise 2)
13	1	1	0	0	0	0	
15	1	− 1/2	1/2	0	0	1/2	
10	1	− 1/2	− 1/2	0	0	1/2	
20	− 1	0	0	1/2	1/2	− 1/2	
18	− 1	0	0	− 1/2	1/2	− 1/2	
35	− 1	0	0	0	− 1	0	

3. The F values and significance levels for the comparisons are:

(1) $F = 18.58$, signif. $< .01$.

(2) $F = .02$, not signif.

(3) $F = 1.14$, not signif.

(4) $F = .18$, not signif.

(5) $F = 15.53$, signif. $< .01$

Notice that the means in this experiment are not, strictly speaking, independent, as they are all based on responses by the same subjects. The status of these F tests is therefore somewhat uncertain, as compared to those for independent means.

5. A possible set of comparisons is given by

Group	Mean	1	2	3
1	73.1	1	0	0
2	82.0	− 1/3	1	0
3	70.0	− 1/3	− 1/2	1
4	69.0	− 1/3	− 1/2	− 1
		− .567	12.500	1.000

The only significant comparison is the second, where the F value of 5.268 is significant for 1 and 36 degrees of freedom ($\alpha = .05$).

7. $F = \dfrac{18.58 + .02 + 1.14 + .18 + 15.53}{5} = 7.09$, significant beyond the .01 level for 5 and 45 degrees of freedom.

9. The means and comparisons are

	Mean	$\hat{\psi}_1$	$\hat{\psi}_2$	$\hat{\psi}_3$
Constant reward	12.0	1	0	0
Frequent reward	11.0	$-1/3$	1	0
Infrequent	16.0	$-1/3$	$-1/2$	1/2
Never	16.6	$-1/3$	$-1/2$	$-1/2$
		-2.533	-5.300	$-.300$

For 1 and 16 degrees of freedom, and with an MS error equal to 3.45, the first two comparisons are significant beyond the .05 level. The third comparison is not significant.

11. The probability of at least one Type I error is, for three independent tests, .143. If $\alpha = .017$, then the probability of at least one Type I error is less than or equal to .051.

13. Since each comparison must be orthogonal to the general mean, and hence have weights that sum to zero, there are $J - 1$ rather than J comparisons.

15. For $K = 3$ groups, and 27 degrees of freedom, the value of the studentized range statistic is approximately $q = 3.49$. Then, for MS error $= 19.84$ and $n = 10$ the Tukey HSD value is about 4.92. Hence the differences between Mean 2 and Mean 1, and between Mean 2 and Mean 3 prove to be significant at the .05 level.

17. Instead of an F test, a t test with the appropriate sign would be used for each comparison, and a directional test employed.

19.

Source	SS	df	MS	F	p
Between groups					
comparison 1	405.0	1	405.0	10.13	<.01
comparison 2	2,880.0	1	2,880.0	72.0	<<.01
Other comparisons	27.0	2	13.5	0.34	—
Error					
(within groups)	1,800.0	45	40.0		
Total	5,112.0	49			

21. For Comparisons 1 and 2, the weighted sum of products is

$$\frac{(4.25)(0)}{5} + \frac{(-1.5)(2.75)}{6} + \frac{(-1.75)(-1.75)}{7} + \frac{(-1)(-1)}{4} = 0$$

For 1 and 3 the weighted sum of products is

$$\frac{(4.25)(0)}{5} + \frac{(-1.5)(0)}{6} + \frac{(-1.75)(1)}{7} + \frac{(-1)(-1)}{4} = 0.$$

Finally for Comparisons 2 and 3 are the sum of products is

$$\frac{(0)(0)}{5} + \frac{(2.75)(0)}{6} + \frac{(-1.75)(1)}{7} + \frac{(-1)(-1)}{4} = 0,$$

showing that these three comparisons are pairwise orthogonal.

23. The table of differences between all pairs of the six means is as follows:

	− 1.4	−.6	.8	3.0	5.4	6.4
− 1.4	0	0.8	2.2	4.4	6.8*	7.8*
−.6		0	1.4	3.6	6.0	7.0*
0.8			0	2.2	4.6	5.6
3.0				0	2.4	3.4
5.4					0	1.0
6.4						0

(* indicates significance at the .05 level)

For the first layer of this table, in which $k = 6$ and the degrees of freedom are 24, the value of the HSD is 7.02. Then for the second layer where $k = 5$ the HSD value is 6.70, and for the third layer and $k = 4$ the HSD $= 6.26$. The values for the fifth and six layers, where $k = 3$ and 2 respectively, are 5.67 and 4.69.

CHAPTER 12

1. The row, column, and cell means were as follows:

	B_1	B_2	B_3	Row mean
A_1	80	78	70	76
A_2	73.5	75	64.5	71
A_3	68.5	72	66.5	69
col. mean	74	75	67	

3. For rows

est. $\alpha_1 = 3.44$, est. $\alpha_2 = -3.44$.

For columns

est. $\beta_1 = 12.44$, est. $\beta_2 = -11.47$, est. $\beta_3 = -0.97$.

For interaction

est. $(\alpha\beta)_{11} = -2.10$, est. $(\alpha\beta)_{12} = -1.36$, est. $(\alpha\beta)_{13} = 3.46$

est. $(\alpha\beta)_{21} = 2.10$, est. $(\alpha\beta)_{22} = 1.36$, est. $(\alpha\beta)_{23} = -3.46$

5. In terms of the effects, as stated in Exercise 1, and for $n = 4$ observations in each of nine cells, the sums of squares are

$$\text{SS rows or SS A} = Kn \sum_{j=1}^{3} a_j^2 = 12(16 + 1 + 9) = 312$$

$$\text{SS cols or SS B} = Jn \sum_{k=1}^{3} b_k^2 = 12(4 + 9 + 25) = 456$$

$$\text{SS A} \times \text{B interaction} = n \sum_{j}^{3} \sum_{k}^{3} (ab)_{jk}^2 = 4(4 + 1 + 1$$
$$+ .25 + 1 + 2.25 + 6.25 + 0 + 6.25) = 88.$$

Then

$$\text{SS error} = \text{SS total} - \text{SS A} - \text{SS B} - \text{SS A} \times \text{B} = 1,174.$$

The summary table is:

Source	SS	df	MS	F	p
A (rows)	313	2	156	3.59	<.05
B (cols)	456	2	228	5.24	<.05
A × B int.	88	4	22	0.51	—
Error	1,174	27	43.48		
Total	2,030	35			

7. The estimated effects are shown in the following table:

est. β_k =	3.06	2.39	−5.45	est. α_j =
est. $(\alpha\beta)_{jk}$ =	7.19	−1.39	−5.80	−10.78
	5.19	3.11	−8.30	1.72
	−12.39	−1.72	14.11	9.06

9. p(at least one Type I error) $= 1 - (.95)^3 = .1426$, for $\alpha = .05$
p(at least one Type I error) $= 1 - (.99)^3 = .0297$, for $\alpha = .99$.

A small value of α is to be preferred, since it will provide a reasonable value for the probability of at least one Type I error.

11. The eight orthogonal comparisons among the nine cells are as follows:

	A_1B_1	A_2B_1	A_3B_1	A_1B_2	A_2B_2	A_3B_2	A_1B_3	A_2B_3	A_3B_3	$\hat{\psi}$	SS $(\hat{\psi})$
$A^{(1)}$	2	−1	−1	2	−1	−1	2	−1	−1	−7	5.44
$A^{(2)}$	0	1	−1	0	1	−1	0	1	−1	−1	0.33
$B^{(1)}$	2	2	2	−1	−1	−1	−1	−1	−1	6.5	4.69
$B^{(2)}$	0	0	0	1	1	1	−1	−1	−1	−6.5	14.08
$AB^{(11)}$	4	−2	−2	−2	1	1	−2	1	1	−5	1.39
$AB^{(12)}$	0	0	0	2	−1	−1	−2	1	1	26	112.67
$AB^{(21)}$	0	2	−2	0	−1	1	0	−1	1	22	80.67
$AB^{(22)}$	0	0	0	0	1	−1	0	−1	1	3	4.5

SS A = 5.44 + 0.33 = 5.77

SS B = 4.69 + 14.08 = 18.77

SS A × B = 1.39 + 112.67 + 80.67 + 4.5 = 199.23

SS total = 284.28

SS error = 60.51

13.

Source	SS	df	MS	F	p
Between columns (A)	44.67	2	22.34	1.23	not signif.
Between rows (B)	284.67	2	142.34	7.83	<.05
Error (interaction)	72.66	4	18.11		
Total	402.00	8			

15.

Source	SS	df	MS	F	p
Between columns	508.17	2	254.09	0.12	—
Between rows	3,952.67	3	1,317.56	0.64	—
Error (interaction)	12,299.83	6	2,049.97		
Total	16,760.67	11			

Here, interaction effects are assumed to be absent.

17. For nine groups (cells) and $N - 9 = 9$ degrees of freedom for error, the studentized range statistic is equal to 5.60, for alpha $= .05$. The value of the Tukey HSD is then

$$\text{HSD} = q_{(9,9)} \sqrt{\text{MS error}/n} = 5.60 \sqrt{6.72/2}$$

$$= 10.26 .$$

However, among the nine groups only one difference equals or exceeds this HSD value: this is the difference of 10.5 between Group A_2B_2 and Group A_3B_3. Therefore, only this difference is significant at or beyond the .05 level.

19. The estimate of the second-order interaction effect, or est. $(\alpha\beta\gamma)_{jkm}$ may be found from the other estimates of main and interaction effects by taking

$$\text{est. } \alpha\beta\gamma_{jkm} = \text{est. } \alpha\beta\gamma_{jkm} \text{ cell effect } - \text{ est. } (\alpha\beta)_{jk} - \text{ est. } (\alpha\gamma)_{jm} - \text{ est. } (\beta\gamma)_{km}$$

$$- \text{ est. } \alpha_j - \text{ est. } \beta_k - \text{ est. } \gamma_m$$

$$= \bar{y}_{jkm} - \bar{y}_{jk\cdot} - \bar{y}_{j\cdot m} - \bar{y}_{\cdot km} + \bar{y}_{j\cdot\cdot} + \bar{y}_{\cdot k\cdot} + \bar{y}_{\cdot\cdot m} - \bar{y}_{\cdot\cdot\cdot}.$$

21. The estimated effects are shown in the tables below: For A (rows), B (columns), and A × B interaction (cells):

0.1042	0.7708	−3.0625	2.1875	−0.5208
2.6042	−2.9792	1.4375	−1.0625	−1.2708
−2.7083	2.2083	1.6250	−1.1250	1.7917

−3.1042	−0.2708	−0.4375	3.8125

For A (rows), C (columns), and A × C interaction (cells):

−0.1042	0.1042	−0.5208
0.1458	−0.1458	−1.2708
−0.0417	0.0417	1.7917

−.6458	.6458

For B (rows), C (columns) and B × C interaction (cells):

0.0625	−0.0625	−3.1042
−0.2708	0.2708	−0.2708
0.0625	−0.0625	−0.4375
0.1458	−0.1458	3.8125

−0.6458	0.6458

For A × B × C interaction (cells):

	B_1	B_2	B_3	B_4
A_1	0.1875	0.0208	0.1875	−0.3958
A_2	0.1875	−0.2292	−0.3125	0.3542
A_3	−0.3750	0.2083	0.1250	0.0417

$$C_1$$

	B_1	B_2	B_3	B_4
A_1	−0.1875	−0.0208	−0.1875	0.3958
A_2	−0.1875	0.2292	0.3125	−0.3542
A_3	0.3750	−0.2083	−0.1250	−0.0417

$$C_2$$

CHAPTER 13

1.

Source	SS	df	MS	F	p
Between managers	731.49	7	104.50	39.43	<<.01
Within managers	190.90	72	2.65		
Total	922.39	79			

Yes, the managers definitely appear to account for variation in employee satisfaction.

3.

Source	SS	df	MS	F	p
Between stretches	10.17	6	1.70	0.10	—
Within stretches	1,049.60	63	16.67		
Total	1,059.77	69			

The coverage of the highway appears to be evenly spaced. Since est. $\sigma_A^2 = 0$, there is no variability attributable to stretches to be inferred from this study.

5.

Source	SS	df	MS	F	p
Between stimuli	162	7	23.14	13.23	<.01
Within stimuli	14	8	1.75		
Total	176	15			

The hypothesis of no between-stimuli variance can be rejected.

7. Here, H_0: $\rho_I \leq .50$ is equivalent to the H_0: $\theta \leq 1$. Then since $n = 10$, we would reject the hypothesis for $\alpha = .05$ if $F = \dfrac{\text{MS between}}{\text{MS within } (1 + 10)} \geq 2.4$, for 5 and 54 degrees of freedom. However, this ratio is only .32, so H_0 is not rejected.

9. The ANOVA summary table is as follows:

Source	SS	df	MS	F	p
Blocks	249.42	9	27.71		
Within blocks					
Factor A	240.95	5	48.19	21.04	<.01
A × Blocks	102.88	45	2.29		
Total	593.25	59			

11.

Source	SS	df	MS	F	p
Between columns	31,141.19	3	10,390.4	17.80	<.01
Between rows	460.06	2	230.03	0.39	—
Interaction	3,499.72	6	583.29	1.87	not signif.
Error	7,489.33	24	312.06		

Here, although interaction is not significant at the .05 level, the F is, nevertheless, substantially more than 1.00, and would be significant somewhere between .25 and .10. Thus, interaction might be pooled with error, and both rows and columns tested against a pooled error of 366.30 with 30 degrees of freedom. This makes the F for both rows and columns larger than before, although that for rows is still less than 1.00.

13. Since we are assuming that circularity is true, the MS error term in our Tukey HSD calculation is taken to be the MS residual (or MS A by blocks term). For these data this term is equal to 2.29. Then, using Table XI (Appendix E), we find that for alpha $= .05$, and with six groups (levels of A) and $(5)(9) = 45$ degrees of freedom, the approximate value of $q = 4.22$. Therefore,

$$\text{HSD} = 4.22 \sqrt{2.29/10} = 2.02 \ .$$

15.

Source	SS	df	MS	F	p
Between subjects	131.29	7	18.76		
Within subjects					
Lists	111.08	2	55.54	10.87	<.01
Subjects by lists	71.59	14	5.11		
Total	313.96				

Yes, the lists are significantly different.

17. Here we form the ratio $.10 = \theta/(1 + \theta)$ and find that $\theta = 1/9$. Then the F ratio is

$$\frac{3.81}{1 + (4/9)} = 2.64.$$

This is larger than the value required for significance with $\alpha = .05$, for 6 and 21 degrees of freedom.

19.

Source	Fixed effects		Random effects		Mixed	
	F	p	F	p	F	p
Factor A	2.98	<.025	1.29	not signif.	1.29	not signif.
Factor B	6.12	<.01	2.64	<.05	6.12	<.01
Interaction	2.32	<.01	2.32	<.01	2.32	<.01

21. est. $\sigma^2_{groups} = \dfrac{.59 - .40}{10} = .009$

est. $\sigma^2_{somat.} = .175$, est. $\sigma^2_{int.} = 0$

est. $\sigma^2_Y = .584$

The proportion of variance accounted for by somatotype $= .30$.

23. Since $\rho_I = .5$, $\theta = 1$. Then power $= \text{prob}\left[F \geq \dfrac{F_{(.05)}}{1 + 4}\right] = \text{prob}(F \geq .42)$

We find that in a distribution with 33 and 10 degrees of freedom the probability of F greater than $(1/.42)$ or 2.38 is between .90 and .95. Hence the power is between .90 and .95.

25. Estimate of average $\alpha^2_j = \dfrac{22.02 - 5.82}{18} = .9$

Estimate of σ^2_{AB} $= \dfrac{5.82 - .81}{2} = 2.5$

Estimate of σ^2_B $= \dfrac{1.36 - .81}{6} = .09$

Estimate of proportion of σ^2_Y due to tasks $= .21$

Estimate of proportion of σ^2_Y due to classes $= .02$

Estimate of proportion of σ^2_Y due to interaction $= .58$

Estimate of proportion of σ^2_Y unaccounted for $= .19$

27. When the data in Exercise 9 are set up as in Table 13.19.1 of the text, we have

			Treatments				Block eff. b_k
Blocks	A_1	A_2	A_3	A_4	A_5	A_6	
1	5	14	8	10	11	6	−0.7500
2	7	10	7	9	12	5	−1.4167
3	11	9	10	11	14	6	0.4167
4	9	10	6	13	15	7	0.2500
5	13	12	7	14	16	11	2.4167
6	7	9	8	6	11	5	−2.0833
7	10	11	8	12	13	8	0.5833
8	4	8	5	7	9	4	−3.5833
9	14	13	11	15	17	12	3.9167
10	9	9	8	11	14	9	0.2500
treat. eff. a_j	−0.85	0.75	−1.95	1.05	3.45	−2.45	

Then

$$\sum_j \sum_k a_j\, b_k\, y_{jk} = 17.9979 \ .$$

Furthermore, we know that

SS blocks $= 249.42$, SS treatments (A) $= 240.95$,

SS A $\times$ blocks $=$ SS residual $= 102.88$,

so that

$$\text{SS non-add} = \frac{10 \times 6 \times 17.9979^2}{249.42 \times 240.95} = 0.32.$$

Then

SS remainder $=$ SS residual $-$ SS non-add $= 102.88 - 0.32 = 102.56$

and

MS remainder $= 102.56/(45 - 1) = 2.33$.

Thus, $F = .32/2.33 = 0.14$ with 1 and 44 degrees of freedom.

Hence, the hypothesis of additivity is definitely *not* rejected. According to this evidence it is quite safe to assume no interaction effects exist for blocks by treatments. One consequence is that blocks can be tested for significance by use of SS residual (or, if you will, SS A by blocks), which now represents only error.

CHAPTER 14

1. See Section 14.13.

3. $y' = 92.38 + 12.31x$. (Graph is not shown.)

5.

Source	SS	df	MS	F	p
Linear regression	37,157.14	1	37,157.14	158.59	$<<.01$
Deviations from linear regression	13,589.26	58	234.30		
Total	50,746.40	59			

7. Here the value of SP_{xy} total $= 25.245$, and that for SS_x total $= 60.0$ and SS_y total $= 30.250$. Hence,

$$b_{y.x} = (25.245)/(60) = 0.4208 = \text{est. } \beta_{y.x}.$$

The SS_y regression on $x = 0.4208^2 \times 60 = 10.62$, so that

$$\text{est. } \sigma_{y.x} = \sqrt{\frac{\text{SS deviations from regression}}{N - 2}}$$

$$= \sqrt{(30.25 - 10.62)/46} = 0.6533.$$

The confidence limits are then given by Eq. 14.16.1 to be

$$\text{est. } \beta_{y.x} \pm \frac{(\text{est. } \sigma_{y.x}) \, t_{(\alpha/2, N-2)}}{\sqrt{SS_x} \text{ total}}$$

or

$$0.4208 \pm (0.6533 \times 2.014)/\sqrt{60}.$$

The final confidence limits for $\beta_{y.x}$ are therefore .2509 and .5907.

9.

Source	SS	df	MS	F	p
Regression on x	30,945.03	1	30,945.03	199.65	$<<.01$
Deviations from reg.	12,864.90	83	155.00		
Total	43,809.93	84			

Here, in order to carry out the test of significance we assume that within each x value the true distribution of y values is normal, with the same variance. The x values may be regarded here as fixed, with a sampling of cases having taken place within that x value. No special assumption is then necessary about the distribution of x values.

11. $r_{AB} = .831$, $r_{AC} = .588$, $r_{BC} = .568$

13. $S_{z_A \cdot z_B} = \sqrt{1 - (.831)^2} = .556$

$S_{z_B \cdot z_C} = \sqrt{1 - (.568)^2} = .823$

$S_{z_C \cdot z_A} = \sqrt{1 - (.588)^2} = .809$

15. $r^2_{AB} = .69$, $r^2_{AC} = .35$, $r^2_{BC} = .32$, proportions of variance accounted for in one variable by the other.

17. $x' = 67.64 + .083y$ (the figure is not shown). The appropriate measure of scatter is $S^2_{x \cdot y} = S^2_x(1 - r^2_{xy}) = 1.58$, $S_{y \cdot x} = 1.26$.

19. $r_{xy} = .68$. The t value of 6.29 is significant beyond the .01 level for 46 degrees of freedom.

21. The confidence limits for the predicted proficiency, under the population regression equation, are about 84.17 and 120.21.

23. The difference between the correlations is not significant.

25. The table of intercorrelations among the X and Y variables is as follows:

	X_1	X_2	X_3	Y	Mean	SD
X_1	1	.7726	.4246	.6073	33.09	7.61
X_2		1	.3818	.4753	29.51	6.56
X_3			1	.7935	10.71	3.78
Y				1	30.77	9.83

The correlations between X_1 and X_3, and between X_2 and X_3 are significant only beyond the .05 level. The others reach significance beyond the .01 level.

CHAPTER 15

1. $r_{12 \cdot 3} = \dfrac{.38 - (.45 \times -.17)}{\sqrt{[1 - (.45)^2][1 - (.17)^2]}} = .5187$

$r_{23 \cdot 1} = \dfrac{-.17 - (.38 \times .45)}{\sqrt{[1 - (.38)^2][1 - (.45)^2]}} = -.4128$.

3. If the three correlations shown were all true, then it would also have to be true that

$r_{12 \cdot 3} = \dfrac{-.75 - (-.75)(-.75)}{1 - (-.75)^2} = -3.$

This cannot be true of any correlation, of course. If $r_{12} = r_{13} = r_{23} = r$, then if $r_{12 \cdot 3}$ is to be no smaller than -1,

$\dfrac{r - r^2}{1 - r^2} \geq -1$ so that $r \geq -1/2$.

5. If we let

$x_{B \cdot A} = x_B - [\bar{x}_B - b_{B \cdot A}(x_A - \bar{x}_A)]$

and

$x_{C \cdot A} = x_C - [\bar{x}_C - b_{C \cdot A}(x_A - \bar{x}_A)]$

then the adjusted data are as follows:

	$X_{B \cdot A}$	$X_{C \cdot A}$
1	−1.5008	0.7039
2	−0.7273	−0.1633
3	0.6281	−0.9406
4	1.4016	1.1922
5	−3.1141	−1.2297
6	1.4992	0.7039
7	−1.2430	−1.5852
8	1.4992	−1.2961
9	1.4016	−0.8078
10	−1.1141	−3.2297
11	1.4992	−1.2961
12	0.6281	−0.9406
13	−1.2430	1.4148
14	−0.3719	1.0594
15	0.7570	6.4148

The correlation between the adjusted $x_{B \cdot A}$ and the adjusted $x_{C \cdot A}$ values is then about .175. As a check we can also find the the value of the partial correlation, $r_{BC \cdot A}$, which is also .175:

$$r_{BC \cdot A} = \frac{.5675 - .8311(.5884)}{\sqrt{[1 - (.8311)^2][1 - (.5884)^2]}} = .175$$

7. Here, $B_2 = .4701$ and $B_3 = .5299$, so that we have

$$z_1' = .4701 (1.9) + .5299 (-1.2) = .2573 .$$

9. First we find

$$r_{y3 \cdot 1} = .7447, \; r_{y2 \cdot 1} = .0121, \; r_{23 \cdot 1} = .0935$$

so that

$$r_{y3 \cdot 12} = \frac{r_{y3 \cdot 1} - r_{y2 \cdot 1} (r_{23 \cdot 1})}{\sqrt{[1 - (r_{y2 \cdot 1})^2][1 - (r_{23 \cdot 1})^2]}}$$

$$= .7469 .$$

11. The data yield the correlation matrix, means, and standard deviations as follow:

	B	C	A	Mean	SD
B	1	.5675	.8311	8.5333	2.4998
C		1	.5884	6.8000	2.6128
A			1	7.3333	2.3851

The standardized regression weights may then be found:

$$B_B = .7335 \text{ and } B_C = .1721 .$$

The unstandardized weights are therefore

$$b_B = (2.3851)(.7335)/(2.4998) = .6998$$

and

$$b_C = (2.3851)(.1721)/2.6128 = .1571 .$$

Then,

$$x'_A = 7.3333 + .6998(x_B - 8.5333) + .1571(x_C - 6.8000)$$

or, on simplifying, the equation becomes

$$x'_A = .293 + .700x_B + .157x_C .$$

13. Given the correlations and the standardized regression weights found in Exercise 11, we find that

$$R^2_{A \cdot BC} = (.7335)(.8311) + (.1721)(.5884)$$

$$= .71.$$

Then since there are $K = 2$ predictors and $N = 15$ case, the significance may be tested by taking

$$F = \frac{.71 (15 - 2 - 1)}{(1 - .71)(2)}$$

$$= 14.69$$

which, for 2 and 12 degrees of freedom, is significant beyond the .01 level.

15. The normal equations for this multiple regression problem are:

$$B_1 + .7726B_2 + .4246B_3 = .6073$$

$$.7726B_1 + \quad B_2 + .3818B_3 = .4753$$

$$.4246B_1 + .3818B_2 + \quad B_3 = .7935 .$$

If we use the method of determinants to solve these equations, the results are the following standardized regression weights:

$$B_1 = .3839, B_2 = -.0726, B_3 = .6582 .$$

The corresponding unstandardized weights are then

$$b_1 = .3839(9.83)/(7.61) \quad = 0.4959$$

$$b_2 = (-.0726)(9.83)/(6.56) = -0.1088$$

$$b_3 = .6582(9.83)/(3.78) \quad = 1.7117$$

giving the regression equation

$$y' = 30.77 + 0.4959(x_1 - 33.09) - 0.1088(x_2 - 29.51) + 1.7117(x_3 - 10.71)$$

or

$$y' = -0.7610 + 0.4959x_1 - 0.1088x_2 + 1.7117x_3 .$$

17. Several approaches can be taken to this problem. In terms of the partial correlations $r_{y1} = .6073$, $r_{y2 \cdot 1} = .0121$, and $r_{y3 \cdot 12} = .7501$, we can find the value of the squared multiple correlation by solving

$$(1 - R^2_{y \cdot 123}) = (1 - r^2_{y1})(1 - r^2_{y2 \cdot 1})(1 - r^2_{y3 \cdot 12})$$

$$= .276$$

so that the value of R^2 is about .72. In terms of part correlations, we have $r_{y(2\cdot1)} = .0096$ and

$$r_{y(3\cdot12)} = r_{y3\cdot12} \sqrt{[1 - r^2_{y1}][1 - r^2_{y2\cdot1}]}$$

$$= .5959.$$

Then,

$$R^2_{y\cdot123} = (.6073)^2 + (.0096)^2 + (.5959)^2 = .72$$

just as before.

19. By use of the sweep-out method the successive matrices are found to be:

SWP(1) =

			B_j	$r^2_{y(j\cdot1)}$
.31	.27	.22	.22	
.9039	.2337	.4282		.2028
.2337	.9271	−.2894		.0903
.4282	−.2894	.9516		

$$R^2_{y1} = .0484$$

SWP(2)

		B_j	$r^2_{y(j\cdot12)}$
.3501	.3669	.3669	
.2585	.4737	.4737	
.8667	−.4001		.1847
−.4001	.7488		

$$R^2_{y\cdot12} = .2512$$

SWP(3)

	B_j
.5285	.5285
.5930	.5390
−.4616	−.4616
.5640	

$$R^2_{y\cdot123} = .4360$$

21. The matrices that result after the correlation matrix have been swept for each variable are

SWP(1)

				B_j	$r^2_{y(j\cdot1)}$
.15	.11	−.18	.26	.26	
.9775	−.1465	.2370	−.1890		.0365
−.1465	.9879	.0898	.3914		.1551
.2370	.0898	.9676	−.1532		.0243
−.1890	.3914	−.1532	.9324		

$$R^2_{y\cdot1} = .0676$$

SWP(12)

			B_j	$r^2_{y(j\cdot12)}$
.1325	−.2164	.2890	.2890	
−.1499	.2425	−.1934	.1934	
.9659	.1253	.3631		.1365
.1253	.9101	−.1074		.0127
.3631	−.1074	.8959		

$$R^2_{y\cdot12} = .1041$$

SWP(123)

$$
\begin{bmatrix}
-.2336 & .2392 \\
.2619 & -.1370 \\
.1297 & .3759 \\
.8938 & -.1545 \\
-.1545 & .7594
\end{bmatrix}
$$

	B_j	$r^2_{y(j \cdot 123)}$
	.2392	
	-.1370	
	.3759	
		.0267

$$R^2_{y \cdot 123} = .2406$$

SWP(1234)

$$
\begin{bmatrix}
.1988 \\
-.0918 \\
.3983 \\
-.1728 \\
.7327
\end{bmatrix}
$$

	B_j
	.1988
	-.0918
	.3983
	-.1728

$$R^2_{y \cdot 1234} = .2673.$$

CHAPTER 16

1. In terms of dummy variables, the data are as follow:

Case	X_1	X_2	X_3	Y
1	1	0	0	26
2	1	0	0	19
3	1	0	0	24
4	1	0	0	27
5	0	1	0	22
6	0	1	0	29
7	0	1	0	30
8	0	1	0	23
9	0	0	1	26
10	0	0	1	13
11	0	0	1	20
12	0	0	1	25
13	0	0	0	25
14	0	0	0	32
15	0	0	0	33
16	0	0	0	29

The matrix of correlation is then

	X_1	X_2	X_3	Y	Mean	SD
X_1	1	-.3333	-.3333	-.1381	0.25	0.4330
X_2		1	-.3333	.0945	0.25	0.4330
X_3			1	-.4869	0.25	0.4330
Y				1	25.1875	4.9651

$$SS_y \text{ total} = 16(4.9651)^2 = 394.44$$

3. After inclusion of X_1 in the regression equation for predicting Y, the proportion of variance accounted for is

$$r_{y1}^2 = R_{y\cdot 1}^2 = .0191.$$

The proportion of variance due both to X_1 and X_2 is

$$R_{y\cdot 12}^2 = .0217 .$$

Thus, the gain in proportion of variance accounted for due to X_2 is

$$R_{y\cdot 12}^2 - R_{y\cdot 1}^2 = r_{y(2\cdot 1)}^2 = .0217 - .0191 = .0026 .$$

Finally, the proportion of variance explained by X_1, X_2, and X_3 is

$$R_{y\cdot 123}^2 = .4099.$$

so that the gain in proportion of variance due to X_3 equals

$$R_{y\cdot 123}^2 - R_{y\cdot 12}^2 = r_{y(3\cdot 12)}^2 = .4099 - .0217 = .3882.$$

5. The correlation matrix is as follows:

	X_1	X_2	X_3	Y
X_1	1	$-.3321$	$-.3705$	$-.5080$
X_2		1	$-.4183$	$-.1062$
X_3			1	.4258
Y				1

This leads to a squared multiple correlation coefficient

$$R_{y\cdot 123}^2 = .3523.$$

The corresponding ANOVA table is

Source	SS	df	MS	F	p
Between	89.94	3	29.98	3.26	<.05
Within	165.38	18	9.19		
Total	255.32	21			

Here, the sample proportion of Y variance due to membership in the levels of Factor A is given by

$$R_{y\cdot A}^2 = SS_y \text{ between}/S_y \text{ total} = .3523 ,$$

which agrees, of course, with the R-square value found above.

7. In the multiple regression analysis it turns out that the proportion of variance due to variable X_1 is .258 with an additional .085 due to the addition of X_2 to the regression equation. Then the proportion of variance accounted for by X_1 and X_2 jointly is .343.

For the first comparison of Exercise 21, Chapter 11, the proportion of varaince explained is given by

$$SS_y(\hat{\psi}_1)/SS_y \text{ total} = 65.88/255.32 = .258 .$$

The equivalent proportion for the second comparison is .085, so that the variance explained either by the first or the second comparison is $.258 + .085 = .343$, agreeing with the values from the multiple regression analysis.

9. Below is the ANOVA summary table for the data in Exercise 8, along with the sample proportions of variance accounted for by Factor A, Factor B, and by the A × B interaction. Note the agreement with Exercise 8.

Source	SS	df	MS	F	p
Factor A	2,889.06	1	2889.06	104.8	<.01
Factor B	18.06	1	18.06	0.65	—
A × B Inter.	663.06	1	663.06	24.06	<.01
Error	330.76	12	27.56		
Total	3,900.94	15			

The proportions of variance accounted for are

$$R^2_{y \cdot A} = R^2_{y \cdot 1} = (2,889.06)/3,900.94 = .7406,$$

$$R^2_{y \cdot B} = R_{y \cdot 12} - R_{y \cdot 1} = (18.06)/3,900.94 = .0046,$$

and

$$R^2_{y \cdot A \times B} = R^2_{y \cdot 123} - R^2_{y \cdot 12} = (663.06)/3,900.94 = .1700.$$

11. The hierarchical analysis based on effects coding yields the following:

$$R^2_{y \cdot A} = R^2_{y \cdot 1} = .0024$$

so that the

$$SS_y \, A = R^2_{y \cdot A} \, (SS_y \, \text{total}) = 1.06.$$

For B, after adjustment for A, we have

$$R^2_{y(B \cdot A)} = R^2_{y \cdot 12} - R^2_{y \cdot 1} = .1079,$$

leading to

$$SS_y \, B \text{ beyond } A = R^2 \, (SS_y \, \text{total}) = 64.11.$$

Finally, the proportion of variance due to interaction, over and above A and B effects, is

$$R^2_{y(A \times B \cdot A,B)} = R_{y \cdot 123} - R^2_{y \cdot 12} = .4301,$$

so that the sum of squares for interaction is

$$SS \, A \times B \text{ beyond } A \text{ and } B = (.4301)(440.94) = 189.64.$$

The major difference between these SS values and those found in Exercise 10 is the shade larger SS B in the previous regression solution.

13. (In this exercise it is important to remember that each mean shown in a cell is actually based on $n = 3$ cases. Therefore, the row, column, and cell *sums* are actually 3 times the summed values obtained from the table itself. Consequently, the actual row sums are 270 for A_1, 417 for A_2, 387 for A_3, and 321 for A_4. Starting with B_1, the column sums are 267, 267, 246, 237, 216, and 162.) let us use the symbol a to stand for one of the four age values associated with the rows, or Factor A, and also use the symbol b to stand for one of the six deprivation

levels associated with Factor B. Then we can calculate the respective totals, and the SS values for a, b, and y, as well as the SP values for a and y and b and y:

$T_a = 180$, $T_b = -180$, and $T_y = 1,395$.

The sum of squares for variable a is

$$SS_a = [4^2 + 3^2 + 2^2 + 1^2]18 - (180^2/72) = 90.$$

Similarly the sum of squares of the column values b turns out to be

$$SS_b = [(-5)^2 + \cdots + (0)^2]12 - (-180^2/72) = 210.$$

The sums of products are

$$SP_{ay} = [4(270) + 3(417) + 2(387) + 1(321)] - [(180 \times 1,395)/72]$$

$$= -61.5$$

and

$$SP_{by} = [-5(267) - 4(267) - \cdots - 0(162)] - [(-180 \times 1,395)/72]$$

$$= -343.5 .$$

The SS for regression of Y on variable a is then

$$SS_y \text{ regression on } a = (SP_{ay})^2/SS_a = 42.03$$

and the equivalent term for variable b is

$$SS_y \text{ regression on } b = (SP_{by})^2/SS_b = 561.87 .$$

The completed analysis is summarized as follows:

Source	SS	df	MS	F	p
Regression on Factor A	42.03	1	42.03	1.16	NS
Regression on Factor B	561.87	1	561.87	15.55	<.01
Deviations	2,493.53	69	36.14		
Total	3,087.43	71			

Here the experimental design creates a correlation of 0 between variables a and b, and we are assuming no interaction between Factors A and B.

15. In Exercise 14, three trends turned out to be significant: the first degree, the second degree, and the fourth degree. The regression weight for each trend is found by dividing the actual comparison value $\hat{\psi}_g$ by the sum of the squared weights for that trend:

$$b_g = \hat{\psi}_g/nw_g$$

These regression weights turn out to be

$$b_1 = 510/70 = 7.2857$$

$$b_2 = -77.1/84 = -0.9179$$

$$b_4 = 93.5/23 = 3.3393 .$$

In addition, the overall mean $\bar{y} = 48.6$.

Our curvilinear regression equation for any group j is then

$$y_j' = 48.6 + 7.2851(c_{j1}) - 0.9179(c_{j2}) + 3.3393(c_{j4}) \,,$$

where the c_{jg} are the orthogonal polynomial weights applied to group j for the significant trends g (here, 1, 2, and 4). The predicted and actual values are

Hours	8	12	16	20	24	28
predicted	10.92	17.64	51.66	66.24	61.36	83.78
actual	11.1	18.5	48.2	69.7	60.5	83.6

17.

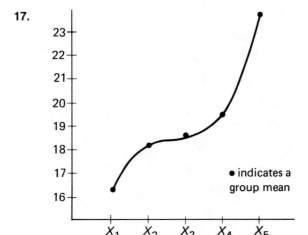

x-value	y'-value
19.5	16.35
39.5	18.26
59.5	18.43
79.5	19.41
99.5	23.78

• indicates a group mean

19. In the correlation problem model, X values are not fixed in advance, and thus planned comparisons among population means are meaningless in and of themselves. Furthermore, under the assumption of a bivariate normal population, most often made in a problem in correlation, there can be no curvilinear regression.

21. For this example, the ANOVA summary table would look something like this:

Source	SS	df	MS	F	p
between	(41,242.40)	(5)			
Quardratic	707.67	1	707.67	4.02	<.05
Other trends	40,534.73	4	10,133.68	57.58	<<.01
Error	9,504.00	54	176		
Total	50,746.40	59			

23. Here the proportion of variance accounted for by linear regression is $r_{y1}^2 = 7,744/21,390 = .362$. However, the proportion of variance accounted for by curvilinear trends is only $R_{y(G \cdot 1)}^2 = 336/21,390 = .016$. The sample relation is decidedly linear, and only minimally curvilinear.

25. Here the proportions of variance accounted for by the different trends and interactions are as follows:

$$r^2_{y \cdot A \text{ lin}} = 42.03/3,097.43 = .0136$$

$$r^2_{y \cdot A \text{ quad}} = 630.13/3,097.43 = .2034$$

and

$$r^2_{y \cdot A \text{ other}} = 55.23/3,093.43 = .0179$$

For Factor B:

$$r^2_{y \cdot B \text{ lin}} = 561.87/3,097.43 = .1814$$

$$r^2_{y \cdot B \text{ quad}} = 72.32/3,097.43 = .0233.$$

and

$$r^2_{y \cdot B \text{ other}} = 17.95/3,097.43 = .0058$$

For trend interactions we then have

$$r^2_{y \cdot A \text{ lin} \times B \text{ lin}} = 67.13/3,097.43 = .0217$$

$$r^2_{y \cdot A \text{ quad} \times B \text{ lin}} = 156.87/3,097.43 = .0506$$

$$r^2_{y \cdot A \text{ lin} \times B \text{ quad}} = 6.01/3,097.43 = .0019$$

$$r^2_{y \cdot A \text{ quad} \times B \text{ quad}} = 95.57/3,097.43 = .0312 .$$

The proportion of variance due to all other interactions is then

$$R^2_{Y \cdot \text{other } A \times B} = 369.78/3,097.43 = .119.$$

CHAPTER 17

1.

Source	SS	df	MS	F	p
Adj. means	559.62	3	186.54	86.18	$\ll .01$
Adj. error	43.21	19	2.27		
Adj. total	602.83	22			

There appears to be a strong relation between authoritarianism and success, even when anxiety is controlled as a factor. The proportion of total Y variance accounted for in the sample by the adjusted means is $559.62/667.96 = .84$. Observe that the denominator is taken to be the unadjusted SS total of 667.96.

3. The test for homogeneity of regression may be summarized as follows:

Source	SS	df	MS	F	p
Between reg.	5.24	3	1.75	0.74	
Remainder	37.97	16	2.37		—
Adj. error	43.21	19			

Since the F value is less than 1.00, we can be quite comfortable with the assumption of homogeneity of regression here.

5. Ignoring the variable X we get the following ANOVA table:

Source	SS	df	MS	F	p
Columns	4,994.13	2	2,497.07	96.66	$<<.01$
Remainder (error)	1,457.60	57	25.57		
Total	6,451.73	59			

Both columns and the "error" sums of squares are larger here than in the ANCOVA above, because of the influence of the uncontrolled variability due to the differing levels of skill of the subjects. (Note that the term *error* here is, however, much greater than in the original problem of Section 12.7, where effects of norm groups and of interaction were both removed from the error sum of squares.)

7. The regression equation used to find the adjusted group means is here

adj. $\bar{y}_j = \bar{y}_j - b_w (\bar{x}_j - \bar{x})$,

where

$b_w = (SP_{xy}$ within$)/SS_x$ within $= -.9236$.

The adjusted means are then approximately

adj. $\bar{y}_1 = 20.84$;   adj. $\bar{y}_2 = 26.46$:   adj. $\bar{y}_3 = 36.70$.

9. The test for homogeneity of regression is summarized below:

Source	SS	df	MS	F	p
Between reg.	61.93	5	12.39	1.04	NS
Remainder	571.27	48	11.9		
Adj.err.	633.2	53			

Since the F value in this test fails to reach significance even for the .25 level, it appears safe to conclude that the populations represented by the cells have homogeneous regressions of Y on X.

11. The two analyses of covariance are summarized as follows: For Y_3 with Y_1 as covariate,

Source	SS	df	MS	F	p
Adj. Groups	76.95	1	76.95	10.30	<.01
Adj. error	127.03	17	7.47		
Adj. total	203.98	18			

Again for Y_3, but with Y_2 as covariate,

Source	SS	df	MS	F	p
Adj. Groups	9.63	1	9.63	0.86	—
Adj. error	190.00	17	11.18		
Adj. total	199.63	18			

As compared with the original SS between for Y_3, the adjusted values are both larger, but only that adjusted for Y_1 is significant.

13. The relevant SSCP matrices here are as follows (with X listed first throughout)

Factor A:
$$\begin{bmatrix} 50.81 & 46.93 \\ & 47.15 \end{bmatrix}$$

Factor B:
$$\begin{bmatrix} 167.70 & 115.04 \\ & 517.15 \end{bmatrix}$$

Factor C:
$$\begin{bmatrix} 28.17 & 79.44 \\ & 224.07 \end{bmatrix}$$

A × B int.:
$$\begin{bmatrix} 210.74 & 56.91 \\ & 68.19 \end{bmatrix}$$

A × C int.:
$$\begin{bmatrix} 17.33 & -32.11 \\ & 74.93 \end{bmatrix}$$

B × C int.:
$$\begin{bmatrix} 55.11 & 97.11 \\ & 235.15 \end{bmatrix}$$

A × B × C int.:
$$\begin{bmatrix} 168.89 & 18.39 \\ & 154.85 \end{bmatrix}$$

Within cells:
$$\begin{bmatrix} 873.33 & 158.00 \\ & 486.17 \end{bmatrix}$$

The ANCOVA summary table is then

Source	SS	df	MS	F	p
adj. A	30.29	2	15.15	1.16	NS
adj. B	474.12	2	237.06	18.11	<.01
adj. C	211.04	1	211.04	16.12	<.01
adj. A × B	54.17	4	13.54	1.03	NS
adj. A × C	85.72	2	42.86	3.27	.05
adj. B × C	193.64	2	96.82	7.40	<.01
adj. A × B × C	153.58	4	38.40	2.93	<.05
adj. error	458.08	35	13.09		

After adjustment for variable X, Factors B and C seem to show the strongest relationships with variable Y, although the significant adjusted B × C interaction suggests that one should attempt to predict particular adjusted B effects only given information about C, and *vice versa*.

15. See Section 17.15.

CHAPTER 18

1. The chi-square value of about 9.18 is not significant for 8 degrees of freedom. The hypothesis that true $p = .5$ is not rejected.
3. The value of chi-square is 74.00, with 8 degrees of freedom. This is significant beyond the .01 level, so that we may safely reject the hypothesis that the stated distribution characterizes the population of girls.
5. Here, $\chi^2 =$ about 202, which, for 8 degrees of freedom, is very significant. The null hypothesis of equal likelihood in any half-century may be rejected.
7. The value of V_C' is .46. The significance is tested by chi-square.
9. Prediction of income from ethnic balance: $\lambda_A = .298$; prediction of ethnic balance from income: $\lambda_B = .427$.
11. Occupation from interest pattern: $\lambda_B = .256$; interest pattern from occupation: $\lambda_A = .238$.
13. The chi-square value of about 10.9 is not significant for 6 degrees of freedom.
15. Since the mean of each sample is exactly 6, then $F = 0$. However, samples do differ in other ways, and these are the things reflected in the chi-square test.
17. No strong relationship is suggested, since the proportional reduction in error in predicting parental dominance, given the culture, is only about .07.
19. The two relevant tables here are

1	5	6
5	1	6
6	6	

with probability of $\dfrac{\binom{6}{1}\binom{6}{5}}{\binom{12}{6}} = .039$

and

0	6	6
6	0	6
6	6	

with probability of $\dfrac{\binom{6}{0}\binom{6}{6}}{\binom{12}{6}} = .001.$

21. The chi-square value of about 14.8 is not quite significant at the .05 level, for 9 degrees of freedom.
23. The chi-square value corrected for continuity is about .40. This is not significant. We do not reject the hypothesis of independence between identical twinship and sex of infant. Here one assumes that every entry in the table is independent of every other (note that each set of twins accounts for only one entry).

CHAPTER 19

1. For these data the mean is 37.3692, and the estimated standard deviation for the population is 15.5672. After conversion of each of the real class limits of the distribution into standardized values, using these estimates of the population and standard deviation, we then can find proportion in the corresponding class interval of z values in a normal distribution. These are shown below, along with cumulative relative frequencies for the sample distribution $F_S(x)$ as well as for the theoretical normal distribution, or $F_T(x)$.

z interval	$F_S(x)$	$F_T(x)$
1.81 up	1.00	1.00
1.36–1.81	.9766	.9648
0.91–1.36	.8972	.9131
0.46–0.91	.8037	.8186
0.01–0.46	.6682	.6772
−0.44–0.01	.5047	.5039
−0.89–0.44	.3224	.3300
−1.34–0.89	.1963	.1867
−1.79–(−1.34)	.1028	.0901
−2.24–(−1.79)	.0467	.0367

then the largest absolute difference between corresponding $F_S(x)$ and $F_T(x)$ values is

$$D_K = |.8037 - .8186| = .0149.$$

When this is compared with the required value of .093 as given in Table XIII, Appendix E, it is easily seen that this result is not significant at the .05 level. Hence, the null hypothesis of a normal population distribution is *not* rejected.

3. The cumulative relative frequencies for the two distributions are shown below:

Class interval	$F_1(x)$	$F_2(x)$
73–79	1.00	1.00
66–72	1.00	.9836
59–65	.9766	.9180
52–58	.8972	.7596
45–51	.8037	.6120
38–44	.6682	.4426
31–37	.5047	.2896
24–30	.3224	.1639
17–23	.1963	.0984
10–16	.1028	.0492
3– 9	.0467	.0109

The maximum absolute difference between the cumulative relative frequencies of the two samples is then

$$D_K = .6682 - .4426 = .2256,$$

so that the chi-square approximation of Eq. 19.1.3 is

$$\chi^2 = 4(.2256)^2[(214) \times (183)/(214 + 183)] = 20.08.$$

For 2 degrees of freedom this exceeds the value required for significance at the .01 level. The two populations seem to be different in their distributions of this variable.

5. Here a z value of 2.3 permits us to reject the null hypothesis beyond the .05 level, two-tailed.

7. The chi-square value of about 52 with 4 degrees of freedom is very significant.

9. The H value of 12.59 is significant beyond the .01 level for 3 degrees of freedom.

11. For B and C, the z value of −2.21 is significant beyond the .05 level. For A and C the z value of −1.87 is not quite significant. Technically, these three tests are not independent of each other, and the probability of at least one Type I error is about .15 when α = .05 for a single test.

13. The H value of 1.56 is not significant for 4 degrees of freedom.

15. For $U' = 35.5$, $E(U) = 40.5$, and $\sigma_U = 11.32$, the z value of $-.44$ is not significant.

17. The Q value of 16.4 is significant beyond the .01 level for 4 degrees of freedom. The six puzzles do appear to differ significantly in difficulty.

19. The t value of 7.06 is significant far beyond the .01 level for 31 degrees of freedom.

21. Here, $N_1 = N_2 = 10$, and the number of runs is 8. The probability of exactly eight runs is

$$p(8) = 2\frac{\binom{9}{3}\binom{9}{3}}{\binom{20}{10}} = .076$$

Since this is greater than .05, we know immediately that the probability of eight or fewer runs is also greater than .05, so that the result is not significant. We can also find $E(R) = 11$, $\sigma_R^2 = 4.74$, $\sigma_R = 2.18$, so that $z = -1.38$. This is clearly not significant.

23. If we score $y \geq 6$ as a "1," and $y < 6$ as a "0," then the Q obtained is 13.57.

25. In terms of the γ measure, $S_+ = 1,973$, $S_- = 781$, so that $\gamma = .43$. The corresponding value of tau is .306.

27. Here $N_1 = 10$, $N_2 = 12$. There are 10 runs, with $E(F) = 11.9$. The probability of 10 runs is

$$\frac{2\binom{9}{4}\binom{11}{4}}{\binom{22}{10}} = .129$$

Since the probability of exactly 10 runs is about .13, the probability of 10 or fewer runs is obviously $>.05$. Hence the result is not significant.

29. For Items II and III, $\gamma = .47$; for Items I and III, $\gamma = .07$; for Items I and II, $\gamma = -.18$.

31. The chi-square value of 35.51 with 9 degrees of freedom is significant beyond the .01 level. The gamma measure then has a value of $-.45$. There is a fairly negative ordinal association.

33. There are six inversions in order between pairs, sot that $\tau = .73$.

35. Here the τ value is about .76.

APPENDIX A

1. $x_1 + x_2 + x_3 + x_4 + x_5$

3. $x_1^3 + x_2^3 + x_3^3 + x_5^3 + x_6^3 + x_7^3 + x_8^3$

5. $\displaystyle\sum_{i=1}^{3} 4\left(\sum_{j=1}^{2} x_{ij}\right) = 4\sum_{i=1}^{3}(x_{i1} + x_{i2}) = 4\left(\sum_{i}^{3} x_{i1} + \sum_{i}^{3} x_{i2}\right)$

$$= 4[(x_{11} + x_{21} + x_{31}) + (x_{12} + x_{22} + x_{32})]$$

7. $(x_1^2 g_1 + x_2^2 g_2 + x_3^2 g_3 + x_4^2 g_4)/(g_1 + g_2 + g_3 + g_4)$.

9. $\displaystyle\sum_{i=1}^{5}\left(\sum_{j=2}^{4} x_{ij}^2 + \sum_{j=2}^{4} y_i\right) = \sum_{i=1}^{5}(x_{i2}^2 + x_{i3}^2 + x_{i4}^2 + 3y_i) = 3(y_1 + \cdots + y_5) + x_{12}^2 + \cdots$

$$+ x_{52}^2 + x_{13}^2 + \cdots + x_{53}^2 + x_{14}^2 + \cdots + x_{54}^2$$

11. $\left(\sum_{i=1}^{4} x_i\right)/N$

13. $\sum_{i=2}^{4} x_i f(x_i)$

15. $3 \sum_{i=1}^{N} (x_1 + y_i)^2$

17. $15 \left(\sum_{i=1}^{4} x_i\right)^2$

19. $4 \sum_{i=2}^{4} \frac{(x_i - 5)^2}{y_i}$.

21. $\left(\sum_{i=1}^{4} x_i\right)^2 \left(\sum_{i=1}^{4} y_i^2\right)$

23. $\sum_{i=1}^{5} x_i + 210c$

25. 0

27. $N\sqrt{(a + b)c}$

29. $\text{var}(X) = \dfrac{\sum x_i^2}{N} - \bar{x}^2$ where $\bar{x} = \left[\sum_{i=1}^{N} x_i\right]\bigg/ N$

31. -3

33. -26

35. $47/23 = 2.04$

APPENDIX B

1. .72

3. 3.1

5. 21.66

7. $E(X^2)$

9. $E[X - E(X)]^2 = \text{var}(X)$

11. $\dfrac{E(X)}{10} + 3.5$

13. $E(X^2) - E^2(X) = \text{var}(X)$

15. 17

17. $E(aX) = \sum_i a x_i p(x_i) = a \sum_i x_i p(x_i) = aE(X)$

19. $E(X - E(X)) = E(X) - E(E(X)) = E(X) - E(X) = 0$

21. *(1)* -8.25
     (2) -9.22
     (3) 3.50

23. Since the variables are independent, $E(XY) = E(X)E(Y)$, so $E(XY) = (3.1)\,(115.60) = 358.36$.

References and Suggestions for Further Reading

PROBABILITY THEORY

Derman, C., Gleser, L. J., & Olkin, I. (1973). *A guide to probability theory and applications*. New York: Holt, Rinehart, and Winston.

de Finetti, B. (1974). *Theory of probability: A critical introductory treatment* (Vol. I). London: Wiley.

Feller, W. (1968). *An introduction to probability theory and its applications* (Vol. 1, 3rd ed.). New York: Wiley.

Kyburg, H. E., & Smokler, H. E. (1964). *Studies in subjective probability*. New York: Wiley.

Parzen, E. (1960). *Modern probability theory and its applications*. New York: Wiley.

Ross, S. (1972). *Introduction to probability models*. New York: Academic Press.

Thompson, W. A. (1969). *Applied probability*. New York: Holt, Rinehart, and Winston.

MEASUREMENT AND RELATED MATTERS

Coombs, C. H., Dawes, R. M., & Tversky, A. (1970). *Mathematical psychology: An elementary introduction*. Englewood Cliffs, NJ: Prentice-Hall.

Krantz, D. H., Luce, R. D., Suppes, P., & Tversky, A. (1971). *Foundations of measurement*. New York: Academic Press.

Stevens, S. S. (1951). *Handbook of experimental psychology*. New York: Wiley.

Thrall, R. M., Coombs, C. H., & Davis, R. L. (1959). *Decision processes*. New York: Wiley.

Torgerson, W. (1958). *Theory and methods of scaling*. New York: Wiley.

Tufte, E. R. (1983). *The visual display of quantitative information*. Cheshire, CT: Graphics Press.

STATISTICAL INFERENCE IN GENERAL

Brunk, H. D. (1965). *An introduction to mathematical statistics* (2nd ed.). Boston: Ginn and Co.

Cramér, H. (1946). *Mathematical methods of statistics*. Princeton, NJ: Princton University Press.

Dixon, W. J., & Massey, F. J. (1969). *Introduction to statistical analysis* (3rd ed.). New York: McGraw-Hill.

Freund, J. E. (1971). *Mathematical statistics* (2nd ed.). Englewood Cliffs, NJ: Prentice-Hall.

Harnett, D. L. (1975). *Introduction to statistical methods* (2nd ed.). Reading, MA. Addison-Wesley.

Hodges, J. L., & Lehmann, E. L. (1970). *Basic concepts of probability and statistics* (2nd ed.). San Francisco: Holden-Day.

Hoel, P. G. (1971). *Introduction to mathematical statistics* (4th ed.). New York: Wiley.

Hogg, R. V., & Craig, A. T. (1970). *Introduction to mathematical statistics* (3rd ed.). New York: Macmillan.

Huff, D. (1954). *How to lie with statistics*. New York: Norton.

Kendall, M. G., & Stuart, A. (1958/1961/1966). *The advanced theory of statistics*. (Vols. I–III). London: Charles Griffin.

Lippman, S. A. (1971). *Elements of probability and statistics*. New York: Holt, Rinehart, and Winston.

Mood, A. M., Graybill, F. A., & Boes, D. C. (1974). *Introduction to the theory of statistics* (3rd ed.). New York: McGraw-Hill.

Tate, R. F. & Klett, G. W. (1959). Optimal confidence intervals for the variance of a normal distribution. *Journal of the American Statistical Association, 54,* 674–682.

Tukey, J. W. (1977). *Exploratory data analysis*. Reading, MA: Addison-Wesley.

Wilks, S. S. (1962). *Mathematical statistics*. New York: Wiley.

DECISION THEORY

Brown, R. V., Kahr, A. S., & Peterson, C. (1974). *Decision analysis for the manager*. New York: Holt, Rinehart, and Winston.

Chernoff, H., & Moses, L. E. (1959). *Elementary decision theory*. New York: Wiley.

Lindley, D. V. (1971). *Making decisions*. London: Wiley.

Luce, R. D., & Raiffa, H. (1957). *Games and decisions*. New York: Wiley.

Novick, M. R., & Jackson, P. H. (1974). *Statistical methods for educational and psychological research*. New York: McGraw-Hill.

Phillips, L. D. (1974). *Bayesian analysis for social scientists*. New York: Thomas Y. Crowell,

Schlaifer, R. (1969). *Analysis of decisions under uncertainty*. New York: McGraw-Hill.

Winkler, R. L. (1972). *An introduction to Bayesian inference and decision*. New York: Holt Rinehart, and Winston.

TOPICS IN ANALYSIS OF VARIANCE AND EXPERIMENTAL DESIGN

Box, G. E. P. (1953). Non-normality and tests on variances. *Biometrika, 40,* 318–335.

Box, G. E. P (1954a). Some theorems on quadratic forms applied in the study of analysis of variance problems: I. Effects of inequality of variance in the one-way classification. *Annals of Mathematical Statistics, 25,* 290–302.

Box, G. E. P. (1954b). Some theorems on quadratic forms applied in the study of analysis of variance problems: II. Effects of inequality of variance and of correlation of errors in the two-way classification. *Annals of Mathematical Statistics 25*, 484–489.

Bryant, J. L. & Paulson, A. S. (1976). An extension of Tukey's method of comparisons to experimental designs with random concomitant variables. *Biometrika, 63*, 631–638.

Cochran, W. G., & Cox, G. M. (1957). *Experimental designs* (2nd ed.). New York: Wiley.

Cohen, J. (1977). *Statistical power analysis for the behavioral sciences* (rev. ed.). New York: Academic Press.

Cook, T. D., & Campbell, D. T. (1979). *Quasi-experimentation: Design and analysis issues for field settings*. Chicago: Rand-McNally.

Cox, D. R. (1958). *Planning of experiments*. New York: Wiley.

Edwards, A. L. (1985). *Experimental design in psychological research* (5th ed.). New York: Harper and Row.

Feingold, M., & Korsog, P. E. (1986). The correlation and dependence between two F statistics with the same denominator. *The American Statistician, 40*, 218–220.

Feldt, L. S. (1958). A comparison of three experimental designs employing a concomitant variable. *Psychometrika, 23*, 335–353.

Fisher, R. A. (1966). *The design of experiments* (8th ed.). Edinburgh: Oliver and Boyd.

Glass, G. V., & Hakstian, A. R. (1969). Measures of association in comparative experiments: Their development and interpretation. *American Educational Research Journal, 6*, 404–413.

Graybill, R. A. (1961). *An introduction to linear statistical models* (Vol I.). New York: McGraw-Hill.

Guenther, W. C. (1964). *Analysis of variance*. Englewood Cliffs, NJ: Prentice-Hall.

Herr, D. G. (1986). On the history of ANOVA in unbalanced factorial designs: The first 30 years. *American Statistician, 40*, 265–270.

Huitema, B. E. (1980). *The analysis of covariance and alternatives*. New York: Wiley.

Huynh, H., & Feldt, L. S. (1970). Conditions under which mean square ratios in repeated measures designs have exact F distributions. *Journal of the American Statistical Association, 65*, 1582–1589.

Huynh, H., & Feldt, L. S. (1976). Estimation of the Box corrector for degrees of freedom from sample data in randomized block and split-plot designs. *Journal of Educational Statistics, 1*, 69–82.

Huynh, H., & Feldt, L. S. (1980). Performance of traditional F tests in repeated measures designs under covariance heterogeneity. *Communications in Statistics—Theoretical Methods, A9(1)*, 61–74.

Huynh, H., & Mandeville, G. K. (1979). Validity conditions in repeated measures designs. *Psychological Bulletin, 86*, 964–973.

Keppel, G. (1982). *Design and analysis: A researcher's handbook* (2nd ed.). Englewood Cliffs, NJ: Prentice-Hall.

Keuls, M. (1952). The use of the studentized range in connection with an analysis of variance. *Euphytica, 1*, 112–122.

Kirk, R. E. (1982). *Experimental design: Procedures for the behavioral sciences*. (2nd ed.). Belmont, CA: Brooks-Cole.

Mauchley, J. W. (1940). Significance tests for sphericity of a normal n-variate distribution. *Annals of Mathematical Statistics, 11*, 204–209.

Mendenhall, W. (1968). *Introduction to linear models and the design and analysis of experiments*. Belmont, CA: Wadsworth.

Myers, J. L. (1979). *Fundamentals of experimental design* (3rd ed.). Boston: Allyn and Bacon.

Neter, J., & Wasserman, W. (1974). *Applied linear statistical models*. Homewood, IL: Richard D. Irwin.

Newman, D. (1939). The distribution of the range in samples from a normal population, expressed in terms of an independent estimate of standard deviation. *Biometrika, 31,* 20–30.

Overall, J. E., Lee, P. M., & Hornick, C. W. (1981). Comparison of two strategies for analysis of variance in nonorthogonal designs. *Psychological Bulletin, 90,* 367–375.

Overall, J. E., & Spiegel, D. K. (1969). Concerning least squares analysis in experimental data. *Psychological Bulletin, 72,* 311–322.

Overall, J. E., Spiegel, D. K., & Cohen, J. (1975). Equivalence of orthogonal and nonorthogonal analysis of variance. *Psychological Bulletin, 82,* 182–186.

Rouanet, H., & Lepine, D. (1970). Comparisons between treatments in a repeated measurement design: ANOVA and multivariate methods. *Bristish Journal of Mathematics and Statistics for Psychology, 23,* 147–163.

Scheffé, H. (1959). *The analysis of variance.* New York: Wiley.

Snedecor, G. W., & Cochran, W. G. (1980). *Statistical methods* (7th ed.). Ames, IA: Iowa State College Press.

Tukey, J. W. (1949). One degree of freedom for nonadditivity. *Biometrics, 5,* 232–242.

Vaughn, G. M., & Corballis, M. C. (1969). Beyond tests of significance: Estimating strength of effects in selected ANOVA designs. *Psychological Bulletin, 72,* 204–213.

Ward, J., & Jennings, E. (1973). *Introduction to linear models.* Englewood Cliffs, NJ: Prentice-Hall.

Winer, B. J. (1971). *Statistical principles in experimental design* (2nd ed.). New York: McGraw-Hill.

CORRELATION AND REGRESSION, MULTIVARIATE METHODS

Bennett, S., & Bower, D. (1976). *An introduction to multivariate techniques for the social and behavioral sciences.* New York: Wiley.

Blalock, H. M. (Ed.) (1970). *Causal models in the social sciences.* Chicago: Aldine Press.

Carroll, J. B. (1961). The nature of the data, or how to choose a correlation coefficient. *Psychometrika, 26,* 347–372.

Cliff, N. (1987). *Analyzing multivariate data.* Orlando, FL: Harcourt-Brace-Jovanovich.

Cohen, J., & Cohen, P. (1983). *Applied multiple regression: Correlation analysis for the behavioral sciences* (2nd ed.). New York: Erlbaum.

Draper, N. R., & Smith, H. (1981). *Applied regression analysis* (2nd ed.). New York: Wiley.

Green, P. E., & Carroll, J. D. (1976). *Mathematical tools for applied multivariate analysis.* New York: Academic Press.

Guilford, J. P., & Fruchter, B. (1978). *Fundamental statistics in psychology and education* (6th ed.). New York: McGraw-Hill.

Harris, R. J. (1985). *A primer of multivariate statistics* (2nd ed.). New York: Academic Press.

Horst, P. (1963). *Matrix algebra for social scientists.* New York: Holt, Rinehart, and Winston.

Konishi, S. (1981). Normalizing transformations of some statistics in multivariate analysis. *Biometrika, 68,* 647–651.

Lewis, D. (1960). *Quantitative methods in psychology.* New York: McGraw-Hill.

Marascuilo, L. A., & Levin, J. R. (1983). *Multivariate statistics in the social sciences.* Monterey, CA: Brooks-Cole.

Morrison, D. F. (1976). *Multivariate statistical methods* (2nd ed.). New York: McGraw-Hill.

Mosteller, F., & Tukey, J. W. (1977). *Data analysis and regression.* Reading, MA: Addison-Wesley.

Nunnally, J. C. (1978). *Psychometric thoery.* (2nd ed.) New York: McGraw-Hill.

Overall, J. E. and Klett, C. J. (1972). San Franscisco, CA: McGraw-Hill.

Pedhazer, E. (1982). *Multiple regression in behavioral research* (2nd ed.). New York: Holt, Rinehart, and Winston.

Speed F. M., Hocking, P. R., & Hackney, O. P. (1978). Methods of analysis of linear models with unbalanced data. *Journal of the American Statistical Association, 73,* 105–112.

Tabachnick, B. G., & Fidell, L. S. (1983). *Using multivariate statistics.* New York: Harper and Row.

Timm, N. H. (1975). *Multivariate analysis, with applications in education and psychology.* Belmont, CA: Brooks-Cole.

QUALITATIVE AND ORDINAL METHODS

Bishop, Y. M. M., Fienberg, S. E., & Holland, P. W. (1975). *Discrete multivariate analysis: Theory and practice.* Cambridge, MA: M. I. T. Press.

Bradley, J. V. (1968). *Distribution-free statistical tests.* Englewood, NJ: Prentice-Hall.

Cochran, W. G. (1950). The comparison of percentages in matched samples. *Biometrika, 37,* 256–266.

Cohen, J. (1960). A coefficient of agreement for nominal scales. *Educational and Psychological Measurement, 20,* 37–46.

Fienberg, S. E. (1980). *The analysis of cross-classified categorical data* (2nd ed.). Cambridge, MA:M. I. T. Press.

Fisher, R. A. (1934). Statistical methods for research workers (5th ed.). Edinburgh: Oliver and Boyd.

Fisher, R. A. (1940). The precision of discriminant functions. *Annals of Eugencis, 10,* 422–429.

Fleiss, J. L. (1971). Measuring nominal scale agreement among many raters. *Psychological Bulletin, 76,* 378–382.

Friedman, M. The use of ranks to avoid the assumption of normality implicit in the analysis of variance. (1937). *Journal of the American Statistical Association, 32,* 675–701.

Gibbons, J. D. (1971). *Nonparametric statistical inference.* New York: McGraw-Hill.

Gilula, Z. (1986). Grouping and association in contingency tables: An exploratory canonical correlation approach. *Journal of the American Statistical Association, 81,* 773–779.

Goodman, L. A. (1978). *Analyzing qualitative/categorical data.* Cambridge, MA: Abt Books.

Goodman, L. A. (1984). *The analysis of cross-classified data having ordered categories.* Cambridge, MA: Harvard University Press.

Goodman, L. A., & Kruskal, W. H. (1979). *Measures of association for cross-classifications.* New York: Springer-Verlag.

Haberman, S. J. (1981). Tests for independence in two-way contingency tables based on canonical correlation and on linear by linear interaction. *Annals of Statistics, 9,* 1178–1186.

Kendall, M. G. (1963). *Rank correlation methods* (3rd ed.). London: Griffin.

Kennedy, J. J. (1983). *Analyzing qualitative data.* New York: Praeger.

Kolmogorov, A. (1941). Confidence limits for an unknown distribution function. *Annals of Mathematical Statistics, 12,* 461–463.

Kruskal, W. H. (1958). Ordinal measures of association. *Journal of the American Statistical Association, 53,* 814–861.

Kruskal, W. H., & Wallis, W. A. (1952). Use of ranks in one-criterion variance analysis. *Journal of the American Statistical Association, 47,* 583–621.

Mann, H. B., & Whitney, D. R. (1947). On a test of whether one of two variables is stochastically larger than the other. *Annals of Mathematical Statistics, 18,* 50–60.

Maxwell, A. E. (1961). *Analysing qualitative data.* London: Methuen.

McNemar, Q. (1975). *Psychological statistics* (5th ed.). New York: Wiley.

Siegel, S. (1956). *Nonparametric methods for the behavioral sciences*. New York: McGraw-Hill.

Wald, A., & Wolfowitz, J. (1940). On a test whether two samples are from the same population. *Annals of Mathematical Statistics, 11,* 147–162.

Walsh, J. E. (1962, 1967). *A handbook of nonparametric statistics* (2 vols.) New York: Van Nostrand.

Wilcoxon, F. (1949). *Some rapid approximate statistical procedures*. Stamford, CT: American Cyanamid.

Wilson, E. B., & Hilferty, M. M. (1931). The distribution of chi-square. *Proceedings of National Academy of Sciences, 17,* 684–688.

COMPUTER PACKAGES AND PROGRAMS

Bock, R. D., & Yates, G. (1973). *MULTIQUAL: Log-linear analysis of nominal and ordinal data by the method of maximum likelihood*. Chicago: National Educational Resources.

Dixon, W. J. (ed.) (1985). *BMDP statistical software manual*. Berkeley, CA: University of California Press.

Fay, R. E., & Goodman, L. A. (1975). *ECTA program: Description for users*. Chicago: University of Chicago.

Hull, C. H., & Nie, N. H. (1981). *SPSS update 7–9*. New York: McGraw-Hill.

Nie, N. H., Hull, C. H., Jenkins, J. G., Steinbrenner, K., & Bent, D. H. (1975). *Statistical package for the social sciences*. (2nd ed.). New York: McGraw-Hill.

Ryan, T. A., Joiner, B. L., & Ryan, B. F. (1976). *MINITAB student handbook*. Duxbury, MA: Duxbury Press.

SAS Institute, Inc. (1982). *SAS user's guide* (1982 edition). Cary, NC: Author.

SPSS Inc. (1986). *SPSS^x user's guide* (2nd ed.). Chicago: Author.

MATHEMATICAL RESOURCES AND STATISTICAL TABLES

Korn, G. A., & Korn, T. M. (1968). *Mathematical handbook for scientists and engineers* (2nd ed.). New York: McGraw-Hill.

Noble, B., & Daniel, J. W. (1977). *Applied linear algebra* (2nd ed.). Englewood Cliffs, NJ: Prentice-Hall.

Pearson, E. S., & Hartley, H. O. (1966). *Biometrika tables for statisticians*, (Vol. 1, 3rd ed.). Cambridge: Cambridge University Press.

RAND Corporation. (1955). *A million random digits, with 100,000 normal deviates*. New York: Free Press.

Glossary of Symbols

CONVENTIONAL MATHEMATICAL SYMBOLS

SET THEORY

A	a set, or event (1.3)
$A = \{a,b,c,d\}$	set A includes the following members; a set specified by listing (1.4)
$a \leq b$	a is less than or equal to b (1.5)
$a \geq b$	a is greater than or equal to b (3)
$a < b$	a is less than (but not equal to) b (2.1)
$a > b$	a is greater than (but not equal to) b (2.1)
$\varnothing$	the empty or "impossible" event; the null set (1.3)
$A \cup B$	union of sets or events A and B (1.3)
$A \cap B$	intersection of sets or events A and B (1.3)
$\overline{A}$	complement of set A; the event "not A" (1.3)
$A - B$	difference between sets A and B (15.20)
$A = B$	sets A and B are equal or equivalent (1.5)
(a,b)	ordered pair of elements, a from set A and b from set B (Appendix C)

MATRIX THEORY

a, b, x, y, etc.	symbols for vectors of values (C.4)
A, B, X, Y, etc.	symbols for matrices of values (D.1)
a′	transpose of a vector **a** (C.4)
A′	transpose of a matrix **A** (D.1)
I	the square identity matrix (D.1)
$\mathbf{A}^{-1}$	the inverse of a square matrix **A,** such that $\mathbf{AA}^{-1} = \mathbf{A}^{-1}\mathbf{A} = \mathbf{I}$ (D.1)
deter. **A** or $|\mathbf{A}|$	determinant of square matrix **A** (D.2)

MATHEMATICAL ANALYSIS

$y = f(x),\ y = G(x),$ etc.	y is functionally related to x (2.18)
$y = f(x,z)$	y is a function of the two variables x and z (Appendix C)

∞	an infinite value (6.1)
Σ	sum of (see Appendix A)
$\int_a^b f(x)\,dx$	area under the curve generated by the function $y = f(x)$, as cut off by the x interval with limits a and b (2.18)
$N!$	factorial of the interger N (3.5)
$\binom{N}{r}$	number of unordered combinations of N things taken r at a time, $0 \le r \le N$; a binomial coefficient (3.7)
e	mathematical constant equal approximately to 2.7182818; base of the natural system of logarithms (3.18)
$\log_e m$	logarithm to the base e for the value m (14.21)
π	mathematical constant, equal approximately to 3.14159265; ratio of the circumference of a circle to its diameter (6.1)

STATISTICAL SYMBOLS USED IN THIS TEXT

a_j	sample effect of the treatment appearing as the jth such treatment in the experiment (10.3, 13.2)
a_0	a constant in the linear model for predicting y from x (14.1)
A_0	a constant in the linear model for predicting z_Y from z_X (14.4)
$(a_0, a_1, \cdots, a_J)$	weights given to variables $(X_0, X_1, \cdots, X_J)$ in a linear model for predicting Y (10.2)
$[ab]_{jk}$	the effect of membership in the *cell* defined by A_j and B_k (12.2)
$(ab)_{jk}$	the interation effect of the cell defined by A_j and B_k (12.4)
$|AD|$	the average absolute deviation (4.15)
α	(small Greek alpha) probability of rejecting H_0 when it is true; probability of Type I error (7.6)
$\alpha_0 = \mu$	constant "effect" in analysis of variance model under least squares criterion (10.5)
$\alpha_j = \mu_j - \mu$	true effect associated with the jth treatment applied; a fixed effect (10.5)
est. α_j	estimated effect for treatment j (10.7)
α_{FW}	the familywise Type I error rate (11.13)
α_{PC}	the Type I error rate per comparison (11.13)
$[\alpha\beta]_{jk}$	true cell effect for the population given treatments A_j and B_k (12.4)
$(\alpha\beta)_{jk}$	true interaction effect for A_j and B_k (12.4)
$b_h = \dfrac{\psi_h}{nw_h}$	weight given to the orthogonal polynomial coefficient representing a trend of degree h in estimation of a curvilinear prediction function (16.15)
b_k	sample effect of level B_k (12.2)
$b_{y\cdot 1}$	linear regression coefficient for prediction of Y from X_1 (16.15)
$b_{Y\cdot X}$	sample raw score coefficient of linear regression of Y on X (14.2)
$b_{12\cdot 3\cdots K}$	regression coefficient for weighting X_2 in the prediction of X_1 from $X_2, \cdots, X_K$ (15.11)
B_j	the standardized partial regression coefficient for variable X_j in the prediction of Y from $(X_1, \cdots, X_j, \cdots, X_K)$ (15.11)
B_j^*	the population standardized partial regression weight for X_j in the prediction of Y from $H = (X_1, \cdots, X_K)$ (15.18)

$B_{Y \cdot X}$	standardized sample regression coefficient for predicting z_y from z_x (14.4)
$B_{12 \cdot 3}$	standardized regression coefficient applied to variable z_2 in predictions of z_1 from z_2 and z_3 (15.11)
$B_{13 \cdot 2}$	standardized regression coefficient applied to variable z_3 in predictions of z_1 from z_2 and z_3 (15.11)
β	(small Greek beta) probability of failing to reject H_0 when it is false; probability of Type II error (7.6)
β_k	true fixed effect associated with the treatment or factor level k (12.4); also any true unstandardized partial regression weight in the prediction of Y from $H = (X_1, \cdots, X_K)$ (15.11)
$\beta_{Y \cdot X}$	unstandardized regression coefficient for prediction of z_y from z_x in the population (14.12)
$\beta_1, \beta_2, \cdots, \beta_{J-1}$	population regression coefficients for linear and curvilinear trends between Y and X (16.14)
$1 - \beta$	power of a statistical test against some given true alternative to the null hypothesis (7.8)
c_j	constant applied to a sample or population mean corresponding to treatment j in a comparison (11.2)
C_{AB}	coefficient of contingency (18.8)
$(c_1, \cdots, c_n)$	set of n constant weights figuring in some linear combination of n random variables (Appendix C)
cf	cumulative frequency (4.2)
χ^2	(small Greek chi, squared) random variable chi-square; also the Pearson chi-square statistic (9.1, 18.1)
$\chi^2 = \sum\limits_j \dfrac{(f_j - m_j)^2}{m_j}$	Pearson χ^2 statistic in a test of goodness of fit (18.1)
$\chi^2_{(\nu)}$	chi-square variable with ν degrees of freedom (9.1)
$\chi^2_{(N-1;\alpha/2)}$; $\chi^2_{(N-1;1-\alpha/2)}$	values in the distribution of χ^2 with $N - 1$ degrees of freedom, cutting off the upper $\alpha/2$ and $1 - \alpha/2$ proportion of cases, respectively (9.5)
$\mathrm{cov}(X,Y)$	covariance of the two random variables X and Y (see Appendix B)
$\mathrm{cov}(\bar{y}_1, \bar{y}_2)$	covariance of pairs of sample means (8.16)
d_i	deviation of the score for observation i from the grand mean (4.3)
$\bar{D}$	the mean difference between values in a set of matched pairs (8.16); also the mean of diagonal values (13.23)
$D_i = (y_{i1} - y_{i2})$	difference between the scores for members of matched pair i from among N such pairs (8.16)
δ	(small Greek delta) noncentrality parameter for noncentral t or, as δ^2, noncentral F (8.11, 10.21)
ΔX	width of an arbitrarily small interval of values for random variable X (2.18)
e	the error in a measurement (6.5)
e_{ij}	random error associated with the ith observation made under treatment j (10.5)
e_{ijk}	random error associated with the ith observation in the treatment combination j, k (12.4)
$\bar{e}$	mean error in entire sample (10.10)

$\bar{e}_j$	mean error in level or group A_j (10.10)
$E(X)$	expectation of random variable X (see Appendix B)
$E(u\|H_0)$	expected loss, given a decision rule and truth of hypothesis H_0 (7.3)
$\hat{\epsilon}$	(small Greek epsilon) estimated value of Box correction to degrees of freedom (13.23)
f	frequency of a given measurement or event class (2.3)
f_j	actual frequency in class A of attribute A (18.1)
f_{j+}	marginal frequency in row class A_j in a two-dimensional contingency table (18.3)
f_{+k}	marginal frequency in column class B in a two-dimensional contingency table (18.3)
$f(a) = f(X = a)$	probability density for random variable X at the value $X = a$ (2.18)
$\max_k \cdot f_{\cdot k}$	the largest marginal frequency among the rows of a table (15.12)
F	F ratio computed from a sample; the random variable F (9.8)
F_α	value of F required for significance at the α level, one-tailed, for a given number of degrees of freedom (11.12)
$F_{(\alpha;\nu_1,\nu_2)}$	value cutting off the upper α proportion of cases in an F distribution with ν_1 and ν_2 degrees of freedom (9.9)
F_g	the F ratio associated with comparison $\hat{\psi}_g$ (11.9)
$F_{(\nu_1;\nu_2)}$	random variable distributed as F with ν_1 and ν_2 degrees of freedom (9.9)
$F(a) = p(X \le a)$	cumulative probability of random variable X at the value a (2.19)
G	the set of significant trends in a problem involving orthogonal polynomials (16.15)
G^2	the likelihood ratio test statistic under the log-linear model (18.14)
$G = H - \{X_j\}$	set notation for a reduced set of predictors in a regression equation for predicting Y (15.20)
γ	(small Greek gamma) coefficient of predictive association between sets of ordered classes (19.13)
H	the entire set of predictors $(X_1, \cdots, X_K)$ in a multiple regression equation for predicting variable Y (15.20)
H:	indicator of a statistical hypothesis (7.1)
H_0:	hypothesis actually being tested; the ''null'' hypothesis (7.1)
H_1:	hypothesis to be entertained if H_0 is rejected; the alternative hypothesis (7.1)
i	class interval size in some frequency distribution (2.5); also running subscript, ordinarily indicating the ith observation among some N distinct observations (Appendix A)
j	running subscript indicating the jth treatment group or factor level (Appendix A)
J	number of different experimental treatments or groups in an experiment (Appendix A)

k	an arbitrarily small positive number (4.18); also a running subscript, ordinarily indicating the kth treatment group or block of observations (Appendix A)
K	in a two-factor experiment, number of treatment groups or levels of the second factor; number of "blocks" of observations in an experiment (12.1)
K	(capital Greek kappa) an index of interjudge agreement (18.13)
$L(x_1, \cdots, x_N|\theta)$	likelihood of a particular set of N sample values, given the value of the parameter (or set of parameters) θ (5.5)
λ_{AB}	(small Greek lambda) symmetric measure of predictive association in a contingency table (18.9)
λ_B	(small Greek lambda) asymmetric measure of predictive association for a contingency table (18.9)
m	the intensity parameter in a Poisson process (3.18)
m_j	the expected frequency in category j under some null hypothesis (18.1)
$m(o)$	measured amount of some property possessed by object o (2.1)
m_{jk}	the expected frequency in cell jk under the hypothesis of independence of row and column attributes (18.3)
$\hat{m}_{jk}$	the estimate of the expected frequency in cell jk under the hypothesis of independence (18.3)
MS	mean square in the analysis of variance (10.10)
$\mu = E(X)$	(small Greek mu) mean of the probability distribution of the random variable X (4.9)
$\mu_G = E(G)$	mean of the sampling distribution for some statistic G (5.3)
μ_j	true mean of the potential population of observations made under treatment j (10.5)
μ_{jk}	mean of the potential population of observations made under the treatment combination j, k (12.4)
μ_0	value of the population mean specified by H_0 (7.9)
μ_1	true value of the mean of the population; a value of the mean covered by the alternative hypothesis H_1 (7.9)
n	size of any one of several sample groups containing the same number of observations (12.1)
n_j	number of sample observations in treatment group j (10.8)
n_{jk}	number of sample observations in treatment combination j, k (12.1)
$n(A)$	the frequency of event A in the sample space (1.6)
N	total number of trials in a simple experiment or of observations in a given sample; the total number of elementary events in a sample space (1.6)
ν	(small Greek nu) degrees of freedom parameter for a t or chi-square variable (8.2, 9.1)
ν_1, ν_2	number of degrees of freedom for numerator and denominator, respectively, for an F ratio; the parameters of the F distribution (9.8)
$\mathbb{O} = (o_1, \cdots, o_N)$	a set of objects of measurement (2.1)

ω^2	(small Greek omega, squared) population index showing the relative or proportional reduction in the variance of Y given the X status or value for an observation (8.12)
est ω^2	sample estimate of the proportional reduction in variance of Y given X (8.15)
p	probability of a given event class; probability of a "success" in a single Bernoulli trial (2.13)
$p(A)$	probability associated with a particular event A in a probability function (1.4)
$p(B\|A) = \dfrac{p(A \cap B)}{p(A)}$	conditional probability of event B given the event A (1.12)
$p(x)$	the probability that random variable X takes on value $X = x$ (2.14)
$p(X_j)$	in the fixed-effects model, the relative frequency in the grand population of those receiving treatment j (10.5)
$p(X = r; N,p)$	binomial probability for $X = r$, given the parameters N and p (3.10)
P	sample proportion of "successes" in sampling from a Bernoulli process (4.9)
$\mathscr{P}(H)$	personal probability associated by the experimenter with the hypothesis H (7.4)
φ^2	(small Greek phi) index of mean square contingency for a sample contingency table (18.8)
Φ^2	(capital Greek phi) index of mean square contingency for a two-dimensional population table (18.8)
ψ	(small Greek psi) value of a particular comparison among population means (11.2)
$\hat{\psi}$	value of a particular comparison among sample means (11.2)
$\hat{\psi}_g$	value of a particular comparison g on a set of sample means (11.9)
q	the Studentized range statistic (11.17)
$q = 1 - p$	probability of a "failure" on a single Bernoulli trial (2.13)
Q	proportion of cases cut off by a given value on the upper tail of a sampling distribution (8.4)
Q,Y	Yule's indices for a fourfold table (18.8)
$Q = 1 - P$	proportion of sample "failures" in sampling from a Bernoulli process (4.9)
r	the number of successes out of N trials of a Bernoulli process (3.11)
r_S	Spearman rank-correlation coefficient (16.11)
r_{xy}	sample correlation coefficient between X and Y (14.4)
$r_{(Y',Y)}$	correlation between the predicted and actual values of Y (13.18)
$r_{1(2 \cdot 3)}$	part correlation between X_1 and X_2, with X_3 variance removed from x_2 (15.4)
$r_{12 \cdot 3}$	partial correlation between Variables 1 and 2, with 3 held constant (15.2)
$r_{y \cdot h}^2$	the proportion of variance accounted for in the data by any trend h (16.14)
$R_{y \cdot G}$	the multiple correlation between Y and the reduced set of predictors, $G = H - \{X_j\}$ (15.20)

$R_{y \cdot H}$	the multiple correlation between Y and the full set of predictor variables, H (15.20)
$R_{y \cdot 12 \cdots K}$	the multiple correlation for Y and the set of predictors $X_1, \cdots, X_K$ (15.14)
$R_{1 \cdot 23}$	multiple correlation of Variable 1 with Variables 2 and 3 (15.14)
$R^2_{y \cdot A}$	the proportion of y variance accounted for by membership in levels of Factor A (10.17)
$R^2_{y \cdot G}$	proportion of variance accounted for by a set of predictor variables G (15.20)
$R^2_{y \cdot H}$	the proportion of variance in Y explained by the set of variables in H (15.20)
ρ_I	(small Greek rho, sub I) intraclass correlation coefficient for a population (13.5)
ρ_{XY}	population correlation coefficient between variables X and Y (14.19, Appendix B)
$P_{y \cdot 1 \cdots K}$	(capital Greek rho) the population multiple correlation for Y predicted from $X_1, \cdots, X_K$ (15.18)
$s = \sqrt{s^2}$	corrected standard deviation for a sample (5.9)
$s^2 = \dfrac{N}{N-1} S^2$	corrected variance; the unbiased estimator of the population variance from a sample (5.9)
s^2_D	corrected variance based on the difference for each of N matched pairs of observations (8.16)
S	standard deviation computed for a sample of data (4.12)
$S_{Y \cdot X}$	sample standard error of estimate for predictions of Y from X (14.8)
$S_{z_Y \cdot z_X}$	sample standard error of estimate for predictions of z_Y from z_X (14.4)
S_+, S_-	number of agreements and disagreements, respectively, about the ordering of pairs of objects in two rank orders (19.11)
$S = \sqrt{(J-1)F_\alpha}$	constant determining the width of confidence intervals for post hoc comparisons. Scheffé method (11.6)
S^2	sample variance (4.11)
$S^2_{y \cdot 1 \cdots K}$	sample variance of estimate in the prediction of Y from $X_1, \cdots, X_K$ (15.15)
S^2_z	variance of z or standardized scores calculated from a sample of data (4.17)
$S^2_{z_1 \cdot z_2 \cdots z_K}$	sample variance of estimate for z_1 values predicted from $z_2, \cdots, z_K$ via a multiple regression equation (15.15)
$S^2 = (J-1)F_\alpha$	Scheffé criterion for evaluating an F in post hoc comparisons of J means (11.6)
$\mathscr{S}$	sample space for a particular simple experiment (1.3)
$SE(u \| D)$	subjective expected loss associated with a particular decision rule D by the experimenter (7.4)
SS	sum of squares in the analysis of variance (10.8)
SSCP	the matrix of sums of squares (SS) and sums of products (SP) values (14.2)
SWP$(x_k, \cdots)$	symbol for a matrix of correlations (or variance–covariance matrix) "swept" for one or more variables $x_k, \cdots$ (15.27)
σ	(small Greek sigma) standard deviation of a random variable (4.16)
$\sigma_{\text{diff.}}$	standard error of the difference between two means (8.7)

est. $\sigma_{\text{diff.}}$	estimated error of the difference between two means (8.7)
σ_M	true standard error of the mean, given samples of size N from some population (5.8)
est. σ_M	sample estimate of the standard error of the mean (5.9)
est. σ_{M_D}	estimated standard error of the mean difference between N matched pairs (8.16)
$\sigma_{Y \cdot X}$	true standard error of estimate for predictions of Y from X in some population (14.12)
$\sigma_{(Z_1 - Z_2)}$	variance of the difference between Z values for pairs of independent samples (14.21)
σ^2	variance of a random variable (4.16)
σ_A^2	variance of the distribution of random effects representing Factor A (13.2)
σ_{AB}^2	variance of the random interaction effects for the Factors A and B (13.10)
σ_B^2	variance of the random effects representing Factor B (13.10)
σ_e^2	variance of random errors, e (10.9)
σ_G^2	variance of the sampling distribution of some statistic G (5.3)
σ_M^2	variance of the sampling distribution of the mean (5.7)
σ_P^2	variance of a sampling distribution of sample proportions, P (4.16)
$\sigma_{y \cdot 12 \cdots K}^2$	the true variance of multiple estimate for predicting Y from $X_1, \cdots, X_K$ (15.18)
σ_Y^2	variance of the random variable Y; "marginal" variance of Y in a joint distribution of (x,y) events (see Appendices B and C)
$\sigma_{Y\|X}^2$	variance of the conditional distribution of Y, given the value of X (8.12, Appendix C)
$\sigma_Y^2 - \sigma_{Y\|X}^2$	reduction in the variance of Y afforded by specification of X (8.12)
σ_z^2	variance of a random variable transformed to z or standardized values (4.17)
σ_0^2	value of the population variance specified by the hypothesis H_0 (9.4)
$\boldsymbol{\Sigma}$	(capital Greek sigma, boldface) the population variance–covariance matrix for two or more variables (C.3)
$\hat{\boldsymbol{\Sigma}}$	the sample estimate of the population variance–covariance matrix (D.6)
t	a random variable following the Student distribution of t with ν degrees of freedom (8.2)
$t_{(\alpha/2;\nu)}$	value of t in a distribution with ν degrees of freedom, cutting off the upper $\alpha/2$ proportion of cases (8.9)
t_B	value of t required under the Bonferroni criterion for significance (11.14)
$t_{(o)}$	true amount of some property possessed by object o (2.1)
$t = \dfrac{\bar{x} - E(\bar{x})}{\text{est. } \sigma_M}$	t ratio based on the mean of a single sample (8.2)
$t = \dfrac{\bar{x}_1 - \bar{x}_2 - E(\bar{x}_1 - \bar{x}_2)}{\text{est. } \sigma_{\text{diff.}}}$	t ratio based on the difference between means of two samples (8.7)
T	total number of potential observations in a finite population (5.11)
T_{jk}	total of the Y values in cell jk (12.6)
T_y	total of Y values for N cases (10.13)

$T_j = T_{j.}$	total of the Y values in level A_j (10.13, 12.6)
$T_k = T_{.k}$	total of the Y values in level B_k (12.6)
τ	(small Greek tau) Kendall's coefficient of rank-order agreement (19.11)
θ	(small Greek theta) general symbol for a population parameter (5.5)
θ_0	value of θ specified by the null hypothesis H_0 (13.6)
$\theta = \sigma_A^2/\sigma_e^2$	ratio of the variance of effects due to Factor A to the error variance (13.5)
$u, u_i, u_{ij},$ etc.	"effects" in the log-linear model (18.14)
$u(H_0\|H_1)$	loss associated with the decision to accept H_0 when H_1 is in fact true (7.4)
$u(H_1\|H_0)$	loss associated with the decision to accept H_1 when H_0 is in fact true (7.4)
v_{gj}	an indicator variable equal to 1 when treatment g is selected and labeled as group j, and 0 otherwise (13.2)
$\mathbf{V}$	a set of vectors $(\mathbf{V}_0, \cdots, \mathbf{V}_{J-1})$ forming an orthogonal basis for a set of vectors $\mathbf{X}$ (Appendix C)
V_C	Cramér's index of association in a contingency table (18.10)
$\text{var}(X)$	variance of the random variable X (see Appendix B)
$\text{var}(\bar{y}_1 - \bar{y}_2)$	variance of the difference between means (8.7)
$\text{var}(\hat{\psi})$	variance of a comparison among sample means (11.2)
w_g	weighting factor used in obtaining the sum of squares for comparison $\hat{\psi}_g$ (11.9)
$w(b\|a)$	conditional density for $Y = b$ given that $X = a$ (Appendix C)
x	a particular value which random variable X can assume (2.14); also midpoint of a class interval (4.2)
(x,y)	a joint event, consisting of a value for variable X paired with a value for variable Y (Appendix C)
x_{i+}, x_{+i}	marginal frequencies in an $I \times I$ table (18.13)
x_{i+}, x_{+j}	marginal frequencies in an $I \times J$ table (18.14)
$\bar{x}$ or $\bar{y}$	the sample mean (4.2)
$\bar{x}_G$	the geometric mean (4.2)
$\bar{x}_H$	the harmonic mean (4.2)
X	a random variable; the independent variable in an experiment (2.14)
$\|X - \mu\|$	absolute deviation (disregarding sign) of the value of X from the value of μ (4.18)
$\mathbf{X}$	a set of m vectors each of dimension n (Appendix C)
y_{ij}	score associated with the observation i in experimental group j (10.5)
y_{ijk}	score associated with the ith observation in the treatment combination j,k (12.2)
y'	raw score on Y predicted from the value of X (14.1)
$\bar{y}_{jk}$	mean in the cell for A_j and B_k (12.2)
$\bar{y}_{j.}$	mean of the Y values in Group A_j (12.2)
$\bar{y}_{.k}$	mean of the y values in Group B (12.2)
$\bar{y}_{..}$	grand mean for a set of N cases (12.2)

Y	a random variable; ordinarily, the dependent variable in an experiment (8.12)
$z_{\text{diff.}}$	standardized value for the difference between two means (8.7)
z_X, z_Y	standardized scores corresponding to particular values of X and Y, respectively (14.4)
$z_{(1-\alpha/2)}$	standardized value cutting off the lower $1 - \alpha/2$ proportion of cases in a normal distribution (8.13)
$\lvert z \rvert$	absolute value of a standardized score (4.18)
$\bar{z}$	mean of a set of standardized values (4.17)
z'_Y	predicted standardized value on variable Y (14.4)
$z = \dfrac{X - E(X)}{\sigma}$	standardized score or value corresponding to a particular value of X, relative to a population distribution (4.17)
$z = \dfrac{X - \bar{x}}{S}$	standardized score or value corresponding to a sample value X, relative to a sample distribution (4.17)
$z_M = \dfrac{\bar{x} - E(\bar{x})}{\sigma_M}$	standardized value corresponding to a particular value of sample $\bar{x}$ in a sampling distribution of means (5.8)
Z	value corresponding to r_{XY} in the Fisher r- to- Z transformation (14.21)
ζ	(small Greek zeta) value in the Fisher r- to- Z transformation corresponding to the population correlation ρ (14.21)

Index